Handbook of
Membrane Channels

Handbook of Membrane Channels

MOLECULAR AND CELLULAR PHYSIOLOGY

Edited by

Camillo Peracchia

Department of Physiology
University of Rochester
School of Medicine and Dentistry
Rochester, New York

Academic Press

San Diego   New York   Boston   London   Sydney   Tokyo   Toronto

This book is printed on acid-free paper. ∞

Academic Press, Inc.
A Division of Harcourt Brace & Company
525 B Street, Suite 1900, San Diego, California 92101-4495

United Kingdom Edition published by
Academic Press Limited
24–28 Oval Road, London NW1 7DX

Library of Congress Cataloging-in-Publication Data

Handbook of membrane channels : molecular and cellular physiology /
  edited by Camillo Peracchia.
      p.   cm.
   Includes bibliographical references and index.
   ISBN 0-12-550640-6
   1. Ion channels-Handbooks, manuals, etc.   2. Cell membranes-
-Handbooks, manuals, etc.   3. Mitochondrial membranes--Handbooks,.
manuals, etc.   4. Gap junctions (Cell biology)--Handbooks, manuals,
etc.   I. Peracchia, Camillo.
   QH603.I54H36 1994
   574.87'5--dc20   94-2019
                                                        CIP

PRINTED IN THE UNITED STATES OF AMERICA
94   95   96   97   98   99   QW   9   8   7   6   5   4   3   2   1

This book is dedicated to Professor Shosaku Numa
whose pioneering work on the molecular genetics of membrane channels
has had profound impact on the field.
His premature death has left a large vacuum
in membrane research.

Contents

21

Inhibitory Ligand-Gated Ion Channel Receptors: Molecular Biology and Pharmacology of GABA$_A$ and Glycine Receptors

Robert J. Vandenberg and Peter R. Schofield

PART

F

Cyclic-Nucleotide-Sensitive Channels

22

Hyperpolarization-Activated (i_f) Current in Heart

Dario DiFrancesco

23

Selectivity and Single-Channel Properties of the cGMP-Activated Channel in Amphibian Retinal Rods

Vincent Torre and Anna Menini

PART

G

Gap Junction Channels

24

Molecular Models of Channel Interaction and Gating in Gap Junctions

Camillo Peracchia, Ahmed Lazrak, and Lillian L. Peracchia

25

Molecular Biology and Electrophysiology of Cardiac Gap Junctions

Eric C. Beyer and Richard D. Veenstra

26

Gap Junction Proteins in the Lens

Lisa Ebihara

PART

H

Channels of the MIP Family

27

The MIP Transmembrane Channel Gene Family

Ana B. Chepelinsky

PART

I

Calmodulin-Regulated Channels

28

Calmodulin-Sensitive Channels

Yoshiro Saimi, Kit-Yin Ling, and Ching Kung

PART

J

Channels of Unicellular Organisms

29

Microbial Channels

Boris Martinac, Xin-Liang Zhou, Andrzej Kubalski, Sergei Sukharev, and Ching Kung

---

SECTION

II

CHANNELS OF INTRACELLULAR MEMBRANES

PART

A

Channels of Calcium Stores

30

Ryanodine Receptor/Ca^{2+}-Release Channel of Excitable Tissues

Gerhard Meissner

Contributors

Numbers in parentheses indicate the pages on which the authors' contributions begin.

David J. Adams (255), Department of Molecular and Cellular Pharmacology, University of Miami School of Medicine, Miami, Florida 33101

E. X. Albuquerque (269), Department of Pharmacology, and Experimental Therapeutics, University of Maryland School of Medicine, Baltimore, Maryland 21201, and Labóratorio de Farmacologia Molecular II, Instiudo de Biofisica Carlos Chagas Filho, Universitade Federal de Rio de Janeiro, Rio de Janeiro, Brazil 21941

M. Alkondon (269), Department of Pharmacology, and Experimental Therapeutics, University of Maryland School of Medicine, Baltimore, Maryland 21201

Bruce P. Bean (199), Department of Neurobiology, Harvard Medical School, Boston, Massachusetts 02115

Ted Begenisich (17), Department of Physiology, University of Rochester Medical Center, Rochester, New York 14642

Francisco Bezanilla (29), Department of Physiology, Jerry Lewis Neuromuscular Research Institute, University of California at Los Angeles, Los Angeles, California 90024

Eric C. Beyer (379), Departments of Pediatrics and Cell Biology and Physiology, Washington University School of Medicine, St. Louis, Missouri 63110

Ilya Bezprozvanny[1] (511), Division of Cardiology, University of Connecticut Health Center, Farmington, Connecticut 06032

Ana B. Chepelinsky (413), Regulation of Gene Expression Section, Laboratory of Molecular and Developmental Biology, National Eye Institute, National Institutes of Health, Bethesda, Maryland 20892

Mark G. Darlison (303), Institut für Zellbiochemie und Klinische Neurobiologie, Universitäts-Krankenhaus Eppendorf, Universität Hamburg, D-20246 Hamburg, Germany

Dario DiFrancesco (335), Universita di Milano, Dipartimento di Fisiologia e Biochimica Generali, Elettrofisiologia, 20133 Milano, Italy

Lisa Ebihara (403), Department of Pharmacology, Columbia University, New York, New York 10032

Barbara E. Ehrlich (511), Departments of Medicine and Physiology, University of Connecticut, Farmington, Connecticut 06032, and The Marine Biological Laboratory, Woods Hole, Massachusetts 02543

Julio M. Fernandez (557), Department of Physiology and Biophysics, Mayo Foundation, Rochester, Minnesota 55905

Fabio Franciolini (245, 255), Instituto Biologie Cellulare, Università di Perugia, 06123 Perugia, Italy

Alan L. Goldin (121), Department of Microbiology and Molecular Genetics, University of California, Irvine, Irvine, California 92717

Rainer Greger (229), Physiologisches Institut der Universität, Albert-Ludwigs-Universität Freiburg, D-79104 Freiburg, Germany

T. Higuti (549), Faculty of Pharmaceutical Sciences, Tokushima University, Tokushima 770, Japan

I. Inoue (549), Institute for Enzyme Research, Tokushima University, Tokushima 770, Japan

---

[1] Present address: Department of Molecular and Cellular Physiology, Beckman Center, Stanford University Medical Center, Stanford, CA 94305-5426.

Robert S. Kass (187), Department of Physiology, University of Rochester, Rochester, New York 14642

C. Kentros (41), Department of Physiology and Biophysics, New York University Medical Center, New York, New York 10016

Kathleen W. Kinnally (529), Department of Biological Sciences, State University of New York at Albany, Albany, New York 12222

Douglas S. Krafte (115), Vascular and Biochemical Pharmacology, Sterling Winthrop, Inc., Collegeville, Pennsylvania 19426

Andrzej Kubalski (447), Department of Cell Biology, Nencki Institute of Experimental Biology, 02-093 Warsaw, Poland

Ching Kung (447, 435), Laboratory of Molecular Biology, and Department of Genetics, University of Wisconsin, Madison, Madison, Wisconsin 53706

Ramon Latorre (79), Centro de Estudios Científicos de Santiago, and Departamento de Biologia, Facultad de Ciencias, Universidad de Chile, Santiago 9, Chile

D. Lau (41), Department of Physiology and Biophysics, New York University Medical Center, New York, New York 10016

Ahmed Lazrak (361), Department of Physiology, School of Medicine and Dentistry, University of Rochester, Rochester, New York 14642

Kit-Yin Ling (435), Laboratory of Molecular Biology, University of Wisconsin, Madison, Madison, Wisconsin 53706

Boris Martinac (447), Department of Pharmacology, The University of Western Australia, Newland, Western Australia 6009, Australia

Gerhard Meissner (465), Departments of Biochemistry and Biophysics, University of North Carolina at Chapel Hill, Chapel Hill, North Carolina 27599

Anna Menini (345), Istituto di Cibernetica e Biofisica, Consiglio Nazionale delle Ricerche, 16146 Genova, Italy

Isabelle M. Mintz (199), Department of Neurobiology, Harvard Medical School, Boston, Massachusetts 02115

Edward Moczydlowski (137), Department of Pharmacology, Yale University School of Medicine, New Haven, Connecticut 06510

Jonathan R. Monck (557), Department of Physiology and Biophysics, Mayo Foundation, Rochester, Minnesota 55905

J. G. Montes (269), Department of Pharmacology and Experimental Therapeutics, University of Maryland School of Medicine, Baltimore, Maryland 21201

H. Moreno (41), Department of Physiology and Biophysics, New York University Medical Center, New York, New York 10016

Yasuo Mori (163), Department of Pharmacology and Cell Biophysics, University of Cincinnati College of Medicine, Cincinnati, Ohio 45267, and Department of Molecular Genetics, Kyoto University Faculty of Medicine, Kyoto 606, Japan

Andres F. Oberhauser (557), Department of Physiology and Biophysics, Mayo Foundation, Rochester, Minnesota 55905

Camillo Peracchia (361), Department of Physiology, School of Medicine and Dentistry, University of Rochester, Rochester, New York 14642

Lillian L. Peracchia (361), Department of Physiology, School of Medicine and Dentistry, University of Rochester, Rochester, New York 14642

E. F. R. Pereira (269), Department of Pharmacology, and Experimental Therapeutics, University of Maryland School of Medicine, Baltimore, Maryland 21201

Eduardo Perozo (29), Jules Stein Eye Institute, and Department of Chemistry and Biochemistry, University of California at Los Angeles, Los Angeles, California 90024

Isaac N. Pessah (475), Department of Molecular Biosciences, School of Veterinary Medicine, University of California, Davis, Davis, California 95616

Anna Petris (245), Instituto Biologie Cellulare E Molecolare, Università di Perugia, 06123 Perugia, Italy

Paul J. Pfaffinger (5), Division of Neuroscience, Baylor College of Medicine, Houston, Texas 77030

B. Rudy (41), Department of Physiology and Biophysics, New York University Medical Center, New York, New York 10016

Yoshiro Saimi (435), Laboratory of Molecular Biology, University of Wisconsin, Madison, Madison, Wisconsin 53706

Laurent Schild (137), Institute of Pharmacology and Toxicology, Université de Lausanne, CH-1005 Lausanne, Switzerland

Peter R. Schofield (317), The Garvan Institute of Medical Research, St. Vincent's Hospital, Sydney, Darlinghurst NSW 2010, Australia

P. Serodio (41), Department of Physiology and Biophysics, New York University Medical Center, New York, New York 10016

N. Vivienne Shen (5), Division of Neuroscience, Baylor College of Medicine, Houston, Texas 77030

Peter Shrager (105), Department of Physiology, University of Rochester Medical Center, Rochester, New York 14642

Trevor J. Shuttleworth (495), Department of Physiology, University of Rochester, School of Medicine and Dentistry, Rochester, New York 14627

Enrico Stefani (29), Department of Molecular Physiology and Biophysics, Baylor College of Medicine, Houston, Texas 77030

Sergei Sukharev (447), Laboratory of Molecular Biology, University of Wisconsin, Madison, Madison, Wisconsin 53706

Tsutomu Tanabe (177), Howard Hughes Medical Institute, Department of Cellular and Molecular Physiology, Yale University School of Medicine, New Haven, Connecticut 06536

Peter E. R. Tatham (557), Department of Physiology, University College London, London WC1E 6JJ, United Kingdom

Henry Tedeschi (529), Department of Biological Sciences, State University of New York at Albany, Albany, New York 12222

Ligia Toro (29), Department of Molecular Physiology and Biophysics, Baylor College of Medicine, Houston, Texas 77030

Vincent Torre (345), Dipartimento di Fisica, Universita di Genova, 16146 Genova, Italy

Carlo Trequattrini (245), Instituto Biologie Cellulare E Molecolare, Università di Perugia, 06123 Perugia, Italy

Robert J. Vandenberg (317), The Garvan Institute of Medical Research, St. Vincent's Hospital, Sydney, Darlinghurst NSW 2010, Australia

Richard D. Veenstra (379), Department of Pharmacology, State University of New York, Health Science Center, Syracuse, New York 13210

E. Vega-Saenz de Miera (41), Department of Physiology and Biophysics, New York University Medical Center, New York, New York 10016

M. Weiser (41), Department of Physiology and Biophysics, New York University Medical Center, New York, New York 10016

David S. Weiss (213), University of South Florida College of Medicine, Department of Physiology and Biophysics, Tampa, Florida 33612

Martin Wilson (287), Section of Neurobiology, Physiology, and Behavior, Division of Biological Sciences, University of California, Davis, Davis, California 95616

Jin V. Wu (105), Department of Physiology, University of Rochester Medical Center, Rochester, New York 14642

Xin-Liang Zhou (447), Laboratory of Molecular Biology, University of Wisconsin, Madison, Madison, Wisconsin 53706

Ildiko Zimanyi (475), Novo Nordisk Entoech, Inc., Davis, California 95616

Preface

This year marks the tenth anniversary of publications that have opened new avenues in our approach to channel research. Much to the credit of Shosaku Numa, Masaharu Noda, and their co-workers, the discovery of the amino acid sequence of acetylcholine receptor subunits and voltage-dependent sodium channels and the subsequent expression of normal and genetically mutated channel proteins in easily accessible expression systems have opened the way to an unprecedented wealth of discoveries on channel structure and function at the molecular level.

While the fifties, sixties, and seventies will be remembered as the decades of channel biophysics, as the development of the voltage clamp allowed the recording of multiple-channel electrical activity to become a routine exercise, the eighties and nineties will be recognized as the decades of molecular physiology of channels, as the development of the patch clamp by Erwin Neher and Bert Sakmann and the merging of molecular biology and biophysics have provided the means for studying channels at the molecular level. Although it is clear that a complete understanding of the structural basis of channel behavior will only derive from future crystallographic data of channel protein conformation, it is unquestionable that these new approaches have enabled channel research to achieve a depth and a new level of selectivity, such that it is no longer mesmerizing to open an issue of *Science* or *Nature* and find that a question which has been puzzling generations of scientists has been answered by a molecular geneticist who replaced a single amino acid residue of the channel protein and a biophysicist who expressed the mutated protein in a frog's egg.

It is primarily as a testament to the wealth of discoveries generated by the merging of these powerful approaches that this book has been conceived. Although several books on channel biophysics are available, a thorough account of this new direction in channel research is needed and timely. This book is intended to fill this need by bringing together world leaders in their respective fields and a "wealth of goods" including exciting new findings, provocative hypotheses, creative molecular models, and ingenious methodologies. Although our goal has been to ensure that each subject is covered in terms of channel biophysics, pharmacology, molecular biology, and physiological regulation, and that all the major channel families are included, the vastness and mobility of the field have discouraged us from an attempt to make it a more comprehensive, yet potentially cumbersome, publication.

This book provides a useful reference on the state of the art of channel research for a large group of biomedical scientists and students including cell and molecular biologists, biophysicists, physiologists, neuroscientists, pharmacologists, biochemists, and developmental biologists. Reference made in many chapters to channel malfunction in disease states will undoubtedly call the attention of research-oriented clinicians as well.

I express my sincere gratitude to the authors for their keen interest in this publication, to the Academic Press editors for their efficiency and professionalism, and to my wife, Lillian, for her invaluable assistance.

Camillo Peracchia

CHANNELS OF PLASMA MEMBRANES

Potassium Channels

Conservation of K⁺ Channel Properties in Gene Subfamilies

N. VIVIENNE SHEN AND PAUL J. PFAFFINGER

Ion channel studies now involve a rich interaction of molecular biological, biochemical, and biophysical approaches. Each technique provides a different view of the channel, allowing a unique perspective on channel biology. This review focuses on analyses involving molecular biological approaches and ideas, and extends these ideas to include results obtained using more classical biophysical techniques. The important ideas and methods required to analyze K^+ channels properly are described and are applied to the analysis of K^+ channel protein structural and functional evolution. The molecular analysis of K^+ channel proteins builds on the tremendous knowledge base accumulated over years of functional studies. The goal of molecular analysis is to determine specific molecular correlates to the functional features previously identified, as well as to provide new and unexpected insights into K^+ channels and their function.

I. What Is Homology?

The identification of sequence information about K^+ channels immediately raises questions about the relationships among these channels, and their relationships to other types of channels and proteins. How are such relationships defined, and how is significance judged? How is the knowledge that certain proteins are related interpreted? What relationships are produced by the similarity of specific functionally important structures? Are proteins closely related because they have diverged only recently or because they occupy similar or identical physiological niches?

A. Homology and Similarity

Relationships between sequences are often described in terms of homology and similarity. However, how these terms are used is often unclear. Classical evolutionary studies define homology as the descent of two structures from a common ancestor. For example, a bat wing and a rat forelimb are homologous since both evolved from a common ancestral forelimb. A bat wing is not homologous to a dragonfly wing, since no common ancestor can be defined, although the bat and dragonfly wings are related in function. Homology is a useful term to describe the existence of an evolutionary relationship between two different structures. Molecular biologists have usurped this term to describe two genes that share a common ancestral gene. The term can be used in two manners. The first is to refer to genes in different organisms that derive from a single gene found in a common ancestral organism; these are orthologous sequences. The other is to refer to two genes in a single organism that derive, through gene duplication, from

5

a common ancestral gene; these are paralogous sequences. The possible relationships between genes can be much more complex, however, than those between animals with which the term homology originates.

In its classical evolutionary study origins, homology is a declaration of a relationship and not a measure (Patterson, 1988). Similarity and identity are measures that correctly define homologous relationships. Therefore, stating that two genes are "25% homologous" is technically incorrect unless the sequences are partially derived from a common ancestor through recombination, sharing 25% of the residues.

Identity is the most easily understood measure of similarity between sequences. Identity counts the number of residues of a protein that remained unchanged as the protein evolved. These residues are defined by aligning sequences to maximize the number of amino acid residue identities and counting all residues that are identical in the two proteins. The percentage identity is then defined by the fraction of residues that are identical. However, determining identity is difficult when insertions or deletions alter the sequence of the protein. These changes complicate determining the correct amino acids to compare between hypothetically homologous sequences. In addi-

tion, some amino acids can be changed with very little consequence to the overall function of the protein. Such slight changes should not be considered indicative of weak relationships between proteins.

To provide a more complete measure of sequence relationships, a measure of percentage similarity was introduced. This measure takes into account the probabilities of certain amino acid substitutions occurring in homologous sequences when determining whether the sequences are related. This measure becomes subjective in the assignment of these substitution probabilities. No *a priori* rules for the protein of interest can be determined; therefore, the probabilities are assigned based on database analysis of conservative substitution patterns in genes that are homologous. These probabilities are then incorporated into a matrix that is used in algorithms that search for relationships between sequences.

B. Genes Can Undergo Novel Transformations

A major complication in comparing sequences is that genes are not stable entities; they can undergo dramatic changes such as gene duplication, exon shuffling, and other rearrangements (Fig. 1). Gene duplication is a poorly understood genetic process through

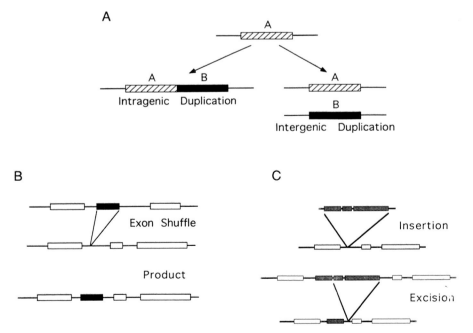

FIGURE 1 Mechanisms for rapid gene evolution. (A) Gene duplications produce multiple copies of a sequence. Intragenic duplication creates internally repeated structures. Intergenic duplication creates a duplicate copy of a gene that can evolve separately. (B) Exon shuffling allows a functional domain to be moved into a new environment, creating a novel gene product. (C) Mobile DNA can transfer sequences between genes and organisms during insertion/excision cycles.

which sequences can be duplicated in the genome. In intragenic gene duplication, a single gene has a repeat structure and, thus, internal homology because of the internal repeating of a sequence. For example, the Na^+ and Ca^{2+} channel genes (see Chapter 10) are tetrameric repeats of an ancestral gene that is homologous to the gene for voltage-gated K^+ channels (reviewed by Jan and Jan, 1992). Thus, K^+ channel genes are not compared with the whole Na^+ or Ca^{2+} channel sequence but to a single repeat of these much larger genes. If the duplication is not intragenic, a new copy of the entire gene can be created. This new duplicate gene is then able to follow a separate evolutionary path. Gene duplication is the basis of gene homology.

In exon shuffling, large contiguous sequences are transferred from one gene to another. These exons are often found to encode single functional domains that can act autonomously from the parent protein. In this case, the concept of gene homology is replaced by the concept of domain homology. The comparisons become especially confusing when several domains, each from different parent genes, are combined into a single new gene.

Other factors that confuse the comparison of sequences are pseudogenes and mobile DNA elements, such as transposable elements and viruses. Pseudogenes are nonfunctional sequences that are copies of functional genes. These sequences may represent rapidly mutating versions of genes that can later be turned into functional genes through recombination or other unknown processes. Mobile DNA can rapidly transfer pieces of the genome within a single organism's genome or between genomes of evolutionarily distinct organisms. Such transfers theoretically introduce tremendous rapid changes in sequences, and blur the evolutionary relationships that normally exist between genes.

Homology analyses can be conducted at several levels. The simplest approach is a statistical analysis comparing the primary sequences of proteins to determine whether sequence similarities can be detected. The second type of analysis takes advantage of structural or domain information about the protein to evaluate its homology to other proteins. Finally, functional information about the role of the protein and the corresponding mechanism provides further information about possible homologies.

II. Primary Sequence Homology

The determination of primary sequence homology is a process that has largely been left to large computer program packages, with algorithms that are inaccessi-

ble to the researcher. However, the researcher should have some understanding of the analyses that are being undertaken to make sense of the information provided by the computer and to avoid making nonsensical comparisons using incorrect algorithms or parameters.

The raw material of sequence homology is the determination of the coding sequences for proteins of interest using molecular biological techniques. These DNA sequences provide the deduced amino acid sequence of the protein. This amino acid sequence, the primary sequence, is the fundamental unit of analysis for most comparisons between proteins. This linear array of information is ideal for computer algorithms and is generally compared quite effectively with sequences of other known proteins, providing a rich mixture of possible relationships between the protein of interest and other sequences.

A. Programs to Determine Primary Sequence Similarity

Analysis of sequence homology is essentially a problem of statistics. The statistical question addresses the similarity of two sequences. Because of the large amount of information contained in protein sequences, estimated at 4.16 bits per amino acid residue, for two sequences to be very similar over reasonable stretches and not be homologous is virtually impossible (Gribskov and Devereux, 1991). For example, 20 amino acid identities between two sequences requires specifying the same 83 bits of information. The odds of this event occurring by chance is 2^{-83} or 10^{-25}. The probabilities are unchanged, whether the identities are contiguous or located at 20 independent sites aligned without gaps. Thus, the probabilities for even reasonably weak amino acid sequence similarities to occur by chance soon become unreachable.

Two types of analysis are most commonly undertaken: (1) comparisons between two sequences that are suspected to be homologous and (2) comparison of a sequence of interest against an entire database of protein sequences to detect similarities that are significantly far from the mean. Comparisons between two sequences are undertaken most often with dot matrix analyses or dynamic programming algorithms (Gribskov and Devereux, 1991). These programs, although extremely effective for detecting similarities in such binary comparisons, are unsuited for database searches because they are either labor intensive or computation intensive. In dot matrix methods, the sequences to be compared are written out vertically and horizontally. In the basic dot matrix, all amino acid identities between the sequences are recorded by a dot at the intersection of the vertical and horizontal sequences, creating a two-dimensional matrix. The number of comparisons done is equal to the size of the matrix (N^2) if the sequences are the same length. Regions of high identity are revealed by diagonal stretches with a high density of dots. These comparisons are made easier by filtering the output using windowing techniques (Gribskov and Devereux, 1991). Dynamic programming algorithms perform a similar comparison, but the program records the stretches of sequence that produce the highest percentage identity when aligned. The entire alignment is then reconstructed with gaps inserted where necessary. Again, N^2 comparisons are needed.

Database programs dramatically speed up the search for homologies by creating a look-up table to identify shared "words," short contiguous stretches of sequence, between the sequence of interest and the database (Gribskov and Devereux, 1991). This method increases the efficiency of the algorithm to approximately N comparisons, where N is the size of the database. The FASTA program preforms an identity matching search (Pearson, 1990), whereas the newer BLAST program preforms a sophisticated similarity search with an algorithm optimized to take advantage of the full computational strength of the processor (Altschul *et al.*, 1990).

B. Defining Homology from Sequence Similarity

When is a primary sequence similar enough to conclude homology? This issue, like many in statistics, has no absolute answer. The best response is that alignments that are obvious, such as those among *Shaker* subfamily K$^+$ channels, are clearly homologies. For weaker similarities, a common measure is the Z score, which records how many standard deviations beyond the mean a particular alignment is, when the mean is determined by analysis against the entire database or against randomized sequences (Gribskov and Devereux, 1991). Z scores above 6 are unlikely to occur by chance and, thus, are likely to represent true homologies. Z scores between 3 and 6 may indicate significant homologies. Various different measures are used by other programs to estimate the probable significance of a particular alignment.

Statistical measures of less stringent similarities cannot prove that these alignments are not homologies, but that information from such an analysis alone is unable to make the determination of homology. Identifying these homologies requires more information about the structure of the protein and presence of conserved sequence motifs. This information is often obtainable by analyses based on tests of conserved structures.

C. K$^+$ Channel Gene Cloning and Primary Sequence Homology

The first K$^+$ channel to be cloned was the *Drosophila* K$^+$ channel *Shaker* (Kamb *et al.*, 1987; Tempel *et al.*, 1987; Pongs *et al.*, 1988). This gene was identified by classical genetic techniques as a result of the *Shaker* mutation, causing leg shaking under anesthesia. Chromosome walking provided a means to identify the gene with no structural or sequence information available. After cloning of *Shaker*, a wealth of information about K$^+$ channels was suddenly available; strategies were designed to clone more members of the K$^+$ channel gene family. Two directions were taken that took advantage of the concept of homology: the first was to look for evolutionarily conserved *Shaker* genes in other species; the other was to attempt to cross-hybridize to different K$^+$ channel genes in *Drosophila*. The first approach rapidly showed that the *Shaker* gene was evolutionarily conserved from *Aplysia* to humans. The high conservation among the identified clones, often greater than 70% identity, left little doubt that these genes from other species were *Shaker* gene homologs. However, as researchers continued using these approaches, they began to identify additional genes in some species that also had >70% identity to *Shaker*. Extensive screening in *Drosophila* clearly showed that only one *Shaker* gene exists. Eventually cloning from several mammalian species and other invertebrates demonstrated that mammals contained multiple *Shaker* type genes and that the invertebrates only have a single *Shaker* gene. This observation suggests the following hypothesis: a single *Shaker* gene was duplicated several times during evolutionary stages between invertebrates and mammals to produce a large increase in the number of mammalian *Shaker* genes (see Strong *et al.*, 1993).

D. Sequence Homology to Define Subfamilies

Low stringency screening in *Drosophila* eventually identified other genes—*Shab*, *Shal*, and *Shaw*—with similarities to *Shaker* in the 40% range. These similarities were clustered among the transmembrane domains and, in particular, showed conserved S4 domains and P domains with high similarities. Based on these sequence correlates, researchers concluded that homologous K$^+$ channel genes had been cloned, a conclusion that was confirmed by functional expression in *Xenopus* oocytes. This gene homology, a 40% identity, was much weaker than the identity found among the multiple *Shaker* type genes in mammals. Thus, these genes were concluded to encode different kinds of K⁺ channels (Butler *et al.*, 1989).

Interestingly, cloning in mammals soon showed that K⁺ channel gene duplication was occurring in all these channel systems (reviewed by Salkoff *et al.*, 1992). For *Shab* and *Shal*, multiple genes with greater than 75% identity to *Drosophila* clones and to each other were cloned. For *Shaw*, mammalian genes with about 55% sequence identity to the *Drosophila* clone were found. Cloning in the invertebrate *Aplysia* revealed that only single genes for these different types of K⁺ channels were present, suggesting that this extensive duplication of various K⁺ channel genes also occurred during the transition from invertebrates to mammals (Pfaffinger *et al.*, 1991).

As cloning continues, more genes are being identified with even lower homology to the original *Shaker* gene (Chandy, 1991). These clones are being identified by a combination of techniques, including low stringency screening, expression cloning, and gene mutation isolation by chromosome walking. In most cases, the complexity of gene duplications for these other types of K⁺ channels has not been examined, so whether they also show extensive gene duplication in mammals remains unknown.

III. Structural Homology

Structural homology refers to the tendency of related proteins to assume similar three-dimensional shapes. As more results from structural analyses of proteins become available, researchers are learning that relationships between proteins can often be understood best by understanding the structures into which they are folded. Many proteins with similar enzymatic functions at first glance appear to have little in common. Closer analysis, including detailed structural work, often shows that the proteins assume similar folded structures and that the amino acid residues that are responsible for enzymatic activity are, in fact, completely conserved. This observation has motivated searches for similar structural motifs.

A. Sources of Structural Homology

Protein structure results from the combined influence of the folding of individual structural elements, such as α helices or β sheets, into the secondary structure of the protein and from the incorporation of these elements into a three-dimensional relationship called the tertiary structure. This folding is the direct consequence of the primary sequence as well as a result of the active folding of the polypeptide chain by chaperonin molecules. A strong homology analysis between two sequences often involves showing how they can fold into similar three-dimensional structural arrangements.

The tertiary structure is often broken up into a collection of structural units called domains. Individual structural domains are often moved between different genes during exon shuffling, thus creating novel homologous relationships (Craik *et al.*, 1983). The likely explanation for this shuffling is that each domain can fold and function as an independent entity. Domain homologies are often best seen as local blocks of sequence that are highly conserved within an array of nonconserved residues. Programs that produce visual outputs, such as the dot matrix programs described previously, can readily reveal the presence of such domains. Once a domain has been located, it can be screened against an entire database of sequences to locate other sequences containing this motif and, thus, provide likely clues to the function of the domain. Identification of specific domains or motifs provides strong evidence for sequence homology, even when the similarity of the remaining sequence is very low. Indeed, certain types of protein homologies can be identified by the conserved location of only a few amino acid residues.

B. Programs to Determine Structural Homology

Five main types of analysis have been performed most often to evaluate the conservation of structural features. The first is a prediction program, such as the Chou and Fasman (1978) method, to identify secondary structural elements that may be conserved. Such analysis, however, generally results in only about 50% success in predicting the placement and type of specific secondary structure. This type of analysis is generally best applied when comparing already suspected domain homologies for characteristic strong secondary features.

The second analysis is of the hydropathy pattern of the protein, using an algorithm such as the Kyte–Doolittle (1982) method. This analysis seeks to determine the parts of the protein that are shielded from the aqueous environment and, thus, are likely to reside within the globular protein or in the lipid bilayer. Hydropathy analysis can show whether two proteins have similar overall shapes or topologies.

The third analysis is a pattern analysis that seeks conserved relationships between amino acid residues that are correlated with the presence of particular motifs. The most common strategy used is a search with the sequence of interest against a database of known sequence motifs using PROSITE (Bairoch, 1989). These pattern searches are very useful in highlighting important structural features that are otherwise difficult to identify.

A fourth type of analysis begins with a database search using the sequence of a suspected domain, identified by dot matrix analysis or other techniques, to identify possible homologous domains in other proteins. If other proteins share domains with significant similarity to the protein of interest, then the functional importance of the suspected domain can often be discerned. In addition, if crystallographic information is available for any proteins of suspected homology, an analysis can be made of the ability of the domain of interest to fold into a similar structure.

Finally, an important analysis involves comparing the topological organization of the protein to that of other proteins of suspected homology. Often such comparisons reveal a characteristic topological relationship of sequence motifs among proteins that otherwise are not suspected homologs. This type of analysis is particularly important in establishing gene superfamily organization, as in the ABC family (Jan and Jan, 1992).

C. Defining Homology by Structural Similarity

Many analyses of structural similarity are best suited for establishing the homology of distantly related proteins that are difficult to compare at the level of the amino acid sequence. An excellent example of this type of homology comparison is between cyclic-nucleotide-gated channels and K^+ channels (Jan and Jan, 1990). However, information about specific domains is also invaluable in determining specific residue changes that are critical in the evolution of proteins into different types, or subfamilies, of genes. The presence of a specific domain type can be critical to inclusion of a protein into the homologous class. For example, the presence of the homeobox domain uniquely classifies otherwise weakly related proteins into the family of homeobox-containing proteins (reviewed by McGinnis and Krumlauf, 1992).

Because domains are often swapped between proteins, distinguishing whether small domains with similar functions result from sequence conservation, convergent evolution, or transfer between proteins by mechanisms such as exon shuffling is difficult. An interesting example of this problem concerns the inactivation mechanisms of voltage-gated K^+ channels. Several *Shaker* type channels, all known *Shal* channels, and the mammalian Kv3.4 (Raw3) *Shaw* type channel (see Chapter 4) show rapid inactivation (Tempel *et al.*,

1987; Stuhmer *et al.*, 1989; Wei *et al.*, 1990; Pfaffinger *et al.*, 1991; Schroter *et al.*, 1991). *Shaker* type K^+ channels and Kv3.4 have an inactivation particle present on the N terminus that functions as an inactivation ''ball,'' connected to the channel by a ''chain'' whose length controls the rate of inactivation (Hoshi *et al.*, 1990; Rettig *et al.*, 1992). However, *Shal* type channels inactivate rapidly without an N-terminal inactivation particle (Baldwin *et al.*, 1991). This result seems to indicate that two independent mechanisms for inactivation have been conserved in different K^+ channel subfamilies. The complication in this analysis is that not all *Shaker* channels have an N-terminal inactivation ball, and that *Shaw* type channels from invertebrates do not have N-terminal inactivation balls (see Fig. 2). Sequence comparison of N-terminal regions reveals that only very general properties, rather than specific amino acid sequences, are conserved. Therefore, the inactivation particle may not be evolutionarily conserved but may be a product of convergent evolution. In fact, the true conserved feature in the channels may be the receptor for the inactivation ball, which appears to be an integral part of the permeation pathway (Hoshi *et al.*, 1990). Additional experiments are needed to better understand the similarities and differences of the inactivation mechanisms of the different ion channels.

D. K^+ Channel Homology Domains to Define Subfamilies

Given this background of modern molecular genetics, how did K^+ channel genes evolve, what are the fundamental units of K^+ channel genes, how are these units deployed, and what other nonfundamental units are being added to and subtracted from these genes to produce the genetic diversity found in modern animals?

Analyses of K^+ channel sequences have determined several distinguishing structural elements that are present in K^+ channels (see Fig. 3). The domains that have been identified to date are inactivation ''ball and chain''; the T1 assembly domain; the transmembrane domains S1, S2, S3, S5, and S6; the S4 voltage sensor; and the P region. Although not all these domains have been identified in all K^+ channels to date, and certainly some are absent from some channels, the order of these domains is conserved among all channels studied to date. Some of these elements are also present in distantly related ion channels. This conservation suggests that these structures play a critical role in the function of a K^+ channel. This section describes how these structures are identified and how they are distinguishing features of K^+ channels. The

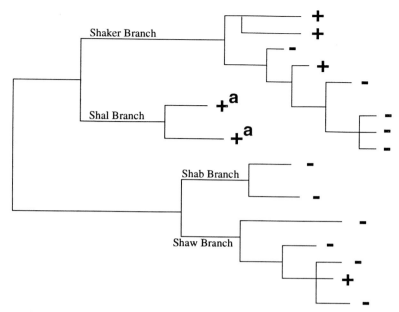

FIGURE 2 A K⁺ channel family tree shows that N-terminal fast inactivation appears to have evolved more than once in K⁺ channels. The N-terminal inactivation mechanism in a *Shaw* clone cannot be connected to the *Shaker* N-terminal inactivators by any common N-terminal inactivating line. In addition, the split between the *Shaker* and *Shal* branches divides N-terminal and non-N-terminal fast inactivation mechanisms. +, Fast inactivation, N-terminal; +ᵃ, fast inactivation, not N-terminal; −, no fast inactivation. Tree structure based on Strong *et al.* (1993).

"ball and chain" inactivation region was described already.

1. T1 Domain

K⁺ channels are assembled from individual subunit proteins that aggregate to form a functional tetramer. Several experiments now suggest that this assembly is driven and controlled by sequences found in an N-terminal domain called the T1 domain. Peptides produced in bacteria that contain the T1 domain show self-binding activity in filter blot assays. Switching the T1 domain from K⁺ channels of one subfamily with that of another produces novel functional channels after coexpression in *Xenopus* oocytes (Li *et al.*, 1992). Other work shows, biochemically, that subunit protein assembly is driven by the T1 domain and that

the T1 domain is required for assembly (Pfaffinger and Chen, 1992; Shen *et al.*, 1993). Assembly specificity was seen by showing that combinations of *Shaker* and *Shaw* type proteins assemble into nonheteromultimerizing tetramers when expressed together, suggesting that the T1 domain contains some essential sequences for controlling the heteromultimerization of subunit proteins and, possibly, for maintaining the subfamily-specific gene organization of K⁺ channels proteins.

2. Hydropathy Plot

K⁺ channels are membrane proteins. To conduct ions across the membrane, some part of the protein must be a transmembrane domain. To predict these transmembrane domains, a hydrophobicity scale is used, the most common of which is the Kyte and Doolittle (1982) scale. Each amino acid is assigned a hydrophobicity number. Since transmembrane domains are often α helices, and at least 19 amino acids are needed to span the membrane as an α helix, a "window" of 19 amino acids is often used to identify possible transmembrane domains from the primary sequence. A hydropathy index, which is the mean value of the amino acids within the window, is calculated. The more positive the number, the more hydrophobic the region. The window is then moved

B C T1 S1 S2 S3 S4 S5 P S6

FIGURE 3 K⁺ channel domain organization. The organization of identified K⁺ channel functional domains is shown. B, Inactivation ball; C, chain region; T1, tetramerization domain 1; S1, S2, S3, S5, S6, hydrophobic segments hypothesized to be transmembrane α-helices; S4, amphipathic helix hypothesized to be the voltage sensor; P, novel hydrophobic segment hypothesized to be the pore-lining segment. S1–S6 and P domains have been identified in all K⁺ channels. Ball, chain, and T1 domains have only been identified in a few K⁺ channels to date.

progressively along the sequence; the indices are plotted against residue number. Transmembrane α helical domains are generally considered to be present when stretches of at least 19 amino acids with a high hydropathy index are identified. Additional evidence is obtained if the region shows a strong tendency to form an α helix by Chou–Fasman analysis.

When hydropathy plots are carried out on K⁺ channel subunit sequences, a distinct central region of multiple hydrophobic peaks is seen (see Fig. 4). In every K⁺ channel studied to date, six large hydrophobic regions containing the requisite number of amino acids for a transmembrane α helix are found. These domains have been named S1–S6 and are the main candidates for the transmembrane domains. In addition, a smaller hydrophobic region that is too small to constitute a classical transmembrane α helix, called the P domain, is found between the fifth and sixth putative transmembrane domains. All K⁺ channel sequences that can be classified into the S4 gene superfamily have these six putative transmembrane domains and one P region.

Other voltage-gated channels, the Na⁺ and Ca²⁺ channels, contain four repeats of this hydrophobicity pattern. This result provides some evidence for structural homology among these voltage-gated channels. This homology implies that functional K⁺ channels are tetramers. Supporting this hypothesis are experiments performed by MacKinnon (1991) and Pfaffinger and Chen (1992), in which K⁺ channels are shown to assemble with an N of 4.

3. S4 Domain

The S4 transmembrane domain has been identified as a unique structural element and is the most conserved of the putative transmembrane α helices among a wide variety of ion channel proteins. S4 is an amphipathic helix in which every third or fourth residue is positively charged. This amphipathic helix is proposed to be part of the voltage-sensing apparatus of the channel that mediates channel activation (see Chapter 3). This hypothesis has been supported by mutagenesis of charged residues in the S4 regions of Na⁺ and K⁺ channels (Stuhmer et al., 1989; Papazian et al., 1991). These results are not clear cut, however, suggesting that other motifs within the K⁺ channel protein may also contribute to voltage-dependent activation.

The S4 motif has also been found in the cGMP-dependent channel of photoreceptors and in other cyclic-nucleotide-gated ion channels (Jan and Jan, 1990). These channels are known not to be voltage gated. What other roles could the S4 domain play? Jan and Jan (1990) suggest that the S4 domain may be required for close packing of the protein core of this channel. This suggestion has led to the proposal that S4 is a key domain for a variety of channel proteins. Thus, these proteins are classified as members of the S4 superfamily of ion channel proteins.

Although S4 is a very conserved transmembrane α helix, the charge motif has not been conserved as completely in the *Shaw* subfamily of K⁺ channels (see Chapter 4). Among *Shaker*, *Shal*, and *Shab*, the charged residues in S4 are nearly identical in number and configuration, but this is not the case in *Shaw* type K⁺ channels. Across species, the number of charges ranges from six in some mammalian *Shaw* channels to two in *Caenorhaloditis elegans* (reviewed by Salkoff et al., 1992). A small number of charges correlates with *Shaw* K⁺ channels that have a shallower I–V relationship, showing less voltage-dependent activation and slower activation rates than *Shaker* type K⁺ channels. On the other hand, the mammalian RKv3.4 (Raw 3) protein contains all six charged amino acid residues and shows strong and fast voltage-dependent activation (Schroter et al., 1991).

4. P Domain

The P domain, found in the linker region between S5 and S6, is the most conserved domain among K⁺ channel clones. Various experiments point to the P domain as constituting a critical region of the transmembrane ion channel (see Chapter 2) (Hartmann et al., 1991; Kavanaugh et al., 1991; Yellen et al., 1991). The core of the P domain is a 19-amino-acid stretch of conserved amino acids, generally numbered 1–19, independent of the actual clone sequence numbering (Fig. 5). Within this sequence are a number of virtually invariant amino acids. A TMTT motif (9–12) is found in the middle and, near the C terminus of this region, a GYGN sequence (14–17) is found. Tyrosine 15 is found in every K⁺ channel P region sequence except *eag* (Warmke et al., 1991), and has been hypothesized

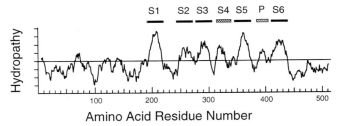

FIGURE 4 The hydropathy profile for K⁺ channels shows a characteristic pattern of hydrophobic segments. This Kyte–Doolittle profile for the *Shaker* type clone AKv1.1a (Pfaffinger et al., 1990) shows S1–S6 and the P domain as peaks of hydrophobicity.

K⁺ channel P Domain Evolution

1	2	3	4	5	6	7	8	9	10	11	12	13	14	15	16	17	18	19
T	A	F	W	W	A	I	V	T	M	T	T	V	G	Y	G	D	M	Y
D[1]	G[3]	L	Y	Y[4]	T	V	I[2]	S	L	S	S	L		F		N	I[2]	V
A	S	V		F	L	L	T	C									V	H
V	C	M			S	G											L	A
L						T												T
I						M												R
E																		K

FIGURE 5 P domain sequence conservation is seen in the alignment of all published P domain sequences from S4 type K⁺ channels. A segment of 19 amino acids can be aligned. The upper sequence is the highest consensus sequence. All bold residues are found in more than one K⁺ channel subfamily. Amino acid residues 9–17 are particularly conserved; only number 13 has an alternative residue that is found in more than one subfamily. In rare cases, a particular amino acid residue is found only in all members of a single subfamily; these residues are footnoted as (1) in all *Shaker* type channels, (2) in all *Shab* type channels, (3) in all *Shaw* type channels, and (4) in all *Shal* type channels.

to form the narrowest portion of the pore. Residues 1 and 19 are thought to be located at the external mouth of the pore and show more amino acid variability than other positions in the P domain. These and other residues outside this 19-amino-acid motif appear to play an important role in the pharmacology of K⁺ channels. Within the 19-amino-acid stretch is a threonine (position 11) that seems to be important for internal tetraethylammonium (TEA) binding.

The structure of the P region is believed to be conserved among ion channels of the S4 superfamily. To form the entire pore region, this sequence would need to span the membrane twice to preserve the cytoplasmic orientation of the N- and C-terminal ends of the subunit proteins. However, the number of amino acids is not even sufficient to span the membrane once as an α helix. Several groups of investigators have produced models that predict ways in which this P region might fold to span all or part of the membrane twice; models induce α/β barrel structures, β sheets, or an extended conformation (Guy and Conti, 1990; Bogusz and Busath, 1992; Durrell and Guy, 1992).

E. Unique Homology Domains That Define Subfamilies

The previous sections detail how evolutionary conservation patterns result in the variety of molecularly distinct but related ion channels. However, another evolutionary process leading to diversity is evident that involves the addition of novel features to ion channels. The clearest examples of the addition of novel domains are the putative cyclic-nucleotide-binding region of the *eag* K⁺ channel (Warmke *et al.*, 1991) and the putative Ca²⁺ regulatory region of *slo* K⁺ channels (Atkinson *et al.*, 1991). The presence of these domains provides a new level of categorization that is independent of primary sequence homology in the remainder of the sequence.

IV. Functional Homology

The oldest homology analysis between proteins is functional homology: proteins that perform similar chores are hypothesized to be related structurally. This idea is often embellished by more exacting biochemical and pharmacological analyses of the relationship. However, all functional comparisons on their own suffer from the difficulty of detecting convergent evolution, that is, certain proteins can be modified from very different precursors to conduct the same type of activity. A possible example of this occurred with the cloning of the min-K⁺ gene family. This class of single transmembrane proteins appears to perform the function of passing ions across the membrane much as voltage-gated K⁺ channels do. However, no homology between these two proteins is apparent (Takumi *et al.*, 1988). Functional homology comparisons also fail to identify the unexpected relationships between different genes that become so apparent when sequence comparisons are made. One such example is the evidence that cyclic-nucleotide-gated channels in photoreceptors and olfactory epithelium are homologous of the voltage-gated ion channels (Jan and Jan, 1990).

A. Defining Functional Homologies for K⁺ Channels

Classical functional analyses rely on testing proteins derived from their native expression environment. This approach is limited to the often difficult problem of determining the exact subunit composition of the protein of study. In addition, the protein is often only obtainable in extremely limited quantities and is difficult or impossible to manipulate in interesting ways. Much success has been obtained by expressing cloned genes in heterologous expression systems. In this way, the subunit composition of the protein can be known more precisely, the protein can be obtained in larger quantities, and manipulation of the DNA sequence allows a variety of novel and interesting functional tests. A particularly powerful strategy has been the testing of chimeras, in which a presumed homologous domain in one protein is substituted for natural sequence. This ability allows a direct test of the homologous nature of these sequences and a determination of how the domain has changed in the two proteins.

Structure and function are related; therefore, knowledge about protein function can reveal much about protein structure. Certainly many of the homology arguments made by a functional analysis of voltage-gated ion channels have been upheld by sequence analysis. Indeed, the relationships among voltage-gated K⁺ channels, Na⁺ channels, and Ca²⁺ channels were predicted long before sequence information was available. Although these homologies were predicted, the exact nature of these relationships was a complete surprise that awaited the cloning of the principal genes.

Functional analyses have played important roles in understanding the relationships between K⁺ channel proteins. These experiments have been motivated by an interest in relating specific K⁺ channel clones to precloning classifications of ion channels, and to understanding the functional similarities and differences among the cloned K⁺ channels (Hille, 1992). Whole-cell K⁺ currents, single-channel conductances, and pharmacological properties are some of the characteristics that have been studied. In addition, experiments have tested the ability of different K⁺ channel proteins subunits to heteromultimerize in forming ion channels.

V. Integrating Homology Analyses

A real strength in obtaining true homology relationships and in using these relationships to better understand ion channels lies in the ability to combine results from different homology analyses and to begin to integrate these results into a complete picture of the relationships among these ion channel proteins.

A. Integrating Sequence and Structural Homology Analyses

Often, determination of sequence homologies requires information about the parts of the sequence that should be compared. This type of analysis was required to compare cyclic-nucleotide-gated channels with K⁺ channels (Jan and Jan, 1990). Hydropathy analysis revealed that these proteins had similar hydropathy patterns, indicative of a similar folding pattern and a possible homologous relationship. This hydropathy analysis revealed the probable location of homologous transmembrane segments. Comparison of each of these putative transmembrane segments revealed that these proteins were, in fact, homologous. This analysis was not possible by low stringency alone, since the sequences of the entire proteins diverged too much, but by focusing on the most conserved regions—the transmembrane domains—the relationship was revealed. This interplay between a global structural analysis, such as a hydropathy plot, and more specific motif or landmark analysis is the basis for grouping of proteins into families, such as the S4 channels, the ionotrophic receptors, the metabotrophic receptors, and the ABC proteins.

B. Integrating Protein Analyses and Functional Analyses

The hypothesis that channel types homologous at a level lower than 50%, and thus classified as members of different subfamilies, were forming independent channel types was tested by examining the ability of different channel subunit proteins to form functional heteromultimeric channels. Several groups independently showed that different *Shaker* type K⁺ channels can heteromultimerize to form channels with unique properties (Christie *et al.*, 1990; Isacoff *et al.*, 1990; Ruppersberg *et al.*, 1990). Covarrubias *et al.* (1991) showed that K⁺ channel subunits from different subfamilies do not appear to heteromultimerize functionally. When two K⁺ channel genes from different subfamilies are coexpressed in oocytes, they find two distinct channel populations. Thus, heteromultimerization of K⁺ channel subunit proteins serves as an independent means for classifying channel genes into the appropriate subfamilies.

The integration of structural and functional analyses in the establishment of protein homologies often involves the functional analysis of mutagenesis experiments. Two basic mutagenesis strategies are often used: (1) point mutagenesis of specific highly conserved amino acid residues, and (2) construction of chimeras by artificially swapping domains of suspected homology.

These integrative approaches have provided the experimental evidence that the P region lines or is near the pore of the K⁺ channel. In one set of experiments, the effects of blockade by the K⁺ channel blockers charybdotoxin and TEA on K⁺ currents were changed by single point mutations in the sequence between S5 and S6. Point mutations in this P region also changed the single-channel conductance properties of the channel (reviewed by Pongs, 1992). By far the strongest evidence to date is from studies on MKv3.1 and RKv2.1 K⁺ channels that are chimeric in this region (Hartmann *et al.*, 1991). The MKv3.1 has a higher single-channel conductance than the RKv2.1 K⁺ channel, of which Hartmann *et al.* (1991) took advantage in their chimeric K⁺ channel pore studies. The putative pore region from one K⁺ channel was cloned into another K⁺ channel with very different conductance and TEA binding properties. These investigators showed that they could alter the K⁺ channel conductance from that of the wild-type to that of the donor K⁺ channel by cloning in the donor P region.

VI. Conclusions

An overall analysis of the results from homology comparisons of ion channel proteins suggests that an ancestral gene encoding a protein similar to modern K⁺ channels diverged into the large number of different types of ion channels that exist currently. This ancestral protein contains an S4 domain and a P region similar to those of modern ion channels. These features define this entire class of proteins as members of the S4 superfamily. The ancestral protein diverged in two directions by intragenic duplications and intergenic duplications. Intragenic duplication produced the ancestor of the Ca^{2+} and Na^+ channels. Intergenic duplication produced the K⁺ channels, the cyclic-nucleotide-gated channels, the Ca^{2+}-gated K⁺ channels, and others. These gene duplications form the basis for the different families of modern ion channel proteins. These homologies are best seen as structural homologies among proteins with similar hydrophobicity patterns and similar topological organization of domains such as S4 and the P region. Within voltage-gated K⁺ channels, a series of additional gene duplications produced genes encoding

proteins with easily discernible primary sequence similarity. In the invertebrates, these genes produce multiple transcripts capable of producing unique protein products. Functional analyses suggest that these invertebrate genes are independent channel systems and, thus, should be considered members of different subfamilies. The process of gene duplication is clearly evident in the switch from invertebrates to vertebrates and, eventually, mammals, in which the different subfamily genes have been duplicated, producing additional diversity. Now, however, the genes of a single subfamily are capable of heteromultimerizing and, thus, of forming a single ion channel system. At all levels of diversification, up to the subfamily gene duplications, the ion channel genes have evolved to form independent channel systems. Since a mechanism of diversification at the subfamily level was likely to be present even in ancient invertebrates, for example, alternative splicing, diversification at the subfamily level may be critical to the formation of native K⁺ channels with the needed properties. However, equally plausible is the idea that genes in the same subfamily are in the process of diverging into independent channel systems. Identification of channels from the same subfamily that cannot heteromultimerize would strongly support the latter hypothesis.

References

Altschul, S. F., Gish, W., Miller, W., Myers, E. W., and Lipman, D. J. (1990). Basic local alignment search tool. *J. Mol. Biol.* **215,** 403–410.

Atkinson, N. S., Robertson, G. A., and Ganetzky, B. (1991). A component of calcium-activated potassium channels encoded by the *Drosophila slo* locus. *Science* **253,** 551–555.

Bairoch, A. (1989). PROSITE: A dictionary of protein sites and patterns. *In* "EMBL Biocomputing Technical Document 4". EMBL, Heidelberg, Germany.

Baldwin, T. J., Tsaur, M.-L., Lopez, G. A., Jan, Y. N., and Jan, L. Y. (1991). Characterization of a mammalian cDNA for an inactivating voltage-sensitive K⁺ channel. *Neuron* **7,** 471–483.

Bogusz, S., and Busath, D. (1992). Is a β-barrel model of the K⁺ channel energetically feasible? *Biophys. J.* **62,** 19–21.

Butler, A., Wei, A., Baker, K., and Salkoff, L. (1989). A family of putative potassium channel genes in *Drosophila*. *Science* **243,** 943–947.

Chandy, K. G. (1991). Simplified gene nomenclature. *Nature (London)* **352,** 26.

Chou, P. Y., and Fasman, G. D. (1978). Predictions of the secondary structure of proteins from their amino acid sequence. *Adv. Enzymol.* **47,** 45–147.

Christie, M. J., North, R. A., Osborne, P. B., Douglass, J., and Adelman, J. P. (1990). Heteropolymeric potassium channels expressed in *Xenopus* oocytes from cloned subunits. *Neuron* **4,** 405–411.

Covarrubias, M., Wei, A., and Salkoff, L. (1991). *Shaker, Shal, Shab,* and *Shaw* express independent K⁺ current systems. *Neuron* **7,** 763–773.

Craik, C. S., Rutter, W. J., and Fletterick, R. (1983). Splice junctions:

Association with variations in protein structure. *Science* **220**, 1125–1129.

Durell, S. R., and Guy, H. R. (1992). Atomic scale structure and functional models for voltage-gated potassium channels. *Biophys. J.* **62**, 238–250.

Gribskov, M., and Devereux, J. (1991). *"Sequence Analysis Primer."* Stockton Press, New York.

Guy, H. R., and Conti, F. (1990). Pursuing the structure and function of voltage-gated channels. *Trends Neurosci.* **13**, 201–206.

Hartmann, H. A., Kirsch, G. E., Drewe, J. A., Taglialatela, M., Joho, R. H., and Brown, A. M. (1991). Exchange of conduction pathway between two related K$^+$ channels. *Science* **251**, 942–944.

Hille, B. (1992). *"Ionic Channels of Excitable Membranes,"* 2d Ed. Sinauer Associates, Sunderland, Massachussetts.

Hoshi, T., Zagotta, W. N., and Aldrich, R. W. (1990). Biophysical and molecular mechanisms of *Shaker* potassium channel inactivation. *Science* **250**, 533–538.

Isacoff, E. Y., Jan, Y. N., and Jan, L. Y. (1990). Evidence for the formation of heteromultimeric potassium channels in *Xenopus* oocytes. *Nature (London)* **345**, 530–534.

Jan, L. Y., and Jan, Y. N. (1990). A superfamily of ion channels. *Nature (London)* **345**, 672.

Jan, L. Y., and Jan, Y. N. (1992). Tracing the roots of ion channels. *Cell* **69**, 715–718.

Kamb, A., Iverson, L. E., and Tanouye, M. A. (1987). Molecular characterization of *Shaker*, a *Drosophila* gene that encodes a potassium channel. *Cell* **50**, 405–415.

Kavanaugh, M. P., Varnum, M. D., Osborne, P. B., Christie, M. J., Busch, A. E., Adelman, J. P., and North, R. A. (1991). Interaction between tetraethylammonium and amino acid residues in the pore of cloned voltage-gated potassium channels. *J. Biol. Chem.* **266**, 7583–7587.

Kyte, J., and Doolittle, R. F. (1982). A simple method for displaying the hydropathic character of a protein. *J. Mol. Biol.* **157**, 105–132.

Li, M., Jan, Y. N., and Jan, L. Y. (1992). Specification of subunit assembly by the hydrophilic amino-terminal domain of the *Shaker* potassium channel. *Science* **257**, 1225–1230.

MacKinnon, R. (1991). Determination of the subunit stoichiometry of a voltage-activated potassium channel. *Nature (London)* **350**, 232–235.

McGinnis, W., and Krumlauf, R. (1992). Homeobox genes and axial patterning. *Cell* **68**, 283–302.

Papazian, D. M., Timpe, L. C., Jan, Y. N., and Jan, L. Y. (1991). Alteration of voltage-dependence of *Shaker* potassium channel by mutations in the S4 sequence. *Nature (London)* **349**, 305–310.

Patterson, C. (1988). Homology in classical and molecular biology. *Mol. Biol. Evol.* **56**, 603–625.

Pearson, W. R. (1990). Rapid and sensitive sequence comparison with FASTP and FASTA. *Meth. Enzymol.* **183**, 63–98.

Pfaffinger, P. J., and Chen, X. (1992). Biochemical analysis of potassium channel proteins synthesized by *in vitro* translation. *Biophys. J.* **61**, A256 (Abstract).

Pfaffinger, P. J., Furukawa, Y., Zhao, B., Dugan, D., and Kandel,

E. R. (1991). Cloning and expression of an *Aplysia* K$^+$ channel and comparison with native *Aplysia* K$^+$ currents. *J. Neurosci.* **11**, 918–927.

Pongs, O. (1992). Structural basis of voltage-gated K$^+$ channel pharmacology. *Trends Pharmacol. Sci.* **13**, 359–365.

Pongs, O., Kecskemethy, N., Muller, R., Krah-Jentgens, I., Baumann, A., Kiltz, H. H., Canal, I., Llamazares, S., and Ferrus, A. (1988). *Shaker* encodes a family of putative potassium channel proteins in the nervous system of *Drosophila*. *EMBO J.* **7**, 1087–1096.

Rettig, J., Wunder, F., Stocker, M., Lichtinghagen, R., Mastiaux, F., Beckh, S., Kues, W., Pedarzani, P., Schroter, K. H., Ruppersberg, J. P., Veh, R., and Pongs, O. (1992). Characterization of a *Shaw*-related potassium channel family in rat brain. *EMBO J.* **11**, 2473–2486.

Ruppersberg, J. P., Schroter, K. H., Sakmann, B., Stocker, M., Sewing, S., and Pongs, O. (1990). Heteromultimeric channels formed by rat brain potassium channel proteins. *Nature (London)* **345**, 535–537.

Salkoff, L., Baker, A., Butler, A., Covarrubias, M., Pak, M. D., and Wei, A. (1992). An essential "set" of K$^+$ channels conserved inflies, mice, and humans. *Trends Neurosci.* **15**, 161–166.

Schroter, K. H., Ruppersberg, J. P., Wunder, F., Rettig, J., Stocker, M., and Pongs, O. (1991). Cloning and functional expression of a TEA-sensitive A-type potassium channel from rat brain. *FEBS Lett.* **278**, 211–216.

Shen, N. V., Chen, X., Boyer, M. M., and Pfaffinger, P. J. (1993). Deletion analysis of K$^+$ channel assembly. *Neuron.* **11**, 67–76.

Strong, M., Chandy, K. G., and Gutman, G. A. (1993). Molecular evolution of voltage-sensitive ion channel genes: On the origins of electrical excitability. *Mol. Biol. Evol.* **10**, 221–242.

Stuhmer, W., Conti, F., Suzuki, H., Wang, X. D., Noda, M., Yahazi, N., Kubo, H., and Numa, S. (1989). Structural parts involved in activation and inactivation of the sodium channel. *Nature (London)* **339**, 597–603.

Takumi, T., Ohkubo, H., and Nakanishi, S. (1988). Cloning of a membrane protein that induces a slow voltage-gated potassium channel. *Science* **242**, 1042–1045.

Tempel, B. L., Papazian, D. M., Schwarz, T. L., Jan, Y. N., and Jan, L. Y. (1987). Sequence of a probable potassium channel component encoded at the *Shaker* locus of *Drosophila*. *Science* **237**, 770–775.

Warmke, J., Drysdale, R., and Ganetzky, B. (1991). A distinct potassium channel polypeptide encoded by the *Drosophila eag* logus. *Science* **252**, 1560–1562.

Wei, A., Covarrubias, M., Butler, A., Baker, K., Pak, M., and Salkoff, L. (1990). K$^+$ current diversity is produced by an extended gene family conserved in *Drosophila* and mouse. *Science* **248**, 599–603.

Yellen, G., Jurman, M. E., Abramson, T., and MacKinnon, R. (1991). Mutations affecting internal TEA blockade identify the probable pore-forming region of a K$^+$ channel. *Science* **251**, 939–942.

Permeation Properties of Cloned K⁺ Channels

TED BEGENISICH

I. Introduction

A. Ion Permeation

Transporting an ion across the lipid bilayer of cell membranes is not a trivial undertaking. The low dielectric constant of the hydrocarbon chains of membrane lipids represents an energy barrier to ion movement of ~40 kcal/mol (Parsegian, 1969). Thus, unaided, only one ion in 10^{14} has enough energy to pass through the bilayer. Two types of membrane proteins have evolved to catalyze ion permeation: "pumps" and "channels." Pumps provide relatively small ion flux but exhibit extremely high ion discrimination. The aqueous pores of ion channel proteins can deliver a very large ion flux but must sacrifice some selectivity.

A major thrust of modern biophysical research is to acquire an understanding at the molecular level of the process of ion permeation. One of the best studied permeation systems is the pore in voltage-gated potassium channels. This chapter summarizes some of these studies and interprets these results in terms of molecular and mathematical models of the pore structure.

A first step toward a molecular understanding of ion movement through channel pores is the gathering of permeation data. A catalog of the permeation properties of an ion channel pore includes (1) the dependence of ion current on ion concentration and transmembrane voltage; (2) selectivity, that is, the ability of the pore to discriminate among different ions; and (3) structural information obtained by probing the pore with blocking ions that occlude the pore and inhibit the flow of permeant ions. Determining the level of complexity of the permeation process, as reflected in the number of ions that can occupy the pore simultaneously, is also important. Thus, some pores may be almost always empty, others may be occupied by no more than one ion, and still others may allow simultaneous multi-ion occupancy. Various types of experiments provide data that can discriminate among these various pore types; these studies have been discussed several times (Hille and Schwarz, 1978,1979; Begenisich, 1982,1987).

This chapter reviews the available data on ion permeation in cloned K⁺ channels, with special emphasis on experiments employing site-specific mutagenesis. From these data, a picture of the channel pore is composed. Finally, a mathematical model for permeation is developed and computations from the model are compared with the experimental data. This model is not intended to be the end of the study of permeation in these channels; instead, it is intended as a tool for organizing the available data and for suggesting new

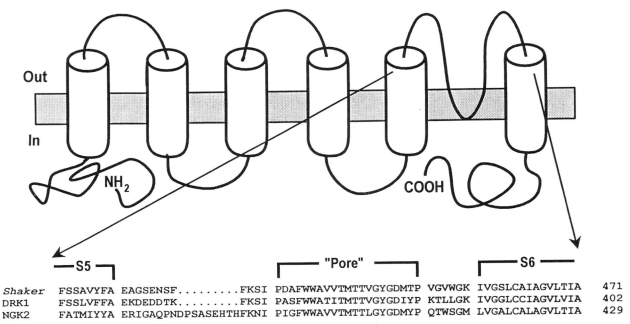

	S5		"Pore"		S6	
Shaker	FSSAVYFA EAGSENSF.........FKSI	PDAFWWAVVTMTTVGYGDMTP	VGVWGK	IVGSLCAIAGVLTIA	471
DRK1	FSSLVFFA EKDEDDTK.........FKSI	PASFWWATITMTTVGYGDIYP	KTLLGK	IVGGLCCIAGVLVIA	402
NGK2	FATMIYYA ERIGAQPNDPSASEHTHFKNI	PIGFWWAVVTMTTLGYGDMYP	QTWSGM	LVGALCALAGVLTIA	429

FIGURE 1 Putative folding of *Shaker* K$^+$ channel protein (after Hartmann *et al.*, 1991). In the sequence alignments in the S5–S6 linker, dots have been added to *Shaker* and DRK1 to match the longer NGK2 segment. Numbers at the right apply to the last residue in the aligned sequences.

experiments to further illuminate the ion permeation process.

B. Cloned Channels

Beginning with the *Drosophila Shaker* gene, an unreasonably large number of K$^+$ channels has been cloned, including many mammalian homologs of the *Drosophila* channels (reviewed by Jan and Jan, 1992). The voltage-gated types of these channels can be grouped into four large families named for the *Drosophila* genes *Shaker*, *Shab*, *Shaw*, and *Shal*. Functional channels encoded by these cDNAs can be expressed in a variety of *in vitro* systems and appear to form homotetramers (MacKinnon, 1991; Liman *et al.*, 1992). The amino acid sequences of these channels suggest six membrane-spanning domains (S1–S6; Fig. 1). The linker region connecting S5 and S6 is highly conserved among all these channels.

An alignment of the S5–S6 linker sequence of representatives of three of the K$^+$ channel classes is illustrated in Fig. 1. Shown in this figure are *Shaker* H4[1] (Kamb *et al.*, 1988), DRK1 (Frech *et al.*, 1989) from the *Shab* subfamily (also known as Kv2.1; Chandy, 1991),

and NGK2 (Yokoyama *et al.*, 1989; also known as Kv3.1) from the *Shaw* subfamily. One 21-amino-acid segment in the S5–S6 linker region (*Shaker* P430 to P450; marked as "Pore" in Fig. 1) is extremely highly conserved and, as will be discussed subsequently, constitutes part of the K$^+$ channel pore (MacKinnon and Yellen, 1990; Hartmann *et al.*, 1991; Yellen *et al.*, 1991; Yool and Schwarz, 1991).

Until direct evidence is available for the three-dimensional structure of K$^+$ channels, information on the structural and functional organization of the ion permeation pathway will have to be obtained by more indirect methods. One approach is finding the residues that, when mutated, alter such permeation properties as pore block, current–voltage relationship, and pore selectivity. A common concern in mutagenesis is the possibility of inducing long-range structural changes. Consequently, this chapter concentrates as much as possible on mutations with effects that are localized, at least to the extent that they do not produce changes in channel gating properties. Table I summarizes site-specific mutations in the pore region of *Shaker* channels that have been shown to alter some pore property without significant alteration in channel gating. Mutations in related cloned channels, as well as *Shaker* mutations that alter gating but nevertheless may have implications for pore structure, are discussed in the text.

---

[1] *Shaker* H4 is essentially identical to one of the *Shaker* splice variants called *Shaker* B (Timpe *et al.*, 1988).

II. Probing the Structure of K$^+$ Channel Pores

A. Charybdotoxin

Scorpion venoms contain many proteins that interact with ion channels. One of these is a small (37 amino acids) protein from *Leiurus quinquestriatus* called charybdotoxin (CTX). CTX (and its many isoforms), applied extracellularly, blocks K$^+$ channels with a very high affinity (Miller *et al.*, 1985; Gimenez-Gallego *et al.*, 1989; Oliva *et al.*, 1991). Considerable evidence exists that this toxin blocks Ca^{2+}-activated K$^+$ channels by physically occluding the pore. The strongest such evidence is that *external* CTX block is antagonized by increasing the *internal* concentration of permeant ions K$^+$ and Rb$^+$ but not by increasing the concentration of impermeant ions (MacKinnon and Miller, 1988). External tetraethylammonium (TEA) ions compete with CTX for blocking both Ca^{2+}-activated (Miller, 1988) and *Shaker* (unpublished observations in MacKinnon *et al.*, 1990) K$^+$ channel pores. Therefore, TEA and CTX are both useful molecules for probing the pore of cloned K$^+$ channels.

Site-specific mutagenesis (see Table I) of residues in the *Shaker* S5–S6 linker region have identified four "hot spots" for CTX binding, two near each end of the linker region, at positions 425, 431, 449, and 451. Mutations of these hot spots inhibit CTX binding by factors of 100 to >2,000. Two other residues near the N terminus of the linker (Glu 422 and Lys 427) have small effects on CTX affinity, probably through an electrostatic mechanism. At the other end of the linker is another site that also has only a small effect on the CTX–channel interaction (Gly 452).

TABLE I **Mutations near *Shaker* Pore That Express Ionic Currents and Have No Gating Effects**

Mutation	Effects[a]	Reference
E422D,Q,K	Electrostatic (small) effect on CTX block	MacKinnon and Miller (1989)
E422K	No significant effect on TEA$_o$ block or on K$^+$ current–voltage relationship	MacKinnon and Yellen (1990)
F425G	"Hot spot" for CTX	Goldstein and Miller (1992)
K427R,N,E	Electrostatic effect on CTX block	MacKinnon *et al.* (1990)
K427E	No significant effect on TEA$_o$ block or on K$^+$ current–voltage relationship	MacKinnon and Yellen (1990)
D431E,N,K	"Hot spot" for CTX	MacKinnon *et al.* (1990)
D431E,N,K	Small (electrostatic?) effect on TEA$_o$ block; (N) produces no significant modification of K$^+$ current–voltage relationship; current–voltage relationship with (K) not done	MacKinnon and Yellen (1990)
F433Y,S	Alters NH$_4^+$/K$^+$ and (much smaller) Rb$^+$/K$^+$ permeabilities; increases inward NH$_4^+$ and Rb$^+$ currents; no significant effect on K$^+$ current–voltage relationship	Yool and Schwarz (1991)
T439S	No effect on TEA$_o$ or TEA$_i$ block	Yellen *et al.* (1991)
M440I	"Hot spot" for TEA$_i$	Choi *et al.* (1993)
T441S	"Hot spot" for TEA$_i$; no effect on K$^+$ current–voltage relationship	Yellen *et al.* (1991); Choi *et al.* (1993)
T441S	Alters NH$_4^+$/K$^+$ and (much smaller) Rb$^+$/K$^+$ permeabilities; increases inward NH$_4^+$ and Rb$^+$ currents; no significant effect on K$^+$ current–voltage relationship	Yool and Schwarz (1991)
T449K,Q,Y	"Hot spot" for CTX	MacKinnon *et al.* (1990)
T449K,V,Y,Q,R	"Hot spot" for TEA$_o$; (K,Q,R) linearizes K$^+$ current–voltage relationship; (VY) = wild type	MacKinnon and Yellen (1990)
T449Y,F,V,I,S,W	"Hot spot" for TEA$_o$	Heginbotham and MacKinnon (1992)
V451K	"Hot spot" for CTX	MacKinnon *et al.* (1990)
G452R,E	Small (electrostatic?) effect on CTX	MacKinnon *et al.* (1990)
G452R	"Hot spot" for TEA$_o$	MacKinnon and Yellen (1990)

[a] Abbreviations: CTX, charybdotoxin; TEA, tetraethylammonium.

B. TEA

Several studies have identified residues that alter the ability of both internal and external TEA to block the pore of K$^+$ channels. A modest reduction in the affinity of external TEA for *Shaker* channels is produced by mutations at Asp 431; extremely large changes are produced by mutations at 449. Indeed, the type of amino acid at position 449 almost certainly determines much of the TEA sensitivity in the many "wild-type" channels (MacKinnon and Yellen, 1990; Kavanaugh *et al.*, 1991). Channels with a tyrosine at 449 have a very high sensitivity to external TEA, and the block has little voltage dependence (equivalent to a binding site at a position about 5% into the membrane voltage). The voltage dependence of the block increases to an equivalent of 20% of the membrane voltage when a threonine occupies position 449 (Heginbotham and MacKinnon, 1992). All four channel subunits appear to contribute to the aromatic binding site formed with tyrosine at 449 (Heginbotham and MacKinnon, 1992; Kavanaugh *et al.*, 1992).

Mutations of residues 440 and 441 in *Shaker* substantially alter the affinity of the channel for internal TEA (Yellen *et al.*, 1991; Choi *et al.*, 1993). These residues are located nearly halfway into the 21-amino-acid P region (Fig. 1). The voltage dependence of block by internal TEA is equivalent to binding of this ion at a site that experiences 75–85% of the membrane voltage (Kirsch *et al.*, 1991; Taglialatela *et al.*, 1991; Yellen *et al.*, 1991; Choi *et al.*, 1993). Most of this voltage dependence appears in the "on" rate constant of the blocking mechanism (Kirsch *et al.*, 1991).

Mutagenesis of DRK1, NGK2, and a DRK1/NGK2 chimera suggests that residues 438 and 443 (*Shaker* numbering) can alter internal TEA binding (Kirsch *et al.*, 1992a). These mutations also produce large changes in single-channel conductance and in channel gating (Kirsch *et al.*, 1992a,b; De Biasi *et al.*, 1993) and thus demonstrate that separating permeation and gating properties is not always possible.

III. Mutagenic Alterations of Pore Selectivity

The most fundamental permeation property is selectivity; channels are categorized according to which ions pass through most readily. Therefore, current understanding of the mechanism of ion permeation will be improved substantially by knowing which amino acids are important for ion selectivity. Yool and Schwarz (1991) found that mutations at three locations in the *Shaker* P region produced changes in relative ion selectivity. Two of these positions had no significant effect on gating properties and are included in Table I. Mutations of Phe 433 or Thr 441 to serine increased the NH$_4^+$ and, to a much smaller extent, the relative Rb$^+$ permeability (based on reversal potential measurements in bi-ionic conditions). These changes are specific for only certain amino acids: replacing Phe 433 with tyrosine instead of serine produced little or no alteration in NH$_4^+$ or Rb$^+$ permeability.

In their study of DRK1, NGK2, and the DRK1/NGK2 chimeras, Kirsch *et al.* (1992a,b) found that mutations of the hydrophobic amino acid valine (443 in *Shaker*) produced profound changes in pore selectivity. These mutations also altered single-channel conductance without much effect on channel gating (Kirsch *et al.*, 1992a,b).

IV. Mutagenic Alterations of the Current–Voltage Relationship

A. Mutations in the P Region

Under a normal (high internal, low external) K$^+$ concentration gradient, the open channel current–voltage relationship of *Shaker* is strongly outwardly rectifying (Iverson *et al.*, 1988; Timpe *et al.*, 1988). As the external concentration is increased, the rectification is reduced; the current–voltage relationship becomes linear at about 40 mM (Timpe *et al.*, 1988; Pérez-Cornejo, 1992). In symmetrical 100 mM K$^+$ concentrations, rectification of *Shaker* becomes inwardly directed (MacKinnon and Yellen, 1990; Isacoff *et al.*, 1991). This inward rectification is an intrinsic part of the K$^+$ permeation process and is not the result of block by intracellular Mg^{+2} ions (MacKinnon and Yellen, 1990) as it is in some other cells (Horie *et al.*, 1987; Matsuda *et al.*, 1987; Vandenberg, 1987).

As described earlier, position 449 is an important location: it is a hot spot for both CTX and external TEA. Mutations of the wild-type threonine to lysine, arginine, or glutamine (but *not* tyrosine) decrease inward current more than outward current, and tend to linearize the channel current–voltage relationship (MacKinnon and Yellen, 1990). The result with neutral glutamine is quite significant, since it shows that the linearization is not caused entirely by an electrostatic mechanism. The mutations described earlier that alter *Shaker* channel selectivity also produce alterations in the current carried by Rb$^+$ and NH$_4^+$. Mutations of Phe 433 or Thr 441 to serine substantially increase inward NH$_4^+$ current and, to a lesser extent, Rb$^+$ current. Again these effects are selective for certain amino acids; replacement of Phe 433 with a tyrosine induces only small changes in ion current.

Significant work has also been done with DRK1 and NGK2 channels. In 20 mM external K$^+$ and 120 mM internal K$^+$, DRK1 rectification is strongly outward (Kirsch *et al.*, 1992a). In symmetrical 120 mM K$^+$ concentrations, the single-channel current–voltage relationships of DRK1 and NGK2 are very different than those of *Shaker*. The DRK1 current–voltage relationship is nearly linear and NGK2 currents are about twice as large as those of *Shaker* (Kirsch *et al.*, 1992b). An idealized representation of these observations is illustrated in Fig. 2. The very large differences in the current–voltage relationships in Fig. 2 are surprising given the similarity of the P region sequences (Fig. 1). Perhaps residues outside the P region contribute to channel permeation properties.

B. Amino Acids Outside the P Region

Just outside the P region in *Shaker* channels is Val 451. As described earlier, this position is a hot spot for CTX; replacing the wild-type uncharged valine with the positively charged lysine essentially eliminates CTX binding (MacKinnon *et al.*, 1990). Wild-type DRK1 channels contain a lysine at the equivalent

position (see Fig. 1). The very strong outward rectification of these channels is reduced substantially by replacing the lysine with glutamine (Kirsch *et al.*, 1992a).

Researchers have long known that the internal TEA binding site includes a hydrophobic component (Armstrong, 1971). Mutation of Thr 469 in S6 (see Fig. 1) to more hydrophobic amino acids substantially increases the interaction of hydrophobic TEA analogs with the *Shaker* pore (Choi *et al.*, 1993). This result certainly indicates that residues of S6 are not far from the internal end of the pore. Thr 469 is likely to be close enough to form part of the pore structure involved in internal TEA binding.

Kirsch *et al.* (1992a,b) studied the permeation properties of DRK1, NGK2, and a chimeric channel made by inserting a piece of the NGK2 sequence into a DRK1 host. This chimeric channel has a P region consisting of the first 7 residues from DRK1 and the last 14 from NGK2. The current–voltage relationships of these three channels are quite different, especially with Rb$^+$ as the charge carrier. The chimeric P region differs from NGK2 by only 2 of 21 residues, yet this channel has permeation properties that are very different from those of NGK2. Replacing these two amino acids in the chimera with their NGK2 equivalents produces a channel with many but not all of the permeation properties appropriate for NKG2; several significant differences remain between this reversion mutant and native NGK2, especially with Rb$^+$ as the charge carrier (Kirsch *et al.*, 1992b). Such findings suggest that amino acids outside the 21-amino-acid P region are involved in ion permeation in K$^+$ channels.

This same conclusion was reached by Isacoff *et al.* (1991) in a study designed to identify the receptor for the cytoplasmic inactivation gate in *Shaker* channels. These investigators found that mutations of several residues in the S4–S5 linker (see Fig. 1) altered the current–voltage relationship measured with symmetrical isotonic K$^+$ concentrations. The amino acid sequence of this linker region is shown in Fig. 3 for *Shaker*, DRK1, and NGK2 channels. Mutations at two locations produced substantial effects. Replacement of Leu 385 in the S4–S5 linker (see Fig. 3) with alanine reduced the currents at all voltages, that is, the current–voltage relationship was not linearized; indeed, inward rectification was increased. Replacement with

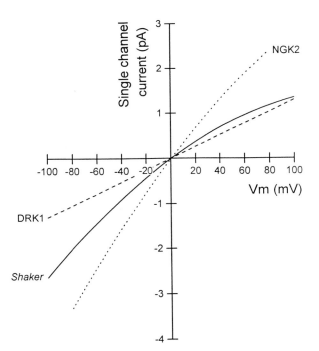

FIGURE 2 Current–voltage relationships of three cloned K$^+$ channels. These smooth curves are idealized from patch-clamp data. *Shaker* data from MacKinnon and Yellen (1990) were recorded with symmetrical 100 mM K$^+$ concentrations. DRK1 and NKG2 data from Kirsch *et al.* (1992b), recorded in symmetrical 120 mM K$^+$, were scaled down by 20% to facilitate comparison with the *Shaker* data.

	-- S4 --	--- linker ---	-- S5 --	
Shaker	SRHSKGLQ	ILGRTLKASMRELG	LLIFFLFI	405
DRK1	ARHSTGLQ	SLGFTLRRSYNELG	LLILFLAM	336
NGK2	TRHFVGLR	VLGHTLRASTNEFL	LLIIFLAL	354

FIGURE 3 Amino acids in the S4–S5 linker. Leucine 385 and serine 392 are underlined.

valine produced very small effects. Ser 392 also appears to influence permeation; replacement with a cysteine or an alanine produces a large reduction of channel currents (and an increase in inward rectification). Note that these mutations produced effects on the channel gating properties, as they were intended to do. Thus, the effects of these site-specific mutations must be considered with caution, since unintended long-range structural changes could have occurred.

As seen in Fig. 3, substantial homology exists in this S4–S5 linker region among the three types of channels; the important residues at *Shaker* 385 and 392 are identical. However, several differences are also apparent. For example, in *Shaker*, three positively charged residues (arg 387, Lys 390, Arg 394) and one negative charge (Glu 395) are found. Fewer positive charges are seen in DRK1 [Arg 321 (*Sh* 390), Arg 322 (*Sh* 391), and Glu 326 (*Sh* 395)], and even fewer in NGK2 [Arg 339 (*Sh* 390) and Glu 344 (*Sh* 395)].[2]

Mutations were made in *Shaker* at three of the positions containing charged groups (Isacoff *et al.*, 1991). The current–voltage relationship was not altered by replacing Lys 390 or Arg 394 with neutral glutamine. The conservative replacement of Glu 395 with aspartate was also without effect on the current–voltage relationship. Unfortunately, because of effects on gating, the current–voltage relationship of the potentially interesting mutant E395Q (Glu → Gln) was not examined.

Although large differences are apparent in the number of charged residues in *Shaker*, DRK1, and NGK2 K⁺ channels, insufficient information is available to assign a role to these residues. The possibility exists that the differences in positive charges among the three channels (especially at *Sh* 387) might contribute to the differences in permeation properties.

V. Multi-ion Nature of K⁺ Channels

Ion permeation and block of many types of ion channels, especially K⁺ channels, is a complex process that appears to reflect the simultaneous presence of several ions in the pore. The first such evidence was obtained by Hodgkin and Keynes (1955) with K⁺ channels from *Sepia*. These authors also showed that this type of behavior was consistent with single-file pores containing more than one ion at a time. Hille and Schwarz (1978) analyzed many of the properties possible in multi-ion pores. Several reviews of these

properties and their applicability to ion channels are available (Hille and Schwarz, 1979; Begenisich, 1982; Begenisich and Smith, 1984; Hille, 1992). The conclusion of these various studies and analyses is that at least some types of K⁺ channels simultaneously contain at least three ions.

Very few tests for multi-ion properties have been done in cloned K⁺ channels, but the available data suggest that the pores of these channels can simultaneously accommodate more than a single ion. Newland et al. (1992) showed that internal and external TEA are mutually antagonistic blockers of currents in *Shaker* and in two other kinds of K⁺ channels (RBK1 and RBK2). In other words, the presence of one of these blocking ions at one end of the pore has an influence on the ion at the other end. Pérez-Cornejo (1992) found that Cs⁺ ions block *Shaker* channels as though this ion bound to a site 130% across the membrane field—a nonsensical result for a one-ion pore. In multi-ion pores, the apparent voltage dependence of block by an ion such as Cs⁺ includes not only the voltage-dependent movement of Cs⁺ to its binding site but also any required voltage-dependent movement of permeant ions. In one-ion pores, permeability ratios (usually determined under bi-ionic conditions) may be functions of voltage but not of ion concentration. In *Shaker*, the apparent Rb⁺/K⁺ permeability is ~0.9 with high internal K⁺ and low external Rb⁺ concentrations. The ratio is reduced to ~0.6 when the ion concentrations are reversed (Pérez-Cornejo, 1992). This concentration-dependent permeability ratio is another property expected of multi-ion pores.

VI. Structural Model of the Pore

The mutagenesis data discussed earlier and summarized in Table I provide important clues to the structural and functional organization of the ion permeation pathway of K⁺ channels. Residues Phe 425, Asp 431, Thr 449, and Val 451 must be accessible from the external surface since these positions are hot spots for external CTX and external TEA. Position 441 is a hot spot for *internal* TEA. Consequently, as illustrated in Fig. 1, the 21-amino-acid P region probably "loops" into the membrane so Tyr 441 is near the inner membrane surface and residues 425, 431, 449, and 451 are near the external surface. The results of the mutagenesis studies suggest that certain amino acids strongly influence permeation or pore block (Table I), so they are likely to be part of the aqueous pore lining. Residues Phe 425, Asp 431, Phe 433, Thr 441, Thr 449, and Val 451 in the 21-amino-acid P region are all hot spots for some permeation property and are likely to

---

[2] Some additional positive charge may be contributed by His336 (*sh* 387).

point into the ion pathway. As discussed earlier, Val 443 is also likely to be an important part of the pore structure. Thr 439 may point away from the pore, since replacement of this residue with a serine has no effect on block by internal or external TEA (see Table I), yet the nearby position 441 is a hot spot for internal TEA and for other permeation properties.

The 21-amino-acid P region is too short to cross the lipid bilayer twice in an α-helical conformation. Consequently, modeling efforts have considered this segment to be in the more elongated β-barrel conformation. Durrell and Guy (1992) described an 8-stranded, antiparallel barrel that produces a rather short pore structure. In this segment, amino acids Thr 439, Thr 441, Tyr 445, and Asp 447 point into the pore. The model includes an inner vestibule formed by some of the residues in the S4–S5 linker region. This model has two apparent weaknesses: (1) the apparently nonfunctional Thr 439 (see Table I) points into the aqueous pore and (2) the narrow pore region in this model is quite short. As argued by Durrell and Guy (1992), the necessary 60–75% of the membrane voltage (see previous discussion) may drop across this short structure, but this suggestion certainly pushes the limits of such considerations.

Other analyses of possible barrel structures for the "core" pore region have been described by Bogusz and Busath (1992) and Bogusz *et al.* (1992). Two classes of structures were considered, based on whether an even or odd number of amino acids is found in the "hairpin turn" of the necessary loop. The structures with an even number of residues in the turn seem less likely than their odd-numbered cousins, since the former have important amino acids (Asp 431, Phe 433, or Thr 449) pointing away from the pore. The odd-numbered class can be further subdivided into two groups. In one group, an ion approaching the pore from the external surface would encounter amino acids in the order Thr 449, Asp 431, Asp 447. In the other group of structures, the order is Asp 431, Thr 449, Phe 433, Asp 447. The former arrangement seems less likely than the latter since mutations of Asp 431 have very little effect on external TEA binding, even when the negatively charge aspartate is replaced by positively charged lysine. Consequently, Asp 431 is likely to be best positioned as far as possible from the interior of the pore.

With these considerations in mind, a speculative model of the K⁺ channel pore was constructed (Fig. 4). This view relies heavily on the β-barrel model of the P region (Bogusz *et al.*, 1992). In this longitudinal cross section, the individual amino acids appear as top and bottom pairs. This is the two-dimensional representation of the 4-fold symmetry of these pro-

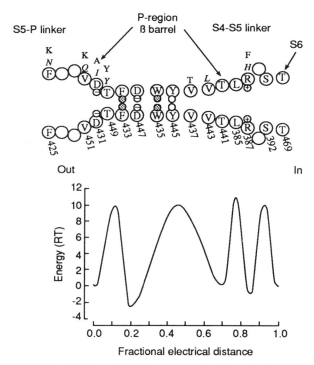

FIGURE 4 (*Top*) Longitudinal cross section through *Shaker* pore. The circled amino acids are those of wild-type *Shaker*. The different DRK1 and NGK2 residues are shown above the corresponding *Shaker* location; those of NGK2 are in italics. Charged amino acids are identified by small circles containing "plus" or "minus" signs. Large polar residues (only tyrosine in this figure) include a small empty circle. Large nonpolar amino acids (phenylalanine and tryptophan) are identified with small shaded circles. The residue numbers are for the *Shaker* sequence. The order appears jumbled (cf., Fig. 1) because two of the four channel subunits in this β barrel are visible in this cross section (Bogusz *et al.*, 1992). (*Bottom*) K⁺ ion free energy as a function of electrical distance. The energy minima were placed as described in text. For simplicity, energy maxima were located equidistant between adjacent minima. RT = 583 cal/mol.

teins (at least as expressed in *in vitro* systems). The third and fourth identical amino acids would project in front of and behind the plane of the picture.

Using this diagram to suggest some elements of the ion permeation pathway is tempting. Although the mutagenesis data (see Table I) identify Asp 431 as a hot spot for CTX, so it faces the aqueous pore, this residue does not seem to be very involved in external TEA block or K⁺ permeation. Consequently, in the model in Fig. 4, it is located as part of the outer vestibule, not part of the narrow ion-selective region. Position 449 is a hot spot for external TEA (as well as for CTX). With tyrosines at this position, the voltage dependence of TEA block is equivalent to only 5% of the membrane voltage. Thus, Thr 449 is drawn in the model in Fig. 4 in the narrow portion of the outer vestibule, just at the entrance of the narrow pore region. Perhaps the larger voltage dependence of external TEA block with the wild-type Thr 449 occurs because this ion moves past position 449 further into the pore, perhaps as far as Asp 447. The narrow part of the pore region is formed by residues Phe 433, Asp 447, Trp 435, Tyr 445, Thr 437, Val 443, and Thr 441. The sequential positions of the amino acids and the three-dimensional structure of this narrow part of the pore are constrained by the β-barrel model (see Bogusz et al., 1992, for an accurate three-dimensional representation). This model does not, however, define the positions of amino acids away from this narrow region. This uncertainty is reflected in the figure by the presence of large circles without an associated amino acid or residue number.

To include residues in the S4–S5 linker (Isacoff et al., 1991) and the voltage dependence of internal TEA block (at Thr 441), an inner vestibule is drawn in Fig. 4 using residues (Leu 385 and Ser 392) from the S4–S5 linker, which appear to be involved in permeation. Arg 387 is included to leave open the possibility that charged groups may also be located in this vestibule. Also included in the inner vestibule is Thr 469 of S6, since block by hydrophobic TEA analogs is affected by mutations of this residue (see previous discussion). To date, few structural constraints on this region are available (but see the model by Durrell and Guy, 1992). This uncertainty is reflected in the figure by the presence of empty circles.

VII. Mathematical Model for Ion Permeation

As described earlier, a considerable amount of structural (mutagenesis and modeling) and functional information is available on ion permeation in K⁺ channels. Nevertheless, the available data are not sufficient to allow construction of an unambiguous mathematical description of ion permeation. Indeed, the nature of the appropriate mathematical formalism is not even clear. Two general approaches have been used: (1) continuum diffusion theory and (2) a more discrete approach using transition state (or Eyring rate) theory. Each approach has its advantages and limitations (and associated advocates and detractors) that have been addressed in many published reports (see Cooper et al., 1985, 1988a,b; Levitt, 1986) and will not be repeated here, except to note that a potentially quite useful extension of the continuum theory has been published (Chen et al., 1992). This latter work also contains an excellent historical description of the continuum approach. Despite the preceding qualifications, describing a mathematical model for permeation through K⁺ channels may, nevertheless, be useful. Such a model

may help illuminate the permeation process or may suggest new experiments. Although either mathematical formalism could be used for this purpose, the Eyring rate theory approach is presented here.

In Eyring rate theory, ion movement is modeled as "jumps" from site to site over intervening energy barriers. As a consequence the first parameter to consider is the number of ion binding sites. The available data on the multi-ion properties of K⁺ channels suggest that at least three K⁺ ions can occupy the pore simultaneously (Hille and Schwarz, 1978; Begenisich and De Weer, 1980; Hille, 1992); therefore, there may be three binding sites for K⁺ ions. This conclusion is not entirely model independent, but is probably a reasonable starting point.

After deciding on the number of sites, their locations within the pore must be established. For the purpose of computing ion movement in an Eyring model, the positions within the membrane electric field, not the spatial locations, are required. Assigning K⁺ binding sites to locations important for block by TEA ions might seem appropriate. In this view, the outermost K⁺ binding site would be at position 449. This position is certainly important for external TEA block but, as discussed earlier, TEA appears to bind at residue 449 (at 5% of the membrane field) *only* if a tyrosine is at this position. Otherwise TEA (and perhaps K⁺) appears to move further into the pore to bind at a site 20% into the field. Consequently, as illustrated in Fig. 4, the first ion binding site (energy minimum) is positioned at about 20% into the electric field and is associated with Asp 447. The outermost energy barrier is then associated with Thr 449 and Phe 433. Partial dehydration may be necessary before K⁺ ions can get past the large Phe 433.

The innermost barrier may involve residues outside the P region and is drawn in Fig. 4 near Ser 392 and Leu 385, mutations of which alter the K⁺ current–voltage relationship as described earlier. Associating the innermost K⁺ ion binding site with the internal TEA hot spot at position 441, and placing it at 85% through the electric field is then convenient. The very large hydrophobic Trp 435 might represent another dehydration energy barrier.[3] The remaining energy minimum has been placed at 70% of the field.

Recall that data from *Shaker* and a few other cloned channels indicate that these channels contain a multi-ion pore. As discussed earlier for Cs⁺ ions, using the measured voltage dependence of block to assign a position for the ion binding site may be misleading.

---

[3] This barrier may also include an electrostatic "image" barrier (Parsegian, 1969; Levitt, 1978; Jordan, 1982).

This condition may also apply to TEA block but, in the absence of direct data, the preceding assignments of ion binding sites will have to be sufficient.

One of the first goals of such a mathematical model of the pore is reproduction of the wild-type K⁺ ion current–voltage relationship. The inward rectification of *Shaker* currents in symmetrical K⁺ concentrations (see preceding discussions and Fig. 2) suggests that some asymmetry must exist in the mathematical pore properties. Although such asymmetries could be provided by the depths of the outer and inner wells, these measures generally have less effect on rectification than asymmetries in barrier heights (Begenisich and Cahalan, 1980b).

Charged residues in the inner and outer vestibules could also contribute to a rectifying current–voltage relationship in symmetrical K⁺ concentrations by accumulating or depleting K⁺ ions at the entrances to the pore. The results of mutagenesis in the outer vestibule suggest that charged residues Glu 422, Lys 427, and Asp 431 in *Shaker* do not play significant roles in the K⁺ current–voltage relationship. Mutagenesis of positively charged residues Lys 390 and Arg 394, which may be in the inner vestibule, was also without significant effect on the *Shaker* current–voltage relationship. Although testing the involvement of Arg 387 and Glu 395 would still be useful, discounting the role played by charges in the inner or outer vestibules in the rectification of the *Shaker* current–voltage relationship seems reasonable (but see discussion for DRK1).

All these speculations are given concrete form in the simulations presented in Fig. 5. The curve labeled *Shaker* in this figure is the result of computing the single-channel current–voltage relationship from Eyring rate theory with methods described by Begenisich and Cahalan (1980a). The positions and energies of the barriers and wells are those presented in Fig. 4 and are given explicitly in Fig. 5. The model current–voltage relationship is a reasonable representation of the experimental data (cf., Fig. 2).

As discussed earlier, very few differences are apparent in the amino acid sequences of *Shaker*, NGK2, and DRK1 in the 21-amino-acid P region (see Figs. 1, 4). The T449Y (Thr → Tyr) mutation in *Shaker* does not alter the K⁺ current–voltage relationship (see Table I), so the presence of tyrosine instead of the *Shaker* threonine at this location in NGK2 and DRK1 cannot account for the large differences in the current–voltage relationships of these channels. The sequences of these three channels also differ at position 431 (see Figs. 3, 4). As summarized in Table I, mutations of Asp 431 produce small effects on external TEA block, and the mutation D431N (Asp → Asn) has no signifi-

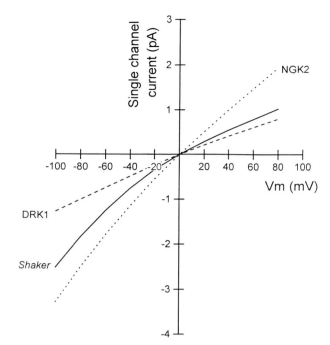

FIGURE 5 Simulations of the current–voltage relationships of cloned K⁺ channels. For all calculations, the internal and external K⁺ concentrations were 100 m*M*. *Shaker*—Barriers (left to right in the lower part of Fig. 4): 9.5, 10, 11, 10 RT; wells: −2.5, 0, −1 RT. NGK2—Same as *Shaker*, except the third barrier was 9 RT. DRK1—Same as *Shaker*, but with a 45-mV external surface potential.

cant effect on the *Shaker* current–voltage relationship. Another difference appears at position 437, where *Shaker* contains a valine and DRK1 a threonine. A V437T (Val → Thr) mutation was made in the DRK1/NGK2 chimera described earlier (Kirsch *et al.*, 1992a,b) and had little effect on K⁺ current. The only remaining difference between NGK2 and *Shaker* and DRK1 is the leucine instead of valine at position 443. This conservative substitution might be expected to be insignificant, but the V443L (Val → Leu) mutation in DRK1 produces an increase in current that approaches that of NGK2 channels (Kirsch *et al.*, 1992a). The third energy barrier in Fig. 4 is located near this residue, so investigating how changes in this barrier would affect the computed current relationship was reasonable. As illustrated in Fig. 5, reducing this barrier height from 11 RT (the *Shaker* value) to 9 RT produced a current–voltage relationship (curve labeled NGK2) that is a reasonable representation of the experimental data (cf., Fig. 2).

The mutagenesis data indicate that residues in the P region cannot account for the small currents and the linearity of the DRK1 current–voltage relationship. Therefore, the amino acids responsible for this difference must lie elsewhere, perhaps in the outer or inner vestibule. One interesting candidate is the positively charged lysine in DRK1 at position 451. Perhaps this positive charge depletes K^+ in the outer vestibule and subsequently lowers current. Indeed, as discussed earlier, replacing lysine with glutamine in DRK1 channels substantially reduces outward rectification (Kirsch *et al.*, 1992a). A simulation of the effects of adding a $+45$-mV external surface charge is presented in Fig. 5 (curve labeled DRK1). As seen in this figure, the effects are the reduction of current and the linearization of the current–voltage relationship, much like the experimental data (cf., Fig. 2).

VIII. Limitations

The mathematical model presented here has several obvious limitations. Little information is available on cloned K^+ channels with which to place constraints on the depths of the energy wells. The outer well was chosen as -2.5 RT to produce a reasonably linear current–voltage relationship with 40 mM external K^+, as required by the experimental data (see previous discussion). No similar data are available for the effects of internal K^+ concentration, so use of a -1 RT inner well is arbitrary. The barriers used in the computations were adjusted to produce a reasonable simulation of the experimental data. If, however, new data show that different well depths are required for the ion binding sites, these barriers will need to be adjusted accordingly. Using site-specific mutants, identifying the amino acids that constitute these sites might also be possible.

To minimize the number of adjustable model parameters, the barriers were chosen to be symmetrical in shape. This choice may be reasonable, and certainly has been the choice in many previous Eyring permeation models. However, the data of Kirsch *et al.*, (1991) show that internal TEA^+ appears to experience an asymmetrical barrier. No such direct information is available for the barriers that K^+ ions must cross, but this possibility must be considered.

The choice of three sites of K^+ ion occupancy was based primarily on data from K^+ channels that have not been cloned (see Hille, 1992). All the cloned K^+ channels have amazing similarity in their respective P regions, but this point must be addressed directly with appropriate data from cloned channels.

Another element in the modeling was, except for DRK1, the lack of a role for charges in the inner and outer pore vestibules. Experiments designed specifically to determine whether surface potentials are involved in ion permeation in cloned K^+ channels would be of great value. Replacing Phe 425 of *Shaker*

with the lysine of DRK1 would test the speculation that this residue is responsible for the small currents and the relatively linear current–voltage relationship of DRK1. Examining in more detail the charged residues in the S4–S5 linker for involvement in permeation would also be useful.

As described earlier, the speculative structural and mathematical models presented here have obvious limitations. Nevertheless, these models have been useful in suggesting additional data necessary for a further understanding of ion permeation. With these additional data, refined structural and mathematical models can be constructed that may suggest further valuable experiments. Eventually, more direct structural (e.g., crystallographic) data may be obtained, as well as results from advanced modeling techniques (e.g., molecular dynamic simulations), which hopefully will lead to a clearer molecular picture of ion permeation through K^+ channels.

Acknowledgments

I thank David Busath for numerous extensive and useful conversations about many of the topics in this work. I also thank Dr. Busath and Arthur M. Brown and Sherrill Spires for critical reading of the manuscript. Dr. Brown also generously provided details of some interesting results prior to publication. This work was supported in part by U. S. Public Health Services Grant NS-14138.

References

Armstrong, C. M. (1971). Interaction of tetraethylammonium ion derivatives with the potassium channels of giant axons. *J. Gen. Physiol.* **58,** 413–437.

Begenisich, T. (1982). Ion permeation through channels in excitable cells. How many ions per pore? *Membranes Transport* **2,** 373–378.

Begenisich, T. (1987). Molecular properties of ion permeation through sodium channels. *Annu. Rev. Biophys. Biophys. Chem.* **16,** 247–263.

Begenisich, T., and Cahalan, M. D. (1980a). Sodium channel permeation in squid axons. I: Reversal potential experiments. *J. Physiol.* **307,** 217–242.

Begenisich, T., and Cahalan, M. D. (1980b). Sodium channel permeation in squid axons. II: Non-independence and current–voltage relations. *J. Physiol.* **307,** 243–257.

Begenisich, T., and De Weer, P. (1980). Potassium flux ratio in voltage-clamped squid giant axons. *J. Gen. Physiol.* **76,** 83–98.

Begenisich, T., and Smith, C. (1984). Multi-ion nature of potassium channels in squid axons. *Curr. Top. Membranes Transport* **22,** 353–369.

Bogusz, S., and Busath, D. (1992). Is a β-barrel model of the K^+ channel energetically feasible? *Biophys. J.* **62,** 19–21.

Bogusz, S., Boxer, A., and Busath, D. (1992). An SS1-SS2 β-barrel structure for the voltage-activated potassium channel. *Protein Eng.* **5,** 285–293.

Chandy, K. G. (1991). Simplified gene nomenclature. *Nature (London)* **352,** 26.

Chen, D. P., Barcilon, V., and Eisenberg, R. S. (1992). Constant fields and constant gradients in open ionic channels. *Biophys. J.* **61,** 1372–1393.

Choi, K. L., Mossman, C., Aubé, and Yellen, G. (1993). The internal quaternary ammonium receptor site of *Shaker* potassium channels. *Neuron* **10,** 533–541.

Cooper, K. E., Jakobsson, E., and Wolynes, P. G. (1985). The theory of ion transport through membrane channels. *Prog. Biophys. Mol. Biol.* **46,** 51–96.

Cooper, K. E., Gates, P. Y., and Eisenberg, R. S. (1988a). Surmounting barriers in ionic channels. *Q. Rev. Biophys.* **21,** 331–364.

Cooper, K. E., Gates, P. Y., and Eisenberg, R. S. (1988b). Diffusion theory and discrete rate constants in ion permeation. *J. Membrane Biol.* **106,** 95–105.

De Biasi, M., Hartmann, H. A., Drewe, J. A., Taglialatela, M., Brown, A. M., and Kirsch, G. E. (1993). Inactivation determined by a single site in K$^+$ pores. *Pflügers Arch.* **422,** 354–363.

Durrell, S. R., and Guy, H. R. (1992). Atomic scale structure and functional models of voltage-gated potassium channels. *Biophys. J.* **62,** 238–250.

Frech, G. C., VanDonagen, A. M. J., Schuster, G., Brown, A. M., and Joho, R. H. (1989). A novel potassium channel with delayed rectifier properties isolated from rat brain by expression cloning. *Nature (London)* **340,** 642–645.

Gimenez-Gallego, G., Navia, M. A., Reubeb, J. P., Katz, G. M., Kaczorowski, G. J., and Garcia, M. L. (1989). Purification, sequence and model structure of CTX, a protein selective inhibitor of Ca^{2+}-activated K$^+$ channels. *Proc. Natl. Acad. Sci. USA* **85,** 3329–3337.

Goldstein, S. A. N., and Miller, C. (1992). A point mutation in a *Shaker* K$^+$ channel changes its charybdotoxin binding site from low to high affinity. *Biophys. J.* **62,** 5–7.

Hartmann, H. A., Kirsch, G. E., Drewe, J. A., Taglialatela, Joho, R. H., and Brown, A. M. (1991). Exchange of conduction pathways between two related K$^+$ channels. *Science* **251,** 942–944.

Heginbotham, L., and MacKinnon, R. (1992). The aromatic binding site for tetraethylammonium ion potassium channels. *Neuron* **8,** 483–491.

Hille, B. (1992). "Ionic Channels of Excitable Membranes." Sinauer Associates, Sunderland, Massachusetts.

Hille, B., and Schwarz, W. (1978). Potassium channels as multi-ion single-file pores. *J. Gen. Physiol.* **72,** 409–442.

Hille, B., and Schwarz, W. (1979). K$^+$ channels in excitable cells as multi-ion pores. *Brain Res. Bull.* **4,** 159–162.

Hodgkin, A. L., and Keynes, R. D. (1955). The potassium permeability of a giant nerve fibre. *J. Physiol.* **128,** 61–88.

Horie, M., Irasawa, H., and Noma, A. (1987). Voltage-dependent magnesium block of adenosine-triphosphate-sensitive potassium channels in guinea-pig ventricular cells. *J. Physiol.* **387,** 251–272.

Isacoff, E. Y., Jan, N. J., and Jan, L. Y. (1991). Putative receptor for the cytoplasmic inactivation gate in the *Shaker* K$^+$ channel. *Nature (London)* **353,** 86–90.

Iverson, L. E., Tanouye, M. A., Lester, H. A., Davidson, N., and Rudy, B. (1988). A-type potassium channels expressed from *Shaker* locus cDNA. *Proc. Natl. Acad. Sci. USA* **85,** 5723–5727.

Jan, L. Y., and Jan, Y. N. (1992). Structural elements involved in specific K$^+$ channel functions. *Annu. Rev. Physiol.* **54,** 537–555.

Jordan, P. C. (1982). Electrostatic modeling of ion pores: Energy barriers and electric field profiles. *Biophys. J.* **39,** 157–164.

Kamb, A., Tseng-Crank, J., and Tanouye, M. A. (1988). Multiple products of the *Drosophila Shaker* gene may contribute to potassium diversity. *Neuron* **1,** 421–430.

Kavanaugh, M. P., Varnum, M. D., Osborne, P. B., Christie, M. J., Busch, A., Adelman, J. P., and North, R. A. (1991). Interaction between tetraethylammonium and amino acid residues in the pore of cloned voltage-dependent potassium channels. *J. Biol. Chem.* **266,** 7583–7587.

Kavanaugh, M. P., Hurst, R. S., Yakel, J., Varnum, M. D., Adelman, J. P., and North, R. A. (1992). Multiple subunits of a voltage-dependent potassium channel contribute to the binding site for tetraethylammonium. *Neuron* **8,** 493–497.

Kirsch, G. E., Taglialatela, M., and Brown, A. M. (1991). Internal and external TEA block in single cloned K$^+$ channels. *Am. J. Physiol.* **261,** C583–590.

Kirsch, G. E., Drewe, J. A., Hartmann, H. A., Taglialatela, M., de Biasi, M., Brown, A. M., and Joho, R. H. (1992a). Differences between the deep pores of K$^+$ channels determined by an interacting pair of nonpolar amino acids. *Neuron* **8,** 499–505.

Kirsch, G. E., Drewe, J. A., Taglialatela, M., Joho, R. H., de Biasi, M., Hartmann, H. A., and Brown, A. M. (1992b). A single nonpolar residue in the deep pore of related K$^+$ channels acts as a K$^+$:Rb$^+$ conductance switch. *Biophys. J.* **62,** 136–144.

Levitt, D. G. (1978). Electrostatic calculations for an ion channel. I. Energy and potential profiles and interactions between ions. *Biophys. J.* **22,** 209–219.

Levitt, D. G. (1986). Interpretation of biological ion channel flux data: Reaction-rate versus continuum theory. *Annu. Rev. Biophys. Biophys. Chem.* **15,** 29–57.

Liman, E. R., Tytgat, J., and Hess, P. (1992). Subunit stoichiometry of a mammalian K$^+$ channel determined by construction of multimeric cDNAs. *Neuron* **9,** 861–871.

MacKinnon, R. (1991). Determination of the subunit stoichiometry of a voltage-activated potassium channel. *Nature (London)* **350,** 232–235.

MacKinnon, R., and Miller, C. (1988). Mechanism of charybdotoxin block of the high-conductance, Ca^{2+}-activated K$^+$ channel. *J. Gen. Physiol.* **91,** 335–349.

MacKinnon, R., and Miller, C. (1989). Mutant potassium channels with altered binding of charybdotoxin, a pore-blocking peptide inhibitor. *Science* **245,** 1382–1385.

MacKinnon, R., and Yellen, G. (1990). Mutations affecting TEA blockade and ion permeation in voltage-activated K$^+$ channels. *Science* **250,** 276–279.

MacKinnon, R., Heginbotham, L., and Abramson, T. (1990). Mapping the receptor site for charybdotoxin, a pore-blocking potassium channel inhibitor. *Neuron* **5,** 767–771.

Matsuda, H., Saigusa, A., and Irisawa, H. (1987). Ohmic conductance through the inwardly rectifying K channel and blocking by internal Mg^{2+}. *Nature (London)* **325,** 156–159.

Miller, C. (1988). Competition for block of a Ca^{2+}-activated K$^+$ channel by charybdotoxin and tetraethylammonium. *Neuron* **1,** 1003–1006.

Miller, C., Moczydlowski, E., Latorre, R., and Phillips, M. (1985). Charybdotoxin, a protein inhibitor of single Ca^{2+}-activated K$^+$ channels from mammalian skeletal muscle. *Nature (London)* **313,** 316–318.

Newland, C. F., Adelman, J. P., Tempel, B. L., and Almers, W. (1992). Repulsion between tetraethylammonium ions in cloned voltage-gated potassium. *Neuron* **6,** 975–982.

Oliva, C., Folander, K., and Smith, J. S. (1991). Charybdotoxin is not a high affinity blocker of *Shaker* K$^+$ channels in *Xenopus* oocytes. *Biophys. J.* **59,** 450a.

Parsegian, A. (1969). Energy of an ion crossing a low dielectric membrane: Solutions to four relevant electrostatic problems. *Nature (London)* **221,** 844–846.

Pérez-Cornejo, G. P. (1992). "Multi-ion Nature of the *Shaker* K$^+$ Channel," M. S. Thesis. University of Rochester, Rochester, New York.

Taglialatela, M., VanDonagen, A. M. J., Drewe, J. A., Joho, R. H.,

Brown, A. M., and Kirsch, G. E. (1991). Patterns of internal and external tetraethylammonium block in four homologous K$^+$ channels. *Mol. Pharmacol.* **40,** 299–307.

Timpe, L. C., Schwarz, T. L., Tempel, B. L., Papazian, D. M., Jan, Y. N., and Jan, L. Y. (1988). Expression of functional potassium channels from *Shaker* cDNA in *Xenopus* oocytes. *Nature (London)* **31,** 143–145.

Vandenberg, C. A. (1987). Inward rectification of a potassium channel in cardiac ventricular cells depend on internal magnesium ions. *Proc. Natl. Acad. Sci. USA* **84,** 2560–2564.

Yellen, G., Jurman, M. E., Abramson, T., and MacKinnon, R. (1991). Mutations affecting internal TEA blockade identify the probable pore-forming region of a K$^+$ channel. *Science* **251,** 939–942.

Yokoyama, S., Imoto, K., Kawamura, T., Higashida, H., Iwabe, N., Miyata, T., and Numa, S. (1989). Potassium channels from NG108-15 neuroblastoma-glioma hybrid cells: Primary structure and functional expression from cDNAs. *FEBS Lett.* **259,** 37–42.

Yool, A. J., and Schwarz, T. L. (1991). Alteration of ionic selectivity of a K$^+$ channel by mutation of the H5 region. *Nature (London)* **349,** 700–704.

3

Gating Currents of Cloned Shaker K$^+$ Channels

ENRICO STEFANI, LIGIA TORO,
EDUARDO PEROZO,
AND FRANCISCO BEZANILLA

I. Introduction

Voltage-dependent ion channels are integral membrane proteins that respond to changes in membrane potential with rapid variations in membrane permeability to ions. The sequence of events leading to the opening of ion channel pores is thought to require the movement of elements that sense the membrane electric field. This movement of charged elements generates displacement currents called gating currents (Armstrong and Bezanilla, 1973; Schneider and Chandler, 1973; Bezanilla, 1985). Ionic current measurements mainly reflect the channel open state, whereas gating currents give additional information about the channel transitions that occur among many closed states prior to channel opening. Thus, ionic currents in conjunction with gating current measurements may provide insight into the molecular basis of the different conformations of the channel protein during voltage activation.

Cells have a mixed population of a variety of voltage-dependent proteins. Therefore, gating currents from a single channel type are difficult to measure in isolation, since contaminating gating components from other membrane proteins (e.g., channels and transporters) cannot be eliminated completely by

pharmacological tools or by changes in the ionic composition of the solutions. Nevertheless, squid axon Na$^+$ and K$^+$ gating currents have been characterized extensively and have provided initial insight into the transitions among states and the interactions between gating elements (Bezanilla, 1985). For example, the steady-state relationship between charge movement and conductance indicated the presence of many closed states before channel opening. Additionally, the similarity in the time constant values of ionic current deactivation and OFF charge relaxation discredited a model of independent and equal gating subunits. Another important finding was the relationship between the fast inactivation of Na$^+$ currents and the slow charge recovery (charge immobilization), suggesting the presence of a different pathway for the return from the inactivated to the closed state (Armstrong and Bezanilla, 1977; White and Bezanilla, 1985; Spires and Begenisich, 1989).

In this chapter, our previous studies on gating currents of *Shaker* K$^+$ cannels are extended, using *Xenopus laevis* oocytes as a heterologous expression system that synthesize a homogeneous population of exogenous protein when injected with foreign RNA (Dascal, 1987). In addition, oocytes have the enormous advantage that capacity and ionic current measurements can be performed in control oocytes before complementary messenger RNA (cRNA) injection. Further, improvements in the preparation and handling of cRNA have made possible the superexpression of channel protein which, in conjunction with the development of a fast and low-voltage noise clamp (cut-open oocyte vaseline clamp; Taglialatela *et al.*, 1992), allows measurements of gating currents, even in the absence of subtracting pulses. Another development that has significantly improved the characterization of gating currents is a point mutation in the pore region of *Shaker* K$^+$ channels (W434F; Trp → Phe) that eliminates ion conduction and allows the direct measurement of charge movement in the presence of "normal" conducting ion and the absence of channel blockers (Perozo *et al.*, 1993). These improvements in the clamp technique and channel expression have made possible the direct recording of unsubtracted gating currents with a high signal-to-noise ratio, revealing features of protein movement under the electric field.

II. Methods

A. Oocyte Preparation and Injection of cRNAs

Xenopus laevis oocytes (stage V–VI) were used. Oocytes were treated with collagenase (200 U/ml; GIBCO, St. Louis) in a Ca^{2+}-free solution, 1 day before injection of cRNA, to remove the follicular layer. Oocytes were microinjected with 40–60 nl cRNA at 1 μg/μl (40–60 ng) and maintained at 19°C in an amphibian saline solution supplemented with 50 μg/ml gentamycin and 2.5 mM pyruvate, for 2–6 days. All the recordings reported were performed at room temperature (22°C), 2–7 days after cRNA injection.

B. cRNA Preparation

cDNA encoding *Shaker* H4 (*Sh*H4) K$^+$ channels, which is nearly identical to the *Shaker* B clone, was used (Tempel *et al.*, 1987; Kamb *et al.*, 1988). The following clones were used: *Shaker* wild-type (*Sh*H4); *Shaker* with inactivation removed, which has the deletion Δ6–46 (*Sh*H4-IR), and the corresponding nonconducting counterparts (*Sh*H4-W434F and *Sh*H4-IR-W434F); (Perozo *et al.*, 1993). Conditions have been determined for the synthesis of cRNA, allowing us to record large ionic and gating currents reliably and consistently. Several ratios of capping nucleotide (GpppG) to GTP (i.e., 0.6/0.2 mM, 0.6/1.0 mM, 1/0.4 mM) have been examined. A ratio of 0.6 mM/1 mM was found to be optimal, and GpppG had to be dissolved in Tris-EDTA (TE) buffer. Another factor that increased expression

in the oocytes was the nature of the 5'-untranslated leader of the cRNA (Fu *et al.*, 1991). In this respect, a construct of the *Sh*H4-IR K$^+$ channel made by R. MacKinnon (Dept. Neurobiology, Harvard Medical School) in which bases 1–223 of the original 5'-untranslated leader were deleted gave high expression.

C. Solutions

Solutions were made by mixing stock isotonic solutions of the main cation (220 mOsm)—glutamate (Glu), methanesulfonate (MES), *n*-methylglucamine (NMG), and tetraethylammonium (TEA)—containing 10 mM HEPES at pH 7.0. The isotonic solutions used were NaCl, KCl, NaMES, K-MES, K-Glu, TEA-MES, NMG-Glu, TEA-Glu (all 110 mM) and Ca(MES)$_2$ (74 mM) EGTA (1 mM) was added to solutions for intracellular perfusion. A typical solution for recording gating currents was made by mixing NaMES and Ca(MES)$_2$ to obtain NaMES (107 mM), Ca(MES)$_2$ (2 mM) HEPES (10 mM), abbreviated NaMESCa2. Standard solutions for intracellular micropipettes were 3 M KCl or 3 M NaMES, 10 mM NaCl, and 10 mM Na$_2$EGTA buffered to pH 7.0 with 10 mM HEPES. Solutions were replaced in the top and guard compartments. Solution replacement in the top chamber can be achieved in less than 1 sec by a fast perfusion system ($\sim$ 1 ml/sec) and can be done at a slower rate in the guard compartment.

D. Recording of Macroscopic Ionic and Gating Currents

Experiments were performed with the cut-open oocyte clamp technique (COVG; Bezanilla *et al.*, 1991, Taglialatela *et al.*, 1992) and conventional patch-clamp techniques.

E. Cut-Open Oocyte Clamp

This method is a low-noise fast clamp that allows control of the internal solution. Defolliculated occytes were mounted in a Perspex® chamber specifically designed to allow the compartmentalization of the oocyte membrane using two thin vaseline rims. This procedure established three electrically independent pools: (1) a lower pool (I) to inject current and to have access to the cell interior, (2) a middle pool (GS) to connect an electronic guard shield, and (3) an upper pool (P) in which the oocyte membrane to be clamped was exposed. Connections between pools and the electronic components were made via 1 M NaMES agar bridges and nonpolarizable Ag–AgCl electrodes immersed in 1 M NaCl.

The dynamics of the system was further improved by actively clamping the GS compartment. For a scheme of the circuit diagram and experimental compartments, see Taglialatela *et al.* (1992) and Perozo *et al.* (1992). The electronic system consisted of three independent active voltage clamps: two clamps controlled pool P and pool GS to the command potential while the third clamped to ground the interior of the oocyte membrane facing pool P. Membrane currents arising from the voltage-clamped area of the oocyte membrane facing the upper compartment P (radius 350 μm, capacitance 20–60 nF) were recorded. The circuit consisted of a current to voltage (I–V) transducer with variable resistances (0.1–2 MΩ) in the feedback loop. The upper compartment P had two electrodes. One of the electrodes in pool P was connected to the negative input of the I–V transducer, and the other to the output of the same amplifier via the feedback resistance. The command pulse was fed into the positive input of the I–V transducer. GS was also clamped to the command potential with two separate electrodes; ideally, no current should flow between GS and P since both pools were clamped at the same potential. A conventional microelectrode voltage clamp was used to clamp to ground the interior of the oocyte membrane facing compartment P. The output of this clamp amplifier was connected to the lower pool via a low resistance electrode (I). The potential difference across the clamped oocyte membrane was differentially recorded between the intracellular microelectrode and an external one positioned close to the external oocyte surface. Recorded pulse potentials closely followed the command pulse. The clamp system had capacity compensation at the negative input of the I–V converter to prevent amplifier saturation during large capacity transients. The surface of the oocyte membrane facing pool I was opened to inject current intracellularly by superfusing K-MES with saponin (0.1%) for 1–3 min. Oocyte opening was detected by a dramatic fall in series resistance, at which time the saponin-containing solution was exchanged with the intracellullar solution.

An IBM-compatible personal computer was used for data acquisition. Analog signals were filtered at 1/4th the sampling frequency (10–50 kHz). Oocytes membrane potential was generally clamped to −90 mV holding potential (HP). Linear capacity and resistive components were digitally subtracted by scaled control currents obtained with small negative (P/−4) or positive (P/4) pulses of 1/4 test-pulse amplitude from negative (−120 to −150 mV) or positive (20 mV) subtracting holding potentials (SHP), respectively. The adequacy of the subtracting protocol was tested using different SHPs and comparing subtracted and unsubtracted gating charge records (see Fig. 5).

F. Patch Clamp

The vitelline membrane was removed manually after incubation in a hyperosmotic solution (NaMESCa$_2$ + 300 mM sucrose). Membrane currents were recorded in cell-attached or -excised patch configuration. In cell-attached mode, the bath solution was K-MESCa0 with 1 mM Na$_2$EGTA that should bring the oocyte membrane potential close to 0 mV. Activity was recorded with Axopatch 200A patch amplifier (Axon Instruments, Burlingame, California).

III. Results and Discussion

A. Control Oocytes Have Undetectable or Very Small Nonlinear Capacity Currents

Records in Fig. 1A illustrate the linearity of the oocyte membrane and the virtual absence of gating currents and/or nonlinear capacity components (<5 nA) in noninjected oocytes and in the explored voltage range (−140 to +50 mV). For comparison, peak gating currents from oocytes expressing *Shaker* K$^+$ channels can measure, at 0 mV, up to 10 μA (see subsequent figures). In some batches of oocytes (10 of more than 100), small (5–25 nA) nonlinear currents were recorded that were blocked by 100 μM ouabain and were suppressed by replacing external Na$^+$ with K$^+$ or TEA (Fig. 1B,C); these currents have been attributed to the Na$^+$/K$^+$ pump (Rakowski, 1993). Because of their small size, these nonlinear currents, when present, cannot significantly contaminate the gating charge records. This assumption has been confirmed by demonstrating that K$^+$ gating current records were not altered when recorded using Na$^+$-free solutions and ouabain.

Records in Fig. 1D show unsubtracted traces in another control oocyte in the same solution. The membrane capacity was rapidly charged with a time constant of 30 μsec. The capacity transient smoothly decayed to baseline without detectable voltage- and time-dependent nonlinear capacity components. For comparison (in Fig. 5B), the capacity transient in an oocyte injected with *Sh*H4 noninactivating and nonconducting clone (*Sh*H4-IR-W434F) cRNA showed a complex time course because of the voltage-dependent conformational changes of the channel protein that are thought to generate gating currents. These experiments indicate that the oocyte expression system for measurements of gating currents is excellent, provided that controls for nonlinear components are carried out. This control becomes critical when small gating currents are studied in batches of oocytes showing substantial Na$^+$/K$^+$ pump nonlinear cur-

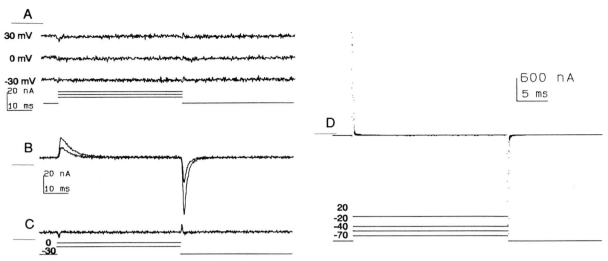

FIGURE 1 Linear and nonlinear transient components in control uninjected oocytes. (A) Subtracted records (HP = −90 mV, P/−4 and SHP = −120 mV) in external NaMESCa2. (B) Nonlinear transient currents to −30- and 0-mV pulse potentials in subtracted records (HP = −90 mV, P/−4 and SHP = −120 mV). The external solution was NaMESCa2 and the bottom solution was K-Glu EGTA. (C) Same pulses and conditions as in B, but after the addition of 100 μM ouabain to the external solution. (D) Capacity transients in unsubtracted records to various potentials (numbers at left) from HP = −90 mV in external NaMESCa2 and bottom K-Glu EGTA. Temperature, 22°C.

rents. In these cases, external Na^+-free solution or the addition of ouabain (100 μM) was adequate to eliminate contamination by pump currents, since K^+ gating currents were not modified whereas the nonlinear transporter currents were eliminated.

B. Gating and Ionic Currents in Shaker Inactivating (ShH4) and Noninactivating (ShH4-IR) K^+ Channels

Figure 2 shows ionic and gating currents measured from ShH4 (A,B) and ShH4-IR (C,D) channels. ShH4 ionic currents (*left*, Fig. 2A) display the typical characteristics of fast-inactivating K^+ current, with threshold activation around −50 mV and fast inactivation in the millisecond time range. On the other hand, as previously described for *Shaker* B K^+ channels (Hoshi *et al.*, 1990), ShH4 K^+ channels without amino acids 6–46 in the N terminus (Δ6–46, ShH4-IR) do not show inactivation, and the outward K^+ current is maintained during the depolarizing pulse.

Shaker B K^+ gating currents have been previously characterized after eliminating K^+ ionic currents by replacing the cytoplasmic oocyte solution with NMG-Glu using an internal perfusion system (Bezanilla *et al.*, 1991). To record charge movement in the presence of K^+, the mutation W434F, which has been shown to eliminate ionic conduction in ShH4-IR clones without major effects on gating current voltage dependence and kinetics, was used (Perozo *et al.*, 1993). The results

are shown in Fig. 2, which illustrates gating currents from ShH4 with the W434F mutation in normal inactivating channels (ShH4-W434F; Fig. 2B) and with inactivation removed (ShH4-IR-W434F; Fig. 2D). The salient feature in comparing ionic and gating currents is that charge movement can be detected at more negative potentials than ionic currents, indicating the presence of closed transitions before channel opening.

ON gating currents elicited during depolarizing pulses (Q_{on}) in inactivating (Fig. 2B) and noninactivating (Fig. 2D) K^+ channels share the same properties. For small depolarizations, the gating current rises very quickly and then relaxes with an apparent single exponential time constant, whereas for larger depolarizations, a slow rising phase appears that becomes more prominent with hyperpolarizing prepulses (see subsequent discussion). The main decay time constant becomes progressively slower for depolarizations up to nearly −30 mVs; thereafter, more positive potentials increase the rate of gating current decay. In a tetrameric channel model (MacKinnon, 1991b), in which four identical and independent gating subunits undergo a single voltage-dependent structural change for channel opening, gating currents should rise instantaneously and then decay following a single exponential. The recorded slow rising phase discards the hypothesis that identical gating subunits with two states respond to voltage in an independent manner.

The OFF gating currents of the mutant and the wild-type channels have fast return for small depolar-

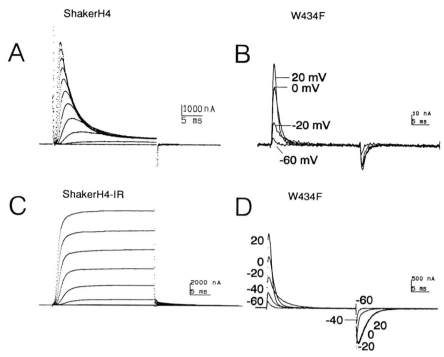

FIGURE 2 Ionic and gating currents in inactivating (*Sh*H4) and noninactivating (*Sh*H4-IR) K+ channels. (A) Outward inactivating currents in *Sh*H4 K+ channel clone. Pulses from −80 to 60 mV in 10-mV increments from HP = −90 mV. Records are unsubtracted. Solutions were NaMESCa2 (external) and K-Glu EGTA (internal). (B) Gating currents from *Sh*H4-W434F clone in the same solutions used in A. HP = −90 mV, P/−4 and SHP = −120 mV. Numbers are pulse potential values. (C) Outward noninactivating currents in *Sh*H4-IR K+ channel clone. Pulses from −70 to 20 mV in 10-mV increments from HP = −90 mV. Records were subtracted; P/−4 and SHP = −120 mV. Solutions were NaMESCa2 (external) and K-Glu EGTA (internal). (D) Gating currents from *Sh*H4-IR-W434F clone in the same solutions used in D. HP = −90 mV, P/−4 and SHP = −120 mV. Numbers are pulse potential values. Temperature, 22°C.

izing pulses (from −80 up to −50 mV). With larger depolarizations that should populate the open state, the OFF charge displays a rising phase and a longer time to relaxation. These findings are consistent with the idea that the transition from the open state to the last closed state carries less charge than the other transitions, as suggested from single channel studies (Zagotta and Aldrich, 1990).

The primary difference between inactivating and noninactivating channels is the amount of charge returned during repolarization to the holding potential. In the noninactivating channel, the same amount of charge moved during the depolarizing pulses returns during the sampling time (23 msec) after the end of the pulse (Figs. 2D, 3A), whereas in the inactivating channel only about 1/4th of the charge returns during the same time (Figs. 2B, 3B); the remaining charge in this clone has a much slower return. These results indicate that the same region of the N terminus that blocks the channel and induces inactivation is also

responsible for "immobilization" of the gating charge; these results support the correspondence between fast inactivation of the ionic current and charge immobilization. A similar conclusion was obtained in studies performed in squid axon Na+ ionic and gating currents. The OFF/ON ratio of the Na+ gating current time integral was reduced to about 30% when the depolarizing pulse length was increased from 0.3–1 msec to 10–20 msec, which increased the population of the open state (Bezanilla and Armstrong, 1975; Armstrong and Bezanilla, 1977; Meves and Vogel, 1977). In contrast, in DRK1 K+ channels, the OFF gating charge, as in the squid axon delayed rectifier (Almers and Armstrong, 1980; White and Bezanilla, 1985), did not show any charge immobilization, even when 125-msec pulses to +15 mV were used (Tagliatela and Stefani, 1993).

In summary, both inactivating and noninactivating nonconducting mutants reproduced the gating charge steady-state and kinetic properties of the correspond-

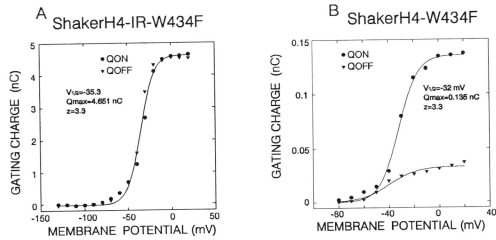

FIGURE 3 Charge–voltage relationship in noninactivating (ShH4-IR-W434F) (A) and inactivating (ShH4-IR-W434F) (B) nonconducting K$^+$ channels. Filled circles are Q$_{ON}$ values during 25-msec pulses; filled triangles are Q$_{OFF}$ values integrated for 23 msec after the end of the pulse. HP = −90 mV, P/−4 and SHP = −120 mV for depolarizing pulses and P/4, SHP = −130 mV for hyperpolarizing pulses. Temperature, 22°C.

ing conducting *Shaker* K$^+$ clones previously described using NMG-Glu as the internal solution (Bezanilla *et al.*, 1991). Further, W434F, which is a mutant of the putative pore-forming region, renders the K$^+$ channel nonconducting even though it undergoes the normal voltage-sensitive conformational changes such as charge immobilization in response to membrane depolarization. Thus, the normal kinetics of the gating charge movements strongly indicate that the mutant channel, except for a region in the pore, must maintain the normal structure.

C. Subtracted and Unsubtracted Gating Charge Records

Large gating currents (1–10 μA at 0 mV), like those illustrated in Fig. 4, could be recorded 2–4 days after injection of *Sh*H4-IR cRNA. The high expression of

these channels in conjunction with the high time resolution of the clamp allowed the recording of unsubtracted gating currents after a large fraction of the linear membrane capacity had been charged. Figure 4 shows subtracted (A) and unsubtracted (B) gating currents from the same oocyte expressing *Sh*H4-IR-W434F channels. The unsubtracted records have an initial off scale capacity transient (few sampling points taken at 50 μsec interval) that is followed by the gating charge. Note the similarity between the subtracted and unsubtracted gating charge records. Both records show (1) ON charge rising phase with a plateau and OFF rising phase for large depolarizations (in the unsubtracted records, the ON rising phase merges with the decay of the capacity transient, but a plateau phase is clearly discernible), (2) ON charge relaxation that is voltage dependent (the decay time constant became progressively slower up to pulses to −40 mV and

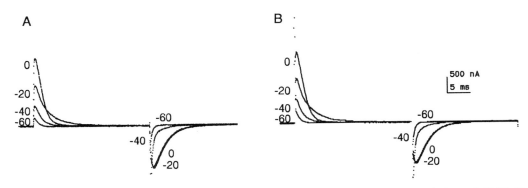

FIGURE 4 Subtracted (A) and unsubtracted (B) gating currents in noninactivating K$^+$ channel (*Sh*H4-IR-W434F). Records were obtained from the same oocyte in external NaMESCa2 and internal K-Glu EGTA. Numbers are pulse potentials from HP = −90 mV. (A) P/−4 and SHP = −120 mV. Temperature, 22°C.

faster with further depolarization), and (3) OFF gating charge with a slow relaxation component for pulses more depolarized than −40 mV and a prominent rising phase for larger depolarizations. These recordings clearly show that subtracted and unsubtracted gating currents are practically indistinguishable, thus confirming that negative subtracting pulses, from SHP = 120 mV, generally used in gating current subtracted records are adequate. Another observation from these records, as mentioned before, is that gating charge characteristics in the nonconducting *Sh*H4-IR-W434F clone are indistinguishable from those of the conducting clone *Sh*H4-IR, strengthening the notion that this mutation does not change the folding of the intramembranous channel protein.

D. Comparison of Gating Charge from Cut-Open Oocyte and Macropatches

Figure 5 compares gating current recordings obtained with the COVG with those obtained with cell-attached macropatches. Gating currents recorded with both methods are very similar. The right panel shows expanded traces that illustrate the ON rising phase and the charge decay with the superimposed fit to a single decaying exponential function.

With both techniques, the ON gate had the same kinetics and voltage dependence at all potentials. The OFF gate for stimulating potentials more negative than channel activation (−40 mV), from HP = −90 mV, was also practically identical with both techniques. However, when larger depolarizations (> −40 mV) were applied, the OFF gate was quite variable with the macropatch technique. On excision, the OFF gate was much slower with the macropatches; in some cases, the charge returned so slowly that it was difficult to separate it from baseline noise. In contrast, with COVG, the OFF charge remains practically invariant during the recording period. Records by Stühmer *et al.* (1991) and Schoppa *et al.* (1992) with excised macropatches showed a very small and slow OFF charge in the presence or absence of internal

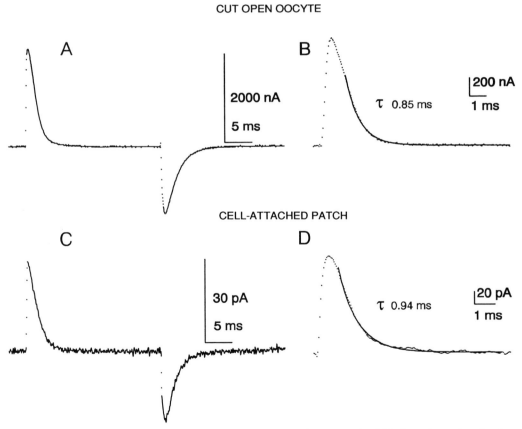

FIGURE 5 Gating current records in cut-open oocyte (A,B) and macropatch (C,D) clamp in noninactivating K⁺ channels (*Sh*H4-IR-W434F). Records were subtracted (P/−4, SHP = −120 mV, HP = −90 mV) for a pulse to 0 mV. (A,B) Traces were obtained in external NaMESCa2 and internal K-Glu EGTA. (C,D) Traces were obtained in external NaMESCa2. (B,D) Expanded traces with superimposed fitted single exponential to the decay phase of the gating currents. Temperature, 22°C.

TEA, respectively. The small OFF charge in the presence of internal TEA can be explained by the observation that internal TEA induces "charge immobilization" (Bezanilla *et al.*, 1991). However, this observation cannot explain the results of studies by Schoppa *et al.* (1992) that were done in the absence of internal TEA. These data are consistent with the smaller (although variable) gating charge tail that we observed in excised macropatch recordings.

In summary, these results indicate that the ON K$^+$ gating currents from *Sh*H4-IR-W434F have the same properties using COVG and cell-attached macropatches, and are as previously described (Bezanilla *et al.*, 1991; Perozo *et al.*, 1992). At present, no explanation is available for the variability and changes with time of OFF gating currents with the excised macropatch technique.

E. Hyperpolarizing Prepulses Enhance ON Gating Charge Rising Phase and Delay K$^+$ Ionic Currents

The ON gating charge rising phase for larger depolarizing pulses was attributed to initial slow forward rates among the closed states, as a consequence of interacting gating elements. A prediction from this model is that, although negative prepulses should increase the gating charge rising phase and delay ionic currents, more positive conditioning prepulses

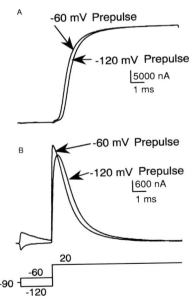

FIGURE 6 Effect of conditioning prepulses on ionic and gating currents. Ionic (A) and gating (B) currents recorded in *Sh*H4-IR and *Sh*H4-IR-W434F, respectively. Solutions were NaMESCa2 (external) and K-Glu EGTA (internal). P/-4, SHP = -120 mV; HP = -90 mV. Temperature, 22°C.

should have opposite effects (Cole–Moore effect; Cole and Moore, 1960). Figure 6 shows one of these experiments with ionic current (A; *Sh*H4-IR) and gating current (B; *Sh*H4-IR-W434F) recordings. In Fig. 6B, from HP = -90 mV, a prepulse to -120 mV elicited a small inward gating charge that was followed by a large outward gating charge for a test pulse to 20 mV. A prepulse to -60 mV that should bring the system closer to the open state generates a small ON gating charge that relaxes during the prepulse. Once that initial charge at -60 mV has moved, the pulse to 20 mV elicits a larger and faster ON gating charge practically lacking a rising phase. This behavior was associated with a faster onset of the ionic current (Fig. 6A). Similar experiments conducted in DRK1 clones had these general features, but with different voltage and time dependence of the installation of the ON charge rising phase (Taglialatela and Stefani, 1993). Thus, this behavior seems to be a general feature, as it was initially described by Cole and Moore (1960) in K$^+$ ionic currents of the squid axon, and argues in favor of the presence of initial slow transitions.

F. Internal TEA Immobilizes OFF Charge Movement

Internal TEA has been shown to induce charge immobilization (Bezanilla *et al.*, 1991). Those experiments were performed using subtraction protocols. To eliminate possible artifacts due to subtraction, equivalent experiments are performed recording unsubtracted gating current, which is now possible because of the high expression of these channels in conjunction with the high time resolution of the clamp.

Experimental evidence supports the view that internal TEA blockage of K$^+$ currents requires the channel to be in its open conformation (Armstrong, 1969). In *Sh*H4-IR, internal TEA has previously been shown not only to block ion conduction in the inner mouth of the channel but to directly or indirectly interact with the gating particles to produce OFF gating charge immobilization as the channel opens (Bezanilla *et al.*, 1991; Perozo *et al.*, 1992). In these experiments, the action of internal TEA in the nonconducting *Sh*H4-W434F mutant was explored to confirm its effect on charge immobilization and, further, as a probe for voltage-dependent channel conformation. A voltage- and time-dependent charge immobilization by internal TEA would indicate that the channel undergoes voltage- and time-dependent conformational changes to allow TEA to enter the mouth of the channel, even in the absence of ion conduction, a property conferred by the W434F mutation.

Figure 7 shows unsubtracted gating currents elic-

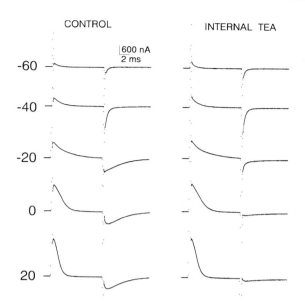

CONTROL   INTERNAL TEA

|600 nA
2 ms

-60

-40

-20

0

20

FIGURE 7 Charge immobilization induced by internal tetraethyl-ammonium (TEA) in unsubstracted gating current records. Gating current in *Sh*H4-IR-W434F to different pulse potentials (numbers at left) from HP = −90 mV. (*Left*) In external NaMESCa2 and internal K-Glu EGTA. (*Right*) Same cell with the same external solution, but 12 min after replacing the internal solution with 120 m*M* TEA-MES and 1 m*M* BAPTA. Temperature, 22°C.

ited with pulses to different potentials under control conditions (*left*) and in the presence of internal TEA (*right*). The sampling points at the beginning and end of the pulses correspond to off-scale capacity transients. Under control conditions, the time course of charge movement has the same kinetic features described previously in subtracted records (Bezanilla *et al.*, 1991; Perozo *et al.*, 1992). Gating currents elicited by small pulses do not have well-defined ON and OFF rising phases (−60 and −40 mV). As the depolarizing pulses were increased, ON gating currents showed a discernible rising phase (0 and 20 mV) and OFF gating currents became slower (−20 mV), with a prominent OFF rising phase (0 and 20 mV). The ON gating current rising phase was explained by positive cooperativity among the gating subunits, whereas the OFF gating current that developed for larger pulses was attributed to the progressive population of the open state, with increasing depolarization and a weakly voltage-dependent transition in the return from the open to the closed states.

In the presence of internal TEA (*right*), the ON charge remained unmodified at all potentials with respect to the control. The OFF charge also remained unmodified for potentials more negative than −40 mV that would not have elicited ionic currents. However, for more positive potentials, the OFF

charge was greatly reduced in size (immobilized). This result suggests that, although the channel is not able to conduct ions, it can still undergo the series of conformational changes that lead to the open state, allowing the entry of TEA into the internal mouth of the channel where it induces charge immobilization. Although internal TEA has been shown to increase the degree of charge immobilization in *Shaker* K⁺ channels (Bezanilla *et al.*, 1991), under identical conditions isotonic internal TEA does not immobilize the OFF charge in DRK1 K⁺ channels (Taglialatela and Stefani, 1993). This difference in the effect of internal TEA on charge immobilization could arise from differences in the TEA binding site located in the internal mouth of the pore (Hartmann *et al.*, 1991; Taglialatela *et al.*, 1991; Yellen *et al.*, 1991; Kirsch *et al.*, 1992; for review, see MacKinnon, 1991a) or in the coupling mechanisms between the pore and the gating machinery of the channels.

G. Basic Model of Channel Activation and Inactivation

Figure 8 is a schematic representation of a model that explains many of the properties of ionic and gating current records in the inactivating *Shaker* K⁺ channel. The channel can have several closed states (C_1 to C_5) preceding the open state (O). Depolarization favors the occupancy of states on the right whereas hyperpolarization favors the states on the left. At negative potentials, the channel is closed and the gating charges are in the resting state (R). On depolarization, charges can move to the active state (A). When all four gating charges have moved, the channel opens by moving from C_5 to O without a significant charge displacement. This model is in accordance with a C_5 to O transition that is weakly voltage dependent. In the open state (O), the channel can be inactivated (blocked) by the ball peptide, resulting in the inactivated state I_1 that is progressively populated as the channel opens during depolarization. On repolarization, only a small fraction of channels can return to the closed states from I_1 through the open state (Demo and Yellen, 1991; Ruppersberg *et al.*, 1991). The majority of the channels will move from I_1 to I_3. These last transitions are associated with the return of about 1/4th of the charge on repolarization and correspond to the charge movement that rapidly returns, after repolarization, from a depolarizing inactivating step. The rate-limiting step of charge return is the slow $I_3 \to C_4$ transition that reflects the OFF rate of the ball peptide from the channel inner mouth. This slow transition limits the charge return and is associated with the slow charge recovery (charge immobilization)

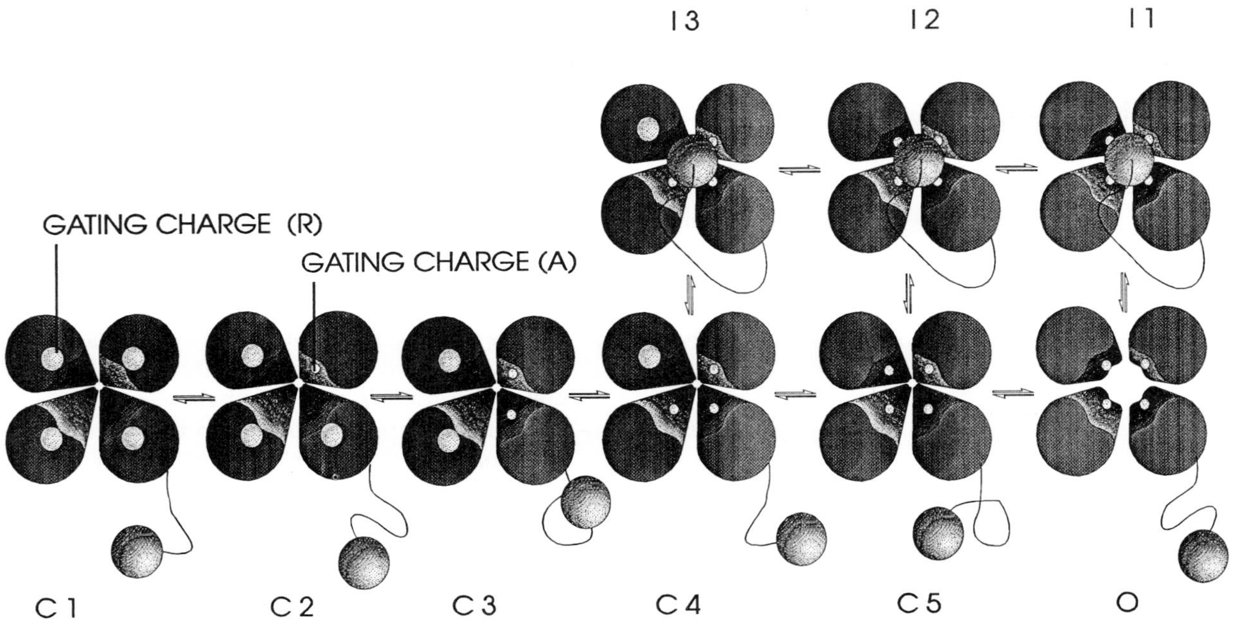

I 3   I 2   I 1

GATING CHARGE (R)

GATING CHARGE (A)

C 1   C 2   C 3   C 4   C 5   O

FIGURE 8 Scheme of channel states in an inactivating K^+ channel. The cytoplasmic side of the channel is represented and, for simplicity, one "ball and chain" structure for the channel is depicted. In the resting state (R), the charge elements are closer to the cytoplasmic surface whereas during activation (A) the charge elements are displaced to the center of the channel protein. When all the charges have moved (C_5) the channel opens (O) without significant charge displacement. The channel in the open state can be inactivated (blocked) by the "ball" peptide, resulting in the inactivated states (I_1, I_2, and I_3). Ball peptide-induced inactivation prevents charge return and, on repolarization only, one-fourth of the charge can be recovered rapidly. The subsequent slow recovery of charge is slowed by the rate limiting step of the transition $I_3 \rightarrow C_4$.

after repolarization. Thus, a small fraction of charge (about 1/4) returns rapidly on repolarization (I_1 to I_3 transition). The remaining charge remains "immobilized" and slowly returns because of the slow $I_3 \rightarrow C4$ transition. Consequently, the second phase of slow charge return (Armstrong and Bezanilla, 1977) and the recovery from inactivation are governed by the rates of this transition. That these processes depend on the membrane potential suggests that the ball peptide can sense a fraction of the electric field.

IV. Summary and Conclusions

Ionic and gating currents in *Shaker* K^+ channels expressed in *Xenopus laevis* oocytes were studied. The cut-open oocyte voltage-clamp technique with active guards were used to further improve the frequency response. In most cases, control oocytes had undetectable nonlinear capacity transients; however, in few cases, small (5–20 nA) nonlinear components blocked by ouabain (100 μM) and sensitive to external Na^+ were detected. When present, these currents, attributed to the Na^+/K^+ pump, did not distort gating current records since they could be eliminated selectively

with ouabain or Na^+-free solutions. Ionic and gating currents in inactivating and noninactivating *Shaker* K^+ channels were studied to investigate the mechanisms of channel activation and inactivation. The high channel expression and the speed of the clamp facilitated measuring gating currents in the absence of subtracting pulses. Further, a single point mutation (W434F) that obliterated ionic conduction without affecting gating current properties was used. Thus, gating currents were recorded in the presence of normal internal K^+ concentration. Gating currents in this mutant had the same general properties as gating currents from the equivalent clones with normal ion conduction, recorded in the absence of internal and external K^+. Gating currents had the following salient properties: (1) the turn-on of the gating current shows a rising phase, suggesting that the hypothetical activating subunits are not independent; (2) the more negative position of the charge–voltage curve relative to the conductance–voltage curve and the charge displacement by hyperpolarizing prepulses indicates that a large fraction of the voltage dependence occurs in the transitions between closed states; (3) the open-to-closed transition is weakly voltage dependent, as suggested by the presence of a rising phase in the

OFF gating currents; (4) in K$^+$ channels showing fast inactivation, the OFF gating charge is partially immobilized for large depolarizing pulses, whereas in mutant channels lacking inactivation the charge is recovered quickly at the end of the pulse; and (5) internal TEA mimics the inactivation particle, inducing charge immobilization without affecting the ON charge. These results eliminate the hypothesis of equal and independent gating subunits with two positions, but suggest the presence of interactions among the gating subunits and are consistent with a model for K$^+$ channel activation in which an early slower and/or less voltage-dependent transition between very deep closed states is followed by more voltage-dependent transitions between later closed states, and, finally, by a voltage-independent transition between the last closed state and the open state. Further, a mutation (W434F) in the proposed pore region between transmembrane segments S5 and S6 completely abolishes ion conduction without affecting the voltage dependence of gating charge of the channel. Although the conduction pathway in this mutant is not functional, the channel can still undergo the closed–open conformation transition in response to voltage changes, since internal TEA induces OFF charge immobilization for large depolarization that would have opened the channel.

Acknowledgments

Supported by United States Public Health Services Grants GM30376 (F.B.), GM50550 (E.S.), and HL47382 (L.T.). E. Perozo was supported by the Pew Charitable Trust.

References

Almers, W., and Armstrong, C. M. (1980). Survival of K$^+$ permeability and gating currents in squid axons perfused with K$^+$-free media. *J. Gen. Physiol.* **75**, 61–78.

Armstrong, C. M. (1969). Inactivation of the potassium conductance and related phenomena caused by quaternary ammonium ion injection in squid axons. *J. Gen. Physiol.* **54**, 553–575.

Armstrong, C. M., and Bezanilla, F. (1973). Currents related to movement of the gating particles of the sodium channels. *Nature (London)* **242**, 459–461.

Armstrong, C. M., and Bezanilla, F. (1977). Inactivation of the sodium channel. II. Gating current experiments. *J. Gen. Physiol.* **70**, 567–590.

Bezanilla, F. (1985). Gating of sodium and potassium channels. *J. Membrane Biol.* **88**, 97–111.

Bezanilla, F., and Armstrong, C. M. (1975). Kinetic properties and inactivation of the gating currents of sodium channels in squid axon. *Philosol. Trans. Roy. Soc. Lond.* **270**, 449–458.

Bezanilla, F., Perozo, E., Papazian, D. M., and Stefani, E. (1991). Molecular basis of gating charge immobilization in *Shaker* potassium channels. *Science* **254**, 679–683.

Cole, K. S., and Moore, J. W. (1960). Potassium ion current in the squid giant axon: Dynamic characteristics. *Biophys. J.* **1**, 161–202.

Dascal, N. (1987). The use of *Xenopus* oocytes for the study of ion channels. *CRC Crit. Rev. Biochem. Mol. Biol.* **22**, 317–387.

Demo, S. D., and Yellen, G. (1991). The inactivation gate of the *Shaker* K$^+$ channel behaves like an open-channel blocker. *Neuron* **7**, 743–753.

Fu, L., Ye, R., Browder, L. W., and Johnston, R. N. (1991). Translational potentiation of messenger RNA with secondary structure in *Xenopus*. *Science* **251**, 807–810.

Hartmann, H. A., Kirsch, G. E., Drewe, J. A., Taglialatela, M., Joho, R. H., and Brown, A. M. (1991). Exchange of conduction pathways between two related K$^+$ channels. *Science* **251**, 942–944.

Hoshi, T., Zagotta, W. N., and Aldrich, R. W. (1990). Biophysical and molecular mechanisms of *Shaker* potassium channel inactivation. *Science* **250**, 533–538.

Kamb, A., Tseng-Crank, J., and Tanouye, M. A. (1988). Multiple products of the *Drosophila Shaker* gene may contribute to potassium channel diversity. *Neuron* **1**, 421–430.

Kirsch, G. E., Drewe, J. A., Hartmann, H. A., Taglialatela, M., de Biasi, M., Brown, A. M., and Joho, R. H. (1992). Differences between the deep pores of K$^+$ channels determined by an interacting pair of nonpolar amino acids. *Neuron* **8**, 499–505.

MacKinnon, R. (1991a). Using mutagenesis to study potassium channel mechanisms. *J. Bioenerg. Biomembr.* **23**, 647–663.

MacKinnon, R. (1991b). Determination of the subunit stoichiometry of a voltage-activated potassium channel. *Nature (London)* **350**, 232–235.

Meves, H., and Vogel, W. (1977). Slow recovery of sodium current and gating current from inactivation. *J. Physiol.* **267**, 395–410.

Perozo, E., MacKinnon, R., Bezanilla, F., and Stefani, E. (1993). Gating currents from a non-conducting mutant reveal open-closed conformations in *Shaker* K$^+$ channels. *Neuron* **11**, 353–358.

Perozo, E., Papazian, D. M., Stefani, E., and Bezanilla, F. (1992). Gating currents in *Shaker* K$^+$ channels. Implications for activation and inactivation models. *Biophys. J.* **62**, 160–171.

Rakowski, R. F. (1993). Charge movement by the Na$^+$/K$^+$ pump in *Xenopus* oocytes. *J. Gen. Physiol.* **101**, 117–144.

Ruppersberg, J. P., Frank, R., Pongs, O., and Stocker, M. (1991). Cloned neuronal I$_k$(A) channels reopen during recovery from inactivation. *Nature (London)* **353**, 657–660.

Schneider, M. F., and Chandler, W. K. (1973). Voltage-dependent charge movement in skeletal muscle: A possible step in excitation-contraction coupling. *Nature (London)* **242**, 244–246.

Schoppa, N. E., McCormack, K., Tanouye, M. A., and Sigworth, F. J. (1992). The size of gating charge in wild-type and mutant *Shaker* potassium channels. *Science* **255**, 1712–1715.

Spires, S., and Begenisich, T. (1989). Pharmacological and kinetic analysis of K$^+$ channel gating currents. *J. Gen. Physiol.* **93**, 263–283.

Stühmer, W., Conti, F., Stocker, M., Pongs, O., and Heinemann, S. H. (1991). Gating currents of inactivating and non-inactivating potassium channels expressed in *Xenopus* oocytes. *Pflügers Arch.* **418**, 423–429.

Taglialatela, M., and Stefani, E. (1993). Gating currents of the cloned delayed rectifier K$^+$ channel DRK1. *Proc. Natl. Acad. Sci. USA* **90**, 4758–4762.

Taglialatela, M., vanDongen, A. M. J., Drewe, J. A., Joho, R. H., Brown, A. M., and Kirsch, G. E. (1991). Patterns of internal and external tetraethylammonium block in four homologous K$^+$ channels. *Mol. Pharmacol.* **40**, 299–307.

Taglialatela, M., Toro, L., and Stefani, E. (1992). A novel voltage clamp to record small, fast currents from ion channels expressed in *Xenopus* oocytes. *Biophys. J.* **61**, 78–82.

Tempel, B. L., Papazian, D. M., Schwarz, T. L., Jan, Y. L., and

Jan, L. Y. (1987). Sequence of a probable potassium channel component encoded at *Shaker* locus of *Drosophila*. *Science* **237**, 770–775.

White, M. M., and Bezanilla, F. (1985). Activation of squid axon K$^+$ channels. *J. Gen. Physiol.* **85**, 539–554.

Yellen, G., Jurman, M. E., Abramson, T., and MacKinnon, R. (1991). Mutations affecting internal TEA blockade identify the probable pore-forming region of a K$^+$ channel. *Science* **251**, 939–942.

Zegotta, W. N., and Aldrich, R. W. (1990). Voltage-dependent gating of *Shaker* A-type potassium channels in *Drosophila* muscle. *J. Gen. Physiol.* **95**, 29–60.

Shaw-*Related* K$^+$ *Channels* *in Mammals*

E. VEGA-SAENZ de MIERA, M. WEISER,
C. KENTROS, D. LAU, H. MORENO,
P. SERODIO, AND B. RUDY

I. Introduction

Twenty-five transcripts from genes related to the *Drosophila Shaker* gene (the *Sh* gene family) have been identified in mammals by molecular methods. *Sh* transcripts are thought to encode subunits of tetrameric voltage-gated K$^+$ channels. However, the relationship between native K$^+$ channels and the molecular components identified by cloning, and thereby the physiological significance of this molecular diversity, remain to be elucidated.

Sh genes and their products can be divided into four subfamilies. Compelling evidence suggests that subunits of the same subfamily, but not of different subfamilies, form heteromultimeric channels *in vitro*; thus, each gene subfamily is postulated to encode components of an independent channel system. The potential significance of heteromultimer formation demonstrates the importance of studying the genes of one subfamily as a group when the physiological role of any of these genes is being pursued. This chapter focuses on the characterization of the *Shaw*-related or *Sh*III subfamily in mammals, emphasizing the work done to understand the relationship between cloned components and native K$^+$ channels. This work illustrates how molecular biology may contribute to the discovery of new, previously unidentified K$^+$ channels.

II. Molecular Diversity and Neuronal Function

Diversity is an old and recurring theme in the organization and function of the nervous system. The early neuroanatomists recognized that the large variety of neuronal forms must be associated with functional specialization. The application of intracellular recording techniques to the nervous system several decades later revealed a similar diversity of intrinsic electrophysiological properties of neurons. This theme is re-emerging in the study of the molecules responsible for the generation and transmission of signals in neurons.

Large numbers of protein components of ligand- and voltage-gated ion channels as well as neurotransmitters and neuropeptide receptors (which modulate the activity of ion channels through intermediary second messenger cascades) are being discovered by molecular cloning (Betz, 1990,1992; Cockroft *et al.*, 1990; Buck and Axel, 1991; Catterall, 1992; Gasic and Hollmann, 1992; Hille, 1992; Nakanishi, 1992; Neher, 1992; Role, 1992; Sommer and Seeburg, 1992; Wisden and Seeburg, 1992; Jessell and Kandel, 1993; Sargent, 1993). This diversity suggests a degree of functional specificity well beyond that previously considered and that the power of generating variation by gene duplication and divergence played a role in the gener-

41

ation of complexity in the nervous system of animals with elaborate nervous systems. These studies will influence our views on the organization and function of the nervous system.

Much of the cloning work has been done in the absence of prior isolation of native proteins. Therefore, the exact relationship between the molecular components identified by cloning and the native channels or receptors is, in most cases, not known. Therefore, the physiological significance of the molecular diversity that is being discovered remains to be elucidated.

One notable example is the large number of subunits of γ-aminobutyric acid (GABA$_A$) receptors that has been cloned (reviewed by Seeburg et al., 1990; Wisden and Seeburg, 1992; see also Chapter 21). Since GABA$_A$ receptors are thought to be pentameric, the variety of receptors that could be generated from different combinations of these subunits is truly staggering, suggesting variations in synaptic inhibition that were not previously expected. New views on neuronal integration are likely to emerge from understanding the physiological significance of this molecular diversity. Although the problem is complex, progress is being made in determining which combinations of subunits may exist in vivo and where they might be present (Wisden and Seeburg, 1992), in establishing functional differences between receptors with different subunit combinations reconstituted in model expression systems (Seeburg et al., 1990; Sigel et al., 1990; Verdoorn et al., 1990), and in determining the possible roles of specific isoforms in brain function (Korpi et al., 1993). A similar problem exists for the components of K$^+$ channels that have been identified by cloning methods.

III. Families of K$^+$ Channel Genes

K$^+$ channels, a group of ligand- and voltage-gated channels that conduct K$^+$ ions preferentially, are present in most, if not all, eukaryotic cells. These channels play key roles in a wide range of cellular functions including excitability, secretion, differentiation, mitogenesis, and osmotic regulation (e.g., Thompson and Aldrich, 1980; Klein et al., 1982; Llinas, 1984,1988; Peterson and Maruyama, 1984; Adams and Galvan, 1986; Kaczmareck and Levitan, 1987; Levitan, 1988; Lewis and Cahalan, 1988; Rudy, 1988; Latorre et al., 1989; Grinstein and Smith, 1990; Hille, 1992; Dubois and Rouzaire-Dubois, 1993). Among the ion channels, K$^+$ channels constitute a group exhibiting particularly large functional diversity (Rudy, 1988; Latorre et al., 1989; Hille, 1992); K$^+$ channel types vary in their re-

sponse to factors, such as intracellular Na$^+$ and Ca^{2+} concentrations and the membrane potential, that determine the opening and closing of the channel. K$^+$ channels also vary in kinetics, permeability properties, and sensitivity and type of response to toxins, drugs, and modulating factors such as second messenger cascades. This diversity allows a fine tuning of resting membrane potentials and action potential waveforms, and a modulation of firing and neurotransmitter-secretion patterns (Klein et al., 1982; Kaczmareck and Levitan, 1987; Rudy, 1988; Levitan, 1988; Llinas, 1988; McCormick, 1990; Baxter and Byrne, 1991; Hille, 1992). Since K$^+$ channels are frequently targets of second messenger cascades, their diversity influences the type of cellular response to different types of stimuli (Kaczmareck and Levitan, 1987; Levitan, 1988; Lopez-Barneo et al., 1988; Rudy, 1988; Hille, 1992). Alterations in excitability may arise by differential spatiotemporal expression of K$^+$ channels.

Molecular cloning of K$^+$ channel components is an enterprise that began with the cloning of the Shaker locus in Drosophila (Baumann et al., 1987; Kamb et al., 1987; Papazian et al., 1987; Tempel et al., 1987). Subsequent work has revealed a molecular diversity of K$^+$ channel components that parallels the functional diversity of K$^+$ channels in vivo. Over 30 different cDNAs corresponding to transcripts from several gene families have already been discovered (Fig. 1). Of these families, the one consisting of homologs of the Drosophila Shaker gene (called here the Shaker or Sh gene family) encoding putative subunits of voltage-gated K$^+$ channels (reviewed in Jan and Jan, 1990a; Perney and Kaczmareck, 1991; Rehm and Tempel, 1991; Rudy et al., 1991a; Pongs, 1992) has been studied most extensively and is the subject of this review.

A. Sh Proteins: Putative Subunits of Voltage-Gated K$^+$ Channels

Nearly 20 Sh genes have been identified in mammals. Based on sequence similarities and, hence, probable evolutionary relationships, the family is divided into four groups or subfamilies (reviewed by Jan and Jan, 1990a; Perney and Kaczmareck, 1991; Rudy, et al., 1991a; Pongs, 1992; Salkoff et al., 1992; see also Chapter 1) designated here as ShI, ShII, ShIII and ShIV. A member of a subfamily in mammals is also more similar to one of four related Drosophila genes (Shaker, Shab, Shaw, and Shal; Butler et al., 1989; Wei et al., 1990) than to a mammalian member of a different subfamily, suggesting that these subfamilies are derived from precursors that existed prior to the divergence of mammals and insects. ShI mammalian genes are thought to be homologs of Shaker, ShII of

S4 Superfamily

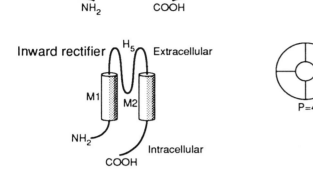

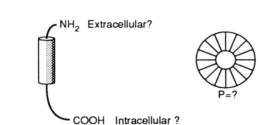

FIGURE 1 Families of K⁺ channel subunits. Three different types of protein subunits of K⁺ channels are shown. (*Top*) K⁺ channel proteins of the S4 superfamily include those of the *Sh* and *eag* families, which are components of voltage-gated K⁺ channels, as well as those of the *slo* family, which are components of calcium-activated K⁺ channels. This superfamily also includes cyclic nucleotide-gated channels and the α subunits of voltage-gated Na⁺ and Ca²⁺ channels (Jan and Jan, 1990b). K⁺ channel proteins of the S4 superfamily and cyclic nucleotide-gated channel proteins contain six membrane spanning domains (including an S4 domain) and a pore domain (with an H5-like sequence on K⁺ channels); four such polypeptides form one functional channel. α subunits of voltage-gated Na⁺ and Ca²⁺ channels consist of four homologous domains, each resembling a K⁺ channel or cyclic nucleotide-gated channel protein (Catterall, 1988). (*Middle*) Inward rectifier K⁺ channel proteins, also proposed to form tetrameric channels (Kubo *et al.*, 1993), contain two membrane spanning domains and one H5-like pore domain. (*Bottom*) The I_{SK} protein expresses slowly activating voltage-dependent K⁺ currents and contains a single membrane spanning domain and no H5 domain. How many such subunits form a functional channel is not known (Takumi *et al.*, 1988).

Shab, *Sh*III of *Shaw*, and *Sh*IV of *Shal*. Multiples genes of each subfamily have been described in mammals (*Sh*I, 7; *Sh*II, 2; *Sh*III, 4; and *Sh*IV, 3), indicating that, in the chordate line (Deuterostomia), the precursors underwent extensive duplication and subsequent variation, leading to a large number of closely related genes. In addition, two *Sh* cDNAs isolated from rat libraries may define two additional subfamilies (Drewe *et al.*, 1992). Moreover, mammalian *Sh*III genes encode more than one product by the process of alternative splicing (Luneau *et al.*, 1991a,b; Rudy *et al.*, 1992; Vega-Saenz, de Miera *et al.*, 1992), leading to the existence of 25 known *Sh* transcripts. All these transcripts are expressed in brain, and some are expressed in other tissues as well.

Each *Sh* transcript induces the expression of voltage-dependent K⁺ currents with characteristic voltage dependence and kinetic and pharmacological properties when introduced in heterologous expression systems such as *Xenopus* oocytes and cells in culture (Jan and Jan, 1990a; Perney and Kaczmareck, 1991; Rudy *et al.*, 1991a; Pongs, 1992; Salkoff *et al.*, 1992). Evidence strongly suggests that the channels underlying these currents are homomultimers (Jan and Jan, 1990a; Perney and Kaczmarek, 1991; Rudy *et al.*, 1991a; Pongs, 1992, Salkoff *et al.*, 1992), probably tetramers of *Sh* proteins (MacKinnon, 1991a; Liman *et al.*, 1992). Therefore, the proteins encoded by *Sh* transcripts are likely to be subunits of voltage-gated K⁺ channels found in neurons and other cells. One of the current challenges is understanding the functional roles of the cloned products, beginning with the identification of native voltage-gated K⁺ channels that contain these components.

B. Heteromultimer Formation and Channel Subunit Composition

Compelling evidence exists that *in vitro* coexpression of different subunits from the same subfamily, *but not from different subfamilies*, results in the formation of electrophysiologically distinct heteromultimeric channels (Christie *et al.*, 1990; Isacoff *et al.*, 1990; K. McCormack *et al.*, 1990; Ruppersberg *et al.*, 1990; Covarrubias *et al.*, 1991; Weiser *et al.*, 1993). The significance of heteromultimer formation in determining channel composition *in vivo* is still not fully appreciated. Other factors, such as interaction with non-*Sh* subunits and posttranslational modifications, may also contribute to the functional properties of native channels containing *Sh* proteins (Rehm and Lasdunzky, 1988; Rehm *et al.*, 1989a,b; Rudy *et al.*, 1988,1991a; Warmke *et al.*, 1991; Zhong and Wu, 1991). Further, the influence of heteromultimer formation, interaction with non-*Sh* subunits, and posttranslational modifications could vary from cell to cell. K⁺ channel diversity arising from these factors may contribute to the individualities in electrophysiological properties of single neurons in the nervous system.

Given the number of subunits available and of factors that could determine channel composition *in vivo*, the task of understanding the relationship between the diversity of molecular components and the functional diversity of K$^+$ channels is far from simple. These problems are symptomatic of the challenge that follows the isolation of cloned gene products: the right pieces are available but we must discover what they do. This chapter focuses on work that addresses this problem for the products of the *Sh*III or *Shaw*-related subfamily of K$^+$ channel genes.

IV. Toward an Understanding of the Functional Role of Mammalian *Shaw*-Related Proteins

A. Four *Shaw-Related* Genes Encode at Least 11 Transcripts

As in the other mammalian *Sh* subfamilies the *Shaw*-related or *Sh*III subfamily consists of several genes formed by gene duplication throughout animal evolution. However, this group is the only class of *Sh* genes in mammals for which evidence exists of alternative splicing also contributing to the generation of diversity. Researchers have isolated 11 different *Sh*III cDNAs from libraries derived from rodent and human mRNA (Yokoyama *et al.*, 1989; T. McCormack *et al.*, 1990,1991; Luneau *et al.*, 1991a,b; Rudy *et al.*, 1991b,1992; Sen *et al.*, 1991; Schroter *et al.*, 1991; Ghanshani *et al.*, 1992; Vega-Saenz de Miera *et al.*, 1992). Based on sequence analysis, these cDNAs are thought to represent transcripts from four different genes that generate more than one product by alternative splicing (Fig. 2).

Various names for the four *Sh*III genes and their products are shown in Table I. This table also includes the corresponding names in a standardized nomenclature proposed by several investigators in the field (Chandy *et al.*, 1991) and the name assigned to each gene by the Human Genome Nomenclature Committee. The nomenclature of Chandy *et al.* will be implemented throughout this chapter.

1. Evidence for the Existence of Four *Sh*III Genes

The nucleotide sequences of the cDNAs thought to be products of independent genes differ throughout the length of the cDNAs, suggesting that they are derived from transcripts of distinct genes rather than by alternative splicing. The analysis of genomic clones (see subsequent discussion) and the locations of *Sh*III genes in human and mouse chromosomes (Table II) are further evidence that the eleven transcripts are derived from four independent genes.

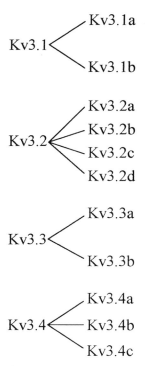

FIGURE 2 *Sh*III K$^+$ channel subunits. Four mammalian *Shaw*-related genes KV3.1, KV3.2, KV3.3, and KV3.4 encode 11 different K$^+$ channel subunit isoforms by alternative splicing.

TABLE I   Nomenclature of *Shaw*-Related K$^+$ Channel Transcripts in Mammals

Name used in this chapter[a]	Other names	Gene symbol[b]
KV3.1a	NGK2, RKShIIIB, KShIIIB, KV3.1β	KCNC1
KV3.1b	KV4, KV3.1α, Raw2	
KV3.2a	RKShIIIA, (R)KShIIIA.1	KCNC2
KV3.2b	KShIIIA.3	
KV3.2c	—	
KV3.2d	KShIIIA.2, Raw1	
KV3.3a	KShIIID.1	KCNC3
KV3.3b	KShIIID.2	
KV3.4a	Raw3	KCNC4
KV3.4b	(H)(R)KShIIIC	
KV3.4c	—	

[a] Nomenclature based on Chandy *et al.* (1991).
[b] Symbol assigned to the gene by the human genome nomenclature committee.

TABLE II   **Chromosomal Localization of *Sh*III Genes in Mice and Humans**

Gene	Chromosomal location (mice)	Chromosomal location (human)
KCNC1	7 (a)	11p15 (b,c)
KCNC2	10 (a)	12 (c); 19 (a)
KCNC3	10 (a); 7 (d)	19q13.3-4 (a,e)
KCNC4	NA	1p21 (e,f)

Mapping references are (a) Haas *et al.*, 1993; (b) Ried *et al.*, 1993; (c) Grissmer *et al.*, 1992; (d) D. Lau, V. Clarke, P. D'Eustachio, and B. Rudy, unpublished observations; (e) Ghanshani *et al.*, 1992; (f) Rudy *et al.*, 1991b.

NA, not available. The discrepancies in the locations of KCNC3 in mouse and KCNC2 in human remain to be clarified.

2. Alternative Splicing Generates Carboxyl-End *Sh*III Variants

Putative alternatively spliced transcripts of each gene have identical nucleotide sequences from the assumed starting ATG up to a point of divergence near, but prior to, the in-frame stop codon, predicting protein products with different carboxyl ends (Fig. 3). Mature transcripts from each gene diverge at the same point in the sequence, following an AG, the dinucleotide characteristic of the exonic portion of donor or 5' splice junctions (Mount, 1982). These sequence relationships suggest that the different transcripts arise by alternative splicing of a primary transcript rather than by transcription from separate but highly homologous genes. In two cases, KV3.1 (Luneau *et al.*, 1991a) and KV3.2 (Luneau *et al.*, 1991b), support for the alternative splicing mechanism comes from analysis of rat genomic DNA in Southern blots. Probes derived from the sequence preceding the point of divergence detected a single restriction fragment for each gene in Southern blots of rat genomic DNA hybridized under high stringency conditions. This result indicates that the genes identified by those probes are represented only once in the rat genome.

B. Patterns of Alternative Splicing and Genomic Structure

Given the results of partial genomic analysis of the KV3.1 (Luneau *et al.*, 1991a), KV3.3 (Ghanshani *et al.*, 1992), and KV3.2 (D. Lau and B. Rudy, unpublished observations) genes, as well as analysis of cDNAs thought to be derived from incompletely processed RNAs (E. Vega-Saenz de Miera and B. Rudy, unpublished observations), all *Sh*III genes apparently have a similar structure. This hypothesis is consistent with the idea that all four genes arose by duplication of a single ancestral gene. This possibility is further borne out by the patterns of alternative splicing of *Sh*III genes (Fig. 4).

As depicted in Fig. 4, the amino end of the proteins is encoded in a single exon (exon 1) that is constitutively spliced in the four genes to a downstream exon (exon 2) encoding most of the membrane core portion of the polypeptides,[1] from the beginning of the S1 domain to a point in the carboxyl end region following the S6 domain. Evidence that the amino end is encoded in a single exon comes from the analysis of rodent genomic clones of KV3.3 (Ghanshani *et al.*, 1992) and KV3.2 (D. Lau and B. Rudy, unpublished results). Exon 2 has been identified in genomic clones of KV3.1 (Luneau *et al.*, 1991a), KV3.2 (D. Lau and B. Rudy, unpublished results), and KV3.3 (Ghanshani *et al.*, 1992). The data are consistent with the presence of two sites of alternative splicing in *Sh*III genes: the first at the end of exon 2 and the second at the end of exon 3. In two of the four *Sh*III genes, KV3.1 (Luneau *et al.*, 1991a) and KV3.3 (Vega-Saenz de Miera *et al.*, 1992), the sequence immediately following exon 2 is either read through, generating KV3.1a and KV3.3b, or spliced out and replaced by exons present downstream to generate KV3.1b and KV3.3a. In KV3.1a and KV3.3b, the nucleotide sequence following the AG at the end of exon 2 starts with GT, the dinucleotide characteristic of the intronic portion of 5' splice sites (Mount, 1982), supporting the idea that the divergent region of these transcripts is derived from the sequence following the splice site.

Genomic DNA analysis provides direct evidence for this "splice/don't splice" alternative splicing mechanism (McKeown, 1992). Utilizing polymerase chain reaction (PCR) of rat genomic DNA, Luneau *et al.* (1991a) demonstrated that the sequence encoding the divergent 3' end of KV3.1a is contiguous with exon 2. These investigators also isolated a rat genomic clone that contains two exons encoding the divergent sequence of KV3.1b (exon 3 and exon 4). The 5' end of exon 3 of the KV3.1 gene starts with the sequence of KV3.1b immediately following the point of divergence and extends 188 bp in the 3' direction. Exon 4, separated from exon 3 by an intron, contains the remainder of the 3' end of KV3.1b. In the case of the KV3.3 gene, Ghanshani *et al.* (1992) isolated a mouse genomic clone

---

[1] As described in Section IV,G, *Sh*III proteins (like other *Sh* proteins) consist of a core membrane region with six putative (S1, S2, S3, S4, S5, and S6) membrane-spanning domains flanked by putatively intracellular amino and carboxyl domains.

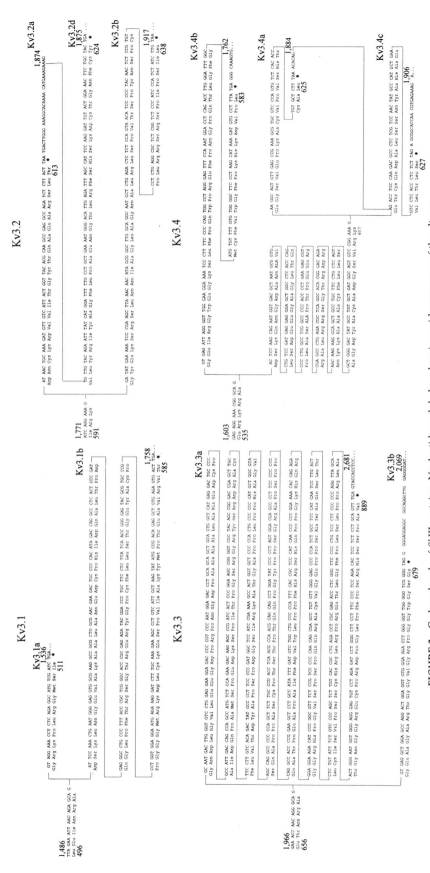

FIGURE 3 Carboxyl ends of *ShIII* proteins. Nucleotide and deduced amino acid sequences of the alternatively spliced 3' ends of transcripts from four *ShIII* genes. In all cases, the nucleotide and predicted amino acid sequences diverge after an AG. The first in-frame stop condons are indicated with asterisks. The alternatively spliced carboxyl ends contain serines and threonines within consensus sequences for phosphorylation by various protein kinases (see Table IV).

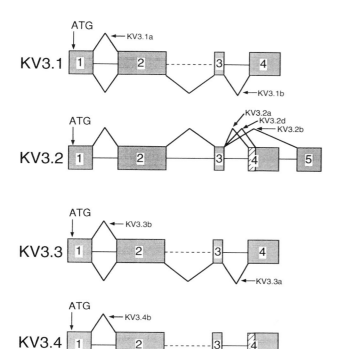

FIGURE 4 Genomic structure and patterns of alternative exon usage of four *Sh*III genes. In this schematic depiction of the structure of the coding regions of four *Sh*III genes, exons are indicated by numbered filled boxes, and introns are indicated by lines. Broken lines in KV3.1, KV3.3, and KV3.4 indicate sequences that are read through to generate one alternatively spliced transcript or spliced out and replaced by downstream exons to generate other mature transcripts. Exon 4 contains an internal acceptor site in KV3.2 and KV3.4. The region of the exon preceding this acceptor site is cross-hatched. Lengths of exons and introns are not necessarily to scale.

tion to the end of exon 2 in KV3.1 and KV3.3 (indicated with an arrow in Fig. 11) by 165 bp. Moreover, this additional sequence shows significant similarity (Fig. 5) to exon 3 of the other genes. Therefore, alternative splicing of KV3.2 may take place at the end of exon 3 rather than at the end of exon 2.

No direct information is available on how the known KV3.2 variants are generated. However, sequence analysis suggests that the KV3.2a and KV3.2b variants arise from the use of alternative exons that follow exon 3 (Fig. 4). The number and position of these exons in genomic DNA might differ from that shown in Fig. 4. The sequence of KV3.2d, immediately following the splicing point, is found in KV3.2a 95 bp downstream (see Fig. 3). After this new overlap, the two sequences are identical. Thus, KV3.2d may arise by use of an internal acceptor site in the same exon used to generate KV3.2a, as suggested in Fig. 4. Luneau *et al.* (1991b) isolated a partial cDNA that may correspond to a fourth alternatively spliced KV3.2 transcript (KV3.2c). However, these investigators were unable to amplify the 5′ end of this cDNA with primers derived from the 5′ end of KV3.2a. Further characterization is required to understand whether this is a *bone fide* transcript of the KV3.2 gene and how it is generated.

Both alternative splice sites of *Sh*III genes (at the end of exon 2 and at the end of exon 3) described so far appear to be used to generate KV3.4 variants, although no genomic evidence exists in this case. KV3.4b, which has only been found in humans, diverges from KV3.4a and KV3.4c at the same point as KV3.1 and KV3.3 variants. Further, the nucleotide sequence of KV3.4b immediately after the divergence starts with GT, as in KV3.1a and KV3.3b. This observation is consistent with the hypothesis that KV3.4b arises by reading through the sequence immediately following exon 2 of the KV3.4 gene. KV3.4a and KV3.4c, which have only been identified in rat, have an identical 204-bp sequence following their divergence from KV3.4b. This sequence also shows significant sequence similarity with the sequence thought to derive from exon 3 in KV3.1b, KV3.2, and KV3.3a transcripts (Fig. 5). Thus, the alternative splicing in this case is likely to take place at the end of exon 3. The relationship between KV3.4a and KV3.4c is similar to the relationship between KV3.2a and KV3.2d. The sequence of KV3.4c immediately following the splice point is found 62 bp downstream in KV3.4a. After this new overlap, the two sequences are identical. These differences suggest an internal acceptor splicing mechanism, as in the case of KV3.2a and KV3.2d.

Except for KV3.3b and KV3.4c (which have not been tried) all alternatively spliced *Sh*III transcripts

in which the divergent sequence of KV3.3b is contiguous with the sequence of exon 2. Although the exons encoding the carboxyl end of KV3.3a have not been identified, the structure of the KV3.3 gene in this region is suspected to be similar to that of the KV3.1 gene, based mainly on the sequence similarities between the divergent sequence of KV3.1b and KV3.3a transcripts (Vega-Saenz de Miera *et al.*, 1992), particularly in the region postulated to be encoded by exon 3 (Fig. 5).

In contrast to the KV3.1 and KV3.3 alternatively spliced variants, none of the divergent KV3.2 carboxyl ends identified to date begin with GT, suggesting that KV3.2 variants arise by a different splicing mechanism. Sequence comparisons also suggest that the site of alternative splicing in KV3.2 is different than that in KV3.1 and KV3.3. The region shared by all KV3.2 transcripts extends beyond a site equivalent in posi-

A

```
                        ATTCCAAACTGAATGGGAGGTGGCGAAGGCCGCGCTGGCGAACGAAGAC   1554
Kv3.1
Kv3.2       TGTTATCAGGTG-CGA--GTACAGGAA-T---                               1647
Kv3.3   GCAATGACTTGGGTGTCCTGGAGGAGGAG-CC--CGT-CC-----A--CCCT--AGCA-T--A----TC-T--G---   2061
Kv3.4             -C-----G-A-----T--C-    -T--T--G-T----T-CG-T--G--G          1665
```

```
        TGCCCCCACATAGACCAG         GCCCTCACTCCGATGAGGGCCTGCCCTTTACCCGCTCGGGCACCCGCGA   1622
Kv3.1
Kv3.2                     CCGG-AT-AT-A--TCCG--AA-G--C---A--C-GA-----TA-T---A-A--          1700
Kv3.3   -----GC---T------CCA       --A-GT----A--A--CAAGAGC--GA-C--T   C-C--A-G---G--G     2129
Kv3.4   G-AG-TGG-C-CAC----CCCCTGGCCTCG----C---C--T--A--C-T-GAG--C-G-GA---A-----A--G--   1745
```

```
        GAGATACGGACCC         TGCTTCCTCTTATCAACCGGGAGTACGCGTGCCCGCCTGGTGGAGGAATGAGAAAG   1692
Kv3.1
Kv3.2   C-A-A--A--AGAGGGGAAACA--T-----G--GA-G--A--T--T--A-----G-TT--A------C--G--AG      1780
Kv3.3   TC-C--A-C-GGGACCGAGCT------TG-CA--    --C--T--CCCTT-C--A---CTCC---CC----AG       2203
Kv3.4   C---A--AAGAAGGCAGCTGCC----------GC-CAGTG-T-----C--T--C--TG-TGA--CA-T-   -CCAG--AG   1822
```

B

```
        DSKLNGEVAKAALANEDCPHIDQ         ALTPDEGLPFTRSGTRERYGP   CFLLSTGEYACPPGGGMRK   564
Kv3.1
Kv3.2   VLSGD--TGSEP                    P-S-P-R--IR--S--DKNRRGET----T--D-T-ASD--I--     593
Kv3.3   GNDLGVLEEG-PRP--DP-A----H----A---P   -MS-EDKS-I-PGS  -G--SRDRA---VT  D--PS-D-SI--   734
Kv3.4   ---Q--D  -N-V-SD-EGAGLT-PLAS-P--E-RRALR-----D-NKKAAA----A-D---A D-SV--            607
```

FIGURE 5 Sequence of putative exon 3 of four *Shill* genes. (A) The nucleotide sequence of exon 3 of the KV3.1 gene is compared with the last 165 bp of the divergent end of KV3.3a, and the first 204 bases of the divergent ends of KV3.4a and KV3.4c. In all cases, the sequence shown is preceded by AG. The sequences are aligned to maximize long stretches of identity. Nucleotides identical to those in KV3.1 are shown with a dash. Gaps required for optimized alignment of the sequences are shown as blanks. (B) Alignment of the amino acid sequences encoded by the putative exon 3 of each gene. Amino acid residues identical to those in KV3.1 are shown with a dash. Gaps required for optimized alignment of the sequences are shown as blanks.

identified here are found in brain poly(A)-selected mRNA and express currents in *Xenopus* oocytes (Yokoyama *et al.*, 1989; T. McCormack *et al.*, 1990; Luneau *et al.*, 1991a,b; Rudy *et al.*, 1991b,1992; Schröter *et al.*, 1991; Vega-Saenz de Miera *et al.*, 1992; Weiser *et al.*, 1994). These results support the notion that these cDNAs represent real transcripts rather than the result of priming of unprocessed pre-mRNA.

1. Alternative Splicing of 5′ UTR

The previous section discussed alternative splicing of *Sh*III genes 3′ to the putative initiation codon. However, Kentros *et al.* (1992) have isolated cDNA clones of KV3.2 containing two different 5′ untranslated regions (5′UTRs) as well, which they have designated α and β. Northern blot analysis of rat brain poly(A) RNA with probes specific to these two 5′ UTRs shows hybridization to the major bands seen with coding

region probes, whereas PCR experiments with primers specific to both α and β show that both 5′ UTRs can be associated with more than one of the various 3′ coding region, alternatively spliced variants of KV3.2. These experiments suggest that both 5′ UTRs are not simply the result of unspliced RNA. Alternative splicing of the 5′ UTR was also reported for KV1.1 a Shaker-related gene (Tempel *et al.*, 1988).

C. Multiple Transcripts of Each ShIII Gene Are Found in Brain mRNA

Northern blot analysis of rat brain poly(A) RNA shows that all members of the *Sh*III subfamily have multiple transcripts (Fig. 6). Specifically, KV3.1 probes hybridize to bands of approximately 10.5, 8, and 4.5 kb (Luneau *et al.*, 1991a; Weiser *et al.*, 1994), and KV3.2 probes hybridize to bands of approxi-

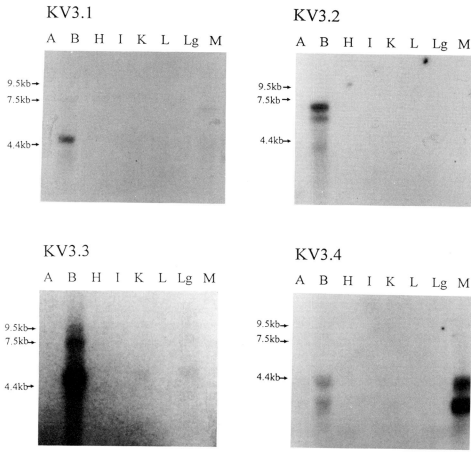

FIGURE 6 Tissue expression of four *Sh*III genes. Northern blot analysis of adult rat poly(A) RNA (3 μg per lane) from adrenal gland (A), brain (B), heart (H), intestine (I), kidney (K), liver (L), lung (Lg), and skeletal muscle (M) hybridized with probes specific to the KV3.1, KV3.2, KV3.3, and KV3.4 K+ channel genes. The blots were exposed to X-ray film at −70°C with two intensifying screens for 2 days (KV3.1), 4 days (KV3.2, KV3.3), or 7 days (KV3.4). The positions of three RNA size markers (9.5, 7.5, and 4.4 kb) are indicated.

mately 7, 6, and 4 kb (Luneau *et al.*, 1991b; Rudy *et al.*, 1992; Weiser *et al.*, 1994). Interestingly, the 6-kb band is much weaker in thalamic RNA (Rudy *et al.*, 1992). KV3.3 probes hybridize to bands of approximately 8.5 and 5.5 kb (Vega-Saenz de Miera *et al.*, 1992; Weiser *et al.*, 1994) and KV3.4 probes to bands of 4.5 and 3.5 kb (Weiser *et al.*, 1994). Little data on the nature of the multiple transcripts of *Sh*III genes detected in Northern blots is currently available. For two of the members of this subfamily, KV3.2 (Luneau *et al.*, 1991b; Rudy *et al.*, 1992) and KV3.3 (Vega-Saenz de Miera *et al.*, 1992), probes specific to two 3' coding alternatively spliced products produce banding patterns that are similar to each other. However, Luneau *et al.* (1991a) showed that for KV3.1, probes specific to KV3.1a hybridized to the 8-kb band whereas probes specific to the other alternatively spliced product, KV3.1b, hybridized to the 4.5-kb band. Nevertheless, both probes still hybridized to a relatively weak 10.5-kb band.

The meaning of the complex banding patterns just discussed remains unknown. Moreover, since the size of the coding regions of the cDNAs is less than 2.7 kb for all cases, the size of the transcripts seen in Northern blots is difficult to account for. The most obvious explanation for these phenomena would be incomplete splicing of the RNAs. However, for most (if not all) genes described to date, unspliced heterologous nuclear RNA is relatively short-lived and is certainly not the major RNA species, as would be the case for a number of these transcripts. Alternatively, these could be mature transcripts and the discrepancy in size from that of the coding regions could be the result of unusually large untranslated regions. The multiple bands could be explained by use of different polyadenylation sites and/or of as yet undefined alternative splice sites. Alternative splicing of 5' UTRs is a possibility for which some evidence was discussed in the previous section.

The functional significance (if any) of these multiple transcripts from each *Sh*III gene is still unclear, but various possibilities exist. First, the mRNAs themselves may have a distinct subcellular localization. Such localization of mRNAs has been documented in the nervous system (Steward and Banker, 1992) and has been shown for some cases to depend on the 3' untranslated region of the RNA. Second, 5' untranslated regions have been shown in various systems (Kozak, 1992; Melefors and Hentze, 1993) to confer sequence-specific regulation of translation, which may indeed be the case for these transcripts. Yet another possibility is the regulation of mRNA stability (Saini *et al.*, 1990), a phenomenon that has been shown to depend on untranslated regions as well as length

of the poly(A) tail. The preceding observations raise some interesting unanswered questions about the potential role of posttranscriptional regulation of *Sh*III K$^+$ channel gene expression.

D. Electrophysiological Properties of ShIII Channels in Xenopus Oocytes

Xenopus oocytes injected with *Sh*III transcripts express voltage-dependent K$^+$ currents that are absent in uninjected oocytes (Yokoyama *et al.*, 1989; T. McCormack *et al.*, 1990; Luneau *et al.*, 1991a,b; Rudy *et al.*, 1991b; Schröter *et al.*, 1991; Vega-Saenz de Miera *et al.*, 1992). The currents expressed by an alternatively spliced transcript of each of the four *Sh*III genes are shown in Fig. 7. The currents were recorded with a two-microelectrode voltage clamp that had sufficient time resolution to resolve the overall kinetic features of these currents.

A notable feature of the channels expressed by *Sh*III transcripts is their voltage dependence, which is nearly identical for all transcripts of this subfamily investigated to date (see Fig. 8A for representative conductance–voltage curves). Membrane potentials more positive than −10 mV are required to activate significant *Sh*III currents, a more depolarized value than that required to activate channels of the other subfamilies. The pharmacological properties of *Sh*III channels are also very similar. These channels are highly sensitive to both 4-aminopyridine (4-AP) and tetraethylammonium (TEA) and are not blocked by peptide toxins that block other *Sh* channels (see Table III).

*Sh*III currents do show variability in their kinetic properties. KV3.1 and KV3.2 transcripts express very similar slowly rising, delayed-rectifier type currents that inactivate extremely slowly. KV3.3 and KV3.4 transcripts, however, express A-type inactivating currents that are significantly different from each other. For instance, although macroscopic inactivation rates change with voltage in both cases, they are an order of magnitude faster in KV3.4 currents than in KV3.3 currents (Fig. 7). KV3.4 currents also rise faster than those carried by all other *Sh*III channels (Fig. 7).

KV3.3 and KV3.4 currents also differ in the voltage-dependence of steady-state inactivation. As a result of these differences, KV3.3 currents conduct in the steady state over a much broader window of potentials than KV3.4 currents (Fig. 8B). Vega-Saenz de Miera *et al.* (1992) suggested that the differences in the voltage dependence of steady-state inactivation of KV3.3 and KV3.4 currents could be explained by differences in the coupling between inactivation and channel opening, with KV3.4 channels inactivating

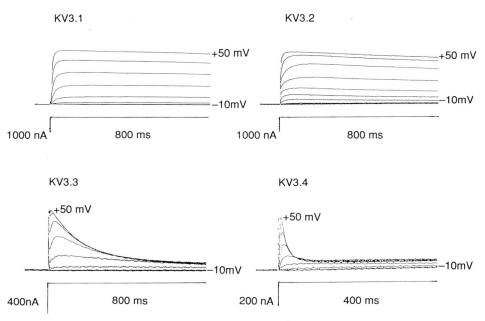

FIGURE 7 Currents expressed by one alternatively spliced variant of each of four *Sh*III genes in *Xenopus* oocytes. Shown are the currents recorded under voltage-clamp during depolarizing pulses from −60 to +50 mV in 10-mV increments delivered at a rate of one every 60 sec from a holding potential of −100 mV. Currents have been leak subtracted. The alternatively spliced variants used were KV3.1b, KV3.2a, KV3.3a, and KV3.4b. Note that the current increment decreases at high voltages. Modified with permission from Weiser *et al.* (1994).

from closed states much more readily than KV3.3 channels.

A provocative study by Ruppersberg *et al.* (1991a) showed that inactivated KV3.4a channels are able to conduct ions during repolarization. Apparently, these channels are unable to enter the resting closed state when the membrane is hyperpolarized until their inactivation is first removed, resulting in a long-lasting current (at voltages much more negative than those required for activation) while they are recovering from inactivation. KV3.3 channels, which inactivate more slowly, may or may not behave in a similar fashion. Such behavior could drastically affect the role played by these channels *in vivo* (see Section IV,F).

KV3.1a channels have also been expressed in human embryonic kidney cells (Critz *et al.*, 1993). Significantly, the overall properties of the currents are very similar to those seen in *Xenopus* oocytes, although a more detailed comparison of the KV3.1a currents in the two systems has not been reported. The electrophysiological and pharmacological properties of the channels formed by products of the four *Sh*III genes are summarized in Table III. Overall, *Sh*III channels display less functional variation than *Sh*I channels, which vary widely in voltage dependence, kinetics, and pharmacological properties (Stühmer *et al.*, 1989a).

Another aspect shared by all *Sh*III currents is a decline in conductance at membrane potentials more positive than +20 mV, although this feature is not quite as evident in KV3.1 currents (T. McCormack *et al.*, 1990; Rudy *et al.*, 1991b; Schröter *et al.*, 1991; Vega-Saenz de Miera *et al.*, 1992). In single-channel recordings, KV3.2a channels display a fast flickering at voltages more positive than +20 mV. This flickering produces a decline of the average *single-channel current* (Fig. 9) and is reminiscent of the effects of fast open-channel blockers on a number of channels (Hille, 1992). Based on observations such as those shown in Fig. 9, we proposed that the decline in conductance of *Sh*III channels results from fast voltage-dependent open-channel block by some intracellular component, perhaps Na^+ or Mg^{2+} (Rudy *et al.*, 1991b; Vega-Saenz de Miera *et al.*, 1992). These ions have been shown to produce similar block of other voltage-gated K^+ channels, but at more positive potentials (Linsdell *et al.*, 1990; Lopatin and Nichols, 1993). Rettig *et al.* (1992) showed that a similar blockage of channels expressed by a KV3.4 construct lacking the N-terminal 28 residues is abolished in inside-out patches on removal of Mg^{2+} from the bath solution.

The significance of so many different alternatively spliced transcripts from *Sh*III genes is quite baffling, since none of the alternatively spliced transcripts

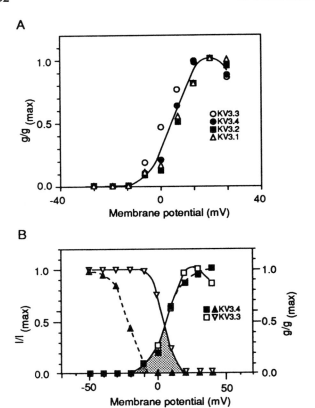

isoform-specific channel localization, mobility, and other forms of modulation.

E. Functional Consequences of Heteromultimer Formation between ShIII Proteins

Formation of heteromultimeric channels results in a large increase in the potential number of different channels that a given set of *Sh* subunits could form. This property leads to the formation of channels with subtle, but often physiologically significant, differences (e.g., Po *et al.*, 1993). For example, if 10 *Sh*III subunits are capable of forming heteromultimeric channels, assuming the channels are tetramers 715 channels with differing subunit composition could be formed.[2] Most studies investigating heteromultimer formation have used experiments in which different pairs of cRNAs are co-injected in *Xenopus* oocytes or coexpressed in other cell types. Heteromultimer formation is suggested by the fact that, when two different subunits are coexpressed, currents are obtained that cannot be the algebraic sum of the currents expressed by each channel independently.

In general, heteromultimeric *Sh*I channels have properties that are intermediate between those of the corresponding homomultimers, although some properties might be closer to those of one or the other homomultimer (Isacoff *et al.*, 1990; K. McCormack *et al.*, 1990; Ruppersberg *et al.*, 1990; MacKinnon *et al.*, 1991). Since all *Sh*III channels have similar voltage dependencies, this parameter was expected not to change in heteromultimers. This expectation has been confirmed experimentally; when two different *Sh*III cRNAs are injected into oocytes, the total current has a voltage dependence similar to that of the *Sh*III homomultimers (H. Moreno, E. Vega-Saenz de Miera, and B. Rudy, unpublished observations). Moreover, and perhaps not surprisingly, the currents recorded in oocytes co-injected with KV3.1 and KV3.2 cRNAs, which produce similar currents in a homomultimer channel, are similar to the currents of KV3.1 or KV3.2 alone.

Oocytes co-injected with *Sh*III cRNAs that express currents that inactivate at different rates when injected alone expressed currents suggestive of heteromultimer formation (Weiser *et al.*, 1994). One of the most interesting results from these co-injection experiments is the observed changes when KV3.4 cRNA is co-injected with KV3.1 cRNA. These oocytes had fast inactivating currents that were several-fold larger than those seen in oocytes injected with the same amount of KV3.4 cRNA alone (Fig. 10). This result is contrary

FIGURE 8 (A) Conductance–voltage relationship of four *Sh*III currents. Plots of average normalized conductance (g/g_{max}; conductance at the indicated potential divided by the maximum conductance) as a function of voltage. A single curve through the data of the four channels was drawn by hand. (B) Comparison of the conductance–voltage and steady-state inactivation–voltage relationships of KV3.3 and KV3.4 currents. Plots of normalized conductance (g/g_{max}; squares) and steady-state inactivation (I/I_{max}; triangles) as a function of voltage. Note that, in the steady state, KV3.3 channels conduct over a much broader window of potentials than KV3.4 channels (see area filled with dots and bars, respectively).

tested to date (KV3.1a, Yokoyama *et al.*, 1989; KV3.1b, Luneau *et al.*, 1991a; KV3.2a, T McCormack *et al.*, 1990; KV3.2b, Luneau *et al.*, 1991a; KV3.2d, H. Moreno and B. Rudy, unpublished observations; KV3.4a, Schroter *et al.*, 1991; KV3.4b, Rudy *et al.*, 1991b) expresses currents in *Xenopus* oocytes that are noticeably different from those of any other transcript from the same gene. However, more in-depth analysis of the channels expressed by these transcripts may reveal previously overlooked subtle differences. Alternatively, these channels may respond differently to various second messenger systems (see Section IV,G). Alternative splicing may thus allow for isoform-specific differences in the modulation of channel function by neurotransmitter and neuropeptide-induced cellular responses. The divergent C termini could be used as well in protein–protein interactions, thus conferring

---

[2] The number of different subunit combinations can be calculated from the equation $[p + (n - 1)]!/[p!(n - 1)!]$, where p is the aggregation number and n is the number of subunits.

TABLE III Electrophysiological and Pharmacological Properties of Mammalian *ShIII* Channels in *Xenopus* Oocytes[a,b]

Gene	Activation			Inactivation			g (pS)	Pharmacology—IC$_{50}$ (mM)					
	$V_{.5}$ (mV)	k (mV)	Δt (msec)	$V_{.5}$ (mV)	k (mV)	τ (msec)		TEA	4AP	Quinine	DTX	CTX	MCDP
KV3.1	+10	8	24	—	—	—	16	0.150	0.6	1	NB	NB	NB
KV3.2	+10	7	23	—	—	—	16	0.150	0.9	ND	NB	NB	NB
KV3.3	+7	6	25	+5.2	6.1	240	14	0.140	1.2	ND	ND	ND	ND
KV3.4	+14	8.5	9	−20	7.2	20	12	0.200	0.6	0.5	NB	ND	NB

[a] Data derived from Yokoyama et al. (1989), T. McCormack et al. (1990), Rudy et al. (1991b), Rettig et al. (1992), Vega-Saenz de Miera et al. (1992), and H. Moreno and B. Rudy (unpublished observations).

[b] Abbreviations: For activation: $V_{.5}$, membrane potential at which the conductance is half maximal; k, slope of normalized g/V curve; Δt, time for the current to rise to 90% of its final value for noninactivating currents and time to peak for inactivating currents, in both cases at +40 mV. For inactivation: $V_{.5}$, membrane potential of a 1-sec (for KV3.4) and 2-sec (for KV3.3) prepulse producing 50% inactivation; k, slope of the steady-state inactivation vs. voltage curve; τ, time constant of inactivation at +30 mV. g, single channel conductance in physiological K$^+$ concentrations, in cell-attached patches. For pharmacology: IC$_{50}$ (mM), concentrations blocking 50% of the current. TEA, tetraethylammonium; 4AP, 4-aminopyridine; DTX, dentrotoxin; CTX, charybdotoxin; MCDP, mast cell degranulating peptide; NB, no block; ND, not determined.

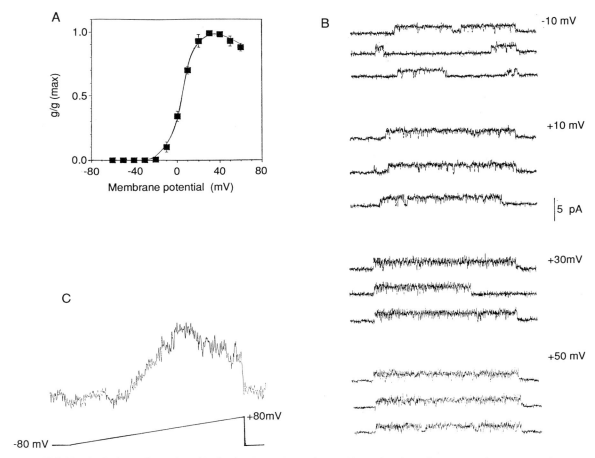

FIGURE 9 Voltage-dependent block of KV3.2a channels. (A) Normalized conductance–voltage curve of KV3.2a channels obtained from macroscopic currents in *Xenopus* oocytes. (B) Representative single KV3.2a channel records at four different voltages in a cell-attached patch on an oocyte injected with KV3.2a cRNA. (C) Ensemble average of 32 sweeps obtained from the same patch used in B during a voltage ramp from −80 to +80 mV.

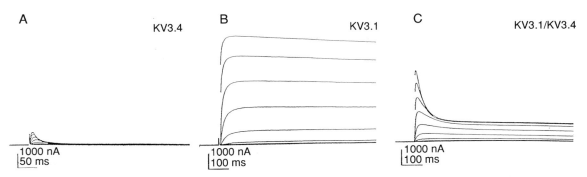

FIGURE 10 KV3.1–KV3.4 heteromultimeric channels in *Xenopus* oocytes. (A) Currents recorded in a representative oocyte injected with KV3.4b cRNA alone. (B) Currents recorded in a representative oocyte injected with KV3.1b cRNA alone. (C) Currents recorded in a representative oocyte injected with the same amount of KV3.4b cRNA as the oocyte in A plus the same amount of KV3.1b cRNA as the oocyte in B. The currents shown here are representative of several similar experiments. For example, the average transient current at +40 mV of 8 oocytes injected with KV3.4b cRNA alone was 300 nA; that in oocytes co-injected with KV3.1b and KV3.4b was 1800 nA. Modified with permission from Weiser *et al.* (1994).

to what might be expected from the algebraic sum of two independent currents and is consistent with the formation of heteromultimeric channels containing one or more KV3.4 subunits and the remaining KV3.1 subunits (Weiser *et al.*, 1994). The amplification of the currents is caused by both an increase in single channel conductance in the heteromultimeric channels (compared with that of KV3.4 channels) and the fact that apparently less than four KV3.4 subunits are sufficient to impart fast inactivating properties on the channel. Similar studies with *Sh*I subunits have also shown that the presence of a single inactivating subunit is sufficient to impart inactivating properties on the resultant channels (e.g., Isacoff *et al.*, 1990; K. McCormack *et al.*, 1990; Ruppersberg *et al.*, 1990; MacKinnon *et al.*, 1991). This behavior can be understood in terms of the N-terminal "ball and chain" model of inactivation (Hoshi *et al.*, 1990; Zagotta *et al.*, 1990; see Section IV,G). The "ball and chain" of a single inactivating subunit is sufficient to inactivate the channel, although the rate of inactivation increases with the number of inactivating subunits (MacKinnon *et al.*, 1991), as was also the case for KV3.1–KV3.4 heteromultimers (Weiser *et al.*, 1994). Similar results have been obtained in oocytes co-injected with KV3.4 and KV3.2 or KV3.3 (H. Moreno, unpublished observations).

F. Role of *Sh*III Channels in Excitability

Voltage-dependent K⁺ currents modulate various important aspects of neuronal excitability including membrane resting potential, action potential waveform, firing patterns, neurotransmitter release, and postsynaptic potentials. The extent of the involvement of a given K⁺ channel in these processes is dependent on the number of channels and their conductance, voltage dependence, kinetics, and subcellular localization, as well as on the balance of other currents (Hille, 1992). Further, these properties are often altered by cellular responses to neurotransmitters and neuropeptides, making K⁺ channels important mediators of the longer term effects of synaptic transmission (Levitan, 1988; Kaczmareck and Levitan, 1987; Rudy, 1988).

As discussed in the preceding sections, transcripts from the four *Sh*III genes express K⁺ channels in *Xenopus* oocytes that start activating at highly depolarized potentials. Heteromultimer formation allows for the existence of channels with this characteristic possessing a wide range of inactivation kinetics. Particularly interesting is the finding that small numbers of KV3.4 subunits might be sufficient to impart fast inactivating properties on channels composed mainly of

other *Sh*III subunits (see Section IV,E). Assuming that native *Sh*III channels have properties similar to those seen in oocytes, certain predictions can be made about the kinds of electrical activity native *Sh*III channels may or may not be involved in, and what their effects might be on neuronal excitability.

Voltage-gated K⁺ channels can affect fast Na⁺ action potential waveforms primarily by their influence on the repolarization of the spike. *Sh*III channels could easily be active at the depolarized potentials (+10 to +30 mV) characteristic of these spikes. In fact, Perney *et al.* (1992) suggested that the physiological role of KV3.1 channels is to shorten the duration of action potentials in fast-firing neurons. In support of their hypothesis, these investigators reported studies with AtT20 cells transfected with the *ras* oncogene, which caused a 20-fold reduction in the width of action potentials relative to untransfected cells and was associated with the induction of KV3.1 mRNA (Hemmick *et al.*, 1992).

Although it is conceivable that KV3.1 channels (or KV3.2 channels, which express similar currents) act to shorten the duration of action potentials in fast-firing neurons, this hypothetical function must be reconciled with the relatively slow activation kinetics of these channels (see Section IV,D) and the short duration of neuronal fast action potentials (<1 msec at 35–37°C). Of course, channel density could make up for the small percentage of these channels that would be activated during a spike and would provide sufficient current to repolarize the membrane. Alternatively, KV3.1 or KV3.2 channels could activate much faster during an action potential as a result of having been exposed to depolarized voltages below those necessary to produce measurable current. KV3.1 and KV3.2 channels could easily be activated, however, during action potentials of longer duration.

The spike frequency of trains of action potentials depends mainly on the membrane potential and resistance between spikes. These parameters are governed in large part by K⁺ channels that are activated at voltages below the threshold of action potentials. However, the voltage dependence of *Sh*III channels precludes their involvement in subthreshold phenomena such as regulation of spike frequency or latency to first spike. In this context, the work of Ruppersberg *et al.* (1991a), described in Section IV,D, provides an interesting caveat to this generalization. If inactivated KV3.4 channels are indeed able to conduct ions at negative potentials while they recover from inactivation, they could have quite interesting effects on firing patterns. Specifically, the first spike in a train could open and then quickly inactivate these channels. On repolarization of the membrane, the current through

```
Kv3.3  MLSSVCVWSFSGRQGTRKQHSQPAPTPQPPESSPPPLLPPPQQQCAQPGTAASPAGAPLSCGPGGRRAEP   70
Kv3.4  MISSVCVSSYRGRKSGNKPPSKTCLKEE                                            28


Kv3.1       MGQ   GDESERIVINVGGTRHQTYRSTLRTLPGTRLAWLAEPDAHSHFDYDPRA           52
Kv3.2       --K    IENN--VIL-------E------K-------L--SSEPQGDCLTAAGDKLQPLP       58
Kv3.3  CSGLPAVA--RHGGG-GD-GK-------V--E---------------G-T--E-AAR-----GT        134
Kv3.4       -AK    -EA--K-I--------E-----------------D--GGGRPES-GGG            80


Kv3.1                                      DEFFFDRHPGVFAHILNYYRTGKLHCPAD       81
Kv3.2  PPLSPPPRPPPLSPVPSGCFEGGAGNCSSHGGNGSDHPGGGR-----------YV-------------  128
Kv3.3                                   -------------YV-------------        163
Kv3.4                 AGSSGSSGGGGGC-----------YV-------------               121


Kv3.1  VCGPLYEEELAFWGIDETDVEPCCWMTYRQHRDAEEEALDSFGGAPLDNSADDADADGPGDSGDGEDELEM  151
Kv3.2  -----F---------------------------I-ETPD-IGGDPGD-E-LG-KRLGI--AAGL       198
Kv3.3  -----F----G----------A-----------------EAPDSSGN-NANAGGAHDAGL-D-AGAGG    233
Kv3.4  -----F----T---------------------I-ESPDGGGGGAGPGDEAGD-ERELALQRLG        191

                                                             ↓          S 1
Kv3.1  TKRLALSDSPDGRP              GGFWRRWQPRIWALFEDPYSSRYARYVAFASLFFILV        202
Kv3.2  GGPDGK                      S-R--KL---M-----------A--FI----------      241
Kv3.3  GGLDGAGGELKRLCFQDAGGGAGGPAGGPGGAG-TW-------V-----------A-------------I  303
Kv3.4  PHEGGSGPGAG                 S-GC-G----M-----------A--V-----------      239

              ▲                      ▲                          S 2
Kv3.1  SITTTFCLETHERFNPIVNKTEIE    NVRNGTQVRYYREAETEAFLTYIEGVCVVWFTFEFLMRVV    265
Kv3.2  ----------A--IVK----        P-I---SAVLQY-I--DPA---V------------V-I-    302
Kv3.3  ----------G-IH-S---VTQASPIPGAPPE-I-N-   -V---P----V---------------T     369
Kv3.4  ----------A--IDR-V---H      R-G-I-S--FR--V---PI---------M---L---V-I-    302

                      S 3                          S 4  •          •
Kv3.1  FCPNKVEFIKNSLNIIDFVAILPFYLEVGLSGLSSKAAKDVLGFLRVVRFVRILRIFKLTRHFVGLRVLG  335
Kv3.2  -S---L-----L-----------------C------------------------------------    372
Kv3.3  ---D----L-S------C-----------------------------------------------    439
Kv3.4  C--DTLD-V--L--------------------------R-------------------------    372

          •         •        •  S 5                              H 5
Kv3.1  HTLRASTNEFLLLIIFLALGVLIFATMIYYAERIGAQPNDPSASEHTHFKNIPIGFWWAVVTMTTLGYGD  405
Kv3.2  ----------------------------V----------x---Q---------------------     442
Kv3.3  --------------------------D-D-ILG-N--Y-------------------------        509
Kv3.4  --------------------------R-S--RGND--D------------------------        442

                     S 6                          ★
Kv3.1  MXPQTWSGMLVGALCALAGVLTIAMPVPVIVNNFGMYYSLAMAKQKLPKKKKKHIPRPPQLGSPNYCK    473
Kv3.2  -------------------------------------------------R-R-----PA-LAS--TF--  510
Kv3.3  ---K-------------------------------------------N---------P-------PD    579
Kv3.4  ---K-----------------------------------------------R---V------E--I---  510


Kv3.1                                               SVVNSPHHSTQSDTCPL         490
Kv3.2                                               TEL-MACN-------LG         527
Kv3.3  PPPPPPPHPHHGSGGISPPPPITPPSMGVTVAGAYPPGPHTHPGLLRGGAGGLGIMGL-PLPAPGEP---  649
Kv3.4                                               -EET--RD--Y---S-P         527

                ↓
Kv3.1  AQEE ILEINRA                                                          501
Kv3.2  KENR L--H--SVLSGDDSTGSEPPLSPPERLPIRRSSTRDKNRRGETCFLLTTGDYTCASDGGIRK    593
Kv3.3  ---- VI-T---                                                          660
Kv3.4  -R--GMV-RK--                                                          539
```

FIGURE 11 Comparison of the constant region of K⁺ channel proteins encoded by four *Sh*III genes. Predicted amino acid sequences of the constant region (the sequences preceding the alternatively spliced carboxyl ends) of proteins encoded by *Sh*III transcripts. The sequences are aligned to maximize long stretches of identity. Amino acids identical to those in KV3.1 are shown with a dash. Gaps required for optimal alignment are shown as blanks. Note that the second methionine of KV3.3 and KV3.4 has been aligned with the starting methionines of KV3.1 and KV3.2. The sequences between the first and second methionines of KV3.3 and KV3.4 are important for channel inactivation. The first arrow marks the end of the sequences encoded by exon 1 and the second arrow marks the end of the sequences encoded by exon 2. The S1–S6 and H5 domains are overlined. Proline-rich sequences in KV3.2 and KV3.3 are underlined. Two putative *N*-glycosylation sites in all *Sh*III proteins in the linker between the S1 and S2 domains are indicated with triangles. KV3.3 has an additional putative *N*-glycosylation site in the S5–H5 linker (also indicated with a triangle). The positively charged residues in S4 have been

KV3.4 channels recovering from inactivation could prevent the next spike or delay its onset.

Both the pre- and the postsynaptic sides of synaptic transmission can be modulated by K⁺ channels. The amount of neurotransmitter release from the presynaptic terminal is dependent on the amount of Ca^{2+} entry through voltage-gated Ca^{2+} channels. Activation of K⁺ channels in the presynaptic terminal can limit the magnitude or duration of local depolarization and, thus, Ca^{2+} entry and transmitter release. On the other side of the synapse, activation of K⁺ channels could limit the spread of the depolarization caused by receptor activation. *Sh*III channels could play a role in both of these processes, provided that presynaptic potentials or excitatory postsynaptic potentials (EPSPs) reach voltages sufficient to activate these channels. This event would be more likely in areas devoid of large numbers of K⁺ channels that activate at lower voltages.

The vagueness of the preceding discussion underscores the importance of studies of *Sh*III channels that are aimed at elucidating the electrophysiological properties and subcellular localization of K⁺ channels containing *Sh*III subunits *in vivo*. The state of present work on these problems and future directions for the field are discussed in later sections of this chapter.

G. Structural and Functional Domains in ShIII Proteins

The overall structure of mammalian *Shaw*-related or *Sh*III proteins (Yokoyama *et al.*, 1989; T. McCormack *et al.*, 1990,1991; Luneau *et al.*, 1991a,b; Rudy *et al.*, 1991b,1992; Schröter *et al.*, 1991; Vega-Saenz de Miera *et al.*, 1992; Ried *et al.*, 1993) is similar to that of all other *Sh* proteins (reviewed by Jan and Jan, 1990a;1992; Perney and Kaczmareck, 1991; Rudy *et al.*, 1991a; Pongs, 1992) and other channel proteins of the S4 superfamily (see Fig. 1). A comparison of the constant region (prior to divergence of the carboxyl ends) of the protein products of the four *Sh*III genes is shown in Fig. 11. *Sh* proteins are several hundred amino acids in length and are characterized by a core membrane region consisting of six putative membrane-spanning domains designated S1, S2, S3, S4, S5, and S6 flanked by intracellular amino and carboxyl domains of variable lengths (see Figs. 3, 11). Domains S1–S3 and S5–S6 are hydrophobic and are thought to span the membrane as α helices. The S4 domain is amphipathic, but is also believed to be membrane spanning. The topology of the S1–S6 domains is not known, but models have been proposed based on the idea that less conserved, highly hydrophobic domains are likely to interact with lipid whereas highly conserved domains may be involved in protein–protein interactions and may be located in the interior of the protein (Guy, 1989). In addition to having a similar overall structure, *Sh*III proteins and *Sh* proteins of the other subfamilies share significant amino acid sequence similarity (~40% sequence identity in a core region that includes all the membrane portion of the polypeptides). Details of the structure of *Sh* proteins that are not addressed here can be found in several reviews (Jan and Jan, 1990a;1992; McKinnon, 1991; Miller, 1991; Rudy *et al.*, 1991a; Pongs, 1992).

As is the case for *Sh* proteins of the other subfamilies, *Sh*III proteins are most similar in the core membrane region (which corresponds roughly to the sequence encoded in exon 2 of *Sh*III genes; see Section IV,B). In this region, *Sh*III proteins share ~85% amino acid identity, compared with ~70% *overall* amino acid identity, indicating that the regions preceding and following the core region are less conserved (see Figs. 3, 11). Most of the differences among *Sh*III proteins are localized to parts of the amino end region preceding the first membrane-spanning domain, the carboxyl domain following the membrane portion of the polypeptide, and the S1–S2 and S2–S3 linkers.

The conservation of the core region among *Sh*III proteins is somewhat higher than that among different *Sh*I proteins (75–80% identity). This conservation may explain why the channels composed of *Sh*III proteins display less functional variability than those composed of *Sh*I proteins (see Section III,D). The core regions of *Sh*II proteins are even more conserved than those of *Sh*III proteins (Frech *et al.*, 1989; Hwang *et al.*, 1992). *Sh* proteins also share several structural and functional domains, the most characteristic being the H5 or pore domain and the S4 or voltage-sensor domain.

1. H5 or Pore (P) Domain

Between the fifth and the sixth membrane-spanning domains is a highly conserved and somewhat hydrophobic sequence knows as H5 (or SS1–SS2 by some authors; Jan and Jan, 1990a; Perney and Kaczmareck, 1990; Rudy *et al.*, 1991a; Pongs, 1992; see also Chapter 2). H5 domains are present not only

---

boxed. The leucines in the *Sh* leucine heptad repeat adjacent to the S4 domain are indicated with a filled circle. The fourth residue in the repeat is phenylalanine in *Sh*III proteins, but leucine in all other *Sh* proteins. Residues in the H5 domain that were found by mutagenesis to be important in determining pore properties of KV3.1 are boxed. A cluster of positively charged residues after the S6 domain is indicated with a star.

in the K$^+$ channels of the S4 superfamily (the *Sh, eag,* and *slo* families) but also in inward rectifier channels that otherwise bear no resemblance to the S4-containing K$^+$ channels (Ho *et al.*, 1993; Kubo *et al.*, 1993). The only known exception is the I$_{SK}$ channel, which lacks a clear H5 domain. Mutations of the H5 domain in *Drosophila* and mammalian *Sh* proteins affect ion selectivity, open-channel conductance, and blockade by TEA, providing strong evidence that this domain contributes to the formation of the channel pore (MacKinnon and Yellen, 1990; Yellen *et al.*, 1991; Yool and Schwarz, 1991; Hartmann *et al.*, 1991; Kirsch *et al.*, 1992a,b). In one of these experiments (Hartmann *et al.*, 1991; Kirsch *et al.*, 1992a,b), the H5 domain of KV2.1 (a mammalian *Shab*-related protein) was replaced with the H5 domain from a mammalian *Sh*III protein (KV3.1), yielding a chimeric protein expressing channels with permeability properties characteristic of KV3.1 channels, including single-channel conductance and internal TEA sensitivity. Na$^+$ and Ca^{2+} channels contain a domain in an analogous position (but with a different amino acid sequence), known as SS1–SS2, that participates in the regulation of ion permeability in these channels (Guy and Conti, 1990; Heinemann *et al.*, 1992).

In the accepted topology of *Sh* proteins, the H5 domain begins at the extracellular side of the membrane. The H5 domain of each subunit has been proposed to form an antiparallel β hairpin loop that enters and exits the membrane so the pore is a hole in the middle of an eight-stranded β barrel (Guy, 1989; Guy and Conti, 1990). Consistent with this hypothesis, mutagenesis studies demonstrate that the H5 domain is accessible from both sides of the membrane. Mutations of amino acid residues near the beginning and the end of the H5 domain affect channel block by external TEA, whereas mutations of residues near the middle of H5 affect mainly blockade by internal TEA (Yellen *et al.*, 1991).

Residue 19 in the H5 domain of all mammalian *Sh*III proteins is a tyrosine (Fig. 11). In *Drosophila Shaker* channels, a tyrosine in this position conferred high sensitivity to external TEA (MacKinnon and Yellen, 1990). All *Sh*III channels are blocked by submillimolar concentrations of external TEA (see Table III). Tyr 19 probably contributes to the sensitivity of *Sh*III channels to external TEA, but is clearly not sufficient since KV2.1 channels have a tyrosine in this position but are about 20-fold less sensitive to external TEA than *Sh*III channels. Moreover, the KV2.1 chimera with a KV3.1 H5 domain described earlier (KV2.1–KV3.1 chimera) was less sensitive to external TEA than KV3.1 channels (Hartmann *et al.*, 1991).

Mutagenesis studies, particularly those with the KV2.1–KV3.1 chimera (Hartmann *et al.*, 1991; Kirsch

et al., 1992a,b), have been useful in identifying other residues that contribute to the pore properties of *Sh*III subunits. These residues include a leucine and a valine in the middle of the KV3.1 H5 domain that are replaced by valine and isoleucine, respectively, in KV2.1 (Fig. 12A). These residues have been shown to influence the K$^+$ conductance, mean open time, and relative K$^+$/Rb$^+$ conductance of KV3.1 channels. These residues are also important in determining the sensitivity of KV3.1 channels to internal TEA blockade, which is significantly lower in these channels (Taglialatela *et al.*, 1991; Kirsch *et al.*, 1992a,b). KV3.2 proteins are identical to KV3.1 in the region of the H5 domain; however, KV3.3 and KV3.4 have a lysine two positions C-terminal to Tyr 19, whereas KV3.1 and KV3.2 have a glutamine at this position. Substitution of this glutamine for lysine in the KV2.1–KV3.1 chimera, a change resulting in an increase in the charge close to the putative external mouth of the channel, reduced the single-channel conductance to a degree similar to the difference in single-channel conductance between KV3.1 (or KV3.2) and KV3.3 or KV3.4 channels (see Table III). This substitution also reduced the external TEA sensitivity 5-fold; however, KV3.3 and KV3.4 channels have external TEA sensitivities that are not very different from those of KV3.1 and KV3.2 (see Table III). Other residues, or the structure of the pore itself, may explain these differences. Indeed, a triple mutation was required (V8I, L13V, Q21K) to restore the external TEA sensitivity of the KV2.1–KV3.1 chimera to that of KV2.1 channels (Kirsch *et al.*, 1992a,b).

Peculiar to mammalian *Sh*III proteins is the presence of a 9-residue insert just prior to the beginning of the H5 domain that results in elongation of the linker between the S5 and the H5 domains relative to all other *Sh* channels (Fig. 12A). *Sh*III proteins expressing inactivating and noninactivating channels differ in the sequence of this insert (Fig. 11). The insert is nearly identical in KV3.1 and KV3.2 proteins that express delayed-rectifier type currents, but different in KV3.3 and KV3.4 that express inactivating channels.

2. S4 Domain

S4 domains, located between the third (S3) and the fourth (S5) hydrophobic segments, are characterized by the repetition of a motif consisting of two neutral residues (usually hydrophobic, except toward the carboxyl end of S4) and one positively charged residue. In *Sh* proteins, the positively charged residue is usually arginine, except toward the carboxyl end of S4 where some lysines are located. S4 domains are also present in a number of other voltage-gated channels including K$^+$ channels of the *eag* family, Na$^+$ and Ca^{2+} channels,

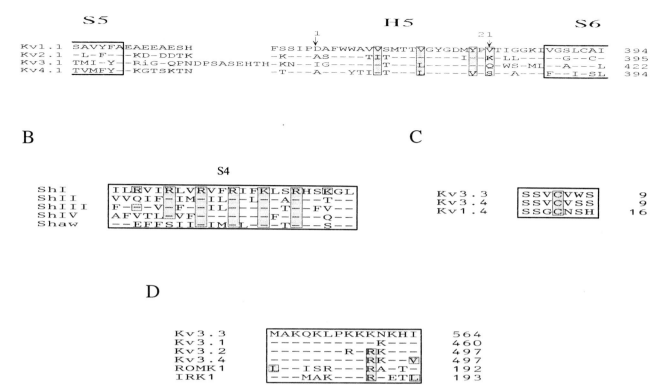

FIGURE 12 (A) Comparison of the sequence between the S5 and S6 domains (including the H5 domain) of mammalian *Sh* proteins of different subfamilies. Note the 9-residue insert in KV3.1. Other *Sh*III proteins have a similar insert (see Fig. 11). The sequences are aligned to maximize long stretches of identity. Gaps required for optimal alignment are shown as blanks. The amino acids in the H5 region have been arbitrarily numbered from 1 to 21. Residues in the H5 domain that were found by mutagenesis to be important in determining pore properties of KV3.1 are boxed (see text). (B) Comparison of the sequence of the S4 domains of mammalian *Sh* proteins and the *Drosophila Shaw* protein. Positively charged residues have been boxed. (C) Conserved sequence containing a cysteine residue at the amino end of KV3.3, KV3.4, and KV1.4 proteins. (D) A region rich in positively charged residues that follows the S6 domain of the protein products of the four *Sh*III genes is compared with a similar sequence in two inward rectifier channels. In A, C, and D the numbers on the right represent the position of the last residue shown relative to the starting methionine. In A, B, and D a dash represents an amino acid identical to that of the sequence shown in the top row.

and some channels that are thought not to be primarily voltage gated, such as cyclic-nucleotide-activated channels (Kaupp *et al.*, 1989; Dhallan *et al.*, 1990; Jan and Jan, 1990b) and Ca²⁺-activated K⁺ channels of the *slo* family (Atkinson *et al.*, 1991). The number of repetitions of the sequence motif X–X–R/K, where X is a hydrophobic or neutral residue, in the S4 domain is characteristic of the proteins of each *Sh* subfamily: seven in *Sh*I proteins, five in *Sh*II proteins, six in mammalian *Sh*III proteins, and five in *Sh*IV proteins (Fig. 12B). Curiously, the *Drosophila Shaw* protein contains a distorted S4 domain containing four positively charged residues, one negatively charged residue, and one serine instead of one of the arginines (Fig. 12B).

Mutagenesis experiments on Na⁺ and *Sh* K⁺ channels support the notion that the S4 domain acts as a voltage sensor (Noda *et al.*, 1986; Stühmer *et al.*, 1989b; Liman *et al.*, 1991; Papazian *et al.*, 1991; see also Chapter 3). However, the mechanism of voltage-dependent gating and the role of the S4 domain in this mechanism is far from understood and might be quite complex, involving structures outside the S4 domain (Liman *et al.*, 1991; K. McCormack *et al.*, 1991). For example, clearly not all positive charges are equivalent; in addition, for some positions, the nature of the positively charged residue (arginine or lysine) is important. Considering these results, it is not surprising that no clear-cut correlation exists between the reported slope of the conductance–voltage relationship (gating charge) of *Sh* channels and the number of positive charges in the S4 domain. Nevertheless, the fact that the structure of the S4 domains is constant for each subfamily is intriguing.

Other potentially important domains of *Sh*III proteins may or may not be found in *Sh* proteins of other subfamilies, or even in all *Sh*III proteins. Starting at the amino end, several of these domains are described in the following sections.

3. Amino End "Ball and Chain" Domains in Inactivating *Sh*III Proteins

KV3.3 and KV3.4 proteins, which express inactivating channels, differ from those that do not inactivate by an N-terminal insert of 78 residues in KV3.3 and 28 residues in KV3.4 (see Fig. 11). At the C-terminal end of these inserts is a methionine that can be aligned with the starting methionine of KV3.1 and KV3.2 to produce proteins with similar sequence (Fig. 11).

In *Drosophila Shaker* channels, the N termini are involved in a fast inactivation process known as N-inactivation. In addition, another inactivation process known as C-inactivation is dependent on structures toward the C terminus of the protein (perhaps between H5 and S6; Hoshi *et al.*, 1991). Deletions of the

N terminus in *Shaker* channels results in removal of the N-inactivation (Hoshi *et al.*, 1990; Zagotta *et al.*, 1990). The N terminus is hypothesized to act as a "ball and chain" (Hoshi *et al.*, 1990; Zagotta *et al.*, 1990). The tethered "ball" is proposed to produce inactivation by occluding the pore of open channels (Armstrong and Bezanilla, 1977; Hoshi *et al.*, 1990; Zagotta *et al.*, 1990).

The presence of clear N-terminal inserts in inactivating KV3.3 and KV3.4 channels suggests that these channels also inactivate by a "ball and chain" mechanism. In support of this hypothesis, deletion of the 78 residues of the N-terminal insert of KV3.3 and the 28 residues of the N-terminal insert of KV3.4 results in the expression of channels that do not inactivate (Fig. 13; see also Rettig *et al.*, 1992). Whether the differences in inactivation properties of the channels formed by KV3.3 and KV3.4 proteins (Section IV,D) are the results of the differences in their amino ends still remains to be determined.

4. A Domain for the Regulation of Channel Activity by the Redox Potential

The inactivating inserts of KV3.3 and KV3.4 contain a cysteine residue surrounded by a very conserved sequence (Fig. 12C). A very similar sequence is found

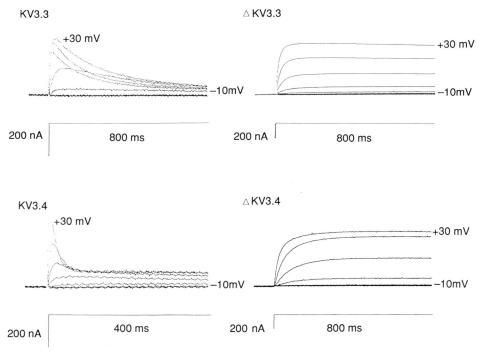

FIGURE 13 Deletion of the N-terminal inserts of KV3.3 and KV3.4 removes inactivation. ΔKV3.3 and ΔKV3.4 correspond to constructs in which the first 78 amino acids of KV3.3a and the first 28 amino acids of KV3.4b, respectively, have been deleted. Shown are the currents recorded under voltage clamp during depolarizing pulses from −60 to +30 mV in 10-mV increments, delivered at a rate of one every 60 sec from a holding potential of −100 mV. Currents have been leak subtracted.

in KV1.4, a fast-inactivating *Shaker*-related mammalian channel (Fig. 12C). The conservation of this sequence is even more notable since otherwise very little amino acid sequence conservation is seen in the N-termini of these proteins (Stühmer *et al.*, 1989a; Rudy *et al.*, 1991b; Schroter *et al.*, 1991; Vega-Saenz de Miera *et al.*, 1992).

Ruppersberg *et al.* (1991b) showed that exposure of the cytoplasmic surface of KV1.4 or KV3.4 channels to air, presumably to oxygen (in inside-out excised membrane patches), resulted in removal of the inactivation process. In contrast to the effects of N-terminal deletions and other treatments, this inactivation removal was reversible. Inactivation was restored if the patch was reinserted into the cell or by exposure to reduced glutathione or dithiothreitol (DTT; Ruppersberg *et al.*, 1991b). The effects of oxidation were not seen in channels expressed from a KV1.4 protein in which the cysteine was replaced with serine (Ruppersberg *et al.*, 1991b). A model to explain how the redox state of this cysteine may modulate channel inactivation was proposed by these investigators.

Similarly, exposure of oocytes expressing KV1.4, KV3.3, or KV3.4 channels to high micromolar concentrations of external H_2O_2 resulted in a reversible removal of inactivation (Vega-Saenz de Miera and Rudy, 1992). Concomitant with inactivation removal there was a significant increase in current magnitude that was particularly large in KV3.4 injected oocytes. The effects of H_2O_2 were specific to these three channels. No effect was detected in oocytes expressing channels that lack the conserved cysteine-containing sequence, including noninactivating KV3.2 channels, inactivating *Drosophila Shaker* channels (Vega-Saenz de Miera and Rudy, 1992), and inactivating KV4.1 and KV4.2 channels (of the *ShIV* subfamily) (P. Serodio and B. Rudy, unpublished observations). Effects similar to those of H_2O_2 were seen on exposure to *N*-ethyl maleimide, a reagent that alkylates free sulfhydryls; in this case the effects were, as expected, irreversible.

5. Proline-Rich Domains

KV3.2 and KV3.3 proteins have proline-rich areas (underlined in Fig. 11) in regions of the protein that are thought to be cytoplasmic. One such sequence is found in KV3.2 in the N-terminal side of the S1 domain, and one is found in the C-terminal region of KV3.3 after the S6 domain. In addition, the inactivating N-terminal insert of KV3.3 is also proline rich.

T. McCormack *et al.* (1990) first observed one such proline-rich insert in KV3.2a that had the sequence PPLSPPPRPPPLSP. These investigators suggested that the serines in this sequence might be sites of cytoplasmic O-glycosylation, based on similarities with the sequences surrounding serines in other proteins that are known to undergo cytoplasmic O-glycosylation.

More recently, researchers have found that proline-rich sequences such as those seen in KV3.2 and KV3.3 are binding sites for SH3 domains, a module present in a number of proteins believed to mediate protein–protein interactions that are important in signal transduction (Olivier *et al.*, 1993; Ren *et al.*, 1993). The proline-rich areas in KV3.2 and KV3.3 contain stretches that fit the proposed consensus sequences for SH3-domain binding. Our laboratory is currently investigating the interaction of the proline-rich domains of KV3.2 and KV3.3 with SH3-domain-containing proteins.

6. Subunit Association Domains

The amino acid sequence of the N terminal region of *ShIII* proteins that follows the N-terminal inserts in KV3.3 and KV3.4 is extremely well conserved (see Fig. 11). The sequences are also very similar to the equivalent sequence in the *Drosophila Shaw* protein (Vega-Saenz de Miera *et al.*, 1992; see also Table VI). In analogous positions of other *Sh* proteins, the amino acid sequence is very well conserved within members of the same subfamily, but shows relatively little conservation among members of different subfamilies (Stühmer *et al.*, 1989a; Jan and Jan, 1990a; Perney and Kaczmareck, 1990; Baldwin *et al.*, 1991; Rudy *et al.*, 1991a; Hwang *et al.*, 1992; Pongs, 1992). In *ShIII* proteins, the region of similarity is broken into two conserved areas by the polyproline insert of KV3.2 (see Fig. 11).

Li *et al.* (1992) and Shen and Pfaffinger (1992) showed that the conserved area prepared *in vitro* can form tetramers or bind to *in vitro* translated subunits of the same subfamily but not of different subfamilies. These and other elegant experiments provide evidence that the self-aggregating conserved area includes domains (hereafter called subunit recognition domains) that are important for both subunit recognition and channel formation, and may be the basis for the formation of heteromultimeric channels by subunits of the same subfamily. Additional evidence that these subunit recognition domains are crucial for the assembly of appropriate subunits was provided by Li *et al.* (1992), who used chimeric proteins to show that these domains were sufficient to determine coassembly of subunits into functional channels.

7. *N*-Glycosylation Sites in the Linker between the S1 and S2 Domains

All *ShIII* proteins have two putative *N*-glycosylation sites in the linker between the S1 and the S2 domains that are thought to face the extracellular surface of the membrane (indicated with triangles in Fig. 11). In

KV3.3, another putative *N*-glycosylation site is found in the linker between the S5 and H5 domains (see Fig. 11). Whether these sites are indeed glycosylated in *Sh*III proteins, and the functional consequences, if any, of this glycosylation, are not yet known. Most *Sh*I proteins also contain a putative *N*-glycosylation site in the S1–S2 linker, whereas *Sh*II proteins have an *N*-glycosylation site in the S3–S4 linker.

8. Disruption of the Leucine-Heptad Repeat in *Sh*III Proteins

When the sequence of the S4 domain, the linker between the S4 and S5 domains, and the beginning of the S5 domain of *Sh* proteins is drawn in α-helical form, one side of the α helix is particularly well conserved (Fig. 14). In particular, conservation of one arm of the helix (number 1 in Fig. 14), consisting of the third arginine in S4 and five leucine residues, spaced 7 amino acids apart, is apparent (K. McCormack *et al.*, 1989,1991). Mutagenesis of these residues leads to profound changes in voltage-dependent gating (K. McCormack *et al.*, 1991; Schoppa *et al.*, 1992). K. McCormack *et al.* (1991) proposed that the leucines in the heptad repeat mediate hydrophobic interactions that are crucial to the conformational changes involved in voltage-dependent gating.

Interestingly, in all *Sh*III proteins the fourth leucine of the leucine heptad repeat is substituted by phenyl-

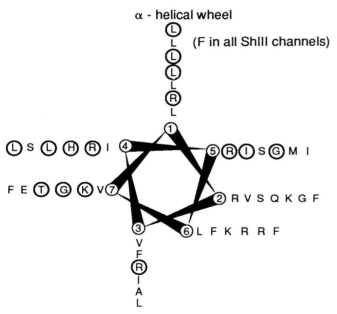

FIGURE 14 Helical representation of the S4–S5 region of the KV1.1 protein. The sequence of the KV1.1 protein, starting at the second residue in S4 (a leucine) and ending in the last leucine of the leucine heptad repeat (located in the S5 domain), has been projected onto an α-helical wheel. Residues that are conserved among all mammalian *Sh* proteins have been circled.

alanine. Researchers have suggested that this substitution might explain why *Sh*III channels require very positive potentials for activation (T. McCormack *et al.*, 1990; Rudy *et al.*, 1991a,b; Vega-Saenz de Miera *et al.*, 1992). This hypothesis has only been tested indirectly. Rettig *et al.* (1992) reported that replacing the phenylalanine with leucine and the next residue in the sequence (a leucine in all *Sh*III proteins) with glycine (making this area similar to *Shaker*-related proteins) produced a 16-mV negative shift in the voltage-dependence of a KV3.4 channel. The interpretation of their results is not straightforward since two residues were changed. Further, the interpretation is complicated by the fact that Rettig *et al.* (1992) utilized a KV3.4 cDNA with a spontaneous mutation in the S4 domain (H364Q), which itself produced a large positive shift in the voltage dependence of this channel relative to normal KV3.4 channels. Nevertheless, this result is encouraging and stimulates interest in further mutagenesis studies.

9. Cluster of Positively Charged Residues After the S6 Domain

All *Sh*III proteins possess a cluster of positively charged residues just after the S6 domain region (indicated with a star in Fig. 11). A similar sequence is found (see Fig. 12D) in the two inward rectifier K$^+$ channels IRK1 (Kubo *et al.*, 1993) and ROMK1 (Ho *et al.*, 1993). These similarities are intriguing. Perhaps these structural similarities are related to the rectification seen in *Sh*III channels (see Section IV,D).

10. Carboxyl Ends

*Sh*III proteins become particularly variable after the common cluster of positive charges found after the S6 domain. Moreover, all *Sh*III proteins terminate in divergent sequences derived by alternative splicing. Nevertheless, some short stretches of highly conserved amino acid sequence are seen, particularly in the portions of the protein thought to be encoded by exon 3. As an example, consider the sequence PLAQEEI (residues 489–495) in KV3.1, PLAQEEV (residues 648–654) in KV3.3, and PPAREEG in KV3.4 (residues 526–532) (see Fig. 11).

11. Putative Phosphorylation Sites

Several sites that fit the consensus sequence for protein kinase phosphorylation are found in regions of *Sh*III proteins that are thought to be intracellular. Some of these sites are found in the products of all the genes, some are found only in proteins encoded by a single gene, and some are specific to putative alternatively spliced isoforms. The consensus sites for three kinases are listed in Table IV. None of these

TABLE IV Putative Sites for Protein Kinase Phosphorylation in *ShIII* Channels[a]

			KV3.1	KV3.2	KV3.3	KV3.4
Putative sites for protein kinase C	Sites common to all *ShIII* channels	Threonine	21, 25, 73, 108, 337	21, 25, 155, 374	103, 107, 155, 190, 441	49, 53, 113, 148, 374
	Gene-specific sites	Serine	185	224	286	222
		Threonine	152	205, 564	16, 657	9, 35, 515
		Serine			11, 89	
	Sites specific to an alternatively spliced version		KV3.1a serine 510, KV3.1b serine 503, threonine 539, 585	KV3.2d threonine 609	KV3.3a serine 731, 769	KV3.4a, KV3.4c serine 604
Putative sites for protein kinase A	Sites common to all *ShIII* channels	Threonine	21, 25, 342	21, 25, 379	103, 107, 446	49, 53, 379
		Serine	341	378	445	378
	Gene-specific sites	Threonine		574	16	520
		Serine		563, 564		15, 519
	Sites specific to an alternatively spliced version			KV3.2a threonine 613, KV3.2d serine 613, KV3.2b serine 615	KV3.3a threonine 736, 852, 865, serine 713, 885, 886	KV3.4a, KV3.4c serine 541, 578, KV3.4c threonine 609
Putative sites for casein kinase II	Sites common to all *ShIII* channels	Threonine	99	146	181	139
		Serine	341	378	445	378
	Gene-specific sites	Threonine	483	17, 520, 546, 579	120	45, 520
		Serine	44, 130	414, 541, 553, 564		515
	Sites specific to an alternatively spliced version		KV3.1b threonine 527, serine 550	KV3.2a threonine 605	KV3.3a serine 697	KV3.4a, KV3.4c threonine 568, serines 552, 594, KV3.4c threonine 609

[a] The deduced amino acid sequences of *ShIII* channels contain several intracellular sites that fit the consensus sequences of protein kinase A (R-[R/K]-X-[T/S] >> R-X$_2$-[S/T] = R-X-[S/T]; Kemp and Pearson, 1990; Kennelly adn Krebs, 1991; some of these sites may be phosphorylated as well by cGMP-dependent protein kinase); protein kinase C ([R/K$_{1-3}$]-X$_{1-20}$-[S/T]-X-[R/K] or [S/T]-X-[R/K]; Kemp and Pearson, 1990; Kennelly and Krebs, 1991; sites where the S or T is separated by more than one residue from the C-terminal basic residue have not been included); and casein kinase II ([S/T]-X$_2$-[D/E]; (Marin *et al.*, 1986; additional putative sites for this kinase are not included; see Kennelly and Krebs, 1991). The numbers represent the positions of the amino acids that are putatively phosphorylated. We have grouped these sites into three different categories: (1) sites common to all *ShIII* channels, that is, sites that are in a similar position when all *ShIII* polypeptides are aligned to maximize aa identity; (2) gene-specific sites, that is, sites only found in the proteins encoded by a given *ShIII* gene; and (3) sites specific to alternatively spliced versions, which are found in the region of divergence of the alternatively spliced products. For protein kinase A, the numbers in italics correspond to sites with a better consensus sequence (R-[R/K]-X-[T/S]; Kennelly and Krebs, 1991). The influences of residues near the putative recognition sequence motif (Kemp and Pearson, 1990) were not considered.

sites has been shown to be phosphorylated, although Critz *et al.* (1993) reported inhibition of a KV3.1 channel by phorbol esters that activate protein kinase C and we have seen inhibition of KV3.2 currents by cAMP-dependent phosphorylation (H. Moreno and B. Rudy, unpublished observations).

Since alternatively spliced variants express nearly identical currents in *Xenopus* oocytes (see Section IV,D), and since they also appear to have similar expression patterns in brain (see Section IV,H), the existence of transcript-specific putative phosphorylation sites is especially interesting.

H. Tissue and Cellular Distribution of ShIII Transcripts

Studies of the distribution of *Sh*III mRNAs utilizing Northern blot and RNAse protection analysis, as well as *in situ* hybridization histochemistry (Drewe *et al.*, 1992; Perney *et al.*, 1992;[3] Rudy *et al.*, 1992; Weiser *et al.*, 1994), have revealed a number of interesting features of the tissue distribution of *Sh*III transcripts in rat. In addition, these studies have provided important clues about which subunit combinations might occur *in vivo*. Although the specific localization of *Sh*III proteins remains to be determined, the mRNA localizations suggest where to begin searching for native channels containing *Sh*III proteins using electrophysiological methods.

A Northern blot of mRNA isolated from various rat tissues hybridized with *Sh*III gene-specific probes is shown in Fig. 6. The probes used in this experiment differentiate among the products of different genes but not among alternatively spliced variants of the same gene. KV3.1, KV3.2, and KV3.3 are expressed mainly in the brain. Low levels of KV3.1 transcripts (mainly the 8-kb band) are also seen in skeletal muscle, whereas low levels of KV3.3 mRNAs are found in kidney and lung. According to Ghanshani *et al.* (1992), KV3.3 transcripts are also present in mouse liver, thymus, and heart. In contrast, KV3.4 transcripts are more abundant in skeletal muscle than in brain. KV3.1 (KV3.1a only), KV3.3, and KV3.4 products have also been identified in PC12 pheochromocytoma cells (Vega-Saenz de Miera *et al.*, 1991,1992; Ried *et al.*, 1993). KV3.1 transcripts were also found in T lymphocytes, in which they are thought to be a primary component of the L-type K$^+$ channel and in spleen (Grissmer *et al.*, 1992).

Quantitative analysis (Weiser *et al.*, 1994) demonstrated that, in the adult rat brain, KV3.1 (KV3.1b) and KV3.3 transcripts were the most abundant *Sh*III mRNAs. They appeared to be present in amounts similar to KV1.2 mRNAs, a *Shaker*-related transcript (Chandy *et al.*, 1991). KV3.1 (KV3.1b) and KV3.3 mRNAs were ~3 times more abundant than KV3.2 mRNAs, ~6 times more abundant than KV3.4 transcripts, and about 1/6 as abundant as Na$^+$ channel α subunit mRNAs.

1. ShIII Gene Expression during Development

Little is known about the expression of *Sh*III genes during development. Perney *et al.* (1992) found that low levels of KV3.1 transcripts were detected in the rat brain as early as embryonic day 19. By postnatal day 3 expression was detected in the cortex and in the hippocampus. A large increase in expression of these transcripts was seen, however, between the 7th and 14th day after birth. In fact, postnatal day 14 rats exhibited a pattern of expression in brain similar to that seen in the adult rat. KV3.2 mRNAs were also detectable in postnatal day 1 rat brains, but increase several-fold between day 10 and 15 (C. Kentros, unpublished observations).

2. Specific but Overlapping Expression of ShIII mRNAs in the CNS

In situ hybridization histochemistry has been used to study the distribution of *Sh*III mRNAs in the central nervous system (CNS) of adult rats (Drewe *et al.*, 1992; Perney *et al.*, 1992; Rettig *et al.*, 1992; Rudy *et al.*, 1992; Weiser *et al.*, 1994). The work by Perney *et al.* (1992) on KV3.1 and the work by Weiser *et al.* (1994) on KV3.1, KV3.2, KV3.3, and KV3.4 include, in addition to low-resolution X-ray autoradiography (which provides regional localization), a higher cellular-resolution microscopic analysis of emulsion-dipped sections. The data from these two studies form the basis for the descriptive analysis presented here.

As is the case with neurotransmitters, receptors, and other molecular markers, *Sh*III mRNAs were distributed in a nonhomogeneous fashion throughout the CNS. In fact, some neuronal types (e.g., most neurons in the cerebral cortex, the caudate–putamen, the amygdala, the epithalamus, the hypothalamus, the substantia nigra compacta, and a few structures in the brain stem such as the inferior olive) did not appear to express significant amounts of any *Sh*III mRNAs. Interestingly, the main neuronal population of the caudate–putamen appears to express few *Sh* transcripts of the other subfamilies as well (Weiser *et al.*, 1994). Of those tested, only probes for KV1.4 of

---

[3] KV3.1α and KV3.1β in Perney *et al.* (1992) correspond to KV3.1b and KV3.1a, respectively, in Weiser *et al.* (1994) and in this chapter.

the *Sh*I subfamily (Sheng *et al.*, 1992) and KV4.2 of the *Sh*IV subfamily (Tsaur *et al.*, 1992) labeled the caudate–putamen strongly. Only weak labeling was seen with *Sh*II probes (Drewe *et al.*, 1992; Hwang *et al.*, 1992).

Based on the distribution of hybridization signals as well as on cell size and morphology of labeled cells in Nissl-counterstained sections, Perney *et al.* (1992), Rudy *et al.* (1992), and Weiser *et al.* (1994) concluded that in the CNS *Sh*III mRNAs were present mainly in neurons and not glia. In all cases, hybridization signals were seen in the larger diameter cells with characteristics of neuronal somata. Moreover, highly sensitive RNAse protection assays with RNA from glial cells cultured from rat neonate optic nerve and from C6 glioma cells did not detect KV3.1 transcripts (Perney *et al.*, 1992).

In the rat CNS, each *Sh*III gene was found to exhibit a unique pattern of expression. However, a significant degree of overlap was seen in the distribution of KV3.1 and KV3.3 mRNAs. Moreover, KV3.4 transcripts were only present, albeit at lower levels, in several of the neuronal populations that also expressed KV3.1 and/or KV3.3 mRNAs. The expression patterns of transcripts of the four *Sh*III genes in adult rat brain are illustrated diagramatically in Fig. 15.

KV3.1, KV3.3, and KV3.4 exhibited patterns of expression more similar to each other than that of KV3.2 transcripts, thus defining two trends in the expression of *Sh*III genes in the CNS. Many neurons expressing KV3.2 mRNAs expressed these transcripts predominantly (relative to the other *Sh*III transcripts), if not exclusively (e.g., in thalamic relay neurons of many nuclei of the dorsal thalamus, and in neurons of the optic layer of the superior colliculus and the locus coeruleus). At the same time, many regions that expressed KV3.1, KV3.3, or KV3.4 mRNAs prominently—such as the cerebellar cortex, the spinal cord, the reticular thalamic nucleus, the inferior colliculus, and many nuclei in the brain stem—appeared to express little or no KV3.2 transcripts.

Other interesting features of *Sh*III mRNA expression were the differences between dorsal and ventral thalamus (Fig. 15). KV3.2 mRNAs were most abundant in thalamic relay neurons throughout the dorsal thalamus, but were weakly expressed in the reticular thalamic nucleus, a structure of the ventral thalamus. On the other hand, KV3.1 and KV3.3 mRNAs were much more abundant in the neurons of the reticular thalamic nucleus than in thalamic relay neurons of the dorsal thalamus. These two parts of the thalamus have been demonstrated to have different embryological origins (Rose, 1942) and are functionally distinct, since only the nuclei in the dorsal thalamus project to the cerebral cortex (Jones, 1985; Steriade *et al.*, 1990).

3. Transcript-Specific Expression

The study by Weiser *et al.* (1994) used gene-specific probes that did not distinguish among the alternatively spliced variants of each gene. In two cases, KV3.1 (Perney *et al.*, 1992) and KV3.2 (Rudy *et al.*, 1992), *in situ* hybridization with probes specific to alternatively spliced variants produced spatially identical labeling patterns, suggesting that the variants of each gene are colocalized to the same neurons. Perney *et al.* (1992), however, noted differences in the developmental expression of KV3.1a and KV3.1b transcripts in the rat brain. KV3.1a transcripts were more abundant prenatally and at earlier postnatal periods, whereas the reverse was true in the adult brain.

4. *Sh*III mRNA Expression Distinguishes among Subsets of Interneurons in the Cortex, Hippocampus, and Caudate Putamen

*Sh*III probes produced patchy labeling in the cerebral cortex and the hippocampus (Perney *et al.*, 1992; Weiser *et al.*, 1994). Higher magnification studies in these areas showed that these signals were caused by hybridization in a small number of neurons that were morphologically indistinguishable from other cells in the area. In addition, *Sh*III probes labeled a unique subset of neurons in the caudate–putamen (Weiser *et al.*, 1994). Therefore, subsets of neurons in these regions apparently can be distinguished by their expression of *Sh*III mRNAs.

Based on distribution and cell morphology, Perney *et al.* (1992) and Weiser *et al.* (1993) concluded that the cells expressing *Sh*III mRNAs in the neocortex were local circuit neurons which, in the rat cortex, represent a small percentage of the total neuronal population. KV3.2 mRNAs were present in neurons located in layers IV–VI, resembling the distribution of D1 dopamine receptors and somatostatin binding (Weiser *et al.*, 1994). KV3.1 mRNAs and, to a lesser degree, KV3.3 and KV3.4 mRNAs were present in cells located throughout layers II–VI, being more numerous in layer II and in the superficial parts of layers III and IV. Thus, in X-ray autoradiograms, probes specific for KV3.2 mRNAs produced a band localized toward deep layers of the cortex, whereas KV3.1, KV3.3, and KV3.4 probes produced a double band: one in superficial layers, which was more intense, and one in deeper layers (Perney *et al.*, 1992; Weiser *et al.*, 1994).

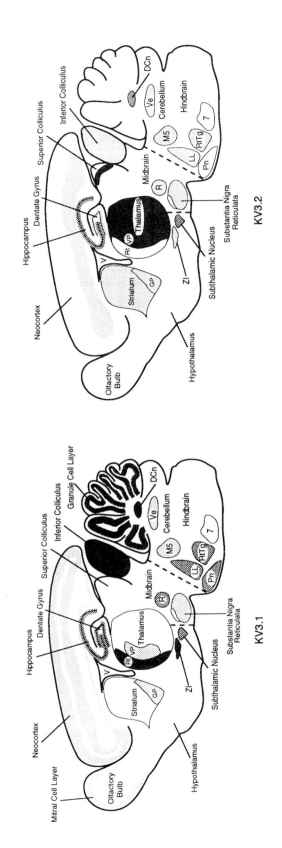

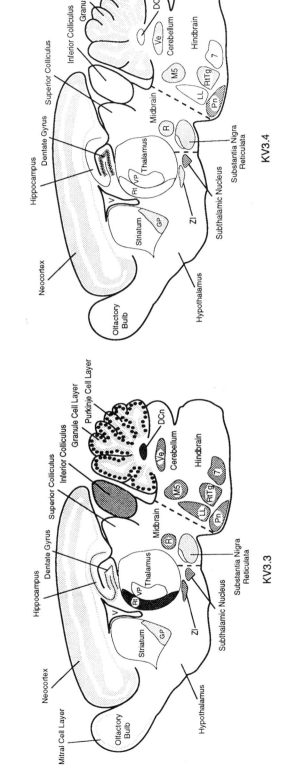

*Sh*III mRNAs were weakly to moderately expressed in the pyramidal and granule cells of the hippocampus (Fig. 15). In addition, KV3.1, KV3.2, and KV3.3 mRNAs were also expressed in specific subsets of interneurons (Perney *et al.*, 1992; Weiser *et al.*, 1994). The differences in labeled interneuron distribution were more notable in the CA1 and hilar regions. The distribution of cells expressing KV3.2 mRNAs was similar to that reported for somatostatin immunoreactivity (Weiser *et al.*, 1994). Labeled cells were numerous in the stratum oriens in the CA1 region as well as in the hilus. In contrast, the distribution of cells hybridizing with KV3.1 and KV3.3 probes resembled that for GABA or cholecystokinin (Perney *et al.*, 1992; Weiser *et al.*, 1994). In the CA1 region, labeled cells were seen along the stratum radiatum–pyramidalis border and along the stratum oriens–pyramidalis border. In the area dentata, KV3.1- and KV3.3-labeled cells were more numerous in the borders of the granular cell layer than in the hilus proper. Subsets of cortical and hippocampal interneurons have been described based on differences in morphology (DeFelipe and Jones, 1988; Zilles, 1990) or in the expression of various neurotransmitters, neuropeptides, and receptors (Somogyi *et al.*, 1984; Douglas and Martin, 1990; Lin *et al.*, 1986; Sloviter and Nilaver, 1987; Demeulemeester *et al.*, 1988; Woodson *et al.*, 1989; Zilles *et al.*, 1990).

In the caudate–putamen, *Sh*III probes labeled a very small number of neurons (Weiser *et al.*, 1994) that were described to be a subset of small to medium-sized cells distinct from the large cholinergic local circuit neurons. Weiser *et al.* (1994) hypothesized that the *Sh*III mRNA-containing neurons were a subpopulation of the principal GABAergic neurons that comprise over 90% of the cells in the caudate–putamen, or perhaps yet another distinct subpopulation of interneurons.

5. Coexpression of *Sh*III Transcripts in the Same Neuronal Populations Reveals a High Potential for Heteromultimer Formation

Based on compelling evidence from overlapping labeling with multiple probes in easily identifiable neuronal populations, Weiser *et al.* (1994) concluded that neurons in several CNS areas are likely to coexpress transcripts of more than one *Sh*III gene. These investigators argued that heteromultimer formation is a potentially significant feature of the channels formed by subunits of the *Shaw* subfamily and raised the interesting possibility that the subunit composition of heteromultimers could vary in different neurons since the ratio of overlapping signals changed from one neuronal population to another.

The neuronal populations that may coexpress transcripts of distinct *Sh*III genes include the mitral cells of the olfactory bulb; the principal neurons of the reticular thalamic nucleus, the lateral leminiscus, the reticulo-tegmental nucleus of the pons, and the trigeminal motor nucleus, which were labeled with

FIGURE 15 Distribution of *Sh*III mRNAs in the rat brain. The levels of expression of transcripts from four different *Sh*III genes, based on *in situ* hybridization histochemistry studies by Perney *et al.* (1992) and Weiser *et al.* (1994), are represented by different grades of shading. KV3.1 and KV3.3 mRNAs were most prominent in the cerebellar cortex, but are also abundant in the reticular thalamic nucleus (Rt), the inferior colliculus, and several nuclei in the brain stem, including the nuclei of the lateral lemniscus (LL), the reticulo-tegmental nucleus of the pons (RtTg), the pontine nuclei (Pn), the trigeminal motor nuclei (M5), and the vestibular nuclei (Ve). These RNAs were also expressed prominently in the cochlear nucleus and spinal cord (not shown). Moderate to low expression was seen in the mitral cells of the olfactory bulb and in thalamic relay neurons of some nuclei of the dorsal thalamus (such as the ventral posterior complex, VP). Hybridization signals with KV3.2 probes were strongest in thalamic relay neurons throughout the dorsal thalamus. KV3.2 signals were also prominent on neurons of the optic layer of the superior colliculus. Moderate to weak hybridization was seen in the locus coeruleus and a few other structures in the brain stem. KV3.4 mRNAs were the least abundant *Sh*III RNA in the central nervous system. They were seen mainly in areas that also contain KV3.1 and/or KV3.3 transcripts, usually at higher levels.

All four *Sh*III mRNAs are expressed in the giant neurons of the gigantocellular part of the red nucleus (R) and in the majority of the neurons of the subthalamic nucleus. All were present as well in parts of the septum, zona incerta (ZI), substantia nigra reticulata, globus pallidus (GP), reticulo-tegmental nucleus of the pons (RtTg), and oculomotor nucleus. KV3.1, KV3.3 mRNAs, and, to a lesser degree, KV3.2 mRNAs were expressed in the large neurons of deep cerebellar nuclei (DCn). Each *Sh*III mRNA had a characteristic distribution in the hippocampus. KV3.1, KV3.2, and KV3.3 mRNAs, but not KV3.4 mRNAs, were expressed weakly in the pyramidal cell layers of the CA1–CA3 fields. KV3.1, KV3.4, and, to a lesser degree, KV3.3 mRNAs were expressed in granule cells of the dentate gyrus. KV3.1, KV3.2, and KV3.3 mRNAs also were expressed in distinct subsets of hippocampal interneurons. See text and original references for further details.

KV3.1 and KV3.3 probes; and neurons of the pontine nuclei and some cerebellar Purkinje cells, which were labeled with KV3.1, KV3.3, and KV3.4 probes. Coexpression of KV3.1 and KV3.3 is also likely in the inferior colliculus and the reticular nuclei of the hindbrain, although the complex cytoarchitecture of these areas made it more difficult for Weiser *et al.* (1994) to assess whether cells were co-labeled.

Other neuronal populations appeared to express all *Sh*III mRNAs, albeit at different levels. These groups included the principal neurons of the globus pallidus, the subthalamic nucleus, the diagonal band of Broca, the substantia nigra pars reticulata, the zona incerta, and the giant neurons of the gigantocellular part of the red nucleus (see Fig. 15). In addition, thalamic relay neurons of some nuclei of the dorsal thalamus were found to coexpress KV3.2, KV3.1, and KV3.3 mRNAs.

6. Correlation between Expression Patterns of *Sh*III mRNAs and Other Neuronal Markers

Many, but by no means all, neurons expressing *Sh*III transcripts, particularly KV3.1, KV3.3, and KV3.4 mRNAs (e.g., interneurons in the cerebral cortex and the hippocampus; the neurons of the reticular thalamic nucleus, globus pallidus, zona incerta, inferior colliculus, nuclei of the lateral lemniscus, and the substantia nigra pars reticulata; and cerebellar Purkinje cells) are GABAergic, a feature first noted by Perney *et al.* (1992) for KV3.1 mRNAs. Although this extensive correlation is intriguing, note that some important GABAergic systems, notably the main neuronal population of the caudate–putamen and the olfactory bulb, do not appear to express these mRNAs. Moreover, some cells that express the mRNAs prominently, such as granule cells in the cerebellum and thalamic relay neurons, are not GABAergic but do receive GABAergic input. Perhaps channels containing these subunits mediate the effects of putative $GABA_B$ receptors.

The resemblance between the distribution of KV3.2 mRNAs in the cerebral cortex and the localization of somatostatin receptors is also interesting. Somatostatin has been shown to modulate a voltage-dependent K^+ current in cultured rat neocortical neurons that resembles KV3.2 currents in *Xenopus* oocytes in voltage dependence and kinetics (Wang *et al.*, 1989).

Other than these similarities, no obvious common characteristics of the neurons expressing any or all *Sh*III transcripts were found. Such a correlation would have provided strong hints about the functional role of *Sh*III subunits. However, the lack of correlation is not necessarily surprising. There are multiple receptors for most neurotransmitters and neuropeptides, each acting on different second messenger cascades. In addition to this divergence, there is a convergence of different neurotransmitter and neuropeptide receptors eliciting the same second messenger response. Further, a given voltage-gated K^+ channel could play different roles in different neurons or even in different places in the same neuron. In some cells, this channel may mediate the response to one neurotransmitter, whereas in another cell it could mediate the response to another neurotransmitter acting on the same second messenger system. Moreover, the same channel could be affected by more than one kinase and, hence, mediate the effect of a different second messenger system in a third group of cells, and play a role as a voltage-gated channel independent of second messenger modulation in another group of cells.

I. Searching for Candidate Native Sh III K^+ Channels

One approach to identifying native channels containing cloned K^+ channel subunits is to search for *in vivo* currents that resemble the currents expressed by these subunits in heterologous expression systems. Using such a strategy, investigators have found some attractive candidates, although adequate methods to prove unequivocally that the candidate channels contain a given subunit are unavailable. For example, Pardo *et al.* (1992) found that an A current recorded in rat cultured hippocampal neurons resembled the currents expressed *in vitro* by KV1.4 transcripts, of the *Shaker*-related subfamily, in voltage dependence, pharmacology, and kinetics, including a slow recovery from inactivation. Moreover, the native current was greatly reduced when the cell was perfused with K^+-free solution, returning to normal when the external K^+ was restored. Among several *Sh* channels tested in *Xenopus* oocytes, only KV1.4 channels were similarly affected by K^+-free solutions (Pardo *et al.*, 1992).

Another example is the N-type K^+ channel in T lymphocytes, which resemble the channels expressed in *Xenopus* oocytes by KV1.3 transcripts, also of the *Shaker*-related subfamily (Grissmer *et al.*, 1990), in many macroscopic and microscopic properties. Moreover, KV1.3 channels in *Xenopus* oocytes were the only known cloned K^+ channel blocked by margatoxin, a peptide derived from the venom of the scorpion *Centruroides margaritus*, which also blocks the N-type K^+ channel in lymphocytes (Novick *et al.*, 1991; Leonard *et al.*, 1992).

A different and interesting situation is encountered in the case of channels of the *Shal*-related subfamily.

A *Shal*-related transcript, KV4.2, expresses currents in *Xenopus* oocytes that resemble the low voltage-activating A currents that have been recorded in several neuronal types (Baldwin *et al.*, 1991; Pak *et al.*, 1991) and those seen in *Xenopus* oocytes injected with rat brain poly(A) mRNA (Rudy *et al.*, 1988). KV4.2 currents activate and inactivate near neuronal resting potentials, are insensitive to TEA but are blocked by millimolar concentrations of 4-AP, and have macroscopic time constants of inactivation that do not change with the membrane potential. However, KV4.2 currents in *Xenopus* oocytes recover very slowly from inactivation unlike typical low voltage-activating A currents found in neurons or those seen in oocytes injected with rat brain mRNA. Antisense hybrid-arrest experiments demonstrate that KV4.2 subunits indeed constitute a main component of the low voltage-activating A current that is expressed in *Xenopus* oocytes injected with rat brain mRNA (Serodio *et al.*, 1992). Serodio *et al.* (1992) suggested that additional proteins encoded in total rat brain mRNA (perhaps other channel subunits or modifying enzymes) are responsible for the differences between KV4.2 channels and those expressed from brain mRNA. Native channels might be similarly modified by these putative additional proteins. Chabala *et al.* (1992) also obtained evidence for the existence of modifying factors for another *Shal*-related transcript, KV4.1, that is much less abundant in brain than KV4.2 (P. Serodio and B. Rudy, unpublished observations). The situation encountered with these *Sh* channels is reminiscent of that of the rat brain Na$^+$ channel α subunit which, when expressed in *Xenopus* oocytes, has inactivation properties that are different from those of native Na$^+$ channels. Co-injection with rat brain Na$^+$ channel β subunits results in the expression of channels that more closely resemble native Na$^+$ channels (Isom *et al.*, 1992).

Regarding the identification of native K$^+$ channels containing *Sh*III subunits, Grissmer *et al.* (1992) proposed that the L-type K$^+$ channel in T lymphocytes is probably a homomultimer of KV3.1 subunits. These workers found that macroscopic KV3.1 currents in oocytes were indistinguishable from those of the L-type channel in T cells expressing KV3.1 RNAs. To eliminate differences resulting from different solutes across the membrane, Grissmer *et al.* (1992) made all their recordings in excised patches using identical solutions across the patch. The single-channel conductance of KV3.1 channels in *Xenopus* oocytes under the recording conditions used (27 pS) was also found to be very similar to that of L-type K$^+$ channels in mouse and human T lymphocytes (21–27 pS). Moreover, KV3.1 RNA was not found in cells containing little or no L-type channels, such as the EL4 mouse T cell line (Grissmer *et al.*, 1992).

Another native K$^+$ channel proposed to contain *Sh*III subunits (Vega-Saenz de Miera and Rudy, 1992) is the O$_2$-sensitive K$^+$ channel present in type I cells of the carotid body (Lopez-Barneo *et al.*, 1988; Lopez-Lopez *et al.*, 1989; Urena *et al.*, 1989; Ganfornina and Lopez-Barneo, 1991,1992a,b). Lopez-Barneo and colleagues hypothesized that the modulation of the activity of this K$^+$ channel by changes in arterial O$_2$ tension is one of the initial triggering events of the sensory response to changes in pO$_2$ in the carotid body (Lopez-Barneo *et al.*, 1988; Lopez-Lopez *et al.*, 1989; Urena *et al.*, 1989; Ganfornina and Lopez-Barneo, 1991, 1992a,b). The Po$_2$-sensitive current in type I cells of the carotid body is a TEA-sensitive, voltage-dependent, transient K$^+$ current that activates when the membrane potential is more positive than -20 mV. Of all cloned K$^+$ channels, only two *Sh*III channels—KV3.3 and KV3.4—express transient, TEA-sensitive, voltage-dependent K$^+$ currents requiring voltages more positive than -20 mV to activate in *Xenopus* oocytes. The kinetics of the pO$_2$-sensitive current, especially its time course of inactivation, are particularly similar to the currents expressed by KV3.3 mRNAs (KV3.4 currents inactivate much more rapidly). However, heteromultimers of KV3.4 and other *Sh*III proteins have intermediate inactivation properties (see Section IV,E); therefore, heteromultimers of KV3.4 and KV3.3 or KV3.4 with *Sh*III proteins expressing noninactivating channels could also produce currents similar to those of the pO$_2$-sensitive channel. Interestingly, PCR experiments demonstrate that KV3.3 mRNAs are present in the rabbit carotid body (E. Vega-Saenz de Miera and B. Rudy, unpublished observations). The pO$_2$-sensitive channel has been well characterized at the single-channel level (Ganfornina and Lopez-Barneo, 1991,1992a,b). Microscopic analysis of KV3.3 channels (or other *Sh*III channels) in heterologous expression systems may allow a more detailed comparison with the native channel.

Ito *et al.* (1992) recorded oocytes injected with KV3.1a cRNA with long (5 sec) depolarizing pulses to characterize the slow inactivation of KV3.1a currents and to compare these currents with a slowly inactivating current found in NG108-15 cells (Robbins and Sims, 1990) expressing KV3.1a mRNAs (Yokoyama *et al.*, 1989). Although the NG108-15 currents showed some resemblance to KV3.1a currents in oocytes, they began activating at more negative voltages and are much less sensitive to external TEA.

Given the results of mRNA expression studies, one would expect to find *Sh*III channels in several CNS neurons. However, the most notable feature is the general lack of resemblance of *in vitro ShIII* channels to typical K^+ currents seen in somatic recordings of neurons. For example, KV3.2 mRNAs appear to be most abundant in thalamic relay neurons. Two delayed rectifier currents, I_{K1} and I_{K2}, and a low voltage-activating transient current have been described in acutely dissociated thalamic relay neurons (Hugenard and Prince, 1991). None of these currents shows properties identical to those of KV3.2 channels in *Xenopus* oocytes (or those expected of any KV3.2 heteromultimeric combination). I_{K1}, which accounts for only 2% of the total outward current, is similar to KV3.2 channels in voltage dependence and sensitivity to TEA, but is proposed to be Ca^{2+} activated and is more sensitive to 4-AP, whereas I_{K2} activates at more negative voltages and is less sensitive to TEA.

Attempts at similar comparisons between the K^+ currents recorded in cerebellar Purkinje cells (Bossu *et al.*, 1988; Gähwiler and Llano, 1989; Hirano and Ohmori, 1986) and the currents expressed in oocytes by the *ShIII* transcripts present in these cells (KV3.3 >> KV3.4 > KV3.1) present difficulties that may explain why we are unable to find many clear-cut examples of *in vivo* currents in neurons expressing *ShIII* mRNAs similar to those of *in vitro ShIII* channels. The isolation of individual components of the total K^+ current is often tailored to the goals of a particular investigation. If the native *ShIII* currents in a given cell do not constitute a major portion of the total K^+ current, they could be masked by currents with overlapping electrophysiological and pharmacological properties. For example, none of the currents described in cultured cerebellar Purkinje cells is very similar to the currents expressed in oocytes by KV3.3 RNAs (which are abundant in cerebellar Purkinje cells) or by mixtures of KV3.3 and KV3.4 or KV3.1 RNAs. However, KV3.3 currents could account for some of the inactivating component of the "maintained," TEA-sensitive outward current seen in Purkinje cells in organotypic cultures, or some of the slow components of the transient current (Gähwiler and Llano, 1989). Because of the difficulties in isolating K^+ current components, electrophysiological experiments specifically designed to search for native channels with properties similar to those of *ShIII* channels *in vitro* might be required before considering other reasons for the apparent lack of correspondence between native and oocyte currents.

At least three additional explanations are possible for the apparent lack of oocyte-like *ShIII* currents in neurons expressing *ShIII* mRNAs. First, voltage-clamp studies from neurons often use dissociated embryonic or early postnatal tissue, which may not express *ShIII* subunits. Second, native channels containing *ShIII* proteins might have, at least in some cell populations, electrophysiological properties different from those of *in vitro ShIII* channels because of as yet unknown factors (additional subunits, posttranslational modifications, membrane lipid composition, etc.). Finally, *ShIII* channels might be more prominent in neuronal processes or terminals not accessible to the electrophysiological methods used.

J. Mammalian-Like ShIII Genes Might Be a Recent Evolutionary Acquisition

The coding sequence of one *ShIII* transcript (KV3.1a) has been sequenced in rodents (Yokoyama *et al.*, 1989; Ried *et al.*, 1993) and in humans (Ried *et al.*, 1993). In addition, sequence information is available on exon 2 of the human KV3.2 (D. Lau, E. Han, E. Vega-Saenz de Miera, and B. Rudy, unpublished) and KV3.3 (Lee and Roses, GenBank accession number Z11585) genes which can be compared with the corresponding sequences from KV3.2 and KV3.3 rodent cDNAs (T. McCormack *et al.*, 1990, 1991; Luneau *et al.*, 1991b; Ghanshani *et al.*, 1992; Rudy *et al.*, 1992; Vega-Saenz de Miera *et al.*, 1992). The sequence of the constant region (prior to the divergent 3' ends) of KV3.4 in human (Rudy *et al.*, 1991b) and in rat (Schroter *et al.*, 1991) is also available. This partial comparative information can be used for a preliminary assessment of the degree of conservation of *ShIII* proteins throughout mammalian evolution (Table V). In all cases, the predicted proteins are highly conserved. For example, only a single amino acid difference is seen in the 511 residues of KV3.1a in rat and human. The degree of conservation of *ShIII* proteins is similar to that seen in the histones, the most conserved proteins known, and is indicative of very strong selection for the *differences* in structure and function of the protein products of the four *ShIII* genes. Conservation throughout the length of the protein is not required for assembly into functional channels since different *ShIII* proteins (with 70–80% amino acid identity) can form functional heteromultimeric channels. The conservation of the amino domain of KV3.4 could be associated with preserving the inactivation properties determined by this domain (Section IV,G). However, the consequences of the conservation of other portions of polypeptides that are not conserved among different members of the subfamily (i.e., the sequence C-terminal to the sixth membrane-spanning domain in KV3.1a, and the sequence of the linkers between the first and the second and the second and the third membrane-spanning domains in the four genes) are less clear, but may signify that these regions are also

TABLE V Fraction of Identical Amino Acids in Analogous Positions in Pairs of Human and Rodent *Sh*III K⁺ Channel Proteins[a]

	KCNC2 (exon 2)	KCNC3 (exon 2)	KCNC4	Kcnc1	Kcnc2	Kcnc3	Kcnc4
KCNC1	.80	.82	.75	.99	.76	.77	.76
KCNC2 (exon 2)		.79	.84	.78	.99	.79	.82
KCNC3 (exon 2)			.78	.82	.76	.99	.78
KCNC4				.75	.74	.74	.97
Kcnc1					.76	.77	.76
Kcnc2						.72	.75
Kcnc3							.73

[a] Human genes are capitalized, rodent genes are lower cased.

important for specific roles of the protein products of each *Sh*III gene.

The functions that are being preserved are probably not merely the electrophysiological properties of the channels. For example, strong interspecific conservation is seen in KV3.1 and KV3.2 proteins, but both express currents that are very similar in voltage dependence, kinetics, and pharmacology in *Xenopus* oocytes. Additional comparative studies may shed some light on the nature of the functions that are being preserved. Alternatively, these proteins could be functionally identical, but their expression could be spatiotemporally distinct. Thus, both genes could be maintained only to allow for the expression of certain K⁺ channels through mutually exclusive developmental pathways. However, the preliminary comparative analysis shows that the *differences* between the two proteins are conserved among species, implying that areas with gene-specific sequence encode functionally distinct domains. Information on the conservation of the divergent 3′ ends, now almost completely lacking (except for the carboxyl end of KV3.1a), may also be enlightening. The conservation seen to date suggests that, after duplication and subsequent variation, each one of the *Sh*III gene products seen in mammals has been incorporated into an important function that cannot be substituted by another member of the same subfamily.

As described in the introduction, mammalian products of a given *Sh* subfamily are more similar to one of the four *Shaker*-related genes in *Drosophila* than to a mammalian product of a different subfamily, indicating that precursor genes to each subfamily existed prior to the divergence of chordates and arthropods. *Sh*III proteins are more similar to the products of the *Drosophila Shaw* gene (Butler *et al.*, 1989) than to other

Drosophila or mammalian *Sh* proteins (Yokoyama *et al.*, 1989; T. McCormack *et al.*, 1990, 1991; Luneau *et al.*, 1991a,b; Rudy *et al.*, 1991a,b,1992; Schroter *et al.*, 1991; Ghanshani *et al.*, 1992; Vega-Saenz de Miera *et al.*, 1992). However, the percentage of amino acid identity seen between mammalian *Sh*III proteins and *Shaw* (49–56%) is less than that seen between mammalian and fly homologs in the other three subfamilies (70–80%) (Yokoyama *et al.*, 1989; T. McCormack *et al.*, 1990; Luneau *et al.*, 1991a,b; Rudy *et al.*, 1991b; Schroter *et al.*, 1991; Ghanshani *et al.*, 1992; Vega-Saenz de Miera *et al.*, 1992). Perhaps even more significant is the fact that, although *Sh*III proteins are more similar to *Shaw* in certain regions of the protein, they are more similar to members of other *Sh* subfamilies in the so-called S4 domain and in the sequences near this domain (Table VI). Consistent with this feature, the channels expressed by mammalian *Sh*III cDNAs in *Xenopus* oocytes differ considerably more from those expressed by *Shaw* than do members of the other subfamilies from their *Drosophila* homologs (compare mammalian *Sh*III currents described in Section IV,D with those expressed by *Shaw*; (Wei *et al.*, 1990).

The mechanism of evolution of mammalian-like *Sh*III subfamily genes may have been different from that of the genes in the other subfamilies. Assuming that mammalian-like *Sh*III genes are the homologs of the *Shaw* gene, several mechanisms can be postulated. If the hypothetical precursor of mammalian-like *Sh*III genes and *Shaw* resembled mammalian-like *Sh*III genes, *Shaw* might have arisen by accelerated divergence in the evolutionary line leading to arthropods. Alternatively, if the precursor resembled *Shaw*, mammalian-like *Sh*III genes may have evolved by parallel evolution or gene conversion events. The

TABLE VI   **Amino Acid Conservation of Several Domains of *Sh* Proteins**[a]

Fraction of identical residues in KV3.2 and *Sh*K[+] channels of several subfamilies

Protein domains	KV3.2/*Shaw*	KV3.2/KV2.1	KV3.2/KV1.1
NH$_2$ domain[b]	0.67	0.47	0.34
S1	0.55	0.36	0.55
S2	0.55	0.50	0.41
S3	0.43	0.48	0.38
S4	0.45	0.50	0.64
S4–S5 linker	0.50	0.64	0.57
S5	0.55	0.45	0.45
H5	0.76	0.71	0.81
S6	0.91	0.68	0.82

Fraction of identical residues in rat and *Drosophila Sh*I and *Sh*II proteins

Protein domains	KV1.1/*Shaker*	KV2.1/*Shab*
NH$_2$ domain[b]	0.77	0.67
S1	0.77	0.73
S2	0.77	0.77
S3	0.81	0.71
S4	1.00	0.91
S4–S5 linker	0.93	0.86
S5	0.95	0.77
H5	0.86	0.71
S6	1.00	0.77

[a] The upper panel compares domains of KV3.2 proteins with the corresponding domains of *Shaw* (the fly homolog) and mammalian proteins of other subfamilies. The lower panel compares the same domains in fly and mammalian proteins of *Shaker* and *Shab* subfamilies.

[b] Refers to the area of the amino domain containing subunit recognition domains (see Section IV,G).

mosaic-like similarities of mammalian *Sh*III proteins and *Sh* genes of different subfamilies (Table VI) suggest gene conversion events. Whatever the mechanism, mammalian-like *Sh*III genes appear to be more specific to species in the evolutionary line leading to mammals. This factor may be important in considering possible functional roles of *Sh*III proteins, and justifies further studies on the evolution of these genes.

V. Summary and Perspectives

We have presented a discussion of the salient features of a subfamily of genes encoding putative subunits of voltage-gated K[+] channels; the *Shaw*-related or *Sh*III subfamily. Four *Sh*III genes have been identified, each generating multiple transcripts by alternative splicing. A total of 11 different *Sh*III cDNAs has been cloned. The various alternatively spliced transcripts from *Sh*III genes express voltage-gated K[+] channels in heterologous expression systems that activate at depolarized potentials more positive than −10mV. Transcripts of two genes (KV3.1 and KV3.2) express delayed-rectifier type currents whereas transcripts of the two other genes (KV3.3 and KV3.4) express A-type inactivating currents. *Sh*III currents can exhibit a wide range of inactivation properties as a result of the formation of heteromultimeric channels among different *Sh*III subunits. The specific but overlapping distribution of some *Sh*III mRNAs in rat brain, particularly the products of the KV3.1, KV3.3, and KV3.4 genes, lends credence to the idea of a system of heteromultimeric *Sh*III channels in the mammalian CNS. KV3.2 transcripts, which are expressed in many neurons that do not express transcripts of the other genes, may constitute a second independent channel system. Studies described in this chapter, particularly those localizing *Sh*III mRNAs to various tissues and neuronal populations in the CNS as well as those identifying candidate *Sh*III currents *in vivo*, represent important first steps in determining the exact relationship between cloned *Sh*III K[+] channel subunits and their *in vivo* correlates.

Perhaps the most important tools needed to pursue the identification of native *Sh*III channels in neurons and to gain insight into their functional role are subunit-specific antibodies. These antibodies are necessary to determine the localization of *Sh*III proteins to different regions of the neuronal membrane. The mRNA expression studies discussed here suggest (see also Weiser *et al.*, 1994) the interesting possibility that native *Sh*III channels may vary in subunit composition from one neuronal population to another, with one subunit type dominating in one case and another dominating in another case. If we accept this possibility then, by extension, subunit composition in a given cell may change as a function of time, for example, throughout development or in response to activity or specific stimuli. In this manner, channel subunit composition could contribute to neuronal plasticity. Subunit-specific antibodies would be extremely useful to test these ideas. Antibodies could also be used to determine subunit composition of native channels immunoprecipitated from membrane extracts.

The identification of native *Sh*III channels to date has relied on finding similarities between the physiological properties of native channels and those of *Sh*III channels expressed in heterologous expression systems. Additional electrophysiological studies in heterologous expression systems, including determining the effects of second messenger cascades, will provide more clues that can be used to identify native *Sh*III channels and to understand the functional role of *Sh*III proteins *in vivo*. The goal of identifying native *Sh*III channels would be greatly facilitated, however, by the successful implementation of techniques that can specifically eliminate the currents expressed by a given *Sh*III channel. Such techniques would include novel channel-specific toxins as well as molecular techniques such as antisense hybrid arrest (Lotan, 1992) and gene knockout techniques in transgenic mice.

Studies on the regulation of expression of K⁺ channel subunits in the nervous system could illuminate the longer term processes involved in its normal and pathological functions. Such studies would be without physiological context if we are uncertain of the nature of the currents produced by these subunits *in vivo*. This difficulty alone justifies the somewhat daunting task of identifying native *Sh* channels.

Moreover, the identification of channel proteins by cloning, in conjunction with approaches that allow the identification of the *in vivo* channels containing those proteins, can lead to the discovery of new, previously undescribed channels. This possibility is not surprising considering the current limitations of electrophysiological techniques, including the existence of electrophysiologically inaccessible neuronal processes and pre- and postsynaptic terminals. Studies with cloned channels *in vitro* may also lead to the discovery of new forms of K⁺ channel function and modulation, as in the discovery by Ruppersberg *et al.* (1991b), described earlier in this chapter, that K⁺

channels may be modulated by the oxidation of cysteine residues.

Note added in proof: (1) Attali *et al.* (1993a) have proposed that the I_{SK} protein may not be a channel but rather an activator of endogenous K⁺ or Cl⁻ channels present in *Xenopus* oocytes. (2) Sheng *et al.* (1993) and Wang *et al.* (1993) have shown coprecipitation of two different proteins of the Shaker subfamily from solubilized membrane extracts in the presence of an antibody against only one of the proteins. These results are strong evidence that heteromultimeric channels exist *in vivo*. (3) Phorbol esters that activate protein kinase C have been shown to block inactivation of KV3.4 channels presumably as a result of phosphorylation of serines or threonines in the N-terminal insert mediating N-inactivation of these channels (Vyas *et al.*, 1993). (4) The existence of mRNA isoforms generated by alternatively-splicing has now been demonstrated as well for a mammalian Shaker homolog (Attali *et al.*, 1993b) and for a mammalian Shab homolog (D. Lau and B. Rudy, unpublished observations).

Acknowledgments

This research was supported by NIH Grant NS30989, NSF Grant IBN 9209523 and a Grant-In-Aid from the American Heart Association. We would like to thank Earl A. Bueno and Daniel J. Podberesky for their outstanding assistance in the preparation of this chapter.

References

Adams, P. R., and Galvan, M. (1986). Voltage-dependent currents of vertebrate neurons and their role in membrane excitability. *Adv. Neurol.* **44,** 137–170.

Armstrong, C. M., and Benzanilla, F. (1977). Inactivation of sodium channel. II. Gating current experiments. *J. Gen. Physiol.* **70,** 567–590.

Atkinson, N. S., Robertson, G. A., and Ganetzky, B. (1991). A component of calcium-activated potassium channels encoded by the *Drosophila slo* locus. *Science* **253,** 551–554.

Attali, B., Guillemare, E., Lesage, F., Honore, E., Romey, G., Lazdunski, M., and Barhanin, J. (1993a). The protein IsK is a dual activator of K⁺ and Cl⁻ channels. *Nature (London)* **365,** 850–852.

Attali, B., Lesage, F., Ziliani, P., Guillemare, E., Honore, E., Waldmann, R., Hugnot, J. P., Mattei, M. G., Lazdunski, M., and Barhanin, J. (1993b). Multiple mRNA isoforms encoding the mouse cardiac Kv1.5 delayed rectifier K⁺ channel. *J. Biol. Chem.* **268,** 24283–24289.

Baldwin, T. J., Tsaur, M. L., Lopez, G. A., Jan, Y. N., and Jan, L. Y. (1991). Characterization of a mammalian cDNA for an inactivating voltage-sensitive K channel. *Neuron* **7,** 471–483.

Baumann, A., Krah-Jentgens, I., Muller, R., Muller-Holtkamp, F., Seidel, R., Kecskemethy, N., Casal, J., Ferrus, A., and Pongs, O. (1987). Molecular organization of the maternal effect region of the *Shaker* complex of *Drosophila*: Characterization of an IA channel transcript with homology to vertebrate Na⁺ channel. *EMBO J.* **6,** 3419–3429.

Baxter, D. A., and Byrne, J. H. (1991). Ionic conductance mechanisms contributing to the electrophysiological properties of neurons. *Curr. Opin. Neurobiol.* **1,** 105–112.

Betz, H. (1990). Ligand-gated ion channels in the brain: The amino acid receptor superfamily. *Neuron* **5,** 383–392.

Betz, H. (1992). Structure and function of inhibitory glycine receptors. *Q. Rev. Biophys.* **25,** 381–394.

Bossu, J. L., Dupont, J. L., and Feltz, A. (1988). Potassium currents in rat cerebellar Purkinje neurons maintained in culture in L-15 (Leibovitz) medium. *Neurosci. Lett.* **89,** 55–62.

Buck, L., and Axel, R. (1991). A novel multigene family may encode odorant receptors: A molecular basis for odor recognition. *Cell* **65,** 175–187.

Butler, A., Wei, A., Baker, K., and Salkoff, L. (1989). A family of putative potassium channel genes in *Drosophila*. *Science* **243,** 943–947.

Catterall, W. A. (1988). Structure and function of voltage-sensitive ion channels. *Science* **242,** 50–61.

Catterall, W. A. (1992). Cellular and molecular biology of voltage-gated sodium channels. *Physiol. Rev. (Suppl.)* **72,** S15–S48.

Chabala, L. D., Bakry, N., and Covarrubias, M. (1992). Modulation of kinetic properties of a cloned mammalian A-type K$^+$ channel by brain poly (A)$^+$ mRNA. *Soc. Neurosci. Abst.* **18,** 922.

Chandy, K. G., Williams, C. B., Spencer, R. H., Aguilar, B. A., Ghanshani, S., Tempel, B. L., and Gutman, G. A. (1990). A family of three mouse potassium channel genes with intronless coding regions. *Science* **247,** 973–975.

Chandy, K. G., Douglas, J., Gutman, G. A., Jan, L., Joho, R., Kaczmarek, L., McKinnon, D., North, R. A., Numa, S., Philipson, L., Rivera, A. B., Rudy, B., Salkoff, L., Swanson, R., Steiner, D., Tanouye, M., and Tempel, B. L. (1991). Simplified gene nomenclature. *Nature (London)* **352,** 26.

Christie, M. J., North, R. A., Osborne, P. B., Douglass, J., and Adelman, J. P. (1990). Heteropolymeric potassium channels expressed in *Xenopus* oocytes from cloned subunits. *Neuron* **4,** 405–411.

Cockcroft, V. B., Osguthorpe, D. J., Barnard, E. A., Friday, A. E., and Lunt, G. G. (1990). Ligand-gated ion channels: Homology and diversity. *Mol. Neurobiol.* **4,** 129–169.

Covarrubias, M., Wei, A., and Salkoff, L. (1991). *Shaker, Shal, Shab,* and *Shaw* express independent K$^+$ current systems. *Neuron* **7,** 763–773.

Critz, S. D., Wible, B. A., Lopez, H. S., and Brown, A. M. (1993). Stable expression and regulation of rat brain K$^+$ channels. *J. Neurochem.* **60,** 1175–1178.

DeFelipe, J., and Jones, E. G. (1988). "Cajal on the Cerebral Cortex." Oxford University Press, New York.

Demeulemeester, H., Vandesande, F., Orban, G. A., Brandon, C., and Vanderhaeghen, J. J. (1988). Heterogeneity of GABAergic cells in cat visual cortex. *J. Neurosci.* **8,** 988–1000.

Dhallan, R. S., Yau, K. W., Schrader, K. A., and Reed, R. R. (1990). Primary structure and functional expression of a cyclic nucleotide-activated channel from olfactory neurons. *Nature (London)* **347,** 184–187.

Douglas, R. J., and Martin, K. A. C. (1990). Neocortex. In "The Synaptic Organization of the Brain" (G. M. Shepherd, ed.), 3d Ed., pp. 389–438. Oxford University Press, New York.

Drewe, J. A., Verma, S., Frech, G., and Joho, R. H. (1992). Distinct spatial and temporal expression patterns of K$^+$ channel mRNAs from different subfamilies. *J. Neurosci.* **12,** 538–548.

Dubois, J. M., and Rouzaire-Dubois, B. (1993). Role of potassium channels in mitogenesis. *Prog. Biophys. Mol. Biol.* **59,** 1–21.

Frech, G. C., VanDongen, A. M. J., Schuster, G., Brown, A. M., and Joho, R. H. (1989). A novel potassium channel with delayed rectifier properties isolated from rat brain by expression cloning. *Nature (London)* **340,** 642–645.

Gähwiler, B. H., and Llano, I. (1989). Sodium and potassium conductances in somatic membranes of rat Purkinje cells from organotypic cerebellar cultures. *J. Physiol.* **417,** 105–122.

Ganfornina, M. D., and Lopez-Barneo, J. (1991). Single K$^+$ channels in membrane patches of arterial chemoreceptor cells are modulated by O$_2$ tension. *Proc. Natl. Acad. Sci. USA* **88,** 2927–2930.

Ganfornina, M. D., and López-Barneo, J. (1992a). Potassium channel types in arterial chemoreceptor cells and their selective modulation by oxygen. *J. Gen. Physiol.* **100,** 401–426.

Ganfornina, M. D., and López-Barneo, J. (1992b). Gating of O$_2$-sensitive K$^+$ channels of arterial chemoreceptor cells and kinetic modifications induced by low pO$_2$. *J. Gen. Physiol.* **100,** 427–455.

Gasic, G. P., and Hollmann, M. (1992). Molecular neurobiology of glutamate receptors. *Annu. Rev. Physiol.* **54,** 507–536.

Ghanshani, S., Pak, M., McPherson, J. D., Strong, M., Dethlefs, B., Wasmuth, J. J., Salkoff, L., Gutman, G. A., and Chandy, K. G. (1992). Genomic organization, nucleotide sequence, and cellular distribution of a *Shaw*-related potassium channel gene, *Kv3.3,* and mapping of *Kv3.3* and *Kv3.4* to human chromosome 19 and 1. *Genomics* **12,** 190–196.

Grinstein, S., and Smith, J. D. (1990). Ca^{2+}-independent cell volume regulation in human lymphocytes. *J. Gen. Physiol.* **95,** 97–120.

Grissmer, S., Dethlefs, B., Wasmuth, J. J., Goldin, A. L., Gutman, G. A., Cahalan, M. D., and Chandy, G. (1990). Expression and chromosomal localization of a lymphocyte K$^+$ channel gene. *Proc. Natl. Acad. Sci. USA* **87,** 9411–9415.

Grissmer, S., Ghanshani, S., Dethlefs, B., McPherson, J. D., Wasmuth, J. J., Gutman, G. A., Cahalan, M. D., and Chandy, G. (1992). The *Shaw*-related potassium channel, *Kv3.1,* on human chromosome 11, encodes the Type *l* K$^+$ channel in T cells. *J. Biol. Chem.* **267,** 20971–20979.

Guy, H. R. (1989). Models of voltage- and transmitter-activated membrane channels based on their amino acid sequences. In "Monovalent Cations in Biological Systems" (C. A. Pasternak, ed.), pp. 31–58. CRC Press, Boca Raton, Florida.

Guy, H. R., and Conti, F. (1990). Pursuing the structure and function of voltage-gated channels. *Trends Neurosci.* **13,** 201–206.

Haas, M., Ward, D. C., Lee, J., Roses, A. D., Clarke, V., D'Eustachio, P. D., Lau, D., Vega-Saenz de Miera, E., and Rudy, B. (1993). Localization of Shaw-related K$^+$ channel genes on mouse and human chromosomes. *Mammalian Genome* **4,** 711–715.

Hartmann, H. A., Kirsch, G. E., Drewe, J. A., Taglialatela, M., Joho, R. H., and Brown, A. M. (1991). Exchange of conductance pathways between two related K$^+$ channels. *Science* **251,** 942–944.

Heinemann, S. H., Terlau, H., Stühmer, W., Imoto, K., and Numa, S. (1992). Calcium channel characteristics conferred on the sodium channel by single mutations. *Nature (London)* **356,** 441–443.

Hemmick, L. M., Perney, T. M., Flamm, R. E., Kaczmarek, L. K., and Birnberg, N. C. (1992). Expression of the H-*ras* oncogene induces potassium conductance and neuron-specific potassium channel mRBAs in the AtT20 cell line. *J. Neurosci.* **12,** 2007–2014.

Hille, B. (1992). "Ionic Channels of Excitable Membranes," 2d Ed. Sinauer Associates, Sunderland, Massachusetts.

Hirano, T., and Ohmori, H. (1986). Voltage gated and synaptic currents in rat Purkinje cells in dissociated cell cultures. *Proc. Natl. Acad. Sci. USA* **83,** 1945–1949.

Ho, K., Nichols, C. G., Lederer, W. J., Lytoon, J., Vassilev, P. M., Kanazirska, M. V., and Hebert, S. C. (1993). Cloning and expression of an inwardly rectifying ATP-regulated potassium channel. *Nature (London)* **362,** 31–38.

Hoshi, T., Zagotta, W. N., and Aldrich, R. W. (1990). Biophysical and molecular mechanisms of *Shaker* potassium channel inactivation. *Science* **250,** 533–538.

Hoshi, T., Zagotta, W. N., and Aldrich, R. W. (1991). Two types of inactivation in *Shaker* K channels: Effects of alterations in the carboxy-terminal region. *Neuron* **7**, 547–556.

Huguenard, J. R., and Prince, D. A. (1991). Slow inactivation of TEA-sensitive K⁺ current in acutely isolated rat thalamic relay neurons. *J. Neurophysiol.* **66**, 1316–1328.

Hwang, P. M., Glatt, C. E., Bredt, D. S., Yellen, G., and Snyder, S. H. (1992). A novel K⁺ channel with unique localizations in mammalian brain: Molecular cloning and characterization. *Neuron* **8**, 473–482.

Isacoff, E. Y., Jan, Y. N., and Jan, L. Y. (1990). Evidence for the formation of heteromultimeric potassium channels in *Xenopus* oocytes. *Nature (London)* **345**, 530–534.

Isom, L. L., De Jongh, K. S., Patton, D. E., Reber, B. F. X., Offord, J., Charbonneau, H., Walsh, K., Goldin, A. L., and Catterall, W. A. (1992). Primary structure and functional expression of the β_1 subunits of the rat brain sodium channel. *Science* **256**, 839–842.

Ito, Y., Yokoyama, S., and Higashida, H. (1992). Potassium channels cloned from neuroblastoma cells display slowly inactivating outward currents in *Xenopus* oocytes. *Proc. R. Soc. London B.* **248**, 95–101.

Jan, L. Y., and Jan, Y. N. (1990a). How might the diversity of potassium channels be generated? *Trends Neurosci.* **13**, 415–419.

Jan, Y. N., and Jan, L. Y. (1990b). A superfamily of ion channels. *Nature (London)* **345**, 672.

Jan, L.-Y. and Jan, Y.-N. (1992). Structural elements involved in specific K⁺ channel functions. *Ann. Rev. Physiol.* **54**, 537–555.

Jessell, T. M., and Kandel, E. R. (1993). Synaptic transmission: A bidirectional and self-modifiable form of cell–cell communication. *Cell* **72**/*Neuron* **10**, 1–30.

Jones, E. G. (1985). "The Thalamus." Plenum Press, New York.

Kaczmareck, L. K. and Levitan, I. B. (1987). "Neuromodulation: The Biochemical Control of Neuronal Excitability." Oxford University Press, New York.

Kamb, A., Iverson, L., and Tanouye, M. A. (1987). Molecular characterization of *Shaker*, a *Drosophila* gene that encodes a potassium channel. *Cell* **50**, 405–413.

Kaupp, U. B., Niidome, T., Tanabe, T., Terada, S., Bönigk, W., Stühmer, W., Cook, N. J., Kangawa, K., Matsuo, H., Hirose, T., Miyata, T., and Numa, S. (1989). Primary structure and functional expression from complementary DNA of the rod photoreceptor cyclic GMP-gated channel. *Nature (London)* **342**, 762–766.

Kemp, B., and Pearson, R. B. (1990). Protein kinase recognition sequence motifs. *Trends Biochem. Sci.* **15**, 342–346.

Kennelly, P. J., and Krebs, E. G. (1991). Consensus sequences as substrate specificity determinants for protein kinases and protein phosphatases. *J. Biol. Chem.* **266**, 15555–15558.

Kentros, C., Weiser, M., Vega-Saenz de Miera, E., Morel, K., Baker, H., and Rudy, B. (1992). Alternative splicing of the 5'-untranslated region of a gene encoding K⁺ channel components. *Soc. Neurosci. Abst.* **18**, 1093.

Kirsch, G. E., Drewe, J. A., Hartmann, H. A., Taglialatela, M., de Biasi, M., Brown, A. M., and Joho, R. H. (1992a). Differences between the deep pores of K⁺ channels determined by an interacting pair of nonpolar amino acids. *Neuron* **8**, 499–505.

Kirsch, G. E., Drewe, J. A., Taglialatela, M., Joho, R. H., DeBiasi, M., Hartmann, H. A., and Brown, A. M. (1992b). A single nonpolar residue in the deep pore of related K⁺ channels acts as a K⁺ : Rb⁺ conductance switch. *Biophys. J. (Disc.)* **62**, 136–144.

Klein, M., Camardo, J., and Kandel, E. R. (1982). Serotonin modulates a specific potassium current in the sensory neurons that show presynaptic facilitation in Aplysia. *Proc. Natl. Acad. Sci. USA* **79**, 5713–5717.

Korpi, E. R., Kleingoor, C., Kettenmann, H., and Seeburg, P. H. (1993). Benzodiazepine-induced motor impairment linked to point mutation in cerebellar GABA_A receptor. *Nature (London)* **361**, 356–359.

Kozak, M. (1992). Regulation of translation in eukaryotic systems. *Annu. Rev. Cell Biol.* **8**, 197–225.

Kubo, Y., Baldwin, T. J., Jan, Y. N., and Jan, L. Y. (1993). Primary structure and functional expression of a mouse inward rectifier potassium channel. *Nature (London)* **362**, 127–133.

Latorre, R., Oberhauser, A., Labarca, P., and Alvarez, O. (1989). Varieties of calcium-activated potassium channels. *Annu. Rev. Physiol.* **51**, 385–399.

Leonard, R. J., Garcia, M. L., Slaughter, R. S., and Reuben, J. P. (1992). Selective blockers of voltage-gated K⁺ channels depolarize human T lymphocytes: Mechanism of the antiproliferative effect of charybdotoxin. *Pro. Natl. Acad. Sci. USA* **89**, 10094–10098.

Levitan, I. B. (1988). Modulation of ion channels in neurons and other cells. *Annu. Rev. Neurosci.* **11**, 119–136.

Lewis, R. S., and Cahalan, M. D. (1988). The plasticity of ion channels: parallels between the nervous and immune systems. *Trends Neurosci.* **11**, 214–218.

Li, M., Jan, Y. N., and Jan, L. Y. (1992). Specification of subunit assembly by the hydrophilic amino-terminal domain of the *Shaker* potassium channel. *Science* **257**, 1225–1230.

Liman, E. R., Hess, P., Weaver, F., and Koren, G. (1991). Voltage-sensing residues in S4 region of a mammalian K⁺ channel. *Nature (London)* **353**, 752–756.

Liman, E. R., Tytgat, J., and Hess, P. (1992). Subunit stoichiometry of mammalian K⁺ channel determined by construction of multimeric cDNAs. *Neuron* **9**, 861–871.

Lin, C. S., Lu, S. M., and Schmechel, D. E. (1986). Glutamic acid decarboxylase and somatostatin immunoreactives in rat visual cortex. *J. Comp. Neurol.* **244**, 369–383.

Linsdell, P., Forsythe, I. D., and Stanfield, P. R. (1990). Rectification of unitary A-current in cultured rat locus coeruleus neurons is the result of a voltage-dependent block by internal Mg^{2+} and Na⁺. *J. Physiol. (London)* **430**, 125P.

Llinas, R. (1984). Comparative electrobiology of mammalian central neurons. *In* "Brain Slices" (R. Dingledine, ed.), pp. 7–24. Plenum Press, New York.

Llinas, R. (1988). The intrinsic electrophysiological properties of mammalian neurons: Insights into central nervous system function. *Science* **242**, 1654–1664.

Lopatin, A. N., and Nichols, C. G. (1993). Block of outward rectifier DRK1 potassium channels expressed in *Xenopus* oocytes by Mg^{2+}. *J. Physiol. (London)* **473**, 45 p.

Lopez-Barneo, J., Lopez-Lopez, J. R., Urena, J., and Gonzalez, C. (1988). Chemotransduction in the carotid body: K⁺ current modulated by PO_2 in type I chemoreceptor cells. *Science* **241**, 580–582.

Lopez-Lopez, J., Gonzalez, C., Urena, J., and Lopez-Barneo, J. (1989). Low pO_2 selectively inhibits K channel activity in chemoreceptor cells of the mammalian carotid body. *J. Gen. Physiol.* **93**, 1001–1015.

Lotan, I. (1992). Hybrid arrest technique to test for functional roles of cloned cDNAs and to identify homologies among ion channel genes. *Meth. Enzymol.* Ion Channels. Vol. **207**, 605–612.

Luneau, C. J., Williams, J. B., Marshall, J., Levitan, E. S., Oliva, C., Smith, J. S., Antanavage, J., Folander, K., Stein, R. B., Swanson, R., Kaczmarek, L., and Buhrow, S. A. (1991a). Alternative splicing contributes to K channel diversity in the mammalian central nervous system. *Proc. Natl. Acad. Sci. USA* **88**, 3932–3936.

Luneau, C. J., Wiedmann, R., Smith, J. S., and Williams, J. B.

(1991b). *Shaw*-like rat brain potassium channel cDNAs with divergent 3' ends. *FEBS Lett.* **288**, 163–167.

MacKinnon, R. (1991a). Determination of the subunit stoichiometry of a voltage-activated potassium channel. *Nature (London)* **350**, 232–235.

MacKinnon, R. (1991b). New insights into the structure and function of potassium channels. *Curr. Opin. Neurobiol.* **1**, 14–19.

MacKinnon, R., and Yellen, G. (1990). Mutations affecting TEA blockade and ion permeation in voltage-activated K+ channels. *Science* **250**, 276–279.

MacKinnon, R., Zagotta, W. N., and Aldrich, R. W. (1991). How many balls does it take to inactivate a K+ channel? *Biophys. J.* **59**, 404a.

Marin, O., Meggio, F., Marchiori, F., Borin, G., and Pinna, L. A. (1986). Site specificity of casein kinase-2 (TS) from rat liver cytosol. A study with model peptide substrates. *Eur. J. Biochem.* **160**, 239–244.

McCormack, K., Campanelli, J. T., Ramaswami, M., Mathew, M. K., Tanouye, M. A., Iverson, L. E., and Rudy, B. (1989). Leucine-zipper motif update. *Nature (London)* **340**, 103.

McCormack, K., Lin, J. W., Iverson, L., and Rudy, B. (1990). *Shaker* K+ channel subunits form heteromultimeric channels with novel functional properties. *Biochem. Biophys. Res. Commun.* **171**, 1361–1371.

McCormack, K., Tanouye, M. A., Iverson, L. E., Lin, J., Ramaswami, M., McCormack, T., Campanelli, J. T., Mathew, M. K., and Rudy, B. (1991). A role for hydrophobic residues in the voltage-dependent gating of Shaker K+ channels. *Proc. Natl. Acad. Sci. USA* **88**, 2931–2935.

McCormack, T., Vega-Saenz de Miera, E., and Rudy, B. (1990). Molecular cloning of a member of a third class of *Shaker*-family K+ channel genes in mammals. *Proc. Natl. Acad. Sci. USA* **87**, 5227–5231.

McCormack, T., Vega-Saenz de Miera, E., and Rudy, B. (1991). Molecular cloning of a member of a third class of Shaker-family K+ channel genes in mammals. Correction. *Proc. Natl. Acad. Sci. USA* **88**, 4060.

McCormick, D. (1990). Membrane properties and neurotransmitter actions. *In* "The Synaptic Organization of the Brain" (G. M. Shepherd, ed.), 3d Ed., pp. 32–66. Oxford University Press, New York.

McKeown, M. (1992). Alternative mRNA splicing. *Ann. Rev. Cell Biol.* **8**, 133–155.

Melefors, Ö., and Hentze, M. W. (1993). Translational regulation by mRNA/protein interactions in eukaryotic cells. *BioEssays* **15**, 85–89.

Miller, C. (1991). *Annus mirabilis* of K channels. *Science* **252**, 1092–1096.

Mount, S. M. (1982). A catalogue of splice junction sequences. *Nucleic Acids Res.* **10**, 459–472.

Nakanishi, S. (1992). Molecular diversity of glutamate receptors and implications for brain function. *Science* **258**, 597–603.

Neher, E. (1992). Ion channels for communication between and within cells. *Neuron* **8**, 605–612.

Noda, M., Ikeda, T., Kayano, T., Suzuki, H., Takeshima, H., Kurasaki, M., Takahashi, H., and Numa, S. (1986). Existence of distinct sodium channel messenger RNAs in rat brain. *Nature (London)* **320**, 188–192.

Novick, J., Leonard, R. J., King, V. F., Schmalhofer, W., Kaczorowski, G. L., and Garcia, M. L. (1991). Purification and characterization of two novel peptidyl toxins directed against K+ channels from venom of new world scorpions. *Biophys. J.* **59**, 78a.

Pak, M. D., Baker, K., Covarrubias, M., Butler, A., Ratcliffe, A., and Salkoff, L. (1991). *mShal*, a subfamily of A-type K+ channels cloned from mammalian brain. *Proc. Natl. Acad. Sci. USA* **88**, 4386–4390.

Papazian, D. M., Schwarz, T. L., Tempel, B. L., Jan, Y. N., and Jan, L. Y. (1987). Cloning of genomic and complementary DNA from *Shaker*, a putative potassium channel gene from *Drosophila*. *Science* **237**, 749–753.

Papazian, D. M., Timpe, L. C., Jan, Y. N., and Jan, L. Y. (1991). Alteration of voltage dependence of *Shaker* potassium channel by mutations in the S4 sequence. *Nature (London)* **349**, 305–310.

Pardo, L. A., Heinemann, S. H., Terlau, H., Ludewig, U., Lorra, C., Pongs, O., and Stühmer, W. (1992). Extracellular K+ specifically modulates a rat brain K+ channel. *Proc. Natl. Acad. Sci. USA* **89**, 2466–2470.

Perney, T. M., and Kaczmarek, L. K. (1991). The molecular biology of K channels. *Curr. Opin. Cell Biol.* **3**, 663–670.

Perney, T. M., Marshall, J., Martin, K. A., Hockfield, S., and Kaczmarek, L. K. (1992). Expression of the mRNAs for the Kv3.1 potassium channel gene in the adult and developing rat brain. *J. Neurophysiol.* **68**, 756–766.

Peterson, O. H., and Maruyama, Y. (1984). Ca-activated K channels and their role in secretion. *Nature (London)* **307**, 693–696.

Po, S., Roberds, S., Snyders, D. J., Tamkun, M. M., and Bennett, P. B. (1993). Heteromultimeric Assembly of Human Potassium Channels. Molecular Basis of a Transient Outward Current? *Circulation Research.* **72**, 1326–1336.

Pongs, O. (1992). Molecular biology of voltage-dependent potassium channels. *Physiol. Rev. (Suppl.)* **72**, S69–S88.

Rehm, H., and Lazdunski, M. (1988). Purification and subunit structure of a putative K+-channel protein identified by its binding properties for dendrotoxin I. *Proc. Natl. Acad. Sci. USA* **85**, 4919–4923.

Rehm, H., and Tempel, B. L. (1991). Voltage-gated K+ channels of the mammalian brain. *FASEB J.* **5**, 164–170.

Rehm, H., Newitt, R. A., and Tempel, B. L. (1989a). Immunological evidence for a relationship between the dendrotoxin-binding protein and the mammalian homologue of the *Drosophila Shaker* K+ channel. *FEBS Lett.* **249**, 224–228.

Rehm, H., Pelzer, S., Cochet, C., Chambaz, E., Tempel, B. L., Trautwein, W., Pelzer, D., and Lazdunski, M. (1989b). Dendrotoxin-binding brain membrane protein displays a K+ channel activity that is stimulated by both cAMP-dependent and endogenous phosphorylations. *Biochemistry* **28**, 6455–6460.

Ren, R., Mayer, B. J., Cicchetti, P., and Baltimore, D. (1993). Identification of a ten-amino-acid proline-rich SH3 binding site. *Science* **259**, 1157–1161.

Rettig, J., Wunder, F., Stocker, M., Lichtinghagen, R., Mastiaux, F., Beckh, S., Kues, W., Pedarzani, P., Schröter, K. H., Ruppersberg, J. P., Veh, R., and Pongs, O. (1992). Characterization of *Shaw*-related potassium channel family in rat brain. *EMBO J.* **11**, 2473–2486.

Ried, T., Rudy, B., Vega-Saenz de Miera, E., Lau, D., Ward, D. C., and Sen, K. (1993). Localization of a highly conserved potassium channel gene (NGK2-KV4; KCNC1) to chromosome 11p15. *Genomics* **15**, 405–411.

Robbins, J., and Sims, J. A. (1990). A transient outward current in NG108-15 neuroblastoma × glioma hybrid cells. *Pflügers Arch.* **416**, 130–137.

Role, L. W. (1992). Diversity in primary structure and function of neuronal nicotinic acetylcholine receptor channels. *Curr. Opin. Neurobiol.* **2**, 254–262.

Rose, J. E. (1942). The ontogenetic development of the rabbit's diencephalon. *J. Comp. Neurol.* **77**, 61–129.

Rudy, B. (1988). Diversity and ubiquity of K channels. *Neurosci.* **25**, 729–750.

Rudy, B., Hoger, J. H., Lester, H. A., and Davidson, N. (1988). At least two mRNA species contribute to the properties of rat brain A-type potassium channels expressed in *Xenopus* oocytes. *Neuron* **1**, 649–658.

Rudy, B., Kentros, C., and Vega-Saenz de Miera, E. (1991a). Families of potassium channel genes in mammals: Toward an understanding of the molecular basis of potassium channel diversity. *Mol. Cell. Neurosci.* **2**, 89–102.

Rudy, B., Sen, K., Vega-Saenz de Miera, E., Lau, D., Ried, T., and Ward, D. C. (1991b). Cloning of a human cDNA expressing a high voltage-activating, TEA-sensitive, type-A K⁺ channel gene which maps to chromosome 1 band p21. *J. Neurosci. Res.* **29**, 401–412.

Rudy, B., Kentros, C., Weiser, M., Fruhling, D., Serodio, P., Vega-Saenz de Miera, E., Ellisman, M. H., Pollock, J. A., and Baker, H. (1992). Region-specific expression of a K⁺ channel gene in brain. *Proc. Natl. Acad. Sci. USA* **89**, 4603–4607.

Ruppersberg, J. P., Schröter, K. H., Sakmann, B., Stocker, M., Sewing, S., and Pongs, O. (1990). Heteromultimeric channels formed by rat brain potassium-channel proteins. *Nature (London)* **345**, 535–537.

Ruppersberg, J. P., Frank, R., Pongs, O., and Stocker, M. (1991a). Cloned neuronal I_K(A) channels reopen during recovery from inactivation. *Nature (London)* **353**, 657–660.

Ruppersberg, J. P., Stocker, M., Pongs, O., Heinemann, S. H., Rainer, F., and Koenen, M. (1991b). Regulation of fast inactivation of cloned mammalian I_K (A) channels by cysteine oxidation. *Nature (London)* **352**, 711–714.

Saini, K. S., Summerhayes, I. C., and Thomas, P. (1990). Molecular events regulating messenger RNA stability in eukaryotes. *Mol. Cell. Biochem.* **96**, 16–23.

Salkoff, L., Baker, K., Butler, A., Covarrubias, M., Pak, M. D., and Wei, A. (1992). An essential "set" of K⁺ channels conserved in flies, mice and humans. *Trends Neurosci.* **15**, 161–166.

Sargent, P. B. (1993). The diversity of neuronal nicotinic acetylcholine receptors. *Annu. Rev. Neurosci.* **16**, 403–443.

Schoppa, N. E., McCormack, K., Tanouye, M. A., and Sigworth, F. J. (1992). The size of gating charge in wild-type and mutant *Shaker* potassium channels. *Science* **255**, 1712–1715.

Schröter, K. H., Ruppersberg, J. P., Wunder, F., Rettig, J., Stocker, M., and Pongs, O. (1991). Cloning and functional expression of a TEA-sensitive A-type potassium channel from rat brain. *FEBS Lett.* **278**, 211–216.

Seeburg, P. H., Wisden, W., Verdoorn, T. A., Pritchett, D. B., Werner, P., Herb, A., Luddens, H., Sprengel, R., and Sakmann, B. (1990). The GABA_A receptor family: Molecular and functional diversity. *Cold Spring Harbor Symp. Quant. Biol.* **55**, 29–40.

Sen, K., Vega-Saenz de Miera, E., Chiu, N., Lin, J. W., Lau, D., and Rudy, B. (1991). Characterization of the *Shaw*-like class of K⁺ channel genes in mammals. *Biophys. J.* **59**, 2a.

Serodio, P., Vega-Saenz de Miera, E., Lau, D., and Rudy, B. (1992). Molecular components of low-voltage-activating A currents. *Soc. Neurosci. Abst.* **18**, 78.

Shen, N. V., and Pfaffinger, P. J. (1992). Identification of a K⁺ channel subfamily specific oligomerization domain. *Soc. Neurosci. Abst.* **18**, 77.

Sheng, M., Liao, Y., Jan, Y.-N., and Jan, L.-Y. (1993). Presynaptic A-current based on heteromultimeric K⁺ channels detected *in vivo*. *Nature (London)* **365**, 72–75.

Sheng, M., Tsaur, M. L., Jan, Y. N., and Jan, L. Y. (1992). Subcellular segregation of two A-type K⁺ channel proteins in rat central neurons. *Neuron* **9**, 271–284.

Sigel, E., Baur, R., Trube, G., Mohler, H., and Malherbe, P. (1990). The effect of subunit composition of rat brain GABA_A receptors on channel function. *Neuron* **5**, 703–711.

Sloviter, R. S., and Nilaver, G. (1987). Immunocytochemical localization of GABA-, cholecystokinin-, vasoactive intestinal polypeptide-, and somatostatin-like immunoreactivity in the area dentata and hippocampus of the rat. *J. Comp. Neurol.* **256**, 42–60.

Sommer, B., and Seeburg, P. H. (1992). Glutamate receptor channels: Novel properties and new clones. *Trends Pharmacol. Sci.* **13**, 291–296.

Somogyi, P., Hodgson, A. J., Smith, A. D., Nunzi, M. G., Gorio, A., and Wu, J. Y. (1984). Different populations of GABAergic neurons in the visual cortex and hippocampus of cat contain somatostatin- or cholecystokinin-immunoreactive material. *J. Neurosci.* **4**, 2590–2603.

Steriade, M., Jones, E. G., and Llinas, R. R. (1990). "Thalamic Oscillations and Signaling." John Wiley and Sons, New York.

Steward, O., and Banker, G. A. (1992). Getting the message from the gene to the synapse: Sorting and intracellular transport of RNA in neurons. *Trends Neurosci.* **15**, 180–186.

Stühmer, W., Ruppersberg, J. P., Schröter, K. H., Sakmann, B., Stocker, M., Giese, K. P., Perschke, A., Baumann, A., and Pongs, O. (1989a). Molecular basis of functional diversity of voltage-gated potassium channels in mammalian brain. *EMBO J.* **8**, 3235–3244.

Stühmer, W., Conti, F., Suzuki, H., Wang, X., Noda, M., Yagi, N., Kubo, N., and Numa, S. (1989b). Structural parts involved in activation and inactivation of the sodium channel. *Nature (London)* **339**, 597–603.

Taglialatela, M., VanDongen, A. M. J., Drewe, J. A., Joho, R., Brown, A. M., and Kirsch, G. E. (1991). Pattern of internal and external tetraethylammonium block in four homologous K⁺ channels. *Mol. Pharmacol.* **40**, 299–307.

Takumi, T., Ohkuba, H., and Nakanishi, S. (1988). Cloning of a membrane protein that induces a slow voltage-gated potassium current. *Science* **242**, 1042–1045.

Tempel, B. L., Papazian, D. M., Schwarz, T. L., Jan, Y. N., and Jan, L. Y. (1987). Sequence of a probable potassium channel component encoded at *Shaker* locus of *Drosophila*. *Science* **237**, 770–775.

Tempel, B. L., Jan, Y. N., and Jan, L. Y. (1988). Cloning of a probable potassium channel gene from mouse brain. *Nature (London)* **332**, 837–839.

Thompson, S. H., and Aldrich, R. W. (1980). Membrane potassium channels. *In* "The Cell Surface and Neuronal Function" (C. W. Cotman, G. Poste, and G. L. Nicholson, eds), Vol. 6, pp. 49–85. Elsevier/North Holland, Amsterdam,

Tsaur, M. L., Sheng, M., Lowenstein, D. H., Jan, Y. N., and Jan, L. Y. (1992). Differential expression of K⁺ channel mRNAs in the rat brain and down-regulation in the hippocampus following seizures. *Neuron* **8**, 1055–1067.

Urena, J., Lopez-Lopez, J., Gonzalez, C., and Lopez-Barneo, J. (1989). Ionic currents in dispersed chemoreceptor cells of the mammalian carotid body. *J. Gen. Physiol.* **93**, 979–999.

Vega-Saenz de Miera, E., and Rudy, B. (1992). Modulation of K⁺ channels by hydrogen peroxide. *Biochem. Biophys. Res. Commun.* **186**, 1681–1687.

Vega-Saenz de Miera, E., Chiu, N., Sen, K., Lau, D., Lin, J. W., and Rudy, B. (1991). Toward an understanding of the molecular composition of K⁺ channels: Products of at least nine distinct *Shaker* family K⁺ channel genes are expressed in a single cell. *Biophys. J.* **59**, 197a.

Vega-Saenz de Miera, E., Moreno, H., Fruhling, D., Kentros, C., and Rudy, B. (1992). Cloning of ShIII (*Shaw*-like) cDNAs encoding a novel high-voltage-activating, TEA-sensitive, type-A K⁺ channel. *Proc. R. Soc. London B* **248**, 9–18.

Verdoorn, T. A., Draguhn, A., Ymer, S., and Seeburg, P. H. (1990). Functional properties of recombinant rat GABA_A receptors depend upon subunit composition. *Neuron* **4**, 919–928.

Vyas, T. B., Wei, A., Salkoff, L., and Covarrubias, M. (1993). Fast inactivation of a human K$^+$ channel is regulated by protein phosphorylation. *Soc. Neurosci. Abstr.* **19,** 712.

Wang, H. L., Bogen, C., Reisine, T., and Dichter, M. (1989). Somatostatin-14 and somatostatin-28 induce opposite effects on potassium currents in rat neocortical neurons. *Proc. Natl. Acad. Sci. USA* **86,** 9616–9619.

Wang, H., Kunkel, D. D., Martin, T. M., Schwartzkroin, P. A., and Tempel, B. L. (1993). Heteromultimeric K$^+$ channels in terminal and juxtaparanotal regions of neurons. *Nature (London)* **365,** 75–79.

Warmke, J. W., Drysdale, R. A., and Ganetzky, B. (1991). A distinct potassium channel polypeptide encoded by the *Drosophila eag* locus. *Science* **252,** 1560–1562.

Wei, A., Covarrubias, M., Butler, A., Baker, K., Pak, M., and Salkoff, L. (1990). K$^+$ current diversity is produced by an extended gene family conserved in *Drosophila* and mouse. *Science* **248,** 599–603.

Weiser, M., Vega-Saenz de Miera, E., Kentros, C., Moreno, H., Franzen, L., Hillman, D., Baker, H., and Rudy, B. (1994). Differential expression of *Shaw*-related K$^+$ channels in the rat central nervous system. *J. Neurosci.* **14,** 949–972.

Wisden, W., and Seeburg, P. H. (1992). GABA$_A$ receptor channels: from subunits to functional entities. *Curr. Opin. Neurobiol.* **2,** 263–269.

Woodson, W., Nitecka, L., and Ben-Ari, Y. (1989). Organization of the GABAergic system in the rat hippocampal formation: a quantitative immunocytochemical study. *J. Comp. Neurol.* **280,** 254–271.

Yellen, G., Jurman, M., Abramson, T., and McKinnon, R. (1991). Mutations affecting internal TEA blockade identify the probable pore-forming region of a K$^+$ channel. *Science* **251,** 939–941.

Yokoyama, S., Imoto, K., Kawamura, T., Higashida, H., Iwabe, N., Miyata, T., and Numa, S. (1989). Potassium channels from NG108-15 neuroblastoma-glioma hybrid cells. *FEBS Lett.* **259,** 37–42.

Yool, A., and Schwarz, T. L. (1991). Alteration of ionic selectivity of a K$^+$ channel by mutation of the H5 region. *Nature (London)* **349,** 700–704.

Zagotta, W. N., Hoshi, T., and Aldrich, R. W. (1990). Restoration of inactivation in mutants of *Shaker* potassium channels by a peptide derived from ShB. *Science* **250,** 568–571.

Zhong, Y., and Wu, C. F. (1991). Alteration of four identified K$^+$ currents in *Drosophila* muscle by mutations in eag. *Science* **252,** 1562–1564.

Zilles, K. (1990). Anatomy of the neocortex: cytoarchitecture and myeloarchitecture. *In* "The Cerebral Cortex of the Rat" (B. Kolb and R. C. Tees, eds.), pp. 77–112. MIT Press, Cambridge, Massachusetts.

Zilles, K., Wree, A., and Dausch, N. D. (1990). Anatomy of the neocortex: neurochemical organization. *In* "The Cerebral Cortex of the Rat (B. Kolb and R. C. Tees, eds.), pp. 113–150. MIT Press, Cambridge, Massachusetts.

CHAPTER

5

Molecular Workings of Large Conductance (Maxi) Ca^{2+}-Activated K^+ Channels

RAMON LATORRE

I. Introduction

This chapter discusses the possible mechanisms that mediate the activity of a class of calcium-activated potassium [K(Ca)] channels. These K(Ca) channels possess a large unitary conductance (≈200 pS in symmetrical 100 m*M* KCl). Large conductance K(Ca) channels are ubiquitously distributed in different types of cells and tissues, and play an important role in such diverse physiological processes as repolarization of the action potential in neurons (e.g., Pennefather *et al.*, 1985) and regulation of arterial tone (Brayden and Nelson, 1992).

Despite their large conductance, this type of channel is also characterized by the exquisite ability to discriminate between monovalent ions. K(Ca) channels are essentially impermeant to Na^+ and Cs^+ and conduct K^+ about 10-fold more effectively than Rb^+, an ion that is often used as a K^+ substitute in radioisotope flux studies. K(Ca) channels are activated by cytoplasmic Ca^{2+}, but the amount of Ca^{2+} necessary to increase the fraction of time the channel spends in the open configuration (open probability, P_o) is extremely variable. For example, the internal Ca^{2+} concentration that produces a P_o of 0.5 at 0 mV applied voltage can be as low 10 n*M* in pig pancreas acinar cells (Maruyama *et al.*, 1983) and as high as 100 μM in rat plasma muscle membrane channels incorporated into planar bilayers (Moczydlowski and Latorre, 1983). Most of the large conductance, also called "maxi" or "BK," K(Ca) channels are voltage dependent. P_o increases with depolarizing potentials; the voltage needed to change P_o *e*-fold ranges from 8 (Lang and Ritchie, 1990) to ~39 mV (Markwardt and Isenberg, 1992). Pharmacologically, maxi K(Ca) channels are characterized by their sensitivity to the scorpion toxin charybdotoxin (ChTX; Miller *et al.*, 1985). However, note that not all maxi K(Ca) are sensitive to this toxin (Reinhardt *et al.*, 1989; Adelman *et al.*, 1992) and, most importantly, that ChTX also blocks K^+ channels that are *not* activated by Ca^{2+} and some Ca^{2+}-activated K^+ channels of small conductance. (For reviews on K^+ channels and toxins, the reader should consult Moczydlowski *et al.*, 1988; Strong, 1990; Garcia *et al.*, 1991.) Regarding toxin specificity, another scorpion toxin, iberiotoxin (IbTX; Galvez *et al.*, 1990), has proved to be more specific for maxi K(Ca) channels than ChTX (Garcia *et al.*, 1991).

Maxi K(Ca) channels have been very resilient to the cloning efforts of several research teams, but two groups (Atkinson *et al.*, 1991; Adelman *et al.*, 1992) took advantage of the existence of the *Drosophila* mutant *slowpoke* (Elkins *et al.*, 1986). These investigators found that the *slowpoke* locus encodes a polypeptide similar to voltage-dependent K⁺ channels. Moreover, Adelman *et al.* (1992) were able to express a maxi K(Ca) channel by injecting RNA transcribed from complementary DNA derived from the *slowpoke* locus, beginning the molecular biological study of this type of channel. Apparently, the structure of maxi K(Ca) channels have been conserved from fly to mouse (see Butler *et al.*, 1993).

In this chapter, some of the gating kinetic models that have been proposed to explain the manner in which K(Ca) channels open and close are discussed. Whenever possible, the data are presented in terms of the most simple and plausible physical explanations that can provide information about the molecular nature of the Ca^{2+} and voltage dependence of maxi K(Ca) channels. Since the three-dimensional structure of K⁺ channels is unknown, but that of some Ca^{2+}-binding proteins has been resolved, the conformational changes brought about by Ca^{2+} binding to the latter are discussed in an effort to infer details about the possible mechanisms of activation by Ca^{2+} of maxi K(Ca) channels. The data on ion conduction are reviewed, and an explanation based on these data is proposed for the remarkable combination of very large conductance and finely tuned selectivity in favor of K⁺ over other cations. The subsequent section is dedicated to studies of channel structure with the aid of channel blockade by ions and toxins. Since most K⁺ channels appear to have evolved with a common pattern for their ion conduction system, the behavior of

maxi K(Ca) channels when confronted with blocking ions is compared with that of K⁺ channels of well-known primary structure. This exercise is important since in some cases, such as the *Shaker* K⁺ channel, the crucial amino acid residues that interact with blocking ions such as tetraethylammonium (TEA) are known (e.g., MacKinnon, 1991; see Chapter 2). Evidence is mounting that not only internal Ca^{2+}, but also most permeant cations, can modify channel activity. Modifications of gating by external permeant cations and a blocker such as Ba^{2+} are also presented. Most of the evidence described in this chapter is derived from single-channel data and, since the article describes a rather personal account of the workings of K(Ca) channels, no attempt has been made to review the literature of K(Ca) channels comprehensively. Readers interested in variety, physiological roles, and metabolic regulation may consult several review articles (Levitan, 1988; Rudy, 1988; Latorre *et al.*, 1989; McManus, 1991; Garcia *et al.*, 1991; Toro and Stefani, 1991,1993).

II. Channel Gating

Figure 1 shows the most salient features of K(Ca) channel activation by depolarizing voltages and internal Ca^{2+}. These two ways of activating a channel lead to the same result: an increase in P_o. However, this class of channels shows spontaneous shifts into kinetic modes of lower or higher open probability at constant voltage and $[Ca^{2+}]$, as well as into a fast flickering mode (Methfessel and Boheim, 1983; Moczydlowski and Latorre, 1983; McManus and Magleby, 1988). When constructing kinetic models, most studies disregard the presence of subconductance states because they represent a small percentage of the total time the channel spends in the open configuration. A characterization of the subconductance states of a maxi K(Ca) channel indicates the presence of at least six conductance states (Stockbridge *et al.*, 1991). The detailed study of the characteristics of these substates can prove useful in the selection between different gating mechanisms. The studies discussed here consider only channels with stable P_o and disregard the presence of conductance substates.

A. Equilibrium Behavior

Activation by Ca^{2+} of maxi K(Ca) channels has been discussed in detail by McManus (1991), so only the most salient features are mentioned here. Data for the probability of opening, as a function of internal Ca^{2+} are usually fit with the relationship:

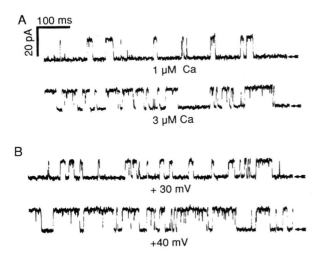

FIGURE 1 Effect of cytoplasmic Ca^{2+} and voltage on a maxi K(Ca) channel from skeletal muscle incorporated into a planar lipid bilayer membrane. (A) Records obtained at +40 mV and at the indicated $[Ca^{2+}]$. Open probabilities were 0.22 and 0.66 at 1 and 3 μM Ca^{2+}, respectively. (B) Records obtained at 3 μM internal Ca^{2+} and at the indicated voltages. Open probabilities were 0.29 and 0.66 at 30 and 40 mV, respectively. Redrawn with permission from Moczydlowski and Latorre (1983).

$$P_o = P_o{}^{max}[Ca]^N/([Ca]^N + K_d), \qquad (1)$$

where N is the Hill coefficient and K_d is the apparent dissociation constant. In the extreme case of a reaction with infinite cooperativity, N is the actual number of Ca^{2+} binding sites in the channel protein since, in this case, the binding can be represented by the reaction scheme (Cantor and Schimmel, 1971)

Closed channel + NCa^{2+} ⇌ Open channel–Ca_N
(R-1)

and, hence, Eq. 1. However, this behavior is not observed in practice and N is always the lower limit of the number of sites involved in the reaction. In other words, N is the minimum number of Ca^{2+} ions needed to fully activate a K(Ca) channel. For maxi K(Ca) channels, N is on average larger than 2 (McManus, 1991), but a Hill coefficient as large as 5.3 has been reported (Carl and Sanders, 1989).

For the skeletal muscle K(Ca) channel, Golowash *et al.* (1986) found a marked effect of internal Mg^{2+} on N, despite the fact that this cation is unable to activate the channel when it is the only divalent cation present in the internal solution (Latorre *et al.*, 1982). The Hill coefficient increases 2.1- to 3-fold when the internal $[Mg^{2+}]$ is raised from 0 to 10 mM; effects on N can be appreciable at physiological $[Mg^{2+}]$ (i.e., 1–2 mM; Golowasch *et al.*, 1986; Squire and Petersen, 1987; Oberhauser *et al.*, 1988). Magnesium appears to act allosterically by binding to a site with a divalent cation selectivity that is different from that of the Ca^{2+} binding site. Whereas the effectiveness of divalent cations in activating maxi K(Ca) channels—measured as the ability of a given cation to increase P_o at constant voltage—follows the sequence

$$Ca^{2+} > Cd^{2+} > Sr^{2+} > Mn^{2+} > Fe^{2+} > Co^{2+}$$
$$\gg Mg^{2+}, Ba^{2+}, Ni^{2+}, Pb^{2+},$$

in the presence of contaminant amounts of Ca^{2+}, this ability to increase P_o is changed to

$$Ca^{2+} > Cd^{2+} > Mn^{2+} > Fe^{2+} > Co^{2+} > Ni^{2+}$$
$$> Sr^{2+} > Mg^{2+}$$

(Oberhauser *et al.*, 1988). Note that Ni^{2+} and Mg^{2+} activate the channel in the presence of contaminant amounts of Ca^{2+}, although they do not activate it in the absence of Ca^{2+}. Actually, Ni^{2+} under these conditions is more effective than Sr^{2+}. Another interesting observation is that Ca^{2+} and Sr^{2+} do not modify N, although they are potent activators. Cd^{2+}, Mn^{2+}, and Co^{2+} are able to increase N in a concentration-dependent fashion and are also able to activate in the absence of Ca^{2+}. Thus, some cations appear to bind only to the modulator site (e.g., Mg^{2+}) and others to

the Ca^{2+} sites and to the modulator site (e.g., Co^{2+}). Mechanistically, Mg^{2+} and the other cations able to modify N could expose previously inaccessible Ca^{2+} binding sites (Fig. 2), change the extent of cooperativity between Ca^{2+} binding sites, or modify the coupling between Ca^{2+} binding and channel opening.

The apparent dissociation constant, K_d, that appears in Eq. 1 is highly variable for different channels; variations in Ca^{2+} sensitivity of maxi K(Ca) channels of the same tissue are also found (McManus, 1991). For example, the internal $[Ca^{2+}]$ necessary to produce P_o of 0.5 at 0 mV is 10 nM in mouse lacrimal gland (Findlay, 1984) and 20–100 μM in rat pancreas β cells (Tabcharani and Misler, 1989). In the former case, the channel should be fully activated at physiological concentrations of internal Ca^{2+} whereas in the latter case, an activated channel should be difficult or even impossible to find under physiological conditions. A different behavior was found for a K(Ca) channel of bovine parathyroid cells (Jia *et al.*, 1988). For this maxi K(Ca) channel, Ca^{2+} behaves as an activator up to an internal $[Ca^{2+}]$ of 160 nM, above this $[Ca^{2+}]$, the divalent cation inhibits channel activity.

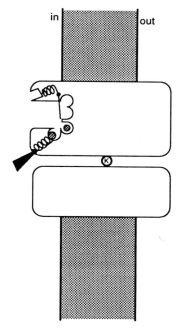

FIGURE 2 Maxi K(Ca) channel model, according to the main features described in the text. Accessible Ca^{2+} binding sites are located in an activation cleft. Filled circles represent bound Ca^{2+}. Only two sites were drawn for the sake of simplicity. Calcium binding lowers the activation energy of opening. Hidden Ca^{2+} binding sites that are unmasked only when Mg^{2+} binds to the site are represented with a rectangular shape. Calcium binding in this model is voltage dependent.

The activity of maxi K(Ca) channels is extremely sensitive to the internal pH. At constant $[Ca^{2+}]$, intracellular acidification induces a decrease in P_o and a shift of the activation curve to the right along the voltage axis. This displacement ranges from 23.5 to 80 mV per pH unit (Cook *et al.*, 1984; Christensen and Zeuthen, 1987; Cornejo *et al.*, 1989; Kume *et al.*, 1990; Copello *et al.*, 1991; Laurido *et al.*, 1991). Because of the extremely high sensitivity of maxi K(Ca) channels to protons, some of the differences in Ca^{2+} sensitivities between different maxi K(Ca) channels could be the result of measurements of P_o under different pH conditions.

Increasing the cytoplasmic proton concentration increases the apparent dissociation constant for Ca^{2+} binding without affecting N. A fit to the P_o vs $[H^+]$ data with a Hill equation indicates that the minimum number of proton binding sites is 1. Although researchers generally agree that protons act by weakening Ca^{2+} binding (i.e., K_d increases), whether protons compete for the same sites with Ca^{2+} (e.g., Kume *et al.*, 1990) or exert their action allosterically by binding to a different site (Laurido *et al.*, 1991) is not clear. The kinetic arguments given by Laurido *et al.* (1991) and arguments based on known molecular properties of Ca^{2+} sites in Ca^{2+}-binding proteins (see subsequent discussion) favor a clearly defined proton binding site that is structurally different from the activator sites.

B. Voltage Gating or Ion Gating?

The probability of opening is voltage dependent: P_o increases as the membrane potential is increased. The relationship between P_o and membrane potential (V) can be fit reasonable well by the expression

$$P_o = P_o^{max}\{1 + \exp[(-A)(V - V_o)]\}^{-1}, \quad (2)$$

where A (also called the slope parameter) is the parameter that contains the channel voltage sensitivity and V_o is the potential at which P_o is 0.5. For a two-state channel, $A = z\delta FV/RT$, where z is the effective number of charges ("gating particles") that move an electrical distance δ down the electrical field during channel activation. For channels such as maxi K(Ca) channels that contain multiple kinetic states (McManus and Magleby, 1991), the slope of the activation curve on a logarithmic ordinate only becomes constant and *independent* of the number of states at infinitesimal values of P_o (Almers, 1978, see Chapter 3). Therefore, caution must be taken in interpreting A in terms of gating charges. The parameter A ranges from 8 mV^{-1} in a rat brain channel incorporated into planar bilayers and in channels of a pituitary cell line (Reinhardt *et al.*, 1989; Lang and Ritchie, 1990) to 39 mV^{-1} in myocyte maxi K(Ca) channels (Markwardt and Isenberg,

1992). Therefore, maxi K(Ca) channels are less voltage dependent than the classical delayed-rectifier K$^+$ channel (2–10 mV^{-1}; Rudy, 1988) and the A-type K$^+$ channels. The voltage dependence of P_o has the consequence that the apparent K_d (Eq. 1) increases with more positive potentials. Voltage-dependent binding constants can be explained using the analysis of Woodhull (1973) for the case of ion channel blockers. Using this analogy, K_d can be expressed as

$$K_d(V) = K_d(0)\exp(-z\delta FV/RT), \quad (3)$$

where $K(0)$ is the zero voltage dissociation constant. An issue that has not been resolved is the molecular origin of the voltage dependence of maxi K(Ca) channels. Two interpretations are possible: (1) voltage modulates the movement of charged amino acid residues in the protein-forming channel and (2) the Ca^{2+}-binding reaction(s) is voltage dependent. In the latter case, Ca^{2+} binds to sites located within the electrical field of the membrane (Fig. 2). Voltage-dependent Ca^{2+}-binding reactions were first used by Wong *et al.* (1982) to interpret the translation along the voltage axis of the P_o vs V curves that occurred as internal $[Ca^{2+}]$ was varied. A voltage-dependent Ca^{2+} binding (Eq. 3a) an infinite cooperative (R-1) demands that (Wong *et al.*, 1982):

$$V_o = -(N/A)\ln\{K_d(0)[Ca^{2+}]\} \quad (3a)$$

Moreover, Wong *et al.* (1982) showed that a linear relation between V_o and $\ln[Ca^{2+}]$ is also obtained for the more general case that Ca^{2+} binds sequentially (R-3) to the sites present in the channel-forming protein. For the case of a maxi K(Ca) channel from skeletal muscle incorporated into neutral phospholipid bilayers, this prediction of the model was put to test by Moczydlowski *et al.* (1985) who found that it holds in the $[Ca^{2+}]$ range from 0.001 to 2 mM. In addition, the gating kinetics data of Moczydlowski and Latorre (1983) were adequately described by assuming that all the maxi K(Ca) channel voltage dependence resided in voltage-dependent binding reactions (see subsequent discussion). The same type of model was used by Markwardt and Isenberg (1992). In this case, channel gating was studied by Ca^{2+} concentration jumps, the best fit to the equilibrium data was obtained by assuming the existence of three identical binding sites that acted independently and were located at an electrical distance of 0.31 from the cytoplasmic side.

Ion gating has been questioned by Pallota (1985), who reported that after treatment with the protein-modifying agent *N*-bromoacetamide (NBA), the open probability, albeit much lower than that of the untreated channels, became insensitive to internal Ca^{2+} in the range of 1 nM to 100 μM. Under these conditions, channel open probability exhibited a voltage dependence similar to that of channels not exposed to NBA. These results suggested that the Ca^{2+}-dependent component of the channel could be removed without much alteration of its voltage dependence. Does NBA really uncouple the Ca^{2+}- from the voltage-dependent components of the channel? Apparently not, since Cornejo *et al.* (1987) showed that, although NBA exposure virtually abolished maxi K(Ca) activity in the presence of low [Ca^{2+}], this characteristic could be partially restored at sufficiently high [Ca^{2+}]. More recently, Salomao *et al.* (1992) showed that trypsin, like NBA, reduces but does not abolish maxi K(Ca) sensitivity to internal Ca^{2+}. After trypsin treatment and at high internal [Ca^{2+}], the channel voltage dependence is the same as that observed before protease treatment. The data of Cornejo *et al.* (1987) and Salomao *et al.* (1992), although favoring the voltage-dependent Ca^{2+}-binding hypothesis, do not prove that it is correct. The channel activation parameters (K_d and/or N) that are altered by these protein-modifying agents remain unknown.

The *slowpoke* maxi K(Ca) channel has an amino acid stretch in which every third amino acid is positively charged (arginine). The positive charges are separated by hydrophobic residues (Adelman *et al.*, 1992). This α-helical region (S4) is present in voltage-dependent K$^+$ channels and has been implicated in voltage-sensitive gating functions of K$^+$ channels (e.g., Jan and Jan, 1992). Nevertheless, the S4 region of *slowpoke* contains three arginine residues whereas the S4 region of the *Shaker* K$^+$ channel contains seven positively charged residues (five arginines and two lysines). This design for the S4 transmembrane domain is conserved in the maxi K(Ca) channels of mouse brain and skeletal muscle (Butler *et al.*, 1993). Mutations of the positively charged residues in the *slowpoke* S4 region to neutral or negatively charged residues will be crucial in determining the involvement of these charges in channel activation.

C. Kinetic Models Compatible with Single-Channel Activity

A channel binding N Ca^{2+} ions should have, in principle, a minimum of $N + 1$ closed and open states. From a thermodynamic standpoint, none of the possible states can be excluded. The approach followed by most authors (including the most daring) is not to include some states with a low probability of occurrence in the kinetic scheme. For example, in the case of K(Ca) channels, it seems reasonable to assume that a closed-to-open transition with no bound Ca^{2+} is very unlikely. On the other hand, the number of states that are experimentally accessible is determined by the effective time resolution of the

system. For example, with a 30-μsec time resolution, McManus and Magleby (1991) found five exponential components in the dwell-time distribution for the closed states and three exponential components in the dwell-time distribution of the open states for the maxi K(Ca) channel of rat skeletal muscle. The number of closed states is in agreement with the Hill coefficient ($N \sim 3.7$). The major features of the equilibrium and kinetic data can be described by Scheme R-2.

$$C \rightleftharpoons CCa^{2+} \rightleftharpoons CCa_2^{2+} \rightleftharpoons CCa_3^{2+} \rightleftharpoons CCa_4^{2+}$$
$$\updownarrow \qquad \updownarrow \qquad \updownarrow$$
$$OCa_2^{2+} \rightleftharpoons OCa_3^{2+} \rightleftharpoons OCa_4^{2+}$$
(R-2)

Linear models in which only one pathway leads to the open state are excluded since McManus and Magleby (1985,1991) found an inverse relationship between the durations of adjacent open and closed intervals. This result is consistent with Scheme R-2 if the duration of the closed states decrease from C to CCa$_4^{2+}$ and the duration of the open states increases from OCa$_2^{2+}$ to OCa$_4^{2+}$. At 1 μM Ca^{2+}, using a maximum likelihood method to estimate the rate constants, McManus and Magleby (1991) found that the mean lifetime decreases 3.16×10^4-fold as the states progress from C to CCa$_3^{2+}$ and the lifetime increases 6.1-fold as the states progress from OCa$_2^{2+}$ to OCa$_4^{2+}$. An important observation is that, for Scheme R-2, the second order rate constants for Ca^{2+} binding become increasingly faster as the reaction progresses toward the right, suggesting a cooperative interaction among the binding sites. When Scheme R-2 is constrained for independent and identical Ca^{2+}-binding sites (e.g., Markwardt and Isenberg, 1992), the normalized likelihood is 4×10^{19}-fold smaller than for the unconstrained cooperative model. On this basis, McManus and Magleby (1991) have rejected identical and independent Ca^{2+}-binding sites for Scheme R-2. According to Scheme R-2, the Ca^{2+} dependence of the maxi K(Ca) channel arises because an increase in cytoplasmic [Ca^{2+}] drives the reaction toward closed states of shorter duration and open states of longer duration.

1. Physical Models for Channel Gating

The analysis of McManus and Magleby (1991) also allowed the identification of the preferential gating pathway

$$C \rightleftharpoons CCa^{2+} \rightleftharpoons CCa_2^{2+} \rightleftharpoons CCa_3^{2+} \rightleftharpoons OCa_3^{2+}$$
$$\rightleftharpoons OCa_4^{2+},$$
(R-3)

which encompasses most of the models used previously to explain the kinetic behavior of maxi K(Ca) channels obtained with less effective time resolution (Methfessel and Boheim, 1982; Moczydlowski and Latorre, 1983; Singer and Walsh, 1987; Christensen and Zeuthen, 1987; Sheppard *et al.*, 1988). Two of these less complex kinetic models are worth discussing here, since they offer specific mechanisms for gating. Based on their kinetic results, Methfessel and Boheim (1982) proposed a model in which Ca^{2+} binds to two sites: an activation site, where Ca^{2+} binding leads to channel opening, and a site on a blocking particle, where Ca^{2+} binding prevents the particle from blocking the channel (activation-blockade mechanism). Methfessel and Boheim *assumed* that the voltage-dependent step is the open/closed conformational change. In their model, Ca^{2+}-binding constants are voltage independent. On the other hand, Moczydlowski and Latorre (1983), using the same kinetic model (i.e., $C-CCa^{2+}-OCa^{2+}-OCa_2^{2+}$), found that, as demanded by the model, the unconditional mean open time τ_o, was directly proportional to $[Ca^{2+}]$ whereas the mean closed time τ_c, was inversely proportional to $[Ca^{2+}]$. When the Ca^{2+} dependence of the mean times was studied at different voltages, the results obtained only made sense under the assumption that the rate constants for opening and closing, for the reaction $CCa^{2+} \rightleftharpoons OCa^{2+}$ were voltage independent. Inasmuch as the shapes of the curves that describe the relationships between the mean times and $[Ca^{2+}]$ are voltage dependent, all the voltage dependence must reside in the equilibrium dissociation constants. The dissociation constants were found to be exponential functions of voltage (see also Markwardt and Isenberg, 1992). This finding, as discussed earlier (see Eq. 3), can be explained by assuming that the binding sites "sense" the electric field (Fig. 2). Consistent with this view, Markwardt and Isenberg (1992) found that Ca^{2+} binding to a maxi K(Ca) channel is rate limiting for the time course of activation.

D. Fast Inactivation in Maxi K(Ca) Channels

In general, maxi K(Ca) channels originate sustained macroscopic current, so their level of activity is considered to be a reflection of the membrane potential difference and the cytoplasmic $[Ca^{2+}]$ (i.e., channels are noninactivating; Latorre *et al.*, 1989; McManus, 1991; but see Pallota, 1985; Ikemoto *et al.*, 1989). However, a report by Solaro and Lingle (1992) provides strong evidence for the presence of rapid inactivation in K(Ca) channels of rat adrenal chromaffin cells. This maxi K(Ca) channel (275 pS in symmetrical 140 mM KCl) shares most of the typical features

that characterize large unitary conductance K(Ca) channels: it is activated by depolarizing voltages and it is blocked by micromolar amounts of externally added TEA. However, in contrast with most maxi K(Ca) channels, the chromaffin cell channel shows a fast inactivation process (Fig. 3A) that resembles that found in A-type K^+ channels (e.g., Hoshi *et al.*, 1990). Fast inactivation proceeds in an internal $[Ca^{2+}]$- and voltage-dependent fashion (Fig. 3B,C). These two characteristics, Solaro and Lingle (1992) argued, are not intrinsic properties of the inactivation gate but are consequences of a coupling that exists between inactivation and $[Ca^{2+}]$- and voltage-dependent activation processes. Note that, at high internal $[Ca^{2+}]$ ($[Ca^{2+}]_i$) or at high depolarizing voltages, the time constant for inactivation saturates (Fig. 3B,C), suggesting that the transition from the open to the inactivated state(s) is independent of $[Ca^{2+}]_i$ and voltage. Moreover, the fact that fast inactivation can be re-

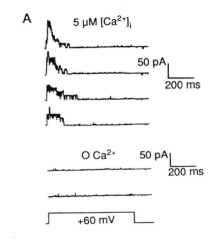

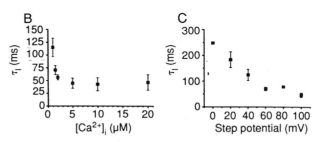

FIGURE 3 Fast-inactivating maxi K(Ca) channel from rat adrenal chromaffin cells. (A) 60-mV depolarizing pulses were applied to a membrane patch containing several maxi K(Ca) channels in the presence of the indicated $[Ca^{2+}]$. Channels inactivate with a time constant of about 40 msec (cf., Fig. 1). No channel activity can be discerned after removal of the internal Ca^{2+}. Traces are consecutive. (B,C) Inactivation time constant, τ_i, as a function of internal $[Ca^{2+}]$ (B) and voltage (C). Note that at high $[Ca^{2+}]$ and high depolarizing voltages, τ_i remains rather constant. Redrawn with permission from Solaro and Lingle (1992).

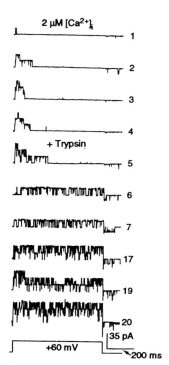

FIGURE 4 Internally applied trypsin removes fast inactivation of a maxi K(Ca) channel from rat adrenal chromaffin cells. Records were taken at 3-sec intervals. Redrawn with permission from Solaro and Lingle (1992).

moved by exposing the internal side of the membrane to trypsin (Fig. 4) is consistent with the hypothesis that the "ball and chain" model, proposed originally for Na⁺ channel fast inactivation (Armstrong and Bezanilla, 1977), may also be applicable to this particular maxi K(Ca) channel.

E. Interaction of Inactivating Peptide with Maxi K(Ca) Channels

Hoshi et al. (1990) identified the inactivation gate in the *Shaker* A-type K⁺ channel as the cytoplasmic domain of the channel-forming protein located in the N terminus. In terms of the "ball and chain" model, the first 20 amino acids conform the ball that "hangs" on a chain formed by the subsequent 60 or so residues. The ball is able to swing into and occlude the pore once the pore opens. The synthesized *Shaker* B inactivating peptide ("ball peptide") is able to interact with noninactivating maxi K(Ca) channels (Foster et al., 1992; Toro et al., 1992) and with trypsin-treated chromaffin maxi K(Ca) channels (Solaro and Lingle, 1992) from the cytoplasmic side only. The ball peptide produces inhibition of channel activity. For noninactivating K(Ca) channels the changes in channel gating kinetics induced by the peptide can be most simply interpreted

in terms of a voltage-dependent binding of the peptide to the internal mouth, hindering ion flow through the channel. The interaction between the ball peptide and noninactivating K(Ca) channels appears to be rather specific, since a point mutation (L7E; Leu → Glu) renders the peptide unable to promote channel inhibition. Interestingly, a similar specificity is found for *Shaker* B type K⁺ channels, in which the same point mutation of the ball peptide eliminates inactivation (Zagotta et al., 1990). These results suggest that the ball peptide receptor has been conserved throughout the evolution of K⁺ channels. The difference between inactivating and noninactivating maxi K(Ca) channels may be a consequence of a variable N terminus encoded by multiple genes and/or a single gene giving rise to multiple protein products because of alternative splicing. Diversity originating through these two mechanisms is found in the K⁺ channels of *Drosophila* (Salkoff et al., 1992).

In this regard, the lack of fast inactivation in *slowpoke* (Adelman et al., 1992) and in the maxi K(Ca) channel cloned from mouse brain and skeletal muscle (*mslo*; Butler et al., 1993) can be explained on the basis of the lack of identity between the N terminus of these two channels and the N terminus of *Shaker* K⁺ channels. Moreover, in the two cloned maxi K(Ca) channels there is a negatively charged residue (glu in *mslo*, asp in *slowpoke*) in position 7. As discussed earlier, replacement of the leucine existing in position 7 of the wild type inactivating gate of *Shaker* K⁺ channels by an acidic amino acid, eliminates the ability of the peptide to block maxi K(Ca) channels. The receptor for the inactivating gate have been localized between the fourth and fifth transmembrane segments of the *Shaker* K⁺ channel (Isacoff et al., 1991). Sequence alignment of this cytoplasmic stretch of amino acids with the corresponding segments in *slowpoke* and *mslo* reveals about 40% identity of the ball peptide receptor of *Shaker* K⁺ and that of maxi K(Ca) channels.

F. Gating Modifiers

1. Up Regulators

a. Endothelin Endothelin is a vasoactive peptide containing 21 amino acids and two disulfide bonds. The action of endothelin-1 on maxi K(Ca) channels has been characterized in a channel from smooth muscle cells of porcine coronary artery (Hu et al., 1991). At 1 nM concentration, endothelin increases P_o ~4-fold. Above 1 nM, endothelin is a down regulator. Endothelin appears to be specific for certain types of maxi K(Ca) channels, since no potentiation induced by the drug can be determined if the maxi K(Ca) channels are from rabbit portal vein (Hu et al., 1991).

b. Niflumic acid Trifluoro-methyl-3-phenylam-ino-2-nicotinic acid (niflumic acid) is known for its inhibitory action on Ca^{2+}-activated Cl^- channels. Niflumic acid to a final concentration of 20 μM greatly increases P_o of a maxi K(Ca) channel from coronary smooth muscle incorporated into planar bilayers (Toro *et al.*, 1993). Niflumic acid appears to modify gating only from the external side of the channel.

c. Activators Isolated from **Desmodium adscendens;** *a medicinal herb* Three organic compounds present in the crude extract of a medicinal herb used in Ghana to treat ailments related to smooth muscle contraction have proved to be potent activators of maxi K(Ca) channels (McManus *et al.* 1993). The compounds were identified as triterpenoid glycosides [dehydrosoyasaponin (DHS-I), soyasaponin I, and soyasaponin III]. In particular, DHS-I at nanomolar concentrations increases P_o, modifies gating only when added to the internal side, and through allosteric interactions, is able to increase the rate of dissociation of ChTX. DHS-I is unable to activate maxi K(Ca) channels in the absence of internal Ca^{2+} implying that the receptor for this compound is likely to be structurally different from the binding sites for Ca^{2+}.

2. Down Regulators

a. ATP ATP decreases P_o of rabbit aortic, rabbit tracheal, and pig coronary artery maxi K(Ca) channels with a K_d in the range of 0.2–0.6 mM (Gelband *et al.*, 1990; Groschner *et al.*, 1991). ATP shifts the P_o vs voltage curve to the right along the voltage axis without changing the channel voltage dependence; 0.5 mM ATP shifted V_o of the aortic smooth muscle channel by about 36 mV (Gelband *et al.*, 1990). However, reports that ATP does not modify gating of two maxi K(Ca) channels, also from smooth muscle membrane, indicate the existence of maxi K(Ca) channel subtypes (Williams *et al.*, 1988; Standen *et al.*, 1989). ATP does not affect the gating of a skeletal muscle maxi K(Ca) channel (I. B. Levitan and R. Latorre, unpublished observations).

b. Angiotensin II Angiotensin II is one of the most potent vasoconstrictors of the coronary vessels. Toro *et al.* (1990) showed that angiotensin II greatly inhibits maxi K(Ca) channels from coronary smooth muscle incorporated into lipid bilayers. Angiotensin II binds to the maxi K(Ca) channel with a K_d of 58 nM and a Hill coefficient of 2.2. The decrease in P_o is caused by a shortening of the open and a lengthening of the closed times. Toro *et al.* (1990) argued that modifications of gating by angiotensin II are the result of

a direct interaction between the vasoconstrictor and the maxi K(Ca) channel.

c. Thromboxane A_2 Thromboxane A_2, a metabolite of arachidonic acid, is a potent vasoconstrictor. When added to the external side in the concentration range of 50–150 nM, U46619, a thromboxane agonist, decreases P_o and this inhibition of channel activity can be explained on the basis of a shortening of mean open and a lengthening of the mean closed time (Scornik and Toro, 1992). The U46619 effect can be reversed by increasing the internal (Ca^{2+}) or by addition of SQ29548, a potent thromboxane competitive antagonist.

3. Special Case

Dendrotoxin I, a 60-residue peptide belonging to the dendrotoxin family of mamba snake neurotoxins, is a potent inhibitor of some types of voltage-dependent K^+ channels. Added to the cytoplasmic side of a skeletal muscle maxi K(Ca) channel, dendrotoxin I promotes the appearance of a long-lasting conductance substate with 66% of the current of the fully open state (Lucchesi and Moczydlowski, 1990). The toxin appears to interfere with K^+ permeation without affecting the Ca^{2+} dependence or the voltage dependence of gating. The lack of effect of external K^+ on internal dendrotoxin action makes a blocking mechanism unlikely. Lucchesi and Moczydlowski (1990) proposed that, by binding to a unique internal binding site (K_d = 90 nM), dendrotoxin I stabilizes one of the many normally occurring substate conformations of the channel.

G. Lessons from Ca^{2+}-Binding Proteins

1. Divalent Cation Selectivity

The divalent selectivity sequence found for the maxi K(Ca) channel of skeletal muscle follows the same order as that found for Ca^{2+}-binding proteins such as calmodulin, troponin C, and parvalbumin. However, in contrast with the maxi K(Ca) channel, troponin binds Pb^{2+} better than Mn^{2+} (Fuchs, 1971) and parvalbumin binds Cd^{2+} more tightly than Ca^{2+} (Cave *et al.*, 1979). Note that the potency of divalent cation to open the small conductance K(Ca) channel of human erythrocytes follows the sequence Pb^{2+}, $Cd^{2+} > Ca^{2+} \geq Co^{2+} > Mg^{2+}$, Fe^{2+} (Leinders *et al.*, 1992). This sequence resembles the selectivity sequence of the Ca^{2+} site in parvalbumin (Cave *et al.*, 1979). Perhaps in small conductance K(Ca) channels the divalent binding sites are closely related to those found in the parvalbumin, whereas the sites in maxi

K(Ca) channels resemble the Ca^{2+}-binding sites of calmodulin more ($Ca^{2+} > Cd^{2+} > Hg^{2+} > Sr^{2+} > Mn^{2+} > Zn^{2+} > Pb^{2+} > Co^{2+} > Mg^{2+}$, Ni^{2+}, Ba^{2+}; Chao *et al.*, 1984) or troponin C ($Ca^{2+} > Cd^{2+} > Sr^{2+} > Pb^{2+} > Mn^{2+} >> Mg^{2+}$, Ni^{2+}, Zn^{2+}, Ba^{2+}, Co^{2+}). In Ca^{2+}-activated K^+ channels, as is the case for Ca^{2+}-coordinating sites of the troponin superfamily, the genes that encode these sites may have evolved from an ancestral gene that encodes a single Ca^{2+}-binding domain. These analogies must be viewed with caution, since information about the molecular structure of Ca^{2+}-binding sites in K(Ca) channels is still lacking. Note also that the ability of a certain cation to increase the probability of opening is measured, *not* direct binding of the divalent cation to the channel protein. Some divalent cations may bind to the sites well, but are unable to promote the conformational change that leads to channel opening.

2. Calcium Binding

The widely different degrees of Ca^{2+} binding in maxi K(Ca) channels can be the result of structurally different Ca^{2+}-binding sites and/or of modification of the primary protein structure in the neighborhood of the sites. Falke *et al.* (1991) reported that point mutations of the Ca^{2+}-binding site of the *Escherichia coli* D-galactose receptor can promote dramatic cation specificity changes. This Ca^{2+} site shares key structural elements with the EF-hand class of Ca^{2+} sites. Falke *et al.* (1991) changed the cavity size and the charge by replacing a glutamine (142) with asparagine, (N), glutamate (E), or aspartate (D). Even the rather conservative mutation Q142N produces large changes in binding characteristics. The mutant Q142N, although it binds Ca^{2+} almost as well as the wild-type protein, binds Mg^{2+} 10-fold less and has increased affinity for trivalent cations. The preference for Ca^{2+} is lost in the mutant Q142E and the mutant Q142D binds Ca^{2+} ~500-fold less that the wild-type protein. In these last two cases, the selectivity for trivalent over divalent cations is highly increased. Another interesting example of how differences in Ca^{2+}-binding affinities may arise is given in the work of Pantoliano *et al.* (1988). In this case, amino acids with acidic chains replaced uncharged amino acids in the vicinity of a weak Ca^{2+}-binding site in the bacterial serine protease. These changes increased the affinity of the Ca^{2+}-binding site 2 to 4-fold. Therefore, altering the electrostatic potential around the metal ion binding site in proteins changes divalent cation binding affinity.

3. Magnesium Binding

Troponin C and the parvalbumins, but not calmodulin, bind Mg^{2+} (Klee *et al.*, 1980; Leavis and Gergely,

1984). However, no evidence exists for positive cooperativity in these proteins regarding Ca^{2+} binding induced by Mg^{2+}, as found in some maxi K(Ca) channels. Positive cooperativity is found, however, in the dimeric crayfish muscle and sandworm muscle Ca^{2+}-binding proteins. In the latter case, Ca^{2+} binding becomes strongly cooperative at physiological levels of Mg^{2+} (Cox *et al.*, 1977).

4. Proton Modulation

In troponin C, the number and the affinity for Ca^{2+} of the Ca^{2+}-binding sites remain essentially constant in the pH range of 5–8 (Fuchs, 1974). These results suggest that Ca^{2+}-binding sites in this protein contain solely carboxyl groups that are completely dissociated at pH $\geq$ 5. This conclusion is supported by the fact that the most common ligands at the Ca^{2+}-binding sites are the peptide carbonyl groups and the carboxyl groups of aspartate and glutamate residues, both with pK values $\leq$ 4 (Einspahr and Bugg, 1978). Moreover, the *slowpoke* channel monomer has at least one sequence with possible similarity to a calcium-binding domain (Adelman *et al.*, 1992). However, this sequence is not conserved in *mslo* (Butler *et al.*, 1993). On the basis of these results protons are unlikely to alter maxi K(Ca) channel activity by competing with Ca^{2+} for the Ca^{2+}-binding sites. The effect of protons on channel activity is clearly seen at pH values several units above the pK of glutamic and aspartic acid.

III. Selectivity, Ion Conduction, and Blockade

A. Ion Selectivity in Maxi K(Ca) Channels

Customarily, ion selectivity is defined as the ability of a transporter to discriminate between different ionic species. In the case of ion channels, ion selectivity sequences are built on the basis of two different measurements: (1) permeability ratios obtained from reversal potential measurements, ideally under bi-ionic conditions, and (2) single-channel conductance ratios obtained by measuring the channel conductance under symmetrical conditions. Table I shows that in maxi K(Ca) channels these two different types of experiment produce different selectivity sequences. Conductance in symmetrical solutions follows the sequence $K^+ > Tl^+ > Rb^+ > NH_4^+$ (Eisenman *et al.*, 1986). On the other hand, permeability ratios generate the sequence $Tl^+ > K^+ > Rb^+ > NH_4^+ >> Na^+$, Li^+, Cs^+ (Blatz and Magleby, 1984; Yellen, 1984a; Eisenman *et al.*, 1986; Smart, 1987; Mayer *et al.*, 1990). Note, for example, that reversal potential measurements show Rb^+ to be almost as "permeant" as K^+,

TABLE 1   K⁺(Ca²⁺) Channel Selectivity Parameters[a]

Ion (X)	g^b (pS)	g^c (pS)	$g_X/g_K{}^b$	$g_X/g_K{}^c$	$g_X/g_K{}^d$	$P_X/P_K{}^b$	$P_X/P_K{}^e$	$P_X/P_K{}^f$	$P_X/P_K{}^c$	$P_X/P_K{}^d$
Tl⁺	130		0.41			1.3	1.20			
K⁺	320	23.5	1.00	1.00	1.00	1.0	1.00	1.00	1.00	1.00
Rb⁺	22	11.4	0.07	0.48	0.70	0.7	0.67	0.83	0.71	0.96
NH₄⁺	56	14.2	0.18	0.60	1.20	0.1	0.11		0.12	0.17
Cs⁺					0.10					0.07
Na⁺							<0.01	<0.03		
Li⁺							<0.02			

[a] Permeability ratios were calculated from reversal potentials and defined as $P_X/P_K = \exp(-FV_r/RT)$ where V_r is the measured reversal potential.

[b] Eisenman *et al.* (1986). Single-channel conductance, g, are the values obtained at zero voltage and were obtained in symmetrical single-ion solutions at 300 mM concentration.

[c] Christophersen (1991). Small conductance K(Ca) channel. Single-channel conductances were obtained at 130 mM concentration.

[d] Grissmer *et al.* (1993). Small conductance K(Ca) channel. Data obtained from macroscopic currents at 160 mM concentration.

[e] Blatz and Magleby (1984).

[f] Yellen (1984a). A $P_{Na}/P_K = 0.014$–0.033 was reported by Gitter *et al.* (1987) for maxi K(Ca) channels from apical membranes of cell from renal cortical duct.

but the single-channel conductance in symmetrical solution of these two ions shows that the K(Ca) channel conducts K⁺ ~14.5-fold better than Rb⁺. Moreover, a comparison of permeability and conductance ratios for Tl⁺ and K⁺ shows that the channel conducts K⁺ better, but that Tl⁺ has a permeability coefficient about 1.3-fold larger than that of K⁺. The large differences between these two measures of ionic selectivity are well-known consequences of the fact that ions do not move independently of one another in the pore (e.g., Hille, 1992; see Chapter 2). The higher permeabilities for Tl⁺ and Rb⁺ relative to their single-channel conductance suggest that these ions bind more tightly in the conduction pathway than K⁺. The small conductance K(Ca) channel of the red cell membrane ($g_K = 23.5$ pS) shows a permeability ratio K:Rb = 1.4 and a conductance ratio for the same two cations of ~2 (Christophersen, 1991). Thus, despite their large conductance, maxi K(Ca) channels are able to discriminate better between K⁺ and Rb⁺ than small K(Ca) channels. That conductance cannot be equated with the ability of a given channel to discriminate between different ions is also exemplified by a K(Ca) of small conductance present in T lymphocytes (Grissmer *et al.*, 1993). This 35 pS channel shows a P_{Cs}/P_K of 0.07 and a ratio of macroscopic conductances between these two ions of 0.1 (Table I). The T lymphocyte channel is essentially impermeable to Na⁺. On the other hand, maxi K(Ca) channels are essentially impermeable to Na⁺, Li⁺, and Cs⁺ and, in this regard, are not very different from other K⁺-selective channels (Latorre, 1986; Hille, 1992; but see Reuter and

Stevens, 1980). Hille (1992) pointed out that a channel through which Rb⁺ with a radius, r, of 0.148 nm can pass, but Cs⁺ ($r = 0.169$ nm) cannot, should have a narrow region with a diameter of 0.296–0.338 nm. The experimental evidence indicates that this filter is somewhat deformable since blockade of maxi K(Ca) channels by external Cs⁺ is relieved at high negative voltages (Cecchi *et al.*, 1987).

Now that the primary structure of two maxi K(Ca) channels [*slowpoke*; Atkinson *et al.*, 1991; Adelman *et al.*, 1992; and *mslo* (actually a family of maxi K(Ca) channels); Butler *et al.*, 1993; see subsequent discussion] is available, these observations are not surprising, since the amino acid sequence between the S5 and S6 regions (P, also called H5), which corresponds to an integral part of the ion conduction pore in *slowpoke* and *mslo* (Adelman *et al.*, 1992; Butler *et al.*, 1993), shows a great similarity with the P region of other K⁺ channels (Fig. 5). However, when comparing maxi K(Ca) with other K⁺ channels, some qualitative and quantitative differences regarding channel selectivity should be mentioned. Kirsch *et al.* (1992) found that the ratio g_{Rb}/g_K is 1.5 and 0.7 in two delayed-rectifier K⁺ channels (Kv2.1 and Kv3.1, respectively). This ratio becomes 0.3 if a chimeric channel is built by replacing the P region of Kv2.1 with that of Kv3.1. A specific residue in P appears to be responsible for the differences in K⁺ and Rb⁺ conductances in these two different delayed rectifiers. Replacing a valine residue located toward the internal pore side (position 443 in Fig. 5; position 374 in Kirsch *et al.*, 1992) in Kv2.1 by an isoleucine present at position 442 in Kv3.1 changes

									441								449		
Sh	F	W	W	A	V	V	T	M		T	V	G	Y	G	D	M	T
Kv2.1	F	W	W	A	T	I	T	M		T	V	G	Y	G	D	I	Y
slo	V	Y	F	L	I	V	T	M	S		T	V	G	Y	G	D	V	Y
mslo	V	Y	L	L	M	V	T	M	S		T	V	G	Y	G	D	V	Y

FIGURE 5 Alignment of the deduced amino acid sequences of the pore (P) region of several K⁺-selective channels. *Shaker* and Kv2.1 are voltage-dependent channels (Schwarz *et al.*, 1988; Frech *et al.*, 1989; Kirsch *et al.*, 1992). *slo* is the *slowpoke* channel described in the text (Atkinson *et al.*, 1991; Adelman *et al.*, 1992). *mslo* is the mouse maxi K(Ca) channel (Butler *et al.*, 1993). Residues labeled 441 and 449 indicate positions in the *Shaker* K⁺ channel at which point mutations alter the internal and external tetraethylammonium (TEA) binding, respectively.

g_{Rb}/g_K from 1.5 to 0.2. As shown in Fig. 5, Val 443 is conserved in *slowpoke*; therefore, a conductance ratio g_{Rb}/g_K similar to that found in Kv2.1 is expected. This ratio has not been measured in slowpoke, but the ratio g_{Rb}/g_K is 0.07 in the skeletal muscle maxi K(Ca) channel (see Table I), a value much lower than that for any of the mutants in position 443 tested by Kirsch *et al.* (1992) or Taglialatela *et al.* (1993). Assuming that the P region is conserved in all maxi K(Ca) channels, these results suggest that, in this type of channel, more than one amino acid residue are modulating K⁺/Rb⁺ selectivity or that this ion selection is regulated by multi-ion occupancy of the pore. The degree of ion occupancy of K(Ca) channels may be different from that of small conductance K⁺ channels such as Kv2.1 or Kv3.1 (5–24 pS).

B. Maxi K(Ca) Channels Are Multi-Ion Channels

Several experimental criteria have been used to determine the degree of ion occupancy of maxi K(Ca) channels (Vergara *et al.*, 1984; Yellen, 1984b; Eisenman *et al.*, 1986; Cecchi *et al.*, 1987; Neyton and Miller, 1988a,b; Turnheim *et al.*, 1989; Villarroel, 1989; Neyton and Pelleschi, 1991). Maxi K(Ca) channels can be considered part of the class of multi-ion pores based on several observations.

(1) The single-channel conductance vs. [K⁺] relationship cannot be described by a simple single-ion mechanism based on Michaelis–Menten behavior (Vergara *et al.*, 1984; Moczydlowski *et al.*, 1985; Turnheim *et al.*, 1989; Villarroel, 1989; Nomura *et al.*, 1990). This test of the degree of channel occupancy has limitations since, in general, the $g_{K(Ca)}$ vs [C] curve is not obtained at constant ionic strength when [C] is changed. Changes in ionic strength may accompany changes in surface potential that could distort the channel conductance vs permeant ion concentration

relationship (see subsequent discussion). For multi-ion pores, conductance is expected to increase up to a certain concentration, reach a maximum, and then decrease at higher concentrations (Hille and Schwarz, 1978). Even at [K⁺] as high as 3 M, no maximum has been obtained for the $g_{K(Ca)}$ vs [K⁺] relationship, but when Rb⁺ is used as the permeant ion, a maximum is obtained for the skeletal muscle maxi K(Ca) channel (Villarroel *et al.*, 1988a; Villarroel, 1989).

(2) Increasing external K⁺ relieves blockade by internal Na⁺ by increasing the rate of Na⁺ *exit* from the skeletal muscle K(Ca) channel (Yellen, 1984b). A single-ion channel model predicts that, in the presence of two competing ions, only the apparent Na⁺ *blocking* rate should be affected by alterations in [K⁺] since, in this case, the site can be occupied by permeant ion (K⁺) or by the blocker (Na⁺).

(3) External Cs⁺ blockade of maxi K(Ca) channels has two characteristics that are incompatible with a single-ion pore (Cecchi *et al.*, 1987). First, increasing external [K⁺] enhances external Cs⁺ blockade. Second, for a fast block, the ratio i/i_o, of the single-channel current in the presence of the blocker, B, to that in the absence of blocker is given by the expression

$$i/i_o = \{1 + [B]/K_d\}^{-1}, \qquad (4)$$

where K_d is the dissociation constant for the reaction

$$\text{Open channel} + \text{B} \rightleftharpoons \text{Blocked channel} \qquad (\text{R-4})$$

If the site of blockade is located at a fractional distance, δ, down the electric field, the site "feels" a fraction δV of the applied voltage; hence, K_d is described by Eq. 3. For channels that cannot be occupied by more than one ion at any given time, $z\delta$ cannot be >1. Cecchi *et al.* (1987) found that, for external Cs⁺ blockade, $z\delta$ can reach values as high as 1.5.

(4) Vergara and Latorre (1983) showed that internal Ba²⁺ is able to block maxi K(Ca) channels in a voltage-dependent manner, only when the channel is in the open configuration. Binding of Ba²⁺ to the pore walls is very tight [$K_d(0) \sim 3\mu M$ at 100 mM K⁺; Vergara and Latorre, 1983]; because this ion is very similar in size to K⁺, these investigators suggested that Ba²⁺ binds to site(s) governing the passage of K⁺ through the channel (Vergara and Latorre, 1983). Neyton and Miller (1988a) found that increasing external [K⁺] in the range of 0–1 mM slows the rate of Ba²⁺ (applied internally) exit from *and* the rate of Ba²⁺ association with the skeletal muscle K(Ca) channel, that is, the binding of a blocking ion such as Ba²⁺ is *stabilized* by the permeant ion (cf., internal Na⁺ blockade). This finding can be easily rationalized by assuming that the maxi K(Ca) pore can simultaneously hold the divalent and the monovalent ion (Neyton and Miller, 1988a).

In the absence of external K$^+$, Ba^{2+} can dissociate toward the internal or the external side (Fig. 6A), but it is easier for a Ba^{2+} ion to leave the channel toward the external than toward the cytoplasmic side. Since binding of K$^+$ stabilizes Ba^{2+} in its blocking position, the site to which K$^+$ binds has been termed the "lock-in" site. Neyton and Miller (1988a) localized the lock-in site toward the external side of the conduction pathway (Fig. 6A) with a selectivity, measured as the

ability of a given ion to keep Ba^{2+} locked, of Tl$^+$ > Rb$^+$ > K$^+$ > Cs$^+$ > NH$_4^+$. Externally located K$^+$-binding sites were first postulated for this type of channel by Marty (1983). However, increasing the external [K$^+$] in the range of 100–1000 mM has the effect of speeding up Ba^{2+} dissociation from the channel, suggesting the existence of an additional low-affinity K$^+$ site (Neyton and Miller, 1988b). A third K$^+$-binding site located toward the internal side of the pore has been postulated, since internal K$^+$ is also able to lock Ba^{2+} in its blocking site (Fig. 6B).

(5) Maxi K(Ca) channels show "anomalous mole fraction behavior" (Eisenman *et al.*, 1986; Villaroel, 1989), that is, a minimum in single-channel conductance, when the composition of a two-ion mixture (e.g., NH$_4^+$/Rb$^+$, NH$_4^+$/K$^+$) is varied at constant total ion concentration.

Much needed information about K$^+$ fluxes through maxi K(Ca) channels is the flux ratio exponent, since this exponent is a direct measure of the number of ions occupying the channel single-filing region (Hodgkin and Keynes, 1952). The flux ratio exponent, n, has been obtained for the small conductance K(Ca) channels of red blood cells (Vestegaard-Vogin *et al.*, 1985). n was found to be ~2.7, indicating that three K$^+$ ions can dwell in the pore simultaneously. Since occupancy is a function of permeant ion concentration, n is expected to change as [K$^+$] is varied. n for the red blood cell channel is essentially constant over a wide range of [K$^+$], demonstrating that ion occupancy of the channel does not decrease as the permeant ion concentration of the media is lowered. For this to occur, the ions must be tightly bound to their sites in the conduction system.

C. Model for Ion Conduction: A Multi-Ion Pore with Charged Vestibules

As stated in the introduction, maxi K(Ca) channels possess both a large conductance that is near the limit theoretically expected for a pore (Latorre, 1986; Hille, 1992) and a high cation selectivity (see preceding discussion). To reconcile these apparently contradictory features, models of ion conduction through maxi K(Ca) channels that view the pore with the shape of an hourglass have been proposed. In these models, the channel vestibules are negatively charged because strategically located charged groups in this region exert long-range electrostatic forces to control ion access to the channel conduction pore (Dani, 1986). For example, Imoto *et al.* (1988) found that a ring of negatively charged residues lining the entry to the nicotinic acetylcholine receptor channel enhances the conduc-

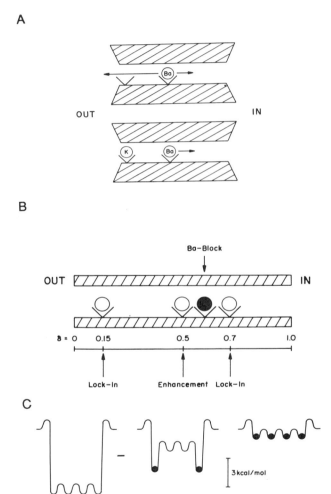

FIGURE 6   Barium ion can be locked in the maxi K(Ca) channel pore by external and internal K$^+$. (A) In the absence of external K$^+$, Ba^{2+} can exit toward the external side of the channel more easily. External K$^+$ increases the mean blocked time ("lock-in" effect). (B) At least four ion binding sites are needed to explain the effects of internal and external K$^+$ (○) on Ba^{2+} (●) blockade. (C) Qualitative effects of ion occupancy on the free-energy profile model for the maxi K(Ca) channel. For more details see text. Redrawn with permission from Neyton and Miller (1988b).

tion of Na$^+$ through this cation-selective pore. In other words, the channel conduction machinery "senses" all or part of the negative electrostatic potential that arises as a consequence of the negative charges located in the vestibules. The negative surface potential causes local accumulation of cations and a depletion of anions in the vestibules, with a subsequent increase in channel conductance.

To determine whether negative charges residing in the maxi K(Ca) channel-forming protein are involved in ion conduction, MacKinnon and Miller (1989) and MacKinnon *et al.* (1989) measured single-channel conductance as a function of [K$^+$] in the absence and presence of external trimethyloxonium (TMO; Fig. 7A). TMO is an esterifying agent that methylates carboxylic groups. Figure 7B demonstrates that the effectiveness of TMO in lowering channel conductance is strongly dependent on ionic strength. At low ionic strength, when the electrostatic forces are large, TMO treatment reduces channel conductance 2-fold

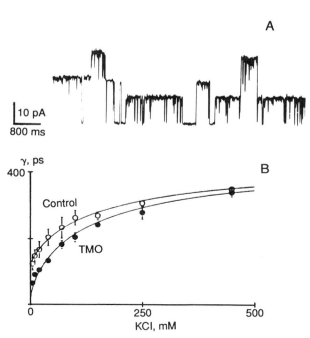

FIGURE 7 Effect of externally added trimethyl oxonium (TMO) on channel conductance. (A) After TMO treatment and perfusion with a TMO-free buffer of a single-channel membrane, a second maxi K(Ca) channel was incorporated into the bilayer. The two channels can be differentiated by their current amplitudes. TMO-treated channel; 12 pA; untreated channel, 19 pA. Records were taken in symmetrical 150 m*M* KCl. (B) Maxi K(Ca) channel conductance as a function of KCl concentration before (open circles) and after (closed circles) treatment with TMO. Redrawn with permission from MacKinnon and Miller (1989) and MacKinnon *et al.* (1989).

whereas at high ionic strength the effect is barely discernible. This result is consistent with the presence of surface charges in the external vestibule of the channel. This local negative surface charge raises the local concentration of the current carrying cations near the channel mouth.

Net negative charges in the vestibules explain the high conductance obtained at low ion concentration (Fig. 7); multi-ion occupancy of the pore increases the rate of ion transport at high permeant ion concentrations. For example, in a doubly occupied channel, the entry of the second ion effectively decreases the energy barrier for the exit by providing an electrostatic repulsive force on the first ion (Andersen and Procopio, 1980; Latorre, 1986). However, the results obtained using Ba^{2+} as a probe of multi-ion occupancy of maxi K(Ca) channels (Neyton and Miller, 1988b) strongly suggest the existence of at least four ion binding sites able to bind K$^+$ with high affinity, all of which can be occupied simultaneously (Fig. 6C). Binding to the sites in this model is so tight that the channel only begins to conduct ions when the repulsive interactions become strong enough; this event occurs when the channel is occupied by at least three ions. Assuming that the ions move in single file as a concerted stack (e.g., Hodgkin and Keynes, 1955), a conduction mechanism for the maxi K(Ca) channel was proposed by Schumaker (1992). In this model, a channel with m ion binding sites can be associated with m or $m - 1$ ions (Fig. 8). An ion from the solution may join the "stack" of $m - 1$ ions at the vacant terminal position and an ion can leave from the other pore end (Fig. 8A). The stack of $m - 1$ ions can also go from the first $m - 1$ site to the last $m - 1$ site within the conduction pore; hence the name "shaking stack" was determined for this particular model (Fig. 8B). The model is able to account for the conductance vs [K$^+$] curve and can explain nicely the high monovalent cation selectivity shown by maxi K(Ca) channels. In this model, assuming that the ions interact independently with the pore walls and that all the energy barriers that separate binding sites are equal in height (Fig. 9), the total energy for the stack movement is $n\Delta G$, where n is the number of barriers (3 if there are 4 sites; Fig. 9) and ΔG is the barrier height. Therefore, any difference in translocation energy between two ions will be amplified by n; the ratio of the translocation rate constants for the stack of, for example, ions 1 and 2 becomes $\{\exp[-(\Delta G_1 - \Delta G_2)/kT]\}^n$. Thus, the longer the stack, the more selective the channel becomes, despite the fact that the difference in chemical properties of two different ions can be quite small.

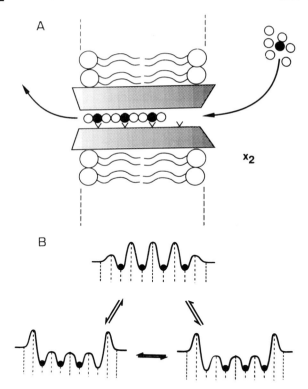

FIGURE 8 The shaking stack model for ion conduction. (A) In agreement with the Ba^{2+} blockade results, the channel is pictured as containing four K^+ binding sites, three of which are occupied by K^+ (closed circles) in this particular case. Open circles are water molecules. (B) Potential profiles for the three possible kinetic states of the shaking stack model. Reproduced from the *Biophysical Journal* (1992). **63**, 1032–1044, by © permission of the Biophysical Society.

D. Blocking Ions as Probes of Pore Structure

1. Probing Structure with Quaternary Ammonium Ions

Quaternary ammonium ions have proven to be rather specific blockers of K^+ channels. Armstrong (1969,1971) used these organic cations to investigate the gross architecture of the squid axon delayed-rectifier K^+ channel. By measuring the effect of these

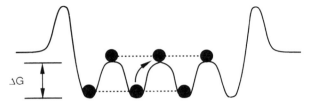

FIGURE 9 Ion stack translocation, assuming that ions interact independently with energy barriers of height ΔG. Reproduced from the *Biophysical Journal* (1992). **63**, 1032–1044, by © permission of the Biophysical Society.

cations on the macroscopic currents, this investigator proposed that the delayed rectifier contains, at its inner end, a mouth wide enough to accommodate TEA.

Large organic cations such as TEA and nonyltrimethylammonium are also able to block maxi K(Ca) channels (Blatz and Magleby, 1984; Vergara *et al.*, 1984; Yellen, 1984a; Iwatsuki and Petersen, 1985; Smart, 1987; Villarroel *et al.*, 1988b; Bokvist *et al.*, 1990; Carl *et al.*, 1990; Lang and Ritchie, 1990). TEA added to the internal side of a K(Ca) channel induces a decrease in channel conductance. This fast blockade is characterized by K_d = 27–60 mM and $z\delta$ = 0.3 (Blatz and Magleby, 1984; Vergara *et al.*, 1984; Yellen, 1984a; Villarroel *et al.*, 1988b). Adding TEA to the external side of the channel promotes a more potent fast blockade, that is less voltage dependent. In this case, K_d = 0.14–0.29 mM and $z\delta$ = 0.13–0.20 (but see Benham *et al.*, 1985; Wong and Adler, 1986). Thus, the internal affinity of the channel for TEA is 155-fold lower than the external affinity. Some other important differences exist between internal and external TEA blockade that suggest different molecular structures for these two vestibules. The external TEA site is specific for TEA; increasing or decreasing the side chains by one methylene group decreases potency ~400-fold, indicating that the fit with the quaternary ammonium ion is tight (Villarroel *et al.*, 1988). In the case of internal blockade, blocking potency increases as the compound is made more hydrophobic.

The study of the effect of voltage on the effect of different quaternary ammonium compounds applied to the cytoplasmic side of the channel shows that the $z\delta$ values are very similar for all the organic ions, despite their large differences in binding constants. This result indicates that all quaternary ammonium compounds block the channel from the internal side by binding to a site located at an electrical distance equivalent to ~30% of the voltage drop within the channel from the internal side.

This situation is radically changed if divalent quaternary (*bis*-quaternary) ammonium ions are used (Villarroel *et al.*, 1988; see also Miller, 1982). *bis*-Quaternary ammonium (bisQA) ions are of the type $(CH_3)_3-N^+-(CH_2)_n-N^+-(CH_3)_3$. Figure 10 shows the $z\delta$ parameter as a function of chain length. For short hydrocarbon chains, the $z\delta$ parameter is larger than the average value found for monovalent QA compounds; $z\delta$ reaches a constant value of about 0.26 for bisQ5 and bisQ6 and increases again for bixQ$_n$ ions with longer chains. In other words, for bisQ$_n$ compounds with the charges separated by a very short or a very long hydrocarbon chain, $z\delta$ has a value that is

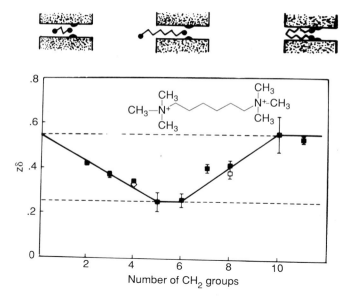

FIGURE 10 For *bis*-quaternary ammonium (bis-QA) ions, the voltage dependence ($z\delta$) of blockade is a function of the length of the hydrocarbon chain of the QA. (*Bottom*) The lower broken line is drawn at the mean $z\delta$ obtained for monovalent quaternary ammonium ions such as tetraethylammonium (TEA). The upper broken line is $2z\delta$. The open diamond is 2-butene, *N,N,N,N,N',N'*-hexamethyl-1,4-*bis*-ammonium and the open square is 1,8-*bis*-guanidinium-*n*-octane. (*Top*) A possible interpretation for the divalent QA ion interaction with its receptor in the maxi K(Ca) channel external mouth. For more details see text. Redrawn with permission from Villarroel *et al.* (1988) and Miller (1982).

nearly twice the value obtained for monovalent QAs; and for bisQ5 and bisQ6, this value is very close to the average value obtained for monovalent QA ions (i.e., 0.3).

Assuming that one of the charges is always interacting with a site located at $\delta \sim 0.3$ (Fig. 10, inset), a measurement of the voltage dependence of a given bisQ$_n$ ion facilitates determination of δ for the other charge since, in this case, the voltage dependence for the bisQ$_n$ ion is given by the sum of the $z\delta$ values for each charge. A decrease in $z\delta$ is expected as the chain length increases, until it reaches a constant value when the second charge is completely outside the field. Because bisQ5 behaves as a monovalent ion, the second charge of this compound is likely to be outside the field. The length of this compound is 0.96 nm; therefore, $\sim$30% of the voltage drop occurs over a distance of $\sim$1 nm. Compounds larger than bisQ7 apparently can block in a bent conformation, as suggested by Miller (1982) on the basis of its conformation in aqueous solution (Dufourcq *et al.*, 1972; see also Alvarez *et al.*, 1983). For bisQ10, both charges appear

to interact with the site since this compound shows a $z\delta$ of 0.55 (Fig. 10, inset).

A similar strategy for studying the gross architecture of the internal channel mouth was followed by Nomura *et al.* (1990) for rat brain synaptosomal maxi K(Ca) channel. These authors used alkylmonoamine ($z = +1$) and alkyldiamino ($z = +2$) blockers containing different numbers of carbon atoms. Nomura *et al.* (1990) found that the monovalent blockers blocked by binding to a site located at $\delta = 0.25$, suggesting that these compounds interact with the internal TEA receptor described by Villarroel *et al.* (1988b). However, regardless of the hydrocarbon chain length separating the two charges of the alkyldiamino compounds, the $z\delta$ was approximately constant and equal to 0.46–0.51. One possible interpretation of these results is that the internal mouth of the brain maxi K(Ca) channel is wider than that of the skeletal muscle channel, allowing the two charges of the long diaminoalkyl molecule to interact with the binding site.

Several conclusions can be drawn from the results obtained with quaternary ammonium ions, that suggest features of the gross architecture of the maxi K(Ca) channel. The channel can be viewed as an hourglass with two large vestibules that are able to accommodate TEA, communicating via a narrow tunnel in which ion discrimination must take place.[1] The internal mouth contains a hydrophobic region that is able to accommodate the long hydrocarbon tails of ions such as nonyltrimethylammonium. The external vestibule is specific for TEA, is not hydrophobic in nature, and appears to select quaternary ammonium ions by their sizes. The analysis of the data obtained with the bisQ$_n$ compounds provided an estimate of the physical distance from the place where the voltage drop begins to the blocking site; thus, a relationship was obtained between physical and electrical distances for the internal mouth of the channel (but see Nomura *et al.*, 1990).

a. Molecular Characteristics of TEA Binding Sites in slowpoke   In *Shaker* K⁺ channels, two amino acid

---

[1] Alcayaga *et al.* (1989) determined the number of water molecules (Γ) that are coupled to each ion transported across the narrow region of a maxi K(Ca) channel. For multi-ion channels, Γ represents the average number of water molecules between ions. $\Gamma = 2$ at 20 mM KCl. Although the channel occupancy at this [K⁺] is unknown, given the results of Neyton and Miller (1988b), the maxi K(Ca) channel should contain more than one ion in its conduction system. Therefore, the pore single-filing length cannot be much shorter than 1.1 nm (double-occupied pore) and cannot be longer than 3.5 nm (all four sites occupied by K⁺).

residues located in the P region are able to influence the extracellular binding of TEA. These residues are located at positions 431 and 449 and correspond to aspartate and threonine, respectively. Both residues face the extracellular side of P (MacKinnon, 1991). In particular, mutations involving residue 449 (Fig. 5) affect external TEA inhibition dramatically. A change from threonine to tyrosine changes the K_d from ~30 mM to 0.65 mM; a further reduction is obtained (K_d = 0.35) if a phenylalanine is located at position 449 (Heginbotham and MacKinnon, 1992). Note that, in contrast with *Shaker*, *slowpoke* has a tyrosine at position 449, explaining the high TEA affinity showed by this maxi K(Ca) channel (Adelman *et al.*, 1992) and suggesting that a tyrosine or a phenylalanine should exist in most maxi K(Ca) channels. In *slowpoke*, residue 449 is also crucial for external TEA binding. Adelman *et al.* (1992) showed that an expressed *slowpoke* channel containing a valine at position 449 is essentially TEA resistant. The *Shaker* K$^+$ channel possesses a threonine in position 441. This residue is important for the binding of internal TEA, which limits the pore length to a stretch of 8 amino acid residues (441 to 449). Yellen *et al.* (1991) found that, in the *Shaker* K$^+$ channel, the mutations Thr 441 to ser reduces internal TEA affinity 10-fold. In both *slowpoke* and *mslo* a serine replaces Thr 441. The *slowpoke* channel binds internal TEA very weakly (Adelman *et al.*, 1992).

2. Probing Channel Structure with Ba^{2+}

As stated in Section III,B, the use of Ba^{2+} has been decisive in determining the degree of occupancy of a maxi K(Ca) channel. In this section, some findings that have led to the conclusion that an ion can get trapped in the channel once it closes are highlighted. How these findings can be interpreted in terms of a physical model (Miller *et al.*, 1987) is discussed also. The basic result is that Ba^{2+} can be trapped within the channel by closing the channel with a strong hyperpolarizing potential applied during a blocked interval. Trapped Ba^{2+} cannot leave the channel until after the channel opens by the application of a depolarizing potential. The distribution of first latency times (the time necessary for a blocked channel to become unblocked after the depolarizing pulse) agrees with that of blocked times obtained under equilibrium conditions. Conversely, if the channel is closed during an unblocked period, the channel is in the unblocked state on depolarization. These results strongly suggest that a closed blocked channel must go through the open-blocked configuration before Ba^{2+} can leave the channel, and that the site is inaccessible to the divalent

cation when the channel is closed.[2] The results suggest that some sort of gate exists on the cytoplasmic side of the channel that prevents the exit of Ba^{2+} from the closed channel. The preceding discussion mentioned that, in the presence of high external [K$^+$] (lock-in site occupied), Ba^{2+} cannot dissociate toward the cytoplasmic side of the pore. Currently, whether once the channel has closed the divalent cation cannot exit toward the external site because the lock-in site is occupied or because of the existence of a gating region located on the external side of the pore is unclear.

Channel blockade by Ba^{2+} not only hinders the passage of ions but also modifies the channel gating reaction (Miller *et al.*, 1987; Neyton and Pelleschi, 1991). When Ba^{2+} occupies the channel pore, more energy is required to close the channel.[2] Interestingly, the modulatory effect of Ba^{2+} on channel gating is regulated by external K$^+$. The Ba^{2+}-blocked open state is stabilized at K$^+$ concentrations greater than 5 mM (Miller *et al.*, 1987) whereas, in the absence of external permeant ions, stabilization of the closed-blocked configuration of the channel occurs (Neyton and Pelleschi, 1991). Thus, occupancy of the external lock-in site not only is able to control the Ba^{2+} blocking reaction, but also appears to play an important role in regulating the gating of K(Ca) channels. Neyton and Pelleschi (1991) have proposed that occupancy of this site modulates gating by speeding the rate of channel opening.

---

[2] The most simple model able to account for the kinetic behavior of Ba^{2+} blockade is (Miller *et al.*, 1987)

$$
\begin{array}{ccc}
\text{Closed} & \underset{}{\overset{K_1}{\rightleftharpoons}} & \text{Open} \\
{\alpha[\text{Ba}]} \Big\Updownarrow {\beta} & & \\
\text{Closed–blocked} & \underset{}{\overset{K_2}{\rightleftharpoons}} & \text{Open–blocked}
\end{array}
$$

where K_1 and K_2 are the equilibrium constants for the gating reactions for the opening of the unblocked and blocked channels, respectively, and α and β are the association and dissociation rate constants for the blocking reaction. Since $P_o = 1/(1 + K_1)$, the open-blocked probability P_{ob} is given by $P_{ob} = 1/(1 + K)$, and the unblocking rate $k_{off} = \beta P_{ob}$

$$k_{off} = \beta\Theta[\Theta - 1 + 1/P_o]^{-1},$$

where $\Theta = K_2/K_1$. Θ can be obtained by plotting k_{off} against P_o, leaving Θ as an adjustable parameter. In terms of channel gating, $\Theta > 1$ implies that the open-blocked configuration of the channel is stabilized (*i.e.*, more energy is required to close a blocked channel than an unblocked channel). On the other hand, the closed-blocked configuration is stabilized where $\Theta < 1$.

FIGURE 11 Primary structures of iberiotoxin (IbTX) and charybdotoxin (ChTX). Rectangles enclose portions of the sequence that are regions of identity in the two toxins.

IV. Toxins That Block Maxi K⁺(Ca²⁺) Channels

A. Understanding the Molecular Mode of Action of Charybdotoxin

A small component of the venom of the scorpion *Leiurus quinquestriatus* var. *hebraeus*, CTX at nanomolar concentrations is able to block several types of K⁺-selective channels (Miller *et al.*, 1985; Castle and Strong, 1986; Leneveu and Simonneau, 1986; Gimenez-Gallego *et al.*, 1988; for reviews, see Moczydlowski *et al.*, 1988; Strong, 1990; Garcia *et al.*, 1991). ChTX is a 37-amino-acid peptide that contains several charged residues, conferring a net charge of +5 on the molecule (Sugg *et al.*, 1990; Fig. 11). The three-dimensional structure of this toxin has been solved by proton NMR (Lambert *et al.*, 1990; Bontems *et al.*, 1991a,b,1992). The molecule has a slightly oblong shape with axes of ~2 and 2.5 nm. ChTX consists of a β sheet linked to an extended strand by a disulfide bond and to an α helix by two such bonds.

When ChTX is added to the external side of most of maxi K(Ca) channels, the normal channel activity is interrupted for periods that can last several tens of seconds, in which the channel remains quiescent (Fig. 12B). These silent periods of activity have been interpreted as blocked time intervals. Detailed kinetic analysis of active and blocked periods indicates that ChTX inhibits maxi K(Ca) channel activity by binding, in a bimolecular fashion, to a receptor located in the external vestibule, and thus blocking the passage of K⁺ through the conduction pore (Miller *et al.*, 1985; MacKinnon and Miller, 1988; Miller, 1988). The dissociation constant is 10 nM when the external medium is 150 mM NaCl (Anderson *et al.*, 1988). Toxin blockade is voltage dependent. Anderson *et al.* (1988) showed that only the *dissociation rate* constant (k_{off}) is voltage dependent. Depolarization speeds up dissociation of the toxin. Moreover, an increase in internal K⁺ relieves channel blockade promoted by the toxin. Again, this effect is due exclusively to an increase in k_{off}. These experimental results, in conjunction with the fact that k_{off} voltage dependence is lost if internal K⁺ is

replaced by Na⁺, an impermeant ion, led MacKinnon and Miller (1988) to propose that toxin binding is intrinsically voltage independent. The effect of voltage on toxin potency is actually the result of voltage-dependent binding of K⁺ to a site located in the conduction pathway and near the external mouth of the channel. When K⁺ is occupying this site, the positively charged ChTX is electrostatically flushed away from its blocking position (Fig. 13).

ChTX binding to the external channel mouth is strongly dependent on external ionic strength; the effect is essentially the same whether K⁺, Na⁺, or arginine is used to change the ionic strength (Anderson *et al.*, 1988). These observations, in conjunction with the fact that only the ChTX association rate constant is affected by this experimental *maneuver*, suggest nonspecific electrostatic contributions to the ChTX–channel interactions. In other words, the nega-

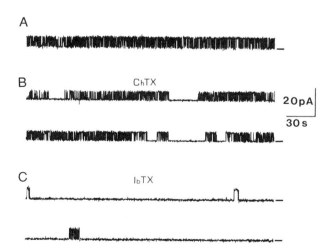

FIGURE 12 Maxi K(Ca) channel blockade induced by ChTX and IbTX under similar experimental conditions. In both cases, the internal medium contained 300 mM K⁺ and the external medium 300 mM Na⁺; the records were obtained at 0 mV. (A) Control. (B) IbTX was added to the external side to a final concentration of 17 nM. (C) In a different single-channel membrane, but under otherwise identical experimental conditions, ChTX was added to a final concentration of 18 nM. Redrawn with permission from Candia *et al.* (1992).

tive local surface potential created by the protein car-boxyl groups located in the external mouth (MacKinnon *et al.*, 1989) increases the concentration of positively[3] charged toxin in the vicinity of the ChTX binding site.

To obtain structural information about toxin–channel interaction, Park *et al.* (1991) expressed a gene encoding ChTX in *E. coli.* The gene was expressed as a cleavable fusion protein and ChTX was obtain by enzymatic cleavage of ChTX followed by oxidation of the cysteine residues and synthesis of the N-terminal pyroglutamate. The final primary structure of the toxin is the same as that shown in Fig. 11. The genetically engineered ChTX blocks skeletal muscle maxi K(Ca) channels with the same characteristics as the native toxin. Modification of ChTX using site-directed mutagenesis has allowed the identification of the positively charged residues that make direct contact with the protein surface that forms the external mouth of the channel (Park and Miller, 1992a,b). Replacement of all eight charged residues by neutral ones (glutamine) showed that only three of them are important for toxin binding: Arg 25, Lys 27, and Arg 34 (see Fig. 11). Replacement of these residues by glutamine increases the dissociation constant. In the case of the K27Q (Lys → Glu), K_d changed from 15 nM (wild-type ChTX) to 7500 nM. The change in affinities is mainly the result of a large increase in the dissociation rate, with only mild effects on the association rate constant. The mutations Z1Q (Glx → Gln), F2A (Phe → Ala), M29L (Met → Leu), and Y36N (Tyr → Asn) also have a destabilizing effect on toxin binding, whereas mutations V5E (Val → Glu), V16E (Val → Glu), and L20N (Leu → Asn) are innocuous (Stampe *et al.*, 1992). These results indicate that these amino acid residues make intimate contact with the toxin receptor located in the external mouth of the channel. In the two-dimensional NMR structure of

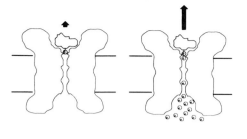

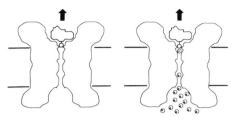

FIGURE 13 Interaction of wild-type ChTX (A) and mutant (K27N or K27Q) ChTX (B) with K[+]. (A) Increasing internal [K[+]] largely increases the k_{off} of wild-type ChTX (represented by the size of the arrow) because K[+], by binding to an external site, electrostatically flushes away the toxin. (B) k_{off} is not modified by alterations in internal K when lysine 27 is changed to an asparagine or to a glutamine. Reprinted with permission from Park and Miller (1992a). © Cell Press.

CTX, Arg 25, Lys 27, and Arg 34 lie close together whereas the rest of the positively charged residues are at the opposite end of the molecule (Bontems *et al.*, 1991, 1992). Moreover, residues Glx 1, Phe 2, Met 29, and Tyr 36 are on the same "hot" surface of ChTX. Park and Miller (1992a) proposed that the three crucial charged residues form part of an area of the ChTX molecule that makes direct contact with the channel mouth. The observation that the mutation K27Q is the only one that completely eliminates the voltage dependence of toxin blockade provides further assurance that this interpretation is correct, and adds one more twist to the molecular mechanism of toxin action. As depicted in Fig. 13, Lys 27 would be the amino acid that makes close contact with the pore, and the voltage dependence of the blockade disappears because neutralization of this residue removes the electrostatic interaction between the toxin and K[+] moving along the pore (Park and Miller, 1992b).

B. Iberiotoxin Blockade of Maxi K(Ca) Channels Is Highly Selective

Despite the fact that ChTX is a high affinity blocker of maxi K(Ca) channels, its specificity is poor. ChTX also blocks the apamin-sensitive channel of guinea pig hepatocyte (Abia *et al.*, 1986), a small conductance

---

[3] ChTX contains 8 positively charged residues (4 Lys, 3 Arg, and 1 His). Taking into account the negatively charged residues, the peptide net charge is +5. The concentration of a pentavalent ion X near a negatively charged surface relates to the bulk concentration through equation $[X]_{surface} = [X]_{bulk}\exp(-5\Phi_5 F/RT)$, where Φ_5 is the surface potential. Hence, even a very small electrostatic potential in the external mouth should, in principle, promote a large increase in toxin concentration in the neighborhood of the ChTX binding site. However, in ChTX the charge is spread all over the toxin molecule (i.e., it is not a point charge), making it difficult to predict quantitatively the effect of the surface potential since the "effective" valence of the toxin is, in general, less than the net charge. For example, in the extreme case that the charges are separated by distances that largely exceed the solution Debye length (~1 nm in 0.1 M monovalent salt), the multivalent ion will behave as a monovalent one (Alvarez *et al.*, 1983).

K(Ca) channel of *Aplysia* neurons (Hermann and Erxleben, 1987), a small conductance K(Ca) channel (35 pS in symmetrical 160 mM KCl) and voltage-dependent type n and n' K⁺ channels of T lymphocytes (Sands *et al.*, 1989; Grissmer *et al.*, 1993), and large conductance cation channels present in the vacuolar membrane of some plants (Weiser and Bentrup, 1991). On the other hand, some maxi K(Ca) channels are not blocked by ChTX. Reinhardt *et al.* (1989) found that a brain maxi K(Ca) channel is not blocked by ChTX; Adelman *et al.* (1992) reported that the *slowpoke* channel is also insensitive to this toxin. The lack of selectivity of ChTX prompted a search for new and more specific toxins for maxi K(Ca) channels. Galvez *et al.* (1990) found that the venom of the scorpion *Buthus tamulus* was able to modulate ChTX binding to smooth muscle sarcolemmal membranes. Fractionation of this venom and purification of the active component isolated a peptide displaying 68% identity with ChTX, but possessing four more negatively charged (Asp) and one less positively charged (H21F; His → Phe) residue (iberiotoxin, IbTX; Fig. 11). Therefore, the toxin carries, a net charge of +1. The three-dimensional structure of IbTX has been determined and appears very similar to that of ChTX (Johnson and Sugg, 1992).

ITX shows a high selectivity for maxi K(Ca) channels. In GH₃ cells, IbTX does not alter A-type K⁺, Ca²⁺, or Na⁺ currents but completely blocks the Ca²⁺-activated K⁺ current present in this type of cell. Moreover, voltage-dependent K⁺ channels of lymphocytes are unaffected by IbTX (Garcia *et al.*, 1991).

In the presence of external ITX, channel activity is interrupted by closed periods that can last several minutes (Fig. 12C). A kinetic analysis of single-channel records in the presence of IbTX has provided strong evidence that IbTX blocks maxi K(Ca) channels by a mechanism very similar to that of ChTX (Giangiacomo *et al.*, 1991; Candia *et al.*, 1992): (1) the toxin binding reaction is bimolecular; (2) TEA alleviates blockade in a strictly competitive manner; and (3) internal K⁺ enhances the IbTX dissociation rate. A fitting to the data predicts a $k_{off} = 6 \times 10^{-5}$ sec⁻¹ in the absence of external K⁺ (Giangiacomo *et al.*, 1991). This results means that, under these experimental conditions, the toxin would dwell in the channel for an average of 4.6 hr. The K_d for IbTX (~1 nM) is about 10-fold smaller than that of ChTX. This difference is mainly the result of very low dissociation rate constant for IbTX (0.004 sec⁻¹ vs 0.08 sec⁻¹ for ChTX). Not all maxi K(Ca) channels bind IbTX more tightly than ChTX. The maxi K(Ca) channel of the apical membrane of rabitt proximal convoluted tubule although binds ChTX with a K_d similar to the skeletal muscle

channel (~10 nM), binds IbTX with a K_d of 550 nM (Tauc *et al.*, 1993). As in the case of CTX (Anderson *et al.*, 1988), the second order IbTX–channel association rate constant is decreased as the external ionic strength is increased. Although qualitatively the effect of ionic strength is the same for both toxins, quantitatively ChTX binding to the channel is much more sensitive to ionic strength variations than IbTX binding. Because ChTX has a much higher net positive charge than IbTX, its local concentration should be a much steeper function of ionic strength.[2]

Although less well characterized, limbatotoxin isolated from the scorpion *Centuroides limbatus* appears to be as potent and specific for maxi K(Ca) channels as IbTX (Giangiacomo *et al.*, 1993; M. L. Garcia, personal communication). On the other hand, a minor component of the venom of *Leiurus quinquestriatus* var. *hebraeus* showed substantial homology with ChTX, with differences observed at eight positions (Lucchesl *et al.*, 1989). This toxin (ChTX-Lq2) is about 40-fold less potent than ChTX. The lower blocking affinity of ChTX is a consequence of a faster dissociation rate and a smaller association rate constant.

C. Tetrandine

The alkaloid 6,6',7,12-tetramethoxy-2,2'-dimethylberbamam (tetrandine) is used in China for the treatment of cardiac ailments. Tetrandine blocks a maxi K(Ca) channel in nerve terminals of rat neurohypophysis with a K_d of 0.21 μM (Wang and Lemos, 1992). Tetrandine may become a powerful tool for studying K(Ca) channels, because it appears to block selectively those maxi K(Ca) channels insensitive to charybdotoxin.

Acknowledgments

I thank M. L. Garcia, M. Garcia-Calud, I. Levitan, O. McManus, C. Miller, E. Moczydlowski, L. Toro, and E. Stefani for providing manuscripts before publication. The generosity of and the many interesting discussions on maxi K(Ca) channels held with Ligia Toro and Enrico Stefani are cheerfully acknowledged. This work was supported by Grants 863-1991 and 1940277 from Fondo Nacional de Investigaciones, Chile, and a group of Chilean private companies (COPEC, CMPC, ENERSIS).

References

Abia, A., Lobaton, C. D., Moreno, A., and Garcia-Sancho, J. (1986). *Leiurus quinquestriatus* venom inhibits different kinds of Ca²⁺-dependent K⁺ channels. *Biochim. Biophys. Acta* **856**, 403–407.

Adelman, J. P., Shen, K.-Z., Kavanaugh, M. P., Warren, R. A., Wu, Y.-N., Lagrutta, A., Bond, C. T., and North, R. A. (1992). Calcium-activated potassium channels expressed from cloned complementary DNAs. *Neuron* **9**, 209–216.

Alcayaga, C., Cecchi, X., Alvarez, O., and Latorre, R. (1989). Streaming potential measurements in Ca^{2+}-activated K^+ channels from skeletal and smooth muscle: Coupling of ion and water fluxes. *Biophys. J.* **55**, 367–371.

Almers, W. (1978). Gating currents and charge movements in excitable membranes. *Rev. Physiol. Biochem. Pharmacol.* **82**, 96–190.

Alvarez, O., Brodwick, M., Latorre, R., McLaughlin, A., McLaughlin, S., and Szabo, G. (1983). Large divalent cations and electrostatic potentials adjacent to membranes: Experimental results with hexamethonium. *Biophys. J.* **44**, 333–342.

Andersen, O. S., and Procopio, J. (1980). Ion movements through the gramicidin channel. *Acta Physiol. Scand. Suppl.* **481**, 27–35.

Anderson, C. S., MacKinnon, R., Smith, C., and Miller, C. (1988). Charybdotoxin block of single Ca^{2+}-activated K^+ channels. Effects of channel gating, voltage, and ionic strength. *J. Gen. Physiol.* **91**, 317–332.

Armstrong, C. M. (1969). Inactivation of the potassium conductance and related phenomena caused by quaternary ammonium ions injected in squid axons. *J. Gen. Physiol.* **54**, 553–575.

Armstrong, C. M. (1971). Interaction of tetraethylammonium ions derivatives with the potassium channel of giant axon. *J. Gen. Physiol.* **58**, 413–437.

Armstrong, C., and Bezanilla, F. (1977). Inactivation of the sodium channel. II. Gating current experiments. *J. Gen. Physiol.* **70**, 567–590.

Atkinson, N. S., Robertson, G., and Ganetzky, B. (1991). A component of calcium-activated potassium channels encoded by the *Drosophila slow* locus. *Science* **253**, 551–555.

Benham, C. D., Bolton, T. B., Lang, R. J., and Takewaki, T. (1985). The mechanism of action of Ba^{2+} and TEA on single Ca^{2+}-activated K^+ channels in arterial and intestinal smooth muscle cell membranes. *Pflügers Arch.* **403**, 120–127.

Blatz, A. L., and Magleby, K. L. (1984). Ion conductance and selectivity of single calcium-activated potassium channels in cultured rat muscle. *J. Gen. Physiol.* **84**, 1–23.

Bokvist, K., Rorsman, P., and Smith, P. (1990). Block of ATP-regulated and Ca^{2+}-activated K^+ channels in mouse pancreatic β-cells by external tetraethylammonium and quinine. *J. Physiol. (London)* **423**, 327–342.

Bontems, F., Roumestand, C., Boyot, P., Gilquin, B., Doljansky, Y., Menez, A., and Toma, F. (1991a). Three-dimensional structure of natural charybdotoxin in aqueous solution by ^{1}H-NMR. Charybdotoxin possesses a structural motif found in other scorpion toxins. *Eur. J. Biochem.* **196**, 19–28.

Bontems, F., Roumestand, C., Gilquin, B., Menez, A., and Toma, F. (1991b). Refined structure of charybdotoxin: Common motifs in scorpions toxins and insect defensins. *Science* **254**, 1521–1523.

Bontems, F., Gilquin, B. B., Roumestand, C., Menez, A., and Toma, F. (1992). Analysis of side-chain organization on a refined model of charybdotoxin: Structural and functional implications. *Biochemistry* **31**, 7756–7764.

Brayden, J. E., and Nelson, M. T. (1992). Regulation of arterial tone by activation of calcium-dependent potassium channels. *Science* **256**, 532–535.

Butler, A., Tsunoda, S., McCobb, D. P., Wei, A., and Salkoff, L. (1993). *mslo*, a complex mouse gene encoding "maxi" calcium-activated potassium channels. *Science* **261**, 221–224.

Candia, S., Garcia, M. L., and Latorre, R. (1992). Mode of action of iberiotoxin, a potent blocker of the large conductance Ca^{2+}-activated K^+ channel. *Biophys. J.* **63**, 583–590.

Cantor, C. R., and Schimel, P. R. (1980). "Biophysical Chemistry." Freeman, San Francisco.

Carl, A., and Sanders, K. M. (1989). Ca^{2+}-activated K^+ channels of canine colonic myocites. *Am. J. Physiol.* **257**, C470–C480.

Carl, A., McHale, N. G., Publicover, N. G., and Sanders, K. M. (1990). Participation of Ca^{2+}-activated K^+ channels in electrical activity of canine gastric smooth muscle. *J. Physiol. (London)* **429**, 205–221.

Castle, N. A., and Strong, P. N. (1986). Identification of two toxins from scorpion (*Leiurus quinquestriatus*) venom which block distinct classes of calcium-activated potassium channel. *FEBS Lett.* **209**, 117–121.

Cave, A., Daures, M.-F., Parello, J., Saint-Yves, A., and Sempere, R. (1979). NMR studies of primary and secondary sites of parvalbumins using two paramagnetic probes Gd(III) and Mn(II). *Biochemie* **61**, 755–765.

Cecchi, X., Wolff, D., Alvarez, O., and Latorre, R. (1987). Mechanisms of Cs^+ blockade in a Ca^{2+}-activated K^+ channel from smooth muscle. *Biophys. J.* **52**, 707–716.

Chao, S-H., Suzuki, Y., Zysk, J. R., and Cheung, W. Y. (1984). Activation of calmodulin by various metal cations as a function of ionic radius. *Mol. Pharmacol.* **26**, 75–82.

Christensen, O., and Zeuthen, T. (1987). Maxi K^+ channels in leaky epithelia are regulated by intracellular Ca^{2+}, pH, and membrane potential. *Pflügers Arch.* **408**, 249–259.

Christophersen, P. (1991). Ca^{2+}-activated K^+ channel from human erythrocyte membranes: Single channel rectification and selectivity. *J. Membr. Biol.* **119**, 75–83.

Cook, D. L., Ikeuchi, M., and Fujimoto, W. Y. (1984). Lowering pH inhibits Ca^{2+}-activated K^+ channels in pancreatic β cells. *Nature (London)* **311**, 269–271.

Copello, J., Segal, Y., and Reuss, L. (1991). Cytosolic pH regulates maxi K^+ channels in necturus gallbladder epithelial cells. *J. Physiol. (London)* **434**, 577–590.

Cornejo, M., Guggino, S. E., and Guggino, W. B. (1987). Modification of Ca^{2+}-activated K^+ channels in cultured medullary thick ascending limb cells by N-bromoacetamide. *J. Membr. Biol.* **99**, 147–155.

Cornejo, M., Guggino, S. E., and Guggino, W. B. (1989). Ca^{2+}-activated K^+ channels from cultured renal medullary thick ascending limb cells: Effect of pH. *J. Membr. Biol.* **110**, 49–55.

Cox, J. A., Wlodzimiers, W., and Stein, E. A. (1977). Regulation of calcium-binding by magnesium. *In* "Proceedings of the International Symposium on Calcium Binding Proteins and Calcium Function in Health and Disease" (R. H. Wasserman, ed.), pp. 266–269. Elsevier/North Holland, New York.

Dani, J. (1986). Ion-channel entrances influences permeation. Net charge, size, shape and binding considerations. *Biophys. J.* **49**, 607–618.

Dufourcq, J., Clin, B., and Lemanceau, B. (1972). NMR study of ganglion blocking and curare-like dimethoniums. Conformation in aqueous solutions. *FEBS Lett.* **22**, 205–209.

Einspahr, H., and Bugg, C. E. (1978). Crystal structures of calcium complexes of amino acids, peptides and related models. *In* "Proceedings of the International Symposium on Calcium Binding Proteins and Calcium Function in Health and Disease" (R. H. Wasserman, ed.), pp. 13–20. Elsevier/North Holland, New York.

Eisenman, G., Latorre, R., and Miller, C. (1986). Multi-ion conduction and selectivity in the high-conductance Ca^{2+}-activated K^+ channel from skeletal muscle. *Biophys. J.* **50**, 1025–1034.

Elkins, T., Ganetzky, B., and Wu, C-F. (1986). A *Drosophila* mutant that eliminates a calcium-dependent potassium current. *Proc. Natl. Acad. Sci. USA* **83**, 8415–8419.

Falke, J. J., Snyder, E. E., Thatcher, K. C., and Voertler, C. S. (1991). Quantitating and engineering of an EF-hand-like Ca^{2+} binding site. *Biochemistry* **30**, 8690–8697.

Findlay, I. (1984). A patch-clamp study of potassium channels and whole-cell currents in acinar cells of the mouth lacrimal gland. *J. Physiol.* (*London*) **350**, 179–195.

Foster, Ch. D., Chung, S., Zagotta, W. N., Aldrich, R. W., and Levitan, I. B. (1992). A peptide derived from the *Shaker* B K⁺ channel produces short and long blocks of reconstituted Ca²⁺-dependent K⁺ channels. *Neuron* **9**, 229–236.

Frech, G. C., VanDongen, A. M., Schuster, G., Brown, A. M., and Joho, R. H. (1989). A novel potassium channel with delayed rectifier properties isolated from rat brain by expression cloning. *Nature* (*London*) **340**, 642–645.

Fuchs, F. (1971). Ion exchange properties of the calcium receptor site of troponin. *Biochim. Biophys. Acta* **245**, 221–229.

Fuchs, F. (1974). Chemical properties of the calcium receptor of troponin as determined from binding studies. *In* "Calcium Binding Proteins" (W. Drabikowski, H. Strzelecka-Golaszewska, and E. Carafoli, eds.), pp. 1–37. Elsevier Scientific, Amsterdam.

Galvez, A., Gimenez-Gallego, G., Reuben, J., Roy-Contancin, L., Feingenbaum, P., Kaczorowski, G. J., and Garcia, M. L. (1990). Purification and characterization of a unique, potent, peptidyl-probe for the high-conductance calcium-activated potassium channel from the venom of the scorpion *Buthus tamulus*. *J. Biol. Chem.* **265**, 11083–11090.

Garcia, M. L., Galvez, A., Garcia-Calvo, M., King, V. F., Vazquez, J., and Kaczorowski, G. (1991). Use of toxins to study potassium channels. *J. Biomembr. Bioenerg.* **23**, 615–646.

Gelband, C. H., Silberberg, S. D., Groschner, K., and van Breemen, C. (1990). ATP inhibits smooth muscle Ca²⁺-activated K⁺ channels. *Proc. R. Soc. London B.* **242**, 23–28.

Giangiacomo, K. M., Garcia, M. L., and McManus, O. (1991). Mechanism of iberiotoxin block of the large conductance calcium-activated potassium channel from bovine aortic smooth muscle. *Biochemistry* **31**, 6719–6727.

Giangiacomo, K., Sugg, E. E., Garcia-Calvo, M., Leonard, R. J., McManus, O. B., Kaczorowski, G. J., and Garcia, M. L. (1993). Synthetic charybdotoxin-iberiotoxin chimeric peptides define toxin binding sites on calcium-activated and voltage-dependent potassium channel. *Biochemistry* **32**, 2363–2370.

Gimenez-Gallego, G., Navia, M. A., Reuben, J. P., Katz, G. M., Kaczorowski, G. J., and Garcia, M. L. (1988). Purification, synthesis, and model structure of charybdotoxin, a potent selective inhibitor of calcium-activated potassium channels. *Proc. Natl. Acad. Sci. USA* **85**, 3329–3333.

Gitter, A. H., Beyenbach, K. W., Christine, Ch. W., Gross, P., Minuth, W. W., and Frömter, E. (1987). High-conductance K⁺ channel in apical membranes of principal cells cultured from rabbit renal cortical collecting duct anlagen. *Pflügers Arch.* **408**, 282–290.

Golowasch, J., Kirkwood, A., and Miller, C. (1986). Allosteric effects of Mg²⁺ on the gating of Ca²⁺-activated K⁺ channels from mammalian skeletal muscle. *J. Exp. Biol.* **124**, 5–13.

Grissmer, S., Nguyen, A. N., and Cahalan, M. D. (1993). Calcium activated potassium channels in resting and activated human T lymphocytes. *J. Gen. Physiol.* **102**, 601–630.

Groschner, K., Silberberg, S. D., Gelband, C. H., and Van Breemen, C. (1991). Ca²⁺-activated K⁺ channels in airway smooth muscle are inhibited by cytoplasmic adenosine triphosphate. *Pflügers Arch.* **417**, 517–522.

Heginbotham, L., and MacKinnon, R. (1992). The aromatic binding site for tetraethylammonium ion on potassium channels. *Neuron* **8**, 483–491.

Herman, A., and Erxleben, C. (1987). Charibdotoxin selectivity blocks small Ca-activated K channels in *Aplysia* neurons. *J. Gen. Physiol.* **90**, 27–47.

Hille, B. (1992). "Ionic Channels of Excitable Membranes," 2d Ed. Sinhauer Associates, Sunderland, Massachussetts.

Hille, B., and Schwarz, W. (1978). Potassium channels as multi-ion single-file pores. *J. Gen. Physiol.* **72**, 409–442.

Hodgkin, A. L., and Keynes, R. D. (1955). The potassium permeability of a giant nerve fiber. *J. Physiol.* (*London*) **128**, 61–68.

Hoshi, T., Zagotta, N., and Aldrich, R. W. (1990). Biophysical and molecular mechanisms of *Shaker* potassium channel inactivation. *Science* **280**, 533–538.

Hu, S., Kim, H. S., and Jeng, A. Y. (1991). Dual action of endothelin-1 on the Ca²⁺-activated K⁺ channel in smooth muscle cells of porcine coronary artery. *Eur. J. Pharmacol.* **194**, 31–36.

Ikemoto, Y., Ono, K., Yoshida, A., and Akaike, N. (1989). Delayed activation of large conductance Ca²⁺-activated K channels in hippocampal neurons of rat. *Biophys. J.* **56**, 207–212.

Imoto, K., Busch, C., Sakmann, B., Konno, M., Nakai, T., Bujo, J., Mori, H., Fukuda, Y., and Numa, S. (1988). Rings of negatively charged amino acids determine the acetylcholine receptor channel conductance. *Nature* (*London*) **335**, 645–648.

Iwatsuki, N., and Petersen, O. H. (1985). Action of tetraethylammonium on calcium-activated potassium channels in pig acinar cells studied by patch-clamp single-channel and whole-cell recording. *J. Membr. Biol.* **86**, 139–144.

Jan, L. Y., and Jan, Y. N. (1992). Structural elements involved in specific K⁺ channel functions. *Annu. Rev. Physiol.* **54**, 537–555.

Jia, M., Ehrenstein, G., and Iwasa, K. (1988). Unusual calcium-activated potassium channels of bovine parathyroid cells. *Proc. Natl. Acad. Sci. USA* **85**, 7236–7239.

Johnson, B. A., and Sugg, E. E. (1992). Determination of the three-dimensional structure of iberiotoxin in solution by ¹H nuclear magnetic resonance spectroscopy. *Biochemistry* **31**, 8151–8159.

Kirsch, G. E., Drewe, J., Taglialatela, M., Joho, R. H., Hartmann, H. A., DeBiasi, M., and Brown, A. M. (1992). A single nonpolar residue in the deep pore of related K⁺ channels acts as a K⁺ : Rb⁺ conductance switch. *Biophys. J.* **62**, 136–144.

Klee, C., Crouch, T., and Richman, P. (1980). Calmodulin. *Annu. Rev. Biochem.* **49**, 489–515.

Kume, H., Takagi, K., Satake, T., Tokuno, H., and Tomita, T. (1990). Effect of intracellular pH on calcium-activated potassium channels in rabbit tracheal smooth muscle. *J. Physiol.* (*London*) **424**, 445–457.

Lambert, P., Kuroda, H., Chino, N., Waranabe, T., Kimura, T., and Sakakibara, S. (1990). Solution synthesis of charibdotoxin (ChTX) a K⁺ channel blocker. *Biochim. Biophys. Res. Commun.* **170**, 684–690.

Lang, D. G., and Ritchie, A. K. (1990). Tetraethylammonium blockade of apamin-sensitive and insensitive Ca²⁺-activated K⁺ channels in a pituitary cell line. *J. Physiol.* (*London*) **425**, 117–132.

Latorre, R. (1986). The large calcium-activated potassium channel. *In* "Ion Channel Reconstitution" (C. Miller, ed.), pp. 431–467. Plenum Press, New York.

Latorre, R., Vergara, C., and Hidalgo, C. (1982). Reconstitution in planar lipid bilayers of a Ca²⁺-dependent K⁺ channel from transverse tubule membranes isolated from rabbit skeletal muscle. *Proc. Natl. Acad. Sci. USA* **79**, 805–809.

Latorre, R., Oberhauser, A., Labarca, P., and Alvarez, O. (1989). Varieties of calcium-activated potassium channels. *Annu. Rev. Physiol.* **51**, 385–399.

Laurido, C., Candia, S., Wolff, D., and Latorre, R. (1991). Proton modulation of a Ca²⁺-activated K⁺ channel from rat skeletal muscle incorporated into planar bilayers. *J. Gen. Physiol.* **98**, 1025–1043.

Leavis, P. C., and Gergely, J. (1984). Thin filaments proteins and thin filament regulation of vertebrate muscle contraction. *CRC Crit. Rev. Biochem.* **16,** 235–305.

Leinders, T., van Kleef, R. G. D. M., and Vijverberg, H. P. M. (1992). Distinct metal ion binding sites on Ca^{2+}-activated K^{2+} channels in inside-out patches of human erythrocytes. *Biochim. Biophys. Acta* **1112,** 75–82.

Leneveu, E., and Simonneau, M. (1986). Scorpion venom inhibits selectively Ca^{2+}-activated K^+ channels in situ. *FEBS Lett.* **209,** 165–168.

Levitan, I. B. (1988). Modulation of ion channels in neurons and other cells. *Annu. Rev. Neurosci.* **11,** 119–136.

Lucchesi, K., and Moczydlowski, E. (1990). Subconductance behavior in a maxi Ca^{2+}-activated K^+ channel induced by dendrotoxin-I. *Neuron* **2,** 141–148.

Lucchesi, K., Ravindran, A., Young, H., and Moczydlowski, E. (1989). Analysis of the blocking activity of charybdotoxin homologs and iodinated derivatives against Ca^{2+}-activated K^+ channels. *J. Membr. Biol.* **109,** 269–281.

MacKinnon, R. (1991). Using mutagenesis to study potassium channel mechanisms. *J. Bioenerg. Biomembr.* **23,** 647–663.

MacKinnon, R., and Miller, C. (1988). Mechanism of charybdotoxin block of the high-conductance, Ca^{2+}-activated K^+ channel. *J. Gen. Physiol.* **91,** 335–349.

MacKinnon, R., and Miller, C. (1989). Functional modification of a Ca^{2+}-activated K^+ channel by trimethyloxonium. *Biochemistry* **28,** 8087–8092.

MacKinnon, R., Latorre, R., and Miller, C. (1989). Role of surface electrostatics in the operation of a high-conductance Ca^{2+}-activated K^+ channel. *Biochemistry* **28,** 8092–8099.

Markwardt, F., and Isenberg, G. (1992). Gating of maxi K^+ channels studied by Ca^{2+} concentration jumps in excised inside-out multichannel patches (Myocytes from guinea pig urinary bladder). *J. Gen. Physiol.* **99,** 841–862.

Marty, A. (1983). Blocking of large unitary calcium-dependent potassium currents by internal sodium ions. *Pflügers Arch.* **396,** 179–181.

Maruyama, Y., Petersen, O. H., Flanagan, P., and Pearson, G. T. (1983). Quantification of Ca^{2+}-activated K^+ channels under hormonal control in pig pancreas acinar cells. *Nature (London)* **305,** 228–232.

Mayer, E. A., Loo, D. D. F., Snape, W. J., Jr., and Sachs, G. (1990). The activation of calcium and calcium-activated potassium channels in mammalian colonic smooth muscle by substance P. *J. Physiol. (London)* **420,** 47–71.

McManus, O. (1991). Calcium-activated potassium channels: Regulation by calcium. *J. Bioenerg. Biomem.* **23,** 537–560.

McManus, O. B., and Magleby, K. L. (1985). Inverse relationship of the duration of adjacent open and shut intervals in Cl and K channels. *Nature (London)* **317,** 625–627.

McManus, O. B., and Magleby, K. L. (1988). Kinetic states and modes of single large-conductance calcium-activated potassium channels in cultured rat skeletal muscle. *J. Physiol.* **402,** 79–120.

McManus, O. B., and Magleby, K. L. (1991). Accounting for the Ca^{2+}-dependent kinetics of single large-conductance Ca^{2+}-activated K^+ channels in rat skeletal muscle. *J. Physiol. (London)* **443,** 739–777.

McManus, O. B., Harris, G. H., Giangiacomo, K. M., Feigenbaum, P., Reuben, J. P., Addy, M. E., Burka, J. F., Kaczorowski, G. J., and Garcia, M. L. (1993). An activator of calcium-dependent potassium channels isolated from a medicinal herb. *Biochemistry* **32,** 6128–6133.

Methfessel, C., and Boheim, G. (1983). The gating mode of single calcium-dependent potassium channel is described by an activation/blockade mechanism. *Biophys. Struct. Mech.* **9,** 35–60.

Miller, C. (1982). Bis-quaternary ammonium blockers as structural probes of the sarcoplasmic reticulum K^+ channel. *J. Gen. Physiol.* **79,** 861–891.

Miller, C. (1988). Competition for block of a Ca^{2+}-activated K^+ channel by charybdotoxin and tetraethylammonium. *Neuron* **1,** 1003–1006.

Miller, C., Moczydlowski, E., Latorre, R., and Phillips, M. (1985). Charybdotoxin, a protein inhibitor of single Ca^{2+}-activated K^+ channels from mammalian skeletal muscle. *Nature (London)* **313,** 316–318.

Miller, C., Latorre, R., and Reisin, I. (1987). Coupling of voltage-dependent gating and Ba^{2+} block in the high-conductance, Ca^{2+}-activated K^+ channel. *J. Gen. Physiol.* **90,** 427–449.

Moczydlowski, E., Alvarez, O., Vergara, C., and Latorre, R. (1985). Effect of phospholipid surface charge on the conductance and gating of a Ca^{2+}-activated K^+ channel in planar lipid bilayers. *J. Membr. Biol.* **83,** 273–282.

Moczydlowski, E., and Latorre, R. (1983). Gating kinetics of Ca^{2+}-activated K^+ channels from rat muscle incorporated into planar lipid bilayers: Evidence for two voltage-dependent Ca^{2+} binding reactions. *J. Gen. Physiol.* **82,** 511–542.

Moczylowski, E., Lucchesi, K., and Ravindran, A. (1988). An emerging pharmacology of peptide toxins targeting against potassium channels. *J. Membr. Biol.* **105,** 95–111.

Neyton, J., and Miller, C. (1988a). Potassium blocks barium permeation through a calcium-activated potassium channel. *J. Gen. Physiol.* **92,** 549–567.

Neyton, J., and Miller, C. (1988b). Discrete Ba^{2+} block as a probe of ion occupancy and pore structure in the high-conductance Ca^{2+}-activated K^+ channel. *J. Gen. Physiol.* **92,** 569–586.

Neyton, J., and Pelleschi, M. (1991). Multi-ion occupancy alters gating in high-conductance, Ca^{2+}-activated K^+ channels. *J. Gen. Physiol.* **97,** 641–665.

Nomura, K., Naruse, K., Watanabe, K., and Sokabe, M. (1990). Aminoglycoside blockade of Ca^{2+}-activated K^+ channel from rat brain synaptosomal membranes incorporated into planar bilayers. *J. Membr. Biol.* **115,** 241–251.

Oberhauser, A., Alvarez, O., and Latorre, R. (1988). Activation by divalent cations of a Ca^{2+}-activated K^+ channel from skeletal muscle membrane. *J. Gen. Physiol.* **92,** 67–86.

Pallotta, B. S. (1985). N-Bromoacetamide removes a calcium-dependent component of channel opening from calcium-activated potassium channels in rat skeletal muscle. *J. Gen. Physiol.* **86,** 601–611.

Pantoliano, M. W., Whitlow, M., Wood, J. F., Rollence, M. L., Finzel, B. C., Gilliland, G. L., Poulos, T. L., and Bryan, P. N. (1988). The engineering of binding affinity at metal binding sites for the stabilization of proteins: Subtilisin a test case. *Biochemistry* **27,** 8311–8317.

Park, C-S., and Miller, C. (1992a). Mapping function to structure in a channel-blocking peptide: Electrostatic mutants of charybdotoxin. *Biochemistry* **31,** 7749–7755.

Park, C-S., and Miller, C. (1992b). Interaction of charybdotoxin with permeant ions inside the pore of a K^+ channel. *Neuron* **9,** 307–313.

Park, C-S., Hausdorff, S. F., and Miller, C. (1991). Design, synthesis, and functional expression of a gene for charybdotoxin, a peptide blocker of K^+ channels. *Proc. Natl. Acad. Sci. USA* **88,** 2046–2050.

Pennefather, P., Lancaster, B., Adams, P. R., and Nicoll, R. A. (1985). Two distinct Ca-dependent K currents in bullfrog sympathetic ganglion cells. *Proc. Natl. Acad. Sci. USA* **82,** 3040–3044.

Reinhart, P. H., Chung, S., and Levitan, I. B. (1989). A family of calcium-dependent potassium channels from rat brain. *Neuron* **2**, 1031–1041.

Reuter, H., and Stevens, C. F. (1980). Ion conductance and ion selectivity of potassium channels in snail neurons. *J. Membr. Biol.* **57**, 103–118.

Rudy, B. (1988). Diversity and ubiquity of K$^+$ channels. *Neurosci.* **25**, 729–749.

Salkoff, L., Baker, K., Butler, A., Covarrubias, M., Pak, M. D., and Wei, A. (1992). An essential 'set' of K$^+$ channels conserved in flies, mice and humans. *Trends Neurosci.* **15**, 161–166.

Salomao, L., Wark, G., Dubinsky, W. P., and Schultz, S. T. (1992). Effect of trypsin on a Ca^{2+}-activated K$^+$ channel reconstituted into planar phospholipid bilayers. *Am. J. Physiol.* **31**, C971–C974.

Sands, S. B., Lewis, R. S., and Cahalan, M. D. (1989). Charybdotoxin blocks voltage-gated K$^+$ channels in human and murine T-lymphocytes. *J. Gen. Physiol.* **93**, 1061–1074.

Schumaker, M. F. (1992). Shaking stack model of ion conduction through the Ca^{2+}-activated K$^+$ channel. *Biophys. J.* **63**, 1032–1044.

Scornik, F. S., and Toro, L. (1992). U46619, a thromboxane A$_2$ agonist, inhibits K$_{Ca}$ channel activity from pig coronary artery. *Am. J. Physiol.* **262**, C708–C713.

Schwarz, T. L., Tempel, B. L., Papazian, D. M., Jan, Y. N., and Jan, L. Y. (1988). Multiple potassium channels are produced by alternative splicing at *Shaker* locus in *Drosophila*. *Nature* (*London*) **331**, 137–142.

Sheppard, D. N., Giraldez, F., and Sepulveda, F. V. (1988). Kinetics of voltage and Ca^{2+} activation and Ba^{2+} blockade of a large conductance K$^+$ channel from *Necturus* enterocytes. *J. Membr. Biol.* **105**, 65–75.

Singer, J. J., and Walsh, J. V., Jr. (1987). Characterization of calcium-activated potassium channels in single smooth muscle cells using the patch-clamp technique. *Pflügers Arch.* **408**, 98–111.

Smart, T. G. (1987). Single calcium-activated potassium channels recorded from cultured rat sympathetic neurones. *J. Physiol.* **389**, 337–360.

Solaro, C. R., and Lingle, C. J. (1992). Trypsin-sensitive, rapid inactivation of a calcium-activated potassium channel. *Science* **257**, 1694–1698.

Squire, L. G., and Petersen, O. H. (1987). Modulation of Ca^{2+}- and voltage-activated K$^+$ channels by internal Mg^{2+} in salivary acinar cells. *Biochim. Biophys. Acta* **899**, 171–175.

Stampe, P., Kolmakova-Partensky, L., and Miller, C. (1992). Mapping hydrophobic residues of the interaction surface of charybdotoxin. *Biophys. J.* **62**, 8–9.

Standen, N. B., Quayle, J. M., Davies, N. W., Brayden, J. E., Huang, Y., and Nelson, M. T. (1989). Hyperpolarizing vasodilators activate ATP-sensitive K$^+$ channels in arterial smooth muscle. *Science* **245**, 177–180.

Strockbridge, L. L., French, A. S., and Man, S. F. P. (1991). Subconductance states in calcium-activated potassium channels from canine airway smooth muscle. *Biochim. Biophys. Acta* **1064**, 212–218.

Strong, P. N. (1990). Potassium channel toxins. *Pharmacol. Therapeut.* **46**, 137–162.

Sugg, E. E., Garcia, M. L., Reuben, J. P., Patchett, A. A., and Kaczorowski, G. J. (1990). Synthesis and structural characterization of charybdotoxin, a potent peptidyl inhibitor of the high conductance Ca^{2+}-activated K$^+$ channel. *J. Biol. Chem.* **265**, 18745–18748.

Tabcharani, J. A., and Misler, S. (1989). Ca^{2+}-Activated K$^+$ channels in rat pancreatic islet β-cells: permeation, gating and blockade by cations. *Biochim. Biophys. Acta* **982**, 62–72.

Taglialatela, M., Drewe, J., Kirsch, G. E., De Biasi, M., Hartmann, H., and Brown, A. M. (1993). Regulation of K$^+$/Rb$^+$ selectivity and internal TEA blockade by mutation at a single site in K$^+$ pores. *Pflügers Arch.* **423**, 104–112.

Tauc, M., Congar, P., Poncet, V., Merot, J., Vita, C., and Poujeol, P.(1993). Toxin pharmacology of the large conductance Ca^{2+}-activated K$^+$ channel in the apical membrane of rabbit proximal convoluted tubule in primary culture. *Pflugers Arch.* **425**, 126–133.

Toro, L., and Stefani, E. (1991). Calcium-activated K$^+$ channels: Metabolic regulation. *J. Bioenerg. Biomembr.* **23**, 561–576.

Toro, L., and Stefani, E. (1993). Modulation of maxi calcium-activated K channels: Role of ligands, phosphorylation, and G-proteins. *In* "Handbook of Experimental Pharmacology" (B. Dickey and L. Birnbaumer, eds.), Vols. I and II. Springer-Verlag, New York.

Toro, L., Amador, M., and Stefani, E. (1990). ANG II inhibits calcium-activated potassium channels from coronary smooth muscle in lipid bilayers. *Am. J. Physiol.* **258**, H912–H915.

Toro, L., Stefani, E., and Latorre, R. (1992). Internal blockade of a Ca^{2+}-activated K$^+$ channel by *Shaker* B inactivating "Ball" peptide. *Neuron* **9**, 237–245.

Toro, L., Ottolia, M., Olcese, R., and Stefani, R. (1993). Niflumic acid activated large conductance K(Ca) channels. *Biophys. J.* **64**, A2 (*Abstract*).

Turnheim, K., Constantin, J., Chan, S., and Schultz, S. G. (1989). Reconstitution of a calcium-activated potassium channel in basolateral membranes of rabbit colonocytes into planar lipid bilayers. *J. Membr. Biol.* **112**, 247–254.

Vergara, C., and Latorre, R. (1983). Kinetics of Ca^{2+}-activated K$^+$ channels from rabbit muscle incorporated into planar bilayers. Evidence for a Ca^{2+} and Ba^{2+} blockade. *J. Gen. Physiol.* **82**, 543–568.

Vergara, C., Moczydlowski, E., and Latorre, R. (1984). Conduction, blockade and gating in a Ca^{2+}-activated K$^+$ channel incorporated into planar lipid bilayers. *Biophys. J.* **45**, 73–76.

Vestegaard-Bogind, B., Stampe, P., and Chistophersen, P. (1985). Single-file diffusion through the Ca^{2+}-activated K$^+$ channel of human red blood cells. *J. Membr. Biol.* **88**, 67–75.

Villarroel, A. (1989). "Mechanism of Ion Conduction in the Large Calcium-Activated Potassium Channel." Ph.D. Thesis. University of California, Los Angeles.

Villarroel, A., Alvarez, O., and Eisenman, G. (1988a). A maximum conductance occurs for the large Ca-activated K channel at high Rb concentration. *Biophys. J.* **53**, 259a.

Villarroel, A., Alvarez, O., Oberhauser, A., and Latorre, R. (1988b). Probing a Ca^{2+}-activated K$^+$ channel with quaternary ammonium ions. *Pflügers Arch.* **413**, 118–126.

Wang, G., and Lemos, J. R. (1992). Tetrandin blocks a slow, large-conductance Ca^{2+}-activated potassium channel besides inhibiting a non-inactivating Ca^{2+} current isolated nerve terminals of the rat neurohypophysis. *Pflügers Arch.* **421**, 558–565.

Weiser, Th., and Bentrup, F.-W. (1991). Charybdotoxin blocks cation-channels in the vacuolar membrane of suspension cells of *Chenopodium rubrum* L. *Biochim. Biophys. Acta* **1066**, 109–110.

Williams, D. C., Jr., Katz, G. M., Roy-Contancin, L., and Reuben, J. P. (1988). Guanine 5'-monophosphate modulates gating of high-conductance Ca^{2+}-activated K$^+$ channels in vascular smooth muscle. *Proc. Natl. Acad. Sci. USA.* **85**, 9360–9364.

Wong, B. S., and Adler, M. (1986). Tetraethylammonium blockade of calcium-activated potassium channels in clonal anterior pituitary cells. *Pflügers Arch.* **407**, 279–284.

Wong, B. S., Lecar, H., and Adler, M. (1982). Single calcium-dependent potassium channels in clonal anterior pituitary cells. *Biophys. J.* **39**, 313–317.

Woodhull, A. M. (1973). Ionic blockade of sodium channels in nerve. *J. Gen. Physiol.* **61,** 687–708.

Yellen, G. (1984a). Ion permeation and blockade in Ca^{2+}-activated K$^+$ channels of bovine chromaffin cells. *J. Gen. Physiol.* **84,** 157–186.

Yellen, G. (1984b). Relief of sodium block of Ca^{2+}-activated channels by external cations. *J. Gen. Physiol.* **84,** 187–199.

Yellen, G., Jurman, M. E., Abramson, J., and MacKinnon, R. (1991). Mutations altering TEA blockade identify the probable pore forming region of a K$^+$ channel. *Science* **251,** 939–942.

Zagotta, W. N., Hoshi, T., and Aldrich, R. W. (1990). Restoration of inactivation in mutants of *Shaker* potassium channels by a peptide derived from SbB. *Science* **250,** 568–571.

Sodium Channels

Ionic Channels of Myelinated Axons

PETER SHRAGER AND JIN V. WU

I. Ionic Channels at Nodes and Paranodes

The evolution of myelin has facilitated high propagation velocities in axons of small diameter. The proper functioning of these fibers depends on the integrity of the myelin sheath as well as on a distinct and heterogeneous distribution of ionic channels. With the development of methods for voltage-clamp measurements at nodes of Ranvier (Dodge and Frankenhaeuser, 1958,1959), evidence emerged rapidly for a high density of Na^+ channels at these sites and, in the case of amphibian axons, voltage-dependent K^+ channels as well. However, the myelin sheath, so effective in insulating the internodal axolemma to allow saltatory conduction, was also an impenetrable barrier to electrophysiologists seeking information about ion channel distributions all along the fiber. In recent years, several methods have been developed to overcome that limitation. This chapter focuses on new approaches that have significantly altered scientist's views on the topography and role of ionic channels in axons and glia of myelinated axons.

Early work was suggestive of a regional specialization of the myelinated axolemma. Rosenbluth (1976) studied freeze-fracture replicas and found large intramembranous particles concentrated at the nodal axolemma at densities of about $1500/\mu m^2$ (Tao-Cheng and Rosenbluth, 1980). Quick and Waxman (1977) and

Waxman and Quick (1978) showed that ferric ion ferrocyanide preferentially stained the nodal membrane of peripheral and central fibers. The first evidence using probes specific for ionic channels was derived by Ritchie and Rogart (1977), who compared [^{3}H]saxitoxin (STX) binding to rabbit sciatic nerves that were intact or homogenized, with the rationale that internodal channels would only be accessible in the latter case. Binding was indistinguishable within experimental error in the two preparations, leading to the conclusion that virtually all axonal Na^+ channels were localized at the nodes of Ranvier, with a density of $12,000/\mu m^2$. However, Ritchie and Rang (1983) later showed that, in this same species, [^{3}H]STX binding rose during Wallerian degeneration, leading to the speculation that rabbit Schwann cells, which proliferate during this process, may express Na^+ channels. Direct evidence for such an event was obtained by the Ritchie laboratory. Chiu *et al.* (1984) and Shrager *et al.* (1985) measured whole-cell and single-channel Na^+ currents in membranes of Schwann cells cultured from neonatal rabbits. This work led to a caveat in interpreting the STX binding results as well as to an interesting speculation for the role of Schwann cell channels. If, in fact, most of the toxin binds to Schwann cell channels, then the contribution of axonal sites may be masked sufficiently that internodal binding sites are not apparent. Correspondingly, Na^+

channels have not been detected in frog Schwann cells to date (Shrager, 1988) and saturable binding was not demonstrable in homogenized frog sciatic nerve (Ritchie and Rogart, 1977). On the other hand, the Ritchie laboratory has advanced the hypothesis that Schwann cells might serve as local sites for synthesis of Na$^+$ channels that might later be transferred to axons (Shrager *et al.*, 1985). Evidence for this idea is discussed subsequently. Finally, the issue of which Schwann cells in peripheral nerves express Na$^+$ channels has been debated rather vigorously in the literature (Chiu, 1987; Howe and Ritchie, 1990).

More direct attempts to measure internodal ionic channels have used acute and chronic demyelination. Chiu and Ritchie (1981,1982) applied lysolecithin to nodes or internodes under gap voltage clamp and, in each case, recorded voltage-dependent K$^+$ currents (but no Na$^+$ currents) at critical times, presumably as the paranodal or internodal membrane was abruptly exposed. Grissmer (1985) repeated the experiments on frog internodes, but blocked almost all K$^+$ currents with internal Cs$^+$. Under these conditions, a small transient Na$^+$ current was observed. Chiu and Schwarz (1987) used the same technique on rabbit internodes and recorded very large Na$^+$ currents. In one experiment, a peak current of 90 nA was obtained from a 100-μm long segment, corresponding to a minimum of about 30 Na$^+$ channels/μm^2. These authors measured capacitative currents after lysolecithin-induced myelin damage that were 10-fold larger than those expected from axonal dimensions, and attributed these Na$^+$ currents to Schwann cell membranes that had fused to the axolemma.

II. Patch Voltage Clamp Recording of Macroscopic Currents from the Internodes of Axons

The loose patch-clamp technique has been used to better define the distribution of ionic channels in paranodal and internodal regions (Shrager, 1987,1988,1989). The circuitry was based on the bridge design of Stuhmer *et al.* (1983), but was modified to allow a more accurate compensation for the seal resistance. Frog and rat sciatic nerves were demyelinated by a surgical intraneural injection of lysolecithin (1 μl, 1%; Hall and Gregson, 1971). This compound damaged and vesiculated myelin at several points along an internode. Macrophages were recruited to the site and removed all remaining myelin and myelin debris by phagocytosis. Complete demyelination required about 1 wk, but loose patch recordings were possible just 1 day postinjection. Over the second week,

Schwann cells proliferated, adhered to the axon, and began the process of remyelination. Beginning at about 4 wk in the frog (2 wk in the rat), new nodes of Ranvier appeared in regions that previously were internodal, forming a sequence of short internodes. The following discussion centers on *Xenopus* axons. Results for rat fibers were very similar, and are summarized later. Macroscopic Na$^+$ currents were recorded from demyelinated internodes at all stages tested, from 1 day to 5 mo postinjection (Fig. 1). Are these Na$^+$ channels present normally, or do they only appear after demyelination? If the latter, are they newly synthesized or do they migrate from existing nodes of Ranvier? A number of experiments have been designed to answer these questions.

At paranodes, the amplitude of peak Na$^+$ currents increased about 3-fold over the first 6 days. This result may reflect some lateral diffusion of channels from the node after myelin disruption. However, at internodal sites, Na$^+$ currents were constant over the entire 5-mo period (Shrager, 1987,1988). Thus, internodal Na$^+$ channels were present prior to lysolecithin exposure or reached a steady level within 24 hr, when myelin damage is confined to a few discrete loci. From estimates of axolemmal area in the loose-patch pipette, the density of Na$^+$ channels in the internode is calculated to be about 4% of that at the node. This number

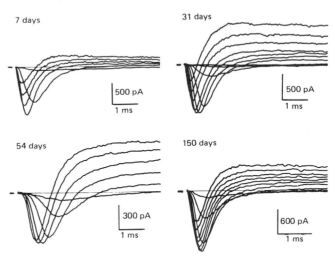

FIGURE 1 Loose patch-clamp recording of internodal ionic currents from *Xenopus* axons. Fibers at 7 and 31 days were exposed to lysolecithin once. In the other cases, the sciatic nerve was surgically re-injected with a low concentration of lysolecithin to re-expose the axolemma. Holding (V_h) and test (V_t) potentials (mV relative to rest) follow. 7 days: $V_h = -20$; $V_t = 40, 60, 80, 100, 120, 140$. 31 days: $V_h = 0$; $V_t = 30, 40, 50, 60, 70, 80, 90, 110, 130, 150$. 54 days: $V_h = -20$; $V_t = 40, 50, 60, 70, 80, 90$. 150 days: $V_h = -40$; $V_t = 70, 80, 90, 100, 110, 120, 130, 140, 150, 160$. Reproduced with permission from the *Journal of Physiology* (Shrager, 1988).

may, at first, appear to be insignificant. However, the best estimates of nodal density are in the range 2000–3000/μm^2 (Hille, 1992), suggesting that the internodal density is on the order of 80–120/μm^2. This number is similar to estimates made in unmyelinated fibers of the same caliber, and therefore may be adequate to support conduction if the demyelinated region can be depolarized sufficiently to activate them. Finally, since the internode is about 1000 times the length of the node, these results suggest that, of the *total* Na$^+$ channels in the axolemma of a myelinated axon, most are internodal and only 2–3% are found at the nodes.

Voltage-dependent K$^+$ currents were also detected at internodal sites. These currents were only partially blocked by 7 mM tetraethylammonium ion (TEA$^+$), a level sufficient to block all delayed-rectifier current at frog nodes (Hille, 1967). The loose patch clamp is inadequate for studying slowly activating currents; K$^+$ channels are much better analyzed with single-channel recording. Experiments with gigohm seal patch-clamp procedures have been conducted in several laboratories and are described in detail in a later section.

At internodal sites, Na$^+$ currents were well behaved kinetically, suggesting adequate potential control, but when the patch pipette was applied to paranodal zones, results were often quite different. Figure 2 illustrates traces recorded from "transition nodes," sites often found at the periphery of the injected zone at which the myelin was damaged and retracted on only one side, exposing the paranodal region. A depolarization of 73 mV from the resting level produced a normal ionic current, but increasing the pulse height by just 1 mV resulted in a large transient outward current. This pattern was then changed little by an

additional 6-mV increase. These currents are interpreted as follows. At the patch site (paranodal), the density of Na$^+$ channels is low, but just outside this zone is the nodal axolemma, with a very high concentration of channels. During depolarization, patch Na$^+$ channels are activated and an inward current develops. This current spreads internally and a portion of it reaches the node, depolarizing it. If this voltage change is sufficient, the nodal channels open, producing a large inward current. This current also spreads; a fraction of it enters the pipette and is recorded as outward current through the patch. The extremely sharp voltage dependence (zero to full activation over 1 mV) argues against the outward transient current resulting from an inactivating K$^+$ channel. In support of this interpretation, these records were obtained *only* at paranodal sites and never at internodal zones. Thus, the results provide evidence for a sharp gradient in Na$^+$ channel density at nodes of Ranvier. An interesting point is that this pattern was encountered only a few days postinjection and persisted for up to 5 mo (the longest times tested), long after the formation of nodes of Ranvier during remyelination. Thus, if lateral diffusion of Na$^+$ channels from nodes occurs, it must involve only a fraction of the nodal channels. Subsequently, researchers have shown in a computational model that such lateral diffusion would not be helpful in restoring conduction in demyelinated or remyelinating fibers (Hines and Shrager, 1991).

Results of studies on mammalian axons were similar to those in *Xenopus* in most details. Experiments were made more difficult by the smaller diameter of rat sciatic axons, particularly after demyelination. Voltage-dependent Na$^+$ currents were recorded from internodes (Fig. 3) and their peak amplitude did not change over the first week postinjection, again pointing to their presence in the normal axon. At paranodes, transient outward currents reflective of a sharp

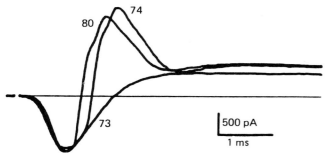

FIGURE 2 Ionic currents recorded a few μm from a half-demyelinated node, 57 days after injection of lysolecithin. The holding potential was equal to V_{rest}; the patch was depolarized by the amount shown next to each trace (mV). The Ringer solution contained 7.4 mM tetraethylammonium (TEA). Reproduced with permission from the *Journal of Physiology* (Shrager, 1988).

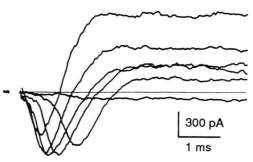

FIGURE 3 Loose patch-clamp recording of internodal currents from a rat internode, 3 days after exposure to lysolecithin. Relative to rest (mV): $V_h = -30$; $V_t = 40, 60, 70, 80, 90, 100$. Reproduced with permission from *Brain Research* (Shrager, 1989).

gradient in Na⁺ channel density were recorded (Fig. 4). As in the frog, the currents were recorded well into the period of remyelination, suggesting again that lateral diffusion of these channels is limited. Two results argue against significant fusion of Schwann cell membranes with the axolemma as a source of Na⁺ channels, as was postulated by Chiu and Schwarz (1987). First, electron micrographs provide no evidence for membrane fusion, and no unusual folding of the axolemma to accommodate the larger area. Demyelinated axons are of small diameter, particularly in the rat. Second, in many axons myelin is loosened, but not vesiculated, by lysolecithin treatment prior to removal by macrophages. Chiu and Schwarz (1987) calculated that, in the rabbit, 30 times more total Na⁺ channels are found in internodes than in nodes. In the rat, the estimated ratio is 20–25, which is considered in good agreement considering the uncertainties in both measurements.

In an additional test of the origin of internodal Na⁺ channels, *Xenopus* axons have been examined after transection of the sciatic nerve. Whereas in the mammal the distal stump degenerates in 1–2 days, in the frog it survives for over 1 wk with conduction intact (Wang, 1985). Rubinstein and Shrager (1990) cut all branches of the sciatic nerve in one leg near the spinal cord, and then injected lysolecithin at a more distal site in either the peroneal or tibial branch; the other site served as a control for degeneration. The nerve was dissected 6 days later and was tested with the loose patch clamp. Na⁺ currents, with peak amplitudes similar to those of uncut axons, were recorded from demyelinated internodes. These Na⁺ channels could not have been synthesized in response to the demyelination since transection preceded demyelination. These studies were carried out over the entire period of recovery by dividing it into 6-day segments. Remyelination, including formation of functional nodes of Ranvier, occurred on schedule, despite the loss of continuous communication with the cell body. Thus, these events are under local control.

Therefore, the internodal axolemma is likely normally to contain a significant density of Na⁺ channels. No evidence is available that these channels are activated or play any role in normal conduction. However, in demyelinated fibers, these channels may be involved in action potential propagation. This possibility has been tested in two ways. First, the loose patch clamp was used to record ionic currents generated by a propagating impulse that invaded a demyelinated region. The pipette was sealed to a demyelinated site in the usual way, and the patch was held at the resting potential. The nerve was stimulated several millimeters proximal to the patch. Since the patch clamp only controlled the external potential, a propagating signal could depolarize the patch internally. If Na⁺ channels were activated, an all-or-none transient inward current should result. An example is given in Fig. 5, which illustrates results at a transition node. Two traces are superimposed, differing only slightly in stimulus amplitude. The initial biphasic response is the stimulus artifact, and differs negligibly in the two traces. The higher stimulus, however, produces a second biphasic current. The initial outward component is likely to be largely capacitative, and the later large inward current results primarily from the activation of Na⁺ channels. In Fig. 6, the patch site was a developing new node of Ranvier in an axon 37 days postinjection. The upper trace represents the compound action potential; the more delayed components result from remyelinating axons. The lower record is the patch current, and corresponds in time to the latter remyelinating fibers. The large inward current suggests that a significant accumulation of Na⁺ channels has already occurred at this site. Such patches could also be depolarized by pulses to the clamp in

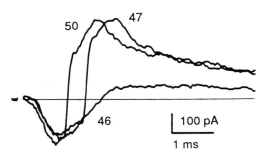

FIGURE 4 Ionic currents a few μm from a half-demyelinated rat node, 13 days after surgery. The patch was held at rest and depolarized by the amounts shown (mV). Reproduced with permission from *Brain Research* (Shrager, 1989).

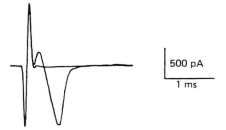

FIGURE 5 Loose patch-clamp recording of membrane current in response to a propagating impulse in *Xenopus*, 4 days after lysolecithin exposure. Half-demyelinated node (paranode) at the proximal end of the demyelinated zone. Two traces are superimposed; the stimulus amplitude differs by only a small amount. The initial biphasic current represents the stimulus artifact. Reproduced with permission from the *Journal of Physiology* (Shrager, 1987).

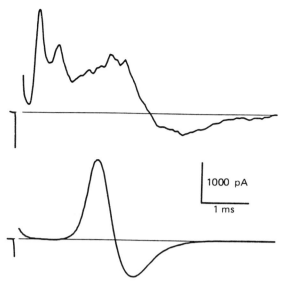

FIGURE 6 Currents from a new node of Ranvier, forming in a remyelinating *Xenopus* axon 37 days after surgery. The top trace is the compound action potential, recorded about 5 mm distal to the patch. The lower trace represents all-or-none current at the loose patch pipette in response to external stimulation, 9 mm proximal to the site. The time scale applies to both records. The current scale is for the lower trace only. Reproduced with permission from the *Journal of Physiology* (Shrager, 1988).

the usual manner, allowing an assessment of the maximum Na^+ current available. The combined results thus provided a measure of the fractional activation of Na^+ channels at specific sites by propagating signals. At very early stages of remyelination, when Schwann cells are just beginning to elaborate their first lamellae, conduction is boosted by Na^+ channel activation, but may be decremental over the initial segment of the internode (Shrager, 1988).

III. Internodal Channel Function Analyzed by Optical Recording of Propagating Action Potentials in Demyelinated Fibers

In multiple sclerosis, a demyelinating disease of the central nervous system (CNS), patients often experience an exacerbating–remitting course. Recovery from periods of markedly increased neurological deficits may occur over a period of weeks. Remyelination in the CNS is rare and incomplete; thus, how axonal conduction recovers so rapidly remains unclear. To better judge the requirements for successful conduction in demyelinated axons, optical recording of propagating signals in single axons has been employed. The preparation was identical to that used for loose-patch recording, except that axons were stained

with a voltage-sensitive dye. The fractional change in transmitted light was measured by imaging a node or a 15-μm length of demyelinated axon on a photodiode. By signal averaging 8–64 sweeps, action potentials and subthreshold signals at these sites could be recorded. We have attempted to evaluate the possible role of internodal Na^+ channels in recovery from demyelination. Very briefly summarized, conduction is blocked in completely demyelinated internodes, but can be restored at low velocity by a very minimal level of remyelination. In fact, the first incomplete wrap of newly proliferating Schwann cells may suffice to allow the most proximal segment of the demyelinated region to reach threshold. Conduction is then continuous through the demyelinated zone.

Optical signals have been recorded at numerous sites along axons at various stages of demyelination and remyelination. Conduction has also been modelled mathematically, using a computer simulation that allowed independent specification of axolemmal and Schwann cell–myelin layers (Hines and Shrager, 1991). Two points are of particular interest here. First, the model Na^+ channel density in the internodal axolemma that reproduced the experimental results was in the range measured with the loose patch clamp (2–6% of the nodal density). Second, with this internodal Na^+ channel distribution, conduction was restored with the most minimal Schwann cell association. The optical experiments, combined with the model calculations, thus support the idea that internodal Na^+ channels play an important role in the restoration of conduction during early recovery from demyelination. Despite their low density relative to the nodal region, these channels allow signal propagation through the demyelinated lesion at much earlier times than would be possible in their absence.

IV. Single-Channel Recording from Demyelinated Paranodes and Internodes

Although recording and analyzing internodal Na^+ currents by using the loose patch clamp has been possible, the corresponding characterization of K^+ channels requires other approaches. Two primary limitations of the loose-patch system interfere with K^+ current recording. First, a slow, possibly electrostrictive, current develops during strong depolarizations and interferes with K^+ currents. Second, many important classes of K^+ channels are metabolically modulated or are relatively voltage insensitive and are not easily resolved with macroscopic current measurements. Thus, several laboratories have used gigohm seal single-channel patch-clamp recording techniques to analyze K⁺ channels. The preparation utilized by Vogel's laboratory consists of frog or rat sciatic axons with myelin retracted enzymatically to expose the nodal and paranodal axolemma. These investigators found voltage-gated Na⁺ channels and delayed-rectifier K⁺ channels with fast, intermediate, and slow deactivation kinetics, a Ca^{2+}-activated K⁺ channel with a single channel conductance of 132 pS, and a 44-pS ATP-sensitive K⁺ channel (Jonas *et al.*, 1989,1991).

We have applied single-channel analysis to *Xenopus* internodes demyelinated by lysolecithin exposure and macrophage phagocytosis. The main advantage of this approach over the one used by Vogel's group is that accessibility is extended over the entire internode and is not restricted to the node ± 50 μm, as is the case for enzymatic retraction. Figure 7 illustrates a demyelinated internode with a patch pipette in the initial cell-attached position. The patch clamp was used in the excised inside-out configuration (Hamill *et al.*, 1981) to allow good potential control and manipulation of the internal solution. Except where noted or modified, solutions contained symmetrical KCl concentrations of 115 mM. Four distinct classes of K⁺ channels have been characterized in demyelinated internodes. A K⁺-selective channel with a single-channel conductance of 125 pS was inhibited by application of 2 mM ATP to the cytoplasmic surface. Activation of this channel was insensitive to voltage. Figure 8A illustrates inward currents at −60 mV in the absence (*top*) and presence (*bottom*) of 2 mM ATP. Membrane potential independence is shown in the response to a voltage ramp in Fig. 8B. The frequency of channel openings is relatively constant over the voltage range tested (±100 mV). This channel may be

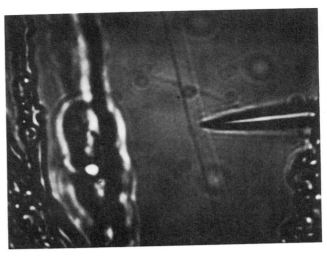

FIGURE 7 Light micrograph of a demyelinated *Xenopus* axon, 6 days after injection of lysolecithin. At this stage, myelin removal is complete and Schwann cell attachment has not yet begun. The axon diameter is 7 μm. A patch pipette is sealed to the axon surface.

important in maintenance of the resting potential. In metabolic deficit, if [ATP]$_i$ declines, this channel would be activated, helping hold V_m at a strong negative level despite the possible loss of the electrogenic component of the Na⁺/K⁺ ATPase.

A Ca^{2+}-activated K⁺ channel [K(CA)] has been characterized in internodes. The channel is activated by both [Ca^{2+}]$_i$ and V_m depolarization. At +40 mV, half activation is achieved at [Ca^{2+}]$_i$ = 10^{-5} M. Records at this potential are illustrated in Fig. 9 for a patch containing two K(Ca) channels. [Ca^{2+}]$_i$ for each trace is given on the left. At 10^{-7} M, the channels are closed; at 2×10^{-3} M, they are open most of the

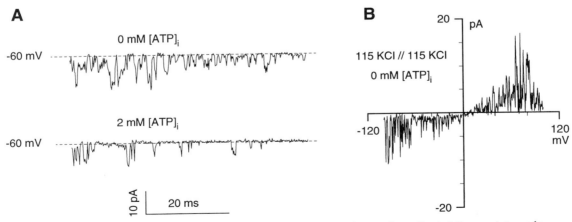

FIGURE 8 An ATP-inhibited K⁺ channel in an inside-out patch from a demyelinated *Xenopus* internode. The pipette and bath solutions are symmetrical 115 mM KCl. (A) Current recorded in the absence (upper trace) and presence (lower trace) of ATP. (B) Current recorded in response to a voltage ramp from −100 to +100 mV over 600 msec.

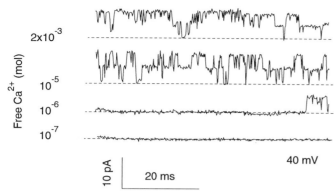

FIGURE 9 Outward currents from a Ca²⁺-activated K⁺ channel at several levels of $[Ca^{2+}]_i$ (given at the left of each trace). The K⁺ concentration in pipette and bath solutions is 115 mM KCl. $V_m = +40$ mV.

time. These channels were often found in clusters of 4–6 per patch; the single-channel conductance was 235 pS. The role of internodal K(Ca) channels in excitation is unknown, but a number of intriguing possibilities exist. Lev Ram and Grinvald (1987) demonstrated increased $[Ca^{2+}]_i$ in rat optic nerve in response to depolarization, suggesting the presence of voltage-gated Ca^{2+} channels in the axolemma. Subsequent activation of K(Ca) channels could increase the afterhyperpolarization that follows strong repetitive activity, and thus could render the axon less excitable (Bostock and Grafe, 1985).

Two additional internodal K⁺ channels have been characterized, but are mentioned here only briefly. A 176-pS K⁺ channel had an outwardly rectifying unitary current in symmetrical solutions but an ohmic ensemble-averaged I–V curve. This channel had a weak voltage dependence of the open probability: 220 mV per *e*-fold change. The nearly symmetrical

TABLE I Ionic Channels in Myelinated Axons

Channel type	Conductance (pS)	Nodal/paranodal presence	Internodal presence	Voltage dependence (activation)	Blocker(s)	Reference
Na⁺	7–11	+ + + + + + + + + + + + +	+	> -70 mV	TTX, LA	Shrager (1987, 1988, 1989) Jonas *et al.* (1989)
$K^+_{DR}(F)$	30	+		-60 to -40 mV	TEA	Jonas *et al.* (1989)
$K^+_{DR}(I)$	23	+ + + +		-70 to -40 mV	TEA, DTX	Chiu and Ritchie (1982)
	28		+ + +	-41 mV, *e* fold/8 mV	TEA, 4-AP	Shrager (1987, 1988) Jonas *et al.* (1989) Wu *et al.* (1993a,b)
$K^+_{DR}(S)$	7	+ + +		> -90 mV	TEA	Jonas *et al.* (1989)
K^+_{Ca}	132	+ +		> -50 mV	TEA	Jonas *et al.* (1991)
	235		+ + + +	> -24 mV, *e* fold/20 mV	TEA	Wu *et al.* (1993a,b)
K^+_{ATP}	44	+			TEA, Glib.	Jonas *et al.* (1991)
	125		+	independent		Wu *et al.* (1993a,b)
K^+_{BG}	176		+ +	< 0 mV, *e* fold/220 mV		Wu *et al.* (1993a,b)
Cl^-_{IR}	50		+ +	independent	SITS, DIDS	Wu and Shrager (1994)
Cl^-_{OR}	33–65	+ +		independent		Strupp and Grafe (1991)
	50		+ + +	> -45 mV, *e* fold/42 mV		Wu and Shrager (1994)
Cl^-_{MAX}	300–400		+	closed > 40 mV		Wu and Shrager (1994)

Abbreviations: 4-AP, 4-aminopyridine; DIDS, 4,4'-diisothiocyanatostilbene-2,2'-disulfonic acid; DTX, dendrotoxin; Glib, glibenclamide; LA, local anesthetics; SITS, 4-acetamido-4'-isothiocyanatostilbene-2,2'-disulfonic acid; TEA, tetraethylammonium; TTX, tetrodotoxin.

macroscopic I–V curve of the background channel suggests that this channel may play a role in maintaining the axonal resting potential. A delayed-rectifier K^+ channel with a single-channel conductance of 28 pS has been found. This channel is blocked internally by 2 mM 4-aminopyridine and by 10 mM TEA. The potential for half activation of this channel is −41 mV and its voltage sensitivity is 8 mV per e-fold change in open probability. Delayed rectifiers are possible targets for pharmacological intervention in demyelinating disease, since they have little influence on the resting potential but may strongly alter the threshold for action potential generation (Davis *et al.*, 1990).

V. Summary

Table I summarizes results on the distribution and role of ionic channels in myelinated nerve fibers. Clearly, the heterogeneity in species and geography of these channels is far greater than was originally apparent from the early gap voltage-clamp measurements on nodes. Several important questions remain, and are being avidly investigated. First what role, if any, do the internodal channels play in normal nondemyelinated axons. The periaxonal space between axon and Schwann cell–myelin constitutes the immediate extracellular environment of the internodal axon. This space is very narrow (typically 100–200 Å), and its ionic composition is thus easily altered. Some of the axonal internodal channels may by involved in maintaining homeostasis in this region, which would be important for stability of the resting potential. In this regard, the fact that the classical model of myelin as electrically completely isolating the internodal axolemma has been challenged is of considerable interest (Barrett and Barrett, 1982; Funch and Faber, 1984; Blight, 1985; Bostock and Grafe, 1985; Shrager, 1993). If current pathways do exist between the internodal axon and the extracellular medium, the ionic channels in this region might play important roles in excitability in normal axons. The second interesting question concerns the mechanism by which ionic channels are spatially segregated along the axon. Particularly striking is the distribution of Na^+ channels, which are concentrated about 25-fold at nodes of Ranvier. Researchers have suggested that cytoskeletal association is important; evidence linking Na^+ channels with ankyrin reveals important progress toward elucidating the mechanism involved (Srinivasan *et al.*, 1988). Finally, the ability to influence ion channel function may lead to new therapeutic procedures in demyelinating disease. Current trials with 4-aminopyridine are encour-

aging (Davis *et al.*, 1990), but this drug is too nonspecific and is of too low an affinity. Biophysical experiments aimed at identifying compounds that preferentially modulate internodal channels in ways that improve conduction may ultimately produce better drugs.

Acknowledgments

Work in the authors' laboratory has been supported by Grant NS-17965 from the National Institutes of Health, Grant RG-1774 from the National Multiple Sclerosis Society, and a Messersmith Fellowship.

References

Barrett, E. F., and Barrett, J. N. (1982). Intracellular recording from vertebrate myelinated axons: Mechanism of the depolarizing afterpotential. *J. Physiol.* **323**, 117–144.

Blight, A. R. (1985). Computer simulation of action potentials and afterpotentials in mammalian myelinated axons: The case for a lower resistance myelin sheath. *Neurosci.* **15**, 13–31.

Bostock, H., and Grafe, P. (1985). Activity-dependent excitability changes in normal and demyelinated rat spinal root axons. *J. Physiol.* **365**, 239–257.

Chiu, S. Y. (1987). Sodium currents in axon-associated Schwann cells from adult rabbits. *J. Physiol.* **386**, 181–203.

Chiu, S. Y., and Ritchie, J. M. (1981). Evidence for the presence of potassium channels in the paranodal region of acutely demyelinated mammalian single nerve fibres. *J. Physiol.* **313**, 415–437.

Chiu, S. Y., and Ritchie, J. M. (1982). Evidence for the presence of potassium channels in the internode of frog myelinated nerve fibres. *J. Physiol.* **322**, 485–501.

Chiu, S. Y., and Schwarz, W. (1987). Sodium and potassium currents in acutely demyelinated internodes of rabbit sciatic nerves. *J. Physiol.* **391**, 631–649.

Chiu, S. Y., Shrager, P., and Ritchie, J. M. (1984). Neuronal-type Na^+ and K^+ channels in rabbit cultured Schwann cells. *Nature (London)* **311**, 156–157.

Davis, F. A., Stéfoski, D., and Rush, J. (1990). Orally administered 4-aminopyridine improves clinical signs in multiple sclerosis [see comments]. *Ann. Neurol.* **27**, 186–192.

Dodge, F., and Frankenhaeuser, B. (1958). Membrane currents in isolated frog nerve fibre under voltage clamp conditions. *J. Physiol.* **143**, 76–90.

Dodge, F., and Frankenhaeuser, B. (1959). Sodium currents in the myelinated nerve fibre of *Xenopus laevis* investigated with the voltage clamp technique. *J. Physiol.* **148**, 188–200.

Funch, P. G., and Faber, D. S. (1984). Measurement of myelin sheath resistances: Implications for axonal conduction and pathophysiology. *Science* **225**, 538–540.

Grissmer, S. (1985). Sodium currents in the isolated voltage-clamped frog internode. *J. Physiol.* **372**, 53P.

Hall, S. M., and Gregson, N. A. (1971). The *in vivo* and ultrastructural effects of injection of lysophosphatidyl choline into myelinated peripheral nerve fibres of the adult mouse. *J. Cell Sci.* **9**, 769–789.

Hamill, O. P., Marty, A., Neher, E., Sakmann, B., and Sigworth, F. J. (1981). Improved patch-clamp techniques for high-resolution current recording from cells and cell-free membrane patches. *Pflügers Arch. (Eur. J. Physiol.)* **391**, 85–100.

Hille, B. (1967). The selective inhibition of delayed potassium currents in nerve by tetraethylammonium ion. *J. Gen. Physiol.* **50**, 1287–1302.

Hille, B. (1992). "Ionic Channels of Excitable Membranes." Sinauer Associates, Sunderland, Massachusetts.

Hines, M., and Shrager, P. (1991). A computational test of the requirements for conduction in demyelinated axons. *Restor. Neurol. Neurosci.* **3**, 81–93.

Howe, J. R., and Ritchie, J. M. (1990). Sodium currents in Schwann cells from myelinated and non-myelinated nerves of neonatal and adult rabbits. *J. Physiol.* **425**, 169–210.

Jonas, P., Brau, M. E., Hermsteiner, M., and Vogel, W. (1989). Single-channel recording in myelinated nerve fibers reveals one type of Na channel but different K channels. *Proc. Natl. Acad. Sci. USA* **86**, 7238–7242.

Jonas, P., Koh, D. S., Kampe, K., Hermsteiner, M., and Vogel, W. (1991). ATP-sensitive and Ca-activated K channels in vertebrate axons: Novel links between metabolism and excitability. *Pflügers Arch.* (*Eur. J. Physiol.*) **418**, 68–73.

Lev-Ram, V., and Grinvald, A. (1987). Activity-dependent calcium transients in central nervous system myelinated axons revealed by the calcium indicator Fura-2. *Biophys. J.* **52**, 571–576.

Quick, D. C., and Waxman, S. G. (1977). Specific staining of the axon membrane at nodes of Ranvier with ferric ion and ferrocyanide. *J. Neurol. Sci.* **31**, 1–11.

Ritchie, J. M., and Rang, H. P. (1983). Extraneuronal saxitoxin binding sites in rabbit myelinated nerve. *Proc. Natl. Acad. Sci. USA* **80**, 2803–2807.

Ritchie, J. M., and Rogart, R. B. (1977). Density of sodium channels in mammalian myelinated nerve fibers and nature of the axonal membrane under the myelin sheath. *Proc. Natl. Acad. Sci. USA* **74**, 211–215.

Rosenbluth, J. (1976). Intramembranous particle distribution at the node of Ranvier and adjacent axolemma in myelinated axons of the frog brain. *J. Neurocytol.* **5**, 731–745.

Rubinstein, C. T., and Shrager, P. (1990). Remyelination of nerve fibers in the transected frog sciatic nerve. *Brain Res.* **524**, 303–312.

Shrager, P. (1987). The distribution of sodium and potassium channels in single demyelinated axons of the frog. *J. Physiol.* **392**, 587–602.

Shrager, P. (1988). Ionic channels and signal conduction in single remyelinating frog nerve fibres. *J. Physiol.* **404**, 695–712.

Shrager, P. (1989). Sodium channels in single demyelinated mammalian axons. *Brain Res.* **483**, 149–154.

Shrager, P. (1993). Axonal coding of action potentials in demyelinated nerve fibers. *Brain Res.* **619**, 278–290.

Shrager, P., Chiu, S. Y., and Ritchie, J. M. (1985). Voltage-dependent sodium and potassium channels in mammalian cultured Schwann cells. *Proc. Natl. Acad. Sci. USA* **82**, 948–952.

Srinivasan, Y., Elmer, L., Davis, J., Bennett, V., and Angelides, K. (1988). Ankyrin and spectrin associate with voltage-dependent sodium channels in brain. *Nature* (*London*) **333**, 177–180.

Strupp, M., and Grafe, P. (1991). A chloride channel in rat and human axons. *Neurosci. Lett.* **133**, 237–240.

Stuhmer, W., Roberts, W. M., and Almers, W. (1983). The loose patch clamp. *In* "Single-Channel Recording" (B. Sakmann and E. Neher, eds.), pp. 123–132. Plenum Press, New York.

Tao-Cheng, J. H., and Rosenbluth, J. (1980). Nodal and paranodal membrane structure in complementary freeze-fracture replicas of amphibian peripheral nerves. *Brain Res.* **199**, 249–265.

Wang, G. K. (1985). The long-term excitability of myelinated nerve fibres in the transected frog sciatic nerve. *J. Physiol.* **368**, 309–321.

Waxman, S. G., and Quick, D. C. (1978). Intra-axonal ferric ion-ferrocyanide staining of nodes of Ranvier and initial segments in central myelinated fibers. *Brain Res.* **144**, 1–10.

Wu, J. V., and Shrager, P. (1994). Three types of chloride channels in the internodes of demyelinated frog axons. *J. Neurosci. Res.* In press.

Wu, J. V., Rubinstein, C. T., and Shrager, P. (1993a). Single channel characterization of multiple types of K+ channels in the internode of frog axons. *Biophys. J.* **64**, 100a.

Wu, J. V., Rubinstein, C. T., and Shrager, P. (1993b). Single channel characterization of multiple types of potassium channels in demyelinated Xenopus axons. *J. Neurosci.* **13**, 5153–5163.

Cardiac Sodium Channel Expression in Xenopus Oocytes

DOUGLAS S. KRAFTE

I. Introduction

Xenopus laevis oocytes provide one type of heterologous expression system in which exogenous proteins can be translated, inserted into plasma membranes, and quantitatively studied. Largely because of the technical ease with which an oocyte can be impaled with microelectrodes, this system has proven particularly amenable to the electrophysiological assessment of ion channel function (for review, see Dascal, 1987; Sigel, 1990). Initial studies were primarily directed toward expression of ion channels and neurotransmitter receptors after injection of RNA derived from brain tissue (Sumikawa *et al.*, 1981; Gundersen *et al.*, 1983). More recently, channels have been expressed from cRNA of cloned genes as well as tissue RNA derived from nonneuronal sources. Ion channels of cardiac origin fall into this latter category and can also be expressed in *Xenopus* oocytes (Krafte *et al.*, 1991), although in certain instances modifications of standard expression techniques are necessary. This chapter gives a brief introduction to sodium channel function in the heart and describes the methodology required to achieve reliable cardiac ion channel expression in *Xenopus* oocytes. In addition, the physiological and pharmacological properties of cardiac Na$^+$ channels expressed in oocytes from tissue-derived RNA and cRNA are discussed and compared.

II. Sodium Channel Function in the Heart

After stimulation of the myocardium, Na$^+$ conductance is activated that drives the upstroke of the cardiac action potential and provides the majority of current that supports impulse conduction. The magnitude of this conductance, resulting from flow of Na$^+$ ions through voltage-dependent Na$^+$ channels, determines how quickly excitation will spread throughout the heart. Soon after Na$^+$ channels open, they begin to close through a process known as inactivation (see Chapter 8). The inactivation process is relatively fast ($\tau \leq 5$ msec) but, at certain voltages, the probability of finding any one channel in the inactivated state is less than 1. Since the voltage range in which this phenomenon occurs overlaps with the voltages at which channels activate or open, the probability of finding an open Na$^+$ channel several hundred milliseconds after a stimulus is finite. Simply stated, Na$^+$ channels can be open during the plateau of the cardiac action potential and, therefore, can influence action potential duration in addition to upstroke velocity. Prolongation of action potential duration can affect cardiac refractoriness and contractility, the latter by allowing longer periods of time for Ca^{2+} to enter the cell.

In addition to direct effects on cardiac action potentials, Na^+ flow can also affect $[Ca^{2+}]_i$ since the Na^+ and Ca^{2+} electrochemical gradients are coupled. One of the major pathways used by a myocardial cell to regulate intracellular Ca^{2+} is electrogenic Na^+/Ca^{2+} exchange, taking advantage of the Na^+ gradient to pump Ca^{2+} out of the cell (Kimura *et al.*, 1987). Consequently, even transient increases in $[Na^+]_i$, as a result of channel opening, can potentially increase $[Ca^{2+}]_i$, thereby enhancing cardiac contractility. Conversely, decreases in $[Na^+]_i$ due to blockade of Na^+ channels can potentially decrease $[Ca^{2+}]_i$ and depress cardiac contractility.

Since voltage-dependent Na^+ channels play a central role in regulating cardiac rhythm and contractility, significant research efforts have been focused on this channel, from both physiological and pharmacological perspectives. Results also indicate that subtypes of Na^+ channels may exist in the heart (George *et al.*, 1992). In particular, one channel type appears to be active during the action potential plateau because of a lack of inactivation (Saint *et al.*, 1992). The effects of this channel on myocardial properties would be analogous to those discussed earlier for classical voltage-dependent channels for which the probability of inactivation is <1.

Cardiac Na^+ channels have several characteristics that allow them to be distinguished from Na^+ channels found in nerve and innervated skeletal muscle, the most discriminating of which are pharmacological properties. Cardiac channels are 100-fold less sensitive to block by tetrodotoxin (TTX; Cohen *et al.*, 1981) and are 100-fold more sensitive to block by Cd^{2+} and Zn^{2+} (Frelin *et al.*, 1986) than channels found in nerve. In addition, molecules such as DPI 201-106 selectively slow inactivation of cardiac Na^+ channels (see Scholtysik, 1989). Biophysical properties of Na^+ channels tend to be similar among the different channel types, but cardiac Na^+ channels inactivate more slowly than those present in noncardiac tissues (Colatsky, 1980). This "fingerprint" of the cardiac Na^+ channel has proven useful in experimentally confirming expression of these channels in *Xenopus* oocytes.

Alternative methods to investigate channels can often provide new insights into function. Heterologous expression systems, particularly the *Xenopus* oocyte, have played an important role in these studies. The oocyte system becomes particularly useful when expressing different subtypes of channels derived from cDNA clones and in studies of structure–function relationships (e.g., Pongs, 1992).

III. Method to Express Cardiac Sodium Channels from Tissue RNA

Standard techniques for ion channel expression in *Xenopus* oocytes involve intracellular injection of RNA followed by incubation at 21°C for several days (Soreq and Seidman, 1992). Expression of sufficient numbers of channels or receptors for quantitative study can be achieved in as little as 24 hr using cRNAs encoding a specific protein (e.g., Na^+ channel α subunits) or as long as 4–10 days for other types of channels, particularly when using tissue-derived RNA. Incubation at 21°C, however, does not lead to significant expression of cardiac Na^+ channels after injection of total RNA (i.e., not enriched by oligo-dT chromatography). If oocytes are incubated at higher temperatures for defined periods, however, reliable expression of cardiac channels can be achieved. Figure 1 illustrates the effects of incubation at 30°C on the expression of cardiac Na^+ channels in *Xenopus* oocytes assessed by measuring maximum current amplitudes. Incubation of cells at 30°C for defined periods causes a 20-fold increase in channel expression relative to incubation at room temperature. Figure 1 also illustrates that uninjected oocytes are not induced to express Na^+ channels after incubation at 30°C, provided voltage-dependent Na^+ channels are not present in the control population (Krafte and Volberg, 1992). The ideal protocol to mimic translation and processing of Na^+ channels in mammalian cells would be to incubate oocytes contin-

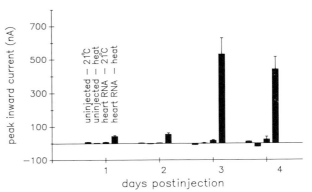

FIGURE 1 Expression of cardiac Na^+ channels under different incubation protocols. Oocytes were injected with 250 μg total guinea pig heart RNA and then incubated for 4 days. Na^+ currents were measured using standard two-microelectrode voltage-clamp techniques. Currents were measured in oocytes incubated at 21°C for 4 days or under a modified protocol in which oocytes were incubated overnight at 30°C on the day of injection and two days later. As controls, uninjected oocytes were carried through the same protocols.

uously at 37°C; however, oocytes will not tolerate prolonged incubation at this temperature. Consequently, the empirically derived protocol illustrated in Fig. 1 was found to be the most reliable procedure for reproducible expression of cardiac channels. Incubation of *Xenopus* oocytes at temperatures above 21°C may also prove useful for expression of ion channels from noncardiac sources.

The application of molecular biological techniques to the study of ion channels has identified genes for many different channels and membrane receptors. With respect to cardiac Na$^+$ channels, cDNAs have been cloned from rat heart (Rogart *et al.*, 1989) and human heart (Gellens *et al.*, 1992). These cDNAs encode the α subunit of the Na$^+$ channel multisubunit complex; *in vitro* transcripts direct expression of functional voltage-dependent Na$^+$ channels in *Xenopus* oocytes. When using cRNA, the 30°C incubation protocol noted earlier need not be used. Reliable expression can be achieved over short periods of time ($\geq$24 hr). These results indicate that the temperature dependence of expression for Na$^+$ channels derived from tissue RNA does not arise from an inherent property of the channel α subunit itself, but suggests the presence of an inhibitory element encoded in the total RNA population. To date, this phenomenon has only been observed when expressing cardiac ion channels from tissue-derived RNA.

IV. Electrophysiological Properties of Expressed Cardiac Sodium Channels

Inward currents can be elicited during voltage steps from -100 mV to potentials ≥ -50 mV 3 days after injection of total heart RNA (250 μg) into *Xenopus* oocytes. When using RNA derived from guinea pig cardiac tissue, current amplitudes average -351 ± 37 nA and peak amplitudes are observed near -10 mV. Steady-state inactivation assessed with 50-msec prepulses is characterized by a $V_{1/2}$ value of -49 ± 0.6 mV and a slope factor of 7.3 ± 0.2 mV. Properties are similar for channels derived from rat and human cardiac RNA, although the amplitude of expressed currents is smaller than that observed for Na$^+$ currents expressed from an equal amount of guinea pig cardiac RNA (Krafte *et al.*, 1991). One possible explanation for this observation is a lower abundance of Na$^+$ channel message in rat and human RNA than in RNA from the guinea pig.

When Na$^+$ channels are expressed in *Xenopus* oocytes from cRNAs, the overall voltage-dependent properties are similar to those described earlier for

tissue RNA derived channels. The threshold for activation ranges from -60 to -40 mV (Cribbs *et al.*, 1990; Gellens *et al.*, 1992; Krafte *et al.*, 1994) and peak current amplitudes are observed from -20 to -10 mV. $V_{1/2}$ values for inactivation of human heart Na$^+$ channels (hH1) range from -57 to -62 mV with slope factors of 7.3 to 7.7 mV (Gellens *et al.*, 1992; Krafte *et al.*, 1994). Inactivation properties for the rat heart Na$^+$ channel (RH-I) have also been reported with a $V_{1/2}$ value of -68.5 mV and a slope factor of 6.2 (Satin *et al.*, 1992a). The differences between these values and those for hH1 may be related to the longer prepulses used in the latter study (5 sec vs 45–50 msec) rather than to an inherent difference between hH1 and RH-I α subunits.

Amplitudes of currents can vary a great deal among Na$^+$ channels expressed from cRNAs, but this difference most likely reflects the particular construct utilized for channel expression (see White *et al.*, 1991) or the amount of cRNA injected into each oocyte. As such, direct comparisons among different channel types are not relevant when discussing properties of Na$^+$ channels expressed from cRNAs.

V. Channel Blockade by Tetrodotoxin

One of the classical properties used to distinguish cardiac Na$^+$ channels from those of neuronal and skeletal muscle is a lower affinity for TTX, a blocker of voltage-dependent Na$^+$ channels. As noted earlier, TTX blocks neuronal and skeletal muscle Na$^+$ channels with high affinity (Pappone *et al.*, 1980; Neumcke and Stampfli, 1982), but requires approximately 100-fold greater concentrations to block cardiac channels (Cohen *et al.*, 1981). Figure 2 illustrates the TTX sensitivity of cardiac Na$^+$ channels expressed from guinea pig, rat, and human heart total RNA. The IC$_{50}$ value for channel block ranged from 0.65 to 1.3 μM. Neuronal channels were blocked by TTX with an IC$_{50}$ value of ~10 nM (Krafte *et al.*, 1991). These values are consistent with predictions based on data obtained from cardiac and neuronal cells.

Cardiac Na$^+$ channels derived from cRNAs have comparable IC$_{50}$ values for TTX block. Human heart Na$^+$ channels are 50% blocked at TTX concentrations of 1.8–5.7 μM (Gellens *et al.*, 1992; Krafte *et al.*, 1994) whereas rat heart channels have an IC$_{50}$ value of 1.5 μM (Cribbs *et al.*, 1990). Conversely, Na$^+$ channels expressed from cRNAs encoding neuronal α subunits have IC$_{50}$ values of ~10 nM (Noda *et al.*, 1986). These data indicate that the binding site for TTX is an integral part of the α subunit and that this site differs in cardiac

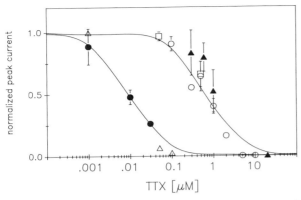

FIGURE 2 Dose–response curves for tetrodotoxin (TTX) block of heart and brain channels. Peak current was normalized to control values and plotted against TTX concentration. Fitted IC_{50} values for guinea pig, rat, human heart, and rat brain Na^+ currents were 651 nM, 931 nM, 1.3 μM, and 9.1 nM, respectively. Fitted curves are shown for guinea pig heart and rat brain only. ●, Rat brain; △, guinea pig brain; ○, guinea pig heart; □, rat heart; ▲, human heart. Reprinted with permission from Krafte et al. (1991).

and neuronal Na^+ channels. The exact nature of the binding site has been elegantly defined using site-directed mutagenesis coupled with expression of Na^+ channels in *Xenopus* oocytes (Satin et al., 1992b). Substitution of a single cysteine with either phenylalanine or tyrosine appears to define the sensitivity for TTX blockade, cysteine being the naturally occurring amino acid in the cardiac Na^+ channel α subunits that have been cloned to date.

VI. Blockade of Expressed Cardiac Sodium Channels by Class I Antiarrhythmic Drugs

Other agents that block Na^+ channels are the Class I antiarrhythmic drugs. These drugs are designed to block cardiac channels and, as a consequence, decrease membrane excitability and slow conduction velocity in the myocardium. The Class I drugs are subdivided into three groups, defined primarily by the rate at which channels recover from block (Rosen and Schwartz, 1991). Representative agents from all three classes have been tested on cardiac Na^+ channels and have been found to be active.

Lidocaine, a drug that shows rapid block and recovery rates, has been tested on both tissue RNA derived and cRNA-derived channels. Lidocaine tonically blocks cardiac Na^+ channels expressed from guinea pig heart RNA with an IC_{50} value of 446 μM (Volberg et al., 1991). When a train of stimuli is applied, however, significant use-dependent block can be observed at concentrations as low as 10 μM (Krafte et al., 1991).

Similar results have been obtained for hH1 channels, with tonic IC_{50} values of 226–521 μM and 60–70% use-dependent block observed at 100 μM (Chahine et al., 1992; Krafte et al., 1994). Satin et al. (1992a) reported approximately 10% tonic block at 100 μM lidocaine concentrations and 33% use-dependent block during a 2-Hz train of stimuli for the RH-I channel. These data indicate similar results for lidocaine block of guinea pig, rat, and human Na^+ channels expressed in the *Xenopus* oocyte.

Our laboratory also investigated block of expressed cardiac Na^+ channels by quinidine, flecainide, and encainide. Channels blocked by quinidine show moderate rates of recovery whereas those blocked by encainide or flecainide recover slowly (Krafte et al., 1994). Quinidine tonically blocks guinea pig and hH1 channels with respective IC_{50} values of 74 μM and 198 μM. Encainide and flecainide are more potent tonic blockers, with IC_{50} values of 30–40 μM (Volberg et al., 1991; Krafte et al., 1994). However, whether channels are expressed from tissue RNA or cRNA, quinidine does not show appreciable use-dependent block at concentrations lower than those that induce tonic block. This result differs from reported data in intact mammalian myocytes, in which use-dependent block is observed at lower concentrations. Whether this phenomenon is a result of expression in *Xenopus* oocytes rather than mammalian cells remains to be determined.

VII. Pharmacological Slowing of Channel Inactivation

Another class of pharmacological agents designed to interact with cardiac Na^+ channels is typified by DPI 201-106 (Buggisch et al., 1985). Rather than block channels, these drugs keep them open longer. As stated at the outset of this chapter, the end result of this effect is to increase cardiac refractoriness and contractility. Figure 3 illustrates the effect of 10 μM DPI 201-106 on hH1 Na^+ currents. In the presence of drug, channel inactivation is slowed and/or removed and channels do not close until the membrane potential is returned to -100 mV. DPI 201-106 is effective on guinea pig cardiac Na^+ channels and hH1 channels. Agents such as DPI 201-106 also reduce peak current amplitude and block channels in a use-dependent manner during a train of stimuli.

A series of polypeptide toxins has also been shown to interact with Na^+ channels to slow inactivation (Narahashi and Herman, 1992). In our laboratory, the scorpion venom from *Leiurus quinquestriatus*, as well as sea anemone toxins such as *Anemonia sulcata* toxin

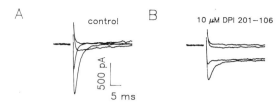

FIGURE 3 Effects of DPI 201-106 on hH1 Na⁺ currents expressed in *Xenopus* oocytes. (A) Control current records elicited by test depolarizations from −100 to −60 mV, −40 mV, −20 mV, and 0 mV. (B) Currents elicited by the same test depolarizations done in A following a 10-min exposure to 10 μM DPI 201-106.

II and anthopleurin A, was shown to slow inactivation of cardiac Na⁺ channels expressed in oocytes. The interaction between these polypeptides and the Na⁺ channel is thought to occur at a site unique from that at which DPI 201-106 interacts with the channel (Scholtysik *et al.*, 1986). Consistent with action at a different site, polypeptide toxins do not reduce current amplitudes as do agents such as DPI 201-106.

VIII. Summary

Expression of cardiac Na⁺ channels in a heterologous system such as the *Xenopus* oocyte will allow channel properties to be examined in greater detail than previously possible. The effects of adding or subtracting different subunits can be examined, as has been reported for neuronal Na⁺ channels (Isom *et al.*, 1992). In addition, expressed channels retain binding sites for Class I antiarrhythmic drugs, DPI 201-106, and a variety of toxins. The structure–function studies initiated by Satin *et al.* (1992b) can now be extended to investigate the sites of interaction between these agents and voltage-dependent Na⁺ channels. Definition of the binding site for Class I antiarrhythmic drugs may give new insight into permeation mechanisms of Na⁺ channels, since these agents block conductance through the channel. Perhaps of more interest from a therapeutic perspective, definition of the binding site for DPI 201-106 and site III neurotoxins may lead to new agents that modify channel inactivation. This type of molecule, at least in principle, offers the potential for improved antiarrhythmic therapy with a concomitant enhancement of contractility. Discovery and development of such agents would expand the repertoire of drugs available for the treatment of impaired myocardial function.

Regardless of the direction taken by such studies, expression of Na⁺ and other ion channels will undoubtedly continue to expand our knowledge of channel structure–function relationships (see Chapter 2).

By capitalizing on this information, we may be able to design a new generation of therapeutic agents that modulate ion channel function for the treatment of different diseases.

Acknowledgments

The author wishes to thank Paul J. Silver for constructive comments and suggestions during the preparation of this chapter.

References

Buggisch, D., Isenberg, G., Ravens, U., and Sholtysik, G. (1985). The role of sodium channels in the effects of the cardiotonic compound DPI 201-106 on contractility and membrane potentials in isolated mammalian heart preparations. *Eur. J. Pharmacol.* **118**, 303.

Chahine, M., Chen, L.-Q., Barchi, R. L., Kallen, R. G., and Horn, R. (1992). Lidocaine block of human heart sodium channels expressed in *Xenopus* oocytes. *J. Mol. Cell. Cardiol.* **24**, 1231–1236.

Cohen, C., Bean, B. P., and Tsien, R. W. (1981). Tetrodotoxin block of sodium channels in rabbit Purkinje fibers: Interaction between toxin binding and channel gating. *J. Gen. Physiol.* **78**, 383–411.

Colatsky, T. J. (1980). Voltage clamp measurements of sodium channel properties in rabbit cardiac Purkinje fibers. *J. Physiol. (London)* **305**, 215–234.

Cribbs, L. L., Satin, J., Fozzard, H. A., and Rogart, R. B. (1990). Functional expression of the rat heart I Na⁺ channel isoform. Demonstration of properties characteristic of native cardiac Na⁺ channels. *FEBS Lett.* **275**, 195–200.

Dascal, N. (1987). The use of *Xenopus* oocytes for the study of ion channels. *CRC Crit. Rev. Biochem.* **22**, 317–387.

Frelin, C., Cognard, C., Vigne, P., and Lazdunski, M. (1986). Tetrodotoxin-sensitive and tetrodotoxin-resistant Na⁺ channels differ in their sensitivity to Cd²⁺ and Zn²⁺. *Eur. J. Pharmacol.* **122**, 245–250.

Gellens, M. E., George, A. L., Jr., Chen, L., Chahine, M., Horn, R., Barchi, R. L., and Kallen, R. G. (1992). Primary structure and functional expression of the human cardiac tetrodotoxin-insensitive voltage-dependent sodium channel. *Proc. Natl. Acad. Sci. USA* **89**, 554–558.

George, A. L., Jr., Knittle, T. J., and Tamkun, M. M. (1992). Molecular cloning of an atypical voltage-gated sodium channel expressed in human heart and uterus: Evidence for a distinct gene family. *Proc. Natl. Acad. Sci. USA* **89**, 4893–4897.

Gundersen, C. B., Miledi, R., and Parker, I. (1983). Voltage-operated channels induced by foreign messenger RNA in *Xenopus* oocytes. *Proc. R. Soc. London B* **220**, 131–140.

Isom, L. L., De Jongh, K. S., Patton, D. E., Reber, B. F. X., Offord, J., Charbonneau, H., Walsh, K., Goldin, A. L., and Catterall, W. A. (1992). Primary structure and functional expression of the beta-1 subunit of the rat brain sodium channel. *Science* **256**, 839–842.

Krafte, D. S., and Volberg, W. A. (1992). Properties of endogenous voltage-dependent sodium currents in *Xenopus laevis* oocytes. *J. Neurosci. Meth.* **43**, 189–193.

Krafte, D. S., Volberg, W. A., Dillon, K., and Ezrin, A. M. (1991). Expression of cardiac Na channels with appropriate physiological and pharmacological properties in *Xenopus* oocytes. *Proc. Natl. Acad. Sci. USA* **88**, 4071–4074.

Krafte, D. S., Davison, K., Dugrenier, N., Estep, K., Josef, K., Barchi, R. L., Kallen, R. G., Silver, P. J., and Ezrin, A. M. (1994).

Pharmacological modulation of human cardiac Na$^+$ channels. *Eur. J. Pharmacol.* **266**, 245–254.

Kimura, J., Miyamae, S., and Noma, A. (1987). Identification of sodium-calcium exchange current in single ventricular cells of guinea pig. *J. Physiol. (London).* **384**, 199–222.

Narahashi, T., and Herman, M. D. (1992). Overview of toxins and drugs as tools to study excitable membrane ion channels: I. Voltage-activated channels. *Meth. Enzymol.* **207**, 620–643.

Neumcke, B., and Stampfli, R. (1982). Sodium currents and sodium current fluctuations in rat myelinated nerve fibers. *J. Physiol. (London)* **329**, 163–184.

Noda, M., Ikeda, T., Suzuki, H., Takeshima, H., Takahashi, T., Kuno, M., and Numa, S. (1986). Expression of functional sodium channels from cloned cDNA. *Nature (London)* **322**, 826–828.

Pappone, P. A. (1980). Voltage-clamp experiments in normal and denervated mammalian skeletal muscle. *J. Physiol. (London)* **306**, 377–410.

Pongs, O. (1992). Molecular biology of voltage-dependent potassium channels. *Physiol. Rev.* **72**, S69–S88.

Rogart, R. B., Cribbs, L. L., Muglia, L. K., Kephart, D. D., and Kaiser, M. W. (1989). Molecular cloning of a putative tetrodotoxin-resistant rat heart Na$^+$ channel isoform. *Proc. Natl. Acad. Sci. USA* **86**, 8170–8174.

Rosen, M. R., and Schwartz, P. J. (1991). The Sicilian gambit, A new approach to the classification of antiarrhythmic drugs based on their action on arrhythmogenic mechanisms. *Circulation* **84**, 1831–1851.

Saint, D. A., Ju, Y-K., and Gage, P. W. (1992). A persistent sodium current in rat ventricular myocytes. *J. Physiol. (London)* **453**, 219–231.

Satin, J., Kyle, J. W., Chen, M., Rogart, R. B., and Fozzard, H. A. (1992a). The cloned cardiac Na channel α-subunit expressed in *Xenopus* oocytes show gating and blocking properties of native channels. *J. Membr. Biol.* **130**, 11–22.

Satin, J., Kyle, J. W., Chen, M., Bell, P., Cribbs, L. L., Fozzard, H. A., and Rogart, R. B. (1992b). A mutant of TTX-resistant cardiac sodium channels with TTX-sensitive properties. *Science* **256**, 1202–1205.

Scholtysik, G. (1989). Cardiac Na$^+$ channel activation as a positive inotropic principle. *J. Cardiovasc. Pharmacol.* **14**, S24–S29.

Scholtysik, G., Quast, U., and Schaad, A. (1986). Evidence for different receptor sites for the novel cardiotonic S-DPI 201-106, ATX II, and veratridine on the cardiac sodium channel. *Eur. J. Pharmacol.* **125**, 111–118.

Sigel, E. (1990). Use of *Xenopus* oocytes for the functional expression of plasma membrane protein. *J. Membr. Biol.* **117**, 201–221.

Soreq, H., and Seidman, S. (1992). *Xenopus* oocyte microinjection: from gene to protein. *Meth. Enzymol.* **207**, 225–265.

Sumikawa, K., Houghton, N., Emtage, J. S., Richards, B. M., and Barnard, E. A. (1981). Active multi-subunit ACh receptors assembled by translation of heterologous mRNA in *Xenopus* oocytes. *Nature (London)* **292**, 862–864.

Volberg, W. A., Krafte, D. S., and Ezrin, A. M. (1991). Effects of Class I antiarrhythmics on guinea pig brain and heart Na currents expressed in *Xenopus* oocytes. *Biophys. J.* **59**, 97a.

White, M. M., Chen, L., Kleinfield, R., Kallen, R. G., and Barchi, R. L. (1991). SkM2, a Na$^+$ channel cDNA clone from denervated skeletal muscle, encodes a tetrodotoxin-insensitive Na$^+$ channel. *Mol. Pharmacol.* **39**, 604–608.

Molecular Analysis of Sodium Channel Inactivation

ALAN L. GOLDIN

I. Introduction

Voltage-sensitive sodium channels are responsible for the initial inward current during the depolarization phase of an action potential. Membrane depolarization activates the channels, causing a voltage-dependent conformational change that increases the permeability to sodium ions. This event is followed by inactivation, in which the channels close and the permeability to sodium ions returns to the resting level. Once the sodium channels have been inactivated, they cannot reopen until a sufficient period of time has elapsed to allow for recovery from inactivation. Therefore, the frequency and the duration of action potentials are at least partially determined by the rate at which sodium channels enter the inactivated state.

The sodium channel complex has been purified from a number of different tissues, including *Electrophorus electricus* (eel) electroplax membranes (Agnew *et al.*, 1978), rat muscle (Barchi *et al.*, 1980), chicken cardiac muscle (Lombet and Lazdunski, 1984), and rat brain (Hartshorne and Catterall, 1981). The electroplax (Agnew *et al.*, 1980, Miller *et al.*, 1983) and chicken heart (Lombet and Lazdunski, 1984) proteins consist of a single large subunit of molecular mass

230–270 kDa. The rat brain, muscle, and heart complexes, however, contain one large subunit (α) of 260 kDa and associated small subunits, β_1 (36 kDa) and β_2 (33 kDa) in the brain (Hartshorne *et al.*, 1982) and one β subunit in rat muscle (Kraner *et al.*, 1985; Casadei *et al.*, 1986) and rat heart (D. Gordon *et al.*, 1988). These proteins are sufficient for sodium channel function, since the complexes from eel and rat brain are functional when reconstituted into phospholipid vesicles (Talvenheimo *et al.*, 1982; Weigele and Barchi, 1982; Rosenberg *et al.*, 1984a,b). Since the eel α subunit makes fully functional sodium channels, a β subunit is not necessary for function.

Using antibodies and protein sequence information, cDNA clones have been isolated that encode the α subunits from eel electroplax membranes (Noda *et al.*, 1984), rat brain (Noda *et al.*, 1986a; Auld *et al.*, 1988; Kayano *et al.*, 1988; Joho *et al.*, 1990), rat skeletal muscle (Trimmer *et al.*, 1989; Kallen *et al.*, 1990), rat cardiac muscle (Rogart *et al.*, 1989), human skeletal and cardiac muscle (Gellens *et al.*, 1992; George *et al.*, 1992), and human brain (Ahmed *et al.*, 1992), as have genomic clones for the α subunit from *Drosophila* (Salkoff *et al.*, 1987; Loughney *et al.*, 1989; Ramaswami and Tanouye, 1989). Based on the predicted amino acid sequence, the α subunit of the sodium channel consists of four homologous domains termed I–IV, as

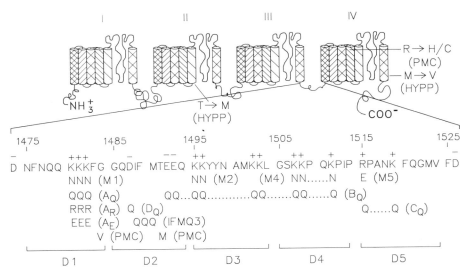

FIGURE 1 Diagram of the sodium channel α subunit showing inactivation mutations. The α subunit of the sodium channel is shown in schematic form; the amino acid sequence of the rat brain sodium channel III–IV linker is presented below the diagram. Mutations M1, M2, M3, and M4 in the rat brain III channel were reported by Moorman *et al.* (1990a). Mutations A_Q, B_Q, C_Q, D_Q, A_R, and A_E and deletion mutations D1, D2, D3, D4, and D5 in the rat brain IIA channel were reported by Patton *et al.* (1992). Mutation IFMQ3 in the rat brain IIA channel was reported by West *et al.* (1992). Mutations in the human skeletal muscle sodium channel gene, which are the cause of paramyotonia congenita (PMC), are changes of glycine to valine at position 1484 (V) and threonine to methionine at position 1491 (M) in the III–IV linker (McClatchey *et al.*, 1992) and alteration of arginine to either histidine (R → H) or cysteine (R → C) in domain IVS4 (Ptacek *et al.*, 1992a). Mutations in the human skeletal muscle sodium channel gene that are the cause of hyperkalemic periodic paralysis (HYPP) are changes of threonine to methionine (T → M) in domain IIS5 (Ptacek *et al.*, 1991a) and methionine to valine (M → V) in domain IVS6 (Rojas *et al.*, 1991). Abbreviations for the amino acid residues are A, Ala; C, Cys; D, Asp; E, Glu; F, Phe; G, Gly; H, His; I, Ile; K, Lys; L, Leu; M, Met; N, Asn; P, Pro; Q, Gln; R, Arg; S, Ser; T, Thr; V, Val; and Y, Tyr.

depicted in Fig. 1. Each of these domains is 40–60% identical to the others at the amino acid level. Hydrophobicity profiles indicate five hydrophobic segments of approximately 20 amino acids each at equivalent positions in the four domains (S1, S2, S3, S5, and S6). In addition, an amphipathic segment of approximately the same size, with positively charged residues at every third position (S4), is also present. To account for these domains, the secondary structure is most often represented with S1 through S6 as membrane-spanning α helical domains (Catterall, 1988). The region between S5 and S6 most likely spans the membrane as the actual pore of the channel, based on results obtained with the potassium channel (Hartmann *et al.*, 1991; Yellen *et al.*, 1991; Yool and Schwarz, 1991; see Chapter 2). The linker regions connecting the four homologous domains are all thought to be located on the cytoplasmic side of the membrane (R. Gordon *et al.*, 1987, 1988; Rossie *et al.*, 1987). The voltage-gated calcium channel appears to have a structure comparable to that of the sodium channel; the voltage-gated potassium channel looks like one of the

four sodium channel domains (reviewed by Jan and Jan, 1989). Since the potassium channel most likely functions as a tetramer (MacKinnon, 1991), the three voltage-gated cation-selective channels are structurally homologous. A more detailed model of the predicted structure of the voltage-gated potassium channel was proposed by Durell and Guy (1992).

The functional role of the β_1 and β_2 subunits in the rat brain sodium channel is unknown. Schmidt *et al.*, (1985) and Schmidt and Catterall (1986) obtained data suggesting that disulfide linkage of the α and β_2 subunits, insertion into the cell-surface membrane, and attainment of a functional conformation are closely related late events in the biogenesis of functional sodium channels in rat brain. In addition, selective removal of the β_1 subunit from purified rat brain sodium channels in detergent solution is accompanied by loss of the ability to reconstitute neurotoxin-activated sodium influx (Messner and Catterall, 1986; Messner *et al.*, 1986), suggesting that a complex of α and β_1 is required for function in this system. The β subunits are not essential for functional expression of the rat brain sodium channel in *Xenopus* oocytes, however. Injection into oocytes of unfractionated polyadenylated RNA from rat brain results in the synthesis of functional sodium channels (Gundersen *et al.*, 1983, 1984). If the RNA is size fractionated using a sucrose gradient to remove RNA that would be likely to encode the β subunits, a high molecular weight RNA fraction is sufficient for the synthesis of functional channels (Sumikawa *et al.*, 1984; Hirono *et al.*, 1985; Goldin *et al.*, 1986; Krafte *et al.*, 1988). These results suggest that the α subunit is sufficient for functional sodium channels in *Xenopus* oocytes.

II. Effects of Low Molecular Weight Proteins on Sodium Channel Inactivation

Since high molecular weight rat brain RNA encodes many proteins in addition to the sodium channel α subunit, the experiments using sucrose gradient fractionated RNA did not prove that the rat brain α subunit is sufficient to constitute functional sodium channels. This hypothesis was proven once cDNA clones encoding sodium channel α subunits were obtained, because injection of *in vitro* transcribed RNA made from the cloned α subunits resulted in functional sodium channels in *Xenopus* oocytes (Noda *et al.*, 1986b; Auld *et al.*, 1988; Suzuki *et al.*, 1988; Trimmer *et al.*, 1989; Cribbs *et al.*, 1990; Joho *et al.*, 1990; White *et al.*, 1991). However, the properties of the α subunit or high molecular weight RNA-induced sodium channels were not identical to those of channels resulting from the injection of rat brain polyadenylated RNA (Auld *et al.*, 1988; Krafte *et al.*, 1988, Trimmer *et al.*, 1989; Moorman *et al.*, 1990b; Zhou *et al.*, 1991). In particular, the macroscopic rate of inactivation was

slower than that observed with unfractionated rat brain RNA-induced channels or *in vivo*. These results suggest that some low molecular weight protein, perhaps a β subunit, is necessary for normal sodium channel inactivation in *Xenopus* oocytes.

Krafte *et al.* (1988) examined the properties of sodium channels resulting from injection into oocytes of high molecular weight rat brain RNA. The slower macroscopic rate of inactivation was reflected in a time constant that was about 2-fold larger than the time constant for inactivation of unfractionated RNA-induced channels (4.1 msec compared with 1.8 msec at −20 mV). However, at the single-channel level, the mean open time was not significantly different (1.12 msec for fractionated RNA compared with 0.93 msec for unfractionated RNA). The slower macroscopic rate of inactivation was a result of the channels continuing to burst throughout the depolarizing interval. Another major difference was that the voltage dependence of steady-state inactivation was shifted 11 mV in the positive direction for high molecular weight RNA-induced sodium channels. All these properties were restored to normal by adding back the low molecular weight RNA fractions to the high molecular weight RNA fraction before injection. These results suggest that some protein(s) encoded in low molecular weight RNA modifies the inactivation properties of α subunit sodium channels in *Xenpous* oocytes.

Experiments using cDNA clones encoding the α subunit confirmed the findings of Krafte *et al.*, (1988). Auld *et al.*, (1988) observed that sodium channels formed after injection of oocytes with RNA transcribed *in vitro* from the rat brain type IIA sodium channel clones inactivated with a slower macroscopic time constant than those resulting from injection of rat brain polyadenylated RNA (Fig. 2). Normal inacti-

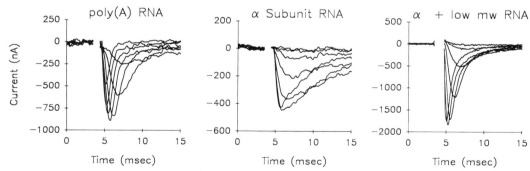

FIGURE 2 The effect of low molecular weight RNA on inactivation of α subunit sodium channels. *Xenopus* oocytes were injected with 70 ng rat brain polyadenylated RNA [poly(A) RNA], 0.7 ng *in vitro* transcribed RNA encoding the rat IIA α subunit (α subunit RNA), or 0.7 ng α subunit RNA with 70 ng low molecular weight rat brain polyadenylated RNA (α + low mw RNA). After 2 days incubation at 20°C, sodium currents were analyzed by two-microelectrode voltage clamping. The traces shown were elicited from a holding potential of −100 mV by depolarizations ranging from −30 to +30 mV in 10-mV increments. Reprinted with permission from Auld *et al.* (1988). © Cell Press.

vation was restored by adding back a low molecular weight fraction from rat brain RNA (Fig. 2). Slower inactivation was also observed for the rat skeletal muscle μl (SkM1) sodium channel α subunit expressed in oocytes by Trimmer *et al.*, (1989); co-injection of rat brain RNA restored the normally rapid kinetics of inactivation to the muscle channel (Zhou *et al.*, 1991). Joho *et al.*, (1990) found that the rat brain type III α subunit induced sodium channels that inactivated even more slowly than the rat type IIA or SkM1 channels in oocytes. Inactivation of rat type III channels occurred with a time course best fit by two exponential terms, with time constants of approximately 20 msec and 200 msec, compared with less than 10 msec for the rat type II and SkM1 channels.

However, not all experiments involving expression of sodium channel α subunits in oocytes have demonstrated slow inactivation. The rat brain type II sodium channel appeared to inactivate more rapidly when examined by Stühmer *et al.*, (1987) using macropatch recording techniques on *Xenopus* oocytes. This difference between the rat brain type II and type IIA sodium channels most likely represents a difference in recording techniques rather than a difference in the two channel subtypes, for the following reasons. First, the rat type II channel differs from the rat type IIA channel at only seven amino acid positions; mutation of each of these residues does not affect inactivation of the channel (Auld *et al.*, 1990). Second, when the rat type II channel was examined, using whole-cell oocyte recording, by Noda *et al.*, (1986b), the macroscopic rate of inactivation did not appear to be very different from that observed for the rat type IIA channel by Auld *et al.*, (1988). Third, single-channel recordings obtained by Stühmer *et al.*, (1987) show bursting rat type II sodium channels, with slowly inactivating ensemble average currents; these results are similar to the results obtained for rat type IIA channels by Krafte *et al.*, (1990). Therefore, the faster inactivation observed by Stühmer *et al.*, (1987) for the rat type II channel most likely resulted from the macropatch recording technique. One plausible explanation is that the very high level of expression of the channels in oocytes used for macropatch recordings resulted in a more rapid time course of inactivation.

The electrophysiological properties of sodium channels resulting from injection of only the α subunit are quite similar to those resulting from injection of high molecular weight rat brain RNA. Krafte *et al.*, (1990) found that sodium channels resulting from injection of the rat type IIA α subunit inactivated slowly, a feature that was reflected in two inactivation time constants of about 2 msec and 18 msec during a depolarization to -10mV. In contrast, rat brain RNA induced sodium channels that inactivated with a single predominant time constant similar to the fast time constant for α subunit channels. At the single-channel level, the mean open times were not significantly different, but the α subunit channels continued to burst throughout the depolarization interval. Sodium channels resulting from injection of only the α subunit also recovered from inactivation more slowly than did channels resulting from injection of rat brain RNA. In addition to the kinetic differences, the voltage dependence of steady-state inactivation was shifted approximately 20 mV in the positive direction for α subunit sodium channels compared with those resulting from injection of rat brain RNA. All these properties were restored to normal by co-injection of low molecular weight rat brain RNA. In addition, co-injection of low molecular weight RNA resulted in a 4-fold increase in the maximal peak amplitude of sodium currents following injection of the same quantity of α subunit RNA (Auld *et al.*, 1988). Therefore, some protein(s) encoded in low molecular weight rat brain RNA modulates the electrophysiological properties and enhances currents from α subunit sodium channels in *Xenopus* oocytes.

The single-channel basis for the slower macroscopic rate of inactivation of α subunit sodium channels has been extensively analyzed by Moorman *et al.*, (1990b) and Zhou *et al.*, (1991). Moorman *et al.*, (1990b) found that single rat type III channels gate in two distinct modes, with fast and slow kinetics. Channels gating in the fast mode were characterized by fewer reopenings, shorter mean open time, and a smaller ratio of open time to pulse duration than channels gating in the slow mode. The slow gating mode appeared to occur in runs. Patches with a single-channel demonstrated both modes of gating. These results indicate that a single channel can shift between the two modes of gating, but that the shift from slow to fast mode does not occur at a very high frequency. In a similar analysis, Zhou *et al.*, (1991) found that rat SkM1 α subunit channels also gated with multiple distinct modes. The two principal modes were comparable to the fast and slow modes observed for the rat type III channels, but these investigators also noted that the data could not be explained without additional modes of gating. In addition, channels gating in the fast mode demonstrated a shorter latency to first opening than channels gating in the slow mode. Coexpression of rat skeletal muscle or brain RNA, or RNA made *in vitro* from a fraction of a rat brain cDNA library, shifted the channels into gating in a predominantly fast mode. These results indicate that some accessory protein(s) encoded in both rat brain and skeletal muscle RNA modulates both the activation and the inactivation gating of sodium channels, and that shifts between the various gating modes can have dramatic effects on the macroscopic kinetics of inactivation.

III. Role of the β_1 Subunit in Sodium Channel Inactivation

The experiments using low molecular weight RNA or fractions of a cDNA library demonstrated that the modulatory factor(s) was a relatively low molecular weight protein(s). However, the factor could be one of the accessory β subunits of the channel or it could be a protein that indirectly affects channel function, for example, a kinase or phosphatase. The functional effects of the β_1 subunit could be tested once a cDNA clone encoding it was isolated by Isom *et al.*, (1992). Based on the predicted amino acid sequence, the rat brain β_1 subunit is a 209 residue protein with a 19-amino acid leader sequence at the N terminus and a single membrane-spanning segment about 45 residues from the C terminus. The protein contains four potential N-linked glycosylation sites in regions predicted to be extracellular, a feature that is consistent with biochemical data indicating three or four N-linked carbohydrate side chains (Messner and Catterall, 1985). RNA hybridizing to the β_1 cDNA was detected at high levels in rat brain and spinal cord, at moderate levels in rat heart, and at low but detectable levels in rat skeletal muscle, consistent with the results of antibody binding experiments performed by Sutkowski and Catterall (1990). When RNA encoding the

β_1 subunit was coexpressed in oocytes with the α subunit, the macroscopic kinetics of inactivation were markedly faster than for the α subunit alone (Fig. 3A). In addition, coexpression of β_1 shifted the voltage dependence of steady-state inactivation in the negative direction by 19 mV (Fig. 3B). Finally, coexpression of β_1 increased the level of current observed from the same quantity of α subunit RNA about 2.5-fold (Fig. 3C). All these results are similar to the effects observed after co-injection of low molecular weight rat brain RNA (Krafte *et al.*, 1990), suggesting that the β_1 subunit does modulate the inactivation properties of the sodium channel.

Rat brain type II, type III, and skeletal muscle type SkM1 sodium channels all demonstrate unusually slow inactivation when the α subunits are expressed alone in *Xenopus* oocytes. Further, some low molecular weight factor(s) from rat brain modulates inactivation of at least the type IIA and SkM1 channels. However, these three sodium channel types do not contain the same accessory β subunits *in vivo*. The sodium channel complex from rat skeletal muscle contains only a single β subunit that is immunologically related to but possibly distinct from the rat brain β_1 subunit (Casadei *et al.*, 1986; Roberts and Barchi, 1987; Sutkowski and Catterall, 1990). The rat brain type III sodium channel is most likely a fetal subtype, since it is expressed at highest levels during fetal and early postnatal stages of development (Beckh *et al.*, 1989). The fetal sodium channel is probably not associated with any β subunits (Scheinman *et al.*, 1989; Sutkow-

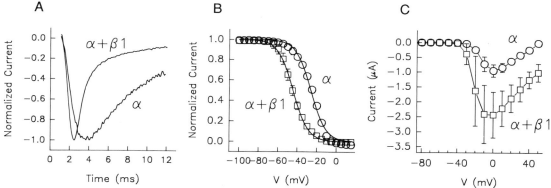

FIGURE 3 The effect of β_1 on inactivation of α subunit sodium channels. Oocytes were injected with 0.25 ng *in vitro* transcribed RNA encoding the rat IIA α subunit alone or 0.25 ng α subunit RNA with 5 ng *in vitro* transcribed rat brain β_1 subunit RNA. After 2 days incubation at 20°C, sodium currents were analyzed by two-microelectrode voltage clamping. (A) The traces shown were elicited by depolarization to −10 mV from a holding potential of −100 mV. (B) The voltage dependence of inactivation was determined using a two-pulse inactivation protocol with an inactivating prepulse of 50 msec and a test pulse to −5 mV. Peak current during the test pulse is plotted against the prepulse potential. The smooth curves represent Boltzmann distributions that best fit the data. (C) Current–voltage relationships were obtained by plotting the current against the depolarizing potential for depolarizations ranging from −80 to +50 mV in 10-mV increments. Reprinted with permission from Isom *et al.* (1992). © AAAS.

ski and Catterall, 1990). Despite this diversity of subunit composition, coexpression of rat or human β_1 with the rat type III or SkM1 α subunit resulted in a significant increase in the rate of fast inactivation compared with either α subunit alone (Cannon *et al.*, 1993; McClatchey *et al.*, 1993; Tong *et al.*, 1993; Bennet *et al.*, 1993; Patton *et al.*, 1994). Therefore, the same β_1 subunit speeds up inactivation when coexpressed with each of these three distinct sodium channel types, indicating that these different channels are all capable of associating with and being modulated by the same β_1 subunit. These results further suggest that the different sodium channel α subunits may associate with comparable β subunits *in vivo*.

IV. "Ball and Chain" Model of Sodium Channel Inactivation

To account for fast sodium channel inactivation, Armstrong and Bezanilla (1977) proposed a mechanical representation that has since been referred to as the "ball and chain" model. In this representation, a portion of the molecule on the cytoplasmic side of the membrane functions as an inactivating particle that occludes the pore of the channel when in the inactivated state. The positively charged particle was proposed to interact with a negatively charged docking site near the mouth of the channel. This model can successfully explain a number of the known electrophysiological characteristics of sodium channel inactivation. Treatment of the inside of a cell with Pronase® removes fast sodium channel inactivation (Armstrong *et al.*, 1973; Rojas and Rudy, 1976; Stimers *et al.*, 1985) which, according to the ball and chain model, would be the result of digesting the inactivating particle. Sodium channel inactivation has little or no intrinsic voltage dependence, but appears to be voltage dependent because of the coupling between activation and inactivation (Goldman and Schauf, 1972; Bezanilla and Armstrong, 1977; Nonner, 1980; Aldrich and Stevens, 1987). According to the model, this effect would be caused by the fact that the inactivating particle cannot interact with the docking site of the channel in the resting state, so some voltage-dependent conformational change associated with activation must first occur. For the same reason, inactivation is delayed following depolarization and is very slow for small depolarizations (Goldman and Schauf, 1972; Bezanilla and Armstrong, 1977). Finally, detecting any component of gating current with the time course of inactivation has been difficult (see Greeff and Forster, 1991, for one exception), but inactivation immobilizes

about two-thirds of the activation gating charge (Armstrong and Bezanilla, 1977; Meves and Vogel, 1977; Nonner, 1980). According to the ball and chain model, this immobilization would result from the channel being unable to return completely to the resting state while the inactivating particle is still associated with the docking site. In addition to being consistent with the mechanical ball and chain representation, all these findings are consistent with the state diagram associated with the model (Armstrong and Bezanilla, 1977).

The ball and chain model was developed based on electrophysiological data and without any knowledge of the molecular structure of the sodium channel. Once cDNA clones encoding voltage-gated ion channels were isolated, it became possible to make predictions concerning the structure of the channel. Hoshi *et al.* (1990) and Zagotta *et al.* (1990) examined the applicability of the ball and chain model to the potassium channel by identifying a region of the channel that might function as an inactivating particle. They demonstrated that the N terminus of the *Shaker* potassium channel appeared to act as the inactivating particle predicted by Armstrong and Bezanilla (1977) for the sodium channel. Deletions within the first 20 residues of the N-terminal region of the channel eliminated fast inactivation, consistent with elimination of the actual particle. In contrast, deletions beyond amino acid 23 resulted in faster inactivation, consistent with shortening the chain by which the particle was connected. Mutation of a single leucine at position 7 eliminated fast inactivation, indicating that this residue is critical in stabilizing the structure of the inactivating particle. Finally, to demonstrate that the particle can inactivate the channel in *trans*, these investigators showed that a peptide consisting of the first 20 amino acids of the channel could inactivate a channel from which the particle had been removed by a deletion. Later studies by Isacoff *et al.* (1991) suggested that the docking site with which the particle interacts includes residues within the cytoplasmic loop between putative transmembrane segments S4 and S5. All these data are consistent with the model that the voltage-gated potassium channel inactivates by a ball and chain type mechanism.

V. Role of the Cytoplasmic Linker between Domains III and IV in Inactivation

Despite the fact that the ball and chain model was originally proposed for the voltage-gated sodium channel, a molecular basis for that model in the sodium channel has been much harder to demonstrate. Neither the N-terminal nor the C-terminal region appears to function as an inactivating particle, because deletions of either end result in functional channels that inactivate normally when expressed in *Xenopus* oocytes (Stühmer *et al.*, 1989a). Instead, the linker connecting domains III and IV has been implicated as being involved in fast inactivation of the sodium channel. The first line of evidence was obtained by Vassilev *et al.* (1988), who showed that antibodies against that region, but not against the I–II or II–III linkers, slowed inactivation of the channel in rat muscle cells. Vassilev *et al.* (1989) later extended their findings to show that the same antibody could inhibit inactivation of single sodium channels in rat brain neurons, and that additional antibodies directed against other regions still had no effect on inactivation.

Stühmer *et al.* (1989a) tested the importance of the cytoplasmic linkers for fast inactivation by constructing channels consisting of noncovalently connected domains. Expression of RNA encoding domains I, II, and III in addition to RNA encoding domain IV resulted in channels that inactivated about 10-fold more slowly than wild-type channels. For depolarizations to -20 mV, the time constant for inactivation was ~ 10 msec (Stühmer *et al.*, 1989a) compared with ~ 1 msec for the wild-type channel (Stühmer *et al.*, 1987); the single-channel mean open time was 5.8 msec (Stühmer *et al.*, 1989a) compared with 0.43 msec for the wild-type channel (Stühmer *et al.*, 1987). On the other hand, expression of RNA encoding domains I and II with RNA encoding domains III and IV resulted in channels that inactivated with kinetics similar to those of wild-type channels. Expression of RNA encoding domain I with RNA encoding domains II, III, and IV resulted in only marginal sodium current.

Patton and Goldin (1991) examined the importance of the length of the sodium channel III–IV linker by constructing a mutation that contained amino acid insertions at both ends of the linker. Mutant NaIII–IVins contained insertions of nine amino acids between IIIS6 and the III–IV linker and three amino acids between the III–IV linker and IVS1. NaIII–IVins channels inactivated with a macroscopic rate that was about 2-fold slower than the rate constant for the wild-type rat type IIA channel. These results suggest that increasing the length of the cytoplasmic linker can decrease the rate of sodium channel inactivation.

Further evidence implicating the III–IV linker in fast sodium channel inactivation was obtained by West *et al.* (1991). Numann *et al.* (1991) demonstrated that phosphorylation of the sodium channel by protein kinase C resulted in decreased currents and slower inactivation in both rat brain neurons and Chi-

nese hamster ovary cells expressing a transfected rat brain type IIA gene. West *et al.* (1991) then identified a specific serine residue in the III–IV linker of the rat type IIA channel that was required for this modulation. Sodium channels containing an alanine instead of a serine at position 1506 in the III–IV linker demonstrated normal electrophysiological functions in Chinese hamster ovary cells, but their inactivation properties were not modulated by protein kinase C activation. All these results are consistent with the hypothesis that the linker connecting domains III and IV is critical for fast sodium channel inactivation.

VI. Effects of Deletions and Charge Neutralization in the III–IV Linker on Inactivation

Since the III–IV linker appeared to be involved in sodium channel inactivation, the positively charged amino acids in this region might be important in stabilizing the inactivating particle in its docking site. Moorman *et al.* (1990a) tested this hypothesis by constructing four mutations that neutralized combinations of positively charged residues in the III–IV linker of the rat type III channel. The mutations consisted of substitutions of asparagine for lysine or glutamate for arginine, and are shown as mutations M1, M2, M4, and M5 in Fig. 1. The effects of the mutations on sodium channel inactivation were examined by expression in *Xenopus* oocytes and by electrophysiological analysis. When the wild-type and mutant channels were examined by two-microelectrode whole-cell voltage clamping, two distinct exponential time constants for fast inactivation were detected. Two of the mutations (M2 and M4) resulted in faster inactivation, which was reflected in a shorter fast time constant and a greater percentage of current inactivating with the fast time constant. Mutant M5 sodium currents were similar to those from the wild-type channel, and mutant M1 did not express functional sodium channels.

The properties of mutants M4 and M5 were then examined more carefully by single-channel analysis. As discussed previously, sodium channels expressed from only the wild-type α subunit in *Xenopus* oocytes continue to burst throughout the depolarizing interval. Records for both mutants demonstrated similar bursting patterns, but with a significant decrease in the burst duration relative to the wild-type channel. This decrease in the burst duration was sufficient to explain the faster macroscopic rate of inactivation for the M4 mutant channels. However, M5 channels displayed macroscopic kinetics of inactivation equivalent to those of the wild-type channel, although these channels had a shorter burst duration at the single-channel level. This seeming contradiction was generated because the mean latency to first opening of the channel was longer for M5 channels than for the wild-type channels. Therefore, neutralization of the positive residues in the III–IV linker affected both activation and fast inactivation of the sodium channel. Somewhat surprisingly, however, neutralization of some of the charged residues did not slow down or prevent inactivation, but resulted in speeding up inactivation by decreasing the burst duration of the channels.

Patton et al. (1992) similarly analyzed the importance of the charged residues in the III–IV linker of the rat type IIA sodium channel. These investigators constructed a series of mutations in which each of the 15 charged residues was changed to glutamine in various combinations (Fig. 1). Mutants A_Q, B_Q, and C_Q resulted in functional channels, as did combinations of A_Q with B_Q or C_Q. However, mutant D_Q did not result in functional sodium channels, nor did the combination of A_Q, B_Q, and C_Q. To avoid inactivation effects resulting from changes in the gating mode, all the mutants were analyzed by coexpression with the β_1 subunit, which causes the channels to gate primarily in the fast mode (Isom et al., 1992). When analyzed with β_1, all the charge neutralization mutants (A_Q, B_Q, and C_Q) inactivated with kinetics similar to the wild-type channel. The only significant difference in kinetics was a slight increase in the amount of maintained current at the end of the pulse for mutant B_Q.

Mutant B_Q includes all the charge neutralizations present in mutants M2 and M4 of Moorman et al. (1990a), yet B_Q did not demonstrate faster inactivation as did M2 and M4. This difference may be the result of the use of glutamines as replacement amino acids in B_Q rather than the use of asparagines in M2 and M4. A more likely explanation, however, is that mutant B_Q was analyzed in the presence of β_1, whereas mutants M2 and M4 were analyzed without β_1. The speeding up of inactivation observed for mutants M2 and M4 may have been the result of increasing the percentage of time the channels spent in the fast gating mode, as was suggested by Zhou et al. (1991). Since mutant B_Q was expressed with the β_1 subunit, those channels would already be gating predominantly in the fast mode. Consistent with this explanation, mutants B_Q and C_Q did demonstrate slightly faster inactivation than the wild-type channel when expressed in the absence of the β_1 subunit (Patton et al., 1992).

Although the kinetics of inactivation were similar for mutants A_Q, B_Q, and C_Q and for the wild-type rat type IIA sodium channel in the presence of β_1, differences were detected in the voltage dependence of activation and inactivation. Mutant B_Q shifted activation in the positive direction, mutant C_Q shifted inactivation in the negative direction, and mutant A_Q shifted inactivation in the positive direction. To determine whether the effects for mutant A_Q were primarily the result of charge alterations, two additional mutants were constructed (Patton et al., 1992). Mutant A_R contains arginines instead of lysines, which conserves the positive charge, and mutant A_E contains glutamates instead of lysines, which replaces positive charges with negative charges (see Fig. 1). Mutant A_R was not significantly different from the wild-type channel, and mutant A_E shifted inactivation in the positive direction to an even greater extent than mutant A_Q. These results are consistent with the hypothesis that these residues interact from the cytoplasmic side with a voltage sensor in the channel.

Since the charge neutralization mutations had very little effect on the kinetics of sodium channel inactivation, Patton et al. (1992) also constructed a series of consecutive deletions (10 amino acids each) to determine whether the III–IV linker was essential for fast inactivation (Fig. 1). Of the five deletions that span the III–IV linker, only deletion mutants 1, 4, and 5 expressed functional channels in oocytes. Deletion 1 completely eliminated fast inactivation, with a time constant of 4.6 sec during a depolarization to -10 mV (Fig. 4). Deletion 4 resulted in markedly slower inactivation, and deletion 5 did not dramatically affect the kinetics of fast inactivation (Fig. 4). The complete elimination of fast inactivation by deletion 1 indicates that the III–IV linker is critical for inactivation. The fact that deletion 5 did not dramatically alter the kinetics of inactivation indicates that the effect was not simply the result of shortening the linker.

The single-channel properties of deletion mutant 1 channels were then compared with those of the wild-type channel to determine the single-channel basis for the lack of inactivation (Fig. 5). For these experiments, the wild-type channel was expressed without the β_1 subunit, so the channel continued to burst for short periods of time. Even with this bursting, the channel completely inactivated within ~10 msec. However, the deletion mutant 1 channel continued to open and close throughout the recording interval. The mean open time in the absence of fast inactivation was distinctly voltage dependent (Fig. 5, inset), suggesting that the final transition from the closed to the open state is a voltage-dependent process. The mutant channel did eventually enter an inactivated state, as shown in the final trace, which probably represents the process of slow inactivation. The single channel conductance for deletion mutant 1 was not significantly different from that of the wild-type channel when determined under comparable ionic condi-

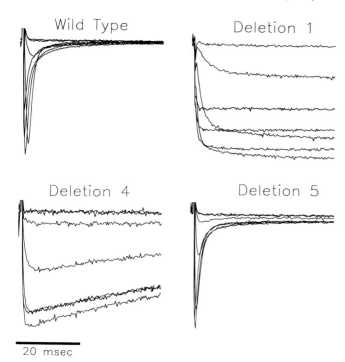

FIGURE 4 Inactivation kinetics of sodium channel deletion mutants. Oocytes were injected with *in vitro* transcribed RNA encoding the wild-type rat IIA sodium channel or deletion mutants 1, 4, or 5 (see Fig. 1). Oocytes were incubated for 2 days at 20°C, at which time sodium currents were analyzed by two-microelectrode voltage clamping. The traces shown were elicited from a holding potential of −100 mV by depolarizations ranging from −50 to +10 mV in 10-mV increments. Reprinted with permission from Patton *et al.* (1992).

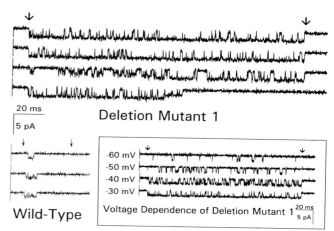

FIGURE 5 Single-channel properties of noninactivating sodium channel deletion mutant. Oocytes were injected with *in vitro* transcribed RNA encoding either the wild-type rat IIA sodium channel or deletion mutant 1 and were incubated for 2 days at 20°C. Outside-out patches were excised, and single-channel openings were analyzed during depolarizations to −30 mV from a holding potential of −100 mV. The inset shows openings obtained during depolarizations to −60, −50, −40, and −30 mV for deletion mutant 1. The recordings from the wild-type channel were obtained using 96 m*M* external sodium, whereas the recordings from deletion mutant 1 were obtained using 115 m*M* external sodium. Reprinted with permission from Patton *et al.* (1992).

tions. However, two distinct classes of openings appear to exist for the deletion mutant 1 channel, as shown in the third current trace. During the first half of that trace, the channel was rapidly flickering between open and closed states, whereas in the last portion of the trace it returned to a state with well-defined openings. These rapid openings may be to a lower conductance state. Alternatively, the apparently smaller current amplitudes might be the result of the rapid fluctuations. These results indicate that deletion of the first 10 amino acids of the III–IV linker selectively eliminates the fast inactivation process in the sodium channel.

VII. Cluster of Hydrophobic Residues in the III–IV Linker Is Critical for Inactivation

The results with the deletions in the III–IV linker demonstrated that some specificity exists in the different regions of the linker. The results with the charge neutralization mutations indicated that none of the charged residues is critical for fast inactivation. To identify specific amino acids that might be essential for fast inactivation, West *et al.* (1992) constructed mutations of nonpolar amino acids in the III–IV linker. One mutation, termed IFMQ3, was a change of three consecutive hydrophobic residues to glutamine (Fig. 1). This mutation resulted in sodium channels that did not demonstrate fast inactivation (Fig. 6). Mutation of individual hydrophobic amino acids elsewhere throughout the III–IV linker did not have any significant effect on inactivation.

To determine which of the three adjacent IFM residues (Ile–Phe–Met) was critical for fast inactivation, West *et al.* (1992) individually mutated each of those residues to glutamine. Mutation of the methionine at position 1490 (M1490Q) did not significantly affect fast inactivation. Mutation of the isoleucine at position 1488 (I1488Q) resulted in sodium channels that inactivated rapidly, but had a distinct amount of noninactivating current. On the other hand, mutation of the phenylalanine at position 1489 (F1489Q) dramatically reduced fast inactivation, almost to the same extent as mutation of the three IFM residues (Fig. 6). The only major difference from the IFMQ3 mutant was a small component of rapidly inactivating current. These results indicate that the phenylalanine at position 1489 is critical for fast sodium channel inactivation.

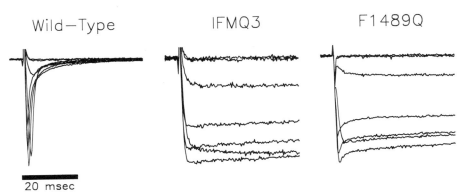

FIGURE 6 A cluster of hydrophobic residues is critical for sodium channel inactivation. Oocytes were injected with *in vitro* transcribed RNA encoding the wild-type rat IIA sodium channel, mutant IFMQ3, or mutant F1489Q (see Fig. 1). After 2 days incubation at 20°C, sodium currents were analyzed by two-microelectrode voltage clamping. The traces shown were elicited from a holding potential of −100 mV by depolarizations ranging from −50 to +10 mV in 10-mV increments. Reprinted with permission from West *et al.* (1992). © Cell Press.

To explain these results, West *et al.* (1992) proposed a modification of the ball and chain model, which they termed the "hinged-lid" model, for fast sodium channel inactivation. This model is based on the hinged-lid structures of allosteric enzymes (Vermersch *et al.*, 1990), and proposes two hinge points at either end of the linker that allow the majority of the linker region to fold over the pore of the channel, as a lid would. Deletion of the first 10 amino acids would prevent that process from occurring, hence removing fast inactivation. On the other hand, the last 10 amino acids would be located in a region of the linker that neither constitutes the lid nor interferes with its movements; therefore, deletion of these residues has no significant effect on fast inactivation. The hydrophobic phenylalanine residue located at position 1489 would be critical for stabilizing the lid in its closed position, so mutation of that residue to glutamine would eliminate fast inactivation. This model shares many of the important functional features of the ball and chain model; the major distinction is that the inactivating particle or lid in the sodium channel is tethered to the channel at both ends.

VIII. Functional Analogy of the Inactivation Regions of Sodium and Potassium Channels

The results discussed in the preceding section are consistent with the idea that fast inactivation in sodium channels occurs by a modified ball and chain or hinged-lid mechanism, similar to the inactivation mechanism of voltage-dependent potassium channels. Patton *et al.* (1993) investigated the functional similarity of fast inactivation in the two types of channels by constructing chimeras in which the sodium channel III–IV linker was attached to the N terminus of a noninactivating potassium channel.

The potassium channel that was used for these studies was the mouse brain channel termed MK1 (Kv1.1), which is a naturally noninactivating homolog of the *Shaker* potassium channel (Tempel *et al.*, 1988; Stühmer *et al.*, 1989b; Chandy *et al.*, 1990). First researchers had to determine whether the MK1 channel could be inactivated by a *Shaker* particle. This possibility was tested by inserting the first 57 amino acids of *Shaker* H4, which includes the inactivating particle, into the N terminus of MK1. This chimeric channel (H4MK1) inactivated even more rapidly than the wild-type *Shaker* H4 channel (Fig. 7), indicating that the MK1 channel does contain a docking site with which the *Shaker* inactivating particle can interact. Then, to test whether the sodium channel III–IV linker could function as an inactivating particle, the III–IV linker was inserted at the N terminus of the noninactivating MK1 clone. This chimeric channel (NaMK1) also displayed fast inactivation (Fig. 7), although with somewhat slower kinetics and more maintained current than observed for the H4–MK1 chimera.

Finally, to determine if fast inactivation in the chimeric potassium channel resulted specifically from the sodium channel linker, the III–IV linker containing the IFMQ3 mutation was placed at the N terminus of the noninactivating MK1 clone. This chimeric channel (NaMK1 IFMQ3) displayed no fast inactivation (Fig-

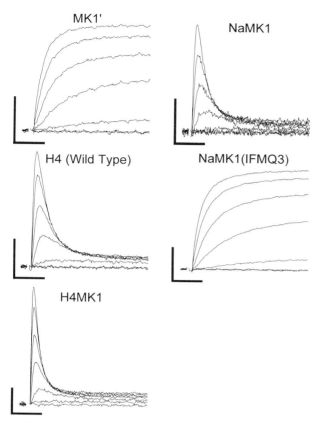

FIGURE 7 Inactivation kinetics of chimeric potassium channels. Oocytes were injected with *in vitro* transcribed RNA encoding a modified MK1 potassium channel, the wild-type *Shaker* H4 potassium channel (Iverson *et al.*, 1988), chimeric channel H4MK1, chimeric channel NaMK1, or chimeric channel NaMK1 (IFMQ3). The modified MK1 channel consists of amino acids 1–4 of *Shaker* H4 followed by amino acids 8 to the end of MK1. This channel is comparable to the wild-type MK1 channel in its electrophysiological properties, but the clone contains a unique restriction site for insertion of exogenous sequences after the fourth amino acid. H4MK1 consists of amino acids 1-57 of *Shaker* H4 and amino acids 8 to the end of MK1. NaMK1 consists of amino acids 1–4 of *Shaker* H4, amino acids 1470 to 1526 of the rat IIA sodium channel (comprising the entire III–IV linker) and amino acids 8 to the end of MK1. NaMK1 (IFMQ3) is comparable to NaMK1, except that the IFM residues at position 1488-1490 have been substituted with glutamines. After 2 days incubation at 20°C, currents were analyzed by two-microelectrode voltage clamping. The traces shown were elicited from a holding potential of −100 mV by depolarizations ranging from −70 mV to +50 mV in 20 mV increments. The current bars represent 1 μA, and the time bars represent 10 ms. Reprinted with permission Patton *et al.* (1993). © Cell Press.

ure 7), demonstrating that the same sequences within the III–IV linker are essential for fast inactivation in the sodium channel and in the chimeric channel. These results suggest that the III–IV linker functions by a similar mechanism to cause fast inactivation in the two channels.

IX. Clinical Relevance of Disorders of Sodium Channel Inactivation

Disorders of sodium channel inactivation have been implicated in the etiology of a number of human neuromuscular diseases including hyperkalemic periodic paralysis (HYPP), paramyotonia congenita (PMC), and myotonia congenita (MC). HYPP is a disease in which increased levels of serum potassium lead to muscle hypoexcitability and paralysis. PMC patients experience cold-induced weakness and paralysis that is aggravated by increased muscle activity. MC is similar to PMC without the temperature-sensitive effects. All these diseases are inherited in an autosomal dominant manner. Electrophysiological studies of cells from patients with these disorders have shown that sodium channel inactivation was slower than normal. In particular, sodium channels in muscle cells from HYPP patients demonstrated an aberrant gating mode in which the channels continued to reopen (Lehmann-Horn *et al.*, 1991; Cannon *et al.*, 1991); this mode is similar to the slow gating mode observed for the sodium channel α subunit expressed in oocytes (Krafte *et al.*, 1990; Moorman *et al.*, 1990b; Zhou *et al.*, 1991).

Since then, researchers have shown that HYPP (Fontaine *et al.*, 1990; Ebers *et al.*, 1991; Ptacek *et al.*, 1991b,c) PMC (Ebers *et al.*, 1991; Ptacek *et al.*, 1991b), and at least one form of MC (Ptacek *et al.*, 1992b) are tightly linked to the human skeletal muscle sodium channel gene on chromosome 17q (George *et al.*, 1991; Wang *et al.*, 1992). Further studies have identified the actual mutations in some families with these diseases. As shown in Fig. 1, two different mutations that result in HYPP are alterations of threonine to methionine (T→M) in domain IIS5 (Ptacek *et al.*, 1991a) and methionine to valine (M→V) in domain IVS6 (Rojas *et al.*, 1991). Four mutations in two distinct regions have been identified to cause PMC (Fig. 1). These mutations include alteration of arginine to either histidine (R→H) or cysteine (R→C) in domain IVS4 (Ptacek *et al.*, 1992a) and alterations of glycine to valine at position 1484 (G→V) and threonine to methionine at position 1491 (T→M) in the III–IV linker (McClatchey *et al.*, 1992). The mechanism by which these mutations affect sodium channel inactivation is currently unknown, but the results suggest that fast inactivation is strongly dependent on the structure of the molecule in regions other than the III–IV linker. These other regions might constitute part of a docking site for the inactivating particle or lid, or they might indirectly affect inactivation, perhaps by shifting the relative proportions of the different gating modes.

X. Future Directions

Currently, strong evidence exists that the cytoplasmic linker between domains III and IV is a critical component in sodium channel inactivation. However, very little data is available concerning the mechanism by which this linker is involved. In particular, the III–IV linker still may not function as the inactivation gate, but may affect inactivation indirectly. These alternatives could be distinguished by determining whether peptides representing portions or all of the III–IV linker can inactivate a channel in which the linker is no longer functional. If the peptides can inactivate the channel, the III–IV linker would be shown to occlude the sodium channel pore in *trans*. This approach is similar to the experiments demonstrating that the potassium channel N terminus functions as an inactivating particle (Zagotta *et al.*, 1990). However, such results may still be inconclusive for the sodium channel, because the mutated III–IV linker in the noninactivating sodium channel might prevent the peptide from interacting with the docking site. If the III–IV linker does function as an inactivating particle, then the regions of the channel that make up the docking site must be identified. Regions that are likely candidates include the S4–S5 segment, analogous to the potassium channel docking site identified by Isacoff *et al.* (1991), and regions IIS5, IVS4, and IVS6, all of which have been shown to contain mutations that cause human neuromuscular disorders.

Finally, the mechanism by which sodium channels shift between gating modes remains completely unknown. Although the β_1 subunit is known to modulate channels by shifting them into a fast gating mode, the mechanism of this interaction has not been identified. Whether other low molecular weight proteins, such as the β_2 subunit, might modify sodium channel function is also unknown. One possibility is that some of the mutations causing human diseases may not directly affect inactivation, but may affect the ability of the channel to be modulated by some accessory subunit such as β_1. For these reasons, many regions of the sodium channel α subunit may be important for fast sodium channel inactivation, but the mechanisms by which the different regions are involved might be completely different.

Acknowledgments

I thank Drs. Dave Patton and Kris Kontis for critical reading of the manuscript. The author is a Lucille P. Markey Scholar. Work in this laboratory is supported by grants from the U.S. National Institutes of Health, the Lucille P. Markey Charitable Trust, the Muscular Dystrophy Association and the National Multiple Sclerosis Society.

References

Agnew, W. S., Levinson, S. R., Brabson, J. S., and Raftery, M. A. (1978). Purification of the tetrodotoxin-binding component associated with the voltage-sensitive sodium channel from *Electrophorus electricus* electroplax membranes. *Proc. Natl. Acad. Sci. USA* **75**, 2606–2610.

Agnew, W. S., Moore, A. C., Levinson, S. R., and Raftery, M. A. (1980). Identification of a large molecular weight peptide associated with a tetrodotoxin binding protein from the electroplax of *Electrophorus electricus*. *Biochem. Biophys. Res. Commun.* **92**, 860–866.

Ahmed, C. M. I., Ware, D. H., Lee, S. C., Patten, C. D., Ferrer-Montiel, A. V., Schinder, A. F., McPherson, J. D., Wagner-McPherson, C. B., Wasmuth, J. J., Evans, G. A., and Montal, M. (1992). Primary structure, chromosomal localization, and functional expression of a voltage-gated sodium channel from human brain. *Proc. Natl. Acad. Sci. USA* **89**, 8220–8224.

Aldrich, R. W., and Stevens, C. F. (1987). Voltage-dependent gating of single sodium channels from mammalian neuroblastoma cells. *J. Neurosci.* **7**, 418–431.

Armstrong, C. M., and Bezanilla, F. (1977). Inactivation of the sodium channel. II. Gating current experiments. *J. Gen. Physiol.* **70**, 567–590.

Armstrong, C. M., Bezanilla, F., and Rojas, E. (1973). Destruction of sodium conductance inactivation in squid axons perfused with pronase. *J. Gen. Physiol.* **62**, 375–391.

Auld, V. J., Goldin, A. L., Krafte, D. S., Marshall, J., Dunn, J. M., Catterall, W. A., Lester, H. A., Davidson, N., and Dunn, R. J. (1988). A rat brain Na$^+$ channel α subunit with novel gating properties. *Neuron* **1**, 449–461.

Auld, V. J., Goldin, A. L., Krafte, D. S., Catterall, W. A., Lester, H. A., Davidson, N., and Dunn, R. J. (1990). A neutral amino acid change in segment IIS4 dramatically alters the gating properties of the voltage-dependent sodium channel. *Proc. Natl. Acad. Sci. USA* **87**, 323–327.

Barchi, R. L., Cohen, S. A., and Murphy, L. E. (1980). Purification from rat sarcolemma of the saxitoxin-binding component of the excitable membrane sodium channel. *Proc. Natl. Acad. Sci. USA* **77**, 1306–1310.

Beckh, S., Noda, M., Lübbert, H., and Numa, S. (1989). Differential regulation of three sodium channel messenger RNAs in the rat central nervous system during development. *EMBO J.* **8**, 3611–3636.

Bennett, P. B., Jr., Makita, N., and George, A. L., Jr. (1993). A molecular basis for gating mode transitions in human skeletal muscle Na$^+$ channels. *FEBS Lett.* **326**, 21–24.

Bezanilla, F., and Armstrong, C. M. (1977). Inactivation of the sodium channel—Sodium current experiments. *J. Gen. Physiol.* **70**, 549–566.

Cannon, S. C., Brown, R. H., Jr., and Corey, D. P. (1991). A sodium channel defect in hyperkalemic periodic paralysis: potassium-induced failure of inactivation. *Neuron* **6**, 619–626.

Cannon, S. C., McClatchey, A. I., and Gusella, J. F. (1993). Modification of the Na$^+$ current conducted by the rat skeletal muscle α subunit by co-expression with a human brain β subunit. *Pflügers Arch.* **423**, 155–157.

Casadei, R. M., Gordon, R. D., and Barchi, R. L. (1986). Immunoaffinity isolation of Na$^+$ channels from rat skeletal muscle—Analysis of subunits. *J. Biol. Chem.* **261**, 4318–4323.

Catterall, W. A. (1988). Structure and function of voltage-sensitive ion channels. *Science* **242**, 50–61.

Chandy, K. G., Williams, C. B., Spencer, R. H., Aguilar, B. A., Ghanshani, S., Tempel, B. L., and Gutman, G. A. (1990). A family of three mouse potassium channel genes with intronless coding regions. *Science* **247**, 973–975.

Cribbs, L. L., Satin, J., Fozzard, H. A., and Rogart, R. B. (1990). Functional expression of the rat heart I Na$^+$ channel isoform. Demonstration of properties characteristic of native cardiac Na$^+$ channels. *FEBS Lett.* **275**, 195–200.

Durell, S. R., and Guy, H. R. (1992). Atomic scale structure and functional models of voltage-gated potassium channels. *Biophys. J.* **62**, 238–250.

Ebers, G. C., George, A. L., Jr., Barchi, R. L., Ting-Passador, S. S., Kallen, R. G., Lathrop, G. M., Beckmann, J. S., Hahn, A. F., Brown, W. F., Campbell, R. D., and Hudson, A. J. (1991). Paramyotonia congenita and hyperkalemic periodic paralysis are linked to the adult muscle sodium channel gene. *Ann. Neurol.* **30**, 810–816.

Fontaine, B., Khurana, T. S., Hoffman, E. P., Bruns, G. A. P., Haines, J. L., Trofatter, J. A., Hanson, M. P., Rich, J., McFarlane, H., Yasek, D. M., Romano, D., Gusella, J. F., and Brown, R. H., Jr. (1990). Hyperkalemic periodic paralysis and the adult muscle sodium channel α-subunit gene. *Science* **250**, 1000–1002.

Gellens, M. E., George, A. L., Jr., Chen, L., Chahine, M., Horn, R., Barchi, R. L., and Kallen, R. G. (1992). Primary structure and functional expression of the human cardiac tetrodotoxin-insensitive voltage-dependent sodium channel. *Proc. Natl. Acad. Sci. USA* **89**, 554–558.

George, A. L., Jr., Ledbetter, D. H., Kallen, R. G., and Barchi, R. L. (1991). Assignment of a human skeletal muscle sodium channel α-subunit gene (*SCN4A*) to 17q23.1-25.3. *Genomics* **9**, 555–556.

George, A. L., Jr., Komisarof, J., Kallen, R. G., and Barchi, R. L. (1992). Primary structure of the adult human skeletal muscle voltage-dependent sodium channel. *Ann. Neurol.* **31**, 131–137.

Goldin, A. L., Snutch, T., Lübbert, H., Dowsett, A., Marshall, J., Auld, V., Downey, W., Fritz, L. C., Lester, H. A., Dunn, R., Catterall, W. A., and Davidson, N. (1986). Messenger RNA coding for only the α subunit of the rat brain Na channel is sufficient for expression of functional channels in *Xenopus* oocytes. *Proc. Natl. Acad. Sci. USA* **83**, 7503–7507.

Goldman, L., and Schauf, C. L. (1972). Inactivation of the sodium current in *Myxicola* giant axons. Evidence for coupling to the activation process. *J. Gen. Physiol.* **59**, 659–975.

Gordon, D., Merrick, D., Wollner, D. A., and Catterall, W. A. (1988). Biochemical properties of sodium channels in a wide range of excitable tissues studied with site-directed antibodies. *Biochemistry* **27**, 7032–7038.

Gordon, R. D., Fieles, W. E., Schotland, D. L., Hogue-Angeletti, R., and Barchi, R. L. (1987). Topographical localization of the C-terminal region of the voltage-dependent sodium channel from *Electrophorus electricus* using antibodies raised against a synthetic peptide. *Proc. Natl. Acad. Sci. USA* **84**, 308–312.

Gordon, R. D., Li, Y., Fieles, W. E., Schotland, D. L., and Barchi, R. L. (1988). Topological localization of a segment of the eel voltage-dependent sodium channel primary sequence (AA 927–938) that discriminates between models of tertiary structure. *J. Neurosci.* **8**, 3742–3749.

Greeff, N. H., and Forster, I. C. (1991). The quantal gating charge of sodium channel inactivation. *Eur. Biophys. J.* **20**, 165–176.

Gundersen, C. B., Miledi, R., and Parker, I. (1983). Voltage-operated channels induced by foreign messenger RNA in *Xenopus* oocytes. *Proc. R. Soc. London B* **220**, 131–140.

Gundersen, C. B., Miledi, R., and Parker, I. (1984). Messenger RNA from human brain induces drug- and voltage-operated channels in *Xenopus* oocytes. *Nature (London)* **308**, 421–424.

Hartmann, H. A., Kirsch, G. E., Drewe, J. A., Taglialatela, M., Joho, R. H., and Brown, A. M. (1991). Exchange of conduction pathways between two related K$^+$ channels. *Science* **251**, 942–944.

Hartshorne, R. P., and Catterall, W. A. (1981). Purification of the saxitoxin receptor of the sodium channel from rat brain. *Proc. Natl. Acad. Sci. USA* **78**, 4620–4624.

Hartshorne, R. P., Messner, D. J., Coppersmith, J. C., and Catterall, W. A. (1982). The saxitoxin receptor of the sodium channel from rat brain—Evidence for two nonidentical subunits. *J. Biol. Chem.* **257**, 13888–13891.

Hirono, C., Yamagishi, S., Ohara, R., Hisanaga, Y., Nakayama, T., and Sugiyama, H. (1985). Characterization of mRNA responsible for induction of functional sodium channels in *Xenopus* oocytes. *Brain Res.* **359**, 57–64.

Hoshi, T., Zagotta, W. N., and Aldrich, R. W. (1990). Biophysical and molecular mechanisms of *Shaker* potassium channel inactivation. *Science* **250**, 533–538.

Isacoff, E. Y., Jan, Y. N., and Jan, L. Y. (1991). Putative receptor for the cytoplasmic inactivation gate in the *Shaker* K$^+$ channel. *Nature (London)* **353**, 86–90.

Isom, L. L., DeJongh, K. S., Patton, D. E., Reber, B. F. X., Offord, J., Charbonneau, H., Walsh, K., Goldin, A. L., and Catterall, W. A. (1992). Primary structure and functional expression of the β_1 subunit of the rat brain sodium channel. *Science* **256**, 839–842.

Iverson, L. E., Tanouye, M. A., Lester, H. A., Davidson, N., and Rudy, B. (1988). A-type potassium channels expressed from *Shaker* locus cDNA. *Proc. Natl. Acad. Sci. USA* **85**, 5723–5727.

Jan, L. Y., and Jan, Y.-N. (1989). Voltage-sensitive ion channels. *Cell* **56**, 13–25.

Joho, R. H., Moorman, J. R., VanDongen, A. M. J., Kirsch, G. E., Silberberg, H., Schuster, G., and Brown, A. M. (1990). Toxin and kinetic profile of rat brain type III sodium channel expressed in *Xenopus* oocytes. *Mol. Brain Res.* **7**, 105–113.

Kallen, R. G., Sheng, Z.-H., Yang, J., Chen, L., Rogart, R. B., and Barchi, R. L. (1990). Primary structure and expression of a sodium channel characteristic of denervated and immature rat skeletal muscle. *Neuron* **4**, 233–242.

Kayano, T., Noda, M., Flockerzi, V., Takahashi, H., and Numa, S. (1988). Primary structure of rat brain sodium channel III deduced from the cDNA sequence. *FEBS Lett.* **228**, 187–194.

Krafte, D. S., Snutch, T. P., Leonard, J. P., Davidson, N., and Lester, H. A. (1988). Evidence for the involvement of more than one mRNA species in controlling the inactivation process of rat brain Na channels expressed in *Xenopus* oocytes. *J. Neurosci.* **8**, 2859–2868.

Krafte, D. S., Goldin, A. L., Auld, V. J., Dunn, R. J., Davidson, N., and Lester, H. A. (1990). Inactivation of cloned Na channels expressed in *Xenopus* oocytes. *J. Gen. Physiol.* **96**, 689–706.

Kraner, S. D., Tanaka, J. C., and Barchi, R. L. (1985). Purification and functional reconstitution of the voltage-sensitive sodium channel from rabbit T-tubular membranes. *J. Biol. Chem.* **260**, 6341–6347.

Lehmann-Horn, F., Iaizzo, P. A., Hatt, H., and Franke, C. (1991). Altered gating and conductance of Na$^+$ channels in hyperkalemic periodic paralysis. *Pflügers Arch.* **418**, 297–299.

Lombet, A., and Lazdunski, M. (1984). Characterization, solubilization, affinity labeling and purification of the cardiac Na+ channel using *Tityus* toxin γ. *Eur. J. Biochem.* **141**, 651–660.

Loughney, K., Kreber, R., and Ganetzky, B. (1989). Molecular analysis of the *para* locus, a sodium channel gene in Drosophila. *Cell* **58**, 1143–1154.

MacKinnon, R. (1991). Determination of the subunit stoichiometry of a voltage-activated potassium channel. *Nature (London)* **350**, 232–235.

McClatchey, A. I., Cannon, S. C., Slaugenhaupt, S. A., and Gusella, J. F. (1993). The cloning and expression of a sodium channel β1-subunit cDNA from human brain. *Hum. Molec. Genet.* **2**, 745–749.

McClatchey, A. I., Van den Bergh, P., Pericak-Vance, M. A., Raskind, W., Verellen, C., McKenna-Yasek, D., Rao, K., Haines, J. L., Bird, T., Brown, R. H., Jr., and Gusella, J. F. (1992). Temperature-sensitive mutations in the III–IV cytoplasmic loop region of the skeletal muscle sodium channel gene in paramyotonia congenita. *Cell* **68**, 769–774.

Messner, D. J., and Catterall, W. A. (1985). The sodium channel from rat brain—Separation and characterization of subunits. *J. Biol. Chem.* **260**, 10597–10604.

Messner, D. J., and Catterall, W. A. (1986). The sodium channel from rat brain—Role of the β1 and β2 subunits in saxitoxin binding. *J. Biol. Chem.* **261**, 211–215.

Messner, D. J., Feller, D. J., Scheuer, T., and Catterall, W. A. (1986). Functional properties of rat brain sodium channels lacking the β1 or β2 subunit. *J. Biol. Chem.* **261**, 14882–14890.

Meves, H., and Vogel, W. (1977). Inactivation of the asymmetrical displacement current in giant axons of *Loligo forbesi*. *J. Physiol.* **267**, 377–393.

Miller, J. A., Agnew, W. S., and Levinson, S. R. (1983). Principal glycopeptide of the tetrodotoxin/saxitoxin binding protein from *Electrophorus electricus*: Isolation and partial chemical and physical characterization. *Biochemistry* **22**, 470–476.

Moorman, J. R., Kirsch, G. E., Brown, A. M., and Joho, R. H. (1990a). Changes in sodium channel gating produced by point mutations in a cytoplasmic linker. *Science* **250**, 688–691.

Moorman, J. R., Kirsch, G. E., VanDongen, A. M. J., Joho, R. H., and Brown, A. M. (1990b). Fast and slow gating of sodium channels encoded by a single mRNA. *Neuron* **4**, 243–252.

Noda, M., Shimizu, S., Tanabe, T., Takai, T., Kayano, T., Ikeda, T., Takahashi, H., Nakayama, H., Kanaoka, Y., Minamino, N., Kangawa, K., Matsuo, H., Raftery, M. A., Hirose, T., Inayama, S., Hayashida, H., Miyata, T., and Numa, S. (1984). Primary structure of *Electrophorus electricus* sodium channel deduced from cDNA sequence. *Nature (London)* **312**, 121–127.

Noda, M., Ikeda, T., Kayano, T., Suzuki, H., Takeshima, H., Kurasaki, M., Takahashi, H., and Numa, S. (1986a). Existence of distinct sodium channel messenger RNAs in rat brain. *Nature (London)* **320**, 188–192.

Noda, M., Ikeda, T., Suzuki, H., Takeshima, H., Takahashi, T., Kuno, M., and Numa, S. (1986b). Expression of functional sodium channels from cloned cDNA. *Nature (London)* **322**, 826–828.

Nonner, W. (1980). Relations between the inactivation of sodium channels and the immobilization of gating charge in frog myelinated nerve. *J. Physiol.* **299**, 573–603.

Numann, R., Catterall, W. A., and Scheuer, T. (1991). Functional modulation of brain sodium channels by protein kinase C phosphorylation. *Science* **254**, 115–118.

Patton, D. E., and Goldin, A. L. (1991). A voltage-dependent gating transition induces use-dependent block by tetrodotoxin of rat IIA sodium channels expressed in *Xenopus* oocytes. *Neuron* **7**, 637–647.

Patton, D. E., West, J. W., Catterall, W. A., and Goldin, A. L. (1992). Amino acid residues required for fast Na+-channel inactivation. Charge neutralizations and deletions in the III–IV linker. *Proc. Natl. Acad. Sci. USA* **89**, 10905–10909.

Patton, D. E., West, J. W., Catterall, W. A., and Goldin, A. L. (1993). A peptide segment critical for sodium channel inactivation functions as an inactivation gate in a potassium channel. *Neuron* **11**, 967–974.

Patton, D. E., Isom, L. L., Catterall, W. A., and Goldin, A. L. (1994). The adult rat brain β1 subunit modifies activation and inactivation gating of multiple sodium channel α subunits. *J. Biol. Chem.* (in press).

Ptacek, L. J., George, A. L., Jr., Griggs, R. C., Tawil, R., Kallen, R. G., Barchi, R. L., Robertson, M., and Leppert, M. F. (1991a). Identification of a mutations in the gene causing hyperkalemic periodic paralysis. *Cell* **67**, 1021–1027.

Ptacek, L. J., Trimmer, J. S., Agnew, W. S., Roberts, J. W., Petajan, J. H., and Leppert, M. (1991b). Paramyotonia congenita and hyperkalemic periodic paralysis map to the same sodium-channel gene locus. *Am. J. Hum. Genet.* **49**, 851–854.

Ptacek, L. J., Tyler, F., Trimmer, J. S., Agnew, W. S., and Leppert, M. (1991c). Analysis in a large hyperkalemic periodic paralysis pedigree supports tight linkage to a sodium channel locus. *Am. J. Hum. Genet.* **49**, 378–382.

Ptacek, L. J., George, A. L., Jr., Barchi, R. L., Griggs, R. C., Riggs, J. E., Robertson, M., and Leppert, M. F. (1992a). Mutations in an S4 segment of the adult skeletal muscle sodium channel cause paramyotonia congenita. *Neuron* **8**, 891–897.

Ptacek, L. J., Tawil, R., Griggs, R. C., Storvick, D., and Leppert, M. (1992b). Linkage of atypical myotonia congenita to a sodium channel locus. *Neurology* **42**, 431–433.

Ramaswami, M., and Tanouye, M. A. (1989). Two sodium-channel genes in Drosophila: Implications for channel diversity. *Proc. Natl. Acad. Sci. USA* **86**, 2079–2082.

Roberts, R. H., and Barchi, R. L. (1987). The voltage-sensitive sodium channel from rabbit skeletal muscle. Chemical characterization of subunits. *J. Biol. Chem.* **262**, 2298–2303.

Rogart, R. B., Cribbs, L. L., Muglia, L. K., Kephart, D. D., and Kaiser, M. W. (1989). Molecular cloning of a putative tetrodotoxin-resistant rat heart Na+ channel isoform. *Proc. Natl. Acad. Sci. USA* **86**, 8170–8174.

Rojas, C. V., Wang, J., Schwartz, L. S., Hoffman, E. P., Powell, B. R., and Brown, R. H., Jr. (1991). A met-to-val mutation in the skeletal muscle Na+ channel α-subunit in hyperkalaemic periodic paralysis. *Nature (London)* **354**, 387–389.

Rojas, E., and Rudy, B. (1976). Destruction of the sodium conductance inactivation by a specific protease in perfused nerve fibres from *Loligo*. *J. Physiol.* **262**, 501–531.

Rosenberg, R. L., Tomiko, S. A., and Agnew, W. S. (1984a). Reconstitution of neurotoxin-modulated ion transport by the voltage-regulated sodium channel isolated from the electroplax of *Electrophorus electricus*. *Proc. Natl. Acad. Sci. USA* **81**, 1239–1243.

Rosenberg, R. L., Tomiko, S. A., and Agnew, W. S. (1984b). Single-channel properties of the reconstituted voltage-regulated Na channel isolated from the electroplax of *Electrophorus electricus*. *Proc. Natl. Acad. Sci. USA* **81**, 5594–5598.

Rossie, S., Gordon, D., and Catterall, W. A. (1987). Identification of an intracellular domain of the sodium channel having multiple cAMP-dependent phosphorylation sites. *J. Biol. Chem.* **262**, 17530–17535.

Salkoff, L., Butler, A., Wei, A., Scavarda, N., Giffen, K., Ifune, C., Goodman, R., and Mandel, G. (1987). Genomic organization and deduced amino acid sequence of a putative sodium channel gene in Drosophila. *Science* **237**, 744–748.

Scheinman, R. I., Auld, V. J., Goldin, A. L., Davidson, N., Dunn, R. J., and Catterall, W. A. (1989). Developmental regulation of sodium channel expression in the rat forebrain. *J. Biol. Chem.* **264**, 10660–10666.

Schmidt, J. W., and Catterall, W. A. (1986). Biosynthesis and processing of the α subunit of the voltage-sensitive sodium channel in rat brain. *Cell* **46**, 437–445.

Schmidt, J. W., Rossie, S., and Catterall, W. A. (1985). A large intracellular pool of inactive Na channel α subunits in developing rat brain. *Proc. Natl. Acad. Sci. USA* **82**, 4847–4851.

Stimers, J. R., Bezanilla, F., and Taylor, R. E. (1985). Sodium channel activation in the squid giant axon. Steady state properties. *J. Gen. Physiol.* **85**, 65–82.

Stühmer, W., Methfessel, C., Sakmann, B., Noda, M., and Numa, S. (1987). Patch clamp characterization of sodium channels expressed from rat brain cDNA. *Eur. Biophys. J.* **14**, 131–138.

Stühmer, W., Conti, F., Suzuki, H., Wang, X., Noda, M., Yahagi, N., Kubo, H., and Numa, S. (1989a). Structural parts involved in activation and inactivation of the sodium channel. *Nature (London)* **339**, 597–603.

Stühmer, W., Ruppersberg, J. P., Schrote, K. S., Sakmann, B., Stocker, M., Giese, K. P., Perschke, A., Baumann, A., and Pongs, O. (1989b). Molecular basis of functional diversity of voltage gated potassium channels in mammalian brain. *EMBO J.* **8**, 3235–3244.

Sumikawa, K., Parker, I., and Miledi, R. (1984). Partial purification and functional expression of brain mRNAs coding for neurotransmitter receptors and voltage-operated channels. *Proc. Natl. Acad. Sci. USA* **81**, 7994–7998.

Sutkowski, E. M., and Catterall, W. A. (1990). β1 Subunits of sodium channels. Studies with subunit-specific antibodies. *J. Biol. Chem.* **265**, 12393–12399.

Suzuki, H., Beckh, S., Kubo, H., Yahagi, N., Ishida, H., Kayano, T., Noda, M., and Numa, S. (1988). Functional expression of cloned cDNA encoding sodium channel III. *FEBS Lett.* **228**, 195–200.

Talvenheimo, J. A., Tamkun, M. M., and Catterall, W. A. (1982). Reconstitution of neurotoxin-stimulated sodium transport by the voltage-sensitive sodium channel purified from rat brain. *J. Biol. Chem.* **257**, 11868–11871.

Tempel, B. L., Jan, Y.-N., and Jan, L. Y. (1988). Cloning of a probable potassium channel gene from mouse brain. *Nature (London)* **332**, 837–839.

Tong, J., Potts, J. F., Rochelle, J. M., Seldin, M. F., and Agnew, W. S. (1993). A single β₁ subunit mapped to mouse chromosome 7 may be a common component of Na channel isoforms from brain, skeletal muscle and heart. *Biochem. Biophys. Res. Commun.* **195**, 679–685.

Trimmer, J. S., Cooperman, S. S., Tomiko, S. A., Zhou, J., Crean, S. M., Boyle, M. B., Kallen, R. G., Sheng, Z., Barchi, R. L., Sigworth, F. J., Goodman, R. H., Agnew, W. S., and Mandel, G. (1989). Primary structure and functional expression of a mammalian skeletal muscle sodium channel. *Neuron* **3**, 33–49.

Vassilev, P. M., Scheuer, T., and Catterall, W. A. (1988). Identification of an intracellular peptide segment involved in sodium channel inactivation. *Science* **241**, 1658–1661.

Vassilev, P., Scheuer, T., and Catterall, W. A. (1989). Inhibition of inactivation of single sodium channels by a site-directed antibody. *Proc. Natl. Acad. Sci. USA* **86**, 8147–8151.

Vermersch, P. S., Tesmer, J. J. G., Lemon, D. D., and Quiocho, F. A. (1990). A pro to gly mutation in the hinge of the arabinose-binding protein enhances binding and alters specificity. *J. Biol. Chem.* **265**, 16592–16603.

Wang, J., Rojas, C. V., Zhou, J., Schwartz, L. S., Nicholas, H., and Hoffman, E. P. (1992). Sequence and genomic structure of the human adult skeletal muscle sodium channel α subunit gene on 17q. *Biochem. Biophys. Res. Commun.* **182**, 794–801.

Weigele, J. B., and Barchi, R. L. (1982). Functional reconstitution of the purified sodium channel protein from rat sarcolemma. *Proc. Natl. Acad. Sci. USA* **79**, 3651–3655.

West, J. W., Numann, R., Murphy, B. J., Scheuer, T., and Catterall, W. A. (1991). A phosphorylation site in the Na⁺ channel required for modulation by protein kinase C. *Science* **254**, 866–868.

West, J. W., Patton, D. E., Scheuer, T., Wang, Y., Goldin, A. L., and Catterall, W. A. (1992). A cluster of hydrophobic amino acid residues required for fast Na⁺-channel inactivation. *Proc. Natl. Acad. Sci. USA* **89**, 10910–10914.

White, M. M., Chen, L., Kleinfield, R., Kallen, R. G., and Barchi, R. L. (1991). SkM2, a Na⁺ channel cDNA clone from denervated skeletal muscle, encodes a tetrodotoxin-insensitive Na⁺ channel. *Mol. Pharmacol.* **39**, 604–608.

Yellen, G., Jurman, M. E., Abramson, T., and MacKinnon, R. (1991). Mutations affecting internal TEA blockade identify the probable pore-forming region of a K⁺ channel. *Science* **251**, 939–942.

Yool, A. J., and Schwarz, T. L. (1991). Alteration of ionic selectivity of a K⁺ channel by mutation of the H5 region. *Nature (London)* **349**, 700–704.

Zagotta, W. N., Hoshi, T., and Aldrich, R. W. (1990). Restoration of inactivation in mutants of *Shaker* potassium channels by a peptide derived from ShB. *Science* **250**, 568–571.

Zhou, J., Potts, J. F., Trimmer, J. S., Agnew, W. S., and Sigworth, F. J. (1991). Multiple gating modes and the effect of modulating factors on the μI sodium channel. *Neuron* **7**, 775–785.

Unitary Properties of the Batrachotoxin-Trapped State of Voltage-Sensitive Sodium Channels

EDWARD MOCZYDLOWSKI
AND LAURENT SCHILD

I. Introduction

A. Na⁺ Channel Isoforms

The voltage-sensitive Na^+ channel and its relatives that conduct Ca^{2+} and K^+ are integral membrane proteins that share primary responsibility for electrical excitability of nerve, muscle, and heart. As such, these channels have stimulated efforts of electrophysiologists and molecular biologists to elucidate their function and molecular properties. Voltage-sensitive Na^+ channels contain a large glycosylated α subunit. Molecular cloning studies to date have revealed a family of Na^+ channel α subunits that are the products of at least five distinct genes in the rat. The proteins encoded by these rat genes range from 1820 to 2019 amino acid residues, corresponding to a protein molecular mass of 208–229 kDa (Numa and Noda, 1986; Kayano et al., 1988; Rogart et al., 1989; Trimmer et al., 1989). Na^+ channels in mammalian brain have also been found to contain smaller β_1 and β_2 (33–36 kDa) subunits (Catterall, 1988). Mammalian muscle Na^+ channels similarly contain a 38-kDa subunit (Roberts and Barchi, 1987) that is analogous to β_1 of brain. Reconstitution of purified α subunits and studies of cloned and expressed α subunits have shown that this subunit alone is capable of forming functional channels that are blocked by the guanidinium toxins

tetrodotoxin (TTX) and saxitoxin (STX) (for review, Trimmer and Agnew, 1989). Researchers have found that β_1 subunit appears to play a role in the assembly of functional channels and in the modulation of gating kinetics (Isom et al., 1992).

The various Na^+ channel isoforms exhibit differential tissue-specific expression and developmental regulation. These isoforms also account for pharmacological subtypes of Na^+ currents that have been distinguished on the basis of sensitivity to TTX, STX, μ-conotoxin, certain divalent cations such as Cd^{2+} and Zn^{2+}, and other ligands. Rat brain Na^+ channels I, II, and III, and the $\mu1$ (or SkM1) isoform of adult skeletal muscle are all blocked by nanomolar concentrations of TTX and STX (Noda et al., 1986; Suzuki et al., 1988; Trimmer et al., 1989). In contrast, the rat cardiac Na^+ channel (RH1 or SkM2) that is also expressed in denervated skeletal muscle (Kallen et al., 1990) is about 100-fold less sensitive to these guanidinium toxins (Cribbs et al., 1990; White et al., 1991). The small μ-conotoxin peptide produced by the marine snail *Conus geographus* is an effective blocker of the skeletal muscle $\mu1$ Na^+ channel (Moczydlowski et al., 1986; Trimmer et al., 1989) but has low affinity for other isoforms. The cardiac RH1 isoform is also characterized by greater sensitivity to block by the group IIB divalent cations Cd^{2+} and Zn^{2+} (Frelin et al., 1986; Ravindran et al., 1991). Primary sequences of the analogous Na^+ channel isoforms cloned from human skeletal and cardiac muscle have been reported (Gellens *et al.*, 1992; George *et al.*, 1992b). Isolation of a unique type of Na$^+$ channel cDNA from adult human heart has also been described (George *et al.*, 1992a). This isoform, called hNa$_v$2.1 is expressed in heart and uterus and has only moderate similarity (<50%) to other isoforms; this result is suggestive of a distinct gene family.

At the most basic level, three questions about Na$^+$ channel function have long been at the heart of this field of investigation since it was firmly placed on a quantitative foundation by Hodgkin and Huxley (1952). How does the channel open and close? How do ions selectively pass through the pore? How do drugs and toxins affect function? Each of these deceivingly simple questions motivates whole fields of research that are more broadly described as gating, permeation, and pharmacology. This chapter reviews these issues based on a perspective that derives from work on a drugged form of the channel—the batrachotoxin-activated or batrachotoxin-modified Na$^+$ channel.

B. Pharmacology of Batrachotoxin

Batrachotoxin (BTX) is an unusual steroidal alkaloid that is isolated from the skin of small colorful frogs of the South American rain forest that belong to the genus *Phyllobates* (Tokuyama *et al.*, 1969; Albuquerque *et al.*, 1971; Albuquerque and Daly, 1976). The pharmacological history of BTX extends from its primitive use as a poison in blow dart hunting by native South Americans to modern applications in Na$^+$ channel biophysics. Figure 1 shows the structures of several BTX homologs as well as a photograph of *Phyllobates aurotaenia*, the frog species that was the original source of samples of BTX for structural chemistry (Tokuyama *et al.*, 1969). Of the four structures shown in Fig. 1, batrachotoxinin A, batrachotoxin, and homobatrachotoxin are natural homologs; however, batrachotoxinin A is about 500-fold less toxic than the others. The structure of the synthetic 20-α-*p*-bromobenzoate derivative of batrachotoxinin A was deduced by X-ray crystallography (Tokuyama *et al.*, 1968). Current understanding of the mechanism of action of BTX and its availability for the study of Na$^+$ channels owes much to the work of Daly and collaborators (Daly *et al.*, 1987). Nearly every publication dealing with BTX cites the gift of this toxin from Daly (Laboratory of Bioorganic Chemistry, U.S. National Institutes of Health). The surprising discovery of homobatrachotoxin in the feathers of the toxic pitohui bird of New Guinea (Dumbacher *et al.*, 1992) suggests that the oc-

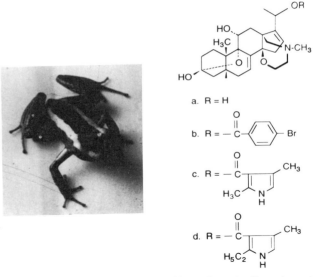

FIGURE 1 Source and structure of batrachotoxin. Batrachotoxin (c) was originally isolated from the skin of the Colombian poison arrow frog *Phyllobates aurotaneia* (*left*), which also contains the natural homologs batrachotoxinin A (a) and homobatrachotoxin (d). The synthetic derivative (b) is the *O*-*p*-bromobenzoate of batrachotoxinin A. The photograph is by Charles Myers (American Museum of Natural History). The structures are reproduced with permission from Albuquerque *et al.* (1971). Copyright © 1971 by the AAAS.

currence of BTX in nature may be more extensive than previously suspected.

In practically all preparations of excitable cells that have been tested, BTX causes a time- and dose-dependent depolarization of the resting membrane potential that is the result of persistent activation of voltage-sensitive Na$^+$ channels. Although the Na$^+$ channel is the functionally relevant target of BTX, BTX has also been reported to block voltage-dependent Ca^{2+} channels with a $K_{0.5}$ of 40 nM (Romey and Lazdunski, 1982). The binding site recognized by BTX on Na$^+$ channels has been named receptor site 2 by Catterall (1988), whose work has shown that this site also binds various plant alkaloids such as veratridine, aconitine, and grayanotoxin, as judged by competitive binding interactions (Catterall, 1975; Catterall *et al.*, 1981). BTX and other alkaloids acting at receptor site 2 have profound effects on many aspects of Na$^+$ channel function. These effects include supression of fast and slow inactivation, a hyperpolarizing shift in the voltage-dependence of activation, reduction in unitary conductance, alteration of ionic selectivity, and reduction of the blocking affinity of many local anesthetic drugs (Khodorov, 1985). In exerting these effects on Na$^+$ channel function, BTX is the most potent and effective agent, in terms of removal of time-dependent inactivation, that is known to act at receptor site 2.

In view of this profound alteration of "normal" function, why has the BTX-modified form of the Na^+ channel been chosen as the focus of such extensive investigation? One reason is that the alterations induced by BTX appear to reflect changes in function that are also exhibited by normal unadulterated Na^+ channels. For example, a small fraction of normal Na^+ channels in muscle and heart exhibit long-lived states of high opening probability that indicate a supression of inactivation (Patlak and Ortiz, 1985,1986). Single Na^+ channels in cell membrane patches have also been found to exhibit modal shifts to states of high opening probability that are reminiscent of BTX-activated channels (Nilius, 1988a; Kirsch and Brown, 1989; Ukomadu *et al.*, 1992). Discrete subconductance events of single Na^+ channels have been documented in a variety of preparations that may be related to the lower unitary conductance state induced by BTX (Patlak, 1988; Chinn and Narahashi, 1989; Nilius *et al.*, 1989). Heterogeneity in the ionic selectivity of normal Na^+ channels from *Electrophorus* electrolytes has been inferred from patch-clamp studies (Shenkel and Sigworth, 1991). Differences in the binding affinity of local anesthetic molecules to various states of the channel also constitute an aspect of normal function according to the "modulated receptor hypothesis" (Hille, 1992). Such phenomena are only a few examples that emphasize that the Na^+ channel is a conformationally active protein. Alterations in function and ligand binding that occur during turnover among closed, open, and inactivated states are extraordinarily complex and often cannot be derived unambiguously from normal Na^+ channels observed under physiological conditions.

BTX is an important pharmacological tool for studying the Na^+ channel primarily because it greatly stabilizes the open conformation of the channel. At membrane voltages more positive than about -70 mV, BTX-modified Na^+ channels exhibit a high open-state probability. In this voltage range, BTX-activated Na^+ channels are effectively locked into a well-defined open state, permitting ligand interactions with the open channel to be studied at equilibrium. For biophysical studies of ligand interactions, this situation greatly reduces the complexity of interpreting voltage-pulse experiments that cycle the channel among numerous intermediate states, each of which may exhibit altered affinity and function with respect to any given ligand. The use of BTX to trap an open channel conformation that is normally brief and elusive may be compared with analogous strategies in enzymology or protein crystallography in which inhibitors are used to trap conformational intermediates. Such an approach is often necessary in systems in which conformational changes that accompany turnover must be understood.

II. Gating Behavior

A. Preferential Binding of BTX to the Open State

The site that binds BTX appears to be coupled to structural domains of the Na^+ channel protein that are involved in voltage-dependent gating. This coupling is implied by the alteration of the gating kinetics of BTX-bound channels and the observation that the binding of BTX itself is state dependent. BTX binds very slowly, if at all, to the normal closed or resting state of Na^+ channels. Various electrophysiological studies have shown that BTX modification is favored by repetitive depolarization (Khodorov, 1985). For example, perfusion of a squid axon for up to 1 hr with 5 μM BTX has no effect on Na^+ current when the axon membrane is held clamped at -150 mV (Tanguy and Yeh, 1991). However, BTX modification in this preparation can readily be achieved by repetitive depolarization to $+80$ mV, which permits the exposure of BTX to the open state. In squid, close examination of the state dependence of BTX binding indicates that binding rapidly occurs to the normal open state and at a lower rate to the slow inactivated state of channels lacking fast inactivation by treatment with Pronase®, chloramine-T, or *N*-bromoacetamide (Tanguy and Yeh, 1991). Once bound, BTX has a slow rate of dissociation. The half-time for dissociation of a tritiated derivative of BTX is 60 min at 36°C in rat brain synaptosomes (Catterall *et al.*, 1981). This rate is similar to that estimated from the rate of "spontaneous disappearance" of BTX-modified channels in planar bilayers (Green *et al.*, 1987a), an event that is thought to correspond to BTX dissociation.

The radioligand [³H]batrachotoxinin-A 20-α-benzoate ([³H]BTX-B) has been developed for use in analyzing interactions between the BTX-binding site and other sites of drug/toxin interaction with Na^+ channels (Brown *et al.*, 1981). Binding of [³H]BTX-B to rat brain synatopsomal membranes in the presence of *Leiurus* scorpion toxin is well described by a single class of sites with a K_d of 82 nM (Catterall *et al.*, 1981). Positive heterotropic binding interactions (mutually stabilizing) occur between the BTX-binding site and sites for other ligands that enhance channel opening, including scorpion and sea anemone polypeptide toxins (Catterall *et al.*, 1981), pyrethroid insecticides (Brown *et al.*, 1988), and brevetoxin (Catterall and Gainer, 1985). In contrast, negative heterotropic (mutually destabilizing) binding interactions have been observed between [³H]BTX-B binding and sites for various channel blockers such as cocaine and other local anesthetics (Creveling *et al.*, 1983; Postma and Catterall, 1984; McNeal *et al.*, 1985; Reith *et al.*, 1986), anticonvulsants such as carbamazepine and diphenylhydantoin (phenytoin) (Willow and Catterall, 1982), and the guanidinium toxins STX and TTX (Brown, 1986). Thus, state-dependent binding of BTX is also manifested by allosteric binding interactions with a multitude of other drug/toxin binding sites that are known to mediate, block, or modulate channel gating.

B. Gating of BTX-Modified Na⁺ Channels

The slow rate of BTX dissociation permits the BTX-trapped state of Na⁺ channels to be isolated and studied for long periods of time. The stable BTX-bound form of the channel avoids kinetic complexities that occur with reversible modifiers, for example, dissociation of drug-bound states and interconversion with the unmodified channel. With this advantage, the gating behavior of BTX-modified Na⁺ channels has been extensively studied at the macroscopic and single-channel levels. Typical examples of macroscopic Na⁺ currents recorded before and after BTX modification in a mouse–rat neuroblastoma–glioma cell line (Huang *et al.*, 1982) and in squid giant axon (Tanguy and Yeh, 1988) are shown in Fig. 2. Such experiments show that the normal fast inactivation process is essentially abolished by BTX, as judged by the absence of significant decay in the maximal level of the activated current (Fig. 2B,D).

In addition to fast time-dependent inactivation elicited by brief depolarizations (~10 msec), normally gating Na⁺ channels also undergo a slow inactivation process that occurs when the membrane voltage is held at depolarized levels for seconds to minutes (Rudy, 1978). This slow inactivation process has also been reported to be abolished by BTX in neuroblastoma cells (Huang *et al.*, 1982). However, BTX-modified Na⁺ currents in rat cardiac cells exhibit about 25% residual slow inactivation (Huang *et al.*, 1987). In GH₃ pituitary cells, Wang and Wang (1992) have described a form of slow inactivation that occurs when BTX-modified channels are exposed to a 5-sec depolarizing prepulse. In contrast to the steady-state voltage dependence of normal fast inactivation, which can be approximated by a Boltzmann function of voltage, inactivation of BTX-modified Na⁺ current in GH₃ cells

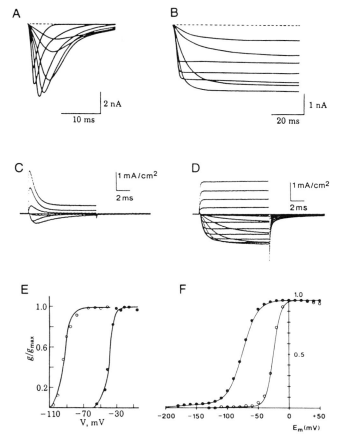

FIGURE 2 Effect of batrachotoxin (BTX) on macroscopic Na⁺ currents in neuroblastoma cells and squid axon. (A) Normal Na⁺ currents from a mouse–rat neuroblastoma–glioma cell in response to a series of step depolarizations in the range of −45 to +15 mV. (B) BTX-modified Na⁺ currents from a neuroblastoma cell. Depolarization steps were from −90 to −25 mV. (C) Normal Na⁺ currents from squid giant axon. Depolarization steps were in the range of −40 to +60 mV. (D) BTX-modified Na⁺ currents from squid giant axon. Depolarization steps were from −40 to +60 mV. (E) Voltage dependence of activation in neuroblastoma cells described by relative Na⁺ conductance (*g*/*g*ₘₐₓ) before (●) and after (○) modification by BTX. (F) Voltage dependence of activation (*g*/*g*ₘₐₓ) in squid giant axon before (○) and after (●) modification by BTX. (A, B, E) Reproduced with permission from Huang *et al.* (1982). (C, D) Reproduced from Tanguy and Yeh, *The Biophysical Journal* (1988), **54**, 719–730, by copyright permission of the Biophysical Society. (F) Reproduced from Tanguy and Yeh, *The Journal of General Physiology* (1991), **97**, 499–159, by copyright permission of the Rockefeller University Press.

exhibits a biphasic voltage dependence with a minimum near −70 mV, at which 45% of the current is inactivated (Wang and Wang, 1992). The relationship of this latter phenomenon to normal fast or slow inactivation is unclear, but the authors have suggested that BTX-modified channels may exhibit a residual form of inactivation related to a certain population of spontaneous closing events observed in single BTX-modified channels.

In addition to the virtual absence of fast inactivation, the kinetics of macroscopic BTX-modified Na⁺ currents differ from normal Na⁺ currents in other important ways. Whereas the rising phase of normal Na⁺ channel activation is slightly sigmoidal, accounted for by m^3 in the Hodgkin–Huxley model (Hodgkin and Huxley, 1952), BTX-modified Na⁺ currents activate with a nearly mono-exponential time course (Huang *et al.*, 1982,1987). Also, the voltage dependence of activation is shifted to a voltage range that is ~50 mV more negative than that of unmodified Na⁺ currents. This effect is illustrated in Fig. 2(E,F) by plots of relative conductance vs. voltage for neuroblastoma cells (Huang *et al.*, 1982) and squid axon (Tanguy and Yeh, 1991). Generally, the steepness of the voltage dependence or apparent gating charge (q) reflected in the conductance–voltage relationship is not significantly different from that of unmodified Na⁺ current (Fig. 2E); however, results in squid axon show a ~2-fold reduction in the steepness of the macroscopic voltage dependence of activation after BTX modification (Fig. 2F; Tanguy and Yeh, 1991).

To determine whether BTX alters the movement of gating charge that accompanies Na⁺ channel opening, Tanguy and Yeh (1988) measured the ON gating current that remains in squid axon after various durations of a conditioning prepulse. These investigators found that the conditioning pulse led to an immobilization of ~57% of the total gating charge for both normal and BTX-modified current. However, the onset of charge immobilization was about three times faster in BTX-modified than in unmodified axons. These experiments also showed that the voltage dependence of total gating charge movement (Q_{ON} vs V) was not significantly different in normal and BTX-modified axons. The phenomenon of gating charge immobilization is thought to be closely linked to inactivation (Armstrong and Bezanilla, 1977). In support of this view, the treatment of squid axons with agents such as Pronase® or chloramine-T abolishes fast inactivation and concomitantly eliminates charge immobilization (Tanguy and Yeh, 1988). Since BTX removes inactivation while retaining gating charge immobilization, BTX is proposed to uncouple these two processes. According to the "ball and chain" model of inactivation, one way to explain these observations is that BTX interferes with the binding of the inactivation particle to its receptor but mimics this state with respect to immobilization of gating charge (Tanguy and Yeh, 1988; Correa *et al.*, 1992).

Further insight into the gating process of BTX-modified Na⁺ channels has been gained by analysis of opening and closing fluctuations of single channels isolated in patch pipettes or planar bilayers. Figure 3 shows a comparison of steady-state current records from single BTX-modified Na⁺ channels from squid optic nerve and rat skeletal muscle recorded in planar bilayers. This comparison illustrates that single BTX-modified Na⁺ channels from diverse sources exhibit a characteristic voltage dependence of steady-state opening probability. At holding voltages more positive than about −60 mV, the channel dwells in the open state nearly all the time. In this voltage range, the opening probability is essentially voltage independent. Several studies have found that single BTX-modified Na⁺ channels do exhibit occasional brief closing events at positive voltage, resulting in a maximal open-state probability of ~0.95 (Huang *et al.*, 1984; Moczydlowski *et al.*, 1984a; Keller *et al.*, 1986). However, higher P_{open} values approaching 1.00 are found in some preparations (Behrens *et al.*, 1989; Correa *et al.*, 1992). In a narrow voltage range more negative than −60 mV, BTX-modified Na⁺ channels exhibit rapid closing fluctuations (Fig. 3). In this range P_{open} declines to zero as a steep function of voltage that closely matches the voltage range of activation found macroscopically (e.g., Fig. 2E,F). The dependence of P_{open} on voltage can generally be described by the Boltzmann function

$$P_{open} = \{1 + \exp[-qF(V - V_0)/RT]\}^{-1}, \qquad (1)$$

where q is the apparent gating charge and V_0 is the voltage at which channels are open 50% of the time. However, in some cases, better fits of P_{open} vs V have been found with a more complicated function containing two exponential terms (Behrens *et al.*, 1989; Correa *et al.*, 1992). Using Eq. 1 for purposes of comparison, values of V_0 and q derived from single-channel analysis are quite similar among BTX-modified Na⁺ channels from different species or tissue subtypes, and in native or artificial membranes (Table I). In planar bilayers, in the absence of divalent cations, and for 0.1–0.5 M symmetrical NaCl, average V_0 values range from −83 to −111 mV and q lies in the range of 3.7 to 4.0. Such P_{open} measurements are complicated by the fact that BTX-modified Na⁺ channels exhibit spontaneous shifts in open-state probability (Chabala *et al.*, 1991). Some preparations are also characterized by a rather large range of V_0 and q values that describes channel-to-channel variation; for example, V_0 varies from −77 to −129 and z varies from 1.2 to 6.5 for canine brain Na⁺ channels (Chabala *et al.*, 1991).

With the ultimate goal of deriving a kinetic pathway of the major gating transitions in BTX-modified Na⁺ channels, several groups of workers have analyzed dwell time distributions of closed- and open-state

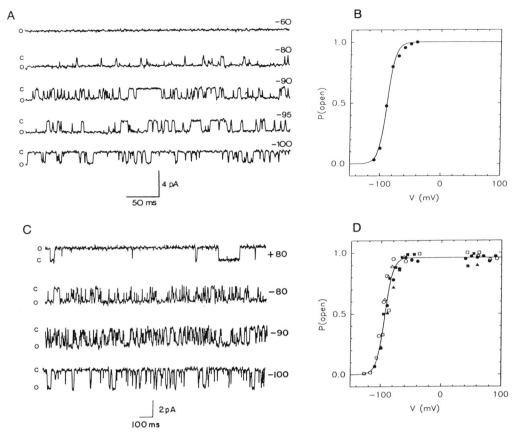

FIGURE 3 Voltage-dependent activation of single batrachotoxin (BTX)-modified Na$^+$ channels from squid neurons and rat muscle in planar lipid bilayers. (A) Recordings of a single BTX-modified Na$^+$ channel from squid optic nerve in the presence of 0.2 M symmetrical NaCl. c, Closed current level; o, open current level. Holding voltage (mV) is given at the right. (B) Open-state probability vs voltage relationship for BTX-modified Na$^+$ channel from squid. Data are fit to Eq. 1 with $q = 3.7$ and $V_0 = -88.8$ mV. (C) Records of a BTX-modified Na$^+$ channel from rat skeletal muscle. (D) Open-state probability vs voltage relationship for a single rat muscle Na$^+$ channel modified by BTX. Data are fit to a modified form of Eq. 1 with $q = 3.8$, $V_0 = -94$ mV, and $P_{max} = 0.96$. (A, C) Reproduced from Behrens *et al.*, *The Journal of General Physiology* (1989), **93**, 23–41, by copyright permission of the Rockefeller University Press. (C, D) Reproduced from Moczydlowski *et al.*, *The Journal of General Physiology* (1984), **84**, 665–686, by copyright permission of the Rockefeller University Press.

TABLE I Gating Parameters of Batrachotoxin-Modified Na$^+$ Channels from Various Preparations[a]

NA$^+$ channel preparation	V_0 (mV)	q	Reference
Rat muscle, bilayer	−94	3.8	Moczydlowski *et al.* (1984a)
Rat brain, bilayer	−91.5 ± 10.1	3.9 ± 0.9	French *et al.* (1986)
Eel electroplax, bilayer	−83 ± 18	4.0 ± 0.4	Shenkel *et al.* (1989)
Canine brain, bilayer	−111 ± 11	3.7 ± 1.2	Chabala *et al.* (1991)
Squid ganglion, bilayer	−88.8	3.7	Behrens *et al.* (1989)
Squid axon, patch clamp	−59 to −80	3.5 to 4.1	Correa *et al.* (1992)
Neuroblastoma, whole cell	−85	4.5	Huang *et al.* (1982)

[a] V_0 and q values were obtained from fits to Eq. 1 for data obtained at 22°C in all cases except patch recording of squid axon, which was at 15°C. Error estimates indicate the standard deviation.

events as a function of voltage. Using patch recording in NG108-15 neuroblastoma cells, Huang *et al.* (1984) found that open-state distributions at all voltages can be fit by a single-exponential distribution and that the corresponding closing rate constant, β, leading away from the single open state, monotonically increases as a function of voltage from -40 to -100 mV. Dwell-time distributions of closed-state events can be fit by a single exponential in the gating range, but a sum-of-two-exponentials distribution is required for closed-state populations more positive than -70 mV. The gating of BTX-modified Na^+ channels in neuroblastoma cells can be interpreted on the basis of a linear C_1–C_2–O model (Scheme 1, Fig. 4), but a C_1–O–C_2 kinetic model cannot be excluded (Huang *et al.*, 1984).

Analysis of dwell-time distributions in planar bilayers is hindered by the lower resolution of brief events that is a result of heavy filtering required when recording from a large membrane area. Technical improvements in forming small bilayers suitable for routine channel incorporation promise to improve this situation (Wonderlin *et al.*, 1990). Despite lower resolution, bilayer studies of gating fluctuations in single BTX-modified Na^+ channels from mammalian sources have generally confirmed the patch-clamp results of Huang *et al.* (1984). BTX-modified Na^+ channels from rat brain and muscle exhibit single-exponential dwell-time distributions for the open state at all voltages and a sum-of-two-exponentials distribution for the closed state at voltages more positive than -70 mV (Moczydlowski *et al.*, 1984a; French *et al.*, 1986; Keller *et al.*, 1986).

Using purified Na^+ channels from rat brain in bilayers, Keller *et al.* (1986) obtained a fit to a C_1–C_2–O

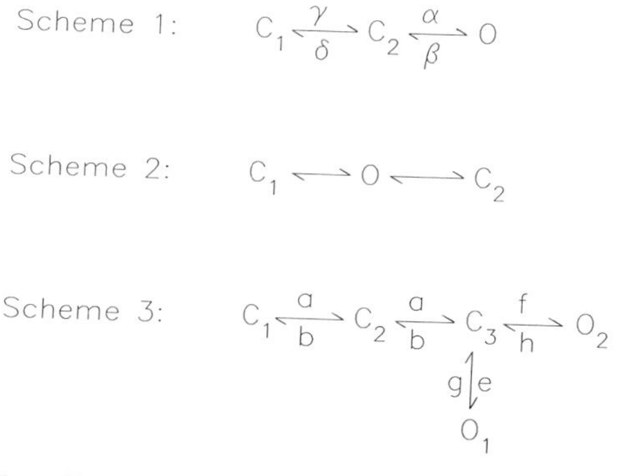

Scheme 1: $\quad C_1 \underset{\delta}{\overset{\gamma}{\rightleftarrows}} C_2 \underset{\beta}{\overset{\alpha}{\rightleftarrows}} O$

Scheme 2: $\quad C_1 \rightleftharpoons O \rightleftharpoons C_2$

Scheme 3: $\quad C_1 \underset{b}{\overset{a}{\rightleftarrows}} C_2 \underset{b}{\overset{a}{\rightleftarrows}} C_3 \underset{h}{\overset{f}{\rightleftarrows}} O_2$
$\qquad\qquad\qquad\qquad\quad g \updownarrow e$
$\qquad\qquad\qquad\qquad\quad O_1$

FIGURE 4 Kinetic models used to describe voltage-dependent gating of batrachotoxin (BTX)-modified Na^+ channels. Single-channel data at 20–23°C from neuroblastoma (Huang *et al.*, 1984) and rat brain (Keller *et al.*, 1986) Na^+ channels have been interpreted according to Scheme 1 or 2. Data at 0°C from squid axon (Correa *et al.*, 1992) have been modeled according to Scheme 3.

kinetic scheme that was very similar to that described for neuroblastoma cells (Huang *et al.*, 1984). Covariance analysis was used to demonstrate the absence of correlation between durations of successive open or closed events, as expected for a system that exhibits a single open state. A fit to the model of Scheme 1 with rate constants expressed as exponential functions of voltage was able to simulate the dependence of open-state probability on applied voltage in the gating range of -120 to -70 mV. However, this scheme cannot explain the saturation at positive voltages of the mean open time and P_{open} at a value less than 1.0. To explain this discrepancy, Keller *et al.* (1986) proposed that the rate constants of Scheme 1 are nonexponential functions of voltage that saturate at positive applied voltage. Such a dependence of gating transitions on voltage could be accounted for by a charged gating domain that moves out of the membrane electric field (Keller *et al.*, 1986).

For BTX-modified Na^+ channels from squid optic nerve recorded in planar bilayers, open and closed state histograms are approximated by single-exponential distributions (Behrens *et al.*, 1989). Open-state probability for this species of Na^+ channel appears to reach 1.00 at -50 mV. The mean open time for this channel increases with voltage in an exponential fashion from -110 to -50 mV; the mean closed time decreases with positive voltage in a similar fashion in the gating region, but appears to saturate at voltages more positive than -70 mV.

A higher resolution description of the squid axon Na^+ channel was obtained by patch-clamp recording of single BTX-modified Na^+ channels by the cut-open axon technique at low temperature (Correa *et al.*, 1992). At 5°C and 530 mM symmetrical NaCl, V_0 and q values for P_{open} vs V data fit to Eq. 1 range from -79 to -86 mV and 3.0 to 3.4, respectively. As temperature is increased, the voltage dependence of opening probability is not significantly affected, but the V_0 midpoint of the gating reaction increases by $+10$ mV from 0 to 14°C. This result indicates that the open state is stabilized by low temperature. This temperature dependence can be interpreted as a decrease in entropy during channel opening, meaning that open states are more ordered than closed states. Analysis of dwell-time distributions collected at 0°C showed that open events are best described by two major exponential components with a minor third component, whereas closed state events are described by three exponentials at all voltages. To limit the number of possible models describing the data, adjacent interval analysis was used to show a lack of correlation of the mean duration of open events with the duration of adjacent closed events, suggesting that all open states originate from a common closed state. Statistical comparisons of different models led to consideration of the preferred gating model depicted as Scheme 3 in Fig. 4. To interpret the results in a manner compatible with the gating scheme that has emerged from studies of the normal squid Na^+ channel (Vandenberg and Bezanilla, 1991), Correa *et al.* (1992) proposed that the O_2 open state of Scheme 3 (Fig. 4) represents BTX-induced modification of a normal inactivated (I) state to a conducting state (O_2). This interpretation has the advantage that it can explain why BTX modification keeps the process of gating charge immobilization intact. BTX is envisioned to bind to the open state of the normal channel and to keep the channel open in the inactivated state by preventing binding of the inactivation particle to its receptor, while retaining normal immobilization of gating charge.

C. Mode Behavior

The preceding discussion of gating kinetics hinges on the assumption that dwell-time populations can be collected from a single channel that reflect a true steady state. In actuality, several laboratories have described nonstationary behavior of BTX-modified Na^+ channels in planar bilayers (Moczydlowski *et al.*, 1984a; French *et al.*, 1986; Chabala *et al.*, 1991). An example of such behavior for a single canine brain Na^+ channel from the work of Chabala *et al.* (1991) is shown in Fig. 5. This particular channel exhibited spontaneous increases and decreases in open-state

probability (Fig. 5A) that would be accompanied by corresponding changes in open and closed dwell-time distributions. Measurements of P_{open} averaged over selected intervals during the observation suggested that the channel made a reversible change between two different P_{open}–voltage relationships with similar gating charges (q) but shifted by ~16 mV along the voltage axis (Fig. 5B). In addition to such transitions and channel-to-channel variation in q and V_0, Chabala *et al.* (1991) also observed rare shifts of a single channel to states with low apparent gating charge (e.g., a shift from $q = 3.0$ to 1.2). Similar examples of nonstationary gating have been observed in single-channel recordings of Na^+, Ca^{2+}, and K^+ channels from intact cells under near-physiological conditions and have been attributed to modal patterns of gating (McManus and Magleby, 1988; Patlak and Ortiz, 1986; Pietrobon and Hess, 1990). Such behavior is problematic when deriving a canonical gating scheme. If such mode shifts occur during the collection of closed or open state histograms, the predicted effect would be to smear dwell-time distributions with extra components belonging to a collection of distinct modes. To avoid this effect, one strategy is selecting data sets that do not contain obvious mode shifts. However, if mode changes are not kinetically well separated, many subtle shifts in P_{open} can lead to histograms that approximate a continuum of states rather than a small set of energetically stables states. If such spontaneous gating shifts also underlie macroscopic current measurements, the macroscopic time course may not represent

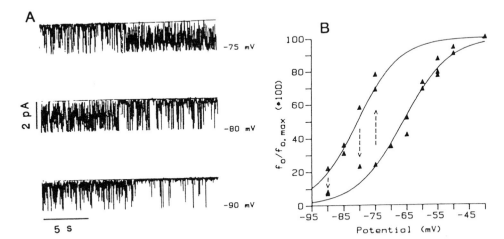

FIGURE 5 Spontaneous shifts in steady-state gating behavior of a single batrachotoxin (BTX)-modified Na^+ channel. (A) Current records of a BTX-modified Na^+ channel from canine brain in the presence of 1.0 M symmetrical NaCl. Openings are downward. The records illustrate abrupt shifts in fractional open time, f_0, from 0.24 to 0.69 at −75 mV, 0.58 to 0.23 at −80 mV, and 0.22 to 0.07 at −90 mV. (B) Fractional open time data for the channel in A can be described as a shift between two activation gating curves. The data are fit to Eq. 1 with $q = 3.5$ and $V_0 = −82$ or −66 mV. Reproduced from Chabala *et al.*, *The Journal of General Physiology* (1991), **98**, 197–224, by copyright permission of the Rockefeller University Press.

the kind of homogeneous kinetic behavior that is usually assumed. Rather, the measurement would reflect a distribution of channels cycling among a set of interconverting modes. The predicted consequence of this effect would be an anomalous broadening in the macroscopic voltage dependence of a gating process that no longer reflects an intrinsic molecular gating charge (Chabala *et al.*, 1991). Although assessing the significance of mode behavior is premature, such considerations have inspired attempts to describe the gating of BTX-activated Na^+ channels by continuum models (Rubinson, 1992).

D. Effect of Surface Charge on Gating

Since voltage-dependent activation of BTX-modified Na^+ channels can be studied in planar bilayers in the absence of fast inactivation (Kreuger *et al.*, 1983), this is an ideal system for investigating the effect of surface charge on activation gating. An influence of surface charge in modulating gating of normal Na^+ channels was first proposed by Frankenhauser and Hodgkin (1957), who found that the voltage range of activation in squid axon was shifted to more positive voltages by increasing external Ca^{2+}. According to current theory for this effect, external Ca^{2+} acts to lower the external surface potential by screening or by binding to negative charges on the surface of the protein or lipid (for review Hille, 1992). A reduction in external negative surface potential alters the electric field across the membrane to produce an effect equivalent to hyperpolarization. This effect leads to a depolarizing shift in the observed activation gating curve since a stronger depolarization is required to open channels. This hypothesis is supported by experiments showing that an increase in external ionic strength causes a depolarizing shift of activation gating similar to that induced by Ca^{2+} (Hille *et al.*, 1975). However, comparison of the ability of various divalent cations to shift Na^+ channel gating in frog nodes reveals a specific order of effectiveness: $Zn^{2+} > Ni^{2+} > Co^{2+} > Mn^{2+} > Ca^{2+} > Mg^{2+} > Ba^{2+}$, Sr^{2+} (Hille *et al.*, 1975). Since various divalent cations do not exhibit exactly the same effectiveness, as expected for a screening mechanism responsive to ionic strength alone, chemically selective binding of divalent cations must also be involved.

Other classical experiments have shown that the underlying mechanism of gating shifts is more complicated than a simple shift in the transmembrane electric field. Such a mechanism would be expected to shift all gating parameters equally; however, external Zn^{2+} and La^{3+} preferentially affect channel opening over channel closing kinetics (Gilly and Armstrong, 1982;

Armstrong and Cota, 1991). In contrast to the effect of Zn^{2+} and La^{3+}, external Ca^{2+} does appear to cause equal shifts of gating parameters in frog muscle (Hahin and Campbell, 1983). Thus, studies of macroscopic Na^+ currents have led to the notion that surface charge effects on voltage-dependent processes are mediated by at least three types of phenomena: screening, binding, and direct modulation of particular steps in the gating process.

The preceding concepts generated from classical studies at the macroscopic level have been further refined at the microscopic level by studies on single BTX-modified Na^+ channels from mammalian brain in planar bilayers of defined lipid composition (Cukierman *et al.*, 1988; Cukierman and Krueger, 1990, 1991; Cukierman, 1991a). In uncharged bilayers formed from neutral lipids, an increase in external ionic strength by [NaCl] was observed to shift the P_{open} vs V relationship in the positive direction (Fig. 6A; Cukierman, 1991a). A similar increase in internal [NaCl] shifted the P_{open} vs V gating curve in the negative direction (Fig. 6C). These gating shifts are due to a change in the midpoint (V_0) of the gating curve with no significant effect on the apparent gating charge q. Analysis of the dwell-time distribution of closed- and open-state events showed that the shifts produced by NaCl were the result of a roughly equivalent effect on opening and closing rate constants without a significant change in voltage dependence. Thus, in this assay, changes in the internal and external ionic strength modulated activation gating in accordance with the simplest version of surface charge theory. An analysis describing the shift of V_0 as a function of external or internal ionic strength according to the Gouy–Chapman theory of planar surface charge found an external surface charge density of 1 e-/533 $Å^2$ and an internal surface charge density of 1 e-/1231 $Å^2$ for this behavior (Fig. 6C,D). Comparison of gating shifts in neutral bilayers with those containing 70% negatively charged phosphatidylserine (PS) also showed that an extra amount of surface charge was sensed in a negatively charged bilayer. The magnitude of the extra shift induced by PS was equivalent to the membrane electrostatic potential at a distance of 48 Å and 28 Å from the plane of the membrane on the outer and inner side of the channel, respectively. The fact that not all the lipid surface potential is sensed by the gating process suggests that some insulating distance exists between the voltage sensor and the lipid portion of the bilayer (Cukierman, 1991a).

Examination of the effect of divalent cations on gating of BTX-modified Na^+ channels in bilayers has also reinforced classical observations. External Ca^{2+} and Ba^{2+} shifted the midpoint of activation gating

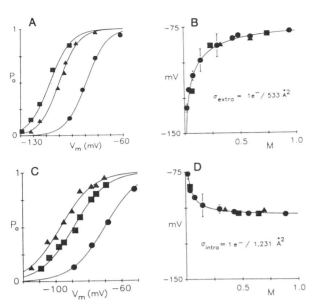

FIGURE 6 Effect of surface potential on Na$^+$ channel gating as demonstrated by dependence on ionic strength. (A) Open-state probability vs voltage relationship for single batrachotoxin (BTX)-modified Na$^+$ channels from rat brain in a neutral lipid bilayer in the presence of 150 mM NaCl inside and 18 (■), 38 (▲), or 150 (●) mM NaCl outside. Data are fit to Eq. 1 with q and V_0 values of 3.2 and −117 mV (■); 3.4 and −110 (▲); and 3.2 and −89 mV (●). (B) V_0 values from experiments in A are plotted against external salt concentration. Data are fit to a Gouy–Chapman model of surface potential with a planar density of extracellular negative surface charge of 1 e$^-$/533 Å^2. (C) Results of experiments analogous to those in A with 150 mM NaCl ouside and 20 (●), 150 (■), or 600 (▲) mM NaCl inside. Data are fit to Eq. 1 with q and V_0 values of 2.5 and −70 mV (●); 2.4 and −89 mV (■); and 2.3 and −97 mV (▲). (D) V_0 values from experiments in D are plotted against intracellular salt concentration. Data are fit using a planar density of negative intracellular surface charge of 1 e$^-$/1231 Å^2. Reproduced from Cukierman, *The Biophysical Journal* (1991), **60**, 845–855, by copyright permission of the Biophysical Society.

in the positive direction along the voltage axis and internal Ca^{2+} or Ba^{2+} shifted V_0 in the negative direction (Cukierman *et al.*, 1988; Chabala *et al.*, 1991). Similar to the results obtained with NaCl, the magnitude of the shifts induced by Ca^{2+} and Ba^{2+} indicates that the density of negative protein surface charge sensed by the gating process is greater on the outside than on the inside of the channel (Cukierman *et al.*, 1988; Chabala *et al.*, 1991). External divalent cations were found to exert a greater effect on the channel opening reaction than on the channel closing reaction. The observed mean dwell time of closed state events (τ_C) increased an average of 2.8-fold, whereas the mean dwell time of open-state events decreased 2.0-fold on addition of 5 mM external Ba^{2+} in the presence of 150 mM NaCl (Cukierman and Krueger, 1990). This effect is reminiscent of the asymmetric effect of Zn^{2+}

on Na$^+$ channel gating in squid axon (Gilly and Armstrong, 1982). In contrast, Behrens *et al.* (1989) found essentially similar shifts in τ_C and τ_O by addition of external Ca^{2+} in studies of BTX-modified Na$^+$ channels from squid optic nerve.

For rat brain Na$^+$ channels, the relative ability of a series of divalent cations to shift activation gating is different for the outside and the inside of the channel. The order of effectiveness for the depolarizing shift that occurs with external divalents is Ca^{2+} > Ba^{2+} > Sr^{2+} > Mg^{2+} (Cukierman and Krueger, 1990). The hyperpolarizing shift that occurs with internal divalents follows the sequence of effectiveness: Mg^{2+} > Ca^{2+} > Ba^{2+} > Sr^{2+} (Cukierman and Kruger, 1991). The differential sensitivity of the Na$^+$ channel to internal and external Mg^{2+} points to the involvement of distinct types of divalent cation binding sites with different chemical specificities. When the effect of 5 mM internal Ba^{2+} on gating kinetics was studied, a greater effect on channel closing than on channel opening was found: τ_C increased 3-fold compared with a 2-fold decrease in τ_C.

The effects of divalent cations on gating of single BTX-activated Na$^+$ channels reinforce the view that surface electrostatic phenomena involved in gating modulation are complex. The underlying processes at work appear to include (1) surface potential screening that is not dependent on ionic species, (2) reduction of surface potential by chemically selective binding of divalent cations at one or more sites, and (3) a component not strictly related to surface potential in which specific binding of certain divalent cations exhibits preferential effects on channel opening or closing kinetics. This final effect may involve divalent cation binding sites that exert allosteric effects on the gating process. Since a given divalent cation could act by all three mechanisms, assessing the relative contributions of the various processes is difficult. However, one possible approach involves selective neutralization of negatively charged residues that participate in surface charge modulation or elimination of divalent cation binding sites by site-specific mutagenesis of cloned Na$^+$ channels. If such mutant channels exhibit defective gating modulation by ionic strength or divalent cations, resolving some of the individual electrostatic components that contribute to the gating process may ultimately be possible.

E. Effect of Scorpion Toxin and Other Modifiers

In addition to BTX, a number of other toxins and protein modification reagents are known to modify the gating kinetics of voltage-dependent Na$^+$ channels

(Strichartz *et al.*, 1987). For example, a class of small protein toxins called α scorpion toxins slows the inactivation process; a similar class of β scorpion toxins binds at a different external site and also affects activation kinetics. Since the binding of scorpion toxins is synergistic with that of BTX, O'Leary and Krueger (1989) studied the effect of α scorpion toxin from *Leiurus quinquestriatus* (α-LqqTX) on single BTX-modified Na$^+$ channels from rat brain in planar bilayers. These researchers found that, in the presence of both BTX and α-LqqTX, Na$^+$ channels exhibited an abnormally high open-state probability in the voltage range of -120 to -80 mV where BTX-modified Na$^+$ channels are normally closed by hyperpolarization. Such channels modified by both BTX and α-LqqTX are fully open, even at -135 mV. Whether such channels do eventually close at a more negative voltage range than -135 mV is not yet known. This experiment is difficult since planar bilayers are unstable at high voltage. In any case, the binding of α scorpion toxin appears to further stabilize the open state of the BTX-modified Na$^+$ channel without significantly affecting unitary conductance or block by STX and Ca^{2+} (O'Leary and Krueger, 1989). Treatment of single BTX-activated Na$^+$ channels with the inactivation modifiers trypsin or chloramine-T also shifts the voltage dependence of activation to more negative voltages, by -22.4 mV on average for trypsin and -12.9 mV for chloramine-T (Cukierman, 1991b). These hyperpolarizing shifts of activation gating do not appear to involve a change in gating charge, and are the result of equal effects on the mean dwell time of open- and closed-state events, thus mimicking a generalized surface charge effect.

III. Ion Permeation

A. *Differences between Normal and BTX-Modified Na$^+$ Channels*

Macroscopic measurements of the reversal potential of normal and BTX-modified Na$^+$ currents indicate that BTX alters ionic selectivity. For example, Na$^+$ channel currents in neuroblastoma cells normally exhibit a relative permeability sequence of Na$^+(1.0) >$ K$^+(0.1) >$ Cs$^+(0.01)$. Treatment with BTX shifts the reversal potential by -30 to -40 mV, indicating a decrease in ionic selectivity with a permeability sequence of Na$^+(1.0) >$ K$^+(0.38) >$ Cs$^+(0.14)$ (Huang *et al.*, 1982). Similarly, in frog nodes of Ranvier, macroscopic Na$^+$ currents normally exhibit the following permeability sequence in the absence of BTX: Na$^+(1.0) \sim$ Li$^+(0.93) >$ NH$_4^+$ $(0.16) >$ K$^+(0.086) >$

Rb$^+(0.012) >$ Cs$^+(0.013)$ (Hille, 1992). BTX-modified currents in the same preparation are less selective: Na$^+(1.0) \sim$ Li$^+(1.0) >$ NH$_4^+$ $(0.5) >$ K$^+(0.2-0.4) >$ Rb$^+(0.25) >$ Cs$^+(0.15)$ (Khodorov and Revenko, 1979; Khodorov, 1985).

At the single-channel level, Shenkel *et al.* (1989) found a selectivity ratio P_K/P_{Na} of 0.024 for trypsin-activated Na$^+$ channels from *Electrophorus electricus*. BTX-modified channels from the same preparation exhibited lower selectivity ($P_K/P_{Na} = 0.17$) under bi-ionic conditions of 400 mM NaCl (external) and 400 mM KCl (internal). Trypsin-activated Na$^+$ channels from *Electrophorus* also had a higher unitary conductance of 44.8 pS, compared with 20.0 pS for BTX-modified channels in symmetrical 400 mM NaCl. BTX modification of Na$^+$ channels from squid giant axon also reduced unitary conductance to 10 pS, compared with 25 pS for normal channels (Correa *et al.*, 1991). However, the selectivity for certain ions does not appear to be significantly affected by BTX. In the case of Li$^+$, for example, measurements on single BTX-modified channels (Garber, 1988) and normal Na$^+$ currents (Hille, 1992) both reported P_{Li}/P_{Na} values near 1.0. Likewise, measurements in squid axon indicated the same selectivity for NH$_4^+$/Na$^+$ in normal and BTX-modified channels ($P_{NH4}/P_{Na} = 0.34-0.37$; Correa *et al.*, 1991).

The range of values for P_K/P_{Na} reported for single BTX-modified Na$^+$ channels by different laboratories varies widely from 0.07 to 0.30 (Krueger *et al.*, 1983; Hartshorne *et al.*, 1985; Moczydlowski *et al.*, 1984a; Green *et al.*, 1987a; Recio-Pinto *et al.*, 1987; Duch *et al.*, 1988). This variation in reported values of P_K/P_{Na} can be attributed to the species of Na$^+$ channel studied, the exact ionic conditions of assay, and the orientation of the Na$^+$ and K$^+$ gradients used to make the measurement. For BTX-activated Na$^+$ channels from rat skeletal muscle, Garber (1988) observed a P_K/P_{Na} of 0.26 for bi-ionic conditions of Na$^+$ (out)/K$^+$ (in) and a P_K/P_{Na} of 0.12 for K$^+$ (out)/Na$^+$ (in). For human brain Na$^+$ channels activated by BTX, Duch *et al.* (1988) observed a P_K/P_{Na} of 0.30 for conditions in which K$^+$ was intracellular and a P_K/P_{Na} of 0.18 when Na$^+$ was intracellular. This orientation-dependent selectivity for Na$^+$ and K$^+$ was not observed for BTX-modified Na$^+$ channels from canine brain, for which P_K/P_{Na} is in the range of 0.14–0.18 (Green *et al.*, 1987a); however, strictly bi-ionic conditions were not examined in this case.

Despite differences in unitary conductance and ionic selectivity relative to the native channel, BTX-modified Na$^+$ channels are a valuable preparation for investigating the mechanism of ion permeation. The gating characteristics of BTX-modified channels permit well-resolved conductance measurements on single channels. In planar bilayers, such measurements can be made in the virtual absence of divalent cations and in membranes of defined lipid composition. For example, a comparison of conductance behavior in bilayers composed of phosphatidylethanolamine or phosphatidylserine indicates that the conduction pathway of rat brain Na$^+$ channels is electrostatically isolated from phospholipid head groups (Worley et al., 1992). Unitary current–voltage relationships and the dependence on permeant ion concentration can be related to the energetics of ion conduction. When compared with particular physical models of ion permeation, such information can provide the basis for rejecting or refining models that do not conform to observed behavior.

B. Interpreting the Dependence of Unitary Conductance on [Na$^+$]

The unitary conductance, g, of BTX-modified Na$^+$ channels from rat skeletal muscle in neutral bilayers was initially characterized as a Michaelis–Menten function of Na$^+$ concentration (Moczydlowski et al., 1984a):

$$g = g_{max} [Na^+]/\{K_{0.5} + [Na^+]\}, \qquad (2)$$

where g_{max} is the maximal conductance at saturation and $K_{0.5}$ is equivalent to the Na$^+$ concentration at which $g = g_{max}/2$. The behavior of rat muscle Na$^+$ channels is approximately described by Eq. 2 in the range of 3–500 mM Na$^+$, with $g_{max} = 21$ pS and $K_{0.5} = 6.5$–8 mM (Moczydlowski et al., 1984a; Garber and Miller, 1987). Similarly, BTX-modified channels from rat brain and squid optic nerve in neutral lipid membranes have also been interpreted on the basis of Eq. 2, using $g_{max} = 31.8$ pS and $K_{0.5} = 40.5$ for rat brain (Worley et al., 1992) and $g_{max} = 23$ pS and $K_{0.5} = 11$ mM for squid (Behrens et al., 1989). In contrast, the [Na$^+$] dependence of unitary conductance for canine brain Na$^+$ channels does not correspond to a Michaelis–Menten function when analyzed in terms of [Na$^+$] in the bulk solution (Green et al., 1987a). However, the conductance behavior can be described as a Michaelis–Menten dependence on local [Na$^+$], [Na$^+$]$_l$, at the entrance of the channel. For this approach, [Na$^+$]$_l$ is expressed as a Boltzmann distribution with respect to electrostatic surface potential:

$$[Na^+]_l = [Na^+]_b \exp(-\Psi_e F/RT). \qquad (3)$$

In this equation, [Na]$_b$ is [Na$^+$] in the bulk solution, Ψ_e is the electrostatic surface potential at the surface of the channel, and $RT/F = 25.4$ mV at 22°C. Although

estimating the electrostatic surface potential at the surface of a membrane protein is difficult, Green et al. (1987a) assumed an infinite planar charge distribution that allows the relationship between Ψ_e, surface charge density (σ), and ionic strength to be calculated from Gouy–Chapman theory (Green and Anderson, 1991). This approach was used to obtain values of g_{max} (45 pS), $K_{0.5}$ (1.5 M), and σ (1 e$^-$/263 Å^2).

The differing interpretations of Green et al. (1987a) for canine brain Na$^+$ channels and of other groups working with rat brain (Worley et al., 1992), rat muscle (Moczydlowski et al., 1984a; Garber and Miller, 1987), and squid optic nerve (Behrens et al., 1989) appear to result from actual differences in Na$^+$ channel properties among species as well as the range of [Na$^+$] on which conclusions are based (Naranjo and Latorre, 1993). For example, comparison of rat muscle and canine heart Na$^+$ channels under identical conditions in planar bilayers showed that conductance–[Na$^+$] behavior differs slightly but significantly between these two isoforms (Ravindran et al., 1992). In addition, when conductance measurements are made over a wider range of [Na$^+$], departure from Michaelis–Menten behavior is obvious at high [Na$^+$], because conductance continues to increase without reaching saturation at 3000 mM Na$^+$ (Ravindran et al., 1992; Worley et al., 1992).

The biophysical interpretation of the conductance vs [Na$^+$] relationship for the BTX-modified Na$^+$ channel concerns two difficult issues: the influence of protein surface charge and the contribution of multiple-ion occupancy. Mutagenesis experiments have implicated the linker region between presumed membrane-spanning domains S5 and S6 of voltage-dependent Na$^+$ channels as a structure that is intimately associated with the external vestibule of the channel itself (Heinemann et al., 1992). Examination of the primary sequence of this S5–S6 linker region for each of the four internally homologous repeats (I–IV) reveals an excess of aspartate and glutamate over lysine and arginine residues. This primary structure suggests that a net negative charge is associated with the channel vestibule (Green and Andersen, 1991). Such a distribution of negative charge would be expected to result in an accumulation of permeant cations near the vestibule that would enhance current through the channel, particularly at low ionic strength when little charge screening occurs. However, because the three-dimensional structure of the channel is unknown, estimation of the magnitude of this effect requires assumptions about the location and geometry of the negative charge distribution with respect to the pore. The planar charge distribution that is often used to model the effect of surface charge on ion conduction is mathematically advantageous because the Gouy–Chapman theory can be used to obtain an explicit dependence of surface potential on ionic strength and distance from the plane of the membrane (Green and Andersen, 1991). However, because of its assumption of an infinite plane, this particular model predicts that the surface concentration of cations will approach a nonzero limiting value at low ionic strength. This prediction gives rise to a predicted conductance plateau in the limit of low [Na$^+$] (Green and Andersen, 1991; Latorre *et al.*, 1992).

Since the surface distribution of negative charge on a real channel protein is clearly different than that of an infinite plane, conductance–[Na$^+$] behavior is not necessarily expected to exhibit a limiting conductance value at low [Na$^+$]. Numerical computations of electrostatic effects at the mouth of a charged channel-like structure, according to the Possion–Boltzmann equation, predict zero conductance in the limit of low [Na$^+$] (Dani, 1986; Cai and Jordan, 1990). According to the latter approach, the data of Green *et al.* (1987a) can be modeled as a channel with single-ion occupancy for Na$^+$, with $g_{max} = 51.7$ pS, $K_{0.5} = 2.5$ M, and a local surface charge density of 1 e$^-$/67 Å^2 spread at the innermost entrance of a funnel-shaped vestibule (Fig. 7A,B; Cai and Jordan, 1990). Based on conductance data alone, this prediction suggests that a single-ion channel coupled with a negatively charged vestibule can account for the dependence of g on [Na$^+$] for BTX-modified Na$^+$ channels.

However, additional data from the literature suggest that Na$^+$ channels may be capable of multiple-ion occupancy, which also appears to be the case for other members of the family of voltage-dependent ion channels (Begenisich, 1987; Tsien *et al.*, 1987; Yellen, 1987). For native Na$^+$ channels, this evidence includes permeability ratios for certain monovalent cations that are dependent on the internal ion concentration (Chandler and Meves, 1965; Cahalan and Begenisich, 1976; Begenisich and Cahalan, 1980) and complex interactions between permeant ions and blocking ions that are consistent with simultaneous occupancy of the channel by more than one ion (e.g., Wang, 1988). Considering this possibility, Ravindran *et al.* (1991) analyzed permeation of Na$^+$, Li$^+$, and K$^+$ through BTX-modified Na$^+$ channels of rat skeletal muscle on the basis of an Eyring-type, 3-barrier–2-site model with double-ion occupancy (3B2S model). When examined over the range of 0.5 to 3000 mM Na$^+$, the dependence of unitary conductance on [Na$^+$] displays biphasic behavior that can be interpreted as evidence for the existence of two binding sites for Na$^+$. According to this interpretation, the unoccupied channel can bind one Na$^+$ ion with a high affinity for Na$^+$ of

~3.7 mM and a second Na$^+$ ion with a much lower affinity of ~4900 mM. By neglecting the possible contribution of vestibule surface charge, this model was able to simulate the unitary current–voltage behavior over the range of 0.5 to 3000 mM Na$^+$ (Fig. 7C). The best-fit energy profile for Na$^+$ is based on a relatively symmetrical energy barrier structure with respect to the center of the membrane (Fig. 7D). To simulate the effects of double occupancy, this model includes a term for ion–ion repulsion that raises the energy of the well and peak adjacent to an occupied site. The basic conclusion of such modeling work is that a repulsive interaction between two Na$^+$ ions could theoretically produce the distinctive biphasic dependence on [Na$^+$] (Ravindran *et al.*, 1992).

The 3B2S double-occupancy model of ion conduction through the BTX-modified Na$^+$ channel was also extended to include permeation of Li$^+$ and K$^+$ (Ravindran *et al.*, 1992). The dependence of unitary conductance on the mole fraction of Li$^+$ or K$^+$ in mixtures with Na$^+$ at constant ionic strength (206 or 2006 mM) followed monotonic behavior with no evidence of a minimum or an "anomalous mole fraction effect." This latter type of mole fraction behavior has been observed previously in several other types of ion channels and is considered to be strong evidence for multiple occupancy (Neher, 1975; Hille, 1992). Despite the absence of anomalous mole fraction behavior, Eyring barrier profiles for Li$^+$ and K$^+$ were found that simulate observed current–voltage behavior of single Na$^+$ channels in these experiments. A particular success of this approach is that the 3B2S double-occupancy model for Na$^+$ and Li$^+$ can account for the low unitary conductance in pure Li$^+$ (~10 pS), which is about half that of pure Na$^+$ (21 pS), and the observed permeability ratio for P_{Li}/P_{Na} of ~1.0. Garber (1988) previously found that a 4B3S single-occupancy model could not satisfactorily account for this behavior.

In summary, research into the nature of ion permeation through single BTX-modified Na$^+$ channels has provided two alternative explanations summarized by the data and models in Fig. 7. In the view presented by Green *et al.* (1987a) and refined by Cai and Jordan (1990), the channel is occupied by no more than one weakly binding Na$^+$ ion, and negatively charged vestibules are responsible for a strong enhancement of unitary current at low ionic strength. In the minimal model of Ravindran *et al.* (1992), two Na$^+$ binding sites with ion–ion repulsion are sufficient to describe permeation without consideration of the effect of vestibule surface charge. Although the issues are far from resolved, the real truth probably lies somewhere between these extremes. Independent information on the [Na$^+$] dependence of TTX and STX association

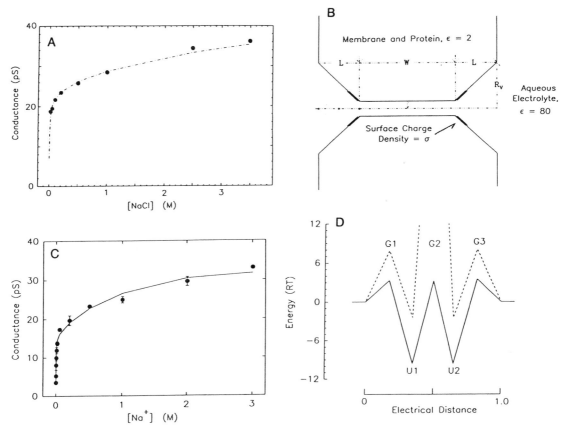

FIGURE 7 Two different models to explain the dependence of unitary conductance on Na$^+$ concentration. (A) Unitary conductance data vs [NaCl] for batrachotoxin (BTX)-modified Na$^+$ channels from canine brain, obtained by Green *et al.* (1987a). Data points are fit according to the model of Cai and Jordan (1990) illustrated in B. (B) The Na$^+$ channel is assumed to bind only one ion at a time. The drawing depicts a water-filled channel separated by two funnel-shaped vestibules. The model assumes the dimensions $L = 20$ Å, $R_v = 15$ Å, $W = 20$ Å, and channel radius = 3 Å. A surface charge density of 1 e$^-$/76 Å^2 is placed symmetrically at each entrance. The protein–membrane dielectric constant is $\varepsilon = 2$ and that of the aqueous channel is $\varepsilon = 80$. (C) Unitary conductance data vs. [Na$^+$] for BTX-modified Na$^+$ channels from rat muscle, obtained by Ravindran *et al.* (1992). Data are fit to an Eyring energy barrier model illustrated in D. (D) The energy profile of the channel is represented by three barriers, G1–G3, and two sites for Na$^+$, U1 and U2. G1 is on the intracellular side. The channel can be occupied by one or two Na$^+$ ions. In the singly occupied state, transport rates are determined by the solid line profile. Na$^+$ entry to or exit from the doubly occupied state is determined by the dashed profile, which simulates the effect of ion–ion repulsion. (A, B) Reproduced from Cai and Jordan, *The Biophysical Journal* (1990), **57,** 883–891, by copyright permission of the Biophysical Society. (C, D) Reproduced from Ravindran *et al.*, *The Biophysical Journal* (1992), **61,** 494–508, by copyright permission of the Biophysical Society.

rates (Green *et al.*, 1987a; Ravindran and Moczydlowski, 1989) and the effects of carboxyl group modification (Worley *et al.*, 1986) suggest that negatively charged surface residues do affect functions related to permeation. A study by Naranjo and Latorre (1993) on conduction through Na$^+$ channels of toad skeletal muscle provides a more compromising view. These authors conclude that an energy barrier profile with double-ion occupancy and asymmetrical negative surface charge at outer and inner vestibules is necessary to adequately describe permeation.

IV. Pharmacology of Block

A. Tetrodotoxin, Saxitoxin, and μ-Conotoxin

The heterocyclic guanidinium toxins, TTX and STX, as well as the 22-residue peptide μ-conotoxin, are mutually competitive ligands that permit discrimination of Na$^+$ channel isoforms on the basis of toxin binding affinity. These molecules are also important tools for investigating blocking interactions at an external site that appears to be intimately associated with the outer mouth of the channel. The BTX-modified form of the open channel makes it is possible to resolve discrete blocking events induced by these toxins in single Na^+ channels, that can range from seconds to minutes, depending on the toxin homolog (Krueger *et al.*, 1983; Cruz *et al.*, 1985; Green *et al.*, 1987b; Guo *et al.*, 1987). Stochastic analysis of such individual blocking and unblocking events indicates that Na^+ channel inhibition is caused by a reversible reaction between one toxin molecule and a single binding site on the channel. Such studies have led to a descriptive model for the guanidinium toxin receptor of Na^+ channels, which is now being pursued at the amino acid level of the channel protein, as outlined in the following discussion.

The simplest mechanism that is often proposed for channel block involves a direct occlusion of the pore: the toxin enters and physically blocks the channel. Identification of block by occlusion is normally based on the voltage dependence of the blocking reaction. One of the characteristic features of TTX/STX block of BTX-modified Na^+ channels in planar bilayers is its strong voltage dependence, the block being enhanced at negative membrane potentials (Krueger *et al.*, 1983; French *et al.*, 1984). The usual model to explain such a voltage-dependent binding reaction assumes that the voltage dependence arises from a direct effect of the transmembrane electric field on the blocking molecule as it enters and dissociates from the pore. This model was originally proposed to explain voltage-dependent block of Na^+ channels by small blocking ions such as H^+ and Ca^{2+} (Woodhull, 1973). A salient feature of the Woodhull mechanism is the direct relationship between the voltage dependence of block and the charge of the blocking molecule.

Direct occlusion or plugging of the Na^+ channel by TTX/STX is a classical hypothesis that evolved from early electrophysiological studies and considerations of toxin structure (Kao and Nishiyama, 1965; Hille, 1975b). Since guanidinium ions are weakly permeant through Na^+ channels and the larger TTX and STX molecules possess one or two guanidinium moieties, respectively, the concept of a toxin guanidinium group partly stuck into the narrow pore of the channel is very appealing. The finding that alkali cations inhibit binding of TTX/STX with an ionic selectivity sequence similar to permeation (Henderson *et al.*, 1974; Reed and Raftery, 1976; Weigele and Barchi, 1978) further suggested that the site of occlusion must be part of the channel's "selectivity filter" (Hille, 1975a). However, the tight binding affinity of guanidinium toxins in the nanomolar range implies that specific protein–ligand interactions are involved in achieving

such binding energy. Doubts about the ability of a narrow plugging site to avidly recognize the distinct configuration of polar groups of both TTX and STX led Kao and Walker (1982) to argue that the toxin receptor lies at "the outside surface of the membrane, close to, but not inside, the sodium channel." This perspective changed the debate about the mechanism of toxin block to a question of depth. How deep is the toxin binding site? Does the toxin act like a "plug" or a "lid?"

The ability of single BTX-modified Na^+ channels to reveal individual durations of toxin block encouraged attempts to answer this question, but studies of the voltage dependence of block have not provided much illumination. Analysis according to Woodhull (1973) indicates that the divalent cation STX and the monovalent cation TTX exhibit the same voltage dependence, equivalent to a singly charged blocking ion that senses ~65% of the transmembrane voltage drop (French *et al.*, 1984; Moczydlowski *et al.*, 1984b; Green *et al.*, 1987b). Similar studies of a collection of natural derivatives of STX with net charges ranging from -1 to $+2$ reveal the same lack of dependence on toxin charge (Moczydlowski *et al.*, 1984b; Guo *et al.*, 1987). Likewise, polycationic μ-conotoxins, small proteins that apparently bind to an overlapping site, exhibit similar voltage-dependent blocking kinetics (Cruz *et al.*, 1985). Since charged toxin molecules do not appear to directly sense applied voltage (as judged by the criterion that the voltage-dependence must be proportional to toxin change), researchers proposed that the apparent voltage dependence of TTX/STX block arises from a voltage-dependent conformational change of the binding site (Moczydlowski *et al.*, 1984b) or a voltage-dependent closure of the channel subsequent to toxin binding (Green *et al.*, 1987b). Perhaps the most important finding from this work is that voltage dependence of block does not always correlate with the valence of the blocker, implying that allosteric mechanisms may play a role in this phenomenon.

Interpretation of the voltage dependence of TTX/STX block of single BTX-modified Na^+ channels is also complicated by apparently conflicting observations in other types of assays. Blockade of normal macroscopic Na^+ currents by guanidinium toxins usually does not exhibit a dependence on holding potential or pulse potential, unless repetitive pulsing is used to reveal the use-dependent block that occurs in some preparations, particularly in cardiac Na^+ channels (Cohen *et al.*, 1981; Lonnendonker, 1989). Rando and Strichartz (1986) addressed the question of whether the apparent voltage dependence of STX block is a phenomenon induced by BTX modification. These investigators found that STX block of normal Na^+ currents in frog nodes of Ranvier did not vary with changes in holding voltage or repetitive depolarization. However, after BTX modification, repetitive depolarization caused a time-dependent relief of STX block, consistent with the relief of STX block observed with increasing positive voltage in bilayer studies. However, this effect was only observed at voltages more positive than the range of channel activation, leading to the conclusion that STX only binds in a voltage-dependent manner to the BTX-modified form of the open channel (Rando and Strichartz, 1986; Strichartz et al., 1986). A similar investigation of the effect of TTX on BTX-modified Na^+ currents, measured by the whole-cell voltage-clamp technique in cardiac cells, led to the different conclusion that TTX block is state dependent but not voltage dependent (Huang et al., 1987).

In addition to differences in the phenomenon of voltage-dependent block observed at the whole-cell and single-channel level, evidence suggests that the effect is unique to certain modifiers. For example, voltage-dependent block by the STX derivative C1 is observed for single Na^+ channels modified by veratridine (Garber and Miller, 1987), but not for TTX block of single Na^+ channels modified by the alkaloid germitrine (Dugas et al., 1989). The concern that voltage-dependent block by TTX/STX may be peculiar to planar bilayer assays prompted an examination of TTX block of single BTX-modified Na^+ channels in neuroblastoma cells by cell-attached patch recording (McPherson and Huang, 1991). Reassuringly, this study did find voltage-dependent block by TTX similar to that observed in planar bilayers. Although more work is needed, the safest conclusion currently seems to be that certain modifiers such as BTX reveal voltage-dependent kinetics of TTX/STX binding that are absent in other states of the channel.

The implication from bilayer work, that TTX and STX do not directly sense the applied electric field, has led to the suggestion that the toxins bind at a superficial site and block by inducing a conformational change that closes the channel (Green et al., 1987b). One important test of the proximity of a channel blocker to the pore is the demonstration of a *trans* effect, that is, relief of block on one side of the channel by permeant ions or a charged blocker added to the opposite side (e.g., Newland et al., 1992). In the case of block of BTX-modified Na^+ channels by TTX/STX, reported tests for a *trans* effect have failed. No significant relief of toxin block is generated by increasing internal $[Na^+]$ (Moczydlowski et al., 1984a) or by simultaneous occupancy of the channel by internal blockers such as tetraethylammonium (Green et al., 1987a) or methylguanidinium ions (Moczydlowski et al., 1986). These negative results are in stark contrast

with the finding that block of the Na^+ channel by an internal local anesthetic, cocaine, is relieved by increasing external $[Na^+]$ (Wang, 1988). The lack of a demonstrated *trans* effect certainly does not rule out an occlusion mechanism for guanidinium toxin block, but does make proving such a hypothesis difficult.

B. Surface Charge at the Mouth and Block by Divalent Cations

A great deal of evidence supports the presence of fixed negative charges at the external surface of Na^+ channels, including effects of ionic strength and divalent cations on activation gating (discussed earlier), as well as the effect of chemical modification of carboxyl groups on single channels (Worley et al., 1986). Using Na^+ channels from canine brain, Green et al. (1987a) obtained an estimate of the magnitude of surface charge density at the guanidinium toxin binding site. By measuring the rate constants of association for the divalent toxin STX^{2+} relative to that of the monovalent toxin TTX^+ as a function of external $[NaCl]$, the Gouy–Chapman theory of planar surface charge was used to estimate an apparent surface charge density of $1 \ e^-/300 \ \text{Å}^2$ in the vicinity of the TTX/STX binding site. Application of the same approach to the TTX-insensitive Na^+ channel isoform from canine cardiac muscle yielded a similar value for negative surface charge density (Ravindran and Moczydlowski, 1987; Cai and Jordan, 1990). These results imply that negative charge is a conserved feature in these two channel isoforms and that a difference in negative surface charge in the vicinity of the toxin receptor is not the basis for the different affinities for TTX/STX.

Evidence for the importance of specific chemical interactions, such as hydrogen bonds between functional groups of guanidinium toxins and the receptor site, is abundant from structure–activity studies of the blocking potency of TTX and STX derivatives (Kao, 1986; Strichartz, 1984; Guo et al., 1987). In particular, faster dissociation rates of TTX/STX observed directly in single-channel recordings of toxin-insensitive Na^+ channels from cardiac muscle imply that the low binding affinity is due to the absence or perturbation of particular bonding interaction(s) (Guo et al., 1987). This possibility has been confirmed by mutagenesis experiments using cloned cDNAs encoding skeletal muscle and cardiac isoforms, which demonstrate that the substitution of a single cysteine in place of a tyrosine residue can fully account for low affinity for TTX/STX in heart Na^+ channels (Satin et al., 1992; Backx et al., 1992). A clue leading to this discovery was provided by studying the block of Na^+ channel isoforms by external divalent cations.

Patch-clamp recordings of single Na^+ channels have shown that, in addition to shifting the voltage range of activation gating to more positive voltage, external Ca^{2+} ions cause a reduction in unitary Na^+ current (Yamamoto *et al.*, 1984; Weiss and Horn, 1986; Sheets *et al.*, 1987; Nilius, 1988b). Such a reduction in conductance is characteristic of a fast block, when flickering between the open and blocked current levels is so rapid that the recording reflects a filtered average because of the limited bandwidth of the amplifier (Hille, 1992). However, another mechanism that could produce a lower current is the screening of surface charge by Ca^{2+}, which could decrease the local concentration of Na^+ near the external mouth of the channel. Because of their similar effects, evaluating the relative contribution of block and screening mechanisms in the inhibition of unitary channel currents by divalent cations can be difficult. This difficulty is discussed by Green and Andersen (1991) in reference to the effect of Ba^{2+} on BTX-activated Na^+ channels from canine brain (Green *et al.*, 1987a). When added to the external side of the channel, Ba^{2+} causes a weakly rectifying current reduction that is suggestive of a voltage-dependent block. However, when Ba^{2+} is added symmetrically to both sides, a uniform 70% reduction in current occurs at positive and negative voltage. Such an effect would be expected for surface charge screening, but a weak block and binding to surface charges are probably also involved. The contribution of block is more clearly demonstrated by the effect of external Ca^{2+} and Zn^{2+} on BTX-activated Na^+ channels from brain, where these divalent cations produce a reduction in unitary current that is enhanced with hyperpolarization (Worley *et al.*, 1986; Green *et al.*, 1987a). Numerous investigators have interpreted this effect according to a fast Woodhull-type block in which the site for the divalent cation is located about 20% of the way along the membrane field from the external side (see Ravindran *et al.*, 1991).

Evidence for specific blocking site(s) for divalent cations has been obtained by quantitative comparison of blocking ion selectivity for BTX-modified Na^+ channels from rat skeletal muscle and canine heart (Ravindran *et al.*, 1991). The apparent equilibrium dissociation constant for fast voltage-dependent block ranged from 8 to 90 mM for the following divalent cations, according to the affinity sequence: $Co^{2+} \sim Ni^{2+} > Mn^{2+} > Ca^{2+} > Mg^{2+} > Sr^{2+} > Ba^{2+}$. Divalent cations with the smallest ionic radii exhibited higher affinity, suggesting that electrostatic attraction predominates over dehydration energy in binding to the blocking site. The participation of surface charge screening in modulating this blocking reaction can be inferred from the shape of Ca^{2+} titration curves. The

increase in ionic strength as $CaCl_2$ is added to titrate the block results in a deviation from a Langmuir binding isotherm. This effect can be simulated by Gouy–Chapman theory to predict the decrease in negative surface potential as $CaCl_2$ is increased. This analysis suggests that the density of negative surface charge in the vicinity of the Ca^{2+} blocking site is about 1 e$^-$/400 Å^2 for the muscle and the cardiac Na^+ channel (Ravindran *et al.*, 1991). This value is very similar to that previously found for the TTX/STX binding reaction (Green *et al.*, 1987b; Ravindran and Moczydlowski, 1989), suggesting a common location of the blocking sites for divalent cations and guanidinium toxins. A correlation has also been suggested between the fraction of Na^+ channels blocked by Ca^{2+} and the effect of Ca^{2+} in shifting the voltage sensitivity of activation gating (Armstrong and Cota, 1991). However, when four alkaline earth cations are compared, blocking affinity follows the sequence $Ca^{2+} > Mg^{2+} > Sr^{2+} > Ba^{2+}$ whereas gating shift follows the sequence $Ca^{2+} > Ba^{2+} > Sr^{2+} > Mg^{2+}$ (Cukierman and Krueger, 1990). This difference implies that different sites mediate the effect on gating and blocking.

Additional evidence for an overlapping location of a divalent cation blocking site and the site for TTX/STX has been obtained from competition studies. Experiments with single BTX-activated Na^+ channels from brain have shown that external Ca^{2+} and Zn^{2+}, acting as fast voltage-dependent blockers, also lower the apparent association rate of STX in a manner suggestive of binding competition (Worley *et al.*, 1986; Krueger *et al.*, 1986; Green *et al.*, 1987b). However, the necessity of using relatively high concentrations of divalent cations in these latter experiments introduces the possibility of surface potential screening, making it difficult to discern true binding competition. This problem was circumvented by the finding that mammalian cardiac Na^+ channels display higher sensitivity than other isoforms to group IIB divalent cations such as Cd^{2+} and Zn^{2+} (DiFrancesco *et al.*, 1985; Frelin *et al.*, 1986; Visentin *et al.*, 1990). To illustrate this difference, Fig. 8 shows a comparison of the effect of external Zn^{2+} on BTX-modified Na^+ channels from rat muscle and canine heart in planar bilayers (Ravindran *et al.*, 1991; Schild *et al.*, 1991). The rat muscle channel exhibits a fast block that appears as a lower unitary current in the presence of 10 mM Zn^{2+}, whereas the cardiac channel displays a discrete flickering block in the presence of 40 μM Zn^{2+}.

In planar bilayers, the blocking affinity of canine heart Na^+ channels by Cd^{2+} and Zn^{2+} is, respectively, 46-fold and 110-fold greater than that of rat muscle Na^+ channels (Ravindran *et al.*, 1991; Schild *et al.*, 1991). Since the voltage dependence of these blocking

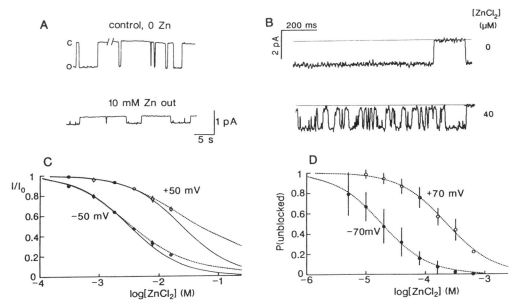

FIGURE 8 Differential sensitivity of muscle and cardiac Na$^+$ channels to block by external Zn^{2+}. (A) Records (opening downward) at -50 mV of a single batrachotoxin (BTX)-modified Na$^+$ channel from rat skeletal muscle showing the fast blocking effect of 10 mM external Zn^{2+}. (B) Records (opening downward) at -70 mV of a single BTX-modified Na$^+$ channel from canine heart showing discrete flickering block induced by 40 μM Zn^{2+}. (C) Zn^{2+} sensitivity of rat muscle Na$^+$ channel is illustrated by titration of the relative unitary current, I/I_0, vs external [ZnCl$_2$]. Data are described by a fit to one-site inhibition (solid curves) by Zn^{2+} with $K_d = 3.2$ mM at -50 mV and $K_d = 24$ mM at $+50$ mV. Dashed line fits include a correction for negative surface charge. (D) Zn^{2+} sensitivity of canine cardiac Na$^+$ channel illustrated by titration of unblocked probability vs external [ZnCl$_2$]. Data at low [ZnCl$_2$] are described by a Hill equation with $K_d = 18$ μM, $n = 1.1$ at -70 mV and $K_d = 220$ μM, $n = 1.2$ at $+70$ mV. (A, C, D) Reproduced from Ravindran *et al.*, *The Journal of General Physiology* (1991), **97**, 89–115, by copyright permission of the Rockefeller University Press. (B) Reproduced from Schild *et al.*, *The Journal of General Physiology* (1991), **97**, 117–142, by copyright permission of the Rockefeller University Press.

reactions is similar for the two isoforms (~20% of the field), this result suggests that block by Cd^{2+} or Zn^{2+} occurs at structurally analogous sites in the two Na$^+$ channel isoforms, but that a difference in amino acid sequence alters the binding affinity. The high affinity of the cardiac Na$^+$ channel for Zn^{2+} ($K_d = 67$ μM at 0 mV) permits Zn^{2+}–STX competition to be examined at low Zn^{2+} concentrations that have insignificant screening effects. Single-channel experiments with the BTX-activated Na$^+$ channel from heart provided clear evidence of binding competition between Zn^{2+} and STX (Schild and Moczydlowski, 1991), strengthening the notion that the Zn^{2+} blocking site is a subsite of the STX binding site. Since binding sites for group IIB divalent cations in proteins often contain cysteine residues, the effect of the sulfhydryl alkylating reagent iodoacetamide was also studied. Modification of single heart Na$^+$ channels by iodoacetamide resulted in abolition of high-affinity Zn^{2+} block and a simultaneous reduction in affinity for STX (Schild and Moczydlowski, 1991). Independent evidence for

Zn^{2+}–STX competition and a cysteine residue in the binding site has been obtained from biochemical studies of [^{3}H]STX binding to Na$^+$ channels in plasma membranes isolated from sheep heart (Doyle *et al.*, 1993). A particularly important clue to the location of the Zn^{2+} blocking site in the cardiac Na$^+$ channel is the finding that Zn^{2+} block corresponds to a small subconductance state in single BTX-modified channels (Schild *et al.*, 1991). To explain the small Na$^+$ current flowing in the Zn^{2+}-occupied substate, this observation implies that Zn^{2+} binds somewhere within the external vestibule but does not completely occlude the single-filing region of the channel.

C. Mapping Blocking Sites by Mutational Analysis

The feasibility of analyzing functional consequences of mutations of cloned Na$^+$ channels in *Xenopus* oocytes has allowed identification of protein domains that are associated with external blocking sites.

The first advance in this direction came with the finding that the substitution of Glu 387 with glutamine in the rat brain type II isoform abolished sensitivity to block by TTX and STX without seriously affecting macroscopic Na$^+$ current (Noda *et al.*, 1989). The conserved Glu 387 residue is located within the S5–S6 extracellular linker region of the internally homologous domain I of Na$^+$ channels. This location is in the analogous region that has been identified to form part of the pore in voltage-dependent K$^+$ channels (Yellen *et al.*, 1991). More detailed mutational studies have also identified a number of other residues in the SS2 portion of the S5–S6 linker in each of the I–IV domains that have a large influence on blocking affinity for the guanidinium toxins (Terlau *et al.*, 1991). These residues are generally located in two homologous positions in each of the I–IV domains; five of eight are negatively charged glutamate or aspartate residues. Since several of these mutations also decrease unitary conductance, these two clusters of predominantly negatively charged groups are thought to form two charged rings that line part of the extracellular mouth. This hypothesis is supported by the finding that two of the non-negatively charged residues in these ring locations, Lys 1422 and Ala 1714, control the relative permeability of monovalent and divalent inorganic cations (Heinemann *et al.*, 1992).

Complementary results have been obtained by an approach based on pharmacological differences between Na$^+$ channels isoforms. The higher blocking affinity of the cardiac isoform for Zn^{2+} and Cd^{2+}, and the abolition of Zn^{2+} block by alkylation of a sulfhydryl group, led to the suggestion that a unique cysteine residue in the rat heart isoform (RH1), Cys 374, was involved in this effect (Schild and Moczydlowski, 1991). Cys 374 is also located in the S5–S6 linker region and is only two residues away from the position of the glutamate residue originally implicated in TTX/STX binding by Noda *et al.* (1989). The involvement of Cys 374 was verified by the demonstration that substitution of this residue with tyrosine reduced sensitivity to block by Cd^{2+} and concomitantly enhanced sensitivity to TTX/STX (Satin *et al.*, 1992). Similarly, Backx *et al.* (1992) showed that the converse tyrosine to cysteine mutation in the μ1 isoform from rat skeletal muscle enhanced sensitivity to block by external Cd^{2+} and reduced affinity for TTX/STX. These results have led to the proposal that a major energetic contribution to the high affinity of TTX/STX binding found in neuronal and skeletal muscle isoforms is a π-electron–cation interaction between an aromatic tyrosine/phenylalanine residue and a toxin guanidinium group in TTX or STX (Satin *et al.*, 1992).

D. Local Anesthetics

Inhibition of normal resting macroscopic Na$^+$ current by local anesthetics such as lidocaine is characterized by a reduction in current after a single depolarizing stimulus. In addition, many local anesthetics induce a further accumulative reduction of Na$^+$ current in response to frequent repetitive stimulation. The former type of inhibition has been termed "tonic" or resting inhibition; the latter effect has been termed "phasic" or "use-dependent" inhibition (e.g., Lee-Son *et al.*, 1992). This complex blocking behavior is understood to reflect changes in binding affinities for local anesthetics that depend on voltage-dependent conformational states of the channel that occur during the activation and inactivation process (Hille, 1992). Since BTX modification alters channel gating and greatly stabilizes an open form of the channel at depolarized membrane potentials, it is not surprising that the blocking pharmacology of BTX-modified channels by local anesthetics differs from that of the normal channel studied in the absence of an activator (Khodorov, 1985).

Permanently charged quaternary ammonium derivatives of local anesthetics, such as the lidocaine derivative QX-314, do not readily cross lipid bilayers and have been used to demonstrate an internal site of action (Frazier *et al.*, 1970; Strichartz, 1973). Addition of QX-314 to the internal side of single BTX-modified Na$^+$ channels causes an apparent reduction in unitary conductance with increased noise that is characteristic of fast unresolved blocking events. Related organic cations such as tetraethylammonium and short-chain alkylguanidinium compounds have been shown to induce a similar fast block when millimolar concentrations of drug are added to the internal side (Moczydlowski *et al.*, 1986; Green *et al.*, 1987a). The fast block induced by these compounds is markedly enhanced at positive voltage. Analysis according to a Woodhull (1973) model implies that the blocking reaction senses 34–59% of the membrane field, depending on the blocker (Moczydlowski *et al.*, 1986; Wang, 1988). The more hydrophobic analogs of local anesthetics exhibit longer dwell times in the channel, permitting resolution of discrete blocking events. Examples of molecules belonging to this latter class of compounds include bupivicaine, cocaine, and long-chain derivatives of *n*-alkyl-trimethylammonium (Wang, 1988; Wang *et al.*, 1991). At +50 mV, cocaine exhibits a mean blocked duration of ~400 msec on single BTX-activated channels from rabbit muscle and has been used to examine the effect of Na$^+$ on the kinetics of the blocking reaction (Wang, 1988). Although internal Na^+ appears to have no effect, an increase in external Na^+ relieves block by internal cocaine. Kinetic analysis indicates that relief of internal cocaine block by external Na^+ is the result of a lower association rate and a faster dissociation rate of cocaine. The fact that the binding rates of an internal blocker can be modulated by external permeant ions suggests that the cocaine binding site is in the inner mouth of the channel and is able to sense occupancy of the pore by Na^+, possibly by electrostatic repulsion.

The stereospecificity of local anesthetics has been used to probe the topology of the receptor site on normal and BTX-modified Na^+ channels. For normal Na^+ channels, the I isomer of the chiral local anesthetic RAC421 is more potent than the corresponding II isomer (Yeh, 1980). Surprisingly, this stereospecificity is reversed in BTX-modified Na^+ channels, for which the II isomer of RAC421 is more potent (Wang, 1990). As suggested by Lee-Son *et al.* (1992), a change in stereospecificity of local anesthetics in BTX-modified Na^+ channels can be explained by two sites of action of local anesthetics or by an allosteric interaction with BTX that changes the affinity and stereospecificity of drug binding. An increase in the dissociation rate of [^{3}H]BTX-B induced by various local anesthetics has been documented in radioligand binding studies (Postma and Catterall, 1984). This result implies that BTX and local anesthetics simultaneously bind to distinct sites that interact allosterically in a destabilizing fashion.

In addition to the block induced by QX-314 and cocaine that is enhanced by depolarization, certain local anesthetic drugs such as procaine and benzocaine induce a different kind of long-lived blocking event in single BTX-modified channels that is enhanced by hyperpolarization in the range of $+60$ to -60 mV (Moczydlowski *et al.*, 1986). This blocking effect is currently poorly understood, but appears to be related to a residual form of inactivation in BTX-modified Na^+ channels, as described by Wang and Wang (1992).

V. Summary

The use of BTX to analyze the behavior of voltage-sensitive Na^+ channels activated by this toxin has greatly enhanced understanding of various aspects of channel function. In particular, the importance of surface charge phenomena has been clearly defined and elementary processes that are likely to play direct roles in ion permeation and selectivity have been identified. The molecular level has been reached with the mapping of protein residues involved in the binding of Na^+ channel blockers such as TTX, STX, Cd^{2+}, and Zn^{2+}. Many important problems remain, including the location of binding sites for BTX and local anesthetics and the nature of allosteric interactions between channel domains involved in gating and those involved in drug binding.

References

Albuquerque, E. X., and Daly, J. W. (1976). Batrachotoxin, a selective probe for channels modulating sodium conductances in electrogenic membranes. *In* "The Specificity and Action of Animal, Bacterial and Plant Toxins" (P. Cuatrecsas, ed.), pp. 297–338. Chapman and Hall, London.

Albuquerque, E. X., Daly, J. W., and Witkop, B. (1971). Batrachotoxin: Chemistry and pharmacology. *Science* **172**, 995–1002.

Armstrong, C. M., and Bezanilla, F. (1977). Inactivation of the sodium channel. II. Gating current experiments. *J. Gen. Physiol.* **70**, 567–590.

Armstrong, C. M., and Cota, G. (1991). Modification of sodium channel gating by lanthanum. Some effects that cannot be explained by surface charge theory. *J. Gen. Physiol.* **96**, 1129–1140.

Backx, P. H., Yue, D. T., Lawrence, J. H., Marban, E., and Tomaselli, G. F. (1992). Molecular localization of an ion-binding site within the pore of mammalian sodium channels. *Science* **257**, 248–251.

Begenisich, T. (1987). Molecular properties of ion permeation through sodium channels. *Annu. Rev. Biophys. Biophys. Chem.* **16**, 247–263.

Begenisich, T. B. and Cahalan, M. D. (1980). Sodium channel permeation in squid axons. I. Reversal potential experiments. *J. Physiol. (London)* **307**, 217–242.

Behrens, M. I., Oberhauser, A., Bezanilla, F., and Latorre, R. (1989). Batrachotoxin-modified sodium channels from squid optic nerve in planar bilayers. Ion conduction and gating properties. *J. Gen. Physiol.* **93**, 23–41.

Brown, G. B. (1986). ^{3}H-Batrachotoxinin-A benzoate binding to voltage-sensitive sodium channels: Inhibition by the channel blockers tetrodotoxin and saxitoxin. *J. Neurosci.* **6**, 2064–2070.

Brown, G. B., Tieszen, S. C., Daly, J. W., Warnick, J. E., and Albuquerque, E. X. (1981). Batrachotoxinin-A 20-α-benzoate: A new radioactive ligand for voltage-sensitive sodium channels. *Cell. Mol. Neurobiol.* **1**, 19–40.

Brown, G. B., Gaupp, J. E., and Olsen, R. W. (1988). Pyrethroid insecticides: Stereospecific allosteric interaction with the batrachotoxinin-A benzoate binding site of the mammalian voltage-sensitive sodium channels. *Mol. Pharmacol.* **34**, 54–59.

Cahalan, M., and Begenisich, T. (1976). Sodium channel selectivity. Dependence on internal permeant ion concentration. *J. Gen. Physiol.* **68**, 111–125.

Cai, M., and Jordan, P. C. (1990). How does vestibule surface charge affect ion conduction and toxin binding in a sodium channel? *Biophys. J.* **57**, 883–891.

Catterall, W. A. (1975). Activation of the action potential Na^+ ionophore of cultured neuroblastoma cells by veratridine and batrachotoxin. *J. Biol. Chem.* **250**, 4053–4059.

Catterall, W. A. (1988). Structure and function of voltage-sensitive ion channels. *Science* **242**, 50–61.

Catterall, W. A., and Gainer, M. (1985). Interaction of brevetoxin A, with a new receptor site on the sodium channel. *Toxicon* **23**, 497–504.

Catterall, W. A., Morrow, C. S., Daly, J. W., and Brown, G. B. (1981). Binding of batrachotoxinin A 20-α-benzoate to a receptor site associated with sodium channels in synaptic nerve ending particles. *J. Biol. Chem.* **256**, 8922–8927.

Chabala, L. D., Urban, B. W., Weiss, L. B., Green, W. N., and Andersen, O. S. (1991). Steady-state gating of batrachotoxin-modified sodium channels. Variability and electrolyte-dependent modulation. *J. Gen. Physiol.* **98**, 197–224.

Chandler, W. K., and Meves, H. (1965). Voltage clamp experiments on internally perfused giant axons. *J. Physiol. (London)* **180**, 788–820.

Chinn, K., and Narahashi, T. (1989). Temperature-dependent subconducting states and kinetics of deltamethrin-modified sodium channels of neuroblastoma cells. *Pfluegers Arch.* **413**, 571–579.

Cohen, C. J., Bean, B. P., Colatsky, T. J., and Tsien, R. W. (1981). Tetrodotoxin block of sodium channels in rabbit purkinje fibers. Interactions between toxin binding and channel gating. *J. Gen. Physiol.* **78**, 383–411.

Correa, A. M., Latorre, R., and Bezanilla, F. (1991). Ion permeation in normal and batrachotoxin-modified Na^+ channels in the squid giant axons. *J. Gen. Physiol.* **97**, 605–625.

Correa, A. M., Bezanilla, F., and Latorre, R. (1992). Gating kinetics of batrachotoxin-modified Na^+ channels in the squid giant axon. Voltage and temperature effects. *Biophys. J.* **61**, 1332–1352.

Creveling, C. R., McNeal, E. T., Daly, J. W., and Brown, G. B. (1983). Batrachotoxin-induced depolarization and [^{3}H]batrachotoxinin-A 20-α-benzoate binding in a vesicular preparation from guinea pig cerebral cortex. Inhibition by local anesthetics. *Mol. Pharmacol.* **23**, 350–358.

Cribbs, L. L., Satin, J., Fozzard, H. A., and Rogart, R. B. (1990). Functional expression of the rat heart I Na channel isoform: Demonstration of properties characteristic of native Na channels. *FEBS Lett.* **275**, 195–200.

Cruz, L. J., Gray, W. R., Olivera, B. M., Zeikus, R. D., Kerr, L., Yoshikami, D., and Moczydlowski, E. (1985). *Conus geographus* toxins that discriminate between neuronal and muscle sodium channels. *J. Biol. Chem.* **260**, 9280–9288.

Cukierman, S. (1991a). Asymmetric electrostatic effects on the gating of rat brain sodium channels in planar lipid bilayers. *Biophys. J.* **60**, 845–855.

Cukierman, S. (1991b). Inactivation modifiers of Na^+ currents and the gating of rat brain Na^+ channels in planar lipid membranes. *Pfluegers Arch.* **419**, 514–521.

Cukierman, S., and Krueger, B. K. (1990). Modulation of sodium channel gating by external divalent cations: Differential effects on opening and closing rates. *Pfluegers Arch.* **416**, 360–367.

Cukierman, S., and Krueger, B. K. (1991). Effects of internal divalent cations on the gating of rat brain Na^+ channels reconstituted in planar lipid bilayers. *Pfluegers Arch.* **419**, 559–565.

Cukierman, S., Zinkand, W. C., French, R. J., and Krueger, B. K. (1988). Effects of membrane surface charge and calcium on the gating of rat brain sodium channels in planar bilayers. *J. Gen. Physiol.* **92**, 431–447.

Daly, J. W., Myers, C. W., and Whittaker, N. (1987). Further classification of skin alkaloids from neotropical poison frogs (Dendrobatidae), with a general survey of toxic/noxious substances in the Amphibia. *Toxicon* **25**, 1023–1095.

Dani, J. A. (1986). Ion channel entrances influence permeation: Net charge, size, shape and binding considerations. *Biophys. J.* **54**, 767–776.

Di Francesco, D., Ferroni, A., Visentin, S., and Zaza, A. (1985). Cadium-induced blockade of the cardiac fast Na channels in calf Purkinje fibers. *Proc. R. Soc. London B* **233**, 475–484.

Doyle, D. D., Guo, Y., Lustig, S. L., Satin, J., Rogart, R. B., and Fozzard, H. A. (1993). Divalent cation competition with ^{3}H-saxitoxin binding to tetrodotoxin-resistant and -sensitive sodium

channels. A two-site structural model of ion/toxin interaction. *J. Gen. Physiol.* **101**, 153–182.

Duch, D. S., Recio-Pinto, E., Frenkel, C., and Urban, B. W. (1988). Human brain sodium channels in bilayers. *Mol. Brain Res.* **4**, 171–177.

Dugas, M., Honerjager, P., and Masslich, U. (1989). Tetrodotoxin block of single germitrine-activated sodium channels in cultured rat cardiac cells. *J. Physiol. (London)* **411**, 611–626.

Dumbacher, J. P., Beehler, B. M., Spande, T. F., Garraffo, H. M., and Daly, J. W. (1992). Homobatrachotoxin in the genus *Pitohui*: Chemical defense in birds? *Science* **258**, 799–801.

Frankenhaeuser, B., and Hodgkin, A. L. (1957). The action of calcium on the electrical properties of squid axons. *J. Physiol. (London)* **137**, 218–244.

Frazier, D. T., Narahashi, T., and Yamada, M. (1970). The site of action and active form of local anesthetics. II. Experiments with quaternary compounds. *J. Pharmacol. Exp. Ther.* **171**, 45–51.

Frelin, C., Cognard, C., Vigne, P., and Lazdunski, M. (1986). Tetrodotoxin-sensitive and tetrodotoxin-resistant Na^+ channels differ in their sensitivity to Cd^{2+} and Zn^{2+}. *Eur. J. Pharmacol.* **122**, 245–250.

French, R. J., Worley, J. F., and Krueger, B. K. (1984). Voltage-dependent block by saxitoxin of sodium channels incorporated into planar lipid bilayers. *Biophys. J.* **45**, 301–310.

French, R. J., Worley, J. F., Blaustein, M. B., Romine, W. O., Tam, K. K., and Krueger, B. K. (1986). Gating of batrachotoxin-activated sodium channels in lipid bilayers. In ''Ion Channel Reconstitution'' (C. Miller, ed.), pp. 363–383. Plenum Press, New York.

Garber, S. S. (1988). Symmetry and asymmetry of permeation through toxin-modified Na^+ channels. *Biophys. J.* **54**, 767–776.

Garber, S. S., and Miller, C. (1987). Single Na^+ channels activated by veratridine and batrachotoxin. *J. Gen. Physiol.* **89**, 459–480.

Gellens, M. E., George, A. L., Chen, L., Chahine, M., Horn, R., Barchi, R. L., and Kallen, R. G. (1992). Primary structure and functional expression of the human cardiac tetrodotoxin-insensitive voltage-dependent sodium channel. *Proc. Natl. Acad. Sci. USA* **89**, 554–558.

George, A. L., Knittle, T. J., and Tamkun, M. M. (1992a). Molecular cloning of an atypical voltage-gated sodium channel expressed in human heart and uterus: Evidence for a distinct gene family. *Proc. Natl. Acad. Sci. USA* **89**, 4893–4897.

George, A. L., Komisarof, J., Kallen, R. G., and Barchi, R. L. (1992b). Primary structure of the adult human skeletal muscle voltage-dependent sodium channels. *Ann. Neurol.* **31**, 131–137.

Gilly, W. F., and Armstrong, C. M. (1982). Slowing of sodium channel opening in squid axon by extracellular zinc. *J. Gen. Physiol.* **79**, 935–964.

Green, W. N., and Andersen, O. S. (1991). Surface charges and ion channel function. *Annu. Rev. Physiol.* **53**, 341–359.

Green, W. N., Weiss, L. B., and Andersen, O. S. (1987a). Batrachotoxin-modified sodium channels in planar bilayers. Ion permeation and block. *J. Gen. Physiol.* **89**, 841–872.

Green, W. N., Weiss, L. B., and Andersen, O. S. (1987b). Batrachotoxin-modified sodium channels in planar lipid bilayers. Characterization of saxitoxin- and tetrodotoxin-induced channels closures. *J. Gen. Physiol.* **89**, 873–903.

Guo, X., Z. Uehara, A. Ravindran, S. H. Bryant, S. Hall, and E. Moczydlowski. (1987). Kinetic basis for insensitivity to tetrodotoxin and saxitoxin in sodium channels of canine heart and denervated rat skeletal muscle. *Biochemistry* **26**, 7546–7556.

Hahin, R., and Campbell, D. T. (1983). Simple shifts in the voltage-dependence of sodium channel gating caused by divalent cations. *J. Gen. Physiol.* **82**, 785–805.

Hartshorne, R. P., Keller, B. U., Talvenheimo, J. A., Catterall,

W. A., and Montal, M. (1985). Functional reconstitution of the purified brain sodium channel in planar lipid bilayers. *Proc. Natl. Acad. Sci. USA* **82**, 240–244.

Heinemann, S. H., Terlau, H., Stuhmer, W., Imoto, K., and Numa, S. (1992). Calcium channel characteristics conferred on the sodium channel by single mutations. *Nature (London)* **356**, 441–443.

Henderson, R., Ritchie, J. M., and Strichartz, G. R. (1974). Evidence that tetrodotoxin and saxitoxin act at a metal cation binding site in the sodium channel of nerve membrane. *Proc. Natl. Acad. Sci. USA* **71**, 3936–3940.

Hille, B. (1975a). Ionic selectivity, saturation and block in sodium channels: A four barrier model. *J. Gen. Physiol.* **66**, 535–560.

Hille, B. (1975b). The receptor for tetrodotoxin and saxitoxin. A structural hypothesis. *Biophys. J.* **15**, 615–619.

Hille, B. (1992). "Ionic Channels of Excitable Membranes." Sinauer Associates, Sunderland, Massachusetts.

Hille, B., Woodhull, A. M., and Shapiro, B. I. (1975). Negative surface charge near sodium channels of nerve: Divalent ions, monovalent ions, and pH. *Phil. Trans. R. Soc. London B* **270**, 301–318.

Hodgkin, A. L., and Huxley, A. F. (1952). A quantitative description of membrane current and its application to conduction and excitation in nerve. *J. Physiol. (London)* **117**, 500–544.

Huang, L.-Y. M., Moran, N., and Ehrenstein, G. (1982). Batrachotoxin modifies the gating kinetics of sodium channels in internally perfused neuroblastoma cells. *Proc. Natl. Acad. Sci. USA* **79**, 2082–2085.

Huang, L.-Y., M., Moran, N., and Ehrenstein, G. (1984). Gating kinetics of batrachotoxin-modified sodium channels in neuroblastoma cells determined from single-channel measurements. *Biophys. J.* **45**, 313–322.

Huang, L.-Y. M., Yatani, A., and Brown, A. M. (1987). The properties of batrachotoxin-modified cardiac Na channels, including state-dependent block by tetrodotoxin. *J. Gen. Physiol.* **90**, 341–360.

Isom, L. L., De Jongh, K. S., Patton, D. E., Reber, B. F. X., Offord, J., Charbonneau, H., Walsh, K., Goldin, A. L., and Catterall, W. A. (1992). Primary structure and functional expression of the β1 subunit of the rat brain sodium channel. *Science* **256**, 839–842.

Kallen, R. G., Sheng, Z.-H., Yang, J., Chen, L., Rogart, R. B., and Barchi, R. L. (1990). Primary structure and expression of a sodium channel characteristic of denervated and immature rat skeletal muscle. *Neuron* **4**, 233–242.

Kao, C. Y., and Nishiyama, A. (1965). Actions of saxitoxin on peripheral neuromuscular systems. *J. Physiol. (London)* **180**, 50.

Kao, C. Y. (1986). Structure-activity relations of tetrodotoxin, saxitoxin and analogues. *Ann. NY Acad. Sci.* **479**, 52–67.

Kao, C. Y., and Walker, S. E. (1982). Active groups of saxitoxin and tetrodotoxin as deduced from actions of saxitoxin analogues on frog muscle and squid axon. *J. Physiol. (London)* **323**, 619–637.

Kayano, T., Noda, M., Flockerzi, V., Takahashi, H., and Numa, S. (1988). Primary structure of rat brain sodium channel III deduced from the cDNA sequence. *FEBS Lett.* **228**, 187–194.

Keller, B. U., Hartshorne, R. P., Talvenheimo, J. A., Catterall, W. A., and Montal, M. (1986). Sodium channels in planar lipid bilayers. Channel gating kinetics of purified sodium channels modified by batrachotoxin. *J. Gen. Physiol.* **88**, 1–23.

Khodorov, B. I. (1985). Batrachotoxin as a tool to study voltage-sensitive sodium channels of excitable membranes. *Prog. Biophys. Mol. Biol.* **45**, 57–148.

Khodorov, B. I., and Revenko, S. V. (1979). Further analysis of the mechanisms of action of batrachotoxin on the membrane of myelinated nerve. *Neurosci.* **4**, 1315–1330.

Kirsch, G. E., and Brown, A. M. (1989). Kinetic properties of single sodium channels in rat heart and rat brain. *J. Gen. Physiol.* **93**, 85–99.

Krueger, B. K., Worley, J. F., and French, R. J. (1983). Single sodium channels from rat brain incorporated into planar lipid bilayer membranes. *Nature (London)* **303**, 172–175.

Krueger, B. K., Worley, J. F., and French, R. J. (1986). Block of sodium channels in planar lipid bilayers by guanidinium toxins and calcium. Are the mechanisms of voltage-dependence the same? *Ann. N.Y. Acad. Sci.* **479**, 257–268.

Latorre, R., Labarca, P., and Naranjo, D. (1992). Surface charge effects on ion conduction in ion channels. *Meth. Enzymol.* **207**, 471–501.

Lee-Son, S., Wang, G. K., Concus, A., Crill, E., and Strichartz, G. (1992). Stereoselective inhibition of neuronal sodium channels by local anesthetics. Evidence for two sites of action? *Anesthesiol.* **77**, 324–335.

Lonnendonker, U. (1989). Use-dependent block of sodium channels in frog myelinated nerve by tetrodotoxin and saxitoxin at negative holding potentials. *Biochim. Biophys. Acta* **985**, 153–160.

McManus, O. B., and Magleby, K. L. (1988). Kinetic states and modes of single large-conductance calcium-activated potassium channels in cultured rat skeletal muscle. *J. Physiol. (London)* **402**, 79–120.

McNeal, E. T., Lewandowski, G. A., Daly, J. W., and Creveling, C. R. (1985). [³H]Batrachotoxinin A 20-α-benzoate binding to voltage-sensitive sodium channels: a rapid and quantitative assay for local anesthetic activity in a variety of drugs. *J. Med. Chem.* **28**, 381–388.

McPherson, D. R., and Huang, L.-Y. M. (1991). Voltage-dependent tetrodotoxin binding to single batrachotoxin-modified Na channels recorded from intact neuroblastoma cells. *Neurosci. Kett.* **131**, 201–204.

Moczydlowski, E., Garber, S. S., and Miller, C. (1984a). Batrachotoxin-activated Na⁺ channels in planar lipid bilayers: competition of tetrodotoxin block by Na⁺. *J. Gen. Physiol.* **84**, 665–686.

Moczydlowski, E., Hall, S., Garber, S. S., and Miller, C. (1984b). Voltage-dependent blockade of muscle Na⁺ channels by guanidinium toxins: effect of toxin charge. *J. Gen. Physiol.* **84**, 687–704.

Moczydlowski, E., Uehara, A., and Hall, S. (1986). Blocking pharmacology of batrachotoxin-activated sodium channels. *In* "Ion Channel Reconstitution" (C. Miller, ed.), pp. 405–428. Plenum Press, New York.

Naranjo, D., and Latorre, R. (1993). Ion conduction in substates of the BTX-modified Na⁺ channel from toad skeletal muscle. *Biophys. J.* **64**, 1038–1050.

Newland, C. F., Adelman, J. P., Tempel, B. L., and Almers, W. (1992). Repulsion between tetraethylammonium ions in cloned voltage-gated potassium channels. *Neuron* **8**, 975–982.

Nilius, B. (1988a). Modal gating behavior of cardiac sodium channels in cell-free membrane patches. *Biophys. J.* **53**, 857–862.

Nilius, B. (1988b). Calcium block of guinea-pig heart sodium channels with and without modification by the piperazinylindole DPI-201-106. *J. Physiol. (London)* **399**, 537–558.

Nilius, B., Vereecke, J., and Carmeliet, E. (1989). Different conductance states of the bursting Na channel in guinea-pig ventricular myocytes. *Pfluegers Arch.* **413**, 242–248.

Noda, M., Ikeda, T., Suzuki, H., Takeshima, H., Takahashi, T., Kuno, M., and Numa, S. (1986). Expression of functional sodium channels from cloned cDNA. *Nature (London)* **322**, 826–828.

Noda, M., Suzuki, H., Numa, S., and Stuhmer, W. (1989). A single point mutation confers tetrodotoxin and saxitoxin insensitivity on the sodium channel II. *FEBS Lett.* **259**, 213–216.

Numa, S., and Noda, M. (1986). Molecular structure of sodium channels. *Ann. N.Y. Acad. Sci.* **479,** 338–355.

O'Leary, M. E., and Krueger, B. K. (1989). Batrachotoxin and α-scorpion toxin stabilize the open state of single voltage-gated sodium channels. *Mol. Pharmacol.* **36,** 789–795.

Patlak, J. B. (1988). Sodium channel subconductance levels measured with a new variance-mean analysis. *J. Gen. Physiol.* **92,** 413–430.

Patlak, J. B., and Ortiz, M. (1985). Slow currents through single sodium channels of the adult rat heart. *J. Gen. Physiol.* **86,** 89–104.

Patlak, J. B., and Ortiz, M. (1986). Two modes of gating during late Na$^+$ channel currents in frog sartorius muscle. *J. Gen. Physiol.* **87,** 305–326.

Pietrobon, D., and Hess, P. (1990). Novel mechanism of voltage-dependent gating in L-type calcium channels. *Nature (London)* **346,** 651–655.

Postma, S. W., and Catterall, W. A. (1984). Inhibition of binding of [^{3}H]batrachotoxinin A 20-α-benzoate to sodium channels by local anesthetics. *Mol. Pharmacol.* **25,** 219–227.

Rando, T. A., and Strichartz, G. R. (1986). Saxitoxin blocks batrachotoxin-modified sodium channels in the node of Ranvier in a voltage-dependent manner. *Biophys. J.* **49,** 785–794.

Ravindran, A., and E. Moczydlowski. (1989). Influence of negative surface charge on toxin binding to canine heart Na channels in planar bilayers. *Biophys. J.* **55,** 359–365.

Ravindran, A., Schild, L., and Moczydlowski, E. (1991). Divalent cation selectivity for external block of voltage-dependent Na$^+$ channels prolonged by batrachotoxin. Zn^{2+} induces discrete substates in cardiac Na$^+$ channels. *J. Gen. Physiol.* **97,** 89–115.

Ravindran, A., Kwiecinski, H., Alvarez, O., Eisenman, G., and Moczydlowski, E. (1992). Modeling ion permeation through batrachotoxin-modified Na$^+$ channels from rat skeletal muscle with a multi-ion pore. *Biophys. J.* **61,** 494–508.

Recio-Pinto, E., Duch, D. S., Levinson, S. R., and Urban, B. W. (1987). Purified and unpurified sodium channels from eel electroplax in planar lipid bilayers. *J. Gen. Physiol.* **90,** 375–395.

Reed, J. K., and Raftery, M. A. (1976). Properties of the tetrodotoxin binding component in plasma membranes isolated from *Electrophorus electricus. Biochemistry* **15,** 944–953.

Reith, M. E. A., Kim, S. S., and Lajtha, A. (1986). Structural requirements for cocaine congeners to interact with [^{3}H]batrachotoxinin-A 20-α-benzoate binding sites on sodium channels in mouse brain synaptosomes. *J. Biol. Chem.* **261,** 7300–7305.

Roberts, R. H., and Barchi, R. L. (1987). The voltage-sensitive sodium channels from rabbit skeletal muscle. Chemical characterization of subunits. *J. Biol. Chem.* **262,** 2298–2303.

Rogart, R. B., Cribbs, L. L., Muglia, L. K., Kephart, D. D., and Kaiser, M. W. (1989). Molecular cloning of a putative tetrodotoxin-resistant rat heart Na$^+$ channel isoform. *Proc. Natl. Acad. Sci. USA* **86,** 8170–8174.

Romey, G., and Lazdunski, M. (1982). Lipid-soluble toxins thought to be specific for Na$^+$ channels block Ca^{2+} channels in neuronal cells. *Nature (London)* **297,** 79–80.

Rubinson, K. A. (1992). Steady-state kinetics of solitary batrachotoxin-treated sodium channels. Kinetics on a bounded continuum of polymer states. *Biophys. J.* **61,** 463–479.

Rudy, B. (1978). Slow inactivation of the sodium conductance in squid giant axons. Pronase resistance. *J. Physiol. (London)* **283,** 1–21.

Satin, J., Kyle, J. W., Chen, M., Bell, P., Cribbs, L. L., Fozzard, H. A., and Rogart, R. B. (1992). A mutant of TTX-resistant cardiac sodium channels with TTX-sensitive properties. *Science* **256,** 1202–1205.

Schild, L., and Moczydlowski, E. (1991). Competitive binding interaction between Zn^{2+} and saxitoxin in cardiac Na$^+$ channels. Evidence for a sulfhydryl group in the Zn^{2+}/saxitoxin binding site. *Biophys. J.* **59,** 523–537.

Schild, L., Ravindran, A., and Moczydlowski, E. (1991). Zn^{2+}-Induced subconductance events in cardiac Na$^+$ channels prolonged by batrachotoxin. Current-voltage behavior and single channel kinetics. *J. Gen. Physiol.* **97,** 117–142.

Sheets, M. F., Scanley, B. E., Hanck, D. A., Makielski, J. C., and Fozzard, H. A. (1987). Open sodium channel properties of single canine cardiac Purkinje cells. *Biophys. J.* **52,** 13–22.

Shenkel, S., and Sigworth, F. J. (1991). Patch recordings from the electrocytes of *Electrophorus electricus*. Na currents and P_{Na}/P_K variability. *J. Gen. Physiol.* **97,** 1013–1041.

Shenkel, S., Cooper, E. C., James, W., Agnew, W. S., and Sigworth, F. J. (1989). Purified, modified eel sodium channels are active in planar bilayers in the absence of activating neurotoxins. *Proc. Natl. Acad. Sci. USA* **86,** 9592–9596.

Strichartz, G. (1973). The inhibition of sodium currents in myelinated nerve by quaternary derivatives of lidocaine. *J. Gen. Physiol.* **62,** 37–57.

Strichartz, G. (1984). Structural determinants of the affinity of saxitoxin for neuronal sodium channels. Electrophysiological studies on frog peripheral nerve. *J. Gen. Physiol.* **84,** 281–305.

Strichartz, G., T. Rando, S. Hall, J. Gitschier, L. Hall, B. Magnani, and C. Hansen-Bay. (1986). On the mechanism by which saxitoxin binds to and blocks sodium channels. *Ann. NY Acad. Sci.* **479,** 96–112.

Strichartz, G., Rando, T., and Wang, G. K. (1987). An integrated view of the molecular toxinology of sodium channel gating in excitable cells. *Annu. Rev. Neurosci.* **10,** 237–267.

Suzuki, H., Beckh, S., Kubo, H., Yahagi, N., Ishida, H., Kayano, T., Noda, M., and Numa, S. (1988). Functional expression of cloned cDNA encoding sodium channel III. *FEBS Lett.* **228,** 195–200.

Tanguy, J., and Yeh, J. Z. (1988). Batrachotoxin uncouples gating charge immobilization from fast Na inactivation in squid giant axons. *Biophys. J.* **54,** 719–730.

Tanguy, J., and Yeh, J. Z. (1991). BTX modification of Na channels in squid axons. I. State dependence of BTX action. *J. Gen. Physiol.* **97,** 499–519.

Terlau, H., Heinemann, S. H., Stuhmer, W., Pusch, M., Conti, F., Imoto, K., and Numa, S. (1991). Mapping the site of block by tetrodotoxin and saxitoxin of sodium channel II. *FEBS Lett.* **293,** 93–96.

Tokuyama, T., Daly, J., Witkop, B., Karle, I. L., and Karle, J. J. (1968). The structure of batrachotoxinin A, a novel steroidal alkaloid from the Columbian arrow poison frog, *Phyllobates aurotaenia. J. Am. Chem. Soc.* **90,** 1917–1918.

Tokuyama, T., Daly, J., and Witkop, B. (1969). The structure of batrachotoxin, a steroidal alkaloid from the columbian arrow poison frog, *Phyllobates aurotaenia*, and partial synthesis of batrachotoxin and its analogs and homologs. *J. Am. Chem. Soc.* **91,** 3931–3938.

Trimmer, J. S., and Agnew, W. S. (1989). Molecular diversity of voltage sensitive Na channels. *Annu. Rev. Physiol.* **51,** 401–418.

Trimmer, J. S., Cooperman, S. S., Tomiko, S. A., Zhou, J., Crean, S. M., Boyle, M. B., Kallen, R. G., Sheng, Z., Barchi, R. L., Sigworth, F. J., Goodman, R. H., Agnew, W. S., and Mandel, G. (1989). Primary structure and functional expression of a mammalian skeletal muscle sodium channel. *Neuron* **3,** 33–49.

Tsien, R. W., Hess, P., McCleskey, E. W., and Rosenberg, R. L. (1987). Calcium channels: Mechanisms of selectivity, permeation and block. *Annu. Rev. Biophys. Biophys. Chem.* **16,** 265–290.

Ukomadu, C., Zhou, J., Sigworth, F. J., and Agnew, W. S. (1992). μI Na$^+$ channels expressed transiently in embryonic kidney cells: Biochemical and biophysical properties. *Neuron* **8,** 663–676.

Vandenberg, C. A., and Bezanilla, F. (1991). A sodium channel gating model based on single channel, macroscopic ionic and gating currents in the squid giant axon. *Biophys. J.* **60,** 1511–1533.

Visentin, S., Zaza, A., Ferroni, A., Tromba, C., and Di Francesco, C. (1990). Sodium current block caused by group IIb cations in calf Purkinje fibers and in guinea-pig ventricular myocytes. *Pfluegers Arch. (Eur. J. Physiol.)* **417,** 213–222.

Wang, G. K. (1988). Cocaine-induced closures of single batrachotoxin-activated Na$^+$ channels in planar lipid bilayers. *J. Gen. Physiol.* **92,** 747–765.

Wang, G. K. (1990). Binding affinity and stereoselectivity of local anesthetics in single batrachotoxin-activated Na$^+$ channels. *J. Gen. Physiol.* **96,** 1105–1127.

Wang, G. K., and Wang, S.-Y. (1992). Inactivation of batrachotoxin-modified Na$^+$ channels in GH3 cells. Characterization and pharmacological modification. *J. Gen. Physiol.* **99,** 1–20.

Wang, G. K., Simon, R., and Wang, S.-Y. (1991). Quaternary ammonium compounds as structural probes of single batrachotoxin-activated Na$^+$ channels. *J. Gen. Physiol.* **98,** 1005–1024.

Weigele, J. B., and Barchi, R. L. (1978). Saxitoxin binding to the mamalian sodium channel: competition by monovalent and divalent cations. *FEBS Lett.* **95,** 49–53.

Weiss, R. E., and Horn, R. (1986). Single-channel studies of TTX-sensitive and TTX-resistant sodium channels in developing rat muscle reveal different open channel properties. *Ann. N.Y. Acad. Sci.* **479,** 152–161.

White, M. M., Chen, L., Kleinfield, R., Kallen, R. G., and Barchi, R. L. (1991). SkM2, a Na$^+$ channel cDNA clone from denervated skeletal muscle encodes a tetrodotoxin-insensitive Na$^+$ channel. *Mol. Pharmacol.* **39,** 604–608.

Willow, M., and Catterall, W. A. (1982). Inhibition of binding of [^{3}H]batrachotoxinin A 20-α-benzoate to sodium channels by the anticonvulsant drugs diphenylhydantoin and carbamazepine. *Mol. Pharmacol.* **22,** 627–635.

Wonderlin, W. F., Finkel, A., and French, R. J. (1990). Optimizing planar lipid bilayer single-channel recordings for high resolution with rapid voltage steps. *Biophys. J.* **58,** 289–297.

Woodhull, A. (1973). Ionic blockage of sodium channels in nerve. *J. Gen. Physiol.* **61,** 687–708.

Worley, J. F., French, R. J., and Krueger, B. K. (1986). Trimethyloxonium modification of single batrachotoxin-activated sodium channels in planar bilayers. Changes in unit conductance and block by saxitoxin and calcium. *J. Gen. Physiol.* **87,** 327–349.

Worley, J. F., French, R. J., Pailthorpe, B. A., and Krueger, B. K. (1992). Lipid surface charge does not influence conductance or calcium block of single sodium channels in planar bilayers. *Biophys. J.* **61,** 1353–1363.

Yamamoto, D., Yeh, J. Z., and Narahashi, T. (1984). Voltage-dependent calcium block of normal and tetramethrin-modified single sodium channels. *Biophys. J.* **45,** 337–344.

Yeh, J. Z. (1980). Blockage of sodium channels by stereoisomers of local anesthetics. *In* "Molecular Mechanisms of Anesthesia" (B. R. Fink, ed.), Vol. 2, pp. 35–44. Raven Press, New York.

Yellen, G. (1987). Permeation in potassium channels: Implications for channel structure. *Annu. Rev. Biophys. Biophys. Chem.* **16,** 227–246.

Yellen, G., Jurman, M. E., Abramson, T., and MacKinnon, R. (1991). Mutations affecting internal TEA blockade identify the probable pore-forming region of a K$^+$ channel. *Science* **251,** 939–942.

Calcium
Channels

10

Molecular Biology of Voltage-Dependent Calcium Channels

YASUO MORI

I. Introduction

Voltage-dependent calcium channels play important roles in the regulation of diverse cellular functions including membrane excitability, muscle contraction, and synaptic transmission and other forms of secretion (Tsien and Tsien, 1990). Four types of calcium channels (L-, N-, P-, and T-type) have been distinguished by their functional properties such as time- and voltage-dependent kinetics, single-channel conductance, and pharmacology (Table I) (Hess 1990; Tsien *et al.*, 1991). L-Type calcium channels are high-voltage activated, are sensitive to dihydropyridine (DHP), and show a single-channel conductance of 22–27 pS. These channels are found in virtually all excitable tissues and in many nonexcitable cells. The channels trigger excitation–contraction coupling in skeletal muscle, heart, and smooth muscle and control hormone or transmitter release from endocrine cells and some neurons. N-Type calcium channels are also high-voltage activated, but differ pharmacologically from L-type channels by being resistant to DHP and blocked by ω-conotoxin (ω-CgTX). N-Type channels often exhibit a smaller single-channel conductance than L-type channels and tend to inactivate with depolarized holding potentials. N-Type channels have

been shown to be largely restricted to neurons and to be involved in neurotransmitter release. P-Type calcium channels, first identified in Purkinje cells (Llinás *et al.*, 1989; Hillman *et al.*, 1991), are found in a variety of neurons (Regan *et al.*, 1991). These channels are high-voltage activated calcium channels that are selectively blocked by ω-agatoxin-IVA (ω-Aga-IVA) (Mintz *et al.*, 1992a) or funnel-web spider toxin (FTX) (Llinás *et al.*, 1989), but are insensitive to DHP and ω-CgTx. P-Type channels play an essential role in inducing long-term depression. T-Type calcium channels are known as low-voltage activated calcium channels and are called "fast" because they inactivate rapidly. These channels are sensitive to Ni^{2+}, amiloride, and octanol but are resistant to DHP and ω-CgTx. T-Type channels have a single-channel conductance of ~8 pS, and have been implicated in repetitive firing and pacemaker activity in heart and neurons.

To elucidate molecular mechanisms that underlie diverse cellular functions, distinction of multiple types of calcium channels is a prerequisite. Several studies have illustrated that the classification of calcium channels may be more complicated than previously described, and that properties of calcium channels of a certain type may vary more than previously appreciated (Aosaki and Kasai, 1989; Plummer *et al.*, 1989; Swandulla *et al.*, 1991; Artalejo *et al.*, 1992;

TABLE I Functional Classification of Voltage-Dependent Calcium Channels

Type	Properties	Function
T-Type	Low-voltage activated, blocked by low Ni^{2+} and octanol, inactivates rapidly	Pacemaker activity and repetitive firing in heart and neurons
L-Type	High-voltage activated, blocked by dihydropyridine antagonists and low Cd^{2+}	Excitation–contraction coupling, excitation–secretion coupling in endocrine cells and some neurons
N-Type	High-voltage activated, blocked by ω-conotoxin and low Cd^{2+}	Triggering neurotransmitter release
P-Type	High-voltage activated, blocked by ω-Aga-IVA, insensitive to dihydropyridine and ω-conotoxin	Triggering neurotransmitter release, calcium spike in some neurons, induction of long-term depression

Hillyard *et al.*, 1992; Mintz *et al.*, 1992b; Minlieff and Beam, 1992). The extent and rate of inactivation varies considerably among N-type channels in various neuronal cells (Tsien and Tsien, 1990). ω-CgTx, a selective blocker of N-type calcium channels in neurons, blocks both inactivating and noninactivating current components (Aosaki and Kasai, 1989; Plummer *et al.*, 1989; Plummer and Hess, 1991; Artalejo *et al.*, 1992). Moreover, N- and L-type channels are barely distinguishable on the basis of channel mean open time in the absence of DHP agonists and of unitary conductance (20 and 25 pS, respectively; Plummer *et al.*, 1989; Swandulla *et al.*, 1991).

Over the last 5 years, rapid advances have been made in molecular biological approaches to studying calcium channels. The approaches have been proven to be a powerful aid to the understanding of the diversity of voltage-dependent calcium channels. This chapter assembles the available facts on molecular cloning, tissue location, and recombinant expression of voltage-dependent calcium channels. Classification of calcium channels is also addressed, primarily from a structural perspective.

II. Structural Diversity and Differential Expression of Calcium Channels

Biochemical and molecular biological approaches began with studies of the L-type calcium channel (the DHP receptor) complex in skeletal muscle (Campbell *et al.*, 1988). The DHP receptor has been purified from skeletal muscle by several laboratories (Hosey *et al.*, 1987; Leung *et al.*, 1987; Morton and Froehner, 1987; Takahashi *et al.*, 1987; Tanabe *et al.*, 1987; Vaghy *et al.*, 1987). The receptor is composed of the α_1 subunit (170-kDa polypeptide), the α_2 subunit (140-kDa polypeptide), the β subunit (55-kDa polypeptide), and the

γ subunit (33-kDa polypeptide). A stoichiometric ratio of 1:1:1:1 was obtained for the α_1, α_2, β and γ subunits, suggesting that all four subunits are integral components of the DHP receptor (Leung *et al.*, 1988). The δ subunit has been observed as a small protein of 24–33 kDa, disulfide-linked to α_2 subunits to form α_2/δ (Takahashi *et al.*, 1987). The subunit composition of calcium channel complexes is not the same among tissues (Takahashi and Catterall, 1987; Ahlijanian *et al.*, 1990). One form of L-type calcium channel in brain contains α_1-like subunits in association with α_2/δ- and β-like subunits. The brain ω-CgTx-sensitive N-type calcium channel contains components homologous to the α_1 (Abe and Saisu, 1987; Cruz *et al.*, 1987; Marqueze *et al.*, 1988; McEnery *et al.*, 1991), α_2 (Abe and Saisu, 1987; McEnery *et al.*, 1991), and β (Glossmann and Striessnig, 1990; McEnery *et al.*, 1991; Sakamoto and Campbell, 1991) subunits of the DHP receptor. An additional polypeptide of 110 kDa (McEnery *et al.*, 1991) or 100 kDa (Ahlijanian *et al.*, 1990) is present in N-type or L-type calcium channels, respectively, from brain. Other polypeptides of 36 kDa, 28 kDa (Saisu *et al.*, 1991), and 58 kDa (Takahashi *et al.*, 1991) associated with ω-CgTx receptors have also been reported. Using FTX to form an affinity gel, a protein with an apparent molecular mass of 90–100 kDa was isolated from brain (Cherksey *et al.*, 1991).

A. α_1 Subunit

Molecular biological studies have revealed an even greater diversity among calcium channels than functional studies did, because of multiple genes and alternative splicing. Evidence suggests that α_1 subunits are encoded by a gene family comprising at least six distinct genes (Table II; Snutch *et al.*, 1990; Niidome *et al.*, 1992). Amino acid sequences of the six classes of calcium channels have been deduced by cDNA

TABLE II **Molecular Classification of Voltage-Dependent Calcium Channels**

Numa class	Snutch class	Perez–Reyes class	Primary tissue location	Functional class
SK	—	1	Skeletal muscle	L-Type
C	C	2	Heart, smooth muscle	L-Type
BIV	D	3	Brain, pancreas	L-Type
BI	A	4	Brain	P-Type
BII	E	—	Brain	?
BIII	B	5	Brain	N-Type

cloning and sequencing (Fig. 1). All the α_1 subunits share general structural features with voltage-dependent sodium channels; thus, calcium channels apparently have the same transmembrane topology proposed for sodium channels (Numa and Noda, 1986). Calcium channels contain four repeating units of homology, each of which has one positively charged segment (S4) that probably represents a voltage-sensing region (Stühmer et al., 1989) and five hydrophobic segments (S1, S2, S3, S5, and S6). The conserved charged residues in segments S2 and S3 (Numa and Noda, 1986) are also retained. The glutamic acid residues in the SS1–SS2 region, which may be critical for ion selectivity of calcium channels (Heinemann et al., 1992), are conserved among the α_1 subunits. Previous cDNA expression studies suggested that the function of the sodium channel could be manifested by the large subunit alone (Noda et al., 1986). Thus, the striking structural similarity found between the α_1 subunit of calcium and sodium channels implies that the α_1 subunit itself may be a voltage-dependent channel protein. cDNA expression studies of calcium channels support this notion (see Section III).

The cDNA of the skeletal muscle DHP receptor (class Sk calcium channel) was first of all the α_1 subunits cloned (Tanabe et al., 1987; Grabner et al., 1991). Using immunochemical techniques, this channel was detected only in skeletal muscle (Morton and Froehner, 1989). A prominent transcript of the class Sk channel was detected only in RNA preparations from skeletal muscle by Northern blot analysis (Ellis et al., 1988). However, successful isolation of class Sk channel cDNAs from brain, pancreatic β cell-derived HIT cells, and ovary implies that this class cannot be considered exclusively a skeletal muscle gene (Perez-Reyes et al., 1990). The class Sk channel is expressed in a developmentally regulated fashion, being in-

duced on myogenic differentiation (Morton and Froehner, 1989; Varadi et al., 1989). Splice variants of this class, which differ in the region between S3 and S4 of repeat IV (Perez-Reyes et al., 1990), and, interestingly, cDNAs encoding a two-motif isoform of the class Sk channel have been isolated (Malouf et al., 1992).

The class C calcium channel is a ubiquitous calcium channel. cDNAs encoding the class C channel were isolated from heart (Mikami et al., 1989), lung (Biel et al., 1990), aorta (Koch et al., 1990), brain (Snutch et al., 1991), skin fibroblast cell line (Soldatov, 1992), ovarian cells, and hamster pancreatic β cell-derived HIT cells (Perez-Reyes et al., 1990). Expression of this class in intestine, stomach, spinal cord, pituitary, adrenal gland, liver, kidney, testes, and spleen was shown by Northern blot analyses and analyses using polymerase chain reaction (PCR; Koch et al., 1990; Perez-Reyes et al., 1990; Snutch et al., 1991). The class C channel mRNA was also found in early skeletal muscle myotubes (Chaudhari and Beam, 1991). Numerous isoforms of the class C channel are generated by alternative splicing at different sites (Biel et al., 1990; Koch et al., 1990; Perez-Reyes et al., 1990; Snutch et al., 1991; Diebold et al., 1992; Soldatov, 1992). The sites of structural variation as a result of alternative splicing are located in the N-terminal portion, the region between segment S5 and S6 of repeat I (IS5 and IS6, respectively), IS6, two regions in the vicinity of IIS6, the region between repeats I and II (I–II loop), the II–III loop, IIIS2, IVS3 and its vicinity, and the C-terminal region. The variability within IVS3 is generated by developmentally regulated, mutually exclusive splicing mechanisms (Diebold et al., 1992). Differential expression of these variants was shown by PCR analysis of RNAs from different tissues (Snutch et al., 1991).

Class BIV (or class D) calcium channel cDNAs were isolated from brain, human neuroblastoma IMR32 cell line, pancreatic islet cells, hamster pancreatic β cell-derived HIT cells, and ovarian cells (Perez-Reyes et al., 1990; Hui et al., 1991; Seino et al., 1991; Williams et al., 1992a). Therefore, the BIV channel is called the neuroendocrine type (Perez-Reyes et al., 1990). Expression of this class in pancreatic β cells (Seino et al., 1992) and in central nervous system (Chin et al., 1992) was shown by in situ hybridization. Multiple types of the class BIV channel isoform are generated by alternative splicing (Perez-Reyes et al., 1990; Williams et al., 1992a). Variable regions of this channel are located in IS6, the I–II loop, IVS3 and its vicinity, and the C-terminal region.

Until recently, molecular identification of calcium channels was limited to L-type calcium channels. The

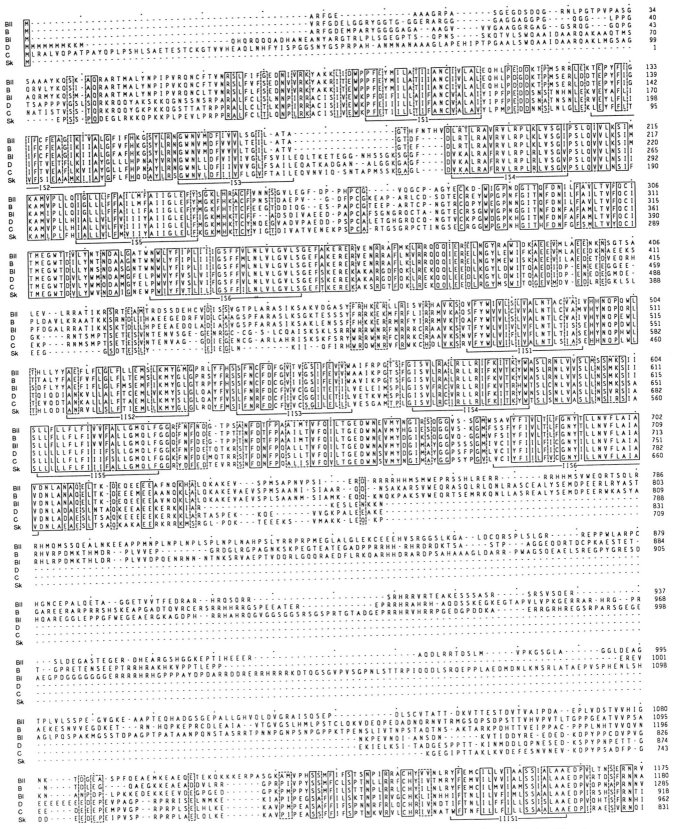

FIGURE 1 Alignment of amino acid sequences of the different calcium channel α_1 subunits. The six sequences compared (from top to bottom) are rabbit brain class BII calcium channel (Niidome *et al.*, 1992); rat brain calcium channel rbB-I (class BIII); rabbit brain class BI calcium channel (Mori *et al.*, 1991); human brain DHP-sensitive calcium channel α_{1D} (class BIV; Williams *et al.*, 1992a); rabbit cardiac muscle DHP-sensitive calcium channel (class C; Mikami *et al.*, 1989); and rabbit skeletal muscle dihydropyridine (DHP)-

sensitive calcium channel (class Sk; Tanabe *et al.*, 1987). BI and BII represent BI-2 and BII-2 sequences, respectively. Sets of six identical residues at one position are boxed; sets of six identical or conservative residues (Numa and Noda, 1986) are boxed with dashed lines. The numbers of the amino acid residues are given at the right of the individual lines. Segments S1–S6 in each repeat are shown.

class BI (or class A) calcium channel cDNA was the first cloned non-L-type calcium channel cDNA (see Section III; Mori *et al.*, 1991; Starr *et al.*, 1991). The presence of multiple isoforms of this class, which is generated by alternative RNA splicing, has been reported (Mori *et al.*, 1991). The sites of variation are located in the II–III loop, the vicinity of IVS6, and the C-terminal region. The class BI channel is distributed widely in the brain, and is abundant in the cerebellum (Mori *et al.*, 1991). Blotting analysis of RNA from the cerebella of mutant mice with different types of cerebellar degeneration suggested that this channel is expressed in Purkinje cells and granule cells. Expression of this class in the heart (Mori *et al.*, 1991; Starr *et al.*, 1991) and pituitary (Starr *et al.*, 1991) was also demonstrated.

Class BII (or class E) calcium channel cDNA was also isolated from brain (Niidome *et al.*, 1992). Two isoforms, possibly generated by alternative splicing, differ from each other in the C-terminal sequence. The spatial distribution of the class BII channel in the brain is different from that of the class BI channel. This class is abundant in cerebral cortex, hippocampus, and corpus striatum.

Class BIII (or class B) calcium channel cDNAs were cloned from brain (Dubel *et al.*, 1992; Fujita *et al.*, 1993) and human neuroblastoma IMR32 cell line (Williams *et al.*, 1992b). Two isoforms that differ in the C-terminal region are generated by alternative RNA splicing (Williams *et al.*, 1992b). Polyclonal antisera generated against a peptide from the class BIII sequence selectively immunoprecipitate high-affinity ω-CgTx binding sites from forebrain membranes (Dubel *et al.*, 1992). The class BIII channel mRNA is present in all the regions of the central nervous system, including the spinal cord (Dubel *et al.*, 1992; Williams *et al.*, 1992b; Fujita *et al.*, 1993). This class is expressed also in calcitonin-secreting C cells and undifferentiated and differentiated rat pheochromocytoma PC12 cells (Dubel *et al.*, 1992). cDNAs encoding different classes of calcium channels have been isolated from the electric organ of the marine ray *D. ommata* (Horne *et al.*, 1991).

The α_1 subunits of voltage-dependent calcium channels can be classified into two subfamilies, according to the degree of amino acid sequence homology between calcium channel pairs (Fig. 2) and the phylogenetic tree (Fujita *et al.*, 1993). One subfamily is the DHP-sensitive L-type calcium channel subfamily, consisting of the class Sk, class C, and class BIV channel; the other is the neuronal calcium channel subfamily, consisting of the class BI, BII, and BIII channels. The regions corresponding to the four internal repeats and the short segment between repeats III and IV (III–IV loop) are relatively well conserved. However,

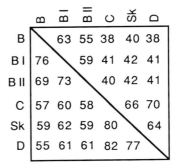

FIGURE 2 Homology matrix for the calcium channel α_1 subunits. The calculations are based on the alignment in Fig. 1. Gaps are counted as one substitution, regardless of their length. The percentage identity of each pair of the calcium channels is shown on the upper right side of the diagonal; the percentage homology calculated for identical plus conservative residues (Numa and Noda, 1986) is shown on the lower left side. BI and BII represent BI-2 and BII-2 sequences, respectively.

the remaining regions, all of which are assigned to the cytoplasmic side of the membrane, are less well conserved. The amino acid sequence of the putative cytoplasmic region between repeats II and III (II–III loop) and the C-terminal region are highly divergent among calcium channels.

B. α_2/δ Subunit

Cloning and sequencing of the cDNA of the skeletal muscle α_2 subunit revealed that it is a glycosylated hydrophilic protein without homology to other known protein sequences (Fig. 3; Ellis *et al.*, 1988). The α_2 subunit contains three hydrophobic segments that may constitute transmembrane domains. Northern blot and immunoblot analyses suggest that the α_2 subunit is present in a variety of tissues (Morton and Froehner, 1989; Varadi *et al.*, 1989). The α_2 subunit is expressed in myoblasts and is induced on myogenic differentiation (Morton and Froehner, 1989; Varadi *et al.*, 1989). No significant differences were observed in the patterns of DNA blot analysis of genomic DNA under conditions of high and low stringency (Ellis *et al.*, 1988), suggesting that significant homology exists between the skeletal muscle α_2 subunit and those of other tissues. Indeed, the cDNA of the α_2 subunit isolated from brain encoded a splice variant of the skeletal muscle α_2 subunit (Williams *et al.*, 1992a). N-Terminal sequence analysis of the δ subunits revealed identical sequences that match those of the C-terminal portion of the α_2 subunit (De Jongh *et al.*, 1990; Jay *et al.*, 1991). This result supports a model of the α_2 subunit in which the same polypeptide is processed into

```
                                            MAAGRPLAWTLTLWQAWLILIGPSSE          -1

EPFPSAVTIKSWVDKMQEDLVTLAKTASGVHQLVDIYEKYQDLYTVEPNNARQLVEIAARDIEKLLSNRSKALVRLALEAEKVQAAHQWREDFASNEVVY     100
                                                                  *
YNAKDDLDPEKNDSEPGSQRIKPVFIDDANFRRQVSYQHAAVHIPTDIYEGSTIVLNELNWTSALDDVFKKNREEDPSLLWQVFGSATGLARYYPASPWV     200
          *                                                 *
DNSRTPNKIDLYDVRRRPWYIQGAASPKDMLILVDVSGSVSGLTLKLIRTSVSEMLETLSDDDFVNVASFNSNAQDVSCFQHLVQANVRNKKVLKDAVNN     300
                                                                                              *
ITAKGITDYKKGFSFAFEQLLNYNVSRANCNKIIMLFTDGGEERAQEIFAKYNKDKKVRVFTFSVGQHNYDRGPIQWMACENKGYYYEIPSIGAIRINTQ     400
                    *
EYLDVLGRPMVLAGDKAKQVQWTNVYLDALELGLVITGTLPVFNITGQFENKTNLKNQLILGVMGVDVSLEDIKRLTPRFTLCPNGYYFAIDPNGYVLLH     500
                      *                        *
PNLQPKPIGVGIPTINLRKRRPNVQNPKSQEPVTLDFLDAELENDIKVEIRNKMIDGESGEKTFRTLVKSQDERYIDKGNRTYTWTPVNGTDYSSLALVL     600
                                                                  *              *
PTYSFYYIKAKIEETITQARYSETLKPDNFEESGYTFLAPRDYCSDLKPSDNNTEFLLNFNEFIDRKTPNNPSCNTDLINRVLLDAGFTNELVQNYWSKQ     700
                                                *              *
KNIKGVKARFVVTDGGITRVYPKEAGENWQENPETYEDSFYKRSLDNDNYVFTAPYFNKSGPGAYESGIMVSKAVEIYIQGKLLKPAVVGIKIDVNSWIE     800
                                                            *
NFTKTSIRDPCAGPVCDCKRNSDVMDCVILDDGGFLLMANHDDYTNQIGRFFGEIDPSLMRHLVNISVYAFNKSYDYQSVCEPGAAPKQGAGHRSAYVPS     900
*                                                                     *        *
IADILQIGWWATAAAWSILQQFLLSLTFPRLLEAADMEDDDFTASMSKQSCITEQTQYFFDNDSKSFSGVLDCGNCSRIFHVEKLMNTNLIFIMVESKGT    1000
                                                            *                *
CPCDTRLLIQAEQTSDGPDPCDMVKQPRYRKGPDVCFDNNVLEDYTDCGGVSGLNPSLWSIIGIQFVLLWLVSGSRHCLL                        1080
                                                    *
```

FIGURE 3 Amino acid sequence of the rabbit skeletal muscle calcium channel α_2 subunit (Ellis *et al.*, 1988). Positive numbering starts with the proposed N-terminal residue of the mature protein. Three putative transmembrane regions are underlined. Potential *N*-glycosylation sites are identified by asterisks.

two chains that remain attached through disulfide linkages.

C. β Subunit

Four distinct β subunit genes (designated CaB1, CaB2, and CaB3; Fig. 4) have been reported to date. As for other subunits, the cDNA encoding the β subunit was first cloned from skeletal muscle (Ruth *et al.*, 1989). An analysis of the β subunit sequence reveals the absence of a typical membrane region and weak similarity to some isoforms of actin, keratin, glial fibrillary acidic protein, myosin, and tropomyosin. Several potential cyclic AMP-dependent phosphorylation sites are present in the β subunit. cDNAs encoding splice variants of this class (CaB1) have been isolated from brain (Pragnell *et al.*, 1991; Williams *et al.*, 1992a). Class CaB2 β subunit cDNAs were isolated from heart and brain (Hullin *et al.*, 1992; Perez-Reyes *et al.*, 1992). The existence of splice variants was also demonstrated for class CaB2 (Hullin *et al.*, 1992; Perez-Reyes *et al.*, 1992). Recently, cDNA of the fourth β subunit (β4) was isolated from brain (Castellano *et al.*, 1993). β4 is abundantly expressed in cerebellum. Class CaB3 cDNAs were isolated from heart (Hullin *et al.*, 1992). Northern blot analysis revealed the expression of class CaB2 in heart, brain, aorta, lung, trachea, and two insulin-secreting tumor cell lines HIT (hamster) and RIN5mF (rat) (Hullin *et al.*, 1992; Perez-Reyes *et al.*, 1992); similar studies revealed class CaB3 in heart, skeletal muscle, brain, aorta, trachea, and lung (Hullin *et al.*, 1992).

D. γ Subunit

The primary structure of the γ subunit, deduced by sequencing cDNAs cloned from skeletal muscle, indicates that the γ subunit is a glycosylated protein with four putative transmembrane domains (Fig. 5; Bosse *et al.*, 1990; Jay *et al.*, 1990). No amino acid sequence of other known proteins showed significant homology to that of this subunit. Expression of γ subunit mRNA was also detected in lung (Jay *et al.*, 1990).

E. Other Polypeptides Associated with Calcium Channels

Syntaxin (Synaptocanalin) was found to be associated with ω-CgTx-sensitive calcium channels. This protein has significant homology to epimorphine, which is predominantly expressed in the nervous system (Bennett *et al.*, 1992; Morita *et al.*, 1992).

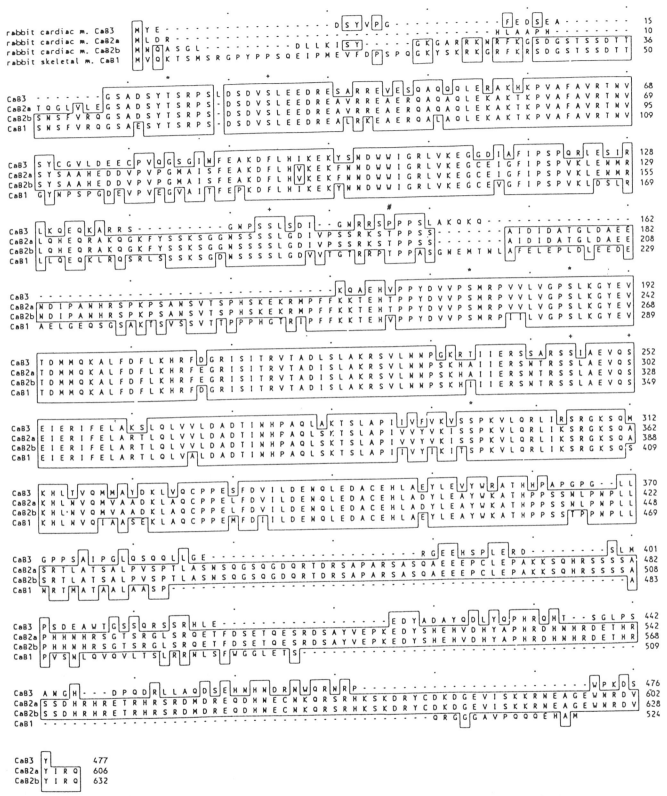

FIGURE 4 Alignment of amino acid sequences of the rabbit β subunits. Regions of sequence identity are boxed. Gaps in the sequences are represented by dashes. The potential sites for phosphorylation are indicated above the line with the following symbols: cAMP-dependent protein kinase (#), protein kinase C (*), casein kinase II (+). Reprinted from Hullin *et al.* (1992) by permission of Oxford University Press.

```
                .                .             .              .
MSPTEAPKVRVTLFCILVGIVLAMTAVVSDHWAVLSPHMENHNTTCEAAH      50
                                                  *
                .                .             .              .
FGLWRICTKRIALGEDRSCGPITLPGEKNCSYFRHFNPGESSEIFEFTTQ     100
                            *
                .                .             .              .
KEYSISAAAISVFSLGFLIMGTICALMAFRKKRDYLLRPASMFYVFAGLC     150
                .                .             .              .
LFVSLEVMRQSVKRMIDSEDTVWIEYYYSWSFACACAAFVLLFLGGISLL     200
                .                .
LFSLPRMPQNPWESCMDAEPEH                                 222
```

FIGURE 5 Amino acid sequence of the skeletal muscle calcium channel γ subunit (Bosse *et al.*, 1990). Four putative transmembrane regions are underlined. Potential *N*-glycosylation sites are indicated by asterisks.

III. Heterologous Expression of Calcium Channels

Expression of mRNAs or cDNAs in foreign cells has proven essential to relating the molecular classification of calcium channels to their functional classification. This form of functional reconstitution is termed heterologous expression. Several expression systems have been developed for calcium channels, the most favored of which is based on injection of mRNAs into *Xenopus* oocytes. This method provides both decisive proof of successful cloning and a useful tool for studying structure–function relationships of various types of ion channels. Techniques of expression in mammalian cells also offer useful ways to study the functional properties of calcium channels. These two kinds of system have been carefully compared in another review (Nargeot *et al.*, 1992).

A. Xenopus *Oocytes*

Successful expression of plasma membrane proteins was first reported for nicotinic acetylcholine receptors in *Xenopus* oocytes (Sumikawa *et al.*, 1981). Proteins expressed after microinjection of crude mRNA retain the pharmacological and electrophysiological properties observed in their original tissue.

Cardiac calcium channels were expressed by microinjecting rat heart poly(A) + RNA into oocytes (Dascal *et al.*, 1986; Moorman *et al.*, 1987; Lory *et al.*, 1990). Because endogenous chloride currents activated by Ca^{2+} influx severely limit the functional detection of expressed calcium channels, extracellular Ca^{2+} was replaced by Ba^{2+} (40 m*M*) and Cl^- was replaced by methanesulfonate. The observed inward Ba^{2+} current was modulated by DHP, isoproterenol, acetylcholine, and cyclic AMP, and corresponded to

that of L-type channels in native cardiac cells. Single-channel properties, that is, conductance (18–25 pS with 110 m*M* Ba^{2+} as the charge carrier), activation threshold (-20 to -10 mV), and long openings in the presence of Bay-K 8644 also corresponded to features of calcium channels in native tissue (Moorman *et al.*, 1987).

Smooth muscle calcium channels were expressed in oocytes by injecting poly(A) + RNA from rat myometrium (Fournier *et al.*, 1989). The expressed channels were partially sensitive to DHP and, in contrast with the expressed cardiac channels, displayed a half-inactivation time of ~400 msec at +10 mV.

The calcium channel activity obtained after injection of poly(A) + RNA from rat brain was insensitive to both DHP and ω-CgTx (Kaneko and Nomura, 1987; Leonard *et al.*, 1987). Several results showed that calcium channels expressed by injection of rat brain poly(A) + RNA were blocked by FTX, suggesting that the RNA preparation contains mRNA species encoding P-type channels (Lin *et al.*, 1990). Human brain RNA and *Torpedo* electric lobe RNA induced N-type channel currents that were blocked by ω-CgTx (Gundersen *et al.*, 1988; Umbach and Gundersen, 1987).

Skeletal muscle calcium channels of vertebrates have been poorly expressed in oocytes (Nargeot *et al.*, 1992). As described earlier, injection of poly(A) + RNA into oocytes can create calcium channels that resemble those of native tissues. However, multiple types of ion channels and subtypes of calcium channels are expressed, reflecting the expression of multiple channels in the cells of origin. One way to identify the expression of a given calcium channel subtype is to selectively inhibit a given RNA translation in *Xenopus* oocytes using antisense oligodeoxynucleotide sequences (Lotan *et al.*, 1989; Snutch *et al.*, 1990,1991). Oligonucleotides complementary to an mRNA sequence specifically inactivate this mRNA in oocytes by an RNase H-like degradation mechanism (Dash *et al.*, 1987). The oligonucleotide designed to hybridize with the DHP receptor mRNA suppressed the expression of cardiac calcium channel currents but not the expression of voltage-dependent Na^+ and K^+ currents (Lotan *et al.*, 1989; Slish *et al.*, 1989).

To reconstitute a particular calcium channel subtype, the subtype-specific mRNA should be injected into oocytes. This specificity has become possible by *in vitro* transcription of cDNAs using the bacteriophage SP6 promotor (Mishina *et al.*, 1985). The technique has provided an effective approach to determining what roles the individual subunits, domains, or amino acids play in the operation of calcium channels.

In *Xenopus* oocytes, mRNA derived from class C calcium channel cDNA isolated from heart directed the formation of a functional DHP-sensitive calcium channel current (Mikami *et al.*, 1989). The Ba^{2+} current carried by the channel was high-voltage activated, was increased in amplitude several-fold by BAY-K 8644, and was virtually abolished by nifedipine and Cd^{2+}. The properties of a splice variant of the class C channel from lung were similar to those of the class C variant from heart (Biel *et al.*, 1990). These results indicate that the α_1 subunit alone is sufficient to exhibit L-type calcium channel activity. Co-injection of the skeletal muscle α_2 subunit enhanced the amplitude of the peak inward current approximately 3-fold. Current amplitude enhancement was also observed with coexpression of the class CaB1 β subunit from skeletal muscle (Singer *et al.*, 1991; Wei *et al.*, 1991; Itagaki *et al.*, 1992), the class CaB2 β subunit from brain (Perez-Reyes *et al.*, 1992) and heart (Hullin *et al.*, 1992), the class CaB3 β subunit from heart (Hullin *et al.*, 1992), and the γ subunit from skeletal muscle (Singer *et al.*, 1991; Wei *et al.*, 1991).

The two class B1 (class A) channel isoforms (BI-1 and BI-2) were expressed in oocytes (Mori *et al.*, 1991). Both isoforms were high-voltage activated; insensitive to Ni^{2+}, nifedipine, and ω-CgTx; and inhibited moderately by Bay-K 8644 and strongly by FTX and low concentrations of Cd^{2+}, with the half-blocking concentrations of 0.5 μM. The single-channel slope conductance of the BI channel was 16 pS, with 110 mM Ba^{2+} as the charge carrier. Thus, the properties of the BI channel, as well as its tissue distribution, suggest that the BI channel represents the P-type channel. Oocytes injected with the BI-specific mRNA alone showed only small inward Ba^{2+} current in 40 mM external Ba^{2+}, but when the α_2 subunit and the class CaB1 β subunit from skeletal muscle were coexpressed with the BI channels, the Ba^{2+} current increased by two orders of magnitude. Coexpression with only the α_2 or the β subunit increased the current

much less, the latter being more effective, whereas coexpression with the skeletal muscle γ subunit had no significant effect.

The class BIV (class D) channel expressed in oocytes is a DHP-sensitive L-type channel (Williams *et al.*, 1992a). Coexpression with the neuronal type splice variant of the class CaB1 β subunit was necessary for the functional expression of the BIV channel, whereas the neuronal type splice variant of the α_2 subunit played an accessory role that potentiated channel activity. The BIV channel was reversibly blocked by ω-CgTx. This observation, in conjunction with the tissue distribution of class C and BIV channels in neurons, suggests that class BIV represents the neuronal L-type channel that is reversibly inhibited by ω-CgTx (Aosaki and Kasai, 1989) and that class C represents the neuronal L-type channel that is insensitive to ω-CgTx (Plummer *et al.*, 1989).

The effect of coexpression of auxiliary subunits on macroscopic current properties and pharmacological sensitivities of class C and BI channel has not been reported. However, some work has shown that auxiliary subunits have pronounced effects on macroscopic characteristics such as drug sensitivity, kinetics, and voltage dependence of activation and inactivation of class C calcium channels (Singer *et al.*, 1991; Wei *et al.*, 1991; Hullin *et al.*, 1992; Perez-Reyes *et al.*, 1992). The presence of the β subunit was required for cAMP-mediated increase in class C channel activity (Klöckner *et al.*, 1992). Different combinations of multiple α_1 and β subunits may constitute heterologous calcium channel complexes and may contribute to the functional heterogeneity of calcium channels.

B. Mammalian Cells

As mentioned earlier, the class Sk channel-specific mRNA has failed to generate L-type calcium channel activity in oocytes (T. Tanabe and Y. Mori, unpublished results). Skeletal muscle myotubes from mice with muscular dysgenesis (*mdg*) have constituted a useful expression system for various classes of calcium channels. *mdg* is a fatal autosomal recessive mutation (Glyecksohn-Waelsch, 1963) that is manifested in skeletal muscle as a failure of excitation–contraction (E–C) coupling (Powell and Fambrough, 1973) and an absence of slow DHP-sensitive L-type calcium current (Beam *et al.*, 1986). Blot hybridization analysis of genomic DNA and skeletal muscle RNA suggested that the *mdg* mutation alters the structural gene for the skeletal muscle DHP receptor (Tanabe *et al.*, 1988). Both functional defects of dysgenic myotubes were restored by microinjection of cDNA for the class Sk channel (the skeletal muscle DHP receptor). The cardiac L-type calcium channel that activates more rapidly than that of skeletal muscle, was also expressed in myotubes from *mdg* mice by microinjection of cDNA for the class C channel from heart (Tanabe *et al.*, 1990). E-C coupling restored by injection of the class C cDNA did not require Ca^{2+} entry, in contrast to the activity restored by injection of cDNA for the class Sk channel. Thus, the behavior of the expressed calcium channels mirrors the physiological situation in skeletal and cardiac muscle.

The skeletal muscle L-type channel was stably expressed in murine L-cells (Perez-Reyes *et al.*, 1989). Since neither the α_2 nor the β subunit was expressed in this cell line (Kim *et al.*, 1990), the class CaB1 β subunit cDNA from skeletal muscle was transfected to examine its functional roles. Coexpression of the β subunit accelerated activation kinetics and increased drug binding sites, but did not increase currents (Lacerda *et al.*, 1991; Varadi *et al.*, 1991).

In Chinese hamster ovary (CHO) cells, the class C channel cDNA from smooth muscle alone was sufficient for stable expression of functional calcium channels (Bosse *et al.*, 1992). The single-channel conductance was 26 pS, with 80 mM Ba^{2+} as the charge carrier, corresponding to 24.6 pS in vascular smooth muscle.

The class BIII (class B) channel induced the recombinant expression of N-type channel activity when transiently coexpressed in human embryonic kidney (HEK 293) cells with the class CaB1 β subunit and the α_2 subunit from brain. The recombinant channel was irreversibly blocked by ω-CgTx but was insensitive to DHP (Williams *et al.*, 1992b). N-Type channels were also expressed by microinjecting class BIII cDNA into myotubes from *mdg* mutant mice (Fujita *et al.*, 1993). The expressed channel gave rise to both an inactivating and a noninactivating current, suggesting that L- and N-type channels cannot be distinguished by inactivation kinetics.

IV. Conclusions

To date, molecular biological studies of calcium channels have confirmed earlier functional classification and have helped eliminate ambiguities in functional categorization, for example, inactivation kinetics of N-type channels and ω-CgTx sensitivities of N- and L-type channels in neurons.

Two subfamilies of α_1 subunit, defined from the molecular point of view, correspond to two classes of channels—L-types and non-L-types—that can be distinguished by DHP sensitivity (Table II). The L-type subfamily consists of Sk, C, and BIV class channels. The expressed Sk and C class channels mirror physiological behaviors in skeletal and cardiac muscle, respectively. The class BIV channel is the neuroendocrine calcium channel that is reversibly blocked by ω-CgTx. On the other hand, the non-L-type subfamily consists of BI, BII, and BIII class channels. The presence of the BI channel in cerebellar Purkinje cells and in granule cells, as well as its pharmacological and electrophysiological properties, suggests that the BI channel represents the P-type calcium channel of neurons. The BIII calcium channel is the N-type calcium channel of neurons. Because the voltage-sensing region S4 (Stühmer *et al.*, 1989), the pore-forming region between S5 and S6 (Heinemann *et al.*, 1992), and the region between repeats III and IV that is involved in the inactivation process (Stühmer *et al.*, 1989) are highly conserved among BI, BII, and BIII class channels, the class BIII channel is likely to be a high-voltage activated non-L-type channel.

What direction should future studies of voltage-dependent calcium channels take? One possibility is dismantling a calcium channel molecule into structural elements that are responsible for specific functions, such as Ca^{2+} selectivity, voltage sensing, and gating. Another possibility is searching for other proteins associated with calcium channels and trying to integrate these proteins into subcellular structures, that is, dendritic spines and presynaptic active zones in neurons or triads in myotubes. This would be the most direct approach to elucidating molecular mechanisms of memory in brain, neurotransmitter release, and excitation–contraction coupling.

Acknowledgments

I would like to thank Keiji Imoto and Haruyuki Atomi for critically reading the manuscript, Tsutomu Tanabe for helpful discussion, and Emiko Mori for her help during the production of this manuscript.

References

Abe, T., and Saisu, H. (1987). Identification of the receptor for ω-conotoxin in brain. *J. Biol. Chem.* **262**, 9877–9882.

Ahlijanian, M. K., Westenbroek, R. E., and Catterall, W. A. (1990). Subunit structure and localization of dihydropyridine-sensitive calcium channels in mammalian brain, spinal cord and retina. *Neuron* **4**, 819–832.

Aosaki, T., and Kasai, H. (1989). Characterization of two kinds of high-voltage-activated Ca-channel currents in chick sensory neurons. *Pflügers Arch.* **44**, 150–156.

Artalejo, C. R., Perlman, R. L., and Fox, A. P. (1992). ω-Conotoxin GVIA blocks a Ca^{2+} current in bovine chromaffin cells that is not the "classic" N type. *Neuron* **8**, 85–95.

Beam, K. G., Knudson, C. M., and Powell, J. A. (1986). A lethal mutation in mice eliminates the slow calcium current in skeletal muscle. *Nature (London)* **320**, 168–170.

Bennet, M. K., Calakos, N., and Scheller, R. H. (1992). Syntaxin: a synaptic protein implicated in docking of synaptic vesicles at presynaptic active zones. *Science* **257**, 255–259.

Biel, M., Ruth, P., Bosse, E., Hullin, R., Stühmer, W., Flockerzi, V., and Hofmann, F. (1990). Primary structure and functional expression of a high voltage-activated calcium channel from rabbit lung. *FEBS Lett.* **269**, 409–412.

Bosse, E., Regulla, S., Biel, M., Ruth, P., Meyer, H. E., Flockerzi, V., and Hofmann, F. (1990). The cDNA and deduced amino acid sequence of the γ subunit of the L-type calcium channel from rabbit skeletal muscle. *FEBS Lett.* **267**, 153–156.

Bosse, E., Bottlender, R., Kleppisch, T., Hescheler, J., Welling, A., Hofmann, F., and Flockerzi, V. (1992). Stable and functional expression of the calcium channel α_1 subunit from smooth muscle in somatic cell lines. *EMBO J.* **11**, 2033–2038.

Campbell, K. P., Leung, A. T., and Sharp, A. H. (1988). The biochemistry and molecular biology of the dihydropyridine-sensitive calcium channel. *Trends Neurosci.* **11**, 425–430.

Castellano, A., Wei, X., Birnbaumer, L., and Perez-Reyes, E. (1993). Cloning and expression of a neuronal calcium channel β subunit. *J. Biol. Chem.* **268**, 3450–3455.

Chaudhari, N., and Beam, K. G. (1991). mRNA for the cardiac calcium channel is expressed during development of skeletal muscle. *Soc. Neurosci. Abstr.* **17**, 772.

Cherksey, B. D., Sugimori, M., and Llinás, R. R. (1991). Properties of calcium channels isolated with spider toxin, FTX. *Ann. N.Y. Acad. Sci.* **635**, 80–89.

Chin, H., Smith, M. A., Kim, H.-L., and Kim, H. (1992). Expression of dihydropyridine-sensitive brain calcium channels in the rat central nervous system. *FEBS Lett.* **299**, 69–74.

Cruz, L. J., Johnson, D. S., and Olivera, B. M. (1987). Characterization of the ω-conotoxin target. Evidence for tissue-specific heterogeneity in calcium channel types. *Biochemistry* **26**, 820–824.

Dascal, N., Snutch, T. P., Lübbert, H., Davidson, N., and Lester, H. A. (1986). Expression and modulation of voltage-gated calcium channels after RNA injection in *Xenopus* oocytes. *Science* **231**, 1147–1150.

Dash, P., Lotan, I., Knapp, M., Kandel, E. R., and Goelet, P. (1987). Selective elimination of mRNAs *in vivo*: Complementary oligonucleotides promote RNA degradation by an RNase H-like activity. *Proc. Natl. Acad. Sci. USA* **84**, 7896–7900.

De Jongh, K. S., Warner, C., and Catteral, W. A. (1990). Subunits of purified calcium channels. *J. Biol. Chem.* **265**, 14738–14741.

Diebold, R. J., Koch, W. J., Ellinor, P. T., Wang, J.-J., Muthuchamy, M., Wieczorek, D. F., and Schwartz, A. (1992). Mutually exclusive exon splicing of the cardiac calcium channel a1 subunit gene generates developmentally regulated isoforms in the rat heart. *Proc. Natl. Acad. Sci. USA* **89**, 1497–1501.

Dubel, S. J., Starr, T. V. B., Hell, J., Ahlijanian, M. K., Enyeart, J. J., Catterall, W. A., and Snutch, T. P. (1992). Molecular cloning of the α-1 subunit of an ω-conotoxin-sensitive calcium channel. *Proc. Natl. Acad. Sci. USA* **89**, 5058–5062.

Ellis, S., Williams, M. E., Ways, N. R., Brenner, R., Sharp, A. H., Leung, A. T., Campbell, K. P., McKenna, E., Koch, J. K., Hui, A., Schwartz, A., and Harpold, M. M. (1988). Sequence and expression of mRNAs encoding the α_1 and α_2 subunits of a DHP-sensitive calcium channel. *Science* **241**, 1661–1664.

Fournier, F., Honoré, E., Brûlé, G., Mironneau, J., and Guilbault, P. (1989). Expression of Ba currents in *Xenopus* oocyte injected with pregnant rat myometrium mRNA. *Pflügers Arch.* **413**, 682–684.

Fujta, Y., Mynlieff, M., Dirksen, R. T., Kim, M.-S., Niidome, T., Nakai, J., Friedrich, T., Iwabe, N., Miyata, T., Furuichi, T., Furutama, D., Mikoshiba, K., Mori, Y., and Beam, K. G. (1993). Primary structure and functional expression of the ω-conotoxin-sensitive N-type calcium channel from rabbit brain. *Neuron* **10**, 585–598.

Glossmann, H., and Striessnig, J. (1990). Molecular properties of calcium channels. *Rev. Physiol. Biochem. Pharmacol.* **114**, 1–105.

Gluecksohn-Waelsch, S. (1963). Lethal genes and analysis of differentiation. In higher organisms lethal genes serves as tools for studies of cell differentiation and cell genetics. *Science* **142**, 1269–1276.

Grabner, M., Friedrich, K., Knaus, H.-G., Sriessing, J., Scheffaur, F., Staudinger, F., Koch, W. J., Schwartz, A., and Glossmann, H. (1991). Calcium channels from *Cyprinus carpio* skeletal muscle. *Proc. Natl. Acad. Sci. USA* **88**, 727–731.

Gundersen, C. B., Umbach, J. A., and Swartz, B. E. (1988). Barbiturates depress currents through human brain calcium channels studied in *Xenopus* oocytes. *J. Pharmacol. Exp. Ther.* **247**, 824–829.

Heinemann, S. H., Terlau, H., Stühmer, W., Imoto, K., and Numa, S. (1992). Calcium channel characteristics conferred on the sodium channel by single mutations. *Nature (London)* **356**, 441–443.

Hess, P. (1990). Calcium channels in vertebrate cells. *Annu. Rev. Neurosci.* **13**, 337–356.

Hillman, D., Chen, S., Aung, T. T., Cherksey, B., Sugimori, M., and Llinás, R. (1991). Localization of P-type calcium channels in the central nervous system. *Proc. Natl. Acad. Sci. USA* **88**, 7076–7080.

Hillyard, D. R., Monje, V. D., Mintz, I. M., Bean, B. P., Nadasdi, L., Ramachandran, J., Miljanich, G., Azimi-Zoonooz, A., McIntosh, J. M., Cruz, L. J., Imperial, J. S., and Olivera, B. M. (1992). A new conus peptide ligand for mammalian presynaptic Ca^{2+} channels. *Neuron* **9**, 69–77.

Horne, W. A., Inman, I. P., Schwarz, T. L., and Tsien, R. W. (1991). Neuronal calcium channel cDNAs from a marine ray. *Biophys. J.* **59**, 338a.

Hosey, M. M., Barhanin, J., Schmid, A., Vandaele, S., Ptasienski, J., O'callahan, C., Cooper, C., and Lazdunski, M. (1987). Photoaffinity labelling and phosphorylation of a 165 kilodalton peptide associated with dihydropyridine and phenylalkylamine-sensitive calcium channels. *Biochem. Biophys. Res. Commun.* **147**, 1137–1145.

Hui, A., Ellinor, P. T., Krizanova, O., Wang, J.-J., Diebold, R. J., and Schwartz, A. (1991). Molecular cloning of multiple subtypes of a novel rat brain isoform of the α_1 subunit of the voltage-dependent calcium channel. *Neuron* **7**, 35–44.

Hullin, R., Singer-Lahat, D., Friechel, M., Biel, M., Dascal, N., Hoffmann, F., and Flockerzi, V. (1992). Calcium channel β subunit heterogeneity: Functional expression of cloned cDNA from heart, aorta and brain. *EMBO J.* **11**, 885–890.

Itagaki, K., Koch, W. J., Bodi, I., Klöckner, U., Slish, D. F., and Schwartz, A. (1992). Native-type DHP-sensitive calcium channel currents are produced by cloned rat aortic smooth muscle and cardiac α_1 subunits expressed in *Xenopus laevis* oocytes and are regulated by α_2- and β-subunits. *FEBS Lett.* **297**, 221–225.

Jay, S. D., Ellis, S. B., McCue, A. E., Williams, M. E., Vedvick, T. S., Harpold, M. M., and Campbell, K. P. (1990). Primary structure of the γ subunit of the DHP-sensitive calcium channel from skeletal muscle. *Science* **248**, 490–492.

Jay, S. D., Sharp, A. H., Kahl, S. D., Vedvick, T. S., Harpold, M. M., and Campbell, K. P. (1991). Structural characterization of the dihydropyridine-sensitive calcium channel α_2-subunit and the associated δ peptides. *J. Biol. Chem.* **266**, 3287–3293.

Kaneko, S., and Nomura, Y. (1987). Cyclic AMP facilitates slow-inactivating Ca^{2+} channel currents expressed by *Xenopus* oocyte after injection of rat brain mRNA. *Neurosci. Lett.* **83,** 123–127.

Kim, H. S., Wei, X., Ruth, P., Perez-Reyes, E., Flockerzi, V., Hofmann, F., and Birnbaumer, L. (1990). Studies on the structural requirements for the activity of the skeletal muscle dihydropyridine receptor/slow Ca^{2+} channel. *J. Biol. Chem.* **265,** 11858–11863.

Klöckner, U., Itagaki, K., Bodi, I., and Schwartz, A. (1992). β-subunit expression is required for cAMP-dependent increase of cloned cardiac and vascular calcium channel currents. *Pflügers Arch.* **420,** 413–415.

Koch, W. J., Ellinor, P. T., and Schwartz, A. (1990). cDNA cloning of a dihydropyridine-sensitive calcium channel from rat aorta. *J. Biol. Chem.* **265,** 17786–17791.

Lacerda, A. E., Kim, H. S., Ruth, P., Perez-Reyes, E., Flockerzi, V., Hofmann, F., Birnbaumer, L., and Brown, A. M. (1991). Normalization of current kinetics by interaction between the α_1 and β subunits of the skeletal muscle dihydropyridine-sensitive Ca^{2+} channel. (1991). *Nature (London)* **352,** 527–530.

Leonard, J. P., Nargeot, J., Snutch, T. P., Davidson, N., and Lester, H. A. (1987). Ca channels induced in *Xenopus* oocytes by rat brain mRNA. *J. Neurosci.* **7,** 875–881.

Leung, A. T., Imagawa, T., and Campbell, K. P. (1987). Structural characterization of the 1,4-dihydropyridine receptor of the voltage-dependent Ca^{2+} channel from rabbit skeletal muscle. *J. Biol. Chem.* **262,** 7943–7946.

Leung, A. T., Imagawa, T., Block, B., Franzini-Armstrong, C., and Campbell, K. P. (1988). Biochemical and ultrastructure of the 1,4-dihydropyridine receptor from rabbit skeletal muscle. *J. Biol. Chem.* **263,** 994–1001.

Lin, J.-W., Rudy, B., and Llinás, R. (1990). Funnel-web spider venom and a toxin fraction block calcium current expressed from rat brain mRNA in *Xenopus* oocytes. *Proc. Natl. Acad. Sci. USA* **87,** 4538–4542.

Llinas, R., Sugimori, M., Lin, J-W., and Cherksey, B. (1989). Blocking and isolation of a calcium channel from neurons in mammals and cephalopods utilizing a toxin fraction (FTX) from funnel-web spider poison. *Proc. Natl. Acad. Sci. USA* **86,** 1689–1693.

Lory, P., Rassendren, F. A., Richard, S., Tiaho, F., and Nargeot, J. (1990). Characterization of voltage-dependent calcium channels expressed in *Xenopus* oocytes injected with mRNA from rat heart. *J. Physiol.* **429,** 95–112.

Lotan, I., Goelet, P., Gigi, A., and Dascal, N. (1989). Specific block of calcium channel expression by a fragment of dihydropyridine receptor cDNA. *Science* **243,** 666–669.

Malouf, N. N., McMahon, D. K., Hainsworth, C. N., and Kay, B. K. (1992). A two-motif isoform of the major calcium channel subunit in skeletal muscle. *Neuron* **8,** 899–906.

Marqueze, B., Martin-Moutot, N., Leveque, C., and Couraud, F. (1988). Characterization of the ω-conotoxin-binding molecule in rat brain synaptosomes and cultured neurons. *Mol. Pharmacol.* **34,** 87–90.

McEnery, M. W., Snowman, A. M., Sharp, A. H., Adams, M. E., and Snyder, S. H. (1991). Purified ω-conotoxin GVIA receptor of rat brain resembles a dihydropyridine-sensitive L-type calcium channel. *Proc. Natl. Acad. Sci. USA* **88,** 11095–11099.

Mikami, A., Imoto, K., Tanabe, T., Niidome, T., Mori, Y., Takeshima, H., Narumiya, S., and Numa, S. (1989). Primary structure and functional expression of the cardiac dihydropyridine-sensitive calcium channel. *Nature (London)* **340,** 230–233.

Minlieff, M., and Beam, K. G. (1992). Pharmacological analysis of voltage-dependent calcium channels in mouse motoneurons. *J. Neurophysiol.* **68,** 85–92.

Mintz, I. M., Venema, V. J., Swiderek, K. M., Lee, T. D., Bean, B. P., and Adams, M. E. (1992a). P-type calcium channels blocked by the spider toxin ω-Aga-IVA. *Nature (London)* **355,** 827–829.

Mintz, I. M., Adams, M. E., and Bean, B. P. (1992b). P-type calcium channels in rat central and peripheral neurons. *Neuron* **9,** 85–95.

Mishina, M., Kurosaki, T., Tobimatsu, T., Morimoto, Y., Noda, M., Yamamoto, T., Terao, M., Lindstrom, J., Takahashi, T., Kuno, M., and Numa, S. (1984a). Expression of functional acetylcholine receptor from cloned cDNAs. *Nature (London)* **307,** 604–608.

Mishina, M., Tobimatsu, T., Imoto, K., Tanaka, K., Fujita, Y., Fukuda, K., Kurasaki, M., Morimoto, Y., Takahashi, H., Hirose, T., Inayama, S., Takahashi, T., Kuno, M., and Numa, S. (1985). Location of functional regions of acetylcholine receptor α-subunit by site-directed mutagenesis. *Nature (London)* **313,** 364–369.

Moorman, J. R., Zhou, Z., Kirsch, G. E., Lacerda, A. E., Caffrey, J. M., Lam, D. M.-K., Joho, R. H., and Brown, A. M. (1987). Expression of single calcium channels in *Xenopus* oocytes after injection of mRNA from rat heart. *Am. J. Physiol.* **250,** H985–H991.

Mori, Y., Friedrich, T., Kim, M.-S., Mikami, A., Nakai, J., Ruth, P., Bosse, E., Hofmann, F., Flockerzi, V., Furuichi, T., Mikoshiba, K., Imoto, K., Tanabe, T., and Numa, S. (1991). Primary structure and functional expression from complementary DNA of a brain calcium channel. *Nature (London)* **350,** 398–402.

Morita, T., Mori, H., Sakimura, K., Mishina, M., Sekine, Y., Tsugata, A., Odani, S., Horikawa, H. P. M., Saisu, H., and Abe, T. (1992). Synaptocanalin I, a protein associated with brain ω-conotoxin-sensitive calcium channels. *Biomed. Res.* **13,** 357–364.

Morton, M. E., and Froehner, S. C. (1987). Monoclonal antibody identifies a 200-kDa subunit of the dihydropyridine-sensitive calcium channel. *J. Biol. Chem.* **262,** 11904–11907.

Morton, M. E., and Froehner, S. C. (1989). The α_1 and α_2 polypeptides of the dihydropyridine-sensitive calcium channel differ in developmental expression and tissue distribution. *Neuron* **2,** 1499–1506.

Nargeot, J., Dascal, N., and Lester, H. A. (1992). Heterologous expression of calcium channels. *J. Membr. Biol.* **126,** 97–108.

Niidome, T., Kim, M.-S., Friedrich, T., and Mori, Y. (1992). Molecular cloning and characterization of a novel calcium channel from rabbit brain. *FEBS Lett.* **308,** 7–13.

Noda, M., Ikeda, T., Suzuki, H., Takeshima, H., Takahashi, Y., Kuno, M., and Numa, S. (1986). Expression of functional sodium channels from cloned cDNA. *Nature (London)* **322,** 826–828.

Numa, S., and Noda, M. (1986). Molecular structure of sodium channels. *Ann. N.Y. Acad. Sci.* **479,** 338–355.

Perez-Reyes, E., Kim, H. S., Lacerda, A. E., Horne, W., Wei, X., Rampe, D., Campbell, K. P., Brown, A. M., and Birnbaumer, L. (1989). Induction of calcium currents by the expression of the α_1-subunit of the dihydropyridine receptor from skeletal muscle. *Nature (London)* **340,** 233–236.

Perez-Reyes, E., Wei, X., Castellano, A., and Birnbaumer, L. (1990). Molecular diversity of L-type calcium channels. *J. Biol. Chem.* **265,** 20430–20436.

Perez-Reyes, E., Castellano, A., Kim, H. S., Bertrand, P., Baggstrom, E., Lacerda, A. E., Wei, X., and Birnbaumer, L. (1992). Cloning and expression of a cardiac/brain β subunit of the L-type calcium channel. *J. Biol. Chem.* **267,** 1792–1797.

Plummer, M. R., and Hess, P. (1991). Reversible uncoupling of inactivation in N-type calcium channels. *Nature (London)* **351,** 657–659.

Plummer, M. R., Logothetis, D. E., and Hess, P. (1989). Elementary properties and pharmacological sensitivities of calcium channels in mammalian peripheral neurons. *Neuron* **2**, 1453–1463.

Powell, J. A., and Fambrough, D. M. (1973). Electrical properties of normal and dysgenic mouse skeletal muscle in culture. *J. Cell. Physiol.* **82**, 21–38.

Pragnell, M., Sakamoto, J., Jay, S. D., and Campbell, K. P. (1991). Cloning and tissue specific expression of the brain calcium channel β-subunit. *FEBS Lett.* **291**, 253–258.

Regan, L. J., Sah, D. W., and Bean, B. P. (1991). Ca²⁺ channels in rat central and peripheral neurons: High-threshold current resistant to dihydropyridine blockers and ω-conotoxin. *Neuron* **6**, 269–280.

Ruth, P., Rohrkasten, A., Biel, M., Bosse, E., Regulla, S., Meyer, H. E., Flockerzi, V., and Hofmann, F. (1989). Primary structure of the β subunit of the DHP-sensitive calcium channel from skeletal muscle. *Science* **245**, 1115–1118.

Saisu, H., Ibaraki, K., Yamaguchi, T., Sekine, Y., and Abe, T. (1991). Monoclonal antibodies immunoprecipitating ω-conotoxin-sensitive calcium channel molecules recognize two novel proteins localized in the nervous system. *Biochem. Biophys. Res. Commun.* **181**, 59–66.

Sakamoto, J., and Campbell, K. P. (1991). A monoclonal antibody to the β subunit of the skeletal muscle dihydropyridine receptor immunoprecipitates the brain ω-conotoxin GVIA receptor. *J. Biol. Chem.* **266**, 18914–18919.

Seino, S., Chen, L., Seino, M., Blondel, O., Takeda, J., Johnson, J. H., and Bell, G. I. (1992). Cloning of the α₁ subunit of a voltage-dependent calcium channel expressed in pancreatic β cells. *Proc. Natl. Acad. Sci. USA* **89**, 584–588.

Singer, D., Biel, M., Lotan, I., Flockerzi, V., Hofmann, F., and Dascal, N. (1991). The roles of the subunits in the function of the calcium channel. *Science* **253**, 1553–1557.

Slish, D. F., Engle, D. B., Varadi, G., Lotan, I., Singer, D., Dascal, N., and Schwartz, A. (1989). Evidence for the existence of a cardiac specific isoform of the α₁ subunit of the voltage dependent calcium channel. *FEBS Lett.* **250**, 509–514.

Snutch, T. P., Leonard, J. P., Gilbert, M. M., Lester, H. A., and Davidson, N. (1990). Rat brain expresses a heterogeneous family of calcium channels. *Proc. Natl. Acad. Sci. USA* **87**, 3391–3395.

Snutch, T. P., Tomlinson, W. J., Leonard, J. P., and Gilbert, M. M. (1991). Distinct calcium channels are generated by alternative splicing and are differentially expressed in the mammalian CNS. *Neuron* **7**, 45–57.

Soldadov, N. M. (1992). Molecular diversity of L-type calcium channel transcripts in human fibroblasts. *Proc. Natl. Acad. Sci. USA* **89**, 4628–4632.

Starr, T. V. B., Prystay, W., and Snutch, T. P. (1991). Primary structure of a calcium channel that is highly expressed in the rat cerebellum. *Proc. Natl. Acad. Sci. USA* **88**, 5621–5625.

Stühmer, W., Conti, F., Suzuki, H., Wang, X., Noda, M., Yahagi, N., Kubo, H., and Numa, S. (1989). Structural parts involved in activation and inactivation of the sodium channel. *Nature (London)* **339**, 597–603.

Sumikawa, K., Houghton, M., Emtage, J. S., Richards, B. M., and Barnard, E. A. (1981). Active multi-subunit ACh receptor assembled by translation of heterologous mRNA in *Xenopus* oocytes. *Nature (London)* **292**, 862–864.

Swandulla, D., Carbone, E., and Lux, H. D. (1991). Do calcium channel classifications account for neuronal calcium channel diversity? *Trends Neurosci.* **14**, 46–51.

Takahashi, M., and Catterall, W. A. (1987). Identification of an a subunit of dihydropyridine-sensitive brain calcium channels. *Science* **236**, 88–91.

Takahashi, M., Segar, M. J., Jones, J. F., Reber, B. F. X., and Catterall, W. A. (1987). Subunit structure of dihydropyridine-sensitive calcium channels from skeletal muscle. *Proc. Natl. Acad. Sci. USA* **84**, 5478–5482.

Takahashi, M., Arimatsu, Y., Fujita, S., Fujimoto, Y., Kondo, S., Hama, T., and Miyamoto, E. (1991). Protein kinase C and Ca²⁺/calmodulin-dependent protein kinase II phosphorylate a novel 58-kDa protein in synaptic vesicles. *Brain Res.* **551**, 279–292.

Tanabe, T., Takeshima, H., Mikami, A., Flockerzi, V., Takahashi, H., Kangawa, K., Kojima, M., Matsuo, H., Hirose, T., and Numa, S. (1987). Primary structure of the receptor for calcium channel blockers from skeletal muscle. *Nature (London)* **328**, 313–318.

Tanabe, T., Beam, K. G., Powell, J. A., and Numa, S. (1988). Restoration of excitation-contraction coupling and slow calcium current in dysgenic muscle by dihydropyridine receptor complementary DNA. *Nature (London)* **336**, 134–139.

Tanabe, T., Mikami, A., Numa, S., and Beam, K. G. (1990). Cardiac-type excitation-contraction coupling in dysgenic skeletal muscle injected with cardiac dihydropiridine receptor cDNA. *Nature (London)* **344**, 451–453.

Tsien, R. W., and Tsien, R. Y. (1990). Calcium channels, stores, and oscillations. *Annu. Rev. Cell Biol.* **6**, 715–760.

Tsien, R. W., Ellinor, P. T., and Horne, W. A. (1991). Molecular diversity of voltage-dependent Ca²⁺ channels. *Trends Pharmacol. Sci.* **12**, 349–354.

Umbach, J. A., and Gundersen, C. B. (1987). Expression of an ω-conotoxin-sensitive calcium channel in *Xenopus* oocytes injected with mRNA from *Torpedo* electric lobe. *Proc. Natl. Acad. Sci. USA* **84**, 5464–5468.

Vaghy, P. L., Striessnig, J., Miwa, K., Knaus, H.-G., Itagaki, K., McKenna, E., Glossmann, H., and Schwartz, A. (1987). Identification of a novel 1,4-dihydropyridine- and phenylalkylamine-binding polypeptide in calcium channel preparations. *J. Biol. Chem.* **262**, 14337–14342.

Varadi, G., Orlowski, J., and Schwartz, A. (1989). Developmental regulation of expression of the α₁ and α₂ subunits mRNAs of the voltage-dependent calcium channel in a differentiating myogenic cell line. *FEBS Lett.* **2**, 515–518.

Varadi, G., Lory, P., Schultz, D., Varadi, M., and Schwartz, A. (1991). Acceleration of activation and inactivation by the β subunit of the skeletal muscle calcium channel. *Nature* **352**, 159–162.

Wei, X., Perez-Reyes, E., Lacerda, A. E., Schuster, G., Brown, A. M., and Birnbaumer, L. (1991). Heterologous regulation of the cardiac Ca²⁺ channel α₁ subunit by skeletal muscle β and γ subunits. *J. Biol. Chem.* **266**, 21943–21947.

Williams, M. E., Feldman, D. H., McCue, A. F., Brenner, R., Velicelebi, G., Ellis, S. B., and Harpold, M. M. (1992a). Structure and functional expression of α₁, α₂, and β subunit of a novel human neuronal calcium channel subtype. *Neuron* **8**, 71–84.

Williams, M. E., Brust, P. F., Feldman, D. H., Patthi, S., Simerson, S., Maroufi, A., McCue, A. F., Velicelebi, G., Ellis, S. B., and Harpold, M. M. (1992b). Structure and functional expression of an ω-conotoxin-sensitive human N-type calcium channel. *Science* **257**, 389–395.

Structure and Function of Skeletal Muscle and Cardiac Dihydropyridine Receptors

TSUTOMU TANABE

I. Introduction

Voltage-dependent calcium channels play important roles in the regulation of a variety of cellular functions, including membrane excitability, muscle contraction, synaptic transmission, and secretion. At least four types (termed L, T, N, and P) of calcium channels have been distinguished by their electrophysiological and pharmacological properties (Bean, 1989; Tsien and Tsien, 1990; Bertolino and Llinás, 1992). Studies combining molecular biology and electrophysiology have provided evidence that the diversity of calcium channel types derives largely from differences in the pore-forming subunit of these channels (α_1 subunit). In addition, the other subunits associated with α_1 can modify the channel properties. Further, investigators have shown that the diversity of α_1 and of the other subunits originates not only from differences in the genes encoding these subunits but also from alternative splicing of their RNAs. Several types of calcium channels are known to be coexpressed in single cells; each cell apparently uses these channels for different purposes. The discussion in this chapter focuses on work dealing with the structure–function relationships of skeletal and cardiac L-type calcium channels.

II. Subunit Composition of Calcium Channels

The skeletal muscle L-type calcium channel protein was the first voltage-dependent calcium channel to be purified and cloned. This channel protein was purified using dihydropyridine (DHP) calcium channel blockers as ligands. The L-type channel is composed of multimeric subunits α_1 (DHP receptor), α_2/δ, β, and γ (Campbell et al., 1988; Catterall et al., 1988). All the cDNAs encoding each subunit have been cloned (Tanabe et al., 1987; Ellis et al., 1988; Ruth et al., 1989; Bosse et al., 1990; Jay et al., 1990). Using cDNA probes from the skeletal muscle DHP receptor as well as from the other subunits, homologs of the skeletal muscle subunit cDNAs were cloned from several different tissues (Mikami et al., 1989; Biel et al., 1990; Koch et al., 1990; Grabner et al., 1991; Hui et al., 1991; Mori et al., 1991; Pragnel et al., 1991; Snutch et al., 1991; Starr et al., 1991; Dubel et al., 1992; Hullin et al., 1992; Niidome et al., 1992; Perez-Reyes et al., 1992; Soldatov, 1992; Seino et al., 1992; Williams et al., 1992a,b). These experiments identified the presence of heterogeneity in each subunit in several tissues. The heterogeneity arises from the presence of multiple genes encoding each subunit and from the mechanism of alternative RNA splicing.

Expression studies of these cDNAs identified the different types of channel molecules, including the P-like channel and the N channel as well as the L channels (Tanabe *et al.*, 1988; Mikami *et al.*, 1989; Perez-Reyes *et al.*, 1989; Biel *et al.*, 1990; Mori *et al.*, 1991; Williams *et al.*, 1992a,b). Although the subunit composition is only clearly known for the skeletal muscle L channel, investigators have shown that skeletal subunits can facilitate the expression and/or modify the channel properties of heterogeneous α_1 subunits (Mikami *et al.*, 1989; Biel *et al.*, 1990; Lacerda *et al.*, 1991; Mori *et al.*, 1991; Singer *et al.*, 1991; Varadi *et al.*, 1991; Wei *et al.*, 1991; Perez-Reyes *et al.*, 1992; Williams *et al.*, 1992a,b). This result suggests similar subunit composition not only in L channels of other tissues but also in other types of channels. However, the effect of coexpression of these subunits is different for each α_1 subunit, perhaps because each α_1 subunit interacts with the specific subunits of its own tissue, and skeletal subunits can mimic, but not represent exactly, the same interaction and modification. Determining the actual subunit composition of a particular type of channel is most difficult because subunits of one type of channel can modify the properties of other types of channel and several different types of calcium channel molecules are coexpressed in many different cell types.

III. Molecular Structure of the Skeletal Muscle and Cardiac DHP Receptor

The primary structure of the rabbit skeletal muscle and cardiac DHP receptors was deduced by cloning and sequencing the cDNAs (Tanabe *et al.*, 1987; Mikami *et al.*, 1989). These receptors are composed of 1873 and 2171 amino acid residues (including the initiating methionine), respectively. The level of amino acid sequence homology between these two DHP receptors is 66%. The DHP receptors contain four internal repeats (I–IV) with homologous sequences. The regions corresponding to the four internal repeats are highly conserved whereas the remaining regions, all of which are assigned to the cytoplasmic side of the membrane, are less well conserved, except for the short segment between repeats III and IV. These receptors are homologous in amino acid sequence with the voltage-dependent sodium channels (Numa and Noda, 1986), particularly in the regions constituting the four internal repeats (Fig. 1). This observed structural similarity suggests that the genes encoding the DHP receptors and sodium channels arose from a common ancestor. Thus, investigators infer that voltage-dependent ion channels represent a family of evolutionarily and structurally related gene products.

Each internal repeat I–IV has five hydrophobic segments (S1, S2, S3, S5, and S6) and one positively charged segment (S4), all of which exhibit predicted secondary structure. Segments S1, S2, and S3 generally contain a few charged residues. S3 has a few negative charges and S2 a few negative and positive charges, whereas S1 is not uniform in charge. Segment S4 contains five or six arginine or lysine residues located at every third position (except for the basic residue closest to the carboxyl end of segment S4 in repeats III and IV, residing at the fourth position), which are interrupted primarily by nonpolar residues. The unique structure of segment S4 in all repeats is strikingly well conserved between the two DHP receptors. The positive charges present in this segment, many of which presumably form dipoles, are most likely to represent the voltage sensor. These charged residues would move outward in response to depolarization, causing conformational changes and possible rearrangement of ion pairs. Evidence supporting this event has been reported for sodium channels (Stühmer *et al.*, 1989) and potassium channels (Papazian *et al.*, 1991). Segments S5 and S6 are highly hydrophobic regions without any charged residues. A region called SS1 and SS2 between S5 and S6 is postu-

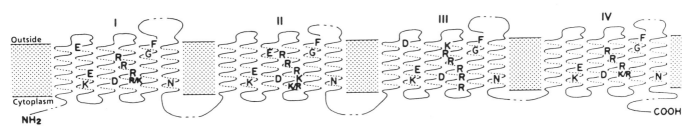

FIGURE 1 Structural characteristic common to calcium channels and sodium channels. The four units of homology (repeat I to IV) spanning the membrane, which are assumed to surround the ionic channel, are displayed linearly. The putative transmembrane α-helical segments S1 to S6 (from left to right) in each repeat are illustrated, and characteristic amino acid residues are indicated. Reproduced with permission from Tanabe *et al.* (1987). Copyright © 1987 Macmillan Magazines Ltd.

lated to partly span the membrane as a hairpin (Guy and Conti, 1990). This region in potassium and sodium channels has been shown to constitute the channel lining region (channel pore) (MacKinnon and Yellen, 1990; Hartmann *et al.*, 1991; Yellen *et al.*, 1991; Yool and Schwarz, 1991; Heinemann *et al.*, 1992). The structural similarity of this region suggests that it plays the same role for calcium channels.

IV. Functional Expression of the DHP Receptor

A. Expression in Xenopus Oocytes

The skeletal muscle DHP receptor is not successfully expressed in the *Xenopus* oocyte system, but the cardiac DHP receptor can be expressed in this system (Mikami *et al.*, 1989). mRNA specific for the cardiac DHP receptor was synthesized by transcription *in vitro* of the cloned cDNA and was injected into *Xenopus* oocytes. Using Cl^--free Ba^{2+} solution, inward current could be measured. This current was unaffected by addition of tetrodotoxin (30 μM) or by replacement of both Na^+ and K^+ with Ba^{2+}, but was completely suppressed by addition of Cd^{2+} (0.2 mM). The peak inward current was increased several-fold by 5 μM

BAY-K 8644, whereas it was decreased to 30–80% of the control value by 1 μM nifedipine and was virtually abolished by 10 μM nifedipine. These results collectively indicate that inward current is a Ba^{2+} current through a DHP-sensitive calcium channel and that cardiac DHP receptor alone is sufficient to direct the formation of a functional L-type calcium channel in *Xenopus* oocytes. Functional expression of smooth muscle L-type calcium channel (Biel *et al.*, 1990), brain non-L-, non-N-type calcium channel (Mori *et al.*, 1991), and brain L-type calcium channel (Williams *et al.*, 1992a) has also been reported in this system.

B. Expression in Dysgenic Myotubes

1. Mutant Mouse Skeletal Muscle

Muscular dysgenesis (*mdg*) is a fatal autosomal recessive mutation of mice (Fig. 2; Gluecksohn-Waelsch, 1963). Mice homozygous for this mutation only survive until birth, at which time they die of respiratory arrest due to complete paralysis of their skeletal muscle. This mutation eliminates excitation-contraction (E–C) coupling and DHP-sensitive slow L-type calcium current from skeletal muscle (Powell and Fambrough, 1973; Klaus *et al.*, 1983; Beam *et al.*, 1986; Bournaud and Mallart, 1987; Rieger *et al.*, 1987).

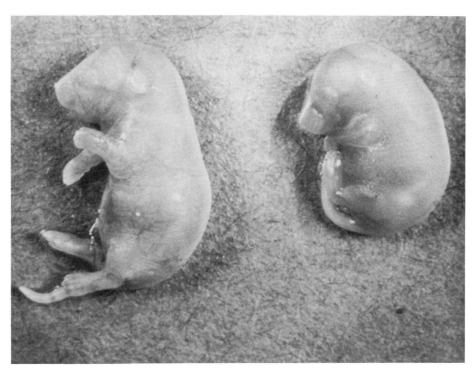

FIGURE 2 18-day-old mouse normal embryo (*left*) and embryo with muscular dysgenesis (*right*).

The muscular dysgenesis phenotype is transmitted as a single recessive defect and has been maintained in highly inbred colonies of mice. Analysis of genomic DNA and skeletal muscle RNA indicates that the *mdg* mutation is associated with alterations of the structural gene for the skeletal muscle DHP receptor (Tanabe *et al.*, 1988). Further, analysis using antibodies against rabbit skeletal muscle proteins showed that the DHP receptor was specifically eliminated in the skeletal muscle of neonatal dysgenic mice (Knudson *et al.*, 1989). Thus, genetic and biochemical analyses suggest that the primary defect in muscular dysgenesis is the alteration and prevention of expression of the skeletal muscle DHP receptor gene. The structural characteristics of the DHP receptor molecule and the biological, biochemical, molecular biological, and electrophysiological analyses of dysgenic muscle suggest that the DHP receptor in the transverse tubule membrane of skeletal muscle functions both as the voltage sensor for E–C coupling and as the L-type calcium channel. An involvement of the T-tubular DHP receptor in E–C coupling was suggested previously on the basis of the observed effect of DHP on charge movement and myoplasmic Ca^{2+} transients (Rios and Brum, 1987).

2. Expression of DHP Receptor

a. E–C Coupling An expression plasmid carrying the entire protein coding sequence of the rabbit skeletal muscle DHP receptor cDNA (pCAC6) or the cardiac DHP receptor cDNA (pCARD1) was injected into nuclei of dysgenic myotubes in primary culture (Tanabe *et al.*, 1988, 1990a). pCAC6-injected and pCARD1-injected dysgenic myotubes displayed spontaneous and electrically evoked contractions. Thus, skeletal muscle and cardiac DHP receptors restored depolarization–contraction coupling, which is missing in dysgenic myotubes. However, the nature of this coupling is different for the two kinds of DHP receptor. Dysgenic myotubes injected with pCAC6 underwent electrically evoked contractions in normal rodent Ringer's solution, in Ca^{2+}-free Ringer, and in 0.5 mM Cd^{2+}-containing Ringer (Fig. 3A), as do normal myotubes in which Ca^{2+} entering across the sarcolemma is not necessary for E–C coupling (Armstrong *et al.*, 1972; Chiarandini *et al.*, 1980). On the other hand, pCARD1-injected dysgenic myotubes displayed electrically evoked contractions in normal rodent Ringer, but not in Ca^{2+}-free Ringer or in 0.5 mM Cd^{2+}-containing Ringer (Fig. 3B). Thus, unlike the E–C coupling that is restored in dysgenic myotubes by injection of pCAC6, depolarization–contraction coupling restored in dysgenic myotubes by injection of pCARD1 requires the voltage-dependent entry of Ca^{2+}. An examination

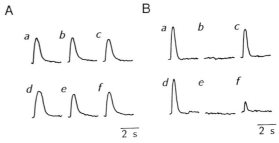

FIGURE 3 Comparison of electrically evoked contractions in dysgenic myotubes that were injected with pCAC6 (A) or pCARD1 (B). The cells were first bathed in normal rodent Ringer (traces a,d), then exposed to either Ca^{2+}-free (trace b) or 0.5 mM Cd^{2+}-containing Ringer (trace e), and finally returned to normal Ringer (traces c,f). Recovery of electrically evoked contraction in the pCARD1-injected cell after exposure to Cd^{2+} was incomplete. However, after being returned to culture medium and maintained for ~30 min in the tissue culture incubator, the cell regained the ability to contract vigorously. The blockade of contraction in pCARD1-injected myotubes by brief exposures to lower concentrations of Cd^{2+} was readily reversible. For the method, see Tanabe *et al.* (1988). Reproduced with permission from Tanabe *et al.* (1990a). Copyright © 1990 Macmillan Magazines Ltd.

using caffeine-treated myotubes suggested the involvement of Ca^{2+}-induced release of Ca^{2+} from the sarcoplasmic reticulum (SR; Endo, 1977) in depolarization–contraction coupling in pCARD1-injected myotubes. Thus, depolarization–contraction coupling in pCARD1-injected dysgenic myotubes appears to mimic coupling in cardiac cells (Fabiato, 1985; Beuckelmann and Wier, 1988; Näbauer *et al.*, 1989): Ca^{2+} entry is required and the Ca^{2+} that enters triggers additional Ca^{2+} release from the SR.

b. Charge Movement The skeletal muscle DHP receptor is postulated to be the voltage sensor that gives rise to the intramembrane current, termed charge movement (Schneider and Chandler, 1973). To test this hypothesis, intramembrane charge movement in dysgenic myotubes was measured before and after injection with expression plasmid carrying skeletal muscle DHP receptor cDNA (Adams *et al.*, 1990). Immobilization-resistant charge movement was measured using the prepulse protocol, which led to complete inactivation of both the sodium current and the T-type calcium current without affecting the L-type calcium current. The immobilization-resistant charge movement in dysgenic myotubes is found to be deficient, but is fully restored after injection of an expression plasmid carrying the skeletal muscle DHP receptor cDNA, providing evidence that the DHP receptor is the voltage sensor for E–C coupling in skeletal muscle.

c. L-Type Current Dysgenic myotubes that had been injected with pCAC6 or pCARD1 and had been observed to contract expressed L-type calcium currents, but the nature of these currents was very different (Tanabe *et al.*, 1990a). In pCAC6-injected myotubes, the rate of activation was slow, similar to that observed in normal skeletal muscle (Sánchez and Stefani, 1978; Donaldson and Beam, 1983). In contrast, the L-type calcium current in pCARD1-injected myotubes activated more rapidly, as does the L-type current in cardiac muscle (Isenberg and Klöckner, 1982; Lee and Tsien, 1982; Fig. 4).

3. Dual Function of the Skeletal Muscle DHP Receptor

Researchers have shown that the lack of E–C coupling and L-type calcium current and deficient intramembrane charge movement in dysgenic myotubes can be restored by microinjection of the expression plasmid carrying the skeletal muscle DHP receptor cDNA. These results provide evidence that the skeletal muscle DHP receptor in the T-tubule membrane has a dual role, acting as the voltage sensor for E–C coupling and as the L-type calcium channel. The cardiac DHP receptor expressed in skeletal muscle is able to function as an L-type calcium channel but not as a voltage sensor that directly controls release of Ca^{2+} from the SR.

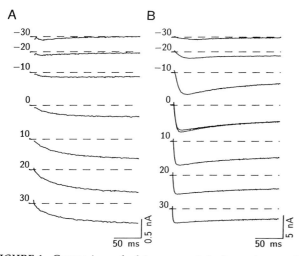

FIGURE 4 Comparison of calcium currents in dysgenic myotubes injected with pCAC6 (A) or pCARD1 (B). Calcium current was measured using the whole-cell variant of the patch-clamp technique (Hamill *et al.*, 1981). The test potential (mV) is indicated next to each current trace. Reproduced with permission from Tanabe *et al.* (1990a). Copyright © 1990 Macmillan Magazines Ltd.

V. Functional Region of the DHP Receptor

A. Functional Difference between Skeletal Muscle and Cardiac DHP Receptor

Injection of an expression plasmid carrying the skeletal muscle DHP receptor cDNA restores both E–C coupling and skeletal L-type calcium current. The restored coupling resembles normal skeletal muscle E–C coupling, which does not require entry of extracellular Ca^{2+}. In contrast, injection into dysgenic myotubes of an expression plasmid carrying the cardiac DHP receptor cDNA produces cardiac-like L-type current and cardiac-type E–C coupling, which does require entry of extracellular Ca^{2+}. To investigate the molecular basis for these differences in calcium currents and in E–C coupling, various chimeric DHP receptor cDNAs were expressed in dysgenic myotubes.

B. Regions Critical for Skeletal-Type E–C Coupling

Depolarization causes a molecular rearrangement of the voltage sensor, which is postulated to gate calcium flow across the SR. If the DHP receptor is responsible for controlling the release of Ca^{2+} from the SR and the coupling is an electromechanical mechanism (Chandler *et al.*, 1976), some regions of the DHP receptor might interact directly or indirectly through a putative linking protein with the foot region of the ryanodine receptor (Fleischer and Inui, 1989) to gate the calcium-release channel. The main differences in primary structure of the skeletal muscle and the cardiac DHP receptors reside in the large, putative cytoplasmic regions, that is, in the N- and C-terminal regions as well as in the regions linking repeats I and II and repeats II and III. To identify the region that is responsible for this interaction, cDNAs encoding chimeras with one or more of the large, putative cytoplasmic regions of the cardiac DHP receptor replaced with corresponding regions of the skeletal muscle DHP receptor were expressed in dysgenic myotubes (Fig. 5). Examination of the electrically evoked contraction of these myotubes showed that the putative cytoplasmic region linking repeats II and III of the skeletal muscle DHP receptor is a major determinant for skeletal-type E–C coupling (Tanabe *et al.*, 1990b). Further, analysis of calcium current and intramembrane charge movement indicated that the slow activation of the skeletal muscle L-type channel is not an obligatory consequence of the role of the DHP receptor as the voltage sensor for E–C coupling. In addition, participation in direct skeletal-type E–C coupling does

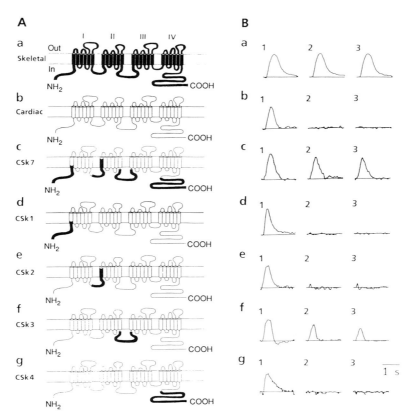

FIGURE 5 (A) Schematic representation of the structures of the skeletal muscle (a), cardiac (b), and chimeric (c–g) DHP receptors. For each DHP receptor, the four units of homology (repeat I–IV) are displayed linearly and the six putative transmembrane segments (S1–S6 from left to right) in each repeat are shown as cylinders. The darkly shaded areas indicate regions of the skeletal muscle DHP receptor. Note that the junctional sequences common to the skeletal muscle and cardiac DHP receptors are not darkly shaded and that the exchanged portion of segment S1 of repeat II is very similar in amino acid sequence in the two DHP receptors. (B) Comparison of electrically evoked contractions in dysgenic myotubes injected with pCAC6 (a), pCARD1 (b), pSkC7 (c), pSkC1 (d), pSkC2 (e), pSkC3 (f), or pSkC4 (g). Contractions were recorded initially in normal rodent Ringer (trace 1), then exposed to either Ca^{2+}-free (trace 2) or 0.5 mM Cd^{2+}-containing Ringer (trace 3). Reproduced with permission from Tanabe *et al.* (1990b). Copyright © 1990 Macmillan Magazines Ltd.

not necessarily cause a DHP receptor to lose its efficiency as a calcium channel (Adams *et al.*, 1990).

C. Regions Critical in Determining Activation Kinetics

For sodium channels and calcium channels, kinetic analysis of current has suggested that, during activation, the channel undergoes several distinct conformational changes before reaching the open state (Hodgkin and Huxley, 1952; Armstrong, 1981; Kostyuk *et al.*, 1981; Sánchez and Stefani, 1983; Keynes *et al.*, 1990). Based on the structural characteristics of these channels, investigators have postulated that the

distinct conformational transitions inferred from kinetic analyses may be equated with conformational changes of the individual structural repeats. To test this hypothesis, cDNAs encoding chimeric calcium channels in which one or more of the four repeats of the skeletal muscle DHP receptor were replaced with the corresponding repeats derived from the cardiac DHP receptor were expressed in dysgenic myotubes (Tanabe *et al.*, 1991). The current produced by the expression of pSkC15, in which repeat I is skeletal and repeats II–IV are cardiac, activated slowly, resembling the current produced by pCAC6. In contrast, the current produced by pSkC11, in which repeat I is cardiac and repeats II–IV are skeletal, activated rapidly, resembling the current for pCARD1 (Fig. 6). The measured currents could be approximated by a single exponential function, providing a simple measure (τ_{act}) of the activation kinetics for different constructs. Because the rate of activation varies with test potential, τ_{act} was determined in each myotube expressing a cDNA for a test potential at or just above the potential that elicited maximal inward current. For any given plasmid, τ_{act} shows a considerable range of values, but the data clearly fall into two groups with almost no overlap. The only consistent structural feature that distinguishes these two groups of chimeric plasmids is repeat I (Fig. 7). For all chimeric plasmids in which repeat I is of skeletal muscle origin, τ_{act} is large, whereas for those in which repeat I is of cardiac origin, τ_{act} is small. Thus, a single repeat—the first—deter-

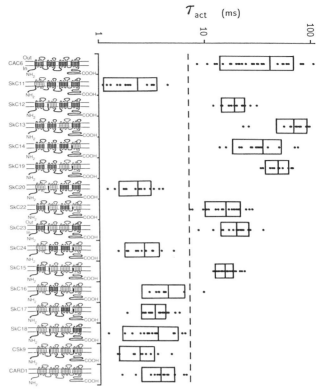

FIGURE 7 Time constant of activation (τ_{act}) of calcium current expressed in dysgenic myotubes injected with expression plasmids carrying the skeletal muscle (pCAC6), cardiac (pCARD1), and chimeric (pSkC11–pSkC20, pSkC22–pSkC24, and pCSk9) DHP receptor cDNAs. Structure of the chimeric DHP receptor is represented schematically (see Fig. 5). Individual data points are plotted as filled circles (overlapping values are plotted as single points); the short horizontal line indicates the mean and the box indicates $\pm$ SD. Note that values of τ_{act} for all the constructs with repeat I of skeletal muscle origin were greater than 7 msec (indicated by dashed horizontal line), and that values of τ_{act} for all the constructs with repeat I of cardiac origin were less than 7 msec (with the exception of one data point). Reproduced with permission from Tanabe *et al.* (1991). Copyright © 1991 Macmillan Magazines Ltd.

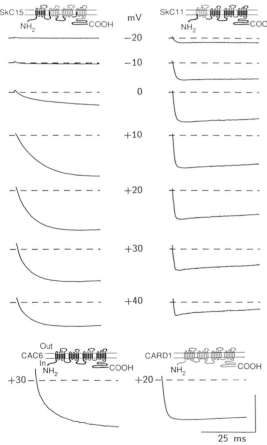

FIGURE 6 Calcium currents expressed in dysgenic myotubes injected with pSkC15 (*left*) and pSkC11 (*right*). The insets on the bottom left and right, respectively, illustrate calcium currents recorded from a pCAC6-injected myotubes and pCARD1-injected myotubes. Test potentials are indicated next to the current traces. Structure of the chimeric DHP receptor is represented schematically (see Fig. 5). Reproduced with permission from Tanabe *et al.* (1991). Copyright © 1991 Macmillan Magazines Ltd.

mines whether the calcium channel has slow (skeletal-like) or rapid (cardiac-like) activation. One simple mechanism that can account for the results is that each of the four repeats can independently and reversibly interconvert between resting and activated states, and that the conformational transition of repeat I might be slow or the identity of repeat I might affect the interconversion of other repeats.

VI. Conclusion

Skeletal muscle and cardiac DHP receptors function as essential components of E–C coupling in their respective tissues. In skeletal muscle, the DHP receptor primarily functions as the voltage sensor that controls the release of Ca^{2+} from the SR in response to transverse tubular depolarization without the need for entry of Ca^{2+} across the sarcolemma. In contrast, the cardiac muscle DHP receptor functions as an L-type calcium channel, and Ca^{2+} entry through this channel triggers the subsequent large release of Ca^{2+} from the SR. We have investigated the molecular basis of these differences in the function of the DHP receptor molecule in each tissue. The results obtained indicate that the putative cytoplasmic region between repeats II and III of the skeletal muscle DHP receptor is an important determinant of skeletal muscle-type E–C coupling, and that repeat I determines whether the chimeric calcium channel shows slow (skeletal muscle-type) or rapid (cardiac-type) activation.

Acknowledgments

The author thanks the many collaborators who contributed to the work described in this chapter. The author is an investigator of the Howard Hughes Medical Institute.

References

Adams, B. A., Tanabe, T., Mikami, A., Numa, S., and Beam, K. G. (1990). Intramembrane charge movement restored in dysgenic skeletal muscle by injection of dihydropyridine receptor cDNAs. *Nature (London)* **346**, 569–572.

Armstrong, C. M. (1981). Sodium channels and gating currents. *Physiol. Rev.* **61(3)**, 644–683.

Armstrong, C. M., Bezanilla, F. M., and Horowicz, P. (1972). Twitches in the presence of ethylene glycol bis (β-aminoethyl ether)-N,N'-tetraacetic acid. *Biochim. Biophys. Acta* **267**, 605–608.

Beam, K. G., Knudson, C. M., and Powell, J. A. (1986). A lethal mutation in mice eliminates the slow calcium current in skeletal muscle cells. *Nature (London)* **320**, 168–170.

Bean, B. P. (1989). Classes of calcium channels in vertebrate cells. *Annu. Rev. Physiol.* **51**, 367–384.

Bertolino, M., and Llinás, R. R. (1992). The central role of voltage-activated and receptor-operated calcium channels in neuronal cells. *Annu. Rev. Pharmacol. Toxicol.* **32**, 399–421.

Beuckelmann, D. J., and Wier, W. G. (1988). Mechanism of release of calcium from sarcoplasmic reticulum of guinea-pig cardiac cells. *J. Physiol.* **405**, 233–255.

Biel, M., Ruth, P., Bosse, E., Hullin, R., Stühmer, W., Flockerzi, V., and Hofmann, F. (1990). Primary structure and functional expression of a high voltage activated calcium channel from rabbit lung. *FEBS Lett.* **269(2)**, 409–412.

Bosse, E., Regulla, S., Biel, M., Ruth, P., Meyer, H. E., Flockerzi, V., and Hofmann, F. (1990). The cDNA and deduced amino acid sequence of the γ subunit of the L-type calcium channel from rabbit skeletal muscle. *FEBS. Lett.* **267(1)**, 153–156.

Bournaud, R., and Mallart, A. (1987). An electrophysiological study of skeletal muscle fibres in "muscular dysgenesis" mutation of the mouse. *Pflügers Arch.* **409**, 468–476.

Campbell, K. P., Leung, A. T., and Sharp, A. H. (1988). The biochemistry and molecular biology of the dihydropyridine-sensitive calcium channel. *Trends Neurosci.* **11(10)**, 425–430.

Catterall, W. A., Seagar, M. J., and Takahashi, M. (1988). Molecular

properties of dihydropyridine-sensitive calcium channels in skeletal muscle. *J. Biol. Chem.* **263(8)**, 3535–3538.

Chandler, W. K., Rakowski, R. F., and Schneider, M. F. (1976). Effects of glycerol treatment and maintained depolarization on charge movement in skeletal muscle. *J. Physiol.* **254**, 285–316.

Chiarandini, D. J., Sánchez, J. A., and Stefani, E. (1980). Effects of calcium withdrawal on mechanical threshold in skeletal muscle fibers of the frog. *J. Physiol.* **303**, 153–163.

Donaldson, P. L., and Beam, K. G. (1983). Calcium currents in a fast-twitch skeletal muscle of the rat. *J. Gen. Physiol.* **82**, 449–468.

Dubel, S. J., Starr, T. V. B., Hell, J., Ahlijanian, M. K., Enyeart, J. J., Catterall, W. A., and Snutch, T. P. (1992). Molecular cloning of the α₁ subunit of an ω-conotoxin-sensitive calcium channel. *Proc. Natl. Acad. Sci. USA* **89**, 5058–5062.

Ellis, S. B., Williams, M. E., Ways, N. R., Brenner, R., Sharp, A. H., Leung, A. T., Campbell, K. P., McKenna, E., Koch, W. J., Hui, A., Schwartz, A., and Harpold, M. M. (1988). Sequence and expression of mRNAs encoding the α₁ and α₂ subunits of a DHP-sensitive calcium channel. *Science* **241**, 1661–1664.

Endo, M. (1977). Calcium release from the sarcoplasmic reticulum. *Physiol. Rev.* **57**, 71–108.

Fabiato, A. (1985). Simulated calcium current can both cause calcium loading in and trigger calcium release from the sarcoplasmic reticulum of a skinned canine cardiac Purkinje cell. *J. Gen. Physiol.* **85**, 291–320.

Fleischer, S., and Inui, M. (1989). Biochemistry and biophysics of excitation–contraction coupling. *Annu. Rev. Biophys. Biophys. Chem.* **18**, 333–364.

Glucksohn-Waelsch, S. (1963). Lethal genes and analysis of differentiation. In higher organisms lethal genes serve as tools for studies of cell differentiation and cell genetics. *Science* **142**, 1269–1276.

Grabner, M., Friedrich, K., Knaus, H., Striessnig, J., Scheffauer, F., Staudinger, R., Koch, W. J., Schwartz, A., and Glossmann, H. (1991). Calcium channels from *Cyprinus carpio* skeletal muscle. *Proc. Natl. Acad. Sci. USA* **88**, 727–731.

Guy, H. R., and Conti, F. (1990). Pursuing the structure and function of voltage-gated channels. *Trends Neurosci.* **13(6)**, 201–206.

Hamill, O. P., Marty, A., Neher, E., Sakmann, B., and Sigworth, F. J. (1981). Improved patch-clamp techniques for high-resolution current recording from cells and cell-free membrane patches. *Pflügers Arch.* **391**, 85–100.

Hartmann, H. A., Kirsch, G. E., Drewe, J. A., Taglialatela, M., Joho, R. H., and Brown, A. M. (1991). Exchange of conduction pathways between two related K⁺ channels. *Science* **251**, 942–945.

Heinemann, S. H., Terlau, H., Stühmer, W., Imoto, K., and Numa, S. (1992). Calcium channel characteristics conferred on the sodium channel by single mutations. *Nature (London)* **356**, 441–443.

Hodgkin, A. L., and Huxley, A. F. (1952). A quantitative description of membrane current and its application to conduction and excitation in nerve. *J. Physiol.* **117**, 500–544.

Hui, A., Ellinor, P. T., Krizanova, O., Wang, J. J., Diebold, R. J., and Schwartz, A. (1991). Molecular cloning of multiple subtypes of a novel rat brain isoform of the α₁ subunit of the voltage-dependent calcium channel. *Neuron* **7**, 35–44.

Hullin, R., Singer-Lahat, D., Freichel, M., Biel, M., Dascal, N., Hofmann, F., and Flockerzi, V. (1992). Calcium channel β subunit heterogeneity: Functional expression of cloned cDNA from heart, aorta and brain. *EMBO J.* **11(3)**, 885–890.

Isenberg, G., and Klöckner, U. (1982). Calcium currents of isolated bovine ventricular myocytes are fast and of large amplitude. *Pflügers Arch.* **395**, 30–41.

Jay, S. D., Ellis, S. B., McCue, A. F., Williams, M. E., Vedvick, T.

S., Harpold, M. M., and Campbell, K. P. (1990). Primary structure of the γ subunit of the DHP-sensitive calcium channel from skeletal muscle. *Science* **248**, 490–492.

Keynes, R. D., Greeff, N. G., and Forster, I. C. (1990). Kinetic analysis of the sodium gating current in the squid giant axon. *Proc. R. Soc. London B* **240**, 411–423.

Klaus, M. M., Scordilis, S. P., Rapalus, J. M., Briggs, R. T., and Powell, J. A. (1983). Evidence for dysfunction in the regulation of cytosolic Ca^{2+} in excitation–contraction uncoupled dysgenic muscle. *Dev. Biol.* **99**, 152–166.

Knudson, C. M., Chaudhari, N., Sharp, A. H., Powell, J. A., Beam, K. G., and Campbell, K. P. (1989). Specific absence of the α_1 subunit of the dihydropyridine receptor in mice with muscular dysgenesis. *J. Biol. Chem.* **264(3)**, 1345–1348.

Koch, W. J., Ellinor, P. T., and Schwartz, A. (1990). cDNA cloning of a dihydropyridine-sensitive calcium channel from rat aorta. *J. Biol. Chem.* **265(29)**, 17786–17791.

Kostyuk, P. G., Krishtal, O. A., and Pidoplichko, V. I. (1981). Calcium inward current and related charge movements in the membrane of snail neurones. *J. Physiol.* **310**, 403–421.

Lacerda, A. E., Kim, H. S., Ruth, P., Perez-Reyes, E., Flockerzi, V., Hofmann, F., Birnbaumer, L., and Brown, A. M. (1991). Normalization of current kinetics by interaction between the α_1 and β subunits of the skeletal muscle dihydropyridine-sensitive Ca^{2+} channel. *Nature (London)* **352**, 527–530.

Lee, K. S., and Tsien, R. W. (1982). Reversal of current through calcium channels in dialysed single heart cells. *Nature (London)* **297**, 498–501.

MacKinnon, R., and Yellen, G. (1990). Mutations affecting TEA blockage and ion permeation in voltage-activated K$^+$ channels. *Science* **250**, 276–279.

Mikami, A., Imoto, K., Tanabe, T., Niidome, T., Mori, Y., Takeshima, H., Narumiya, S., and Numa, S. (1989). Primary structure and functional expression of the cardiac dihydropyridine-sensitive calcium channel. *Nature (London)* **340**, 230–233.

Mori, Y., Friedrich, T., Kim, M., Mikami, A., Nakai, J., Ruth, P., Bosse, E., Hofmann, F., Flockerzi, V., Furuichi, T., Mikoshiba, K., Imoto, K., Tanabe, T., and Numa, S. (1991). Primary structure and functional expression from complementary DNA of a brain calcium channel. *Nature (London)* **350**, 398–402.

Näbauer, M., Callewaert, G., Cleemann, L., and Morad, M. (1989). Regulation of calcium release is gated by calcium current, not gating charge, in cardiac myocytes. *Science* **244**, 800–803.

Niidome, T., Kim, M. S., Friedrich, T., and Mori, Y. (1992). Molecular cloning and characterization of a novel calcium channel from rabbit brain. *FEBS Lett.* **308**, 7–13.

Numa, S., and Noda, M. (1986). Molecular structure of sodium channels. *Ann. N.Y. Acad. Sci.* **479**, 338–355.

Papazian, D. M., Timpe, L. C., Jan, Y. N., and Jan, L. Y. (1991). Alteration of voltage-dependence of *Shaker* potassium channel by mutations in the S4 sequence. *Nature (London)* **349**, 305–310.

Perez-Reyes, E., Kim, H. S., Lacerda, A. E., Horne, W., Wei, X., Rampe, D., Campbell, K. P., Brown, A. M., and Birnbaumer, L. (1989). Induction of calcium currents by the expression of the α_1-subunit of the dihydropyridine receptor from skeletal muscle. *Nature (London)* **340**, 233–236.

Perez-Reyes, E., Castellano, A., Kim, H. S., Bertrand, P., Baggstrom, E., Lacerda, A. E., Wei, X., and Birnbaumer, L. (1992). Cloning and expression of a cardiac/brain β subunit of the L-type calcium channel. *J. Biol. Chem.* **267(3)**, 1792–1797.

Powell, J. A., and Fambrough, D. M. (1973). Electrical properties of normal and dysgenic mouse skeletal muscle in culture. *J. Cell Physiol.* **82**, 21–38.

Pragnell, M., Sakamoto, J., Jay, S. D., and Campbell, K. P. (1991).

Cloning and tissue specific expression of the brain calcium channel β-subunit. *FEBS. Lett.* **29(2)**, 253–258.

Rieger, F., Bournaud, R., Shimahara, T., Garcia, L., Pinçon-Raymond, M., Romey, G., and Lazdunski, M. (1987). Restoration of dysgenic muscle contraction and calcium channel function by co-culture with normal spinal cord neurons. *Nature (London)* **330**, 563–566.

Rios, E., and Brum, G. (1987). Involvement of dihydropyridine receptors in excitation–contraction coupling in skeletal muscle. *Nature (London)* **325**, 717–720.

Ruth, P., Röhrkasten, A., Biel, M., Bosse, E., Regulla, S., Meyer, H. E., Flockerzi, V., and Hofmann, F. (1989). Primary structure of the β subunit of the DHP-sensitive calcium channel from skeletal muscle. *Science* **245**, 1115–1118.

Sánchez, J. A., and Stefani, E. (1978). Inward calcium current in twitch muscle fibres of the frog. *J. Physiol.* **283**, 197–209.

Sánchez, J. A., and Stefani, E. (1983). Kinetic properties of calcium channels of twitch muscle fibres of the frog. *J. Physiol.* **337**, 1–17.

Schneider, M. F., and Chandler, W. K. (1973). Voltage dependent charge movement in skeletal muscle: A possible step in excitation-contraction coupling. *Nature (London)* **242**, 244–246.

Seino, S., Chen, L., Seino, M., Blondel, O., Takeda, J., Johnson, J. H., and Bell, G. I. (1992). Cloning of the α_1 subunit of a voltage-dependent calcium channel expressed in pancreatic β cells. *Proc. Natl. Acad. Sci. USA* **89**, 584–588.

Singer, D., Biel, M., Lotan, I., Flockerzi, V., Hofmann, F., and Dascal, N. (1991). The roles of the subunits in the function of the calcium channel. *Science* **253**, 1553–1557.

Snutch, T. P., Tomlinson, W. J., Leonard, J. P., and Gilbert, M. M. (1991). Distinct calcium channels are generated by alternative splicing and are differentially expressed in the mammalian CNS. *Neuron* **7**, 45–57.

Soldatov, N. M. (1992). Molecular diversity of L-type Ca^{2+} channel transcripts in human fibroblasts. *Proc. Natl. Acad. Sci. USA* **89**, 4628–4632.

Starr, T. V. B., Prystay, W., and Snutch, T. P. (1991). Primary structure of calcium channel that is highly expressed in the rat cerebellum. *Proc. Natl. Acad. Sci. USA* **88**, 5621–5625.

Stühmer, W., Conti, F., Suzuki, H., Wang, S., Noda, M., Yahagi, N., Kubo, H., and Numa, S. (1989). Structural parts involved in activation and inactivation of the sodium channel. *Nature (London)* **339**, 597–603.

Tanabe, T., Takeshima, H., Mikami, A., Flockerzi, V., Takahashi, H., Kangawa, K., Kojima, M., Matsuo, H., Hirose, T., and Numa, S. (1987). Primary structure of the receptor for calcium channel blockers from skeletal muscle. *Nature (London)* **328**, 313–318.

Tanabe, T., Beam, K. G., Powell, J. A., and Numa, S. (1988). Restoration of excitation–contraction coupling and slow calcium current in dysgenic muscle by dihydropyridine receptor complementary DNA. *Nature (London)* **336**, 134–139.

Tanabe, T., Mikami, A., Numa, S., and Beam, K. G. (1990a). Cardiac-type excitation–contraction coupling in dysgenic skeletal muscle injected with cardiac dihydropyridine receptor cDNA. *Nature (London)* **344**, 451–453.

Tanabe, T., Beam, K. G., Adams, B. A., Niidome, T., and Numa, S. (1990b). Regions of the skeletal muscle dihydropyridine receptor critical for excitation–contraction coupling. *Nature (London)* **346**, 567–569.

Tanabe, T., Adams, B. A., Numa, S., and Beam, K. G. (1991). Repeat I of the dihydropyridine receptor is critical in determining calcium channel activation kinetics. *Nature (London)* **352**, 800–803.

Tsien, R. W., and Tsien, R. Y. (1990). Calcium channels, stores and oscillations. *Annu. Rev. Cell Biol.* **6**, 715–760.

Varadi, G., Lory, P., Schultz, D., Varadi, M., and Schwartz, A. (1991). Acceleration of activation and inactivation by the β subunit of the skeletal muscle calcium channel. *Nature (London)* **352,** 159–162.

Wei, X., Perez-Reyes, E., Lacerda, A. E., Schuster, G., Brown, A. M., and Birnbaumer, L. (1991). Heterologous regulation of the cardiac Ca^{2+} channel $α_1$ subunit by skeletal muscles β and γ subunits. *J. Biol. Chem.* **226,** 21943–21947.

Williams, M. E., Feldman, D. H., McCue, A. F., Brenner, R., Velicelebi, G., Ellis, S. B., and Harpold, M. M. (1992a). Structure and functional expression of $α_1$, $α_2$ and β subunits of a novel human neuronal calcium channel subtype. *Neuron* **8,** 71–84.

Williams, M. E., Brust, P. F., Feldman, D. H., Patthi, S., Simerson, S., Maroufi, A., McCue, A. F., Velicelebi, G., Ellis, S. B., and Harpold, M. M. (1992b). Structure and functional expression of an ω-conotoxin-sensitive human N-type calcium channel. *Science* **257,** 389–395.

Yellen, G., Jurman, M. E., Abramson, T., and MacKinnon, R. (1991). Mutations affecting internal TEA blockade identify the probable pore-forming region of a K^+ channel. *Science* **251,** 939–942.

Yool, A. J., and Schwarz, T. L. (1991). Alteration of ionic selectivity of a K^+ channel by mutation of the H5 region. *Nature (London)* **349,** 700–704.

Molecular Pharmacology of Cardiac L-Type Calcium Channels

ROBERT S. KASS

I. Introduction

Voltage-dependent calcium channels underlie depolarization-induced calcium influx in many cell types. Therefore, these channels control a diverse range of cellular functions from activation of contractile proteins to exocytotic secretion. In the cardiovascular system, the pioneering work of Reuter (1967) showed that calcium channels are key to maintaining cardiac electrical and mechanical activity (Reuter, 1979; Kass et al., 1987). Thus, that drugs that modulate calcium channel activity—calcium channel blockers—have been found to be powerful therapeutic tools with which to treat a variety of cardiovascular diseases including hypertension, angina, and some forms of cardiac arrhythmias is not surprising. More recently, therapeutic applications of these drugs have been expanded to the treatment of congestive heart failure, cardiomyopathy, atherosclerosis, and cerebral and peripheral vascular disorders (Triggle, 1991). In addition to the major contributions calcium channel blockers have made to the treatment and management of clinical disorders, these drugs have been equally important in the development of our understanding of the molecular properties of calcium channels.

At least four calcium channel subtypes (P, T, N,

and L) have now been identified based on their pharmacological and/or biophysical properties (Bean, 1989; Hess, 1990; Tsien et al., 1991). Both T- and L-type channels have been studied in the heart (Pelzer et al., 1990), but the L-type channel is the target of the most extensively developed calcium channel pharmacology. The drugs that have received the most attention belong to three distinct chemical classes: (1) phenylalkylamines (verapamil, D-600), (2) benzothiazepines [(+)cis-diltiazem], and (3) the 1,4-dihydropyridines (PN 200-110, nitrendipine, nifedipine, nisoldipine). These drugs bind to distinct but allosterically coupled receptors on the channel protein (Glossmann and Striessnig, 1990). Of all the chemical compounds that interact with calcium channels, the dihydropyridines (DHPs) have proven to be the most useful in molecular studies of the L-type channel, because these drugs bind to the channel with highest affinity and specificity (Janis and Triggle, 1990).

Calcium channel antagonists in general, and the DHP derivatives in particular, regulate calcium influx by modulating channel gating (Bean, 1984; Hess et al., 1984; Sanguinetti and Kass, 1984; Sanguinetti et al., 1986). Drug-induced gating changes resemble modes of gating that can occur under drug-free conditions (Artalejo et al., 1990; Pietrobon and Hess, 1990), supporting the view that molecular perturbations induced by these compounds promote indigenous conformational changes of the channel proteins. Since researchers have shown that the α_1 subunit contains the specific binding sites for all three major classes of calcium channel blockers just described (Catterall, 1988; Catterall *et al.*, 1988; Hosey and Lazdunski, 1988; Glossmann and Striessnig, 1990), drug-induced conformational changes of the α_1 subunit are very likely to underlie these gating changes.

The goal of this chapter is to review the electrophysiological and biochemical studies that have emphasized the molecular aspects of the mechanism of action of calcium channel blockers. Although three major classes of calcium channel antagonists have been identified, this chapter focuses on the molecular aspects of L-type calcium channel pharmacology of phenylalkylamine and DHP compounds.

A. Materials and Methods

Although this is a review chapter, some of the basic methodology that will be referenced in the next sections is worth mentioning. Membrane currents are described that were measured in single as well as multicellular cardiac and smooth muscle preparations. In multicellular experiments, ionic currents were measured with a two-microelectrode (Purkinje fiber) or sucrose-gap (ventricular muscle preparations) voltage-clamp arrangement. Readers are referred to earlier reports, such as that of Kass and Bennett (1985), for discussion of these approaches.

More recent experiments are described that used patch-clamp procedures in the single-channel or whole-cell configuration. In these experiments, extra- and intracellular solutions were designed to focus on calcium channel current activity, and blockers or ion substitution were employed to eliminate or minimize potassium and sodium channel currents. Readers are referred to any of several patch-clamp studies for details (Hess *et al.*, 1984; Kass and Arena, 1989). Patch-clamp methods are as described by Hamill *et al.* (1981) for the whole-cell or single-channel configuration.

1. Drugs

Most of the compounds described in the chapter can be obtained from sources described in the literature. Some of the charged and tertiary DHP compounds are presented here because of their limited availability. Amlodipine was dissolved in water as a concentrated stock solution. Amlodipine concentrations were chosen as previously described (Burges *et al.*, 1985,1987; Kass and Arena, 1989). SDZ-207-180

was a gift of Sandoz (CH-4002 Basel, Switzerland), nisoldipine was a gift from Miles Laboratories (New Haven, Connecticut), and amlodipine was a gift from Pfizer Central Research (Sandwich, England).

2. Voltage Protocols

The voltage protocols referenced in the chapter are described in detail and illustrated by Kass and Arena (1989). Drug onset was measured by applying a depolarizing "train" protocol in which the holding potential was changed from -80 to -40 or -50 mV and test voltage pulses were applied once every 5 sec. Test voltages were either 0 mV (Na^+ currents), $+10$ mV (Ca^{2+} and Ba^{2+} currents), or $+40$ to $+60$ mV (Cs^+ and K^+ currents); test pulse duration was 200 msec. Recovery was measured using train protocols in which the holding potential was changed from -40 to -80 mV and pulses were applied to the same test potentials, but test pulse duration was 20 or 40 msec to minimize pulse-induced inactivation. In experiments that required recording currents from potentials negative to -60 mV, 50- to 100-msec prepulses were applied to -40 mV to inactivate sodium channel and T-type calcium channel currents (Bean, 1985; Marchetti and Brown, 1988). Thus, in this chapter, current referred to as Ca^{2+} channel current (I_{Ca}) corresponds to L-type Ca^{2+} channel current according to the terminology suggested by Nilius *et al.* (1985). Sodium channels were also blocked by 10–50 μM tetrodotoxin (TTX; Behring Diagnostics, La Jolla, California) and by replacement of NaCl by Tris-Cl in some experiments.

II. Calcium Channel Antagonists: Defining the Basic Categories

Lindner (1960) and Haas and Hartfelder (1962) were among the first to report that verapamil and prenylamine, drugs developed as coronary vasodilators, caused negative ionotropic effects in the myocardium that distinguished them from drugs that had been classically defined as vasodilators, including nitroglycerine and papaverine (Henry, 1980). Fleckenstein and colleagues (Fleckenstein, 1977,1983,1988) were the first to recognize that the actions of these compounds involved inhibition of transmembrane calcium movements, and used the term "calcium antagonists" to classify these organic compounds as a new group of potentially important therapeutic drugs. The term "calcium antagonist" referred to the reversible "sealing" of putative "calcium channels" in the membranes of smooth and cardiac muscle cells, and to the fact that the inhibitory actions of the drugs are antagonized by the concentration of extracellular calcium ions.

This early work was crucial to the probing and identification of the now identified L-type, or DHP-sensitive, calcium channel proteins in heart, smooth muscle, skeletal muscle, and neuronal cells (Triggle and Janis, 1984). Many organic compounds have been found to inhibit the activity of these important channel proteins. The best studied of these compounds are the phenylalkylamines, the benzodiazepines, and the DHPs.

Biochemical analysis, specifically the binding of radiolabeled ligands to cardiac and smooth muscle membrane fragments, was used to probe the basic molecular properties of the sites on these membranes to which calcium channel antagonists bind. Three distinct but allosterically coupled sites were identified: site 1 (DHP binding site), site 2 (phenylalkylamine binding site), and site 3 (benzodiazepine binding site; Glossmann *et al.*, 1984; Fig. 1). The most potent of these drugs are the DHPs which bind with high affinity and stereospecificity to the membranes of cardiac and smooth muscle cells (Triggle and Janis, 1984). Linking all these sites was the binding and modulation by divalents cations, notably calcium (Glossmann *et al.*, 1984). Thus, the term "calcium antagonist" appeared to be well suited to these compounds; however, the mode of action of the drugs in controlling the activity of calcium channels required functional (electrophysiological) analysis. This work began with the characterization of the phenylalkylamine verapamil and its methoxy derivative D600, and has since focused more on the actions of the DHPs.

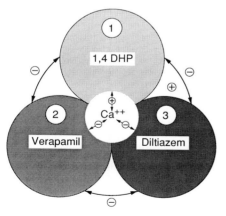

FIGURE 1 Interaction of three distinct receptor sites with calcium on the L-type calcium channel. Reprinted with permission from Glossmann *et al.* (1984).

A. Phenylalkylamines and Calcium Channel Regulation

Early evidence in favor of specific calcium channel inhibition by the calcium antagonists was provided by experiments in which contractile activity and action potentials were measured simultaneously in muscle strips. One of the first reports of separation of sodium- from calcium-dependent electrical activity was that of Kohlhardt *et al.* (1972), in which the actions of verapamil, D600 (methoxy derivative of verapamil), and ion substitution were compared on electrical and mechanical activity of cat right ventricular trabeculae (Fig. 2). The results of these studies indicated a differential susceptibility of Na^+ and Ca^{2+}-dependent electrical activity and led to the hypothesis that these drugs (the calcium antagonists) could selectively block calcium entry pathways in the myocardium.

The question of ion channel selectivity of D600 was again addressed by Kass and Tsien (1975), who compared the effects of this drug to those of Mn^{2+} and La^{2+} in calf cardiac Purkinje fibers. Although this work did support the view that D600 was a potent blocker of calcium channel currents, it also provided evidence that each of these organic or inorganic blockers affected at least one other channel in the Purkinje fiber: the delayed-rectifier outward current. Thus, these experiments demonstrated multiple effects of these calcium antagonists in the heart.

Despite the slight nonspecific effects on other channels, the mechanism of action of the phenylalkylamines was of great interest because of their potential in the therapeutic approach to cardiovascular problems. One of the key characteristics reported for inhibition of electrical and mechanical activity by these drugs was the marked dependence of drug activity on stimulation rate (Bayer *et al.*, 1975; Fig. 3). Inhibition appeared to be accelerated by stimulation frequency and relieved at negative diastolic potentials (Ehara and Kaufmann, 1978).

These properties were investigated in detail by McDonald *et al.* (1980, 1984) and Trautwein *et al.* (1987), who used sucrose-gap voltage-clamp procedures to study the effects of the drug on membrane currents in muscle strips (Fig. 4). D600 block of calcium channel current did not occur unless the cell membrane was depolarized and repetitive depolarizing pulses were applied. Then, current was blocked in a pulse-dependent manner, and block could be relieved at negative voltages, as previously reported. The rate of removal of block was accelerated as membrane potential was made more negative.

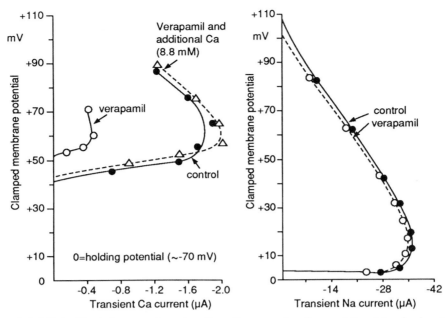

FIGURE 2 The influence of verapamil on the current–voltage relationships for heart calcium (*left*) and sodium (*right*) currents. Note that the voltage and current axes are inverted from the usual presentation. Reprinted with permission from Kohlhardt *et al.* (1972).

III. Identifying the Phenylalkylamine Binding Site

A. Electrophysiological Evidence for an Intracellular Location

The effects of D600 and a related compound, AQA 39, were interpreted as evidence for channel state-dependent interactions. Researchers proposed that these drugs preferentially blocked open channels and that channel availability was shifted in the hyperpolarizing direction by this state- (or voltage-) dependent action. Although these experiments provided clear evidence for a role of membrane potential in the regulation of drug action, they could not be used to assay the location of the site to which this family of drugs binds because the compounds are tertiary and contain neutral as well as charged forms at physiological pH. Ionized, but not neutral, drugs are restricted in their access to channel-associated binding sites (Hille, 1977a; Hondeghem and Katzung, 1977) and are very useful in identifying the location of these sites relative to the cell surface. To address the issue of the binding site location, Hescheler *et al.* (1982) applied a quaternary phenylalkylamine to the intracellular surface of heart cells by dialysis with a patch-recording pipette. The results of this study indicated that the quaternary form of D600 (D800) blocked calcium channels from the intracellular membrane surface, consistent with drug access gained via open channels for extracellular application of tertiary compounds (such as D600).

B. Biochemical Evidence: An Intracellular Site on the α_1 Subunit

The location of the phenylalkylamine receptor site has also been investigated using a combination of photoaffinity labeling and immunoprecipitation with several sequence-specific antibodies. The results of this work restrict the site of photolabeling to the transmembrane helix S6 of domain IV and the beginning of the long intracellular C-terminal tail (Striessnig *et al.*, 1990). In conjunction with the functional studies of Hescheler *et al.* (1982), these data provide strong evidence for an intracellular location of the phenylalkylamine binding site.

Catterall and Striessnig (1992) have proposed that the binding of phenylalkylamines to a receptor site formed in part by sequences near the intracellular end of the transmembrane segment IVS6 could inhibit channel activity by actually occluding the channel pore. Thus, in this view, the actions of these drugs result from an actual physical block of channel conduction pathways and not by regulation of indigenous channel gating.

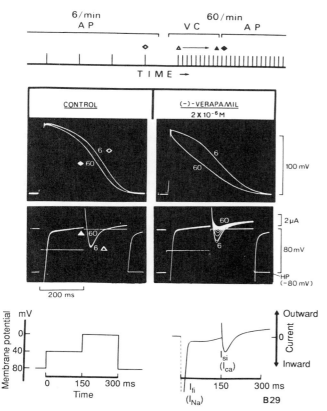

FIGURE 3 Evidence for frequency-dependent effects of verapamil. Action potential and membrane currents measured at 6/min and 60/min in the absence (*left*) and presence (*right*) of 20 μM verapamil. The effects of the drug on the action potential plateau and on the transmembrane calcium current are more pronounced at the higher stimulation frequency. Reprinted with permission from Bayer *et al.* (1975).

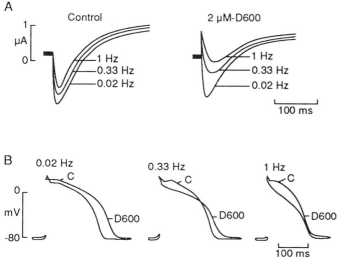

FIGURE 4 Rate-dependent effects of D600 (2 μM) on membrane current (A) and action potentials (B) recorded in cat ventricular muscle. Reprinted with permission from McDonald *et al.* (1984).

IV. Dihydropyridine Modulation of L-Type Channels

Structure–activity studies indicated that the DHPs were the most potent calcium channel antagonists (Janis and Triggle, 1990) and that the mode of action of these compounds might differ from that of the phenylalkylamines (Rodenkirchen *et al.*, 1982). Two points were of great interest: (1) the DHPs had a high degree of tissue selectivity for smooth muscle over heart muscle cells and (2) the use dependence of these drugs differed greatly from that of the phenylalkylamines (Bayer *et al.*, 1982).

Some of the differences in electrical effects between the phenylalkylamines and the DHPs were found to be caused by a higher selectivity of the DHPs for calcium channel activity. The DHP nisoldipine was found to inhibit calcium channel currents selectively and not to affect the delayed-rectifier outward current in the Purkinje fiber at drug concentrations as high as 10 μM (Kass, 1982). Further, addition of D600 after calcium channels were completely blocked by nisoldipine still inhibited delayed rectification, providing evidence that inhibition of the outward current was a direct effect of D600 and not a secondary effect of block of calcium channel current (Kass, 1982,1984).

However, the most important and interesting property of the DHPs is the unique voltage dependence of the actions of these drugs. Bean (1984) and Sanguinetti and Kass (1984) found that the inhibitory actions of DHP compounds were greatly enhanced by depolarization. Importantly, the influence of membrane potential on DHP inhibition of calcium channel activity contrasted with the role of voltage in regulating the phenylalkylamines because channel openings were not needed to induce DHP block of the channels (Sanguinetti and Kass, 1984; Fig. 5). Instead, marked modulation of drug–channel interactions was caused by changes in steady-state holding potential alone. Thus, several investigators postulated that the drugs bound preferentially to inactivated channels and that this state dependence was the mechanism underlying the well-known tissue selectivity of DHPs for smooth muscle over heart muscle cells. However, despite the appeal of this explanation, some groups reported distinction in tissue selectivity that could not be explained entirely by voltage dependence (Bean *et al.*, 1986; Hermsmeyer *et al.*, 1988).

Interest in the mechanism of action of DHP antagonist regulation of calcium channel activity was further enhanced by the discovery of a series of closely related DHPs that actually stimulated channel activity, and did so in a voltage dependent manner (Sanguinetti *et*

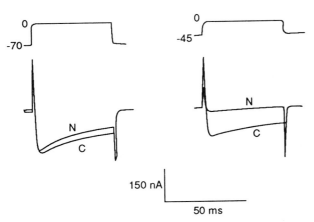

FIGURE 5 Membrane potential modulates the inhibitory actions of nisoldipine in isolated calf Purkinje fibers. Calcium channel currents measure once every 30 sec from holding potentials of −70 mV (*left*) and −40 mV (*right*) in the absence (C) and presence (N) of nisoldipine (200 n*M*). Current recorded at the depolarized holding potential is completely blocked, but the current at −70 mV is not affected. Reprinted with permission from Sanguinetti and Kass (1984).

al., 1986). Most importantly, these drugs were shown to modulate channel activity by controlling channel gating (Hess *et al.*, 1984). This type of channel control was later shown to be consistent with the mechanism by which antagonist compounds inhibit channel activity (Kass, 1990). Consequently, understanding the molecular basis for DHP action is likely to yield important information about the mechanisms underlying gating of these important channels.

V. Calcium Channels as Dihydropyridine Receptors

Because of the high affinity of DHPs for calcium channels, the initial strategy for the isolation and purification of L-type calcium channels was to isolate and purify the protein components as isolated DHP receptors with the chance that the purified components would constitute functional calcium channels (Hosey and Lazdunski, 1988). This approach was applied to skeletal muscle t-tubules, the richest source of L-type calcium channels, and yielded a purified DHP receptor that consisted of five subunits: α_1, α_2, β_2, γ, and δ (Borsotto *et al.*, 1984; Curtis and Catterall, 1984; Vaghy *et al.*, 1987; Catterall, 1988; Glossmann and Striessnig, 1988). Antibodies raised against one component were capable of immunoprecipitating the others, suggesting that all five subunits were involved in forming a structure for native calcium channel function (Vandaele *et al.*, 1987; Sharp and Campell, 1989).

Reconstitution experiments showed that the purified DHP receptor could function as a voltage-dependent calcium channel (Curtis and Catterall, 1984; Flockerzi *et al.*, 1986), but functional roles of individual subunits could not be specified.

Amino acid analysis of the purified rabbit skeletal muscle DHP receptor α_1 subunit was used successfully as a probe to clone the receptor subunit and to deduce its primary structure by sequencing its cDNA (Tanabe *et al.*, 1987). The α_1 subunit is a large hydrophobic polypeptide that has sequence characteristics similar to those of the voltage-gated sodium channel (containing four internally homologous domains with six putative transmembrane segments designated S1 through S6), strongly suggesting that this subunit is the calcium channel pore itself (Catterall, 1988; Fig. 6). The cardiac DHP receptor was then cloned using the skeletal muscle α_1 subunit as a probe; its sequence showed 66% homology with the skeletal muscle subunit and the same structural resemblance to the voltage-gated sodium channel (Mikami *et al.*, 1989). Expression of L-type channel activity was first demonstrated by injection of skeletal α_1 cDNA into myotubes of dysgenic mice (Tanabe *et al.*, 1988) and then by stable transfection of mouse L cells with α_1 cDNA (Perez-Reyes *et al.*, 1989). Functional channel activity could be measured after injection of cardiac α_1 subunit mRNA into *Xenopus* oocytes (Mikami *et al.*, 1989). Coexpression of the cardiac α_1 and skeletal β_2 subunits doubled the expressed currents, strongly suggesting that subunits other than α_1 were likely to have important functional roles in channel activity (Mikami *et al.*, 1989). The genes for the remaining skeletal muscle DHP receptor subunits have now also been cloned (Ellis *et al.*, 1988; Jay *et al.*, 1989; Ruth *et al.*, 1989; Bosse *et al.*, 1990), providing tools for functional studies of individual subunit combinations and for determining the specific functional roles of molecular components of the channel. With this structural information in hand, it is now possible to begin to determine the molecular components that underlie important functional characteristics of the L-type channel in heart and other tissues.

VI. A Unique Role for Calcium in the Mechanism of Action of Calcium Channel Antagonists?

Since the discovery of organic calcium channel blockers by Fleckenstein (reviewed by Fleckenstein, 1988), researchers have known that an interrelationship exists between the blocking activity of these drugs and divalent ion concentration. Divalent ions

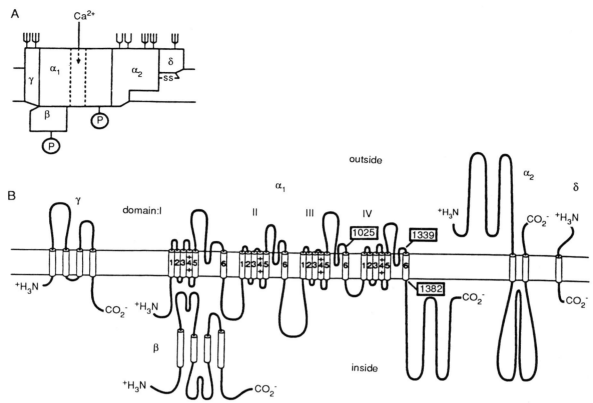

FIGURE 6 Subunit structure of skeletal muscle calcium channels. (A) A model of the subunit structure based on biochemical studies. (B) Transmembrane folding models of the channel subunits. Reprinted with permission from Catterall (1991).

inhibit the binding of radiolabeled phenlyalkylamines to membrane-bound L-type channels, but high-affinity DHP labeling of L-type channels in brain, cardiac, or smooth muscle membranes depends on the presence of divalent ions (Glossmann and Striessnig, 1988, 1990) as does the binding of DHPs to the purified DHP receptor (Flockerzi *et al.*, 1986). Binding of calcium also has important regulatory roles in the permeability and gating properties of drug-free native L-type channels. Calcium influx and the high calcium selectivity of skeletal muscle and heart L-type channels is best explained by a multiple-ion (at least two) binding model in which ion–ion repulsion is seen as the key to overcoming strong intrapore cation binding (Tsien *et al.*, 1987). Divalent ion binding has been shown to induce protein conformational changes that, in turn, could affect the binding of other ions (notably H^+; Prod'hom *et al.*, 1989; Pietrobon and Hess, 1990). Inactivation of cardiac L-type channels has also been shown to be influenced by calcium and is enhanced by calcium entry (Kass and Sanguinetti, 1984; Yue *et al.*, 1990). Thus, the binding of calcium to at least one

regulatory site is very likely to change the conformation of the α_1 subunit so DHP–receptor interactions are modified and, possibly, permeation and gating are controlled in the absence of drug binding.

The structural predictions of a close physical relationship between the DHP binding site and a regulatory calcium binding site (Babitch, 1990; Regulla *et al.*, 1991) suggest a molecular mechanism for this interrelationship between drug action and divalent ion binding. This view has been strengthened by a report that positive allosteric regulators of DHP binding increase the calcium affinity of the L-type calcium channel and, simultaneously, alter DHP receptor kinetics. However, the location of this putative regulatory calcium binding site remains to be determined, since conflicting evidence has been presented for extracellular (Ebata *et al.*, 1990) or cytoplasmic (Babitch, 1990; Regulla *et al.*, 1991) sites. Clearly, knowledge of the drug- and calcium-binding domains on the α_1 subunit will contribute to our understanding of the molecular mechanisms of channel gating, as well as of the basis for allosteric drug interactions.

VII. Probing the Dihydropyridine Binding Site Location

A. Electrophysiological Experiments

Most DHP derivatives used in electrophysiological studies, including PN 200-110, nitrendipine, nisoldipine, and nifedipine, have pk_a values less than 3.5 (Rodenkirchen *et al.*, 1982). Thus, at physiological pH, these are neutral molecules that are not restricted by the lipid bilayer of the cell membrane when accessing the DHP binding site (Hille, 1977a,b). Therefore, voltage-dependent DHP modulation of calcium channels in whole-cell or single-channel experiments in which these drugs are used is not sufficient to rule out or prove intracellular or extracellular access to the drug binding site (Hess *et al.*, 1984; Kokubun *et al.*, 1987).

Kass and Arena (1989) carried out experiments in which amlodipine, a tertiary DHP (pk_a of 8.6) that is 94% ionized at physiological pH, was used to block calcium channels and probe the location of the DHP binding site. The strategy of these experiments was to compare the block and unblock of I_{Ca} by neutral and charged amlodipine, and to use this information to probe the location of the DHP receptor binding site in or near the calcium channel. Different forms of the drug were determined by external pH. The data presented earlier were used to determine H^+-induced changes in channel activity in the absence of drug.

Amlodipine had previously been shown by others (Burges *et al.*, 1985) to compete with nitrendipine binding sites; thus, it was not surprising that the neutral form of amlodipine (pH_o 10.0) blocked I_{Ca} in a voltage-dependent manner that closely resembled the behavior of other neutral DHP compounds (Fig. 7). Block was promoted by depolarization; changes in holding potential in the absence of additional pulsing were sufficient to cause block to develop. Most im-

portantly, block by neutral amlodipine is rapidly relieved when membrane potential is returned to negative values in a manner that, again, resembles the behavior of other previously investigated neutral DHPs. Ionized amlodipine applied externally also blocked I_{Ca} in a voltage-dependent manner, but block onset was slower than with the neutral drug form and the development of block appeared to be influenced by depolarizing pulses. Most importantly, however, was the finding that recovery from block by ionized amlodipine (pH 7.4) was very slow and very incomplete, suggesting that access to the DHP binding site might be restricted for the ionized but not the neutral form of the drug (Fig. 8).

The recovery from block was used as an assay for access of the drug-bound receptor by external H^+. After blocking channels in pH 7.4, recovery from block was assayed by returning the membrane potential to -80 mV. In pH 7.4, recovery was always slow and very incomplete. Then, with membrane potential fixed at -80 mV, external pH was rapidly changed from 7.4 to 10.0 and recovery of current was monitored. Drug-blocked channels were shown to recover rapidly from block when external pH was increased (Fig. 9), demonstrating that the drug-bound receptor could be titrated by external pH. The same effects of external pH were found even if channels had been completely blocked by amlodipine in pH 7.4. Thus, open channels appear to be unnecessary for the drug-bound receptor to be titrated by external H^+.

These results resemble those obtained for local anesthetic block of sodium channels because the drug–receptor complex can be titrated by external H^+. Like the sodium channel local anesthetic receptor (Hille, 1977a,b; Schwarz *et al.*, 1977), the DHP receptor binding site might be accessible to ionized drugs only from an intracellular pathway. In this model, at pH 7.4 externally applied amlodipine accesses the DHP receptor via the small fraction of drug (6%) that is

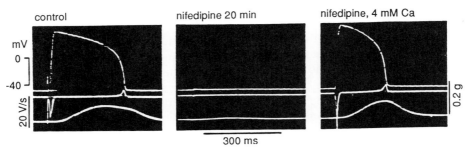

FIGURE 7 Inhibition of electrical and mechanical activity in depolarized guinea pig papillary muscle by nifedipine, and antagonistic effects of calcium. Nifedipine (1.4 μM) completely inhibits contraction and Ca^{2+}-dependent action potential (*middle*). This inhibition is reversed by elevating extracellular calcium (*right*).

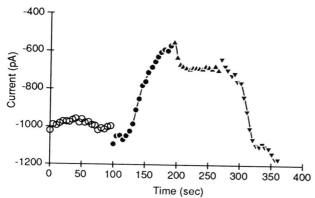

FIGURE 8 Influence of pH on amlodipine block of calcium channels. Current amplitude is plotted against time during application of depolarizing trains in the absence (○) and presence (●) of amlodipine at pH 7.4. At pH 7.4, channels recover little current (▲), but recovery protocols applied in pH 10.0 cause most channels to recover from drug block (▼). Thus, the drug-bound channel is titrated by extracellular hydrogen ions. Reprinted with permission from Kass and Arena (1989).

neutral under these conditions. To test more directly for parallels between the DHP receptor binding site in calcium channels and the sodium channel local anesthetic binding site, experiments had to be carried out with quaternary DHP derivatives.

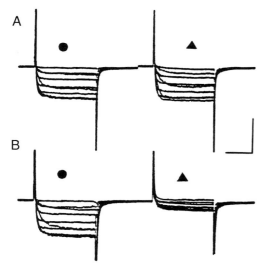

FIGURE 9 Block of calcium channel currents in GH₄C₁ cells by external (*bottom*) but not internal (*top*) SDZ-207-180, a permanently charged dihydropyridine. Currents shown were measured in 10-mV increments from a −30-mV holding potential. (A) Currents were measured after applying a depolarizing train protocol, immediately (*left*) and 3 min after (*right*) establishing whole-cell conditions with a 500 nM SDZ-207-180-containing patch pipette. (B) Currents were recorded in the same cell 5 min after establishing whole-cell conditions (*left*) and then immediately after exposing the cell to 500 nM external SDZ 207-180 (*right*). Reprinted with permission from Kass et al. (1991).

However, unlike sodium channels, external but *not* internal application of a fully charged DHP compound SDZ 207-180 blocked calcium channel current in a voltage-dependent manner (Kass *et al.*, 1991). Further, these results also contrasted with the work of Heschler *et al.* (1982), who identified an intracellular site for regulation of the same type of calcium channels by phenylalkylamines. Although additional experiments with a series of structurally related quaternary DHPs are needed to improve the electrophysiological resolution of the binding site location, these experiments clearly demonstrate that permanently charged DHPs, *which cannot cross the lipid bilayer*, can block L-type channels from the extracellular surface and do not require channel openings for this action. Thus an intracellular DHP binding domain is unlikely.

B. Biochemical Experiments: Binding and Subunit-Specific Antibodies

The location of the binding domain for DHPs has also been investigated by two groups using biochemical techniques with isolated and purified channel protein subunits. In the first published study, the DHP binding site on the rabbit skeletal muscle α₁ subunit was investigated using tritiated azidopine and nitrendipine as ligands (Regulla *et al.*, 1991). Several peptides were found to be labeled using this approach, but the two major labeled peptides were shown to be located on the putative cytosolic domain of the calcium channel. Interestingly, this domain adjoins a possible calcium binding site (Babitch, 1990), suggesting the possibility of conformation-dependent interactions between DHP and calcium binding.

In another approach, Catterall and colleagues (Nakayama *et al.*, 1991; Striessnig *et al.*, 1991) used a combination of radioligand binding and peptide-specific antibodies raised against targeted fragments of the L-type channel α₁ subunit to also identify the DHP binding domain on the channel protein. This approach yielded two additional sites that appeared to be located on the extracellular domain of the channel. The postulated binding site locations on the assembled channel protein are presented in Fig. 10. The electrophysiological data summarized earlier remain the principal means of distinguishing between these two possibilities. Additional work is planned in the future to use probes specifically designed to improve the electrophysiological estimate of the binding site location. In the model of the Catterall and Striessnig (1992), the action of DHPs is to control channel gating by causing allosteric interactions among channel subunits, thus inhibiting current by controlling the open-

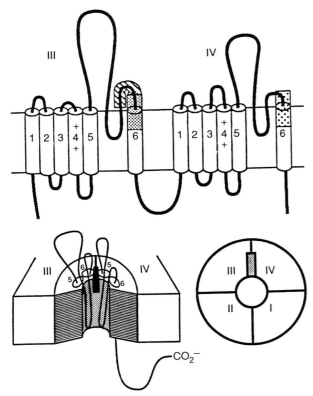

FIGURE 10 A domain interface model for dihydropyridine binding and action on calcium channels. (*Top*) Transmembrane folding model of part of the α_1 subunit. Shaded and dotted rectangles indicate extracellular sites detected by photolabeling. (*Bottom*) Proposed binding cleft for dihydropyridines formed at the extracellular end of the interface between domains III and IV is illustrated in vertical (*left*, black bar) and horizontal (*right*, shaded) cross-sectional views of the α_1 subunit. Reprinted with permission from Catterall and Striessnig (1992).

ing and closing of channels and not by occluding the channel pore.

Although calcium binding sites are clearly important to channel function and modulation by DHPs and other drugs, clearly more work must be done to clarify the structural basis of the activity of these drugs and to determine the roles, if any, of channel subunits in causing the important regulatory actions of the calcium channel antagonists.

VIII. Summary

Calcium channel antagonists are unique drugs that have gained prominence from their major contributions to clinical medicine and to the basic science of calcium channel proteins. Because of the multiple roles of calcium ions in cell function, communication, and biochemical regulation, the control of calcium entry by these drugs has provided therapeutic tools for

a wide variety of clinical disorders. In addition, the high degree of selectivity and great sensitivity of these drugs for the L-type channel has allowed the cloning and molecular fingerprinting of the channel protein. Work is now directed at combining the approaches of molecular biology and electrophysiology to provide further information about the molecular basis of tissue selectivity of these drugs, and to allow the design of structurally related drug molecules that will be able to target and control specific regions of the channel protein.

References

Artalejo, C. R., Ariano, M. A., Perlman, R. L., and Fox, A. P. (1990). Activation of facilitation calcium channels in chromaffin cells by D1 dopamine receptors through a cAMP/protein kinase A-dependent mechanism. *Nature (London)* **348**, 239–242.

Babitch, J. (1990). Channel hands. *Nature (London)* **346**, 321–322.

Bayer, R., Hennekes, R., Kaufmann, R., and Mannhold, R. (1975). Ionotropic and electrophysiological actions of verapamil and D600 on mammalian myocardium: I. Pattern of ionotropic effects of the racemic compounds. *Naunyn-Schmiedeberg's Arch. Pharmacol.* **290**, 49–68.

Bayer, R., Kaufmann, R., Mannhold, R., and Rodenkirchen, R. (1982). The action of specific Ca antagonists on cardiac electrical activity. *Prog. Pharmacol.* **5**, 53–85.

Bean, B. P. (1984). Nitrendipine block of cardiac calcium channels: High-affinity binding to the inactivated state. *Proc. Natl. Acad. Sci. USA* **81**, 6388–6392.

Bean, B. P. (1985). Two kinds of calcium channels in canine atrial cells. Differences in kinetics, selectivity, and pharmacology. *J. Gen. Physiol.* **86**, 1–30.

Bean, B. P. (1989). Classes of calcium channels in vertebrate cells. *Annu. Rev. Physiol.* **51**, 367–384.

Bean, B. P., Sturek, M., Puga, A., and Hermsmeyer, K. (1986). Calcium channels in muscle cells isolated from rat mesenteric arteries: Modulation by dihydropyridine drugs. *Circ. Res.* **59**, 229–235.

Borsotto, M., Barhanin, J., Norman, R. I., and Lazdunski, M. (1984). Purification of the dihydropyridine receptor of the voltage-dependent calcium channel from skeletal muscle transverse tubule using (+)[3H]PN200-110. *Biochem. Biophys. Res. Commun.* **122**, 1357–1366.

Bosse, E., Regulla, S., Biel, M., Ruth, P., Meyer, H. E., Flockerzi, V., and Hofmann, F. (1990). The cDNA and deduced amino acid sequence of the gamma subunit of the L-type calcium channel from rabbit skeletal muscle. *FEBS Lett.* **267**, 153–156.

Burges, R. A., Carter, A. J., Gardiner, D. F., and Higgins, A. J. (1985). Amlodipine, a new dihydropyridine calcium channel blocker with slow onset and long duration of action. *Br. J. Pharmacol.* **85**, 281P.

Burges, R. A., Gardiner, D. G., Gwilt, M., Higgins, A. J., Blackburn, K. J., Campbell, S. F., Cross, P. E., and Stubbs, J. K. (1987). Calcium channel blocking properties of amlodipine in vascular smooth muscle and cardiac muscle *in Vitro*: Evidence for voltage modulation of vascular dihydropyridine receptors. *J. Cardiovasc. Pharmacol.* **9**, 110–119.

Catterall, W. A. (1988). Structure and function of voltage-sensitive ion channels. *Science* **242**, 50–60.

Catterall, W. A. (1991). Functional subunit structure of voltage-gated calcium channels. *Science* **253**, 1499–1500.

Catterall, W. A., and Striessnig, J. (1992). Receptor sites for Ca channel antagonists. *Trends Pharmacol. Sci.* **13**, 256–262.

Catterall, W. A., Seagar, M. J., and Takahashi, M. (1988). Molecular properties of dihydropyridine-sensitive calcium channels in skeletal muscle. *J. Biol. Chem.* **263**, 3535–3538.

Curtis, B. M., and Catterall, W. A. (1984). Purification of the calcium antagonist receptor of the voltage-sensitive calcium channel from skeletal muscle transverse tubules. *Biochemistry* **23**, 2113–2118.

Ebata, H., Mills, J. S., Nemcek, K., and Johnson, J. D. (1990). Calcium binding to extracellular sites of skeletal muscle calcium channels regulates dihydropyridine binding. *J. Biol. Chem.* **265(1)**, 177–182.

Ehara, T., and Kaufmann, R. (1978). The voltage- and time-dependent effects of (minus)-verapamil on the slow inward current in isolated cat ventricular myocardium. *J. Pharmcol. Exp. Thera.* **207**, 49–55.

Ellis, S. B., Williams, M. E., Ways, N. R., Brenner, R., Sharp, A. H., Leung, A. T., Campbell, K. P., McKenna, E., Koch, W. J., Schwartz, A., and Harpold, M. M. (1988). Sequence and expression of mRNAs encoding the α_1 and α_2 subunits of a DHP-sensitive calcium channel. *Science* **241**, 1661–1663.

Fleckenstein, A. (1977). Specific pharmacology of calcium in myocardium, cardiac pacemakers, and vascular smooth muscle. *Annu. Rev. Pharmacol. Toxicol.* **17**, 149–166.

Fleckenstein, A. (1983). Calcium Antagonism in Heart and Smooth Muscle. Wiley, New York.

Fleckenstein, A. (1988). Calcium antagonists: Pharmacology and clinical research. *Ann. N.Y. Acad. Sci.* **522**, 1–15.

Flockerzi, V., Oeken, H. J., Hofmann, F., Pelzer, D., Cavalie, A., and Trautwein, W. (1986). Purified dihydropyridine-binding site from skeletal muscle T-tubules is a functional calcium channel. *Nature (London)* **323**, 66–68.

Glossmann, H., and Striessnig, J. (1988). Calcium channels. *Vit. Horm.* **44**, 155–328.

Glossmann, H., and Striessnig, J. (1990). Molecular properties of calcium channels. *Rev. Physiol. Biochem. Pharmacol.* **114**, 1–105.

Glossmann, H., Ferry, D. R., Goll, A., Striessnig, J., and Zernig, G. (1984). Calcium channels: Introduction into their molecular pharmacology. *In* "Cardiovascular Effects of Dihydropyridine-Type Calcium Antagonists and Agonists" (A. Fleckenstein, C. Van Breemen, R. Gross, and F. Hoffmeister, eds.), pp. 113–139. Springer-Verlag, Heidelberg.

Haas, H., and Hartfelder, G. (1962). Alpha-isopropyl-alpha-(n-methylhomoveratryl-gamma-aminopropyl)-3,4-dimethoxy-phenylacetonitrol, eine Substanz mit coronargefässerweiternden Eigenschaften. *Arzneim. Forsch.* **12**, 549–558.

Hamill, O. P., Marty, A., Neher, E., Sakmann, B., and Sigworth, F. J. (1981). Improved patch-clamp techniques for high-resolution current recording from cells and cell-free membrane patches. *Pflügers Arch.* **391**, 85–100.

Henry, P. D. (1980). Comparative pharmacology of calcium antagonists: Nifedipine, verapamil, and diltiazem. *Am. J. Cardiol.* **46**, 1047–1058.

Hermsmeyer, K., Sturek, M., and Rusch, N. J. (1988). Calcium channel modulation by dihydropyridines in vascular smooth muscle. *Ann. N.Y. Acad. of Sc.* **522**, 25–31.

Hescheler, J., Pelzer, D., Trube, G., and Trautwein, W. (1982). Does the organic calcium channel blocker D600 act from inside or outside on the cardiac cell membrane? *Pflügers Arch.* **393**, 287–291.

Hess, P. (1990). Calcium channels in vertebrate cells. *Annu. Rev. Neurosci.* **13**, 337–356.

Hess, P., Lansman, J. B., and Tsien, R. W. (1984). Different modes of gating behavior favoured by dihydropyridine agonists and antagonists. *Nature (London)* **311**, 538–544.

Hille, B. (1977a). Local anesthetics: Hydrophilic and hydrophobic pathways for the drug–receptor reaction. *J. Gen. Physiol.* **69**, 497–515.

Hille, B. (1977b). The pH-dependent rate of action of local anesthetics on the node of Ranvier. *J. Gen. Physiol.* **69**, 475–496.

Hondeghem, L. M., and Katxung, B. G. (1977). Time and voltage dependent interaction of antiarrhythmic drugs with cardiac sodium channels. *Biochim. Biophys. Acta* **472**, 373–398.

Hosey, M. M., and Lazdunski, M. (1988). Calcium channels: Molecular pharmacology, structure and regulation. *J. Membr. Biol.* **104**, 81–105.

Janis, R. A., and Triggle, D. J. (1990). "The Calcium Channel: Its Properties, Function, Regulation, and Clinical Relevance." CRC Press, Cleveland.

Jay, S. D., Ellis, S. B., McCue, A. F., Williams, M. E., Vedvick, T. S., Harpold, M. M., and Campbell, K. P. (1989). Primary structure of the gamma subunit of the DHP-sensitive calcium channel from skeletal muscle. *Science* **248**, 490–492.

Kass, R. S. (1982). Nisoldipine: A new, more selective calcium current blocker in cardiac Purkinje fibers. *J. Pharmacol. Exp. Ther.* **223**, 446–456.

Kass, R. S. (1984). Delayed rectification in the cardiac Purkinje fiber is not activated by intracellular calcium. *Biophys. J.* **45**, 837–839.

Kass, R. S. (1990). Calcium antagonists: Cellular mechanisms of action. *Jap. Pharmacol. Ther.* **18**, 253–265.

Kass, R. S., and Arena, J. P. (1989). Influence of pHo on calcium channel block by amlodipine, a charged dihydropyridine compound: Implications for location of the dihydropyridine receptor. *J. Gen. Physiol.* **93**, 1109–1127.

Kass, R. S., and Bennett, P. B. (1985). Microelectrode voltage clamp: The cardiac Purkinje fiber. In "Voltage and Patch Clamping with Microelectrodes" (T. G. Smith, Jr., H. Lecar, S. J. Redman, and P. W. Gage, eds.), pp. 171–191. American Physiological Society, Bethesda, Maryland.

Kass, R. S., and Sanguinetti, M. C. (1984). Calcium channel inactivation in the cardiac Purkinje fiber. Evidence for voltage- and calcium-mediated mechanisms. *J. Gen Physiol.* **84**, 705–726.

Kass, R. S., and Tsien, R. W. (1975). Multiple effects of calcium antagonists on plateau currents in cardiac Purkinje fibers. *J. Gen. Physiol.* **66**, 169–192.

Kass, R. S., Arena, J. P., and Chin, S. (1991). Block of L-type calcium channels by charged dihydropyridines. Sensitivity to side of application and calcium. *J. Gen. Physiol.* **98**, 63–75.

Kass, R. S., Arena, J. P., and Wiener, R. S. (1987). Membrane potential and nisoldipine block of calcium channels in the heart: Interactions with channel gating. *In* "Nisoldipine 1987" (P. Hugenholtz and J. Meyer, eds.), pp. 13–27. Springer-Verlag, Berlin.

Kohlhardt, M., Bauer, B., Krause, H., and Fleckenstein, A. (1972). Differentiation of the transmembrane Na and Ca channels in mammalian cardiac fibres by the use of specific inhibitors. *Pflügers Arch.* **335**, 309–322.

Kokubun, S., Prod'hom, B., Becker, C., Porzig, H., and Reuter, H. (1987). Studies on Ca channels in intact cardiac cells: Voltage-dependent effects and cooperative interactions of dihydropyridine enantiomers. *Mol. Pharmacol.* **30**, 571–584.

Lindner, E. (1960). Phenyl-propyl-diphenyl-propyl-amin, eine neue Substanz mit coronargefässerweiternder Wirkung. *Arzneim. Forsch.* **10**, 569–573.

Marchetti, C., and Brown, A. M. (1988). Protein kinase activator 1-oleoyl-2-acetyl-*sn*-glycerol inhibits two types of calcium currents in GH3 cells. *Am. J. Physiol.* **23**, C206–C210.

McDonald, T. F., Pelzer, D., and Trautwein, W. (1980). On the mechanism of slow calcium channel block in heart. *Pflügers Arch.* **385**, 175–179.

McDonald, T. F., Pelzer, D., and Trautwein, W. (1984). Cat ventricular muscle treated with D600: Characteristics of calcium channel block and unblock. *J. Physiol.* **352**, 217–241.

Mikami, A., Imoto, K., Tanabe, T., Niidome, T., Mori, Y., Takeshima, H., Narumiya, S., and Numa, S. (1989). Primary structure and functional expression of the cardiac dihydropyridine-sensitive calcium channel. *Nature (London)* **340**, 230–233.

Nakayama, H., Taki, M., Striessnig, J., Glossmann, H., Catterall, W. A., and Kanaoka, Y. (1991). Identification of 1,4-dihydropyridine binding regions within the a1 subunit of skeletal muscle Ca_{2+} channels by photoaffinity labeling with diazepine. *Proc. Natl. Acad. Sci. U.S.A.* **88**, 9203–9207.

Pelzer, D., Pelzer, S., and McDonald, T. F. (1990). Properties and regulation of calcium channels in muscle cells. *Rev. Physiol. Biochem. Pharmacol.* **114**, 108–207.

Perez-Reyes, E., Kim, H. S., Lacxerda, A. E., Horne, W., Wei, X., Rampe, D., Campbell, K. P., Brown, A. M., and Birnbaumer, L. (1989). Induction of calcium currents by the expression of the subunit of the dihydropyridine receptor from skeletal muscle. *Nature (London)* **340**, 233–236.

Pietrobon, D., and Hess, P. (1990). Novel mechanism of voltage-dependent gating in L-type calcium channels. *Nature (London)* **346**, 651.

Prod'ham, B., Pietrobon, D., and Hess, P. (1989). Interactions of protons with single-open L-type calcium channels. Location of protonation site and dependence of proton-induced current fluctuations on concentration and species of permeant ion. *J. Gen. Physiol.* **94**, 23–42.

Regulla, S., Schneider, T., Nastainczyk, W., Meer, H. W., and Hofmann, F. (1991). Identification of the site of interaction of the dihydropyridine channel blockers nitrendipine and azidopine with the calcium-channel α_1 subunit. *EMBO J.* **10**, 45–49.

Reuter, H. (1967). The dependence of slow inward current in Purkinje fibres on the extracellular calcium-concentration. *J. Physiol.* **192**, 479–492.

Reuter, H. (1979). Properties of two inward membrane currents in the heart. *Annu. Rev. Physiol.* **41**, 413–424.

Rodenkirchen, R., Bayer, R., and Mannhold, R. (1982). Specific and non-specific Ca antagonists: A structure-activity analysis of cardiodepressive drugs. *Prog. Pharmacol.* **5**, 9–23.

Ruth, P., Rohrkasten, A., Biel, M., Bosse, E., Regulla, S., Meyer, H. E., Flockerzi, V., and Hofmann, F. (1989). Primary structure of the β subunit of the DHP-sensitive calcium channel from skeletal muscle. *Science* **245**, 1115–1118.

Sanguinetti, M. C., and Kass, R. S. (1984). Voltage-dependent block of calcium channel current in the calf cardiac Purkinje fiber by dihydropyridine calcium channel antagonists. *Circ. Res.* **55**, 336–348.

Sanguinetti, M. C., Krafte, D. S., and Kass, R. S. (1986). Bay K8644: Voltage-dependent modulation of Ca channel current in heart cells. *J. Gen. Physiol.* **88**, 369–392.

Schwarz, W., Palade, P. T., and Hille, B. (1977). Local anesthetics. Effect of pH on use-dependent block of sodium channels in frog muscle. *Biophys. J.* **20**, 343–368.

Sharp, A. H., and Campell, K. P. (1989). Characterization of the 1,4-dihydropyridine receptor using subunit-specific polyclonal antibodies. Evidence for a 32,000-Da subunit. *J. Biol. Chem.* **264**, 2816–2825.

Striessnig, J., Glossmann, H., and Catterall, W. A. (1990). Identification of a phenylalkylamine binding region within the α_1 subunit of skeletal muscle Ca^{2+} channels. *Proc. Natl. Acad. Sci. USA* **87**, 9108–9112.

Striessnig, J., Murphy, B. J., and Catterall, W. A. (1991). Dihydropyridine receptor of L-type Ca channels: Identification of binding domains for H-PN200-110 and H-azidopine within the alpha1 subunit. *Proc. Natl. Acad. Sci. USA* **88**, 10769–10773.

Tanabe, T., Takeshima, H., Mikami, A., Flockerzi, V., Takahashi, H., Kangawa, K., Kojima, M., Matsuo, H., Hiose, T., and Numa, S. (1987). Primary structure of the receptor for calcium channel blockers from skeletal muscle. *Nature (London)* **328**, 313–318.

Tanabe, T., Beam, K. G., Powell, J. A., and Numa, S. (1988). Restoration of excitation–contraction coupling and slow calcium current in dysgenic muscle by dihydropyridine receptor complementary DNA. *Nature (London)* **336**, 134–139.

Trautwein, W., Cavalie, A., Flockerzi, V., Hoffman, F., and Pelzer, D. (1987). Modulation of calcium channel function by phosphorylation in guinea pig ventricular cells and phospholipid bilayer membranes. *Cir. Res.* **61**(Suppl. I), 117–123.

Triggle, D. J. (1991). Calcium-channel drugs: Structure–function relationships and selectivity of action. *J. Cardiovasc. Pharmacol.* **18**, S1–S6.

Triggle, D. J., and Janis, R. A. (1984). The 1,4-dihydropyridine receptor: A regulatory component of the Ca channel. *J. Cardiovasc. Pharmacol.* **6**, S949–S955.

Tsien, R. W., Hess, P., McCleskey, E. W., and Rosenberg, R. L. (1987). Calcium channels: Mechanisms of selectivity, permeation, and block. *Annu. Rev. Biophys. and Biophys. Chem.* **16**, 265–290.

Tsien, R. W., Ellinor, P. T., and Horne, W. A. (1991). Molecular diversity of voltage-dependent Ca channels. *Trends Pharmacol. Sci.* **12**, 349–354.

Vaghy, P. L., Striessnig, J., Miwa, K., Knaus, H. G., Itagaki, K., McKenna, E., Glossmann, H., and Schwartz, A. (1987). Identification of a novel 1,4-dihydropyridine and phenyl-alkylamine-binding polypeptide in calcium channel preparations. *J. Biol. Chem.* **262**, 14337–14342.

Vandaele, S., Fosset, M., Galizzi, J. P., and Lazdunski, M. (1987). Monoclonal antibodies that coimmunoprecipitate the 1,4-dihydropyridine and phenylalkylamine receptors and reveal the Ca^{2+} channel structure. *Biochemistry* **26**, 5–9.

Yue, D. T., Backx, P. H., and Imredy, J. P. (1990). Calcium-sensitive inactivation in the gating of single calcium channels. *Science* **21**, 1735–1738.

13

Pharmacology of Different Types of Calcium Channels in Rat Neurons

BRUCE P. BEAN AND ISABELLE M. MINTZ

I. Introduction

Voltage-dependent Ca^{2+} channels are present in the cell membranes of almost all excitable cells. They contribute to many cellular functions, from contraction of smooth and cardiac muscle to synaptic transmission between neurons. In the last decade, researchers have identified multiple types of Ca^{2+} channels in mammalian cells. The existence of distinct kinds of Ca^{2+} channels was first recognized because of differences in voltage dependence, kinetics, and single-channel properties. Different types of Ca^{2+} channels also differ in their sensitivity to various drugs and toxins, so pharmacology has become particularly valuable in recognizing distinct classes of channels, especially the various types of Ca^{2+} channels in neurons, which seem to contain the greatest variety of channels (reviewed by Bean, 1989; Hess, 1990; Tsien *et al.*, 1991; Miller, 1992).

The existence of multiple types of Ca^{2+} channels has also become clear from molecular biology (see Chapter 10; reviewed by Tsien *et al.*, 1991; Miller, 1992; Snutch and Reiner, 1992). A great deal of on-going work is focused on cloning new Ca^{2+} channel types and correlating the channels identified by cloning with those expressed in differentiated cells. In some cases (for example, Ca^{2+} channels in cardiac muscle, skeletal muscle, and smooth muscle), correlating cloned channels with native channels has been relatively straightforward (see Chapter 11). In neurons, however, our understanding is still incomplete. Clearly pharmacological distinctions between different Ca^{2+} channel types will be crucial in establishing links between cloned channels and the native channels expressed in neurons.

This chapter reviews the pharmacology of voltage-dependent Ca^{2+} channels in mammalian cells. The emphasis is on Ca^{2+} channels in neurons and on drugs and toxins that have been useful in making distinctions between different kinds of Ca^{2+} channels.

We begin with an overview of the emerging classification of Ca^{2+} channels. Each section then describes the actions of drugs and toxins that have helped lead to the classification.

II. Overview

Recordings of native channels in normal cells have distinguished four types of voltage-dependent Ca^{2+} channels—T-, L-, N-, and P-type channels—to date, with hints of a fifth type. Of these, L- and N-type channels can be clearly correlated with particular cloned channels. The correlation of P-type channels with cloned channel types is still ambiguous (Tsien *et al.*, 1991), and T-type channels have not yet been identified by cloning.

T-Type Ca^{2+} channels can be activated by small depolarizations from the resting potential, so they are also known as low-voltage activated or low-threshold Ca^{2+} channels. These channels are easily distinguished from the various high-threshold Ca^{2+} channel types by their very different voltage dependence, kinetics, and single-channel conductance, but no potent selective blocker for T-type Ca^{2+} channels is currently available. L-Type, N-type, and P-type channels are all so-called high-threshold channels, requiring larger depolarizations to be activated. Because differences in voltage dependence, kinetics, and single-channel properties are relatively subtle, pharmacology has played a large role in distinguishing among them. L-Type Ca^{2+} channels can be potently blocked by dihydropyridine drugs. L-Type channels produce most or all of the current in most muscle cells and a significant fraction of the current in most neurons and neuroendocrine cells. N-Type Ca^{2+} channels are potently blocked by the snail toxin ω-conotoxin; these channels seem to be restricted to neurons and neuroendocrine cells. Similarly, P-type Ca^{2+} channels have been found only in neurons to date; these channels have been identified by high-affinity block by ω-agatoxin-IVA (ω-Aga-IVA), a toxin from spider venom. In some neurons, a fraction of high-threshold Ca^{2+} channel current exists that does not correspond to L-type, N-type, or P-type properties and could represent a fifth type of channel.

As will be discussed in more detail, L-type channels at least should be regarded as a class or family of channels. Clear differences exist between L-type channels in different tissues, in kinetics and voltage dependence as well as in molecular structure (see Chapters 10 and 11). The other classes of channels have not yet been studied as thoroughly as L-type channels on any level. Each type might turn out to comprise distinct members with strong similarities but also some differences.

III. T-Type Channels

T-Type Ca^{2+} channels were originally distinguished from other Ca^{2+} channels by their much smaller single-channel current, lower threshold of activation, faster inactivation, and disappearance (because of steady-state inactivation) at moderately depolarized holding potentials (Carbone and Lux, 1984,1987; Fox et al., 1987a,b). T-Type Ca^{2+} channels have been described in many kinds of excitable cells including neurons, neuroendocrine cells, skeletal muscle, cardiac muscle, and smooth muscle. However, within a given tissue, these channels tend to be

restricted to only a small fraction of cells. Broadly speaking, a correlation may exist between T-type Ca^{2+} channels and cells that are capable of spontaneous electrical pacemaking activity (see Bean, 1989). For example, in the heart, T-type Ca^{2+} channels are especially prominent in cells of the sinoatrial node (Hagiwara et al., 1988); in the mature brain, they are prominent in thalamic (Coulter et al., 1989a; Huguenard and Prince, 1992) and hypothalamic (Akaike et al., 1989) neurons. However, T-type Ca^{2+} channels are not restricted entirely to cells capable of spontaneous activity. For example, these channels can be found in individual cardiac atrial and ventricular cells that are quiescent in the absence of external stimulation.

The pharmacology of T-type Ca^{2+} channels is not yet well developed. Several drugs have been described that show some selectivity for blocking T-type Ca^{2+} channels over other types (see Herrington and Lingle, 1992), but none is both potent and highly selective. Amiloride, best known as a blocker of epithelial non-voltage-dependent Na^{+} channels, is able to inhibit T-type Ca^{2+} channels in cardiac muscle and in neurons (Tang et al., 1988). Amiloride is not very potent—~1 mM is required to inhibit T-type current nearly completely—and it is not perfectly selective—slightly higher concentrations can inhibit high-threshold Ca^{2+} channels as well. Nonetheless, this drug has proved useful in helping separate components of T-type current from those carried by other channels.

Perhaps the most intriguing drugs that interact with T-type Ca^{2+} channels are petit mal anticonvulsants such as ethosuximide (Coulter et al., 1989b). These drugs partially inhibit T-type Ca^{2+} current in thalamic relay neurons and are relatively selective for T-type current over components of high-threshold current in the same neurons. In contrast, ethosuximide has almost no effect on T-type current in GH$_3$ cells (Herrington and Lingle, 1992), suggesting selectivity for different subtypes of T-type channels. Testing ethosuximide against T-type channels in various types of central and peripheral neurons and in muscle will be interesting.

IV. L-Type Channels

A. Dihydropyridine Blockers

L-Type Ca^{2+} channels have come to be defined by their sensitivity to dihydropyridine blockers such as nifedipine, nitrendipine, and nimodipine. L-Type Ca^{2+} channels appear to pass most or all of the voltage-dependent calcium current in skeletal muscle, cardiac muscle, and smooth muscle (T-type channels contribute a small fraction in some individual cell types). In neurons, L-type channels are almost always present but typically contribute only a minor fraction of the overall high-threshold current, ranging from ~5% in rat sympathetic neurons to ~30% in rat and guinea pig CA3 hippocampal neurons (Regan *et al.*, 1991; Mintz *et al.*, 1992a).

Obvious differences exist in the kinetics and voltage dependence of the L-type current in different cell types. For example, L-type current in cardiac cells inactivates relatively rapidly (although more slowly than T-type channels) whereas L-type current in neurons or the GH_3 pituitary cell line inactivates very slowly or not at all. The L-type Ca^{2+} current in skeletal muscle activates very slowly (time constant of tens to hundreds of milliseconds) compared with that in other cells (time constant of several milliseconds). Nonetheless, even before the structure of the channels was known, regarding these channels as members of the same family seemed reasonable since all could be blocked potently by dihydropyridine drugs (as well as because the L-type channels in different cells share the property of a large single-channel conductance, another early defining characteristic of L-type channels). This concept has been justified by results from molecular biological studies. Structurally distinct subunits are encoded by mRNA expressed in different tissues. At least nine different forms of the main Ca^{2+} channel subunit (the α_1 subunit) exist, encoded by three different genes; some gene products produce different subunits by alternative splicing (Snutch and Reiner, 1992; Hullin *et al.*, 1993). Clear homologies exist among the various subunits, however, justifying the idea that they form a distinct class of Ca^{2+} channels. All subunits that have been functionally expressed are inhibited by dihydropyridines, justifying the use of dihydropyridine sensitivity as a defining feature of L-type channels.

Experiments with L-type channels in cardiac and smooth muscle cells have been crucial in elucidating the mode of action of dihydropyridine blockers (see Chapter 12). Figure 1 shows Ca^{2+} channel current elicited by a step from −70 to −20 mV in an atrial cardiac myocyte. As in most muscle cells, the current is virtually all L-type current. In this case, the current was nearly completely blocked by 1 μM nimodipine. The existence in many cardiac and smooth muscle cells of a single population of L-type channels has been very helpful in characterizing drug actions on them.

Dihydropyridine drugs block L-type Ca^{2+} channel current more potently when the current is elicited from depolarized holding potentials, when the chan-

Rat atrial myocyte

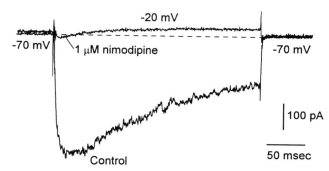

FIGURE 1 Nimodipine block of L-type current in a rat atrial myocyte. Internal solution: 115 mM CsCl, 5 mM MgATP, 0.3 mM GTP (Tris salt), 5 mM creatine phosphate (diTris salt), 10 mM BAPTA, 10 mM HEPES, pH adjusted to 7.4 with CsOH. External solution: 10 mM BaCl$_2$, 160 mM tetraethylammonium(TEA) Cl, 10 mM HEPES, 3 μM tetrodotoxin, pH adjusted to 7.4 with TEAOH. Adapted by permission of Elsevier Publishing Co., Inc. from Bean, *American Journal of Hypertension*, **4**, 4068–4115. Copyright © 1991 by American Journal of Hypertension, Inc.

nels are partially inactivated, than from hyperpolarized holding potentials, when the channels are in the noninactivated "resting" state (Bean, 1984; Sanguinetti and Kass, 1984; Uehara and Hume, 1985). This result can be interpreted to reflect high-affinity binding to the inactivated state of the channel but much weaker binding to the resting state of the channel. For nitrendipine block of cardiac L-type channels, the half-blocking concentration is ~0.4 nM for currents elicited from a depolarized holding potential but ~700 nM for current elicited from a negative holding potential (Bean, 1984).

The voltage dependence of dihydropyridine block is a complicating factor in their use as tools to block L-type channels, since relatively high concentrations are needed to block at negative holding potentials. However, concentrations of drug apparently can be found that can block L-type channels completely, even at negative holding potentials, yet retain selectivity for L-type channels. Nimodipine is apparently more potent than nitrendipine or nifedipine in blocking at negative holding potentials. At 1 μM, nimodipine blocks cardiac L-type current completely, even from negative holding potentials (e.g., Fig. 1). At 1 μM, nimodipine has almost no effect on P-type Ca^{2+} channels in rat cerebellar Purkinje neurons or N-type Ca^{2+} channels in rat sympathetic neurons (Mintz *et al.*, 1992a; I. M. Mintz and B. P. Bean, unpublished results), suggesting a high degree of specificity.

The most detailed studies of dihydropyridine action on L-type Ca^{2+} channels have been carried out in cardiac and smooth muscle. Since L-type Ca^{2+} channels in neurons are (at least in part) different gene products (Snutch and Reiner, 1992; Hullin *et al.*, 1993) and show a number of functional differences, notably a difference in the extent and kinetics of inactivation, determining how similar dihydropyridine block of L-type channels is in neuronal and muscle cells is important. This question is still not clearly answered. Quantitative experiments in neurons are difficult because the L-type current is always mixed with that from other channel types. The experiment in Fig. 2 shows nimodipine applied to a rat sensory neuron. Figure 2 shows the effects of applying two different concentrations of nimodipine to a rat dorsal root ganglion neuron while recording overall high-threshold Ca^{2+} channel current composed of L-type, N-type, P-type, and perhaps other channel components. Only about 20% of the calcium channel current elicited by depolarizing the cell membrane from -100 to 0 mV (*top*) was inhibited by 300 nM nimodipine. Although blocking only a fraction of the current, 300 nM nimodi-

pine produced nearly saturating effects, since increasing the nimodipine concentration to 3 μM had only a small additional effect. Figure 2 (*bottom*) shows the effects of the two nimodipine concentrations on currents elicited from a depolarized holding potential (-30 mV) in the same cell. Nimodipine blocked about the same amount of absolute current elicited from the two holding potentials, although the fractional block of current elicited from -30 mV is greater because the control current is smaller than that elicited from -100 mV. A simple interpretation is that L-type Ca^{2+} channels contribute about the same amount of current from the two holding potentials and 300 nM *and* 3 μM nimodipine effectively block this current at both holding potentials, but other kinds of channels contribute substantial current from -100 mV and much less from -30 mV.

Since both concentrations of nimodipine apparently produce nearly saturating block of L-type current from both holding potentials, the experiment in Fig. 2 does not address the important question of whether the potency of dihydropyridine block is sensitive to holding potential in neurons, as it is in cardiac cells. However, this sensitivity is likely, since McCarthy and Tanpiengco (1992) showed that very low (nanomolar) concentrations of nimodipine are capable of inhibiting L-type Ca^{2+} channels in depolarized rabbit sensory neurons. However, a detailed examination of this issue remains to be done.

Although the evidence suggests that 1 μM nimodipine may be sufficient to block L-type Ca^{2+} channels with minimal effects on other channel types, many experiments have been done using much higher concentrations of dihydropyridine blockers that definitely had nonspecific blocking effects on non-L-type Ca^{2+} channels as well as on Na^+ and K^+ channels. When blockers are used in excess of 10 μM, significant nonspecific effects can result. At 10 μM, nitrendipine inhibits cardiac T-type current by ~50% (Bean, 1985), neuronal N-type current by 10–25% (Regan *et al.*, 1991), and cardiac Na^+ current by 10–50% (Yatani and Brown, 1985). Cardiac K^+ channels can be similarly affected by similar concentrations of nifedipine (Hume, 1985). Interestingly, P-type Ca^{2+} channels in cerebellar Purkinje neurons are almost completely unaffected by 10 μM nitrendipine (Regan *et al.*, 1991), suggesting that Na^+ channels are more sensitive to "spill-over" effects of high concentrations than P-type Ca^{2+} channels.

Although dihydropyridine drugs are the most frequently used selective blockers of L-type channels, two other classes of clinically used Ca^{2+} channel blockers are effective blockers of L-type channels: phenylalkylamines, typified by verapamil (see Chapter 12),

Rat DRG neuron

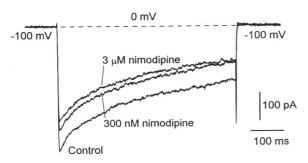

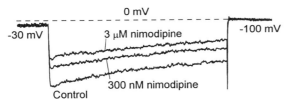

FIGURE 2 Nimodipine inhibition of Ca^{2+} channel current in a rat dorsal root ganglion neuron. (*Top*) Partial block by 300 nM and 3 μM nimodipine of current elicited by a step from -100 to 0 mV. (*Bottom*) In the same cell, partial block by 300 nM and 3μM nimodipine when the holding voltage was -30 mV. Internal solution: 115 mM CsCl, 5 mM MgATP, 0.3 mM GTP (Tris salt), 5 mM creatine phosphate (diTris salt), 10 mM BAPTA, 10 mM HEPES, pH adjusted to 7.4 with CsOH. External solution: 10 mM BaCl$_2$, 160 mM tetraethylammonium(TEA) Cl, 10 mM HEPES, 3 μM tetrodotoxin, pH adjusted to 7.4 with TEAOH. Adapted by permission of Elsevier Publishing Co., Inc. from Bean, *American Journal of Hypertension*, **4**, 4065–4115. Copyright © 1991 by American Journal of Hypertension, Inc.

and benzothiazepines, typified by diltiazem. Both possess high-affinity binding sites on L-type Ca^{2+} channels that are distinct from the high-affinity dihydropyridine binding sites. However, both verapamil and diltiazem show complicated use-dependent block, even at low stimulation rates, so arriving at steady-state block in reasonable amounts of time is difficult or impossible. Because relatively few studies have been done with verapamil and diltiazem in cells other than cardiac muscle, the degree of specificity for L-type channels over other Ca^{2+} channel types is not yet very well established.

B. Dihydropyridine Agonists

Some dihydropyridine molecules have been found to enhance rather than inhibit L-type Ca^{2+} channels. Most experiments have been done with Bay-K 8644, a mixture of two optical isomers, and with another dihydropyridine molecule, (+)-(S)-202-791. The primary mode of action of the dihydropyridine agonists is to increase the probability of channel opening by inducing or stabilizing unusually long-lasting openings (Hess *et al.*, 1984; Nowycky *et al.*, 1985). Channel activity is more dramatically affected at low depolarizations, at which the probability of opening is lowest in controls, than at high depolarizations. The result is a shift in the voltage dependence of activation to more negative voltages. Figure 3 shows results from an experiment in which (+)-(S)-202-791 was applied to a cardiac myocyte (from rabbit atrium) that had only L-type Ca^{2+} current. Under control conditions, depolarization to −30 mV activated only a small current (115 pA); in the presence of drug, this current was enhanced 14-fold. For a larger depolarization, to 0 mV, current was enhanced 3-fold.

This experiment also shows the action of dihydropyridine agonists in dramatically slowing the deactivation kinetics of tail currents on repolarization of the membrane. The slowing of L-type tail currents is very useful in serving as a means of identifying components of L-type current in neurons, where L-type current is only one of several components (Plummer *et al.*, 1989; Regan *et al.*, 1991). Under control conditions, the deactivation of tail currents is comparable for L-type, N-type, and P-type channels, although the deactivation of L-type channels is somewhat faster (Regan *et al.*, 1991). On application of Bay-K 8644 or (+)-(S)-202-791, tail current carried by L-type channels is slowed several-fold, so a new slow tail current is induced that can be identified unambiguously as resulting from L-type channels.

Figure 4 shows the effect of 1 μM (+)-(S)-202-791 applied to a sensory neuron from the dorsal root gan-

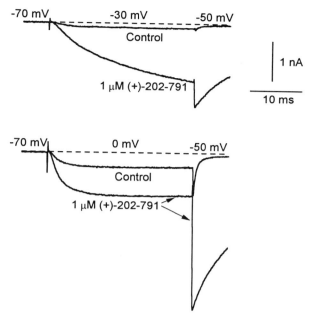

Rabbit atrial myocyte

FIGURE 3 Enhancement of L-type Ca^{2+} channel current in a rabbit atrial myocyte by the dihydropyridine agonist (+)-202-791. Currents activated by depolarizations to −30 or 0 mV are shown before and 1 min after addition of 1 μM (+)-202-791. Internal solution: 115 mM CsCl, 5 mM MgATP, 0.3 mM GTP (Tris salt), 5 mM creatine phosphate (diTris salt), 10 mM BAPTA, 10 mM HEPES, pH adjusted to 7.4 with CsOH. External solution: 2 mM BaCl$_2$, 160 mM tetraethylammonium(TEA) Cl, 10 mM HEPES, 3 μM tetrodotoxin, pH adjusted to 7.4 with TEAOH. Cell W58D.

glion of rat. The drug produced a large enhancement in the current elicited by a depolarization to −40 mV (*top*) as well as a dramatic slowing of the deactivation of the current after repolarization to −60 mV. These effects are very similar to those seen for agonist enhancement of a pure population of L-type channels in cardiac muscle. The small control current (160 pA) at −40 mV may represent a mixture of channel types, but the enhanced current (730 pA) must be at least 80% L-type current induced by the drug. Actually, the control current is likely to be mainly L-type current since this current preferentially activates at small depolarizations; the drug-enhanced current may, therefore, represent nearly pure L-type current. Figure 4 (*bottom*) shows that (+)-(S)-202-791 significantly *inhibits* the current activated by a larger depolarization to −10 mV in the same cell. Similar inhibition by Bay-K 8644 has been observed in various neurons (e.g., Boll and Lux, 1985; Carbone *et al.*, 1990). This effect is most likely an inhibition of N-type channels, similar to that produced by "spill-over" of high concentrations of dihydropyridine blockers, but somewhat more po-

Rat DRG neuron

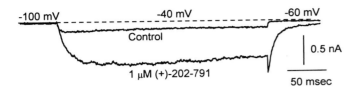

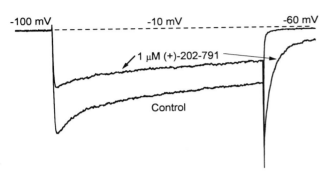

FIGURE 4 Effects of (+)-202-791 on Ca^{2+} channel currents in a rat sensory neuron. Internal solution: 115 mM CsCl, 5 mM MgATP, 0.3 mM GTP (Tris salt), 4 mM cyclic AMP, 5 mM creatine phosphate (diTris salt), 10 mM BAPTA, 10 mM HEPES, pH adjusted to 7.4 with CsOH. External solution: 3 mM BaCl$_2$, 160 mM tetraethylammonium(TEA) Cl, 10 mM HEPES, pH adjusted to 7.4 with TEAOH. Cell W53D.

tent. Although the current at -10 mV is inhibited, the drug induces a slow component of the tail current after repolarization to -60 mV, presumably reflecting the population of L-type channels modified by the drug.

V. N-Type Channels: ω-Conotoxin

N-Type Ca^{2+} channels in neurons were first distinguished from L-type channels by different characteristics of voltage dependence, kinetics, and single-channel behavior in recordings from chick sensory neurons in tissue culture. Subsequent work in a variety of peripheral and central neurons has shown that the differences in kinetics and single-channel size can be even more subtle than those originally recognized. Probably the most convincing evidence that distinct current components are carried by L-type and N-type channels has come from their different pharmacology. The most effective blocker of N-type channels known to date is a peptide toxin isolated from the venom of the sea snail *Conus geographus* (Olivera *et al.*, 1985; McCleskey *et al.*, 1987). This toxin, called ω-conotoxin, is now believed to be a highly selective blocker of N-type Ca^{2+} channels (Aosaki and Kasai, 1989; Plummer *et al.*, 1989; Regan *et al.*, 1991).

Figure 5 summarizes the effect of ω-conotoxin on high-threshold Ca^{2+} channel current in rat dorsal root ganglion neurons. Saturating concentrations of the toxin (>1 μM) block about half the overall high-threshold current; the half-maximal dose of toxin for blocking this fraction of current is ~30 nM. ω-Conotoxin appears to be highly selective for blocking N-type channels over L-type channels. Figure 6 shows an experiment in which a substantial L-type current was isolated in a rat dorsal root ganglion

A   Rat DRG neuron

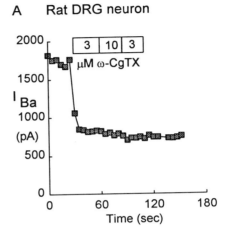

B

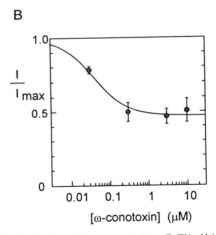

FIGURE 5 Partial block by ω-conotoxin ω-CgTX of high-threshold Ca^{2+} channels in rat DRG neurons. (*Top*) 3 μM ω-CgTX produces saturating but incomplete block of high-threshold current. At 12-msec pulse from -70 to -20 mV was delivered every 5 sec; current was measured at 6–8 msec. (*Bottom*) Dose–response of ω-CgTX block of high-threshold current. Observations (mean ± SEM) from 4–6 cells for each concentration. Internal solution: 108 mM Cs-methanesulfonate, 4.5 mM MgCl$_2$, 9 mM EGTA, 9 mM HEPES, 4 mM MgATP, 0.3 mM GTP (Tris salt), 14 mM creatine phosphate (diTris salt), and 50 U/ml creatine phosphokinase (Sigma Type I), adjusted to pH 7.4 with CsOH. External solution: 5 mM BaCl$_2$, 160 mM TEA Cl, and 10 mM HEPES, adjusted to pH 7.4 with TEAOH. 3 μM tetrodotoxin (TTX) added to the external solution. Redrawn with permission from Regan *et al.* (1991). Copyright © 1991 by Cell Press.

Rat DRG neuron

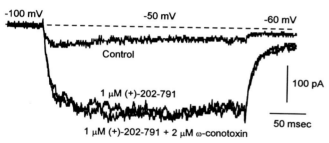

A Cerebellar Purkinje neuron

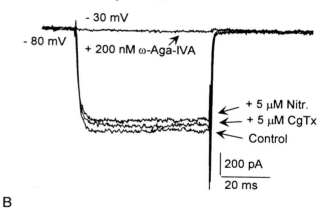

FIGURE 6 ω-Conotoxin (ω-CgTX) does not block (+)-202-791-enhanced L-type current in a rat DRG neuron. The small current elicited by a step from −100 to −50 mV was greatly enhanced by 1 μM (+)-202-791. Cumulative addition of 2 μM ω-CgTX had no effect on the enhanced current. Solutions as in Fig. 4.

B

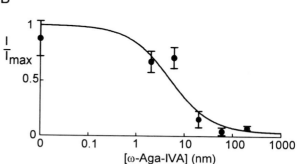

FIGURE 7 ω-Agatoxin-IVA (ω-Aga-IVA) block of high-threshold current in cerebellar Purkinje neurons. (A) Complete block by ω-Aga-IVA of high-threshold Ca^{2+} current resistant to ω-conotoxin and nitrendipine in a cerebellar Purkinje neuron. High-threshold current in control, in the presence of ω-conotoxin (5 μM), ω-conotoxin plus nitrendipine (5 μM), and ω-conotoxin, nitrendipine, and ω-Aga-IVA (200 nM). Internal solution: 108 m*M* Cs-methanesulfonate, 4 m*M* $MgCl_2$, 9 m*M* EGTA, 9 m*M* HEPES, 4 m*M* MgATP, 14 m*M* creatine phosphate (Tris salt), 1 m*M* GTP (Tris salt), pH 7.4. External solution: 5 m*M* $BaCl_2$, 160 m*M* TEACl, 0.1 m*M* EGTA, 10 m*M* HEPES, pH 7.4. Reprinted with permission from Mintz *et al.* (1992a). Copyright © 1992 by Cell Press. (B) Collected dose–response data for block of high-threshold current in Purkinje neurons by ω-Aga-IVA. Reprinted with permission from Mintz *et al.* (1992b). Copyright © 1992 Macmillan Magazines Ltd.

neuron, using (+)-(S)-202-791 to enhance the small current activated by a depolarization to −50 mV. ω-Conotoxin had no effect on this current through L-type channels. The lack of effect of ω-conotoxin on L-type current has been reported in chick and rat sensory neurons (Aosaki and Kasai, 1989; Regan *et al.*, 1991) and in rat sympathetic neurons (Plummer *et al.*, 1989). In all cases, N-type current can be blocked completely with no effect on components of L-type current, identified by its enhancement by Bay-K 8644 or (+)-(S)-202-791.

Despite a preponderance of reports that ω-conotoxin does not block Bay-K 8644-enhanced tail currents, one report claims that the toxin does have some effect in some motor neurons (Mynlieff and Beam, 1992). Interestingly, a cloned neuronal L-type channel was reported to have a slight sensitivity to ω-conotoxin (Williams *et al.*, 1992a), although the block was about 3 orders of magnitude weaker (half-maximal dose >10 μM) than for N-type Ca^{2+} channels in neurons. Consistent with the idea that N-type Ca^{2+} channels can be identified by high-affinity ω-conotoxin block, when an N-type Ca^{2+} channel was cloned and expressed, it was found to be blocked very potently by ω-conotoxin (Williams *et al.*, 1992b).

VI. P-Type Channels: ω-Aga-IVA

Cerebellar Purkinje neurons possess an unusual high-threshold current that has very little contribution from L-type or N-type Ca^{2+} channels. As illustrated in Fig. 7A, dihydropyridines and ω-conotoxin block only a very small fraction of the high-threshold current in rat cerbellar Purkinje neurons. These observations support an earlier suggestion that Ca^{2+} channels in the dendrites of cerebellar Purkinje cells are phar-

macologically different from L-type and N-type channels. The name "P-type" channel was proposed for the predominant Ca^{2+} channels in Purkinje neurons (Llinas *et al.*, 1989).

A 48-amino-acid peptide toxin from the venom of *Agelenopsis aperta*, called ω-Aga-IVA, is a highly effective blocker of P-type Ca^{2+} channels in cerebellar Purkinje neurons (Mintz *et al.*, 1992a,b). Figure 7A shows an example in which 200 nM ω-Aga-IVA completely blocked the high-threshold current remaining in a rat Purkinje neuron after addition of nitrendipine and ω-conotoxin. ω-Aga-IVA is very potent, inhibiting P-type Ca^{2+} channels with a K_d of several nM (Fig. 7B).

ω-Aga-IVA is highly selective for blocking P-type

channels but not other channel types. At 100 nM, well above saturation for P-type channel block (Fig. 8A), this toxin has no effect on T-type current in peripheral neurons such as dorsal rat ganglion neurons (Fig. 8B) or in hippocampal or Purkinje neurons. This toxin is also completely ineffective in blocking Bay-K 8644-enhanced L-type current in neurons (Fig. 8C) and does not block L-type current in rat cardiac myocytes.

Rat sympathetic neurons primarily have current carried by N-type Ca^{2+} channels; ω-Aga-IVA has no effect on the current in sympathetic neurons from the superior cervical ganglion (Fig. 8D).

P-Type Ca^{2+} channels in Purkinje neurons have a somewhat different voltage dependence than L-type and N-type channels (Regan, 1991; Regan et al., 1991). For example, with solutions using 5 mM Ba^{2+} as the charge carrier, P-type Ca^{2+} channel current begins to activate at about -60 mV and reaches a peak at -30 or -20 mV (e.g., Fig. 9). In contrast, under the same ionic conditions, N-type current begins to activate near -40 mV and peaks at -10 or 0 mV. L-Type current has a somewhat shallower voltage dependence than either N-type or P-type current; it begins to activate near -50 mV and peaks near -10 mV. These differences are not dramatic, but they are consistent when examined under the same recording conditions. Figure 9 shows that, although the predominant P-type current in a particular Purkinje neuron reaches a peak at -30 mV, the much smaller current that remains in saturating concentrations of ω-Aga-IVA (a mixture of N-type and L-type current) peaks at -15 mV. This behavior is similar to the voltage dependence of the overall high-threshold current in rat sensory neurons or sympathetic neurons (Regan et al., 1991), where N-type and L-type current in combination make up a majority of the overall current.

Once block of P-type channels by ω-Aga-IVA has developed, it reverses only very slowly with washout of the toxin. However, recovery can be enhanced dramatically by a short series of large depolarizations. Figure 10 illustrates this effect: 200 nM ω-Aga-IVA was applied to a rat cerebellar Purkinje neuron and

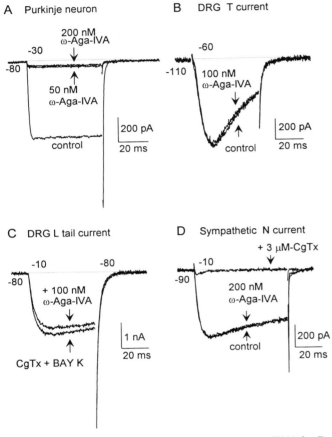

FIGURE 8 Discrimination of ω-agatoxin-IVA (ω-Aga-IVA) for P-type channels. (A) 50 nM ω-Aga-IVA maximally inhibits the high-threshold current in Purkinje neuron with no further effect of 200 nM. Current was elicited by a step from -80 to -30 mV. (B) T-type current in a rat DRG neuron is not blocked by 100 nM ω-Aga-IVA. The current was elicited by a step from -110 to -60 mV; it has the rapid inactivation and slow deactivation typical of T-type channels and was eliminated by holding at -70 mV. (C) Neuronal Bay-K 8644-enhanced L-type current is insensitive to 100 nM ω-Aga-IVA. High-threshold currents were recorded in a rat DRG neuron in the continuous presence of ω-conotoxin (2 μM) and Bay-K 8644 (1 μM). Current was elicited by a step from -80 to -10 mV. ω-Aga-IVA slightly reduced the current elicited at -10 mV and had no effect on the slowly deactivating tail current at -80 mV induced by Bay-K 8644. (D) Ca^{2+} channel current in a rat sympathetic neuron was unaffected by ω-Aga-IVA (200 nM) but was abolished by ω-conotoxin (3 μM). Solutions as in Fig. 7. Redrawn with permission from Mintz et al. (1992a). Copyright © 1992 by Cell Press.

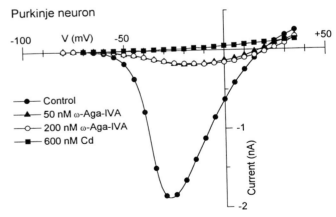

FIGURE 9 Voltage dependence of P-type current in a rat cerebellar Purkinje neuron. About 80% of the overall high-threshold current was blocked by ω-agatoxin-IVA (ω-Aga-IVA). Solutions as in Fig. 7. Control, ●; 50 nM ω-Aga-IVA, ▲; 200 nM ω-Aga-IVA, ○; 600 nM Cd^{2+}, ■.

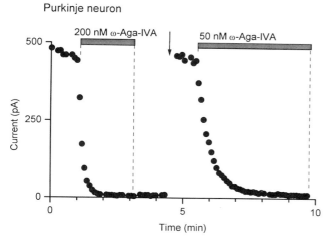

FIGURE 10 Depolarizing pulses alleviate ω-agatoxin-IVA (ω-Aga-IVA)-induced current block of P-type current in a cerebellar Purkinje neuron. Application of 200 n*M* ω-Aga-IVA produced complete inhibition of the current, with very little recovery in 1 min following removal of the toxin. Data points represent currents elicited by 60-msec steps from −90 to −30 mV, applied every 6 sec. After 1 min wash-out, a burst of 25 60-msec steps from −80 to +90 mV (repeated at 1-sec intervals within the burst) was delivered (arrow). This series of short depolarizations produced complete relief of block. The current was then reblocked by addition of 50 n*M* ω-Aga-IVA. Solutions as in Fig. 7.

inhibited the high-threshold current nearly completely. Washing for 1 min produced almost no recovery of the current. Then a series of 25 60-msec depolarizations to +90 mV was delivered (at 1 Hz) and resulted in complete recovery of the current (which was then reblocked by 50 n*M* ω-Aga-IVA). Large depolarizations greatly accelerate the unbinding of toxin from the channels. Quantitatively, the unbinding rate of toxin is accelerated by a factor of about 10^4 at +90 compared with −90 or −80 mV. Assuming that the ON rate for binding of toxin is not affected, this result predicts that the K_d for toxin block changes from a few n*M* at negative holding potentials to >10 μ*M* at very depolarized potentials, so 50–200 n*M* toxin would produce complete block at negative voltages but essentially no block at very depolarized potentials. Consistent with this hypothesis, strong depolarizations can temporarily reverse toxin block even in the continuous presence of toxin (Mintz *et al.*, 1992a). The molecular basis of the striking voltage dependence of block is not yet clear.

ω-Aga-IVA was originally identified by Adams as a potentially interesting Ca²⁺ channel blocker based on its ability to inhibit depolarization-induced Ca²⁺ entry into rat brain synaptosomes which, like the high-threshold Ca²⁺ channel current in Purkinje neurons, is affected very little by blockers of L-type or N-

type Ca²⁺ channels (Mintz *et al.*, 1992b). Since only a tiny fraction of rat brain synaptosomes would originate from Purkinje neurons, the ability of ω-Aga-IVA to inhibit ~70% of the synaptosomal Ca²⁺ flux suggested the existence of ω-Aga-IVA-sensitive channels in cells other than Purkinje neurons.

In fact, electrophysiological experiments have shown that many other types of neurons in addition to Purkinje neurons do possess P-type Ca²⁺ channels, as assayed by sensitivity to block by ω-Aga-IVA. Examples of several central neurons that possess P-type channels (as well as N-type and L-type channels) are shown in Fig. 11. In these neurons (an excitatory pyra-

A Hippocampal CA1

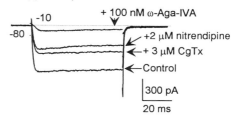

B Visual cortex

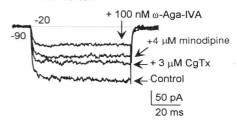

C Spinal cord

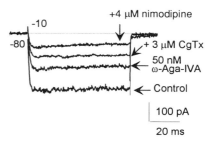

FIGURE 11 Components of high-threshold Ca²⁺ channel current sensitive to ω-agatoxin-IVA (ω-Aga-IVA), ω-conotoxin (ω-CgTx), and dihydropyridine blockers. (A) Hippocampal neuron. Currents were elicited by 60-msec depolarizations from −80 to −10 mV, delivered every 7 sec. (B) Visual cortical neuron. Steps from −90 to −20 mV (60-msec duration) were delivered every 6 sec. Drugs were applied with sequential addition of 3 μ*M* ω-conotoxin, 2 μ*M* nitrendipine, or 4 μ*M* nimodipine, and 100–200 n*M* ω-Aga-IVA. (C) Spinal interneuron. Sequential application of 50 n*M* ω-Aga-IVA, 3 μ*M* ω-conotoxin, and 4 μ*M* nimodipine. Current was elicited every 6 sec by a 60-msec step from −80 to −10 mV. Solutions as in Fig. 7. Reprinted with permission from Mintz *et al.* (1992a). Copyright © 1992 by Cell Press.

midal neuron from the hippocampal CA1 region, a pyramidal neuron from the visual cortex, and an inhibitory interneuron from the spinal cord), ω-Aga-IVA, ω-contoxin, and dihydropyridine durgs block a fraction of the overall high-threshold current. Interestingly, in each of the neurons in Fig. 11, current remains even in the combined presence of all three blockers. Assuming that each blocker produces complete inhibition of its target channels (as can be demonstrated in at least some cases), other channels in addition to L-type, N-type, and P-type channels are contributing to the current.

VII. Coexistence of Channel Types and Undefined Channel Types

Most rat neurons have at least three or four distinct types of Ca^{2+} channels that contribute to the overall current, but large differences are seen in different types of neurons in the relative contributions of different channel types. As already mentioned, only a relatively small fraction of neurons possesses T-type Ca^{2+} channels. Using dihydropyridines to identify L-type current, ω-conotoxin to identify N-type current, and ω-Aga-IVA to identify P-type current, the contribution of these channel types to overall high-threshold current can be assessed (Table I). Virtually every neuron possesses at least some L-type current and at least some N-type current, although in sympathetic neurons the component of L-type current is small and in Purkinje neurons both L-type and N-type channels

contribute only small fractions. With the exception of sympathetic neurons and some CA3 pyramidal neurons, most neurons possess a component of P-type current. As illustrated in Fig 11, in many neurons, application of saturating concentrations of nimodipine, ω-conotoxin, and ω-Aga-IVA still leaves some high-threshold Ca^{2+} channel current unblocked.

The relative contribution of L-type, N-type, and P-type channels to the overall high-threshold current varies widely among different types of neurons. Table I shows the contribution of the current components sensitive to ω-Aga-IVA, ω-conotoxin, or dihydropyridines in a variety of neurons. Purkinje neurons and sympathetic neurons are unusual because they possess nearly pure types of current: 90% P-type in Purkinje neurons and 90% N-type in sympathetic neurons. All other cell types possess significant contributions from two or more channel types, most often including significant contributions from L-type, N-type, and P-type channels as well as from the still-undefined channel types that make up the current that is resistant to combined application of ω-Aga-IVA, ω-conotoxin, and dihydropyridines.

VIII. Future Directions

Perhaps the most pressing immediate issue is relating the distinct components of current in central neurons to particular types of channels identified by cloning. The correlation is fairly clear for at least some N-type and L-type channels, but remains un-

TABLE I   **Pharmacological Components of High-Threshold Ca^{2+} Channel Current**[a,b]

| Neuron type | Percentage of control current inhibited | | | |
	ω-Aga-IVA	ω-Conotoxin GVIA	Nitrendipine or nimodipine	Remaining
Purkinje	92 ± 2 (29)	5 ± 1 (19)	5 ± 2 (18)	0 (8)
Hippocampal CA1	26 ± 4 (20)	37 ± 3 (21)	19 ± 3 (20)	Present
Hippocampal CA3	14 ± 4 (13)	21 ± 2 (10)	36 ± 6 (8)	25 ± 6 (6)
Visual cortex	32 ± 4 (11)	32 ± 7 (6)	23 ± 6 (6)	Present
Spinal cord	45 ± 6 (7)	34 ± 3 (3)	18 ± 5 (3)	Present
Rat DRG	23 ± 4 (15)	43 ± 3 (19)	18 ± 2 (19)	18 ± 4 (10)
Rat sympathetic	0 (7)	93 ± 4 (6)	7 ± 4 (6)	2 ± 2 (6)

[a] Reproduced with permission from Mintz et al. (1992a). Copyright © 1992 by Cell Press.
[b] Fraction of high-threshold current contributed by pharmacologically defined P-, N-, and L-type Ca^{2+} channels in different types of neurons. P-type, N-type, and L-type currents were defined as the percentage of control current (mean ± SEM, number of cells in parentheses) sensitive to ω-Aga-IVA (100–200 nM), ω-conotoxin (3–5 μM), and dihydropyridine blockers (2–5 μM nitrendipine or 2–4 μM nimodipine). Measurements of current blocked by each drug were pooled independently of its order in the sequence of multiple applications. The last column reports current remaining unblocked in the combined presence of all three blockers.

clear for P-type channels and for the channels that remain after blocking L-type, N-type, and P-type channels. As more is learned about all the classes of channels, distinct members of each family are likely to emerge, perhaps with distinguishing pharmacological characteristics. Pharmacology with new toxins or small synthetic compounds should be useful in making further connections between native and cloned channels. Because of the complex mixtures of channel types in most neurons, studies of the pharmacology of individual channel types can be expected to gradually shift to systems using expression of individual channel types.

Much remains to be done in developing Ca^{2+} channel pharmacology using small synthetic compounds. Only for L-type channels are potent, selective synthetic blockers available. Developing such blockers for N-type and P-type channels would be of great interest. A good screening assay for blockers of P-type channels would be inhibition of high K^+-stimulated $^{45}Ca^{2+}$ flux into rat brain synaptosomes, the assay by which Adams identified ω-Aga-IVA as a potential P-type channel blocker (Mintz *et al.*, 1992b).

References

Akaike, N., Kostyuk, P. G., and Osipchuk, Y. V. (1989). Dihydropyridine-sensitive low-threshold calcium channels in isolated rat hypothalamic neurones. *J. Physiol. (London)* **412**, 181–195.

Aosaki, T., and Kasai, H. (1989). Characterization of two kinds of high-voltage-activated Ca-channel currents in chick sensory neurons. Differential sensitivity to dihydropyridines and omega-conotoxin GVIA. *Pflügers. Arch.* **414**, 150–156.

Bean, B. P. (1984). Nitrendipine block of cardiac calcium channels: High-affinity binding to the inactivated state. *Proc. Natl. Acad. Sci. USA* **81**, 6388–6392.

Bean, B. P. (1985). Two types of calcium channels in canine atrial cells. Differences in kinetics, selectivity, and pharmacology. *J. Gen. Physiol.* **86**, 1–30.

Bean, B. P. (1989). Classes of calcium channels in vertebrate cells. *Annu. Rev. Physiol.* **51**, 367–384.

Bean, B. P. (1991). Pharmacology of calcium channels in cardiac muscle, vascular muscle, and neurons. *Am. J. Hypertension* **4**, 406S–411S.

Boll, W., and Lux, H. D. (1985). Action of organic antagonists on neuronal calcium currents. *Neurosci. Lett.* **56**, 335–339.

Carbone, E., and Lux, H. D. (1984). A low-voltage-activated, fully inactivating Ca channel in vertebrate sensory neurons. *Nature (London)* **310**, 501–502.

Carbone, E., and Lux, H. D. (1987). Single low-voltage-activated calcium channels in chick and rat sensory neurones. *J. Physiol. (London)* **386**, 571–601.

Carbone, E., Formenti, A., and Pollo, A. (1990). Multiple actions of Bay K 8644 on high-threshold Ca channels in adult rat sensory neurons. *Neurosci. Lett.* **111**, 315–320.

Coulter, D. A., Huguenard, J. R., and Prince, D. A. (1989a). Calcium currents in rat thalamocortical relay neurones: Kinetic properties of the transient, low-threshold current. *J. Physiol. (London)* **414**, 587–604.

Coulter, D. A., Huguenard, J. R., and Prince, D. A. (1989b). Characterization of ethosuximide reduction of low-threshold calcium current in thalamic relay neurons. *Ann. Neurol.* **25**, 582–593.

Fox, A. P., Nowycky, M. C., and Tsien R. W. (1987a). Single-channel recordings of three types of calcium channels in chick sensory neurones. *J. Physiol. (London)* **394**, 173–200.

Fox, A. P., Nowycky, M. C., and Tsien R. W. (1987b). Kinetic and pharmacological properties distinguishing three types of calcium currents in chick sensory neurones. *J. Physiol. (London)* **394**, 149–172.

Hagiwara, N., Irisawa, H., and Kameyama, M. (1988). Contribution of two types of calcium channels to the pacemaker potentials of rabbit sinoatrial node cells. *J. Physiol. (London)* **395**, 233–252.

Herrington, J., and Lingle, C. J. (1992). Kinetic and pharmacological properties of low voltage-activated Ca^{2+} current in rat clonal (GH$_3$) pituitary cells. *J. Neurophysiol.* **68**, 213–232.

Hess, P. (1990). Calcium channels in vertebrate cells. *Annu. Rev. Neurosci.* **13**, 337–356.

Hess, P., Lansman, J. B., and Tsien, R. W. (1984). Different modes of calcium channel gating behaviour favoured by dihydropyridine agonists and antagonists. *Nature (London)* **311**, 538–544.

Huguenard, J. R., and Prince, D. A. (1992). A novel Ty-type current with slowed inactivation contributes to prolonged Ca^{2+}-dependent burst firing in GABAergic neurons of rat thalamic reticular nucleus. *J. Neurosci.* **12**, 3804–3817.

Hullin, R., Biel, M., Flockerzi, V., and Hofmann, F. (1993). Tissue specific expression of calcium channels. *Trends Cardiovasc. Med.* **3**, 48–53.

Hume, J. R. (1985). Comparative interactions of organic Ca^{++} channel antagonists with myocardial Ca^{++} and K^+ channels. *J. Pharmacol. Exp. Ther.* **234**, 134–140.

Llinas, R., Sugimori, M., Lin, J. W., and Cherksey, B. (1989). Blocking and isolation of a calcium channel from neurons in mammals and cephalopods utilizing a toxin fraction (FTX) from funnel-web spider poison. *Proc. Natl. Acad. Sci. USA* **86**, 1689–1693.

McCarthy, R., and Tanpiengco, P. (1992). Multiple types of high-threshold calcium channels in rabbit sensory neurons: High-affinity block of neuronal L-type by nimodipine. *J. Neurosci.* **12**, 2225–2234.

McCleskey, E. W., Fox, A. P., Feldman, D. H., Cruz, L. J., Olivera, B. M., Tsien, R. W., and Yoshikami, D. (1987). Omega-conotoxin: direct and persistent blockade of specific types of calcium channels in neurons but not muscle. *Proc. Natl. Acad. Sci. USA* **84**, 4327–4331.

Miller, R. J. (1992). Voltage sensitive Ca^{2+} channels. *J. Biol. Chem.* **267**, 1403–1406.

Mintz, I. M., Adams, M. E., and Bean, B. P. (1992a). P-type calcium channels in central and peripheral neurons. *Neuron* **9**, 85–95.

Mintz, I. M., Venema, V. J., Swiderek, K., Lee, T., Bean, B. P., and Adams, M. E. (1992b). P-type calcium channels blocked by the spider toxin w-Aga-IVA. *Nature (London)* **355**, 827–829.

Mynlieff, M., and Beam, K. G. (1992). Characterization of voltage-dependent calcium currents in mouse motorneurons. *J. Neurophysiol.* **68**, 85–92.

Nowycky, M. C., Fox, A. P., and Tsien, R. W. (1985). Long-opening mode of gating of neuronal calcium channels and its promotion by the dihydropyridine calcium agonist Bay K 8644. *Proc. Natl. Acad. Sci. USA* **82**, 2178–2182.

Olivera, B. M., Gray, W. R., Zeikus, R., McIntosh, J. M., Varga, J., Rivier, J., de Santos, V., and Cruz, L. J. (1985). Peptide neurotoxins from fish-hunting cone snails. *Science* **230**, 1338–1343.

Plummer, M. R., Logothetis, D. E., and Hess, P. (1989). Elementary properties and pharmacological sensitivities of calcium channels in mammalian peripheral neurons. *Neuron* **2**, 1453–1463.

Regan, L. J. (1991). Voltage-dependent calcium currents in Purkinje cells from rat cerebellar vermis. *J. Neurosci.* **11**, 2259–2269.

Regan, L. J., Sah, D. W., and Bean, B. P. (1991). Ca²⁺ channels in rat central and peripheral neurons: High-threshold current resistant to dihydropyridine blockers and omega-conotoxin. *Neuron* **6**, 269–280.

Sanguinetti, M. C., and Kass, R. S. (1984). Voltage-dependent block of calcium channel current in the calf cardiac Purkinje fiber by dihydropyridine calcium channel antagonists. *Circ. Res.* **55**, 336–348.

Snutch, T. P., and Reiner, P. B. (1992). Ca²⁺ channels: Diversity of form and function. *Curr. Opin. Neurobiol.* **2**, 247–253.

Tang, C. M., Presser, F., and Morad, M. (1988). Amiloride selectively blocks the low threshold (T) calcium channel. *Science* **240**, 213–215.

Tsien, R. W., Ellinor, P. T., and Horne, W. A. (1991). Molecular diversity of voltage-dependent Ca²⁺ channels. *Trends Pharmacol. Sci.* **12**, 349–354.

Uehara, A., and Hume, J. R. (1985). Interactions of organic calcium channel antagonists with calcium channels in single frog atrial cells. *J. Gen. Physiol.* **85**, 621–647.

Williams, M. E., Feldman, D. H., McCue, A. F., Brenner, R., Velicelebi, G., Ellis, S. B., and Harpold, M. M. (1992a). Structure and function of α1, α2, and β subunits of a novel human neuronal calcium channel subtype. *Neuron* **8**, 71–84.

Williams, M. E., Brust, P. F., Feldman, D. H., Saraswathi, P., Simerson, S., Maoufi, A., McCue, A. F., Velicelebi, G., Ellis, S. B., and Harpold, M. M. (1992b). Structure and functional expression of an ω-conotoxin-sensitive human N-type calcium channel. *Science* **257**, 389–395.

Yatani, A., and Brown, A. M. (1985). The calcium channel blocker nitrendipine blocks sodium channels in neonatal rat cardiac myocytes. *Circ. Res.* **56**, 868–875.

Chloride Channels

Voltage-Dependent Kinetics of the Fast Chloride Channel

DAVID S. WEISS

I. Introduction

In 1985, Blatz and Magleby described a voltage-dependent Cl⁻ channel in cultured myotubes that they called the fast Cl⁻ channel because of its rapid gating kinetics. The fast Cl⁻ channel has since been described in cultured neurons as well (Blatz, 1991). The fast Cl⁻ channel is active at resting membrane potentials and, thus, may contribute to the resting Cl⁻ conductance in these tissues, although this function remains to be demonstrated.

The fast Cl⁻ channel has several features that make it amenable to a detailed analysis of voltage-dependent ion channel gating kinetics. (1) Patches with only a single active fast Cl⁻ channel can be obtained from cultured myotubes. The presence of multiple active channels, especially when the gating kinetics are complex, can greatly compromise the analysis. (2) The fast Cl⁻ channel does not inactivate. For channels that inactivate, the membrane potential must be pulsed and less data can be collected than from a single fast Cl⁻ channel. (3) The channel remains stable for long periods of time, also increasing the amount of data that can be collected. In some cases, over 1×10^6 open and shut intervals have been analyzed for a single active fast Cl⁻ channel. These large data sets

increase the ability to distinguish between putative gating mechanisms.

This chapter describes the voltage-dependent gating of the fast Cl⁻ channel and presents a gating mechanism that can account for the kinetics of the fast Cl⁻ channel.

II. Voltage Dependence of Single Fast Chloride Channels

A. Single-Channel Currents

Figure 1 shows single-channel currents, at several membrane potentials, from an inside-out patch of membrane (Hamil et al., 1981) obtained from a cultured rat myotube (Barrett et al., 1981). Note how the frequency of channel opening (channels open downward) increases with depolarization of the membrane (Fig. 1A). Figure 1B shows the single-channel currents expanded so the individual openings and closings can be discerned more easily.

B. Voltage Dependence of P_{open}

To quantify the voltage dependence, the percentage of time the channel spends open (P_{open}) can be

A

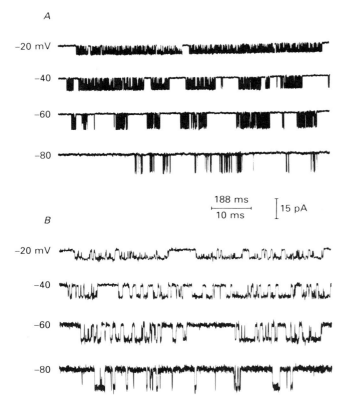

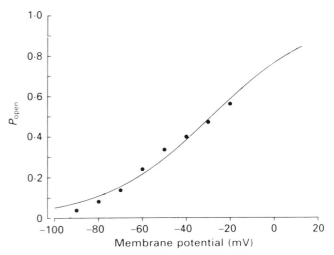

FIGURE 2 Depolarization increases P_{open} as shown in a plot of the fraction of time spent open, P_{open}, against membrane potential. The line is a least-squares fit with Eq. 1 to the data.

FIGURE 1 Currents through a single fast Cl⁻ channel in an excised membrane patch from cultured rat skeletal muscle. Data are presented for four different membrane potentials on slow (A) and fast (B) time bases. Inward (downward) current steps indicate channel opening.

determined at each membrane potential (total open time divided by total observation time), as shown in Fig. 2. The relationship between membrane potential and P_{open} can be described by a Boltzmann distribution (Hodgkin and Huxley, 1952; Hille, 1984) as

$$P_{open} = B \exp(qV/kT), \qquad (1)$$

where B is a constant, q is the effective gating charge, V is the membrane potential, k is the Boltzmann constant, and T is the absolute temperature. The effective gating charge, q, is defined as

$$q = zed, \qquad (2)$$

where z is the valency of the charge, e is the value of an electronic charge (1.6×10^{-19} C), and d is the fractional distance the charge moves in the membrane field. Since z and d cannot be determined separately in these experiments, they are combined into an effective gating charge.

The solid line in Fig. 2 is the best fit of Eq. 1 to the

filled circles. [Actually, $P_{open}/(1-P_{open})$ versus voltage was fitted because this weighted the data in a manner comparable to fits to be presented later.] The effective gating charge was 1.24 electronic charges. The mean ± SD of the effective gating charge from five single fast Cl⁻ channels was 1.65 ± 0.28.

Note that systematic deviations occur between the fitted line and the data in Fig. 2. Namely, the fitted curve overestimates P_{open} at hyperpolarized and depolarized potentials and underestimates P_{open} at the intermediate membrane potentials. These systematic deviations were observed in all five channels and were also observed by Blatz (1991) in studies of the voltage dependence of the fast Cl⁻ channel in cultured neurons. Some deviation is expected, since Eq. 1 assumes a two-state model of channel gating. As will be demonstrated shortly, the assumption of two kinetic states is an oversimplification.

Nevertheless, an effective gating charge of 1.24 implies that 1.24 charges must move through the entire membrane field for the channel to open and close. Since the effective gating charge includes d, 2.48 charges can move halfway through the membrane field ($2.48 \times 0.5 = 1.24$). Concerning the sign of the effective gating charge, voltage-dependent gating could result from a net positive charge moving in the membrane field from in to out or a net negative charge moving in the membrane field from out to in.

The most useful purpose of determining the effective gating charge in Fig. 2 is that it serves as a means of comparing the voltage dependence of different channels. For example, the sodium channel in squid optic lobe has an effective gating charge of 5.4 (Behrens *et al.*, 1989) and is much more voltage sensitive than the fast Cl^- channel. The effective gating charge of the dimeric Cl^- channel in electroplax is 1.1 (Hanke and Miller, 1983), a value very similar to that of the fast Cl^- channel.

C. Voltage Dependence of the Mean Open and Mean Shut Time

As mentioned, the net effective gating charge determined in Fig. 2 is the net amount of charge moved in an open → closed → open transition. This charge could result from a voltage dependence of channel opening, a voltage dependence of channel closing, or both. Figure 3 demonstrates that both channel opening and channel closing are voltage dependent. The filled circles in Fig. 3 are the mean shut times and the open circles are the mean open times plotted as a function of voltage. The solid and dashed lines are linear regressions to these semilogarithmic plots, implying a relationship of the form

$$Y = B \exp(AV), \tag{3}$$

where Y is the mean open or mean shut time, V is the membrane potential, B is a constant, and A is a measure of the voltage sensitivity expressed as the fractional change in Y per millivolt change in membrane potential. The inverse of A is the number of millivolts required for an e-fold (2.72-fold) change in Y. The effective gating charge for channel closing can

be calculated from the voltage dependence of the mean open time and the effective gating charge for channel opening can be calculated from the voltage dependence of the mean shut time, using the relationship

$$q = AkT. \tag{4}$$

For the data in Fig. 3, the effective gating charge for channel opening was 1.00 and the effective gating charge for channel closing was −0.23. The net effective gating charge for opening and closing would be 1.00 − (−0.23) or 1.23, which is in good agreement with 1.24 determined in Fig. 2. Thus, channel opening and channel closing are both voltage dependent, but channel opening is more voltage sensitive.

D. Voltage Dependence of the Number of Observed Open and Shut States

Thus far, a two-state gating model has been assumed. Evidence indicates that the fast Cl^- channel can exist in as many as eight kinetic states (Blatz and Magleby, 1986; Weiss and Magleby, 1990). Knowing whether the number of observed kinetic states changes with voltage is useful. The potential number of kinetic states would not be expected to be voltage dependent, because this condition would assume an infinite effective gating charge in one of the transition pathways (Weiss and Magleby, 1992). However, a voltage dependence of the rates could allow a previously undetected state to be detected.

Information on the minimum number of open and closed states can be determined from the minimum number of exponential components required to describe distributions of open and shut dwell times (Colquhoun and Sigworth, 1983). Figure 4A is a distribution of open dwell times that was described by the sum of two exponential components (solid line). Figure 4B is a shut dwell-time distribution that was described by the sum of six exponential components (solid line). The dashed lines plot the individual components that sum to give the solid line.

Figure 5 plots the number of exponential components required to fit the open (Fig. 5A) and shut (Fig. 5B) dwell-time distributions for five fast Cl^- channels (plotted as different symbols) as a function of voltage. Although some deviations in the estimated number of states were seen, no consistent voltage dependence was detected in the number of exponentials required to fit the distributions. Thus, a kinetic model with two open states and six shut states should be sufficient at all voltages.

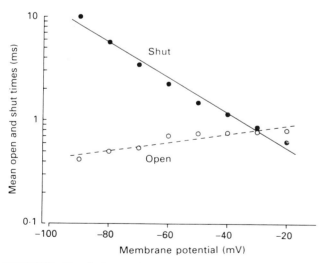

FIGURE 3 Depolarization decreases mean shut time and increases mean open time. The lines are a linear regression to the logarithm of the data (Eq. 3).

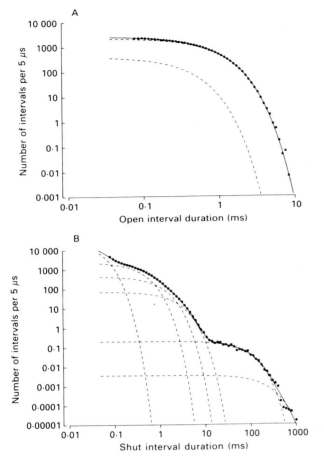

FIGURE 4 Distributions of open and shut dwell times at −40 mV. The continuous lines are the maximum likelihood fitted sums of two exponential components for the open distribution (A) and of six exponential components for the shut distribution (B). The individual components are indicated by the dashed lines.

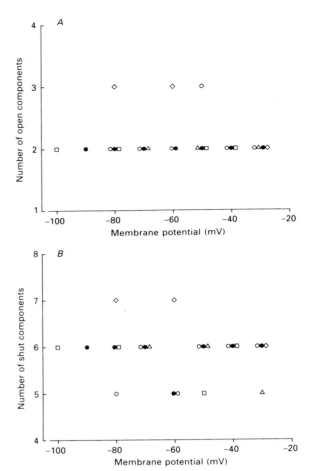

FIGURE 5 The number of kinetic states appears to be independent of membrane potential. Open and shut dwell-time distributions were obtained at the indicated membrane potentials for five channels. Each distribution was fitted with sums of exponentials to determine the number of significant components. (A) The number of open exponential components are plotted against membrane potential. (B) A similar plot is presented for the number of shut components. The channels typically enter at least two open and six shut states over the examined range of membrane potentials.

III. Kinetic Stability

Ion channels can frequently change their kinetic properties over time (Patlak *et al.*, 1979; Moczydlowski and Latorre, 1983; Hess *et al.*, 1984; McManus and Magleby, 1988). As discussed in Section VI, the fast Cl⁻ channel can also display different gating modes. An instability in gating or alternative gating modes can introduce errors into the analysis.

One method by which to test for kinetic stability is to divide the data into segments and to examine the kinetics of these segments of data individually. For the channel shown in Figs. 1 and 2, the data at −40 mV were divided into 11 consecutive groups of 50,000 intervals each. Figure 6 shows the 11 consecutive open (Fig. 6A) and shut (Fig. 6B) dwell-time distri-

butions. The solid lines are the best fit sum of two and five exponential components for the open and shut distributions, respectively. Only five components, not six, were required to describe the shut interval distribution since fewer intervals were being fitted in these subsets of data and the longest component was no longer detected.

The open bars in Fig. 7 plot the mean ± SD of the parameters (time constants and areas) of the fitted sums of exponentials in Fig. 6. The hatched bars in Fig. 7 are from an analysis similar to that in Fig. 6, but with data simulated by a seven-state kinetic model (Scheme I without C_8). The variability for the experi-

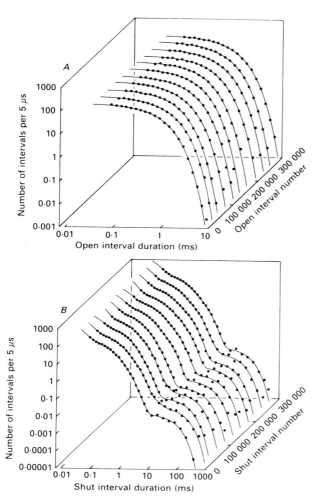

FIGURE 6 Testing for kinetic stability with consecutive dwell-time distributions. The 585,000 intervals collected at −40 mV and analyzed for Fig. 4 were divided into 11 consecutive groups of 50,000 intervals, binned into frequency histograms, and plotted on log–log coordinates. (A) The open dwell-time distributions were fitted (maximum likelihood) with sums of two exponential components. (B) The shut dwell-time distributions were fitted with five (continuous lines).

$$C_8 \rightleftharpoons C_7 \rightleftharpoons C_6 \rightleftharpoons C_5 \rightleftharpoons C_4 \rightleftharpoons C_3$$

$$\Updownarrow \qquad \Updownarrow$$

$$O_2 \qquad O_1$$

SCHEME I

In Scheme I, the voltage dependence of channel gating would arise from a voltage dependence of the transition rates between these eight kinetic states. The following section addresses the form of this voltage dependence.

A. Transition Rates Depend Exponentially on Voltage

Using the Q-matrix method of Colquhoun and Hawkes (1981), the dwell-time distributions can be determined from any kinetic scheme and transition rates. The Q-matrix method can be incorporated into an optimization routine to determine the most likely set of rate constants for a given scheme from the open and shut dwell-time distributions. This optimization procedure is diagrammed in Fig. 8. First, open and shut dwell-time distributions are constructed from the experimental data, as in Fig. 4. Next, a gating mechanism and some starting set of rate constants are assumed. The Q-matrix method is used to predict the dwell-time distributions from the proposed scheme; these predicted dwell-time distributions are compared with the experimental distributions using maximum likelihood fitting techniques (Colquhoun and Sigworth, 1983). A search routine is employed (see PatternSearch in Colquhoun, 1971) to find the most likely set of rate constants. One hundred to two hundred passes around the loop are usually sufficient to find the most likely set of rate constants.

The method in Fig. 8 was performed on the data at each voltage for Scheme I. Examination of the relationship between each of the rates and voltage should provide insights into the form of the voltage dependence of the transition rates. Figure 9 shows semilogarithmic plots of the 14 rate constants as a function of voltage. In most cases, the data are well described by straight lines, suggesting the exponential relationship between the transition rate and voltage

$$\text{rate}_{i \to j} = B \exp(AV), \qquad (5)$$

where $\text{rate}_{i \to j}$ is the rate in going from state i to state j and the other parameters have the same meaning as in Eq. 3.

mental data was similar to the variability of the simulated data generated with a stable kinetic model. This analysis, and other tests of stability (Weiss and Magleby, 1990), insured that the analyzed data were stable and represented a single kinetic mode.

IV. Voltage-Dependent Gating Mechanisms

Scheme I can account for the gating of the fast Cl⁻ channel at any single membrane potential (Blatz and Magleby, 1986,1989; Weiss and Magleby, 1992).

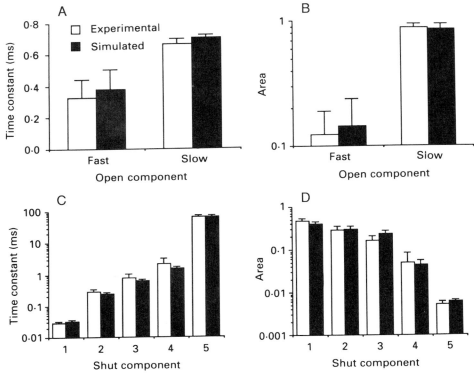

FIGURE 7   The variability in the consecutive dwell-time distributions arises mainly from stochastic variation. (A,B) Plot of experimental and simulated means and standard deviations of the time constants and areas describing the two exponential components fitted to the 11 consecutive open time distributions in Fig. 6A and to the 11 consecutive simulated open time distributions. (C,D) Plot of experimental and simulated means and standard deviations of the time constants and areas describing the six exponential components fitted to the 11 consecutive shut time distributions in Fig. 6B and to the 11 consecutive simulated shut time distributions. Each of the experimental and simulated distributions contained 25,000 intervals. The simulated data were generated using Scheme I with State C8 excluded.

B. Optimizing the Parameters

To optimize the rate constants and their voltage dependence for Scheme I, the open and shut dwell-time distributions at all membrane potentials are fit simultaneously. This procedure is similar to Fig. 8 except Eq. 5 describes each rate constant; thus, two free parameters (B and A) determine each of the 14 transition rates. Each time a rate is changed in the loop, the open and shut dwell-time distributions are determined at each of the experimental membrane potentials; these determinations are all compared with the experimental open and shut dwell-time distributions. The individual likelihoods are then summed to yield a total likelihood value. For a typical data set, convergence of these simultaneous fits requires 1–2 wk continuous computation using an XP accelerator board (Cheshire Engineering, Pasadena, California) housed in a DEC 11/73 computer.

Table I lists the parameters determined for Scheme I using the method just outlined. Figure 10 demon-strates that this model gives an excellent description of the open and shut dwell-time distributions. Dwell-time distributions are shown at three representative membrane potentials. In each case, the solid line is the best fit sum of exponentials and would represent the theoretical best fit (assuming sums of exponentials). The barely visible dotted lines are the predictions of the dwell-time distributions, assuming the relationship between voltage and rates described by the solid lines in Fig. 9, that is, before optimization by simultaneously fitting the data at all membrane potentials. The dashed lines are the predictions of the distributions after optimization (parameters given in Table I).

Figure 11 is a plot of the parameters determined for all five channels. The rate constants (Fig. 11B) are presented (at −50 mV) to be in the range of the experimental membrane potentials. Most of the rates are voltage sensitive (Fig. 11A), implying that most of the transitions between the states move charge in the same or opposite direction of the membrane field.

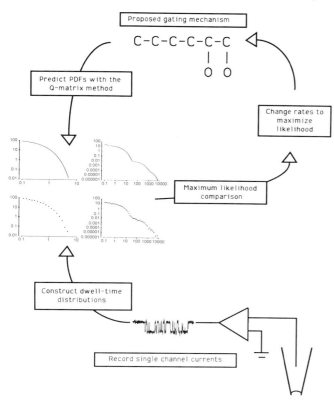

FIGURE 8 Flow chart of the Q-matrix fitting method. Open and shut dwell-time distributions were constructed from experimental data (*bottom*). Dwell-time distributions are predicted with the Q-matrix method of Colquhoun and Hawkes (1981) from the gating scheme to be tested and an initial guess of the rate constants (*top*). The experimental and predicted distributions are compared with maximum likelihood fitting techniques and the loop is repeated using an optimization routine to find the most likely set of rate constants for the proposed scheme.

membrane field between the indicated positions P_i and P_j. P_i and P_j represent the positions of the charge for states i and j; the asterisk represents the position of the charge at the transition state. The formation of the transition state is assumed to be the rate-limiting step between any two states (Glasstone *et al.*, 1941). In Fig. 12A, the transition state between states i and j is formed by moving a positive charge in the same or opposite direction with respect to the membrane field, although in different amounts. Hence, positive and unequal effective gating charges are seen for transitions in either direction between states i and j. Figure 12B is a case in which the effective gating charges are equal and opposite; Fig. 12C is a case in which the transition from state i to state j has a positive effective gating charge whereas the opposite transition (state j to state i) is voltage independent.

C. How the Channel Works

Figure 13 shows the equilibrium occupancies for all eight states in Scheme I as functions of membrane potential. At hyperpolarized membrane potentials, the channel spends most of the time in C_8 and C_7, the two longest-lived shut states. Depolarization drives the channel toward the open states, so at depolarized potentials the channel spends an increasing amount of time in C_4, C_3, O_2, and O_1, yielding the observed depolarization-dependent increase in P_{open}. The increased amount of time spent in these four states also produces the long flickering bursts evident in the data at the more depolarized potentials in Fig. 1.

V. Testing Other Models

Scheme I, with the parameters in Table I, gave excellent descriptions of the open and shut dwell-time distributions. This scheme, however, had a large number of free parameters, so it could be argued that (1) 28 free parameters cannot be defined adequately, (2) models with less than 28 free parameters might describe the data as well, and (3) almost any system can be described with 28 free parameters. Regarding the first argument, proving that an alternative solution (set of rates) does not exist is difficult. Refitting the data with a different starting set of parameters was performed several times for each data set. This process revealed that most of the parameters were well defined whereas a few of the parameters were less well defined. Methods for better defining these parameters are discussed in the summary of this chapter. The second and third arguments can best be ad-

Scheme II includes the effective gating charges for each of the transitions.

$$C_8 \underset{0.00}{\overset{0.85}{\rightleftharpoons}} C_7 \underset{0.22}{\overset{0.39}{\rightleftharpoons}} C_6 \underset{0.63}{\overset{0.27}{\rightleftharpoons}} C_5 \underset{-1.08}{\overset{0.00}{\rightleftharpoons}} C_4 \underset{-1.28}{\overset{0.00}{\rightleftharpoons}} C_3$$

$$\underset{0.00}{\overset{0.00}{\Updownarrow}} \quad \underset{0.00}{\overset{0.00}{\Updownarrow}} \quad \underset{-0.88}{\overset{0.00}{\Updownarrow}}$$

$$O_2 \qquad O_1$$

SCHEME II

At first, that the effective gating charges for transitions between any two states are not equal and opposite might seem strange. For instance, if -0.88 charges are moved going from state C_3 to state O_1, then $+0.88$ charges might be expected to be moved going back to state C_3 from state O_1. Figure 12 shows that this need not be the case. In each of the three scenarios presented, a positive charge is moved through the

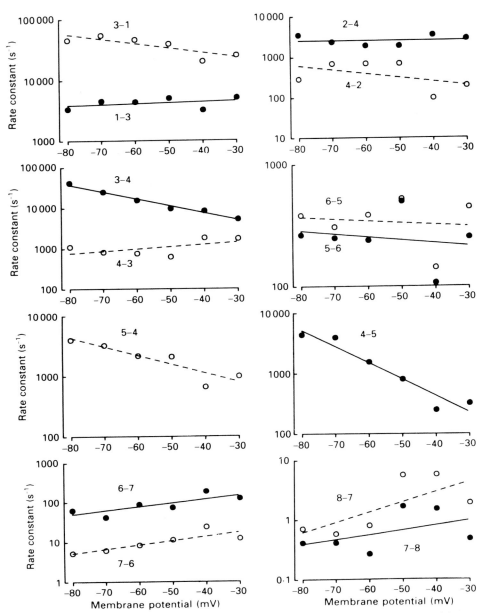

FIGURE 9 Apparent exponential dependence of rate constants on voltage. Estimates of the rate constants are plotted semilogarithmically against membrane potential. The plotted estimates of the rate constants were determined separately for each voltage by fitting the open and shut dwell-time distributions at that voltage with Scheme I. The lines are regression fits to the estimated rate constants. The open symbols and dashed regression lines indicate the forward rate constants for the indicated transition barriers; the filled symbols and continuous regression lines indicate the backward rate constants.

dressed by testing alternative, and simpler, gating mechanisms.

A. Models with Fewer Kinetic States

Schemes similar to Scheme I, but with fewer kinetic states, were tested for their ability to describe the data. Figure 14 summarizes the results from testing schemes with different numbers of open and shut states. The log relative error in Fig. 14 is the difference in the Schwarz criterion between Scheme I and the schemes with fewer kinetic states. The Schwarz criterion (Schwarz, 1978) can be used to rank models and is useful in this case because it applies a heavy penalty for increasing the number of free parameters. The lower the Schwarz criterion, the higher the ranking

TABLE I   **Parameters for Scheme I**

Rate	q^a	B^b
1–3	0.00	10700
2–4	0.00	1090
3–1	−0.876	82600
3–4	−1.28	14500
4–2	0.00	267
4–3	0.00	1880
4–5	−1.08	2020
5–4	0.00	3620
5–6	0.626	458
6–5	0.273	456
6–7	0.224	62.1
7–6	0.389	8.65
7–8	0.00	0.424
8–7	0.848	1.61

[a] Effective gating charge.
[b] Rate at −50 mV.

of the model. Figure 14 shows that decreasing the number of kinetic states lowers the ranking of the model. Thus, six shut and two open states appear necessary to adequately describe the voltage-dependent gating of the fast Cl^- channel.

B. Models with Constrained Voltage Dependence

As discussed in Fig. 12, no reason exists to assume the effective gating charges to be equal and opposite for transitions between any two states. Such a simplifying assumption, however, would place constraints on possible gating mechanisms and would reduce the number of free parameters.

In Scheme III, the effective gating charges are assumed to be equal and opposite for the forward and backward transitions, but the effective gating charge could differ for the seven transition barriers.

$$C_8 \underset{-q_1}{\overset{q_1}{\rightleftharpoons}} C_7 \underset{-q_2}{\overset{q_2}{\rightleftharpoons}} C_6 \underset{-q_3}{\overset{q_3}{\rightleftharpoons}} C_5 \underset{-q_4}{\overset{q_4}{\rightleftharpoons}} C_4 \underset{-q_5}{\overset{q_5}{\rightleftharpoons}} C_3$$

$$O_2 \qquad O_1$$

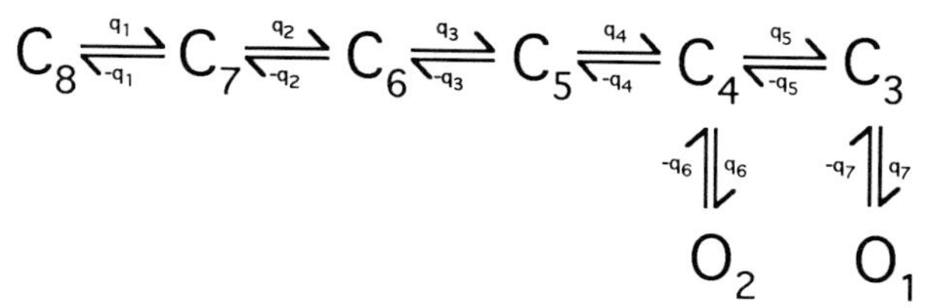

SCHEME III

Scheme IV has the additional constraint that the effective gating charge is the same for all seven transition barriers.

$$C_8 \underset{-q}{\overset{q}{\rightleftharpoons}} C_7 \underset{-q}{\overset{q}{\rightleftharpoons}} C_6 \underset{-q}{\overset{q}{\rightleftharpoons}} C_5 \underset{-q}{\overset{q}{\rightleftharpoons}} C_4 \underset{-q}{\overset{q}{\rightleftharpoons}} C_3$$

$$O_2 \qquad O_1$$

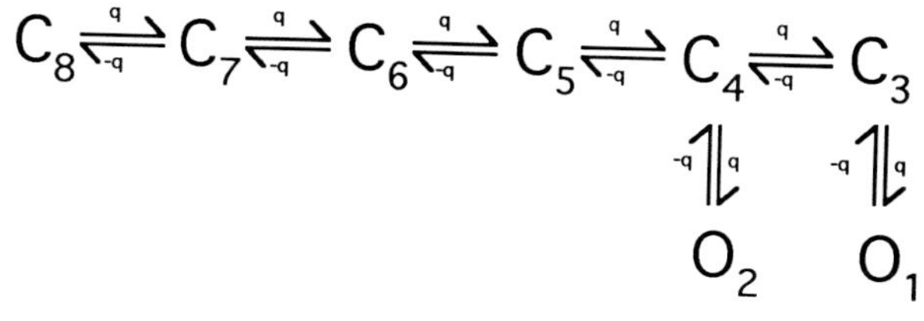

SCHEME IV

Scheme III gives reasonable descriptions of the open and shut dwell-time distributions when assessed visually. Statistical comparison of the likelihood values determined from the fits using the likelihood ratio test (Rao, 1973), however, rank Scheme I much higher. Scheme IV gives a poor description of the dwell-time distributions. Based on these results, Schemes III and IV are considered less likely.

C. Independent Subunits

The gating of voltage-dependent ion channels is often modeled by the movement of independent subunits, or gating particles (Hodgkin and Huxley, 1952; Zagotta and Aldrich, 1990). This model, depicted in Scheme V, was tested for its ability to describe the voltage-dependent gating of the fast Cl^- channel.

$$C_8 \underset{\beta}{\overset{5\alpha}{\rightleftharpoons}} C_7 \underset{2\beta}{\overset{4\alpha}{\rightleftharpoons}} C_6 \underset{3\beta}{\overset{3\alpha}{\rightleftharpoons}} C_5 \underset{4\beta}{\overset{2\alpha}{\rightleftharpoons}} C_4 \underset{5\beta}{\overset{\alpha}{\rightleftharpoons}} C_3$$

$$O_2 \qquad O_1$$

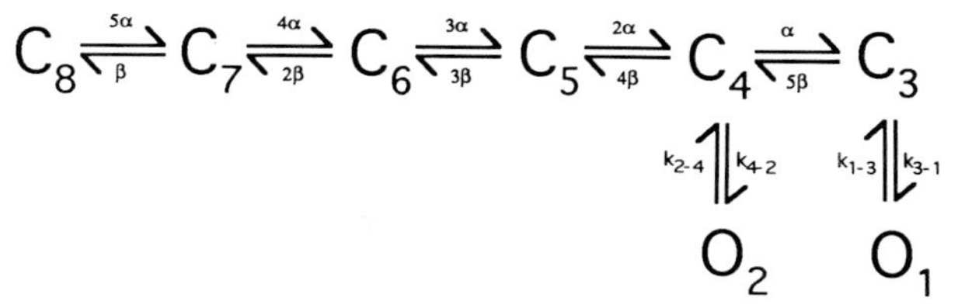

SCHEME V

Figure 15 shows shut dwell-time distributions at two different membrane potentials. The solid line is the best fit with Scheme I and the dashed line is the best description with Scheme V, the model with five identical subunits. As evident from Fig. 15, the model with independent subunits gives significantly worse descriptions of the data than Scheme I and is therefore considered unlikely.

D. Fractal Models

To date, all the schemes that have been presented are Markov models with a discrete number of kinetic states. An alternative model of ion channel gating has been suggested that proposes that the channel can exist in a continuum of kinetic states (Liebovitch *et al.*, 1987). Such a model has been introduced to describe the voltage-dependent gating of a potassium channel found in cultured hippocampal neurons. In the fractal model, the rate constants for leaving the

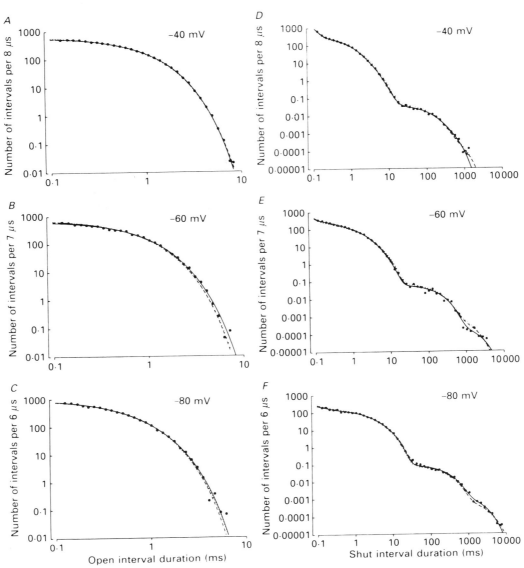

FIGURE 10 Scheme I accounts for the voltage dependence of the fast Cl⁻ channel. The filled circles plot the numbers of observed open (A–C) and shut (D–F) intervals against their durations on double logarithmic coordinates. The continuous, dotted, and dashed lines are dwell-time distributions predicted by Scheme I using different methods to obtain the rate constants. The rate constants for the predictions are given by the plotted symbols in Fig. 9 (continuous lines), regression fits in Fig. 9 (dotted lines), and simultaneous fit with parameters in Table I (dashed lines).

collection of states is given by a fractal scaling equation. The open and shut dwell-time distributions would be described by

$$f(t) = At^{1-D} \exp\{-[A/(2 - D)]t^{2-D}\}, \quad (6)$$

where A is the kinetic setpoint and D is the fractal dimension (Liebovitch and Sullivan, 1987; Liebovitch et al., 1987).

Figure 16 shows open and shut dwell-time distributions for the fast Cl⁻ channel fit by sums of exponentials (solid lines) and by Eq. 6 (dashed lines). The

fractal equation gives a very poor description of the shut dwell-time distribution and is therefore considered unlikely.

VI. Other Kinetic Features of the Fast Chloride Channel

Scheme II can account for the gating of the fast Cl⁻ channel during activity in the normal mode, which accounts for about 99% of the openings and closings.

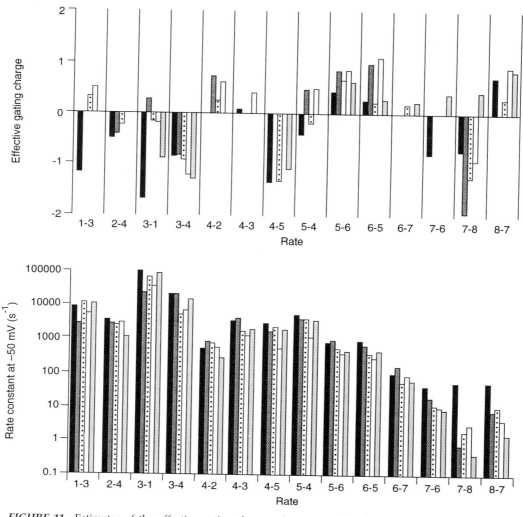

FIGURE 11 Estimates of the effective gating charge, plotted as units of electronic charge, (A) and the values of the rate constants at −50 mV (B) for Scheme I for five different fast Cl⁻ channels. Estimates were made by simultaneous fitting of dwell-time distributions obtained at three to eight different voltages with Scheme I. The bars (from left to right) for each of the indicated transitions plot the results for channels 1–5, respectively. Values of the effective gating charge that could be set to zero without changing the ranking of the fit with the Schwarz criterion are indicated by a plotted effective gating charge of zero.

However, as discussed in this section, the fast Cl⁻ channel can exist in alternative kinetic modes.

A. Buzz Mode

Blatz and Magleby (1986) observed that the fast Cl⁻ channel can occasionally (<1% of the time) enter a mode (buzz mode) consisting of bursts of brief open and shut intervals. Figure 17 shows a transition into this buzz mode at different frequency responses. At low frequency resolution, the buzz mode appears to be a subconductance state. The distribution of open and shut intervals during the buzz mode was de-

scribed by a single-exponential function (Blatz and Magleby, 1986). Thus, during the buzz mode, the channel may make transitions between a single open and a single shut state.

B. Inactive Mode

The fast Cl⁻ channel occasionally enters an inactive mode. This mode is a long-lived closed state that can last seconds or, in some cases, minutes. Although entry into this long-lived shut state(s) is uncommon, the channel can spend a significant portion of its time in this state because of the long lifetime of the state.

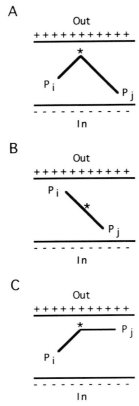

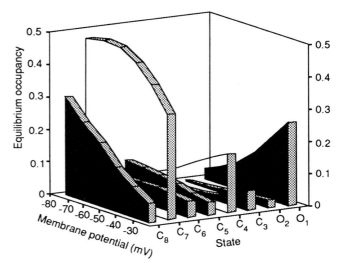

FIGURE 13 Equilibrium occupancy of the kinetic states in Scheme I as a function of voltage, where equilibrium occupancy is the fraction of time spent in the indicated states. Calculations are with the parameters in Table I. Depolarization progressively shifts the activity of the channel from the closed states at the left of Scheme I to the open states at the right of Scheme I.

FIGURE 12 The magnitude and sign of the effective gating charge can be the same or different for forward and backward transitions over the same energy barrier. P_i and P_j are the positions of a positive charge in the electric field of the membrane for states i and j. The asterisk indicates the position of the charge at the assumed transition state of the conformational change, which moves the charge between the two indicated positions. (A) The transitions from P_i to P_j and from P_j to P_i both have positive effective gating charges. (B) The transition from P_i to P_j has a negative effective gating charge and the transition from P_j to P_i has a positive effective gating charge. (C) The transition from P_i to P_j has a negative effective gating charge whereas the transition from P_j to P_i has no effective gating charge.

VII. Summary

This chapter has described the voltage-dependent gating kinetics of the fast Cl^- channel in cultured rat myotubes. A gating mechanism (Scheme I) has been

C. Slow Mode

Finally, the fast Cl^- channel can occasionally enter a mode characterized by somewhat slower gating kinetics. In this slow mode, the mean open time is increased. Figure 18 shows an example of the fast Cl^- channel entering this slow mode. The bottom two traces are shown with an expanded time scale so the kinetics during this mode can be seen in greater detail. In some cases, the fast Cl^- channel remains in this mode for many minutes. Although the gating kinetics have not been analyzed in great detail, it is especially interesting that during this mode the voltage dependence is markedly reduced, if not absent (data not shown).

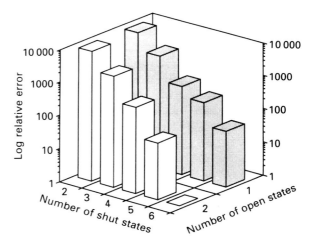

FIGURE 14 Scheme I with six shut and two open states ranks above schemes with fewer numbers of states. The z axis plots the log relative error, which is defined as (Schwarz criterion for the indicated schemes) − (Schwarz criterion for Scheme I). The x and y axes plot the numbers of open and shut states in the examined models. Scheme I with the lowest log relative error (the smallest Schwarz criterion) is the top ranked scheme.

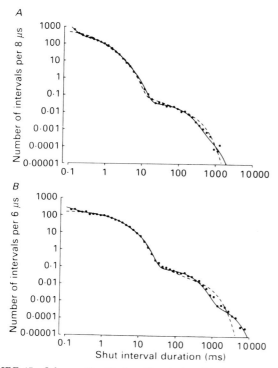

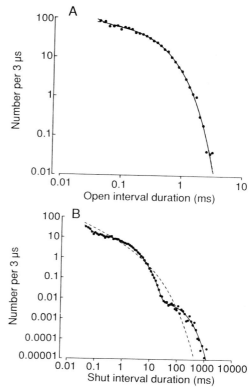

FIGURE 15 Scheme V with five identical and independent subunits gives a worse description of the dwell-time distributions than Scheme I. The filled circles plot closed dwell-time distributions at −40 (A) and −80 mV (B). The dashed lines are the best descriptions of the distributions for Scheme V. The continuous lines are the best descriptions of the distributions with Scheme I. For each model, the rate constants were estimated by the simultaneous fitting of distributions obtained at six different voltages (−30, −40, −50, −60, −70, −80 mV). For Scheme I, the rate constants are defined in Table I. For Scheme V, the values of q and B for the indicated transitions were 1-3: 0.0132 and 2230 s^{-1}; 2-4: −0.133 and 1310 s^{-1}; 3-1: 0.946 and 2260 s^{-1}; 4-2: 0.709 and 472 s^{-1}; α: 0.424 and 6.56 s^{-1}; β: 0.199 and 5.56 s^{-1}.

FIGURE 16 Distributions of open and shut interval durations recorded from the fast Cl^{-} channel fit by the Markov and fractal models. (A) Open times. The continuous line plots the maximum likelihood fit to the open durations with the sum of three exponentials, with time constants (and areas) of 0.013 msec (0.168), 0.208 msec (0.233), and 0.403 msec (0.599). The dashed line plots the maximum likelihood fit to the open times with the fractal continuum model, with $A = 1430$ Hz$^{0.92}$ and $D = 1.08$. (B) Shut times. The continuous line plots the fit to the shut durations with the sum of six exponentials, with time constants (and areas) of 0.024 msec (0.106), 0.362 msec (0.0737), 1.63 msec (0.421), 4.32 msec (0.359), 111 msec (0.0247), and 229 msec (0.0153). The dashed line plots the fit to the shut durations with the fractal continuum model, with $A = 6.22$ Hz$^{0.43}$ and $D = 1.57$. The likelihood ratios of the Markov to fractal model were $e^{3.1}$ for the open distribution and e^{961} for the shut. 7810 open and 7996 shut intervals were fitted and plotted. Reprinted from The *Biophysical Journal* (1988), **54**, 865, by copyright permission of the Biophysical Society.

presented that can account for the gating of the channel across a wide range of membrane potentials. The features of this model are (1) that the channel exists in at least six shut states and two open states at all examined membrane potentials (−10 to −90 mV), (2) that the transition rates between the states depend exponentially on voltage, and (3) that most of the transitions move charge in the same or opposite direction of the membrane field.

Although Scheme I is complex, it currently gives a better description of fast Cl^{-} gating than any other model tested and thus serves as the working hypothesis. As new data and new analysis techniques become available, other mechanisms will be tested for their ability to describe the gating of the fast Cl^{-} channel. For instance, more kinetic information is found in the single-channel data than in the open and shut dwell-

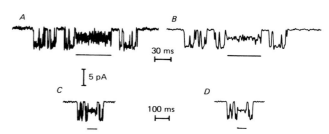

FIGURE 17 Apparent partially conducting state produced by buzz mode at limited frequency response. The horizontal bar indicates activity in the buzz mode. Reprinted with permission from Blatz and Magleby (1986).

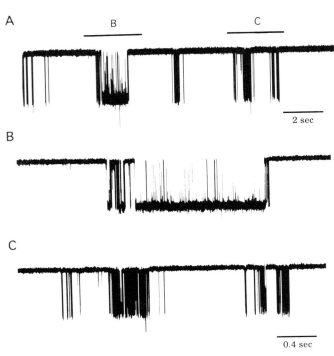

time distributions presented here. Including this other information in the fitting procedure will better define the kinetic parameters and will facilitate the comparison of models (Magleby and Weiss, 1990a,b).

To be complete, the kinetic model should also account for the rarely observed alternative gating modes: buzz mode, inactive mode, and slow mode (Fig. 19). Determining the factors that control the switching between the modes will be interesting because this could be a mechanism of regulating the channel's contribution to the Cl^- conductance of the cell. In addition, once the amino acid sequence of the fast Cl^- channel is known, the working hypothesis for channel gating presented here will be a good starting point for a structure–function analysis.

Acknowledgments

This work was supported by fellowships from the National Institutes of Health to D. S. Weiss (NS 08138 and NS 07044) and Grants from the National Institutes of Health (AR 32805) and the Muscular Dystrophy Association to Karl L. Magleby.

FIGURE 18 Slow mode of the fast Cl^- channel. (A) Stretch of activity showing normal and slow modes. (B,C) Segments in A shown with an expanded time base to show slow (B) and normal (C) modes in greater detail.

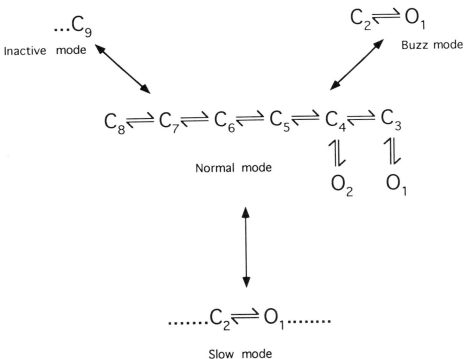

FIGURE 19 Complete kinetic model of the fast Cl^- channel. Normal mode is activity of the type analyzed in Fig. 1–16. The inactive mode is represented as long-lived (seconds or minutes) shut state(s). Buzz mode is shown in Fig. 17 and can be represented by a single open or shut state. The slow mode has not been analyzed in sufficient detail to present a kinetic scheme for activity during this mode.

References

Barrett, J. N., Barrett, E. F., and Dribin, L. B. (1981). Calcium-dependent slow potassium conductance in rat skeletal myotubes. *Dev. Biol.* **82,** 258–266.

Behrens, M. I., Oberhauser, A., Bezanilla, F., and Latorre, R. (1989). Batrachotoxin-modified sodium channels from squid optic nerve in planar bilayers. Ion conduction and gating properties. *J. Gen. Physiol.* **93,** 23–41.

Blatz, A. L. (1991). Properties of single fast chloride channels from rat cerebral cortex neurons. *J. Physiol. (London)* **441,** 1–21.

Blatz, A. L., and Magleby, K. L. (1985). Single chloride-selective channels active at resting membrane potentials in cultured rat skeletal muscle. *Biophys. J.* **47,** 119–123.

Blatz, A. L., and Magleby, K. L. (1986). Quantitative description of three modes of activity of fast chloride channels from rat skeletal muscle. *J. Physiol. (London)* **378,** 141–174.

Blatz, A. L., and Magleby, K. L. (1989). Adjacent interval analysis distinguishes among gating mechanisms for fast chloride channel from rat skeletal muscle. *J. Physiol. (London)* **410,** 561–585.

Colquhoun, D. (1971). "Lectures on Biostatistics." Oxford University Press, London.

Colquhoun, D., and Hawkes, A. G. (1981). On the stochastic properties of single ion channels. *Proc. R. Soc. London B* **211,** 205–235.

Colquhoun, D., and Sigworth, F. J. (1983). Fitting and statistical analysis of single-channel records. *In* "Single-Channel Recording" (B. Sakmann and E. Neher, eds.), pp. 191–263. Plenum Press, New York.

Glasstone, S., Laidler, K. J., and Eyring, H. (1941). "The Theory of Rate Processes." McGraw-Hill, New York.

Hamil, O. P., Marty, A., Neher, E., Sakmann, B., and Sigworth, F. J. (1981). Improved patch-clamp techniques for high-resolution current recording from cells and cell-free membrane patches. *Pflügers Arch.* **391,** 85–100.

Hanke, W., and Miller, C. (1983). Single chloride channels from *Torpedo* electroplax: Activation by protons. *J. Gen. Physiol.* **82,** 23–45.

Hess, P., Lansman, J. B., and Tsien, R. W. (1984). Different modes of Ca^{2+} channel gating behavior favored by dihydrophyridine Ca^{2+} agonists and antagonists. *Nature (London)* **311,** 538–544.

Hille, B. (1984). "Ionic Channels of Excitable Membranes." Sinauer Associates, Sunderland, Massachusetts.

Hodgkin, A. L., and Huxley, A. F. (1952). A quantitative description of membrane current and its application to conduction and excitation in nerve. *J. Physiol. (London)* **117,** 500–544.

Liebovitch, L. S., and Sullivan, J. M. (1987). Fractal analysis of a voltage-dependent potassium channel from cultured mouse hippocampal neurons. *Biophys. J.* **52,** 979–988.

Liebovitch, L. S., Fischbarg, J., Koniarek, J. P., Todorova, I., and Wang, M. (1987). Fractal model of ion channel kinetics. *Biochim. Biophys. Acta* **896,** 173–180.

Magleby, K. L., and Weiss, D. S. (1990a). Identifying gating mechanisms for ion channels by two-dimensional distributions of simulated dwell times. *Proc. R. Soc. London* **241,** 220–228.

Magleby, K. L., and Weiss, D. S. (1990b). Kinetic analysis of single channels using simulation: a general method which resolves the missed-event problem and accounts for noise. *Biophys. J.* **58,** 1411–1426.

McManus, O. B., and Magleby, K. L. (1988). Kinetic states and modes of single large-conductance calcium-activated potassium channels. *J. Physiol. (London)* **402,** 79–120.

Moczydlowski, E., and Latorre, R. (1983). Gating kinetics of Ca^{2+}-activated K$^+$ channels from rat muscle incorporated into planar lipid bilayers. Evidence for two voltage-dependent Ca^{2+} binding reactions. *J. Gen. Physiol.* **82,** 511–542.

Patlak, J. B., Gration, K. A., and Usherwood, P. N. R. (1979). Single glutamate-activated channels in locust muscle. *Nature (London)* **478,** 643–645.

Rao, C. R. (1973). "Linear Statistical Inference and Its Application." Wiley, New York.

Schwarz, G. (1978). Estimating the dimension of a model. *Ann. Stat.* **6,** 461–464.

Weiss, D. S., and Magleby, K. L. (1990). Voltage dependence and stability of the gating kinetics of the fast chloride channel from rat skeletal muscle. *J. Physiol. (London)* **426,** 145–176.

Weiss, D. S., and Magleby, K. L. (1992). Voltage-dependent gating mechanism for single fast chloride channels from rat skeletal muscle. *J. Physiol. (London)* **453,** 279–306.

Zagotta, W. N., and Aldrich, R. W. (1990). Voltage-dependent gating of *Shaker* A-type potassium channels in *Drosophila* muscle. *J. Gen. Physiol.* **95,** 29–60.

Chloride Channels of Colonic Carcinoma Cells

RAINER GREGER

I. Introduction

Colonic carcinoma cell lines such as T84, HT29, and Caco-2 have been established as tools with which to study the function of the colonic mucosa. Studies in the intact colonic mucosa are complicated by the fact that the mucosa contains different types and subtypes of cells, that is, surface and crypt cells, goblet and columnar cells, and so on, as well as by the fact that the submucosa can barely be removed completely.

Attempts to obtain primary cultures from biopsies have generally not been successful. On the other hand, human colonic carcinomas have been used successfully to obtain established cell lines such as the ones just mentioned (Augeron and Laboisse, 1984; Grasset *et al.*, 1985; Mandel *et al.*, 1986). In general, these cultures exhibit properties of the crypt cells, that is, they secrete NaCl, and do not express the functional characteristics of surface cells, namely, the aldosterone-controlled reabsorption of Na^+ (Cartwright *et al.*, 1985; Dharmsathaphorn *et al.*, 1985). Using Ussing chamber-type experiments with cultures of T84 cells grown to confluence, Mandel *et al.* (1986) showed that these cells secrete NaCl by a process that was originally proposed (Greger and Schlatter, 1984; Greger *et al.*, 1984) for the rectal gland of

Squalus acanthias and later was confirmed for many secretory cells and glands (Greger, 1986). The general concept of Cl^- secretion by these cells is shown in Fig. 1.

The primary driving force for Cl^- secretion is provided by the Na^+/K^+ ATPase localized in the basolateral membrane. This pump establishes the ionic gradients, which are inwardly directed for Na^+ and outwardly directed for K^+. Using the favorable gradient for Na^+, Cl^- is taken up by the basolaterally located and loop diuretic-sensitive (Dharmsathaphorn *et al.*, 1985; Franklin *et al.*, 1989; Kubitz *et al.*, 1992) $Na^+/2Cl^-/K^+$ co-transporter. The K^+ taken up by the pump, and by the co-transporter recycles across the basolateral membrane. This event hyperpolarizes the cell to normal voltages of -40 to -80 mV. Na^+ is recycled via the Na^+/K^+ pump. Cl^- leaves the cell via Cl^- channels localized in the luminal membrane because the Cl^- concentration is above equilibrium.

Colonic carcinoma cells have been used as models for Cl^- and mucus secretion in general. The cells usually grow reasonably quickly and form monolayers within 1 wk. These cells have been studied with a broad spectrum of techniques including patch-clamp analysis. More recently, colonic carcinoma cells, because of their expression of the cystic fibrosis transmembrane conductance regulator (CFTR; Breuer *et al.*,

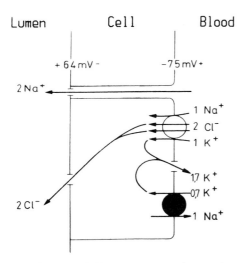

FIGURE 1 Cell model of Cl⁻ secretion in colon carcinoma cells.

1992; Cohn *et al.*, 1992; Sood *et al.*, 1992), have also been examined with respect to the basic mechanisms of Cl⁻ channel regulation.

This chapter reviews the various types of Cl⁻ channels that have been found in colonic carcinoma cells as well as their properties, their pharmacology, and the possible mechanisms of their regulation. Small Cl⁻ channels are probably functionally most relevant. These channels may, in fact, represent the CFTR or may be, in some as yet poorly understood fashion, closely linked to CFTR. Although quite substantial progress has been made over the past few years, interpretations of the mechanisms of stimulation of Cl⁻ secretion are still controversial.

II. Various Types of Chloride Channels

In the first patch-clamp study by Hayslett *et al.* (1987), a Cl⁻ channel of intermediate conductance (30–50 pS) and small conductance (~15 pS) Cl⁻ channels were found. Maxi Cl⁻ channels were found rarely (Hayslett *et al.*, 1987). These channels have never been seen in cell-attached patches. The Cl⁻ channels with 200 to 300-pS conductance appear to be related to the intermediate conductance Cl⁻ channels in the sense that the large channels occur on occasion in excised patches that first contained intermediate conductance Cl⁻ channels. The intermediate conductance Cl⁻ channel has been given the name ICOR (intermediate conductance outwardly rectifying) or ORDIC (outwardly rectifying depolarization-induced Cl⁻ channel); the term ICOR is used in this chapter. Subsequent studies in several laboratories have confirmed the presence of ICOR channels in colonic carcinoma

cells (Halm *et al.*, 1989; Champigny *et al.*, 1990; Halley *et al.*, 1990; Tabcharani *et al.*, 1990). The 30- to 50-pS channel was also found in other epithelia, namely in airway cells (Frizzell *et al.*, 1986a; Welsh, 1986, 1990; Kunzelmann *et al.*, 1989; Hwang *et al.*, 1990; Solc and Wine, 1991). Note that the absolute values of the conductances vary among the various reports for any of the following reasons: (1) Any type of channel will show a biological variation in conductance. For example, the ICOR type channel (cf., subsequent discussion) shows a normal distribution with 95% of all conductances falling in the range of 30–70 pS. (2) The conductance of this channel is nonlinear. At positive clamp voltages it is ~1.5 times larger than at negative voltages. (3) The conductance depends on the ionic concentration on either side of the membrane. Most studies have been done with ~150 mM Cl⁻ on both sides. In cell-attached studies, this consideration is of importance because the Cl⁻ concentration of the cytosol is only 30–50 mM (estimated from measurements of the Cl⁻ concentration dependence of the membrane voltage in HT29 cells with dominating Cl⁻ conductance). Much higher Cl⁻ concentrations are used in experiments in which the channels are reconstituted in lipid bilayers (Bridges *et al.*, 1989; Singh *et al.*, 1991). (4) The conductance is temperature dependent. Most of the cited studies from other laboratories were performed at room temperature, which may result in conductances smaller than those observed at 37°C.

The most frequently observed channel in all these studies has been the one with intermediate conductance and outward rectification (ICOR). Therefore, this channel was believed to be physiologically relevant. This view has been questioned by Cliff and Frizzell (1990), who found that the whole-cell current of T84 cells induced by cAMP was distinctly different from that of the ICOR. Now researchers believe that small Cl⁻ channels are much more relevant.

A. ICOR Channel

1. Properties

The ICOR channel has been seen frequently in all studies of colonic carcinoma cells. It also has been found in many other cells including airway epithelial cells in primary culture (Welsh, 1986; Kunzelmann *et al.*, 1989), fibroblasts (Bear, 1988), lymphocytes (Chen *et al.*, 1989), and keratinocytes (Galietta *et al.*, 1991). A typical recording of this channel from an excised inside-out oriented membrane of an HT29 cell is shown in Fig. 2. The conductance ranges from 30 to 70 pS. With symmetrical Cl⁻ concentrations on both sides of the membrane, the conductance is slightly

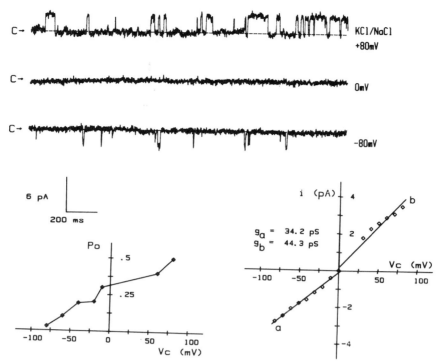

FIGURE 2 ICOR channel in an excised inside/out oriented patch of an HT29 cell. Pipette (extracellular side of membrane) contained KCl–Ringer. The bath contained NaCl–Ringer. Clamp voltage (V_c) refers to the cytosolic side. (*Top*) Typical current traces at different values of V_c. Low pass filtered at 1 kHz. C→ is the zero current level. (*Bottom left*) Voltage dependence of the open probability (P_o) of this channel. (*Bottom right*) Current–voltage relationship. Note that the conductance for positive voltages exceeds that for negative voltages.

larger for the outward than for the inward current. Kinetically, this channel can be described by at least two open and two closed states (Greger *et al.*, 1987; Hayslett *et al.*, 1987; Gögelein, 1988). The open-state probability of this channel is voltage dependent: it increases with positive voltages and decreases with negative voltages. However, if the channel is exposed to a positive clamp voltage for some time, it tends to close. When the channel is then clamped to negative voltages, its activity remains low, but this treatment conditions the channel to become reactivated when clamped to positive voltages again. This complex voltage dependence has been useful in studying the regulation of this channel by cAMP. The voltage dependence is more pronounced at low (room) temperature (Schoumacher *et al.*, 1987; Li *et al.*, 1988; Hwang *et al.*, 1989).

The ion selectivity of the ICOR has been examined by several laboratories (Welsh, 1986; Halm *et al.*, 1989; Kunzelmann *et al.*, 1989). Researchers found that all small halides (I⁻, Br⁻, Cl⁻) are conducted. A selectivity of I⁻ > Br⁻ > Cl⁻ has been reported (Halm *et al.*, 1989; Welsh *et al.*, 1986), but other reports show that all three halides are conducted equally well (Kunzelmann *et al.*, 1989).

The ICOR channel has been found only rarely in intact cells (Kunzelmann *et al.*, 1989). Unlike previous claims (e.g., Frizzell *et al.*, 1986b), pretreatment of the cells with isoproterenol, forskolin, or membrane permeable cAMP did not augment the frequency with which the ICOR channel was observed in cell-attached patches (Kunzelmann *et al.*, 1989,1992a). Other investigators have claimed that a channel similar to the ICOR channel appears in cells swollen by exposure to hypotonic solutions (Solc and Wine, 1991).

In sharp contrast to its rare occurrence in cells, the ICOR is observed very frequently in excised patches. The channel appears with some delay if excision is performed at room temperature (Welsh *et al.*, 1989) and appears instantaneously at 37°C (Kunzelmann *et al.*, 1989,1991). The excision activation is not influenced by the clamp voltage. This activation by excision has been shown to be mediated by the removal of a cytosolic inhibitor rather than by an activation by some constituent of the Ringer-type solution in which the membrane was excised (Kunzelmann *et al.*, 1991).

A component extracted from HT29 and placenta cytosol, but not from red blood cells or skeletal muscle, reversibly inhibits the ICOR channels previously activated by excision. A similar observation was made by Krick *et al.* (1991). This component, named cytosolic inhibitor, has an apparent molecular mass of 0.7–1 kDa, is heat stable, and is amphiphilic (Kunzelmann *et al.*, 1991). Investigators are making a collaborative effort to further purify this inhibitor. The inhibitory activity can be recovered from one peak of a Sephadex column. However, its molecular identity remains still unclear. The cytosolic inhibitor has no influence on K^+ channels but, in addition to its effect on the ICOR, it also inhibits small Cl^- channels that cannot be resolved as single-channel events and, hence, appear as a "baseline current" (Kunzelmann *et al.*, 1992b).

At this stage, the role the ICOR plays in the intact cell is entirely unclear. Several lines of evidence suggest that the ICOR is not responsible for the Cl^- current of the intact colon cell. (1) It is found very rarely cell-attached patches. (2) The macroscopic current of the cell is different from that of the ICOR. (3) The inhibitors of the ICOR (cf., subsequent discussion), such as 5-nitro-2-(3-phenylpropylamino)-benzoate (NPPB), have little effect on Cl^- secretion (Greger *et al.*, 1991) or on the Cl^- current of HT29 monolayers (Fischer *et al.*, 1992b). The possible involvement of the ICOR in regulatory volume decrease (Worrell *et al.*, 1989; Solc and Wine, 1991) and in cAMP-mediated Cl^- secretion (Schoumacher *et al.*, 1987; Li *et al.*, 1988; Hwang *et al.*, 1989) will be discussed in Section III.

2. Pharmacology

The ICOR channel is specifically sensitive to NPPB. An example of this effect and the structure of NPPB are shown in Fig. 3. The block by NPPB is of the flicker type (Dreinhöfer *et al.*, 1988; Champigny *et al.*, 1990; Tilmann *et al.*, 1991). Half-maximal inhibition is obtained at <1 μM. The channel is more sensitive to NPPB from the outside (Tilmann *et al.*, 1991), suggesting that NPPB binds to the external mouth of the channel and that its weaker effect in inside-out patches is due to the fact that NPPB must permeate the membrane to reach its binding site. This feature also explains the finding that the onset of inhibition is delayed in inside-out patches but is instantaneous in outside-out patches. This concept was verified in experiments with macromolecular probes of NPPB. These probes cause inhibition in outside-out patches but not in inside-out ones (Tilmann *et al.*, 1991). ICOR channels can also be blocked by stilbene derivatives such as 4-acetamido-4'-isothiocyanostilbene-2,2'-disulfonic acid (SITS), 4,4'-diisothiocyanestilbene-2,2'-disulfonic acid (DIDS), and 4,4'-dinitrostilbene-

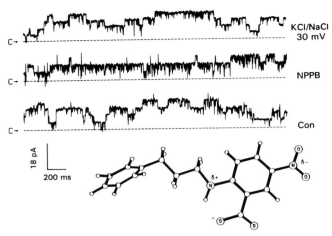

FIGURE 3 Inhibition of ICOR channels by NPPB [5-nitro-2-(3-phenylpropylamino)-benzoate]. (*Top*) Typical current traces in the absence and presence of NPPB (10 μM). Excised inside/out patch from an HT29 cell. Pipette contains KCl–Ringer and bath contains NaCl–Ringer. Low pass filtered at 1 kHz. Con, control. C→ is the zero current level. This patch contains four ICOR channels. NPPB induces a reversible flicker type block. (*Bottom*) Structural formula of NPPB with the anionic carboxylate group and with two partial charges ($\delta+$, $\delta-$).

2,2'-disulfonic acid (DNDS) (Bridges *et al.*, 1989; Tilmann *et al.*, 1991); Singh *et al.*, 1991; by amidine and IAA 94/95 (Tilmann *et al.*, 1991); by high concentrations of substances such as bumetanide (R. Greger, unpublished observations), verapamil (Champigny *et al.*, 1990), and HEPES (Hanrahan and Tabcharani, 1990); by nonsaturated fatty acids (Hwang *et al.*, 1990; Kunzelmann *et al.*, 1991); and even by very high concentrations of aldosterone (J. Disser, personal communication). The mechanisms of drug interaction with the ICOR channel are not clear, although structure–activity relationships have been obtained for some substances (Tilmann *et al.*, 1991; Cabantchik and Greger, 1992). A channel with some properties similar to those of the ICOR channel (its sensitivity to NPPB) has been cloned from MDCK cells (Paulmichl *et al.*, 1992).

B. Small Chloride Channels

Small chloride channels have already been identified in the rectal gland of *Squalus acanthias* (Gögelein *et al.*, 1987b) and in the first report on HT29 cells (Hayslett *et al.*, 1987). The size of these channels has been estimated to be between 11–15 pS, making them still detectably by the patch-clamp technique. The fact that these channels are not as spectacular as the ICOR channel was probably responsible for their neglect until recently.

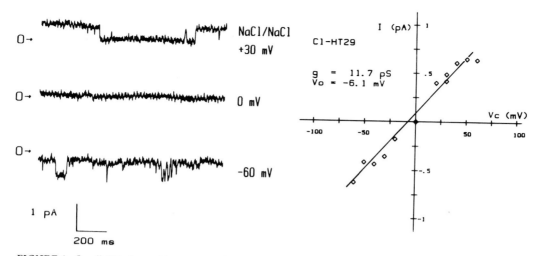

FIGURE 4 Small Cl⁻ channel in an excised inside/out oriented patch of an HT29 cell. Pipette (extracellular side of membrane) and bath contained NaCl–Ringer. Clamp voltage (V_c) refers to the cytosolic side. (*Left*) Typical current traces at different values of V_c. Low pass filtered at 500 Hz. C→ is the zero current level. (*Right*) Current–voltage relationship. Note that the kinetics of this channel appear to be rather slow. Data from Hayslett *et al.* (1987).

1. Properties

Figure 4 shows a small Cl⁻ channel from an HT29 cell. The conditions are equal to those in Fig. 2. The channel has a linear current–voltage relationship. Its conductance is in the range of 7–15 pS. The kinetics appear rather slow, but this measurement may be biased by the filtering that had to be used to increase the signal-to-noise ratio.

Very similar channels have been found in T84 cells (Tabcharani *et al.*, 1990). Their ion selectivity has been determined by Tabcharani *et al.* (1990); a sequence of Br⁻ > Cl⁻ > I⁻ was found. This sequence is identical to the conductance and permeability sequence obtained for the whole-cell Cl⁻ current induced by cAMP (Cliff and Frizzell, 1990). These data have been reproduced by other researchers (Bell and Quinton, 1992; Lin *et al.*, 1992). Very similar conductance and channel behavior was also found in several studies in which CFTR was transfected into cells devoid of CFTR or possessing defective CFTR (Anderson *et al.*, 1991b; Dalemans *et al.*, 1991; Kartner *et al.*, 1991; Tabcharani *et al.*, 1991). Again, on the basis of whole-cell data, results suggest that all three halides (Cl⁻, Br⁻, and I⁻) are conducted almost equally well (Kubitz *et al.*, 1992).

The small Cl⁻ channel has been found in intact cells (Tabcharani *et al.*, 1990; Anderson *et al.*, 1991b) and appears to be up-regulated by cAMP. In excised patches, this channel shows a decrease of open-state probability, but apparently it can be reactivated by

protein kinases A and G in the presence of ATP (Tabcharani *et al.*, 1991; Lin *et al.*, 1992).

Investigators have known for some time that even smaller Cl⁻ channels are present in several other secretory cells including lacrimal gland and pancreatic acinus cells (Marty *et al.*, 1984; Petersen, 1992). Therefore the issue of whether much smaller and, hence, undetectable Cl⁻ channels are also present in HT29 cells has been re-examined. To this end the nystatin patch-clamp method has been modified slightly by reducing the concentration of nystatin in the pipette to less than 50 mg/l (Greger and Kunzelmann, 1991). At this low nystatin concentration, the membrane of the patch is sufficiently permeabilized to permit the recording of the cell voltage, but ion channels present in this membrane are not disturbed. Hence, membrane current and cell voltage can be monitored simultaneously. If a cell is stimulated, for example, by membrane permeable cAMP, the effect on voltage can be verified and the time course of the voltage followed. Simultaneously, Cl⁻ channels would have to appear in the patch, as discrete events or as a superposition of many small and otherwise undetectable events. Their summation would appear as one component of the "baseline current." As shown in Fig. 5 (Kunzelmann *et al.*, 1992a), cAMP induced the expected depolarization and did not induce discrete current events, but increased the baseline current. Data of this kind suggest that cAMP induces small Cl⁻ channels (Kunzelmann *et al.*, 1992a). Obviously, deducing from such

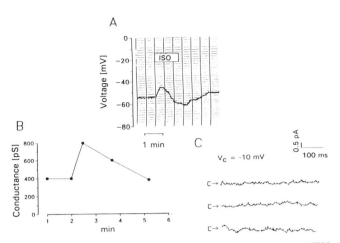

FIGURE 5 Cell-attached nystatin (CAN) recording from an HT29 cell. The pipette was filled with a solution contained 50 mM KCl, 95 mM K-gluconate, 1.6 mM Na$_2$HPO$_4$, 0.4 mM NaH$_2$PO$_4$, 1.6 mM CaCl$_2$, 1.03 mM EGTA, 0.73 mM MgCl$_2$, 5 mM D-glucose, and 20 mg/l nystatin. (A) Voltage recording. Note that isoproterenol (ISO, 10^{-6} M) led to a transient depolarization followed by a hyperpolarization. The voltage change was entirely reversible. (B) Baseline conductance of the patch. Note that the baseline conductance is increased markedly in the presence of ISO. This conductance increase occurs simultaneously with the depolarization, is sustained during the secondary hyperpolarization, and is entirely reversible. (C) High resolution current recording. The clamp voltage (V_c) was −10 mV, corresponding to a net driving force of 70 mV favoring Cl$^-$ exit. Low pass filtered at 200 Hz. C→ is the apparent zero current level. Upper trace taken during control, middle trace taken at the peak of the depolarization and conductance increase, lower trace taken during postexperimental control. Note that no clear single-channel events are apparent from these traces, indicating that the Cl$^-$ channels carrying the 400 pS (Δ conductance in B) must have a very small single-channel conductance.

data how large the single-channel conductance of these channels might be is difficult. A maximal estimate for a given signal-to-noise ratio can be proposed. For example, if the noise band was 0.2 pA at a filtering of 200 Hz, the single event would have to be 0.4 pA to be clearly detectable (twice the noise). To obtain such a current amplitude, a driving force of 80 mV might be applied. Under these conditions, any channel with a conductance ≥5 pS would be detectable. Clearly the filtering is limited by the kinetics of a channel. If the open time constant is only 1 msec, a low pass filtering at 200 Hz would not be advisable. Such a channel would have to have a much larger conductance to be detected. The conductance of small Cl$^-$ channels can also be deduced from noise analysis. Fischer *et al.* (1992b) performed such an analysis in HT29 cells and found that the cAMP-induced current has a very small single-channel conductance of 2–3 pS or less. As discussed later, the Cl$^-$ currents induced in HT29 cells by Ca^{2+}-mobilizing agonists and by cell

swelling are also caused by small and undetectable Cl$^-$ channels (Kubitz *et al.*, 1992; Kunzelmann *et al.*, 1992c).

How do these data compare with those of Tabcharani *et al.* (1990,1991), who found an 8-pS channel in T84 cells? These channels are likely to represent the larger (and, hence, visible) fraction of a Gaussian distribution of the small channels. For any type of channel found in excised or cell-attached patches, the putative role in Cl$^-$ secretion can be deduced from a fairly simple calculation based on the assumptions of (1) macroscopic conductance, (2) the area of the patch, and (3) the average number of active channels observed in a cell-attached patch. For example, cAMP can increase the conductance of an HT29 cell from a few nS to well above 20 nS (Kubitz *et al.*, 1992; Kunzelmann *et al.*, 1992b). The cell surface is assumed to be 1500 μm^2 (based on the mean capacitance and a specific capacitance of 1 μF/cm^2 or on morphological data), and half of it that is, 750 μm^2, represents apical membrane. This surface must accommodate 20 nS of conductance. If the area of the patch is 3–5 μm^2, this area should have a conductance of 107 pS. If the respective channels have an open probability of 0.5, 28 channels with 7.5-pS conductance would have to be activated in each single patch in the presence of cAMP. This prediction is not consistent with results that have been reported. However, an increase in the baseline conductance by some 400 pS has been noted, emphasizing that the preceding calculation is a conservative lower estimate of the number of channels required. 100 channels with 7.5-pS conductance would be necessary to explain these data. To date discrete current levels have not been detected. Therefore, the small (6–10 pS) and possibly nondetectable (<6 pS) Cl$^-$ channels are likely to be of high physiological relevance for Cl$^-$ secretion in colon carcinoma cells.

The small Cl$^-$ channel, as stated earlier, has been shown to be regulated by cAMP (and cGMP)-dependent phosphorylation (Tabcharani *et al.*, 1991; Lin *et al.*, 1992). This issue will be discussed in Section III. This channel has been identified in various cells expressing CFTR, including CHO cells, HeLa cells, Vero cells, and Sf9 cells (Anderson *et al.*, 1991a,b; Dalemans *et al.*, 1991; Kartner *et al.*, 1991; Tabcharani *et al.*, 1991), suggesting that this channel is identical to CFTR. This issue will be discussed in Section IV.

2. Pharmacology

Unfortunately, no specific blocker for the small Cl$^-$ channel in colon carcinoma cells has been identified to date. NPPB does not block this channel (Gögelein *et al.*, 1987b; Gögelein, 1988). The stilbenes such as

DIDS have been examined in a few studies on the cAMP-activated Cl⁻ current, which was reported to be DIDS insensitive (Anderson and Welsh, 1991; Cliff *et al.*, 1992; Cunningham *et al.*, 1992). This result contrasts with the strong inhibitory effect on the ICOR channel, and with the inhibition of Ca^{2+}-mediated Cl^- currents (Anderson and Welsh, 1991; Cunningham *et al.*, 1992). In another study comparing the Cl⁻ currents of normal and cystic fibrosis (CF) airway cells (Chan *et al.*, 1992), all three pathways (Ca^{2+}-, swelling-, and cAMP-induced) were DIDS sensitive. This later report has been confirmed. DIDS at a very high concentration of 0.5 to 1 mM inhibits the Cl⁻ conductances induced by any of the three pathways (Kubitz *et al.*, 1992). Inasmuch as the whole-cell currents correspond to the small conductance Cl⁻ channels (cf., preceding discussion), these channels may be the targets of DIDS. However, DIDS inhibits several other transporters at this high concentration (Cabantchik and Greger, 1992) and might act indirectly, that is, not by an inhibition of the channel itself.

III. Regulation of Chloride Channels

Colon carcinoma cells share the property of colon crypts to secrete Cl⁻ when stimulated by agonists such as isoproterenol, vasoactive inhibitory peptide (VIP), and others (Dharmsathaphorn *et al.*, 1985; Mandel *et al.*, 1986; Anderson and Welsh, 1991; Bajnath *et al.*, 1992; Kunzelmann *et al.*, 1992b, Lohrmann *et al.*, 1992). These hormones mediate their responses via cAMP and cAMP-dependent phosphorylation by protein kinase A (PKA). These cells also possess P_2 (ATP) type receptors, carbachol (CCH) receptors, and receptors for neurotensin (NT) (Kunzelmann *et al.*, 1992c). These receptors induce transient increases in cytosolic Ca^{2+}, which lead to a transient increase in Cl⁻ conductance. A third pathway of increasing the Cl⁻ conductance of colon carcinoma cells is the swelling by hypotonic media (Worrell *et al.*, 1989). Since several other secretory cells such as acinar cells of glands and respiratory epithelial cells in culture share these three general regulatory pathways, the colonic carcinoma cells again serve as a more general model. The role of these cells as a model was even more attractive because colonic crypt cells express the general defect of CF, which is located in the steps distal to the cAMP cascade. The key questions to be answered follow. (1) Which Cl⁻ channels are activated by the various pathways? (2) What is the mechanism of activation? (3) Does any interaction (cross-talk) among the three pathways occur? These issues are discussed in this section, and will set the tone for Section IV,

which addresses the possible role of CFTR and its implication in CF.

A. Hormones Acting via cAMP

This pathway has always attracted the most interest because it appears to be the key mechanism for turning on secretion in the colon physiologically [VIP, Prostaglandin E_2 (PGE)] and pathophysiologically (cholera toxin). The flux data obtained originally by Dharmasathaphorn's group (1985) indicate that cAMP enhances rheogenic Cl⁻ secretion, as discussed in the introduction.

Patch-clamp analysis revealed the presence of the ICOR channels in excised patches from colon carcinoma and airway cells. Researchers have claimed that the ICOR channel is directly under the control of PKA. This relationship has been shown in experiments in which excised patches, ripped off at hyperpolarized clamp voltage, showed ICOR channels when PKA and ATP were added to the cytosolic side. No such effect was seen in time controls (Schoumacher *et al.*, 1987; Li *et al.*, 1988; Hwang *et al.*, 1989). Long-lasting depolarization was used as a control. If a silent channel had been present in the patch, it should also have been activated by depolarization. These studies were also performed in CF respiratory cells (Schoumacher *et al.*, 1987; Li *et al.*, 1988; Hwang *et al.*, 1989), in which researchers found that depolarization-induced activation of ICOR channels was possible in patches excised from normal cells as well as from cells carrying the CF defect. In contrast, activation by PKA was only possible in the patches from normal cells. This result has led to the conclusion that the Cl⁻ channels in CF are basically normal but cannot be opened by PKA-dependent phosphorylation (McCann and Welsh, 1990). Kunzelmann *et al.* (1989) found that excision activation occurred immediately, irrespective of clamp voltage. Hence, no phosphorylation was required for channel activation. Further, patches from CF cells showed the same excision activation of ICOR channels when they were ripped off at hyperpolarized voltages, leading to the conclusion that the ICOR channel apparently is not activated by phosphorylation and is not altered in CF. The difference between these results and those from other studies is not easily explained. The activation of normal channels after addition of PKA might also be explained by spontaneous activation. This possibility is more likely because all other groups worked at room temperature. At 20°C, excision activation is markedly delayed, whereas it is instantaneous at 37°C (Kunzelmann *et al.*, 1989,1991; Welsh *et al.*, 1989).

The finding that small Cl⁻ channels appear when

CFTR is transfected into nonmammalian cells or into mammalian cells not equipped with CFTR (Tabcharani et al., 1991 has shifted the focus of attention completely away from the ICOR to these small Cl⁻ channels. In fact, researchers have shown that the small (8-pS) Cl⁻ channel is opened by PKA phosphorylation and that its activity can be reversed by phosphatase (Tabcharani et al., 1991). Therefore, CFTR is proposed to be the small Cl⁻ channel (see subsequent discussion).

This channel is supposed to have a selectivity sequence of $Br^- > Cl^- > I^-$. To test this feature, the mechanism of cAMP stimulation of HT29 cells was examined with the modified cell-attached nystatin (CAN) technique. A typical experiment (Fig. 5) has already been discussed. Evidently the increased Cl⁻ current was carried by very small (i.e., nondetectable) Cl⁻ channels. A summary of these data is shown in Fig. 6. cAMP addition leads to a depolarization that is paralleled by a marked increase in the Cl⁻ conductance. During the secondary repolarization, a K⁺ conductance is activated in addition to the Cl⁻ conductance. The effect of cAMP is slightly delayed and a maximal effect requires 0.5–3 min to build up. Also, the magnitude of the effect is quite variable. In some cells, the relative increase in conductance is less than 50%; in others the effect is much more marked. In the latter cells, the ion selectivity was examined; cAMP-induced conductance was found not to distinguish among the three tested halides (Kubitz et al., 1992). The discrepancies in the zero current potentials, with Cl⁻ showing the largest and I⁻ showing the smallest depolarization, can be explained by one of two mechanisms: (1) The cAMP regulated channel may have a permeability sequence of $P_{I^-} > P_{Br^-} > P_{Cl^-}$ and a differing conductance sequence of $G_{Cl^-} = G_{Br^-} = G_{I^-}$. Such disparities of permeability and conductance sequences are not uncommon (Gallacher et al., 1984; Gögelein et al., 1987a; Bleich et al., 1990). (2) These changes in voltage may be caused by the selectivity of the $Na^+/2Cl^-/K^+$ co-transporter, which accepts Cl⁻ to a lesser extent than Br⁻ and does not accept I⁻ (Greger, 1985). Consequently, exposure of the cell to a bath containing I⁻ will reduce cytosolic Cl⁻, which will generate a more hyperpolarized voltage (Kubitz et al., 1992). The effects of cAMP are reversible, requiring 5–10 min for the cell to repolarize.

In summary, cAMP up-regulates a Cl⁻ conductance and, shortly thereafter, a K⁺ conductance. These effects occur within 1–3 min. The conductances remain high as long as cAMP is present. The voltage follows sequential conductance changes. An early depolarization is followed by a secondary repolarization.

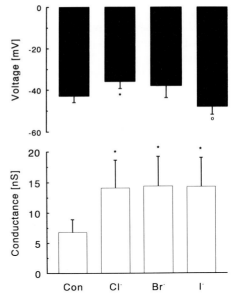

FIGURE 6 Anion selectivity of the cAMP-induced anion conductance in HT29 cells. Whole-cell patch-clamp recordings obtained by mechanical rupture (fast whole-cell) of the membrane in the patch pipette or by high concentrations of nystatin (slow whole-cell). The pipette solution was identical to that described in Fig. 5, except that 1 mM ATP was added and 100 mg/l nystatin was added in the slow whole-cell experiments. The bath contained NaCl–(NaBr–, NaI–) Ringer to which 5×10^{-4} M 8-chloro-thiophenyl-cAMP was added. Fast and slow whole-cell data were pooled to give Mean ± SEM (n = 12) of paired data (all anions were examined in each single experiment). Asterisks indicate significant difference from control (Con). The sequence in which the three anions were examined was permuted. (Top) Voltage. (Bottom) Whole-cell conductance (not corrected for access resistance). Note that the strongest cAMP-induced depolarization is seen with Cl⁻. In the presence of I⁻, the voltage is hyperpolarized significantly (○) with respect to Cl⁻ and Br⁻. The conductances obtained with all three anions are very similar. Data from Kubitz et al. (1992).

The Cl⁻ conductance corresponds to small Cl⁻ channels. Apparently, these channels do not select among the three halides.

B. Hormones Acting via Cytosolic Ca²⁺

These hormones include CCH, NT, and ATP, all of which increase cytosolic Ca²⁺ abruptly (Morris et al., 1990; Leipziger et al., 1991; Fischer et al., 1992a; Nitschke et al., 1993a) with a transient that consists of a rapid spike and a variable secondary component. The spike is caused by IP₃(?)-induced Ca²⁺ release from cytosolic stores. The sustained increase in Ca²⁺ is largely influenced by Ca²⁺ influx across the plasma membrane. The NT transient is very fast; that for ATP is less fast. The Ca²⁺ transient is paralleled by a sudden increase in Cl⁻ conductance that drops while the agonist is still present with a time constant of ~0.5 min for NT and 2 min for ATP. All three agonists also induce a secondary increase in K^+ conductance. A typical CAN experiment for NT is shown in Fig. 7 (Kunzelmann *et al.*, 1992c). NT induces an abrupt depolarization that occurs almost instantaneously ($\leq$1 sec). This very rapid depolarization is paralleled by an increase in Cl^- conductance from 150 to 270 nS. However, no discrete current events can be found in the cell-attached membrane. Therefore, this Cl^- conductance is thought to result from the synchronous operation of many small channels (Kunzelmann *et al.*, 1992c). The voltage re- or even hyperpolarizes while the agonist is still present. At the same time, the conductance drops almost to baseline values.

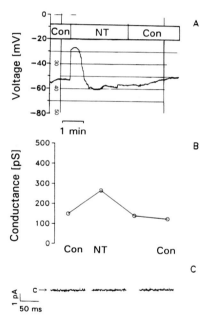

FIGURE 7   Cell-attached nystatin (CAN) recording from an HT29 cell. The pipette was filled with a solution described in Fig. 5. (A) Voltage recording; Con, control. Note that neurotensin (NT, 10^{-8} M) led to an abrupt transient depolarization followed by a hyperpolarization. The voltage change was entirely reversible. (B) Baseline conductance of the patch. Note that the baseline conductance is increased markedly in the presence of NT. This conductance increase occurs simultaneously with the depolarization and is entirely reversible. (C) High resolution current recording. The clamp voltage (V_c) was 0 mV, which corresponds to a net driving force of 60 mV favoring Cl^- exit. Low pass filtered at 200 Hz. C→ is the apparent zero current level. Left trace taken during control, middle trace taken at the peak of the depolarization and conductance increase, right trace taken during postexperimental control. Note that no clear single-channel events are apparent from these traces, indicating that the Cl^- channels carrying the 130 pS (Δ conductance in B) must have a very small single-channel conductance.

Researchers have argued that the Ca^{2+}-mediated Cl^- current of T84 cells is different from the cAMP-mediated current because it is outwardly rectifying (Cliff *et al.*, 1990). Also, this current, unlike the cAMP-induced Cl^- current, was supposedly inhibited by high concentrations (0.5 m*M*) of DIDS (Anderson and Welsh, 1991; Cunningham *et al.*, 1992). Further, in one report, antisense oligoDNA (to CFTR) reduced the cAMP-induced Cl^- current but failed to exert an effect on the Ca^{2+}-induced Cl^- current (Wagner *et al.*, 1992).

The issue of the multiple Cl^- conductances has been examined in HT29 cells. Researchers found that Ca^{2+}-mobilizing agents such as ATP or NT induce a conductance that does not distinguish between Cl^-, Br^-, and I^-, and that this conductance, as well as that induced by cAMP or hypotonic cell swelling, is reduced reversibly by high concentrations of DIDS (0.5–1 m*M*; Kubitz *et al.*, 1992). Further, in studies examining whether the three pathways are additive, investigators found that none of the three pathways is additive with any of the other two. Hence, these data indicate that all three pathways, in some unknown fashion, converge to generate the same Cl^- conductance (Kubitz *et al.*, 1992).

In summary, current knowledge supports the view that the Ca^{2+}-induced currents follow the cytosolic Ca^{2+} transients very closely. The currents are short-lived and collapse almost completely within 1–2 min. The debate continues over whether these currents resemble the operation of Cl^- channels that are different from those opened by cAMP-dependent phosphorylation. The mechanism of Ca^{2+} control of the respective Cl^- channels is also not yet defined. Researchers have suggested that this event is mediated by calmodulin, since calmodulin and calmodulin-kinase inhibitors reduce these currents (Worrell and Frizzell, 1991). In the same vein, that Ca^{2+} itself regulates these Cl^- channels directly appears to be highly unlikely (R. Greger, unpublished observations).

C. Chloride Channels Activated by Hypotonic Cell Swelling

Colonic carcinoma cells such as T84 and HT29 cells respond to cell swelling (Worrell *et al.*, 1989), like many other cells (Hoffmann and Simonsen, 1989; Lang *et al.*, 1990), with regulatory volume decrease (RVD). This RVD response consists of an initial increase in K^+ conductance and a subsequent, even more marked increase in Cl^- conductance. This response is also reflected by voltage changes (Fig. 8). An initial hyperpolarization is followed by a secondary depolarization. When compared with Ca^{2+}-mobiliz-

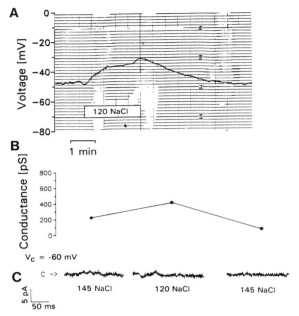

FIGURE 8 Cell-attached nystatin (CAN) recording from an HT29 cell. The pipette was filled with a solution described in legend to Fig. 5. (A) Voltage recording; Con, control. The bath contained the normal NaCl–Ringer (145 m*M* NaCl) or a hypotonic NaCl solution (120 m*M* NaCl). Note that exposure to the hypotonic solution led to a slow and sustained depolarization. The voltage change was entirely reversible. (B) Baseline conductance of the patch. Note that the baseline conductance is increased markedly in the presence of 120 m*M* NaCl. This conductance increase occurs simultaneously with the depolarization and is entirely reversible. (C) High resolution current recording. The clamp voltage (V_c) was − 60 mV, corresponding to a net driving force of ~100 mV favoring Cl⁻ exit. Low pass filtered at 300 Hz. C→ is the apparent zero current level. Left trace taken during control, middle trace taken at the peak of the depolarization and conductance increase, right trace taken during postexperimental control. Note that no clear single-channel events are apparent from these traces, indicating that the Cl⁻ channels carrying the 200 pS (Δ conductance in B) must have a very small single-channel conductance.

ing agents such as ATP, hypotonic cell swelling increases the Cl⁻ conductance very slowly but to a comparable degree. Several studies on T84 cells have shown that the Cl⁻ conductance induced during RVD is outwardly rectifying and prefers I⁻ and Br⁻ over Cl⁻ (Worrell *et al.*, 1989; Solc and Wine, 1991). Also, this conductance appears to be up-regulated by longer lasting depolarization (Worrell *et al.*, 1989; Solc and Wine, 1991). This conductance may reflect many of the properties of the ICOR channel: (1) its outward rectification; (2) its voltage dependence; and (3) its ion selectivity. Solc and Wine (1991) postulated that this current is carried by a channel very similar but not identical to the ICOR type channel. Also, the DIDS inhibition of this conductance has long been claimed

to be peculiar to this pathway. As shown earlier, this does not appear to be the case.

Data (Kubitz *et al.*, 1992; Kunzelmann *et al.*, 1992c) suggest that RVD-induced Cl⁻ channels, like those induced by cAMP or Ca^{2+}, have a very small conductance. The RVD-induced conductance shares ion selectivity with the other two pathways: $G_{Cl^-} = G_{Br^-} = G_{I^-}$. Further, this conductance does not appear to be additive to the cAMP- or Ca^{2+}-induced conductance, and is inhibited by high concentrations of DIDS (Kubitz *et al.*, 1992). The mechanism by which cell swelling induces this conductance is not completely clear. Cell swelling may act via leukotrienes (Christensen and Hoffmann, 1992) or some other membrane-borne mediator, or it may enhance Ca^{2+} (Nitschke *et al.*, 1993b; Warth and Greger, 1993). The Ca^{2+} transients which, in the case of Ca^{2+}-mobilizing agents, parallel the Cl⁻ conductance changes (R. Greger, unpublished observation) do not match the cell swelling-induced Cl⁻ conductance (Nitschke *et al*, 1993b). Cell swelling evokes a triphasic Ca^{2+} activity change with a slow increase, a delayed peak, and a relaxation to a stable plateau. The Cl⁻ conductance increases very slowly to a maximum that is maintained as long as the exposure to hypotonic solution lasts. In these studies, however, the Ca^{2+} transients are monitored over the entire cell, whereas the regulatory process might be influenced by local Ca^{2+} activities close to the membrane. Also, the real Ca^{2+} transient might not be identical to that monitored by 1-[2-(5-carboxyoxazol-2-yl)-6-aminobentofurane-5-oxy]-2-(2′ amino-5′-methylphenoxy)-ethan-*N,N,N′,N′*-tetracetic acid (fura-2).

Researchers had suggested that Ca^{2+} acts directly on the excised membrane (Frizzell *et al.*, 1986a), but this possibility has not been confirmed (Clancy *et al.*, 1990). The Ca^{2+} effect on these Cl⁻ channels may be indirect, since no activation of Cl⁻ channels can be documented in excised patches when the Ca^{2+} activity on the cytosolic side is increased (R. Greger, unpublished data).

In summary, although some studies suggest that the RVD Cl⁻ channel is identical or similar to the ICOR, other data support the view that these channels have small single-channel conductance and may be similar or identical to the Cl⁻ channels induced by Ca^{2+} or cAMP.

IV. Chloride Channels and CFTR

Initially Schulz and Frömter (1968) showed that sweat gland duct Cl⁻ permeability was sharply reduced in patients suffering from CF. After 15 years, this view finally became generally accepted (Quinton, 1983). Since then, researchers have determined that many but by no means all symptoms of CF can be explained by a defect in Cl⁻ channel function. This feature applies specifically to the distal intestine as well, where CF can cause meconium ileus. Initially, investigators claimed that CF is not a defect of the channel itself but a defect in its regulation (Schoumacher *et al.*, 1987; Li *et al.*, 1988; Hwang *et al.*, 1989; McCann and Welsh, 1990). This hypothesis was based on data obtained on the ICOR channel. Since these data have lost their credibility (cf., preceding discussion), this interpretation is now dismissed.

The cloning of CFTR (Riordan *et al.*, 1989) and the subsequent expression of CFTR in systems such as CHO cells, HeLa cells, NIH 3T3 fibroblasts, CF cells, Vero cells, Sf9 insect cells, and *Xenopus* oocytes suggested that CFTR itself is the Cl⁻ channel (Drumm *et al.*, 1990,1991; Rich *et al.*, 1990; Anderson *et al.*, 1991a,b; Dalemans *et al.*, 1991; Kartner *et al.*, 1991; Tabcharani *et al.*, 1991; Cunningham *et al.*, 1992). On the other hand, the structure of CFTR does not compare with that described for any other ion channel to date. Instead, CFTR resembles an ATP-driven pump (Hyde *et al.*, 1990). At this stage, much data supports the view that CFTR is the Cl⁻ channel, but other studies emphasize additional functions of CFTR. The reader is referred to two reviews on this topic (Collins, 1992; Fuller and Benos, 1992).

A. Is CFTR a Chloride Channel?

Many of the symptoms of CF can be explained by defective Cl⁻ secretion. Despite the cloning of CFTR, expressing CFTR in CF tissues or in cells devoid of the mechanisms of cAMP-dependent Cl⁻ transport has been challenging, so studying whether Cl⁻ permeability is increased has been difficult. Several studies in various cell types document this point (Drumm *et al.*, 1990; Rich *et al.*, 1990; Anderson *et al.*, 1991b; Dalemans *et al.*, 1991; Kartner *et al.*, 1991; Tabcharani *et al.*, 1991; Cunningham *et al.*, 1992). These data might be explained if CFTR regulates Cl⁻ channels that are already present in these cells. From this point of view, the insect cell system examined by Kartner *et al.* (1991) is particularly interesting because these cells are very unlikely to contain this type of Cl⁻ channel. Still, CFTR expression in these cells, and subsequent function of cAMP-dependent Cl⁻ channels, might be caused by CFTR acting as some sort of regulator. Much more convincing along these lines are data showing that mutations of CFTR and their expression lead to altered anion selectivity of cAMP-dependent currents (Anderson *et al.*, 1991b). Unfortunately, this study is only presented as a brief note. Why the selected mutations should have had the reported effect on selectivity is not clear. The ultimate proof for the function of CFTR should be provided by its incorporation into lipid bilayers. This manipulation has been performed successfully in two laboratories (Bear *et al.*, 1992; Tilly *et al.*, 1992). In the first study purified CFTR was used; in the second, plasma membrane vesicles containing CFTR were used. Both studies concluded that CFTR must be the Cl⁻ channel. The incorporated channels were even PKA regulated. Additional studies on this important issue will be required to confirm these findings.

The complex structure of CFTR is not reconciled easily with its putative function as a Cl⁻ channel. What is the role of nucleotide binding folds? Why must ATP be hydrolyzed if the driving force for ion movement is provided by the electrochemical gradient for Cl⁻? These questions are currently under intense investigation in several laboratories (Cheng *et al.*, 1991; Rich *et al.*, 1991). Perhaps the current view of the structure of ion channels has been biased too much by information on neuronal channels such as Na⁺, Cl⁻, K⁺, and Ca²⁺ channels. On the other hand, CFTR may also have other (additional) functions. The evidence for this possibility will be reviewed in the following section.

B. CFTR Has Additional Functions

Initially, with the cloning of CFTR (Riordan *et al.*, 1989), it was entirely unclear whether CFTR was a channel or a channel regulator. The complex structure of the molecule was probably the cause for its name of "regulator" rather than "channel." In fact, several findings allude to its function as a regulator rather than as a channel. Bradbury *et al.* (1992) provided evidence that CFTR controls vesicle traffic to the plasma membrane. Further, assuming that all the defects of CF can be explained by a defective Cl⁻ channel seems to be an oversimplification. Compelling evidence suggests that Na⁺ reabsorption is enhanced in CF airway cells and that this event is of pathophysiological relevance (Quinton, 1990). Also, the initial finding that cells transfected with defective ΔF508 CFTR have no cAMP-dependent Cl⁻ permeability has not been confirmed (Dalemans *et al.*, 1991). Further, why should nonsense mutations of CFTR, which prevent expression completely, be clinically more mild than deletions such as ΔF508?

Other data cast some light on another (or additional) role that might be played by CFTR. In normal tissue (HT29 cells), cAMP and ATP or hypotonic cell swelling effects on Cl⁻ conductances are not additive.

On the other hand, cAMP cooperates with Ca^{2+}-mobilizing agents and produces a stable Cl^- conductance, whereas the Ca^{2+}-mobilizing agents produce a transient conductance only (Allert *et al.*, 1992). These data are supported further by other data: If HT29 cells are exposed to forskolin for some time (10–24 hr), they increase their concentration of CFTR, as measured in Western blots (Breuer *et al.*, 1992). This result coincides with an enhanced acute response to cAMP, that is, cAMP induces a larger Cl^- conductance response than in normal or Phorbol-12-mono-myristate-13-acetate (PMA) pretreated cells. This finding may be explained by the interpretation that CFTR is the cAMP-dependent Cl^- channel. However, these same cells also respond much more strongly to activation by ATP, and up-regulate their RVD-induced Cl^- conductance much longer than control cells. Therefore, all three pathways—cAMP, Ca^{2+}, and hypotonic cell swelling—appear to be controlled by the concentration of CFTR. Conversely, PMA suppresses CFTR, and blunts the cAMP response attenuates the ATP response, and increases the collapse of RVD-induced Cl^- conductance. These findings are not compatible with the view that the three major mechanisms to enhance Cl^- conductances act on distinct Cl^- channels, but suggest instead that CFTR stabilizes any Cl^- conductance, irrespective of its cause.

How could CFTR regulate the responses to Ca^{2+}-mobilizing agents and the Cl^- conductance induced in RVD? How can the grossly different time courses of Cl^- conductance activation be explained with rapid (<1 sec) activation by Ca^{2+}-mobilizing agents such as ATP and NT and much slower (30–180 sec) activation by cAMP? cAMP regulation probably occurs on CFTR via PKA-dependent phosphorylation. Since this step is rather slow, it cannot be distal to the Ca^{2+} control of Cl^- channels. To explain these findings, Allert *et al.* (1992) suggested that Ca^{2+} controls the exocytosis of vesicles containing Cl^- channels. This process should be fairly rapid. The transient nature of this conductance response might be explained by endocytosis of this Cl^- conductance-containing membrane. The rate of endocytosis might be regulated by cAMP. The process might be delayed with phosphorylated CFTR and fast with dephosphorylated CFTR or with mutated forms such as $\Delta F508$ CFTR.

This concept predicts that Ca^{2+}-mobilizing agents should enhance the membrane capacitance of these cells. A typical experiment is shown in Fig. 9. Here, a phase-tracking method was used with a dual-channel lock-in amplifier to increase the sensitivity of the measurement. Obviously the membrane capacitance is enhanced abruptly by ATP, an event that is paralleled by abrupt Ca^{2+} mobilization and a respective increase

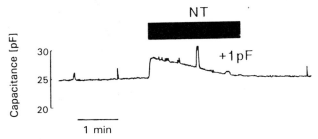

FIGURE 9 Measurement of membrane capacitance changes in an HT29 cell. The membrane capacitance was measured with a dual-channel lock in the amplifier. Note that neurotensin (NT, $10^{-8} M$) induced a rapid and fully reversible increase in the membrane capacitance by 3.5 pF. The capacitance changes were paralleled by the conductance increase, which was approximately 20-fold in this experiment.

in Cl^- conductance. In several series of experiments, with different techniques, agonists such as ATP, NT, and ionomycin have been verified to increase the membrane capacitance by 10–20%. Cell swelling, and probably also cAMP, leads to a much more delayed increase in membrane capacitance.

These data suggest that the Cl^- conductance of HT29 cells may be controlled by exocytosis of Cl^--channel-containing vesicles. cAMP control may act on the length of time this piece of membrane rests in the plasma membrane until it is endocytosed. This concept may not be unique to the colon carcinoma cell; similar mechanisms could be present in the parietal cell of the stomach where cAMP signals (via H_2-receptors) and Ca^{2+} signals (via muscarinic receptors and gastrin receptors) converge on the fusion of vesicles with the tubules of these cells to form the active tubulovesicular system. A "synarchic" mode of interaction between the two messenger systems of Ca^{2+} and cAMP is now increasingly recognized in epithelia (Rasmussen, 1990).

V. Conclusion

Colon carcinoma cells possess Cl^- channels that are probably relevant in Cl^- secretion. Three major pathways can activate these Cl^- channels: hormones acting via cAMP, such as adrenaline; hormones mobilizing cytosolic Ca^{2+}; and regulatory volume decrease. At this stage, two concepts for the function of these pathways have been suggested. One option assumes that each mechanism activates a different type of Cl^- channel. Abundant evidence suggests that the cAMP-induced Cl^- current is mediated by a 6-to-9-pS channel. The RVD-activated channel has been claimed to be of the ICOR type. The ATP-induced Cl^- channel has not yet been identified. The other concept suggests that all pathways converge on the same small or very small conductance Cl⁻ channel because cooperativity can be demonstrated among the three pathways. The two concepts emphasize different functional roles for CFTR. The first hypothesis states that CFTR is the cAMP-regulated Cl⁻ channel. The other assumes that CFTR somehow controls the time an exocytosed piece of Cl⁻ channel-containing membrane spends in the plasma membrane. Future studies will have to reveal how these concepts can be reconciled.

Acknowledgments

The work by the author and his group cited in this chapter has been supported continuously by Deutsche Forschungs Gemeinschaft Grant 480/10). I gratefully acknowledge the expert collaboration of N. Allert, U. Fröbe, H. Gögelein, M. Grolik, C. P. Hansen, J. P. Hayslett, B. Klär, R. Kubitz, K. Kunzelmann, J. Leipziger, R. Nitschke, B. Roch, E. Schlatter, M. Tilmann, and R. Warth.

References

Allert, N., Leipziger, J., and Greger, R. (1992). cAMP and Ca²⁺ act co-operatively on the Cl⁻ conductance of HT29 cells. *Pflügers Arch.* **421,** 403–405.

Anderson, M. P., and Welsh, M. J. (1991). Calcium and cAMP activate different chloride channels in the apical membrane of normal and cystic fibrosis epithelia. *Proc. Natl. Acad. Sci. USA* **88,** 6003–6007.

Anderson, M. P., Berger, H. A., Rich, D. P., Gregory, R. J., Smith, A. E., and Welsh, M. J. (1991a). Nucleoside triphosphates are required to open the CFTR chloride channel. *Cell* **67,** 775–784.

Anderson, M. P., Gregory, R. J., Thompson, S., Souza, D. W., Paul, S., Mulligan, R. C., Smith, A. E., and Welsh, M. J. (1991b). Demonstration that CFTR is a chloride channel by alteration of its anion selectivity. *Science* **253,** 202–205.

Augeron, C., and Laboisse, C. L. (1984). Emergence of permanently differentiated cell clones in a human colonic cancer cell line in culture after treatment with sodium butyrate. *Cancer Res.* **44,** 3961–3969.

Bajnath, R. B., Dekker, K., Vaandrager, A. B., De Jonge, H. R., and Groot, J. A. (1992). Biphasic increase of apical Cl⁻ conductance by muscarinic stimulation of HT29.19A human colon carcinoma cell line: Evidence for activation of different Cl⁻ conductances by carbachol and forskolin. *J. Membr. Biol.* **127,** 81–94.

Bear, C. E. (1988). Phosphorylation-activated chloride channels in human skin fibroblasts. *Fed. Eur. Biochem. Soc. Lett.* **237,** 145–149.

Bear, C. E., Li, C., Kartner, N., Bridges, R. J., Jensen, T. J., Ramjeesingh, M., and Riordan, J. R. (1992). Purification and functional reconstitution of the cystic fibrosis transmembrane conductance regulator (CFTR). *Cell* **68,** 809–818.

Bell, C. L., and Quinton, P. M. (1992). T84 cells anion selectivity demonstrates expression of Cl-conductance affected in cystic fibrosis. *Am. J. Physiol.* **262,** C555–C562.

Bleich, M., Schlatter, E., and Greger, R. (1990). The luminal K⁺ channel of the thick ascending limb of Henle's loop. *Pflügers Arch.* **415,** 449–460.

Bradbury, N. A., Jilling, T., Berta, G., Sorscher, E. J., Bridges, R. J., and Kirk, K. L. (1992). Regulation of plasma membrane recycling by CFTR. *Science* **256,** 530–532.

Breuer, W., Kartner, N., Riordan, J. R., and Cabantchik, Z. I. (1992). Induction of expression of the cystic fibrosis transmembrane conductance regulator. *J. Biol. Chem.* **267,** 10465–10469.

Bridges, R. J., Worrell, R. T., Frizzell, R. A., and Benos, D. J. (1989). Stilbene disulfonate blockade of colonic secretory Cl⁻ channels in planar lipid bilayers. *Am. J. Physiol.* **256,** C902–C912.

Cabantchik, Z. I., and Greger, R. (1992). Chemical probes for anion transporters of mammalian cell membranes. *Am. J. Physiol.* **262,** C803–C827.

Cartwright, C. A., McRoberts, J. A., Mandel, K. G., and Dharmsathaphorn, K. (1985). Synergistic action of cyclic adenosine monophosphate and calcium-mediated chloride secretion in a colonic epithelial cell line. *J. Clin. Invest.* **76,** 1837–1842.

Champigny, G., Verrier, B., and Lazdunski, M. (1990). Ca²⁺ channel blockers inhibit secretory Cl⁻ channels in intestinal epithelial cells. *Biochem. Biophys. Res. Commun.* **171,** 1022–1028.

Chan, H. C., Goldstein, J., and Nelson, D. J. (1992). Alternate pathways for chloride conductance activation in normal and cystic fibrosis airway epithelial cells. *Am. J. Physiol.* **262,** C1273–C1283.

Chen, J. H., Schulman, H., and Gardner, P. (1989). A cAMP-regulated chloride channel in lymphocytes that is affected in cystic fibrosis. *Science* **243,** 657–660.

Cheng, S. H., Rich, D. P., Marshall, J., Gregory, R. J., Welsh, M. J., and Smith, A. E. (1991). Phosphorylation of the R domain by cAMP-dependent protein kinase regulates the CFTR chloride channel. *Cell* **66,** 1027–1036.

Christensen, O., and Hoffmann, E. (1992). Cell swelling activates K⁺ and Cl⁻ channels as well as nonselective, stretch-activated cation channels in Ehrlich ascites tumor cells. *J. Membr. Biol.* **129,** 13–36.

Clancy, J. P., McCann, J. D., Li, M., and Welsh, M. J. (1990). Calcium-dependent regulation of airway epithelial chloride channels. *Am. J. Physiol.* **258,** L25–L32.

Cliff, W. H., and Frizzell, R. A. (1990). Separate Cl⁻ conductances activated by cAMP and Ca²⁺ in Cl-secreting epithelial cells. *Proc. Natl. Acad. Sci. USA* **87,** 4956–4960.

Cliff, W. H., Schoumacher, R. A., and Frizzell, R. A. (1992). cAMP-activated Cl⁻ channels in CFTR-transfected cystic fibrosis pancreatic epithelial cells. *Am. J. Physiol.* **262,** C1154–C1160.

Cohn, J. A., Nairn, A. C., Marino, C. R., Melhus, O., and Kole, J. (1992). Characterization of the cystic fibrosis transmembrane conductance regulator in a colonocyte cell line. *Proc. Natl. Acad. Sci. USA* **89,** 2340–2344.

Collins, F. S. (1992). Cystic fibrosis: molecular biology and therapeutic implications. *Science* **256,** 774–779.

Cunningham, S. A., Worrell, R. T., Benos, D. J., and Frizzell, R. A. (1992). cAMP-stimulated ion currents in *Xenopus* oocytes expressing CFTR cRNA. *Am. J. Physiol.* **262,** C783–C788.

Dalemans, W., Barbry, P., Champigny, G., Jallat, S., Dott, K., Dreyer, D., Crystal, R. G., Pavirani, A., Lecocq, J. P., and Lazdunski, M. (1991). Altered chloride channel kinetics associated with the ΔF508 cystic fibrosis mutation. *Nature (London)* **354,** 526–528.

Dharmsathaphorn, K., Mandel, K. G., Masui, H., and McRoberts, J. A. (1985). Vasoactive intestinal polypeptide-induced chloride secretion by a colonic epithelial cell line. Direct participation of a basolaterally localized Na⁺, K⁺, Cl⁻ co-transport system. *J. Clin. Invest.* **75,** 462–471.

Dreinhöfer, J., Gögelein, H., and Greger, R. (1988). Blocking kinetics of Cl⁻ channels in colonic carcinoma cells (HT29) as revealed by 5-nitro-2-(3-phenyl-propylamino)-benzoic acid (NPPB). *Biochim. Biophys. Acta* **956,** 135–142.

Drumm, M. L., Pope, H. A., Cliff, W. H., Rommens, J. M., Marvin, S. A., Tsui, L. C., Collins, F. S., Frizzell, R. A., and Wilson, J. M. (1990). Correction of the cystic fibrosis defect *in vitro* by retrovirus-mediated gene transfer. *Cell* **62**, 1227–1233.

Drumm, M. L., Wilkinson, D. J., Smit, L. S., Worrell, R. T., Strong, T. V., Frizzell, R. A., Dawson, D. C., and Collins, F. S. (1991). Chloride conductance expressed by ΔF508 and other mutant CFTRs in *Xenopus* oocytes. *Science* **254**, 1797–1799.

Fischer, H., Illek, B., Negulescu, P. A., Clauss, W., and Machen, T. E. (1992a). Carbacol-activated calcium entry into HT-29 cells is regulated by both membrane potential and cell volume. *Proc. Natl. Acad. Sci. USA* **89**.

Fischer, H., Kreusel, K. H., Illek, B., Machen, T. E., Hegel, U., and Clauss, W. (1992b). The outwardly rectifying Cl⁻ channel is not involved in cAMP mediated Cl⁻ secretion in HT29 cells. Evidence for a very-low-conductance Cl⁻-channel. *Pflügers Arch.* **422**, 159–167.

Franklin, C. C., Turner, J. T., and Kim, H. D. (1989). Regulation of Na⁺/K⁺/Cl⁻ cotransport and [³H]bumetanide binding density by phrobol esters in HT29 cells. *J. Biol. Chem.* **264**, 6667–6673.

Frizzell, R. A., Halm, D. R., Rechkemmer, G. R., and Shoemaker, R. L. (1986a). Chloride channel regulation in secretory epithelia. *Fed. Proc.* **45**, 2727–2731.

Frizzell, R. A., Rechkemmer, G. R., and Shoemaker, R. L. (1986b). Altered regulation of airway epithelial cell chloride channels in cystic fibrosis. *Science* **233**, 558–560.

Fuller, C. M., and Benos, D. J. (1992). CFTR! *Am. J. Physiol.* **267**, C267–C287.

Galietta, L. J. V., Barone, V., De Luca, M., and Romeo, G. (1991). Characterization of chloride and cation channels in cultured human keratinocytes. *Pflügers Arch.* **418**, 18–25.

Gallacher, D. V., Maruyama, Y., and Petersen, O. H. (1984). Patch clamp study of rubidium and potassium conductances in single cation channels from mammalian exocrine acini. *Pflügers Arch.* **401**, 361–367.

Gögelein, H. (1988). Chloride channels in epithelia. *Biochim. Biophys. Acta* **947**, 521–547.

Gögelein, H., Greger, R., and Schlatter, E. (1987a). Potassium channels in the basolateral membrane of the rectal gland of *Squalus acanthias*. Regulation and inhibitors. *Pflügers Arch.* **409**, 107–113.

Gögelein, H., Schlatter, E., and Greger, R. (1987b). The "small" conductance chloride channel in the luminal membrane of the rectal gland of the dogfish (*Squalus acanthias*). *Pflügers Arch.* **409**, 122–125.

Grasset, E., Bernabeu, J., and Pinto, M. (1985). Epithelial properties of human colonic carcinoma cell line Caco-2: Effect of secretagogues. *Am. J. Physiol.* **248**, C410–C418.

Greger, R. (1985). Ion transport mechanisms in thick ascending limb of Henle's loop of mammalian nephron. *Physiol. Rev.* **65**, 760–797.

Greger, R. (1986). Chlorid-transportierende Epithelien. *Physiol. Akt.* **2**, 47–58.

Greger, R., and Kunzelmann, K. (1991). Simultaneous recording of the cell membrane potential and properties of the cell attached membrane of HT29 colon carcinoma and CF-PAC cells. *Pflügers Arch.* **419**, 209–211.

Greger, R., and Schlatter, E. (1984). Mechanism of NaCl secretion in the rectal gland of spiny dogfish (*Squalus acanthias*). I. Experiments in isolated *in vitro* perfused rectal gland tubules. *Pflügers Arch.* **402**, 63–75.

Greger, R., Schlatter, E., Wang, F., and Forrest, J. N. Jr. (1984). Mechanism of NaCl secretion in rectal gland tubules of spiny dogfish (*Squalus acanthias*). III. Effects of stimulation of secretion by cyclic AMP. *Pflügers Arch.* **402**, 376–384.

Greger, R., Schlatter, E., and Gögelein, H. (1987). Chloride channels in the luminal membrane of the rectal gland of the dogfish (*Squalus acanthias*). Properties of the "larger" conductance channel. *Pflügers Arch.* **4**, 114–121.

Greger, R., Nitschke, R. B., Lohrmann, E., Burhoff, I., Hropot, M., Englert, H. C., and Lang, H. J. (1991). Effects of arylaminobenzoate-type chloride channel blockers on equivalent short-circuit current in rabbit colon. *Pflügers Arch.* **419**, 190–196.

Halley, D. J. J., Bijman, J., De Jonge, H. R., Sinaasappel, M., Neijens, H. J., and Niermeijer, M. F. (1990). The cystic fibrosis defect approached from different angles—New perspectives on the gene, the chloride channel, diagnosis and therapy. *Eur. J. Pediat.* **149**, 670–677.

Halm, D. R., Rechkemmer, G., Schoumacher, R. A., and Frizzell, R. A. (1989). Biophysical properties of a chloride channel in the apical membrane of a secretory epithelial cell. *Comp. Biochem. Physiol.* **90A**, 597–601.

Hanrahan, J. W., and Tabcharani, J. A. (1990). Inhibition of an outwardly rectifying anion channel by HEPES and related buffers. *J. Membr. Biol.* **116**, 65–77.

Hayslett, J. P., Gögelein, H., Kunzelmann, K., and Greger, R. (1987). Characteristics of apical chloride channels in human colon cells (HT29). *Pflügers Arch.* **410**, 487–494.

Hoffmann, E. K., and Simonsen, L. O. (1989). Membrane mechanisms in volume and pH regulation in vertebrate cells. *Physiol. Rev.* **69**, 315–382.

Hwang, T. C., Lu, L., Zeitlin, L., Gruenert, D. C., Huganir, R., and Guggino, W. B. (1989). Cl⁻ channels in CF: Lack of activation by protein kinase C and cAMP-dependent protein kinase. *Science* **244**, 1351–1353.

Hwang, T. C., Guggino, S. E., and Guggino, W. B. (1990). Direct modulation of secretory chloride channels by arachidonic and other *cis* unsaturated fatty acids. *Proc. Natl. Acad. Sci. USA* **87**, 5706–5709.

Hyde, S., Emsley, P., Hartshorn, M. J., Mimmack, M. M., Gileadi, U., Pearce, S. R., Gallagher, M. P., Gill, D. R., Hubbard, R. E., and Higgins, C. F. (1990). Structural model of ATP-binding proteins associated with cystic fibrosis, multidrug resistance and bacterial transport. *Nature* (*London*) **346**, 362–365.

Kartner, N., Hanrahan, J. W., Jensen, T. J., Naismith, A. L., Sun, S., Ackerley, C. A., Reyes, E. F., Tsui, L. C., Rommens, J. M., Bear, C. E., and Riordan, J. R. (1991). Expression of the cystic fibrosis gene in non-epithelial invertebrate cells produces a regulated anion conductance. *Cell* **64**, 681–691.

Krick, W., Disser, J., Hazama, A., Burckhardt, G., and Frömter, E. (1991). Evidence for a cytosolic inhibitor of epithelial chloride channels. *Pflügers Arch.* **418**, 491–497.

Kubitz, R., Warth, R., Allert, N., Kunzelmann, K., and Greger, R. (1992). Small conductance chloride channels induced by cAMP, Ca²⁺, and hypotonicity in HT29 cells: Ion selectivity, additivity, and stilbene sensitivity. *Pflügers Arch.* **421**, 447–454.

Kunzelmann, K., Pavenstädt, H., and Greger, R. (1989). Properties and regulation of chloride channels in cystic fibrosis and normal airway cells. *Pflügers Arch.* **415**, 172–182.

Kunzelmann, K., Tilmann, M., Hansen, Ch. P., and Greger, R. (1991). Inhibition of epithelial chloride channels by cytosol. *Pflügers Arch.* **418**, 479–490.

Kunzelmann, K., Grolik, M., Kubitz, R., and Greger, R. (1992a). cAMP-dependent activation of small-conductance Cl⁻ channels in HT29 colon carcinoma cells. *Pflügers Arch.* **421**, 230–237.

Kunzelmann, K., Hansen, Ch. P., Grolik, M., Tilmann, M., and Greger, R. (1992b). Regulation of the chloride conductance of colonic carcinoma cells. *In* "Cystic Fibrosis, Basic and Clinical

Research" (N. Hoiby and S. S. Pedersen, eds.), pp. 49–56. Excerpta Medica, Amsterdam.

Kunzelmann, K., Kubitz, R., Grolik, M., Warth, R., and Greger, R. (1992c). Small conductance Cl$^-$ channels in HT29 cells: Activation by Ca^{2+}, hypotonic cell swelling and 9-Br-cGMP. *Pflügers Arch.* **421**, 238–246.

Lang, F., Völkl, H., and Häussinger, D. (1990). General principles in cell volume regulation. *In* "Cell Volume Regulation" (K. W. Beyenbach, ed.), pp. 1–25. Karger, Basel.

Leipziger, J., Nitschke, R., and Greger, R. (1991). Transmitter-induced changes in cytosolic Ca^{2+} activity in HT29 cells. *Cell. Physiol. Biochem.* **1**, 273–285.

Li, M., McCann, J. D., Liedtke, C. M., Nairn, A. C., Greengard, P., and Welsh, M. J. (1988). Cyclic AMP-dependent protein kinase opens chloride channels in normal but not cystic fibrosis airway epithelium. *Nature (London)* **331**, 358–360.

Lin, M., Nairn, A. C., and Guggino, S. E. (1992). cGMP-dependent protein kinase regulation of a chlordie channel in T84 cells. *Am. J. Physiol.* **262**, C1304–C1312.

Lohrmann, E., Cabantchik, Z. I., and Greger, R. (1992). Transmitter induced changes of the membrane potential of HT29 cells. *Pflügers Arch.* **421**, 224–229.

Mandel, K. G., Dharmsathaphorn, K., and McRoberts, J. A. (1986). Characterization of a cyclic AMP-activated Cl$^-$ transport pathway in the apical membrane of a human colonic epithelial cell line. *J. Biol. Chem.* **261**, 704–712.

Marty, A., Tan, Y. P., and Trautmann, A. (1984). Three types of calcium-dependent channel in rat lacrimal glands. *J. Physiol.* **357**, 293–325.

McCann, J. D., and Welsh, M. J. (1990). Regulation of Cl$^-$ and K$^+$ channels in airway epithelium. *Annu. Rev. Physiol.* **52**, 115–135.

Morris, A. P., Kirk, K. L., and Frizzell, R. A. (1990). Simultaneous analysis of cell Ca^{2+} and Ca^{2+}-stimulated chloride conductance in colonic epithelial cells (HT29). *Cell Reg.* **1**, 951–963.

Nitschke, R., Leipziger, J., and Greger, R. (1993a). Agonist induced intracellular Ca^{2+} transients in HT29 cells. *Pflügers Arch.* **423**, 519–526.

Nitschke, R., Leipziger, J., and Greger, R. (1993b). Intracellular Ca^{2+} transients in HT29 cells by hypotonic cell swelling. *Pflügers Arch.* **423**, 274–279.

Paulmichl, M., Li, Y., Wickman, K., Ackerman, M., Peralta, E., and Clapham, D. (1992). New mammalian chloride channel identified by expression cloning. *Nature (London)* **356**, 238–241.

Petersen, O. H. (1992). Stimulus-secretion coupling: cytoplasmic calcium signals and the control of ion channels in exocrine acinar cells. *J. Physiol.* **448**, 1–51.

Quinton, P. M. (1983). Chloride impermeability in cystic fibrosis. *Nature (London)* **301**, 421–422.

Quinton, P. M. (1990). Cystic fibrosis: A disease in electrolyte transport. *FASEB J.* **4**, 2709–2717.

Rasmussen, H. (1990). Stimulus-secretion coupling: General models and specific aspects in epithelial cells. *Meth. Enzymol.* **191**, 661–676.

Rich, D. P., Anderson, M. P., Gregory, R. J., Cheng, S. H., Paul, S., Jefferson, D. M., McCann, J. D., Klinger, K. W., Smith, A. E., and Welsh, M. J. (1990). Expression of cystic fibrosis transmembrane conductance regulator corrects defective chloride channel regulation in cystic fibrosis airway epithelial cells. *Nature (London)* **347**, 358–363.

Rich, D. P., Gregory, R. J., Anderson, M. P., Manavalan, P., Smith, A. E., and Welsh, M. J. (1991). Effect of deleting the R domain on CFTR-generated chloride channels. *Science* **253**, 205–207.

Riordan, J. R., Rommens, J. M., Kerem, B.-S., Alon, N., Rozmahel, R., Grzelczak, Z., Zielenski, J., Lok, S., Plavsic, N., Chou, J.-L., Drumm, M. L., Iannuzzi, M. C., Collins, F. S., and Tsui, L.-C. (1989). Identification of the cystic fibrosis gene: Cloning and characterization of complementary DNA. *Science* **245**, 1066–1073.

Schoumacher, R. A., Shoemaker, R. L., Halm, D. R., Tallant, E. A., Wallace, R. W., and Frizzell, R. A. (1987). Phosphorylation fails to activate chloride channels from cystic fibrosis airway cells. *Nature (London)* **330**, 752–754.

Schulz, I., and Frömter, E. (1968). Mikropunktionsuntersuchungen an Schweißdrüsen von Mukoviszidosepatienten und gesunden Versuchspersonen. *In* "Mucoviscidose Cystische Fibrose II" (A. Windhofer and U. Stephan, eds.), pp. 12–21. Thieme Verlag.

Singh, A. K., Afink, G. B., Venglarik, C. J., Wang, R., and Bridges, R. J. (1991). Colonic Cl channel blockade by three classes of compounds. *Am. J. Physiol.* **260**, C51–C63.

Solc, C. K., and Wine, J. J. (1991). Swelling-induced and depolarization-induced Cl$^-$ channels in normal and cystic fibrosis epithelial cells. *Am. J. Physiol.* **261**, C658–C674.

Sood, R., Bear, C., Auerbach, W., Reyes, E., Jensen, T., Kartner, N., Riordan, J. R., and Buchwald, M. (1992). Regulation of CFTR expression and function during differentiation of intestinal epithelial cells. *EMBO J.* **11**, 2487–2494.

Tabcharani, J. A., Low, W., Elie, D., and Hanrahan, J. W. (1990). Low-conductance chloride channel activated by cAMP in the epithelial cell line T84. *Fed. Eur. Biochem. Soc. Lett.* **270**, 157–164.

Tabcharani, J. A., Chang, X. B., Riordan, J. R., and Hanrahan, J. W. (1991). Phosphorylation-regulated Cl$^-$ channel in CHO cells stably expressing the cystic fibrosis gene. *Nature (London)* **352**, 628–631.

Tilly, B. C., Winter, M. C., Ostedgaard, L. S., O'Riordan, C., Smith, A., and Welsh, M. J. (1992). Cyclic AMP-dependent protein kinase activation of cystic fibrosis transmembrane conductance regulator chloride channels in planar lipid bilayers. *J. Biol. Chem.* **267**, 9470–9473.

Tilmann, M., Kunzelmann, K., Fröbe, U., Cabantchik, Z. I., Lang, H. J., Englert, H. C., and Greger, R. (1991). Different types of blockers of the intermediate conductance outwardly rectifying chloride channel (ICOR) of epithelia. *Pflügers Arch.* **418**, 556–563.

Wagner, J. A., McDonald, T. V., Nghiem, P. T., Lowe, A. W., Schulman, H., Gruenert, D. C., Stryer, L., and Gardner, P. (1992). Antisense oligodeoxynucleotides to the cystic fibrosis transmembrane conductance regulator inhibit cAMP-activated but not calcium-activated chloride currents. *Proc. Natl. Acad. Sci. USA* **89**, 6785–6789.

Warth, R., and Greger, R. (1993). The ion conductances of CFPAC-1 cells. *Cell. Physiol. Biochem.* **3**, 2–16.

Welsh, M. J. (1986). Single apical membrane anion channels in primary cultures of canine tracheal epithelium. *Pflügers Arch.* **407**, S116–S122.

Welsh, M. J. (1990). Abnormal regulation of ion channels in cystic fibrosis epithelia. *FASEB J.* **4**, 2718–2725.

Welsh, M. J., Li, M., and McCann, J. D. (1989). Activation of normal and cystic fibrosis Cl$^-$ channels by voltage, temperature, and trypsin. *J. Clin. Invest.* **84**, 2002–2007.

Worrell, R. T., and Frizzell, R. A. (1991). CaMKII mediates stimulation of chloride conductance by calcium in T84 cells. *Am. J. Physiol.* **260**, C877–C882.

Worrell, R. T., Butt, A. G., Cliff, W. H., and Frizzell, R. A. (1989). A volume-sensitive chloride conductance in human colonic cell line T84. *Am. J. Physiol.* **256**, C1111–C1119.

Structural Information on Background Chloride Channels Obtained by Molecular Genetics

ANNA PETRIS, CARLO TREQUATTRINI, AND
FABIO FRANCIOLINI

I. Introduction

Chloride channels participate in various cellular functions such as regulation of cell volume (Worrel *et al.*, 1989; Strange, 1991; Valverde *et al.*, 1992), transepithelial transport (Gogelein, 1988; Liedtke, 1989; Welsh, 1990), stabilization of membrane potential (Bretag, 1987; Steinmeyer *et al.*, 1991a), and signal transduction (Grenningloh *et al.*, 1987; Schofield *et al.*, 1987). Their importance is underscored by their involvement in the pathogenesis of diseases such as cystic fibrosis and myotonia. A large variety of chloride channels with different repertoires of behavior has been identified (Frizzell, 1987; Gogelein, 1988; Franciolini and Petris, 1990,1992).

This chapter focuses on background chloride channels, which are the anion-selective channels that show a significant probability of being open at resting membrane potential. Background chloride channels derive their name because they are thought to be important in maintaining a sufficiently high resting or background membrane conductance. This conductance stabilizes the membrane potential near the negative limits of the physiological range and, consequently, maintains normal cell excitability. Background chloride channels have intermediate conductance, usually ranging from 10 to 100 pS in symmetrical 150 mM Cl.

Unlike other chloride-selective channels (i.e., glycine and GABA receptors), identification and purification of background chloride channels has been hampered by the lack of a highly specific binding agent. However, using more laborious molecular genetic strategies, the primary structures of several background chloride channels have been obtained.

The plasma membrane chloride channels cloned to date belong to different structural classes: ligand-gated channels (Grenningloh *et al.*, 1987; Schofield *et al.*, 1987), voltage-gated (background) channels (Jentsch *et al.*, 1990; Steinmeyer *et al.*, 1991a; Paulmichl *et al.*, 1992; Thiemann *et al.*, 1992), and, possibly, transporters of the ATP-binding cassette type [if, in fact, the cystic fibrosis transmembrane regulator (Riordan *et al.*, 1989) is a chloride channel (Anderson *et al.*, 1991a,b; Kartner *et al.*, 1991)]. In this chapter, the topological and structural organization of background chloride channels, as derived from a combined molecular genetic and electrophysiological approach, is reviewed. One section is dedicated to the chloride channels of airway epithelia, the malfunctioning of which, as a result of genetic mutations, causes cystic fibrosis.

II. ClC-0: Background Chloride Channel of *Torpedo*

A. General

The electric organ of *Torpedo californica* is an abundant source of the voltage-gated chloride channel (White and Miller, 1979; Miller and White, 1984; for review, see Miller and Richard, 1990). This channel is thought to insure the high conductance of the non-innervated membrane of the electrocyte that is necessary for efficient current generation caused by Na^+ influx through the acetylcholine receptor at the innervated membrane. Despite its abundance, all attempts to identify the channel protein using chemicals that block the *Torpedo* channel (White and Miller, 1979), and presumably bind covalently to it, have failed because of the low binding specificity of the inhibitors used (Taguchi and Kasai, 1980; Jentsch *et al.*, 1989).

B. Topological Organization

In the absence of biochemical data on the protein(s) constituting the channel, Jentsch *et al.* (1990) used a different approach that relied only on function. This method proved successful, since they were able to clone the complete cDNA encoding the chloride channel of *Torpedo marmorata*, isolated by expression cloning in *Xenopus* oocytes. The chloride channel of *Torpedo* (ClC-0) consists of 805 amino acids. Based on hydropathy analysis, the protein shows nine hydrophobic domains of 20–25 amino acids (thus just capable of crossing the membrane as α helices), as

well as two additional hydrophobic domains that are twice as long. Proline or glycine residues are present in the middle of the latter domains. These residues are known to break secondary structures and turn the peptide chains, suggesting that the chains make a sharp U-turn but can still be classified as transmembrane segments. A consensus phosphorylation site for cAMP-dependent kinase is also present in the long intracellular stretch interposed between D12 and D13 (see Fig. 1). Hydropathy analysis showed that the D13 domain is a poor candidate for a transmembrane segment, yet it is functionally important, as shown by the fact that if the *Torpedo* channel is truncated between D12 and D13, the protein becomes nonfunctional in *Xenopus* oocyte. Moreover, D13 is highly conserved in ClC-1, the chloride channel from skeletal muscle of rat (see Section III).

A possible topological organization of the *Torpedo* chloride channel is shown in Fig. 1. When properly arranged on particular domains of the channel protein, charged residues determine the two major channel functions: gating (in voltage-gated channels) and selectivity. Gating is usually caused by a densely charged protein domain that responds to changes in the electric field by moving across the membrane, thus favoring conformational transitions. Although charged residues are present in the *Torpedo* chloride channel, they do not appear to be arranged in a way that would suggest a gating function as observed, for example, in the S4 segment of the sodium channel with its strikingly regular sequence of one charged residue for every two hydrophobic residues (Noda *et al.*, 1984). Similarly, selectivity is also thought to be

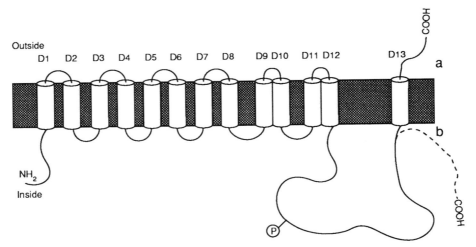

FIGURE 1 Membrane topology of ClC-0. (A) All hydrophobic domains D1–D13 are assumed to cross the plasma membrane as α helices, here shown as cylinders. (B) Domain D13, which displays intermediate hydrophobicity, is assumed not to cross the plasma membrane. P indicates the potential phosphorylation site. Modified with permission from Jentsch *et al.*, (1990). Copyright © 1990 Macmillan Magazines Ltd.

the result of regularly arranged charged (or polar) residues that line the channel pore (for review, see Unwin, 1989).

C. Comparison with Other Chloride Channels

The ClC-0 channel shows no significant similarity to the ligand-gated, chloride-selective GABA- (Schofield *et al.*, 1987) and glycine- (Grenningloh *et al.*, 1987) activated channels, or to voltage-gated Na$^+$ (Noda *et al.*, 1984) or Ca^{2+} (Tanabe *et al.*, 1987; K, Papazian *et al.*, 1987) channels. Certain biophysical characteristics of the cloned ClC-0 (e.g., slow activation by hyperpolarization, current increase with depolarization once activated) seem to indicate that this channel is the chloride channel that has been extensively studied in reconstituted membranes, and is known as the "double-barrelled" chloride channel because of its peculiar kinetics (White and Miller, 1981; Miller, 1982; for review, see Miller and Richard, 1990). Researchers have suggested that the "double-barrelled" chloride channel is a homodimer made of two identical monomers sharing a common gating mechanism. However, the channel also displays an additional gating system that opens and closes each monomer individually in a frequency domain at least one order of magnitude faster than that of the common gating mechanism (Miller and White, 1984). A similar "double-barrelled" chloride channel described in the mammalian kidney (Sansom *et al.*, 1990) displays the same peculiar voltage activation as the cloned chloride channel of *Torpedo*.

Bauer *et al.* (1991) studied ClC-0 expressed in *Xenopus* oocytes injected with cRNA. Their results show that the properties of the newly expressed ClC-0 are almost identical to those reported for the reconstituted native *T. californica* chloride channel (Miller and Richard, 1990). ClC-0 is expressed mostly in clusters in the oocyte membrane. Stocker *et al.* (1990) observed a similar clustering in newly expressed potassium channels. Whether this clustering is an artifact of the oocyte expression system or a natural behavior of channels in the electric organ, for which single-channel data are not available, is unclear. According to Bauer *et al.* (1991), the "double-barrelled" model describes a functional homodimer (composed of two protochannels), which does not imply that it is a homodimer of two proteins encoded by the cloned cDNA. Each of the two protochannels could be an oligomer of two or more identical proteins, and a protein monomer may generate the "double-barrelled" characteristics. A dimeric, or a more general multimeric, channel structure, may be a general phenomenon, because two other "double-barrelled" chloride channels, identified in distal rabbit nephron (Sansom *et al.*, 1990) and in *Aplysia* neurons (Chesnoy-Marchais, 1990), as well as a "triple-barrelled" cardiac potassium channel (Matsuda *et al.*, 1989), have been described. The single polypeptide encoded by the cDNA is able to generate the completely functional "double-barrelled" chloride channel ClC-0. However, currently, no conclusive structural correlation between the cloned channel ClC-0 and the "double-barrelled" chloride channel can be made. Nevertheless, the heterologous expression of a channel with a multibarrelled structure will prove instrumental in future structure–function studies of this class of channel.

III. ClC-1: Background Chloride Channel of Rat Skeletal Muscle

A. General

In skeletal muscle, 70–85% of resting membrane conductance is carried by chloride ions (Bretag, 1987). Since chloride conductance is the largest resting conductance of twitch muscle and chloride ions are distributed almost at equilibrium, chloride channels are thought to stabilize the membrane potential, opposing deviations from the resting state. The importance of chloride channels is obvious in myotonia congenita, in which muscle chloride conductance is unusually low. Consequently, repetitive action potentials are fired (Rudel and Lehmann-Horn, 1985). Similar hyperexcitability can also be induced in isolated muscle by experimental blockage of chloride conductance (Bryant and Morales-Aguilera, 1971).

B. Topological Organization

Steinmeyer *et al.* (1991a) cloned the cDNA encoding a rat skeletal muscle chloride channel by homology screening with the chloride channel from *Torpedo*. These investigators called this channel ClC-1 to distinguish it from ClC-0 and from the mammalian chloride channel ClC-2 (see Section IV). The predicted ClC-1 protein has 994 amino acids and is about 54% identical to ClC-0. Although larger than the *Torpedo* chloride channel, the chloride channel from skeletal muscle has a very similar topology, as suggested by hydropathy analysis, which shows 13 putative membrane-spanning domains. The greatest homology with ClC-0 is in the putative hydrophobic transmembrane α helices and in adjacent regions.

Despite structural and topological similarities, several functional differences exist between ClC-1 and ClC-0, the most conspicuous of which is the lack of slow channel activation by hyperpolarization in ClC-0 (Jentsch *et al.*, 1990), which may result from slow opening of a gate present on both protochannels of the "double-barrelled" channel (Miller, 1982; Miller and White, 1984; Miller and Richard, 1990; Bauer *et al.*, 1991). Whether this difference means that ClC-1 has no "double-barrelled" structure remains to be elucidated in single-channel studies.

C. Developmental Regulation

The tissue distribution of ClC-1, examined by Northern blot analysis, shows that the channel is expressed predominantly in skeletal muscle. Small amounts of ClC-1 mRNA are also detectable in kidney, liver, and heart. In rat muscle, chloride conductance increases markedly during the first weeks after birth (Conte-Camerino *et al.*, 1989). ClC-1 mRNA increases rapidly as will between days 1 and 30 after birth, suggesting an increased channel density as the primary mechanism of postnatal conductance increases (Steinmeyer *et al.*, 1991a). Developmental changes in expression are also known for other muscle channels (Mishina *et al.*, 1986; Trimmer *et al.*, 1990), suggesting similar regulatory mechanisms.

D. ETn Transposon and Myotonia

ClC-1 is expressed predominantly in skeletal muscle, and probably provides the major chloride conductance in this tissue. In fact, its destruction in mouse mutants leads to myotonia (Steinmeyer *et al.*, 1991b). ADR mice provide a good model for studying recessive autosomal myotonia (Rudel, 1990). Steinmeyer *et al.* (1991a) used ClC-1 as a probe in genomic Southern analysis and found aberrant fragments in ADR (homozygous *adr/adr*) mice with several restriction endonucleases. The aberrant fragments of ADR were always larger than the wild-type fragment, suggesting an insertional mutation.

To examine the *adr* mutation in detail, Steinmeyer *et al.* (1991b) screened a complementary DNA library from *adr/adr* skeletal muscle with ClC-1 cDNAs. These authors found four cDNAs, each extending into the D9 transmembrane segment but diverging after the triplet encoding Lys 467. This position corresponds to an exon–intron boundary, suggesting that this exon is spliced to sequences unrelated to the chloride channel. Among the four cDNAs examined, three different splice variants were found that encoded sequences unrelated to the chloride channel. These sequences are nearly identical to known members of the ETn family of mouse transposon (Sonigo, 1987).

In ADR mice, a transposon of this family is inserted into the chloride channel gene (probably into the intron in D9), destroying its coding potential for several membrane-spanning domains. In addition, no recombination occurs between the ClC-1 gene and the *adr* locus. These two factors strongly suggest that the absence of functional chloride channels is the primary cause of mouse myotonia.

IV. ClC-2: Background Chloride Channel from Epithelial and Nonepithelial Cells

A. Topological Organization

Thiemann *et al.* (1992) cloned the cDNA encoding a novel chloride channel, ClC-2, by homology screening to the rat skeletal muscle chloride channel ClC-1. The predicted protein (907 amino acids) has a transmembrane topology similar to that of ClC-0 and ClC-1. Overall homology to the *Torpedo* channel, ClC-0, and the rat muscle channel, ClC-1, is 49 and 55%, respectively. The channel protein has an unusual hydrophobic stretch of about 11 amino acids that follows the putative initiator methionine. Divergence from ClC-0 and ClC-1 also occurs at the C terminus and between domains D12 and D13.

B. Tissue Distribution and Possible Functions

ClC-2 has been isolated from rat heart and brain, but is also expressed in many other tissues such as skeletal muscle (with ClC-1), lung, kidney, pancreas, stomach, intestine, and liver, as well as in cell lines of fibroblastic, neuronal, and epithelial origin including a human pancreatic tumor line expressing the cystic fibrosis phenotype. The channel is closed at physiological voltages and activates slowly after hyperpolarization. Thus, some other mechanism of activation may exist *in vivo*.

The expression of ClC-2 in so many different tissues and cells indicates that this channel may play an important housekeeping role. Expression in the T84 cell line (human colon carcinoma) is especially interesting because these cells have served as a model system for chloride transport. These cells express at least four chloride channels (Worrel *et al.*, 1989; Solc and Wine, 1991; Anderson and Welsh, 1991). ClC-2 closely resembles one of them—the ORDIC (outwardly rectifying depolarization-induced chloride) channel (Ward *et al.*, 1991)—formerly thought to be altered in cystic fibrosis (for review, see Welsh, 1990). However, whether the channel affected in cystic fibrosis can be activated by hyperpolarization, as ClC-2 can, and whether ClC-2 can interact with the product of the cystic fibrosis gene, CFTR (see Section VI), is unclear. The functions of ClC-2, although unidentified, are important because the channel is expressed in cells affected by cystic fibrosis.

V. I_{Cln}: New Structural Motif for Ion Channels

A. Topological Organization

Using expression cloning techniques, Paulmichl *et al.* (1992) isolated and sequenced a cDNA from a canine epithelial cell line (MDCK) encoding a 235-amino-acid protein that gives rise to a chloride-selective outward current (I_{Cln}) when expressed in *Xenopus* oocytes. This putative chloride channel is unusual because hydrophobicity analysis revealed no potential transmembrane helices. The proposed model is shown in Fig. 2, in which two identical 235-amino-acid proteins are arranged as a dimer of four β sheets each. If the channel is a dimer, an eight-stranded antiparallel β barrel might form the pore as in the voltage-dependent anion-selective channel (VDAC) of yeast mitochondrial outer membrane (Blachly-Dyson *et al.*, 1990). Each subunit contains a putative nucleotide-binding site (G-X-G-X-G) that lies near the external mouth of the channel. The expressed channel is reversibly inhibited by cGMP and cAMP and, to a lesser extent, by GTP. However, if only the first of the three glycine residues of the nucleotide

binding site is substituted by alanine, the cAMP sensitivity is lost. The mutation also alters the channel kinetics and confers sensitivity to external calcium concentrations, suggesting that nucleotide- and calcium-binding regions are both located on the external side of the membrane.

B. Comparison with the VDAC Channel

The VDAC channel (or mitochondrial porin) consists of 283 amino acid residues, with 45% charged and polar residues and many stretches of alternating hydrophobic and hydrophilic residues. These stretches could form a β sheet with hydrophobic residues protruding on one side and hydrophilic residues on the other. This possible structure has led to a model of the VDAC channel as a β barrel consisting of a single layer of such a β sheet (Forte *et al.*, 1987).

Based on the effects of site-directed mutagenesis on VDAC selectivity changes, Blachly-Dyson *et al.* (1990) developed a model for the open channel in which each polypeptide contributes 12 β strands and one α helix to form the aqueous pore. The model is also consistent with evidence that VDAC is composed of two identical subunits (Colombini, 1986).

Site-directed mutagenesis experiments showed that, in VDAC and I_{Cln}, channel-specific amino acid residues are crucial for channel function. In VDAC, the replacement of some of the positively charged amino acids lining the channel with negatively charged amino acids (e.g., Lys → Glu) reduces or even reverses channel selectivity. Substitution does not change the selectivity when the amino acids involved do not belong to transmembrane segments (Blachly-Dyson *et al.*, 1990).

Despite their structural similarity, VDAC and I_{Cln} differ in the number of β strands that form the β barrel, as well as in the presence in the VDAC channel of an N-terminal α helix typical of mitochondrial targeting sequences (Roise *et al.*, 1986; von Heijne, 1986). Thus, the I_{Cln} channel may not only represent an independent family of chloride channels, but also display a new structural motif for ion channels.

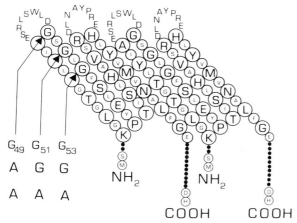

FIGURE 2 Putative channel model of I_{Cln}. Two identical 235-amino-acid proteins are arranged as a dimer of four β sheets each. The large circles represent amino acids facing the inside of the pore, whereas the small circles are amino acids facing the outside. Arrows indicate the experimentally mutated amino acids. Only the proposed transmembrane β sheets and the extracellular loops are shown in detail. Reprinted with permission from Paulmichl *et al.*, (1992). Copyright © 1992 Macmillan Magazines Ltd.

VI. CFTR: Cystic Fibrosis Transmembrane Conductance Regulator

A. General

Cystic fibrosis (CF) is a genetic disease characterized by defective epithelial chloride transport (Boat *et al.*, 1989; Quinton, 1990). CF results from mutations in a single gene that encodes the cystic fibrosis transmembrane conductance regulator (CFTR; Kerem *et al.*, 1989; Riordan *et al.*, 1989; Drumm *et al.*, 1990; Rich *et al.*, 1990). The most frequent mutation (60–70% of CF patients) is the deletion of phenylalanine residue 508 (ΔF508) in the cytoplasmic domain NDB1 (cf., Fig. 3). CFTR can function as a chloride channel (Anderson *et al.*, 1991a) that requires both cAMP-dependent phosphorylation and ATP hydrolysis to open (Anderson *et al.*, 1991b,c). The CFTR encoded by the mutated gene is unable to open in response to phosphorylation by cAMP-dependent protein kinases. This deficiency alters chloride transport, which gives rise to the most severe clinical symptom—accumulation of dense mucus in the lung airways.

B. Topological Organization

The CFTR is a single polypeptide of 1480 amino acids that consists of two structurally similar repeats. Hydropathy profile and predicted secondary structure studies indicate that each repeat consists of six hydrophobic segments long enough to cross the membrane as α helices and a cytoplasmic domain, named the nucleotide binding domain (NBD), that contains a consensus sequence for ATP binding (Fig. 3). The overall symmetry of the two repeats with respect to their topological arrangement within the membrane is remarkable, but amino acid homology is scarce. The strongest sequence identity occurs at the nucleotide binding site. The two repeats are joined by a densely charged cytoplasmic domain (R in Fig. 3) that carries several consensus sequences for protein kinase (A and B) binding and probably has a regulatory function. This structural arrangement shares considerable organization and sequence similarity with several mem-

bers of the ABC (ATP-binding cassette) superfamily of membrane proteins that are involved in transmembrane transport (Hyde *et al.*, 1990).

C. Is the CFTR a Chloride Channel?

The presence of a domain containing several transmembrane α helices in the cloned protein, a feature considered to be the hallmark of ionic channels and transport proteins, in conjunction with the presence of ATP-binding sites, strongly suggests that CFTR is the chloride-conducting pathway of airway epithelia. CFTR is normally activated by ATP-dependent phosphorylation, but loses its sensitivity in the mutated form of cystic fibrosis. However, the observation that the CFTR differed from any known ionic channel but was very similar to members of a family of transport proteins suggested that it might play a regulatory role, rather than a role as an ion pathway.

To demonstrate that the CFTR is the cAMP-activated chloride channel, and not a channel regulator, point mutations of the charged amino acids in the transmembrane segments were made to determine whether changes in conductance or selectivity would follow (Anderson *et al.*, 1991a). Replacement of positively charged lysine with an acidic amino acid changed the selectivity sequence of the channel. This result suggested that the transmembrane domains of the CFTR participate in forming the channel pore and in determining its selectivity. Indeed, replacing only charged residues Lys 95 and Lys 335 on the outer side of the channel markedly affected ion permeation, whereas replacing residues Lys 347 and Lys 1030 on the inner side had negligible effect.

The molecular genetic approach was also used to

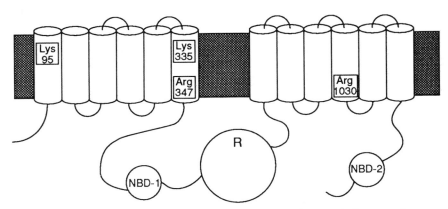

FIGURE 3 Membrane topology of CFTR. The transmembrane α helices are shown as cylinders. The ATP-binding domains (NBD-1 and NBD-2) and the strongly charged R domain that joins the two structurally similar repeats, forming the channel, are also shown. Modified with permission from Anderson *et al.* (1991a). Copyright © 1991 by the AAAS.

delete the R domain (Rich *et al.*, 1991). CFTR with a deleted R domain (CFTRΔR) conducted chloride ions even in the absence of cAMP. However, the basal chloride current increased when cAMP was added. Although this result indicates that the R domain confers some cAMP dependence on the CFTR, the ability of cAMP to further stimulate the chloride current in CFTRΔR suggests that cAMP-sensitive regulatory sites are also present in other parts of the channel protein.

Based on these observations, researchers have proposed that the regulatory R domain physically occludes the channel. This "plug effect" of the R domain can be removed by cAMP-dependent phosphorylation.

D. How Many Functions for CFTR?

CFTR is located in the apical membranes of many secretory cells (Crawford *et al.*, 1991). Epithelial cells show at least two other types of chloride channels, regulated by cell volume (Wagner *et al.*, 1991) and intracellular Ca^{2+} (Worrel *et al.*, 1989), respectively. Valverde *et al.* (1992) reported that volume-regulated chloride channels are generated in a human fibroblast cell line permanently transfected with MDR1 cDNA, the gene that encodes the multidrug-resistance P-glycoprotein. P-Glycoprotein-dependent chloride currents are distinct from those generated by CFTR expression, and have characteristics similar to those of volume-regulated chloride currents of airway epithelial cells (McCann *et al.*, 1989).

P-Glycoprotein is a member of the ABC superfamily of transporters that actively pumps hydrophobic drugs out of cells (Hyde *et al.*, 1990). The chloride currents associated with P-glycoprotein expression are channel-like and not consistent with ATP-dependent chloride transport or with chloride co- or countertransport with a drug molecule. On the basis of these and other observations about P-glycoprotein location in the apical membranes of secretory epithelia, Valverde *et al.* (1992) did not exclude the possibility that P-glycoprotein itself is a chloride channel, or part of one. This hypothesis is consistent with the finding that P-glycoprotein-dependent chloride currents are inhibited by the same drugs that inhibit drug transport by P-glycoprotein. Therefore, P-glycoprotein may provide an example of a single protein involved in both transport and channel function. P-Glycoprotein and CFTR are related to each other, but differ from other known channels (Hyde *et al.*, 1990); they may be members of an independent family of ion channels.

A new function of CFTR was discovered (Bradbury *et al.*, 1992) that could explain the pleiotropic nature of CF. CFTR is involved in the regulation of plasma membrane recycling in pancreatic cell lines. The cystic fibrosis protein, when triggered by cAMP, allows chloride ions and water to leave the cell, and causes vesicles in the cell to fuse with the cell membrane. This behavior suggests that CFTR function goes beyond the regulation of transmembrane chloride conductance. Whatever the mechanism by which CFTR regulates both chloride permeability and plasma-membrane recycling, clearly CFTR participates in the cAMP-dependent regulation of exo- and endocytosis. Moreover, CFTR may act as a chloride channel in cytoplasmic membranes (Barash *et al.*, 1991), thus regulating the trafficking of intracellular membranes and associated proteins.

VII. Conclusion

Several chloride channels have been characterized at the structural level by identifying their primary sequences. Although this approach helps clarify the channel architecture and the topology of its transmembrane segments, this structural understanding does not give satisfying information about channel functions at the molecular level. Moreover, comparison of the structural organization of background chloride channels reveals no unifying features that can explain how these channels are constructed or how they work.

ClC-0, ClC-1, and ClC-2 are the background chloride channels most closely related to each other in their structure. These channels have an overall homology of about 50% and very similar transmembrane topology. Although similarities between ClC-0 and ClC-1 could be expected because of their homologous tissue origin (ClC-1 is predominantly expressed in rat skeletal muscle and ClC-0 was isolated from *Torpedo* electric organ, which is developmentally derived from skeletal muscle), the structural resemblance of ClC-2 to both ClC-0 and ClC-1 is less expected. In fact, ClC-2 is expressed in many tissues and cells of very different origin, suggesting that it might play an important housekeeping role in most cells. However, as stressed in Section IV,B, no information is available about the functions of these channels at the molecular level. The lack of segments similar to the S4 domains of cation-selective voltage-gated channels, which are thought to be the voltage sensor, makes ClC-0, ClC-1, and ClC-2 unlikely candidates for the superfamily of channels to which other voltage-sensitive ion channels belong. At this time, comparison among these channels can only be speculative. The unusual sequence of the I_{Cln} channel from a canine epithelial cell line is even more surprising, and suggests a new class of ion channels not related to other cloned chloride channels. The eight-stranded antiparallel β barrel of this channel (Fig. 2) might form the channel pore, as in the VDAC channel of yeast mitochondrial outer membrane.

CF was the first genetic disease involving an ion channel to be studied at the molecular level. Since the cloning of the CF gene (Riordan et al., 1989) much has been learned about the structure and function of the CFTR protein. Now little doubt remains that CFTR is a chloride channel, although its name reflects the initial uncertainty about the function of the protein.

Collins (1992), in reviewing most of the findings on CFTR, suggested a model in which CFTR is a chloride channel modulated by the cytoplasmic domains NBD-1, NBD-2, and R. The R domain contains four serine residues; in the absence of R-domain phosphorylation, the channel is closed. After cAMP stimulates protein kinase A to phosphorylate one or more serine residue, the NBDs of CFTR bind ATP, which is cleaved to induce a conformational change that opens the channel. Therefore, the channel opens only if the two events (i.e., R-domain phosphorylation and ATP binding to NBDs) occur in sequence.

Unfortunately, this model that accounts for the dual control of CFTR, involving both protein kinase A and ATP, conflicts with the results from other experiments (Arispe et al., 1992) that show that recombinant NBD-1 reconstituted into a planar lipid bilayer has an intrinsic anion channel activity that is blocked by ATP. Moreover, circular dichroism studies of recombinant NBD-1 indicate a substantial proportion of β sheet structure, suggesting that this domain could span the membrane.

Another important point is also under discussion. Researchers have demonstrated that CFTR is involved in the regulation of cAMP-dependent endocytosis and exocytosis (see Section VI,D). In mutant epithelial cells from CF patients, this function of CFTR is defective, as is the function of chloride channel. This result raises the question: Are the symptoms of CF derived from the mutant protein's failure to perform its function as regulator of endocytosis and exocytosis or from its failure to function as a chloride channel? The defective chloride conductance can explain the fact that affected cells secrete less water than normal, but cannot explain the abnormal composition of the mucus proteins in CF patients. A possible answer to this question is that CFTR could act as a chloride channel in cytoplasmic membranes, thus acidifying some cell compartments in which pH-sensitive enzymes modify proteins. In CF, when CFTR does not function, the pH of these compartments is higher than normal, which could change the composition of mucus proteins.

Acknowledgments

We thank Sister Nancy Hutchinson for her help on the English revision of the text. Financial support for this work was provided by Grant Progetto Bilaterale No. 90.01482.CT04 from the Italian Consiglio Nazionale Ricerche.

References

Anderson, M. P., and Welsh, M. J. (1991). Calcium and cAMP activate different chloride channels in the apical membrane of normal and cystic fibrosis epithelia. Proc. Natl. Acad. Sci. USA 88, 6003–6007.

Anderson, M. P., Gregory, R. J., Thompson, S., Souza, D. W., Paul, S., Mulligan, R. C., Smith, A. E., and Welsh, M. J. (1991a). Demonstration that CFTR is a chloride channel by alteration of its anion selectivity. Science 253, 202–205.

Anderson, M. P., Berger, H. A., Rich, D. P., Gregory, R. J., Smith, A. E., and Welsh, M. J. (1991b). Nucleotide triphosphates are required to open the CFTR chloride channel. Cell 67, 775–784.

Anderson, M. P., Rich, D. P., Gregory, R. J., Smith, A. E., and Welsh, M. J. (1991c). Generation of cAMP-activated chloride currents by expression of CFTR. Science 251, 679–682.

Arispe, N., Rojas, E., Hartman, J., Sorscher, E. J., and Pollard, H. B. (1992). Intrinsic anion channel activity of the recombinant first nucleotide binding fold domain of the cystic fibrosis transmembrane regulator protein. Proc. Natl. Acad. Sci. USA 89, 1539–1543.

Barash, J., Kiss, B., Prince, A., Saiman, L., Gruenert, D., and Al-Awqati, A. (1991). Defective acidification of intracellular organelles in cystic fibrosis. Nature (London) 352, 70–73.

Bauer, C. K., Steinmeyer, K., Schwarz, J. R., and Jentsch, T. J. (1991). Completely functional double-barreled chloride channel expressed from a single Torpedo cDNA. Proc. Natl. Acad. Sci. USA 88, 11052–11056.

Blachly-Dyson, E., Peng, S., Colombini, M., and Forte, M. (1990). Selectivity changes in site-directed mutants of the VDAC ion channel: Structural implications. Science 247, 1233–1236.

Boat, T. F., Welsh, M. J., and Beudet, A. L. (1989). Cystic fibrosis. In "The Metabolic Basis of Inherited Disease" (C. R. Scriver, A. L. Beaudet, W. S. Sly, and D. Valle, eds.), pp. 2649–2680. McGraw-Hill, New York.

Bradbury, N. A., Jilling, T., Berta, G., Sorscher, E., Bridges, R. J., and Kirk, K. L. (1992). Regulation of plasma membrane recycling by CFTR. Science 256, 530–532.

Bretag, A. H. (1987). Muscle chloride channels. Physiol. Rev. 67, 618–724.

Bryant, S. H., and Morales-Aguilera, A. (1971). Chloride conductance in normal and myotonic muscle fibres and the action of monocarboxylic aromatic acids. J. Physiol. (London) 219, 367–383.

Chesnoy-Marchais, D. (1990). Hyperpolarization-activated chloride channels in Aplysia neurons. In "Chloride Channels and Carriers in Nerve, Muscle, and Glial Cells" (F. J. Alvarez-Leefmans and J. M. Russel, eds.), p. 367–382. Plenum Press, New York.

Collins, F. S. (1992). Cystic fibrosis: Molecular biology and therapeutic implications. Science 256, 774–779.

Colombini, M. (1986). Voltage gating in VDAC: Towards a molecular mechanism. In "Ion Channel Reconstitution" (C. Miller, ed.), pp. 533–550. Plenum Press, New York.

Conte-Camerino, D., De Luca, A., Mambrini, M., and Vrbova, G. (1989). Membrane ionic conductances in normal and denervated skeletal muscle of the rat during development. Pflügers Arch. 413, 568–570.

Crawford, I., Maloney, P. C., Zeitlin, P. L., Guggino, W. B., Hyde,

S. C., Turley, H., Gatter, K. C., Harris, A., and Higgins, C. F. (1991). Immunocytochemical localization of the cystic fibrosis gene product CFTR. *Proc. Natl. Acad. Sci. USA* **88**, 9262–9266.

Drumm, M. L., Pope, H. A., Cliff, W. H., Rommens, J. M., Marvin, S. A., Tsui, L. C., Collins, F. C., Frizzell, R. A., and Wilson, J. M. (1990). Correction of the cystic fibrosis defect *in vitro* by retrovirus-mediated gene transfer. *Cell* **62**, 1227–1233.

Forte, M., Guy, H. R., and Mannella, C. A. (1987). Molecular genetics of the VDAC ion channel: Structural model and sequence analysis. *J. Bioenerg. Biomembr.* **19**, 341–350.

Franciolini, F., and Petris, A. (1990). Chloride channels of biological membranes. *Biochim. Biophys. Acta* **103**, 247–249.

Franciolini, F., and Petris, A. (1992). Transport mechanism in chloride channels. *Biochim. Biophys. Acta* **1113**, 1–11.

Frizzell, R. A. (1987). Cystic fibrosis: A disease of ion channels? *Trends Neurosci.* **10**, 190–193.

Gogelein, H. (1988). Chloride channels in epithelia. *Biochim. Biophys. Acta* **947**, 521–547.

Grenningloh, G., Rienitz, A., Schmitt, B., Methfessel, C., Zensen, M., Beyreuther, K., Gundelfinger, E. D., and Betz, H. (1987). The strychnine-binding subunit of the glycine receptor shows homology with nicotine acetylcholine receptors. *Nature (London)* **328**, 215–220.

Hyde, S. C., Emsley, P., Hartshorn, M. J., Mimmack, M. M., Gileardi, U., Pearce, S. R., Gallagher, M. P., Gill, D. R., Hubbard, R. E., and Higgins, C. F. (1990). Structural models of ATP-binding proteins associated with cystic fibrosis, multidrug resistance and bacterial transport. *Nature (London)* **346**, 362–365.

Jentsch, T. J., Garcia, A. M., and Lodish, H. F. (1989). Primary structure of a novel 4-acetamido-4'-isothiocyanostilbene-2-2'-disulfonic acid (SITS)-binding membrane protein highly expressed in *Torpedo californica* electroplax. *Biochem. J.* **261**, 155–166.

Jentsch, T. J., Steinmeyer, K., and Schwarz, G. (1990). Primary structure of *Torpedo marmorata* chloride channel isolated by expression cloning in *Xenopus* oocytes. *Nature (London)* **348**, 510–514.

Kartner, N., Hanrahan, J. W., Jensen, T. J., Naismith, A. L., Sun, S., Ackerley, C. A., Reynes, E. F., Tsui, L. C., Rommens, J. M., Bear, C. E., and Riordan, J. R. (1991). Expression of the cystic fibrosis gene in non-epithelial invertebrate cells produces a regulated anion conductance. *Cell* **64**, 681–691.

Kerem, B., Rommens, J. M., Buchanam, J. A., Markievicz, D., Cox, T. K., Chakravarti, A., Buchwald, M., and Tsui, L. C. (1989). Identification of the cystic fibrosis gene: Genetic analysis. *Science* **245**, 1073–1080.

Liedtke, C. M. (1989). Regulation of chloride transport in epithelia. *Annu. Rev. Physiol.* **51**, 143–160.

Matsuda, H., Matsuura, H., and Noma, A. (1989). Triple-barrel structure of inwardly rectifying K^+ channel revealed by Cs^+ and Rb^+ block in guinea-pig heart cells. *J. Physiol. (London)* **413**, 139–157.

McCann, J. D., Li, M., and Welsh, M. J. (1989). Identification and regulation of whole-cell chloride currents in airway epithelium. *J. Gen. Physiol.* **94**, 1015–1036.

Miller, C. (1982). Open-state substructure of single chloride channels from *Torpedo* electroplax. *Phil. Trans. R. Soc. London B* **229**, 401–411.

Miller, C., and White, M. M. (1984). Dimeric structure of single chloride channel from *Torpedo* electroplax. *Proc. Natl. Acad. Sci. USA* **81**, 2772–2775.

Miller, C., and Richard, E. A. (1990). The voltage-dependent chloride channel of *Torpedo* electroplax: Intimations of molecular structure from quirks of single-channel function. *In* "Chloride Channels and Carriers in Nerve, Muscle, and Glial Cells" (F. J. Alvarez-Leefsmans and J. M. Russel, eds.), pp. 383–405. Plenum Press, New York.

Mishina, M., Takai, T., Imoto, K., Noda, M., Takahashi, T., and Numa, S. (1986). Molecular distinction between fetal and adult forms of muscle acetylcholine receptor. *Nature (London)* **321**, 406–411.

Noda, M., Shimizu, S., Tanabe, T., Takai, T., Kayano, T., Ikeda, T., Takahashi, H., Nakayama, H., Kanaoka, Y., Minamino, N., Kangawa, K., Matsuo, H., Raftery, M. A., Hirose, T., Inayama, S., Hayashida, H., Miyata, T., and Numa, S. (1984). Primary structure of *Electrophorus electricus* sodium channel deduced from cDNA sequence. *Nature (London)* **312**, 121–127.

Papazian, D. M., Schwarz, T. L., Tempel, B. L., Jan, Y. N., and Jan, L. Y. (1987). Cloning of genomic and complementary DNA from *Shaker*, a putative potassium channel gene from *Drosophila*. *Science* **237**, 749–753.

Paulmichl, M., Li, Y., Wickman, K., Ackerman, M., Peralta, E., and Clapham, D. (1992). New mammalian chloride channel identified by expression cloning. *Nature (London)* **356**, 238–241.

Quinton, P. M. (1990). Cystic fibrosis: A disease in electrolyte transport. *FASEB J.* **4**, 2709–2717.

Rich, D. P., Anderson, M. P., Gregory, R. J., Cheng, S. H., Paul, S., Jefferson, D. M., McCann, J. D., Klinger, K. W., Smith, A. E., and Welsh, M. J. (1990). Expression of cystic fibrosis transmembrane conductance regulator corrects defective chloride channel regulation in cystic fibrosis airway epithelial cells. *Nature (London)* **347**, 358–363.

Rich, D. P., Gregory, R. J., Anderson, M. P., Manavalan, P., Smith, A. E., and Welsh, M. J. (1991). Effect of deleting the R domain on CFTR-generated chloride channels. *Science* **253**, 205–207.

Riordan, J. R., Rommens, J. M., Kerem, B., Alon, N., Rozmahel, R., Grzelezak, Z., Zielenski, J., Lik, S., Plavsic, N., Chan, J. L., Drumm, M. L., Iannuzzi, M. C., Collins, F. S., and Tsui, L. C. (1989). Identification of the cystic fibrosis gene: Cloning and characterization of complementary DNA. *Science* **245**, 1066–1073.

Roise, D., Horvath, S. J., Tomich, J. M., Richards, J. H., and Schatz, G. (1986). A chemically synthesized pre-sequence of an imported mitochondrial protein can form an amphiphilic helix and perturb natural and artificial phospholipid bilayers. *EMBO J.* **5**, 1327–1334.

Rudel, R. (1990). The myotonic mouse—A realistic model for the study of human recessive generalized myotonia. *Trends Neurosci.* **13**, 1–3.

Rudel, R., and Lehnamm-Horn, F. (1985). Membrane changes in cells from myotonia patients. *Physiol. Rev.* **65**, 310–356.

Sansom, S. C., La, B. Q., and Carosi, S. L. (1990). Double barrelled chloride channels of collecting duct basolateral membrane. *Am. J. Physiol.* **259**, F46–F52.

Schofield, P. R., Darlison, M. G., Fujita, N., Burt, D. R., Stephenson, F. A., Rodriguez, H., Rhee, L. M., Ramachandran, J., Reale, V., Glencorse, T. A., Seeburg, P. H., and Barnard, E. A. (1987). Sequence and functional expression of the $GABA_A$ receptor shows a ligand-gated receptor super-family. *Nature (London)* **328**, 221–227.

Solc, C. K., and Wine, J. J. (1991). Swelling-induced and depolarization-induced Cl channel in normal and cystic fibrosis epithelial cells. *Am. J. Physiol.* **261**, C658–C674.

Sonigo, P. (1987). Nucleotide sequence and evolution of ETn elements. *Proc. Natl. Acad. Sci. USA* **84**, 3768–3771.

Steinmeyer, K., Ortland, C., and Jentsch, T. J. (1991a). Primary structure and functional expression of a developmentally regulated skeletal muscle chloride channel. *Nature (London)* **354**, 301–304.

Steinmeyer, K., Klocke, R., Ortland, C., Gronemeier, M., Jockusch, H., Grunder, S., and Jentsch, T. J. (1991b). Inactivation of muscle chloride channel by transposon insertion in myotonic mice. *Nature (London)* **354**, 304–308.

Stocker, M., Stuhmer, W., Wittka, R., Wang, X., Muller, R., Ferrus, A., and Pongs, O. (1990). Alternative *Shaker* transcripts express either rapidly inactivating or noninactivating K⁺ channels. *Proc. Natl. Acad. Sci. USA* **87**, 8903–8907.

Strange, K. (1991). Volume regulating Cl⁻ loss after Na⁺ pump inhibition in CCT principal cells. *Am. J. Physiol.* **260**, F225–F234.

Taguchi, T., and Kasai, M. (1980). Identification of an anion channel protein from electric organ of *Narke japonica. Biochem. Biophys. Res. Commun.* **96**, 1088–1094.

Tanabe, T., Takeshima, H., Mikami, A., Flockerzi, V., Takahashi, H., Kangawa, K., Kojima, M., Matsuo, H., Horise, T., and Numa, S. (1987). Primary structure of the receptor for calcium channel blockers from skeletal muscle. *Nature (London)* **328**, 313–318.

Thiemann, A., Grunder, S., Pusch, M., and Jentsch, T. J. (1992). A chloride channel widely expressed in epithelial and nonepithelial cells. *Nature (London)* **356**, 57–60.

Trimmer, J., Cooperman, S. S., Agnew, W. S., and Mandel, G. (1990). Regulation of muscle sodium channel transcripts during development and in response to denervation. *Dev. Biol.* **142**, 360–367.

Unwin, N. (1989). The structure of ion channels in membranes of excitable cells. *Neuron* **3**, 665–667.

Valverde, M. A., Diaz, M., Sepulveda, F. V., Gill, D. R., Hyde, S. C., and Higgins, C. F. (1992). Volume-regulated chloride channels associated with the human multidrug-resistance P-glycoprotein. *Nature (London)* **355**, 830–833.

Von Heijne, G. (1986). Mitochondrial targeting sequences may form amphiphilic helices. *EMBO J.* **5**, 1335–1342.

Wagner, J. A., Cozens, A. L., Schulman, H., Gruenert, D. C., Stryer, L., and Gardner, P. (1991). Activation of chloride channels in normal and cystic fibrosis airway epithelial cells by multifunctional calcium/calmodulin-dependent protein kinase. *Nature (London)* **349**, 7983–7962.

Ward, C. L., Krouse, M. E., Gruenert, D. C., Kopit, R. R., and Wine, J. J. (1991). Cystic fibrosis gene expression is not correlated with rectifying Cl- Channels. *Proc. Natl. Acad. Sci. USA* **88**, 5277–5281.

Welsh, M. J. (1990). Abnormal regulation of ion channels in cystic fibrosis epithelia. *FASEB J.* **4**, 2718–2725.

White, M. M., and Miller, C. (1979). A voltage-gated anion channel from electric organ of *Torpedo californica. J. Biol. Chem.* **254**, 10161–10166.

White, M. M., and Miller, C. (1981). Probes of the conduction process of a channel from *Torpedo* electroplax. *J. Gen. Physiol.* **78**, 1–18.

Worrel, R. T., Butt, A. G., Cliff, W. H., and Frizzel, R. A. (1989). A volume-sensitive chloride conductance in human colonic cell line T84. *Am J. Physiol.* **256**, C1111–C1119.

17

Functional Properties of Background Chloride Channels

FABIO FRANCIOLINI AND DAVID J. ADAMS

I. Introduction

A prevailing view of cellular ion homeostasis has assigned chloride ion (Cl⁻) to a thermodynamically passive and unspecialized role. This view arose primarily from studies of the resting Cl⁻ conductance in frog skeletal muscle and red blood cells, in which Cl⁻ is passively distributed across the plasma membrane and equilibrates rapidly.

Chloride ion, one of the most abundant biological anions, accounts for a measurable fraction of the electrical conductance of many biological membranes. Electrophysiological studies using single-channel recording methods (Hamill *et al.*, 1981) have shown that chloride channels exist in many cell types. These channels have been found in the plasma and intracellular membranes of nearly every type of cell examined, from human platelets (Mahaut-Smith, 1990) to the nuclear membrane of rat hepatocytes (Tabares *et al.*, 1991). Although chloride ions do not, in general, play a direct role in membrane excitability as do Na⁺, K⁺, or Ca²⁺ ions, a report has been made of chloride-dependent action potentials in chick skeletal muscle cells grown in tissue culture (Fukuda, 1974), and the chloride conductance of vertebrate skeletal muscle does exhibit a time and voltage dependence (Hutter and Warner, 1972; Warner, 1972).

A wide variety of eukaryotic cells exhibits a *non-equilibrium* distribution of Cl⁻ ions across the plasma membrane, that contributes to cellular processes including pH regulation, volume regulation, generation of the resting membrane potential, and regulation of membrane excitability. Many types of Cl⁻ channel have been described in mammalian cells. This chapter focuses on the characteristics of the background Cl⁻ channels of the plasma membrane. These channels have been characterized only recently for two major reasons. First, experimental difficulties were generated by an inability to clearly isolate and characterize macroscopic Cl⁻ currents, as was possible for voltage-gated Na⁺, K⁺, and Ca²⁺ currents. Whole-cell Cl⁻ currents often exhibit a slow time dependence and weak voltage dependence, and were often considered to be a membrane "leak" current. Second, the presumed physiological irrelevance of Cl⁻ channels caused them to be ignored. During the three decades after the classical description of the action potential in squid axon by Hodgkin and Huxley (1952), membrane excitation and neural transmission seemed to be explained solely by the interplay of Na⁺, K⁺, and Ca²⁺ currents and Cl⁻ currents were assumed to play no part. Therefore, attention was not focused on an experimentally difficult characterization of a Cl⁻ current with no defined function.

The gating properties of most Cl⁻ channels, by which they are active over a wide voltage range and exhibit little time-dependent inactivation, permits the study of background Cl⁻ channels in many cell types. The large number of Cl⁻ channel types reported has been grouped into four categories based on functional properties such as voltage or agonist activation, conductance, and kinetics (Franciolini and Petris, 1990): (1) background Cl⁻ channels, (2) double-barrelled Cl⁻ channels, (3) large conductance ("maxi") Cl⁻ channels, and (4) ligand-gated Cl⁻ channels. This review focuses primarily on the ionic permeation, gating, and pharmacological properties of background Cl⁻ channels. In particular, the ion transport mechanisms that underlie the selectivity and permeability to anions over cations, as well as the channel architecture as derived from molecular biological and electrophysiological approaches, are discussed. Background Cl⁻ channels are defined as the anion-selective channels that show a significant probability of being open at resting membrane potentials, thus contributing to the resting membrane permeability of cells.

Several reviews of chloride channels have appeared describing aspects of background chloride channels (Bretag, 1987; Gögelein, 1988; Vaughan and French, 1989; Blatz, 1990; Franciolini and Petris, 1990,1992).

II. Distribution and Localization

Information about voltage-gated Cl⁻ channel distribution, in contrast to voltage-gated cation and ligand-gated ion channels (see Almers and Stirling, 1984), is limited primarily to the cell membrane (plasmalemma) of vertebrate skeletal muscle. The lack of high-affinity marker ligands for Cl⁻ channels has prevented investigators from directly obtaining information on the distribution of these channels in other cell types. The density of chloride channels in the surface and transverse tubular system (t-system) of amphibian and mammalian skeletal muscle has not been precisely determined, but a pattern is beginning to emerge that suggests that chloride channels in mammals are mainly in the t-system, whereas frog chloride channels are largely in the sarcolemma. Hodgkin and Horowicz (1959), working with single frog muscle fibers, suggested that the rapid response of the membrane potential to changes in external Cl⁻ concentration localizes the Cl⁻ conductance close to the cell surface. By recording membrane conductance of frog muscle fibers before and after glycerol tubular disruption. Eisenberg and Gage (1969) attempted to separate surface and tubular Cl⁻ conductances. These researchers

concluded that all the Cl⁻ conductance resided in the surface membrane. Patch-clamp studies (Woll and Neumcke, 1987; Woll et al., 1987) showed that chloride channels are regularly detected in the sarcolemma of dissociated frog toe muscle fibers.

In mammalian fibers, on the other hand, glycerol disruption has yielded results different from those found in the frog. Dulhunty (1979) concluded, from changes in membrane potential induced by changes in extracellular K⁺ and Cl⁻ in both normal and glycerol-detubulated rat fibers, that most of the fiber's chloride permeability was t-tubular. Further, if reasonable assumptions are made about the ratio of surface to tubular area, that is, the K⁺ conductance of these membranes and the extent of glycerol disruption, the suggestion could be made that surface chloride channel density is low. The low density of surface chloride channels has been confirmed by Chua and Betz (1991), who detected chloride channels in less than 2% of patches from surface membrane of rat fibers. Another study of Cl⁻ channels in human cultured skeletal muscle (myoballs) resulted in calculated channel densities ranging from 0.07 to 0.23 channels/μm^2 (Fahlke et al., 1992). The Cl⁻ current density obtained in this patch clamp study of human myoballs suggests that a higher fraction of Cl⁻ channels is open in cultured than in native muscle membranes.

III. Ionic Permeation

A. Conductance

The unitary conductance of background Cl⁻ channels is as diverse as the number of types of Cl⁻ channels. Therefore, unitary conductance cannot be used as an identifying characteristic of background Cl⁻ channels. Unitary conductances, as measured with the patch-clamp method, span about two orders of magnitude (symmetrical 150 mM Cl⁻), from the 5-pS conductance of the hyperpolarization-activated Cl⁻ channel in mouse astrocytes (Nowak et al., 1987) to the 450-pS conductance of the background (also known as large conductance) Cl⁻ channel of rat Schwann cells (Gray et al., 1984). Indeed, background Cl⁻ channels exhibit the highest single-channel conductance ever recorded in intact cells.

Current-voltage (I–V) relations of background Cl⁻ channels most commonly show significant outward rectification. This effect cannot be explained by an increase in the open probability of the channel at positive voltages, since the I–V plots are obtained from single-channel measurements. Rectification is more likely a result of asymmetries in the electrical profile of the ionic permeation path of the channel; however, no studies have addressed this point specifically.

Another common occurrence in background Cl⁻ channels is the presence of conductance sublevels. Typically, ionic channels gate between two conductance levels, open and closed. Many channels display more complex gating behavior; in particular, anion channels often display one "subconductance" level or more, or conductance levels intermediate to the fully open and closed levels. Subconductance levels have been found in chloride channels from *Torpedo*

electroplax organ, molluscan neurons, rat astrocytes, mouse B lymphocytes, and amphibian skeletal muscle (Fox, 1987). Table I summarizes single-channel conductances of the most representative background Cl⁻ channels, but is not intended to be exhaustive of all the background Cl⁻ channels reported to date.

B. Selectivity

Background Cl⁻ channels are not highly anion selective. Table II gives examples of the poor ion discrimination among these channels as determined from re-

TABLE I **Unitary Conductances of Voltage-Dependent Cl⁻ Channels**

Background Cl⁻ channel type	Main conductance (pS)	$[Cl^-]_i//[Cl^-]_o$	Reference
Human skeletal muscle	31	42//160	Fahlke *et al.* (1992)
	10	42//160	Fahlke *et al.* (1992)
	250	42//160	Fahlke *et al.* (1992)
Nuclear membrane	150	140//140	Tabares *et al.* (1991)
	58	140//140	Tabares *et al.* (1991)
Rat hippocampal neuron	30	150//150	Franciolini and Nonner (1987)
	60	150//150	Franciolini and Petris (1988)
	62	160//160	Shukla and Pockett (1990)
Rat axon	33	150//150	Strupp and Grafe (1991)
	65	150//150	Strupp and Grafe (1991)
Rat skeletal muscle	45	100//100	Blatz and Magleby (1985)
	61	100//100	Blatz and Magleby (1985)
	430	143//143	Blatz and Magleby (1983)
Rat Schwann cell	438	150//150	Gray *et al.* (1984)
Rat cerebral astrocyte	400	140//140	Sonnhof (1987)
Mouse astrocyte	5	150//150	Nowak *et al.* (1987)
Lobster walking leg nerve	15–19	100//100	Lukáes and Moczydlowski (1990)
Aplysia neuron	10–15	—	Chesnoy-Marchais and Evans (1986)
Torpedo electroplax	18	150//150	Miller and White (1984)
Rat aortic smooth muscle	340	150//150	Soejima and Kokubun (1988)
Rat cardiac muscle	400	150//150	Coulombe *et al.* (1987)
Calf cardiac SR membrane	55	100//100	Coronado and Latorre (1982)
Rabbit skeletal SR membrane	200	100//100	Tanifuji *et al.* (1987)
Rabbit urinary bladder epithelium	64	150//150	Hanrahan *et al.* (1985)
	362	150//150	Hanrahan *et al.* (1985)
Renal A6 epithelial cell	360	105//105	Nelson *et al.* (1984)
MDCK epithelial cell	460	150//150	Kolb *et al.* (1985)
Porcine thyroid cell	5.5	140//140	Champigny *et al.* (1990)
Human tracheal epithelial cell	25	145//145	Welsh (1986)
Human T lymphocyte	365	140//140	Schlichter *et al.* (1990)
Mouse B lymphocyte	400	165//165	Bosma (1989)
Mouse pulmonary alveolar type II cells	393	145//145	Schneider *et al.* (1985)
Bovine aortic endothelial cell	382	140//140	Olesen and Bundgaard (1992)

TABLE II Representative Ionic Permeability
Ratios, P_x/P_{Cl}

Ion (X)	λ_x/λ_{Cl}[a]	Large conductance Schwann cell[b]	Small conductance hippocampal neurons[c]	Epithelial cells canine trachea[d]
I⁻	1.0	1.4	1.98	1.75
Br⁻	0.98	1.2	1.46	1.22
Cl⁻	1.0	1.0	1.0	1.0
F⁻	1.38	—	0.46	0.23
NO_3^-	1.07	—	2.35	1.43
SCN⁻	1.16	—	1.44	2.3
SO_4^{2-}	1.92	0.61	<0.01	—
Acetate	1.87	0.39	0.66	—
Propionate	2.13	—	0.5	—
Aspartate	—	<0.03	0.17	—
Glutamate	2.82	<0.03	0.13	—
Benzoate	2.36	—	1.86	—
Na⁺	1.52	0.2	0.25	—
K⁺	1.04	—	0.25	—
Cs⁺	0.99	—	0.35	—

[a] λ_x/λ_{Cl}, Relative hydrated size determined from the inverse of the limiting equivalent conductances (λ) of ions at 25°C in water (Robinson and Stokes, 1959).

[b] Gray et al. (1984).

[c] Franciolini and Nonner (1987).

[d] Li et al. (1990) (cf. Schoppa et al., 1989).

versal potential measurements. Chloride channels pass at least 10 anionic species, including all of the halides. Although, Br⁻, I⁻, NO_3^- and SCN⁻ are more permeant than Cl⁻, the channel is termed as Cl⁻ channel because, under physiological conditions, Cl⁻ is the most relevant anion. Note that, despite the large diversity of Cl⁻ channel conductance (see Table I), the ionic selectivity sequences for these channels are similar and lyotropic: an inverse relationship exists between the crystal diameter of the ions and their permeability ratios, as found for many interactions between anions and proteins. The interaction (or binding) site of these channels typically includes a weak-field-strength site. Such a site would require the ion-exchange equilibrium in the channel site to be dominated by the hydration energy of the permeant ions, rather than by electrostatic interaction between the permeant ions and the binding site. In addition, the permeability of several background Cl⁻ channels to hydrophobic (benzoate, glutamate) anions, as well as the blocking action brought about by a number of aromatic compounds (5-nitro-2-(3-phenylpropylamino) benzoate, anthracene-9-

carboxylate, 9-AC), suggests the presence of a hydrophobic region in the channels.

The selectivity data also provide an estimate of the lower limit for possible channel cross sections. All permeant anions tested can fit through a 6-Å diameter pore, provided that some of the organic anions align their long axis with the channel. This minimal pore size is very close to the dimensions proposed for the endplate channel (6.5 × 6.5 Å; Dwyer et al., 1980), a weakly selective cation channel, but is larger than that of the more selective Na⁺ and K⁺ channels (3.1 × 5.1 Å and 3.3 × 3.3 Å, respectively; Hille, 1992). Another observation made from these ion selectivity experiments is that background Cl⁻ channels show significant permeability to alkali cations as well. This finding contradicts experience with cation channels, of which none has been found to leak anions. For an ion to pass through a channel, it must interact with sites in the pore. Negatively charged binding sites would select against anions and positively charged sites would select against cations. Electrostatics thus would make a channel either cation or anion selective, but not both, since the requirement for a binding site that passes cations would appear to be mutually exclusive with those of one passing anions. The observation that background Cl⁻ channels can be significantly permeable to cations could create a physiological problem. These channels, if permeable to cations under physiological conditions, should destabilize the membrane resting potential (by allowing Na⁺ in) rather than stabilize it, as is actually observed.

C. Transport Mechanism

In the study of cation permeability through background Cl⁻ channels in hippocampal neurons, Franciolini and Nonner (1987) made the following observations. First, although cations were permeant through the open channel, they could not pass in the absence of a permeant anion. For example, when permeant Cl⁻ was replaced with impermeant SO_4^{2-}, no inward cation current could be observed at voltages more negative than the reversal potential predicted by the independence principle from a permeability ratio (P_{cation}/P_{anion}) of 0.25–0.35. Second, the total permeation rate depended to a degree on the species of cation present. Replacement of Na⁺ with Cs⁺ as the major extracellular cation increased the single-channel conductance by approximately 20%. These observations indicated a strong interdependence between anion and cation fluxes in the channel.

To account for these observations, a transport mechanism was proposed in which an "activated complex" is composed of the channel site, a cation,

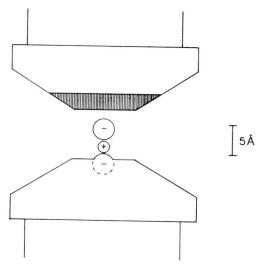

5 Å

FIGURE 1 Schematic diagram of the proposed selectivity filter. The sizes of ions correspond to the crystal radii of Na$^+$ and Cl$^-$ ions. The dark part of the channel wall indicates a hydrophobic moiety. The space between this region and the adsorbed Na$^+$ ion is large enough (5.5 Å) to admit all permeant anions in Table II; the larger organic anions, however, would have to align their long axes with the pore to pass and also require that the width of the channel in the direction vertical to the section shown be ~6.5 Å. The effective radius indicated for the anionic site of the pore is the crystal radius of a Cl$^-$ ion. Ions are shown in the arrangement conforming to the activated complex; the complex may decay by releasing the anion or by releasing the cation–anion pair. Reproduced from the *J. Gen. Phys.* (1987). **90**, 453, 478, by copyright permission of The Rockefeller University Press.

and an anion (Fig. 1). In this model, the channel site is anionic and first binds a cation. The resulting dipole can then bind an anion, forming the activated complex. The complex would decay in one of two ways, either dissociating the anion (anion flux) or dissociating the anion–cation pair (cation flux). Anion flux would then imply anion binding to the cationic (cation-site) dipole, followed by unbinding of the anion of the pair, whereas cation flux would imply either cation binding and unbinding, or cation binding, anion binding to the cationic dipole, and unbinding of the pair from the site. A further study of anion–cation interactions in the pore of rat neuronal background Cl$^-$ channels indicates that anions and cations form mixed complexes of second and higher order while traversing the channel (Franciolini and Nonner, 1994a,b).

This mechanism would also resolve the paradox of membrane stabilization by background Cl$^-$ channels. In contrast to mechanisms in which anions and cations pass independently, the coupled transport mechanism would predict that, under physiological conditions, that is, in the absence of a concentration gradient for cations (since the channel does not discriminate between Na$^+$ and K$^+$), the reversal po-

tential would be set by anions. According to this model, a net influx of positive charge carried by Na$^+$ would require binding of external Na$^+$ to the channel site and movement of internal Cl$^-$ to the channel site to form the activated complex. Since the intracellular Cl$^-$ concentration is generally low, these events are unlikely; thus, the membrane potential is determined primarily by Cl$^-$, independent of cation permeability.

A converse transport mechanism has been proposed for anion–selective amphotericin B channels (Borisova *et al.*, 1986) to explain the dependence of cation permeability on the permeant anion concentration. The binding site for amphotericin B channels, however, is proposed to be cationic, so it first binds an anion, forming a dipole to which cations can bind. As in Cl$^-$ channels, cation permeation can only take place in the presence of a permeant anion. A dependence of cation permeation on anion concentration was not observed for background Cl$^-$ channels from spinal cord neurons (Franciolini and Petris, 1988), in which cationic currents could be observed in the absence of permeant anions. For this channel, independent permeation of cations and anions must be postulated. The simultaneous and independent permeation of anions and cations through membrane channels has been addressed theoretically by Eisenman and Villarroel (1989). These researchers propose the presence of a dipole in the selectivity filter that could present either a cationic or an anionic environment in the channel pore, depending on the orientation of the dipole. Rings of –OH groups, known to be present in several channels, could orientate themselves with the –H of the –OH dipole facing the channel lumen or, because of their ability to rotate freely around the C–O single bond, adopt a conformation in which the partially negative oxygens extend toward the lumen. Although this mechanism could account for independent cation and anion permeation, it does not predict resting membrane potential stability. Since these channels are open at resting membrane potentials, this dipole selectivity filter model would allow Na$^+$ to enter the cell, obviously disrupting membrane potential stability.

IV. Channel Gating

A. Voltage Dependence

Background Cl$^-$ channels can be grouped into three categories based on their voltage sensitivities: (1) channels activated by hyperpolarization, (2) channels activated by depolarization, and (3) channels inactivated by polarization.

1. Hyperpolarization-Activated Channels

Hyperpolarization-activated Cl⁻ currents were first observed in *Aplysia* neurons (Chesnoy-Marchais, 1982). Activation is generally slow, with a time constant of several hundred milliseconds. The gating properties of the channel, analogous to the hyperpolarization-activated (inward rectifier) K⁺ current (Hagiwara *et al.*, 1976), depend on the intracellular Cl⁻ concentration (Chesnoy-Marchais, 1983). Increasing the internal Cl⁻ concentration lowers the activation threshold and accelerates the activation kinetics. Functionally, this behavior would facilitate extrusion of excess intracellular Cl⁻. The hyperpolarization-activated Cl⁻ channel has been identified and characterized at the single-channel level. This channel adopts three equally spaced current levels, with a maximal conductance of 10–15 pS (Chesnoy-Marchais and Evans, 1986), similar to that observed for the double-barreled Cl⁻ channel (Miller and White, 1984).

The double-barreled Cl⁻ channel from *Torpedo* is also activated by hyperpolarization. Steady-state macroscopic conductance, measured in planar lipid bilayers containing on the order of 10^3 Cl⁻ channels, is low at positive voltages and increases with hyperpolarization. The voltage dependence of the channel gating was approximated by a Boltzmann distribution with an equivalent gating charge of ~1.5 and half-maximal activation ($V_{1/2}$) of −50 mV. However, this channel has a complex architecture and has another gating system in addition to a hyperpolarization-activated mechanism. The other gating system shows a positive channel open probability vs voltage ($P_{open}-V$) relationship with an equivalent gating charge of 1.1 and $V_{1/2}$ of −95 mV.

Another background Cl⁻ channel activated by hyperpolarization has been reported by Nowak *et al.* (1987) in cultured mouse astrocytes. The study, carried out on outside-out patches, shows an increase in open probability with hyperpolarization in the range of −40 to −100 mV. This channel displays a low single channel conductance of 5 pS, with a half-conductance sublevel. Note that most hyperpolarization-activated Cl⁻ channels exhibit this double-barreled-like subconductance behavior.

2. Depolarization-Activated Channels

The second and most common type of background Cl⁻ channel is activated by membrane depolarization. The relationship between background Cl⁻ channel open probability and voltage in rat hippocampal neurons is described by a Boltzmann distribution with an equivalent gating charge of 1.6 and $V_{1/2}$ of −45 mV (C. Trequattrini, unpublished). Background Cl⁻ chan-

nels in neurons of rat cerebral cortex exhibited an equivalent gating charge of 1.3, determined from the relationship between membrane potential and channel open probability (Blatz, 1991). Kinetic analysis of these channels showed that most of the observed voltage dependence was due to a decrease in the mean closed intervals with depolarization, whereas the mean open time was relatively independent of the voltage. Similar results were obtained on a background Cl⁻ channel in rat myoballs by Weiss and Magleby (1990). This study showed that the channel is activated by depolarization and has a positive $P_{open}-V$ relationship with a similar equivalent gating charge, and that the major contribution to the increase in P_{open} with positive voltage was a decrease in the mean closed intervals. The only difference between the Cl⁻ channel kinetics exhibited in the two preparations is in the mean open distributions, which could account for the difference in the voltage dependence of channel activation.

3. Polarization-Inactivated Channels

Several large conductance Cl⁻ channels fall into this category, being active at membrane voltages near zero and closing with depolarizing and hyperpolarizing voltages. Gray *et al.* (1984) showed that, in membrane patches excised from rat cultured Schwann cells held between −10 and +20 mV, the large conductance Cl⁻ channel would remain active for long periods of time. If the membrane was polarized outside this range, the channel would close and would remain silent until the potential was again brought within the active range. Similar behavior was observed for the large conductance Cl⁻ channels from rat skeletal muscle (Blatz and Magleby, 1983) and macrophages (Schwarze and Kolb, 1984). These channels closed rapidly (within a few seconds) when the voltage was shifted (>20 mV) from zero. The rate of inactivation is voltage dependent, decreasing with positive voltages. A similar behavior was reported for a 280-pS conductance Cl⁻ channel from amphibian skeletal muscle (Woll and Neumcke 1987; Woll *et al.*, 1987). The voltage range for this channel was somewhat larger than that found for Cl⁻ channels in Schwann cells. The Cl⁻ conductance in amphibian skeletal muscle remained high at depolarized potentials up to +60 mV. In contrast to this kinetic behavior, a large conductance Cl⁻ channel found in cultured aortic endothelial cells closes at membrane potentials more negative than −40 mV (Groschner and Kukovetz, 1992; Olesen and Bundgaard, 1992).

The results from these studies illustrate substantial differences among Cl⁻ channels and their cation-selective counterparts. These data show that the voltage sensitivity of background Cl⁻ channels is much less than that observed for voltage-gated Na⁺ and K⁺ channels. Additionally, the activation time constant for the hyperpolarization-activated Cl⁻ current is several hundred milliseconds, whereas the activation of Na⁺ and K⁺ currents occurs in fractions of milliseconds and milliseconds, respectively. Finally, the equivalent gating charge determined for the background Cl⁻ channels studied was always less than two, compared with an equivalent gating charge of six found for voltage-gated Na⁺ channels.

B. Gating Kinetics

This section is restricted to discussion of two background Cl⁻ channels with gating kinetics that have been described in sufficient detail, namely, the fast Cl⁻ channel found in muscle and neurons (Weiss and Magleby, 1990; Blatz, 1991) and the double-barreled Cl⁻ channel of *Torpedo* electroplax (Miller and White, 1984).

As already mentioned, the fast Cl⁻ channel is active within a wide voltage range that includes the cell resting potential, and it does not inactivate (Blatz and Magleby, 1985). These characteristics are advantageous for kinetic analysis of ionic channels, since this procedure requires large numbers of observed open and closed event durations. For Markovian kinetics, the distribution of these event durations will be the sum of exponential components, each related to one open or closed state. Blatz and Magleby (1986) applied the criterion of maximum likelihood to determine the number of significant exponential components required to fit the distributions of open and closed event durations. Two components were necessary to describe the open interval distribution, and at least five exponential components were required to describe the closed interval distribution. The minimal kinetic model resulting from these data must therefore consist of at least two open and five closed states, as shown in kinetic reaction Scheme I.

$$C1 \rightleftharpoons C2 \rightleftharpoons C3 \rightleftharpoons C4 \rightleftharpoons C5$$
$$\qquad\qquad\qquad\quad \updownarrow \qquad\; \updownarrow$$
$$\qquad\qquad\qquad\;\; O1 \qquad O2$$

SCHEME I

These channels also change, albeit rarely, from this normal kinetic behavior to a different kinetic mode during which the channel switches 10 times faster between the open and closed states. The open and closed interval distributions of the "buzz" mode are both fit by single exponential functions, indicating that a single open and a single closed state prevail in

this mode. A low conductance background Cl⁻ channel from cerebral cortex neurons revealed similar kinetics (Blatz, 1991). The fast neuronal Cl⁻ channel follows closed and open event distributions requiring at least two open and five closed states (see Chapter 14).

The second type of Cl⁻ channel that has been subjected to a great deal of kinetic investigation is the double-barreled Cl⁻ channel of *Torpedo* (White and Miller, 1979). The channel, reconstituted into lipid bilayers, reveals two separate domains of kinetics. Long bursts of activity, during which the current makes transitions among three equally spaced current levels, are interspersed with periods during which the channel is silent. Transitions between the bursting activity and the silent state occur at intervals of seconds, whereas the fast process of "flickering" within the bursts between the three current levels occurs in the millisecond range.

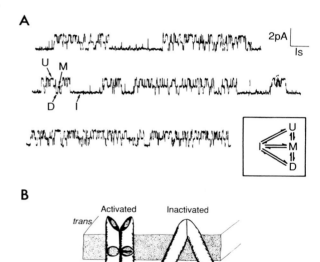

FIGURE 2 Single Cl⁻ channel bursts. (A) A single Cl⁻ channel was inserted into the bilayer and recordings were collected at a holding potential of −90 mV. The four states of the single channel are labeled on the channel record: inactivated state (I) and three substates of the active channel (U, M, and D). (*Inset*) Diagram previously proposed (Miller, 1982) to describe the allowed transitions among the four states. All rate constants in this diagram depend on the voltage and the pH of the *cis* solution. (B) Dimeric Cl⁻ channel complex. This cartoon is drawn to visualize as simply as possible the main conclusions concerning the *Torpedo* Cl⁻ channel structure. The conducting unit is viewed as a dimer of identical protein subunits, or protochannels, in close association spanning the membrane. Each protochannel carries its own "gate" (shown in a cutaway view), which opens and closes on a time scale of milliseconds. A DIDS-reactive inhibitory site on each protochannel is exposed to the *cis*-facing aqueous phase. The dimeric complex can exist in two quaternary structures shown—activated and inactivated. The association of the two subunits in the inactivated conformation leads to occlusion of both protochannels simultaneously. Reprinted with permission from Miller and White (1984).

To account for such kinetic behavior, the channel has been modeled as a dimer of two identical pores (Fig. 2), each opening and closing virtually independently through a fast and local gating mechanism, whereas the aggregate converts between an active and an inactivated state in concert. The channel has two separate and independent gating machineries. The first, which drives the channel between the activated (burst) and the inactivated states, has a voltage dependence in which the channel activates (i.e., increase probability of bursting) on membrane hyperpolarization (solid curve in Fig. 3). The second, the fast gating (opening and closing of individual protochannels between the three current levels), shows a voltage dependence, so as the membrane is depolarized the open probabilities of the protochannels increase (dashed curve in Fig. 3). Both burst probability vs voltage and protochannel open probability vs voltage relationships exhibit two-state Boltzmann behavior, but with opposite voltage dependencies. Therefore, at very negative voltages (< -100 mV), the channels are almost always in the active (bursting) state, but the individual protochannels have low probabilities of being open; thus, the channel as a whole is nonconducting. At positive voltages, the opposite situation occurs; the protochannels would be open, but the slow gating system holds the channel complex in the inactive state. Since, according to the model, the channel would be conducting only when both (slow and fast) gating systems are in the permissive position, the channel open probability will have a bell-shaped function given by the product of the two probabilities.

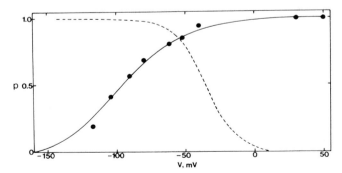

FIGURE 3 Voltage dependence of activation and inactivation gates. Solid curve with points represents individual protochannel activation curve, that is, probability of opening within the burst. Dashed curve represents probability of burst formation. If activation and inactivation are independent processes, the macroscopic conductance–voltage curve will be the product of these two probabilities. Reprinted with permission from Miller and Richard (1990).

V. Channel Pharmacology

Background Cl^- channels are blocked by various agents that can be grouped into three categories: aromatic acids, inorganic cations, and protons. The first group includes anthracene-9-carboxylate (9-AC), diphenylamine-2-carboxylate (DPC), 5-nitro-2-(3-phenylpropylamino)benzoate (NPPB), and disulfonic acid stilbene derivatives including 4-acetamido-4′-isothiocyanostilbene-2,2′-disulfonic acid (SITS) and 4,4′-diisothiocyanostilbene-2,2′-disulfonic acid (DIDS). These compounds have been tested on different preparations and are generally found to act from the extracellular side at micromolar concentrations. An extensive study of blockers of the Cl^- channel in the loop of Henle (Wangemann *et al.*, 1986) demonstrated that the channel could be blocked by both cationic and anionic molecules with a large hydrophobic moiety, and that the external channel mouth is the more susceptible site of blockade.

The molecular mechanism by which these agents block Cl^- channels is not known, but a shared property—hydrophobicity—suggests that hydrophobic domains on the Cl^- channel proteins could be the sites of action. The presence of such hydrophobic regions is consistent with the observed significant permeability of some Cl^- channels to aromatic compounds of smaller size (e.g., benzoate). The observation that hydrophobic NPPB reversibly blocks the outwardly rectifying Cl^- channel in tracheal epithelium when applied to either surface of the membrane, and the flat relationship between NPPB concentration and P_{open} (Li *et al.*, 1990), suggests that NPPB must interact with the membrane lipid to inhibit the Cl^- channel. One review documents the specificity and potency of Cl^- channel blockers in a variety of cell types (Greger, 1990).

Bryant and Morales-Aguilera (1971) demonstrated that the *in vivo* myotonic effects produced in mammals by low doses of certain aromatic monocarboxylic acids were the result of a block of Cl^- conductance, whereas the same aromatic monocarboxylic acids did not block chloride conductance in frog skeletal muscle in the same concentration range. This result indicates a major difference in the Cl^- conductances of amphibian and mammalian channels. Palade and Barchi (1977b) followed up this work with a comprehensive study, in voltage-clamped rat diaphragm muscle fibers, of 25 aromatic carboxylic acids. Of these compounds, 9-AC was found to be the most potent, with a K_i of 1.1×10^{-5} M. The permeability sequence underwent progressive changes with increasing 9-AC concentration, ultimately reversing at high concentrations. These researchers attributed the blocking effect to a change in the ionic selectivity rather than to a direct block. 9-AC is likely to block only one class of chloride channel; the changed selectivity represents conductance through a different (unblocked) class of channel.

More recently, the stereospecific effects of enantiomeric aromatic carboxylic acids on chloride conductances in rat muscle have been elucidated (Bettoni *et al.*, 1987; Conte-Camerino *et al.*, 1988). A series of chiral carboxylic acid derivatives of 2-(*p*-chlorophenoxy)isobutyric acid (CPIB) were synthesized and tested in an effort to find a derivative of clofibric acid that retained cholesterol-lowering activity but was devoid of myotonic side effects. The *R*-(+) derivatives were found to increase Cl$^-$ conductance, and the *S*-(−) and racemic derivatives, like all the previously studied achiral aromatic carboxylic acids, decreased Cl$^-$ conductance. Further, these studies have shown that the concentration–response curve of the *R*-(+) derivatives is actually biphasic, with low concentrations (~1 mM) increasing Cl$^-$ conductance and higher concentrations decreasing Cl$^-$ conductance (Conte-Camerino *et al.*, 1988; De Luca *et al.*, 1992).

The reactive disulfonic stilbenes (DIDS and SITS) do block some background chloride channels including a voltage-gated Cl$^-$ conductance studied macroscopically in squid giant axon and astroglia (Inoue, 1985; Gray and Ritchie, 1986), the low conductance channel of *Torpedo* electroplax and rabbit urinary blad-

der (White and Miller, 1979; Hanrahan *et al.*, 1985), and the large conductance channels of cultured kidney epithelial cells, B lymphocytes, and vascular endothelial cells (Nelson *et al.*, 1984; Bosma, 1989; Groschner and Kukovetz, 1992). Intracellularly or extracellularly applied DIDS reversibly blocks Cl$^-$ channels in a concentrations-dependent manner with high potency (K_i of 10–100 μM) and independent of membrane potential (see Fig. 4; Miller and White, 1984; Kokubun *et al.*, 1991; Groschner and Kukovetz, 1992). The apparent reduction of the unitary current observed in the presence of DIDS is due to extremely rapid and unresolved channel flicker and suggests that this behavior is a consequence of open channel block (Kokubun *et al.*, 1991).

Inorganic cationic blockers constitute another group with a lower potency than the aromatic acids. Zn^{2+} and Cu^{2+} have been found to be the most effective. In frog skeletal muscle, external Zn^{2+} reduces Cl$^-$ conductance with a K_i near 0.1 mM (Hutter and Warner, 1967; Stanfield, 1970; Woodbury and Miles, 1973) and has been shown to reversibly block single-channel Cl$^-$ currents in a voltage-dependent manner when applied to the cytoplasmic face of excised membrane patches (Woll *et al.*, 1987). In contrast, the block of marcoscopic and unitary Cl$^-$ currents by extracellular Zn^{2+} (100 μM) has been reported to be voltage independent in skeletal muscle (Franciolini and Non-

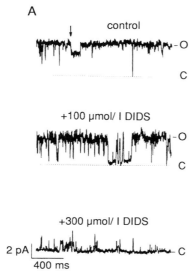

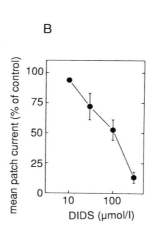

FIGURE 4 Concentration-dependent inhibition of large-conductance Cl$^-$ channels by extracellular DIDS. (A) Representative current traces (filtered at 500 Hz) recorded from a perforated outside-out vesicle at +10 mV in the absence and presence of 100 μM and 300 μM DIDS. Closed state (C) of the channel is indicated in traces as a dashed line. (B) Dose–response curve was obtained with perforated vesicles at +10 mV. The mean patch current amplitude (calculated from 40-sec records), normalized to control (predrug value) is shown as a function of the DIDS concentration. Points represent mean values ± SEM from four experiments. Adapted with permission from Groschner and Kukovetz (1992).

ner, 1987; Woll *et al.*, 1987) and bovine aortic endothelial cells (Shapiro and DeCoursey, 1991; Groschner and Kukovetz, 1992). The direct interaction of Zn^{2+} or H^+ with the Cl^- channel is proposed to prevent the absorption of a cation species that promotes anion permeation and, thus, effectively blocks the channel (Franciolini and Nonner, 1987). The K^+ channel blocker, tetraethylammonium ion (TEA), has been shown recently to block Cl^- channels of lobster walking leg nerves (Lukács and Moczydlowski, 1990) and rat cortical neurons (Sanchez and Blatz, 1992). Externally applied TEA blocked fast Cl^- channels of rat cortical neurons in a voltage-dependent manner, with hyperpolarization favoring block. The dose-response relationship between the fractional Cl^- channel current and external TEA concentration exhibited 50% inhibition at ~10 mM but was incomplete at 50 mM TEA (Sanchez and Blatz, 1992).

The chloride conductance of skeletal muscle is profoundly affected by external pH (Hutter and Warner, 1967; Palade and Barchi, 1977a). The voltage and pH dependence of Cl^- conductance in frog skeletal muscle were first described by Hutter and Warner (1972) using the three-microelectrode voltage-clamp technique. At a pH of 9.8, resting chloride conductance was about 1.8 times higher than at pH 7.4 but, as the membrane was hyperpolarized from -90 mV, the Cl^- conductance fell steeply to about one tenth of the resting value. At pH 5.0, the resting membrane conductance was about one fifth of that at pH 7.4 and remained constant until the membrane was hyperpolarized by >30 mV from the resting potential of -90 mV, after which Cl^- conductance began to increase (Warner, 1972). The effects of low pH on Cl^- conductance in other cell types were qualitatively similar, that is, the conductance turns on at negative potentials and turns off at positive potentials (Shapiro and DeCoursey, 1991). Protons alter the behavior of the gating mechanism of neuronal Cl^- channels, which is manifested as a decrease in P_{open} (Blatz, 1991).

VI. Conclusions

Numerous clues suggest that background Cl^- channels are important in the control of many cellular functions. Maintaining high resting conductance and normal excitability, regulating cell volume, and modulating hormone action are some examples of the functions of background Cl^- channels. This diversity of function is mirrored by a similar diversity in the biophysical characteristics of the channel, such as ionic selectivity sequence, voltage dependence, unitary conductance, kinetic properties, and sensitivity to pharmacological agents. Despite the large amount

of data accumulated on background Cl^- channels, identifying subsets within this class of channels with coherent biophysical features that subserve each specific function is still not possible. The lack of high-affinity ligands for background Cl^- channels has hampered the isolation and purification of the channel protein and, thus, identification of their structure. The use of molecular genetic strategies has provided the primary structure of at least four background Cl^- channels (Jentsch *et al.*, 1990; Steinmeyer *et al.*, 1991; Thiemann *et al.*, 1992): those present in the electric organ of *Torpedo*, rat skeletal muscle, epithelia, and brain (see Chapter 16). The most extensively studied Cl^- channel is the product of the cystic fibrosis gene and is termed the cystic fibrosis transmembrane regulator (CFTR) (for review, see Riordan, 1993). The CFTR protein, heterologously expressed and purified, has been shown to induce Cl^- channels with correct functional properties when reconstituted into planar lipid bilayer membranes. The mutated form, a single amino acid deletion (phenylalanine residue, $\Delta F508$) of the Cl^- channel of airway epithelia, results in a temperature-sensitive defect in protein processing, giving rise to a disease state in which the epithelial Cl^- channels are absent. The primary structure of the CFTR protein is distinctly different from that of known voltage-gated cation (Na^+, K^+, Ca^{2+}) channels in excitable cells, because it contains two motifs of six transmembrane-spanning domains and a cytoplasmic nucleotide-binding domain joined by a regulatory phosphorylation domain. CFTR and other cloned Cl^- channels have structural elements for voltage activation that differ from the S4 helix segment with its regularly spaced positive charges in voltage-dependent cation channels of nerve and muscle. At present, the molecular structure for every type of functional Cl^- channel has not been determined, but future identification of cloned Cl^- channel structures should provide a clearer understanding of the functional properties of background Cl^- channels.

References

Almers, W., and Stirling, C. (1984). Distribution of transport proteins over animal cell membranes. *J. Membr. Biol.* **77**, 169–186.

Bettoni, G., Loiodice, F., Tortorella, V., Conte-Camerino, D., Mambrini, M., Ferrannini, E., and Bryant, S. H. (1987). Stereospecificity of the chloride ion channel: The action of chiral clofibric acid analogues. *J. Med. Chem.* **30**, 1267–1270.

Blatz, A. L. (1990). Chloride channels in skeletal muscle. *In* "Chloride Channels and Carriers in Nerve, Muscle, and Glial Cells" (F. J. Alvarez-Leefmans and J. M. Russell, eds.), pp. 407–420. Plenum Press, New York.

Blatz, A. L. (1991). Properties of single fast chloride channels from rat cerebral cortex neurons. *J. Physiol. (London)* **441**, 1–21.

Blatz, A. L., and Magleby, K. L. (1983). Single voltage-dependent chloride selective channels of large conductance in cultured rat muscle. *Biophys. J.* **43**, 237–241.

Blatz, A. L., and Magleby, K. L. (1985). Single chloride-selective channels active at resting membrane potentials in cultured rat muscle. *Biophys. J.* **47**, 119–123.

Blatz, A. L., and Magleby, K. L. (1986). Quantitative description of three modes of activity of fast chloride channels from rat skeletal muscle. *J. Physiol.* (*London*) **378**, 141–174.

Borisova, M. P., Brutyan, R. A., and Ermishkin, L. N. (1986). Mechanism of anion-cation selectivity of amphotericin B channels. *J. Membr. Biol.* **90**, 13–20.

Bosma, M. M. (1989). Amino channels with multiple conductance levels in a mouse B lymphocyte cell line. *J. Physiol.* (*London*) **410**, 67–90.

Bretag, A. H. (1987). Muscle chloride channels. *Physiol. Rev.* **67**, 618–724.

Bryant, S. H., and Morales-Aguilera, A. (1971). Chloride conductance in normal and myotonic muscle fibres and the action of monocarboxylic aromatic acids. *J. Physiol.* (*London*) **219**, 367–383.

Champigny, G., Verrier, B., Gérard, C., Mauchamp, J., and Lazdunski, M. (1990). Small conductance chloride channels in the apical membrane of thyroid cells. *FEBS Lett.* **259**, 263–268.

Chesnoy-Marchais, D. (1982). A Cl conductance activation by hyperpolarization in *Aplysia* neurons. *Nature* (*London*) **299**, 359–361.

Chesnoy-Marchais, D. (1983). Characterization of a Cl⁻ conductance activated by hyperpolarization in *Aplysia* neurons. *J. Physiol.* (*London*) **342**, 277–308.

Chesnoy-Marchais, D., and Evans, M. G. (1986). Cl channels activated by hyperpolarization in *Aplysia* neurons. *Pflügers Arch.* **407**, 694–696.

Chua, M., and Betz, W. J. (1991). Characterization of ion channels on the surface membrane of adult rat skeletal muscle. *Biophys. J.* **59**, 1251–1260.

Conte-Camerino, D., Mambrini, M., De Luca, A., Tricarico, D., Bryant, S. H., Tortorella, V., and Bettoni, G. (1988). Enantiomers of clofibric acid analogs have opposite actions on rat skeletal muscle chloride channels. *Pflügers Arch.* **413**, 105–107.

Coronado, R., and Latorre, R. (1982). Detection of K⁺ and Cl⁻ channels from calf cardiac sarcolemma in planar lipid bilayer membranes. *Nature* (*London*) **298**, 849–852.

Coulombe, A., Duclohier, H., Coraboeuf, E., and Touzet, N. (1987). Single chloride-permeable channels of large conductance in cultured cardiac cells of new-born rats. *Eur. Biophys. J.* **14**, 155–162.

De Luca, A., Tricarico, D., Wagner, R., Bryant, S. H., Tortorella, V., and Conte-Camerino, D. (1992). Opposite effects of enantiomers of clofibric acid derivative on rat skeletal muscle chloride conductance: Antagonism studies and theoretical modeling of two different receptor site interactions. *J. Pharmacol. Exp. Ther.* **260**, 364–368.

Dulhunty, A. F. (1979). Distribution of potassium and chloride permeability over the surface and T-tubule membranes of mammalian skeletal muscle. *J. Membr. Biol.* **45**, 293–310.

Dwyer, T. M., Adams, D. J., and Hille, B. (1980). The permeability of the endplate channel to organic cations in frog muscle. *J. Gen. Physiol.* **75**, 469–492.

Eisenberg, R. S., and Gage, P. W. (1969). Ionic conductances of the surface and transverse tubular membranes of frog sartorius muscle. *J. Gen. Physiol.* **53**, 279–297.

Eisenman, G., and Villarroel, A. (1989). Ion selectivity of pentameric protein channels. *In* "Monovalent Cations in Biological Membranes" (C. A. Pasternak, ed.), pp. 1–29, CRC Press, Boca Raton, Florida.

Fahlke, Ch., Zachar, E., and Rüdel, R. (1992). Single-channel recordings of chloride currents in cultured human skeletal muscle. *Pflügers Arch.* **421**, 108–116.

Fox, J. A. (1987). Ion channel subconductance states. *J. Membr. Biol.* **97**, 1–8.

Franciolini, F., and Nonner, W. (1987). Anion and cation permeability of a chloride channel in rat hippocampal neurons. *J. Gen. Physiol.* **90**, 453–478.

Franciolini, F., and Nonner, W. (1994). Anion-cation interactions in the pore of neuronal background chloride channels. *J. Gen. Physiol.* (in press).

Franciolini, F., and Nonner, W. (1994). A multi-ion permeation mechanism in neuronal background chloride channels. *J. Gen. Physiol.* (in press).

Franciolini, F., and Petris, A. (1988). Single Cl channels in cultured rat neurons. *Arch. Biochem. Biophys.* **261**, 97–102.

Franciolini, F., and Petris, A. (1990). Chloride channels of biological membranes. *Biochim. Biophys. Acta* **1031**, 247–259.

Franciolini, F., and Petris, A. (1992). Transport mechanisms in chloride channels. *Biochim. Biophys. Acta* **1113**, 1–11.

Fukuda, J. (1974). Chloride spike: A third type of action potential in tissue-cultured skeletal muscle cells from the chick. *Science* **185**, 76–78.

Gögelein, H. (1988). Chloride channels in epithelia. *Biochim. Biophys. Acta* **947**, 521–547.

Gray, P. T. A., and Ritchie, J. M. (1986). A voltage-gated chloride conductance in rat cultured astrocytes. *Proc. R. Soc. London B* **228**, 267–288.

Gray, P. T. A., Bevan, S., and Ritchie, J. M. (1984). High conductance anion-selective channels in rat cultured Schwann cells. *Proc. R. Soc. Lnodon B* **221**, 395–409.

Greger, R. (1990). Chloride channel blockers. *Meth. Enzymol.* **191**, 793–809.

Groschner, K., and Kukovetz, W. R. (1992). Voltage-sensitive chloride channels of large conductance in the membrane of pig aortic endothelial cells. *Pflügers Arch.* **421**, 209–217.

Hagiwara, S., Miyazaki, S., and Rosenthal, N. P. (1976). Potassium current and the effect of cesium on this current during anomalous rectification of the egg cell membrane of a starfish. *J. Gen. Physiol.* **67**, 621–638.

Hamill, O. P., Marty, A., Neher, E., Sakmann, B., and Sigworth, F. J. (1981). Improved patch-clamp techniques for high-resolution current recording from cells and cell-free membrane patches. *Pflügers Arch.* **391**, 85–100.

Hanrahan, J. W., Alles, W. P., and Lewis, S. A. (1985). Single anion-selective channels in basolateral membrane of a mammalian tight epithelium. *Proc. Natl. Acad. Sci. USA* **82**, 7791–7795.

Hille, B. (1992). "Ionic Channel of Excitable Membranes," 2d Ed. Sinauer Associates, Sunderland, Massachusetts.

Hodgkin, A. L., and Horowicz, P. (1959). The influence of potassium and chloride ions on the membrane potential of single muscle fibres. *J. Physiol.* (*London*) **148**, 127–160.

Hodgkin, A. L., and Huxley, A. F. (1952). The components of membrane conductance in the giant axon of *Loligo*. *J. Physiol.* (*London*) **116**, 473–496.

Hutter, O. F., and Warner, A. E. (1967). Action of some foreign cations and anions on the chloride permeability of frog muscle. *J. Physiol.* (*London*) **189**, 445–460.

Hutter, O. F., and Warner, A. E. (1972). The voltage dependence of the chloride conductance of frog muscle. *J. Physiol.* (*London*) **227**, 275–290.

Inoue, I. (1985). Voltage dependent chloride conductance of the squid axon membrane and its blockade by some disulfonic stilbene derivatives. *J. Gen. Physiol.* **85**, 519–537.

Jentsch, T. J., Steinmeyer, K., and Schwarz, G. (1990). Primary structure of *Torpedo marmorata* chloride channel isolated by expression cloning in *Xenopus* oocytes. *Nature* (*London*) **348**, 510–514.

Kokubun, S., Saigusa, A., and Tamura, T. (1991). Blockade of Cl channels by organic and inorganic blockers in vascular smooth muscle cells. *Pflügers Arch.* **418**, 204–213.

Kolb, H.-A., Brown, C. D. A., and Murer, H. (1985). Identification of a voltage-dependent anion channel in the apical membrane of a Cl⁻-secretory epithelium (MDCK). *Pflügers Arch.* **403**, 262–265.

Li, M., McCann, J. D., and Welsh, M. J. (1990). Apical membrane Cl. channels in airway epithelia: Anion selectivity and effect of an inhibitor. *Am. J. Physiol.* **259**, C295–C301.

Lukács, G. L., and Moczydlowski, E. (1990). A chloride channel from lobster walking leg nerves. Characterization of single-channel properties in planar lipid bilayers. *J. Gen. Physiol.* **96**, 707–733.

Mahaut-Smith, M. P. (1990). Chloride channels in human platelets: evidence for activation by internal calcium. *J. Membr. Biol.* **118**, 69–75.

Miller, C. (1982). Open-state substructure of single chloride channels from *Torpedo* electroplax. *Phil. Trans. R. Soc. London* **299**, 401–411.

Miller, C., and Richard, E. A. (1990). The voltage-dependent chloride channel of Torpedo electroplax. *In* "Chloride Channels and Carriers in Nerve, Muscle, and Glial Cells" (F. J. Alvarez-Leefmans and J. M. Russell, eds.), pp. 383–405. Plenum Press, New York.

Miller, C., and White, M. M. (1984). Dimeric structure of single chloride channels from *Torpedo* electroplax. *Proc. Natl. Acad. Sci. USA* **81**, 2772–2775.

Nelson, D. J., Tang, J. M., and Palmer, L. (1984). Single-channel recordings of apical membrane chloride conductance in A6 epithelial cells. *J. Membr. Biol.* **80**, 81–89.

Nowak, L., Asher, P., and Berwald-Netter, Y. (1987). Ionic channels in mouse astrocytes. *J. Neurosci.* **7**, 101–109.

Olesen, S.-P., and Bungaard, M. (1992). Chloride-sensitive channels of large conductance in bovine aortic endothelial cells. *Acta Physiol. Scand.* **144**, 191–198.

Palade, P. T., and Barchi, R. L. (1977a). Characteristics of the chloride conductance in muscle fibers of the rat diaphragm. *J. Gen. Physiol.* **69**, 325–342.

Palade, P. T., and Barchi, R. L. (1977b). On the inhibition of membrane chloride conductance by aromatic carboxylic acids. *J. Gen. Physiol.* **69**, 879–896.

Riordan, J. R. (1993). The cystic fibrosis transmembrane conductance regulator. *Annu. Rev. Physiol.* **55**, 609–630.

Robinson, R. A., and Stokes, R. H. (1959). "Electrolyte Solutions," 2d Ed. Butterworth, London.

Sanchez D. Y., and Blatz, A. L. (1992). Voltage-dependent block of fast chloride channels from rat cortical neurons by external tetraethylammonium ion. *J. Gen. Physiol.* **100**, 217–231.

Schlichter, L. C., Grygorczyk, R., Pahapill, P. A., and Grygorczyk, C. (1990). A large, multiple-conductance chloride channel in normal human T lymphocytes. *Pflügers Arch.* **416**, 413–421.

Schneider, G. T., Cook, D. I., Gage, P. W., and Young, J. A. (1985). Voltage sensitive, high-conductance chloride channels in the luminal membrane of cultured pulmonary alveolar (type II) cells. *Pflügers Arch.* **404**, 354–357.

Schoppa, N., Shorofsky, S. R., Jow, F., and Nelson, D. J. (1989). Voltage-gated chloride currents in cultured canine tracheal epithelial cells. *J. Membr. Biol.* **108**, 73–90.

Schwarze, W., and Kolb, H. A. (1984). Voltage-dependent kinetics

of an anionic channel of large unit conductance in macrophages and myotube membranes. *Pflügers Arch.* **402**, 281–291.

Shapiro, M. S., and DeCoursey, T. E. (1991). Chloride currents in bovine pulmonary artery endothelial cells. *In* "Ion Channels of Vascular Smooth Muscle Cells and Endothelial Cells" (N. Sperelakis and H. Kuriyama, eds.), pp. 327–336. Elsevier Science, Amsterdam.

Shukla, H., and Pockett, S. (1990). A chloride channel in excised patches from cultured rat hippocampal neurons. *Neurosci. Lett.* **112**, 229–233.

Soejima, M., and Kokubun, S. (1988). Single anion-selective channels and its ion selectivity in the vascular smooth muscle cell. *Pflügers Arch.* **411**, 304–311.

Sonnhof, U. (1987). Single voltage-dependent K and Cl channels in cultured rat astrocytes. *Can. J. Physiol. Pharmacol.* **65**, 1043–1050.

Stanfield, P. R. (1970). The differential effects of tetraethylammonium and zinc ions on the resting conductance of frog skeletal muscle. *J. Physiol. (London)* **209**, 213–256.

Steinmeyer, K., Ortland, C., and Jentsch, T. J. (1991). Primary structure and functional expression of a developmentally regulated skeletal muscle chloride channel. *Nature (London)* **354**, 301–304.

Strupp, M., and Grafe, P. (1991). A chloride channel in rat and human axons. *Neurosci. Lett.* **133**, 237–240.

Tabares, L., Mazzanti, M., and Clapham, D. E. (1991). Chloride channels in the nuclear membrane. *J. Membr. Biol.* **123**, 49–54.

Tanifuji, M., Sokabe, M., and Kasai, M. (1987). An anion channel of SR incorporated into planar lipid bilayers: Single-channel behavior and conductance properties. *J. Membr. Biol.* **99**, 103–111.

Thiemann, A., Grander, S., Pusch, M., and Jentsch, T. J. (1992). A chloride channel widely expressed in epithelial and non-epithelial cells. *Nature (London)* **356**, 57–60.

Vaughan, P. C., and French, A. S. (1989). Non-ligand-activated chloride channels of skeletal muscle and epithelia. *Prog. Biophys. Mol. Biol.* **54**, 59–79.

Wangemann, P., Wittner, M., Di Stefano, A., Englert, H. C., Lang, E., Schlatter, E., and Greger, R. (1986). Cl⁻ channel blockers in the thick ascending limb of the loop of Henle. Structure activity relationship. *Pflügers Arch.* **407 (Suppl. 2)**, 5128–5141.

Warner, A. E. (1972). Kinetic properties of the chloride conductance of frog muscle. *J. Physiol. (London)* **227**, 291–312.

Weiss, D. S., and Magleby, K. L. (1990). Voltage dependence and stability of the gating kinetics of the fast chloride channel from rat skeletal muscle. *J. Physiol. (London)* **426**, 145–176.

Welsh, M. J. (1986). An apical-membrane chloride channel in human tracheal epithelium. *Science* **232**, 1648–1650.

White, M. M., and Miller, C. (1979). A voltage-gated anion channel from the electric organ of *Torpedo californica*. *J. Biol. Chem.* **254**, 10161–10166.

Woll, K. H., and Neumcke, B. (1987). Conductance properties and voltage dependence of an anion channel in amphibian skeletal muscle. *Pflügers Arch.* **410**, 641–647.

Woll, K. H., Leibowitz, M. D., Neumcke, B., and Hille, B. (1987). A high-conductance anion channel in adult amphibian skeletal muscle. *Pflügers Arch.* **410**, 632–640.

Woodbury, J. W., and Miles, P. R. (1973). Anion conductance of frog muscle membranes: One channel, two kinds of pH dependence. *J. Gen. Physiol.* **62**, 324–353.

Ligand-Gated Channels

Nicotinic Acetylcholine Receptor of the Mammalian Central Nervous System

J. G. MONTES, M. ALKONDON,
E. F. R. PEREIRA, AND E. X. ALBUQUERQUE

I. Introduction

For several reasons, the nicotinic acetylcholine receptor (nAChR), a member of the family of ligand-gated ion channels, has been one of the most thoroughly studied of all the receptors. The discovery of highly specific snake toxins, notably α-bungarotoxin (α-Bgt), gave the study of the nAChR a decisive push, because use of these toxins facilitated the localization, purification, isolation, and consequently, molecular characterization of this very important transducing element (see Fig. 1). Previously, the most useful antagonist available was the much less specific toxin D-tubocurarine; unfortunately, a natural toxin analogous to α-Bgt was not available for the study of the muscarinic acetylcholine receptor. Another factor that added considerable impetus to the study of the nAChR was the existence of rich stores of nAChR in the form of electric organs of the elasmobranch *Electrophorus* and the teleost fish *Torpedo* (Popot and Changeux, 1984). These natural advantages, as well as advances in experimental techniques applicable to studying the end-plate region of the neuromuscular junction—some of them revolutionary—have secured the nAChR its current position as one of the most studied of all receptors. Subsequently, nAChRs were discovered in anatomical regions other than the neuromuscular junction, including ganglia.

Eventually researchers recognized that nicotine- and α-Bgt-binding sites existed in the central nervous system (CNS), presumably providing evidence for the presence of nicotinic receptors in that region. Soon thereafter, however, the lack of correspondence between binding sites for nicotine or other agonists and those for α-Bgt threatened to compromise any attempt to map the distribution of functional nAChRs in the CNS. Biochemical studies have since verified that α-Bgt-binding proteins that are not necessarily functional nAChRs exist in many neuronal preparations, although in the insect brain the correspondence between functional nAChRs and α-Bgt-binding sites is more nearly complete (Breer *et al.*, 1985), and *in situ* hybridization studies in chick (Couturier *et al.*, 1990) and rat (Séguéla *et al.*, 1993) have demonstrated that a specific subunit (α7) of the nAChR colocalizes with α-Bgt-binding sites in the brain. Ultimately, evidence mounted that a higher degree of structural and functional diversity than anticipated existed in neural preparations. The nAChR of muscle had been successful as a prototype for the study of the neuronal nAChR, but the time had come to recognize the limita-

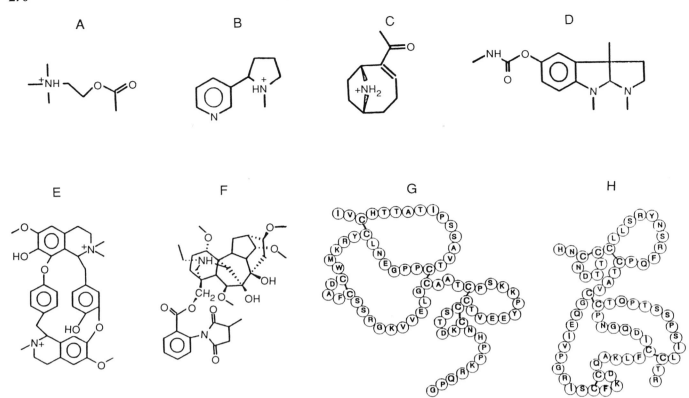

FIGURE 1 Structures of some pharmacological agents important to the study of nicotinic nAChRs. (A) Acetylcholine (ACh). (B) Nicotine. (C) (+)-Anatoxin-a. (D) Physostigmine. (E) D-Tubocurarine. (F) Methyllycaconitine. (G) α-Bungarotoxin (α-Bgt). (H) κ-Bungarotoxin (κ-Bgt). International codes for the amino acids are used in the sequences for the two snake toxins. (A–F) Reprinted with permission from Swanson and Albuquerque (1992). (G) Based on Mebs *et al.*, (1992) and Low (1979), (H); Chiappinelli (1985).

tions of this model as it applied to the more subtle pharmacological and structural properties of the versions of the receptor found in the CNS.

The structure of the muscle nAChR is that of a doughnut-shaped pentamer consisting of subunits arranged in a circular fashion, its aperture corresponding to the cation channel. The five subunits making up the pentamer follow the stoichiometry and order $\alpha\beta\alpha\gamma\delta$ in embryonic muscle and extrajunctional nAChRs, whereas the γ subunit is replaced with an ε subunit in the adult muscle and junctional receptors (Mishina *et al.*, 1986; Changeux, 1990; Witzenmann *et al.*, 1991). In the neural tissues examined to date, however, only α and β (sometimes conservatively termed "non-α") subunits have been found and, although significant homologies exist between the neuronal subunits and their counterparts in muscle, the corresponding subunits in muscle and neurons are not identical. In addition, sufficient heterogeneity has been observed among the neuronal subunits themselves to justify that these neuronal subunits be designated with subscripts or numerical suffixes (we prefer

the numerical suffixes because, unlike subscripts, they cannot be confused with stoichiometric designations). Thus far, seven α subunits ($\alpha2-\alpha8$) and four β subunits ($\beta2-\beta5$) have been identified and designated in neural tissues (reviewed in Deneris *et al.*, 1991; Role, 1992), although much information remains to be uncovered with respect to the stoichiometries and tertiary and quaternary structures of the receptors. In muscle and electric organs, however, only one kind of α subunit has been identified, namely $\alpha1$, which appears to exist in two distinct isoforms (Hartman and Claudio, 1990).

Considerable molecular biological research is currently being conducted on neuronal nAChRs. The most recent approaches emphasize the functional expression of neuronal nAChR subunits in oocytes and other cells, and should be highly informative about the molecular diversity of the CNS nAChR. The pharmacological properties of the reconstituted receptors should provide supplementary information important in assaying the neuronal nAChRs whose structures are still unknown. The electrophysiological methods of patch clamp have been the most effective correlates of pharmacological investigations of the nAChR over the last several years. These techniques are likely to continue to play a very significant role in research for some time.

In the sections that follow, some of the more important molecular pharmacological aspects of the neuronal nAChR are discussed, concentrating on findings related to the CNS of mammals. Some comparisons between neuronal nAChRs and those of the neuromuscular junction and electric organs will be necessary, although readers are referred to the many excellent reviews emphasizing these and other aspects of the nAChR. (An excellent bibliography of reviews, arranged according to categories of specialization, is provided by Hucho, 1986.)

II. Distribution of nAChRs in Different Tissues

The initial observations made on muscle nAChR did not hint at the diversity of structure and function to be discovered later in the nAChRs of other tissues. Indeed, although one can speak of the nAChR in a general sense, a sizeable number of nAChR subtypes exist, even within tissues from the same species. Further, considerable species diversity may occur with respect to nAChRs from homologous tissues.

The nAChR can be subdivided into three general groups predicated on anatomical and functional bases: (1) muscle and muscle-type; (2) ganglion and ganglion-type; and (3) CNS (also called central) neuronal. Nonetheless, developments in molecular pharmacology suggest that the nAChR can be described more simply as belonging to two main groups—muscle and neuronal. The justification for such a grouping is the finding that the $\alpha 1$ subunit is present exclusively in muscle and muscle-type nAChRs, whereas non-$\alpha 1$ subunits are present in the CNS and in ganglia, the chief difference being that one of these subunits, $\alpha 3$, is more prevalent in ganglia than any other α subtype, whereas it appears to be less frequent in the CNS (Role, 1992). Undoubtedly, classification schemes yet to be devised will rely more heavily on structural considerations, and the taxonomy of the nAChR is still evolving.

A. Distribution of Muscle and Muscle-Type nAChRs

The ubiquity of the nAChR at the neuromuscular junction in vertebrates and invertebrates has been confirmed repeatedly in numerous studies (Gerschen-

feld, 1973; Krnjevic, 1975); even cell lines, such as human TE671/RD cells, have been found to contain "muscle-type" nAChRs (i.e., similar to those found in muscle; Lukas, 1990). Among organs derived ontogenically from muscle precursors, piscine electric organs are very rich sources of nAChRs. However, acetylcholine (ACh) is not the only neurotransmitter acting on muscle in all species, as illustrated in insect muscle, in which glutamate serves as the neurotransmitter (Faeder and O'Brien, 1970). This distinction suggests the need for a more extensive survey of invertebrate muscles for the presence of functional nAChRs. The evolution of the nAChR is particularly interesting. The known amino acid sequences of nAChR subunits from a wide range of sources display a great degree of homology to one another.

B. nAChR Distribution in Neuronal Tissues

For some time, researchers have known that muscle nAChRs differ pharmacologically from those in ganglia; in fact, ganglionic nAChRs are, by definition, a subset of neuronal nAChRs. In general, the quaternary ammonium antagonist hexamethonium is better able to competitively block the nAChR in ganglia than the one in muscle, which appears to be more sensitive to decamethonium (see Lukas and Bencherif, 1992). Functional nAChRs have been identified in the sympathetic ganglia of frogs (e.g., Marshall, 1981; Lipscombe, 1988), chickens (e.g., Margiotta *et al.*, 1987; Pinnock *et al.*, 1988; Moss *et al.*, 1989), rats (e.g., Mathie *et al.*, 1987), and of a considerable number of other species.

Functional nAChRs have been found using *in vivo* techniques in a wide range of CNS tissues in a number of animals, including in goldfish brain (Henley *et al.*, 1986), frog pars intermedia (Lamascz *et al.*, 1989), chicken retinal ganglion (Loring *et al.*, 1989) and lateral spiriform nucleus (Sorenson and Chiappinelli, 1990), a number of mammalian brain regions (reviewed in Luetje *et al.*, 1990b), and even the CNS of insects (reviewed in Satelle, 1986). Binding sites for nicotine, ACh, and snake toxins exist in many brain regions in mammals. However, although nAChRs have been found in the CNS of all animals examined to date, muscarinic acetylcholine receptors appear to be even more numerous than the nAChR in neural tissues, at least in vertebrates, and other kinds of receptors, such as *N*-methyl-D-aspartic acid (NMDA) and γ-aminobutyric acid (GABA) receptors, also serve neurotransmission functions. The importance of the nAChR, relative to other types of receptors, in the function of the nervous system remains to be evaluated.

Although the detailed physiological and behavioral

roles played by the nAChRs present in the CNS are yet to be elucidated, the role of nAChRs in animal behavior is manifested in several pathological conditions in which the function and/or number of the nAChRs is altered, as in nicotine addiction (Benwell *et al.*, 1988), Parkinson's disease (Kellar and Wonnacott, 1990), myasthenia gravis (Rash *et al.*, 1976; Albuquerque and Eldefrawi, 1983; Lindstrom, 1985), and Alzheimer's disease, a condition in which the number of neurons expressing nAChRs is considerably reduced (McGeer *et al.*, 1984; Nilsson *et al.*, 1987).

Surveys of brain regions are typically conducted by autoradiographic assay of tissue sections to which radioactively labeled ligands have been added. [^{3}H]Nicotine-binding sites have been found in mouse, rat, bovine, and human brain (discussed by Wonnacott, 1987). Pharmacological, autoradiographic, and biochemical studies have indicated that in the brain, binding sites for [^{3}H]nicotine and [^{3}H]ACh (assayed in the presence of muscarinic antagonists) are probably the same. The inference from earlier investigations was that the αBgt-binding sites were exclusively located on nAChRs. However, this conclusion, at least with respect to functional nAChRs, was initially challenged by the finding that the toxin-binding sites are not necessarily identical to ACh-binding sites (Clarke *et al.*, 1985; Lindstrom *et al.*, 1987) as well as by the isolation and identification of non-nAChR proteins capable of binding α-Bgt (Abood *et al.*, 1980; Wonnacott, 1987). Currently, at least some of the toxin-binding proteins originally thought not to be nAChRs appear to be, in fact, previously unrecognized members of a distinct branch of the nAChR gene family (see discussion by McLane *et al.*, 1993).

Almost all regions of mammalian brain assayed to date appear to bear α-Bgt- and nicotine-binding sites, although varying degrees of regional and species differences in receptor density exist. In mammalian brain, the thalamus is a region particularly rich in nAChRs (Wonnacott, 1987). The distribution of nAChRs generally follows the pattern of cholinergic innervation in the brain, so other regions particularly enriched in nAChRs include the cerebral cortex, the dentate gyrus, the interpeduncular nucleus, and the medial habenula (Luetje *et al.*, 1990b). In Fig. 2, regions and subregions of mammalian brain that have been studied according to sensitivity to two or more nicotinic antagonists are displayed; cholinergic afferents known to stem from one brain region to another are represented by lines and arrows. Figure 2 will be useful when the heterogeneity of the neuronal nAChR is discussed in a subsequent section. The function of the nAChR in nerve tissue, particularly as it relates to presynaptic and postsynaptic localization, is also addressed later.

III. Molecular Pharmacology of Central Neuronal nAChRs

A. Function of Central Neuronal nAChRs

The diverse physiological and behavioral roles of neuronal nAChRs are just beginning to be understood. Although developmental and differentiation studies have been helpful in revealing the physiological function of the neuronal nAChRs of the CNS, these studies have suffered from the perennial attribute of correlations in general, that cause cannot be discerned from effect on the basis of associations alone. Another confounding factor is that functional nAChRs have been demonstrated to exist both presynaptically and postsynaptically in at least some regions of the rat brain (Brown *et al.*, 1983,1984; Mulle *et al.*, 1991). Presynaptic nAChRs appear to be involved in stimulating the release of neurotransmitters such as dopamine; on the other hand, the role of neuronal postsynaptic nAChRs remains less clear. Even if nAChRs are shown to play a role in synapse modeling or maintenance (see Lukas and Bencherif, 1992), how this is achieved at the molecular level would need to be demonstrated.

Cast in more macroscopic terms, very little is known about how the nAChR influences the behavior of animals. This potentially very important aspect of the nAChR can begin to be understood by dissecting the global function of the nAChR into its electrophysiological and biochemical components. Elucidating the contribution made by other types of receptors in affecting behavior is particularly important, especially the relationship of the different types and subtypes of receptors to each other as interconnected rather than as individual elements. The anatomical and microanatomical correlates of the functions of the nAChR and of other neural receptors can be anticipated to become increasingly important to a holistic understanding of the connection between receptor function and animal behavior.

As mentioned earlier, some useful information linking receptor function to animal behavior is already available in the form of knowledge obtained from studies of pathological conditions. Discussing this issue further is not within the scope of this review, so we defer to others (see references related to neurodegenerative disorders) and emphasize instead the properties of the nAChR as seen in individual neurons. Much more sense will be made of the studies

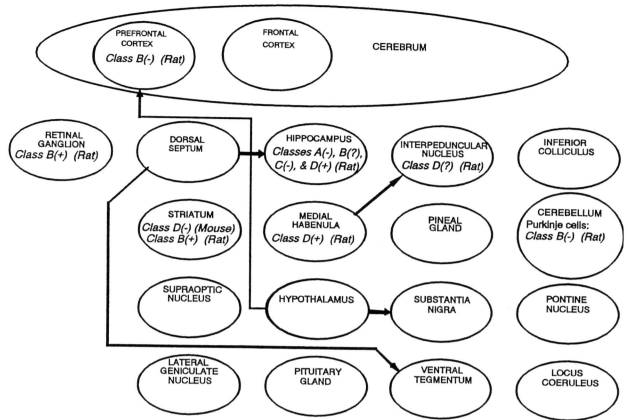

FIGURE 2 Partial distribution of nAChRs in the mammalian central nervous system (CNS). Ovals represent regions of mammalian CNS for which antagonist profiles based on two or more toxins have been determined; other sectors (not shown) also have been found to possess nAChRs, including the thalamus. Profiles of responses to α-bungarotoxin (α-Bgt) and κ-Bgt are given as separate classes (as defined in text and in Table III) for each region for which responses to both these toxins were determined in the same study. The species from which the preparation studied was taken is also given. A plus or minus sign in parentheses after the response class indicates sensitivity (+) or lack of sensitivity (−) to 1 μM mecamylamine. Arrows indicate known cholinergic tracts leading from one brain region to another. Most data were obtained from Lukas and Bencherif (1992). Supplementary data were obtained from Mulle *et al.*, (1991; rat interpeduncular nucleus), Grady *et al.*, (1992; mouse striatum), Alkondon and Albuquerque (1993; rat hippocampus). Cholinergic tract data are from Kasa (1986).

concerning the role of the nAChR in determining or modifying animal behavior once the properties of the neuronal nAChR and other receptors in the nervous system have been more fully revealed at the molecular level.

A major complication in assessing the presence and function of nAChRs in various brain regions is perhaps related to the precise physiological function of these receptors. Nicotine is known to stimulate the release of certain neurotransmitters, including dopamine, from the presynaptic terminals of neurons. The release of dopamine appears to be dependent on Ca^{2+}, at least at lower nicotine concentrations in mouse striatum (Grady *et al.*, 1992), suggesting the involvement of the presynaptic nAChRs in complex physiological processes. In addition, since presynaptic nerve terminals may be major targets of nicotine in the brain (Rapier *et al.*, 1990), and since the nAChRs in brain appear to be structurally and functionally heterogeneous, the central challenge of assigning the newly described nAChR subtypes in individual neurons to their respective cellular regions and functional roles, that is, presynaptic or postsynaptic, remains (e.g., see Mulle *et al.*, 1992). Most of the brain sites that do not bind α-Bgt and that possess a high stereoselective affinity for L-nicotine may represent presynaptic nAChRs (Wonnacott, 1990). The known heterogeneity of the nAChR in the nervous system dictates caution, however, because making correlations between the affinities for the neurotoxins and the origin and location of the nAChRs may be made more difficult by the possibility that more than one structural or functional nAChR subtype can exist in an individual neuron (Vijayaraghavan *et al.*, 1992; Alkondon and Albuquerque, 1993). The report that fractions of rat hippocampal membrane rich in presynaptic nAChRs can be isolated and purified using Percoll gradients (Wonnacott and Thorne, 1990) suggests that soon researchers may be able to quantify the relative frequencies of neuronal nAChRs according to their individual structures and distribution.

The molecular pharmacology of the neuronal nAChR actually begins with the localization and binding studies alluded to earlier (see Section II). However, the earlier binding studies can be related to nAChR function only by implication, thereby highlighting the need for investigations correlating patterns of agonist and antagonist binding in neuronal preparations with electrophysiological responses that result unambiguously from functional nAChRs in these same preparations. In this fashion, the binding studies can be accorded pharmacological relevance to the nAChR.

1. Kinetic Model of the nAChR

A starting point for a discussion of the properties of the nAChR in the CNS, or in any other region for that matter, is a working hypothesis or model describing the kinetics of activation of the receptor ion channel. Years of research on the peripheral nAChR has led to the formulation of a basic scheme for the activation of the receptor (Steinbach, 1968; Adler *et al.*, 1978; Adams, 1981):

$$2A + R \underset{k_{-1}}{\overset{k_1}{\rightleftharpoons}} A + AR \underset{k_{-2}}{\overset{k_2}{\rightleftharpoons}} A_2R \underset{k_{-3}}{\overset{k_3}{\rightleftharpoons}} A_2R^*$$

State
number   1   2   3   4

Scheme I

where two molecules of nicotinic agonist, A, bind to the receptor in two steps (producing states 2 and 3) to yield a non-conducting ligand–receptor complex, A_2R (state 3), which is then free to undergo a transition to the open-state complex, A_2R^* (state 4). The forward and reverse rate constants are shown as k_n and k_{-n}, respectively, where n is the state number. The model presented here is based in part on earlier work on end-plate channels (Katz and Thesleff, 1957), and in part on the assumption that two agonist molecules must bind to the nAChR before it can become activated (Karlin, 1980; Conti-Tronconi and Raftery, 1982). The details of the kinetic model continue to

evolve. This model, widely accepted in its general form, is given here simply as a useful point of departure.

2. Electrophysiological Properties of CNS Neuronal nAChRs

From a set of seven independent studies of the CNS, all but one of them performed in rat neurons, single-channel conductances of neuronal nAChRs can be concluded to be of two or three types, as seen in Table I: a high conductance state of 73 pS, represented here by data from only one source; an intermediate one in the range of 45–50 pS; and a low conductance state of 20–26 pS. The lack of concordance among these conductances and those observed for neuronal nAChRs expressed in oocytes is discussed in a subsequent section. In any case, comparisons among receptors based chiefly on conductances should not be overemphasized because discordant conductance values are often obtained even among neurons from the same preparation (Role, 1992); posttranslational processing of nAChR subunits in the oocytes may result in versions of the original subunits that are not functionally equivalent to those expressed in the cells of their origin (Gehle and Sumikawa, 1991); and experimental conditions, such as the ionic composition of experimental media, are frequently not the same in one study as in another (see Section III,C,2).

Currently, data on open-channel lifetimes for the nAChR ion channels of mammalian central neurons are limited. The few estimates of channel open times of central nAChRs reported to date fall into two general categories, one in the range of ≤ 0.5 msec and the other between 1 and 10 msec. Data for nAChRs expressed in oocytes agree with this pattern in general but, as mentioned before, difficulties arise in making inferences about native nAChRs from expression studies in oocytes. In the native neuronal receptors, only one study has yielded an open time on the order of tenths of milliseconds: a value of 0.1 msec at -80 mV in a single-channel investigation conducted on rat hippocampal neurons by Castro and Albuquerque (1993). Although lifetimes of tenths of milliseconds have been reported for rat neuronal nAChRs expressed in oocytes, as well as in electrophysiological studies of the nAChRs in rat or chick ganglia, the few other studies of native central neuronal nAChRs yielded values in the range of 0.4–10 msec (summarized by Role, 1992; Pereira *et al.*, 1993b). Very fast open times may have been missed at the lower filtering frequencies (typically 3 kHz). In contrast, the 0.1-msec open time was obtained at 7 kHz. Much work remains to be done in this area, and in charac-

TABLE I Conductances of Mammalian Central Neuronal nAChRs

Preparation[a]	Conductance (pS)[b]	Temperature (°C)	Reference
Rat hippocampal neurons[c,d]	20	10	Aracava *et al.*, (1987)
Rat hippocampal neurons[d]	22, 43	22	Pereira *et al.*, (1993a)
Porcine pars intermedia	26	20	Zhang and Feltz (1990)
Rat medial habenula	26	20	Mulle and Changeux (1990)
Rat medial habenula, interpeduncular nucleus	35, 41 (respectively)	?	Mulle *et al.*, (1991)
Rat medial habenula	41	?	Mulle *et al.*, (1992)
Rat retinal ganglion	42–48	22	Ramoa *et al.*, (1990)
Rat retinal ganglion	49	35	Lipton *et al.*, (1987)
Rat medial habenula	50	22	Connolly and Colquhoun (1991)
Rat hippocampal neurons	73	22	Castro and Albuquerque (1993)

[a] All studies were performed using the outside-out configuration and ACh as agonist, unless otherwise noted.

[b] Conductances were determined from single-channel experiments.

[c] Measurements were made using the cell-attached configuration.

[d] Agonist used was (+)-anatoxin-a.

izing other kinetic properties of the ion channels of CNS nAChRs. (For an elementary introduction to electrophysiological methods used in the study of neuronal nAChRs, see Montes *et al.*, 1994.)

3. Pharmacological Properties of CNS Neuronal nAChRs

Electrophysiological studies of the comparative pharmacological properties of a range of nAChR agonists in the CNS of mammals have been confined entirely to rat preparations consisting of medial habenula (Mulle and Changeux, 1990; Mulle *et al.*, 1991), interpeduncular nucleus (Mulle *et al.*, 1991), or hippo-

campal neurons (Alkondon *et al.*, 1992; Zorumski *et al.*, 1992; Alkondon and Albuquerque, 1993). Hierarchies for agonist activation of the receptors differed somewhat among the data for medial habenula and interpeduncular nucleus, except that in both tissues 1,1-dimethyl-4-phenyl-piperazinium (DMPP) was the least active agonist. In hippocampal neurons, three distinct agonist hierarchies have been found (Table II), each one corresponding to a specific type of macroscopic current distinguished from the others by virtue of its decay characteristics and response to various antagonists, as well as by its particular agonist profile. Sample recordings of the three types of current that

TABLE II Hierarchy of Responses of Hippocampal Neuronal nAChRs to Various Agonists According to Type of Evoked Current[a,b]

Type	Rank order of potency								
IA	AnTX	DMPP	(−)Nic	Cyt	ACh	Carb	(+)Nic	Arec	SubCh
II	ACh	AnTX	(−)Nic	DMPP	Carb	Cyt	(+)Nic	SubCh	Arec
III	AnTX	Cyt	(−)Nic	DMPP	ACh	Carb	(+)Nic	Arec	SubCh

[a] Based on Alkondon and Albuquerque (1993).

[b] Nine nicotinic agonists are displayed in rank order of potency in eliciting currents classified as types IA, II, and III on the basis of current decay characteristics and sensitivity to various agonists and antagonists. Symbols used: AnTX, (+)-anatoxin-a; DMPP, 1,1-dimethyl-4-phenyl-piperazinium; (−)Nic, (−)-nicotine; (+)Nic, (+)-nicotine; Cyt, cytisine; ACh, acetylcholine; Carb, carbachol; Arec, arecoline; SubCh, suberyldicholine.

could be evoked in response to 3 mM ACh, and of one hybrid type apparently generated by the simultaneous elicitation of two distinct currents (types IA and II) in the same neuron, are displayed in Fig. 3. Two general conclusions can be drawn from the hierarchies of response to agonists: (1) that arecoline, suberyldicholine, carbachol, and the two stereoisomers of nicotine are relatively weak discriminants of the current types, and (2) that some the agonists studied may have been supernumerary with respect to their use in categorizing individual neurons (and perhaps their receptors) according to the proposed categories. For example, just two of the agonists, ACh and cytisine (Cyt), could be used to sort the currents into the three main categories. Type IA currents are subject to the hierarchy Cyt > ACh, type II currents to ACh ≫ Cyt, and type III currents to Cyt ≫ ACh. Current types IA (the most commonly encountered) and II represent the two extremes of affinity of the receptor for ACh, with EC_{50}s of ~130 μM and 2 μM, respectively. Although only one type of current (IA) could usually be evoked in a given neuron, in some neurons currents (type IB) arising from more than one subtype of receptor were apparent. The varied responses to agonists are strongly indicative of structural and functional heterogeneity of the nAChRs in the CNS.

Also indicative of heterogeneity of the neuronal nAChRs, is that, unlike the nAChR currents in muscle, agonist-evoked nicotinic currents in neuronal nAChRs, including the ganglionic receptor, show inward rectification, a phenomenon that can occur in neurons even in the absence of Mg^{2+} and Ca^{2+} on either side of the membrane (Ifune and Steinbach, 1992; Neuhaus and Cachelin, 1990; Alkondon and Albuquerque, 1991,1993; Sands and Barish, 1992). The degree of rectification can be used as a criterion for classifying the responses of different neurons to agonist, especially when the agonist-evoked currents are recorded using Mg^{2+}-free internal solutions. For example, among the different types of nicotinic currents that could be evoked in hippocampal neurons, the most frequently encountered currents (type IA) showed inward rectification that was highly dependent upon the presence of intracellular Mg^{2+}, whereas a type of current infrequently encountered (type II) exhibited a Mg^{2+}-independent inward rectification (Alkondon and Albuquerque, 1993; Alkondon and Albuquerque, unpublished observations). The fact that different brain regions, or different individual neurons from a particular brain region, can respond dissimilarly to nicotinic antagonists provides additional strong evidence for the structural heterogeneity of the nAChR in the CNS. In addition to nicotinic agonists, the nicotinic antagonists α-Bgt (100 nM), κ-Bgt (100 nM), methyllycaconitine (MLA, 1 nM), mecamylamine (MECA, 1 μm), and dihydro-β-erythroidine (DHβE, 100 nM) have been used as diagnostic tools in categorizing the currents evoked by activation of nAChRs in rat hippocampus, although α-Bgt, κ-Bgt, and MECA alone sufficed to distinguish the different types of responses (Alkondon and Albuquerque, 1993). A workable classification requires a prior selection of threshold concentrations of the antagonists at which a determination of sensitivity or insensitivity to a given antagonist can be made. On the basis of concentrations that were used to discriminate successfully among various types of responses in our studies and in those of other investigators, 200 nM α-Bgt, 200 nM κ-Bgt, and 10 μM MECA have been selected here as the test concentrations for deciding sensitivity to the antagonists. Each antagonist has the possibility of either inhibiting (producing ≥50% inhibition) a response or not inhibiting a response, providing up to four different principal classes of sensitivity to antagonist based only on α-Bgt and κ-Bgt, each with two subclasses based on MECA, for a total of eight possible combinations of receptor responsivity to the antagonists. The four principal classes which, for convenience, are referred to as A, B, C, and D are shown in Table III; if sensitivity to MECA were also used as a criterion, each class shown could be further subdivided into (+) or (−) subclasses, depicting sensitivity or insensitivity to that drug. These classes are indi-

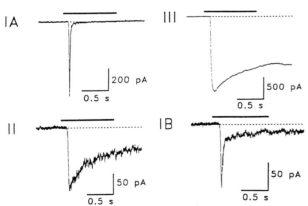

FIGURE 3 Sample recordings of four types of currents evoked by ACh in hippocampal neurons. A sample current recording of each of the four types of current is provided. (upper left) Type IA. (lower left) Type II. (upper right) Type III. (lower right) Type IB, a hybrid type presumed to result from the summation of type IA and type II currents in a single neuron. Horizontal bars indicate duration of 3 mM ACh pulse applied to the neurons. Reprinted with permission from Alkondon and Albuquerque (1993). Diversity of nicotinic acetylcholine receptors in rat hipocampal neurons: I. Pharmacological and functional evidence for distinct structure subtypes. *J. Pharm. Exp. Ther.* **265**, 1455–1473. © Am. Soc. for Pharm. and Exp. Ther.

TABLE III Classes of Neuronal nAChRs According to Sensitivity to α-Bungarotoxin and κ-Bungarotoxin[a]

	α-Bgt	κ-Bgt
Class A	+	+
Class B	−	+
Class C	+	−
Class D	−	−

[a] Four classes based on responses to α- and κ-Bgt were defined according to sensitivity to 200 nM of either of the two toxins.

+, sensitive to toxin; −, insensitive to toxin.

cated in the "brain map" of Fig. 2 according to their location in the CNS of different animal species; responses to MECA are given in parentheses, since this drug has not yet been tested widely enough. Inspection of Fig. 2 with cross-reference to Table III reveals that all four response or sensitivity classes have been discovered in at least one region of the mammalian CNS. Note, however, that Class C, typical of muscle nAChRs, is represented only in one study, which was based rather indirectly on potentials evoked in response to auditory stimuli in adult rat hippocampus *in situ* (Luntz-Leybman *et al.*, 1992). Although a whole-cell current response classifiable as Class C was not seen in work on rat hippocampal neurons, that preparation consisted of cultured fetal neurons (Alkondon and Albuquerque, 1993) rather than *in situ* adult hippocampus as in the other study.

Regardless of the nature of the response, we should caution that in whole-cell patch-clamp and ligand-binding studies, and especially in indirect, more qualitative studies based on secondary responses—such as potentials evoked in response to sensory stimuli—the presence of one kind of response may effectively mask the existence of another. This possibility must be entertained because not every neuron possesses only one functional or structural variant of the nAChR, as has been shown for type IB currents evoked in hippocampal neurons (Alkondon and Albuquerque, 1993). Thus, for example, if two classes of response to toxins can be seen in a particular neuron, and the magnitude of the response belonging to one of the two classes is considerably greater than that of the other, a category may be missed or mislabeled. Based on the criteria in Table III, a low-grade Class C response resulting from a relatively low density of a certain nAChR subtype may not be discernible in a current recording in which a Class A response predominates; similarly, a preponderance of receptor subtypes capable of producing Class D responses

could render the simultaneous presence of any of the other classes difficult to detect. This kind of ambiguity has been invoked as a possible explanation for the previous inability to detect whole-cell currents attributable to α-Bgt-binding components in neuronal membranes. The identity of at least some of these components as functional nAChRs was questioned until recently. Another possibility was that the receptors capable of binding α-Bgt desensitized too rapidly for their currents to be detected against the backdrop of currents that were evoked from other nAChRs and possessed slower decay characteristics (Vijayaraghavan *et al.*, 1992).

In previous work (Alkondon and Albuquerque, 1993), sensitivity to various antagonists was found to be useful in assigning the responses of individual hippocampal neurons to specific categories. Since a series of agonists functioned as well as a series of antagonists in discriminating among the types of responses, the current types that were proposed are likely to encompass sharply demarcated sets of functionally and structurally distinct nAChR subtypes.

Categorizing the types of currents evoked in hippocampal neurons in terms of the separate classes of response to antagonists shown in Table III is possible, especially because each type of current can be matched to a unique pattern of response to the antagonists in hippocampal neurons. Using the same threshold-concentration criteria that were used in defining the four principal classes of response to antagonists, Class A can be equated with type IA currents, Class B with type II currents, and Class D with type III currents; in the brain regions (see Fig. 2), sensitivity to MECA does not appear to correlate unambiguously with either Class B or Class D responses; further, data are insufficient to conclude that all Class A or C responses are insensitive to this compound (Alkondon and Albuquerque, 1993).

Because it is not within the scope of this review to discuss the varied mechanisms responsible for the action of nicotinic antagonists on the nAChRs in neurons and in general, we defer here to other sources of information (e.g., Albuquerque *et al.*, 1988a,b; Swanson and Albuquerque, 1992).

B. Regulation of nAChR Expression and Function

Expression of the nAChR requires the synthesis of the molecular constituents of the receptor in the cell in addition to the proper posttranslational modifications, packaging, delivery, and presentation of the receptor constituents to the cell membrane. Any event interceding at the level at which these processes take place will classify as a potential modulator (or regulator) of nAChR expression of functional nAChRs. Thus, for example, the extent and nature of glycosylation of the N terminus of the nAChR subunits can determine whether the receptor is functional (Covarrubias, 1989; Gehle and Sumikawa, 1991). The stability and rates of transcription of the nucleic acids encoding the subunits of the nAChR will help set the upper limits for the number of receptors available for expression, and the turnover rates for the receptor and its components in the cytoplasm or the cell membrane will set the number of receptors in the membrane at specific levels. The entire regulatory process leading to the expression of the nAChR is much more complex in the case of heterogeneity of the nAChRs in a single cell, because a combination of regulatory processes may operate simultaneously in determining the relative proportions of the different subtypes of nAChR that are expressed in the cell.

The functions of the nAChRs can be regulated via interactions between the nAChRs already formed and a possible multitude of endogenous and exogenous biomolecules. The actions of some of these substances involve more than ligand–receptor interactions, and include such molecular modifications as phosphorylation of the receptors by various kinases. One very important process that regulates nAChR activity is desensitization, through which nAChRs become refractory to their agonists. Desensitization can be brought about by exposure of the nAChRs to high concentrations of agonist, as is dramatically obvious in the case of neuronal nAChRs (Alkondon and Albuquerque, 1993) but not normally apparent in the case of muscle nAChRs (Sherby et al., 1985). Desensitization and its complementary process resensitization are doubtless very important in the physiological and biochemical regulation of nAChR activity. Although little is known or understood about desensitization in central neurons, and even in ganglia, discussing some of the important features recognized to date to operate in the case of the muscle nAChR may be profitable, in anticipation of the potential benefits that may derive from such an effort with respect to future understanding of the nature of desensitization in neuronal nAChRs.

The mechanisms responsible for desensitization, as well as the return to sensitization that ultimately follows, are still uncertain. Desensitization rates, at least for the muscle nAChR, can be grouped into several qualitatively distinct classes (reviewed in Ochoa et al., 1989): (1) ultrafast (less than millisecond time scale), (2) fast (millisecond to second), (3) slow (second to minute), and (4) ultra-slow (minute to hour time scale). These different rate classes suggest the opera-

tion of diverse mechanisms or imply complexity in the details of a singular mechanism. One hypothesis supposes that two or more of these different rates may reflect the transition probabilities connecting a number of alternative, closed states of the agonist-bound receptor with each other and with the bound open state. According to this interpretation, the rates can be incorporated into the widely accepted model shown in Scheme I, with each desensitization rate corresponding to an alternative but inactive conformation of the agonist-bound receptor, as shown in Scheme II for a four-state model implying two distinct desensitization rates (Changeux et al., 1984):

SCHEME II

In this model, either the resting (R) or the active (A) state of the receptor, which may or may not have one or two agonists molecules bound to it, can pass swiftly to the desensitized state I and from there, at another rate, to the alternative desensitized state D, which can also be reached directly from A or R. This model is only one of many that have been proposed to describe the relationship between different states of the nAChR, including desensitized states. Nonetheless, it is a relatively simple model that can serve to illustrate basic concepts and to provide a foundation upon which to construct improved models. Although the requirement for agonist in desensitization has been demonstrated in many studies (reviewed by Ochoa et al., 1989), this model still can be reconciled with the known desensitizing effect of agents other than agonists. Evidence for the existence of the desensitized form, D, of the receptor comes from a number of studies, including a considerable number of kinetics and electrophysiological investigations (reviewed by Ochoa et al., 1989); binding studies with fluorescent agonist (e.g., Heidmann and Changeux, 1980) and the desensitizing agent perhydrohistrionicotoxin (e.g., Albuquerque et al., 1974,1988b; Krodel et al., 1979) reveal the existence of the intermediate state I. An interesting feature of the desensitized forms of the nAChR is that the affinity for agonist follows a hierarchical pattern; the dissociation constants for the R, I, and D states are progressively smaller (i.e., show an increase in affinity for agonist; Boyd and Cohen, 1980; Heidmann and Changeux, 1980; Changeux et al., 1984). The nature of the molecular events accounting for changes in the form of the receptors during desensitization is suggested by the results of cryoelectron microscopic investigations of the resting and desensitized forms of the nAChR of *Torpedo* postsynaptic membranes. These studies indicate that a loss of symmetry in the way the subunits of the receptor are organized may be responsible for desensitization of the receptor (Unwin *et al.*, 1988).

It should be noted that Scheme II, does not take into account that desensitization is responsive to a diverse group of drugs, cofactors, metabolic processes, and a number of experimental and ambient conditions. Among the agents or factors that have been shown to participate directly in desensitization of muscle nAChRs are (1) the agonists themselves, which always, if by definition alone, have the ability to produce desensitization; (2) the concentration of agonist, with higher concentrations producing faster desensitization (e.g., Scubon-Mulieri and Parsons, 1977); (3) temperature, with lower temperatures favoring slower rates of desensitization (Magazanik and Vyskocil, 1975); (4) hyperpolarization, which accelerates desensitization (e.g., Magleby and Pallota, 1981); and (5) protein kinases (Huganir *et al.*, 1986; Hopfield *et al.*, 1988), which are known to participate in the phosphorylation of serine, threonine, or tyrosine residues of the muscle nAChR (reviewed by Huganir and Greengard, 1990). Although no direct biochemical evidence exists for the phosphorylation of neuronal nAChRs, physiological studies indicate that phosphorylation plays a possible regulatory role in the desensitization of neuronal nAChRs (e.g., Margiotta *et al.*, 1987). Modulators of phosphorylation, therefore, also may be viewed as candidate agents influencing desensitization of the nAChR, albeit indirectly, as in the case of substance P (Stallcup and Patrick, 1980; Clapham and Neher, 1984; Role, 1984), calcium ions (Smith *et al.*, 1987), phorbol diesters (Eusebi *et al.*, 1985; Downing and Role, 1987), and cAMP (Albuquerque *et al.*, 1986; Middleton *et al.*, 1986). Phosphorylation may be only one of several alternative means of achieving desensitization (Ochoa *et al.*, 1989; Lukas and Bencherif, 1992).

The study of the desensitization characteristics of central neuronal nAChRs is a relatively recent undertaking. Some of the information comes from nAChR expression studies in oocytes, but more direct evidence comes from electrophysiological measurements conducted on neurons themselves. Rates of desensitization, as assessed from decay half-times, vary widely; in hippocampal neurons, the less frequently encountered types of current (II and III) desensitize more slowly than the predominant type (IA). This behavior is reflected in the fact that each type of current evoked in hippocampal neurons has a characteristic decay profile in response to a saturating concentration of ACh (Alkondon and Albuquerque, 1993).

The number of endogenous substances known to affect nAChR expression or function continues to grow, particularly if the receptor is considered within the broader context of all the different tissues in which it is found. In any case, assessing in advance how many of the discoveries relevant to the nAChRs outside the CNS may be applicable to central neurons is difficult. Note that, on the basis of studies performed in ganglia or central neurons, the range of proper modulators of nAChR activity or number must embrace developmental or ontogenic factors (e.g., Margiotta and Gurantz, 1989; Daubas *et al.*, 1990), innervation by other neurons (e.g., Marshall, 1985), various hormones and growth factors, including nerve growth factor (e.g., Greene and Tischler, 1976), that are known to affect receptor number, and up- or down-regulation of receptor number via a number of agents, including nicotine (e.g., Wonnacott, 1990). Ontogenically related cellular changes are particularly interesting, because they raise the proverbial question of whether a shift in the expression of certain subunits with the age of a cell, either *in vitro* or *in vivo*, is a cause or an effect of cell development and differentiation (Lukas and Bencherif, 1992). Regulation of receptor number constitutes an important way of modifying the physiological role of the nAChR, albeit in a collective fashion.

Another means of achieving regulation of nAChR function may be through other sites on the nAChR that may not be specific for the classical nicotinic agonists. Among the important discoveries is the identification of an ACh-insensitive site on neuronal and muscle nAChRs that recognizes physostigmine as an agonist (Pereira *et al.*, 1993a,b). This site is located in a region of the nAChR subunits that is highly conserved among different tissues and species, and may play a regulatory role in nAChR function.

C. Pharmacologically Relevant Structural Characteristics of Neuronal nAChRs

1. Structure of the Neuronal nAChR α Subunit

Several independent studies relying on different approaches suggest that the neuronal nAChR is a pentamer made up of two α and three β subunits (Anand *et al.*, 1991; Cooper *et al.*, 1991). In its general disposition, therefore, the neuronal nAChR should resemble the muscle nAChR, with the subunits arranged pseudosymmetrically about a central pore. Although the β subunits play a definite role in the pharmacological properties of functional neuronal nAChRs (Luetje and

Patrick, 1991; Papke and Heinemann, 1991), the ago-nist binding sites are present only on the α subunits (Pereira *et al.*, 1993b; reviewed by McLane *et al.*, 1993). A diagrammatic representation of an α subunit of nAChR is provided in Fig. 4. Notable features include the four transmembrane regions, the two ACh bind-ing sites, and the external N and C termini.

Currently, knowledge of which α and β subunits are co-assembled in native neuronal nAChRs is lim-ited. Also, the extent to which functional nAChRs assembled in oocyte studies reflect the structures of the native receptors is unclear. One study has shown that as many as three kinds of subunits (α3, α5, β4) can co-assemble to make functional neuronal nAChRs in chick ciliary ganglion (Vernallis *et al.*, 1993). Func-tional nAChRs identified prior to that study have been characterized as consisting of homomers of α7 or of

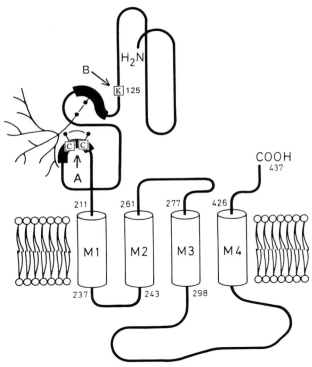

FIGURE 4 Essential structure of the α subunit of the nAChR. The four transmembrane regions—M1, M2, M3, and M4—are shown as cylinders and extramembrane sequences are displayed as thick lines. Requisite constituents of the binding sites for ACh (A) and for physostigmine (B) are indicated with arrows. ACh binding re-quires the presence of two consecutive cysteine residues at the 192 and 193 positions. Physostigmine binding seems to require a single lysine (K) at the 125 position. The branching structure near the physostigmine binding site is a glycosylated region of the subunit. Sulfhydryl bonds are denoted by ●–●. Reprinted with permission from Pereira *et al.*, (1993). Identification and functional characteri-zation of a new agonist site on nicotinic acetylcholine receptors of cultured hippocampal neurons. *J. Pharmacol. Exp. Ther.* **265**, 1474–1491. © Am. Soc. for Pharm. and Exp. Ther.

heteromers of only one kind of α subunit in combina-tion with at least one β2 or β4 subunit (Role, 1992).

2. Structural Heterogeneity of Neuronal nAChRs

The evidence for the structural heterogeneity of the neuronal nAChR, both direct and indirect, comes from many sources, including *in situ* hybridization studies (e.g., Goldman *et al.*, 1987; Heinemann *et al.*, 1990), immunochemical studies (e.g., Patrick and Stallcup, 1977; Swanson *et al.*, 1983; Conti-Tronconi *et al.*, 1985; Sargent *et al.*, 1989; Schroder *et al.*, 1989), and a wide range of functional investigations. Re-search based on functional characteristics is probably the most accessible to a large number of laboratories, and has contributed the most to identifying and quan-tifying the properties of various neuronal nAChRs subtypes. Functional studies reveal the existence of diverse nAChRs in at least one region of rat brain—the hippocampus (Alkondon and Albuquerque, 1991,1993). Data from many sources indicate that pharmacological criteria can be used to categorize the nicotinic responses of individual neurons from a num-ber of brain subregions. One assay for the possible heterogeneity of the nAChR in separate brain regions is to apply the criterion of sensitivity to agonists and antagonist nicotinic currents elicited from neurons, as described in a previous section.

Messenger RNAs (mRNAs) encoding a number of neuronal nAChR subunits have been injected into frog (*Xenopus*) oocytes with the outcome that func-tional nAChRs were expressed in the oocytes. This procedure has allowed the determination of several electrophysiological properties of a variety of nAChRs composed of diverse subunits. The technique has been applied to combinations of mRNAs encoding neuronal α and β nAChR subunits from rat or chick, including α7 or α8 alone (Couturier *et al.*, 1990; Revah *et al.*, 1991; Séguéla *et al.*, 1993; reviewed by Role, 1992). Among the properties that have been determined for neuronal nAChRs in oocyte expression studies are nAChR ion-channel conductances, summarized in Ta-ble IV for mammalian (rat) nAChR subunits studied to date. At first glance, these conductances have no convincing counterparts in the conductances mea-sured directly in neurons (see Table I). However, the experimental solutions used in the whole-cell patch clamping of the oocytes differed substantially in either Ca^{2+} (usually about 2 mM, but sometimes 0) or Na^+ concentration (~115 mM or less) from those typically used in the equivalent experiments performed on neurons (Ca^{2+} usually near 2 mM; Na^+ in excess of 150 mM). Since extracellular Ca^{2+} is known to re-duce conductances through nAChR channels, some of the differences in conductances among reported

TABLE IV Conductances of Rat Neuronal nAChRs Expressed in Oocytes[a]

Preparation	Conductance (pS)	Adjusted conductance (pS)	Temperature (°C)	Reference
Rat $\alpha4\beta2$	13, 22	19, 31	24	Papke *et al.*, (1989)
Rat $\alpha3\beta2$	5, 15	7, 21	24	Papke *et al.*, (1989)
Rat $\alpha3\beta4$	13, 18, 22	19, 26, 31	24	Deneris *et al.*, (1991)
Rat $\alpha2\beta2$	16, 33	23, 47	24	Deneris *et al.*, (1991)
Rat $\alpha2 << \beta2$	12	17	24	Papke *et al.*, (1989)
Rat $\alpha2 >> \beta2$	38	54	24	Papke *et al.*, (1989)

[a] Data from various studies of rat neuronal nAChR subunits expressed as functional nAChRs in *Xenopus* oocytes are summarized with respect to measured conductances. Conductances adjusted for differences in the concentration of external sodium ions between oocyte and neuronal studies are also shown (see text for explanation). The signs $>>$ and $<<$ are used to denote experiments in which an excess or deficit, respectively, of mRNA encoding the specificied α subunit mRNA was injected into the oocytes (9 times greater or 9 times less than the mRNA for the indicated β subunit).

oocyte studies could be reconciled in terms of this effect (Sargent, 1993). However, when comparing ion-channel conductances of nAChRs expressed in oocytes with those for the native receptors in central neurons, note that the main charge carriers through the nAChR ion channels in both types of preparation are monovalent cations and that the concentrations of these ions differed considerably in the two kinds of experiment. Since the currents passing through the receptor ion channels can be expected to obey the Goldman–Hodgkin–Katz current equation, the measured conductances in the oocyte studies could be adjusted to reflect more closely the possible conductances under conditions resembling those under which similar measurements were made on the central neurons (see Table I). In addition to the measured conductance values in oocytes, Table IV provides this kind of adjusted conductances. The rationale for adjusting the conductances follows. The definition of conductance is

$$\gamma = I/V, \qquad (1)$$

where V is the membrane potential, and I is the current, which is almost entirely due to monovalent cations under the experimental conditions in question. If three assumptions are made–(1) the relative permeabilities to monovalent cations are ca. 1.0 and the influence of divalent ions can be ignored, (2) the concentration ratios of internal/external monovalent cations are similar across different preparations, and (3) other interpreparation variables, such as the physicochemical characteristics of the cell membranes in which the receptors are embedded, do not

significantly impact on the electrophysiological characteristics of the receptor ion channels in different preparations—a reasonable estimate of a correction factor can be obtained beginning with the current equation

$$I = Pz^2 \frac{VF^2}{RT} \left(\frac{[C]_i - [C]_o \, e^{-z\frac{VF}{RT}}}{1 - e^{-z\frac{VF}{RT}}} \right), \qquad (2)$$

where P is the permeability of the membrane, z is the charge carried by the ion (unity in this case), V is the membrane potential, F is Faraday's constant, R is the gas constant, T is the temperature, and $[C]_i$ and $[C]_o$ are the internal and external concentrations, respectively, of the charge carrier ion (e.g., Na^+). Dividing Eq. 2 for the neurons by the same equation for the oocytes leads to a ratio of conductances between γ_1 for the neurons and γ_2 for the oocytes, expressed in terms of the external monovalent-cation concentrations, $[C_o]_1$ and $[C_o]_2$, for the neurons and oocytes, respectively:

$$\frac{\gamma_1}{\gamma_2} = \frac{[C_o]_1}{[C_o]_2}. \qquad (3)$$

Using the values for monovalent-cation concentration frequently applicable to each preparation (165 mM for neurons and 115 mM for oocytes), a value of 1.43 for this ratio can be calculated. The adjusted conductances in Table IV were obtained by multiplying the measured oocyte nAChR channel conductances by this factor. The compensated values are considerably closer to those found for neurons (more fall approximately within the range of values for the native receptors in Table I); in fact, the 73-pS conductance seen in rat hippocampal neurons (Castro and Albuquerque, 1993, Table I), believed to correspond to α7-constituted nAChRs (Alkondon and Albuquerque, 1993), could have a counterpart in the 46-pS conductance obtained for chick-brain nAChRs reconstituted as α7 homomers in *Xenopus* oocytes (Revah *et al.*, 1991), when the value of the conductance is adjusted upwards to 66 pS. Perfect agreement must not be expected, since, for example, oocytes probably possess membranes distinctly different from those of neurons and, in addition, may have posttranslation modifications, such as glycosylation, that may result in end-product nAChRs with properties that differ in varying degrees from those of the native nAChRs.

Despite the aforementioned limitations in making comparisons between the nAChRs expressed in oocytes and those native to CNS neurons, researchers have made use of the oocyte data in correlating functional differences observed in neurons with the structural differences that underlie the functional diversity observed in the oocytes (Role, 1992). The most useful comparisons have been those relying on the decay constants of agonist-evoked currents and their dependence of inward rectification on Mg^{2+}, sensitivity to agonists (e.g., as expressed in potency hierarchies), sensitivity to antagonists, and the relative affinities for ACh. In studies in hippocampal neurons, for example, researchers have proposed that the type IA currents, whose rectification is dependent on introcellular Mg^{2+}, were subserved by nAChRs made up at least in part of α7 subunits; the type II currents were subserved by nAChRs consisting of α4 and β2 subunits; and the type III currents were subserved by receptors with an α3β4 composition (Alkondon and Albuquerque, 1993). In contrast to type IA currents, current types II and III displayed a Mg^{2+}-independent rectification. In all studies conducted to date, neuronal nAChRs constituted partly or entirely of α6 or β3 subunits appear to be nonfunctional; not until very recently has a functional neuronal nAChR containing α5 been found (Vernallis *et al.*, 1993). An intriguing possibility to account for the inability of these subunits to form functional receptors is that they may require coexpression with subunits yet to be discovered (Conroy *et al.*, 1992).

IV. Concluding Remarks

In summary, the pharmacological study of the central neuronal nAChRs is in its infancy. Much of the information related here is inferential in nature, relying largely on findings in nonneuronal or neuronal non-CNS receptors. The discoveries that have been made to date in central neuronal preparations, however, indicate that the molecular pharmacology of the CNS nAChRs is far more intricate and complex than that found for the receptors from other anatomical regions. It has been calculated that seven α subunits and three β subunits can combine in heteromeric fashion, according to the $\alpha_2\beta_3$ stoichiometry, to produce 1323 combinations (Role, 1992), each representing a different nAChR with potentially different properties. If other stoichiometries are permitted, the number of possible receptors is much larger. One fundamental question is, To what extent does the neuron avail itself of these opportunities to achieve fine tuning of its functions? Also, if the cell does not produce all hypothetical variants of the nAChR, how does it prevent assembly of otherwise wasteful or undesirable nAChRs? Perhaps, as has been suggested, many subunit combinations may be thermodynamically unfavorable and may be essentially absent in neurons (Vernallis *et al.*, 1993). On the other hand, the existence of at least seven neuronal α subunits makes it quite likely that a minimum of seven structurally dissimilar receptor subtypes is possible, implying the existence of at least the same number of functional variants. If this inference is incorrect, then for what reason do seven different subunits occur in neurons? At present, only three distinct types of currents have been evoked in rat hippocampal neurons (Alkondon and Albuquerque, 1993). Should researchers assume that at least four other current types remain to be identified in that preparation? Does expression of receptor subtypes vary by cellular region, tissue, ontogenically, or in a combination of ways?

Expression studies may be made more relevant to the characterization of the central neuronal nAChRs if the receptors can be expressed in surrogate cells that are more closely related phylogenetically and anatomically to the neurons in the mammalian CNS, thus avoiding some of the pitfalls of oocyte expression studies such as posttranslational and other differences in subunit synthesis and modification between oocytes and neurons, as well as variations in expression caused by differences in membrane composition among diverse species and tissues. One example of this kind of improved approach was reported in which a mouse fibroblast cell line transfected with cDNAs encoding rat α4 and β2 subunits was used to investigate the electrophysiological and pharmacological properties of central neuronal nAChRs (Whiting *et al.*, 1991; Pereira *et al.*, 1993b).

The nature and function of the novel agonist site on the nAChR may have important clinical implications. For instance, investigators have shown that agents such as physostigmine can activate the nAChR even after the receptor has been desensitized by high concentrations of ACh or blocked with α-Bgt (Kuhlmann *et al.*, 1991; Okonjo *et al.*, 1991; Pereira *et al.*, 1993a,b), and that physostigmine can reverse the effects of the intoxication induced by organophosphates (Albuquerque *et al.*, 1988a; Pereira *et al.*, 1991). At least in part, the toxic effects of organophosphates on neuromuscular transmission can be explained by the irreversible blockade of acetylcholinesterase, which would result in enhancement of ACh concentration at the synaptic cleft, resulting in hyperexcitation of nicotinic synapses followed by receptor desensitization. This process could account for the depression observed in subjects chronically exposed to organophosphates. This condition could be treated with drugs that can potentially bypass receptor desensitization and act as antidotes of the intoxication. Another interesting prospect related to the novel agonist site is the possibility that an endogenous molecule may exist for which physostigmine may be a surrogate. If such an agent can be identified, the nature and function of the nAChR must be reassessed.

Since desensitization is still a phenomenon for which mechanisms remain largely in question, and since it can be considered a necessary complementary process to the activation of nAChRs whose full significance is yet to be defined, considerable investigative efforts can be anticipated to be directed at further understanding its nature. Another area that will require special attention is the ontogenic modification of receptor function or, perhaps, the modification of ontogenesis by changes in receptor function. Finally, clarification of the post- and presynaptic functions of the neuronal nAChRs must still be obtained. Are these functions served by essentially identical nAChRs differing only in minor structural details? In other words, How much can function be varied by posttranslational modifications, including phosphorylation and glycosylation? These and many other questions await definitive answers.

Acknowledgments

We are deeply indebted to Barbara Marrow and Mabel A. Zelle for their superb technical assistance and help in the preparation of this manuscript. We also thank Susan Wonnacott and Newton G. Castro for their helpful comments and suggestions. This work was supported in part by U.S. Public Health Service Grants NS25296 and ES05730.

References

Abood, L. G., Reynolds, D. T., and Bidlack, J. M. (1980). Stereospecific ³H-nicotine binding to intact and solubilised rat brain membranes and evidence for its non-cholinergic nature. *Life Sci.* **27**, 1307–1314.

Adams, P. R. (1981). Acetylcholine receptor kinetics. *J. Membr. Biol.* **58**, 161–174.

Adler, M., Albuquerque, E. X., and Lebeda, F. J. (1978). Kinetic analysis of endplate currents altered by atropine and scopolamine. *Mol. Pharmacol.* **14**, 514–529.

Anand, R., Conroy, W. G., Schoepfer, R., Whiting, P., and Lindstrom, J. (1991). Neuronal nicotinic acetylcholine receptors expressed in *Xenopus* oocytes have a pentameric quaternary structure. *J. Biol. Chem.* **266**, 11192–11198.

Albuquerque, E. X., and Eldefrawi, A. T. (1983). "Myasthenia Gravis." Chapman and Hall, London.

Albuquerque, E. X., Kuba, K., and Daly, J. (1974). Effect of histrionicotoxin on the ionic conductance modulator of the cholinergic receptor: A quantitative analysis of the end-plate current. *J. Pharmacol. Exp. Ther.* **189**, 513–524.

Albuquerque, E. X., Deshpande, S. S., Aracava, Y., Alkondon, M., and Daly, J. W. (1986). A possible involvement of cyclic AMP in the expression of desensitization of the nicotinic acetylcholine receptor. *FEBS Lett.* **199**, 113–120.

Albuquerque, E. X., Aracava, Y., Cintra, W. M., Brossi, A., Schönenberger, B., and Deshpande, S. S. (1988a). Structure-activity relationship of reversible cholinesterase inhibitors: Activation, channel blockade and stereospecificity of the nicotinic acetylcholine receptor-ion channel complex. *Brazil. J. Med. Biol. Res.* **21**, 1173–1196.

Albuquerque, E. X., Daly, J. W., and Warnick, J. E. (1988b). Macromolecular sites for specific neurotoxins and drugs on chemosensitive synapses and electrical excitation in biological membranes. *In* "Ion Channels" (T. Narahashi, ed.), pp. 95–162. Plenum Publishing, New York.

Alkondon, M., and Albuquerque, E. X. (1991). Initial characterization of the nicotinic acetylcholine receptors in rat hippocampal neurons. *J. Receptor Res.* **11**, 1001–1021.

Alkondon, M., and Albuquerque, E. X. (1993). Diversity of nicotinic acetylcholine receptors in rat hippocampal neurons: I. Pharmacological and functional evidence for distinct structural subtypes. *J. Pharmacol. Exp. Ther.* **265**, 1455–1473.

Alkondon, M., Pereira, E. F. R., Wonnacott, S., and Albuquerque, E. X. (1992). Blockade of nicotinic currents in hippocampal neurons defines methyllycaconitine as a potent and specific receptor antagonist. *Mol. Pharmacol.* **41**, 802–808.

Aracava, Y., Deshpande, S. S., Swanson, K. L., Rapoport, H., Wonnacott, S., Lunt, G., and Albuquerque, E. X. (1987). Nicotinic acetylcholine receptors in cultured neurons from the hippocampus and brain stem of the rat characterized by single channel recordings. *FEBS Lett.* **222**, 63–70.

Benwell, M. E. M., Balfour, D. J. K., and Anderson, J. M. (1988). Evidence that tobacco smoking increases the density of (−)[³H]-nicotine binding sites in human brain. *J. Neurochem.* **50**, 1243–1247.

Boyd, N. D., and Cohen, J. B. (1980). Kinetics of binding of [³H]acetylcholine and [³H]carbamoylcholine to *Torpedo* postsynaptic membranes: Slow conformational transitions of the cholinergic receptor. *Biochemistry* **19**, 5344–5353.

Breer, H., Kleene, R., and Hinz, G. (1985). Molecular forms and subunit structure of the acetylcholine receptor in the central nervous system of insects. *J. Neurosci.* **5**, 3386–3392.

Brown, D. A., Dochert, R. J., and Halliwell, J. V. (1983). Chemical transmission in the rat interpeduncular nucleus in vitro. *J. Physiol.* (*London*) **341**, 655–670.

Brown, D. A., Dochert, R. J., and Halliwell, J. V. (1984). The action of cholinomimetic substances on impulse conduction in the habenulointerpeduncular pathway of the rat in vitro. *J. Physiol.* (*London*) **353**, 101–109.

Castro, N. G., and Albuquerque, E. X. (1993). Brief-lifetime, fast-inactivating ion channels account for the α-bungarotoxin-sensitive nicotinic responses in hippocampal neurons. *Neurosci. Lett.* **164,** 137–140.

Changeux, J.-P. (1990). Functional architecture and dynamics of the nicotinic acetylcholine receptor: An allosteric ligand-gated ion channel. *Fidia Res. Found. Neurosci. Award Lec.* **4,** 21–168.

Changeux, J.-P., Devillers-Thiery, A., and Chemouilli, P. (1984). Acetylcholine receptor: An allosteric protein. *Science* **225,** 1335–1345.

Chiappinelli, V. A. (1985). Actions of snake venom toxins on neuronal nicotinic receptors and other neuronal receptors. *Pharmac. Ther.* **31,** 1–32.

Clapham, D. E., and Neher, E. (1984). Substance P reduces acetylcholine-induced currents in isolated bovine chromaffin cells. *J. Physiol. (London)* **347,** 255–257.

Clarke, P. B. S., Schwartz, R. D., Paul, S. M., Pert, C. B., and Pert, A. (1985). Nicotinic binding in rat brain: Autoradiographic comparison of [³H]acetylcholine, [³H]nicotine, and [¹²⁵I]α-bungarotoxin. *J. Neurosci.* **5,** 1307–1311.

Connolly, J. G., and Colquhoun, D. (1991). A novel nicotinic acetylcholine receptor in thin slices of rat medial habenula. *J. Physiol. (London)* **438,** 260P.

Conroy, W. G., Vernallis, A. B., and Berg, D. K. (1992). The α5 gene product assembles with multiple acetylcholine receptor subunits to form distinctive receptor subtypes in brain. *Neuron* **9,** 679–691.

Conti-Tronconi, B. M., and Raftery, M. A. (1982). The nicotinic cholinergic receptor: Correlation of molecular structure with functional properties. *Annu. Rev. Biochem.* **51,** 491–530.

Conti-Tronconi, B. M., Dunn, S. M., Barnard, E. A., Dolly, J. O., Lai, F. A., Ray, N., and Raftery, M. A. (1985). Brain and muscle nicotinic receptors are different but homologous proteins. *Proc. Natl. Acad. Sci. USA* **82,** 5208–5212.

Cooper, E., Couturier, S., and Ballivet, M. (1991). Pentameric structure and subunit stoichiometry of a neuronal nicotinic acetylcholine receptor. *Nature (London)* **350,** 235–238.

Couturier, S., Bertrand, D., Matter, J.-M., Hernandez, M.-C., Bertrand, S., Millar, N., Valera, S., Barkas, T., and Ballivet, M. (1990). A neuronal nicotinic acetylcholine receptor subunit (α7) is developmentally regulated and forms a homo-oligomeric channel blocked by α-BTX. *Neuron* **5,** 847–856.

Covarrubias, M., Kopta, C., and Steinbach, J. H. (1989). Inhibitors of asparagine-linked oligosaccharide processing alter the kinetics of the nicotinic acetylcholine receptor. *J. Gen. Physiol.* **93,** 765–783.

Daubas, P., Devillers-Theiry, A., Geoffrey, B., Martinez, S., Bessis, A., and Changeux, J.-P. (1990). Differential expression of the neuronal acetylcholine receptor α2 subunit gene during chick brain development. *Neuron* **5,** 49–60.

Deneris, E. S., Connolly, J., Rogers, S. W., and Duvoisin, R. (1991). Pharmacological and functional diversity of neuronal nicotinic acetylcholine receptors. *Trends Pharmacol. Sci.* **12,** 34–40.

Downing, J. E. G., and Role, L. W. (1987). Activators of protein kinase C enhance acetylcholine receptor desensitization in sympathetic ganglion neurons. *Proc. Natl. Acad. Sci. USA* **84,** 7739–7743.

Eusebi, F., Molinaro, M., and Zani, B. M. (1985). Agents that activate protein kinase C reduce acetylcholine sensitivity in cultured myotubes. *J. Cell Biol.* **100,** 1339–1342.

Faeder, I. R., and O'Brien, R. D. (1970). Responses of perfused isolated leg preparations of the cockroach *Gromphadorhina portentosa* to L-glutamate, GABA, picrotoxin, strychnine, and chlorpromazine. *J. Exp. Zool.* **173,** 203–214.

Gehle, V. M., and Sumikawa, K. (1991). Site-directed mutagenesis of the conserved *N*-glycosylation site on the nicotinic acetylcholine receptor subunits. *Mol. Brain Res.* **11,** 17–25.

Gerschenfeld, H. M. (1973). Chemical transmission in invertebrate central nervous systems and neuromuscular junctions. *Physiol. Rev.* **53,** 1–119.

Goldman, D., Deneris, E., Luyten, W., Kochhar, A., Patrick, J., and Heinemann, S. (1987). Members of a nicotinic acetylcholine receptor gene family are expressed in different regions of the central nervous system. *Cell* **48,** 965–973.

Grady, S., Marks, M. J., Wonnacott, S., and Collins, A. C. (1992). Characterization of nicotinic receptor-mediated [³H]dopamine release from synaptosomes prepared from mouse striatum. *J. Neurochem.* **59,** 848–856.

Greene, L. A., and Tischler, A. S. (1976). Establishment of a noradrenergic clonal line of rat adrenal pheochromocytoma cells which respond to nerve growth factor. *Proc. Natl. Acad. Sci. USA* **73,** 2424–2428.

Hartman, D. S., and Claudio, T. (1990). Coexpression of two distinct muscle acetylcholine receptor α-subunits during development. *Nature (London)* **343,** 372–375.

Heidmann, T., and Changeux, J.-P. (1980). Interaction of a fluorescent agonist with the membrane-bound acetylcholine receptor from *Torpedo marmorata* in the millisecond time range: Resolution of an "intermediate" conformational transition and evidence for positive cooperative effects. *Biochem. Biophys. Res. Commun.* **97,** 889–896.

Heinemann, S., Boulter, J., Deneris, E., Connolly, J., Duvoisin, R., Papke, R., Patrick, J. (1990). The brain nicotinic acetylcholine receptor gene family. *Prog. Brain Res.* **86,** 195–203.

Henley, J. M., Mynlieff, M., Lindstrom, J. M., and Oswald, R. E. (1986). Acetylcholine receptor synthesis in retina and transport to optic tectum in goldfish. *Science* **232,** 1627–1629.

Hopfield, J. F., Tank, D. W., Greengard, P., and Huganir, R. L. (1988). Functional modulation of the nicotinic acetylcholine receptor by tyrosine phosphorylation. *Nature (London)* **36,** 677–680.

Hucho, F. (1986). The nicotinic acetylcholine receptor and its ion channel. *Eur. J. Biochem.* **158,** 211–226.

Huganir, R. L., and Greengard, P. (1990). Regulation of neurotransmitter receptor desensitization by protein phosphorylation. *Neuron* **5,** 555–567.

Huganir, R. L., Delcour, A. H., Greengard, P., and Hess, G. P. (1986). Phosphorylation of the nicotinic acetylcholine receptor regulates its rate of desensitization. *Nature (London)* **321,** 774–776.

Ifune, C. K., and Steinbach, J. H. (1992). Inward rectification of acetylcholine-elicited currents in the pheochromocytoma cells. *J. Physiol. (London)* **457,** 143–165.

Karlin, A. (1980). Molecular properties of nicotinic acetylcholine receptors. *In* "The Cell Surface and Neuronal Function" (C. W. Cotman, G. Poste, and G. L. Nicolson, eds.), pp. 191–260. Elsevier/North-Holland, Amsterdam.

Kasa, P. (1986). The cholinergic systems in brain and spinal cord. *Prog. Neurobiol.* **26,** 211–272.

Katz, B., and Thesleff, S. (1957). A study of the "desensitization" produced by acetylcholine at the motor end-plate. *J. Physiol. (London)* **138,** 63–80.

Kellar, K. J., and Wonnacott, S. (1990). Nicotinic cholinergic receptors in Alzheimer's disease. *In* "Nicotine Psychopharmacology. Molecular, Cellular, and Behavioral Aspects" (S. Wonnacott, M. A. H. Russell, and I. P. Stolerman, eds.), pp. 341–373. Oxford Science Publications, Oxford.

Krnjevic, K. (1975). Acetylcholine receptors in vertebrate CNS. *In* "Handbook of Psychopharmacology" (L. L. Iverson, S. D. Iverson, S. H. Snyder, eds.), Vol. 6, pp. 97–126. Plenum Press, New York.

Krodel, E. K., Beckman, R. A., and Cohen, J. B. (1979). Identification of a local anesthetic binding site in nicotinic postsynaptic membranes isolated from *Torpedo marmorata* electric tissue. *Mol. Pharmacol.* **15,** 294–312.

Kuhlmann, J., Okonjo, K. O., and Maelicke, A. (1991). Desensitization is a property of the cholinergic binding region of the nicotinic acetylcholine receptor, not of the receptor-integral ion channel. *FEBS Lett.* **279,** 216–218.

Lamacz, M., Tonon, M. C., Louiset, E., Cazin, L., Strosberg, D., and Vaudry, H. (1989). Acetylcholine stimulates α-melanocyte-stimulating hormone release from frog pituitary melanotrophs through activation of muscarinic and nicotinic receptors. *Endocrinology* **125,** 707–714.

Lindstrom, J. (1985). Immunobiology of myasthenia gravis, experimental autoimmune myasthenia gravis, and Lambert–Eaton syndrome. *Annu. Rev. Immunol.* **3,** 109–131.

Lindstrom, J., Schopfer, R., and Whiting, P. (1987). Molecular studies of the neuronal nicotinic acetylcholine receptor family. *Mol. Neurobiol.* **1,** 281–337.

Lipscombe, D. (1988). Nicotinic receptors of frog ganglia resemble pharmacologically those of skeletal muscle. *J. Neurosci.* **8,** 3258–3265.

Lipton, S. A., Aizenman, E., and Loring, R. H. (1987). Neural nicotinic acetylcholine responses in solitary mammalian retinal ganglion cells. *Pflügers Arch.* **410,** 37–43.

Loring, R. H., Aizenman, E., Lipton, S. A., and Zigmond, R. E. (1989). Characterization of nicotinic receptors in chick retina using a snake venom neurotoxin that blocks neuronal nicotinic receptor function. *J. Neurosci.* **9,** 2423–2431.

Low, B. A. (1979). The three-dimensional structure of postsynaptic snake neurotoxins: Consideration of structure and function. *In* "Snake Venoms, Handbook of Experimental Pharmacology" (C.-Y. Lee, ed.), Vol. 52, pp. 213–257. Springer-Verlag, Berlin.

Luetje, C. W., and Patrick, J. (1991). Both α- and β-subunits contribute to the agonist sensitivity of neuronal nicotinic acetylcholine receptors. *J. Neurosci.* **11,** 837–845.

Luetje, C. W., Wada, K., Rogers, S., Abramson, S. N., Tsuji, K., Heinemann, S., and Patrick, J. (1990a). Neurotoxins distinguish between different neuronal nicotinic acetylcholine receptor subunit combinations. *J. Neurochem.* **55,** 632–640.

Luetje, C. W., Patrick, J., and Séguéla, P. (1990b). Nicotine receptors in the mammalian brain. *FASEB J.* **4,** 2753–2760.

Lukas, R. J. (1990). Heterogeneity of high-affinity nicotinic [^{3}H]acetylcholine binding sites. *J. Pharmacol. Exp. Ther.* **253,** 51–57.

Lukas, R. J., and Bencherif, M. (1992). Heterogeneity and regulation of nicotinic acetylcholine receptors. *Int. Rev. Neurobiol.* **34,** 25–131.

Luntz-Leybman, V., Bickford, P. C., and Freedman, R. (1992). Cholinergic gating of response to auditory stimuli in rat hippocampus. *Brain Res.* **587,** 130–136.

Magazanik, L. G., and Vyskocil, F. (1975). The effect of temperature on desensitization kinetics at the post-synaptic membrane of the frog muscle fibre. *J. Physiol.* (*London*) **249,** 285–300.

Magleby, K. L., and Pallota, B. S. (1981). A study of desensitization of acetylcholine receptors using nerve-released transmitters in the frog. *J. Physiol.* (*London*) **316,** 225–250.

Margiotta, J. F., and Gurantz, D. (1989). Changes in the number, function, and regulation of nicotinic acetylcholine receptors during neuronal development. *Dev. Biol.* **135,** 326–339.

Margiotta, J. F., Berg, D. K., and Dionne, V. E. (1987). Cyclic AMP regulates the proportion of functional acetylcholine receptors on chicken ciliary ganglion neurons. *Proc. Natl. Acad. Sci. USA* **84,** 8155–8159.

Marshall, L. M. (1981). Synaptic localization of α-bungarotoxin binding which blocks nicotinic transmission at frog sympathetic neurons. *Proc. Natl. Acad. Sci. USA* **78,** 1948–1952.

Marshall, L. M. (1985). Presynaptic control of synaptic channel kinetics in sympathetic neurones. *Nature* (*London*) **317,** 621–623.

Mathie, A., Cull-Candy, S. G., and Colquhoun, D. (1987). Single-channel and whole-cell currents evoked by acetylcholine in dissociated sympathetic neurons of the rat. *Proc. R. Soc. London B* **232,** 239–248.

McGeer, P. L., McGeer, E. G., Suzuki, J., Dolman, C. E., and Nagai, T. (1984). Aging, Alzheimer's disease, and the cholinergic system of the basal forebrain. *Neurology* **34,** 741–745.

McLane, K. E., Dunn, S. J. M., Manfredi, A. A., Conti-Tronconi, B. M., and Raftery, M. A. (1994). The nicotinic acetylcholine receptor as a model for a superfamily of ligand-gated ion channel proteins. *In* "Handbook of Protein and Peptide Design" (P. Carey, ed.). Academic Press, New York. (*in press*).

Mebs, D., Narita, K., Iwanaga, S., Samejima, Y., and Lee, C.-Y. (1972). Purification, properties and amino acid sequence of α-bungarotoxin from the venom of *Bungarus multicinctus*. *Hoppe-Seyler's Z. Physiol. Chem.* **353,** 243–262.

Middleton, P., Jaramillo, R., and Schuetze, S. M. (1986). Forskolin increases the rate of acetylcholine receptor desensitization at rat soleus endplates. *Proc. Natl. Acad. Sci. USA* **83,** 4967–4971.

Mishina, M., Takai, T., Imoto, K., Noda, M., Takahashi, T., Numa, S., Methfessel, C., and Sakmann, B. (1986). Molecular distinction between fetal and adult forms of muscle acetylcholine receptor. *Nature* (*London*) **321,** 406–411.

Montes, J. G., Alkondon, M., Pereira, E. F. R., Castro, N., and Albuquerque, E. X. (1994). Electrophysiological Methods for the Study of Neuronal Nicotinic Acetylcholine Receptor Ion Channels. *In* "Ion Channels of Excitable Cells" (Narahashi, T., ed.), *Methods in Neurosci.* **19,** 121–146. Academic Press, San Diego.

Moss, B. L., Schuetze, S. M., and Role, L. W. (1989). Functional properties and developmental regulation of nicotinic acetylcholine receptors on embryonic chicken sympathetic neurons. *Neuron* **3,** 597–607.

Mulle, C., and Changeux, J.-P. (1990). A novel type of nicotinic receptor in the rat central nervous system characterized by patch-clamp technique. *J. Neurosci.* **10,** 169–175.

Mulle, C., Vidal, C., Benoit, P., and Changeux, J-P. (1991). Existence of different subtypes of nicotinic acetylcholine receptors in the rat habenulo-interpeduncular system. *J. Neurosci.* **11,** 2588–2597.

Mulle, C., Choquet, D., Korn, H., and Changeux, J-P. (1992). Calcium influx through nicotinic receptor in rat central neurons: Its relevance to cellular regulation. *Neuron* **8,** 135–143.

Neuhaus, R., and Cachelin, A. B. (1990). Changes in the conductance of the neuronal nicotinic acetylcholine receptor channel induced by magnesium. *Proc. R. Soc. London B* **241,** 78–84.

Nilsson, L., Adam, A., Hardy, J., Winblad, B., and Nordberg, A. (1987). Do tetrahydroaminoacridine (THA) and physotigmine restore acetylcholine release in Alzheimer brains via nicotinic receptors? *J. Neurol. Trans.* **70,** 357–368.

Ochoa, E. L. M., Chattopadhyay, A., and McNamee, M. G. (1989). Desensitization of the nicotinic acetylcholine receptor: Molecular mechanisms and effect of modulators. *Cell. Mol. Neurobiol.* **9,** 141–177.

Okonjo, K. O., Kuhlmann, J., and Maelicke, A. (1991). A second pathway for the activation of the *Torpedo* acetylcholine receptor. *Eur. J. Biochem.* **200,** 671–677.

Papke, R. L., and Heinemann, S. F. (1991). The role of the β4-subunit in determining the kinetic properties of rat neuronal nicotinic acetylcholine α3-receptors. *J. Physiol.* **440,** 95–112.

Papke, R. L., Boulter, J., Patrick, J., and Heinemann, S. (1989). Single-channel currents of rat neuronal nicotinic acetylcholine receptors expressed in *Xenopus* oocytes. *Neuron* **3,** 589–596.

Patrick, J., and Stallcup, W. (1977). α-Bungarotoxin binding and cholinergic receptor function in a rat sympathetic nerve line. *J. Biol. Chem.* **252,** 8629–8633.

Pereira, E. F. R., Alkondon, M., and Albuquerque, E. X. (1991). Effects of organophosphate (OP) compounds and physostigmine (PHY) on nicotinic acetylcholine receptors (AChR) in the mammalian central nervous system (CNS). *Proc. 1991 Medical Defense Biosci. Rev. U.S. Army Medical Research Institute of Chemical Defense*, 229–233.

Pereira, E. F. R., Alkondon, M., Tano, T., Castro, N. G., Froes-Ferrao, M. M., Rozental, R., Aronstam, R. S., Schrattenholz, A., Maelicke, A., and Albuquerque, E. X. (1993a). A novel agonist binding site on nicotinic acetylcholine receptors. *J. Receptor Res.* **13**, 413–436.

Pereira, E. F. R., Reinhardt-Maelicke, S., Schrattenholz, A., Maelicke, A., and Albuquerque, E. X. (1993b). Identification and functional characterization of a new agonist site on nicotinic acetylcholine receptors of cultured hippocampal neurons. *J. Pharmacol. Exp. Ther.* **265**, 1474–1491.

Pinnock, R. D., Lummis, S. C., Chiappinelli, V. A., and Satelle, D. B. (1988). Kappa-bungarotoxin blocks an alpha-bungarotoxin-sensitive nicotinic receptor in the insect central nervous system. *Brain Res.* **458**, 45–52.

Popot, J.-L., and Changeux, J.-P. (1984). Nicotinic receptor of acetylcholine: Structure of an oligomeric integral membrane protein. *Physiol. Rev.* **64**, 1162–1239.

Ramoa, A. S., Alkondon, M., Aracava, Y., Irons, J., Lunt, G. G., Deshpande, S. S., Wonnacott, S., Aronstam, R. S., and Albuquerque E. X. (1990). The anticonvulsant MK-801 interacts with peripheral and central nicotinic acetylcholine receptor ion channels. *J. Pharmacol. Exp. Ther.* **254**, 71–82.

Rapier, C., Lunt, G. G., and Wonnacott, S. (1990). Nicotinic modulation of [^{3}H]dopamine release from striatal synaptosomes: Pharmacological characterization. *J. Neurochem.* **54**, 937–945.

Rash, J. E., Albuquerque, E. X., Hudson, C. S., Mayer, R. F., and Satterfield, J. R. (1976). Studies of human myasthenia gravis: Electrophysiological and ultrastructural evidence compatible with antibody attachment to the acetylcholine receptor complex. *Proc. Natl. Acad. Sci. USA* **73**, 4584–4588.

Revah, F., Bertrand, D., Galzi, J.-L., Devillers-Thiéry, Mulle, C., Hussy, N., Bertrand, S., Ballivet, M., and Changeux, J.-P. (1991). Mutations in the channel domain after desensitization of neuronal nicotinic receptor. *Nature (London)* **353**, 846–849.

Role, L. W. (1984). Substance P modulation of acetylcholine-induced currents in embryonic chicken sympathetic and ciliary ganglion neurons. *Proc. Natl. Acad. Sci. USA* **81**, 2924–2928.

Role, L. W. (1992). Diversity in primary structure and function of neuronal nicotinic acetylcholine receptor channels. *Curr. Opin. Neurobiol.* **2**, 254–262.

Sands, S. B., and Barish, M. E. (1992). Neuronal nicotinic acetylcholine receptor currents in phaeochromocytoma (PC12) cells: Dual mechanism of rectification. *J. Physiol.* **447**, 467–487.

Sargent, P. B. (1993). The diversity of neuronal nicotinic acetylcholine receptors. *Annu. Rev. Neurosci.* **16**, 403–443.

Sargent, P. B., Pike, S. H., Nadel, D. B., and Lindstrom, J. M. (1989). Nicotinic acetylcholine receptor-like molecules in the retina, retinotectal pathway, and optic tectum of the frog. *J. Neurosci.* **9**, 565–573.

Satelle, D. B. (1986). Insect acetylcholine receptors—biochemical and physiological approaches. *In* "Neuropharmacology and Pesticide Action," (Ford, M. G., Usherwood, P. N., Reay, R. L., and Lunt, G. G., eds.) pp. 445–497. Ellis Horwood Limited, Chichester.

Schroder, H., Zilles, K., Maelicke, A., and Hajos, F. (1989). Immunohisto- and cytochemical localization of cortical nicotinic cholinoceptors in rat and man. *Brain Res.* **502**, 287–95.

Scubon-Mulieri, B., and Parsons, R. L. (1977). Desensitization and recovery at the frog neuromuscular junction. *J. Gen. Physiol.* **69**, 431–447.

Séguéla, P., Wadiche, J., Dineley-Miller, K., Dani, J. A., and Patrick, J. W. (1993). Molecular cloning, functional properties, and distribution of rat brain α_7: A nicotinic cation channel highly permeable to calcium. *J. Neurosci.* **13**, 596–604.

Sherby, S. W., Eldefrawi, A. T., Albuquerque, E. X., and Eldefrawi, M. E. (1985). Comparison of the actions of carbamates and anticholinesterases on the nicotinic acetylcholine receptor. *Mol. Pharmacol.* **27**, 343–348.

Smith, M. M., Merlie, J. P., and Lawrence, J. C. (1987). Regulation of phosphorylation of nicotinic acetylcholine receptors in mouse BC3H myocytes. *Proc. Natl. Acad. Sci. USA* **84**, 6601–6605.

Sorenson, E. M., and Chiappinelli, V. A. (1990). Intracellular recording in avian brain of a nicotinic response that is insensitive to kappa-bungarotoxin. *Neuron* **5**, 307–315.

Stallcup, W. B., and Patrick, J. (1980). Substance P enhances cholinergic receptor desensitization in a clonal nerve cell line. *Proc. Natl. Acad. Sci. USA* **77**, 634–638.

Steinbach, A. B. (1968). A kinetic model for the action of xylocaine on receptors for acetylcholine. *J. Gen. Physiol.* **52**, 162–180.

Swanson, K. L., and Albuquerque, E. X. (1992). Nicotinic acetylcholine receptors and low molecular weight toxins. *In* "Handbook of Experimental Pharmacology" (H. Herken and F. Hucho, eds.), Vol. 102, pp. 611–658. Springer-Verlag, Berlin.

Swanson, L. W., Lindstrom, J., Tzartos, S., Schmued, L. C., O'Leary, D. D. M., and Cowan, W. M. (1983). Immunohistochemical localization of monoclonal antibodies to the nicotinic acetylcholine receptor in chick midbrain. *Proc. Natl. Acad. Sci. USA* **80**, 4532–4536.

Unwin, N., Toyoshima, C., and Kubalek, E. (1988). Arrangement of the acetylcholine receptor subunits in the resting and desensitized states, determined by cryoelectron microscopy of crystallized *Torpedo* postsynaptic membranes. *J. Cell Biol.* **107**, 1123–1138.

Vernallis, A. B., Conroy, W. G., and Berg, D. (1993). Neurons assemble acetylcholine receptors with as many as three kinds of subunits while maintaining subunit segregation among receptor subtypes. *Neuron* **10**, 451–464.

Vijayaraghavan, S., Pugh, P. C., Zhang, Z. W., Rathouz, M. M., and Berg, D. K. (1992). Nicotinic receptors that bind α-bungarotoxin on neurons raise intracellular free Ca^{++}. *Neuron* **8**, 353–362.

Whiting, P., Schoepfer, R., Lindstrom, J., and Priestley, T. (1991). Structural and pharmacological characterization of the major brain nicotinic acetylcholine receptor subtype stably expressed in mouse fibroblasts. *Mol. Pharmacol.* **40**, 463–472.

Witzenmann, V., Brenner, H.-B., and Sakmann, B. (1991). Neural factors regulate AChR subunit mRNAs at rat neuromuscular synapses. *J. Cell Biol.* **114**, 125–141.

Wonnacott, S. (1987). Brain nicotinic binding sites. *Hum. Toxicol.* **6**, 343–353.

Wonnacott, S. (1990). Characterization of nicotine receptor sites in the brain. *In* "Nicotine Psychopharmacology. Molecular, Cellular, and Behavioural Aspects" (S. Wonnacott, M. A. H. Russell, and I. P. Stolerman, eds.), pp. 226–277. Oxford University Press, Oxford.

Wonnacott, S., and Thorne, B. (1990). Separation of pre- and postsynaptic receptors on Percoll gradients. *Biochem. Soc. Trans.* **18**, 885–887.

Zhang, Z. W., and Feltz, P. (1990). Nicotinic acetylcholine receptors in porcine hypophyseal intermediate lobe cells. *J. Physiol. (London)* **422**, 83–101.

Zorumski, C. F., Thio, L. L., Isenberg, K. E., and Clifford, D. B. (1992). Nicotinic acetylcholine currents in cultured postnatal rat hippocampal neurons. *Mol. Pharmacol.* **41**, 931–936.

19

Glutamate-Gated Channels in the Outer Retina

MARTIN WILSON

I. Introduction

Glutamate, the major excitatory transmitter in the brain, was understandably the focus of intense interest throughout the 1980s. As the role of this transmitter in such diverse phenomena as ischemia and memory became established, a wealth of subtleties in the properties of its receptors was uncovered. The 1990s began with a series of spectacular successes in cloning families of receptors as well as remarkable revelations about the diversity of these molecules and their functional design (for review, see Nakanishi, 1992; Sommer and Seeburg, 1992). Few investigators would doubt that research in this field is just beginning, but even if no more clones are acquired, there is ample work to be done describing the functional distinctions between combinations of subunits and mapping their distribution within the nervous system. However, new problems have been generated by the revealed diversity. How is receptor phenotype controlled in cells? This question is not solely one of regulation of gene expression, a problem not peculiar to glutamate receptors, but also involves the rules governing the way in which subunits of multimeric receptors are associated within the cell.

In addition to the questions that exist exclusively at the level of biophysics or molecular biology are those concerned with patterns of receptor expression in the context of neuronal function. Why should a particular neuron express one form of a receptor rather than another? Although this question lies at the heart of understanding membrane channels as elements in neuronal machinery, few answers can be given even for those channels with known functional distinctions. The difficulty in answering this question is not trivial, since a proper answer requires an understanding of the organizing principles and function of the part of the nervous system under consideration; fundamentals that are conspicuously deficient almost everywhere in the central nervous system (CNS).

The retina represents one of the better understood regions of the CNS, in which glutamate is almost certainly the predominant transmitter in the direct signal pathway. Much experimental work has been directed at examining the channel properties and the pharmacology of glutamate receptors in the inner and outer retina (for review, see Massey, 1990), although channel properties are neither completely described nor completely integrated into our understanding of retinal operation. Nevertheless, the properties of some glutamate channels in some cells are known to a point where it is not a completely sterile exercise to ask why these particular channels? The early division of ON and OFF signals in the retina, for example, gives rise to the two classes of bipolar cell whose glutamate channels are so different that it is natural to ask why. Some consideration has already been given to this issue (Falk, 1988); undoubtedly, a deeper understanding will follow a more complete description of these channels.

In a similar vein, the regulation of glutamate channel properties by phosphorylation induced by the action of dopamine (Knapp and Dowling, 1987; Liman *et al.*, 1989) can be placed in its neurobiological context as a component of retinal light/dark adaptation. An interesting comparison with this example is provided by the mammalian hippocampus, in which modulation of glutamate channels apparently occurs through the same mechanism (Greengard *et al.*, 1991; Wang *et al.*, 1991). However, the significance of this mechanism in the function of the hippocampus is completely obscure.

A second and more immediate reason for focusing on the outer retina is that it has provided examples of novel glutamate channels and, as discussed subsequently, it is likely to contain still more unfamiliar channels at glutamatergic synapses. In this chapter, the current understanding of glutamate channels in the outer retina is described, with particular attention to unresolved questions. The discussion is set within the framework of the organization and function of the retina.

A. Outer Retina: Organization and Function

Two general classes of photoreceptor cell, rods, and cones, form the input layer of the vertebrate retina. A large body of evidence suggests that photoreceptors can accumulate (e.g., Marc and Lam, 1981) and release (Copenhagen and Jahr, 1989) an excitatory amino acid. Since photoreceptors hyperpolarize in the presence of light, transmitter is released at its greatest rate in darkness. A convincing demonstration that the synaptic terminal of an individual cone releases glutamate when depolarized used a fluorimetric assay specific for glutamate over aspartate (Ayoub *et al.*, 1989), thereby settling a lingering uncertainty about the identity of the transmitter. Evidence supporting L-glutamate as a transmitter is drawn from work on both rods and cones, but several reports suggest that some cones from primate (Nishimura *et al.*, 1986) and lizard (Engbretson *et al.*, 1988) retina may release γ-aminobutyric acid (GABA). Whether or not these particular suggestions prove correct, available data leaves open the possibility that some photoreceptors in some species may use a transmitter other than glutamate.

Two broad classes of cells are postsynaptic to photoreceptors (Fig. 1): the horizontal cells, characterized by their confinement to the outer retina, and the bipolar cells, which communicate with the inner retina.

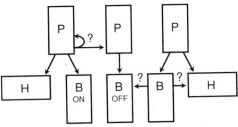

FIGURE 1 Schematic diagram of the outer retina showing known and inferred glutamatergic synapses. Three cell types—photoreceptors (P), bipolar cells (B), and horizontal cells (H) are identified; bipolar cells are further subdivided into ON and OFF cells. Synapses are indicated by an arrow; synapses for which only partial or ambiguous evidence exists are shown with a question mark. As discussed in the text, photoreceptors are responsive to glutamate but the nature of these synapses is unresolved. Bipolar cells are known from electron microscope studies in some retinae to be presynaptic to other bipolar cells (Wong-Riley, 1974; Lasansky, 1980) and to horizontal cells (Lasansky, 1980), although physiological data on these synapses are lacking.

Horizontal cells vary from species to species but have their greatest degree of specialization in teleost fish, in which they occur in several subclasses that have input only from particular subclasses of photoreceptor that contribute to their unique chromatic properties (Stell *et al.*, 1982; Toyoda *et al.*, 1982). Horizontal cells, by antagonizing the locally derived signals in photoreceptors and bipolar cells with signals collected over a wide area, act to remove redundant information and promote efficient coding (Barlow, 1961; Srinivasan *et al.*, 1982). Bipolar cells are two types: ON bipolar cells, also called depolarizing bipolar cells, defined as those that depolarize to the onset of a small centered spot of light or OFF, or hyperpolarizing, bipolar cells that hyperpolarize to the same stimulus. Consistent with glutamate as the photoreceptor transmitter, all second order cells are sensitive to L-glutamate although, as discussed later, the pharmacology of their receptors and the nature of their responses differ considerably.

The outer retina is a device for the reliable and economical encoding of patterns of light but, unlike neurons in many other parts of the brain, neurons of the outer retina generate no action potentials communicating instead with small graded voltages. This behavior might suggest that the outer retina is fundamentally different in construction from the rest of the brain. However, this view receives little support from an examination of its membrane channels which are, in general, commmonplace. The outer retina may be thought of as an ordinary piece of brain with small dimensions and slow signals that have allowed neurons to connect their input dendrites and output terminals directly, without the need for long axons.

Subtleties too numerous to catalog here have been revealed in the anatomy and physiology of the outer retina; these features are incorporated only partially, or in some cases not at all, into our understanding of its function. For example, as shown in Fig. 1, bipolar cells are thought to be presynaptic to horizontal cells and to other bipolar cells at presumed glutamatergic synapses. Because of the absence of any physiological description of these synapses, they have been omitted from further discussion.

B. Glutamate Receptor Cloning and Classification

Until recently, the nomenclature and classification of glutamate channels was determined entirely by pharmacological distinctions. Since the cloning of the first glutamate receptor (Hollman *et al.*, 1989), however, the situation has been shifting toward a classification based on sequence homology. The emerging outline of this classification is that among glutamate receptors are those directly gating an ion channel (the ionotropic receptors) and those that are operated indirectly through second messengers (the metabotropic receptors). The ionotropic receptors comprise the AMPA (d-amino-3-hydroxy-5-methyl-4-isoxazolepropionate) receptors (formerly called non-NMDA or quisqualate/kainate receptors) and the NMDA (*N*-methyl-D-aspartate) receptors, both familiar to physiologists, as well as a few other subunit clones that are not readily recognizable. The metabotropic receptors, in contrast, are all largely unfamiliar to physiologists.

1. AMPA Receptors

Four similar AMPA receptor genes called GluR-A, B, C, and D (Keinänen *et al.*, 1990) or GluR1, 2, 3, and 4 (Boulter *et al.*, 1990) encode subunits that can assemble to form functional homo- or heteromeric channels. All combinations of subunits examined to date are gated by AMPA, kainate, and, of course, L-glutamate (Keinänen *et al.*, 1990; Nakanishi *et al.*, 1990). When pairs of GluR1–GluR4 clones are coexpressed, channels with novel properties not found in either homomeric channel can appear (Boulter *et al.*, 1990; Burnashev *et al.*, 1992). Two additional levels of control multiply the permutations that can be achieved in forming a channel. Alternative splicing can select stretches of 38 amino acids from adjacent exons, giving rise to two variants of the subunit, named flip and flop (Sommer *et al.*, 1990). These two splice alternatives differ because flop variants show a greater degree of desensitization induced by glutamate.

More remarkable is the existence of an RNA editing mechanism that changes a codon for glutamine to one encoding arginine in transcripts of GluR2 but not in the other three, although all share a glutamine codon in that position (Sommer *et al.*, 1991). The amino acid in question lies in the second putative membrane-spanning region and clearly plays a significant role in controlling channel permeability. GluR1, 3, and 4 all express glutamine at this position and give rise to channels, whether homo- or heteromeric, that show both inward and outward rectification and admit divalent cations (Hollman *et al.*, 1991). The presence of arginine contributed by a GluR2 subunit renders the channel impermeable to divalent ions and confers linearity on the current–voltage relationships (Hume *et al.*, 1991; Verdoorn *et al.*, 1991). The spliced and edited transcripts of these four genes are likely to represent the bulk of the messages encoding non-NMDA glutamate transmitter receptors in the brain and the outer retina.

Three very similar clones—GluR5 (Bettler *et al.*, 1990), 6 (Egebjerg *et al.*, 1991), and 7 (Bettler *et al.*, 1992)—are distantly related, having a homology of about 40% with GluR1–4 compared with 70% homology within the group GluR1–4. The functional significance of these subunits is unclear. Although GluR5 and GluR7 show high-affinity kainate binding (Bettler *et al.*, 1992; Sommer *et al.*, 1992); GluR7 does not form functional homomeric channels (Bettler *et al.*, 1992). These subunits may be kainate receptors whose normal function may require coexpression with other subunits. Two other clones, KA1 (Werner *et al.*, 1991) and KA2 (Herb *et al.*, 1992), are only slightly related to GluR1–4 or GluR5–7; these clones also show high-affinity kainate-binding sites but form no functional homomeric channels. Coexpression of GluR5 or 6 with KA2, on the other hand, produces functional channels with properties distinct from those of homomeric GluR5 or 6 (Herb *et al.*, 1992).

2. NMDA Receptors

In 1991, Nakanishi and colleagues (Moriyoshi *et al.*, 1991) expressed a cloned NMDA channel (NMDAR1) with the defining properties of this channel, including voltage-dependent block by Mg^{2+}, permeability to Ca^{2+}, and enhancement by glycine. It has since been shown that alternative splicing can give rise to six variants of this subunit (Sugihara *et al.*, 1992). Although homomeric NMDAR1 channels are clearly functional, heteromeric combinations of NMDAR1 with subunits of different clones, NMDAR2A–D, show enhanced responses to glutamate and NMDA, although these subunits cannot make functional channels by themselves (Kutsuwada *et al.*, 1992; Meguro *et al.*, 1992; Monyer *et al.*, 1992). As do AMPA receptor subunits, heteromeric combinations generate channels with properties that differ in important respects such as response kinetics and sensitivity to glycine from those of homomeric channels.

3. Metabotropic Receptors

Metabotropic glutamate receptors are characterized by their activation with ACPD (*trans*-1-aminocyclopentane-1,3-dicarboxylate) and constitute a highly conserved family of six members: mGluR1 (Houamed *et al.*, 1991; Masu *et al.*, 1991), mGluR2–4 (Tanabe *et al.*, 1992), and mGluR5 and 6 (Nakanishi, 1992). As expected, these receptors show the seven membrane-spanning regions characteristic of those receptors that interact with G proteins and are coupled to second messenger systems, either cAMP (Houamed *et al.*, 1991; Tanabe *et al.*, 1992) or IP$_3$(inositol trisphosphate)/Ca^{2+} (Aramori and Nakanishi, 1992).

C. Where Are GluR Genes Expressed in the Outer Retina?

The cloning of glutamate receptors has allowed the localization of receptor types through the technique of *in situ* hybridization of labeled probes on sections of CNS. Regrettably, few such studies show the retina, although a notable exception (Hughes *et al.*, 1992) examines expression of the AMPA genes GluR1–5 in the rat retina. The most striking results from that study are that all five subunits are expressed somewhere in the retina; as expected, the results are consistent with multiple subunits being expressed per cell. No expression of AMPA receptors was found in photoreceptors which, in the rat, are nearly all rods. The AMPA subunit GluR2 was found to be strongly expressed on ON bipolar cells. This finding is interesting since ON bipolar cells, as discussed later, are notable for a metabotropic glutamate receptor that is thought to be postsynaptic at the photoreceptor synapses. The functional significance of AMPA channels in these cells is unknown but, since the inclusion of GluR2 subunits in a heteromeric AMPA channel renders it impermeable to calcium, some, and perhaps all, AMPA channels in ON bipolar cells must be presumed to exclude calcium.

II. Glutamate-Gated Channel on Photoreceptors?

Electron microscopy of the salamander outer plexiform layer (Lasansky, 1973) shows cone processes that are postsynaptic at rod ribbon synapses. Similarly, in the turtle retina, an electron microscopic Golgi study of cone terminals, supported by physiological evidence, demonstrates synapses between cones (Normann *et al.*, 1984). Therefore, that glutamate iontophoretically applied to isolated cones, in the turtle (Tachibana and Kaneko, 1988) or the salamander (Sarantis *et al.*, 1988) retina, elicits a current is not surprising. However, the functional significance of this current is not at all clear. This current has been viewed as mediating cone–cone interaction (Tachibana and Kaneko, 1988), as generated by autoreceptors to a cone's own transmitter (Sarantis *et al.*, 1988), and as generated by an electrogenic glutamate transporter (Eliasoff and Werblin, 1993).

Several aspects of the current are unusual and call for further examination. Unlike the glutamate responses of second order retinal cells, those of cones show no desensitization over a time course of seconds (Sarantis *et al.*, 1988; Tachibana and Kanebo, 1988; Eliasoff and Werblin, 1993). Although D-aspartate is an effective agonist (Tachibana and Kaneko, 1988; Eliasoff and Werblin, 1993), NMDA and quisqualate are ineffective (Tachibana and Kanebo, 1988; Eliasoff and Werblin, 1993), as in APB (2-amino-4-phosphonobutyric acid; (Eliasoff and Werblin, 1993). The channel blockers kynurenic acid, APV (2-amino-5-phosphonovaleric acid), and CNQX (6-cyano-7-nitroquinoxaline-2,3-dione) are reported to have no effect on the glutamate current (Eliasoff and Werblin, 1993), but the glutamate transporter inhibitors *p*-chloromercuriphenylsulfonic acid (Tachibana and Kaneko, 1988), dihydrokainate, and D,L-threo-β-hydroxyaspartate (Eliasoff and Werblin, 1993) cause partial reduction in the current. The current–voltage relationships are inwardly rectifying; Cl$^-$ concentration at least partially determines the reversal potential, although Na$^+$ ions are also required externally (Sarantis *et al.*, 1988). This evidence suggests that this glutamate current is actually the consequence of an electrogenic transporter rather than an ion channel. Against this interpretation must be set the observation, as shown in Fig. 2, that a measurable noise increase accompanies the glutamate-induced current, the spectrum of which suggests that an ion channel is being opened (Tachibana and Kaneko, 1988). A simple resolution of this paradox, that both an ion channel and an electrogenic pump are involved in generating the glutamate-elicited current, requires that the two components be readily separable. The fact that this does not seem to be the case (Eliasoff and Werblin, 1993) leads to the suspicion that these cells contain a novel channel with unusual pharmacology.

A

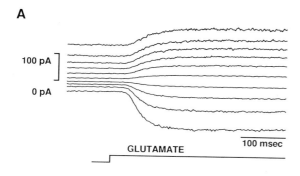

100 pA

0 pA

GLUTAMATE

100 msec

B

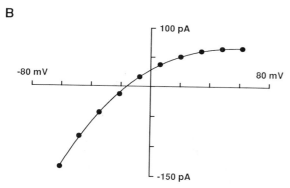

100 pA

-80 mV

80 mV

-150 pA

FIGURE 2 Glutamate-evoked currents in isolated chicken cones. (A) Current records from a whole-cell clamped cone held at voltages from −60 to +60 mV. L-Glutamate (100 μM) was pressure applied to the cone during the step shown. The long latency for the response is attributable to the distance of the drug pipet from the cell. Responses show no desensitization, but are associated with a small but noticeable increase in noise. (B) Current–voltage relationships for the glutamate-elicited current shown in A. The current is strongly inward rectifying and apparently saturates at positive potentials; nevertheless, outward currents and a clear reversal potential are seen. E_{Cl} for this cell was −21 mV, only slightly negative to the measured reversal potential. Data are unpublished results of M. Sarantis, E. Gleason, and M. Wilson.

III. Horizontal Cells

The observation that horizontal cells depolarize when L-glutamate and L-aspartate are applied to the retina (Murakami *et al.*, 1972) originally indicated that an excitatory amino acid was likely to be the photoreceptor transmitter. An inevitable disadvantage of working with the intact retina is that it possesses potent means for sequestering or otherwise disabling its authentic neurotransmitters. Because of this, the concentration of glutamate required to generate responses when applied to the intact retina is alarmingly high (millimolar; see Mangel *et al.*, 1989). Largely as a consequence, the excitatory amino acid—whether

glutamate, aspartate, or some other amino acid—that was the natural transmitter remained unknown for many years (Wu and Dowling, 1978; Shiells *et al.*, 1981). Only with the introduction of methods for the dissociation of the retina into single cells (Kaneko *et al.*, 1976) was it shown that reasonable concentrations of L-glutamate were effective.

Isolated horizontal cells respond strongly to L-glutamate but much less well to D-glutamate or L- or D-aspartate (Lasater and Dowling, 1982; Ishida *et al.*, 1984; Lasater *et al.*, 1984). Quisqualate, kainate, and AMPA are all effective agonists (Ishida *et al.*, 1984; O'Dell and Christensen, 1989a,b), but kainate, as elsewhere (Trussel *et al.*, 1988; Tang *et al.*, 1989), produces responses that show no desensitization. Unlike the AMPA channels expressed in OFF bipolar cells, those of horizontal cells are antagonized weakly by D-*O*-phosphoserine (Slaughter and Miller, 1985b). The effects of this antagonist on expressed GluR has not yet been reported.

Ion substitution experiments indicate that glutamate-gated channels in horizontal cells are nonspecifically permeable to cations (Tachibana, 1985; Hals *et al.*, 1986; O'Dell and Christensen, 1989b). Since two studies (Tachibana, 1985; O'Dell and Christensen, 1989b) report that changes in external calcium concentration can change the reversal potential of the glutamate response, calcium ions are likely to be permeant through these channels, although the extent of permeability remains to be determined. While it is apparent that horizontal cells express AMPA channels, to say whether these channels contain the calcium-excluding subunit GluR2 is not possible. Surprisingly, the current–voltage relationships for glutamate-gated currents that might suggest the composition of these AMPA channels have not been examined in sufficient detail to allow a guess.

The current–voltage relationships for glutamate currents in goldfish horizontal cells are monotonically outwardly rectifying in the presence of external Cs$^+$, which blocks the inward-rectifier potassium channel, but show a negative slope region below −50 mV when Cs$^+$ is absent (Tachibana, 1985; Kaneko and Tachibana, 1985). This curious effect probably originates in the dependence of inward-rectifier activation on the potassium equilibrium potential (Leech and Stanfield, 1981). Pressure application of glutamate is likely to remove the small excess of potassium accumulated just outside the cell membrane in an unstirred layer. The effect of this removal is to shift E_K negative and, thereby, to inactivate the inward potassium current even though the membrane voltage is kept constant. Strong support for this interpretation is provided by experiments in which rapid washing of isolated cells abolishes the effect of glutamate on the inward rectifier (Perlman *et al.*, 1988).

Estimates of the dose–response curve for glutamate on horizontal cells are in many instances not very useful, since they use slow application of agonist. AMPA channels are known to desensitize (Trussel *et al.*, 1988) and, indeed, may do so on a millisecond time scale (Tang *et al.*, 1989), so relatively slow drug applications can be expected to distort the dose–response curve. Some of the most credible data come from a study of stingray H1 horizontal cells in which glutamate was applied rapidly. In this study (O'Dell and Christensen, 1989a), EC_{50} for glutamate was determined to be $\sim$20 μM; the time constant for desensitization, which is concentration dependent, was measured as 7–18 msec when quisqualate was used as the agonist. Two other studies of elasmobranch horizontal cells (Lasater *et al.*, 1984; Shiells *et al.*, 1986) failed to detect any desensitization but, in view of their slow drug application, this is not surprising. As is true for AMPA channels found elsewhere (Mayer and Vyklicky, 1989), treatment with the lectin concanavalin A abolishes desensitization (O'Dell and Christensen, 1989a).

Hill coefficients indicating the minimum number of agonist molecules required to open a channel have been measured for glutamate action on channels in horizontal cells. Reported values lie near 2 (Shiells *et al.*, 1986; O'Dell and Christensen, 1989a), although considerable variability is evident even within the same species (Shiells *et al.*, 1986). Some evidence indicates that a coefficient of 1 is associated with the agonists quisqualate and AMPA, but a coefficient of 2 is associated with glutamate and kainate (O'Dell and Christensen, 1989a). These data might be thought to indicate separate classes of receptor for quisqualate and kainate, but the important finding that homomeric AMPA channels respond to multiple agonists (Boulter *et al.*, 1990) makes this explanation unlikely. In fact, prior to this conclusive demonstration in oocytes, three studies in the retina (Ishida and Neyton, 1985; O'Dell and Christensen, 1989a,b) indicated that quisqualate and kainate bound to the same receptor because responses to different agonists did not add linearly. The actual results of these studies, which show both supra- and sublinear addition of responses, suggest that these AMPA channels have complicated behavior generated by the presence of more than one binding site and rate constants for state changes that are strongly agonist dependent. Understanding these complications can only be properly achieved by studying homogeneous preparations of channels with known subunit stoichiometry, rather than retinal neurons.

NMDA is generally reported to have no effect on horizontal cells (Lasater and Dowling, 1982; Ariel *et al.*, 1984; O'Dell and Christensen, 1989a), but at high concentrations it can act as a kainate antagonist (Ariel *et al.*, 1986; O'Dell, 1989). Several exceptions, both real and apparent, are worth noting. Slaughter and Miller (1983) reported that the first application of NMDA to a horizontal cell in a mudpuppy eyecup caused a transient hyperpolarization, but that subsequent applications were ineffective. A likely explanation for this observation is that the primary effect of NMDA is at the inner retina, where it causes the release and depletion of a second substance that acts on horizontal cells. A similar explanation may apply to a result reported for salamander retina. Consistent with the presence of AMPA channels that are postsynaptic to photoreceptors in horizontal cells, the AMPA channel-specific blocker CNQX (Honoré *et al.*, 1988) abolishes light responses and hyperpolarizes horizontal cells (Yang and Wu, 1989). Nevertheless, glutamate applied in the presence of CNQX produces a large depolarization, prompting the suggestion that horizontal cells contain extrasynaptic NMDA channels (Yang and Wu, 1989). An alternative interpretation is that NMDA receptors on a cell type in the inner retina, perhaps a glycinergic interplexiform cell (Borges and Wilson, 1991; Gilbertson *et al.*, 1991b), are the primary targets of NMDA action.

A convincing case for an NMDA receptor on horizontal cells has been made only for cone horizontal cells of the channel catfish. NMDA acts here with the usual glycine enhancement and the voltage-dependent Mg^{2+} block (O'Dell and Christensen, 1989b; Tachibana and Okada, 1991). Whereas the conductance of NMDA receptors is reported elsewhere to be 40–50 pS (Ascher *et al.*, 1988), these receptors on catfish horizontal cells are thought to have a conductance approximately one order of magnitude less (O'Dell and Christensen, 1989b). Why catfish should have an NMDA receptor while other species apparently do not is unclear.

As described in the next section, APB has been widely used as a selective agonist for the glutamate receptor found in ON bipolar cells. APB also exerts effects on horizontal cells, some of which are direct. APB applied to the light-adapted mudpuppy retina increases the coupling between horizontal cells (Hare and Owen, 1992) in a way that can be reversed by dopamine (Dong and McReynolds, 1989). Since APB is unable to reverse the effects of dopamine, APB has been suggested to act upstream from dopamine release in the inner retina and to affect horizontal cells through this indirect route (Dong and McReynolds, 1991). APB also affects goldfish cone-driven horizontal cells, although in this case the mechanism is direct. In the goldfish retina, APB blocks light responses and causes horizontal cells to hyperpolarize (Nawy *et al.*, 1988). Similar effects are seen in salamander retina (Hare and Owen, 1982) and have been attributed to an inhibition of cone transmitter release. This interpretation cannot be ruled out, but APB acts as an antagonist for glutamate in isolated horizontal cells in fish (Takahashi and Copenhagen, 1992), although remarkably only when intracellular pH is low. Alkalinization causes this antagonism to diminish or disappear. APB has no effect alone, but acts only in the presence of glutamate; because this effect depends on internal pH, it is suggested that APB works through a second messenger in this system.

A recurrent theme in the description of the cellular physiology of the outer retina has been the judicious matching of neuronal properties to the expected signal. A particularly elegant demonstration of this was made in the turtle retina. Rod responses to light are much slower than cone responses. The teleological explanation for this difference is that greater temporal resolution is possible for cones since they operate at higher light intensities at which the random fluctuation in photon arrival is less of a problem. Not only do the responses of rods and cones differ temporally, but the synapses from these photoreceptors to horizontal cells are clearly different in their temporal characteristics. By deconvolving the response waveforms from turtle horizontal cells with those from either rods or cones (Schnapf and Copenhagen, 1982; Copenhagen *et al.*, 1983), it was possible to derive the impulse response for the behavior of these synapses. Cone to H1 AT horizontal cell synapses are roughly one order of magnitude faster than rod synapses to the same cell. The advantage of this arrangement is that the synapse filters out noise frequencies above the useful signal frequency (Baylor and Fettiplace, 1977). Although the rate-limiting steps that generate these impulse response functions are not known, the kinetic properties of glutamate channels may be different at these two synapses. Localized differences in the kinetics of glutamate currents have been demonstrated for chick spinal cord neurons (Trussel *et al.*, 1988), but have not been described for horizontal cells.

Some evidence for a mixed population of glutamate-gated channels on horizontal cells comes from a single-channel study of white bass horizontal cells (Lasater, 1990). In this study, fast (1–2 msec open time) large conductance channels and slow (5.6 msec open time) smaller conductance channels were found. Where these two channel types are located on the horizontal cell membrane is not yet known.

IV. ON Bipolar Cells

The separation of ON and OFF signals first established at the bipolar cell level is preserved for many stages in the visual system. The physiological division in the retina is matched by an antomical segregation in which ON bipolar cells form synapses in the inner half of the inner plexiform layer whereas OFF bipolar cells form synapses in the outer half of the inner plexiform layer (Famiglietti and Kolb, 1976; Famiglietti *et al.*, 1977). The symmetry implied by this arrangement is not complete, since the ON and OFF systems are functionally different. When ON and OFF bipolar cell performance has been compared carefully, ON bipolar cells have been found to respond more slowly than OFF cells (Ashmore and Copenhagen, 1980), a distinction that remains apparent at the level of ganglion cells (Baylor and Fettiplace, 1977) and is likely to constrain the performance of the visual system as a whole.

In addition to the divison of signals into ON and OFF channels, rod and cone signals are kept separate in many retinae. In the mammalian retina, exclusively rod- or exclusively cone-driven bipolar cells are found. Rod bipolar cells are always of the ON type (Dacheux and Raviola, 1986; Dolan and Schiller, 1989), whereas cone bipolar cells may be either ON or OFF. In other vertebrate retinae the separation of rod and cone signals is not precise; bipolar cells are often termed rod or cone dominant (e.g., Saito and Kujiraoka, 1982). In nonmammalian retinae in which the morphology and inputs to bipolar cells have been examined systematically, considerable complexity has been revealed. The turtle, for example, has nine anatomical classes of bipolar cell (Kolb, 1982).

Since photoreceptors may make contact with ON *and* OFF bipolar cells, these two kinds of bipolar cell must differ as a consequence of their different glutamate receptors. This presumption is now known to be true, although the existence of anatomical subtypes of bipolar cell serves notice that the situation may not be simple. This suspicion is reinforced by physiological studies.

A novel idea suggested by Kaneko (1971) was that ON bipolar cells might possess channels closed by photoreceptor transmitter. Further work from many other laboratories supported this interpretation. Direct confirmation of this idea was provided by work on isolated bipolar cells and bipolar cells in retinal slices. Among isolated bipolar cells from the salamander retina, two populations of cells were found (Attwell *et al.*, 1987a,b): one that responsed to glutamate with a conductance increase, presumably the OFF bipolar cells, and one that responded with a conductance decrease, presumably the ON bipolar cells. Although approximately equal numbers of ON and OFF cells were found in that study, a surprisingly large fraction, 67% of cells, showed no response to glutamate. Lasater *et al.*, (1984) similarly found only a minority of isolated skate bipolar cells that responded, although in this instance none showed a conductance decrease. An obvious possibility to explain this lack of response is that glutamate receptors might be mechanically or enzymatically removed during the dissociation procedure. This explanation cannot be ruled out, but the increased success rate enjoyed with perforated-patch recording (Yamashita and Wässle, 1991) over the usual ruptured-patch method (Karschin and Wässle, 1990) indicates that, for ON cells, wash-out of some soluble factor is a major problem. This conclusion is confirmed by the observation that ON bipolar cells in salamander cells in slices show a decline in responsiveness after membrane rupture (Nawy and Jahr, 1990b).

Several points emerged from work on isolated cells. First, it was observed that responses in conductance-decrease cells were much slower than responses in conductance-increase cells. The half-time for the onset of response in a conductance-decrease cell was found to be almost 1 sec (Attwell *et al.*, 1987a). In light of subsequent evidence, these unnaturally slow responses were likely to have been caused by partial wash-out of the second messenger system. Second, not only did a conductance decrease occur during the response, but this decrease was associated with a pronounced noise decrease and a roughly linear relationship between variance and mean current. This result was interpreted to imply (Attwell *et al.*, 1987a,b; Wilson *et al.*, 1987) that less than half the glutamate-closable channels were open, even in the absence of glutamate, and that each open channel contributed a conductance of ~11 pS. A critical assumption in this analysis was that the observed current variance was the result of random fluctuations in the number of channels open around some stationary mean number. Now that a second messenger system is known to gate the channel, as described subsequently, some if not all of the fluctuation is likely to originate at stages earlier than channel opening. Consequently, channel conductance may be smaller than the estimate of 11 pS.

APB, also known as AP4, acts as an agonist for the photoreceptor transmitter at ON bipolar cells (Slaughter and Miller, 1981; Shiells *et al.*, 1981), where the receptor binds L-glutamate in its extended conformation (Slaughter and Miller, 1985a). APB has binding sites throughout the CNS and is thought to act as a presynaptic inhibitor in the spinal cord and elsewhere

(Forsythe and Clements, 1990). One report (Trombley and Westbrook, 1992) described AP4 as inhibiting Ca^{2+} currents in olfactory bulb cells via a G-protein-coupled receptor, although whether the APB receptor of ON bipolar cells are the same as or different from the presynaptic AP4 receptor is still unknown.

APB receptors in the retina are not exclusive to ON bipolar cells, as was originally claimed; nevertheless, the use of this agonist has been important in defining the properties of glutamate receptors on these cells. Much of this work has employed bipolar cells in retinal slices, where responses to glutamate and even light responses (Attwell *et al.*, 1987a) may be recorded fairly readily. Cyclic GMP included in a whole-cell pipette not only reverses the decline of the APB-suppressible current but actually increases it over time (Nawy and Jahr, 1990a). In the slice preparation of salamander, inclusion of cGMP in the pipette augments the inward currents caused by light as a consequence of a diminution in glutamate release by photoreceptors. In dogfish retina, the effect of cGMP is to increase an inward current seen in the dark instead, and thereby to diminish light responses (Shiells and Falk, 1990). Although at first sight these results might seem to conflict, they are, with minor modification of the assumptions, both consistent with the single interpretation that cGMP opens channels and that glutamate binding to the APB receptor causes these channels to close by lowering cGMP concentrations. Since cGMP is well known to open the photocurrent channels of rods (Fesenko *et al.*, 1985; Yau and Baylor, 1989), the two cell types might be supposed to utilize similar transduction machinery, which indeed seems to be the case.

Strong evidence for the interaction of the APB receptor with a G protein is provided by experiments with nucleotide analogs. In slices of salamander retina perfused with Co^{2+} to block transmission, the nonhydrolyzable GTP analog, GTP-γ-S, when included in the patch pipette, causes APB to produce an irreversible effect (Nawy and Jahr, 1990a) that is consistent with the irreversible activation of a G protein. GDP-β-S, a competitive blocker of G-protein activation, prevents APB from closing channels (Nawy and Jahr, 1990a). By analogy with rod transduction, the G protein should activate a phosphodiesterase. Consistent with this idea is the finding that isobutylmethylxanthine (IBMX) a phosphodiesterase inhibitor, reduces or abolishes the ability of APB to close channels (Nawy and Jahr, 1990a; Shiells and Falk, 1990). Although the parallel between this system and phototransduction is appealing, major differences between the two systems are still possible.

Are APB receptors really postsynaptic at photoreceptor synapses? They seem to be since the results with APB generally also hold for the photoreceptor transmitter whose release can be regulated with flashes of light. GTP-γ-S suppresses light responses in dogfish (Shiells and Falk, 1990) and salamander (Nawy and Jahr, 1991) ON bipolar cells, although in neither preparation is the light response completely abolished. A possible explanation for this effect is that, although electrical coupling between bipolar cells (Kujiraoka and Saito, 1986; Borges and Wilson, 1987; Saito and Kujiraoka, 1988; Hare and Owen, 1990) might allow signals to be conducted between the cells, the movement of GTP-γ-S would be more restricted. Dipyridamole, a phosphodiesterase inhibitor, produces a large inward current and reduces the light response, as expected, although again suppression is not complete (Nawy and Jahr, 1991).

The G protein of rod transduction, transducin, is unusual because it is ADP-ribosylated by cholera and pertussis toxins (Van Dop *et al.*, 1984). The A subunit of pertussis toxin in the presence of NAD^+ increases the inward current of dogfish ON biopolar cells, consistent with an inactivation of the G protein. Subunit A of cholera toxin suppresses an inward current in the same way that GTP-γ-S does (Shiells and Falk, 1992a). If a single G protein mediates both of these effects, it resembles tranducin in this respect.

Remarkably little is known about the conduction properties of the channel gated by cGMP. It has been widely reported to be associated with a reversal potential around 0 mV (e.g., Attwell *et al.*, 1987a; Shiells and Falk, 1990). One report suggests that Cl^- is the predominant charge carrier (Miller and Dacheux, 1976), but this is probably not correct. More likely is that the channel is nonspecifically selective for cations. Sodium ions have been shown to carry charge for isolated rat ON bipolar cells (Yamashita and Wässle, 1991), but a careful examination of premeability has not been carried out.

A salient feature of the rod cGMP-gated channels is that removal of external Ca^{2+} greatly increases the channel conductance and linearizes the current–voltage relationships, indicating that Ca^{2+} ions within the pore normally regulate conduction (Zimmerman and Baylor, 1986). In contrast, removal of Ca^{2+} from outside ON bipolar cells does not increase conductance (Shiells and Falk, 1992b) in the dogfish retina. A partial similarity between rod cGMP-gated channels and those of dogfish ON bipolar cells lies in their both being blocked by L-*cis*-diltiazem (Stern *et al.*, 1986; Shiells and Falk, 1992b). The voltage dependence of the block, however, is different in the two sorts of channels, indicating that the block occurs at different sites within the pore.

The molecular identity of the APB receptor at ON bipolar cells is presently unknown, but a reasonable assumption is that it belongs to the family of metabo-

tropic glutamate receptors that have been cloned. The clone mGluR6 found in the retina has been suggested as a likely candidate (Nakanishi, 1992).

ON bipolar cells express protein kinase C (Negishi *et al.*, 1988; Suzuki and Kaneko, 1990). This characteristic has been used to recognize ON bipolar cells when isolated from the retina (Karschin and Wässle, 1990; Yamashita and Wässle, 1991). Protein kinase C, however, is probably not involved in the glutamate-gated pathway, since the kinase inhibitors H7 (Shiells and Falk, 1992a) and staurosporine (Nawy and Jahr, 1991) are without effect on light responses. Similarly, phorbol esters that stimulate protein kinase C also have no effect (Nawy and Jahr, 1991; Yamashita and Wässle, 1991). The function of protein kinase C in these cells remains uncertain.

The elucidation of a mechanism for the action of APB at ON bipolar cells, although incomplete, is a major advance in which the physiologists are ahead of the molecular biologists. However, to suppose that the APB receptor and its associated cGMP-gated channels constitute the only mechanism for glutamate action on these cells would be incorrect. Teleost ON bipolar cells examined in isolated retinae often show little or no conductance change in response to applied glutamate, although this manipulation produces a hyperpolarization (Nawy and Copenhagen, 1990). This observation is consistent with the previous finding that rod and cone inputs to these cells are associated with different reversal potentials (Saito *et al.*, 1979); the cone input has a reversal potential negative to rest and the rod input has a positive reversal potential. Since the effect of rod transmitter was seen to decrease conductance and that of cone transmitter increased conductance, the simultaneous activation of both these receptor types might result in no net conductance change. As expected, the rod transmitter closes channels via the APB receptor so, in the presence of APB, glutamate causes a conductance increase by opening the channels normally gated by cone transmitter (Nawy and Copenhagen, 1987).

Which channel is gated by cone transmitter? Very little is known about this channel and it has not been examined in voltage clamp or in isolated cells. Nevertheless, because intracellular Cs^+ is reported to block this conductance (Nawy and Copenhagen, 1990), the strong suspicion is that this is a K^+ channel. If this assumption is correct, this channel would represent a novel kind of glutamate-gated channel.

Some evidence suggests that cone input may be associated with a positive reversal potential in some fish ON bipolars (Saito and Kujiraoka, 1982), implying that other types of postsynaptic receptor may exist or that rods and cones can share APB receptors.

In retinae other than those of teleost fish, critical

examination of rod and cone input to ON bipolar cells has generally not been carried out. Although salamander ON bipolar cells receive mixed rod and cone input (Hare *et al.*, 1986), Nawy and Jahr (1991) found that these cells have no APB-resistant component to the light response and that glutamate does not activate a different conductance in the presence of APB. These results are strikingly different from those of Hirano and MacLeish (1991), who found that both APB and glutamate activated an outward current associated with a conductance *increase*. These authors report that, unlike Attwell *et al.*, (1987a), who also used the ruptured-patch method on isolated cells, they never saw a conductance decrease in response to applied glutamate or APB. Although the channels gated by cone transmitter may have been missed by other authors, that these channels should be gated by APB which, in fish retina, has, by exclusion, been used to define the cone transmitter receptors is puzzling (Nawy and Copenhagen, 1987).

The synapse between rods and ON bipolar cells is the most critical in the visual system. The ability to detect small numbers of photons reliably depends on the high signal-to-noise transmission of this synapse. By employing a second messenger system in ON bipolar cells, the visual system is apparently duplicating the machinery used for transduction in rods. A key difference is that, whereas rod outer segment channels open in darkness and are closed by light, the opposite is true for the APB-gated channels of ON bipolar cells. Understanding of the APB receptor and its associated pathway lags considerably behind an understanding of rod transduction. Nevertheless, a resaonable assumption is that, as in phototransduction, the gain at the first synapse is high and many channels open for the unbinding of a single receptor. The slowness of the depolarizing bipolar cell relative to the hyperpolarizing bipolar cells, which is presumably a consequence of delays in the second messenger pathway, is not a disadvantage, since rod responses are themselves very slow and, as pointed out by others (Baylor and Fettiplace, 1977), matching the synapse to the expected signal filters out noise.

V. OFF Bipolar Cells

OFF bipolar cells respond to glutamate agonists in a fashion similar to that of horizontal cells. Kainate and quisqualate are effective agonists (Gilbertson *et al.*, 1991). For isolated bipolar cells from the salamander retina, these authors found no responses to NMDA even in the absence of Mg^{2+} and the presence of glycine to potentiate responses (Johnson and

Ascher, 1987). A small depolarizing response of OFF bipolar cells to NMDA was reported for the isolated mudpuppy retina (Slaughter and Miller, 1983), but this result is probably an indirect effect originating in the inner retina where NMDA receptors are widespread. Glutamate-elicited currents in OFF bipolar cells are blocked by CNQX (Gilbertson *et al.*, 1991a). The pharmacological description of these channels leaves no doubt that they are AMPA channels.

Although strong evidence supports the idea that separate sets of channels are postsynaptic at rod and cone synapses for fish ON bipolar cells, studies on teleost OFF bipolar cells suggest that only a single channel type exists for both rod and cone input (Saito and Kaneko, 1983). In this study, a minority of OFF bipolar cells apparently showed a conductance increase in response to light. These experiments were not done in voltage clamp, however, and a voltage-dependent conductance increase could have outweighed the decrease in synaptic conductance. Using transretinal electrical stimulation to elicit photoreceptor transmitter release without transduction, Kaneko and Saito (1983) were able to show a sodium dependence for OFF bipolar cell responses. These investigators estimated the normal reversal potential for the light- and shock-elicited responses to be about $+45$ mV, but the subsequent demonstration of electrical coupling between bipolar cells (Kujiraoka and Saito, 1986; Saito and Kujiraoka, 1988) suggests that this value is probably an overestimate.

To date, the most complete examination of the permeability of glutamate-gated channels in OFF bipolar cells comes from studies of cells isolated from the salamander retina (Gilbertson *et al.*, 1991a). Ion substitution experiments showed that Cl^- ions are impermeant but Na^+ and K^+ are equally permeant. C^+, the chief internal cation used in these experiments, was estimated to be more permeant than K^+ by a factor of ~ 1.2. From a series of experiments in which external Ca^{2+} concentration was varied, it was concluded that this divalent ion must be a charge carrier through these channels (Fig. 3). Based on the measured reversal potentials in different solutions, these investigators estimated a permeability ratio for Ca^{2+} to K^+ of 3.2 to 1, clearly larger than the lower values previously reported for AMPA channels (Mayer and Westbrook, 1987). Calculation of a permeability value for a divalent ion is more complicated than for a monovalent ion (Jan and Jan, 1976; Lewis, 1979) but, more importantly, as argued by Ascher and Nowak (1988), surface charges in the vicinity of the channel mouth can inflate the estimates for divalent permeability. No direct method of measuring these surface charges exists, so inaccuracies could be large. Regardless of the

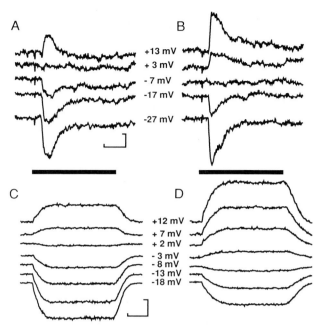

FIGURE 3 Calcium is permeant through AMPA channels of OFF bipolar cells. All panels show current responses from whole-cell clamped bipolar cells isolated from the salamander retina. (A,B) Responses to 200 μM glutamate pressure applied for the duration of the bar shown. (C,D) Responses of another bipolar cell to pressure-applied 20 μM kainate. (A,C) Agonist application in external solution containing 20 mM CaCl$_2$. (B,D) Currents when the external solution was switched to 1 mM CaCl$_2$. For both agonists, lowering Ca^{2+} concentration shifts the reversal potential negative by about 10 mV. Bars; (A,B) 50 msec, 20 pA; (C,D) 50 msec, 50 pA. Reprinted with permission from Gilbertson *et al.* (1991a). Copyright × 1991 by the AAAS.

exact value for permeability, however, clearly Ca^{2+} is significantly permeant, implying that these AMPA channels do not contain the GluR2 gene product. Current–voltage relationships for glutamate current in normal solutions show mild inward *and* outward rectification, as do heteromeric combinations of subunits expressed in embryonic kidney cells (Verdoorn *et al.*, 1991), but unlike the calcium-permeable AMPA channels found in hippocampal cells (Iino *et al.*, 1990) that display very strong inward rectification.

In experiments in which external Mg^{2+} was switched from 0.5 to 53.5 mM (T. A. Gilbertson and M. Wilson, unpublished), the reversal potential charge indicated no permeability for Mg^{2+}. Curiously, however, Mg^{2+} ions exert an effect on desensitization (Gilbertson *et al.*, 1991a). Glutamate (200 μM) elicits responses in these cells that desensitize profoundly with time constants of about 10 msec. The degree of desensitization suggests that these channel subunits exist in the flop edited form (Sommer *et al.*, 1990). Although

Mg^{2+} leaves the transient response to glutamate unaffected, the sustained response is larger in the absence of Mg^{2+} (Fig. 4). Unlike the effect of Mg^{2+} on the NMDA channel (Nowak *et al.*, 1984), the effect on these channels is voltage independent, implying that Mg^{2+} does not have to enter the pore. How Mg^{2+} exerts this effect is still unknown, and has not been reported for channels of known subunit composition examined in an expression system.

VI. Modulation of Glutamate Receptors

In teleost retina, a dopaminergic neuron—an interplexiform cell—feeds back to cells in the outer retina (Dowling, 1992). The release of dopamine orchestrates several changes in the outer retina that are thought to constitute part of dark adaptation (Dowling, 1992). Dopamine, when applied to isolated cone horizontal cells, enhances currents induced by applied glutamate or kainate (Knapp and Dowling, 1987). Like the effect of dopamine in uncoupling horizontal cells (Lasater, 1987), the enhancement of responses is produced by an increase in cAMP that stimulates cAMP-dependent protein kinase. The effect of dopamine is mimicked by the intracellular application of cAMP or the catalytic subunit of cAMP-dependent protein kinase (Liman *et al.*, 1989), which almost certainly phosphorylates one or more subunits of AMPA channels (Wang *et al.*, 1993). Phosphorylation seems to increase neither the number of available glutamate channels nor their conductance (5–10 pS), but instead increases the fraction of channels found in the open state (Knapp *et al.*, 1990), chiefly by increasing the opening rate constant rather than decreasing the closing rate constant.

VII. Outlook and Conclusion

This chapter shows that the disposition and properties of glutamate-gated channels in the outer retina are known at least superficially. The diversity of channel types revealed by cloning and examination of channels in expression systems has opened a new level of resolution for which the appropriate methods are *in situ* hybridization and single-channel analysis, neither of which has yet been widely applied to the retina.

Many important questions attend glutamate-gated channels in retinal cells. An outstandingly interesting problem is the nature of the glutamate channel, if indeed it is a channel, found on cone photoreceptors. A second major issue concerns the APB receptors in the retina, the best-established example of which exists in ON bipolar cells. Although cGMP is clearly

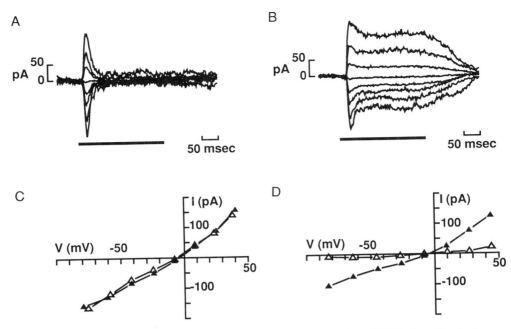

FIGURE 4 External Mg^{2+} enhances desensitization of AMPA channels at OFF bipolar cells of salamander. (A,B) Current responses at a series of voltages for pressure-applied 200 μM glutamate for the same isolated cell. (A) The external solution contained 0.5 mM Mg^{2+}. (B) Mg^{2+} concentration was nominally zero. As shown in the peak (C) and steady-state (D) current–voltage relationships, Mg^{2+} concentration does not affect the peak current but strongly affects the steady-state currents. Reprinted with permission from Gilbertson *et al.* (1991a).

involved in channel opening, the second messenger pathway is incompletely described and the channel itself, which might be expected to contribute minimal noise to postsynaptic responses, is virtually unknown. Evidence exists that other cells in the outer retina have APB receptors (e.g., Takahashi and Copenhagen, 1992). The extent to which these APB receptors and their second messenger pathways resemble those of the ON bipolar cell or the presynaptic APB receptor (Forsythe and Clements, 1990) remains to be determined.

Acknowledgments

I thank Salvador Borges and Matthew Frerking for their help and comments in the preparation of this manuscript.

References

Aramori, I., and Nakanishi, S. (1992). Signal transduction and pharmacological characteristics of a metabotropic glutamate receptor, mGluR1, in transfected CHO cells. *Neuron* **8**, 757–765.

Ariel, M., Lasater, E., Mangel, S., and Dowling, J. (1984). On the sensitivity of H1 horizontal cells of the carp retina to glutamate, aspartate, and their agonists. *Brain Res.* **295**, 179–183.

Ariel, M., Mangel, S. C., and Dowling, J. E. (1986). N-Methyl-D-aspartate acts as an antagonist of the photoreceptor transmitter in the carp retina. *Brain Res.* **372**, 143–148.

Ascher, P., and Nowak, L. (1988). The role of divalent cations in the N-methyl-D-aspartate responses of mouse central neurons in culture. *J. Physiol. (London)* **399**, 247–266.

Ascher, P., Bregestovski, P., and Nowak, L. (1988). N-Methyl-D-aspartate-activated channels of mouse central neurones in magnesium-free solutions. *J. Physiol. (London)* **399**, 207–226.

Ashmore, J. F., and Copenhagen, D. R. (1980). Different postsynaptic events in two types of retinal bipolar cell. *Nature (London)* **288**, 84–86.

Attwell, D., Mobbs, P., Tessier-Lavigne, M., and Wilson, M. (1987a). Neurotransmitter-induced currents in retinal bipolar cells of the axolotl, *Ambystoma mexicanun. J. Physiol. (London)* **387**, 125–161.

Attwell, D., Tessier-Lavigne, M., Wilson, M., and Mobbs, P. (1987b). Bipolar cell membrane currents and signal processing in the axolotl retina. *Neurosci. Res. (Suppl.)* **6**, 5191–5204.

Ayoub, G. S., Korenbrot, J. I., and Copenhagen, D. R. (1989). Release of endogenous glutamate from isolated cone photoreceptors of the lizard. *Neurosci. Res.* **10**, S47–S56.

Barlow, H. B. (1961). Three points about lateral inhibition. *In* "Sensory Communication" (W. A. Rosenblith, ed.), pp. 782–786. MIT Press, Cambridge, Massachusetts.

Baylor, D. A., and Fettiplace, R. (1977). Kinetics of synaptic transfer from receptors to ganglion cells in turtle retina. *J. Physiol. (London)* **271**, 425–448.

Bettler, B., Boulter, J., Hermans-Borgmeyer, I., O'Shea-Greenfield, A., Deneris, E. S., Moll, C., Borgmeyer, U., Hollmann, M., and Heinemann, S. (1990). Cloning of a novel glutamate receptor subunit, GluR5: Expression in the nervous system during development. *Neuron* **5**, 583–595.

Bettler, B., Egebjerg, J., Sharma, G., Pecht, G., Hermans-Borgmeyer, I., Moll, C., Stevens, C. F., and Heinemann, S.

(1992). Cloning of a putative glutamate receptor: A low affinity kainate-binding subunit. *Neuron* **8**, 257–265.

Borges, S., and Wilson, M. (1987). Structure of the receptive fields of bipolar cells in the salamander retina. *J. Neurophysiol.* **58**, 1275–1291.

Borges, S., and Wilson, M. (1991). Dual effect of glycine on horizontal cells of the tiger salamander retina. *J. Neurophysiol.* **66**, 1993–2001.

Boulter, J., Hollmann, M., O'Shea-Greenfield, Hartley, M., Deneris, E., Maron, C., and Heinemann, S. (1990). Molecular cloning and functional expression of glutamate receptor subunit genes. *Science* **249**, 1033–1037.

Burnshev, N., Monyer, H., Seeburg, P. H., and Sakmann, B. (1992). Divalent ion permeability of AMPA receptor channels is dominated by the edited form of a single subunit. *Neuron* **8**, 189–198.

Copenhagen, D. R., and Jahr, C. E. (1989). Release of endogenous excitatory amino acids from turtle photoreceptors. *Nature (London)* **341**, 536–539.

Copenhagen, D. R., Ashmore, J. F., and Schnapf, J. K. (1983). Kinetics of synaptic transmission from photoreceptors to horizontal and bipolar cells in turtle retina. *Vision Res.* **23**, 363–369.

Dacheux, R. F., and Raviola, E. (1986). The rod pathway in the rabbit retina: A depolarizing bipolar and amacrine cell. *J. Neurosci.* **6**, 331–345.

Dolan, R. P., and Schiller, P. H. (1989). Evidence for only depolarizing rod bipolar cells in the primate retina. *Visual Neurosci.* **2**, 421–424.

Dong, C. J., and McReynolds, J. S. (1989). APB increases apparent coupling between horizontal cells in mudpuppy retinal. *Vision Res.* **29**, 541–544.

Dong, C.-J., and McReynolds, J. S. (1991). The relationship between light, dopamine release and horizontal cell coupling in the mudpuppy retina. *J. Physiol. (London)* **440**, 291–309.

Dowling, J. E. (1992). The Charles F. Prentice Medal Award Lecture 1991: Dopamine; a retinal neuromodulator. *Optom. Vision Sci.* **69**, 507–514.

Egebjerg, J., Bettler, B., Hermans-Borgmeyer, I., and Heinemann, S. (1991). Cloning of a cDNA for a glutamate receptor subunit activated by kainate but not AMPA. *Nature (London)* **351**, 745–748.

Eliasof, S., and Werblin, F. (1993). Characterization of the glutamate transporter in retinal cones of the tiger salamander. *J. Neurosci.* **13**, 402–411.

Engbretson, G. A., Anderson, K. J., and Wu, J.-Y. (1988). GABA as a potential transmitter in lizard photoreceptors: Immunocytochemical and biochemical evidence. *J. Comp. Neurol.* **278**, 461–471.

Falk, G. (1988). Signal transmission from rods to bipolar and horizontal cells: A synthesis. *Prog. Retinal Res.* **8**, 255–279.

Famiglietti, E. V., and Kolb, H. (1976). Structural basis for ON- and OFF-center responses in retinal ganglion cells. *Science* **194**, 193–195.

Famiglietti, E. V., Kaneko, A., and Tachibana, M. (1977). Neuronal architecture of on and off pathways to ganglion cells in the carp retina. *Science* **198**, 1267–1269.

Fesenko, E. E., Kolesnikov, S. S., and Lyubarsky, A. L. (1985). Induction by cyclic GMP of cationic conductance in plasma membrane of retinal rod outer segment. *Nature (London)* **313**, 310–313.

Forsythe, I. D., and Clements, J. D. (1990). Presynaptic glutamate receptors depress excitatory monosynaptic transmission between mouse hippocampal neurones. *J. Physiol. (London)* **429**, 1–16.

Gilbertson, T. A., Scobey, R., and Wilson, M. (1991a). Permeation of calcium ions through non-NMDA glutamate channels in retinal bipolar cells. *Science* **251**, 1613–1615.

Gilbertson, T., Borges, S., and Wilson, M. (1991b). The effects of glycine and GABA on isolated horizontal cells from the salamander retina. *J. Neurophysiol.* **66**, 2002–2013.

Greengard, P., Jen, J., Nairn, A. C., and Stevens, C. F. (1991). Enhancement of the glutamate response by cAMP-dependent protein kinase in hippocampal neurons. *Science* **253**, 1135–1138.

Hals, G., Christensen, B. N., O'Dell, T., Christensen, M., and Shingai, R. (1986). Voltage-clamp analysis of currents produced by glutamate and some glutamate analogues on horizontal cells isolated from the catfish retina. *J. Neurophysiol.* **56**, 19–31.

Hare, W. A., and Owen, W. G. (1990). Spatial organization of the bipolar cell's receptive field in the retina of the tiger salamander. *J. Physiol. (London)* **421**, 223–245.

Hare, W. A., and Owen, W. G. (1992). Effects of 2-amino-4-phosphonogutyric acid on cells in the distal layers of the tiger salamander retina. *J. Physiol. (London)* **445**, 741–757.

Hare, W. A., Lowe, J. S., and Owen, G. (1986). Morphology of physiologically identified bipolar cells in the retina of the tiger salamander. *Ambystoma tigrinum. J. Comp. Neurol.* **252**, 130–138.

Herb, A., Burnashev, N., Werner, P., Sakmann, B., Wisden, W., and Seeburg, P. H. (1992). The KA-2 subunit of excitatory amino acid receptors shows widespread expression in brain and forms ion channels with distantly related subunits. *Neuron* **8**, 775–785.

Hirano, A. A., and MacLeish, P. R. (1991). Glutamate and 2-amino-4-phosphonobutyrate evoke an increase in potassium conductance in retinal bipolar cells. *Proc. Natl. Acad. Sci. USA* **88**, 805–809.

Hollman, M., O'Shea-Greenfield, A., Rogers, S., and Heinemann, S. (1989). Cloning by functional expression of a member of the glutamate receptor family. *Nature (London)* **342**, 643–648.

Hollmann, M., Hartley, M., and Heinemann, S. (1991). Calcium permeability of KA-AMPA-gated glutamate receptor channels: Dependence on subunit composition. *Science* **252**, 851–853.

Honoré, T., Davies, S., Drejer, J., Fletcher, E., Jacobsen, P., Lodge, D., and Nielsen, F. (1988). Quinoxalinediones: Potent competitive non-NMDA glutamate receptor antagonists. *Science* **241**, 701–703.

Houamed, K. M., Kuijper, J. L., Gilbert, T. L., Haldeman, B. A., O'Hara, P. J., Mulvihill, E. R., Almers, W., and Hagen, F. S. (1991). Cloning, expression, and gene structure of a G protein-coupled glutamate receptor from rat brain. *Science* **252**, 1318–1321.

Hughes, T. E., Hermans-Borgmeyer, I., and Heinemann, S. (1992). Differential expression of glutamate receptor genes (GluR1-5) in the rat retina. *Visual Neurosci.* **8**, 49–55.

Hume, R. I., Dingledine, R., and Heinemann, S. (1991). Identification of a site in glutamate receptor subunits that controls calcium permeability. *Science* **253**, 1028–1031.

Iino, M., Ozawa, S., and Tsuzuki, K. (1990). Permeation of calcium through excitatory amino acid receptor channels in cultured rat hippocampal neurones. *J. Physiol. (London)* **424**, 151–165.

Ishida, A. T., and Neyton, J. (1985). Quisqualate and I-glutamate inhibit retinal horizontal-cell responses to kainate. *Proc. Natl. Acad. Sci. USA* **82**, 1837–1841.

Ishida, A. T., Kaneko, A., and Tachibana, M. (1984). Responses of solitary retinal horizontal cells from *Carassius auratus. J. Physiol. (London)* **348**, 255–270.

Jan, L. Y., and Jan, Y. N. (1976). L-Glutamate as an excitatory transmitter at the *Drosophila* neuromuscular junction. *J. Physiol. (London)* **262**, 215–236.

Johnson, J. W., and Ascher, P. (1987). Glycine potentiates the NMDA response in cultured mouse brain neurons. *Nature (London)* **325**, 529–531.

Kaneko, A. (1971). Physiological studies of single retinal cells and their morphological identification. *Vision Res.* **3**, 17–26.

Kaneko, A., and Saito, T. (1983). Ionic mechanisms underlying the responses of off-center bipolar cells in the carp retina. II. Studies on responses evoked by transretinal current stimulation. *J. Gen. Physiol.* **81**, 603–612.

Kaneko, A., and Tachibana, M. (1985). Effects of L-glutamate on the anomalous rectifier potassium current in horizontal cells of *Carassius auratus* retina. *J. Physiol.* **358**, 169–182.

Kaneko, A., Lam, D. M. K., and Wiesel, T. N. (1976). Isolated horizontal cells of elasmobranch retinae. *Brain Res.* **105**, 567–572.

Karschin, A., and Wässle, H. (1990). Voltage- and transmitter-gated currents in isolated rod bipolar cells of rat retina. *J. Neurophysiol.* **63**, 860–876.

Keinänen, K., Wisden, W., Sommer, B., Werner, P., Herb, A., Verdoorn, T. A., Sakmann, B., and Seeburg, P. H. (1990). A family of AMPA-selective glutamate receptors. *Science* **249**, 556–560.

Knapp, A. G., and Dowling, J. E. (1987). Dopamine enhances excitatory amino acid-gated conductances in retinal horizontal cells. *Nature (London)* **325**, 437–439.

Knapp, A. G., Schmidt, K. F., and Dowling, J. E. (1990). Dopamine modulates the kinetics of ion channels gated by excitatory amino acids in retinal horizontal cells. *Proc. Natl. Acad. Sci. USA* **87**, 767–771.

Kolb, H. (1982). The morphology of the bipolar cells, amacrine cells and ganglion cells in the retina of the turtle *Pseudemys scripta elegans. Phil. Trans. R. Soc. London B* **298**, 355–393.

Kujiraoka, T., and Saito, T. (1986). Electrical coupling between bipolar cells in carp retina. *Proc. Natl. Acad. Sci. USA* **83**, 4063–4066.

Kutsuwada, T., Kashiwabuchi, N., Mori, H., Sakimura, K., Kushiya, E., Araki, K., Meguro, H., Masaki, H., Kumanishi, T., and Arakawa, M. (1992). Molecular diversity of the NMDA receptor channel. *Nature (London)* **358**, 36–41.

Lasansky, A. (1973). Organization of the outer synaptic layer in the retina of the larval tiger salamander. *Phil. Trans. R. Soc. London B* **265**, 471–489.

Lasansky, A. (1980). Lateral contacts and interactions of horizontal cell dendrites in the retina of the larval tiger salamander. *J. Physiol. (London)* **301**, 59–68.

Lasater, E. M. (1987). Retinal horizontal cell gap junctional conductance is modulated by dopamine through a cyclic AMP-dependent protein kinase. *Proc. Natl. Acad. Sci. USA* **84**, 7319–7323.

Lasater, E. M. (1990). Properties of non-NMDA excitatory amino acid-activated channels in isolated retinal horizontal cells. *J. Neurosci.* **10**, 1654–1663.

Lasater, E. M., and Dowling, J. E. (1982). Carp horizontal cells in culture respond selectively to L-glutamate and its agonists. *Proc. Natl. Acad. Sci. USA* **79**, 936–940.

Lasater, E. M., Dowling, J. E., and Ripps, H. (1984). Pharmacological properties of isolated horizontal and bipolar cells from the skate retina. *J. Neurosci.* **4**, 1966–1975.

Leech, C. A., and Stanfield, P. R. (1981). Inward rectification in frog skeletal muscle fibres and its dependence on membrane potential and external potassium. *J. Physiol. (London)* **319**, 295–309.

Lewis, C. A. (1979). Ion-concentration dependence of the reversal potential and the single channel conductance ion channels at the frog neuromuscular junction. *J. Physiol. (London)* **186**, 417–445.

Liman, E. R., Knapp, A. G., and Dowling, J. E. (1989). Enhancement of kainate-gated currents in retinal horizontal cells by cyclic AMP-dependent protein kinase. *Brain Res.* **481**, 399–402.

Mangel, S. C., Ariel, M., and Dowling, J. E. (1989). D-aspartate potentiates the effects of both L-aspartate and L-glutamate on carp horizontal cells. *Neuroscience* **32**, 19–26.

Marc, R. E., and Lam, D. M.-K. (1981). Uptake of aspartic and glutamic acid by photoreceptors in goldfish retina. *Proc. Natl. Acad. Sci. USA* **78**, 7185–7189.

Massey, S. C. (1990). Cell types using glutamate as a neurotransmitter in the vertebrate retina. *In* "Progress in Retinal Research" (N. Osborne and J. Chader, eds.), pp. 399–425. Pergamon Press, Oxford, England.

Masu, M., Tanabe, Y., Tsuchida, K., Shigemoto, R., and Nakanishi, S. (1991). Sequence and expression of a metabotropic glutamate receptor. *Nature (London)* **349**, 760–765.

Mayer, M. L., and Vyklicky, L. (1989). Concanavalin A selectively reduces desensitization of mammalian neuronal quisqualate receptors. *Proc. Natl. Acad. Sci. USA* **86**, 1411–1415.

Mayer, M. L., and Westbrook, G. L. (1987). Permeation and block of N-methyl-D-aspartic acid receptor channels by divalent cations in mouse cultured central neurones. *J. Physiol. Lond.* **394**, 501–527.

Meguro, H., Mori, H., Araki, K., Kushiya, E., Kutsuwada, T., Yamazaki, M., Kumanishi, T., Arakawa, M., Sakimura, K., and Mishina, M. (1992). Functional characterization of a heteromeric NMDA receptor channel expressed from cloned cDNAs. *Nature (London)* **357**, 70–74.

Miller, R. F., and Dacheux, R. F. (1976). Synaptic organization and ionic basis of on and off channels in mudpuppy retina. I. Intracellular analysis of chloride sensitive electrogenic properties of receptors, horizontal cells, bipolar cells and amacrine cells. *J. Gen. Physiol.* **67**, 639–659.

Monyer, H., Sprengel, R., Schoepfer, R., Herb, A., Higuchi, M., Lomeli, H., Burnashev, N., Sakmann, B., and Seeburg, P. H. (1992). Heteromeric NMDA receptors: molecular and functional distinction of subtypes. *Science* **256**, 1217–1221.

Moriyoshi, K., Masu, M., Ishii, T., Shigemoto, R., Mizuno, N., and Nakanishi, S. (1991). Molecular cloning and characterization of the rat NMDA receptor. *Nature (London)* **354**, 31–37.

Murakami, M., Ohtsu, K., and Ohtsuka, T. (1972). Effects of chemicals on receptors and horizontal cells in the retina. *J. Physiol. (London)* **277**, 899–913.

Nakanishi, N., Schneider, N. A., and Axel, R. (1990). A family of glutamate receptor genes: Evidence for the formation of heteromultimeric receptors with distinct channel properties. *Neuron* **5**, 569–581.

Nakanishi, S. (1992). Molecular diversity of glutamate receptors and implications for brain function. *Science* **258**, 597–603.

Nawy, S., and Copenhagen, D. R. (1987). Multiple classes of glutamate receptor on depolarizing bipolar cells in retina. *Nature (London)* **325**, 56–58.

Nawy, S., and Copenhagen, D. R. (1990). Intracellular cesium separates two glutamate conductances in retinal bipolar cells of goldfish. *Vision Res.* **30**, 967–972.

Nawy, S., and Jahr, C. E. (1990a). Suppression by glutamate of cGMP-activated conductance in retinal bipolar cells. *Nature (London)* **346**, 269–271.

Nawy, S., and Jahr, C. E. (1990b). Time-dependent reduction of glutamate current in retinal bipolar cells. *Neurosci. Lett.* **108**, 279–283.

Nawy, S., and Jahr, C. E. (1991). cGMP-gated conductance in retinal bipolar cells is suppressed by the photoreceptor transmitter. *Neuron* **7**, 677–683.

Nawy, S., Sie, A., and Copenhagen, D. R. (1989). The glutamate analog 2-amino-4-phosphonobutyrate antagonizes synaptic transmission from cones to horizontal cells in the goldfish retina. *Proc. Natl. Acad. Sci. USA* **86**, 1726–1730.

Negishi, K., Kato, S., and Teranishi, T. (1988). Dopamine cells and rod bipolar cells contain protein kinase C-like immunoreactivity in some vertebrate retinas. *Neurosci. Lett.* **94**, 247–252.

Nishimura, Y., Schwartz, M. L., and Rakic, P. (1986). GABA and GAD immunoreactivity of photoreceptor terminals in primate retina. *Nature (London)* **320**, 753–756.

Normann, R. A., Perlman, I., Kolb, H., Jones, J., and Daly, S. J. (1984). Direct excitatory interactions between cones of different spectral types in the turtle retina. *Science* **224**, 625–627.

Nowak, L. Bregestovski, P., Ascher, P., Herbet A., and Prochiantz, A. (1984). Magnesium gates glutamate-activated channels in mouse central neurones. *Nature (London)* **307**, 462–465.

O'Dell, T. J. (1989). Pharmacological characterization of voltage-clamped catfish rod horizontal cell responses to kainic acid. *Brain Res.* **477**, 118–125.

O'Dell, T. J., and Christensen, B. N. (1989a). A voltage-clamp study of isolated stingray horizontal cell non-NMDA excitatory acid receptors. *J. Neurophysiol.* **61**, 162–172.

O'Dell, T. J., and Christensen, B. N. (1989b). Horizontal cells isolated from catfish retina contain two types of excitatory amino acid receptors. *J. Neurophysiol.* **61**, 1097–1109.

Perlman, I., Knapp, A. G., and Dowling, J. E. (1988). Local superfusion modifies the inward rectifying potassium conductance of isolated retinal horizontal cells. *J. Neurophysiol.* **60**, 1322–1332.

Saito, T., and Kaneko, A. (1983). Ionic mechanisms underlying the responses of off-center bipolar cells in the carp retina. I. Studies on responses evoked by light. *J. Gen. Physiol.* **81**, 589–601.

Saito, T., and Kujiraoka, T. (1982). Physiological and morphological identification of two types of on-center bipolar cells in the carp retina. *J. Comp. Neurol.* **205**, 161–170.

Saito, T., and Kujiraoka, T. (1988). Characteristics of bipolar cell coupling in the carp retina. *J. Gen. Physiol.* **91**, 275–287.

Saito, T., Kondo, H., and Toyoda, J. (1979). Ionic mechanisms of two types of on-center bipolar cells in the carp retina. I. The responses to central illumination. *J. Gen. Physiol.* **73**, 73–90.

Sarantis, M., Everett K., and Attwell, D. (1988). A presynaptic action of glutamate at the cone output synapse. *Nature (London)* **332**, 451–453.

Schnapf, J. L., and Copenhagen, D. R. (1982). Differences in the kinetics of rod and cone synaptic transmission. *Nature (London)* **296**, 862–864.

Shiells, R. A., and Falk, G. (1990). Glutamate receptors of rod bipolar cells are linked to a cyclic GMP cascade via a G-protein. *Proc. R. Soc. London B* **242**, 91–94.

Shiells, R. A., and Falk, G. (1992a). The glutamate-receptor linked cGMP cascade of retinal on-bipolar cells is pertussis and cholera toxin-sensitive. *Proc. R. Soc. London B* **247**, 17–20.

Shiells, R. A., and Falk, G. (1992b). Properties of the cGMP-activated channel of retinal on-bipolar cells. *Proc. R. Soc. London B* **247**, 21–25.

Shiells, R. A., Falk, G., and Naghshineh, S. (1981). Action of glutamate and aspartate analogues on rod horizontal and bipolar cells. *Nature (London)* **294**, 592–594.

Shiells, R. A., Falk, G., and Naghshineh, S. (1986). Iontophoretic study of the action of excitatory amino acids on rod horizontal cells of the dogfish retina. *Proc. R. Soc. London B* **227**, 121–135.

Slaughter, M. M., and Miller, R. F. (1981). 2-Amino-4-phosphonobutyric acid: A new pharmacological tool for retina research. *Science* **211**, 182–185.

Slaughter, M. M., and Miller, R. F. (1983). The role of excitatory amino acid transmitters in the mudpuppy retina: An analysis with kainic acid and *N*-methyl aspartate. *J. Neurosci.* **3**, 1701–1711.

Slaughter, M. M., and Miller, R. F. (1985a). Characterization of an extended glutamate receptor of the on bipolar neuron in the vertebrate retina. *J. Neurosci.* **5**, 224–233.

Slaughter, M. M., and Miller, R. F. (1985b). Identification of a distinct synaptic glutamate receptor on horizontal cells in mudpuppy retina. *Nature (London)* **314**, 96–97.

Sommer, B., and Seeburg, P. H. (1992). Glutamate receptor channels: Novel properties and new clones. *Trends Pharmacol. Sci.* **13**, 291–296.

Sommer, B., Keinänen, K., Verdoorn, T. A., Wisden, W., Burnashev, N., Herb, A., Köhler, M., Takagi, T., Sakmann, B., and Seeburg, P. H. (1990). Flip and flop: A cell-specific functional switch in glutamate-operated channels of the CNS. *Science* **249**, 1580–1585.

Sommer, B., Köhler, M., Sprengel, R., and Seeburg, P. H. (1991). RNA editing in brain controls a determinant of ion flow in glutamate-gated channels. *Cell* **67**, 11–20.

Sommer, B., Burnashev, N., Verdoorn, T. A., Keinanen, K., Sakmann, B., and Seeburg, P. H. (1992). A glutamate receptor channel with high affinity for domoate and kainate. *EMBO J.* **11**, 1651–1656.

Srinivasan, M. V., Laughlin, S. B., and Dubs, A. (1982). Predictive coding: A fresh view of inhibition in the retina. *Proc. R. Soc. London B* **216**, 427–459.

Stell, W. K., Kretz, R., and Lightfoot, D. O. (1982). Horizontal cell connectivity in goldfish. *In* "The S-Potential" (B. D. Drujan and M. Laufer, eds.), pp. 51–75. Liss, New York.

Stern, J. H., Kaupp, U. B., and Macleish, P. R. (1986). Control of the light-regulated current in rod photoreceptors by cyclic GMP, calcium, and 1-*cis*-diltiazem. *Proc. Natl. Acad. Sci. USA* **83**, 1163–1167.

Sugihara, H., Moriyoshi, K., Ishii, T., Masu, M., and Nakanishi, S. (1992). Structures and properties of seven isoforms of the NMDA receptor generated by alternative splicing. *Biochem. Biophys. Res. Commun.* **185**, 826–832.

Suzuki, S., and Kaneko, A. (1990). Identification of bipolar cell subtypes by protein kinase C-like immunoreactivity in the goldfish retina. *Visual Neurosci.* **5**, 223–230.

Tachibana, M. (1985). Permeability changes induced by L-glutamate in solitary retinal horizontal cells isolated from *Carassius auratus*. *J. Physiol.* **358**, 153–167.

Tachibana, M., and Kaneko, A. (1988). L-Glutamate-induced depolarization in solitary photoreceptors: A process that may contribute to the interaction between photoreceptors *in situ*. *Proc. Natl. Acad. Sci. USA* **85**, 5315–5319.

Tachibana, M., and Okada, T. (1991). Release of endogenous excitatory amino acids from ON-type bipolar cells isolated from the goldfish retina. *J. Neurosci.* **11**, 2199–2208.

Takahashi, K.-I., and Copenhagen, D. R. (1992). APB suppresses synaptic input to retinal horizontal cells in fish: A direct action on horizontal cells modulated by intracellular pH. *J. Neurophysiol.* **67**, 1633–1642.

Tanabe, Y., Masu, M., Ishii, T., Shigemoto, R., and Nakanishi, S. (1992). A family of metabotropic glutamate receptors. *Neuron* **8**, 169–179.

Tang, C., Dichter, M., and Morad, M. (1989). Quisqualate activates a rapidly inactivating high conductance ionic channel in hippocampal neurons. *Science* **243**, 1474–1477.

Toyoda, J.-I., Kujiraoka, T., and Fujimoto, M. (1982). The opponent color process and interaction of horizontal cells. *In* "The S-potential," (B. D. Drujan and M. Laufer, eds.), pp. 151–160. Liss, New York.

Trombley, P. Q., and Westbrook, G. L. (1992). L-AP4 inhibits calcium currents and synaptic transmission via a G-protein-coupled glutamate receptor. *J. Neurosci.* **12**, 2043–2050.

Trussell, L. O., Thio, L. L., Zorumski, F., and Fischbach, G. D. (1988). Rapid desensitization of glutamate receptors in vertebrate central neurons. *Proc. Natl. Acad. Sci. USA* **85**, 4562–4566.

Van Dop, C., Yamanaka, G., Steinberg, F., Sekura, R. D., Manclark, C. R., Stryer, L., and Bourne, H. R. (1984). ADP-ribosylation of transducin by pertussis toxin blocks the light-stimulated hydrolysis of GTP and cGMP in retinal photoreceptors. *J. Biol. Chem.* **259**, 23–36.

Verdoorn, T. A., Burnashev, N., Monyer, H., Seeburg, P. H., and Sakmann, B. (1991). Structural determinants of ion flow through recombinant glutamate receptor channels. *Science* **252**, 1715–1718.

Wang, L.-Y., Salter, M. W., and MacDonald, J. F. (1991). Regulation of kainate receptors by cAMP-dependent protein kinase and phosphatases. *Science* **253**, 1132–1135.

Wang, L.-Y., Taverna, F. A., Huang, X.-P., MacDonald, J. F., and Hampson, D. R. (1993). Phosphorylation and modulation of a kainate receptor (GluR6) by cAMP-dependent protein kinase. *Science* **259**, 1173–1175.

Werner, P., Voigt, M., Keinänen, K., Wisden, W., and Seeburg, P. H. (1991). Cloning of a putative high-affinity kainate receptor expressed predominantly in hippocampal CA3 cells. *Nature (London)* **351**, 742–744.

Wilson, M., Tessier-Lavigne, M., and Attwell, D. (1987). Noise analysis predicts at least four states of channels closed by glutamate. *Biophys. J.* **52**, 955–960.

Wong-Riley, M. T. T. (1974). Synaptic organization of the inner plexiform layer in the retina of the tiger salamander. *J. Neurocytol.* **3**, 1–33.

Wu, S., and Dowling, J. E. (1978). L-Aspartate: Evidence for a role in cone photoreceptor synaptic transmission in the carp retina. *Proc. Natl. Acad. Sci. USA* **75**, 5202–5209.

Yamashita, M., and Wässle, H. (1991). Responses of rod bipolar cells isolated from the rat retina to the glutamate agonist 2-amino-4-phosphonobutyric acid (APB). *J. Neurosci.* **11**, 2372–2382.

Yang, X.-L., and Wu, S. M. (1989). Effects of CNQX, APB, PDA, and kynurenate of horizontal cells of the tiger salamander retina. *Visual Neurosci.* **3**, 207–212.

Yau, K. W., and Baylor, D. A. (1989). Cyclic GMP-activated conductance of retinal photoreceptor cells. *Annu. Rev. Neurosci.* **12**, 289–327.

Zimmerman, A. L., and Baylor, D. A. (1986). Cyclic GMP-sensitive conductance of retinal rods consists of aqueous pores. *Nature (London)* **321**, 70–72.

Molecular Characterization of GABA-Gated Chloride Ion Channels from Complex and Simple Nervous Systems

MARK G. DARLISON

I. Introduction

Receptors for γ-aminobutyric acid (GABA) can be divided into two major classes. GABA type A (GABA$_A$) receptors, the topic of this chapter, possess an integral chloride-selective channel that is rapidly opened (on the millisecond time scale) on agonist binding. Although how the interaction of GABA with this receptor results in channel opening is not yet clear, researchers generally assume that binding causes a change in the conformation of the protein. In contrast, GABA type B (GABA$_B$) receptors transduce the signal of GABA binding via a guanine nucleotide-binding protein. GABA$_A$ receptors are found not only within the vertebrate central nervous system, where they are widely distributed and play a fundamental role in inhibitory neurotransmission, but also within the muscles and nervous systems of invertebrate species. Indeed, the first evidence in support of GABA acting as an inhibitory neurotransmitter came from studies on the crayfish, in which bovine brain extracts that contained this amino acid (Bazemore et al., 1957) were shown to block the generation of impulses by the stretch receptor neuron.

The observation that GABA$_A$ receptors had a tightly associated binding site for the clinically important benzodiazepine class of compounds led to the isolation and biochemical characterization of this ion channel/receptor complex. Affinity purification of the protein from bovine cerebral cortex, using the immobilized benzodiazepine Ro7-1986/1 (Sigel et al., 1983; Sigel and Barnard, 1984), suggested that the complex was a large hetero-oligomeric glycoprotein that contained two types of polypeptide named α (53,000 daltons) and β (57,000 daltons). Subsequently, two complementary DNA (cDNA) clones were isolated from bovine brain libraries using oligonucleotide probes that were designed on the basis of peptide sequences obtained from the receptor preparation (Schofield et al., 1987). These cDNAs were assigned to encode the α and β polypeptides; indeed, when the corresponding in vitro transcribed RNAs were coexpressed in Xenopus laevis oocytes, GABA-gated chloride ion channels that had many of the properties of native GABA$_A$ receptors formed in the cell membrane. However, the GABA responses of the heterologously expressed receptors were not strongly potentiated by benzodiazepines. Subsequently, Seeburg's group isolated a cDNA for a third type of GABA$_A$ receptor polypeptide, named $\gamma2$, that yielded receptors with a much more typical benzodiazepine pharmacology when it was coexpressed with α- and β-subunit cDNAs (Pritchett et al., 1989b).

Soon after the initial cloning of cDNAs for an α subunit (now called α1) and a β subunit (now called β1), researchers determined that additional sequence-related cDNAs existed that encoded isoforms of these polypeptides. Further, some of the first GABA_A receptor *in situ* hybridization experiments (Wisden *et al.*, 1988) revealed that different α-subunit mRNAs were located in overlapping but nonidentical regions of the vertebrate brain, pointing to the occurrence of GABA_A receptor subtypes that contained different combinations of subunits. Currently, the sequences of 16 vertebrate and 2 invertebrate GABA_A receptor-like polypeptides, each of which is encoded by a separate gene, have been reported. In vertebrates at least, a variety of receptor subtypes is believed to exist, some or all of which probably have different affinities for modulatory compounds, such as the benzodiazepines and certain steroids that occur in brain. Some knowledge is available on the amino acid residues at, or close to, the binding site for the former class of molecules (Pritchett and Seeburg, 1991; Wieland *et al.*, 1992). However, the reason for the multiplicity of GABA_A receptor subtypes that are predicted to exist from the results of a large number of *in situ* hybridization and immunological experiments is not understood, nor

is much known about the subunit compositions and stoichiometries of those subtypes. The differences (which may be pharmacological and/or electrophysiological) among the subtypes that are, presumably, biologically important to a given animal also remain obscure.

In this chapter, the current status of knowledge of the structures and properties of vertebrate GABA_A receptor-like polypeptides is outlined. Results that shed light on the question of receptor subtype subunit composition are discussed as well. Then current knowledge about invertebrate GABA-gated ion channel receptors, which may ultimately be easier to study since the nervous systems in which they occur are considerably simpler than those of vertebrates, is reviewed. Finally, the evolution of this gene family and the possible reason(s) for the apparently large number of vertebrate GABA_A receptor subtypes are addressed. To review all that is known about GABA_A receptors is clearly beyond the scope of this chapter. For areas not covered here (for example, the detailed pharmacologies of putative subtypes), the reader is directed elsewhere (Burt and Kamatchi, 1991; Lüddens and Wisden, 1991; Doble and Martin, 1992; Wisden and Seeburg, 1992; Chapter 21).

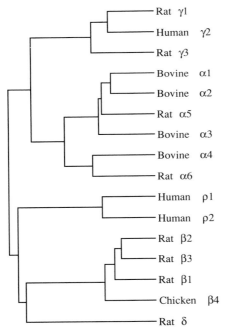

FIGURE 1 Relationships between the sequences of vertebrate GABA-gated ion channel receptor subunits. The mature sequences of the bovine α1, α2, α3, and α4, rat α5, α6, β1, β2, β3, γ1, γ3, and δ, human γ2, ρ1, and ρ2, and chicken β4 subunits (for the original citations, see Burt and Kamatchi, 1991; Cutting *et al.*, 1992; Herb *et al.*, 1992; and references cited therein) were aligned using the computer program GAP (Devereux *et al.*, 1984). The deduced relationships were plotted in the form of a dendrogram.

II. Vertebrate GABA_A Receptor-like Polypeptides

To date, the sequences of six α, three β, three γ, and one δ subunit have been reported for the mammal (see Burt and Kamatchi, 1991, and references cited therein). A fourth β polypeptide that exhibits between 73–77% sequence identity to the rat β1, β2, and β3 subunits has been described in the chicken (Bateson *et al.*, 1991a). That the β4 subunit does not exist in mammalian species seems unlikely since the sequences of other GABA_A receptor polypeptides present in the chicken and the mammal are very similar. For example, the mature sequences of the chicken α1, β2, and γ2 subunits (which probably associate to form a major GABA_A receptor subtype) are 98% (Bateson *et al.*, 1991b), 98% (R. J. Harvey and M. G. Darlison, unpublished data), and 96% (Glencorse *et al.*, 1990) identical to those of their rat counterparts. However, the possibility of species-specific polypeptides, or subtypes, must be considered. The existence of an ε subunit has been alluded to on a number of occasions (see, for example, Schofield, 1989), but the sequence of this has not been reported to date. In addition, cDNAs that encode two different ρ subunits have been isolated from a human retina library (Cutting *et al.*, 1991, 1992).

The classification of receptor subunits as α, β, γ, δ, ε, or ρ types is not founded on any functional criteria but is based solely on sequence similarity. Thus, a mature polypeptide from one subunit class typically displays 30–40% identity to polypeptides from other subunit classes (for example, the rat $\beta 1$ subunit is 38% identical to the rat $\gamma 1$ subunit). In contrast, any two subunits of the same type usually exhibit 60–80% sequence identity (for example, the bovine $\alpha 1$ subunit is 75% identical to the bovine $\alpha 3$ subunit). For comparison, the homologous subunits from two different mammalian species are typically more than 95% identical; even the homologs from mammalian and avian species are at least 90% identical. A dendrogram that depicts the relationships among the different GABA$_A$ receptor-like polypeptides is shown in Fig. 1. This figure shows that α subunits are more closely related to γ subunits, and that β subunits are more similar to the δ subunit.

Hydropathy analysis of any GABA$_A$ receptor polypeptide reveals the presence of four hydrophobic stretches of 20–25 amino acids (three in the middle of each sequence and one very close to the C terminus) which, in the absence of definitive structural data, are presumed to be α-helical membrane-spanning domains (termed M1–M4; Barnard *et al.*, 1987; Schofield *et al.*, 1987). This pattern of hydrophobicity and hydrophilicity is highly reminiscent of those seen in nicotinic acetylcholine receptor (nAChR) and glycine receptor subunits (see Barnard *et al.*, 1987) and of that observed in the sequence of a 5-hydroxytryptamine type 3 (5HT$_3$) receptor polypeptide (Maricq *et al.*, 1991). Further, significant sequence identity between GABA$_A$, nACh, glycine, and 5HT$_3$ receptor subunits is seen. Therefore, the predicted topology of GABA-gated ion channel polypeptides is based on that proposed originally for AChR subunits. Each subunit is assumed to have a 220- to 240-amino acid N-terminal extracellular domain, which contains two highly conserved cysteine residues (separated by 13 amino acids) that are presumed to form a disulfide bond, and a cytoplasmic loop that varies in length in different subunits between M3 and M4. Each subunit also possesses a signal sequence (von Heijne, 1986) at its N terminus that is believed to be removed during insertion of the polypeptide into the membrane.

As in the generally accepted model for the AChR (see Chapter 18), the lining of the chloride-selective channel of GABA$_A$ receptors is thought to be formed by the M2 helix of each subunit. This model also predicts that the binding site determinants for the various agonists, antagonists and modulators of the GABA$_A$ receptor are located in the long N-terminal extracellular domain. In support of this topology, the site-directed mutagenesis of certain residues within the N-terminal segment of α subunits has been shown to have a dramatic effect on the binding of benzodiazepines to heterologously expressed receptors (Pritchett and Seeburg, 1991; Wieland *et al.*, 1992).

Are all 17 sequences ($\alpha 1$–$\alpha 6$, $\beta 1$–$\beta 4$, $\gamma 1$–$\gamma 3$, δ, ε, $\rho 1$ and $\rho 2$) that have been identified by cDNA cloning genuine GABA$_A$ receptor polypeptides? For the α, β, and γ subunits, at least, this seems very likely. First, chemically determined peptide sequences from the purified receptor protein were found to be encoded by the $\alpha 1$- (Schofield *et al.*, 1987), $\alpha 2$- (Levitan *et al.*, 1988), and $\beta 2$- (Ymer *et al.*, 1989) subunit cDNAs. In addition, some of the peptide sequences are found within the predicted amino acid sequences of more than one subunit; for example, the sequence FGSYAYT (single-letter code) that was obtained by trypsin digestion of the gel-eluted α band of the receptor purified from bovine cerebral cortex (Schofield *et al.*, 1987) occurs within the bovine $\alpha 1$ (Schofield *et al.*, 1987), $\alpha 2$, and $\alpha 3$ (Levitan *et al.*, 1988) subunits. Second, a γ subunit clearly seems to be required, in conjunction with an α and a β subunit, to assemble a receptor that exhibits the full range of sensitivities to benzodiazepines (Pritchett *et al.*, 1989b). Further, as discussed in more detail later, the results of *in situ* hybridization (see, for example, Laurie *et al.*, 1992; Wisden *et al.*, 1992) and immunohistochemical (Fritschy *et al.*, 1992) experiments strongly indicate the colocalization of the $\alpha 1$, $\beta 2$, and $\gamma 2$ subunits in many areas of the vertebrate brain.

The δ subunit is also probably a component of a GABA$_A$ receptor, since the mRNA that encodes it has been shown to localize in brain (Shivers *et al.*, 1989) in regions that contain high-affinity binding sites for the GABA$_A$ receptor agonist muscimol. However, the δ-subunit mRNA is found in neuronal populations that are quite distinct, for example, from those that possess the $\gamma 2$-subunit mRNA. Heterologously expressed receptors incorporating the δ subunit instead of the $\gamma 2$ subunit (with an α and a β subunit) are insensitive to benzodiazepines (Shivers *et al.*, 1989). Thus, the δ polypeptide may be a component of a population of benzodiazepine-insensitive GABA$_A$ receptors (see Shivers *et al.*, 1989; Wisden *et al.*, 1992). Since no firm data have been published on the ε subunit, the possible contribution of this polypeptide to GABA$_A$ receptors remains unknown.

Evidence that the two ρ subunits are also components of true GABA$_A$ receptor subtypes is less compelling. In contrast to the situation for the vertebrate α, β, γ, and δ subunits, the $\rho 1$ subunit efficiently forms homo-oligomeric GABA-gated chloride channels (Cutting *et al.*, 1991). However, these receptors are insensitive to the GABA$_A$ receptor antagonist bicuculline and to benzodiazepines (Shimada *et al.*, 1992). Since the ρ-subunit genes are expressed in retina, the encoded polypeptides may represent components of one or more bicuculline-resistant GABA receptor complexes (GABA$_C$ receptors) that have been pharmacologically characterized in visual pathways (see Shimada *et al.*, 1992). The nature of any polypeptides that associate with the ρ1 and ρ2 subunits is currently unknown.

Although, as noted earlier, the detailed pharmacology of GABA$_A$ receptors formed by expressing various combinations of cDNAs will not be discussed in detail here, questioning the function(s) of the different subunit types is appropriate. Of greatest importance, perhaps, is the issue of the polypeptide(s) on which the GABA binding site(s) is/are located. Much conflicting data on this topic is available. For instance, photoaffinity-labeling studies on the receptor purified from bovine brain, using muscimol (Casalotti *et al.*, 1986), indicated that the GABA site resided on a component in the β band, suggesting a β subunit. Consistent with this result is the observation (discussed later) that the expression of an invertebrate (molluscan) GABA$_A$ receptor β-subunit cDNA yields homo-oligomeric GABA-gated channels (Harvey *et al.*, 1991). However, functional vertebrate hetero-oligomeric receptors comprising an α and a β subunit (Schofield *et al.*, 1987; Levitan *et al.*, 1988) or an α and a γ subunit (Verdoorn *et al.*, 1990) can be generated by the coexpression of the appropriate cDNAs, indicating that the GABA binding site may be located on α subunits. In addition, the presence of different α-subunit types in heterologously expressed complexes (Levitan *et al.*, 1988; Sigel *et al.*, 1990) has been shown to produce receptors that have different apparent affinities for GABA. Since most, if not all, GABA$_A$ receptor subunits can form homo-oligomeric receptors, albeit with low efficiency (see, for example, Blair *et al.*, 1988), all subunits are likely to have a GABA site, even if it is only of low affinity on some polypeptides. Perhaps the correct high-affinity agonist sites are formed (within a single subunit or between subunits) only when those combinations of polypeptides that occur in a native receptor subtype associate.

Our knowledge of the determinants of the benzodiazepine binding site on vertebrate GABA$_A$ receptors is greater than that of the channel-activating site. As mentioned previously, the presence of a γ subunit in the receptor complex is a prerequisite for the formation of high-affinity benzodiazepine binding sites (Pritchett *et al.*, 1989b). However, binding studies on heterologously expressed receptors that incorporate a β subunit and the γ2 subunit with different α subunits reveal (Pritchett *et al.*, 1989a) a dramatic difference

in their response to certain benzodiazepines and β-carbolines. For example, α1-subunit-containing receptors have a higher affinity for CL-218872 [3-methyl-6-(3-trifluoromethylphenyl)-1,2,4-triazolo-(4,3-*b*)pyridazine], 2-oxoquazepam, and methyl-β-carboline-3-carboxylate, typical of the so-called benzodiazepine type I receptors (see Doble and Martin, 1992), than receptors containing the α2 or the α3 subunit; the latter pharmacologically resemble benzodiazepine type II receptors. Pritchett and Seeburg (1991) exploited this difference, generated chimeric α1/α3 subunits and, subsequently, point mutants, and showed that a determinant of the benzodiazepine binding site (Gly 200 in rat α1) resides on α subunits. A second residue that appears to be essential for the high-affinity binding of benzodiazepine agonists also has been located on α subunits (His 101 in rat α1; Wieland *et al.*, 1992).

III. Subunit Compositions of Native GABA$_A$ Receptor Subtypes

Functional GABA-gated chloride channels can be assembled from a variety of subunit combinations. If GABA$_A$ receptors are assumed, by analogy with the nAChR and the glycine receptor, to be pentameric (which has not been proven), then a very large number of possible subtypes can be generated from the different vertebrate receptor polypeptides that are currently known (i.e., 6 α, 4 β, 3 γ, and 1 δ), even if each receptor contains an α, a β, and a γ subunit. However, only a small fraction of all the possible combinations are likely to occur *in vivo*. Therefore, defining the number and compositions of native receptor subtypes is extremely important.

Although the number of different GABA$_A$ receptors remains unknown, several groups have begun to address the question of subtype composition by attempting to colocalize subunit mRNAs by *in situ* hybridization or by analyzing receptor polypeptide association by immunohistochemistry or immunoprecipitation with subunit-specific antibodies. For example, Seeburg's group (Laurie *et al.*, 1992; Wisden *et al.*, 1992) has made a thorough study of the locations of 13 GABA$_A$ receptor mRNAs (those that encode the α1–α6, β1–β3, γ1–γ3, and δ subunits) in the adult rat brain. Unfortunately, because many GABA$_A$ receptor genes are expressed widely within the mammalian central nervous system, mRNA distribution patterns frequently overlap. Thus, unequivocal statements about receptor subtype compositions are difficult to make. Nevertheless, certain combinations of subunit mRNAs have been shown to colocalize (see Table I).

TABLE I Possible Vertebrate GABA$_A$ Receptor Subunit Associations and Their Location[a]

Subunits	Species	Method of colocalization	Prominent brain regions	References
α1 and γ2[b]	Chicken	*In situ* hybridization	Optic tectum and cerebellum	Glencorse *et al.* (1991)
α1 and γ2 α2 and γ2 α3 and γ2	Bovine	Immunoaffinity purification and Western blotting	Cerebral cortex	Duggan *et al.* (1991)
α3 and γ2	Rat	Immunofluorescence	Brainstem	Fritschy *et al.* (1992)
α2 and β3	Rat	*In situ* hybridization	Amygdala and hypothalamus	Wisden *et al.* (1992)
α4 and δ	Rat	*In situ* hybridization	Forebrain	Wisden *et al.* (1992)
α5 and β1	Rat	*In situ* hybridization	Hippocampus	Wisden *et al.* (1992)
α1, β1, and γ2	Rat	*In situ* hybridization	Olfactory bulb, hippocampus, and dentate gyrus	Malherbe *et al.* (1990)
α1, β2, and γ2	Rat	*In situ* hybridization	Olfactory bulb and cerebellum	Laurie *et al.* (1992), Wisden *et al.* (1992)
α1, β2/β3,[c] and γ2	Bovine, rat	Immunoprecipitation	Whole brain	Benke *et al.* (1991)
α1, β2/β3, and γ2	Rat	Immunofluorescence	Olfactory bulb, forebrain, and cerebellum	Fritschy *et al.* (1992)
α2, β3, and γ2	Rat	*In situ* hybridization	Spinal cord motor neurons	Persohn *et al.* (1991), Wisden *et al.* (1991)
α3, β1, and γ2	Rat	*In situ* hybridization	Cerebral cortex and hippocampus	Knoflach *et al.* (1992)
α3, β2/β3, and γ2	Rat	Immunofluorescence	Granule cells of the olfactory bulb	Fritschy *et al.* (1992)
α1, α3, β2/β3, and γ2	Rat	Immunofluorescence	Forebrain and cerebellum	Fritschy *et al.* (1992)

[a] Examples of deduced vertebrate GABA$_A$ receptor subunit combinations are given; note that this list is not exhaustive.

[b] γ2 refers to the presence of the γ2S and/or the γ2L subunit since no attempt was made in any of these studies to distinguish between these two γ2 subunit variants (see text).

[c] β2/β3 refers to the presence of the β2 and/or the β3 subunit; the antibody used in these experiments was unable to discriminate between these two β polypeptides.

Further, in view of the observation (Laurie *et al.*, 1992) that relatively high levels of the α1-, α6-, β2-, β3-, γ2-, and δ-subunit mRNAs are found within the granule cells of the cerebellum, at least some neurons may contain more than one GABA$_A$ receptor subtype, thus complicating the mRNA and immunological studies.

One GABA$_A$ receptor subtype on which researchers generally agree comprises α1, β2, and γ2 subunits, although the subunit stoichiometry of this subtype remains unknown. However, the γ2 subunit has been shown to exist in two forms that differ by the absence (γ2S) or presence (γ2L) of 8 amino acids in the presumed intracellular loop region between M3 and M4 (Whiting *et al.*, 1990). This insertion contains a consensus sequence for phosphorylation by protein kinase C that is a substrate for phosphorylation when expressed as part of a fusion protein (Moss *et al.*, 1992). In addition, *in situ* hybridization localization of the γ2S- and γ2L-subunit mRNAs has revealed that,

although the two γ2-subunit transcripts are codistributed throughout the chick brain, certain nuclei (for example, the nucleus isthmi, pars magnocellularis, the nucleus isthmi, pars parvocellularis, and the nucleus solitarius; see Fig. 2) appear to contain predominantly the γ2S- or the γ2L-subunit mRNA (Glencorse *et al.*, 1992). Thus, assuming an "α1β2γ2 subtype" is oversimplifying the actual situation, since α1β2γ2S and α1β2γ2L, as well as α1β2γ2Sγ2L, receptors may occur *in vivo*.

Immunological studies have also contributed to our knowledge of GABA$_A$ receptor subtype polypeptide combinations (see Table I). For example, Fritschy *et al.* (1992) showed, using subunit-specific antibodies and triple immunofluorescence staining, the colocalization of the α1, γ2, and β2 and/or β3 subunits in neurons; unfortunately, the anti-β-subunit antibody used in these studies was unable to discriminate between the β2 and β3 polypeptides. Possible additional

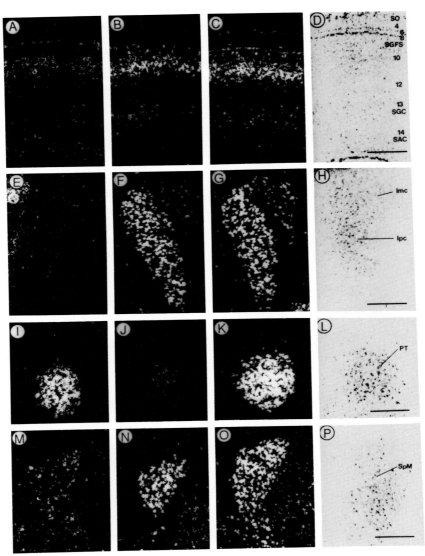

FIGURE 2 *In situ* hybridization localization of two forms of the γ2-subunit mRNA in the chick optic tectum, and in two tectal and two pretectal nuclei. Dark-field photomicrographs of the laminae of the optic tectum (A–C), the nucleus isthmi, pars magnocellularis and the nucleus isthmi, pars parvocellularis (E–G), the nucleus pretectalis (I–K), and the nucleus spiriformis medialis (M–O) are shown. Light-field photomicrographs D, H, L, and P are shown for sections B, E, J, and M, respectively. Sections were hybridized with an oligonucleotide specific for the γ2S-subunit mRNA (A, E, I, and M), an oligonucleotide specific for the γ2L-subunit mRNA (B, F, J, and N), or an oligonucleotide that recognizes both the γ2S- and the γ2L-subunit mRNAs (C, G, K, and O). The layers of the optic tectum are labeled (in D) according to the numerical system of Cajal (see Hunt and Brecha, 1984). Bars: 243 μm (A–D, E–H, M–P) or 309 μm (I–L). Abbreviations: Imc, nucleus isthmi, pars magnocellularis; Ipc, nucleus isthmi, pars parvocellularis; PT, nucleus pretectalis; SAC, stratum album centrale; SGC, stratum griseum centrale; SGFS, stratum griseum et fibrosum superficiale; SO, stratum opticum; SpM, nucleus spiriformis medialis. Reproduced from Glencorse *et al.* (1992) by permission of Oxford University Press.

complexity of GABA$_A$ receptor compositions has been revealed by immunoaffinity purification and immuno-precipitation experiments, which have indicated that more than one type of α subunit can occur in a single receptor molecule. Thus, the coexistence of α1 and α2, α2 and α3 (Duggan *et al.*, 1991), and α1 and α3 (Duggan *et al.*, 1991; McKernan *et al.*, 1991) subunit pairs has been shown in a minority of GABA$_A$ receptors. However, other workers (Lüddens *et al.*, 1991) have suggested that this apparent co-association of different α subunits may be an artifact of receptor aggregation. To complicate matters further, the coexistence of two α subunits (α1 and α3) with a β and a γ subunit on individual neurons has been shown by immunofluorescence staining (Fritschy *et al.*, 1992).

The preceding discussion should illustrate that defining the subunit compositions and stoichiometries of vertebrate GABA$_A$ receptor subtypes will be a long and difficult process. Investigation of the detailed pharmacologies and biological function(s) of the different subtypes that occur *in vivo*, which are prerequisites for the rational design of novel classes of drugs that are specific for particular subtypes and of therapeutic value in humans, is some way off. In view of these difficulties, this laboratory decided approximately three years ago to begin to study GABA$_A$ receptors in an animal with a much less complex nervous system. The fresh water mollusc *Lymnaea stagnalis*, which has been the subject of intensive neurophysiological investigation, was chosen. The nervous system of this organism comprises only 10,000 to 20,000 neurons that are clustered in a small number of major ganglia. Thus, fewer GABA$_A$ receptor subunits and receptor subtypes might be predicted to exist in *Lymnaea* (and, indeed, in other invertebrates) than in vertebrates. In addition, since many molluscan neurons are large and, hence, easily identifiable, electrophysiologically recording from them is a relatively simple matter. Coupled with the amenability of this nervous system to *in situ* hybridization and immunocytochemistry, it should be possible to identify receptors or subtypes on specific neurons and directly compare their electrophysiological and pharmacological properties with those of the corresponding receptor(s) reconstituted in a heterologous system from cloned cDNAs. This approach should also permit an investigation of the biological function(s) of GABA-gated ion channel receptors in this animal.

IV. Invertebrate GABA$_A$ Receptor-like Polypeptides

To date, the complete sequences of only two GABA-gated receptor polypeptides have been reported, those of a *Lymnaea* subunit that exhibits strong similarity (approximately 50% identity) to vertebrate GABA$_A$ receptor β subunits (Harvey *et al.*, 1991) and of a *Drosophila* subunit encoded by the *Rdl* gene (ffrench-Constant *et al.*, 1991; ffrench-Constant and Rocheleau, 1992) that displays a similar level of identity (33–40%) to vertebrate GABA$_A$ and glycine receptor polypeptides (see Darlison, 1992). Both invertebrate subunits

form functional homo-oligomeric (presumably pentameric) GABA-gated channels when *in vitro* transcribed RNAs from the corresponding cDNAs are expressed independently in *Xenopus* oocytes. However, whereas the *Lymnaea* β-subunit receptor is sensitive to the vertebrate GABA$_A$ receptor antagonist bicuculline (Harvey *et al.*, 1991) and the chloride channel blocker picrotoxin (Zaman *et al.*, 1992a), the *Drosophila* receptor is picrotoxin sensitive but bicuculline insensitive (R. H. ffrench-Constant, personal communication).

Although only two invertebrate ionotropic GABA receptor polypeptide sequences have been reported to date, cDNA cloning studies have indicated the existence of several additional subunits. For example, one full-length (M. L. Hutton, R. J. Harvey, F. G. P. Earley, E. A. Barnard, and M. G. Darlison, unpublished data) and two partial (R. J. Harvey, M. L. Hutton, and M. G. Darlison, unpublished data) cDNAs that encode three further *Lymnaea* GABA$_A$ receptor-like subunits have been isolated. In addition, partial clones for two other *Drosophila* GABA$_A$ receptor-like polypeptides have been obtained. Although the sequence encoded by one of these clones is quite dissimilar from the four *Lymnaea* sequences, the other (Sattelle *et al.*, 1991) may encode the species homolog of the *Lymnaea* β subunit (see Darlison, 1992). Thus, at least six different invertebrate GABA$_A$ receptor-like sequences have been identified by molecular cloning approaches. That each invertebrate species possesses its own distinct set of GABA receptor polypeptides seems unlikely. Also doubtful is that the four subunits identified in *Lymnaea* represent the total number of GABA receptor polypeptides in that organism, or that these all associate together in one receptor complex. On the basis of the available evidence, therefore, invertebrates (like vertebrates) are likely to have receptor subtypes.

Since a GABA$_A$ receptor β-like subunit clearly exists in invertebrate species, the question arises as to whether α-like, γ-like, and δ-like polypeptides also occur in these animals. Although there is currently no firm data on this, the existence of α-like subunits seems likely because it has been shown (Harvey *et al.*, 1991) that the *Lymnaea* β subunit can functionally associate with the bovine α1 subunit. The resultant hybrid receptors yield much larger currents in response to a given concentration of GABA than those elicited by the same concentration of agonist from *Lymnaea* β-subunit or bovine α1-subunit homo-oligomeric receptors. However, the occurrence of γ-like subunits is in some doubt. First, as stated earlier, at least one γ subunit is required in a GABA$_A$ receptor complex in vertebrate species for it to display the full range of responses to centrally acting benzodiazepines. However, although invertebrate ionotropic

GABA receptors have an associated benzodiazepine binding site, the pharmacology of this site resembles that of the so-called mammalian peripheral benzodiazepine binding site rather than that of the site that is an integral part of vertebrate GABA$_A$ receptors (see, for example, Lummis and Sattelle, 1986; Robinson *et al.*, 1986). Further, it has been shown (Zaman *et al.*, 1992a) that *Lymnaea* homo-oligomeric β-subunit receptors can be activated directly by benzodiazepines (for example, Ro5-4864, also called 4'-chlorodiazepam) that are selective for the mammalian peripheral site, but not by those that are active at central benzodiazepine binding sites (for example, flunitrazepam and chlordiazepoxide). Second, whereas heterologously expressed vertebrate GABA$_A$ receptors that lack or contain a γ subunit are sensitive or insensitive, respectively, to zinc (Draguhn *et al.*, 1990), GABA receptors on lobster muscle are zinc sensitive (Smart and Constanti, 1982), perhaps suggesting that they do not possess a γ-like polypeptide.

V. Evolutionary Aspects

The structures of three vertebrate GABA$_A$ receptor subunit genes have been described to date, namely those for the mouse δ subunit (Sommer *et al.*, 1990), the human $\beta1$ subunit (Kirkness *et al.*, 1991), and the chicken $\beta4$ subunit (Lasham *et al.*, 1991). These studies, and investigation of other human GABA$_A$ receptor genes (A. A. Hicks and M. G. Darlison, unpublished data), have revealed a very strong conservation of the intron/exon organization, and that the genes themselves are large (usually of the order of 50–100 kb) with the exception of that for the mouse δ subunit, which is 13 kb (Sommer *et al.*, 1990). Typically, the coding region of each gene is split by 8 introns and, hence, is encoded by 9 exons (see Fig. 3). The location of the furthest 5' intron is somewhat variable; however, it usually interrupts the sequence that encodes the N-terminal part of the mature polypeptide, close to the site of cleavage of the signal peptide. The sequence encoding the remainder of the 220- to 240-amino acid putative extracellular domain is split by five introns (the cysteine loop region is encoded by a single exon), the positions of which are highly conserved in the different genes. The next 3' intron is found in the same position in all vertebrate GABA$_A$ receptor subunit genes, where it interrupts the eighth codon of the DNA segment that encodes M2. The furthest 3' intron, the position of which is also variable, splits the sequence that encodes the intracellular loop region between M3 and M4. However, exact comparisons of the positions of this intron are difficult

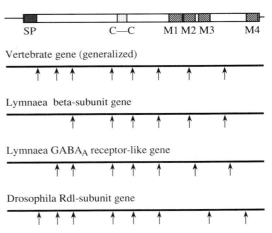

FIGURE 3 Comparison of the organization of vertebrate and invertebrate ionotropic GABA receptor genes. The relative positions of the introns (denoted by arrows) that interrupt the coding sequences of vertebrate GABA$_A$ receptor subunit genes (see Lasham *et al.*, 1991) are compared with those in a *Lymnaea* β-subunit gene (Harvey *et al.*, 1991), a *Lymnaea* GABA$_A$ receptor-like subunit gene (M. L. Hutton, R. J. Harvey, F. G. P. Earley, E. A. Barnard, and M. G. Darlison, unpublished data), and the *Drosophila* GABA$_A$ receptor-like *Rdl*-subunit gene (ffrench-Constant and Rocheleau, 1992). These organizations are shown beneath a schematic representation of a GABA$_A$ receptor subunit. Note that the complete structures of the *Lymnaea* β-subunit gene and the *Lymnaea* GABA$_A$ receptor-like subunit gene have not been determined (see text). SP, signal peptide; C–C, dicysteine loop; M1–M4, four putative membrane-spanning domains.

to make because the lengths of the intracellular domains vary among different GABA$_A$ receptor polypeptides. Hence, the relative positions of this intron are dependent on how any two amino acid sequences are aligned. In view of the evident conservation of gene structure, and the strong sequence similarity of the different receptor polypeptides, the family of GABA$_A$ receptor subunit genes is likely to have arisen by duplication of a common ancestral gene (see Lasham *et al.*, 1991).

Note, however, that a variant of the intron/exon organization of vertebrate GABA$_A$ receptor subunit genes has been observed. In all $\gamma2$-subunit genes that have been investigated, an additional exon of 24 base pairs occurs between the eighth and ninth exons (see Whiting *et al.*, 1990). The corresponding sequence in the primary gene transcript is alternatively spliced to yield two $\gamma2$ subunits, $\gamma2S$ and $\gamma2L$, that lack or contain an extra 8 amino acids, respectively. Similarly, it has recently been found (R. J. Harvey, M. A. Chinchetru, and M. G. Darlison, unpublished data) that the chicken $\beta2$-subunit gene contains an additional exon (of 51 base pairs), in the same relative position as that found in $\gamma2$-subunit genes, that results in two polypeptides (β2S and β2L) that differ by the absence or presence of 17 amino acids.

To date, the complete structure of only one invertebrate ionotropic GABA receptor gene has been reported (ffrench-Constant and Rocheleau, 1992), namely that which encodes the *Drosophila Rdl* subunit. Although the organization of this gene is strikingly similar to that of vertebrate GABA_A receptor polypeptide genes, the relative locations of two introns are significantly different. As in vertebrate genes, the *Rdl*-subunit gene contains an intron in the sequence that encodes the N-terminal portion of the mature polypeptide and five introns that split the sequence that specifies the large extracellular domain (Fig. 3). These introns are all found in the same relative positions as those that occur in vertebrate GABA_A receptor genes. However, no intron is found in the DNA sequence that encodes the M2 domain of the *Rdl* subunit; instead, an intron interrupts the sequence encoding the C-terminal portion of M3. In addition, irrespective of the fact that the intracellular loop of the *Rdl* polypeptide is very long (217 amino acids; ffrench-Constant *et al.*, 1991), the intron that splits the DNA that encodes this region is located much farther 3' relative to the intron in vertebrate GABA_A receptor genes. Indeed, the intron in the *Rdl*-subunit gene occurs only 45 nucleotides 5' of the first codon for the M4 domain; this position is very similar to that found for the only intron that splits the sequence encoding the long intracellular loop of neuronal nAChR subunits (Nef *et al.*, 1988).

Partial structures have been determined for the *Lymnaea* GABA_A receptor β-subunit gene (Harvey *et al.*, 1991) and for one other *Lymnaea* GABA_A receptor-like subunit gene (M. L. Hutton, R. J. Harvey, F. G. P. Earley, E. A. Barnard, and M. G. Darlison, unpublished data). For the β-subunit gene, six intron positions have been established (Fig. 3); the five exons defined by these introns reside on two nonoverlapping genomic clones, each approximately 20 kb in length. Four of the introns interrupt the sequence that encodes the N-terminal extracellular domain, one splits the region that determines the M2 domain, and the sixth disrupts the sequence encoding the intracellular loop region. The relative locations of these introns are either the same as, or very similar to, those found in vertebrate GABA_A receptor genes. For the *Lymnaea* GABA_A receptor-like gene, seven intron positions have been determined (Fig. 3). Five of these split the sequence encoding the extracellular domain in the same relative positions as those found in vertebrate GABA_A receptor genes. One intron interrupts the sequence that specifies the intracellular loop region; exact comparison of the relative position of this intron

with that of the corresponding intron in vertebrate genes is difficult because the intracellular loop region of the *Lymnaea* subunit is significantly shorter (only 67 amino acids) than the equivalent region in vertebrate polypeptides. Nevertheless, its location is more similar to that in vertebrate GABA_A receptor genes and in the *Lymnaea* β-subunit gene than to that in the *Drosophila Rdl*-subunit gene. The seventh intron that has been identified in the *Lymnaea* GABA_A receptor-like gene interrupts the sequence that encodes the M2 segment, but is in a position that is more 3' than the intron that occurs in other GABA_A receptor subunit genes.

The results obtained to date on invertebrate GABA_A receptor-like genes and their encoded polypeptides suggest that these and vertebrate GABA_A receptor subunit genes probably evolved by duplication of a common ancestral gene. However, the positions of the introns that split the DNA sequence that encodes the large extracellular domain, which is believed to contain the binding sites for the various ligands active at these receptors, appear to be more highly conserved than those that interrupt the sequences encoding M2, M3, and the intracellular loop. The significant degree of identity found among the sequences of the extracellular domains of both vertebrate and invertebrate ionotropic GABA receptors, and the conservation of the positions of the introns that interrupt the sequences encoding these domains in different subunits, suggest that this portion of each polypeptide plays a fundamental role in receptor activity. This similarity may also indicate that each of the five exons encodes a specific structural and/or functional domain (see Blake, 1979; Gilbert *et al.*, 1986). However, finding that the intron that is present in the same relative position in the sequences encoding the M2 regions of all vertebrate GABA_A receptor genes analyzed (Sommer *et al.*, 1990; Kirkness *et al.*, 1991; Lasham *et al.*, 1991; A. A. Hicks and M. G. Darlison, unpublished data) is not conserved in two of the three invertebrate GABA_A receptor-like genes that have been studied is somewhat surprising. The reason for this divergence is unclear, but it may reflect a stronger selective pressure on the position of this intron and, perhaps, on its flanking coding sequences in vertebrate than in invertebrate species.

The presence of an extra exon in the same relative position in GABA_A receptor γ2-subunit genes and in the chicken β2-subunit gene suggests that these additional sequences may have been inserted into the basic GABA_A receptor gene structure after the duplication of the presumed ancestral gene. Alternatively, the primordial gene, or one of its very early descendants, may have contained this extra exon and it has been retained or lost during subsequent divergent evolution. If the latter is true, the existence of as yet undetected exons in other GABA$_A$ receptor genes remains an intriguing possibility.

Although all the genes that encode the known vertebrate and invertebrate GABA-gated ion channel polypeptides probably arose by duplication of a common ancestral gene followed by divergent evolution, the nature of the primordial gene remains obscure. One impetus for isolating cDNA clones for ligand-gated ion channels from relatively simple invertebrate species is to determine whether these organisms possess homo-oligomeric receptors. However, regardless of the invertebrate that has been studied (whether a mollusc, an insect, or a nematode), multiple genes for a given receptor (for example, the GABA$_A$ receptor and the nAChR) have always been found. Thus, the ancestral gene for each ion channel/receptor is likely to no longer exist, without related sequences, in any invertebrate. The most obvious interpretation of the dendrogram showing the relatedness of the different vertebrate GABA$_A$ receptor-like polypeptides (Fig. 1) is that the ancestral gene probably duplicated and diverged to give rise to an α-subunit-like and a β-subunit-like gene. The γ- and δ-subunit genes are likely to have arisen later, from the α-subunit and β-subunit genes, respectively. This explanation would coincide with the suggestion, discussed earlier, that invertebrate ionotropic GABA receptors may not have a γ-like subunit. Thus, the original GABA$_A$ receptor subunit may have had some of the properties of what we now recognize as α and β subunits.

VI. Concluding Remarks

One of the most important questions that remains unanswered is, why does the vertebrate brain apparently contain so many different GABA$_A$ receptor subtypes? Early studies on reconstituted α- plus β-subunit receptors (Levitan et al., 1988) indicated that different α subunits imparted different GABA sensitivities on the receptor. Subsequent experiments, performed in the presence of the γ2 subunit (Sigel et al., 1990), also suggested that GABA affinity could be altered by changing the type of α subunit in the complex, although no thorough study on this topic has been done. In situ hybridization studies on the localization of mRNAs in brain suggest that it is possible that a particular neuron could contain more than one GABA$_A$ receptor subtype. If this is indeed the case, and the different subtypes occur together in the postsynaptic membrane, then the effect of the presynaptic release of GABA would depend on the amount of

GABA released as well as on the affinities of the receptor subtypes for the neurotransmitter. Alternatively, the multiplicity of GABA$_A$ receptors in the vertebrate brain might be necessary, at least in part, to permit the differential cellular compartmentalization of subtypes. For example, a particular neuron may have a requirement for two receptors that have different electrophysiological or pharmacological properties, one located on the dendrites and the other on the soma.

Although GABA$_A$ receptors have been shown to have binding sites for a variety of classes of compound (for example, benzodiazepines, barbiturates, β-carbolines, and certain convulsants), these molecules do not naturally occur in the vertebrate brain. However, the sites to which these drugs bind may be used in vivo by as yet undiscovered endogenous modulators. Indeed, that the role of the GABA$_A$ receptor extracellular domain, which comprises more than 1000 amino acid residues, is merely to bind the single amino acid GABA is unlikely. By analogy with the nAChR (see Verrall and Hall, 1992), some of these residues probably play a role in subunit–subunit recognition. However, others may be involved in the binding of endogenous molecules. One class of compounds present in brain that might bind with differing affinity to GABA$_A$ receptor subtypes are steroids some of which (for example, 3α-hydroxy-5α-pregnan-ol-20-one and 3α-pregnan-5α-21-diol-20-one) have been shown to potentiate the response to GABA whereas others (for example, pregnenolone sulfate and dehydroepiandrosterone sulfate) suppress the response (see Lambert et al., 1987; Zaman et al., 1992b, and references cited therein). In this context, note that GABA$_A$ receptors, expressed in a heterologous system from cloned cDNAs, that contain different combinations of subunits can be modulated to different extents by pregnenolone sulfate (Zaman et al., 1992b).

Much debate has taken place over the possible existence of endogenous benzodiazepines that might modulate vertebrate GABA$_A$ receptors in a manner analogous to the mechanism of benzodiazepines used clinically. Although benzodiazepines such as diazepam and nordiazepam, that are found in low concentrations in brain are thought to come indirectly from the food chain, multiple small molecular weight, nonpeptide substances that can act as positive allosteric modulators of GABA$_A$ receptors, and do not appear to be diazepam or nordiazepam, have recently been purified from brain (Rothstein et al., 1992). Evidence in support of the existence of such endogenous molecules comes from studies on benzodiazepines such as the antagonist flumazenil, which is itself devoid of biological activity but can, for example, induce panic attacks in patients with panic disorder (Nutt *et al.*, 1990). This effect may be due to the blockade of the action of naturally occurring anxiolytic benzodiazepines, that are present at lower levels in affected individuals than in control subjects, by the antagonist.

Other results have shown that the *Torpedo* nAChR can be directly activated by the acetylcholinesterase inhibitor (−)physostigmine (or D-eserine) at a site different from the one at which acetylcholine binds (Okonjo *et al.*, 1991). In addition, the expressed *Lymnaea* homo-oligomeric β-subunit GABA$_A$ receptor can be directly gated by certain benzodiazepines, such as Ro5-4864 and diazepam. This effect is insensitive to the vertebrate GABA$_A$ receptor antagonist bicuculline (Zaman *et al.*, 1992a), which does block GABA-evoked currents through this channel (Harvey *et al.*, 1991). These data indicate that ligand-gated ion channels may harbor more than one type of channel-activating site. These results beg the question, do agonists for such a second site occur *in vivo*?

Molecular biology has taught us a great deal about the number and sequences of the different GABA$_A$ receptor polypeptides. It has also provided us with the tools to investigate many aspects of receptor structure, function, and expression. Hopefully, in the near future, these tools will also help us understand why so many GABA$_A$ receptor polypeptides and subtypes exist in brain.

Acknowledgments

The author apologizes to all those whose work on GABA$_A$ receptors, for reasons of space, was not reviewed or cited here. I thank R. J. Harvey for comments on the manuscript.

References

Barnard, E. A., Darlison, M. G., and Seeburg, P. (1987). Molecular biology of the GABA$_A$ receptor: The receptor/channel superfamily. *Trends Neurosci.* **10,** 502–509.

Bateson, A. N., Lasham, A., and Darlison, M. G. (1991a). γ-Aminobutyric acid$_A$ receptor heterogeneity is increased by alternative splicing of a novel β-subunit gene transcript. *J. Neurochem.* **56,** 1437–1440.

Bateson, A. N., Harvey, R. J., Wisden, W., Glencorse, T. A., Hicks, A. A., Hunt, S. P., Barnard, E. A., and Darlison, M. G. (1991b). The chicken GABA$_A$ receptor α1 subunit: cDNA sequence and localization of the corresponding mRNA. *Mol. Brain Res.* **9,** 333–339.

Bazemore, A. W., Elliott, K. A. C., and Florey, E. (1957). Isolation of factor I. *J. Neurochem.* **1,** 334–339.

Benke, D., Mertens, S., Trzeciak, A., Gillessen, D., and Mohler, H. (1991). GABA$_A$ receptors display association of γ2-subunit with α1- and β2/3-subunits. *J. Biol. Chem.* **266,** 4478–4483.

Blair, L. A. C., Levitan, E. S., Marshall, J., Dionne, V. E., and Barnard, E. A. (1988). Single subunits of the GABA$_A$ receptor form ion channels with properties of the native receptor. *Science* **242,** 577–579.

Blake, C. C. F. (1979). Exons encode protein functional units. *Nature (London)* **277,** 598.

Burt, D. R., and Kamatchi, G. L. (1991). GABA$_A$ receptor subtypes: from pharmacology to molecular biology. *FASEB J.* **5,** 2916–2923.

Casalotti, S. O., Stephenson, F. A., and Barnard, E. A. (1986). Separate subunits for agonist and benzodiazepine binding in the γ-aminobutyric acid$_A$ receptor oligomer. *J. Biol. Chem.* **261,** 15013–15016.

Cutting, G. R., Lu, L., O'Hara, B. F., Kasch, L. M., Montrose-Rafizadeh, C., Donovan, D. M., Shimada, S., Antonarakis, S. E., Guggino, W. B., Uhl, G. R., and Kazazian, H. H., Jr. (1991). Cloning of the γ-aminobutyric acid (GABA) ρ_1 cDNA: A GABA receptor subunit highly expressed in the retina. *Proc. Natl. Acad. Sci. USA* **88,** 2673–2677.

Cutting, G. R., Curristin, S., Zoghbi, H., O'Hara, B., Seldin, M. F., and Uhl, G. R. (1992). Identification of a putative γ-aminobutyric acid (GABA) receptor subunit rho$_2$ cDNA and colocalization of the genes encoding rho$_2$ (GABRR2) and rho$_1$ (GABRR1) to human chromosome 6q14–q21 and mouse chromosome 4. *Genomics* **12,** 801–806.

Darlison, M. G. (1992). Invertebrate GABA and glutamate receptors: Molecular biology reveals predictable structures but some unusual pharmacologies. *Trends Neurosci.* **15,** 469–474.

Devereux, J., Haeberli, P., and Smithies, O. (1984). A comprehensive set of sequence analysis programs for the VAX. *Nucleic Acids Res.* **12,** 387–395.

Doble, A., and Martin, I. L. (1992). Multiple benzodiazepine receptors: No reason for anxiety. *Trends Pharmacol. Sci.* **13,** 76–81.

Draguhn, A., Verdorn, T. A., Ewert, M., Seeburg, P. H., and Sakmann, B. (1990). Functional and molecular distinction between recombinant rat GABA$_A$ receptor subtypes by Zn^{2+}. *Neuron* **5,** 781–788.

Duggan, M. J., Pollard, S., and Stephenson, F. A. (1991). Immunoaffinity purification of GABA$_A$ receptor α-subunit iso-oligomers. Demonstration of receptor populations containing α1α2, α1α3, and α2α3 subunit pairs. *J. Biol. Chem.* **266,** 24778–24784.

ffrench-Constant, R. H., and Rocheleau, T. (1992). *Drosophila* cyclodiene resistance gene shows conserved genomic organization with vertebrate γ-aminobutyric acid$_A$ receptors. *J. Neurochem.* **59,** 1562–1565.

ffrench-Constant, R. H., Mortlock, D. P., Shaffer, C. D., MacIntyre, R. J., and Roush, R. T. (1991). Molecular cloning and transformation of cyclodiene resistance in *Drosophila*: An invertebrate γ-aminobutyric acid subtype A receptor locus. *Proc. Natl. Acad. Sci. USA* **88,** 7209–7213.

Fritschy, J.-M., Benke, D., Mertens, S., Oertel, W. H., Bachi, T., and Möhler, H. (1992). Five subtypes of type A γ-aminobutyric acid receptors identified in neurons by double and triple immunofluorescence staining with subunit-specific antibodies. *Proc. Natl. Acad. Sci. USA* **89,** 6726–6730.

Gilbert, W., Marchionni, M., and McKnight, G. (1986). On the antiquity of introns. *Cell* **46,** 151–153.

Glencorse, T. A., Bateson, A. N., and Darlison, M. G. (1990). Sequence of the chicken GABA$_A$ receptor γ2-subunit cDNA. *Nucleic Acids Res.* **18,** 7157.

Glencorse, T. A., Bateson, A. N., Hunt, S. P., and Darlison, M. G. (1991). Distribution of the GABA$_A$ receptor α1- and γ2-subunit mRNAs in chick brain. *Neurosci. Lett.* **133,** 45–48.

Glencorse, T. A., Bateson, A. N., and Darlison, M. G. (1992). Differential localization of two alternatively spliced GABA$_A$ receptor γ2-subunit mRNAs in the chick brain. *Eur. J. Neurosci.* **4,** 271–277.

Harvey, R. J., Vreugdenhil, E., Zaman, S. H., Bhandal, N. S., Usherwood, P. N. R., Barnard, E. A., and Darlison, M. G.

(1991). Sequence of a functional invertebrate GABA_A receptor subunit which can form a chimeric receptor with a vertebrate α subunit. *EMBO J.* **10**, 3239–3245.

Herb, A., Wisden, W., Lüddens, H., Puia, G., Vicini, S., and Seeburg, P. H. (1992). The third γ subunit of the γ-aminobutyric acid type A receptor family. *Proc. Natl. Acad. Sci. USA* **89**, 1433–1437.

Hunt, S. P., and Brecha, N. (1984). The avian optic tectum: A synthesis of morphology and biochemistry. *In* "Comparative Neurology of the Optic Tectum" (H. Vanegas, ed.), pp. 619–648. Plenum Press, New York.

Kirkness, E. F., Kusiak, J. W., Fleming, J. T., Menninger, J., Gocayne, J. D., Ward, D. C., and Venter, J. C. (1991). Isolation, characterization, and localization of human genomic DNA encoding the β1 subunit of the GABA_A receptor (GABRB1). *Genomics* **10**, 985–995.

Knoflach, F., Backus, K. H., Giller, T., Malherbe, P., Pflimlin, P., Möhler, H., and Trube, G. (1992). Pharmacological and electrophysiological properties of recombinant GABA_A receptors comprising the α3, β1 and γ2 subunits. *Eur. J. Neurosci.* **4**, 1–9.

Lambert, J. J., Peters, J. A., and Cottrell, G. A. (1987). Actions of synthetic and endogenous steroids on the GABA_A receptor. *Trends Pharmacol. Sci.* **8**, 224–227.

Lasham, A., Vreugdenhil, E., Bateson, A. N., Barnard, E. A., and Darlison, M. G. (1991). Conserved organization of γ-aminobutyric acid_A receptor genes: Cloning and analysis of the chicken β4-subunit gene. *J. Neurochem.* **57**, 352–355.

Laurie, D. J., Seeburg, P. H., and Wisden, W. (1992). The distribution of 13 GABA_A receptor subunit mRNAs in the rat brain. II. Olfactory bulb and cerebellum. *J. Neurosci.* **12**, 1063–1076.

Levitan, E. S., Schofield, P. R., Burt, D. R., Rhee, L. M., Wisden, W., Köhler, M., Fujita, N., Rodriguez, H. F., Stephenson, A., Darlison, M. G., Barnard, E. A., and Seeburg, P. H. (1988). Structural and functional basis for GABA_A receptor heterogeneity. *Nature (London)* **335**, 76–79.

Lüddens, H., and Wisden, W. (1991). Function and pharmacology of multiple GABA_A receptor subunits. *Trends Pharmacol. Sci.* **12**, 49–51.

Lüddens, H., Killisch, I., and Seeburg, P. H. (1991). More than one alpha variant may exist in a GABA_A/benzodiazepine receptor complex. *J. Receptor Res.* **11**, 535–551.

Lummis, S. C. R., and Sattelle, D. B. (1986). Binding sites for [³H]GABA, [³H]flunitrazepam and [³⁵S]TBPS in insect CNS. *Neurochem. Int.* **9**, 287–293.

Malherbe, P., Sigel, E., Baur, R., Persohn, E., Richards, J. G., and Mohler, H. (1990). Functional characteristics and sites of gene expression of the α_1,β_1,γ_2-isoform of the rat GABA_A receptor. *J. Neurosci.* **10**, 2330–2337.

Maricq, A. V., Peterson, A. S., Brake, A. J., Myers, R. M., and Julius, D. (1991). Primary structure and functional expression of the 5HT_3 receptor, a serotonin-gated ion channel. *Science* **254**, 432–437.

McKernan, R. M., Quirk, K., Prince, R., Cox, P. A., Gillard, N. P., Ragan, C. I., and Whiting, P. (1991). GABA_A receptor subtypes immunopurified from rat brain with α subunit-specific antibodies have unique pharmacological properties. *Neuron* **7**, 667–676.

Moss, S. J., Doherty, C. A., and Huganir, R. L. (1992). Identification of the cAMP-dependent protein kinase and protein kinase C phosphorylation sites within the major intracellular domains of the β_1, γ_2S, and γ_2L subunits of the γ-aminobutyric acid type A receptor. *J. Biol. Chem.* **267**, 14470–14476.

Nef, P., Oneyser, C., Alliod, C., Couturier, S., and Ballivet, M. (1988). Genes expressed in the brain define three distinct neuronal nicotinic acetylcholine receptors. *EMBO J.* **7**, 595–601.

Nutt, D. J., Glue, P., Lawson, C., and Wilson, S. (1990). Flumazenil provocation of panic attacks. Evidence for altered benzodiazepine receptor sensitivity in panic disorder. *Arch. Gen. Psychiat.* **47**, 917–925.

Okonjo, K. O., Kuhlmann, J., and Maelicke, A. (1991). A second pathway of activation of the *Torpedo* acetylcholine receptor channel. *Eur. J. Biochem.* **200**, 671–677.

Persohn, E., Malherbe, P., and Richards, J. G. (1991). *In situ* hybridization histochemistry reveals a diversity of GABA_A receptor subunit mRNAs in neurons of the rat spinal cord and dorsal root ganglia. *Neuroscience* **42**, 497–507.

Pritchett, D. B., and Seeburg, P. H. (1991). γ-Aminobutyric acid type A receptor point mutation increases the affinity of compounds for the benzodiazepine site. *Proc. Natl. Acad. Sci. USA* **88**, 1421–1425.

Pritchett, D. B., Lüddens, H., and Seeburg, P. H. (1989a). Type I and type II GABA_A-benzodiazepine receptors produced in transfected cells. *Science* **245**, 1389–1392.

Pritchett, D. B., Sontheimer, H., Shivers, B. D., Ymer, S., Kettenmann, H., Schofield, P. R., and Seeburg, P. H. (1989b). Importance of a novel GABA_A receptor subunit for benzodiazepine pharmacology. *Nature (London)* **338**, 582–585.

Robinson, T., MacAllan, D., Lunt, G., and Battersby, M. (1986). γ-Aminobutyric acid receptor complex of insect CNS: Characterization of a benzodiazepine binding site. *J. Neurochem.* **47**, 1955–1962.

Rothstein, J. D., Garland, W., Puia, G., Guidotti, A., Weber, R. J., and Costa, E. (1992). Purification and characterization of naturally occurring benzodiazepine receptor ligands in rat and human brain. *J. Neurochem.* **58**, 2102–2115.

Sattelle, D. B., Marshall, J., Lummis, S. C. R., Leech, C. A., Miller, K. W. P., Anthony, N. M., Bai, D., Wafford, K. A., Harrison, J. B., Chapaitis, L. A., Watson, M. K., Benner, E. A., Vassallo, J. G., Wong, J. F. H., and Rauh, J. J. (1991). γ-Aminobutyric acid and L-glutamate receptors of insect nervous tissue. *In* "Transmitter Amino Acid Receptors: Structures, Transduction and Models for Drug Development" (E. A. Barnard and E. Costa, eds.), Fidia Research Foundation Symposium Series Vol. 6, pp. 273–291. Thieme Medical Publishers, New York.

Schofield, P. R. (1989). The GABA_A receptor: Molecular biology reveals a complex picture. *Trends Pharmacol. Sci.* **10**, 476–478.

Schofield, P. R., Darlison, M. G., Fujita, N., Burt, D. R., Stephenson, F. A., Rodriguez, H., Rhee, L. M., Ramachandran, J., Reale, V., Glencorse, T. A., Seeburg, P. H., and Barnard, E. A. (1987). Sequence and functional expression of the GABA_A receptor shows a ligand-gated receptor super-family. *Nature (London)* **328**, 221–227.

Shimada, S., Cutting, G., and Uhl, G. R. (1992). γ-Aminobutyric acid A or C receptor? γ-Aminobutyric acid ρ_1 receptor RNA induces bicuculline-, barbiturate-, and benzodiazepine-insensitive γ-aminobutyric acid responses in *Xenopus* oocytes. *Mol. Pharmacol.* **41**, 683–687.

Shivers, B. D., Killisch, I., Sprengel, R., Sontheimer, H., Köhler, M., Schofield, P. R., and Seeburg, P. H. (1989). Two novel GABA_A receptor subunits exist in distinct neuronal subpopulations. *Neuron* **3**, 327–337.

Sigel, E., and Barnard, E. A. (1984). A γ-aminobutyric acid/benzodiazepine receptor complex from bovine cerebral cortex. Improved purification with preservation of regulatory sites and their interactions. *J. Biol. Chem.* **259**, 7219–7223.

Sigel, E., Stephenson, F. A., Mamalaki, C., and Barnard, E. A. (1983). A γ-aminobutyric acid/benzodiazepine receptor complex of bovine cerebral cortex. Purification and partial characterization. *J. Biol. Chem.* **258**, 6965–6971.

Sigel, E., Baur, R., Trube, G., Möhler, H., and Malherbe, P. (1990). The effect of subunit composition of rat brain GABA$_A$ receptors on channel function. *Neuron* **5,** 703–711.

Smart, T. G., and Constanti, A. (1982). A novel effect of zinc on the lobster muscle GABA receptor. *Proc. R. Soc. London B* **215,** 327–341.

Sommer, B., Poustka, A., Spurr, N. K., and Seeburg, P. H. (1990). The murine GABA$_A$ receptor δ-subunit gene: Structure and assignment to human chromosome 1. *DNA Cell Biol.* **9,** 561–568.

Verdoorn, T. A., Draguhn, A., Ymer, S., Seeburg, P. H., and Sakmann, B. (1990). Functional properties of recombinant rat GABA$_A$ receptors depend upon subunit composition. *Neuron* **4,** 919–928.

Verrall, S., and Hall, Z. W. (1992). The N-terminal domains of acetylcholine receptor subunits contain recognition signals for the initial steps of receptor assembly. *Cell* **68,** 23–31.

von Heijne, G. (1986). A new method for predicting signal sequence cleavage sites. *Nucleic Acids Res.* **14,** 4683–4690.

Whiting, P., McKernan, R. M., and Iversen, L. L. (1990). Another mechanism for creating diversity in γ-aminobutyrate type A receptors: RNA splicing directs expression of two forms of γ2 subunit, one of which contains a protein kinase C phosphorylation site. *Proc. Natl. Acad. Sci. USA* **87,** 9966–9970.

Wieland, H. A., Lüddens, H., and Seeburg, P. H. (1992). A single histidine in GABA$_A$ receptors is essential for benzodiazepine agonist binding. *J. Biol. Chem.* **267,** 1426–1429.

Wisden, W., and Seeburg, P. H. (1992). GABA$_A$ receptor channels: From subunits to functional entities. *Curr. Opin. Neurobiol.* **2,** 263–269.

Wisden, W., Morris, B. J., Darlison, M. G., Hunt, S. P., and Barnard, E. A. (1988). Distinct GABA$_A$ receptor α subunit mRNAs show differential patterns of expression in bovine brain. *Neuron* **1,** 937–947.

Wisden, W., Gundlach, A. L., Barnard, E. A., Seeburg, P. H., and Hunt, S. P. (1991). Distribution of GABA$_A$ receptor subunit mRNAs in rat lumbar spinal cord. *Mol. Brain Res.* **10,** 179–183.

Wisden, W., Laurie, D. J., Monyer, H., and Seeburg, P. H. (1992). The distribution of 13 GABA$_A$ receptor subunit mRNAs in the rat brain. I. Telencephalon, Diencephalon, Mesencephalon. *J. Neurosci.* **12,** 1040–1062.

Ymer, S., Schofield, P. R., Draguhn, A., Werner, P., Köhler, M., and Seeburg, P. H. (1989). GABA$_A$ receptor β subunit heterogeneity: Functional expression of cloned cDNAs. *EMBO J.* **8,** 1665–1670.

Zaman, S. H., Harvey, R. J., Barnard, E. A., and Darlison, M. G. (1992a). Unusual effects of benzodiazepines and cyclodiene insecticides on an expressed invertebrate GABA$_A$ receptor. *FEBS Lett.* **307,** 351–354.

Zaman, S. H., Shingai, R., Harvey, R. J., Darlison, M. G., and Barnard, E. A. (1992b). Effects of subunit types of the recombinant GABA$_A$ receptor on the response to a neurosteroid. *Eur. J. Pharmacol.* **225,** 321–330.

Inhibitory Ligand-Gated Ion Channel Receptors: Molecular Biology and Pharmacology of GABA$_A$ and Glycine Receptors

ROBERT J. VANDENBERG
AND PETER R. SCHOFIELD

I. Introduction

The rapid transmission of information across synapses occurs by the presynaptic release of neurotransmitters that act on postsynaptic cell surface receptors. Whether the signal is excitatory, and mediated by neurotransmitters such as acetylcholine or glutamate, or inhibitory, and mediated by glycine or γ-aminobutyric acid (GABA), the time frame of signal transduction is extremely rapid, occurring within milliseconds. Despite the opposing effects of these two classes of neurotransmitter, excitatory and inhibitory signals are both transduced via a related set of receptor molecules; the ligand-gated ion channel receptors. In this chapter, the development of our understanding of the molecular pharmacology of the glycine- and GABA-gated ion channel receptors is reviewed. Excitatory receptors are referred to only in passing, or when knowledge of these molecules helps us understand the action of the inhibitory receptors (see Chapter 18). The reader is also referred to other reviews of the molecular biology and pharmacology of the inhibitory ligand-gated ion channel receptors (Betz, 1990; Langosch *et al.*, 1990; Burt and Kamatchi, 1991; DeLorey and Olsen, 1992) (see Chapter 20).

II. Pharmacology

GABA was postulated to be an inhibitory neurotransmitter in 1950 (Awapara, 1950; Roberts and Frankel, 1990; Undenfriend, 1950). In the mid-1960s, Aprison first suggested that glycine was a neurotransmitter in the spinal cord (Aprison and Wermann, 1965) and Curtis and colleagues showed that glycine was an inhibitory neurotransmitter in the spinal cord (Curtis *et al.*, 1968). These two molecules have been shown to be the major inhibitory transmitters in the central nervous system. Both operate by increasing the permeability of the neuronal membrane to chloride ions (Coombs *et al.*, 1955). The initial definition of these receptors led to the development of the pharmacology of inhibitory receptors as a central theme of receptor research in the 1960s and 1970s. As in most physiological and pharmacological studies, the identification of high-affinity selective agonists and antagonists has been crucial to the understanding of receptor function.

Glycine receptor (GlyR) agonists are simple amino acids, with a rank order of potency for GlyR activation of glycine $>\beta$-alanine $>$ taurine $>$ L-alanine-, L-serine $>$ proline (Curtis *et al.*, 1968; Fig. 1). Specificity

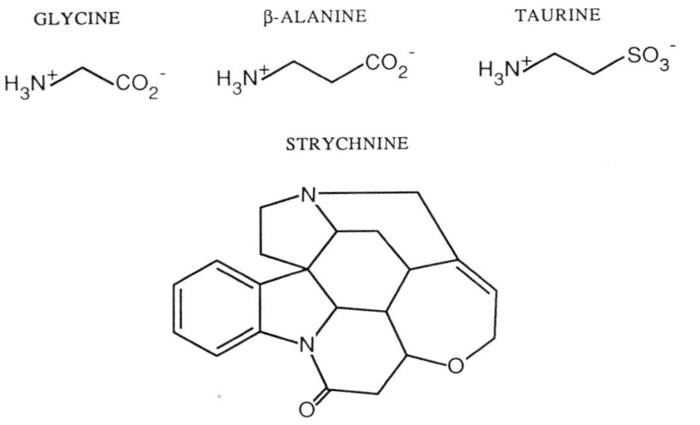

GLYCINE   β-ALANINE   TAURINE   GABA   MUSCIMOL

STRYCHNINE

FIGURE 1 Chemical structures of the prototypical ligands that act as glycine receptor agonists and antagonists.

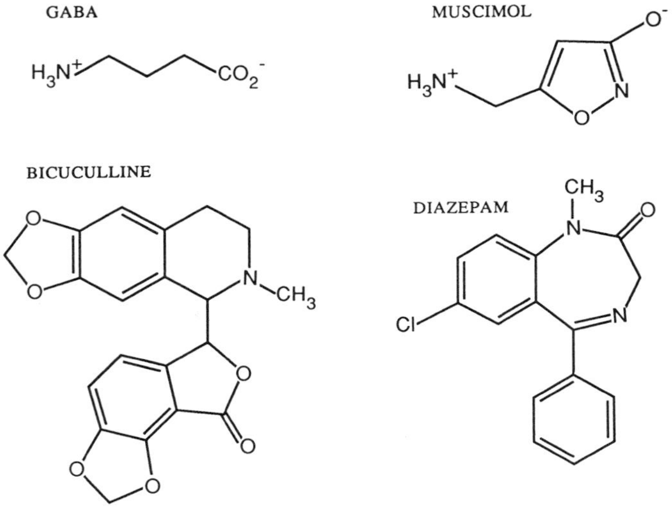

BICUCULLINE

DIAZEPAM

FIGURE 2 Chemical structures of the prototypical ligands that act as GABA$_A$ receptor agonists, antagonists, and allosteric potentiators.

is provided by the amino and carboxyl groups and is limited by the presence of side chains on the central methyl group (Tokutomi et al., 1989). For the GlyR, the prototypical antagonist is the convulsant alkaloid strychnine (Curtis et al., 1971; Fig. 1). Strychnine, isolated from the seeds of the Indian tree Strychnos nux vomica, has been used to poison rats since the sixteenth century, when it was introduced into Germany. Newborn rats are resistant to strychnine poisoning, an observation that may have been made during the Black Plague but was formally reported in the LD$_{50}$ study by Falck (1884). The binding of strychnine to the GlyR may be displaced by various glycinergic agonists. However, modification of the GlyR with acetic anhydride or diazonium tetrazole does not affect strychnine binding, but glycine displacement of strychnine is reduced (Young and Snyder, 1974; Marvison et al., 1986), suggesting that the agonist and antagonist interactions with the receptor are not strictly competitive. The steroid derivative RU 5135 (3α-hydroxy-16-imino-5β-17-aza-androstan-11-one) is an even more potent antagonist of the GlyR, with an affinity of 3 nM (Simmonds and Turner, 1985). However, RU 5135 does not offer good discrimination between GlyR and the GABA$_A$ receptors (GABA$_A$R).

The best characterized selective GABA$_A$R antagonist is the phthalide isoquinoline alkaloid bicuculline, which was first isolated from Dicentra cucullaria (Manske, 1932; Fig. 2). Ligand binding studies indicated that bicuculline, GABA, and the conformationally restricted GABA$_A$R agonist muscimol (Krogsgaard-Larson et al., 1977; Fig. 2) share a common binding site, consistent with bicuculline acting as a competitive antagonist (Curtis et al., 1970). The reader is referred to other reviews for further details (Kerr and Ong, 1992). A more significant class of drugs,

the benzodiazepines, also acts specifically on the GABA$_A$R (Fig. 2). Benzodiazepines were introduced in the 1960s, and Valium® (diazepam), the best known example, became the world's most widely prescribed drug during the 1970s. The benzodiazepines have been widely used as anxiolytics, sedatives, and hypnotics. High-affinity benzodiazepine-binding sites were described in brain (Mohler and Okada, 1977; Squires and Braestrup, 1977) but only in 1978 did Tallman et al. (1978) show that these drugs acted by enhancing GABA$_A$R function. Benzodiazepines allosterically potentiate the action of GABA at its receptor, resulting in an enhanced influx of chloride ions and, thus, enhanced neuronal inhibition (Gottlieb, 1988). The different sensitivities to various benzodiazepines of GABA$_A$Rs expressed in discrete regions of the brain suggested that subtypes of GABA$_A$Rs existed. The triazolopyridazine CL218,872 was the first compound used to distinguish two different classes of benzodiazepine-binding sites in the brain (Klepner et al., 1979). A high-affinity binding site present apparently uniquely in the cerebellum, as well as in other brain regions, was termed the type I benzodiazepine receptor, whereas the lower affinity type II benzodiazepine-binding site was found in fewer areas, including the hippocampus. Sieghart and Karobath (1980) provided evidence of a molecular distinction between these two binding sites by examining the differential effects of CL 218,872 on the [³H]flunitrazepam photoaffinity-labeling of benzodiazepine receptors. Regional differences in the potentiation of benzodiazepine binding by GABA (Leeb-Lundberg and Olsen, 1983) also suggested the existence of distinct receptor subtypes. On the basis of more detailed pharmacological and anatomical investigation, as well as subsequent molecular biological studies (see Section IV and Chapter 20), even these classifications appear oversimplified.

Several other classes of compounds have also been shown to modulate GABA$_A$R function, including the barbiturates, various derivatives of progesterone, and the convulsant picrotoxin. Extensive pharmacological literature exists on the mode of action of these compounds (Johnston *et al.*, 1987; Ticku, 1991; Simmonds, 1991; Kerr and Ong, 1992), but further discussion is limited to the molecular basis of action of GABA and the various benzodiazepines.

III. Receptor Purification

The study of the structural basis for GlyR and GABA$_A$R function was made possible by the purification and subsequent cloning of the subunits of these receptors. For the GlyR and the GABA$_A$R, the purification of solubilized receptor proteins has been achieved using affinity chromatography. 2-Aminostrychnine coupled to agarose beads allowed the one-step purification of a GlyR complex from rat spinal cord containing three proteins of 48 kDa, 58 kDa, and 93 kDa (Pfeiffer *et al.*, 1982). The 48-kDa protein was shown to be the site of photoaffinity labeling with [^{3}H]strychnine (Graham *et al.*, 1983), whereas the 93-kDa protein was shown to be a peripheral component that may play a role in anchoring the GlyR to the postsynaptic cytoskeleton. The GABA$_A$R was purified from bovine cortex by benzodiazepine affinity chromatography. This preparation consisted of α and β subunits of 53 kDa and 57 kDa, respectively; the purified receptor retained its full pharmacological profile (Sigel *et al.*, 1983; Sigel and Barnard, 1984). Photoaffinity labeling of the purified GABA$_A$R subunits with [^{3}H]flunitrazepam revealed that the benzodiazepine binding site was located on the α subunit (Sigel *et al.*, 1983), whereas similar studies using [^{3}H]muscimol indicated that the GABA binding site was located on the β subunit (Casalotti *et al.*, 1986, Deng *et al.*, 1986).

To determine the subunit stoichiometry of the GlyR, Langosch *et al.*, (1988) studied the composition of cross-linked proteins derived from the purified GlyR receptor or from synaptic membrane preparations. Protein complexes with molecular masses of up to 260 kDa were seen, consistent with a subunit structure containing three α subunits (48 kDa) and two β subunits (58 kDa) per receptor. A simplistic gel

analysis of purified GABA$_A$R subunits suggested a tetrameric structure (two α subunits and two β subunits; Mamlaki *et al.*, 1987). The stoichiometry of the excitatory nicotinic acetylcholine receptor (nAChR) has been shown to be $\alpha 2\beta\gamma\delta$ in the electric organ of the ray *Torpedo*; this structure has been visualized by X-ray diffraction studies that have shown that the receptor forms a pseudosymmetric pentameric structure (Brisson and Unwin, 1985; Toyoshima and Unwin, 1990). Since nAChR, GlyR, and GABA$_A$Rs form a superfamily of receptors and share a number of structural and functional similarities (see Section IV), the various GlyRs and GABA$_A$Rs are also likely to be composed of five subunits.

The purification of the GlyR and GABA$_A$R proteins has allowed the protein sequence determinations necessary for cloning the DNA encoding the subunits of both these receptors. Cloning the various subunits has greatly aided the molecular analysis of the function of inhibitory ligand-gated ion channel receptors. In particular, such issues as the molecular structure of the receptors and how these structures regulate ion fluxes; how various agonists, antagonists, and allosteric modulators interact with the receptors, and how these behaviors relate to their functional properties can now be addressed.

IV. Molecular Biology

In 1987, the GABA$_A$R α and β subunits were cloned in the laboratories of Peter Seeburg and Eric Barnard (Schofield *et al.*, 1987). The strychnine-binding α subunit of the GlyR was cloned in Heinrich Betz's laboratory (Grenningloh *et al.*, 1987). Both studies relied on the generation of peptide sequences derived from the purified receptor proteins and the design of appropriate oligonucleotide probes to screen cDNA libraries.

The various ligand-gated ion channel receptor subunits have a conserved transmembrane structure (Grenningloh *et al.*, 1987; Schofield *et al.*, 1987). Each has an N-terminal domain of approximately 220 amino acid residues, containing a highly conserved disulfide-bonded loop motif, four stretches of hydrophobic residues that are thought to traverse the membrane and associate to form the channel structure, and an intracellular domain between the third and fourth transmembrane regions that may play a role in the regulation of receptor function. GlyR and GABA$_A$R form chloride-selective pores; anionic selectivity is thought to be derived from a series of positively charged amino acid residues located at the extracellular and intracellular ends of the transmembrane regions (Grenningloh *et al.*, 1987; Schofield *et al.*, 1987). This structure contrasts with that of the nAChR, in which the equivalent residues are negatively charged to allow the passage of cationic sodium ions (Imoto *et al.*, 1988). Of the four membrane-spanning domains, the second is thought to form the channel lumen. This transmembrane helix is characterized by the preponderance of hydroxylated amino acid residues, such as threonine and serine, which facilitate the passage of hydrated anions or cations, depending on receptor selectivity. Galzi *et al.*, (1992) substituted a number of residues of the nAChR α7 subunit for the corresponding residues of the second transmembrane region of the GlyR α1 subunit for the corresponding residues of the second transmembrane region of the GlyR α1 subunit and were able to generate an nAChR that conducted anions. One of the critical factors that appears to regulate anion selectivity is the presence of a proline residue at the beginning of the transmembrane region, this residue would presumably alter the conformation of the ion channel and, hence, the accessibility of the charged/or hydroxylated residues that may directly interact with the ions.

As highlighted earlier, several lines of pharmacological evidence suggested that GABA$_A$R subtypes existed. In the initial isolation of full-length cDNA clones encoding the GABA$_A$R α subunit, one of the eight cDNAs had a restriction map different from that of the other cDNA clones. On DNA sequence analysis, this clone (designated α2) was found to share ~75% primary sequence identity with the more abundant α1 subunit sequences. The cDNA sequences of each of these receptor subunits reveal that substantial primary sequence similarities exist among the various subunit subtypes within and between receptors. For example, the α subunit subtypes of the GABA$_A$R share about 75% primary sequence identity, whereas the α and β subunits of GABA$_A$R share approximately 35% of their amino acids (Figs. 3,4). This level of sequence conservation is also seen among the subunits of the GABA$_A$R and GlyR. In addition, the subunits of both receptors share about 20% sequence identity with the excitatory nAChR. Thus, the various ligand-gated ion channels were shown to form a superfamily of related proteins (Figs. 3,4).

Large-scale DNA-based homology screening of brain cDNA libraries was undertaken after researchers determined that a receptor superfamily and receptor subunit subtypes existed. This process utilized low-stringency screening with cloned cDNAs, resulting in the isolation of related subunit subtypes, or highly degenerate oligonucleotide probes designed to recognize regions of maximum sequence conservation among the known GABA$_A$R and GlyR subunits. The

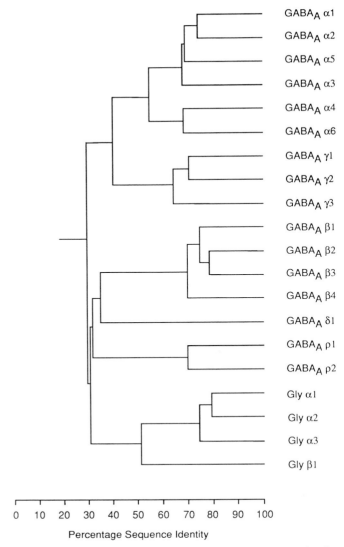

FIGURE 3 The ligand-gated ion channel receptor superfamily. The dendrogram shows the level of amino acid sequence identity that is found on comparison of the various receptor subunits and subunit subtypes. The axis shows the percentage sequence identity.

domain of greatest sequence conservation was a stretch of eight amino acids—Thr–Thr–Val–Leu–Thr–Met–Thr–Thr—that formed part of the second membrane-spanning domain of the receptors. The advantage of screening with a degenerate oligonucleotide probe designed from the conserved sequence was that, since three subunits of the two receptor types contained this motif, this motif was likely to be conserved in additional subunits of the GABA$_A$R and the GlyR.

The use of homology-based screening protocols resulted in the isolation of many new receptor subunits and subunit subtypes. Initially, the bovine GABA$_A$R

α2 and α3 subunits were isolated and characterized (Levitan *et al.*, 1988). After this demonstration of subtype heterogeneity, many other reports followed. The bovine and rat GABA$_A$R α4 subunits (Ymer *et al.*, 1989; Wisden *et al.*, 1991), the rat GABA$_A$R α5 subunit (Khrestchatisky *et al.*, 1989; Malherbe *et al.*, 1990; Pritchett and Seeburg, 1990), and the rat and bovine GABA$_A$R α6 subunits (Lüddens *et al.*, 1990) were all cloned. Similarly, the rat and bovine GABA$_A$R β2 subunits (Ymer *et al.*, 1989) and the rat and bovine GABA$_A$R β3 subunits (Lolait *et al.*, 1989; Ymer *et al.*, 1990) were isolated. More recently, the chicken GABA$_A$R β4 subunit was cloned (Lasham *et al.*, 1991). The chromosomal localization of several of these genes has been undertaken. The α1 subunit gene is located on long arm of chromosome 5 (bands q34–q35; Buckle *et al.*, 1989) whereas the α2 subunit and the β1 subunit gene are located on the short arm of chromosome 4 (bands p12–p13; Buckle *et al.*, 1989; Dean *et al.*, 1991; Kirkness *et al.*, 1991). Unexpectedly, the α3 subunit gene is located on the long arm of the X chromosome, at position Xq28; thus, researchers have speculated that this gene may be a candidate gene for the X-linked form of manic depression (Buckle *et al.*, 1989). Finally, the β3 subunit gene has been mapped to human chromosome 15 (bands q11–q15; Wagstaff *et al.*, 1991b). This location is within the deletions that are observed in Angelman syndrome and Prader-Willi syndrome and suggests a role for GABA and GABA$_A$R function in the pathogenesis of these diseases (Wagstaff *et al.*, 1991a). In addition, these genes are subject to genetic imprinting and differential inheritance, depending on the parental allele that is inherited.

New GABA$_A$R subunits were also isolated by the homology screening approach, most significant of which was the GABA$_A$R γ2 subunit (Pritchett *et al.*, 1989a; Shivers *et al.*, 1989). This novel subunit confers benzodiazepine responsiveness on GABA$_A$Rs. Two other γ subunit subtypes that have also been described are the GABA$_A$R γ1 subunit (Ymer *et al.*, 1990) and the GABA$_A$R γ3 subunit (Knoflach *et al.*, 1991; Wilson-Shaw *et al.*, 1991; Herb *et al.*, 1992). The γ1 subunit gene has been localized to human chromosome 4 between positions 4p14 and 4q21.1 whereas the γ2 subunit gene is located at chromosome 5q31–33.2 (Wilcox *et al.*, 1992). The γ2 and α1 subunit genes are located within 200 kb of each other on the long arm of chromosome 5 (Wilcox *et al.*, 1992). The rat and mouse GABA$_A$R δ subunit genes have been cloned and localized to the short arm of human chromosome 1 (Shivers *et al.*, 1989; Sommer *et al.*, 1990).

Using a related polymerase chain reaction (PCR)-based strategy, another class of GABA$_A$R subunits, ρ, has been isolated from retinal mRNA. The human ρ1 and ρ2 subunits (Cutting *et al.*, 1991,1992) are colocalized on the long arm of human chromosome 6 (bands q14–q21). Finally, invertebrate genes encoding GABA$_A$R and GlyR-like neurotransmitter receptors have also been cloned (Darlison, 1992; see Chapter 20).

Subunits of the GlyR have been cloned, leading to the identification of three α subunit subtypes. The human and rat GlyR α1 and α2 subunits were cloned initially (Grenningloh *et al.*, 1990a; Akagi *et al.*, 1991) and, despite some initial confusion, the α1 subunit gene is located autosomally whereas the α2 subunit gene is located on the X chromosome at position Xp21.2–p22.1 (Siddique *et al.*, 1989; Grenningloh *et al.*, 1990a). A third α subunit, α3 has also been cloned (Kuhse *et al.*, 1990a).

Although the homology-based screening approaches have been enormously successful in the identification of novel subunits and subunit subtypes, the method has not always allowed the cloning of GlyR or GABA$_A$R subunits. For example, the GlyR β subunit was not isolated by homology-based screening and was cloned using oligonucleotide probes based on protein sequence data obtained from the purified subunit (Grenningloh *et al.*, 1990b). Examination of the sequence encoded in the second transmembrane domain reveals several amino acid differences from the invariant sequence motif. These differences precluded the isolation of this subunit using the homology-based screening protocol.

Additional structural heterogeneity is generated by the alternative splicing of various subunit mRNAs of the GABA$_A$R and the GlyR. The alternatively spliced exons have, in general, been small exons located within the intracellular cytoplasmic loop of the receptor subunits. Thus, in the GlyR α1 subunit, an additional eight amino acids are added using an alternative splice acceptor site (Malosio *et al.*, 1991). In the β4 subunit of the GABA$_A$R, a similar situation occurs with the addition of a 4-amino-acid exon by alternative use of splice donor sites (Bateson *et al.*, 1991). Finally, an alternative extracellularly located exon has also been identified in the GlyR α2 subunit (Kuhse *et al.*, 1991) to which a functional role has yet to be ascribed. Although these alternative exons have not yet been shown to affect the properties of the receptor, a novel intracellular exon of eight amino acids in the GABA$_A$R γ2 subunit results in the generation of a novel protein kinase C phosphorylation site (Whiting *et al.*, 1990; Kofuji *et al.*, 1991). Functional modulation of GABA$_A$Rs has already been observed as a result of cAMP-dependent protein phosphorylation (Moss *et al.*, 1992). The presence or absence of this novel exon is

```
                                                                              Q
h GABA_AR α1    1
h GABA_AR β1    0
h GABA_AR γ2    1    Q K M T V W I L L L L S L Y P G F T S Q K S D D D Y E D Y A S N
h GABA_AR δ1    1                                              Q P H H G A R A
h GABA_AR ρ1    1    T E S R M H W P G R E V H E M S K K G R P Q R Q R R E V H E D A H K
h GlyR α1       0
h GlyR β1       1                K E K S S K K G K G K K K Q Y L C P S Q Q

h GABA_AR α1    2    P S L Q D E L K D N T T V F T R I - - L D R L L - - D G Y D N R L R
h GABA_AR β1    1        H S T N E P S N M P Y V K E T V D R L L - - K G Y D I R L R
h GABA_AR γ2   33    K T W V L T P K V P E G D V T V I - L N N L L - - E G Y D N K L R
h GABA_AR δ1    9    M N D I G D Y V G S N L E I S W L P N L D G L M - - E G Y A R N F R R
h GABA_AR ρ1   35    Q V S P I L R R S P D I T K S P L T K S E Q L L R I D D H D F S M R
h GlyR α1       1            A R S A T K P M S P S D F L D K L - - M G R T S G Y D A R I R
h GlyR β1      22    S A E D L A R V P P N S T S N I L N R L - - L V - - - S Y D P R I R

h GABA_AR α1   32    P G L G E R V T E V K T D I F V T S F G P V S D H D M E Y T I D V F
h GABA_AR β1   29    P D F G G P P V D V G M R I D V A S I D M V S E V N M D Y T L T M Y
h GABA_AR γ2   63    P D I G V K P T L I H T D M Y V N S I G P V N A I N M E Y T I D I F
h GABA_AR δ1   41    P G I G G P P V N V A L A L E V A S I D H I S E A N M E Y T M T L Y
h GABA_AR ρ1   69    P G F G G P A I P V G V D V Q V E S L D S I S E V D M D F T M T L Y
h GlyR α1      30    P N F K G P P V N V S C N I F I N S F G S I A E T T M D Y R V N I F
h GlyR β1      51    P N F K G I P V D V V V N I F I N S F G S I Q E T T M D Y R V N I F

h GABA_AR α1   66    F R Q S W K D E R L K F K G P - - M T V - L R L N N L M A S K I R T
h GABA_AR β1   63    F Q Q S W K D K R L S Y S G I - - - P - L N L T L D N R V A D Q L W V
h GABA_AR γ2   97    F A Q M W Y D R R L K F N S T - - I K V - L R L N S N M V G K I W I
h GABA_AR δ1   75    L H R A W R D S R L S Y N H T - - N - E T L G L D S R F V D K L W L
h GABA_AR ρ1  103    L R H Y W K D E R L S F P S T - - N N L S M T F D G R L V K K I W V
h GlyR α1      64    L R Q W N D P R L A Y - N E Y P D - D S L D L D P S M L D S I W K
h GlyR β1      85    L R Q K W N D P R L K L P S D F R G S D A L T V D P T M Y K C L W K

h GABA_AR α1   97    P D T F F H N G K K S V A H N M T M P N K L L R I T E D G T L L Y T
h GABA_AR β1   94    P D T Y F L N D K K S F V H G V T V K N R M I R L H P D G T V L Y G
h GABA_AR γ2  128    P D T F F R N S K K A D A H W I T T P N R M L I R I W N D G R V L Y S
h GABA_AR δ1  106    P D T F I V N A K V C L V H D V T T D N V M L R V Q P D G K V L Y S
h GABA_AR ρ1  135    P D M F F V H S K R S F I H D T T T D N V M L R V Q P D G K V L Y S
h GlyR α1      96    P D L F F A N E K G A H F H E I T T D N K L L R I S R N G N V L Y S
h GlyR β1     119    P D L F F A N E K S A N F H D V T Q E N I L L F I F R D G D V L V S

h GABA_AR α1  131    M R L T V R A E C P M H L E D F P M D A H A C P L K F G S Y A Y T R
h GABA_AR β1  128    L R I T T T A A C M M D L R R Y P L D E Q N C T L E I E S Y G Y T T
h GABA_AR γ2  162    L R I I D A E C Q L Q L H N F P M D E H S C P L E F S S Y G Y P R
h GABA_AR δ1  140    I R I T S T V A C D M D L A K Y P M D E Q E C M L D L E S Y G Y S S
h GABA_AR ρ1  169    L R V T V T A M C N M D F S R F P L D T Q T C S L E I E S Y A Y T E
h GlyR α1     130    I R I T L T L A C P M D L K N F P M D V Q T C I M Q L E S F G Y T M
h GlyR β1     153    M R L S I T L S C P L D L T L F P M D T Q R C K M Q L E S F G Y T T

h GABA_AR α1  165    A E V V Y E W T R E P A R S V V A E D G S R L N Q Y D L L - - - G
h GABA_AR β1  162    D D I E F Y W N G G - - E G A V T G V N K I E L P Q F S I V - - - D
h GABA_AR γ2  196    E E I V Y Q W K R S S V E - - V G D T R S W R L Y Q F S F V - - - G
h GABA_AR δ1  174    E D I V Y Y W S E N - - Q E Q I H G L D R L Q L A Q F T I T - - - S
h GABA_AR ρ1  203    D D L M L Y W K K G - - N D S L K T D E R I S L S Q F L I Q - - - E
h GlyR α1     164    N D L I F E W Q E Q - - - G A V Q V A D G L T L P Q F I L K E E K D
h GlyR β1     187    D D L R F I W Q S G - - - D P V Q L - E K I A L P Q F D I K K E - D

h GABA_AR α1  196    Q T V D S G I V Q - S S T G E Y V V M T T H F H L K R K I G Y F V I
h GABA_AR β1  191    Y K M V S K K V E F - T T G A Y P R L S L S F R L K R N I G Y F I L
h GABA_AR γ2  225    L R N T T E V V K - T T S G D Y V V M S V Y F D L S R R M G Y F T I
h GABA_AR δ1  203    Y R F T T E L M N F K S A G Q F P R L S L H F Q L R R N R G V Y I I
h GABA_AR ρ1  232    F H T T T K L A F Y S S T G W Y N R L Y I N F T L R R H I F F F L L
h GlyR α1     195    L R Y - - C T K H Y N - T G K F T C I E A R F H L E R Q M G Y Y L I
h GlyR β1     216    I E Y G N C T K Y Y K G T G Y Y T C V E V I F T L R R Q M G F Y M M
```

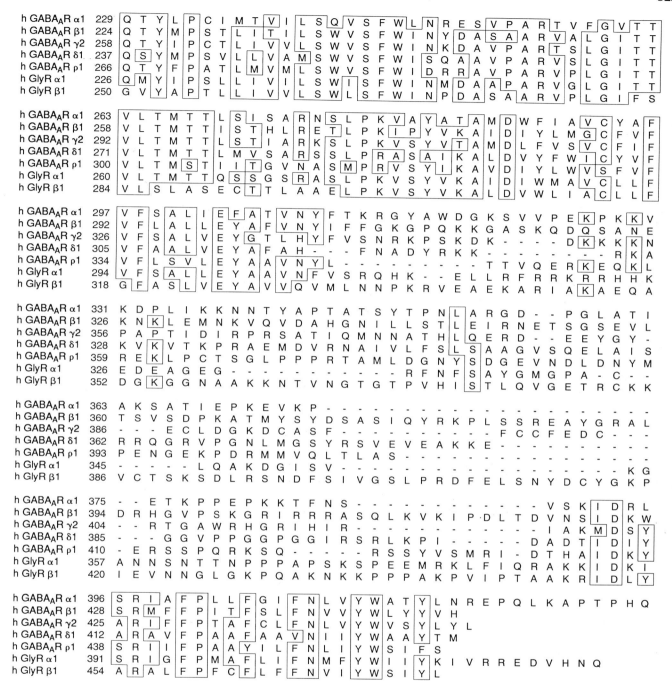

FIGURE 4 Sequence comparison of representative subunits of the ligand-gated ion channel receptor superfamily. Invariant residues are indicated by boxing.

likely to impart additional protein kinase C-regulated functional properties to the GABA$_A$R.

V. Subunits and Functions

The large number of different subunits of the GlyR and GABA$_A$R confers on cells the potential to generate a vast array of receptors. The precise determination of the subunit composition and stoichometry of these receptors *in vivo* has not yet been achieved. One approach that has been used to address this problem is the expression of recombinant receptors containing various subunit combinations in cultured mammalian cells or *Xenopus* oocytes. Measurement of the electrophysiological and pharmacological properties of these recombinant receptors, and comparison with the functional properties observed in brain GABA$_A$Rs or GlyRs, has provided information concerning the role of specific subunits in determining the pharmacological and electrophysiological properties of these receptors.

Each GABA$_A$R receptor subunit, when expressed alone, has the capacity to form a homopentameric receptor that responds to the subunit-specific agonist, although often at reduced efficiency (Blair *et al.*, 1988; Pritchett *et al.*, 1988). The α and β subunits of the GlyR are also capable of forming homomeric glycine-gated ion channels, but only homomeric α subunit receptors produce large currents and are blocked by the antagonist strychnine (Schmieden *et al.*, 1989; Sontheimer *et al.*, 1989). Thus, receptors composed of the homomeric GlyR α1 subunit provide an excellent single-subunit model of native ligand-gated ion channel receptors. Researchers have used this property to facilitate the molecular pharmacological analysis of ligand binding (see Section VI). A similar approach has been taken using the chicken α7 homomeric nAChR (Revah *et al.*, 1991).

To ascertain the functional properties conferred by specific subunits in *Xenopus* oocytes or mammalian cells, various combinations of subunits have been tested. Coexpression of the γ2 subunit of the GABA$_A$R with various α and β subunits generates recombinant receptors that respond to benzodiazepines (Pritchett *et al.*, 1989a). Although the γ2 subunit is essential for the formation of a benzodiazepine-responsive receptor, the presence of a specific α subunit determines the type of benzodiazepine receptor that is formed. Coexpression of the α1 subunit with the β1 and γ2 subunits results in GABA$_A$Rs that show a selectivity for the benzodiazepine-related compounds, such as the β-carbolines, that is characteristic of type I benzodiazepine receptors (Pritchett *et al.*, 1989b; Table I).

TABLE I Specific α Subunits Confer Distinct Pharmacological Properties on Recombinant GABA$_A$ Receptors

α Subunit	Characteristic properties
α1	High affinity for triazolopyridines and 2-oxo-quazepam (Bz agonists); high affinity for β-carbolines (Bz inverse agonists); Bz type I receptor
α2	Binds benzodiazepines; Bz type II receptor
α3	Binds benzodiazepines, enhanced Bz binding in presence of GABA; Bz type II receptor
α4	Does not bind benzodiazepines; high affinity for Ro 15-4513 (alcohol antagonist)
α5	Reduced affinity for Zolpidem® (Bz agonist); Bz type II receptor subtype
α6	Does not bind benzodiazepines; high affinity for Ro 15-4513 (alcohol antagonist)

Receptors containing the α2 and α3 subunits generate GABA$_A$Rs that correspond to type II benzodiazepine receptors (Pritchett *et al.*, 1989b). Moreover, the α3-containing receptors are further distinguished from the α1 or α2-containing receptors by showing a markedly enhanced GABA potentiation of benzodiazepine binding (Pritchett *et al.*, 1989b). The α5 subunit also forms type II benzodiazepine receptors in combination with a β and a γ2 subunit (Pritchett and Seeburg, 1990). However, these particular receptors were characterized by a reduced affinity for the benzodiazepine Zolpidem® (Pritchett and Seeburg, 1991). Thus, the presence of distinct α subunits results in GABA$_A$R that respond selectively to various therapeutic drugs, particularly the benzodiazepines (Table I).

In αxβ2γ2 recombinant GABA$_A$Rs, the presence of the cerebellar α6 subunit or the α4 subunit results in receptors that are insensitive to benzodiazepines, but bind the compound Ro 15-4513 with high affinity (Lüddens *et al.*, 1990; Wisden *et al.*, 1991). Ro15-4513 has been reported to act as an antagonist of alcohol-induced motor incoordination and ataxia. The α4 and α6 subunits are more closely related to each other than to the various other α subunits (Fig. 3), consistent with their common pharmacology, although they are characterized by differential localization in the thalamus and cerebellum, respectively (Lüddens *et al.*, 1990; Wisden *et al.*, 1991).

In addition to conferring benzodiazepine-sensitive pharmacology, the γ2 subunit splice variant that contains the additional 8 amino acid intracellular exon (Whiting *et al.*, 1990; Kofuji *et al.*, 1991) has been shown to confer enhanced ethanol sensitivity (Wafford *et al.*, 1991). Thus, the interaction of ethanol with GABA$_A$Rs is complex, probably involving multiple subunits.

The presence of an α, a β, and a $\gamma2$ subunit appears to be necessary for normal GABA$_A$R function. What role is played by the other subunits? The $\rho1$ subunit, isolated from a retinal cDNA library (Cutting *et al.*, 1991), has been shown to form a bicuculline-insensitive, barbituate-insensitive class of GABA$_A$Rs when expressed in *Xenopus* oocytes (Shimada *et al.*, 1992). This specific pharmacological profile has previously been described in retinal tissue; the bicuculline-insensitive binding site is termed GABA$_C$R (Johnston, 1986). The novel δ subunit, although showing a localization consistent with the high-affinity muscimol-binding site as revealed by *in vitro* autoradiography, has not yet been shown to possess a novel pharmacology.

The approach of testing combinations of GABA$_A$R subunits to mimic the various properties of native receptors has been successful in determining which subunits are responsible for mediating the diverse pharmacological properties seen in GABA$_A$Rs. However, despite these advances in our understanding of the roles of specific subunits in forming functional receptors, little is known of the consequences of receptor subtype heterogeneity for *in vivo* biological function.

VI. Molecular Pharmacology

Site-directed mutagenesis in combination with photoaffinity labeling techniques has been used to identify how various agonists, antagonists, and allosteric modulators specifically interact with the various GABA$_A$R and GlyR subunits.

A. Glycine Receptor Binding Sites

UV light-induced photoaffinity labeling of the GlyR with [^{3}H]strychnine shows specific glycine-inhibitable incorporation into the α subunit. Peptide mapping of the photoaffinity-labeled protein indicates that the strychnine binding site is located between residues 177 and 220 (Ruiz-Gomez *et al.*, 1990). Further refinement of the location and nature of the binding sites has been done using site-directed mutagenesis. Kuhse *et al.*, (1990b) cloned the rat GlyR $\alpha2$ subunit and on expression in *Xenopus* oocytes, observed that this subunit formed GlyRs that were insensitive to strychnine. Since the $\alpha2$ subunit is expressed only in neonatal rats, researchers thought that the properties of this receptor subunit may explain the historical ob-

servation that neonatal rats are resistant (originally described as immune) to the convulsant poison strychnine (Falck, 1884). Comparison of the human and rat GlyR $\alpha2$ subunits revealed only five amino acid differences, the only significant change is located at position 167 of the two $\alpha2$ sequences. The human $\alpha2$ subunit has a glycine residue at this position, whereas the rat $\alpha2$ subunit has a glutamate residue. Mutagenesis of rat $\alpha2$ subunit Glu 167 to glycine generated a GlyR with a 560-fold increase in affinity for strychnine and a 30-fold increase in glycine sensitivity, which are similar to the affinities of strychnine and glycine for the human $\alpha2$ subunit. Thus, this residue may play a role in forming the binding site on the GlyR for the antagonist strychnine and may also be involved in forming part of the glycine-binding site. Further confirmation of this role was obtained by mutating the corresponding glycine residue of the human $\alpha1$ subunit to glutamate, generating a glycine-gated chloride channel with reduced affinity for strychnine (Vandenberg *et al.*, 1992a).

Additional amino acid residues of the GlyR α subunit that are responsible for recognition of strychnine have been identified using alanine-scanning mutagenesis. In this approach, the particular amino acid residue under investigation is changed to alanine, which should remove the capacity of the side chain of the residue to interact with the ligand but would retain any secondary structure that may be present and, thus, would retain the correct tertiary and quaternary structures. This process has identified two tyrosine residues at positions 161 and 202 and a lysine residue at position 200 of the human $\alpha1$ subunit (Vandenberg *et al.*, 1992a). The tyrosine residue at position 161 is next to the glycine/glutamate residue just mentioned. Thus, the lack of a side chain in the glycine at residue 160 may be necessary to allow an interaction between strychnine and Tyr 161. The identification of residues 200 and 202 is consistent with the photoaffinity-labeling experiments; the residue to which the photo-activated strychnine is attached is most likely Tyr 202 (Vandenberg *et al.*, 1992a). Although this list of residues is obviously not complete, at least two separate regions of the extracellular domain of the α subunit are apparently required for strychnine recognition (Fig. 5).

Additional mutational analysis of residues surrounding Lys 200 and Tyr 202 demonstrated that Thr 204 is responsible for binding the agonist glycine (Vandenberg *et al.*, 1992b). Mutation of Thr 204 to alanine or valine did not affect strychnine binding to the receptor, but caused a 3000-fold decrease in the ability of glycine to displace bound strychnine. These studies suggest that the hydroxyl group of the threo-

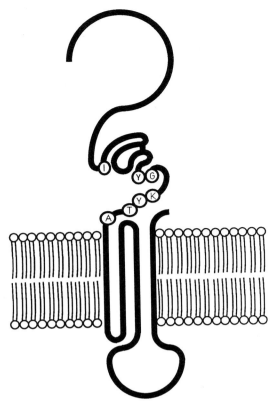

FIGURE 5 Domains of the glycine receptor involved in ligand binding. A single subunit of GlyR is illustrated; residues that are involved in the binding of agonists and antagonists are highlighted. The transmembrane topology of the receptor is indicated. The amino acids illustrated are A, Ala, G, Gly, I, Ile, K, Lys, T, Thr, Y, Tyr.

nine residue is critical for glycine recognition (Fig. 5). A more conservative substitution, serine for threonine, still generated a GlyR with a significant decrease in affinity for glycine, suggesting that not only the presence but also the conformation of the hydroxyl group is important. Note that this threonine residue is conserved among a number of the GABA$_A$R subunits. Since the two agonists glycine and GABA have similar structures, this residue may also be involved in agonist recognition at the GABA$_A$R.

Although all three of the GlyR α subunits, when expressed in *Xenopus* oocytes, form receptors that respond to the three glycinergic agonists glycine, β-alanine, and taurine, the relative potencies and the size of chloride current activated differ for the various subunits. Activation of the homomeric human α1 GlyRs by taurine generates currents of approximately one-fourth the magnitude of the glycine-activated current, whereas activation of the homomeric human α2 GlyRs by taurine generates currents showing only 6%

of the magnitude of glycine activated currents. Schmieden *et al.*, (1992) constructed a series of GlyR α1 subunit point mutants in an attempt to identify amino acid residues responsible for this differential sensitivity to taurine. These investigators noted 11 amino acid residues that were different between the two subunits in the extracellular domain, changed each of these residues in the α1 subunit to the corresponding residue of the α2 subunit, and then analyzed the relative taurine responses. Using this process, these investigators identified two residues, Ile 111 and Ala 212, that appear to be responsible for determining the different taurine responses (Fig. 5). Although these studies demonstrate that these residues are important determinants of agonist sensitivity, whether or not these residues play a direct role in taurine recognition remains unclear.

B. GABA Receptor Binding Sites

The general approach adopted for identifying binding site residues on the GABA$_A$R has been to analyze the functional differences between the expressed α subunit subtypes and then to identify the particular amino acid residues that confer the functional differences on the subunits. As mentioned previously, GABA$_A$Rs formed from α1β2γ2 subunits show benzodiazepine type I receptor specificity whereas GABA$_A$Rs composed of α2β2γ2 or α3β2γ2 subunits have benzodiazepine type II receptor specificity. Pritchett and Seeburg (1991) analyzed the sequence differences among the α1, α2, and α3 subunits. By converting ever smaller regions of the α3 subunit sequence to that of the α1 subunit, these researchers identified a single amino acid residue substitution, a glycine for a glutamate at position 201 of the α1 subunit, that was responsible for conferring type I or type II benzodiazepine receptor specificity (Fig. 6). Substitution of the glutamate for a glycine corresponding to position 201 of the α1 subunit generated a subunit capable of forming a type I benzodiazepine receptor. In a similar study, Seeburg's group analyzed the sequence differences between the α1 (benzodiazepine sensitive) and the α6 (benzodiazepine insensitive) subunits. Through the construction of a series of site-specific mutations, a histidine residue at position 101 of the α1 subunit was identified as forming part of the benzodiazepine-binding site (Wieland *et al.*, 1992). This residue is conserved among the α2, α3, and α5 subunits, but is replaced by an arginine residue in the α6 subunit. Mutation of His 101 of the α1 subunit to arginine generated a subunit with properties similar to those of the α6 subunit (Wieland *et al.*, 1992). Thus, this histidine residue is a major determinant of high-

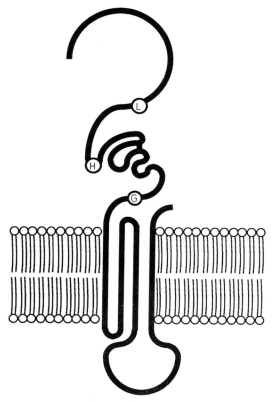

FIGURE 6 Domains of the GABA_A receptor involved in ligand binding. A single subunit of GABA_AR is illustrated; residues that are involved in the binding of agonists, antagonists, and allosteric modulators are highlighted. The transmembrane topology of the receptor is indicated. The amino acids illustrated are G, Gly, H, His, L, Leu.

affinity benzodiazepine binding to the α subunit of GABA_ARs.

Two different amino acid sequences have been published for the rat α1 subunit of the GABA_AR. The sequence published by Sigel *et al.*, (1990) differed from the sequence published by Lolait *et al.*, (1989) by a leucine residue at position 67 (L67) rather than a phenylalanine residue (F67). Sigel *et al.*, (1992) noted that receptors formed with the two different cDNAs in combination with the β2 and γ2 subunits showed a surprising difference in agonist sensitivity. The α1(F67)β2γ2 receptor showed a 200-fold greater affinity for GABA than the α1(L67)β2γ2 receptor. Mutation of the α1 phenylalanine to leucine and expression of the cDNA in combination with the β2 and γ2 subunits generated receptors that exhibited a 200-fold decrease in affinity, not only for GABA but also for the competitive antagonist bicuculline (Fig. 6). As mentioned previously the α, β, and γ subunits are capable of forming homomeric channels that may be gated by GABA.

Therefore, GABA must be capable of binding to, and gating, each subunit. When similar mutations were introduced at the corresponding positions of the β2 and γ2 subunits, little effect was seen on the potencies of GABA or bicuculline. Thus, researchers concluded that residue 67 of only the α subunit is an important determinant of GABA and bicuculline action (Sigel *et al.*, 1992). The lack of an effect of the mutation on the β and γ subunits suggests that this residue of the α subunit does not form part of a structural motif common to all subunits that is required for GABA recognition.

VII. Structural Predictions

The mutagenesis studies on the GlyR have allowed some structural predictions to be made for part of the ligand-binding site. In one region involved in forming the strychnine- and glycine-binding site, the even-numbered residues were identified as being involved in ligand recognition. Lys 200 and Tyr 202 form part of the binding site for the antagonist strychnine, whereas Thr 204 binds the agonist glycine. Mutational analysis of the intervening residues, His 201 and Asn 203, shows that these two residues are not involved in ligand binding. This pattern of alternating residues involved in ligand recognition may be explained if this region of the receptor adopts a β-sheet type structural motif (Fig. 7). In fact, secondary-structure prediction analysis of this sequence as well as of the corresponding region of the GABA_AR subunits suggests that a β sheet would form with a turn at the glycine residue immediately following the threonine residue at position 204. A mechanism for strychnine antagonism of glycine action, based on this structural prediction, has been proposed (Vandenberg *et al.*, 1992b). Although the strychnine- and glycine-binding sites appear to be separable, strychnine bound to the lysine and tyrosine residues provides a steric blockade and prevents glycine binding to Thr 204, thus providing an explanation for the functional antagonism of strychnine (Fig. 7).

The conserved disulfide loop found in the extracellular domain of all subunits of all the ligand-gated ion channel receptors (Grenningloh *et al.*, 1987; Schofield *et al.*, 1987) has been suggested to play a role in ligand recognition (Cockcroft *et al.*, 1990). Computer-aided modeling of the structure of this loop suggests that a rigid, amphiphilic, β hairpin is formed. An aspartate residue at position 11 within the loop is highly conserved and has been suggested to form part of an anionic site on the receptor that interacts with the positively charged amino group of the agonists glycine, GABA, acetylcholine, and serotonin. In the α1

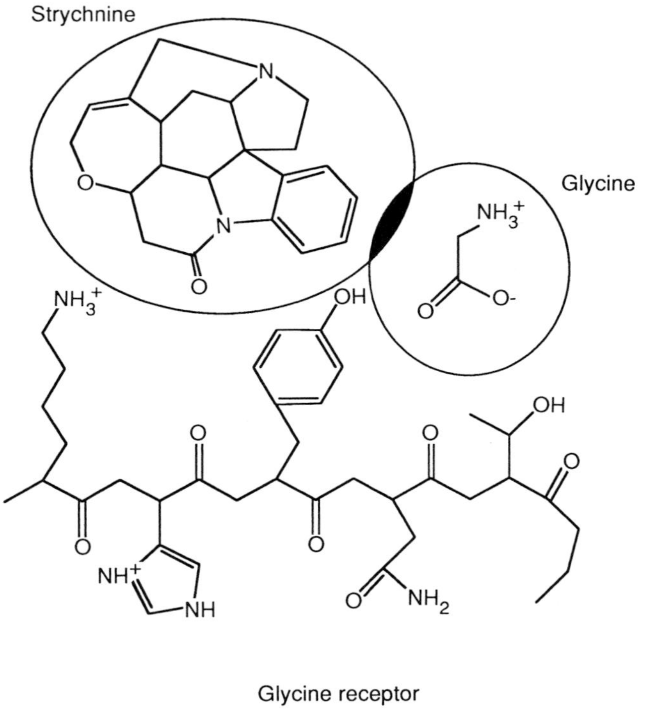

FIGURE 7 Model of the glycine receptor explaining the functional antagonism of strychnine. The amino acid sequence of the receptor illustrated includes the six residues Lys 200–His 201–Tyr 202–Asn 203–Thr 204–Gly 205 that form the agonist and antagonist binding site. The inability of strychnine and glycine to bind to the receptor simultaneously is indicated by the filled domain.

subunit of the GlyR, a lysine residue at position 6 of the loop is predicted to be juxtaposed to the aspartate residue and may form the site of interaction for the carboxyl group of the agonist glycine. The corresponding residue on the GABA$_A$R subunits is histidine, which may also be capable of interacting with the carboxyl group of GABA. Thus, Cockcroft et al., (1990) suggested that this residue may provide the necessary differences to allow the GlyR and GABA$_A$Rs to distinguish between their ligands. This hypothesis may be easily tested using site-directed mutagenesis.

VIII. Agonists, Antagonists, and Allosteric Modulators

The use of site-directed mutagenesis has been useful in determining which residues are important for ligand recognition. However, does this information reveal aspects of the structure of the subunits or how these receptors are activated, antagonized, or allosterically modulated? Although the experimental details are quite limited and are open to a number of interpre-

tations, predictions concerning the structure and function have been made. This chapter is primarily concerned with the inhibitory neurotransmitter-gated ion channels, but it is instructive to consider the studies on the excitatory nAChR also. The identity of ligand-binding sites on the nAChR has been extensively investigated by photoaffinity labeling and site-directed mutagenesis and has been reviewed by Galzi et al., (1991). Dennis et al., (1988) identified residues from three separate regions of the α subunit of the Torpedo nAChR that are involved in antagonist recognition. Alignment of the amino acid sequences of the nAChR α1 subunit with the GlyR and GABA$_A$R α subunits shows that the various residues involved in ligand recognition are located at similar positions within the extracellular domain (Fig. 8). The residues involved in antagonist recognition of the nAChR Tyr 93, Trp 149, Tyr 151, Tyr 190, Cys 192, Cys 193, and Tyr 198) are located in positions similar to those of GlyR residues involved in strychnine recognition (Gly 160, Tyr 161, Lys 200, Tyr 202) (see Fig. 8). Similarly, the two residues of the GABA$_A$R that have been identified as part of the benzodiazepine-binding site (His 101 and Gly 201) correspond to two residues of the nAChR that bind antagonists (Tyr 93 and Tyr 190) (see Fig. 8).

The coincident location of binding sites on the three receptors suggests that all adopt a similar tertiary structure and that these three regions form a cleft that is accessible to ligands. Variations in the key residues

FIGURE 8 Alignment of ligand binding sites of the glycine receptor, the GABA$_A$ receptor, and the nicotinic acetylcholine receptor. Three domains of ligand binding site alignment are observed. Residues involved in agonist, antagonist, or allosteric potentiator binding are boxed.

within the cleft would determine the particular pharmacological specificity of the receptor. However, a conundrum arises since the same residues that bind antagonists on the nAChR also appear to mediate allosteric potentiation at the GABA$_A$R. How does one class of molecule block receptor function and the other enhance it, when both act at the same residues? A mechanism by which this behavior may be achieved is one in which the antagonist locks the receptor molecule so that the conformational changes necessary for activation cannot take place, whereas the allosteric potentiator locks only a portion of the receptor molecule and, in so doing, effects a partial conformational change so, in the presence of agonist, a lower energy is required to achieve receptor activation.

The definition of the ionic basis of neuron inhibition and the pharmacological characterization of the GABA$_A$R and the GlyR (Eccles, 1964) preceded major advances in the elucidation of inhibitory receptor structure and function. The genes encoding the various subunits of the inhibitory GlyR and GABA$_A$R have been cloned, the role of specific subunits has been determined by analysis of recombinant receptors, and several amino acid residues that mediate ligand binding have been identified. These and other structural studies will form the basis for more detailed investigations aimed at understanding the molecular mechanism of drug action, and should provide the information necessary for the development of new receptor subtype-selective compounds for experimental and therapeutic use.

Acknowledgments

The work in the authors' laboratory was supported by the National Health and Medical Research Council (Australia) and the Cystic Fibrosis Foundation (USA).

References

Akagi, H., Hirai, K., and Hishinuma, F. (1991). Cloning of a glycine receptor subtype expressed in rat brain and spinal cord during a specific period of neuronal development. *FEBS Lett.* **281**, 160–166.

Aprison, M. H., and Wermann, R. (1965). The distribution of glycine in cat spinal cord and roots. *Life Sci.* **4**, 2075–2083.

Awapara, J. (1950). Detection and identification of metabolites in tissues by means of paper chromatography. *Fed. Proc.* **9**, 148.

Bateson, A. N., Lasham, A., and Darlison, M. G. (1991). γ-Aminobutyric acid$_A$ receptor heterogeneity is increased by alternative splicing of a novel β-subunit gene transcript. *J. Neurochem.* **56**, 1437–1440.

Betz, H. (1990). Ligand-gated ion channels in the brain: The amino acid receptor superfamily. *Neuron* **5**, 383–392.

Blair, L. A. C., Levitan, E. S., Marshall, J., Dionne, V. E., and Barnard, E. A. (1988). Single subunits of the GABA$_A$ receptor form ion channels with properties of the native receptor. *Science* **242**, 577–579.

Brisson, A., and Unwin, P. N. T. (1985). Quaternary structure of the acetylcholine receptor. *Nature (London)* **315**, 474–477.

Buckle, V. J., Fujita, N., Ryder-Cook, A. S., Derry, J. M. J., Barnard, P. J., Lebo, R. V., Schofield, P. R., Seeburg, P. H., Bateson, A. N., Darlison, M. G., and Barnard, E. A. (1989). Chromosomal localisation of GABA$_A$ receptor subunit genes: Relationship to human genetic disease. *Neuron* **3**, 647–654.

Burt, D. R., and Kamatchi, G. L. (1991). GABA$_A$ receptor subtypes: From pharmacology to molecular biology. *FASEB J.* **5**, 2916–1923.

Casalotti, S. O., Stephenson, F. A., and Barnard, E. A. (1986). Separate subunits for agonist and benzodiazepine binding in the γ-aminobutyric acid receptor complex. *J. Biol. Chem.* **261**, 15013–15016.

Cockcroft, V. B., Osguthorpe, D. J., Barnard, E. A., and Lunt, G. (1990). Modelling of agonist binding to the ligand-gated ion channel superfamily of receptors. *Proteins* **8**, 386–397.

Coombs, J. S., Eccles J. C., and Fatt, P. (1955). The specific ionic conductances and the ionic movements across the motoneuronal membrane that produce the inhibitory post-synaptic potential. *J. Physiol.* **130**, 326–373.

Curtis, D. R., Hosli, L., Johnston, G. A. R., and Johnston, I. H. (1968). Pharmacological study of the depression of spinal neurons by glycine and related amino acids. *Exp. Brain Res.* **6**, 1–18.

Curtis, D. R., Duggan A. E., Felix, D., and Johnston, G. A. R. (1970). GABA, bicuculline and central inhibition. *Nature (London)* **226**, 1222–1224.

Curtis, D. R., Duggan, A. W., and Johnston, G. A. R. (1971). The specificity of strychnine as a glycine antagonist in the mammalian spinal cord. *Exp. Brain Res.* **12**, 547–565.

Cutting, G. R., Lu, L., O'Hara, B. F., Kasch, L. M., Montrose-Rafizadeh, C., Donovan, D. M., Shimada, S., Antonarakis, S. E., Guggino, W. B., Uhl, G. R., and Kazazian, H. H. (1991). Cloning of the γ-aminobutyric acid (GABA) ρ1 cDNA: A GABA receptor subunit highly expressed in the retina. *Proc. Natl. Acad. Sci. USA* **88**, 2673–2677.

Cutting, G. R., Curristin, S., Zoghbi, H., O'Hara, B., Seldin, M. F., and Uhl, G. R. (1992). Identification of a putative γ-aminobutyric acid (GABA) receptor subunit rho$_2$ cDNA and colocalization of the genes encoding rho$_2$ (GABRR2) and rho$_1$ (GABRR1) to human chromosome 6q14-q21 and mouse chromosome 4. *Genomics* **12**, 801–806.

Darlison, M. G. (1992). Invertebrate GABA and glutamate receptors: Molecular biology reveals predictable structures but some unusual pharmacologies. *Trends Neurosci.* **15**, 469–474.

Dean, M., Lucas-Derse, S., Bolos, A., O'Brien, S. J., Kirkness, E. F., Fraser, C. M., and Goldman, D. (1991). Genetic mapping of the β1 GABA receptor gene to human chromosome 4, using a tetranucleotide repeat polymorphism. *Am. J. Hum. Genet.* **49**, 621–626.

DeLorey, T. M., and Olsen, R. W. (1992). γ-Aminobutyric acid$_A$ receptor structure and function. *J. Biol. Chem.* **267**, 16747–16750.

Deng, L., Ransom, R. W., and Olsen, R. W. (1986). [^{3}H]-Muscimol photolabels the GABA receptor site on a peptide subunit distinct from that labelled with benzodiazepine. *Biochem. Biophys. Res. Commun.* **138**, 1308–1314.

Dennis, M., Giraudet, J., Kotzyba-Hibert, F., Goeldner, M., Hirth, C., Chang J-Y., Lazure, C., Chrieten, M., and Changeux, J-P. (1988). Amino acids of the *Torpedo mamorata* acetylcholine receptor α-subunit labelled by a photoaffinity ligand for the acetylcholine binding site. *Biochemistry* **27**, 2345–2351.

Eccles, J. C. (1964). The ionic mechanism of postsynaptic inhibition. *In* "Les Prix Nobel en 1963," pp 261–283. Norstedt, Stockholm.

Falck, F. A. (1884). Ueber den Einfluss des Alters auf die Wirkung des Strychnins. *Arch. Ges. Physiol.* **34**, 530–575.

Galzi, J.-L., Revah, F., Bessis, A., and Changeux, J.-P. (1991). Functional architecture of the nicotinic acetylcholine receptor: From electric organ to brain. *Annu. Rev. Pharmacol.* **31**, 37–72.

Galzi, J-L., Devillers-Thiery, A., Hussy, N., Bertrand, S., Changeux, J-P., and Bertrand, D. (1992). Mutations in the channel domain of a neuronal nicotinic receptor convert ion selectivity from cationic to anionic. *Nature (London)* **359**, 500–505.

Gottlieb, D. I. (1988). GABAergic neurons. *Sci. Am.* **258**, 38–45.

Graham, D., Pfeiffer, F., and Betz, H. (1983). Photoaffinity-labeling of the glycine receptor of rat spinal cord. *Eur. J. Biochem.* **131**, 519–525.

Grenningloh, G., Rienitz, A., Schmitt, B., Methfessel, C., Zensen, M., Beyreuther, K., Gundelfinger, E. D., and Betz, H. (1987). The strychnine-binding subunit of the glycine receptor shows homology with the nicotinic acetylcholine receptors. *Nature (London)* **328**, 215–220.

Grenningloh, G., Schmieden, V., Schofield, P. R., Seeburg, P. H., Siddique, T., Mohandas, T. K., Becker, C.-M., and Betz, H. (1990a). Alpha subunit variants of the human glycine receptor: Primary structures, functional expression and chromosomal localization of the corresponding genes. *EMBO J.* **9**, 771–776.

Grenningloh, G., Pribilla, I., Prior, P., Multhorpe, G., Beyreuther, K., Taleb, O., and Betz, H. (1990b). Cloning and expression of the 58kD β-subunit of the inhibitory glycine receptor. *Neuron* **4**, 963–970.

Herb, A., Wisden, W., Luddens, H., Puia, G., Vicini, S., and Seeburg, P. H. (1992). The third γ subunit of the γ-aminobutyric acid type A receptor family. *Proc. Natl. Acad. Sci. USA* **89**, 1433–1437.

Imoto, K., Busch, C., Sakmann, B., Mishina, M., Konno, T., Nakai, J., Bujo, H., Mori, Y., Fukuda, K., and Numa, S. (1988). Rings of negatively charged amino acids determine the acetylcholine receptor channel conductance. *Nature (London)* **335**, 645–648.

Johnston, G. A. R. (1986). Multiplicity of GABA receptors. *In* "Benzodiazepine/GABA Receptors and Chloride Channels: Structural and Functional Properties" (R. W. Olsen and J. C. Venter, eds.), pp. 57–72. Liss, New York.

Johnston, G. A. R., Kerr D. I. B., and Ong, J. (1987). Stress, steroids and GABA receptors. *In* "Pharmacology" (M. J. Rand and C. Raper, eds.), pp. 121–124. Elsevier, Amsterdam.

Kerr, D. I. B., and Ong, J. (1992). GABA agonists and antagonists. *Med. Res. Rev.* **12**, 593–636.

Khrestchatisky, M., Maclennan, A. J., Chiang, M-Y., Xu, W., Jackson, M. B., Brecha, N., Sternini, C., Olsen, R. W., and Tobin, A. J. (1989). A novel alpha subunit in rat brain GABA_A receptors. *Neuron* **3**, 745–753.

Kirkness, E. F., Kusiak, J. W., Fleming, J. T., Menninger, J., Gocayne, J. D., Ward, D. C., and Venter, J. C. (1991). Isolation, characterization, and localization of human genomic DNA encoding the β subunit of the GABA_A receptor (GABRB1). *Genomics* **10**, 985–995.

Klepner, C. A., Lippa, A. S., Benson, D. I., Sano, M. C., and Beer, B. (1979). Resolution of two biochemically and pharmacologically distinct benzodiazepine receptors. *Pharmacol. Behav.* **11**, 456–462.

Knoflach, F., Rhyner, T., Villa, M., Kellenberger, S., Drescher, U., Malherbe, P., Sigel, E., and Mohler, H. (1991). The γ3-subunit of the GABA_A-receptor confers sensitivity to benzodiazepine receptor ligands. *FEBS Lett.* **293**, 191–194.

Kofuji, P., Wang, J. B., Moss, S. J., Huganir, R. L., and Burt, D. R. (1991). Generation of two forms of the γ-aminobutyric acid_A receptor γ2-subunit in mice by alternative splicing. *J. Neurochem.* **56**, 713–715.

Krogsgaard-Larsen, P., Johnston, G. A. R., Lodge, D., and Curtis, D. R. (1977). A new class of GABA agonist. *Nature (London)* **268**, 53–55.

Kuhse, J., Schmieden, V., and Betz, H. (1990a). Identification and functional expression of a novel ligand binding subunit of the inhibitory glycine receptor. *J. Biol. Chem.* **265**, 22317–22320.

Kuhse, J., Schmieden, V., and Betz, H. (1990b). A single amino acid exchange alters the pharmacology of neonatal rat glycine receptor subunit. *Neuron* **5**, 867–873.

Kuhse, J., Kuryatov, A., Maulet, Y., Malosio, M. L., Schmieden, V., and Betz, H. (1991). Alternative splicing generates two isoforms of the α2 subunit of the inhibitory glycine receptor. *FEBS Lett.* **283**, 73–77.

Langosch, D., Thomas, L., and Betz, H. (1988). Conserved quaternary structure of ligand gated ion channels; The postsynaptic glycine receptor is a pentamer. *Proc. Natl. Acad. Sci. USA* **85**, 7394–7398.

Langosch, D., Becker, C.-M., and Betz, H. (1990). The inhibitory glycine receptor: A ligand-gated chloride channel of the central nervous system. *Eur. J. Biochem.* **194**, 1–8.

Lasham, A., Vreugdenhil, E., Bateson, A. N., Barnard, E. A., and Darlison, M. G. (1991). Conserved organization of γ-aminobutyric acid_A receptor genes: Cloning and analysis of the chicken β4-subunit gene. *J. Neurochem.* **57**, 352–355.

Leeb-Lundberg, L. M. F., and Olsen, R. W. (1983). Heterogeneity of benzodiazepine receptor interactions with GABA and barbiturate receptors. *Mol. Pharmacol.* **23**, 315–325.

Levitan, E. S., Schofield P. R., Burt, D. R., Rhee L. M., Wisden W., Kohler, M., Fujita, N., Rodriguez, H., Stephenson, A., Darlison, M. G., Barnard, E. A., and Seeburg, P. H. (1988). Structural and functional basis for GABA_A receptor heterogeneity. *Nature (London)* **335**, 76–79.

Lolait, S. J., O'Carroll, A.-M., Kusano, K., Muller, J.-M., Brownstein, M. J., and Mahan, L. C. (1989). Cloning and expression of a novel rat GABA_A receptor. *FEBS Lett.* **246**, 145–148.

Lüddens, H., Pritchett, D. B., Kohler, M., Killisch, I., Keinanen, K., Monyer, H., Sprengel, R., and Seeburg P. H. (1990). Cerebellar GABA_A receptor selective for a behavioural alcohol antagonist. *Nature (London)* **346**, 648–651.

Malherbe, P., Sigel, E., Baur, R., Persohn, E., Richards, J. G., and Mohler, H. (1990). Functional expression and sites of gene transcription of a novel α subunit of the GABA_A receptor in rat brain. *FEBS Lett.* **260**, 261–265.

Malosio, M. L., Grenningloh, G., Kuhse, J., Schmieden, V., Schmitt, B., Prior, P., and Betz, H. (1991). Alternative splicing generates two variants of the α1 subunit of the inhibitory glycine receptor. *J. Biol. Chem.* **266**, 2048–2053.

Mamlaki, C., Stephenson, F. A., and Barnard, E. A. (1987). The GABA_A/benzodiazepine receptor is a heterotetramer of homologous α and β subunits. *EMBO J.* **6**, 561–565.

Manske, R. H. F. (1932). The alkaloids of fumaraceous plants II. *Dicentra cucullaria. Can J. Res.* **8**, 265–272.

Marvison, J. C., Vazquez, J., Calvo, M. G., Mayor, F., Gomez, A. R., Valdivieso, F., and Benavides J. (1986). The glycine receptor: Pharmacological studies and mathematical modelling of the allosteric interaction between the glycine- and strychnine-binding sites. *Mol. Pharmacol.* **30**, 590–597.

Mohler, H., and Okada, T. (1977). Benzodiazepine receptor: Demonstration in the central nervous system. *Science* **198**, 849–851.

Moss, S. J., Smart, T. G., Blackstone, C. D., and Huganir, R. L. (1992). Functional modulation of GABA_A receptors by cAMP-dependent protein phosphorylation. *Science* **257**, 661–665.

Pfeiffer, F., Graham, D., and Betz, H. (1982). Purification by affinity chromatography of the glycine receptor of rat spinal cord. *J. Biol. Chem.* **257**, 818–823.

Pritchett, D. B., and Seeburg, P. H. (1990). γ-Aminobutyric acid A receptor α5 subunit creates novel type II benzodiazepine receptor pharmacology. *J. Neurochem.* **54**, 1802–1804.

Pritchett, D. B., and Seeburg, P. H. (1991). γ-Aminobutyric acid type A receptor point mutation increases the affinity of compounds for the benzodiazepine site. *Proc. Natl. Acad. Sci. USA* **88**, 1421–1425.

Pritchett, D. B., Sontheimer, H., Gorman, C. M., Kettenmann, H., Seeburg, P. H., and Schofield, P. R. (1988). Transient expression shows ligand gating and allosteric potentiation of GABA$_A$ receptor subunits. *Science* **242**, 1306–1308.

Pritchett, D. B., Sontheimer, H., Shivers, B. D., Ymer, S., Kettenmann, H., Schofield, P. R., and Seeburg, P. H. (1989a). Importance of a novel GABA$_A$ receptor subunit for benzodiazepine pharmacology. *Nature (London)* **338**, 582–585.

Pritchett, D. B., Luddens, H., and Seeburg, P. H. (1989b). Type I and Type II GABA$_A$-benzodiazepine receptors produced in transfected cells. *Science* **245**, 1389–1392.

Revah, F., Bertrand, D., Galzi, J.-L., Devillers-Thiery, A., Mulle, C., Hussy, N., Bertrand, S., Ballivet, M., and Changeux, J.-P. (1991). Mutations in the channel domain alter desensitization of a neuronal nicotinic receptor. *Nature (London)* **353**, 846–849.

Roberts, E., and Frankel, S. (1950). γ-Aminobutyric acid in brain: Its formation from glutamic acid. *J. Biol. Chem.* **187**, 55–63.

Ruiz-Gomez, A., Morato, E., Garcia-Calco, M., Valdivieso, F., and Mayor, F. (1990). Localization of the strychnine binding site on the 48-kilodalton subunit of the glycine receptor. *Biochemistry* **29**, 7033–7040.

Schmieden, V., Grenningloh, G., Schofield, P. R., and Betz, H. (1989). Functional expression in *Xenopus* oocytes of the strychnine binding 48kd subunit of the glycine receptor. *EMBO J.* **8**, 695–700.

Schmieden, V., Kuhse, J., and Betz, H. (1992). Agonist pharmacology of neonatal and adult glycine receptor α subunits: Identification of amino acid residues involved in taurine action. *EMBO J.* **11**, 2025–2032.

Schofield, P. R., Darlison, M. G., Fujita, N., Burt, D. R., Stephenson, F. A., Rodriguez, H., Rhee, L. M., Ramachandran, J., Reale, V., Glencorse, T. A., Seeburg, P. H., and Barnard, E. A. (1987). Sequence and functional expression of the GABA$_A$ receptor shows a ligand-gated receptor super-family. *Nature (London)* **328**, 221–227.

Shimada, S., Cutting, G., and Uhl, G. R. (1992). γ-Aminobutyric acid A or C receptor? γ-Aminobutyric acid ρ1 receptor RNA induces bicuculline-, barbiturate-, and benzodiazepine-insensitive γ-aminobutyric acid responses in *Xenopus* oocytes. *Mol. Pharmacol.* **41**, 683–687.

Shivers, B. D., Killisch, I., Sprengel, R., Sontheimer, H., Kohler, M., Schofield, P. R., and Seeburg, P. H. (1989). Two novel GABA$_A$ receptor subunits exist in distinct neuronal subpopulations. *Neuron* **3**, 327–337.

Siddique, T., Phillips, K., Betz, H., Grenningloh, G., Warner, K., Hung, W.-Y., Laing, N., and Roses, A. D. (1989). RFLPs of the gene for the human glycine receptor on the X-chromosome. *Nucleic Acids Res.* **17**, 1785.

Sieghart, W., and Karobath, M. (1980). Molecular heterogeneity of benzodiazepine receptors. *Nature (London)* **286**, 285–287.

Sigel, E., and Barnard E. A. (1984). A gamma-aminobutyric acid/benzodiazepine receptor complex from bovine cerebral cortex. *J. Biol. Chem.* **259**, 7219–7223.

Sigel, E., Stephenson, F. A., Mamlaki, C., and Barnard, E. A. (1983). A gamma-aminobutyric acid/benzodiazepine receptor complex of bovine cerebral cortex: Purification and partial characterization. *J. Biol. Chem.* **258**, 6965–6971.

Sigel, E., Baur, R., Trube, G., Mohler H., and Malherbe, P. (1990).

The effect of subunit composition of rat brain GABA$_A$ receptors on channel function. *Neuron* **5**, 703–711.

Sigel, E., Baur, R., Kellenberger, S., and Malherbe, P. (1992). Point mutations affecting antagonist affinity and agonist dependent gating of GABA$_A$ receptor channels. *EMBO J.* **11**, 2017–2023.

Simmonds, M. A. (1991). Modulation of the GABA$_A$ receptor by steroids. *Sem. Neurosci.* **3**, 231–239.

Simmonds, M. A., and Turner, J. P. (1985). Antagonism of inhibitory amino acids by the steroid derivative RU 5135. *Br. J. Pharmacol.* **84**, 631–635.

Sommer, B. M., Poutska, A., Spurr, N. K., and Seeburg, P. H. (1990). The murine GABA$_A$ receptor δ-subunit gene: Structure and assignment to human chromosome 1. *DNA Cell Biol.* **9**, 561–568.

Sontheimer, H., Becker, C.-M., Pritchett, D. B., Schofield, P. R., Grenningloh, G., Kettenmann, H., Betz, H., and Seeburg, P. H. (1989). Functional chloride channels by mammalian cell expression of rat glycine receptor subunit. *Neuron* **2**, 1491–1497.

Squires, R. F., and Braestrup, C. (1977). Benzodiazepine receptors in rat brain. *Nature (London)* **266**, 732–734.

Tallman, J. F., Thomas J. W., and Gallager, D. W. (1978). GABAergic modulation of benzodiazepine binding site sensitivity. *Nature (London)* **274**, 383–385.

Ticku, M. K. (1991). Drug modulation of GABA$_A$ mediated transmission. *Sem. Neurosci.* **3**, 211–218.

Tokutomi, N., Kaneda, M., and Akaike. (1989). What confers specificity on glycine for its receptor site? *Br. J. Pharmacol.* **97**, 353–360.

Toyoshima, C., and Unwin, N. (1990). Three-dimensional structure of the acetylcholine receptor by the cryoelectron microscopy and helical image reconstruction. *J. Cell Biol.* **111**, 2623–2635.

Undenfriend, S. (1950). Identification of γ-aminobutyric acid in brain by isotope derivative method. *J. Biol. Chem.* **187**, 65–69.

Vandenberg, R. J., French, C. R., Barry, P. H., Shine, J., and Schofield, P. R. (1992a). Antagonism of ligand-gated ion channel receptors: Two domains of the glycine receptor α subunit form the strychnine-binding site. *Proc. Natl. Acad. Sci. USA* **89**, 1765–1769.

Vandenberg, R. J., Handford, C. A., and Schofield, P. R. (1992b). Distinct agonist- and antagonist-binding sites on the glycine receptor. *Neuron* **9**, 491–496.

Wafford, K. A., Burnett, D. M., Leidenheimer, N. J., Burt, D. R., Wang, J. B., Kofuji, P., Dunwiddie, T. V., Harris, R. A., and Sikela, J. M. (1991). Ethanol sensitivity of the GABA$_A$ receptor expressed in *Xenopus* oocytes requires eight amino acids contained in the γ2$_L$ subunit. *Neuron* **7**, 27–33.

Wagstaff, J., Knoll, J. H. M., Fleming, J., Kirkness, E. F., Martin-Gallardo, A., Greenberg, F., Graham, Jr., J. M., Menninger, J., Ward, D., Venter, J. C., and Lalande, M. (1991a). Localization of the gene encoding the GABA$_A$ receptor β3 subunit to the Angelman/Prader–Willi Region of human chromosome 15. *Am. J. Hum. Genet.* **49**, 330–337.

Wagstaff, J., Chaillet, J. R., and Lalande, M. (1991b). The GABA$_A$ receptor β3 subunit gene: Characterization of a human cDNA from chromosome 15q11q13 and mapping to a region of conserved synteny on mouse chromosome 7. *Genomics* **11**, 1071–1078.

Whiting P., McKernan R. M., and Iverson L. L. (1990). Another mechanism for creating diversity in γ-aminobutyrate type A receptors: RNA splicing directs expression of two forms of γ2 subunit, one of which contains a protein kinase C site. *Proc. Natl. Acad. Sci. USA* **87**, 9966–9970.

Wieland, H. A., Luddens, H., and Seeburg, P. H. (1992). A single histidine in GABA$_A$ receptors is essential for benzodiazepine agonist binding. *J. Biol. Chem.* **267**, 1426–1428.

Wilcox, A. S., Warrington, J. A., Gardiner, K., Berger, R., Whiting,

P., Altherr, M. R., Wasmuth, J. J., Patterson, D., and Sikela, J. M. (1992). Human chromosomal localization of genes encoding the γ1 and γ2 subunits of the γ-aminobutyric acid receptor indicates that members of this gene family are often clustered in the genome. *Proc. Natl. Acad. Sci. USA* **89,** 5857–5861.

Wilson-Shaw, D., Robinson, M., Gambarana, C., Siegel, R. E., and Sikela, J. M. (1991). A novel γ subunit of the GABA_A receptor identified using the polymerase chain reaction. *FEBS Lett.* **284,** 211–215.

Wisden, W., Herb, A., Wieland H., Keinanen, K., Luddens, H., and Seeburg, P. H. (1991). Cloning, pharmacological characteristics and expression pattern of the rat GABA_A receptor α4 subunit. *FEBS Lett.* **289,** 227–230.

Ymer, S., Schofield, P. R., Draguhn, A., Werner, P., Kohler, M., and Seeburg, P. H. (1989). GABA_A receptor β subunit heterogeneity: Functional expression of cloned cDNAs. *EMBO J.* **8,** 1665–1670.

Ymer, S., Draguhn, A., Wisden, W., Werner, P., Keinanen, K., Schofield, P. R., Sprengel., R., Pritchett, D. B., and Seeburg, P. H. (1990). Structural and functional characterization of the γ1 subunit of GABA_A/benzodiazepine receptors. *EMBO J.* **9,** 3261–3267.

Young, A. B., and Snyder, S. H. (1974). Strychnine binding in rat spinal cord membranes associated with the synaptic glycine receptor: Cooperativity of glycine interactions. *Mol. Pharmacol.* **10,** 790–809.

Cyclic-Nucleotide-Sensitive Channels

Hyperpolarization-Activated (i_f) Current in Heart

DARIO DiFRANCESCO

I. Introduction

Of the ionic currents that contribute to the electrical activity of cardiac myocytes, the so-called "pacemaker" (i_f) current is possibly the one with the most unusual features. This current is activated on hyperpolarization, a property unique among cardiac voltage-gated inward currents; is nonspecifically permeable to monovalent cations, yet has a very low single-channel conductance; and is modulated by the second messenger cAMP but, unlike all other voltage-gated cAMP-dependent currents, its modulation is not mediated by protein kinase A (PKA)-mediated phosphorylation and occurs via direct binding of cAMP to the channel protein.

The functional relevance of these uncommon properties, however, becomes apparent from the role of i_f in cardiac tissue. Pacemaker activity is initiated in sinoatrial (SA) node cells by a slow (diastolic) depolarization phase that develops after repolarization of the action potential. I_f serves a definite purpose here for which its properties are clearly specifically designed: under normal conditions, it begins the diastolic depolarization and controls its rate of development, thereby allowing the onset of spontaneous activity of pacemaker cells. The simplest way in which a depolarizing process can be generated is the activation on

hyperpolarization of a slowly activating inward current (which opposes the hyperpolarizing action of the decaying outward current responsible for action potential repolarization). Even more importantly, the cAMP dependence of i_f is the key mechanism by which neurotransmitters control the rate of development of the diastolic depolarization and, hence, of spontaneous activity. Thus, the properties of i_f are well suited for the generation and neurotransmitter-induced control of pacemaker activity.

This chapter reviews current knowledge of the properties and functional significance of the hyperpolarization-activated i_f current in the mammalian SA node and discusses the role of i_f in the neurotransmitter-induced modulation of cardiac rate.

II. Hyperpolarization-Activated Current of Sinoatrial Node and Pacemaker Activity

A. Properties of i_f in Mammalian Pacemaker Tissue

The hyperpolarization-activated current was characterized for the first time and termed i_f in 1979, in a study of multicellular preparations from the rabbit SA node (Brown *et al.*, 1979a,b). In previous work, an inward current activated on hyperpolarization had been observed in mammalian (Noma and Irisawa, 1976) and amphibian (Brown et al., 1977) pacemaker tissue, but its activation threshold appeared to be too negative for the current to contribute to normal electrical activity. This observation led to the suggestion that the recorded transient could be an artifact (Noma and Irisawa, 1976). In some of the later studies in SA node preparations, the same current was termed i_h (Yanagihara and Irisawa, 1980), although later the i_f terminology was generally agreed on (Irisawa and Hagiwara, 1991).

From the original and subsequent work on the SA node, and from the demonstration that the nodal i_f and the "pacemaker" current in Purkinje fibers were identical currents (DiFrancesco and Ojeda, 1980; DiFrancesco, 1981a), the ionic and kinetic properties of i_f were characterized in some detail. In the SA node and in Purkinje fibers, i_f was shown to be an unspecific inward cation current, normally carried by Na^+ and K^+, and activated on hyperpolarization to voltages within the diastolic range (Brown and DiFrancesco, 1980; Yanagihara and Irisawa, 1980; DiFrancesco, 1981b; DiFrancesco et al., 1986).

As a first observation indicating the presence of a connection between i_f activation threshold and range of diastolic depolarization, note that the diastolic depolarization range is more negative in Purkinje fibers than in SA node cells and that, correspondingly, Purkinje fibers have a more negative i_f activation range than SA node cells. In Fig. 1, action potentials (A), i_f current recordings (B), i_f activation curves, and fully activated current–voltage (I–V) relationships (C) are shown for the two cardiac preparations. Comparison of the two preparations indicates that, whereas the fully activated curves have similar shapes and reversal potentials (E_f between -10 and -15 mV), the activation curve midpoint in the SA node myocyte (-64 mV) is 28 mV more positive than in the Purkinje fiber (-92 mV). The position of the current activation curve is controlled by intracellular cAMP, which is the second messenger that mediates the action of autonomic transmitters on i_f (see subsequent discussion). Thus, the different positions of the i_f activation curves for the two preparations may, at least partly, reflect a different basal cAMP concentration. One way in which this difference has been investigated is discussed in Section II,C,2.

B. Generation of the Diastolic Depolarization

The ionic and kinetic properties of i_f, as summarized in Fig. 1, are appropriate for the generation of a slow depolarizing phase following repolarization of the action potential. The way in which this can be

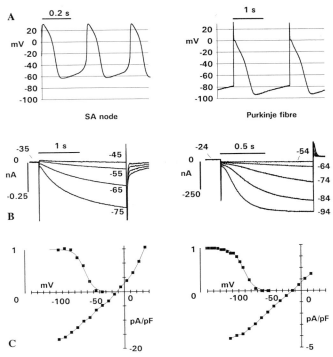

FIGURE 1 A comparison between i_f in the SA node and in the Purkinje fibers. Left panels show action potentials (A), i_f records during hyperpolarizations (B), and the i_f activation curve and fully-activated I/V relation (C) in single SA node cells, whereas right panels show similar recordings from Purkinje fibers. Notice that in both cases i_f is activated in a voltage range comprising the diastolic depolarization range. Also note the more hyperpolarized position of both diastolic depolarization and i_f activation curve in Purkinje fibers. The latter has a half-maximum voltage of -64 mV in the SA node cell (C, left) and -92 mV in the Purkinje fiber (C, right). In both cases the fully activated current is normalized to membrane capacity.

achieved is illustrated in the model reconstruction shown in Fig. 2. The top records show action potentials computed using the DiFrancesco–Noble model (DiFrancesco and Noble, 1985); the bottom records represent the time course of the net ionic current and of some of the components contributing to activity (currents i_{Ca}, i_K, i_f, and i_b). The "background" i_b, or "instantaneous," component is the net sum of all time-independent currents and therefore comprises electrogenic pump and exchange components. In any single cell, of course, the total ionic current (I_{net}) is the determinant, at any time, of the time course of voltage according to the simple relationship $I_{net} = -C \, dV/dt$, where C is cell capacity and V is voltage. A few basic considerations can be made from this relationship. First, a direct correlation exists not so much between current and voltage as between current and voltage derivative; this relationship implies that

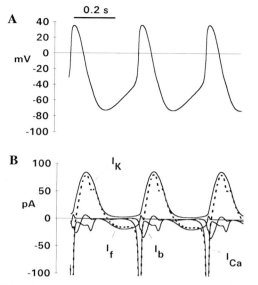

FIGURE 2 Computed spontaneous activity of a single SA node cell generated by the DiFrancesco–Noble model (DiFrancesco and Noble, 1985). (A) action potentials; (B) time course of some of the ionic currents relevant to activity (i_f, i_b, i_{Ca}, and i_K, as indicated). The total membrane ionic current ($= -C \, dV/dt$) is drawn as a broken line. The calculation was performed using the OXSOFT HEART program with parameters as in DiFrancesco (1991) (Appendix B by Noble, Denyer, and DiFrancesco) modified as follows: the mid-point voltage of i_f activation was -64 mV, and the f-conductance was 30 pA/mV (both K and Na components). These parameters yield a spontaneous frequency of about 3 Hz. Note that the net "background" current is essentially zero throughout the diastolic depolarization.

the rate of voltage change can also be controlled by components with slow or no time dependence. For example, simple addition of a constant component I modifies voltage derivative by an amount equal to $-I/C$. Second, very small current changes can bring about relatively large modifications in dV/dt. For example, a typical diastolic depolarization rate of 0.1 V/sec of an SA node cell with mean capacity of about 30 pF (DiFrancesco, 1986; Denyer and Brown, 1990) requires only 3 pA net current; thus, even a minute change of a constant 1 pA, for example, would alter the voltage derivative greatly and would substantially affect spontaneous rate.

The reconstruction of Fig. 2 was made on the basis of experimental data (DiFrancesco, 1991) indicating that the net instantaneous "background" current is outward at voltages of -60 to -65 mV. Since any depolarization requires a net inward current, and normal maximum diastolic potentials (MDP) are typically in the range of -60 to -70 mV, this result implies that the diastolic depolarization cannot be driven by the background current (i_K-decay hypothesis). There-

fore, a direct consequence is that, under normal conditions, i_f activation is the process that initiates the diastolic depolarization (i_f-activation hypothesis). Note that individual time-independent components (for example, the Na^+/Ca^{2+} exchange current) may be *inward* even if the *net* instantaneous component is not. However, since the net current is the determinant of the slope of voltage change (according to the equation given), the inward contribution of background currents is itself unable to initiate a depolarizing process. In the reconstruction of Fig. 2B, i_b is close to zero throughout most of the diastolic depolarization; essentially all the inward current during this phase is supplied by i_f. Thus, the total ionic current (broken line) and i_f have a nearly superimposable time course. A similar conclusion was derived from experimental data in the isolated SA node cell (van Ginneken and Giles, 1991).

C. Autonomous Modulation of Pacemaker Activity and Involvement of i_f

The SA node is the cardiac region receiving the most dense autonomic input and the site of modulation of mammalian heart rate by neurotransmitters. A positive chronotropic effect is exerted by β-adrenergic stimulation; a negative chronotropic effect is generated by muscarinic stimulation. Several ionic components are affected by both adrenergic and cholinergic transmitters; the resulting action on rate thus depends on the action of any individual component. However, experiments indicate that, at low agonist concentrations, one single component may be mainly responsible for the chronotropic action of both β-adrenergic and cholinergic stimulation. For example, low agonist concentrations modify sinus rate by altering the diastolic depolarization only, without affecting other phases of the action potential (DiFrancesco, 1993). As discussed subsequently, at low agonist concentrations i_f modulation appears to be the primary mechanism mediating rate control.

1. β-Adrenergic Modulation

Since i_f was first described in the SA node, it became apparent that its activation was modulated by β-adrenergic stimulation and that this mechanism had an important role in catecholamine-induced rate acceleration (Brown *et al.*, 1979a). Previous work in the Purkinje fibers had indicated that the "pacemaker" current i_{K2}, later re-interpreted and shown to be identical to i_f (DiFrancesco, 1981a), mediated the positive chronotropic effect of catecholamines in the conduction tissue (Hauswirth *et al.*, 1968). In the SA node and the Purkinje fiber, modulation occurs via a shift of the current activation curve to the positive direction, in the absence of a modification of the fully activated current (Tsien, 1974; DiFrancesco *et al.*, 1986). A simple shift of the activation curve could be interpreted in terms of the screening effect of a charged molecule; however, the voltage dependence of the activation variable becomes steeper in the presence of adrenergic stimulation, implying that the activating process must be more complex than that induced by a surface-charge agent (Tsien, 1974; Hart *et al.*, 1980).

A β-adrenergic stimulus accelerates the diastolic depolarization via i_f modulation by increasing adenylyl-cyclase activity, raising the intracellular cAMP concentration and leading to a positive shift of the i_f activation curve; thus, more i_f current will be available for driving the diastolic depolarization process, resulting in an acceleration of the depolarization itself and, hence, of spontaneous rate.

Essentially no chronotropic action is exerted on adult SA node by α-adrenergic modulation (Hewett and Rosen, 1985), although spontaneous activity of other adult and neonatal cardiac preparations is dependent on α-adrenergic stimuli (see Rosen, 1990, for review).

2. Cholinergic Modulation

The notion that acetylcholine (ACh) slows heart rate by muscarinic-induced K^+-permeability increase was first introduced by experiments showing that vagal stimulation causes a K^+-dependent membrane hyperpolarization (Hoffman and Suckling, 1953; Hutter and Trautwein, 1956). This observation led to the well-established view that the process underlying the negative chronotropic action of ACh in pacemaker cells is the activation of a K^+ current ($i_{K,ACh}$), the properties of which have since been studied in detail in intact preparations and in single cells, and have been characterized at the single-channel level (Giles and Noble, 1976; Garnier *et al.*, 1978; Noma and Trautwein, 1978; Sakmann *et al.*, 1983; Soejima and Noma, 1984; Pfaffinger *et al.*, 1985).

However, this view was challenged by the finding that the i_f current is also modulated by ACh. I_f is strongly inhibited by muscarinic stimulation in a way that can account for most of the slowing action of ACh at low doses (DiFrancesco *et al.*, 1989). That other components could be involved in the slowing action had been anticipated by the observation that, for example, slowing induced by moderate vagal stimuli or low ACh concentrations was not accompanied by membrane hyperpolarization (Giles *et al.*, 1989). The possibility that the Ca^{2+} current, also inhibited by ACh in amphibian tissue (Giles and Noble, 1976), could be responsible for the hyperpolarization-

independent slowing action in mammals was not supported by evidence that this component is affected by low ACh doses (Brown *et al.*, 1984).

These unexpected results received a natural explanation with the finding that the i_f current is also under muscarinic control (DiFrancesco and Tromba, 1987). The action of ACh on i_f is qualitatively opposite to that of catecholamines. I_f is inhibited by ACh via a shift of the current activation curve to more negative voltages, which reduces the degree of current available for initiating the diastolic depolarization; further, like catecholamines, ACh does not alter the i_f fully activated I–V relationship (DiFrancesco and Tromba, 1988a). These results are easily accounted for by the hypothesis that ACh acts on i_f via a mechanism opposite to that operating with catecholamines, that is, an inhibition of adenylyl-cyclase and reduction of intracellular cAMP. Indeed, this conclusion was confirmed by experiments based on receptor-independent manipulation of intracellular cAMP levels which demonstrated the involvement of cAMP in the ACh action (DiFrancesco and Tromba, 1988b).

These results stressed the importance of cAMP as second messenger in i_f control by neurotransmitters. In pacemaker cells, cAMP is up-regulated by β-adrenergic agonists and down-regulated by muscarinic agonists; thus, i_f sensitivity to neurotransmitters simply reflects a sensitivity to cAMP. This fact may provide an explanation for the difference, seen in Fig. 1, between the positions of the i_f activation curves in the SA node and the Purkinje fiber. A more hyperpolarized position of the activation curve could be associated, at least partly, with a lower basal level of adenylyl-cyclase activity in the Purkinje fiber than in the SA node. Indeed, in Purkinje fibers, ACh inhibits i_f only after β stimulation, whereas in the SA node it also inhibits basal i_f (Chang *et al.*, 1990).

The finding of ACh-induced i_f modulation also laid the groundwork for re-evaluating the mechanism of action of ACh on cardiac slowing. Researchers could show that the ACh concentrations effective on i_f were much lower than those needed for $i_{K,ACh}$ activation. I_f is half-maximally inhibited by 0.013 μM ACh, a dose some 20-fold lower than that required to half-activate $i_{K,ACh}$ (0.26 μM; DiFrancesco *et al.*, 1989). These observations imply that low doses of ACh slow rhythm by inhibiting i_f and not, as previously thought, by activating a K^+ conductance.

Some experimental findings made in intact tissue provide a natural explanation for these phenomena if the action on i_f of low ACh doses is considered. For example, Hirst *et al.* (1992) reported that, although vagal stimulation slows pacing rate without hyperpolarization of the maximum diastolic potential, slowing induced by superfusion with 10 μM ACh is accompanied by a strong hyperpolarization. To explain their results, these researchers proposed that, as well as activating a K^+ conductance (at this relatively large dose), ACh can also inhibit a "background" Na^+ conductance. This hypothetical ACh-dependent "background" conductance is unnecessary, since these results can be explained simply by i_f blockade at low ACh doses, which is indeed known to slow spontaneous rate without modifying maximum diastolic potential and action potential configuration (DiFrancesco *et al.*, 1989; DiFrancesco, 1993).

Experiments on SA node myocytes indicate that ACh concentrations effective at reducing the Ca current are several-fold higher than those inhibiting i_f (Zaza, Robinson & DiFrancesco, 1992). Thus, ACh concentrations as high as 0.03 μM can be considered as acting specifically on the i_f current.

III. Single-Channel Properties of the i_f Current

Although the "pacemaker" current was one of the first cardiac currents to be analyzed in detail in multicellular preparations (Noble and Tsien, 1968), single pacemaker channels were resolved only several years later. The difficulty of resolving i_f channels was created by the very small elementary current size. Single i_f-channel recording was originally achieved only by a modification of the patch-clamp technique that allowed an improved resolution of single-channel events (DiFrancesco, 1986). This modification consists of using two pipettes on the same cell simultaneously, one in the whole-cell and one in the cell-attached configuration. With this arrangement, transmembrane voltage-clamp steps can be applied through the whole-cell pipette while single-channel activity is recorded with the patch pipette in the absence of capacitative charging of the patch pipette itself and, thus, with greatly improved resolution.

Using this method, the properties of single i_f channels could be investigated. A first notable feature of i_f resulting from the two-pipette patch analysis was the quite small single-channel size. Currents recorded within the activation range were normally lower than 100 fA (1 fA = 10^{-3} pA). Elementary events could not be recorded at voltages more positive than -60 mV, given their exceedingly small amplitude (DiFrancesco, 1986). Therefore, that single i_f channels could not be identified earlier is not surprising.

An example of single i_f channel recording is shown in Fig. 3A. Several traces are superimposed, showing that the current is clearly distributed around discrete

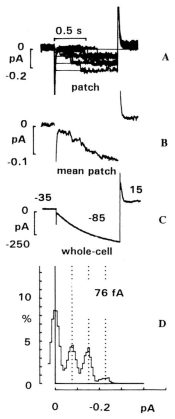

FIGURE 3 Example of single i_f-channel recording from a cell-attached patch of a single SA node myocyte using the two-pipette technique (DiFrancesco, 1986). The current i_f was activated by hyperpolarizations to -85 mV as indicated in panel C (whole-cell record). Eight single i_f channel traces recorded during the same steps are superimposed in panels A and their averages are shown in B. The current distribution during activation is analyzed in the histograms in D (8 fA/bin). As shown by the vertical grids, peaks are concentrated around single channel levels multiple of -76 fA. The patch contained apparently three channels.

levels that are multiples of the elementary single channel size of 76 fA, as apparent from the individual traces and from the histograms of current distribution plotted below each set of records (Fig. 3D). An important feature of the two-pipette method is that it allows simultaneous recording of single-channel and whole-cell currents, making it possible to compare the time course of the whole-cell current (Fig. 3C) with that of the mean-patch current recorded during the same voltage step (Fig. 3B).

The properties of single i_f channels were shown to be consistent with those of the i_f whole-cell current. Channels open in sequence during a hyperpolarization. The maximal number of openings increases at more negative voltages, which reflects the increase in current activation expressed by the i_f activation curve (see, for example, DiFrancesco et al., 1986). The i_f channel conductance, measured by plotting the single-channel I–V relationship, was ~1 pS, one of the smallest recorded by the patch-clamp technique. This result may seem to contrast with the much higher conductance of other nonspecific channels (see, for example, Hille, 1984), but unspecific cation channels with very low conductance are known to exist in sensory neurons (Kaupp, 1991). As discussed subsequently, this characteristic is not the only similarity between i_f and other types of channels in sensory neurons.

Recording of single-channel events also allowed, for the first time, a detailed investigation of the catecholamine-induced control of i_f at the single-channel level (DiFrancesco, 1986). For example, the current activation induced by adrenaline was shown not to be caused by an increased channel conductance but by an increase in the probability of channel opening during hyperpolarization. These results were the single-channel equivalent of those obtained in whole-cell experiments and indicated, as discussed earlier, that adrenaline shifts the current activation curve in the positive direction without altering the fully activated I–V relationship.

IV. Control of i_f by Cyclic AMP

β-Adrenergic and muscarinic stimuli activate and inhibit, respectively, adenylyl-cyclase activity and cAMP production in a variety of tissues (Gilman, 1984). Therefore, the modulation of i_f by autonomic neurotransmitters is very likely to simply reflect a dependence of the current on cAMP. Indeed, a dependence of the cardiac pacemaker current on adenylyl-cyclase activity and cAMP was first proposed in the Purkinje fiber on the basis of experiments indicating a correlation between cAMP production and/or hydrolysis and the extent of i_{K2} current activation (Tsien, 1974). In this section, the basic evidence for the cAMP-induced control of i_f in the SA node is reviewed.

A. Whole-Cell Current Modulation

In intact tissue and in isolated SA node myocytes, i_f responds to several maneuvers that alter the intracellular cAMP concentration, as well as to autonomic transmitters. This result indicates the involvement of cAMP in the control of the i_f current in the SA node. For example, i_f responds to cAMP elevation induced by direct adenylyl-cyclase stimulation by forskolin, by inhibition of cAMP phosphodiesterase (DiFrancesco and Tromba, 1988b), or by direct perfusion of

membrane-permeable cAMP (D. DiFrancesco, unpublished observations). All these manipulations affect i_f by the same mechanism driven by catecholamines, that is, an increase in cAMP induces a rightward shift of the current activation range and leaves unaltered the fully activated I–V relationship. Data derived from studies of Purkinje fibers indicate that the current activation curve does not undergo a simple translation along the voltage axis during activation by catecholamines, but also undergoes a modification of its slope, which becomes steeper (Tsien, 1974). In the isolated SA node, the activation curve protocol is not easily performed because of run-down of the current (DiFrancesco et al., 1986). Information on the possible changes in activation curve steepness induced by catecholamines is lacking. However, data obtained by a simplified protocol indicate that, in the mid-activation range, the i_f activation curve shifts by about +8 mV with 1 μM epinephrine and by about −7 mV with saturating ACh concentrations (DiFrancesco et al., 1989; DiFrancesco and Tortora, 1991).

B. Control of Pacemaker Activity by cAMP

As discussed earlier, i_f changes induced by β-adrenergic and cholinergic transmitters are responsible for rhythm modulation at low agonist concentrations. For example, catecholamines activate i_f by stimulating adenylyl-cyclase activity and cAMP production, resulting in a shift of the current activation curve to more positive voltages or, correspondingly, to a higher probability of single-channel opening. A larger i_f accelerates the development of diastolic depolarization, and thus increases heart rate (Brown et al., 1979b). The opposite set of events characterizes the action of ACh on i_f and on rhythmicity (DiFrancesco et al., 1989). Thus, i_f control by cAMP appears to be the key factor in modulating heart rate. How does cAMP activate i_f channels?

C. Direct cAMP-Induced Activation of i_f Channels

Several voltage-dependent ion channels in cardiac and noncardiac membranes are modulated by cAMP (Levitan, 1985). Of these, possibly the best characterized is the L-type Ca^{2+} channel (Reuter, 1983; Tsien, 1983; Kameyama et al., 1986; Trautwein and Hescheler, 1990; see Chapter 12). Before i_f modulation by cAMP was investigated, all known voltage-gated cAMP-dependent currents had been reported to be activated by cAMP through a phosphorylation process mediated by cAMP-dependent PKA, which thus appeared to be a general mechanism of channel modulation by cAMP. Therefore, researchers assumed that i_f also could be controlled by this mechanism. Surprisingly, the "funny" current turned out to behave in a "funny" way, even in relation to its cAMP dependence. Using macro-patch preparations, which allow recording from inside-out membranes containing a large number of channels (Yatani *et al.*, 1990), researchers showed that i_f activation by cAMP did not involve a phosphorylation process but was the result of a direct cAMP–channel protein interaction (DiFrancesco and Tortora, 1991). Several observations indicated the lack of phosphorylation in the activation process. For example, activation was present during perfusion of the intracellular side of the patch with the holoenzyme and cAMP, but was absent during perfusion with the PKA catalytic subunit; also, i_f was stimulated by cAMP even in the presence of specific and nonspecific inhibitors of PKA. The nature and size of direct macro-patch i_f stimulation by cAMP was in agreement with whole-cell i_f activation by catecholamines. Thus, the macro-patch i_f activation curve is displaced by +11 mV along the voltage axis under the action of saturating cAMP concentrations (>10 μM), whereas the fully activated I–V relationship of macro-patch i_f is unaltered (DiFrancesco and Tortora, 1991). Dose–response analysis further indicates that i_f is half-activated by 0.211 μM cAMP, and that the Hill coefficient is 0.85, suggesting the possibility that cAMP activates i_f channels with a 1:1 stoichiometry.

V. Conclusions

Because of its involvement in the generation and control of the pacemaking process, the i_f current is often referred to as the "pacemaker" current. Indeed, since its first description in the multicellular SA node preparation, i_f appeared to possess unusual features that seemed specifically designed to generate and drive a depolarizing process after action potential repolarization and, thus, to play a key role in driving spontaneous activity.

After more than a decade of detailed investigation, the involvement of i_f in generation of rhythmic behavior has been confirmed. More importantly, its role in the neurotransmitter-induced control of heart rate under normal conditions has been clarified. Not only does i_f mediate the accelaratory action of catecholamines (Brown *et al.*, 1979b), but it also mediates the slowing action of ACh at low concentrations (DiFrancesco *et al.*, 1989).

Several i_f-like currents have been described in a wide variety of cell types, including central and pe-

ripheral neurons, sensory neurons, and smooth muscle myocytes (Barret *et al.*, 1980; Halliwell and Adams, 1982; Mayer and Westbrook, 1983; Bader and Bertrand, 1984; Kubota *et al.*, 1985; Crepel and Penit-Soria, 1986; Benham *et al.*, 1987; Edman *et al.*, 1987; Lacey and North, 1988; Williams *et al.*, 1988; Spain *et al.*, 1991). In cases in which modulation by neurotransmitters has been reported, this effect is mediated by cAMP (Bobker and Williams, 1989; Pape and McCormick, 1989; Tokimasa and Akasu, 1990). Whatever the specific role of i_f-like currents in any of these preparations, these components can clearly produce and maintain a depolarization, as well as control its rate of development according to necessity under the action of modifying agents (hyperpolarization, neurotransmitter).

Neurotransmitters modulate i_f via adenylyl-cyclase and cAMP by a direct activation mechanism that does not involve a phosphorylation process (DiFrancesco and Tortora, 1991), thus allowing a fast metabolism-independent control of heart rate. I_f channels share the property of direct activation mechanism with other cyclic-nucleotide-dependent channels in sensory neurons, such as the light-activated channel and the cation-unspecific channel of olfactory neurons (Kaupp, 1991; see also Chapter 23). I_f channels are unique, however, in being modulated by both voltage and cAMP. Further, since voltage and cAMP are indistinguishable in their action on i_f channels, both modulators may induce similar structural modifications in i_f channels.

Acknowledgments

This work was supported by the Ministero dell' Università e della Ricerca Scientifica e Tecnologica and by the Consiglio Nazionale delle Ricerche.

References

Bader, C. R., and Bertrand, D. (1984). Effects of changes in intra- and extracellular sodium on the inward (anomalous) rectification in salamander photoreceptors. *J. Physiol.* **347,** 611–631.

Barret, E. F., Barret, J. N., and Crill, W. E. (1980). Voltage-sensitive outward currents in cat motoneurones. *J. Physiol.* **304,** 251–276.

Benham, C. D., Bolton, T. B., Denbigh, J. S., and Lang, R. J. (1987). Inward rectification in freshly isolated single smooth muscle cells of the rabbit jejunum. *J. Physiol.* **383,** 461–476.

Bobker, D. H., and Williams, J. T. (1989). Serotonin augments the cationic current I_h in central neurons. *Neuron* **2,** 1535–1540.

Brown, H. F., and DiFrancesco, D. (1980). Voltage-clamp investigations of membrane currents underlying pacemaker activity in rabbit sino-atrial node. *J. Physiol.* **308,** 331–351.

Brown, H. F., DiFrancesco, D., and Noble, S. J. (1979a). Adrenaline action on rabbit sino-atrial node. *J. Physiol.* **290,** 31–32.

Brown, H. F., DiFrancesco, D., and Noble, S. J. (1979b. How does adrenaline accelerate the heart? *Nature (London)* **280,** 235–236.

Brown, H. F., Giles, W., and Noble, S. J. (1977). Membrane currents underlying activity in frog sinus venosus. *J. Physiol.* **271**, 783–816.

Brown, H. F., Kimura, J., Noble, D., Noble, S. J., and Taupignon, A. (1984b). The ionic currents underlying pacemaker activity in rabbit sino-atrial node: Experimental results and computer simulations. *Proc. R. Soc. Lond. B* **222**, 329–347.

Chang, F., Gao, J., Tromba, C., Cohen, I. S., and DiFrancesco, D. (1990). Acetylcholine reverses the effects of β-agonists on the pacemaker current in canine Purkinje fibers but has no direct action. A difference between primary and secondary pacemakers. *Circulation Res.* **66**, 633–636.

Crepel, F., and Penit-Soria, J. (1986). Inward rectification and low threshold calcium conductance in rat cerebellar Purkinje cells. An *in vitro* study. *J. Physiol.* **372**, 1–23.

Denyer, J. C., and Brown, H. F. (1990a). Rabbit isolated sino-atrial node cells: Isolation and electrophysiological properties. *J. Physiol.* **428**, 405–424.

DiFrancesco, D. (1981a). A new interpretation of the pacemaker current in calf Purkinje fibers. *J. Physiol.* **314**, 359–376.

DiFrancesco, D. (1981b). A study of the ionic nature of the pacemaker current in calf Purkinje fibers. *J. Physiol.* **314**, 377–393.

DiFrancesco, D. (1986). Characterization of single pacemaker channels in cardiac sino-atrial node cells. *Nature (London)* **324**, 470–473.

DiFrancesco, D. (1991). The contribution of the hyperpolarization activated current (i_f to the generation of spontaneous activity in rabbit sino-atrial node myocytes. *J. Physiol.* **434**, 23–40.

DiFrancesco, D. (1993). Pacemaker mechanisms in cardiac tissue. *Ann. Rev. Physiol.* **55**, 451–467.

DiFrancesco, D., Ducouret, P., and Robinson, R. B. (1989). Muscarinic modulation of cardiac rate at low acetylcholine concentrations. *Science* **243**, 669–671.

DiFrancesco, D., Ferroni, A., Mazzanti, M., and Tromba, C. (1986). Properties of the hyperpolarizing-activated current (i_f) in cells isolated from the rabbit sino-atrial node. *J. Physiol.* **377**, 61–88.

DiFrancesco, D., and Noble, D. (1985). A model of cardiac electrical activity incorporating ionic pumps and concentration changes. *Phil. Trans. R. Soc. Lond. B* **307**, 353–398.

DiFrancesco, D., and Ojeda, C. (1980). Properties of the current i_f in the sino atrial node of the rabbit compared with those of the current i_{K2} in Purkinje fibers. *J. Physiol.* **308**, 353–367.

DiFrancesco, D., and Tortora, P. (1991). Direct activation of cardiac pacemaker channels by intracellular cyclic AMP. *Nature (London)* **351**, 145–147.

DiFrancesco, D., and Tromba, C. (1987). Acetylcholine inhibits activation of the cardiac pacemaker current, i_f. *Pflügers Arch.* **410**, 139–142.

DiFrancesco, D., and Tromba, C. (1988a). Inhibition of the hyperpolarizing-activated current, i_f, induced by acetylcholine in rabbit sino-atrial node myocytes. *J. Physiol.* **405**, 477–491.

DiFrancesco, D., and Tromba, C. (1988b). Muscarinic control of the hyperpolarizing-activated current i_f in rabbit sino-atrial node myocytes. *J. Physiol.* **405**, 493–510.

Edman, A., Gestrelius, S., and Grampp, W. (1987). Current activation by membrane hyperpolarization in the slowly adapting lobster stretch receptor neurone. *J. Physiol.* **384**, 671–690.

Garnier, D., Nargeot, J., Ojeda, C., and Rougier, O. (1978). The action of acetylcholine on background conductance in frog atrial trabeculae. *J. Physiol.* **274**, 381–396.

Giles, W., Nakajima, T., Ono, K., and Shibata, E. F. (1989). Modulation of the delayed rectifier K⁺ current by isoprenaline in bullfrog atrial myocytes. *J. Physiol.* **415**, 233–249.

Giles, W., and Noble, S. J. (1976). Changes in membrane currents in bullfrog atrium produced by acetylcholine. *J. Physiol.* **261**, 103–123.

Gilman, A. G. (1984). G-proteins and dual control of adenylate-cyclase. *Cell* **36**, 577–579.

Halliwell, J. V., and Adams, P. R. (1982). Voltage clamp analysis of muscarinic excitation in hippocampal neurones. *Brain Res.* **250**, 71–92.

Hart, G., Noble, D., and Shimoni, Y. (1980). Adrenaline shifts the voltage dependence of the sodium and potassium components of i_f in sheep Purkinje fibers. *J. Physiol.* **308**, 34.

Hauswirth, O., Noble, D., and Tsien, R. W. (1968). Adrenaline: Mechanism of action on the pacemaker potential in cardiac Purkinje fibers. *Science* **162**, 916–917.

Hewett, K. W., and Rosen, M. R. (1985). Developmental changes in the rabbit sinus node action potential and its response to adrenergic agonists. *J. Pharmacol. Exp. Ther.* **235**, 308–312.

Hille, B. (1984). Ionic channels of excitable membranes. Sinauer, Sunderland.

Hirst, G. D. S., Bramich, N. J., Edwards, F. R., and Klemm, M. (1992). Transmission at autonomic neuroeffector junctions. *Trends Neurosci.* **15**, 40–46.

Hoffman, B. F., and Suckling, E. E. (1953). Cardiac cellular potentials. Effect of vagal stimulation and acetylcholine. *Am. J. Physiol.* **173**, 312–320.

Hutter, O. and Trautwein, W. (1956). Vagal and sympathetic effects on the pacemaker fibers in the sinus venosus of the heart. *J. Gen. Physiol.* **39**, 715–733.

Irisawa, H. and Hagiwara, N. (1991). Ionic currents in sinoatrial node cells. *J. Cardiov. Electrophysiol.* **2**, 531–540.

Kameyama, M., Hescheler, J., Hofmann, F., and Trautwein, W. (1986). Modulation of Ca current during the phosphorylation cycle in the guinea-pig heart. *Pflügers Arch.* **407**, 123–128.

Kaupp, U. B. (1991). The cyclic-nucleotide-gated channels of vertebrate photoreceptors and olfactory epithelium. *Trends Neurosci.* **14**, 150–157.

Kubota, M., Nakamura, N., and Tsukahara, N. (1985). Ionic conductance associated with electrical activity of guinea-pig red nucleus neurones *in vitro*. *J. Physiol.* **362**, 161–171.

Lacey, M. G., and North, R. A. (1988). An inward current activated by hyperpolarization (I_h) in rat substantia nigra zona compacta neurons *in vitro*. *J. Physiol.* **407**, 18.

Levitan, I. B. (1985). Phosphorylation of ion channels. *J. Memb. Biol.* **87**, 177–190.

Mayer, M. L., and Westbrook, G. L. (1983). A voltage-clamp analysis of inward (anomalous) rectification in mouse spinal sensory ganglion neurones. *J. Physiol.* **340**, 19–45.

Noble, D., and Tsien, R. W. (1968). The kinetics and rectifier properties of the slow potassium current in calf Purkinje fibers. *J. Physiol.* **195**, 185–214.

Noma, A., and Irisawa, H. (1976). Membrane currents in the rabbit sinoatrial node cell as studied by the double microelectrode method. *Pflügers Arch.* **366**, 45–52.

Noma, A., and Trautwein, W. (1978). Relaxation of the ACh-induced potassium current in the rabbit sino-atrial node. *Pflügers Arch.* **377**, 193–200.

Pape, H. C., and McCormick, D. A. (1989). Noradrenaline and serotonin selectively modulate thalamic burst firing by enhancing a hyperpolarization-activated cation current. *Nature (London)* **340**, 715–718.

Pfaffinger, P. J., Martin, J. M., Hunter, D. D., Nathanson, N. M., and Hille, B. (1985). GTP-binding proteins couple cardiac muscarinic receptors to a K channel. *Nature (London)* **317**, 536–538.

Reuter, H. (1983). Calcium channel modulation by neurotransmitters, enzymes and drugs. *Nature (London)* **301**, 569–574.

Rosen, M. R. (1990). Membrane effects of alpha adrenergic catecholamines. *In* "Cardiac Electrophysiology: A textbook" (M. R. Rosen, M. J. Janse, and A. L. Wit, eds.), pp. 847–856. Futura, New York.

Sakmann, B., Noma, A., and Trautwein, W. (1983). Acetylcholine activation of single muscarinic K channels in isolated pacemaker cells of the mammalian heart. *Nature (London)* **303,** 250–253.

Soejima, M., and Noma, A. (1984). Mode of regulation of the ACh-sensitive K channel by the muscarinic receptors in the rabbit atrial cells. *Pflügers Arch.* **400,** 424–431.

Spain, W. J., Schwindt, P. C., and Crill, W. C. (1991). Post-inhibitory excitation and inhibition in layer V pyramidal neurones from cat sensorimotor cortex. *J. Physiol.* **434,** 609–626.

Takahashi, T. (1990). Inward rectification in neonatal rat spinal motoneurones. *J. Physiol.* **423,** 47–62.

Takahashi, T., and Berger, A. J. (1990). Direct excitation of rat spinal motoneurones by serotonin. *J. Physiol.* **423,** 63–76.

Tokimasa, T., and Akasu, T. (1990). Cyclic AMP regulates an inwardly rectifying sodium-potassium current in dissociated bullfrog sympathetic neurones. *J. Physiol.* **420,** 409–429.

Trautwein, W. and Hescheler, J. (1990). Regulation of cardiac L-type calcium current by phosphorylation and G proteins. *Ann. Rev. Physiol.* **52,** 257–274.

Tsien, R. W. (1974). Effects of epinephrine on the pacemaker potassium current of cardiac Purkinje fibers. *J. Gen. Physiol.* **64,** 293–319.

Tsien, R. W. (1983). Ca channels in excitable cell membrane. *Ann. Rev. Physiol.* **45,** 341–358.

van Ginneken A. C. G., and Giles, W. (1991). Voltage clamp measurements of the hyperpolarization-activated inward current i_f in single cells from rabbit sino-atrial node. *J. Physiol.* **434,** 57–83.

Williams, J. T., Colmers, W. F., and Pan, Z. Z. (1988). Voltage- and Ligand-activated inwardly rectifying currents in dorsal raphe neurons *in vitro. J. Neurosci.* **8(9),** 3499–3506.

Yanagihara, K., and Irisawa, H. (1980). Inward current activated during hyperpolarization in the rabbit sinoatrial node cell. *Pflügers Arch.* **385,** 11–19.

Yatani, A., Okabe, K., Codina, J., Birnbaumer, A., and Brown, A. M. (1990). Heart rate regulation by G protein acting on the cardiac pacemaker channel. *Science* **249,** 1163–1166.

Zaza, A., Robinson, R. B., and DiFrancesco, D. (1992). Role of i_f and i_{Ca} in the autonomic modulation of sinoatrial pacemaking. *Circulation,* **86,** I–344.

Selectivity and Single-Channel Properties of the cGMP-Activated Channel in Amphibian Retinal Rods

VINCENT TORRE AND ANNA MENINI

I. Introduction

A major advance in the understanding of the molecular mechanisms underlying phototransduction was the demonstration that the ionic current regulated by light in rod photoreceptors is directly activated by guanosine 3',5'-cyclic monophosphate (cGMP; Fesenko *et al.*, 1985). Soon after this remarkable discovery, ionic channels directly activated by cyclic nucleotides were described in cones (Haynes and Yau, 1985), in olfactory receptor neurons (Nakamura and Gold, 1987), in retinal bipolar cells (Nawy and Jahr, 1990; Shiells and Falk, 1990), and in cells of the pineal gland (Dryer and Henderson, 1991).

The primary sequence of proteins constituting many of these channels is now available (Kaupp *et al.*, 1989; Dhallan *et al.*, 1990,1992; Ludwig *et al.*, 1990; Goulding *et al.*, 1992); the different amino acid sequences show a high degree of homology. Clearly cyclic nucleotide-activated channels form a new superfamily of ionic channels with similar molecular structures and physiological properties (for reviews, see Kaupp, 1991; Kaupp and Altenhofen, 1992; Menini and Anholt, 1994).

In this chapter, results of studies on the ionic selectivity and single-channel properties of the current activated by cGMP in retinal rods of the tiger salamander *Ambystoma tigrinum* are reviewed.

II. Properties of the Macroscopic Current Activated by cGMP

The properties of cGMP-activated channels are usually studied with the patch-clamp technique, excising membrane patches from photoreceptors and adding micromolar concentrations of cGMP to the cytoplasmic side of the membrane. The ionic current flowing through these channels is controlled by cGMP in a highly cooperative way (see Section II,A). Moreover, in the presence of a low concentration of cGMP, the current–voltage (I–V) relationship presents a significant outward rectification, which diminishes after increasing the cyclic nucleotide concentration (see Section II,B). The power spectrum of current fluctuations indicates the presence of a significant component above 2 kHz that is independent of membrane voltage, of cyclic nucleotide concentration, and of the presence of divalent cations (see Section II,C).

345

A. Dependence on cGMP Concentration

When an inside-out membrane patch, excised from the outer segment of a photoreceptor, is exposed to a solution containing micromolar concentrations of cGMP, many channels are often activated (Fesenko *et al.*, 1985; Haynes and Yau, 1985; Zimmerman and Baylor, 1986; Matthews and Watanabe, 1987; Colamartino *et al.*, 1991). As shown in Fig. 1A, at +60 mV and in the presence of a saturating concentration of cGMP (about 100 μM), currents as large as 1–2 nA were measured. Depending on the number of channels in the membrane patch, a small current could sometimes be measured in the presence of very low concentrations of cGMP (Fig. 1B); its amplitude increases steeply with the cGMP concentration. The relationship between activated current and cGMP concentration is reproduced in Fig. 1 (C,D) and can be described by the equation

$$I = I_{max}\frac{g^n}{g^n + K^n},\qquad(1)$$

where I_{max} is the maximal ionic current activated by

cGMP, g is the cGMP concentration, K is the cGMP concentration activating half the maximal current, and n is the Hill coefficient. The experimental points can be fitted with a value between 2 and 3 for n and a value of 10 or 50 μM for K for salamander (Zimmerman and Baylor, 1986; Colamartino *et al.*, 1991) or bovine (Luehring *et al.*, 1990) rods, respectively.

The value between 2 and 3 for n is usually assumed to indicate that the binding of at least three cGMP molecules is necessary to open an ionic channel. The binding of cGMP for the channel, that is, the value of K, has been shown to be regulated by a phosphatase and may change with time (Gordon *et al.*, 1992).

At low concentrations of cGMP, a slow rise in the current is observed (Fig. 1B) that is likely to be caused by a voltage dependence of the opening rate of the channel (Karpen *et al.*, 1988). In membrane patches containing many channels and in the presence of high cGMP concentrations, a time-dependent relaxation of the activated current at the onset and offset of voltage steps is sometimes observed (Fig. 1A). This behavior is not an intrinsic property of the cGMP-activated channel, but is caused by local changes of the ionic

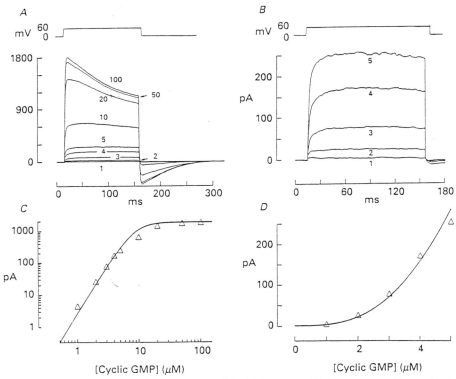

FIGURE 1 Stoichiometry of the activation of the Na$^+$ current. (A) Current recordings activated by 1, 2, 3, 4, 5, 10, 20, 50, and 100 μM cGMP. (B) Current recordings activated by 1, 2, 3, 4, and 5 μM cGMP. Voltage steps were from 0 to +60 mV. (C,D) Relationship between activated current and cGMP concentration on a log–log scale (C) and on a linear plot (D). The continuous line was obtained from Eq. 1, with $n = 3$, $K = 10$ μM, and $I_{max} = 1800$ pA. Reproduced with permission from Colamartino *et al.* (1991).

concentration around the patch. In these patches, the massive ionic flux alters the electrical driving forces across the patch because of hindered diffusion near the membrane patch (Zimmerman *et al.*, 1988). Because of these different effects, the current activated by cGMP in Fig. 1C was measured at late times for cGMP concentration below 10 μM and at early times for higher concentrations.

The ionic channel activated by cGMP in vertebrate photoreceptors also can be activated by other cyclic nucleotides such as cAMP (Tanaka *et al.*, 1989; Furman and Tanaka, 1990; Sanfilippo and Menini, 1993). The maximal current activated by cAMP in a patch immersed in a solution at pH 7.6 is about 40% of that activated by cGMP; the concentration of cAMP activating half the maximal current is about 800 μM, that is, about 80 times that for cGMP. Researchers have shown that lowering the cytoplasmic pH enhances the activation of the channel by cAMP (Sanfilippo and Menini, 1993).

The structural basis for ligand specificity has been analyzed by site-directed mutagenesis experiments in the channel from bovine rods by Altenhofen *et al.* (1991).

B. Current–Voltage Properties

The shape of the I–V relationship of the macroscopic current depends on the concentration of cGMP (Altenhofen *et al.*, 1991; Colamartino *et al.*, 1991), as shown in Fig. 2. In the absence of divalent cations and with saturating concentration of cGMP (100 μM), the I–V relationship is slightly nonlinear, with a small outward rectification, but at low cGMP concentrations the I–V relationship has a strong outward rectification (Fig. 2).

Since the shape of the I–V relationship of the macroscopic current is generated by the I–V relationship of the single channel and by the voltage dependence of the open probability of the channel, these two components must be separated. Analysis of the instantaneous I–V relationship (Zimmerman and Baylor, 1992) suggests that the single-channel current has a small outward rectification that is responsible for the outward rectification observed with saturating concentrations of cGMP (Fig. 2A). Analysis of single-channel activity at positive and negative voltages (Sesti *et al.*, 1994) indicates that the large outward rectification observed in the presence of small concentrations of cGMP (Fig. 2B) is primarily caused by an increase in the open probability at positive voltages. In the presence of a saturating concentration of cGMP, the open probability has already reached a maximum value and is barely affected by membrane voltage.

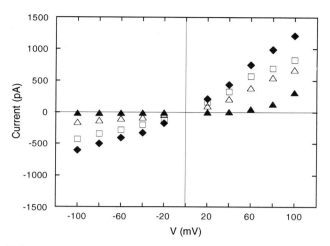

FIGURE 2 I–V relationships in 110 mM NaCl symmetrical solutions (0.2 mM EDTA in the patch pipette and in the bathing solution) in the presence of 5 (▲), 10 (△), 20 (□), and 100 (◆) μM cGMP. Unpublished results of G. Colamartino, A. Menini, and V. Torre.

The large outward rectification of the current in the presence of micromolar amounts of cGMP indicates that outward rectification of the light-sensitive channel in intact rods (Baylor and Nunn, 1986) is largely an intrinsic property of the channel and of low levels of free cGMP in intact rods.

C. Kinetic Properties of cGMP-Induced Fluctuations

Early analysis of current fluctuations induced by cGMP in amphibian vertebrate rods has indicated the existence of two Lorentzian components in the power spectrum, with corner frequencies at 80 and 960 Hz (Matthews, 1987) or 20 and 250 Hz (Haynes *et al.*, 1986). These results were obtained with a recording bandwidth of 0–2000 Hz. When this analysis was repeated at a larger bandwidth, a different picture emerged (Torre *et al.*, 1992; Sesti *et al.*, 1994). The power spectrum of current fluctuations induced by 100 μM cGMP at membrane voltages between -100 and $+100$ mV can be fitted by the sum of at least three Lorentzian components with corner frequencies at ~50, 230, and 4500 Hz (Sesti *et al.*, 1994). The corner frequency of the Lorentzian component at 4500 Hz is caused by the limited bandwidth of the recording system; therefore, the corner frequency of the physiological process originating the high frequency Lorentzian component is likely to be above 4500 Hz. The Lorentzian component of the power spectrum that has a significant energy density at frequencies higher than 2 kHz does not depend on the presence of Ca^{2+} or Mg^{2+} (Torre *et al.*, 1992; Sesti *et al.*, 1994).

The presence of at least three Lorentzian components in the power spectrum suggests the existence of at least three steps in the process leading to the opening of a channel. The Lorentzian component with the cutoff frequency at or above 4500 Hz is likely to represent a rapid flickering between the open and closed states of the channel. The Lorentzian component with a cutoff frequency at about 200 Hz could be associated with the occurrence of bursts; the Lorentzian component with the corner frequency at about 50 Hz could be caused by the same process responsible for the time delay of the activation of the current observed in Fig. 1B.

However, the molecular mechanisms controlling the gating of the channel are largely unknown. A microscopic interpretation of the shape of the power spectrum is, at this moment, rather speculative.

III. Single-Channel Properties

The first measurements of single channels in divalent-free solutions (Zimmerman and Baylor, 1986; Haynes et al., 1986; Matthews, 1986,1987; Matthews and Watanabe, 1987,1988) showed that the cGMP-activated channels are aqueous pores. These experiments reported a great variability of the mean open time ranging from 0.1 to 1 msec and a single-channel conductance between 20 and 25 pS. Experiments obtained at a larger recording bandwidth (Torre et al., 1992) showed that discrete single-channel events were impossible to resolve and that the single-channel conductance is not smaller than 60 pS and the mean open time not greater than 35 μsec (see Section III,A). Very rarely, and only in recordings from the inner segment, another class of cGMP-activated channels exhibiting distinct open/closed transitions was observed (see Section III,B).

A. Outer Segment Recordings

The density of cGMP-activated channels in the plasma membrane of rod outer segments varies from 0.34 to 641 μm^{-2} with a mean value of 126 μm^{-2} (Karpen et al., 1992); membrane patches are usually found to contain many channels. Therefore, to see single-channel openings and to avoid multiple events, the open probability must be reduced by using very small concentrations of cGMP.

Figure 3A illustrates current recordings activated by 2 μM cGMP at a membrane voltage of −60 mV. Evidently, brief current transients typical of single-channel activity can be detected. However, these traces do not show square transitions between an open and a closed state, as would be typical of well-resolved channel openings, but a continuous flickering between the closed and the open state (Fig. 3B). The power spectrum of these fluctuations is shown in Fig. 3C, and the presence of three Lorentzian components is evident. The Lorentzian component with a significant energy above 2 kHz is responsible for the flickering of the channels. As a consequence of this flickering, measuring the single-channel conductance and the mean open time unambiguously is not possible. A theoretical analysis (Torre et al., 1992; Sesti et al., 1994) of the experimental amplitude histograms gives an estimate of the single-channel conductance of at least 60 pS and a mean open time no greater than 35 μsec. The continuous flickering makes it difficult to establish the existence of channel substates with different single-channel conductances.

B. Inner Segment Recordings

Since the density of cGMP-activated channels in the inner segment is much lower, membrane patches can be excised from the inner segment that contain only one cGMP-activated channel (Matthews and Watanabe, 1988). The majority of such patches exhibited channel behavior with flickering properties similar to those seen with outer segment patches. However, some rare recordings showed a very different behavior (Torre et al., 1992; Sesti et al., 1994). In these patches, the application of cGMP induced discrete openings to a clearly defined conductance level, in complete contrast with all the recordings from outer segment patches. At the holding potential of +40 mV, the single-channel conductance was about 42 pS and did not change with the cGMP concentration. Since the Na$^+$ concentration in these experiments had been elevated to 250 mM, the single-channel conductance was estimated to be 25–30 pS at normal Na$^+$ level. The open probability of this channel increased with the cGMP concentration in a cooperative way; the Hill coefficient was between 3 and 4, and the half-activation of this channel occurred at a cGMP concentration of ~18 μM. The properties of this channel appear to be quite similar to those of the channel purified from bovine rods (Cook et al., 1987), subsequently cloned and expressed in oocytes (Kaupp et al., 1989), and of the human rod photoreceptor channel expressed in cell lines (Dhallan et al., 1992). These channels do not show the flickering behavior and have a single-channel conductance between 20 and 25 pS. The simplest interpretation is that the cloned channels from bovine and human rods expressed in oocytes or in cell lines are similar, but not identical, to the channel activated by cGMP in rod outer segments of the

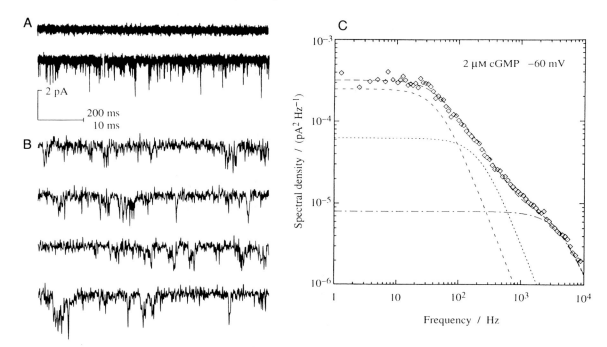

FIGURE 3 (A) Currents recorded with 0 (*top*) or 2 (*bottom*) μM cGMP in the solution bathing the cytoplasmic surface of an excised inside-out patch from the outer segment, at a holding potential of -60 mV (cytoplasmic side). No transient currents were observed in the absence of cGMP (*top*). (B) Raw current recordings with 2 μM cGMP, on a faster time base and filtered at 4 kHz. (C) Difference power spectrum of cGMP-induced current fluctuations. The continuous curve is the sum of three Lorentzian functions (dashed lines) with corner frequencies at 50, 230, and 4400 Hz and zero-frequency spectral densities of 2.5×10^{-4}, 6.25×10^{-5}, and 8×10^{-6} pA^2 Hz^{-1}, giving variances of 0.031, 0.014, and 0.035 pA^2, respectively. The power spectrum was computed as the difference of spectra in the presence and in the absence of cGMP. Reproduced with permission from Torre *et al.* (1992).

tiger salamander. A more thorough analysis and explanation of these differences requires the use of techniques of molecular biology and is beyond the scope of this chapter.

IV. Ionic Selectivity

The ionic selectivity of the channel activated by cGMP in retinal rods has been studied in intact cells and in excised patches. In intact rods, the ionic selectivity was studied by changing ions in the medium bathing the extracellular side of the membrane; the cell was not usually voltage clamped. On the contrary, in excised patches in the inside-out configuration, ions were changed in the medium bathing the cytoplasmic side of the membrane and the current was measured under voltage-clamp conditions. These experiments are largely complementary, but the results obtained with excised patches are easier to interpret because of more controlled experimental conditions. The cGMP-activated channel is permeable to alkali monovalent

cations (see Section IV,A), to a variety of large organic monovalent cations (see Section IV,B), and to divalent cations (see Section IV,C). The current carried by alkali monovalent cations exhibits saturation as the ion activity is increased.

A. Alkali Monovalent Cations

Early studies of the ionic selectivity of the cGMP-activated channel in intact rods showed that this channel does not discriminate very well among alkali monovalent cations (Capovilla *et al.*, 1983; Yau and Nakatani, 1984; Hodgkin *et al.*, 1985). Experiments with excised patches provided a quantitative description of ionic selectivity based on measurements of reversal potentials and conductances (Nunn, 1987; Menini, 1990; Furman and Tanaka, 1990; Luehring *et al.*, 1990).

Figure 4A reproduces currents activated by 100 μM cGMP when the ionic medium bathing the cytoplasmic side of the membrane patch contained 110 mM NaCl, LiCl, KCl, RbCl, or CsCl. The dashed

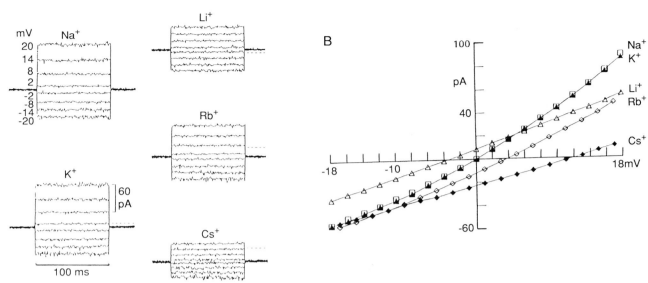

FIGURE 4 (A) Currents activated by 100 μM cGMP under bi-ionic conditions. The solution filling the patch pipette contained 110 mM NaCl, whereas the solution on the cytoplasmic side of the membrane contained 110 mM NaCl, LiCl, KCl, RbCl, or CsCl, as indicated in the figure. Voltage steps of ± 2 mV were given from a holding potential of 0 mV. The dashed lines represent the zero-current level. (B) I–V relationships for the same patch corrected for junction potentials. Reproduced with permission from Menini (1990).

lines in each panel represent the zero-current level. Figure 4B shows the I–V relationships from the experiment illustrated in Fig. 4A. Figure 4 clearly shows that all alkali cations can carry an outward current and, therefore, permeate the channel. The permeability ratio sequence calculated with the Goldman–Hodgkin–Katz potential equation from the measured reversal potentials for alkali cations and for ammonium (NH_4); Menini, 1990; Picco and Menini, 1993) was

$$P_{NH_4}:P_{Li}:P_{Na}:P_K:P_{Rb}:P_{Cs}$$
$$= 2.8:1.14:1:0.98:0.84:0.58, \quad (2)$$

whereas the selectivity ratio sequence obtained for outward currents flowing at $+60$ mV was

$$I_{NH_4}:I_{Na}:I_K:I_{Rb}:I_{Li}:I_{Cs} = 1.5:1:0.85:0.5:0.46:0.15 \quad (3)$$

Sequence 3 is in broad agreement with the selectivity sequence obtained from inward currents exchanging extracellular alkali cations in intact rods (Yau and Nakatani, 1984; Hodgkin *et al.*, 1985). The major discrepancy is the observation that, in intact rods, the ratio between the inward currents carried by Li^+ and Na^+ is 1.4, whereas in excised patches the ratio between the outward currents carried by these cations is 0.46. All these experiments were based on the analysis of macroscopic currents. Whether the various ionic

substitutions also influence the gating of the channel is not known. However, from the analysis of power spectra, Menini *et al.* (1991) concluded that Na^+ to Li^+ substitution did not affect the gating. Experiments on the analysis of the single-channel currents (Sesti *et al.*, 1994) showed that replacing Na^+ with NH_4^+, K^+, or Li^+ did not appreciably change the gating of the channel. The analysis of amplitude histograms indicated that the relationships of single-channel currents at $+60$ mV were

$$NH_4^+:Na^+:K^+:Li^+ = 1.6:1:0.7:0.5. \quad (4)$$

The quantitative agreement between Sequences 3 and 4 suggests that the selectivity sequence of the cGMP-activated channel based on the analysis of macroscopic currents (Furman and Tanaka, 1990; Menini, 1990) is justified. The discrepancy between the experiments with intact rods and excised patches can be explained by an asymmetry in ion transport through the cGMP-activated channel.

B. Organic Monovalent Cations

Experiments in excised patches from retinal rods have shown that at least 13 organic monovalent cations are permeable through the cGMP-activated channel (Picco and Menini, 1993). Figure 5 reproduces currents activated by 100 μM cGMP when some methylated and ethylated ammonium compounds were substituted for Na$^+$ in the medium bathing the cytoplasmic side of the membrane. Methylammonium, dimethylammonium, and ethylammonium are permeable, whereas the other organic cations do not permeate the channel.

Hydroxylammonium, hydrazinium, and methylammonium, which are molecules of very similar shape, size, and molecular weight, permeate the channel with very different permeability ratios relative to Na$^+$: 5.92, 1.99, and 0.6, respectively (Table I). These different permeabilities indicate that the cGMP-activated channel does not discriminate among ions on the basis of the dimensions of cations only, but that chemical interactions are important in determining ion selectivity. Guanidinium and its derivatives formamidinium, aminoguanidinium, acetamidinium, and methylguanidinium are all permeant, with permeability ratios of 1.12, 1.0, 0.63, 0.36, and 0.33, respectively. Estimating the size of all permeant cations from space-filling models gave information on the size of the narrowest region of the pore (see Section VI,D).

Table I reproduces a comparison of permeability ratios, based on the measurement of reversal potential, of alkali and organic cations in the cGMP-activated channel of retinal rods and in other channels (see Section VI,E).

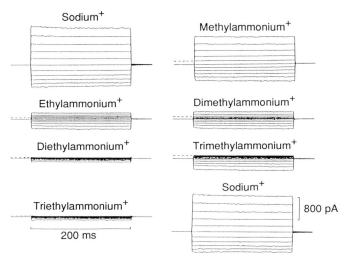

FIGURE 5 Currents activated by 100 μM cGMP in the presence of methylated and ethylated ammonium compounds. The solution filling the patch pipette contained 110 mM NaCl, whereas the solution at the cytoplasmic side contained 110 mM NaCl or the chloride salt of the indicated ammonium compound. Voltage steps of $\pm$ 20 mV were given from a holding potential of 0 mV. The dashed lines represent the zero-current level. Reproduced with permission from Picco and Menini (1993).

C. Divalent Cations

Various divalent cations permeate through the cGMP-activated channel in intact rods (Capovilla *et al.*, 1983; Hodgkin *et al.*, 1985; Menini *et al.*, 1988; Nakatani and Yau, 1988). Divalent cations have a high affinity for the channel and carry a rather small current. The ratios of the estimated maximal inward currents in intact rods were (Menini *et al.*, 1988)

$$Na^+:Sr^{2+}:Ba^{2+}:Ca^{2+}:Mg^{2+}:Mn^{2+}$$
$$= 76.9:4.5:4.1:1:0.7:0, \qquad (5)$$

and the estimated dissociation constants of divalent cations with the channel were (in mM; Menini *et al.*, 1988)

$$K_{Ba}:K_{Sr}:K_{Mg^{2+}}:K_{Ca^{2+}} = 3.9:2.7:2:0.05. \qquad (6)$$

Experiments with membrane patches provided additional information on the permeability of divalent cations. Figure 6 reproduces currents activated by 100 μM cGMP when 110 mM NaCl in the bathing medium was replaced with 73.3 mM SrCl$_2$, CaCl$_2$, BaCl$_2$, MgCl$_2$, or MnCl$_2$.

The ratios of the outward currents carried at + 60 mV by the divalent cations analyzed and by Na$^+$ over the current carried by Ca^{2+} (Colamartino *et al.*, 1991) were

$$Na^+:Sr^{2+}:Ca^{2+}:Ba^{2+}:Mg^{2+}:Mn^{2+}$$
$$= 83.3:1.4:1:0.58:0.33:0.25. \qquad (7)$$

The selectivity sequence 5, obtained in intact rods, and sequence 7, obtained in excised patches in the presence of 100 μM cGMP, are in a broad agreement but do not coincide numerically. The discrepancies between the two sequences can be explained by an asymmetry in ion transport through the channel, as suggested for alkali monovalent cations (Section IV,B) or by more complex mechanisms. Experiments in intact rods (Cervetto *et al.*, 1988; Rispoli and Detwiler, 1990) have suggested that ionic permeability to divalent cations may be modulated by the level of cGMP and of triphosphates (ATP and GTP) and, in particular, may change when the level of cGMP increases. In excised patches in the presence of 10 μM cGMP, a current carried by Ca^{2+} could still be detected but the current carried by Mg^{2+} could not be resolved. Indeed, the half-activation of the current carried by Mg^{2+} occurred at ~85 μM cGMP whereas that for Na$^+$ and Ca^{2+} occurred between 10 and 20 μM cGMP (Colamartino *et al.*, 1991). This result is consistent with the proposed change in permeability of the channel when cGMP increases (Cervetto *et al.*, 1988; Rispoli and Detwiler, 1990). However, an alternative explanation is possible. The appearance of a current carried

TABLE I **Permeability Ratios**[a]

Cation	cGMP[b]	AChR[c]	Na[+d]	K[+e]	Ca[2+f]
Hydroxylammonium	5.92	1.92	0.94	<0.025	—
Ammonium	2.80	1.79	0.16	0.13	1.50
Hydrazinium	1.99	1.32	0.59	<0.029	—
Methylhydroxylammonium	1.15	—	<0.056	—	—
Guanidinium	1.12	1.59	0.13	<0.013	—
Formamidinium	1.00	1.58	0.14	<0.020	—
Aminoguanidinium	0.63	1.37	0.06	—	—
Methylammonium	0.60	1.34	<0.007	<0.021	1.02
Acetamidinium	0.36	1.20	<0.008	—	—
Methylguanidinium	0.33	0.79	<0.010	—	—
Ethanolammonium	0.19	0.72	<0.014	—	—
Ethylammonium	0.16	1.13	—	—	—
Dimethylammonium	0.14	0.87	<0.007	—	0.67
Diethylammonium	<0.003	0.38	—	—	—
Triethylammonium	<0.003	0.09	—	—	—
Trimethylammonium	<0.003	0.36	—	—	0.14
Diethanolammonium	<0.003	0.25	—	—	—
Triethanolammonium	<0.003	0.03	—	—	—
Glucosammonium	<0.019*	0.034	—	—	—
Tetramethylammonium	<0.019*	—	<0.005	—	0.055
Tetraethylammonium	<0.019*	—	<0.008	—	—
Choline	<0.019*	0.13	<0.007	—	—
Lysine	<0.019*	<0.01	—	—	—
Arginine	<0.019*	<0.014	—	—	—
Li	1.14*	0.87	0.93	<0.018	—
Na	1.00*	1.00	1.00	0.01	1.0
K	0.98*	1.11	0.086	1.00	0.8
Rb	0.84*	1.30	<0.012	0.91	0.6
Cs	0.58*	1.42	<0.013	<0.077	0.35

[a] Reproduced with permission from Picco and Menini (1993).
[b] cGMP-activated channel of salamander retinal rods (*Menini, 1990; Picco and Menini, 1993).
[c] nAChR channel of the endplate (Adams *et al.*, 1980; Dwyer *et al.*, 1980).
[d] Na^+ channel of the node of Ranvier (Hille, 1971,1972).
[e] K^+ channel of the node of Ranvier (Hille, 1973).
[f] Ca^{2+} channel of skeletal muscle or of mouse neoplastic B lymphocytes (Fukushima and Hagiwara, 1985; McCleskey and Almers, 1985).

by Mg^{2+} only in the presence of high levels of cGMP may be caused by a change in the ability of cGMP to open ionic channels because of an elevated level of Mg^{2+}.

D. Permeation and Saturation

In excised patches, the outward current at positive voltages increased and saturated when the concentration of the permeating cation was elevated in the me-dium bathing the cytoplasmic side of the membrane (Menini, 1990; Zimmerman and Baylor, 1992). The relationship between the outward current activated by cGMP at +90 mV, I_X, and the ion activity of the permeating cation, $[X^+]_i$, at the cytoplasmic side of the membrane could be fitted by the Michaelis–Menten equation

$$I_X = I_{\max,X}\frac{[X^+]_i}{[X^+]_i + K_{m,X}}, \tag{8}$$

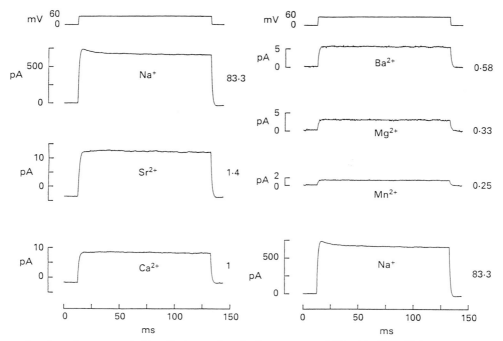

FIGURE 6 Current carried by divalent cations in the presence of 100 μM cGMP. Voltage steps were from 0 to +60 mV. (*Top left, bottom right*) Current recorded in the presence of Na$^+$ at the beginning and at the end of the experiment. Other panels show the current obtained when 110 mM NaCl was replaced with 73.3 mM SrCl$_2$, CaCl$_2$, BaCl$_2$, MgCl$_2$, or MnCl$_2$. Numbers adjacent to each trace are the ratio of the current carried by each cation over the current carried by Ca^{2+}. Reproduced with permission from Colamartino *et al.* (1991).

where $K_{m,X}$ is the concentration activating half-maximal current and $I_{max,X}$ is the maximal current carried by the cation X. $K_{m,X}$ was 249, 203, and 160 mM and the ratio between the maximal current and the current measured in 110 mM NaCl symmetrical solutions, $I_{max,X}/I_{110,X}$, was 3.5, 3, and 0.9 for Na$^+$, K$^+$, and Li$^+$, respectively (Menini, 1990). No evidence for anomalous mole fraction effect was found in mixtures of Na$^+$ and Li$^+$ (Menini, 1990) or of Na$^+$ and Ca^{2+} (Zimmerman and Baylor, 1992). These results indicate the existence of at least one ion binding site along the permeation pathway inside the channel (Menini *et al.*, 1988; Menini, 1990; Zimmerman and Baylor, 1992). Furman and Tanaka (1990) and Rispoli and Detwiler (1990) found evidence for the existence of additional ion binding sites within the channel.

V. Blockage of cGMP-Activated Channels

Divalent cations such as Ca^{2+} and Mg^{2+} are well known to block the cGMP-activated channels (see Section V,A) in intact rods (Hodgkin *et al*, 1985; Lamb and Matthews, 1988) and in excised patches (Fesenko *et al.*, 1985; Haynes *et al.*, 1986; Zimmerman and Bay-

lor, 1986,1992; Colamartino *et al.*, 1991; Su *et al.*, 1991). Large organic cations such as choline and tetraethylammonium (Menini, 1990; Picco and Menini, 1993), protons (Menini and Nunn, 1990), and the compound l-*cis*-diltiazem (Stern *et al.*, 1986; Rispoli and Menini, 1988) also have a powerful blocking effect on the current flowing through cGMP-activated channels (see Section V,B).

A. Divalent Cations

The addition of 1 mM Ca^{2+} or Mg^{2+} to 110 mM NaCl bathing the cytoplasmic side of the membrane reduces the macroscopic current activated by cGMP (Colamartino *et al.*, 1991; Zimmerman and Baylor, 1992). Figure 7 illustrates the voltage dependence of the blocking effect of 1 mM of various divalent cations on the current activated by 100 μM cGMP. The ratio between the currents flowing in the presence and absence of the blocking divalent ion is plotted as a function of membrane voltage for several patches. The sequence of the blockage of divalent cations on the outward Na$^+$ current at +60 mV was

$$Mg^{2+} > Mn^{2+} > Ba^{2+} > Ca^{2+} > Sr^{2+}. \qquad (9)$$

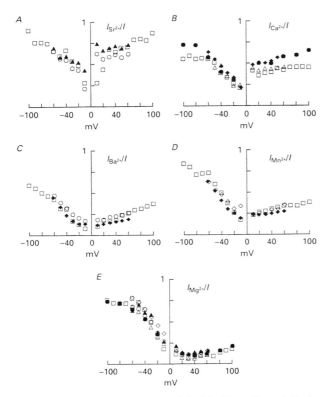

FIGURE 7 Voltage dependence of the blocking effect of divalent cations. The ratio I_X/I is plotted against membrane voltage for Sr^{2+}(A), Ca^{2+}(B), Ba^{2+}(C), Mn^{2+}(D), and Mg^{2+}(E), where I_X is the current flowing in the presence of the blocking ion X and I is the current flowing in its absence. Each symbol represents data from the same patch. Reproduced with permission from Colamartino *et al.* (1991).

sis of the effect of Ca^{2+} on the amplitude histograms of current fluctuations induced by micromolar amounts of cGMP suggested that 1 mM Ca^{2+} reduced the open probability ~3-fold and decreased the single-channel conductance in a voltage dependent way but, in the presence of a saturating concentration of cGMP, the major effect of Ca^{2+} was on the single-channel conductance.

The power spectrum (Fig. 8D) of the fluctuations in the presence of 3 μM cGMP and 1 mM Ca^{2+} had the same shape as that in the presence of 2 μM cGMP and 2 mM EDTA. The shape of the power spectrum was virtually unaltered at membrane voltages between -100 and $+100$ mV at Ca^{2+} concentrations up to 1 mM.

Mg^{2+} ions blocked the current activated by cGMP in a different way. The blocking effect of Mg^{2+} is more voltage dependent; 1 mM cytosolic Mg^{2+} blocked the current quite powerfully at membrane voltages more positive than $+40$ mV, but had a very weak effect at membrane voltages more negative than -40 mV. The presence of millimolar amount of Mg^{2+} altered the shape of the power spectrum of current fluctuations induced by cGMP at positive membrane potentials. The effect of Mg^{2+} on the shape of the power spectrum indicates a more powerful blockage of the low frequency components. The major effect of Mg^{2+} is to reduce the single-channel conductance, but Mg^{2+} has an additional effect on the open probability (Sesti *et al.*, 1994).

B. Organic Cations, Protons, and l-**cis**-diltiazem

Many impermeant organic compounds usually used as substitutes for Na^+, such as choline, tetramethylammonium, tetraethylammonium, lysine, and arginine block the inward current carried by Na^+ at negative membrane potentials (Furman and Tanaka, 1990; Menini, 1990; Picco and Menini, 1993). As do divalent and organic cations, protons have a blocking effect on the cGMP-activated current (Menini and Nunn, 1990). The blocking effect of protons is poorly voltage dependent, and the macroscopic current is reduced by 50% at a pH of ~5.0.

The light-sensitive current in intact rods is blocked by the addition of l-*cis*-diltiazem to the extracellular solution (Stern *et al.*, 1986; Rispoli and Menini, 1988). The same compound at micromolar concentrations blocks cGMP-activated channels in excised patches in a voltage-dependent way (Stern *et al.*, 1986; Haynes, 1992; McLatchie and Matthews, 1992).

At -60 mV, divalent cations still blocked the inward current, but with an efficacy of

$$Ba^{2+} > Ca^{2+} > Sr^{2+} > Mn^{2+} > Mg^{2+}. \quad (10)$$

Several factors are likely to contribute to the voltage dependence of the blocking effect of divalent cations illustrated in Fig. 7; indeed, divalent cations can block the current by reducing the open probability or the single-channel conductance. Another analysis of the blocking effect of Ca^{2+} and Mg^{2+} on single channels showed that these ions have a combined effect on the open probability and on the single-channel conductance (Torre *et al.*, 1992; Sesti *et al.*, 1994).

The addition of 0.5 or 1 mM Ca^{2+} (Fig. 8B,C) to the medium bathing the cytoplasmic side of the membrane reduced but did not abolish the frequency of occurrence of the cGMP-induced single-channel openings seen in the absence of Ca^{2+} (Fig. 8A). Analy-

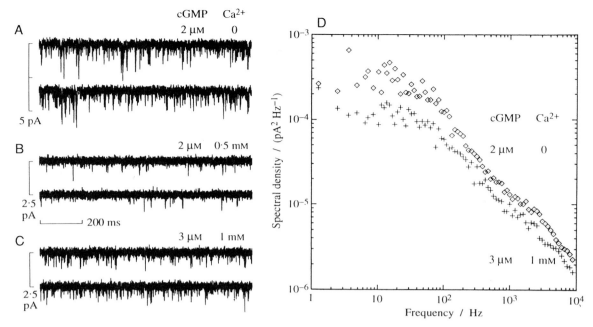

FIGURE 8 cGMP-induced currents and spectra at -60 mV in the presence or absence of Ca^{2+}. (A) 2 μM cGMP and 2 mM EDTA and no added calcium. (B) 2 μM cGMP and 0.5 mM Ca^{2+}. (C) 3 μM cGMP and 1 mM Ca^{2+}. (D) Power spectrum of current fluctuations induced by 2 μM cGMP in the absence of calcium ($\diamond$) and by 3 mM cGMP with 1 mM Ca^{2+} ($+$). Reproduced with permission from Torre *et al.* (1992).

VI. Structure of the Channel

Most electrophysiological results are consistent with a simple model of the cGMP-activated channel. The data suggest a channel structure consisting of a narrow portion, behaving as a one-ion pore, and a large vestibule at the cytoplasmic side, with weakly voltage-dependent binding sites that can be occupied by both permeant and impermeant cations (Menini, 1990). Binding sites may also exist at the extracellular side.

Cook *et al.* (1987) isolated a 63-kDa polypeptide from the bovine retina which, when reconstituted in planar bilayers, behaved as a channel gated by cGMP. Subsequently, this channel was cloned and expressed in oocytes (Kaupp *et al.*, 1989). The properties of the expressed bovine channel are similar, but not identical, to the properties of the cGMP-activated channel that can be analyzed *in situ* in rods of the tiger salamander (Torre *et al.*, 1992; Zagotta *et al.*, 1992). Although some differences may exist between the cGMP-activated channel *in situ* and that expressed in oocytes, data indicate that the isolated protein is indeed the cGMP-activated channel, or a large portion of it.

A. Molecular Structure of the cGMP-Activated Channel

The molecular structure of the cGMP-activated channel has been described in another review (Kaupp, 1991). The channel protein has a single identified binding site for cyclic nucleotides, comprising 80–100 amino acid residues located near the C terminus. Six putative transmembrane segments occur in the sequence. A significant feature of this polypeptide is that its amino acid sequence contains a sequence motif that has been proposed to serve as a voltage sensor in voltage-gated channels (Jan and Jan, 1990; Guy *et al.*, 1991).

B. Number of Subunits

The amino acid sequence of the cGMP-activated channel contains only one binding site for cGMP (Kaupp *et al.*, 1989). The highly cooperative activation of the current by cGMP (Fig. 1) suggests that a functional channel is composed by several monomers. A Hill coefficient of 3 suggests that a channel is formed by at least three such monomers. If the cGMP-activated channel is formed by four monomers, knowing whether the channel requires the binding of four molecules of cGMP to open, or whether it is also conductive, with different properties, when three molecules of cGMP are bound would be interesting.

C. Ion Binding Sites

Analysis of the current carried by different concentrations of monovalent and divalent cations and of mixtures of these ions (Menini *et al.*, 1988; Menini, 1990; Zimmerman and Baylor, 1992) showed that the dependence of current on ionic activity could be described by a simple Michaelis–Menten equation and that no mole fraction effect occurred. Moreover, some impermeant organic cations block the Na^+ inward current. These results have suggested that ion permeation through cGMP-activated channels can be described by assuming that, at least in the narrow region, the channel is a one-ion channel, possibly with other poorly voltage-dependent binding sites in a large inner vestibule (Menini, 1990).

The absence of any evident voltage dependence of the blocking effect of internal pH (Menini and Nunn, 1990) suggests the existence of a site that can be occupied by protons at the cytoplasmic mouth of the channel. Some discrepancies with the ideal one-site channel were found by Furman and Tanaka (1990).

D. Pore Size

By measuring the size of all permeant cations, some information can be obtained on the size of the narrowest region of the pore. The permeant ion of largest dimensions was methylguanidinium, whose two smallest dimensions are 3.8 × 5.9 Å. However, by taking into account the formation of hydrogen bonds within the channel, the effective dimensions could be reduced. Researchers estimated that all the permeant cations will fit in a pore with dimensions of at least 3.8 × 5 Å (Picco and Menini, 1993).

E. Comparison with Other Channels

A comparison of the selectivity for alkali and organic monovalent cations of the cGMP-activated channel of retinal rods and of some voltage-gated Na^+, K^+, and Ca^{2+} channels, as well as the acetylcholine receptor channel, is shown in Table I. A comparison with voltage-gated channels is of some interest, given the proposed structural homology between the cGMP-activated channel and the family of voltage-gated channels (Jan and Jan, 1990; Guy *et al.*, 1991; Kaupp, 1991).

Na^+ and K^+ channels are very selective among ions and exclude methylated cations, whereas the cGMP-activated channel is also permeable to many cations containing methyl groups. The acetylcholine channel is permeable to cations that do not permeate the cGMP-activated channel, including trimethylammonium and choline. The Ca^{2+} channel of skeletal muscle in the absence of divalent cations is also permeable to trimethylammonium and tetramethylammonium.

A comparison of the estimated pore sizes of these channels shows that the pore of the cGMP-activated channel of retinal rods (3.8 × 5 Å) is larger than that of the Na^+ (3 × 5 Å; Hille, 1971) or K^+ (3 Å diameter; Hille, 1973) channels of the node of Ranvier, but smaller than that of the Ca^{2+} channel of skeletal muscle (6 Å diameter or 5.5 × 5.5 Å square; McCleskey and Almers, 1985) or the acetylcholine channel (6.5 × 6.5 Å; Dwyer *et al.*, 1980).

Acknowledgments

This research was supported by the European Community Grant BRA SSS 6961. We thank M. Zanini for helping type the manuscript and L. Giovanelli for checking the English.

References

Adams, D. J., Dwyer, T. M., and Hiue, B. (1980). The permeability of endplate channels to monovalent and divalent metal cations. *J. Gen. Physiol.* **75**, 493–510.
Altenhofen, W., Ludwig, J., Eismann, E., Kraus, W., Bonigk, W., and Kaupp, U. B. (1991). Control of ligand specificity in cyclic nucleotide-gated channels from rod photoreceptors and olfactory epithelium. *Proc. Natl. Acad. Sci. USA* **88**, 9868–9872.
Baylor, D. A., and Nunn, B. J. (1986). Electrical properties of the light-sensitive conductance of salamander rods. *J. Physiol.* **371**, 115–145.
Capovilla, M., Caretta, A., Cervetto, L., and Torre, V. (1983). Ionic movements through light-sensitive channels of toad rods. *J. Physiol.* **343**, 295–310.
Cervetto, L., Menini, A., and Torre, V. (1988). The modulation of the ionic selectivity of the light-sensitive current in isolated rods of the tiger salamander. *J. Physiol.* **440**, 189–206.
Colamartino, G., Menini, A., and Torre, V. (1991). Blockage and permeation of divalent cations through the cyclic GMP-activated channel from tiger salamander retinal rods. *J. Physiol.* **440**, 189–206.
Cook, N. J., Hanke, W., and Kaupp, U. B. (1987). Identification, purification, and functional reconstitution of the cyclic GMP-dependent channel from rod photoreceptors. *Proc. Natl. Acad. Sci. USA* **84**, 585–589.
Dhallan, R. S., Yau, K. W., Schrader, K. A., and Reed, R. R. (1990). Primary structure and functional expression of a cyclic nucleotide-activated channel from olfactory neurons. *Nature (London)* **347**, 184–187.
Dhallan, R. S., Macke, J. P., Eddy, R. L., Shows, T. B., Reed, R. R., Yau, K.-W., and Nathans, J. (1992). Human rod photoreceptor cGMP-gated channel: Amino acid sequence, gene structure, and functional expression. *J. Neurosci.* **12**, 3248–3256.
Dryer, S., and Henderson, D. (1991). A cyclic GMP-activated channel in dissociated cells of the chick pineal gland. *Nature (London)* **353**, 756–758.

Dwyer, T. M., Adams, D. J., and Hille, B. (1980). The permeability of the endplate channel to organic cations in frog muscle. *J. Gen. Physiol.* **75**, 469–492.

Fesenko, E. E., Kolesnikov, S. S., and Lyubarsky, A. L. (1985). Induction by cyclic GMP of cationic conductance in plasma membrane of retinal rod outer segment. *Nature (London)* **313**, 310–313.

Fukushima, Y., and Hagiwara, S. (1985). Currents carried by monovalent cations through calcium channels in mouse neoplastic B lymphocytes. *J. Physiol.* **358**, 255–284.

Furman, R. E., and Tanaka, J. C. (1990). Monovalent selectivity of the cyclic guanosine monophosphate-activated ion channel. *J. Gen. Physiol.* **96**, 57–82.

Gordon, S. E., Brautigan, D. L., and Zimmerman, A. L. (1992). Protein phosphatases modulates the apparent agonist affinity of the light-regulated ion channel in retinal rods. *Neuron* **9**, 739–748.

Goulding, E. H., Ngai, J., Kramer, R. H., Colicos, S., Axel, R., Siegelbaum, S. A., and Chess, A. (1992). Molecular cloning and single-channel properties of the cyclic nucleotide-gated channel from catfish olfactory neurons. *Neuron* **8**, 45–58.

Guy, H. R., Durell, S. R., Warmke, J., Drysdale, R., and Ganetzky, B. (1991). Similarities in amino acid sequences of *Drosophila eag* and cyclic nucleotide-gated channels. *Science* **254**, 730.

Haynes, L. W. (1992). Block of the cyclic GMP-gated channel of vertebrate rod and cone photoreceptors by l-*cis*-diltiazem. *J. Gen. Physiol.* **100**, 783–802.

Haynes, L. W., and Yau, K. W. (1985). Cyclic GMP-sensitive conductance in outer segment membrane of catfish cones. *Nature (London)* **317**, 61–64.

Haynes, L. W., Kay, A. R., and Yau, K. W. (1986). Single cyclic GMP-activated channel activity patches of rod outer segment membranes. *Nature (London)* **321**, 66–70.

Hille, B. (1971). The permeability of the sodium channel to organic cations in myelinated nerve. *J. Gen. Physiol.* **58**, 599–619.

Hille, B. (1972). The permeability of the sodium channel to metal cations in myelinated nerve. *J. Gen. Physiol.* **59**, 637–658.

Hille, B. (1973). Potassium channels in myelinated nerve: Selective permeability to small cations. *J. Gen. Physiol.* **61**, 669–686.

Hodgkin, A. L., McNaughton, P. A., and Nunn, B. J. (1985). The ionic selectivity and calcium dependence of the light-sensitive pathway in toad rods. *J. Physiol.* **358**, 447–468.

Jan, L. Y., and Jan, Y. N. (1990). A superfamily of ion channels. *Nature (London)* **345**, 672.

Karpen, J. W., Zimmerman, A. L., Stryer, L., and Baylor, D. A. (1988). Gating kinetics of the cyclic GMP-activated channel of retinal rods: Flash photolysis and voltage jump studies. *Proc. Natl. Acad. Sci. USA* **85**, 1287–1291.

Karpen, J. W., Loney, D. A., and Baylor, D. A. (1992). Cyclic GMP-activated channels of salamander retinal rods: spatial distribution and variation of responsiveness. *J. Physiol.* **448**, 257–274.

Kaupp, U. B. (1991). The cyclic nucleotide-gated channels of vertebrate photoreceptors and olfactory epithelium. *Trends Neurosci.* **14**, 150–157.

Kaupp, U. B., and Altenhofen, W. (1992). Cyclic nucleotide-gated channels of vertebrate photoreceptor cells and olfactory epithelium. *In* "Sensory Transduction" (D. P. Corey, and S. D. Roper, eds.), pp. 133–150. Rockefeller University Press, New York.

Kaupp, U. B., Niidome, T., Tanabe, T., Terada, S., Bonigk, W., Stuehmer, W., Cook, N. J., Kangawa, K., Matsuo, H., Hirose, T., Miyata, T., and Numa, S. (1989). Primary structure and functional expression from complementary DNA of the rod photoreceptor cyclic GMP-gated channel. *Nature (London)* **342**, 762–766.

Lamb, T. D., and Matthews, H. R. (1988). External and internal actions in the response of salamander retinal rods to altered external calcium concentration. *J. Physiol.* **403**, 473–494.

Ludwig, J., Marglit, T., Eismann, E., Lancet, D., and Kaupp, U. B. (1990). Primary structure of a cAMP-gated channel from bovine olfactory epithelium. *FEBS Lett.* **270**, 24–29.

Luehring, H., Hanke, W., Simmoteit, R., and Kaupp, U. B. (1990). Cation selectivity of the cyclic GMP-gated channel of mammalian rod photoreceptors. *In* "Sensory Transduction" (A. Borsellino, L. Cervetto, and V. Torre, eds.), pp. 169–174. Plenum Press, New York.

Matthews, G. (1986). Comparison of the light-sensitive and cyclic GMP-sensitive conductances of the rod photoreceptor: Noise characteristics. *J. Neurosci.* **6**, 2521–2526.

Matthews, G. (1987). Single-channel recordings demonstrate that cGMP opens the light-sensitive ion channel of the rod photoreceptor. *Proc. Natl. Acad. Sci. USA* **84**, 299–302.

Matthews, G., and Watanabe, S. I. (1987). Properties of ion channels closed by light and opened by guanosine 3',5'-cyclic monophosphate in toad retinal rods. *J. Physiol.* **389**, 691–715.

Matthews, G., and Watanabe, S. I., (1988). Activation of single ion channels from toad retinal rod inner segments by cyclic GMP: Concentration dependence. *J. Physiol.* **403**, 389–405.

McCleskey, E. W., and Almers, W. (1985). The Ca channel in skeletal muscle is a large pore. *Proc. Natl. Acad. Sci. USA* **82**, 7149–7153.

McLatchie, L. M., and Matthews, H. R. (1992). Voltage-dependent block by l-*cis*-diltiazem of the cyclic GMP-activated conductance of salamander rods. *Proc. R. Soc. London B* **247**, 113–119.

Menini, A. (1990). Currents carried by monovalent cations through cyclic GMP-activated channels in excised patches from salamander rods. *J. Physiol.* **424**, 167–185.

Menini, A., and Anholt, R. R. H. (1994). Cyclic nucleotide activated channels. In: "Ion channels and ion pumps: Metabolic and endocrine relationships in biology and clinical medicine." (P. P. Foa and M. F. Walsh, eds.), Vol. 6, pp. 526–548. Springer Verlag, New York.

Menini, A., and Nunn, B. J. (1990). The effect of pH on the cyclic GMP-activated conductance in retinal rods. *In* "Sensory Transduction" (A. Borsellino, L. Cervetto, and V. Torre, eds.), pp. 175–181. Plenum Press, New York.

Menini, A., Rispoli, G., and Torre, V. (1988). The ion selectivity of the light-sensitive current in isolated rods of the tiger salamander. *J. Physiol.* **402**, 279–300.

Menini, A., Picco, C., De Micheli, E., and Conti, F. (1991). Na to Li substitution does not affect the gating of the cGMP-activated channel of retinal rods. *Biophys. J.* **59**, 535a (Abstract).

Nakamura, I., and Gold, G. H. (1987). A cyclic nucleotide-gated conductance in olfactory receptor cilia. *Nature (London)* **325**, 442–444.

Nakatani, K., and Yau, K. W. (1988). Calcium and magnesium fluxes across the plasma membrane of the toad rod outer segment. *J. Physiol.* **395**, 695–729.

Nawy, S., and Jahr, C. E. (1990). Suppression by glutamate of cGMPO-activated conductance in retinal bipolar cells. *Nature (London)* **346**, 269–271.

Nunn, B. J. (1987). Ionic permeability ratios of the cyclic GMP-activated conductance in the outer segment membrane of salamander rods. *J. Physiol.* **394**, 17P.

Picco, C., and Menini, A. (1993). The permeability of the cGMP-activated channel to organic cations in retinal rods of the tiger salamander. *J. Physiol.* **460**, 741–758.

Rispoli, G., and Detwiler, P. B. (1990). Nucleoside triphosphates modulate the light-regulated channel in detached rod outer segments. *Biophys. J.* **57**, 368a (Abstract).

Rispoli, G., and Menini, A. (1988). The blocking effect of 1-*cis*-diltiazem on the light-sensitive current of isolated rods of the tiger salamander. *Eur. Biophys. J.* **16,** 65–71.

Sanfilippo, C., and Menini, A. (1993). Cytosolic acidification enhances the activation by cAMP of the channel in retinal rod outer segments. *Biophys. J.* **64,** A17.

Sesti, F., Straforini, M., Lamb, T. D., and Torre, V. (1994). Gating, selectivity and blockage of single channels activated by cyclic GMP in retinal rods of the tiger salamander. *J. Physiol.* **474,** 203–222.

Shiells, R. A., and Falk, G. (1990). Glutamate receptors of rod bipolar cells are linked to a cyclic GMP cascade via a G protein. *Proc. R. Soc. London B* **242,** 91–94.

Stern, J. H., Kaupp, U. B., and McLeish, P. R. (1986). Control of the light-regulated current in photoreceptors by cyclic GMP, calcium and 1-*cis*-diltiazem. *Proc. Natl. Acad. Sci. USA* **83,** 1163–1167.

Su, K., Furman, R. E., and Tanaka, J. C. (1991). Evidence for calcium binding sites in cGMP-activated channels. *Biophys. J.* **59,** 407a (Abstract).

Tanaka, J. C., Eccleston, J. F., and Furman, R. E. (1989). Photorecep-tors channel activation by nucleotide derivatives. *Biochemistry* **28,** 2776–2784.

Torre, V., Straforini, M., Sesti, F., and Lamb, T. D. (1992). Different channel-gating properties of two classes of cyclic GMP-activated channel in vertebrate photoreceptors. *Proc. R. Soc. London B* **250,** 209–215.

Yau, K. W., and Nakatani, K. (1984). Cation selectivity of light-sensitive conductance in retinal rods. *Nature (London)* **309,** 352–354.

Zagotta, W. N., Middendorf, T. R., Baylor, D., and Aldrich, R. W. (1992). Molecular analysis of cyclic nucleotide-activated channels from salamander. *Biophys. J.* **61,** a52 (Abstract).

Zimmerman, A. L., and Baylor, D. A. (1986). Cyclic-GMP sensitive conductance of retinal rods consists of aqueous pores. *Nature (London)* **321,** 70–72.

Zimmerman, A. L., and Baylor, D. A. (1992). Cation interactions within the cyclic GMP-activated channel of retinal rods from the tiger salamander. *J. Physiol.* **449,** 759–783.

Zimmerman, A. L., Karpen, J. W., and Baylor, D. A. (1988). Hindered diffusion in excised membrane patches from retinal rod outer segments. *Biophys. J.* **54,** 351–355.

Gap Junction Channels

Molecular Models of Channel Interaction and Gating in Gap Junctions

CAMILLO PERACCHIA, AHMED LAZRAK, AND
LILLIAN L. PERACCHIA

I. Introduction

Cell coupling is mediated by channels grouped at discrete cell contact domains known as gap junctions. Each channel results from the extracellular interaction of two hemichannels (connexons) and forms a hydrophilic pathway ~15 nm wide and ~18 nm long, spanning two plasma membranes and a 2 to 3-nm extracellular space (gap). In turn, each connexon is an aggregate of six identical intramembrane proteins (connexins), radially arranged around the channel (reviewed by Peracchia, 1980).

In recent years, ultrastructural and diffraction studies have produced a detailed portrait of gap junction architecture and channel framework. Studies on dye diffusion and metabolite exchange between neighboring cells have clarified the nature, size limit, and charge characteristics of permeant molecules, and have defined the extent of metabolic cooperation among cells. Biochemistry and molecular genetics have provided the means for identifying connexin sequences and for mapping secondary and tertiary structure. Electrophysiology, particularly double whole-cell clamp recording, the development of reliable channel expression systems, and channel reconstitution in artificial membranes have paved the way for defining single-channel attributes such as conduc-

tance, gating kinetics, voltage dependence, and subconductance states, and for clarifying mechanisms of channel regulation and modulation (reviewed by Peracchia, 1991c).

Nevertheless, several crucial aspects of channel structure and regulation are still hypothetical. In the absence of high resolution crystallographic information, the three-dimensional structure of connexins can only be guessed. The parameters that determine connexin–connexin interaction within and across junctional membranes, the structure of the channel lining, and both molecular domains and mechanisms involved in channel regulation and gating are still unclear.

II. Connexin Structure and Interaction

Since the mid-1980s, some 16 members of the connexin family have been sequenced (Paul, 1986; Kumar and Gilula, 1986; Beyer et al., 1987,1990; Gimlich et al., 1988; Ebihara et al., 1989; Zhang and Nicholson, 1989; Fishman et al., 1991; Hoh et al., 1991; Paul et al., 1991; Willecke et al., 1991). A general model of their three-dimensional structure has been drawn. Our knowledge of connexin topology is based on sequence analysis, on evidence for selective binding of site-specific antibodies to gap junctions and on data from limited proteolysis (Zimmer *et al.*, 1987; Goodenough *et al.*, 1988; Hertzberg *et al.*, 1988; Milks *et al.*, 1988; Yancey *et al.*, 1989; Laird and Revel, 1990; Meyer *et al.*, 1990).

Investigators generally agree that connexins span the bilayer four times (TM1–TM4) and have C- and N-termini (C-TER, N-TER) on the cytoplasmic side of the membrane (Fig. 1). Thus, these molecules form two outer loops (OL1, OL2) and one inner loop (IL). Two connexin regions are highly conserved; one spans approximately the first 100 residues of the multiple-alignment (MA) sequence, comprising N-TER (~23 residues), OL1 (~35 residues), TM1 and TM2 (~18 residues each), and the beginning of IL; the other, just about as long, extends from ~MA172 to ~MA269 and contains TM3 (18–20 residues), TM4 (~20 residues), OL2 (44–48 residues), and the beginning of C-TER (Fig. 2). The two remaining regions, most of IL and C-TER, vary considerably not only in sequence but also in length. IL ranges from less than 30 residues (Cx31.1) to over 70 residues (Cx45); C-TER varies from 18 residues (Cx26) to 188 residues (Cx56)(Fig. 2).

A. Extracellular Channel Seal and Connexin–Connexin Interaction

For the formation of functional cell-to-cell channels, at least two elements are deemed necessary: channel sealing and connexon–connexon binding. The channel must be completely insulated from the extracellular medium by a seal of high electrical resistivity to function as an ionic conduit between two cells. Connexon–connexon binding must be sufficiently strong to maintain channel integrity.

In the absence of covalent bonds between connexons, effective channel insulation from the external medium can only be provided by hydrophobic interactions between opposite connexons. A careful analysis of OL1 and OL2 reveals potential clues not only to the nature of the channel seal, but also to the nature of connexon–connexon binding. OL1 and OL2 are similar to each other in terms of –SH groups (three cysteines) and bending or flexible regions (proline or glycine), spaced in almost exactly the same way (Fig. 3), suggesting a similar general configuration that may be important for intra- and inter-loop interaction as well as for the extracellular interaction between connexins. However, the two OLs differ in length because of an extra 8- to 9-residue segment (MA190–MA199) in OL2 (Rahman and Evans, 1991; Bennett *et al.*, 1991)(Fig. 3). Interestingly, this segment is quite hydrophobic in all connexins (Fig. 3), despite being situated outside the membrane, and is directly continuous with the third transmembrane chain (TM3), an amphi-

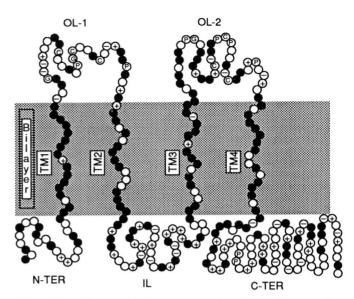

FIGURE 1 General topology of connexins. Mapped here is the rat liver connexin (Cx32). Based on hydrophobicity maps and on data from immunocytochemistry and partial proteolysis, connexins are believed to contain four transmembrane domains (TM), both N (N-TER) and C (C-TER) termini at the cytoplasmic side of the membrane, two outer loops (OL-1 and OL-2), and an inner loop (IL). Filled circles, hydrophobic residues; open circles, polar residues.

philic domain generally believed to line the channel (Nicholson and Zhang, 1988; Milks *et al.*, 1988; Unwin, 1988). One possibility is that this segment provides the extracellular portion of the channel with a high resistance seal by hydrophobically interacting across the gap with the homologous segment of the opposite connexin.

A corollary of the hydrophobic channel-seal hypothesis is that in hemichemicals (nonjunctional connexons), the hydrophobic domains would collapse and interact with each other, creating an external channel gate that would be closed only in free connexons of nonjunctional and intracellular membranes. During gap junction formation, a switch from intra- to interconnexon hydrophobic interaction would occur, causing the opening of cell-to-cell channels and the formation of the hydrophobic seal between the channels and the extracellular medium.

In addition to interacting hydrophobically, opposite connexins may bind to each other via Ca^{2+} bridges. Ca^{2+} involvement in connexon–connexon binding is suggested by previous work on junction splitting (Peracchia, 1977). Splitting was very infrequent in gap junctions of rat livers perfused with hypertonic sucrose at normal $[Ca^{2+}]_o$, but virtually all the junctions became separated when sucrose perfusion at normal $[Ca^{2+}]_o$ was preceded by perfusion

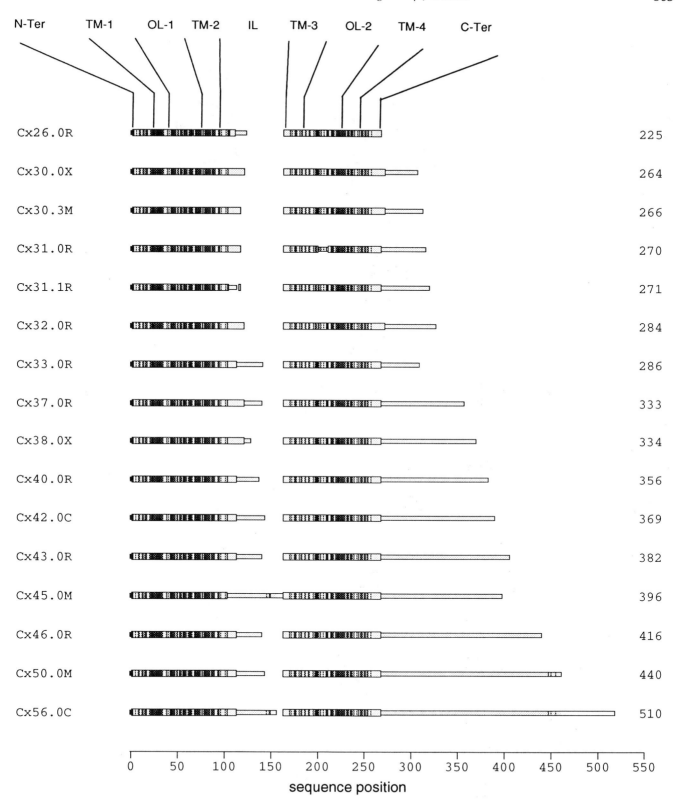

FIGURE 2 Schematic view of multiple sequence alignment (MA) of the 16 members of the connexin family. Note two major areas of similarity: one spanning from the N terminus (N-Ter) to the first 10–12 residues of the inner loop (IL) and the other from the third transmembrane domain (TM-3) to the first 15–20 residues of the C terminus (C-Ter). Most of the inner loop (IL) and the C terminus (C-Ter) varies greatly in length and sequence among connexins.

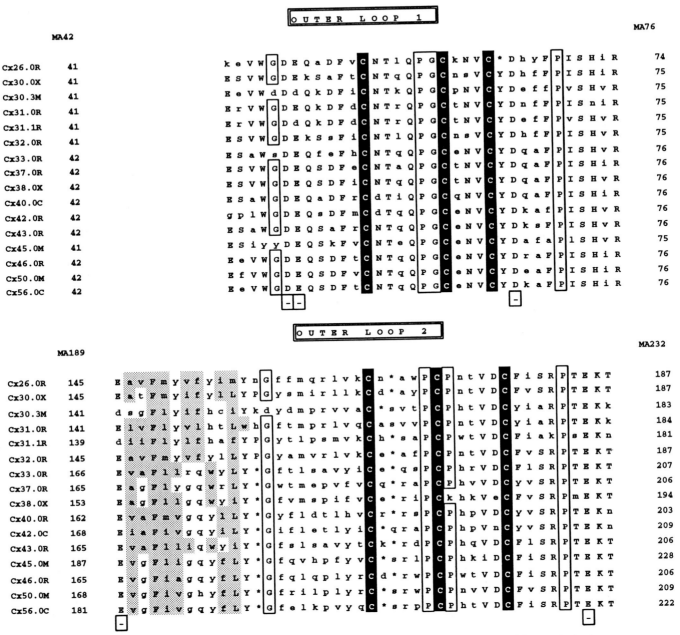

FIGURE 3 Multiple sequence alignment (MA) of connexin outer loops (OL). The two OLs are similar in terms of number and location of cysteine (C), proline (P), and glycine (G) residues, suggesting a similar general configuration, but they differ in length because of an additional segment at the beginning of OL-2. This segment is believed to be mostly extracellular, but is mostly hydrophobic (shaded residues); since it is continuous with TM-3 (the presumed channel-lining chain), OL-2 might represent a site of hydrophobic interaction between connexins across the gap. This site could create a hydrophobic channel seal of high electrical resistivity. Note the conserved negative charges (D or E) in both OLs; they could participate in establishing Ca^{2+} bridges between connexins across the gap.

with EDTA (Ca^{2+}- and Mg^{2+}-free solutions). Consistent with this result is evidence that low $[Ca^{2+}]_o$ prevents the establishment of cell coupling in reassociated cells (Loewenstein, 1966). A reasonable hypothesis is that Ca^{2+} links the carboxyl groups of opposing OLs to each other. Acidic side chains of Asp and Glu ($pK_a < 5$) have a net negative charge at near neutral pH, but lose the charge with severe acidification. This behavior may explain why, in isolated liver gap junctions, the junctional membranes separate with incubation at pH 2 (Zimmer *et al.*, 1987). Indeed, three negative charges are conserved in OL1, and two are conserved in OL2, of all connexins sequenced to date (Fig. 3).

B. Hypothetical Model of Extracellular Interaction between Connexins

As a working hypothesis, a possible model of extracellular connexin–connexin interaction has been proposed that considers the features just mentioned (hydrophobic seal and Ca^{2+} bridges) as well as other important elements. This model is based on the idea that each junctional membrane is unlikely to be a mirror-symmetrical image of the other. In other words, connexins are more likely to be identical than mirror-symmetrical images of their counterparts in a gap junction, because cells have no reason to make two types of connexins, that is, right- and left-handed ones. Mirror symmetry with identical connexins could only occur if OL1 and OL2 were coaxially aligned along the connexon radius, but in this case either TM1 or TM4 would line the channel (Fig. 4). Thus, if TM3 lines the channel and both OL1 and OL2 interact with homologous OLs across the gap, a likely model would place OL1 and OL2 next to each other, radially arranged around the channel with their longitudinal

axes oriented at an ~30° angle to each other (Figs. 5,6). In this model, OL1 and OL2 would still match with OL1 and OL2 of the adjoining membrane. However, opposite connexins would not bind one to one but would be staggered with respect to each other, so each connexin of one membrane would interact with two connexins of the adjoining membrane. In other words, if connexins 1 through 6 of cell 1 were to bind connexins 1a through 6a of cell 2, while OL1 of Cx1 through Cx6 would bind to OL1 of Cx1a through Cx6a, OL2 of Cx1 would bind to OL2 of Cx2a, OL2 of Cx2 to OL2 of Cx3a, and so on (Fig. 7). Indeed, a staggered arrangement would provide a stronger junction than a matched one. Two configurations of the interlocked model are possible. In one, both OLs would have the same N-to-C sequence orientation (Fig. 5), centrifugal with respect to the channel; in the other, only OL2 would have this orientation (Fig. 6).

The conformation of OL1 and OL2 would depend primarily on S–S bonds and on flexible regions (proline and glycine). Intra- and interloop S–S bonds have been demonstrated (Manjunath and Page, 1986; Dupont *et al.*, 1989; John and Revel, 1991; Rahman and Evans, 1991). Consensus data based on analysis of flanking sequences likely to promote S–S bonding caused Rahman and Evans (1991) to suggest the presence of two intraloop and one interloop S–S bonds. Of the three –SH groups per loop, the second (Fig. 3, MA61 and MA217) and third (Fig. 3, MA65 and MA223) are more likely to form the intraloop bond (Fig. 5; see also Rahman and Evans, 1991). Whereas the first of OL1 (Fig. 3, MA54) is likely to form an S–S bond with the first of OL2 (Fig. 3, MA211; Fig. 5). This arrangement would cause TM1 and TM3 to be closer to each other than TM2 and TM4, enabling OL1 and OL2 to diverge by the ~30° angle needed for a staggered connexin–connexin interaction (Fig. 5). The model envisions only one TM per connexin (TM3) lining the channel, which is consistent with the channel size. Since an α helix has a cross-sectional diameter of 0.8–0.9 nm, a channel 1.5–1.8 nm in diameter (circumference of 4.7–5.65 nm) is likely to be enriched by no more than six α-helical chains.

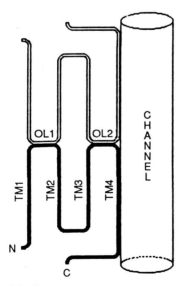

FIGURE 4 Model of one-to-one connexin–connexin interaction. Since the connexins of two interacting cells are most likely identical in configuration, a one-to-one interaction would require mirror symmetry. This model would account for mirror symmetry and one-to-one connexin–connexin interaction, but is unlikely because, contrary to the general hypothesis, it would cause TM-1 or TM-4, rather than TM-3, to line the channel.

III. Channel Gating

Gap junction channels are mostly in an open state, but can close in response to specific changes in the cellular homeostasis. Closure of most of the junctional channels results in electrical and metabolic cell-to-cell uncoupling. Uncoupling is generally believed to be primarily a safety mechanism by which cells can isolate themselves from damaged neighbors, but evi-

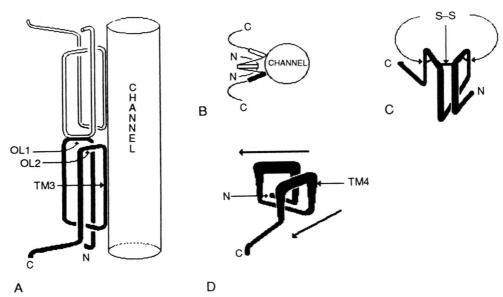

FIGURE 5 Model of staggered (one-to-two) connexin–connexin interaction. In this model, OL-1 and OL-2 of one connexin would interact with homologous OLs of two opposite connexins (A) and vice versa, and would be placed next to each other, radially arranged around the channel with their longitudinal axes forming an~30° angle with each other. This arrangement would depend primarily on S–S bonds (C–C) and bending or flexible regions (P and G). The model would predict that, of the three –SH groups present in each loop, the second and third encountered in the N-to-C (sequence) direction would form an S–S bond, creating an intraloop link (C), whereas the first–SH of OL-1 would form an S–S bond with the first –SH of OL-2 (C). This interaction would place TM-1 and TM-3 closer to each other than TM-2 and TM-4, enabling OL-1 and OL-2 to diverge by the ~30° angle needed for a staggered connexin–connexin interaction. (B) A face view of OL interaction at the gap, along a plane perpendicular to the channel. In this model, both OLs have the same N-to-C sequence orientation (D, arrows), centrifugal with respect to the channel.

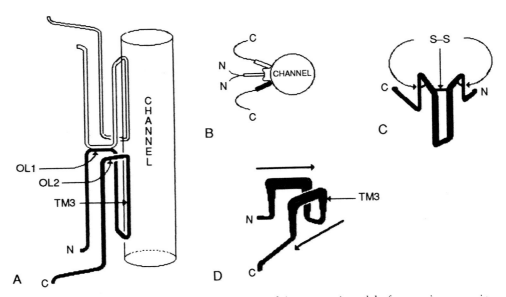

FIGURE 6 Alternative configuration of the staggered (one-to-two) model of connexin–connexin interaction shown in Fig. 5. The only difference between the two is the sequence direction of the OLs. In this model, the two OLs are arranged with opposite N-to-C sequence direction (D, arrows). Obviously, this interaction would cause TM-2 and TM-3 to be closer to each other than TM-1 and TM-4 (A, C, and D) and would involve different residues in the formation of S–S bonds. For other details, see Fig. 5.

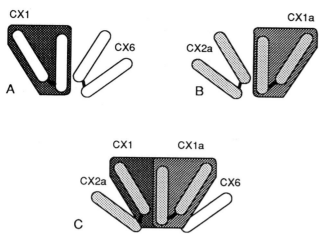

FIGURE 7 Face view of OL interaction in the staggered (one-to-two) model of connexin–connexin binding shown in Fig. 5 and 6. In this diagram, the staggered arrangement of the OLs is viewed along a plane running perpendicular to the channel at the level of the extracellular gap. CX 1 and CX 6 (located below the page) and CX 1a and CX 2a (located above the page) are seen individually in A and B and one on top of the other in D. Note that CX 1 and CX 1a bind to each other via OL-2s only, since OL-1 of CX 1 binds to OL-1 of CX 2a and OL-1 of CX 1a binds to OL-1 of CX 6. This staggered (one-to-two) arrangement would be expected to provide a stronger junction than a matched (one-to-one) configuration.

dence for gap junction channel sensitivity to physiological $[Ca^{2+}]_i$ (Peracchia, 1990b; Lazrak and Peracchia, 1992,1993) indicates that modulation of cell-to-cell coupling may play a role in several Ca^{2+}-mediated regulatory processes. Over the years, several uncoupling agents such as Ca^{2+}, H^+, voltage, cAMP, general anesthetics, and fatty acids have been identified but, other than voltage, whether any of them act directly or via intermediates on the channel gates is unclear.

A. Ca^{2+} Mediates the H^+ Effects on Coupling in Crayfish Axons

In the past, some studies seemed to support a direct effect of H^+ on gap junction channels (Turin and Warner, 1977,1980; Spray *et al.*, 1981), but others did not (Ramón and Rivera, 1987; Peracchia, 1987b; Pressler, 1989). To investigate in more detail the mechanism of low pH-induced uncoupling and to test the possible participation of Ca^{2+}_i in the process, junctional electrical resistance (R_j), pH$_i$, and pCa$^{2+}_i$ in crayfish septate axons uncoupled with an acidifying solution in which NaCl was replaced with Na-acetate (Ac) (Peracchia, 1990b) were monitoried. pH$_i$ and pCa$_i$ were measured with ion-selective electrodes. The Ca^{2+} sensor (ETH 129) contained a neutral carrier that has a logarithmic response to $[Ca^{2+}]$ down to 0.5 nM and is virtually insensitive to other ions (Ammann *et al.*, 1987).

On superfusion with Ac, both R_j and $[H^+]_i$ rapidly increased. Whereas the R_j curve was almost symmetrical, the $[H^+]_i$ curve was asymmetrical: $[H^+]_i$ increased rapidly with Ac but recovered slowly (Fig. 8A). R_j and $[H^+]_i$ curves also differed in peak time, that is, $[H^+]_i$ maxima preceded R_j maxima by 40–90 sec (Fig. 8A). This difference resulted in marked curve hysteresis in the relationship between R_j and $[H^+]_i$. Further, R_j maxima with slow acidification were three times greater than those with fast acidification of the same magnitude.

$[Ca^{2+}]_i$ also increased with Ac by approximately one order of magnitude, but R_j and $[Ca^{2+}]_i$ curves were similar to each other in shape and peak time

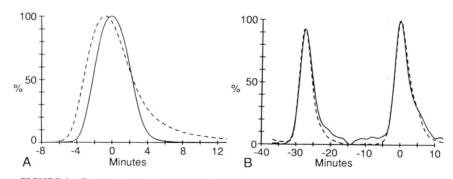

FIGURE 8 Comparison of percentage increase in junctional electrical resistance (R_j, solid lines) and percentage increase in either $[H^+]_i$ (A, dashed line) or $[Ca^{2+}]_i$ (B, dashed line) in crayfish lateral giant axons uncoupled cytosolic acidification induced by superfusion with acetate, for 5 min (A) and (twice) for 3 min (B). Note that whereas R_j and $[H^+]_i$ curves differ from each other in shape and peak time (A), R_j and $[Ca^{2+}]_i$ curves are well matched (B). This result suggests that channel gating with lowered pH$_i$ is a closer function of $[Ca^{2+}]_i$ than of $[H^+]_i$. Reprinted with permission from Peracchia (1991b).

(Fig. 8B), so the relationship between $[Ca^{2+}]_i$ and R_j did not show significant hysteresis; indeed, R_j and $[Ca^{2+}]_i$ curves were virtually superimposed. Both R_j and $[Ca^{2+}]_i$ recovered following a single exponential decay with a time constant (τ) of ~2.3 min. The threshold $[Ca^{2+}]_i$ effective on R_j was in the high nanomolar range (Peracchia, 1990b).

These data suggested that the increase in R_j with lowered pH_i is a closer function of $[Ca^{2+}]_i$ than of $[H^+]_i$. Thus, H_i could simply affect mechanisms involved in the regulation of $[Ca^{2+}]_i$ rather than act directly on the channels. However, at that stage, ruling out any direct action of H^+ seemed premature since the possibility that H^+ increased channel sensitivity to Ca^{2+}, in addition to increasing $[Ca^{2+}]_i$, could not be entirely excluded. This possibility, however, may no longer be valid based on data obtained in Novikoff hepatoma cell pairs (Lazrak and Peracchia, 1992,1993; see Section III,B).

To understand the mechanism of low pH_i-induced $[Ca^{2+}]_i$ increase, R_j and pCa_i were monitored in crayfish septate axons uncoupled with Ac in the presence of drugs that modify Ca^{2+} release from internal stores (caffeine and ryanodine), or with treatments that affect Ca^{2+} entry (Peracchia, 1990a). Addition of 10–30 mM caffeine to Ac solutions more than doubled both R_j and $[Ca^{2+}]_i$ maxima with respect to controls, whereas exposure to caffeine before or before and during Ac significantly reduced both parameters. Caffeine also increased the maximum rate of $[Ca^{2+}]_i$ rise with Ac from ~0.8 to ~1.3 pCa units/min, whereas a substantial rate reduction was observed after caffeine pretreatment. In the absence of Ac, caffeine did not affect R_j, membrane resistance (R_m), or basal $[Ca^{2+}]_i$. In the presence of ryanodine (1–10 μM), both R_j and $[Ca^{2+}]_i$ maxima were approximately one-third of controls, and the maximum rate of $[Ca^{2+}]_i$ rise decreased by approximately 50%. In contrast, different $[Ca^{2+}]_o$ and Ca^{2+} channel blockers did not significantly change the magnitude of R_j and $[Ca^{2+}]_i$ maxima with Ac.

The data indicated that caffeine- and ryanodine-sensitive stores are involved in uncoupling by low pH, and confirmed evidence for an involvement of Ca^{2+} in gating crayfish junctional channels during acidification (Peracchia, 1990b,1991a). In turn, Ca^{2+} may act via a calmodulin (CaM)-like protein (Peracchia and Bernardini, 1984; see Section III,D).

B. Ca^{2+} Mediates the H^+ Effects on Coupling in Novikoff Hepatoma Cells

The sensitivity of cell coupling to acidification has been studied in many mammalian cells, but in most cases coupling was monitored with conventional microelectrodes so the cytosolic composition could not be controlled. Using the double whole-cell clamp technique, G_j was found to decrease in cells exposed to 100% CO_2, but the cells were internally buffered for calcium with EGTA (Burt and Spray, 1988b). EGTA is not as reliable as BAPTA as an intracellular Ca^{2+} buffer for several reasons. First, the Ca-EGTA affinity constant drops by two orders of magnitude with a decrease in pH from 7 to 6 ($K_{Ca-aff} = 1.8 \times 10^6$ and 1.8×10^4 at pH 7 and 6, respectively), whereas that of Ca-BAPTA decreases only slightly with the same pH drop ($K_{ca-aff} = 7 \times 10^6$ and 1.8×10^6, at pH 7 and 6, respectively). Second, calculations of the gradient of $[Ca^{2+}]$ in the vicinity of a channel pore show that EGTA solutions are completely ineffective in buffering $[Ca^{2+}]$ within macromolecular distances of the pore, whereas faster buffers such as BAPTA are quite effective (Marty and Neher, 1985; Stern, 1992).

To test in detail gap junction gating sensitivity to Ca^{2+} and H^+ in mammalian cells, the effects of acidification on junctional conductance (G_j) were studied in Novikoff hepatoma cell pairs internally buffered for Ca^{2+} with either EGTA (a H^+-sensitive Ca^{2+} buffer) or BAPTA (a fast and H^+-insensitive Ca^{2+} buffer), as were the effects of G_j of internal solutions of various pH and $[Ca^{2+}]_i$ values. The Novikoff hepatoma cells constitute a good model system because they express a very common connexin, the rat heart Cx43.

Cytoplasmic acidification caused by exposure to 100% CO_2 had drastically different effects on G_j depending on the intracellular Ca^{2+} buffer used. With 0.5 mM EGTA in the patch pipettes, CO_2 superfusions as short as 2 min had a very large effect on G_j, whereas with 0.5 mM BAPTA even superfusions of 4 min had no effect on G_j. This result indicates that acididification uncouples via an increase in cytosolic Ca^{2+} that is not adequately buffered by EGTA. Indeed, the opposite result would be expected if H^+ rather than Ca^{2+} were acting directly on the channels, because with CO_2 EGTA would bind H^+ and release Ca^{2+} so $[Ca^{2+}]_i$ would increase even more (and $[H^+]_i$ would increase even less) than with BAPTA. Thus, paradoxically, with lowered pH_i, EGTA acts more like an internal source of Ca^{2+} release than like a Ca^{2+} chelator. Based on these data, much if not all of the CO_2 effect on G_j was caused by a low pH_i-induced increase in $[Ca^{2+}]_i$, as in crayfish (Peracchia, 1990b). The obvious test was to measure G_j in cells exposed to different pCa_i and pH_i conditions.

Immediately after the establishment of the double whole-cell configuration, Novikoff hepatoma cell pairs display G_js ranging from 25 to 200 nS but, as in other cells, G_j spontaneously decays with time, even with internal solutions well buffered for Ca^{2+} (1 mM BAPTA) and containing 3 mM ATP. The G_j decay follows a single exponential function with time constants (τ) usually greater than 20 min. To test the effects of different pCa_i and pH_i conditions on G_j, G_j decay curves have been compared among cell pairs exposed to intracellular solutions well buffered for Ca^{2+} (with 1 mM BAPTA) to different pCa conditions (9, 6.9, 6.3, and 6), and for H^+ (with 10 mM HEPES or MES) to pH 7.2 or 6.1. In cell pairs buffered to pCa_i 6.9 or 9, G_j decreased to 10–70% of initial values in ~40 min, following a single exponential curve with a τ of 27.8 ± 6.32 min (mean ± SE; n = 13; Fig. 9). In contrast, in cell pairs buffered to pCa_i 6.3 or 6, G_j decreased to 10–25% of initial values in ~15 min with a τ of 4.87 ± 1.24 min (mean ± SE; n = 7; Fig. 9). The difference between the two groups was statistically significant (P = 0.0178). A further increase in calcium to pCa_i 5.5 resulted in extremely fast uncouplings with τ values of ~20 sec. Acidic pipette solutions affected neither time course nor shape of G_j decay at any pCa tested. At pCa_i 6.9 or greater, τ was 31.7 ± 8.9 (mean ± SE; n = 9) with pH_i 6.1, and 19 ± 3.25 (mean ± SE; n = 4) with pH_i 7.2 (not a statistically significant difference).

The dramatic decrease in time constant with a change in pCa_i from 6.9 to 6.3, regardless of pH_i, indicates that the channel gates are sensitive to nanomolar [Ca^{2+}] and are insensitive to pH_i, at least in the range of 7.2 to 6.1. Lowered pH_i does not seem to change channel sensitivity to Ca^{2+} but, paradoxically, may prevent some channels from closing. This possibility is suggested by evidence for higher residual G_j values at pH 6.1 than at pH 7.2. Data showing lack of synergism between Ca^{2+} and H^+ in reducing G_j contrast with previous evidence from cardiac cells (Burt, 1987; White *et al.*, 1990). Burt (1987) reported no significant change in dye coupling with treatments that increased either [Ca^{2+}]$_i$ or [H^+]$_i$, but significant dye uncoupling with treatments that increased both quantities. However, this study measured neither [Ca^{2+}]$_i$ nor [H^+]$_i$. White *et al.* (1990) reported that G_j was not decreased by reducing pH_i to 5.9–6.0 (with 100% CO_2) in low Ca^{2+} media, but similar acidification in low Ca^{2+} and low Na^+ media, shown in parallel experiments to increase [Ca^{2+}]$_i$ to ~425 nM, reduced G_j to zero. However, this study did not measure [Ca^{2+}]$_i$ and G_j simultaneously, and monitored change in pH_i with neutral carrier pH-sensitive microelectrodes that have been shown to be unreliable in the presence of 100% CO_2 (Peracchia, 1990b).

Early data seemed to support a direct effect of H^+ on G_j (Spray and Bennett, 1985), but some inconsistencies were apparent (Ramón and Rivera, 1987; Peracchia, 1987b). Turin and Warner (1977,1980) showed changes in coupling ratio (α) and pH_i in amphibian embryonic cells, but plotted α against pH_i only for the recoupling phase; thus, possible curve hystereses were not determined. Spray *et al.* (1981) reported a hysteresis in the relationship between pH_i and α, but interpreted it as an artifact of the effects of CO_2 on nonjunctional conductance. The same study reported the absence of hysteresis in the relationship between pH_i and G_j and, thus, concluded that protons act directly on the channel protein. Campos de Carvalho *et al.* (1984) reported that G_j, measured when pH_i decreased, fell along nearly the same Hill relationship as G_j measured when pH_i was recovering, but only with brief acidification; with longer acidification G_j recovered slowly and incompletely despite a normal pH_i recovery rate.

The reported diversity of pH_i sensitivity in different systems and even in the same cells is also hard to reconcile with the idea of a direct H^+ effect on gating. Turin and Warner (1980) showed that, in amphibian embryonic cells, a 50% recovery from CO_2 treatment occurs at pH_i = 6.5 in some experiments and at pH = 6.9 in others. In other cells, the pK of the Hill plot ranged widely from 7.3 to 6.3 (Spray and Bennett, 1985). In addition, a pH_i of 6.6 had only a small effect in mammalian heart fibers (Reber and Weingart, 1982), H^+ affected healing-over in the heart only at pH < 5 (De Mello, 1983), low pH solutions did not change junctional conductance in internally perfused crayfish axons (Johnston and Ramon, 1981; Arellano *et al.*, 1986), and several inconsistencies between pH_i

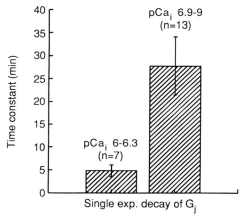

FIGURE 9 The time constant (τ) of G_j decay in Novikoff cell pairs studied by double whole-cell clamp recording is 27.8 ± 6.32 min (mean ± SE, n = 13) with pipette (intracellular) solutions of pCa_i 6.9–9 and 4.87 ± 1.24 min (mean ± SE, n = 7) with pipette solutions of pCa_i 6–6.3 (P = 0.0178). Neither time constant was significantly altered by pH_i in the range 6.1–7.2. This result indicates that the gap junction channels of Novikoff cells are sensitive to nanomolar [Ca^{2+}]$_i$, but are insensitive to pH_i.

and electrical coupling were reported in cells of insect salivary glands (Rose and Rick, 1978). As in crayfish (Peracchia, 1990b), in Novikoff cells Ca^{2+} appears to affect G_j at nanomolar concentrations (Lazrak and Peracchia, 1992,1993; Lazrak et al., 1993).

C. Channel Gating Sensitivity to Nanomolar Calcium

Over the years, various $[Ca^{2+}]_i$ have been reported to induce uncoupling but, although $[Ca^{2+}]_i$ as high as 40–400 μM were effective in ruptured (Oliveira-Castro and Loewenstein, 1971) or internally perfused (Spray et al., 1982) cells, low micromolar to high nanomolar concentrations induced gating in intact cells (Rose and Loewenstein, 1976; Weingart, 1977; Dahl and Isenberg, 1980; Neyton and Trautmann, 1986; Maurer and Weingart, 1987; Noma and Tsuboi, 1987; Veenstra and DeHaan, 1988; Peracchia, 1990b). The weak Ca^{2+} sensitivity of ruptured or perfused cells may follow the partial loss of a cytoplasmic intermediate, possibly a CaM-like protein (Peracchia, 1987b). Indeed, in perforated myocytes, the Ca^{2+} gating sensitivity increases from pCa 5.7 to pCa 7 when CaM is added to the perfusate (H. Sugiura, personal communication). This result is also consistent with evidence for Ca^{2+} insensitivity of channels that were in vitro reconstituted from liver (Young et al., 1987; Harris, 1991) or heart (Shen et al., 1992; Peracchia and Shen, 1993) junction proteins.

In a study of Novikoff hepatoma cells (Lazrak and Peracchia, 1992,1993), G_j was affected by pCa_i 6.3 but not by pCa_i 6.9 (Fig. 9), suggesting that the $[Ca^{2+}]_i$ values at which most Ca^{2+}-driven physiological functions are activated (300–500 nM) may also affect cell coupling. This result indicates not only that Ca^{2+} is a physiological regulator of cell coupling, but also, and more importantly, that Ca^{2+}-induced modulation of cell-to-cell communication may play a role in a variety of Ca^{2+}-mediated cellular activities.

The physiological implications of a high sensitivity of channel gating to Ca^{2+}_i are demonstrated by evidence for simultaneous decrease in G_j, increase in $[Ca^{2+}]_i$, and development of an outward $K^+(Ca^{2+})$ current (I_k) in Novikoff cells exposed to 100 μM ATP (Lazrak et al., 1993). $[Ca^{2+}]_i$, measured with fura-2, increased from 70–100 to 400–500 nM and G_j, measured by double whole-cell clamp with the "perforated-patch" method, decreased by ~30%; G_j minima coincided with I_k maxima, indicating that both phenomena are caused by the same factor, namely, an increase in $[Ca^{2+}]_i$. Since the effect was independent of Ca^{2+}_o, the phenomenon appears to result from ATP-induced activation of $P2_\gamma$ receptors and consequential inositol 1,4,5-trisphosphate ($InsP_3$)-induced Ca^{2+} release from internal stores.

D. Is the Ca^{2+}-Induced Gating Mediated by a CaM-Like Protein?

In 1981, three independent observations suggested the existence of uncoupling intermediates. Johnston and Ramón (1981) report the inability of Ca^{2+} and H^+ to uncouple internally perfused crayfish axons. Peracchia et al. (1981,1983) suggested the participation of CaM-like proteins in the uncoupling mechanism, based on the ability of a CaM blocker (trifluoperazine) to prevent electrical uncoupling in embryonic cells. Herzberg and Gilula (1981) and Welsh et al. (1982) demonstrated the ability of CaM to bind to Cx32 and MIP26 (a lens channel protein).

More recently, calmidazolium and W7, two other specific CaM inhibitors, also inhibited uncoupling in Xenopus embryonic cells (Peracchia, 1984), in crayfish axons (Peracchia, 1987a), and in cardiac (Wojtczak, 1985; Tuganowski et al., 1989) and lens (Gandolfi et al., 1990) cells; internally perfused crayfish axons uncoupled with Ca^{2+} only in the presence of CaM (Arellano et al., 1988). CaM binding to connexins was confirmed through gel overlay (Van Eldick et al., 1985; Zimmer et al., 1987) and immunoelectron microscopy (Fujimoto et al., 1989). In pairs of guinea pig ventricular myocytes in which one cell was voltage clamped and G_j was measured after perforation of the nonjunctional membrane of the partner cell, the gap junction sensitivity to Ca^{2+}-induced gating increased from pCa 5.7 to pCa 7 on perfusion with 10 μM CaM; W7 (but not W5) protected the cells from Ca^{2+}-induced uncoupling (H. Sugiura, personal communication).

The CaM hypothesis is being tested further by studying the effects of blockage of CaM expression with oligonucleotides antisense to CaM mRNAs of Xenopus oocytes. Preliminary data with this approach indicate that, 24–48 hr after oligonucleotide injection, pairs of oocytes coupled by the native connexin (Cx38) have completely lost the capacity to uncouple with 100% CO_2, but regain their gating sensitivity to CO_2 soon after CaM injection (Fig. 10; Peracchia and Peracchia, 1993). Since uncoupling by acidification appears to be mediated by Ca^{2+} (Peracchia, 1990b; Lazrak and Peracchia, 1992,1993; see Section III,A,B), the data confirm a Ca^{2+}–CaM involvement in channel gating.

Gap junction channels are not the only channels for which a direct involvement of Ca^{2+}–CaM has been suggested (see Chapters 28 and 29). Ca^{2+}–CaM, but not Ca^{2+} alone, opened Ca^{2+}-dependent Na^+ and K^+ channels of Paramecium tetraurelia in patch-clamped

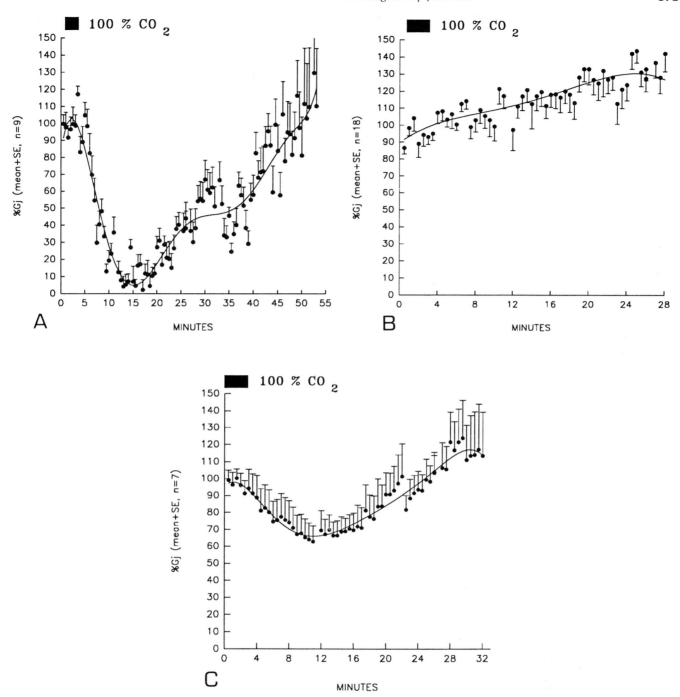

FIGURE 10 Time course of percentage changes in junctional conductance (G_j) in *Xenopus* oocyte pairs (coupled by the native connexin Cx38), following superfusion with solutions bubbled with 100% CO_2. In oocytes injected 24–72 hr earlier with oligonucleotides anti-sense to CaM mRNA (B), CO_2 has virtually no effect on G_j, whereas in uninjected or H_2O-injected oocytes (A), CO_2 causes a complete cell-to-cell uncoupling. A 30–40% recovery of CO_2 gating sensitivity is obtained by injecting CaM into oocytes that have lost gating competency (C).

membranes (Saimi and Ling, 1990; Kink *et al.*, 1991; Preston *et al.*, 1991), as well as Ca^{2+}-activated K^+ channels of rabbit colonocytes in planar bilayers (Liu *et al.*, 1991). In both cases, the Ca^{2+}–CaM effect did not require ATP. Interestingly, in *P tetraurelia*, CaM mutations at the N-terminal lobe abolished the Ca^{2+}-dependent Na^+ currents without affecting the K^+ currents, whereas the reverse was true for CaM mutations at the C-terminal lobe (Kink *et al.*, 1991).

The precise nature of CaM-binding sites is still only partly understood, but structural similarities have been identified in all the CaM receptors studied to date. Small peptides such as β-endorphin, dynorphin, glucagon, mastoparan, and melittin were found to interact strongly ($K_d < 100$ nM) with CaM in a Ca^{2+}-dependent way (Malenick and Anderson, 1982). All were found to contain a basic amphiphilic α helix at least 12 residues long (Malenick and Anderson, 1982; Erickson-Viitanen and DeGrado, 1987). The need for such a structure was soon confirmed by the observation that the CaM-binding domains of myosin light-chain kinase (MLCK: Blumenthal *et al.*, 1985; Blumenthal and Krebs, 1987) and other CaM-activated enzymes are basic amphiphilic α helices. Further, synthetic peptides designed to maximize basicity and amphiphilicity bound CaM with even greater affinity ($K_d < 1$ nM) than CaM-activated enzymes (Erickson-Viitanen and DeGrado, 1987). Interestingly, basicity and amphiphilicity are more important than sequence homology, since CaM-binding sites with similar K_ds vary widely in sequence (Persechini and Kretsinger, 1988; O'Neil and DeGrado, 1990). The importance of basic residues in CaM sites has been demonstrated by evidence for loss of CaM binding in mutants of smooth muscle MLCK in which four basic residues were replaced with acidic residues (Fitzsimons *et al.*, 1992).

On this basis, researchers have scanned (Peracchia, 1988; Girsch and Peracchia, 1991; Peracchia and Shen, 1993) the connexins of rat heart (Cx43R; Beyer *et al.*, 1987), rat liver (Cx32R; Paul, 1986; Kumar and Gilula, 1986), and early *Xenopus* embryo (Cx38X; Ebihara *et al.*, 1989) using the computer program of Erickson-Viitanen and DeGrado (1987). These and the other connexins were found to contain a potential CaM-binding site at the base of the C terminus (Fig. 11). The potential CaM-binding sites are short amphiphilic, mostly basic, sequences (Fig. 12). The degree of

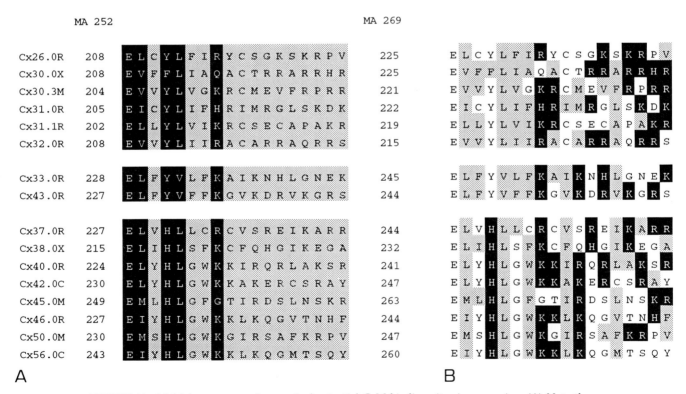

FIGURE 11 Multiple sequence alignment of potential CaM binding sites in connexins. (A) Note three main regions of sequence similarity among connexins. (B) Note the abundance of hydrophobic (shaded) and positively charged (white letters on black) residues, characteristic of CaM binding sites.

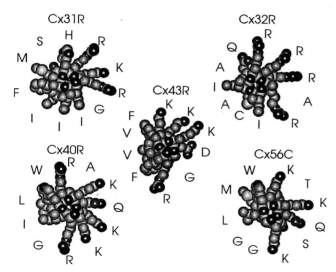

FIGURE 12 Models of axial helical projections (N-to-C sequence orientation) of five potential CaM binding sequences of connexins (see Fig. 11). Note the significant degree of amphiphilicity characterized by the preferential distribution of charged (R, K, and H) or polar (G, Q, S, and T) residues on one side of the helix and of hydrophobic residues (W, L, I, A, M, V, and F) on the other.

amphiphilicity can be evaluated by the helical hydrophobic moment (μ) of the domain (Eisenberg *et al.*, 1982), which is a measure of eccentricity of hydrophobicity in an assumed helix; in regions with high hydrophobic moment, the hydrophobic residues are gathered at one face of the helix and the hydrophilic ones at the other. The potential CaM-binding region of most connexins has a high μ value (Fig. 13), as

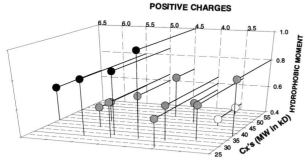

POSITIVE CHARGES

FIGURE 13 Relationship between helical hydrophobic moment and number of positive charges in the potential CaM binding sites (Fig. 11) of connexins. The hydrophobic moment (μ) of a polypeptide chain is the measure of eccentricity of hydrophobicity in an assumed helix; domains with high hydrophobic moment have hydrophobic residues at one face of the helix and hydrophilic ones at the other (see Fig. 12). Note that four connexins (Cx31, 32, 40, and 30, ●) have both a high magnetic moment and numerous positive charges, two (Cx45 and 56, ○) are low in both parameters, and the rest (shaded circles) fall mostly in between. This distribution may predict a range of CaM binding affinities among connexins.

do other known CaM-binding sites (Erickson-Viitanen and DeGrado, 1987), and mostly basic amphiphilicity (Figs. 12 and 13). However, connexins vary from each other in terms of hydrophobic moment and positive charges (Fig. 13). Based on these parameters, the best candidate for CaM binding might be Cx31, followed closely by Cx32, 40, and 30, whereas the worst could be Cx45 and 56; the other 10 connexins fall somewhere between in likelihood. However, extreme caution should be exercised when making predictions of CaM-binding efficiency based on peptide sequence characteristics, because the nature of the environment surrounding the potential CaM-binding peptides within the protein, their interactions with other protein domains, and their accessibility to the cytosol are hard to guess. Indeed, the interaction between CaM and its targets in proteins is still poorly understood. Currently, only some general characteristics of the CaM targets, such as abundance of hydrophobic and basic residues intermixed in a peptide sequence, appear to be common among known CaM-binding proteins.

To test the CaM-binding capacity of some potential CaM-binding sites, four 19-mer peptides—Pep C (MIP26), Cx43-Pep, Cx32-Pep, and Cx38-Pep—were synthesized and tested for CaM binding and conformational change in the presence of Ca^{2+}–CaM (Girsch and Peracchia, 1991; Peracchia and Shen, 1993). The synthetic peptides matched the native peptides (Pep C, 223–242; Cx43-Pep, 227–245; Cx32-Pep, 208–226; Cx38-Pep, 215–233) except for the second residue (leucine or valine), which was substituted with tryptophan to monitor fluorescence emission.

Evidence from spectrofluorometric titration showed that the four peptides bound CaM with 1:1 stoichiometry and K_ds of ~10 nM (Pep C), ~24 nM (Cx43-Pep), ~16 nM (Cx32-Pep), and ~15 nM (Cx38-Pep). The peptides were not affected by Ca^{2+} or H^+ alone. In Ca^{2+}-free solutions containing CaM, the peptides (except Cx38-Pep) underwent a dramatic blue-shift in tryptophan fluorescence emission, indicative of strong hydrophobic interaction, and an increase in circular dichroism (CD) absorption in the α-helical region. Additional fluorescence blue-shift and α-helical CD absorption occurred when Ca^{2+} was added to CaM–Pep solutions. The data suggested a direct interaction between CaM and the synthetic peptides (Peracchia and Girsch, 1989; Girsch and Peracchia, 1991; Peracchia and Shen, 1993), supporting previous evidence for a direct participation of CaM in gap junction function (Peracchia, 1988).

Ca^{2+}-CaM could close the channels by causing conformational changes in connexins or simply by steric hindrance. In the latter case, one CaM molecule per channel would suffice. Alternatively, channel gating could involve a flip–flop mechanism similar to that proposed for several CaM-activated enzymes (Jarrett and Mandhavan, 1991). The flip–flop model suggests the presence of both CaM-binding (basic) peptides and CaM-like (acidic) peptides (targets) in CaM-activated enzymes. The CaM-binding peptide would bind to its target in the absence of Ca^{2+} and would be released from the target by interaction with Ca^{2+}–CaM. In connexins, a possible target could be the inner loop region, in which negatively charged amino acids are abundant.

The molecular mechanism of channel gating is, at present, completely unknown. An interesting hypothesis (Unwin, 1988) suggested a rotation of TM3 as the basis for channel closure. By rotating around its longitudinal axis, TM3 would expose to the channel lumen three phenylalanines normally believed to be in a hydrophobic environment, thus rendering the channel lumen hydrophobic and impermeable to ions. However, of the three phenylalanines believed to be conserved residues in all connexins, only one appears to be so in light of more recent data on additional connexin sequences.

E. Alternative Hypotheses for the Mechanism of Ca^{2+}-Induced Gating

In addition to mechanisms involving direct connexin–CaM interaction, several other ways in which Ca^{2+} could affect channel gating are conceivable. Ca^{2+} could act directly on connexins or via Ca^{2+}–CaM kinases, phosphatases, or proteases. Indeed, phosphorylation of Cx32 by Ca^{2+}–CaM kinase II has been shown to occur, but only in isolated junctions, since intact hepatocytes exposed to the Ca^{2+} ionophore ionomycin did not become phosphorylated (Saez et al., 1990). Further, consensus sequences for Ca^{2+}–CaM kinase II phosphorylation are present in only some of the connexins currently known.

Ca^{2+}-activated proteases are unlikely to be involved because the recovery rate of Ca^{2+}-induced uncoupling is considerably faster than the turnover time of connexins (half-life = ~3 hr; Laird et al., 1991). Phosphatases could also play a role since an increase in $[cAMP]_i$ has been shown to improve coupling (Flagg-Newton et al., 1981; Burt and Spray, 1988a; De Mello, 1988). However, cAMP phosphorylation uncouples the horizontal cells of the retina (McMahon et al., 1989), and data on the effects of protein kinase C activation of coupling are controversial (Chanson et al., 1988; Somogyi et al., 1989).

A direct Ca^{2+} effect on connexins would be expected to involve negative surface charges. However, among connexins, the only intracellular acidic residue

conserved is a C-terminal glutamic acid (MA252). Further, since the channels are not sensitive to Mg_i^{2+} despite its high intracellular concentration, a Ca^{2+} effect at nanomolar concentrations would require Ca^{2+}-specific sites similar to those of Ca^{2+}-modulated proteins (EF hands; Kretsinger, 1975), which are not present in connexins. The absence of an effect of the nearly millimolar $[Mg^{2+}]_i$ also eliminates a possible involvement of membrane phospholipids, since these molecules would not distinguish between Ca^{2+} and Mg^{2+}.

F. Uncoupling Mechanism of General Anesthetics and Fatty Acids

General anesthetics are known to block membrane channels including gap junction channels, but their mechanism of action is still unclear. A direct interaction between anesthetic and amphiphilic domains of channel proteins, possibly, involving hydrophobic interactions and hydrogen bonds, has been suggested (Franks and Lieb, 1982).

In work on the effects of heptanol, halothane, and isoflurane on coupling between crayfish septate axons, evidence was produced for a Ca^{2+}- and H^+-independent uncoupling mechanism potentiated by xanthines (caffeine and theophylline) and inhibited by 4-aminopyridine (4-AP; Peracchia, 1991b). Evidence suggests that anesthetics affect gap junction channels directly, and that xanthines may participate in this scheme by favoring the interaction between anesthetics and connexins. Interestingly, fatty acids are thought to block gap junction channels in a manner similar to anesthetics (Burt, 1989; Giaume et al., 1989), possibly by establishing ionic and hydrophobic interactions with basic amphiphilic domains of connexins (Burt, 1989). CaM-binding sites are reasonable candidates for interaction with anesthetics and fatty acids, not only because they are the principal basic and amphiphilic domains of connexins, but also because fatty acids are known to activate CaM-dependent enzymes directly, in the absence of CaM, by binding to the CaM-binding sites (Tanaka and Hidaka, 1980; Niggli et al., 1981). Another potential target is the primarily hydrophobic domain of OL2 (channel seal?). Indeed, Johnston and Ramón (1981) proposed an extracellular site of action for alkanols because these molecules were effective only when applied externally to internally perfused crayfish axons.

The potentiating effect of xanthines and the inhibitory effects of 4-AP may provide important clues to the mechanisms of general anesthetics and their effects on cardiac rhythm. Indeed, theophylline greatly enhances the arrhythmogenic effect of halothane and other general anesthetics (Koehntop *et al.*, 1977), possibly by enhancing the anesthetic-induced uncoupling. Interestingly, epinephrine enhances the arrhythmogenic action of anesthetics even more than xanthines, and acts synergistically with xanthines on cardiac rhythm (Koehntop *et al.*, 1977). Epinephrine may also act directly on cell-to-cell channels because its effects are not prevented by α and β blockers.

References

Ammann, D., Oesch, U., Buhrer, T., and Simon, W. (1987). Design of ionophores for ion-selective microsensors. *Can. J. Physiol. Pharmacol.* **65,** 879–884.

Arellano, R. O., Ramón, F., Rivera, A., and Zampighi, G. A. (1986). Lowering of pH does not directly affect the junctional resistance of crayfish lateral axons. *J. Membr. Biol.* **94,** 293–299.

Arellano, R. O. Ramón, F., Rivera, A., and Zampighi, G. A. (1988). Calmodulin acts as an intermediary for the effects of calcium on gap junctions from crayfish lateral axons. *J. Membr. Biol.* **101,** 119–131.

Bennett, M. V. L., Barrio, L. C., Bargiello, T. A., Spray, D. C., Hertzberg, E., and Sáez, J. C. (1991). Gap junctions: New tools, new answers, new questions. *Neuron* **6,** 305–320.

Beyer, E. C., Paul, D. L., and Goodenough, D. A. (1987). Connexin43: A protein from rat heart homologous to a gap junction protein from liver. *J. Cell. Biol.* **105,** 2621–2629.

Beyer, E. C., Paul, D. L., and Goodenough, D. A. (1990). Connexin family of gap junction proteins. *J. Membr. Biol.* **116,** 187–194.

Blumenthal, D. K., and Krebs, E. G. (1987). Preparation and properties of the calmodulin-binding protein of skeletal muscle myosin light chain kinase. *Meth. Enzymol.* **139,** 115–126.

Blumenthal, D. K., Takio, K., Edelman, A. M., Charbonneau, H., Titani, K., Walsh, K. A., and Krebs, E. G. (1985). Identification of the calmodulin-binding domain of skeletal muscle myosin light chain kinase. *Proc. Natl. Acad. Sci. USA* **82,** 3187–3191.

Burt, J. M. (1987). Block of intracellular communication: interaction of intracellular H^+ and Ca^{2+}. *Am. J. Physiol.* **253,** C607–C612.

Burt, J. M. (1989). Uncoupling of cardiac cells by doxyl stearic acids: specificity and mechanism of action. *Am. J. Physiol.* **256,** C913–C924.

Burt, J. M., and Spray, D. C. (1988a). Inotropic agents modulate gap junctional conductance between cardiac myocytes. *Am. J. Physiol.* **254,** H1206–H1210.

Burt, J. M., and Spray, D. C. (1988b). Single-channel events and gating behavior of the cardiac gap junction channel. *Proc. Natl. Acad. Sci. USA* **85,** 3431–3434.

Campos de Carvalho, A., Spray, D. C., and Bennett, M. V. (1984). pH dependence of transmission at electrotonic synapses of the crayfish septate axon. *Brain Res.* **321,** 279–286.

Dahl, G., and Isenberg, G. (1980). Decoupling of heart muscle cells: Correlation with increased cytoplasmic calcium activity and with changes of nexus ultrastructure. *J. Membr. Biol.* **53,** 63–75.

De Mello, W. C. (1983). The influence of pH on the healing-over of mammalian cardiac muscle. *J. Physiol.* **339,** 299–307.

De Mello, W. C. (1983). Increase in junctional conductance caused by isoproterenol in heart cell pairs is suppressed by cAMP-dependent protein-kinase inhibitor. *Biochem. Biophys. Res. Commun.* **154,** 509–514.

Dupont, E., El Aoumari, A., Briand, J. P., Fromaget, C., and Gros, D. (1989). Cross-linking of cardiac gap junction connexons by thiol/disulfide exchanges. *J. Membr. Biol.* **108,** 247–252.

Ebihara, L., Beyer, E. C., Swenson, K. I., Paul, D. L., and Goodenough, D. A. (1989). Cloning and expression of a *Xenopus* embryonic gap junction protein. *Science* **243,** 1194–1195.

Eisenberg, D., Weiss, R. M., and Terwilliger, T. C. (1982). The helical hydrophobic moment: A measure of the amphiphilicity of an alpha-helix. *Nature (London)* **299,** 371–374.

Erickson-Viitanen, S., and DeGrado, W. F. (1987). Recognition and characterization of calmodulin-binding sequences in peptides and proteins. *Meth. Enzymol.* **139,** 455–478.

Fishman, G. I., Hertzberg, E. L., Spray, D. C., and Leinwand, L. A. (1991). Expression of connexin43 in the developing rat heart. *Circ. Res.* **68,** 782–787.

Fitzsimons, D. P., Stull, J. T., and Gallagher, P. J. (1992). The role of basic residues in the autoregulation and calmodulin binding of smooth muscle myosin light chain kinase. *Biophys. J.* **61,** A316.

Flagg-Newton, J., Dahl, W., and Loewenstein, W. (1981). Cell junction and cyclic-AMP. 1. Upregulation of junctional membrane permeability and junctional membrane particles by administration of cyclic nucleotide or phosphodiesterase inhibitor. *J. Membr. Biol.* **63,** 105–121.

Franks, N. P., and Lieb, W. R. (1982). Molecular mechanisms of general anaesthesia. *Nature (London)* **300,** 487–492.

Fujimoto, K., Araki, N., Ogawa, K.-S., Kondo, S., Kitaoka, T., and Ogawa, K. (1989). Ultracytochemistry of calmodulin binding sites in myocardial cells by staining of frozen thin sections with colloidal gold-labeled calmodulin. *J. Histochem. Cytochem.* **37,** 249–256.

Gandolfi, S. A., Duncan, G., Tomlinson, J., and Maraini, G. (1990). Mammalian lens inter-fiber resistance is modulated by calcium and calmodulin. *Curr. Eye Res.* **9,** 533–541.

Giaume, C., Randriamampita, C., and Trautmann, A. (1989). Arachidonic acid closes gap junction channels in rat lacrimal glands. *Pflügers Arch.* **413,** 273–279.

Gimlich, R. L., Kumar, N. M., and Gilula, N. B. (1988). Sequence and developmental expression of mRNA coding for a gap junction protein in *Xenopus*. *J. Cell Biol.* **107,** 1065–1073.

Girsch, S. J., and Peracchia, C. (1991). Calmodulin interacts with a C-terminus peptide from the lens membrane protein MIP26. *Curr. Eye Res.* **10,** 839–849.

Goodenough, D. A., Paul, D. L., and Jesaitis, L. (1988). Topological distribution of two connexin32 antigenic sites in intact and split rodent hepatocyte gap junctions. *J. Cell Biol.* **107,** 1817–1824.

Harris, A. L. (1991). Connexin 32 forms ion channels in single artificial membranes. In "Biophysics of Gap Junction Channels" (C. Peracchia, ed.), pp. 273–389. CRC Press, Boca Raton, Florida.

Hertzberg, E. L., and Gilula, N. B. (1981). Liver gap junctions and lens fiber junctions: Comparative analysis and calmodulin interaction. *Cold Spring Harbor Symp. Quant. Biol.* **46,** 639–645.

Hertzberg, E. L., Disher, R. M., Tiller, A. A., Zhou, Y., and Cook, R. (1988). Topology of the Mr 27,000 liver gap junction protein. *J. Biol. Chem.* **263,** 19105–19111.

Hoh, J. H., John, S. A., and Revel, J.-P. (1991). Molecular cloning and characterization of a new member of the gap junction gene family, connexin-31. *J. Biol. Chem.* **266,** 6524–6531.

Jarrett, H. W., and Mandhavan, R. (1991). Calmodulin-binding proteins also have a calmodulin-like binding site within their structure. *J. Biol. Chem.* **266,** 362–371.

John, S. A., and Revel, J.-P. (1991). Connexon integrity is maintained by non-covalent bonds: Intramolecular disulfide bonds link the extracellular domains in rat connexin-43. *Biochem. Biophys. Res. Commun.* **178,** 1312–1318.

Johnston, M. F., and Ramon, F. (1981). Electrotonic coupling in internally perfused crayfish segmented axons. *J. Physiol.* **317,** 509–518.

Kink, J. A., Maley, M. E., Preston, R. R., Ling, K.-Y., Wallen-Friedman, M. A., Saimi, Y., and Kung, C. (1991). Mutations in *Paramecium* calmodulin indicate functional differences between the C-terminal and the N-terminal lobes *in vivo*. *Cell* **62**, 165–174.

Koehntop, D. E., Liao, J.-C., and van Bergen, F. H. (1977). Effects of pharmacologic alterations of adrenergic mechanisms by cocaine, tropolone, aminophylline, and ketamine on epinephrine-induced arrhythmias during halothane-nitrous oxide anesthesia. *Anesthesiology* **46**, 83–93.

Kretsinger, R. H. (1975). Hipothesis: calcium modulated proteins contain EF-hands. *In* "Calcium Transport in Contraction and Secretion" (E. Carafoli, F. Clementi, W., Drabinowski, and A. Margreth, eds.), pp. 469–478. Elsevier, New York.

Kumar, N., and Gilula, N. B. (1986). Cloning and characterization of human and rat liver cDNAs coding for a gap junction protein. *J. Cell Biol.* **103**, 767–776.

Laird, D. W., and Revel, J.-P. (1990). Biochemical and immuno-chemical analysis of the arrangement of connexin43 in rat heart gap junction membranes. *J. Cell Sci.* **97**, 109–117.

Laird, D. W., Puranam, K. L., and Revel, J.-P. (1991). Turnover and phosphorylation dynamics of connexin43 gap junction protein in cultured cardiac myocytes. *Biochem. J.* **273**, 67–72.

Lazrak, A., and Peracchia, C. (1992). Gap junctions of Novikoff hepatoma cells are Ca^{2+}-sensitive but pH-insensitive. *Mol. Biol. Cell*, **3**, A295.

Lazrak, A., and Peracchia, C. (1993). Gap junction gating sensitivity to physiological calcium regardless of pH in Novikoff hepatoma cells. *Biophys. J.* **65**, 2002–2012.

Lazrak, A., Peres, A., Giovannardi, S., and Peracchia, C. (1993). Partial uncoupling, increase in calcium and activation of K(Ca) channels with ATP-induced stimulation of purinergic receptors linked to IP3 turnover. *Proc. 1993 Int. Mtg. Gap Junctions* p. 54. Hiroshima, Japan.

Liu, S., Dubinsky, W. P., Haddox, M. K., and Schultz, S. G. (1991). Reconstitution of isolated Ca^{2+}-activated K^+ channel proteins from basolateral membranes of rabbit colonocytes. *Am. J. Physiol.* **261**, C713–C717.

Loewenstein, W. R. (1966). Permeability of membrane junctions. *Ann. N.Y. Acad. Sci.* **137**, 441–472.

Malenick, D. A., and Anderson, S. R. (1982). Binding of simple peptides, hormones, and neurotransmitters by calmodulin. *Biochemistry* **21**, 3480–3486.

Manjunath, C. K., and Page, E. (1986). Rat heart gap junctions as disulfide-bonded connexon multimers: their depolymerization and solubilization in deoxycholate. *J. Membr. Biol.* **90**, 43–57.

Marty, A., and Neher, E. (1985). Potassium channels in cultured bovine adrenal chromaffin cells. *J. Physiol.* **367**, 117–141.

Maurer, P., and Weingart, R. (1987). Cell pairs isolated from adult guinea pig and rat hearts: Effects of $[Ca^{2+}]_i$ on nexal membrane resistance. *Pflügers Arch.* **409**, 394–402.

McMahon, D. G., Knapp, A. G., and Dowling, J. E. (1989). Horizontal cell gap junctions: Single-channel conductance and modulation by dopamine. *Proc. Natl. Acad. Sci. USA* **86**, 7639–7643.

Meyer, R., Malewicz, B., Baumann, W. J., and Johnson, R. G. (1990). Increased gap junction assembly between cultured cells upon cholesterol supplementation. *J. Cell Sci.* **96**, 231–238.

Milks, L. C., Kumar, N. M., Houghten, R., Unwin, N., and Gilula, N. B. (1988). Topology of the 32-kd liver gap junction protein determined by site-directed antibody localizations. *EMBO J.* **7**, 2967–2975.

Neyton, J., and Trautmann, A. (1986). Acetylcholine modulation of the conductance of intercellular junctions between rat lacrimal cells. *J. Physiol.* **377**, 283–295.

Nicholson, B. J., and Zhang, J. (1988). Multiple protein components

in a single gap junction: Cloning of a second hepatic gap junction protein (Mr 21,000). *In* "Gap Junctions" (E. L. Hertzberg, and R. G. Johnson, eds.), pp. 207–218. Liss, New York.

Niggli, V., Aduryack, E. S., and Carafoli, E. (1981). Acidic phospholipids, unsaturated fatty acids, and limited proteolysis mimic the effect of calmodulin on the purified erythrocyte Ca^{++}-ATPase. *J. Biol. Chem.* **256**, 8588–8592.

Noma, A., and Tsuboi, N. (1987). Dependence of junctional conductance on proton, calcium and magnesium ions in cardiac paired cells of guinea-pig. *J. Physiol.* **382**, 193–211.

O'Neil, K. T., and DeGrado, W. F. (1990). How calmodulin binds its targets: Sequence independent recognition of amphiphilic alpha-helices. *Trends Biochem. Sci.* **15**, 59–64.

Oliveira-Castro, G. M., and Loewenstein, W. R. (1971). Junctional membrane permeability: Effects of divalent cations. *J. Membr. Biol.* **5**, 51–77.

Paul, D. L. (1986). Molecular cloning of cDNA for rat liver gap junction protein. *J. Cell Biol.* **103**, 123–134.

Paul, D. L., Ebihara, L., Takemoto, L. J., Swenson, K. I., and Goodenough, D. A. (1991). Connexin46, a novel lens gap junction protein, induces voltage-gated currents in nonjunctional plasma membrane of *Xenopus* oocytes. *J. Cell Biol.* **115**, 1077–1089.

Peracchia, C. (1977). Gap junctions: Structural changes after uncoupling procedures. *J. Cell Biol.* **72**, 628–641.

Peracchia, C. (1980). Structural correlates of gap junction permeation. *Int. Rev. Cytol.* **66**, 81–146.

Peracchia, C. (1984). Communicating junctions and calmodulin: Inhibition of electrical uncoupling in *Xenopus* embryo by calmidazolium. *J. Membr. Biol.* **81**, 49–58.

Peracchia, C. (1987a). Calmodulin-like proteins and communicating junctions. Electrical uncoupling of crayfish septate axons is inhibited by the calmodulin inhibitor W7 and is not affected by cyclic nucleotides. *Pflügers Arch.* **408**, 379–385.

Peracchia, C. (1987b). Permeability and regulation of gap junction channels in cells and in artificial lipid bilayers. *In* "Cell-to-Cell Communication" (W. C. De Mello, ed.), pp. 65–102. Plenum Press, New York.

Peracchia, C. (1988). The calmodulin hypothesis for gap junction regulation six years later. *In* "Gap Junctions" (E. L. Hertzberg, and R. G. Johnson, ed.), pp. 267–282. Liss, New York.

Peracchia, C. (1990a). Effects of caffeine and ryanodine on low pH_i-induced changes in gap junction conductance and calcium concentration in crayfish septate axons. *J. Membr. Biol.* **117**, 79–89.

Peracchia, C. (1990b). Increase in gap junction resistance with acidification in crayfish septate axons is closely related to changes in intracellular calcium but not hydrogen ion concentration. *J. Membr. Biol.* **113**, 75–92.

Peracchia, C. (1991a). Possible involvement of caffeine- and ryanodine-sensitive calcium stores in low pH-induced regulation of gap junction channels. *In* "Biophysics of Gap Junction Channels" (C. Peracchia, ed.), pp. 13–28. CRC Press, Boca Raton, Florida.

Peracchia, C. (1991b). Effects of the anesthetics heptanol, halothane and isoflurane on gap junction conductance in crayfish septate axons: A calcium- and hydrogen-independent phenomenon potentiated by caffeine and theophylline, and inhibited by 4-aminopyridine. *J. Membr. Biol.* **121**, 67–78.

Peracchia, C. (ed) (1991c). "Biophysics of Gap Junction Channels." CRC Press, Boca Raton, Florida.

Peracchia, C., and Bernardini, G. (1984). Gap junction structure and cell-to-cell coupling regulation: Is there a calmodulin involvement? *Fed. Proc.* **43**, 2681–2691.

Peracchia, C., and Girsch, S. J. (1989). Calmodulin site at the

C-terminus of the putative lens gap junction protein MIP26. *Lens Eye Toxic. Res.* **6**, 613–621.

Peracchia, C., and Peracchia, L. L. (1993). Loss of chemical gating in gap junction of *Xenopus* oocyte pairs injected with oligonucleotides anti-sense to calmodulin mRNA. *Mol. Biol. Cell* **4**, 331a.

Peracchia, C., and Shen, L. (1993). Gap junction channel reconstitution in artificial bilayers and evidence for calmodulin binding sites in MIP26 and connexins from rat heart, liver, and *Xenopus* embryo. *Prog. Cell Res.* **3**, 163–170.

Peracchia, C., Bernardini, G., and Peracchia, L. L. (1981). A calmodulin inhibitor prevents gap junction crystallization and electrical uncoupling. *J. Cell Biol.* **91**, 124a.

Peracchia, C., Bernardini, G., and Peracchia, L. L. (1983). Is calmodulin involved in the regulation of gap junction permeability? *Pflügers Arch.* **399**, 152–154.

Persechini, A., and Kretsinger, R. H. (1988). Toward a model of the calmodulin-myosin light-chain kinase complex: Implications for calmodulin function. *J. Cardiovasc. Pharmacol.* **12**, S1–S12.

Pressler, M. L. (1989). Intracellular pH and cell-to-cell transmission in sheep Purkinje fibers. *Biophys. J.* **55**, 53–65.

Preston, R. R., Kink, J. A., Hinrichsen, R. D., Saimi, Y., and Kung, C. (1991). Calmodulin mutants and Ca^{2+}-dependent channels in *Paramecium*. *Ann. Rev. Physiol.* **53**, 309–319.

Rahman, S., and Evans, W. H. (1991). Topography of connexin32 in rat liver gap junctions. Evidence for an intramolecular disulphide linkage connecting the two extracellular peptide loops. *J. Cell Sci.* **100**, 567–578.

Ramón, F., and Rivera, A. (1987). Gap junction channel modulation: A physiological viewpoint. *Prog. Biophys.* **48**, 127–153.

Reber, W. R., and Weingart, R. (1982). Ungulate cardiac Purkinje fibres: The influence of intracellular pH on the electrical cell-to-cell coupling. *J. Physiol.* **328**, 87–104.

Rose, B., and Loewenstein, W. R. (1976). Permeability of a cell junction and the local cytoplasmic free ionized calcium concentration: A study with aequorin. *J. Membr. Biol.* **28**, 87–119.

Rose, B., and Rick, R. (1978). Intracellular pH, intracellular Ca, and junctional cell–cell coupling. *J. Membr. Biol.* **44**, 377–415.

Saez, J. C., Nairn, A. C., Czernik, A. J., Spray, D. C., Hertzberg, E. L., Greengard, P., and Bennett, M. V. (1990). Phosphorylation of connexin 32, a hepatocyte gap-junction protein, by cAMP-dependent protein kinase, protein kinase C and Ca^{2+}/calmodulin-dependent protein kinase II. *Eur. J. Biochem.* **192**, 263–273.

Saimi, Y., and Ling, K.-Y. (1990). Calmodulin activation of calcium-dependent sodium channels in excised membrane patches of *Paramecium*. *Science* **249**, 1441–1444.

Shen, L., Shrager, S., and Peracchia, C. (1992). Functional characteristics of heart gap junction channels in artificial bilayers. *Biophys. J.* **61**, A506.

Somogyi, R., Batzer, A., and Kolb, H.-A. (1989). Inhibition of electrical coupling in pairs of murine pancreatic acinar cells by OAG and isolated protein kinase C. *J. Membr. Biol.* **108**, 273–282.

Spray, D. C., and Bennett, M. V. L. (1985). Physiology and pharmacology of gap junctions. *Annu. Rev. Physiol.* **47**, 281–303.

Spray, D. C., Harris, A. L., and Bennett, M. V. (1981). Gap junctional conductance is a simple and sensitive function of intracellular pH. *Science* **211**, 712–715.

Spray, D. C., Stern, J. H., Harris, A. L., and Bennett, M. V. (1982).

Gap juctional conductance: Comparison of sensitivities to H and Ca ions. *Proc. Natl. Acad. Sci. USA* **79**, 441–445.

Stern, M. D. (1992). Buffering of calcium in the vicinity of a channel pore. *Cell Calcium* **13**, 183–192.

Tanaka, T., and Hidaka, H. (1980). Hydrophobic regions function in calmodulin-enzyme(s) interaction. *J. Biol. Chem.* **255**, 11078–11080.

Tuganowski, W., Korczynska, I., Wasik, K., and Piatek, G. (1989). Effects of calmidazolium and dibutyryl cyclic AMP on the longitudinal internal resistance in sinus node strips. *Pflügers Arch.* **414**, 351–353.

Turin, L., and Warner, A. E. (1977). Carbon dioxide reversibly abolishes ionic communication between cells of early amphibian embryo. *Nature (London)* **270**, 56–57.

Turin, L., and Warner, A. E. (1980). Intracellular pH in early Xenopus embryos: Its effect on current flow between blastomeres. *J. Physiol.* **300**, 489–504.

Unwin, N. (1988). The structure of ion channels in membranes of excitable cells. *Neuron* **3**, 665–676.

Van Eldick, L. J., Hertzberg, E. L., Berdan, R. C., and Gilula, N. B. (1985). Interaction of calmodulin and other calcium-modulated proteins with mammalian and arthropod junctional membrane proteins. *Biochem. Biophys. Res. Commun.* **126**, 825–832.

Veenstra, R. D., and DeHaan, R. L. (1988). Cardiac gap junction channel activity in embryonic chick ventricle cells. *Am. J. Physiol.* **254**, H170–H180.

Weingart, R. (1977). Action of ouabain on intercellular coupling and conduction-velocity in mammalian ventricular muscle. *J. Physiol.* **264**, 341–365.

Welsh, M. J., Aster, J. C., Ireland, M., and Maisel, H. (1982). Calmodulin binds to chick lens gap junction protein in a calcium-independent manner. *Science* **216**, 642–644.

White, R. L., Doeller, J. E., Verselis, V. K., and Wittenberg, B. A. (1990). Gap junctional conductance between pairs of ventricular myocytes is modulated synergistically by H^+ and Ca^{++}. *J. Gen. Physiol.* **95**, 1061–1075.

Willecke, K., Heynkes, R., Dahl, E., Stutenkemper, R., Hennemann, H., Jungbluth, S., Suchyna, T., and Nicholson, B. J. (1991). Mouse connexin37: Cloning and functional expression of gap junction gene highly expressed in lung. *J. Cell Biol.* **114**, 1049–1057.

Wojtczak, J. A. (1985). Electrical uncoupling induced by general anesthetics: A calcium-independent process? *In* "Gap Junctions" (M. V. L. Bennett, and D. C. Spray, eds.), pp. 167–175. Cold Spring Harbor Laboratory Press, Cold Spring Harbor, New York.

Yancey, S. B., John, S. A., Lal, R., Austin, B. J., and Revel, J.-P. (1989). The 43-kD polypeptide of heart gap junctions: Immunolocalization, topology, and functional domains. *J. Cell Biol.* **108**, 2241–2254.

Young, D.-L., Cohen, Z. A., and Gilula, N. B. (1987). Functional assembly of gap junction conductance in lipid bilayers: Demonstration that the major 27 kDa protein forms the junctional channel. *Cell* **48**, 733–743.

Zhang, J.-T., and Nicholson, B. J. (1989). Sequence and tissue distribution of a second protein of hepatic gap junction, Cx26, as deduced from its cDNA. *J. Cell Biol.* **109**, 3391–3401.

Zimmer, D. B., Green, C. R., Evans, W. H., and Gilula, N. B. (1987). Topological analysis of the major protein in isolated rat liver gap junctions and gap junction-derived single membrane structures. *J. Biol. Chem.* **262**, 7751–7763.

25

Molecular Biology and Electrophysiology of Cardiac Gap Junctions

ERIC C. BEYER AND RICHARD D. VEENSTRA

I. Introduction

Gap junctions are plasma membrane specializations that contain channels that permit the intercellular exchange of ions and small molecules. Gap junction channels are of central importance in electrically excitable tissues such as myocardium, in which they facilitate propagation of action potentials. Electrical coupling of cardiac cells through such low-resistance intercellular pathways was first proposed by Weidmann (1952), who applied linear cable theory to the spread of electrotonic potentials in cardiac Purkinje fibers. Barr *et al.* (1965) subsequently demonstrated that closely apposed regions of adjacent plasma membranes, termed the nexus or gap junction, were the functional sites of electrical contact between heart cells; these investigators showed that disruption of these structures by hypertonic solutions also blocked action potential conduction. In thin-section electron micrographs, the gap junction appeared as a pair of parallel membranes separated by a narrow 20-Å extracellular gap that was spanned by hexagonally arranged protein subunits (Robertson, 1963; Revel and Karnovsky, 1967). Each protein bridge was postulated to consist of a pair of hemichannels (or connexons), one from each cell, that align and join to form a continuous aqueous cell-to-cell passageway insulated from

the extracellular space it traverses (Loewenstein, 1966). The locations and appearance of myocardial gap junctions are depicted in Figure 1.

Much of the information obtained during the first 20 years after identification of the gap junction as the specialized communicating junction has been reviewed and discussed elsewhere (Loewenstein, 1981; Page and Shibata, 1981; De Mello, 1982; Spray and Bennett, 1985). In the last few years, two new technologies have provided valuable information about the molecular properties of gap junction channels of the heart and other tissues. First, adaptation of the two-cell voltage-clamp technique of Spray *et al.* (1981) to pairs of small cells from a variety of tissues has led to the detection of individual gap junction channel currents and has enabled investigators to determine biophysical mechanisms for changes in macroscopic junctional conductance (beginning with the studies of Neyton and Trautmann, 1985,1986; Veenstra and DeHaan, 1986). Second, the molecular cloning of cDNAs for members of a family of gap junction proteins, termed connexins, has provided new information about the primary structure and composition of these channel-forming proteins (reviewed in Beyer *et al.*, 1990). Because of the importance of gap junctions to cardiac physiology, we have concentrated our study on the gap junctions expressed by cells of the cardiovascular system. This review focuses on devel-

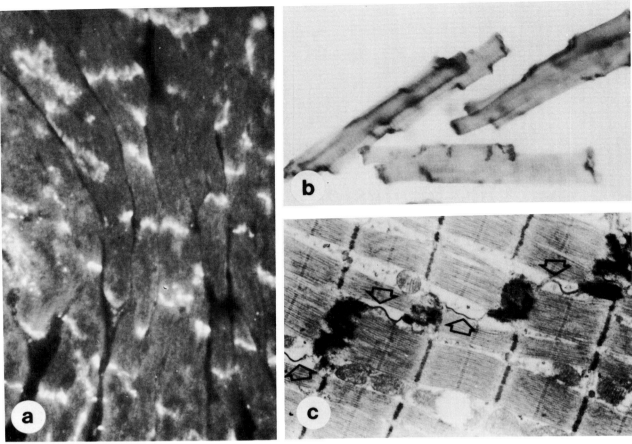

FIGURE 1 Cardiac gap junctions. (a) Immunofluorescent localization of gap junctions in adult rat myocardium (using antiserum against Cx43). (b) Immunoperoxidase localization of gap junctions in isolated canine ventricular myocytes (using antiserum against Cx43). (c) Thin section electron micrograph through the intercalated disk region where two adjacent canine ventricular myocytes join. Gap junctions (arrows) are found both between and within the adhesive elements (fascia adherens). Reprinted by permission of the publisher from Saffitz *et al. Trends in Cardiovascular Medicine*, **2**, 57. Copyright © 1992 by Elsevier Science Publishing Co., Inc.

opments that have enhanced our understanding of the molecular composition and regulation of cardiac gap junction channels with particular emphasis on studies from our laboratories. Other aspects of gap junction biology are discussed in Chapters 24 and 26.

II. Electrophysiology of Cardiac Gap Junctions

A. Channel Permeability, Size, Selectivity, and Conductance

Current flow through an ion channel is governed by certain elementary properties of selective permeability, a limiting electrical conductance, gating, and the electromotive force for ion movement (Hille, 1992). These same basic principles apply to a cardiac

gap junction channel. Applying Ohm's law to an entire gap junction results in the equation

$$I_j = g_j V_j = (\textstyle\sum_i^{\gamma_i} \cdot n_j^i \cdot P_o^i) V_j,$$

where I_j is the macroscopic junctional current, g_j is the macroscopic junctional conductance, V_j is the transjunctional potential, γ_j is the single-channel conductance, n is the number of channels in the entire gap junction, and P_o is the open-state probability for every ith class (or substate) of the channel. The term "gating" refers to the transition of individual channels between different conductance states, namely, the open and closed states of the channel. These alterations are observed as changes in the channel open time. P_o equals the fraction of open channels divided by the total number of channels [$P_o = n_{open}/(n_{open} + n_{closed})$ for every ith class of channel]. The value of V_j will depend on the electrochemical gradients of the permeant ions and the membrane potential (V_m) on each side of the junction. For symmetrical solutions, V_j simply equals the potential difference between adjacent cells ($V_{m2} - V_{m1}$). Each class (or substate) of a channel has a given γ_j associated with it.

1. Permeability

The electric current responsible for cardiac action potential propagation is presumably carried by potassium ions, which are known to permeate the intercalated discs of ventricular trabeculae (Weidmann, 1966). However, gap junction channels are also permeable to a variety of large hydrophilic molecules ranging from tetraethylammonium (TEA) ions to fluorescent dyes (reviewed by Weingart *et al.*, 1975; Loewenstein, 1981; DeMello, 1982). These findings suggest that the junctional channel is an aqueous pore of large diameter and low selectivity. In mammalian cardiac myocytes, tetraglycine-conjugated lissamine rhodamine B-200, a fluorescent dye with a molecular mass of 859 Daltons has the slowest diffusion rate of all permeant tracer molecules tested to date (Imanaga *et al.*, 1987). This result is consistent with the permeability properties of mammalian gap junctions, examined in several cell types and predicts a critical channel diameter of 16 Å (Flagg-Newton and Loewenstein, 1979).

2. Selectivity

For an ion-selective channel, selectivity implies the interaction of a dehydrated ion with fixed dipoles in the narrowest region (selectivity filter) of the channel (Hille, 1992). Since the gap junction channel is an aqueous pore, its selectivity is expected to be low relative to that of other ion channels, but the aqueous nature of the channel also predicts a certain selectivity sequence for the alkali metals ($Cs^+ > Rb^+ > K^+ > Na^+ > Li^+$) based on the hydration shell radius for each ion (Verselis and Brink, 1986). In patch-clamp studies of gap junction channels from earthworm septate axons ($\gamma_j = 100$ pS) and rat lacrimal gland cells ($\gamma_j = 120$ pS), the relative conductance of $Na^+/K^+ = 0.81–0.84$ (Neyton and Trautmann, 1985; Brink and Fan, 1989). In these same studies, the permeability of chloride ions, which have an aqueous mobility nearly identical to that of potassium ions, was lower than expected, as evidenced by a relative Cl^-/K^+ conductance of $0.52–0.67$ (Neyton and Trautmann, 1985; Brink and Fan, 1989). Based on these and previous studies with fluorescent probes, gap junction channels have been proposed to contain fixed anionic sites that can account for this selectivity (Flagg-Newton and Loewenstein, 1979; Brink and

Dewey, 1980). However, none of these studies has been performed in cardiac myocytes. At least some of the gap junction channels characterized for selectivity are formed by different subunit proteins than those in heart cells.

3. Unitary Channel Conductance

Three different single-channel recording techniques have been applied to the measurement of single-channel conductances of gap junctions. The reconstitution of purified proteins from bovine lens and rat liver junctional membranes into artificial lipid bilayers has resulted in the induction of channel activity with conductances ranging from 145 to 360 pS in 0.1 M KCl (Zampighi *et al.*, 1985; Spray *et al.*, 1986; Young *et al.*, 1987; Ehring *et al.*, 1990; Donaldson and Kistler, 1992). A second approach to the recording of gap junction channel activity, by directly patching the junctional membranes of the earthworm septate axon, allows rapid and reversible perfusion of junctional membranes (Brink and Fan, 1989). After blockade of nonjunctional ion-selective channels by the addition of Co^{2+}, Zn^{2+}, Ni^{2+}, Cs^+, and TEA, a 90- to 110-pS channel with characteristics similar to those of other gap junction channels was readily resolved. This approach has provided valuable information about the selectivity (as mentioned earlier) and gating characteristics of presumptive gap junction channels. These approaches have not yet been adapted to cardiac myocytes.

The third and most extensively used approach is the double whole-cell patch-clamp technique, which can be applied to any cell pair of high input resistance and low junctional conductance. Gap junction channel conductance measurements have been obtained with pairs of cells from multiple cardiac sources using this technique, including embryonic chick cardiac myocytes (Veenstra and DeHaan, 1986,1988; Veenstra, 1990), neonatal rat cardiac myocytes (Burt and Spray, 1988a; Rook *et al.*, 1988,1989), adult guinea pig ventricular myocytes (Rudisuli and Weingart, 1989), neonatal hamster myocytes (Wang *et al.*, 1992), and vascular smooth muscle cells (Moore *et al.*, 1991; Christ *et al.*, 1992). An additional advantage of this technique is that junctional currents are readily identifiable as symmetrical signals of equal amplitude and opposite polarity when recorded in the presence of a constant transjunctional potential (V_j; Veenstra and DeHaan, 1986,1988). An example of a cellular preparation and representative junctional current recordings are shown in Fig. 2.

The observation of single gap junction channel activity requires that fewer than 10 channels be open at any given instant in time for any cell pair. Under such

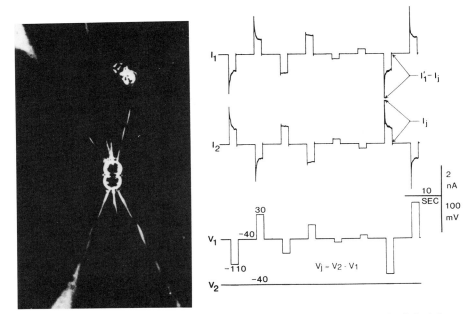

FIGURE 2 Double whole-cell patch-clamp technique. (*Left*) Photomicrograph of chick heart cell pair with patch electrodes in place. (*Right*) Recording protocol for measurement of macroscopic junctional current. I_1 and I_2 are whole-cell currents recorded under voltage-clamp conditions from each cell of an embryonic chick cell pair. V_1 and V_2 are the membrane voltages applied to each cell by separate patch-clamp amplifiers. Transjunctional voltage (V_j) is varied between 100 mV by stepping V_1 from a holding potential of -40 mV to values ranging from -140 to $+60$ mV in 10-mV increments. Each voltage pulse is 2 sec in duration with a 5-sec recovery interval between pulses. V_2 is held constant at -40 mV, so $V_j = V_2 - V_1$. Junctional current (I_j) is observed in both current records as signals of opposite polarity, although a small nonjunctional membrane current (I_1) is also associated with the V_1 voltage step. Hence, I_j is taken as the current signal recorded in I_2 during the V_1 voltage pulse. Macroscopic junctional conductance (G_j) can be calculated by dividing I_j by V_j.

conditions, the symmetrical junctional current signals undergo quantal transitions between distinct current levels that are stable over time. The transitions correspond to the opening or closing of gap junction channels; the stable current levels (states) correspond to the products of unitary currents through the various open channels. Unequal current amplitudes between channel open states represent the different conductances of distinct channel populations or substates of a single population of gap junction channels. Multiple channel records are commonly obtained using this approach, as shown in Fig. 3. In the embryonic chick heart, a 160-pS channel is frequently observed at all ages of the developing heart (Veenstra and DeHaan, 1986,1988; Veenstra, 1990,1991a; Chen and DeHaan, 1992). At least six reproducible channel conductances of approximately 40, 80, 120, 160, 200, and 240 pS were observed in 7-day embryonic chick ventricular myocyte pairs (Chen and DeHaan, 1992).

Direct gap junction channel conductance measurements obtained from neonatal mammalian cardiac ventricular myocytes have produced values ranging from 43 to 60 pS, depending on the composition of the pipette filling solution (Burt and Spray, 1988a; Rook *et al.*, 1988; Wang *et al.*, 1992; Fig. 4). Lower channel conductances of 20–30 pS were also reported for homologous neonatal rat or hamster myocyte pairs (Rook *et al.*, 1988,1989; Wang *et al.*, 1992). Cardiac fibroblasts form gap junction channels with a unitary conductance of 20 pS, and are capable of forming heterologous 30-pS gap junction channels with cardiac myocytes (Rook *et al.*, 1989). Channel conductance measurements are difficult to obtain from adult cardiac myocytes because of their larger membrane surface area (which increases current noise), but have been performed on high input resistance sinus nodal cell pairs (Anumonwo *et al.*, 1992) or by artificially increasing the membrane resistances of adult ventricular myocytes by blocking nonjunctional and junctional membrane conductances with Ba^{2+}, Cd^{2+}, TEA, Cs^+, and heptanol (Rudisuli and Weingart, 1989). Channel conductances of approximately 90, 50, and

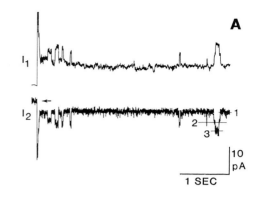

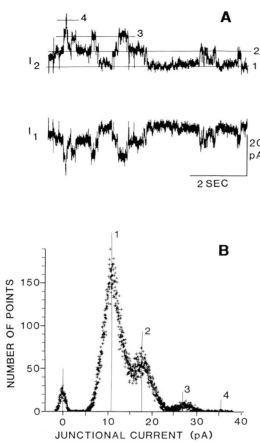

FIGURE 3 Gap junction channel activity from 18-day embryonic chick ventricular cell pairs. (A) A 7-sec current record from a continuous 37-sec 40-mV transjunctional voltage pulse applied to a chick embryo heart cell pair with total junctional conductance (g_j) < 0.9 nS. Gap junction channel openings appear as upward deflections in the I_2 and downward deflections in the I_1 current traces. Thin solid lines in the I_2 trace illustrate the four current levels evident in this record. Low pass filter frequency was 200 Hz and digital sample rate was 1 kHz. (B) Channel amplitude histogram compiled from all 37 sec of the I_j (I_2) recording pictured in A. The zero junctional current level was determined by setting V_j = 0 mV before and after the 40-mV V_j pulse. Current peaks were detected at 10.9, 17.6, 26.5, and 35.5 pA, respectively. Corresponding unitary conductance values for the channel events (transition from one current level to another) were 167 pS (N = 73 channel events), 222 pS (N = 6), and 226 pS (N = 1). Reproduced with permission from Veenstra (1991a).

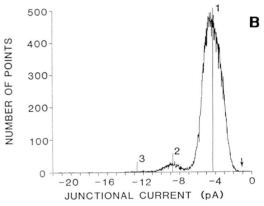

FIGURE 4 Gap junction channel activity from a neonatal hamster ventricular cell pair. (A) A 4-sec gap junction channel recording from a 3-day neonatal hamster ventricular cell pair. Duration of the −90-mV V_j pulse was 42 sec; whole-cell currents were filtered at 200 Hz and digitized at 1 kHz. Channel openings appear as downward deflections in the I_2 and upward deflections in the I_1 current traces. The thin solid lines in the I_2 trace illustrate the major current peaks evident in the channel amplitude histogram. (B) Channel amplitude histogram for the junctional current (I_2) recording described in A. The arrow illustrates the zero I_j level when V_j = 0 mV. The mean of each peak occurred at I_j = −3.5, −8.1, and −11.4 pA, respectively. Current transitions between the first, second, and third peaks have respective γ_j values of 51 (N = 32) and 37 pS (N = 4). Reproduced with permission from Veenstra (1991a).

30 pS were observed in adult rabbit sinus nodal cell pairs (Anumonwo *et al.*, 1992); 37-pS conductances were measured in guinea pig ventricular cell pairs (Rudisuli and Weingart, 1989).

B. Voltage Dependence

Cellular membrane resistance varies with each phase of the cardiac action potential, which can alter the effective coupling ratio (V_{m2}/V_{m1}) between cells. Regulation of g_j by plasma membrane potential is quite rare, being evident only in larval *Drosophila* and *Chironomus* salivary gland cells (Obaid *et al.*, 1983; Verselis *et al.*, 1991). Plasma membrane potential does not appear to regulate g_j in cardiac myocytes from embryonic chick (Veenstra and DeHaan, 1986,1988), adult rat (Metzger and Weingart, 1985; White *et al.*, 1985; Weingart, 1986), or adult guinea pig (Kameyama, 1983; Noma and Tsuboi, 1987; Rudisuli and Weingart, 1989).

Some cardiac gap junctions do appear to be regulated by transjunctional voltages, although their sensitivity is less than that reported for other V_j-dependent gap junctions (Spray and Bennett, 1985). In 7-day embryonic chick cell pairs, g_j declines during prolonged transjunctional voltage steps above ± 30 mV (DeHaan et al., 1989; Veenstra, 1990; Chen and DeHaan, 1992). This time- and voltage-dependent decay in junctional current is faster and more pronounced at larger transjunctional potentials, a pattern consistent with the behavior of all known V_j-dependent gap junctions. Recovery from maximum inactivation is a slow process, occurring with a time constant of 1.1 sec compared with a time constant of approximately 200 msec for inactivation at $V_j = 80$ mV (Veenstra, 1990). If allowed to recover fully between inactivating pulses, the instantaneous I_j–V_j relationship is observed to be linear. Plots of normalized steady-state/instantaneous g_j–V_j relationships are bell-shaped curves, symmetrical about 0 mV, that are adequately defined by a two-state Boltzmann equation as described for other embryonic V_j-dependent gap junctions (Spray et al., 1981). Using criteria identical to those for 7-day heart, examination of the V_j sensitivity of 4-, 14-, and 18-day embryonic chick heart revealed some quantitative differences in the normalized steady-state g_j–V_j relationships (Veenstra, 1991b). A gradual increase is seen in the V_j-insensitive g_j component (g_{min}), an outward shift is seen in the half-inactivation voltage (V_o), and an increase is seen in the slope (A) of the steady-state g_j–V_j relationships with increasing developmental age. The net result of these changes is a significant reduction in the V_j sensitivity of g_j between 4 and 18 days of cardiac development.

Although less fully characterized than in embryonic chick heart, g_j also declines sharply above a V_j of $+50$ mV in newly formed neonatal rat and neonatal hamster myocyte gap junctions (Rook et al., 1988; Wang et al., 1992). The observation of V_j-sensitive mammalian cardiac gap junctions is contrary to earlier findings in paired adult rat and guinea pig myocytes, although V_j gradients of <50 mV were sometimes used (Kameyama, 1983; Metzger and Weingart, 1985; White et al., 1985; Weingart, 1986; Maurer and Weingart, 1987; Noma and Tsuboi, 1987).

1. What Is the Mechanism by Which g_j Is Modulated by V_j?

The conventional Boltzmann model of V_j-dependent gap junctions assumes that two identical gates in series, one for each hemichannel, transit between high (open) and low (closed) conductive states following first order kinetics. Although this model is an oversimplification of the actual gating behavior of

gap junction channels, because some channels do not follow first order kinetics or exist in only two states, the suggestion of bilateral gates is supported by the bilateral symmetry of most g_j–V_j curves. The best evidence for the V_j-dependent gating of cardiac gap junction channels is the ensemble averaging of 5–10 current traces following the onset of a large V_j pulse. Channel current records from neonatal rat, hamster, and embryonic chick cardiac myocytes all reveal closure of individual channels after the onset of repetitive V_j pulses. The signal-averaged channel records result in an exponential decay of I_j that is similar to that of the higher g_j cardiac gap junctions (Rook et al., 1988; Veenstra, 1990,1991b,c). Thus, a V_j-dependent decline in P_o for a single population of gap junction channels appears to account for the time dependence of I_j. This one mechanism may not be the only by which g_j is modulated by V_j. In N2A cells transfected with the gap junction protein Cx37 (discussed subsequently), γ_j values of ~220 and ~120 pS observed at $V_j = 30$ mV were completely absent at 80 mV in favor of lower γ_j states of ~70 and ~40 pS (Reed et al., 1993). Other multiple γ_j channels such as Cx42 may also exhibit V_j-dependent decline in γ_j.

C. Pharmacological Regulation of Cardiac Gap Junction Channels

1. Lipophiles

The first agents shown to block electrical communication were the n-alkanols, 1-heptanol and 1-octanol (Johnston et al., 1980). Both molecules are polar surface-acting agents with general anesthetic properties and are known to modify sodium current kinetics and reduce peak sodium conductance in nerve axons (Haydon and Urban, 1983; Hirche, 1985). In cardiac cell pairs, these agents reversibly block junctional conductance (White et al., 1985; Veenstra and DeHaan, 1986,1988; Burt and Spray, 1988a; Niggli et al., 1989) in a dose-dependent manner (Rudisuli and Weingart, 1989). This decrease in conductance is not caused by reductions in single-channel conductance, since gap junction channel activity recorded in the presence of octanol or heptanol exhibits the same γ_j as control recordings (Burt and Spray, 1988a; Veenstra and DeHaan, 1988; Takens-Kwak, et al., 1992). The reversible effects of the 1-alkanols argues against an alteration in channel number as a mechanism for the blockade of g_j. The favored explanation is a reduction in channel open probability, which is supported by direct observation of a decrease in channel open time of the 160 pS channel in 7-day embryonic chick heart in the presence of 0.12 mM octanol (just prior to complete blockade of g_j; Veenstra and DeHaan, 1989). The 1-alkanols also dramatically reduce inward membrane currents in cardiac myocytes in millimolar concentrations just slightly higher than those used to inhibit g_j, and should not be regarded as specific pharmacological blockers of g_j (Nelson and Makielski, 1991; Takens-Kwak *et al.*, 1992).

The volatile general anesthetics halothane and ethrane also reversibly block cardiac gap junction channels when present in millimolar concentrations, apparently by reductions in the open-state probability of the channel (Burt and Spray, 1989). Unsaturated fatty acids and doxyl stearic acids can also block gap junction-mediated intercellular communication at much lower (micromolar) concentrations than any of the anesthetic compounds mentioned earlier (Aylsworth *et al.*, 1986; Burt, 1989). Arachidonic acid induces closure of gap junction channels between rat lacrimal gland cells and cardiac myocytes (Giaume *et al.*, 1989; Fluri *et al.*, 1990; Massey *et al.*, 1992). These uncoupling agents share the feature of being lipophilic, suggesting that they act via incorporation into the lipid bilayer (Burt, 1989; Niggli *et al.*, 1989; Spray and Burt, 1990). The observation that heptanol, octanol, and halothane also have dose-dependent effects on nonjunctional membrane (slow inward and steady-state) currents in cardiac myocytes raises concerns about the specificity of their gap junction channel-blocking activity (Niggli *et al.*, 1989). Some insights into the possible mechanism of action can be gained from the various fatty acid compounds that block junctional communication with higher potency than the anesthetics. Either the carbon–carbon double bond of an unsaturated fatty acid or the doxyl group of a saturated fatty acid is required for blocking activity to be present (Aylsworth *et al.*, 1986; Burt, 1989). The potency of the various fatty acids was increased by the presence of the polar (carboxyl) head group, longer nonpolar carbon chains, and the *cis*-isomers of unsaturated fatty acids. This observation suggests that ionic interactions at the surface and disruption of the interior of the bilayer are important in the inhibitory action of these agents on junctional communication. The specificity of action of fatty acids on junctional versus nonjunctional conductances has not been thoroughly evaluated under these experimental conditions, although 10–50 μM lysophosphatidylcholine has been noted to induce membrane depolarization and to increase membrane resistance in cardiac tissues (Arnsdorf and Sawicki, 1981; Clarkson and Ten Eick, 1983).

2. ATP

Conditions that include hypoxia, ischemia, and metabolic uncouplers (e.g., 2,4-dinitrophenol) are reported to increase the intercellular resistance of cardiac tissues, suggesting that metabolic state may play a role in determining the total junctional conductance in the heart (DeMello, 1979; Wojtczak, 1979; Kleber *et al.*, 1987; Cascio *et al.*, 1990). Rapid cellular uncoupling occurs 15–20 min after the onset of ischemia in arterially perfused rabbit papillary muscle and is closely associated with extracellular K^+ accumulation and ischemic contracture. Hence, ischemic uncoupling is believed to be correlated with a loss of cellular homeostasis, including an increase in cytosolic Ca^{2+} (Kleber *et al.*, 1987; Cascio *et al.*, 1990). A decrease in cellular ATP, required for active transport processes, leads to increases in intracellular concentrations of H^+, Ca^{2+}, and Na^+ (Baumgarten *et al.*, 1981). However, evidence also suggests that the loss of cellular ATP that occurs during ischemia may directly regulate g_j through specific ligand–receptor interactions. Sugiura *et al.* (1990) used the open-cell system, a miniature cut-end method applied to a cell pair, to demonstrate that ATP concentrations, ranging from 0.1 to 5 mM increase g_j in a reversible, dose-dependent manner (Hill coefficient = 2.6; half-maximal effective concentration of ATP = 0.68 mM). At the single-channel level, γ_j was observed to decrease slightly from 37 to 30 pS on elevation of ATP from 0 or 3 mM to 6 mM in paired adult guinea pig ventricular myocytes (Rudisuli and Weingart, 1989). No mention is made of the effects of 0, 3, or 6 mM ATP on g_j, nor was it possible to examine the reversibility of the effect of ATP on γ_j in this study. Similar experiments have not been conducted in embryonic or neonatal myocyte preparations.

3. Cations

Elevation of cytosolic calcium and hydrogen ion concentration can reduce junctional communication in a variety of tissues (for review, see Loewenstein, 1981; DeMello, 1982; Spray and Bennett, 1985). Exposure of cardiac Purkinje fibers and paired ventricular myocytes to high CO_2 produces an intracellular acidification that increases junctional resistance (Reber and Weingart, 1982; White *et al.*, 1985). Quantitative descriptions of the relationship between g_j and intracellular pH (pH_i) in paired amphibian and teleost blastomeres indicated that g_j was directly modulated by pH_i whereas Ca^{2+} required a 10,000-fold higher concentration to produce similar effects (Spray *et al.*, 1982). Although the pH sensitivity of cardiac gap junctions is much less than that observed in blastomeres and liver hepatocytes (for review, see Spray and Burt, 1990), uncoupling of cardiac myocytes by elevations in intracellular Ca^{2+} ($[Ca^{2+}]_i$) in the absence of acidification required concentrations that elicited sustained contractures (White *et al.*, 1985; Maurer and Weingart, 1987). Based on these findings, lowering pH_i can be concluded to cause electrical uncoupling in the heart. However, if extracellular or intracellular calcium is kept low, the healing-over of damaged tissue and reductions in the incidence of electrical and dye coupling between cardiac myocytes is prevented unless pH is reduced to 5.0 (DeMello, 1983; Burt, 1987). Other studies in *Chironomus* salivary glands and crayfish septate axons indicate that cellular uncoupling correlates more closely with elevations in $[Ca^{2+}]_i$ than with pH_i (Rose and Rick, 1978; Peracchia, 1990a). Further investigations in cardiac Purkinje fibers indicate that the increase in internal longitudinal resistance observed during CO_2 treatment cannot be accounted for by the direct action of protons alone (Pressler, 1989).

Elevation of $[Ca^{2+}]_i$ into the micromolar range causes uncoupling of cardiac gap junctions; extracellular Ca^{2+} is required for the healing-over of myocardial tissues (DeMello *et al.*, 1969; Deleze, 1970; Dahl and Isenberg, 1980). Gap junction channels are permeable to Ca^{2+} (Dunlap *et al.*, 1987; Saez *et al.*, 1989a), but this sensitivity does not preclude modulatory effects of Ca^{2+} on the channel. Evidence that physiological levels of $[Ca^{2+}]_i$ can regulate g_j in cardiac tissues comes from internal dialysis of an "open-cell" or a cell pair with EGTA- and HEPES-buffered solutions (Noma and Tsuboi, 1987; Veenstra and DeHaan, 1988). Noma and Tsuboi (1987) demonstrated that Ca^{2+} and Mg^{2+} cause a fall in g_j with a pCa of 6.6 (0.25 μM), a pMg of 2.5 (3.2 mM), and a Hill coefficient of approximately 3.0 at pH 7.4. In intact cells, pH_i and $[Ca^{2+}]_i$ are interdependent; elevations in the concentration of one cation lead to a concomitant increase in the other (Rose and Rick, 1978; Vaughan-Jones *et al.*, 1983; Peracchia, 1990a,b). Elevation of both H^+ and Ca^{2+} in neonatal cardiac myocytes reduces the incidence of dye coupling of 82%, whereas either ion alone produces negligible effects, suggesting a synergistic action of H^+ and Ca^{2+} in the regulation of g_j (Burt, 1987). Most recently, White *et al.* (1990) conducted a study in which g_j and pH_i were simultaneously monitored in paired adult rat cardiac myocytes. $[Ca^{2+}]_i$ was also monitored in cell suspensions with fura-2 fluorescence during acidification by CO_2 treatment. These procedures demonstrated that increasing either H^+ or Ca^{2+} alone had negligible effects on g_j whereas the combination of $pH_i = 6.0$ and $[Ca^{2+}]_i = 425$ nM produced by exposure to CO_2 essentially blocked g_j. Similarly, the sensitivity of g_j to Ca^{2+} in crayfish septate axons was significantly enhanced by the presence of cyclic AMP-dependent protein kinase activity and low pH (Arellano *et al.*, 1990). Thus, H^+ and Ca^{2+} appear to act synergistically to regulate g_j.

Whatever concentrations of protons and Ca^{2+} are required to alter g_j, all evidence indicates that reductions in g_j by elevations in cation concentration occur by reductions in channel open state without affecting the unitary conductance (Veenstra and DeHaan, 1988; Brink and Fan, 1989). This point is best illustrated by the reversible reductions in channel open-state probabilities and channel event count after exposure of earthworm septate axon gap junction channels to 2 mM Ca^{2+} or pH 6.0 (Brink and Fan, 1989). Channel amplitudes were not affected at all by these treatments.

4. Neurotransmitters and Second Messengers

β-Adrenergic agonists (isoproterenol), cAMP, and its membrane-permeable analogs (dibutyryl and 8-bromo-cAMP) all are reported to increase junctional conductance or permeability in mammalian cardiac tissues (De Mello, 1986,1989; De Mello and van Loon, 1987; Burt and Spray, 1988b). The increase in g_j occurs within a few minutes of β-adrenergic stimulation or application of cAMP derivatives (Burt and Spray, 1988b). The maintenance of electrical coupling is important during catecholamine stimulation if tissue homogeneity is to be maintained, since the effective refractory period of cardiac tissues is reduced by increased heart rate. Contrary to these findings, forskolin, cAMP, and the catalytic subunit of cAMP-dependent protein kinase all failed to produce any significant effects on g_j during perfusion of paired guinea pig ventricular myocytes in the open-cell configuration (Sugiura *et al.*, 1990). Long-term regulatory effects of cAMP identified in other systems (increases in gap junction particles and prolongation of gap junction mRNA half-life) have not yet been examined in the heart (Flagg-Newton *et al.*, 1981; Saez *et al.*, 1989b). The short-term effects on g_j may occur by increasing the open-state probability for the gap junction channel.

Muscarinic agonists, such as carbachol, and cGMP decrease g_j in neonatal rat myocytes with a time course similar to that of cAMP (Burt and Spray, 1988b). This response is thought to be mediated by a cGMP-dependent protein kinase (Spray and Burt, 1990). Although the evidence strongly favors an inhibitory action of protein kinase C on g_j in certain tissues, protein kinase C reportedly increases g_j in mammalian heart (Spray and Burt, 1990). No details have been published yet regarding the nature of this response in cardiac tissues.

In addition to the serine-specific phosphorylation of gap junction proteins, tyrosine-phosphorylating protein kinases (e.g., the oncogene product $pp60^{v\text{-}src}$) typically down-regulate g_j, as determined in a number of fibroblast cell lines (Atkinson and Sheridan, 1988; Azarnia *et al.*, 1988; Hyrc and Rose, 1990).

III. Molecular Biology of Cardiac Gap Junctions

A. Gap Junction Isolation and Biochemistry

In the 1970s, procedures were developed for the isolation of gap junctions from liver (Goodenough and Stoekenius, 1972; Evans and Gurd, 1972). These isolated gap junctions were studied by X-ray diffraction and electron microscopy to develop a low-resolution (25 Å) structural model (Fig. 5; Caspar *et al.*, 1977; Makowski *et al.*, 1977; Unwin and Zampighi, 1980; Makowski, 1988). The model shows that a gap junction plaque is composed of tens to thousands of channels. Each channel is composed of a hexameric structure (connexon) composed of six apparently identical integral membrane subunits (connexins) that surround a central pore. The connexon joins in mirror symmetry with a connexon in the plasma membrane of the adjacent cell. Although they have been studied less thoroughly, cardiac gap junctions are believed to have a generally similar structure, except for larger cytoplasmic domains (Manjunath *et al.*, 1984; Yeager and Gilula, 1992).

Methods were developed for the isolation of myocardial gap junctions (Kensler and Goodenough, 1980; Manjunath *et al.*, 1984) and of lens fiber plasma membranes that contain 5–10% gap junction profiles (Goodenough, 1979; Bloemendal *et al.*, 1972; Alcala

et al., 1975; Broekhuyse *et al.*, 1976; Paul and Goodenough, 1982) as well. SDS-PAGE of these preparations showed that the isolated liver gap junctions are composed primarily of a 27-kDa polypeptide, accompanied by proteolysis fragments, aggregates, and a 21-kDa polypeptide (Henderson *et al.*, 1979; Hertzberg and Gilula, 1979; Finbow *et al.*, 1980). Isolated myocardial gap junctions contain a 43- to 47-kDa polypeptide, cleaved by endogenous proteases to 34-, 32-, and 29-kDa bands (Kensler and Goodenough, 1980; Manjunath *et al.*, 1987). Isolated bovine and ovine lens fiber plasma membranes contain a predominant polypeptide of 26 kDa, called MP26 or MIP26, as well as numerous other polypeptides, notably one of 70 kDa (MP70; Kistler *et al.*, 1985). N-Terminal sequencing of these proteins by Edman degradation demonstrated that liver 27 kDa and 21 kDa (Nicholson *et al.*, 1987), heart 43–47 kDa and its degradation products (Gros *et al.*, 1983; Nicholson *et al.*, 1985), and lens 70 kDa (Kistler *et al.*, 1988) are homologous proteins, whereas lens 26 kDa appears unrelated (Nicholson *et al.*, 1983).

B. Molecular Cloning of Connexin 43

In 1986, Paul cloned cDNA for the rat liver 27-kDa protein by antibody screening of a bacteriophage expression library. Kumar and Gilula (1986) cloned a cDNA for its human counterpart by hybridization screening with an N-terminal oligonucleotide. Both cDNAs encode a polypeptide of 32 kDa. Paul (1986) demonstrated by Northern blotting that mRNA corresponding to this cDNA was expressed in rat liver, brain, stomach, and kidney, but was not detectable in heart or lens.

Beyer *et al.* (1987) isolated a homologous sequence from a rat heart cDNA library by screening with the rat liver cDNA at reduced stringency. The rat heart cDNA encodes a polypeptide of 43 kDa that contains amino acids 43% identical to the protein cloned from rat liver, but also contains two regions with many more identical residues (see subsequent discussion). The N-terminal sequence predicted from the heart clone matches that determined by Edman degradation of the major protein in isolated rat heart gap junctions (Nicholson *et al.*, 1985; Manjunath *et al.*, 1987). Morphological and functional proof that this protein forms cardiac gap junctional channels was provided by immunocytochemistry using specific antipeptide antisera (Beyer *et al.*, 1989; Yancey *et al.*, 1989) and by expression in *Xenopus* oocytes (Swenson *et al.*, 1989; Werner *et al.*, 1989).

Cloning of this second gap junction protein confirmed previous suggestions of a family of related gap

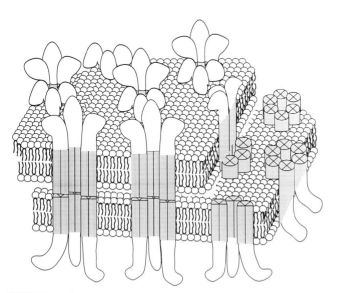

FIGURE 5 Structural model of a gap junction based on electron micrographic and X-ray diffraction studies of isolated liver and heart gap junctions.

junction proteins. Many of these proteins are not uniquely expressed in a single tissue (Paul, 1986; Beyer et al., 1987). The mobilities of these proteins on SDS-PAGE may vary with electrophoresis conditions (Green et al., 1988). Therefore, previous descriptions of gap junction proteins based on tissue of origin or electrophoretic mobility were abandoned in favor of a new operational nomenclature using the generic term connexin (Cx) for the protein family, with an indication of species (as necessary) and a numeric suffix designating the molecular mass in kilodaltons (Beyer et al., 1987,1988,1990). Thus, the 27-kDa protein from rat liver is termed rat connexin32 (Cx32), the 43-kDa protein from rat heart is termed rat connexin43 (Cx43), and so on.

Cx43 or its close homolog has been cloned from Xenopus (Gimlich et al., 1990), chick (Musil et al., 1990a), human (Fishman et al., 1990), mouse (Beyer and Steinberg, 1991), and cow (Lash et al., 1990) as well as rat (Beyer et al., 1987). The sequences are extremely similar: the mammalian Cx43 proteins show ≥97% amino acid identity; the chick protein has 92% identical amino acids relative to the rat; and the Xenopus protein has 87% identical amino acids relative to the rat protein. Many of the substitutions are conservative.

Availability of specific DNA (and antibody) probes for different connexins has allowed determination of the gap junction sequences expressed by different cells and tissues. Nearly all connexins are expressed in multiple locations. Cx43 is not a specific product of cardiac myocytes. Instead, it appears to be the most ubiquitous connexin, and is expressed by cardiac myocytes, myometrial and other smooth muscle cells, ovarian granulosa cells, endothelial cells, lens epithelial cells, pancreatic acinar cells, astrocytes, renal tubular epithelial cells, macrophages, connective tissue cells, fibroblasts and many other tissue culture cell lines, and hepatocellular carcinoma lines (Beyer et al., 1989; Dermietzel et al., 1989,1991; Luke et al., 1989; Larson et al., 1990; Crow et al., 1990; Giaume et al., 1991; Meda et al., 1991; Moore et al., 1991; Spray et al., 1991).

C. Connexin Topological Structure

The amino acid sequences derived from the cloned cDNAs have been used to predict the structures of the connexins. Hydropathy plots of Cx32 and Cx43 predict four hydrophobic domains, with a large C-terminal hydrophilic tail. Three smaller hydrophilic domains separate the hydrophobic regions (Paul, 1986). These data, in conjunction with the results of proteolysis studies of isolated junctions, have been

used to construct very similar topology models for the relationships of these polypeptides to the junctional plasma membrane, assuming that each hydrophobic domain represents a transmembrane segment of the molecule (Beyer et al., 1987; Zimmer et al., 1987). The models have been tested by examining the protease sensitivity of isolated liver gap junctions and by mapping with site-specific antisera by immunocytochemistry. Controlled proteolytic cleavage has demonstrated that the N and C termini of Cx32 face the cytoplasm, and that an additional cytoplasmically accessible proteolytic site is located between the second and third transmembrane segments (Zimmer et al., 1987; Hertzberg et al., 1988). Antisera have been raised against synthetic oligopeptides representing various segments of Cx32 and Cx43, and have been used to map channel topology by electron microscopy (Goodenough et al., 1988; Milks et al., 1988; Beyer et al., 1989; Yancey et al., 1989; Yeager and Gilula, 1992). These studies have confirmed the proteolysis studies, having shown that the N terminus, the C terminus, and a loop in the middle of the protein are all located on the cytoplasmic face of the junctional membrane. To the degree that the harsh experimental conditions do not alter the protein topology, these studies also demonstrate that the predicted extracellular connexin domains can be detected on the extracellular surfaces of the junctional membranes.

D. Connexin43 Biosynthesis/Phosphorylation

The development of specific anticonnexin antibody probes have made possible in vitro studies of connexin biosynthesis by metabolic labeling and immunoprecipitation. The bulk of these studies have been performed on Cx43, and have demonstrated that Cx43 has a short half-life of 1–4 hr in tissue culture cells of different sources (Crow et al., 1990; Musil et al., 1990a; Laird et al., 1991; Larson et al., 1991). These studies have also examined covalent modifications of the connexins (especially Cx43) by phosphorylation.

1. Serine Phosphorylation of Connexin43

Pulse–chase studies have demonstrated that Cx43 is initially synthesized as a 42-kDa polypeptide that subsequently is posttranslationally modified to forms with slightly slower mobility on SDS-PAGE by the addition of phosphate groups onto serine residues (Crow et al., 1990; Filson et al., 1990; Musil et al., 1990a). The exact phosphorylated residues have not been identified, but apparently multiple phosphates are added to each mole of Cx43. Additionally, a serine-rich sequence near the C terminus has multiple potential phosphorylation sites (Beyer et al., 1987). (Other

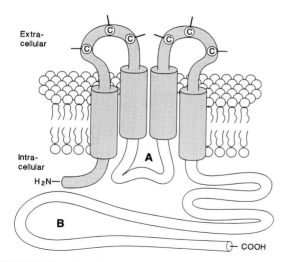

FIGURE 6 Topological model of connexin orientation within the junctional plasma membrane. Shaded regions represent those sequences shared among all connexins. The unshaded predicted cytoplasmic domains (A and B) correspond to the unique connexin-specific regions shown in Fig. 7. Reproduced with permission from Beyer (1993).

connexins have similar sequences, suggesting that they might be similarly modified; see Fig. 4). The kinase(s) responsible for Cx43 phosphorylation and the exact phosphorylated serines have not been determined. However, protein kinase C is strongly implicated in at least some of the phosphorylation events, since treatment of several different cell lines with phorbol esters produces an increase in the phosphorylation of Cx43 and a decrease in intercellular coupling (Brisette *et al.*, 1991; Berthoud *et al.*, 1992; Reynhout *et al.*, 1992).

Musil *et al.* (1990b) have further investigated Cx43 serine phosphorylation. These investigators demonstrated that Cx43 was present in the noncommunicating cell lines L929 and S180, but was not present in cell-surface gap junctional plaques and accumulated intercellularly. Cx43 was incompletely phosphorylated in these cell lines. However, transfection of the S180 cells with the cell-adhesion molecule LCAM (E cadherin) restored gap junctional communication, full phosphorylation of Cx43, and expression in cell-surface gap junctions. These findings suggest a relationship between the ability of cells to fully phosphorylate Cx43 and the ability to form communicating junctions. These results also suggest a hierarchy of events in the formation of intercellular junctions: a primary cell adhesion event is required prior to formation of gap junctions. Similar observations regarding the requirement of cadherin-mediated cell adhesion for development of gap junctions have also been made

in epidermal cells (Jongen *et al.*, 1991). Meyer *et al.* (1992) showed that antibodies against extracellular Cx43 or A-CAM sequences will inhibit gap junction formation by Novikoff hepatoma cells. Musil and Goodenough (1991) extended their original observations in NRK cells by showing the presence of extracellularly accessible Cx43 (in the nonphosphorylated form) and a correlation between phorphorylation of Cx43 and acquisition of insolubility in Triton X-100, suggesting a role for phosphorylation in gap junction formation.

2. Tyrosine Phosphorylation of Connexin 43

A large body of data demonstrating that intercellular communication is abolished in fibroblasts infected with Rous sarcoma virus (RSV; Atkinson *et al.*, 1981; Azarnia *et al.*, 1988) has led several researchers to investigate the effects of pp60^{src} on Cx43-expressing cells. Crow *et al.* (1990) demonstrated that RSV infection uncoupled vole fibroblasts and led to the incorporation of phosphate onto Cx43 tyrosine residues. Filson *et al.* (1990) extended these observations by showing that each molecule of phosphorylated Cx43 contained both phosphoserine and phosphotyrosine, and by showing that the ability of *src* variants to abolish cellular communication correlated with tyrosine phosphorylation of Cx43. Swenson *et al.* (1990) showed that, in a *Xenopus* oocyte expression system, coexpression of pp60$^{v\text{-}src}$ with Cx43 reduced cell–cell coupling and led to tyrosine phosphorylation of Cx43. However, site-directed mutagenesis of Tyr 265 in Cx43 eliminated the tyrosine phosphorylation and the depression of communication.

E. Molecular Cloning of Other Cardiac Connexins

The search for other homologous sequences has identified additional members of the connexin gene family. Beyer (1990) used low stringency hybridization screening of a chick embryo cDNA library with a Cx43 probe to search for connexins expressed in the developing heart and other tissues. Two novel connexin cDNA sequences were identified: chick connexin42 (Cx42) and chick connexin45 (Cx45). These predicted connexin proteins share the homologous transmembrane and extracellular regions noted in other members of this family, and each has its own unique cytoplasmic regions. RNA blots demonstrate that Cx42, Cx43, and Cx45 are each expressed in a number of chick organs, including heart. Blots of total RNA isolated from hearts of chick embryos of different ages demonstrated that the abundance of Cx42 and Cx43 mRNAs varied no more than 2-fold between the embryo and the adult. However, Cx45 mRNA showed a dramatic change, falling 10-fold from the 6-day embryonic heart to the adult. Parallel electrophysiological studies (Veenstra, 1991b) showed a striking change in the voltage dependence of myocardial gap junctions over the same time period. RNA blots, immunoblots, and immunofluorescent localization studies suggested that Cx42 is the major connexin present between chick cardiac myocytes, in contrast to mammals in which Cx43 predominates (Beyer, 1990; El Aoumari *et al.*, 1990; Minkoff *et al.*, 1992; Beyer *et al.*, 1993).

Kanter *et al.* (1992a) used the polymerase chain reaction (PCR) and genomic cloning to isolate mammalian counterparts of chick Cx42 and Cx45. Cx45 appears to be quite highly conserved; 85% of amino acids are identical in dog and chick Cx 45. In contrast, the canine sequence homologous to chick Cx42 encodes a 39,937 dalton protein (Cx40) containing only 70% identical amino acids. Cx40 has also been cloned from rat (Beyer *et al.*, 1992; Haefliger *et al.*, 1992), mouse (Hennemann *et al.*, 1992a), and human (Kanter *et al.*, 1992b). These species also show variability: rodent Cx40 is 83% identical to dog Cx40; human Cx40 is 89% identcal to dog Cx40. In contrast, human, dog, and mouse Cx45 are about 97% identical (Kanter *et al.*, 1992a,b; Hennemann *et al.*, 1992b). Kanter *et al.* (1992a) demonstrated by Northern blotting that Cx40, Cx43, and Cx45 were all expressed by canine ventricular myocytes.

Kanter *et al.* (1992a) raised polyclonal rabbit antisera against unique peptide sequences in the cytoplasmic tails of Cx40, Cx43, and Cx45. These connexin-specific antisera show no cross reactivity, yet all three specifically stain the intercalated disc regions of adult cardiac myocytes and specifically label gap junctions, as determined by electron microscopy (Kanter *et al.*, 1992a,c). To determine whether Cx40 and Cx45 are present in the same junctions as the major cardiac connexin (Cx43), these researchers performed double-label immunohistochemistry using a mouse monoclonal antibody against Cx43 and the polyclonal antisera against Cx40 or Cx45. Analysis of immunoreactive material in isolated myocytes with laser scanning confocal microscopy and double-label immunoelectron microscopy revealed an identical distribution of both proteins in gap junctions. Thus, adult canine myocyte gap junctions contain three distinct connexins. However, regional differences arise in cardiac connexin expression; *in situ* hybridization and immunostaining experiments (Kanter *et al.*, 1992c) suggest that Cx40 is much more abundantly expressed in canine Purkinje fibers than in ventricular myocardium, whereas Cx43 expression differs little between the two locations.

Multiple connexins are coexpressed in other cardiovascular cells. Beyer *et al.* (1992) isolated Cx40 cDNA from a rat aortic smooth muscle library and demonstrated its expression by A7r5 vascular smooth muscle cells. Reed *et al.* (1993) cloned human Cx37 from an umbilical vein endothelial cell library and demonstrated its expression by endothelial cells from a number of sources. Paul *et al.* (1991) found hybridization of a Cx46 probe to adult rat heart RNA, but the cellular source was not identified. Harfst *et al.* (1990) found immunostaining with an antibody against the connexin-related protein MP70 in heart values.

All the identified connexins have a primary structure similar to that determined for Cx32 and Cx43. All contain two homologous regions, one at the beginning of the protein and one in the middle. The other portions of the connexins are unique in each connexin, varying in both length and sequence. Figure 7 shows a representation comparing the primary sequences of several different connexin polypeptides.

Extensive studies mapping sites by proteolysis and immunocytochemistry have not been conducted for the other connexins, but sequence comparison, limited immunocytochemistry, and analogy to the Cx32/Cx43 studies allow some prediction of their structure. All the sequenced connexins share the four hydrophobic domains corresponding to the transmembrane regions in Cx32/Cx43. Comparison of the primary sequences of all the known connexins reveals that the two predicted extracellular domains are the most conserved regions, each containing three invariant cysteines. The transmembrane domans are somewhat less well conserved. The cytoplasmic domains, with the exception of the short N-terminal region, differ markedly among connexins in both sequence and length. These relationships are summarized in Fig. 6 and 7, in which the two cytoplasmic domains are labeled A and B.

F. Studies of Connexin Expression during Cardiac Development

Although electrical coupling through gap junctions is crucial to normal impulse propagation in the mature heart, little is known about the assembly of these intercellular junctions during development. Some descriptive morphological studies have been done (Gros *et al.*, 1978,1979). Beyer (1990) demonstrated that, in the developing chick heart, two gap junction proteins in addition to Cx43 are found: Cx42 and Cx45. Cx45 mRNA is present primarily in early embryos and falls to 10-fold lower levels in the adult heart. Several studies have examined expression of Cx43 in the developing mammalian heart (Fishman *et al.*, 1991a; Fromaget *et al.*, 1990; Gourdie *et al.*, 1992; van Kempen

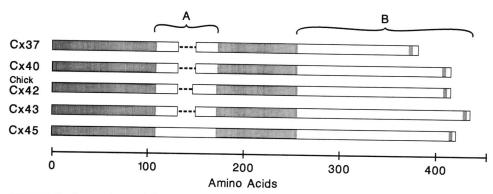

FIGURE 7 Comparison of the primary amino acid sequences of several cloned cardiovascular connexins. Regions containing many identical amino acids are shaded; those that are unique are not shaded. Dashes represent spaces added to optimize alignment. Each connexin contains two stretches of ~100 amino acids that are highly conserved and two unique regions marked A and B. Several connexins also contain a similar, short, serine-rich region near the C terminus.

et al., 1991). Cx43 mRNA is present at the earliest cardiac stage tested (11 days post-coitum in the mouse), but rises 5- to 8-fold by about 1 wk post partum before falling to lower levels in juvenile and adult hearts. Fishman *et al.* (1991a) used immunoblots to demonstrate that Cx43 protein showed parallel, but temporally delayed, changes. van Kempen *et al.* (1991) demonstrated that gap junctions demonstrable by immunostaining with anti-Cx43 are assembled with a similar time course. However, reconciling this peak of connexin expression in neonatal animals with the previous observations that adult myocytes have a much greater total junctional conductance than neonatal myocytes is difficult. Does this conductance represent organization of these junctions into the intercalated disc?

IV. Electrophysiology of Channels Formed by the Cloned Connexins

The availability of cloned connexin DNAs has facilitated more detailed investigation of their functional properties using various expression systems (see Fig. 8). Such expression studies are beginning to address several fundamental questions about gap junction biology. (1) These studies should allow the determination of the biophysical properties of different connexins. Associating the properties identified in the cellular electrophysiology experiments outlined earlier with different connexins should be possible. Do those properties differ for different connexins? For each connexin, the parameters of unitary conductance, pH dependence, V_j dependence, selectivity, and permeability should be known. (2) Does the pharmacological regulation of different connexins in re-

sponse to lipophiles or modulators of protein kinase activity differ? (3) What are the structure–function relationships for the connexin molecules? These experiments should provide a more detailed topological map. They should allow the association of various properties with specific domains within the connexin polypeptides. (4) Are connexins able to mix together to form heteromolecular channels? Two potential mechanisms are possible: each cell of a pair could contribute a different connexin or multiple connexins could mix within a single hexameric hemichannel.

A. Xenopus *Oocyte Expression Studies*

A powerful system has been developed (Dahl *et al.*, 1987) involving the expression of injected gap junction mRNAs in paired *Xenopus* oocytes and the subsequent study of the channels induced between the two cells by voltage-clamp techniques. Several connexins from the cardiovascular system, including Cx43 (Swenson *et al.*, 1989; Werner *et al.*, 1989), Cx37 (Willecke *et al.*, 1991), and Cx40 (Hennemann *et al.*, 1992a), have been shown to be individually sufficient for forming communicating channels.

The *Xenopus* system has been used to begin to determine different biophysical properties of the channels produced from different connexins. For instance, Cx37 and Cx40 form channels that show a pronounced voltage-dependent inactivation, whereas Cx43 shows little such sensitivity. The oocyte system has also been exploited to investigate the possible formation of hybrid junctions composed of two different connexins. In this expression system, Cx43 can make heterologous gap junctions with Cx32 or the endogenous *Xenopus* connexin expressed in the partner cell (Swenson *et al.*, 1989; Werner *et al.*, 1989). However, not all

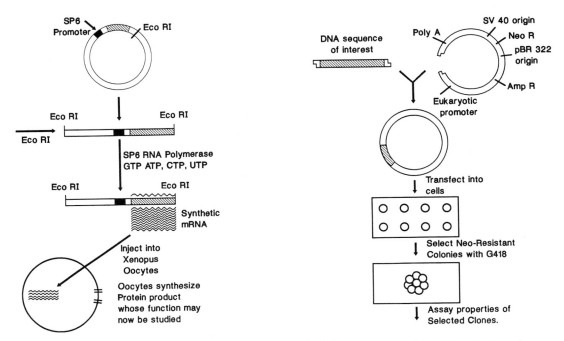

FIGURE 8 Strategies for expression of connexins. (left) *Xenopus* oocytes. (right) Transfection of communication-deficient cells.

connexins will pair successfully to form functional channels (Bruzzone *et al.*, 1991). If two adjacent cells similarly express incompatible connexins *in vivo*, a mechanism for the formation of communication compartments might exist.

The oocyte system is also being exploited to examine mutagenized connexin constructs to determine the functional properties of different portions of the connexin sequences. Studies by Swenson *et al.* (1990) have pinpointed a tyrosine in Cx43 that is the target for pp60^{v-src}. Dunham *et al.* (1992) have produced a truncated mutant of Cx43 (which should have deleted all likely serine phosphorylation sites) that still forms functional channels when expressed in oocytes.

B. Expression of Connexins by Transfection of Communication-Deficient Cells

Functional properties of the cloned connexins have also been investigated by the stable transfection of communication-deficient cells with connexin-containing constructs (Eghbali *et al.*, 1990; Fishman *et al.*, 1990; Moreno *et al.*, 1991a,b; Veenstra *et al.*, 1992). Ideally, this system would utilize small, round, high-resistance cells containing no detectable connexin expression or electrical coupling. These cells should be easily transfectable with connexin sequences, resulting in the establishment of electrical (and possibly dye transfer) communication. Although no cells

perfectly meet all these criteria, several different communication-deficient cell lines have been utilized for such studies (including SKHep-1, RIN, and N2A). Figure 9 illustrates the effects of transfecting Cx43 into the rat insulinoma cell line RIN.

Although the transfected cell system is less amenable to screening multiple connexin mutants than are *Xenopus* oocytes, this system has a number of advantages. This system is probably more relevant to the actual state of gap junctions between small cells in tissues. The system avoids the problems generated by the endogenous amphibian channels. Whole-cell voltage-clamp techniques can be used, allowing more rapid response and greater sensitivity. Single gap junctional channels can be resolved.

Transfection experiments have confirmed the prediction that different connexins have different biophysical properties. Different connexins form channels with different unitary conductances: Cx32 has a unitary conductance of ~120 pS (Eghbali *et al.*, 1990); Cx43 (human or chick) has a single-channel conductance of 45–50 pS (Fishman *et al.*, 1990; Veenstra *et al.*, 1992); Cx45 channels have a unitary conductance of 22–30 pS (Veenstra *et al.*, 1992); chick Cx56 (a lens gap junction protein) channels have a unitary conductance of 100 pS (Rup *et al.*, 1993); Cx42 forms predominantly 160-pS channels (Veenstra *et al.*, 1992); and Cx37 forms predominantly 220-pS channels (Reed *et al.*, 1993). Cx37 and Cx42 show multiple conductance

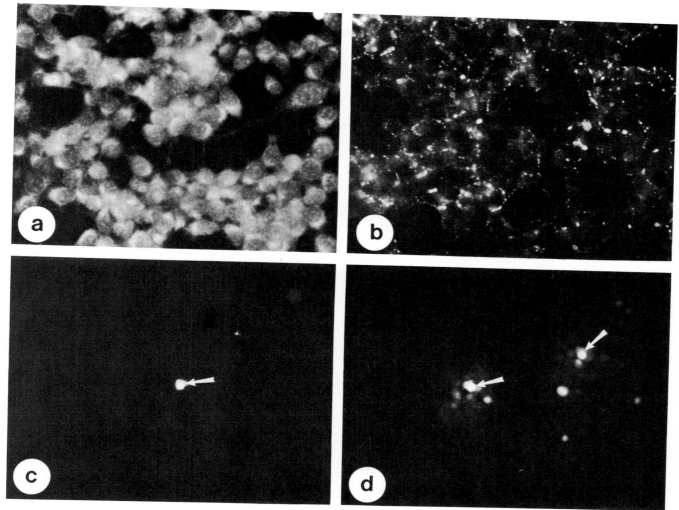

FIGURE 9 Expression of rat Cx43 in transfected RIN cells. (a) Immunofluorescent staining of RIN cells for Cx43. These untransfected parent RIN cells show no specific staining when reacted with anti-Cx43 antibodies and fluorescent secondary reagents. (b) Immunofluorescence with anti-Cx43 in Cx43RIN. RIN cells transfected with Cx43 show bright punctate immunofluorescent staining (characteristic of gap junctions) when reacted with anti-Cx43 antibodies and fluorescent secondary reagents. (c) Lack of dye coupling in parental RIN cell line. A single cell (arrow) was injected with the fluorescent dye Lucifer yellow CH. No passage of dye to neighboring cells is detectable. (d) Dye coupling of Cx43RIN cells. After injection of Lucifer yellow into the cells marked by the arrows, widespread transfer to surrounding cells occurs. This intercellular dye transfer could also be inhibited with heptanol.

states when expressed in the transfected cells (Veenstra *et al.*, 1992; Reed *et al.*, 1993). These results appear to be very similar to the multiple conductances seen by Chen and DeHaan (1992) in chick embryo heart cells. Data from Veenstra *et al.* (1992) comparing the single channels formed by three connexins expressed in the chick embryo heart are shown in Fig. 10.

The gap junctions formed by different connexins also differ in sensitivity to transjunctional voltage (V_j). Cx56 channels appear to be V_j insensitive (Rup *et al.*, 1993). Cx43 gap junctions show only a mild V_j depen-

dence at high transjunctional voltages (Veenstra *et al.*, 1992). Cx45 is very V_j sensitive (Veenstra *et al.*, 1992). Figure 11 compares the V_j dependence (normalized steady-state conductance–voltage relationship) for the three chick embryo cardiac connexins Cx42, Cx43, and Cx45. Different parameters of V_j dependence (including half-inactivation voltage, V_0; voltage-insensitive component, G_{min}; and slope factor, A) differ for each connexin examined. These differences are likely to be consequences of differences among connexins in charged residues sensitive to V_j gradients.

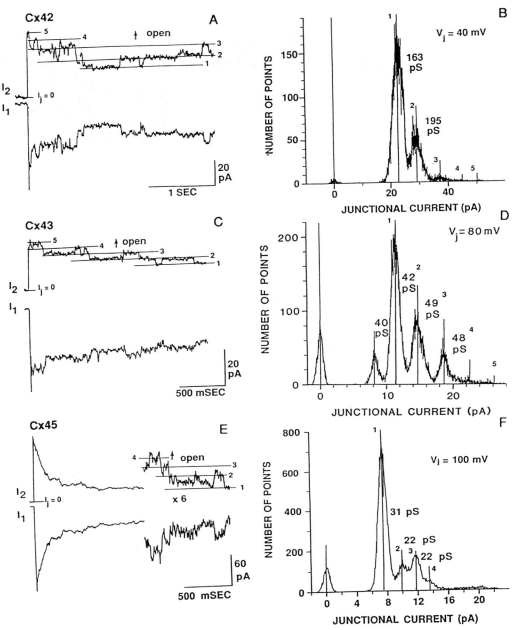

FIGURE 10 Junctional channel currents obtained from connexin cDNA-transfected neuro2A (N2A) cell pairs. (A, C, E) Paired whole-cell currents (I_1 and I_2) obtained during a transjunctional voltage (V_j) pulse to the indicated value for each connexin. In all three cases, the I_2 tracing (which is mirrored in the I_1 tracing) represents channel openings. (B, D, F) The corresponding channel amplitude histograms for each connexin. Corresponding peaks in each histogram to the displayed channel activity are labeled accordingly. Paired whole-cell currents (*left*) were recorded from the onset of a >20-sec V_j pulse applied to one pair from each of the connexin cDNA-transfected N2A cell clones. The break in the connexin42 (Cx42) tracings represents an 8-sec gap in the channel recording. All current tracings were low pass filtered at 125 Hz and digitized at 1 kHz. Channel amplitude histograms compiled from the entire 20-sec I_j recordings are shown (*right*). The area under each peak is proportional to the time spent in each state. Channel events were counted by setting threshold detectors to the valley between adjacent peaks and rescanning the digitized I_2 tracing. Only those transitions having a duration >1 msec were counted as channel events. Only those peaks produced by more than two events of 2% of the total dwell time were used in determining channel amplitudes. Numbers of channel events were 44, 38, and 45 for Cx42, connexin43 (Cx43), and connexin45 (Cx45), respectively. Reproduced with permission from Veenstra *et al.* Copyright © 1992 by the American Heart Association.

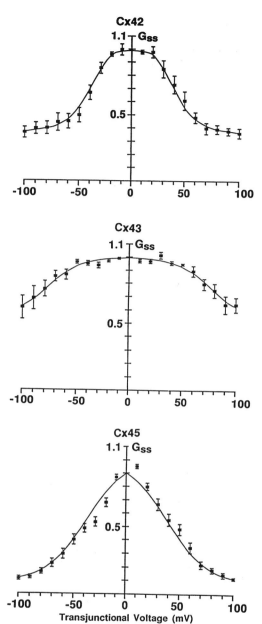

FIGURE 11 Boltzmann relationships demonstrating differences in voltage dependence for embryonic chick cardiac connexin42 (Cx42), connexin43 (Cx43), and connexin45 (Cx45). The filled square and error bars represent the mean ± SEM steady-state junctional conductance/instantaneous junctional conductance (G_{ss}) for each of the indicated connexins. The G_{ss}–transjunctional voltage (V_j) relationship for each connexin was fitted with a theoretical line determined by the Boltzmann relationship $G_{ss} = (1 - G_{min})/\{1 + \exp[A(V_j - V_0)]\} + G_{min}$, where G_{min} is the voltage-insensitive component of G_{ss}, V_0 is the half-inactivation voltage for the voltage-sensitive component of G_{ss}, and A is the slope factor expressing the charge sensitivity of the transition between the high and low conductance states. A is determined by the expression zq/kT, where z is the number of equivalent electrons q, k is Boltzmann's constant, and T is absolute temperature. Reproduced with permission from Veenstra *et al.* (1992). Copyright © 1992 by the American Heart Association.

Further structure–function and mutagenesis studies should elucidate the mechanism of V_j dependence.

The transfection studies have also provided a molecular explanation for the divergent findings observed in studies (discussed earlier) of gap junctions in cardiac myocytes from different mammalian and avian sources. These studies demonstrate that the properties of gap junctions are functions of the connexins expressed, not of the species source of the cells. The predominant channel observed in neonatal mammalian cardiac myocytes is very similar to the cloned/expressed Cx43 channel. Cx43 behaves similarly in transfected cells, whether of chick or human origin (Fishman *et al.*, 1990; Veenstra *et al.*, 1992). Gap junctions from older chick embryo cardiac myocytes closely resemble those expressed from Cx42, the predominant connexin expressed in the chick embryo myocardium (Beyer, 1990; Minkoff *et al.*, 1992; Veenstra *et al.*, 1992; Beyer *et al.*, 1993). The greater V_j dependence of cardiac myocytes from early chick embryos apparently reflects expression of Cx45.

Gap junction channels derived from different connexins also differ in other functional properties. Most gap junction channels are inhibited by cytoplasmic acidification; however, the exact pH required to achieve closure may differ among connexins. Most extreme may be Cx56, whose conductance was unaffected by pH 5 sodium acetate (Rup *et al.*, 1993). Sensitivity to lipophiles may differ; Cx56 channels were unaffected by octanol or heptanol (Rup *et al.*, 1993). Permeability of connexin-specific channels to different sized molecules may differ. Cx43 and Cx37 channels appear to be permeable to ions and fluorescent dyes (such as Lucifer yellow and 6-carboxyfluorescein), whereas Cx45 channels will allow electrical coupling mediated by ions but not passage of the larger dyes (Veenstra *et al.*, 1993).

Soon site-directed mutagenesis experiments and transfection of constructs should elucidate the mechanism of these properties. In the only such study published to date, Fishman *et al.* (1991b) produced mutants of Cx43 with progressive deletions of the carboxyl tail that had different unitary conductances, suggesting that this portion of the protein contributes to determining this property.

V. Conclusion

With the explosive influence of connexin-specific antibody and DNA probes, recombinant expression techniques, and elegant electrophysiological techniques, we shall soon have a detailed molecular understanding of the process of gap junction-mediated intercellular communication. However, the challenge in this field is to address the important biological roles of gap junctions. Can we use these tools to elucidate the contribution of gap junctional channels to such processes as embryonic development and cardiac electrical conduction? Are gap junctions of importance in any disease processes?

Alterations in intercellular coupling have been implicated in the development of re-entrant ventricular arrhythmias (Ursell et al., 1985; Spach and Dobier, 1986). Luke and Saffitz (1991) demonstrated changes in intercellular coupling that might increase anisotropic conduction in infarct border zones. Kanter et al. (1992c,d) demonstrated differences in connexin expression in myocardial tissues with different conduction velocities. Investigations may uncover alterations in connexin expression in diseased hearts, but testable models of the perturbation of cardiac gap junctions should also be possible. Use of information about the molecular structure of connexins may lead to the development of more specific pharmacological agents to modulate cellular electrical coupling. Such drugs would have profound implications for the treatment of arrhythmias or preterm labor.

Acknowledgments

H. Lee Kanter, James G. Laing, and Jeffrey E. Saffitz provided valuable discussions. Research in the authors' laboratories is supported by National Institutes of Health Grants HL45466, HL42220, and EY08368.

References

Alcala, J., Lieska, N., and Maisel, H. (1975). Protein composition of bovine lens cortical fiber cell membranes. Exp. Eye Res. 21, 581–589.

Anumonwo, J. M., Wang, H. Z., Trabka Janik, E., Dunham, B., Veenstra, R. D., Delmar, M., and Jalife, J. (1992). Gap junctional channels in adult mammalian sinus nodal cells. Immunolocalization and electrophysiology. Circ. Res. 71, 229–239.

Arellano, R. O., Rivera, A., and Ramon, F. (1990). Protein phosphorylation and hydrogen ions modulate calcium-induced closure of gap junction channels. Biophys. J. 57, 363–367.

Arnsdorf, M. F. and Sawicki, G. J. (1981). The effects of lysophosphatidyl choline, a toxic metabolite of ischemia, on the components of cardiac excitability in sheep Purkinje fibers. Circ. Res. 49, 16–30.

Atkinson, M. M. and Sheridan, J. D. (1988). Altered junctional permeability between cells transformed by v-ras, v-mos, or v-src. Am. J. Physiol. 255, C674–C683.

Atkinson, M. M., Menko, A. S., Johnson, R. G., Sheppard, J. R., and Sheridan, J. D. (1981). Rapid and reversible reduction of junctional permeability in cells infected with a temperature-sensitive mutant of avian sarcoma virus. J. Cell Biol. 91, 573–578.

Aylsworth, C. F., Trosko, J. E., and Welsch, C. W. (1986). Influence of lipids on gap-junction-mediated intercellular communication between Chinese hamster cells in vitro. Cancer Res. 46, 4527–4533.

Azarnia, R., Reddy, S., Kmiecik, T. E., Shalloway, D., and Loewenstein, W. R. (1988). The cellular src gene product regulates junctional cell-to-cell communication. Science 239, 398–401.

Barr, L., Dewey, M. M., and Berger, W. (1965). Propagation of action potentials and the structure of the nexus in cardiac muscle. J. Gen. Physiol. 48, 797–823.

Baumgarten, C. M., Cohen, C. J., and McDonald, T. F. (1981). Heterogeneity of intracellular potassium activity and membrane potential in hypoxic guinea-pig ventricle. Circ. Res. 49, 1181–1189.

Berthoud, V. M., Ledbetter, M. L. S., Hertzberg, E. L., and Saez, J. C. (1992). Connexin43 in MDCK cells: regulation by a tumor-promoting phorbol ester and Ca^{2+}. Eur. J. Cell Biol. 57, 40–50.

Beyer, E. C. (1990). Molecular cloning and developmental expression of two chick embryo gap junction proteins. J. Biol. Chem. 265, 14439–14443.

Beyer, E. C. (1993). Gap junctions. In "Receptor and Membrane Proteins: Molecular Biology of Membrane Transport" (M. Mueckler and M. Friedlander, eds.). Academic Press, San Diego. (in press).

Beyer, E. C., and Steinberg, T. H. (1991). Evidence that the gap junction protein connexin-43 is the ATP-induced pore of mouse macrophages. J. Biol. Chem. 266, 7971–7974.

Beyer, E. C., Paul, D. L., and Goodenough, D. A. (1987). Connexin43: A protein from rat heart homologous to a gap junction protein from liver. J. Cell Biol. 105, 2621–2629.

Beyer, E. C., Goodenough, D. A., and Paul, D. L. (1988). The connexins, a family of related gap junction proteins. Mod. Cell Biol. 7, 167–175.

Beyer, E. C., Kistler, J., Paul, D. L., and Goodenough, D. A. (1989). Antisera directed against connexin43 peptides react with a 43-kDa protein localized to gap junctions in myocardium and other tissues. J. Cell Biol. 108, 595–605.

Beyer, E. C., Paul, D. L., and Goodenough, D. A. (1990). Connexin family of gap junction proteins. J. Membr. Biol. 116, 187–194.

Beyer, E. C., Reed, K. E., Westphale, E. M., Kanter, H. L., and Larson, D. M. (1992). Molecular cloning and expression of rat connexin40, a gap junction protein expressed in vascular smooth muscle. J. Membr. Biol. 127, 69–76.

Beyer, E. C., Kanter, H. L., Rup, D. M., Westphale, E. M., Reed, K. E., Larson, D. M., and Saffitz, J. E. (1993). Expression of multiple connexins by cells of the cardiovascular system and lens. In "Gap Junctions" (G. Zampighi and J. E. Hall, eds.). pp. 151–155. Elsevier, Amsterdam.

Bloemendal, H., Zweers, A., Vermorken, F., Dunia, I., and Benedetti, E. L. (1972). The plasma membrane of eye lens fibers. Biochemical and structural characterization. Cell Differ. 1, 91–106.

Brink, P. R., and Dewey, M. M. (1980). Evidence for fixed charge in the nexus. Nature (London) 285, 101–102.

Brink, P. R., and Fan, S. F. (1989). Patch clamp recordings from membranes which contain gap junction channels. Biophys. J. 56, 579–593.

Brissette, J. L., Kumar, N. M., Gilula, N. B., and Dotto, G. P. (1991). The tumor promoter 12-O-tetradecanoylphorbol-13-acetate and the ras oncogene modulate expression and phosphorylation of gap junction proteins. Mol. Cell Biol. 11, 5364–5371.

Broekhuyse, R. M., Kuhlmann, E. D., and Stols, A. H. (1976). Lens membrane. II. Isolation and characterization of the main intrinsic peptide (MIP) of bovine lens fiber membranes. Exp. Eye Res. 23, 365–371.

Bruzzone, R., Haefliger, J. A., Gimlich, R. L., and Paul, D. L. (1991). The ability of neighboring cells to form functional intercellular channels may depend on the type of connexin expressed. J. Cell Biol. 115, 4a.

Burt, J. M. (1987). Block of intercellular communication: Interaction of intracellular H^+ and Ca^{2+}. *Am. J. Physiol.* **253**, C607–C612.

Burt, J. M. (1989). Uncoupling of cardiac cells by doxyl stearic acids specificity and mechanism of action. *Am. J. Physiol.* **256**, C913–C924.

Burt, J. M., and Spray, D. C. (1988a). Single-channel events and gating behavior of the cardiac gap junction channel. *Proc. Natl. Acad. Sci. USA* **85**, 3431–3434.

Burt, J. M., and Spray, D. C. (1988b). Inotropic agents modulate gap junctional conductance between cardiac myocytes. *Am. J. Physiol.* **254**, H1206–H1210.

Burt, J. M., and Spray, D. C. (1989). Volatile anesthetics block intercellular communication between neonatal rat myocardial cells. *Circ. Res.* **65**, 829–837.

Cascio, M., Gogol, E., and Wallace, B. A. (1990). The secondary structure of gap junctions. Influence of isolation methods and proteolysis. *J. Biol. Chem.* **265**, 2358–2364.

Caspar, D. L., Goodenough, D. A., Makowski, L., and Phillips, W. C. (1977). Gap junction structures. I. Correlated electron microscopy and X-ray diffraction. *J. Cell Biol.* **74**, 605–628.

Chen, Y. H., and DeHaan, R. L. (1992). Multiple-channel conductance states and voltage regulation of embryonic chick cardiac gap junctions. *J. Membr. Biol.* **127**, 95–111.

Christ, G. J., Moreno, A. P., Melman, A., and Spray, D. C. (1992). Gap junction-mediated intercellular diffusion of Ca^{2+} in cultured human corporal smooth muscle cells. *Am. J. Physiol.* **263**, C373–C383.

Clarkson, C. W. and Ten Eick, R. E. (1983). On the mechanism of lysophosphatidylcholine-induced depolarization of cat ventricular myocardium. *Circ. Res.* **52**, 543–556.

Crow, D. S., Beyer, E. C., Paul, D. L., Kobe, S. S., and Lau, A. F. (1990). Phosphorylation of connexin43 gap junction protein in uninfected and Rous sarcoma virus-transformed mammalian fibroblasts. *Mol. Cell Biol.* **10**, 1754–1763.

Dahl, G., Miller, T., Paul, D., Voellmy, R., and Werner, R. (1987). Expression of functional cell–cell channels from cloned rat liver gap junction complementary DNA. *Science* **236**, 1290–1293.

Dahl, G., and Isenberg, G. (1980). Decoupling of heart muscle cells: Correlation with increased cytoplasmic calcium activity and with changes of nexus ultrastructure. *J. Membr. Biol.* **53**, 63–75.

DeHaan, R. L., Chen, Y. H., and Penrod, R. L. (1989). Voltage dependence of junctional conductance in the embryonic heart. *In* "Molecular and Cellular Mechanisms of Antiarrhythmic Agents" (L. Hondeghem, ed.). Futura Publishing Co., Inc. Mount Kisco, New York. pp. 19–43.

De Mello, W. C. (1979). Effect of 2-4-dinitrophenol on intercellular communication in mammalian cardiac fibers. *Pflugers Arch.* **380**, 267–276.

De Mello, W. C. (1982). Cell-to-cell communication in heart and other tissues. *Prog. Biophys. Mol. Biol.* **39**, 147–182.

De Mello, W. C. (1983). The influence of pH on the healing-over of mammalian cardiac muscle. *J. Physiol. (London)* **339**, 299–307.

De Mello, W. C. (1986). Interaction of cyclic AMP and Ca^{2+} in the control of electrical coupling in heart fibers. *Biochim. Biophys. Acta* **888**, 91–99.

De Mello, W. C. (1989). Effect of isoproterenol and 3-isobutyl-1-methylxanthine on junctional conductance in heart cell pairs. *Biochim. Biophys. Acta* **1012**, 291–298.

De Mello, W. C., and van Loon, P. (1987). Further studies on the influence of cyclic nucleotides on junctional permeability in heart. *J. Mol. Cell Cardiol.* **19**, 763–771.

De Mello, W. C., Motta, G. E., and Chapeau, M. (1969). A study on the healing-over of mammalian cardiac muscle. *Circ. Res.* **24**, 475–487.

Deleze, J. (1970). The recovery of resting potential and input resistance in sheep heart injured by knife or laser. *J. Physiol. (London)* **208**, 547–562.

Dermietzel, R., Traub, O., Hwang, T. K., Beyer, E., Bennett, M. V., Spray, D. C., and Willecke, K. (1989). Differential expression of three gap junction proteins in developing and mature brain tissues. *Proc. Natl. Acad. Sci. USA* **86**, 10148–10152.

Dermietzel, R., Hertberg, E. L., Kessler, J. A., and Spray, D. C. (1991). Gap junctions between cultured astrocytes: Immunocytochemical, molecular, and electrophysiological analysis. *J. Neurosci.* **11**, 1421–1432.

Donaldson, P. and Kistler, J. (1992). Reconstitution of channels from preparations enriched in lens gap junction protein MP70. *J. Membr. Biol.* **129**, 155–165.

Dunham, B., Liu, S., Taffet, S., Trabka Janik, E., Delmar, M., Petryshyn, R., Zheng, S., Perzova, R., and Vallano, M. L. (1992). Immunolocalization and expression of functional and nonfunctional cell-to-cell channels from wild-type and mutant rat heart connexin43 cDNA. *Circ. Res.* **70**, 1233–1243.

Dunlap, K., Takeda, K., and Brehm, P. (1987). Activation of a calcium-dependent photoprotein by chemical signalling through gap junctions. *Nature (London)* **325**, 60–62.

Eghbali, B., Kessler, J. A., and Spray, D. C. (1990). Expression of gap junction channels in communication-incompetent cells after stable transfection with cDNA encoding connexin 32. *Proc. Natl. Acad. Sci. USA* **87**, 1328–1331.

Ehring, G. R., Zampighi, G. A., Horwitz, J., Bok, D., and Hall, J. E. (1990). Properties of channels reconstituted from the major intrinsic protein of lens fiber membrane. *J. Gen. Physiol.* **96**, 631–664.

el Aoumari, A., Fromaget, C., Dupont, E., Reggio, H., Durbec, P., Briand, J. P., Boller, K., Kreitman, B., and Gros, D. (1990). Conservation of a cytoplasmic carboxy-terminal domain of connexin 43, a gap junctional protein, in mammal heart and brain. *J. Membr. Biol.* **115**, 229–240.

Evans, W. H. and Gurd, J. W. (1972). Preparation and properties of nexuses and lipid enriched vesicles from mouse liver plasma membranes. *Biochem. J.* **128**, 691–700.

Filson, A. J., Azarnia, R., Beyer, E. C., Loewenstein, W. R., and Brugge, J. S. (1990). Tyrosine phosphorylation of a gap junction protein correlates with inhibition of cell-to-cell communication. *Cell Growth Differ.* **1**, 661–668.

Finbow, M., Yancey, S. B., Johnson, R., and Revel, J. P. (1980). Independent lines of evidence suggesting a major gap junctional protein with a molecular weight of 26,000. *Proc. Natl. Acad. Sci. USA* **77**, 970–974.

Fishman, G. I., Spray, D. C., and Leinwand, L. A. (1990). Molecular characterization and functional expression of the human cardiac gap junction channel. *J. Cell Biol.* **111**, 589–598.

Fishman, G. I., Hertzberg, E. L., Spray, D. C., and Leinwand, L. A. (1991a). Expression of connexin43 in the developing rat heart. *Circ. Res.* **68**, 782–787.

Fishman, G. I., Moreno, A. P., Spray, D. C., and Leinwand, L. A. (1991b). Functional analysis of human cardiac gap junction channel mutants. *Proc. Natl. Acad. Sci. USA* **88**, 3525–3529.

Flagg Newton, J., and Loewenstein, W. R. (1979). Experimental depression of junctional membrane permeability in mammalian cell culture. A study with tracer molecules in the 300 to 800 Dalton range. *J. Membr. Biol.* **50**, 65–100.

Flagg Newton, J. L., and Loewenstein, W. R. (1981). Cell junction and cyclic AMP: II. Modulations of junctional membrane permeability, dependent on serum and cell density. *J. Membr. Biol.* **63**, 123–131.

Fluri, G. S., Rudisuli, A., Willi, M., Rohr, S., and Weingart, R. (1990). Effects of arachidonic acid on the gap junctions of neonatal rat heart cells. *Pflugers Arch.* **417**, 149–156.

Fromaget, C., el Aoumari, A., Dupont, E., Briand, J. P., and Gros, D. (1990). Changes in the expression of connexin 43, a cardiac gap junctional protein, during mouse heart development. *J. Mol. Cell Cardiol.* **22**, 1245–1258.

Giaume, C., Randriamampita, C., and Trautmann, A. (1989). Arachidonic acid closes gap junction channels in rat lacrimal glands. *Pflugers Arch.* **413**, 273–279.

Giaume, C., Fromaget, C., el Aoumari, A., Cordier, J., Glowinski, J., and Gros, D. (1991). Gap junctions in cultured astrocytes: Single-channel currents and characterization of channel-forming protein. *Neuron* **6**, 133–143.

Gimlich, R. L., Kumar, N. M., and Gilula, N. B. (1990). Differential regulation of the levels of three gap junction mRNAs in Xenopus embryos. *J. Cell Biol.* **110**, 597–605.

Goodenough, D. A. (1979). Lens gap junctions: A structural hypothesis for nonregulated low-resistance intercellular pathways. *Invest. Ophthalmol. Vis. Sci.* **18**, 1104–1122.

Goodenough, D. A., Paul, D. L., and Jesaitis, L. (1988). Topological distribution of two connexin32 antigenic sites in intact and split rodent hepatocyte gap junctions. *J. Cell Biol.* **107**, 1817–1824.

Goodenough, D. A. and Stoeckenius, W. (1972). The isolation of mouse hepatocyte gap junctions. Preliminary chemical characterization and X-ray diffraction. *J. Cell Biol.* **54**, 646–656.

Gourdie, R. G., Green, C. R., Severs, N. J., and Thompson, R. P. (1992). Immunolabelling patterns of gap junction connexins in the developing and mature rat heart. *Anat. Embryol.* **185**, 363–378.

Green, C. R., Harfst, E., Gourdie, R. G., and Severs, N. J. (1988). Analysis of the rat liver gap junction protein: clarification of anomalies in its molecular size. *Proc. R. Soc. Lond. (Biol.).* **233**, 165–174.

Gros, D., Mocquard, J. P., Challice, C. E., and Schrevel, J. (1978). Formation and growth of gap junctions in mouse myocardium during ontogenesis: A freeze-cleave study. *J. Cell Sci.* **30**, 45–61.

Gros, D., Mocquard, J. P., Challice, C. E., and Schrevel, J. (1979). Formation and growth of gap junctions in mouse myocardium during ontogenesis: Quantitative data and their implications on the development of intercellular communication. *J. Mol. Cell Cardiol.* **11**, 543–554.

Gros, D. B., Nicholson, B. J., and Revel, J. P. (1983). Comparative analysis of the gap junction protein from rat heart and liver: Is there a tissue specificity of gap junctions? *Cell* **35**, 539–549.

Haefliger, J. A., Bruzzone, R., Jenkins, N. A., Gilbert, D. J., Copeland, N. G., and Paul, D. L. (1992). Four novel members of the connexin family of gap junction proteins. Molecular cloning, expression, and chromosome mapping. *J. Biol. Chem.* **267**, 2057–2064.

Harfst, E., Severs, N. J., and Green, C. R. (1990). Cardiac myocyte gap junctions: Evidence for a major connexon protein with an apparent relative molecular mass of 70,000. *J. Cell Sci.* **96**, 591–604.

Haydon, D. A., and Urban, B. W. (1983). The action of alcohols and other non-ionic surface active substances on the sodium current of the squid giant axon. *J. Physiol. (Lond)* **341**, 411–427.

Hennemann, H., Schwarz, H. J., and Willecke, K. (1992a). Characterization of gap junction genes expressed in F9 embryonic carcinoma cells: Molecular cloning of mouse connexin31 and -45 cDNAs. *Eur. J. Cell Biol.* **57**, 51–58.

Hennemann, H., Suchyna, T., Lichtenberg Frate, H., Jungbluth, S., Dahl, E., Schwarz, J., Nicholson, B. J., and Willecke, K. (1992b). Molecular cloning and functional expression of mouse connexin40, a second gap junction gene preferentially expressed in lung. *J. Cell Biol.* **117**, 1299–1310.

Hertzberg, E. L., Disher, R. M., Tiller, A. A., Zhou, Y., and Cook, R. G. (1988). Topology of the Mr 27,000 liver gap junction protein. Cytoplasmic localization of amino- and carboxyl termini and a hydrophilic domain which is protease-hypersensitive. *J. Biol. Chem.* **263**, 19105–19111.

Hertzberg, E. L., and Gilula, N. B. (1979). Isolation and characterization of gap junctions from rat liver. *J. Biol. Chem.* **254**, 2138–2147.

Hille, B. (1992). "Ionic Channels of Excitable Membranes" Sinauer Assoc., Inc. Sunderland, Massachusetts.

Hirche, G. A. (1985). Blocking and modifying actions of octanol on Na channels in frog myelinated nerve. *Pflugers Arch.* **405**, 180–187.

Hyrc, K. and Rose, B. (1990). The action of v-*src* on gap junctional permeability is modulated by pH. *J. Cell Biol.* **110**, 1217–1226.

Jongen, W. M., Fitzgerald, D. J., Asamoto, M., Piccoli, C., Slaga, T. J., Gros, D., Takeichi, M., and Yamasaki, H. (1991). Regulation of connexin 43-mediated gap junctional intercellular communication by Ca^{2+} in mouse epidermal cells is controlled by E-cadherin. *J. Cell Biol.* **114**, 545–555.

Kameyama, M. (1983). Electrical coupling between ventricular paired cells isolated from guinea pig heart. *J. Physiol.* **336**, 345–357.

Kanter, H. L., Beyer, E. C., Beau, S. L., and Saffitz, J. E. (1992a). Discrete patterns of connexin expression in canine Purkinje fibers and ventricular muscle. *Circulation* **88**, I–8.

Kanter, H. L., Green, K. G., Beyer, E. C., and Saffitz, J. E. (1992b). Structural and molecular determinants of conduction in canine atrial myocardium. *Circulation* **88**, I–492.

Kanter, H. L., Saffitz, J. E., and Beyer, E. C. (1992c). Cardiac myocytes express multiple gap junction proteins. *Circ. Res.* **70**, 438–444.

Kanter, H. L., Saffitz, J. E., and Beyer, E. C. (1992d). Molecular cloning and characterization of expression of two novel human cardiac gap junction proteins. *Circulation* **88**, I–751.

Kensler, R. W. and Goodenough, D. A. (1980). Isolation of mouse myocardial gap junctions. *J. Cell Biol.* **86**, 755–764.

Kistler, J., Kirkland, B., and Bullivant, S. (1985). Identification of a 70,000-D protein in lens membrane junctional domains. *J. Cell. Biol.* **101**, 28–35.

Kistler, J., Christie, D., and Bullivant, S. (1988). Homologies between gap junction proteins in lens, heart and liver. *Nature (London)* **331**, 721–723.

Kleber, A. G., Riegger, C. B., and Janse, M. J. (1990). Electrical uncoupling and increase of extracellular resistance after induction of ischemia in isolated arterially perfused rabbit papillary muscle. *Circ. Res.* **61**, 271–279.

Kumar, N. M. and Gilula, N. B. (1986). Cloning and characterization of human and rat liver cDNAs coding for a gap junction protein. *J. Cell Biol.* **103**, 767–776.

Laird, D. W., Puranam, K. L., and Revel, J. P. (1991). Turnover and phosphorylation dynamics of connexin43 gap junction protein in cultured cardiac myocytes. *Biochem. J.* **273**, 67–72.

Larson, D. M., Haudenschild, C. C., and Beyer, E. C. (1990). Gap junction messenger RNA expression by vascular wall cells. *Circ. Res.* **66**, 1074–1080.

Larson, D. M., Beyer, E. C., and Haudenschild, C. C. (1991). Biosynthesis of the gap junction protein connexin43 by bovine aortic endothelial cells. *FASEB J.* **5**, A528.

Lash, J. A., Critser, E. S., and Pressler, M. L. (1990). Cloning of a gap junctional protein from vascular smooth muscle and expression in two-cell mouse embryos. *J. Biol. Chem.* **265**, 13113–13117.

Loewenstein, W. R. (1966). Permeability of membrane junctions. *Ann. N. Y. Acad. Sci.* **137**, 441–472.

Loewenstein, W. R. (1981). Junctional intercellular communication: The cell-to-cell membrane channel. *Physiol. Rev.* **61**, 829–913.

Luke, R. A., and Saffitz, J. E. (1991). Remodeling of ventricular conduction pathways in healed canine infarct border zones. *J. Clin. Invest.* **87**, 1594–1602.

Luke, R. A., Beyer, E. C., Hoyt, R. H., and Saffitz, J. E. (1989). Quantitative analysis of intercellular connections by immuno-histochemistry of the cardiac gap junction protein connexin43. *Circ. Res.* **65**, 1450–1457.

Makowski, L. (1988). X-ray diffraction studies of gap junction structure. *Adv. Cell Biol.* **2**, 119–158.

Makowski, L., Caspar, D. L. D., Phillips, W. C., and Goodenough, D. A. (1976). Gap junction structures. II. Analysis of the X-ray diffraction data. *J. Cell Biol.* **74**, 629–645.

Manjunath, C. K., Goings, G. E., and Page, E. (1984). Cytoplasmic surface and intramembrane components of rat heart gap junctional proteins. *Am. J. Physiol.* **246**, H865–H875.

Manjunath, C. K., Nicholson, B. J., Teplow, D., Hood, L., Page, E., and Revel, J. P. (1987). The cardiac gap junction protein (Mr 47,000) has a tissue-specific cytoplasmic domain of Mr 17,000 at its carboxy-terminus. *Biochem. Biophys. Res. Commun.* **142**, 228–234.

Massey, K. D., Minnich, B. N., and Burt, J. M. (1992). Arachidonic acid and lipoxygenase metabolites uncouple neonatal rat cardiac myocyte pairs. *Am. J. Physiol.* **263**, C494–C501.

Maurer, P., and Weingart, R. (1987). Cell pairs isolated from adult guinea pig and rat hearts: Effects ofi on nexal membrane resistance. *Pflugers Arch.* **409**, 394–402.

Meda, P., Chanson, M., Pepper, M., Giordano, E., Bosco, D., Traub, O., Willecke, K., el Aoumari, A., Gros, D., Beyer, E. C., and Spray, D. C. (1991). *In vivo* modulation of connexin 43 gene expression and junctional coupling of pancreatic B-cells. *Exp. Cell Res.* **192**, 469–480.

Metzger, P., and Weingart, R. (1985). Electric current flow in cell pairs isolated from adult rat hearts. *J. Physiol (London)* **366**, 177–195.

Meyer, R. A., Laird, D. W., Revel, J. P., and Johnson, R. G. (1992). Inhibition of gap junction and adherens junction assembly by connexin and A-CAM antibodies. *J. Cell Biol.* **119**, 179–189.

Michalke, W. and Loewenstein, W. R. (1971). Communication between cells of different type. *Nature (London)* **232**, 121–122.

Milks, L. C., Kumar, N. M., Houghten, R., Unwin, N., and Gilula, N. B. (1988). Topology of the 32-kd liver gap junction protein determined by site-directed antibody localizations. *EMBO J.* **7**, 2967–2975.

Minkoff, R., Rundus, V. R., Parker, S. B., Beyer, E. C., and Hertzberg, E. L. (1992). Connexin expression in the developing avian cardiovascular system. *Mol. Biol. Cell* **3**, 294a.

Moore, L. K., Beyer, E. C., and Burt, J. M. (1991). Characterization of gap junction channels in A7r5 vascular smooth muscle cells. *Am. J. Physiol.* **260**, C975–C981.

Moreno, A. P., Eghbali, B., and Spray, D. C. (1991a). Connexin32 gap junction channels in stably transfected cells. Equilibrium and kinetic properties. *Biophys. J.* **60**, 1267–1277.

Moreno, A. P., Eghbali, B., and Spray, D. C. (1991b). Connexin32 gap junction channels in stably transfected cells: Unitary conductance. *Biophys. J.* **60**, 1254–1266.

Musil, L. S. and Goodenough, D. A. (1990). Gap junctional intercellular communication and the regulation of connexin expression and function. *Curr. Opin. Cell Biol.* **2**, 875–880.

Musil, L. S., Beyer, E. C., and Goodenough, D. A. (1990a). Expression of the gap junction protein connexin43 in embryonic chick lens: Molecular cloning, ultrastructural localization, and post-translational phosphorylation. *J. Membr. Biol.* **116**, 163–175.

Musil, L. S., Cunningham, B. A., Edelman, G. M., and Goodenough, D. A. (1990b). Differential phosphorylation of the gap junction protein connexin43 in junctional communication-competent and -deficient cell lines. *J. Cell Biol.* **111**, 2077–2088.

Nelson, W. and Makielski, J. C. (1991). *Circ. Res.* **68**, 977–983.

Neyton, J. and Trautmann, A. (1985). Single channel currents of an intercellular junction. *Nature (London)* **317**, 331–335.

Neyton, J. and Trautmann, A. (1986). Physiological modulation of gap junction permeability. *J. Exp. Biol.* **124**, 993–114.

Nicholson, B., Dermietzel, R., Teplow, D., Traub, O., Willecke, K., and Revel, J. P. (1987). Two homologous protein components of hepatic gap junctions. *Nature (London)* **329**, 732–734.

Nicholson, B. J., Takemoto, L. J., Hunkapiller, M. W., Hood, L. E., and Revel, J. P. (1983). Differences between liver gap junction protein and lens MIP 26 from rat: Implications for tissue specificity of gap junctions. *Cell* **32**, 967–978.

Nicholson, B. J., Gros, D. B., Kent, S. B. H., Hood, L. E., and Revel, J. P. (1985). The Mr 28,000 gap junction proteins from rat heart and liver are different but related. *J. Biol. Chem.* **260**, 6514–6517.

Niggli, E., Rudisuli, A., Maurer, P., and Weingart, R. (1989). Effects of general anesthetics on current flow across membranes in guinea-pig myocytes. *Am. J. Physiol.* **256**, C273–C281.

Noma, A., and Tsuboi, N. (1987). Dependence of junctional conductance on proton, calcium, and magnesium ions in cardiac paired cells of guinea pig. *J. Physiol. (London)* **382**, 193–211.

Obaid, A. L., Socolar, S. J., and Rose, B. (1983). Cell-to-cell channels with two independently regulated gates in series: Analysis of junctional conductance modulation by membrane potential, calsium, and pH. *J. Membr. Biol.* **73**, 691–89.

Page, E., and Shibata, Y. (1981). Permeable junctions between cardiac cells. *Ann. Rev. Physiol.* **43**, 431–441.

Paul, D. L. (1986). Molecular cloning of cDNA for rat liver gap junction protein. *J. Cell Biol.* **103**, 123–134.

Paul, D. L., and Goodenough, D. A. (1983). *In vitro* synthesis and membrane insertion of bovine MP26, an integral protein from lens fiber plasma membrane. *J. Cell Biol.* **96**, 633–638.

Paul, D. L., Ebihara, L., Takemoto, L. J., Swenson, K. I., and Goodenough, D. A. (1991). Connexin46, a novel lens gap junction protein, induces voltage-gated currents in nonjunctional plasma membrane of Xenopus oocytes. *J. Cell Biol.* **115**, 1077–1089.

Peracchia, C. (1990a). Increase in gap junction resistance with acidification in crayfish septate axons is closely related to changes in intracellular calcium but not hydrogen ion concentration. *J. Membr. Biol.* **113**, 75–92.

Peracchia, C. (1990b). Effects of caffeine and ryanodine on low pHi-induced changes in gap junction conductance and calcium concentration in crayfish septate axons. *J. Membr. Biol.* **117**, 79–89.

Pressler, M. L. (1989). Intracellular pH and cell-to-cell transmission in sheep Purkinje fibers. *Biophys. J.* **55**, 53–66.

Reber, W. R., and Weingart, R. (1982). Ungulate cardiac purkinje fibres: the influence of intracellular pH on the electrical cell-to-cell coupling. *J. Physiol. (London)* **328**, 87–104.

Reed, K. E., Westphale, E. M., Larson, D. M., Wang, H. Z., Veenstra, R. D., and Beyer, E. C. (1993). Molecular cloning and functional expression of human connexin37, and endothelial cell gap junction protein. *J. Clin. Invest.* **91**, 997–1004.

Revel, J. P., and Karnovsky, M. J. (1967). Hexagonal array of subunits in intercellular junctions of the mouse heart and liver. *J. Cell Biol.* **33**, C7–C12.

Reynhout, J. K., Lampe, P. D., and Johnson, R. G. (1992). An activator of protein kinase C inhibits gap junction communication between cultured bovine lens cells. *Exp. Cell Res.* **198**, 337–342.

Robertson, J. D. (1963). The occurrence of a subunit pattern in the unit membranes of club endings in Mauthner cell synapses in goldfish brains. *J. Cell Biol.* **19**, 201–221.

Rook, M. B., Jongsma, H. J., and van Gienneken, A. C. G. (1988). Properties of single junctional channels between isolated neonatal rat heart cells. *Am. J. Physiol.* **255,** H770–H782.

Rook, M. B., Jongsma, H. J., and de Jonge, B. (1989). Single channel currents of homo- and heterologous gap junctions between cardiac fibroblasts and myocytes. *Pflugers Arch.* **414,** 95–98.

Rose, B., and Rick, R. (1978). Intracellular pH, intracellular free calcium, and junctional cell–cell coupling. *J. Membr. Biol.* **44,** 377–415.

Rudisuli, A., and Weingart, R. (1989). Electrical properties of gap junction channels in guinea-pig ventricular cell pairs revealed by exposure to heptanol. *Pflugers Arch.* **415,** 12–21.

Rup, D. M., Veenstra, R. D., Wang, H. Z., Brink, P. R., and Beyer, E. C. (1993). Chick connexin56, a novel lens gap junction protein. *J. Biol. Chem.* **268,** 706–712.

Saez, J. C., Connor, J. A., Spray, D. C., and Bennett, M. V. (1989a). Hepatocyte gap junctions are permeable to the second messenger, inositol 1,4,5-trisphosphate, and to calcium ions. *Proc. Natl. Acad. Sci. USA* **86,** 2708–2712.

Saez, J. C., Gregory, W. A., Watanabe, T., Dermietzel, R., Hertzberg, E. L., Reid, L., Bennett, M. V., and Spray, D. C. (1989b). cAMP delays disappearance of gap junctions between pairs of rat hepatocytes in primary culture. *Am. J. Physiol.* **257,** C1–11.

Saffitz, J. E., Hoyt, R. H., Luke, R. A., Kanter, H. L., and Beyer, E. C. (1992). Cardiac myocyte interconnections at gap junctions: Role in normal and abnormal electrical conduction. *Trends Cardiovasc. Med.* **2,** 56–60.

Spach, M. S., and Dolber, P. C. (1986). Relating extracellular potentials and their derivatives to anisotropic propagation at a microscopic level in human cardiac muscle: Evidence for electrical uncoupling of side-to-side fiber connections with increasing age. *Circ. Res.* **58,** 356–371.

Spray, D. C., and Bennett, M. V. (1985). Physiology and pharmacology of gap junctions. *Annu. Rev. Physiol.* **47,** 281–303.

Spray, D. C. and Burt, J. M. (1990). Structure–activity relations of the cardiac gap junction channel. *Am. J. Physiol.* **258,** C195–C205.

Spray, D. C., Harris, A. L., and Bennett, M. V. (1981). Equilibrium properties of a voltage-dependent junctional conductance. *J. Gen. Physiol.* **77,** 77–93.

Spray, D. C., Stern, J. H., Harris, A. L., and Bennett, M. V. (1982). Gap junctional conductance: Comparison of sensitivities to H and Ca ions. *Proc. Natl. Acad. Sci. USA* **79,** 441–445.

Spray, D. C., Saez, J. C., Brosius, D., Bennett, M. V., and Hertzberg, E. L. (1986). Isolated liver gap junctions: Gating of transjunctional currents is similar to that in intact pairs of rat hepatocytes. *Proc. Natl. Acad. Sci. USA* **83,** 5494–5497.

Spray, D. C., Chanson, M., Moreno, A. P., Dermietzel, R., and Meda, P. (1991). Distinctive gap junction channel types connect WB cells, a clonal cell line derived from rat liver. *Am. J. Physiol.* **260,** C513–C527.

Sugiura, H., Toyama, J., Tsuboi, N., Kamiya, K., and Kodama, I. (1990). ATP directly affects junctional conductance between paired ventricular myocytes isolated from guinea-pig heart. *Circ. Res.* **66,** 1095–1102.

Swenson, K. I., Jordan, J. R., Beyer, E. C., and Paul, D. L. (1989). Formation of gap junctions by expression of connexins in Xenopus oocyte pairs. *Cell* **57,** 145–155.

Swenson, K. I., Piwnica Worms, H., McNamee, H., and Paul, D. L. (1990). Tyrosine phosphorylation of the gap junction protein connexin43 is required for the pp60v-src-induced inhibition of communication. *Cell Regul.* **1,** 989–1002.

Takens Kwak, B. R., Jongsma, H. J., Rook, M. B., and Van Ginneken, A. C. (1992). Mechanism of heptanol-induced uncoupling of cardiac gap junctions: A perforated patch-clamp study. *Am. J. Physiol.* **262,** C1531–C1538.

Unwin, P. N. T., and Zampighi, G. (1980). Structure of the junction between communicating cells. *Nature (London)* **283,** 545–549.

Ursell, P. C., Gardner, P. I., Albala, A., Fenoglio, J. J., and Wit, A. L. (1985). Structural and electrophysiological changes in the epicardial border zone of canine myocardial infarcts during infarct healing. *Circ. Res.* **56,** 436–451.

van Kempen, M. J., Fromaget, C., Gros, D., Moorman, A. F., and Lamers, W. H. (1991). Spatial distribution of connexin43, the major cardiac gap junction protein, in the developing and adult rat heart. *Circ. Res.* **68,** 1638–1651.

Vaughan-Jones, R. D., Lederer, W. J., and Eisner, D. A. (1983). Ca^{2+} ions can affect intracellular pH in mammalian cardiac muscle. *Nature (London)* **301,** 522–524.

Veenstra, R. D. (1990). Voltage-dependent gating of gap junction channels in embryonic chick ventricular cell pairs. *Am. J. Physiol.* **258,** C662–C672.

Veenstra, R. D. (1991a). Physiological modulation of cardiac gap junction channels. *J. Cardiovasc. Electrophys.* **2,** 168–189.

Veenstra, R. D. (1991b). Comparative physiology of cardiac gap junction channels. In "Biophysics of Gap Junction Channels" (C. Peracchia, ed.). pp. 131–144. CRC Press, Boca Raton, FL.

Veenstra, R. D. (1991c). Developmental changes in regulation of embryonic chick heart gap junctions. *J. Membr. Biol.* **119,** 253–265.

Veenstra, R. D., and DeHaan, R. L. (1986). Measurement of single channel currents from cardiac gap junctions. *Science* **233,** 972–974.

Veenstra, R. D., and DeHaan, R. L. (1988). Cardiac gap junction channel activity in embryonic chick ventricle cells. *Am. J. Physiol.* **254,** H170–H180.

Veenstra, R. D., and DeHaan, R. L. (1989). Regulation of single channel activity in cardiac gap junctions. In "Cell Interactions and Gap Junctions" (N. Sperelakis and W. C. Cole, eds.). CRC Press, Inc., Boca Raton, Florida. pp. 65–83.

Veenstra, R. D., Wang, H. Z., Westphale, E. M., and Beyer, E. C. (1992). Multiple connexins confer distinct regulatory and conductance properties of gap junctions in developing heart. *Circ. Res.* **71,** 1277–1283.

Veenstra, R. D., Wang, H. Z., Beyer, E. C., and Brink, P. R. (1993). Differential permeability of connexin-specific gap junctions to fluorescent tracers. *Biophys. J.* **64,** A235.

Verselis, V., and Brink, P. R. (1986). The gap junction channel. Its aqueous nature as indicated by deuterium oxide effects. *Biophys. J.* **50,** 1003–1007.

Verselis, V. K., Bennett, M. V. L., and Bargiello, T. A. (1991). A voltage-dependent gap junction in Drosophila melanogaster. *Biophys. J.* **59,** 114–125.

Wang, H. Z., Li, J., Lemanski, L. F., and Veenstra, R. D. (1992). Gating of mammalian cardiac gap junction channels by transjunctional voltage. *Biophys. J.* **63,** 139–151.

Weidmann, S. (1952). The electrical constants of Purkinje fibres. *J. Physiol. (London)* **118,** 348–360.

Weidmann, S. (1966). The diffusion of radiopotassium across intercalated disks of mammalian cardiac muscle. *J. Physiol.* **187,** 323–342.

Weingart, R., Imanaga, I., and Weidmann, S. (1975). Low resistance pathways between myocardial cells. *Recent. Adv. Stud. Cardiac. Struct. Metab.* **5,** 227–232.

Weingart, R. (1986). Electrical properties of the nexal membrane studied in rat ventricular cell pairs. *J. Physiol. (London)* **370,** 267–284.

Werner, R., Levine, E., Rabadan Diehl, C., and Dahl, G. (1989). Formation of hybrid cell–cell channels. *Proc. Natl. Acad. Sci. USA* **86,** 5380–5384.

White, R. L., Spray, D. C., Campos De Carvalho, A. C., Wittenberg,

B. A., and Bennett, M. V. (1985). Some electrical and pharmacological properties of gap junctions between adult ventricular myocytes. *Am. J. Physiol.* **249,** C447–C455.

White, R. L., Doeller, J. E., Verselis, V. K., and Wittenberg, B. A. (1990). Gap junctional conductance between pairs of ventricular myocytes is modulated synergistically by H$^+$ and Ca^{2+}. *J. Gen. Physiol.* **95,** 1061–1075.

Willecke, K., Heynkes, R., Dahl, E., Stutenkemper, R., Hennemann, H., Jungbluth, S., Suchyna, T., and Nicholson, B. J. (1991). Mouse connexin37: Cloning and functional expression of a gap junction gene highly expressed in lung. *J. Cell Biol.* **114,** 1049–1057.

Wojtczak, J. (1979). Contractures and increase in internal longitudi-nal resistance of cow ventricular muscle induced by hypoxia. *Circ. Res.* **44,** 88–95.

Yeager, M., and Gilula, N. B. (1992). Membrane topology and quaternary structure of cardiac gap junction ion channels. *J. Mol. Biol.* **223,** 929–948.

Young, J. D., Cohn, Z. A., and Gilula, N. B. (1987). Functional assembly of gap junction conductance in lipid bilayers: Demonstration that the major 27 kd protein forms the junctional channel. *Cell* **48,** 733–743.

Zimmer, D. B., Green, C. R., Evans, W. H., and Gilula, N. B. (1987). Topological analysis of the major protein in isolated intact rat liver gap junctions and gap junction-derived single membrane structures. *J. Biol. Chem.* **262,** 7751–7763.

Gap Junction Proteins in the Lens

LISA EBIHARA

I. Introduction

The lens is a transparent, highly refractive organ with the primary function of focusing light onto the retina. Factors that contribute to its transparency and high refractive index include (1) its avascularity, (2) the absence of intracellular organelles such as mitochondria and nuclei from mature lens fibers; (3) the narrowness of extracellular spaces between fibers; and (4) the high concentrations of water-soluble proteins called crystallins in lens fiber cells. Cataracts can result from changes in cellular volume, which disrupt fiber packing and alter the width of the intercellular clefts, or from aggregation of intracellular proteins.

The two different cell types in the lens are epithelial cells and fiber cells. The epithelial cells are arranged as a monolayer on the anterior surface of the lens. The epithelial cells divide and differentiate into immature fiber cells at the lens equator. As the lens fibers mature, they elongate and lose their intracellular organelles. The bulk of the lens is composed of mature fibers that extend from pole to pole. All these cells are electrically coupled.

A central problem in lens physiology is the mechanism responsible for maintaining a constant steady-state volume. Since the lens is a syncytial tissue, volume regulation must depend on the transport properties of all the cells in the lens. The transport properties of the epithelial cells are different from those of the fiber cells. Kinsey and Reddy (1965) first demonstrated that active transport of Na^+/K^+ is primarily carried out by the anterior epithelium. The passive properties of the epithelial cells and the fiber cells also differ. Mathias et al. (1979,1981,1985) used a theoretical model for current flow in a spherical syncytium (Eisenberg et al., 1979) to interpret impedance studies of the frog lens. These researchers showed that specific conductance of the inner fiber cell membranes was much smaller that that of the surface cells. Further, they found that the potassium conductance resided almost exclusively in the surface cell membranes and that the sodium and chloride conductances resided mostly in the inner fiber membranes. The spatial localization of active and passive transport properties suggests the possibility of circulating ion fluxes and extracellular and intracellular voltage gradients within the lens. Mathias (1985) proposed a model in which ion flux (primarily sodium) is associated with water flowing radially inward along the intercellular clefts, crossing the inner fiber membranes, and flowing radially outward through gap junctions to the surface (Mathias, 1985; Mathias and Rae, 1985). This circulating fluid flow could bring oxygen and metabolites to the fiber cells and remove waste products from the cells. Experimental evidence for steady-state ion fluxes comes from the work of Robinson and Paterson (1983), who measured outward currents at the equator and inward currents at the anterior and posterior poles of the lens using a vibrating probe.

II. Gap Junctions in the Lens

Morphological studies show that the epithelial cells are interconnected by numerous gap junctions that resemble gap junctions found in other tissues. Lens fibers are also interconnected by numerous gap junctions (see Zampighi et al., 1992, for a review). The epithelial and fiber gap junctions exhibit differences in connexon packing when examined using freeze-fracture electron microscopy (Benedetti et al., 1976; Peracchia, 1978; Goodenough et al., 1980; Lo and Harding, 1986; Miller and Goodenough, 1986). The lens fiber gap junctions are composed of randomly packed gap junctional particles, whereas the epithelial gap junctions tend to form crystalline arrays of particles after chemical fixation.

Controversy exists over whether gap junctions occur between epithelial and fiber cells. Goodenough et al. (1980), Miller and Goodenough (1986), and Lo and Harding (1986) have identified epithelial–fiber gap junctions in the chick lens. These gap junctions do not crystallize following chemical fixation. In contrast, Brown et al. (1990) found only rare crystalline gap junctions between epithelial and fiber cells in 2-month-old adult chicken lens. Although the reason for this discrepancy remains unclear, physiological studies show that the epithelial and fiber cells are electrically and dye-coupled, a result that strongly argues for the presence of gap junctional coupling (Rae and Kuszak, 1983; Miller and Goodenough, 1986).

III. Protein Composition of Lens Fiber Gap Junctions

To determine the protein composition of lens fiber gap junctions, procedures were developed to isolate lens junctional membranes (Bloemendal et al., 1972; Alcala et al., 1975; Broekhuyse et al., 1976; Goodenough, 1979; Paul and Goodenough, 1983). In ungulate lens gap junctional preparations analyzed by SDS-PAGE, the major component migrates as a 26-kDa protein. Several other minor components are also seen, including bands at 18, 70, and 64 kDa. The 26-kDa protein is called MP26 or major intrinsic protein (MIP) of the lens. MP26 was originally thought to be the gap junction-forming protein in the lens

because it was the most abundant protein in lens junctional membrane preparations. In addition, early immunoelectron microscopic studies reported that anti-MP26 antibodies labeled junctional membranes (Bloemendal et al., 1972; Alcala et al., 1975; Maisel, 1977; Kuszak et al., 1978). However, doubts arose about the validity of this hypothesis when other studies demonstrated that anti-MP26 antibodies labeled both junctional and nonjunctional membranes or, in one case, only nonjunctional membranes (Paul and Goodenough, 1983; Fitzgerald et al., 1985; Kistler et al. 1985; Sas et al., 1985). Analysis of the amino acid sequence of MP26 showed that MP26 was unrelated to other members of the connexin family (Nicholson et al., 1983; Gorin et al., 1984). Instead, MP26 is a member of a family of membrane proteins that includes the glycerol facilitator found in the inner plasma membrane of Escherishia coli (Baker and Saier, 1990); nodulin 26, a protein found in plants (Sandal and Marcker, 1988); Drosophila bib gene product, a neurogenic gene found in Drosophila (Rao et al., 1990); and CHIP, a protein found in the plasma membrane of erythrocytes and in apical brush border of the proximal convoluted tubules of the kidney (Agre and Preston 1992; see Chapter 27).

When purified MP26 was reconstituted into lipid vesicles or planar lipid bilayers, it formed large nonselective ionic channels (Peracchia and Girsch, 1985; Zampighi et al., 1985; Ehring et al., 1990; Shen et al., 1991). The MP26 channels had a symmetrical voltage dependence that could be modified by phosphorylation (Ehring et al., 1992). These channels were not blocked by acidification or elevation of calcium (Ehring et al., 1990). To determine whether MP26 could form gap junctional channels, Swenson et al. (1989) expressed Cx32, Cx43, and MP26 in Xenopus oocyte pairs. This study showed that oocyte pairs injected with mRNA for Cx32 or Cx43 and were well coupled. In contrast, injection of mRNA for MP26 failed to induce cell–cell channels in oocyte pairs or nonjunctional channels in solitary oocytes.

The physiological role of MP26 in the lens remains controversial. Ehring et al., (1990) proposed that MP26 forms ion channels in the plasma membrane of lens fibers that can contribute significantly to cell–cell coupling by allowing the flow of current from the cytoplasm of one cell to the cytoplasm of the other cell via the extracellular space. These channels could also function to minimize the extracellular space.

Another candidate for the gap junction-forming protein in the lens is MP20. MP20 is the second most common protein in lens junctional membrane preparations, but is unrelated to MP26 or the connexin family (Louis et al., 1989). Tenbroek et al. (1992) found that MP20 was present in lens fiber gap junctions over a restricted region 0.5 to 1.0 mm into the lens. Biochemical studies show that MP20 is a substrate for cAMP-dependent protein kinase and a receptor for calmodulin (Galvan *et al.*, 1989).

Kistler *et al.* (1985) developed a monoclonal antibody against the 70-kDa protein in purified lens fiber membranes (MP70) that bound exclusively to 18- to 20-nm intercellular fiber junctions in thin sections of lens outer cortex. As part of the aging process, MP70 was shown to be cleaved to a 38-kDa protein, thereby losing the epitope for the monoclonal antibody (Kistler and Bullivant, 1987). The N-terminal sequence of MP70 was closely related to that of connexins, suggesting that MP70 is a member of connexin family of gap junctional proteins (Kistler *et al.*, 1988).

IV. Connexins in the Lens

Most, if not all, gap junctions appear to be composed of a family of closely related proteins called connexins. At least 12 different types of mammalian connexins have been cloned. Three different types of connexins have been identified in the lens: rat Cx46, rat Cx43, and mouse Cx50 (Beyer *et al.*, 1987; Paul *et al.*, 1991; White *et al.*, 1992). The N-terminal amino acid sequence of Cx46 and Cx50 strongly resemble the N-terminal amino acid sequence obtained from MP70 by manual protein sequencing, suggesting that Cx46 and Cx50 are closely related to MP70 (Paul *et al.*, 1991). Northern blot analysis of different rat tissues demonstrated that Cx43 was present in many organs including the heart, kidney, brain, and lens (Beyer *et al.*, 1987). Cx46 mRNA was most abundant in the lens and was found at lower levels in the heart and the kidney (Paul *et al.*, 1991). Cx50 mRNA was found exclusively in the lens (White *et al.*, 1992).

The distribution of different gap junctional proteins in the lens has been examined using antibodies against unique polypeptide regions of the connexins (Fig. 1). Immunolocalization using anti-Cx43 antibodies demonstrated that Cx43 was present in the lens epithelium (Beyer *et al.*, 1989). The antibody did not react with fiber cell gap junctions. In contrast, immunolocalization using anti-Cx46 and anti-Cx50 antibodies and a monoclonal antibody against MP70 showed that all three proteins colocalized to lens fiber gap junctions (Paul *et al.*, 1991; White *et al.*, 1992). No reactivity occurred with intraepithelial gap junctions. The antibodies were also used in Western blotting. The anti-Cx46 antibody recognized a protein that was distinct from MP70 in both rat and bovine lens membranes. In contrast, the antibody raised against Cx50

and the monoclonal antibody against MP70 recognized the same protein. On the basis of these findings, White *et al.* (1992) proposed that Cx50 corresponded to the protein MP70.

Rup *et al.* (1993) used primers corresponding to the N-terminal sequence of Cx46 and a consensus sequence of the second extracellular loop conserved in all connexins to isolate a novel chick lens gap junctional protein called chick Cx56. Northern blot analysis of different chick tissues suggests that Cx56 mRNA is present exclusively in the lens.

Figure 2 shows an alignment of the predicted amino acid sequences of Cx50, Cx46, and Cx56 produced by PILEUP. Hydropathy analysis predicts that the three connexins span the membrane for times, with the N and C termini located in the cytoplasm. The regions of the proteins corresponding to the extracellular and transmembrane domains are highly conserved among all known connexins. Cx56 is more closely related to Cx46 than to any other known connexin. Cx46 and Cx56 share 63% overall identity. In contrast, Cx56 shares only 46% identity with Cx50 and 49% identity with chick Cx43.

V. Functional Studies of Cloned and Native Gap Junctions

The presence of distinct connexins in epithelial and fiber cells suggests that epithelial and fiber gap junctions may have different properties. Evidence in support of this hypothesis comes from dye transfer studies in the chick lens. Schuetze and Goodenough (1982) reported that dye transfer between young lens fibers was reversibly blocked by cytoplasmic acidification up to stage 14 in the chick. Beyond stage 14, dye transfer between the fiber cells became insensitive to acidification. This change is associated with a change in connexin morphology and an increase in Cx56 mRNA levels, suggesting that a switch in connexin isoforms may be responsible (Schuetze and Goodenough, 1982; Rup *et al.*, 1993). At a later stage in development, Miller and Goodenough (1986) showed that acidification caused uncoupling of epithelial cells but not fiber cells. Species variations in pH_i sensitivity have been detected (Bassnett and Duncan, 1988; Mathias *et al.*, 1991; Emptage *et al.*, 1992; Baldo and Mathias, 1992). In the frog lens, Mathias *et al.* (1991) reported that the pH_i sensitivity of the fiber cells was dependent on radial location. The transition zone occurred between 500 and 650 μm into the lens. Gap junctions connecting the peripheral fiber cells were blocked by cytoplasmic acidification or addition of 2,4-dinitrophenol, whereas gap junctions inside this zone were unaf-

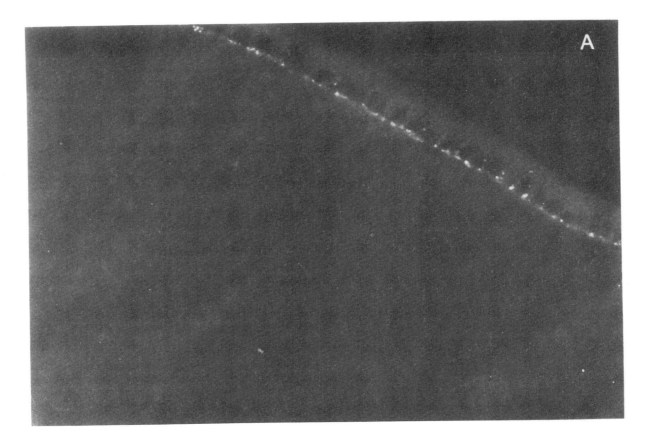

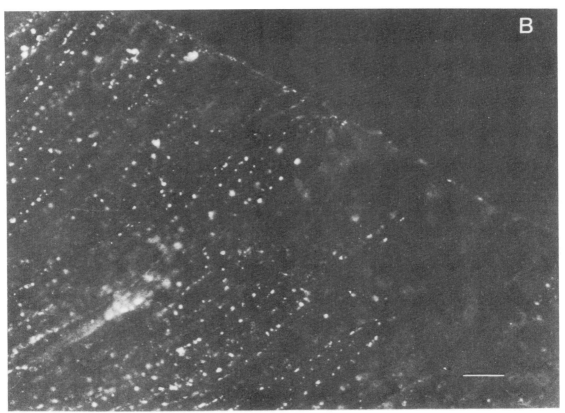

FIGURE 1 Cx46 is expressed by lens fibers. Adjacent frozen section of bovine lens were stained with anti-Cx43 (A) and anti-Cx46(411–416) (B). Cx43 staining is confined to the lens epithelium whereas Cx46 staining is abundant between lens fibers. Reproduced from The *Journal of Cell Biology*, 1991, **115,** 1077–1089, by copyright permission of the Rockefeller University Press.

```
Cx46    MGDWSFLGRL  LENAQEHSTV  IGKVWLTVLF  IFRILVLGAA  AEEVWGDEQS  DFTCNTQQPG  60
Cx56    MGDWSFLGRL  LENAQEHSTV  IGKVWLTVLF  IFRILVLGAA  AEEVWGDEQS  DFTCNTQQPG
Cx50    MGDWSFLGNI  LEEVNEHSTV  IGRVWLTVLF  IFRILILGTA  AEFVWGDEQS  AFRCNTQQPG

Cx46    CENVCYDRAF  PISHIRFWAL  QIIFVSTPTL  IYLGHVLHIV  RMEEKKKERE  EELLRRDNPQ  120
Cx56    CENVCYDKAF  PISHIRFWVL  QIIFVSTPTL  IYLGHVLHIV  RMEEKRKEKE  EELKKRGSVK
Cx50    CENVCYDEAF  PISHIRLWVL  QIIFVSTPSL  MYVGHAVHHV  RMEEKRKDRE  AEELCQQSRS

Cx46    HGRGREPMRT  GSPRDP....  ..........  ..PLRDDRGK  VRIAGALLRT  YVFNIIFKTL  180
Cx56    DNNYPGAATS  GGGSGGGNNF  KDPPIKMGKE  KLPIRDERGR  IRMGGALLRT  YIFNIIFKTL
Cx50    NGGERVPIAP  DQASIRKSSS  SSKGTK....  .........K  FRLEGTLLRT  YVCHIIFKTL

Cx46    FEVGFIAGQY  FLYGFQLQPL  YRCDRWPCPN  TVDCFISRPT  EKTIFVIFML  AVACASLVLN  240
Cx56    FEVGFIVGQY  FLYGFELKPV  YQCSRPPCPH  TVDCFISRPT  EKTIFIIFML  VVASVSLLLN
Cx50    FEVGFIVGHY  FLYGFRILPL  YRCSRWPCPN  VVDCFVSRPT  EKTIFILFML  SVAFVSLFLN

Cx46    MLEIYHLGWK  KLKQGVTNHF  NPDASEVRHK  PLDPLSEAAN  SGPPSV....  ..........  300
Cx56    MLEIYHLGWK  KLKQGMTSQY  ...SLEMPVT  TLTPVMVTGE  SKPVSLPPPA  PPVVVTTTAP
Cx50    IMEMSHLGMK  GIRSAFKRPV  EQPLGEIAEK  SLHSIAVSSI  QKAKGYQ...  ..........

Cx46    ..........  .....SIGLP  PYYTHPACPT  VQGKATG...  .FPGAPLLPA  D.........  360
Cx56    APVLPDTRAV  TPLLAPVTMA  PYYAAAAPRT  RPPSNTASMA  SYPVAPPVPE  NRHRAVTPTP
Cx50    ..........  .LLEEEKIVS  HYFPLTEVGM  VETSPLSAKP  FSQFEEKIGT  GPLADM....

Cx46    .FTVVTLNDA  QGRGHP..VK  HCNGHH.LTT  EQNWASLGAE  PQTPASKPSS  AASS......  420
Cx56    VSTPVTIQRP  YPTPTPAIIN  YFNSKRMLAA  RQNWVNMAAE  QQGKAPSSSA  GSSTPSSVRH
Cx50    ..........  ..........  ...SRSYQET  LPSYAQVGVQ  EVEREEPPIE  EAVEPEV...

Cx46    ..........  .....PHGRK  GLTDSSGSSL  EESALVVT..  .PEGEQALA.  ........TT  500
Cx56    PLPEQEEPLE  QLLPLPAGPP  ITTTNSGSST  SLSGASGSKW  DVEGEEELAE  ERPISATCTT
Cx50    ....GEKKQE  AEKVAPEGQE  TVAVPDRERV  ETPGVGKEDE  KEELQAEKVT  KQGLSAE.KA

Cx46    VEMHSPPLVL  LDP..ERSSK  SSSGRARPGD  LAI*                                560
Cx56    VEMHEPPLLV  DTRRLSRASK  SSSSRARSDD  LAV*
Cx50    PSLCPELTTD  DNRPLSRLSK  .ASSRARSDD  LTI*
```

FIGURE 2 Comparison of the amino acid sequences of Cx46, Cx56, and Cx50. The connexins are optimally aligned to match identical residues, which are shown in boldface.

fected. Researchers have hypothesized that the loss of pH sensitivity may be due to age-related cleavage of the C-terminal region of connexins (Kistler and Bullivant, 1989).

The voltage dependence and kinetic properties of gap junctional channels composed of Cx43 have been characterized in detail using patch-clamp techniques (see Chapter 25). Much less is known about the gating properties of Cx46, Cx50, and Cx56. Attempts to characterize the electrophysiological properties of gap junctional channels in lens fiber cell pairs have been complicated by the elongated geometry of the fiber cells and by difficulties in cell dissociation. To circumvent these problems, Donaldson and Kistler (1992) reconstituted ion channels into planar lipid bilayers using MP70-enriched preparations of lens fiber membranes. Three different classes of channels were observed. One class of channels had a unitary conductance of 2200 pS in 1 M KCl. The behavior of this class of channels was very similar to that of MP26 channels previously described by Ehring *et al.* (1990). A second

class of channels had a unitary conductance of 650 pS and an asymmetrical voltage dependence, and could not discriminate between sodium and potassium. A third class of channels had an unitary conductance of 350 pS and a symmetrical voltage dependence, and could not discriminate between sodium and potassium. The poor ion selectivity and large single-channel conductance of the class II and class III channels are consistent with the idea that they are junctional channels.

Another approach to characterizing the functional properties of lens fiber gap junctions is the expression of cloned lens gap junctional proteins in a heterologous system that permits their characterization. Two different systems have been used to study the functional properties of cloned connexins: (1) *Xenopus* oocyte pairs and (2) communication-deficient cell lines (Dahl *et al.*, 1987; Eghbali *et al.*, 1990). Both systems have advantages and disadvantages. The main advantage of the *Xenopus* oocyte pair system is that exogenous connexins can be expressed rapidly by injecting oocytes with mRNA transcribed from cloned cDNAs or polymerase chain reaction (PCR) products. One disadvantage is that no method is available for studying the single-channel kinetics of the junctional channels. The primary advantage of stably transfected communication-deficient cell lines is that the behavior of single junctional channels can be examined in poorly coupled cell pairs using the dual patch-clamp technique. Further, a mammalian cellular environment may be more appropriate for some of the studies, particularly those involving channel phosphorylation. One disadvantage is the length of time required to establish a stably transfected cell line.

The functional properties of cloned Cx46 and Cx50 channels have been studied in the *Xenopus* oocyte expression system (Paul *et al.*, 1991; White *et al.*, 1992; Ebihara and Steiner, 1993). Cx46 and Cx50 form gap junctional channels that can be gated by transjunctional voltage (V_j) and cytoplasmic acidification. The connexins showed differences in sensitivity to V_j and gating kinetics. The gating kinetics of the Cx50 channels were much faster than those of Cx46 channels. The ability of Cx50, Cx46, and Cx43 to form heterotypic gap junctional channels was tested by pairing oocytes that expressed different connexins (White *et al.*, 1992). Cx50 was capable of forming functional heterotypic channels with Cx46 but not with Cx43. Cx46 could form heterotypic channels with both Cx43 and Cx50.

One unexpected finding was that expression of Cx46 in *Xenopus* oocytes induced the formation of a large time- and voltage-dependent nonjunctional current with properties that are consistent with those expected for a hemichannel current (Paul *et al.*, 1991; Ebihara and Steiner, 1993). Figure 3 shows membrane currents recorded from a solitary Cx46 mRNA-injected oocyte in response to a series of voltage-clamp steps in increments of 10 mV from a holding potential of −40 mV. When voltage-clamp steps were applied to potentials more positive than −15 mV, a large, extremely slowly activating outward current was observed. Following repolarization to −40 mV, the current decayed to baseline over a period of several hundred milliseconds. A current of similar magnitude and kinetics was not observed in noninjected control oocytes or in oocytes injected with mRNA for Cx43, Cx32, or Cx40. The Cx46-induced current is hypothesized to be a hemichannel current based on several criteria. (1) Expression of Cx46 in oocytes induces the formation of both nonjunctional and junctional channels. (2) The Cx46-induced current could be modified by agents that would be expected to alter the properties of gap junctional channels (i.e., protons and calcium). (3) Ion substitution experiments suggest that

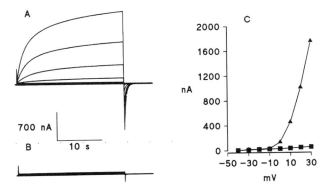

FIGURE 3 Membrane currents in a Cx46 mRNA-injected oocyte (A) and a noninjected control oocyte (B). The whole-cell currents were measured using a two-electrode voltage clamp. A series of depolarizing steps was applied in 10-mV increments between −40 and 30 mV from a holding potential of −40 mV. The interepisode interval was 40 sec. (C) I–V relationships obtained from the data shown in A and B. The current was measured at the end of the 24-sec pulse and was plotted as a function of voltage Cx46, ▲; control, ■. Reproduced from *The Journal of General Physiology* (1993). **102**, 59–74, by copyright permission of the Rockefeller University Press.

the current is poorly selective for cations and anions. (4) The deactivation kinetics of the Cx46-induced current in zero calcium Modified Barth's Solution (MBS) resemble the inactivation kinetics of the gap junctional current, suggesting that the gating properties of the nonjunctional current may account for the voltage- and time-dependent behavior of the gap junctional channels.

Whether Cx46 forms open nonjunctional channels *in vivo* is not clear. Although no reports have been made of a current resembling the Cx46-induced current in lens fiber cells, this Cx46-induced current may have escaped detection because Cx46-induced channels are mostly blocked at physiological concentrations of external calcium (i.e., 1.8 mM). In support of this hypothesis, researchers have shown that when calcium is removed, the lens depolarized and the input conductance of the inner fiber membranes increases by a factor of 10 (Rae *et al.*, 1992). DeVries and Schwarz (1992) reported a hemichannel current in catfish horizontal cells with behavior that is remarkably similar to that of the Cx46-induced current. The hemichannel current is only observed at low external calcium concentrations and can be suppressed by dopamine. Alternatively, Cx46 may be incorrectly modified posttranslationally in the *Xenopus* oocyte, leading to the development of open nonjunctional channels that are not present in the lens. Evidence for this hypothesis comes from the observation that the size of the protein recognized by anti-Cx46 antisera in immunoblots of whole rat lens is 5–10 kDa larger than the translation product in *Xenopus* oocytes (Paul *et al.*, 1991).

The electrophysiological properties of another novel lens connexin protein, chick Cx56, have been investigated by stably expressing the connexin in the communication-deficient cell line N2A (Rup *et al.*, 1993). This study showed that cell lines expressing Cx56 were dye coupled. Further, the Cx56-induced coupling was voltage independent and resistant to uncoupling by heptanol or cytoplasmic acidification. The unique functional properties of Cx56 junctional channels are in agreement with the findings of Miller and Goodenough (1986), who showed that chick fiber gap junctions were insensitive to acidification.

References

Agre, P., and Preston, G. G. (1992). Isolation of the cDNA for erythrocyte integral membrane protein of 28 kilodaltons: Member of an ancient family. *Proc. Natl. Acad. Sci. USA* **88,** 11110–11114.

Alcala, J., Lieska, N., and Maisel, H. (1975). Protein composition of bovine lens cortical fiber cell membranes. *Exp. Eye Res.* **21,** 591–595.

Baker, M. E., and Saier, M. H. (1990). A common ancestor for bovine lens fiber major intrinsic protein, soybean nodulin-26 protein and *E. coli* glycerol facilitator. *Cell* **60,** 185–186.

Baldo, G. J., and Mathias, R. T. (1992). Spatial variations in membrane properties in the intact rat lens. *Biophys. J.* **63(2),** 518–529.

Benedetti, E. L., Dunia, I., Bentzel, C. J., Vermorken, A. J. M., Kibbelaar, M., and Bloemendal, H. (1976). A portrait of plasma membrane specializations in eye lens epithelium fibers. *Biochim. Biophys. Acta* **457,** 353–384.

Beyer, E. C., Paul, D. L., and Goodenough, D. A. (1987). Connexin43: A protein from rat heart homologous to a gap junction protein from liver. *J. Cell Biol.* **105,** 2621–2629.

Beyer, E. C., Kistler, J., Paul, D. L., and Goodenough, D. A. (1989). Antisera directed against connexin43 reacts with a 43kD protein localized to gap junctions in myocardium and other tissues. *J. Cell Biol.* **108,** 595–605.

Bloemendal, H., Zweers, A., Vermorken, F., Dunia, I., and Benedetti, E. L. (1972). The plasma membranes of the eye lens fibres. Biochemical and structural characterization. *Cell Diff.* **1,** 91–106.

Broekhuyse, R., Kuhlman, E., and Stols, A. (1976). Lens membranes II. Isolation and characterization of the main intrinsic polypeptide (MIP) of bovine lens fiber membranes. *Exp Eye Res.* **23,** 365–371.

Brown, H. G., Pappas, G. D., Ireland, M. E., and Kuszak, J. R. (1990). Ultrastructural, biochemical, and immunologic evidence of receptor mediated endocytosis in the crystalline lens. *Invest. Ophthalmol. Vis. Sci.* **32,** 2119–2129.

Dahl, G., Miller, T. Paul, D. L., Voellmy, R., and Werner, R. (1987). Expression of functional cell–cell channels from cloned rat liver gap junction complementary DNA. *Science* **236,** 1290–1293.

DeVries, S. H., and Schwartz, E. A. (1992). Hemi-gap-junction channels in solitary horizontal cells of the catfish retina. *J. Physiol.* **445,** 201–230.

Donaldson, P., and Kistler, J. (1992). Reconstitution of channels from preparations enriched in lens gap junction protein MP70. *J. Membr. Biol.* **129,** 155–165.

Ebihara, L., and Steiner, E. (1993). Properties of a nonjunctional current expressed from a rat connexin46 cDNA in *Xenopus* oocytes. *J. Gen. Physiol.* **102,** 1–16.

Eghbali, B., Kessler, J. A., and Spray, D. C. (1990). Expression of gap junction channels in communication-incompetent cells after stable transfection with cDNA encoding connexin 32. *Proc. Natl. Acad. Sci. USA* **87,** 1328–1331.

Ehring, G. R., Lagos, R., and Zampighi, G. (1991). Phosphorylation modulates the voltage dependence of channels reconstituted from the major intrinsic protein of lens fiber membranes. *J. Membrane Biol.* **126,** 75–88.

Ehring, G. R., Zampighi, G., Horwitz, J., Bok, D., and Hall, J. E. (1990). Properties of channels reconstituted from the major intrinsic protein of lens fiber membranes. *J. Gen. Physiol.* **96,** 631–664.

Eisenberg, R., Barcilon, V., and Mathias, R. T. (1979). Electrical properties of spherical syncytia. *Biophys. J.* **25,** 151–180.

Emptage, N. J., Duncan, G., and Croghan, P. C. (1992). Internal acidification modulateds membrane and junctional resistance in the isolated lens of the frog Rans pipiens. *Exp. Eye Res.* **54 (1),** 33–39.

Fitzgerald, P. G., Bok, D., and Horwitz, J. (1985). The distribution of the main intrinsic membrane polypeptide in ocular lens. *Curr. Eye Res.* **4,** 1203–1217.

Galvan, A., Lampe, P. E., Hur K. C., Howard, J. B., Eccleston, E. D., Arneson, M., and Louis, C. F. (1989). Structural organization of the lens fiber cell plasma membrane protein MP18. *J. Biol. Chem.* **33,** 19974–19978.

Goodenough, D. A. (1979). Lens gap junctions: A structural hypothesis for nonregulated low-resistance intercellular pathways. *Invest. Ophthalmol. Vis. Sci.* **18,** 1104–1122.

Goodenough, D. A., Dick, J. S. B., and Lyon, J. E. (1980). Lens metabolic cooperation: A study of mouse lens permeability visualized with freeze-substitution autoradiographys and electron microscopy. *J. Cell Biol.* **86,** 576–589.

Gorin, M. B., Yancey, S. B., Cline, J., Revel, J-P., and Horwitz, J. (1984). The major intrinsic protein(MIP) of the bovine lens fiber membrane: Characterization and structure based on cDNA cloning. *Cell* **39,** 49–59.

Kinsey, V. E., and Reddy, D. V. N. (1965). Studies on the crystalline lens. XI. The relative role of the epithelium and capsule in transport. *Invest. Ophthalmol. Vis. Sci.* **4,** 104–116.

Kistler, J., and Bullivant, S. (1987). Protein processing in lens intercellular junctions: Cleavage of MP70 to MP38. *Invest. Ophthalmol. Vis. Sci.* **28,** 1687–1692.

Kistler, J., Kirkland, B., and Bullivant, S. (1985). Identification of a 70,000-D protein in lens membrane junctional domains. *J. Cell Biol.* **101,** 28–35.

Kistler, J., Christie, D., and Bullivant, S. (1988). Homologies between gap junction proteins in lens, heart and liver. *Nature (London)* **331,** 721–723.

Kistler, J., and Bullivant S. (1989). Structural and molecular biology of the eye lens membranes. *Critical Reviews in Biochemistry and Molecular Biology.* **24,** 151–181.

Kuszak, J. R., Maisel, H., and Harding, C. V. (1978). Gap junctions of chick lens fiber cells. *Exp. Eye Res.* **27,** 495–498.

Louis, C. F., Hur, K. C., Galvan, A. C., TenBroek, E. M., Jarvis, L. J., Eccleston, E. D., and Howard, J. B. (1989). Identification of an 18,000-dalton protein in mammalian lens fibers. *J. Biol. Chem.* **264,** 19967–19973.

Lo, W.-K., and Harding, C. V. (1986). Structure and distribution of gap junctions in lens epithelium and fiber cells. *Cell Tissue Res.* **244,** 253–263.

Maisel, H. (1977). The nature of the urea-insoluble material of the human lens. *Exp. Eye Res.* **24,** 417–419.

Mathias, R. T. (1985). Steady stage voltages, ion fluxes and volume regulation in syncytial tissues. *Biophys. J.* **48,** 435–448.

Mathias, R. T., and Rae, J. L. (1985). Transport properties of the lens. *Am. J. Physiol.* **249,** C181–C190.

Mathias, R. T., Rae, J. L., and Eisenberg, R. S. (1979). Electrical properties of structural components of the crystalline lens. *Biophys. J.* **25,** 181–201.

Mathias, R. T., Rae, J. L., and Eisenberg, R. S. (1981). The lens as a nonuniform spherical syncytium. *Biophys. J.* **34,** 61–83.

Mathias, R. T., Rae, J. L., Ebihara, L., and McCarthy, R. T. (1985). The localization of transport properties in the frog lens. *Biophys. J.* **48,** 423–434.

Mathias, R. T., Riquelme, G., and Rae, J. L. (1991). Cell to cell communication and pH in the frog lens. *Gen. Physiol.* **98,** 1085–1103.

Miller, T. M., and Goodenough, D. A. (1986). Evidence for two physiologically distinct gap junctions expressed by the chick lens epithelial cell. *J. Cell Biol.* **102,** 194–199.

Moreno, A. P., Campos de Carvalho, A. C., Verselis, V., Eghbali, B., and Spray, D. C. (1991). Voltage-dependent gap junction channels are formed by connexin32, the major gap junction protein of rat liver. *Biophys. J.* **59,** 920–925.

Musil, L. S., and Goodenough, D. A. (1991). Biochemical analysis of connexin43 intracellular transport, phosphorylation, and assembly into gap junctional plaques. *J. Cell Biol.* **115,** 1357–1374.

Nicholson, B. J., Takemoto, L. J., Hunkapillar, L. E., Hood, L. E., and Revel, J. P. (1983). Differences between liver gap junction protein and lens MP26 from rat: Implications for tissue specificity of gap junctions. *Cell* **32,** 967–978.

Paul, D. L., and Goodenough, D. A. (1983). Preparation, characterization, and localization of antisera against bovine MP26, an integral protein from lens fiber plasma membrane. *J. Cell Biol.* **96,** 625–632.

Paul, D. L., Ebihara, L., Takemoto, L. J., Swenson, K. I., and Goodenough, D. A. (1991). Connexin46, a novel lens gap junction protein, induces voltage-gated currents in nonjunctional plasma membrane of *Xenopus* oocytes. *J. Cell Biol.* **115,** 1077–1089.

Peracchia, C. (1978). Calcium effects on gap junction structure and cell coupling. *Nature (London)* **271,** 669–671.

Peracchia, C., and Girsch, S. J. (1985). Functional modulation of cell coupling: Evidence for a calmodulin-driven channel gate. *Am. J. Physiol.* **248,** H765–H782.

Rae, J. L., and Kuszak, J. R. (1983). The electrical coupling of epithelium and fibers in the frog lens. *Exp. Eye Res.* **36,** 317–326.

Rae, J. L., Thomson, R. D., and Eisenberg, R. S. (1982). The effect of 2,4-dinitrophenol on cell to cell communication in the frog lens. *Exp. Eye Res.* **35,** 597–609.

Rae, J. L., Mathias, R. T., Cooper, K., and Baldo, G. (1992). Divalent cation effects on lens conductance and stretch-activated cation channels. *Exp. Eye Res.* **55(1),** 135–144.

Rao, Y., Yan, L. Y., and Jan, Y. N. (1990). Similarity of the product of *Drosophila* neurogenic gene *big brain* to transmembrane channel protein. *Nature (London)* **345,** 163–167.

Robinson, K. R., and Patterson, J. W. (1983). Localization of steady currents in the lens. *Curr. Eye Res.* **2,** 843–847.

Rup, D. M., Veenstra, R. D., Wang, H-Z., Brink, P. R., and Beyer, E. C. (1993). Chick connexin-56, a novel lens gap junction protein. *J. Biol. Chem.* **268,** 706–712.

Sandal, N. N., and Marcker, K. A. (1988). Soybean nodulin 26 is homologous to the major intrinsic protein of the bovine lens fiber membrane. *Nucleic Acids Res.* **16,** 9347–9347.

Sas, D. F., Sas, M. J., Johnson, K. R., Menko, A. S., and Johnson, R. G. (1985). Junctions between lens fiber cells are labeled with a monoclonal antibody shown to be specific for MP26. *J. Cell Biol.* **100,** 216–225.

Schuetze, S. M., and Goodenough, D. A. (1982). Dye transfer between cells of the embryonic chick lens becomes less sensitive to CO_2-treatment with development. *J. Cell Biol.* **92,** 694–705.

Shen, L., Shrager, P., Girsch, S. J., Donaldson, P. J., and Peracchia, C. (1991). Channel reconstitution in liposomes and planar bilayers with HPLC-purified MIP26 of bovine lens. *J. Membr. Biol.* **124,** 21–32.

Swenson, K. I., Jordan, J. R., Beyer, E. C., and Paul, D. L. (1989). Formation of gap junctions by expression of connexins in *Xenopus* oocyte pairs. *Cell* **57,** 145–155.

Tenbroek, E., Arneson, M., Jarvis, L., and Louis, C. (1992). The distribution of fiber cell intrinsic membrane proteins MP20 and connexin46 in the bovine lens. *J. Cell Sci.* **103,** 245–257.

White, T. W., Bruzzone, R., Goodenough, D. A., and Paul, D. L. (1992). Mouse Cx50, a functional member of the connexin family of gap junction proteins, is the lens fiber protein MP70. *Mol. Biol. Cell* **3,** 711–720.

Zampighi, G. A., Hall, J. E., and Kreman, M. (1985). Purified lens junctional protein forms channels in planar lipid films. *Proc. Natl. Acad. Sci. USA* **82,** 8468–8472.

Zampighi, G. A., Simon, S. A., and Hall, J. E. (1992). The specialized junctions of the lens. *Int. Rev. Cytol.* **136,** 185–225.

Channels
of
the
MIP
Family

The MIP Transmembrane
Channel Gene Family

ANA B. CHEPELINSKY

I. Introduction

A. History of the MIP Family

MIP (major intrinsic protein), also called MIP26 and MP26, is the most abundant protein of the ocular lens fiber cell membrane (Broekhuyse et al., 1979; Bloemendal, 1982; Fitzgerald et al., 1983; Paul and Goodenough 1983a; Alcala and Maisel, 1985). The MIP gene is regulated both temporally and spatially during ocular lens development. Expression begins as lens epithelia differentiate into lens fibers, so MIP localizes specifically into the membrane of the primary fibers (Waggoner and Maisel, 1978; Broekhuyse et al., 1979; Yancey et al., 1988; Watanabe et al., 1989). For some time, MIP was considered the gap junction protein of the lens, since lens fiber cell membranes are extremely rich in gap junctions and MIP was initially isolated from lens fiber membranes enriched in gap junctions (Broekhuyse et al., 1976,1979; Fitzgerald et al., 1983; Nicholson et al., 1983; Paul and Goodenough, 1983a,b; Benedetti et al., 1990). Contradictory reports about MIP localization in junctional specializations of the lens fiber membrane, as well as in nonjunctional regions, and the observation that MIP did not form coaxially aligned channels as gap junction proteins do raised doubts about that putative function (Fitzgerald et al., 1985; Paul and Goodenough, 1983a; Sas et al., 1985; Johnson et al., 1988; Zampighi et al., 1989;

Brewer, 1991). The MIP amino acid sequence was first deduced from its cloned cDNA by Gorin et al. (1984). When the amino acid sequences of several gap junction proteins from various tissues were obtained, researchers found that they formed the connexin gene family (Beyer et al., 1988,1990) (see Chapter 25). However, MIP did not belong to this gene family, because no sequence similarity existed between MIP and connexins. In the meantime, several proteins were identified in the lens that belonged to the connexin family and that localized exclusively to gap junctions. Connexin 43 cDNA, initially cloned from a heart cDNA library, is expressed in several tissues; its expression in the lens localized specifically to the lens epithelia, but no connexin 43 is found in lens fibers (Beyer et al., 1987,1989; Musil et al., 1990; Musil and Goodenough, 1991). Connexin 46 was cloned from a lens cDNA library; its expression in the lens is localized to lens fiber gap junctions, and it is expressed at much lower levels in heart and kidney (Beyer et al., 1988; Paul et al., 1991; Tenbroek et al., 1992). When connexin 50 was cloned from a genomic library, researchers found that the coding sequence was identical to that of MP70, a protein known to localize exclusively to the 15 to 17-nm thick lens fiber gap junctions (Kistler et al., 1985; Gruitjers et al., 1987; Kistler and Bullivant, 1989; Zampighi et al., 1989; White et al., 1992) (see Chapter 26).

Moreover, several functional assays indicated that

MIP did not behave like a gap junction protein. Purified MIP forms large channels when incorporated into liposomes (Girsch and Peracchia, 1985a; Gooden *et al.*, 1985a; Peracchia and Girsch, 1985a; Swamy and Abraham, 1992) and voltage-dependent channels when incorporated into planar lipid bilayers (Ehring *et al.*, 1988,1990,1993; Shen *et al.*, 1991). However, MIP expression in *Xenopus* oocyte pairs does not induce cell–cell coupling whereas expression of connexin 32 or 43 does (Swenson *et al.*, 1989). Also, MIP voltage-dependent symmetrical channels differ from the asymmetrical ones formed by connexins 32 and 50 (Ehring *et al.*, 1990,1993; Harris, 1991; Donaldson and Kistler, 1992) (see Chapters 24 and 26).

1. A New Gene Family

After MIP cDNA was cloned (Gorin *et al.*, 1984), several cDNAs were cloned from mammals, plants, *Drosophila*, yeast, and bacteria with unknown or putative functions. These genes aligned with MIP when compared with sequences available in gene banks; it became evident that a new gene family or superfamily was emerging (Baker and Saier, 1990; Rao *et al.*, 1990; Yamamoto *et al.*, 1990; Pao *et al.*, 1991; Preston and Agre, 1991). Since MIP was the first member of the family to be isolated, this group became known as the MIP family. Table I lists the members of the MIP gene family that have been isolated in recent years. Some of the members of the family—MIP, CHIP28, and γTIP—were first isolated as proteins from lens fiber membranes, erythrocyte membrane, and tonoplast membrane, respectively (Broekhuyse *et al.*, 1976; Denker *et al.*, 1988; Johnson *et al.*, 1990). Oligonucleotides synthesized based on the partial amino acid sequence of the protein were used to screen bovine lens, human fetal liver, or plant cDNA libraries, to obtain the respective cDNAs (Gorin *et al.*, 1984; Johnson *et al.*, 1990; Preston and Agre, 1991).

CHIP28 from kidney was obtained by RT-PCR (reverse transcription-polymerase chain reaction) of rat kidney mRNA with oligonucleotides corresponding to erythrocyte CHIP28 sequence and subsequent

TABLE I   **The MIP Family**

Gene	Expression	NPA Repeat[a]	Reference
MIP Major intrinsic protein	Ocular lens fibers	H I S G A H V N P A V T F A F L V Y Y T G A G M N P A R S F A P A I	Gorin *et al.* (1984)
glpF Glycerol facilitator	*Escherichia coli*	G V S G A H L N P A V T I A L W L P L T G F A M N P A R D F G P K V	Muramatsu and Mizuno (1989)
Nod 26 Nodulin 26	Soybean/*Rhizobium* peribacterium membrane	H I S G G H F N P A V T I A F A S P V T G A S M N P A R S L G P A F	Fortin *et al.* (1987)
Pea R7a	Pea root turgor responsive	G I S G G H I N P A V T F G L F L P I T G T G I N P A R S L G A A I	Guerrero *et al.* (1990)
Tob RB7	Tobacco root	N I S G G H L N P A V T L G L A V P F S G G S M N P A R S F G P A V	Yamamoto *et al.* (1990)
At RB7	*Arabidopsis* root	H I S G G H V N P A V T F G I F A A F S G A S M N P A V A F G V A P	Yamamoto *et al.* (1990)
αTIP Tonoplast intrinsic protein	Soybean seeds storage vacuoles	H V S G G H V N P A V T F G A L L P F D G A C M N P A L A F G P S I	Johnson *et al.* (1990)
αTIP Tonoplast intrinsic protein	*Arabidopsis* seeds	N V S G G H V N P A V T F G A L V P F S G A S M N P A R A F G P A L	Hofte *et al.* (1992)
γTIP Tonoplast intrinsic protein	*Arabidopsis* vegetative tissues	N I S G G H V N P A V T F G A F I A F S G A S M N P A V A F G P A V	Hofte *et al.* (1992)
RD28	*Arabidopsis* desiccation responsive	G I S G G H I N P A V T F G L F A P I T G T G I N P A R S F G A A V	Yamaguchi-Shinozaki *et al.* (1992)
bib Big brain	*Drosophila* neural development	H I S G A H I N P A V T L A L C V P Y L N P A R S L G P S F	Rao *et al.* (1990)
CHIP28	Erythrocyte, kidney, various tissues	H I S G A H L N P A V T L G L L L D Y T G C G I N P A R S F G S A V	Preston and Agre (1991)
WCH-CD	Kidney collecting duct	H V S G A H I N P A V T V A C L V Y F T G C S M N P A R S L A P A V	Fushimi *et al.* (1993)

[a] The intramolecular NPA box repeat is indicated for each member of the family.

screening of a rat kidney cDNA library (Deen *et al.*, 1992; Zhang *et al.*, 1993). WCH-CD was obtained by a similar procedure, but the RT-PCR was performed with rat kidney medulla mRNA using oligonucleotides corresponding to the NPA box repeat of the already characterized members of the MIP family (see Section I,B); the RT-PCR product was used to screen a rat kidney cDNA library (Fushimi *et al.*, 1993). Mouse CHIP28 was identified as the growth-factor-induced delayed early response gene DER2 (Lanahan *et al.*, 1992). Nod26 was isolated by screening a nodule-specific cDNA library, devoid of known nodulin sequences, with a probe obtained from *Bradyrhizobium japonicum*-induced nodule polyribosomes precipitated by peribacteroid membrane antisera (Fortin *et al.*, 1987). AtRB7 and TobRB7 were isolated from plant-root-specific cDNA libraries (Yamamoto *et al.*, 1990), and RD28 from a cDNA library specific for genes expressed after plant desiccation (Yamaguchi-Shinozaki *et al.*, 1992).

All the members of the family are intrinsic membrane proteins. Their hydropathy plots are similar because all exhibit six hydrophobic domains that could potentially form transmembrane segments. The members of the family, approximately 250–290 amino acids long, differ mainly at their N and C termini. Their amino acid sequences bear no marked homology to known transport proteins; however, CHIP28, WCH-CD, and γTIP have been found to behave as water channels and GlpF as a glycerol channel in func-

tional assays (Preston *et al.*, 1992a; Fushimi *et al.*, 1993; Maurel *et al.*, 1993).

B. NPA Box Repeat in the MIP Family

Analysis of the amino acid sequences of the members of the MIP family has delineated the presence of a 2-fold repeat in the primary structure of these proteins (Sandal and Marcker, 1990; Pao *et al.*, 1991; Wistow *et al.*, 1991; Reizer *et al.*, 1993). Each repeat shares 30–60% identity with other members of the family (Preston and Agre, 1991). The first repeat corresponds to the sequences encoded by a single exon (exon 1) of the human MIP gene; the second repeat encompasses exons 2–4 (see Fig. 1; Pisano and Chepelinsky, 1991; Wistow *et al.*, 1991). These findings suggest that members of this family may have evolved by gene duplication of a single structural motif, perhaps representing an ancestral monomer capable of forming higher order multimeric structures. The fact that the NPA (Asn-Pro-Ala) box repeat has been conserved throughout evolution from prokaryotes to vertebrates (see Table I) indicates that the MIP family is an ancient family of putative transmembrane channel proteins (Sandal and Marcker, 1990; Pao *et al.*, 1991; Reizer *et al.*, 1993).

1. Structure

All members of the MIP family show six hydrophobic segments that have been considered membrane-spanning domains. However, the presence of an in-

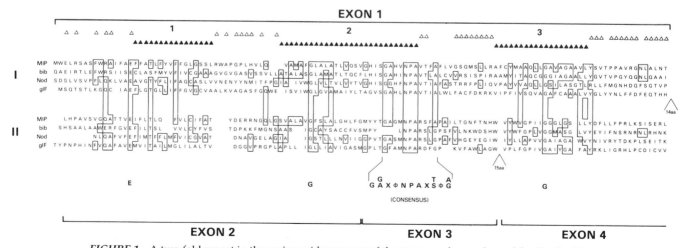

FIGURE 1 A two-fold repeat in the amino acid sequence of the transmembrane channel family. Positions occupied by charged residues in at least one sequence segment are marked with open triangles; long stretches in which all sequences are uncharged are marked with closed triangles and numbered 1–3 for each motif. The NPA box consensus sequence, E, and Gs conserved in each repeat are indicated beneath the alignment. Reprinted with permission from Wistow *et al.* (1991).

tramolecular repeat is inconsistent with that model; the repeated motifs, particularly if they are the result of gene duplication, should have essentially the same tertiary structures and should be inserted into the membrane in a similar way, resulting in a 2-fold axis perpendicular to the membrane (Wistow *et al.*, 1991). Figure 2A shows the generally accepted model of the six transmembrane domains; according to this model, the NPA boxes are on opposite sides of the membranes (Preston and Agre, 1991; Reizer *et al.*, 1993). However, the presence of the internal repeat suggests that both halves of the molecule should be inserted in the membrane in the same orientation (Wistow *et al.*, 1991), which could only be accomplished if each repeat contained an even number of transmembrane domains (Fig. 2B).

The accepted model presented in Fig. 2A is based on three α helices per repeat, of 20–24 amino acids each. MIP contains both α-helix and β-sheet structures (Horwitz and Bok, 1987). The insertion of porin β sheets into membranes has been well characterized; stretches of 8–16 amino acids can be inserted in membranes as β sheets (Jeanteur *et al.*, 1991; Pauptit *et al.*, 1991; Cowan *et al.*, 1992; Nikaido, 1992). If shorter amino acid stretches are included in the membrane, more transmembrane domains can be accommodated. In the model presented in Fig. 2B, an even number of transmembrane domains could be obtained if some β sheets are present rather than the longer α helices presented in Fig. 2A. Alternatively, only two transmembrane helices per repeat could accommodate the structural requirements; the third hydrophobic region might not constitute a transmembrane domain. Figure 1 shows the three stretches of uncharged hydrophobic

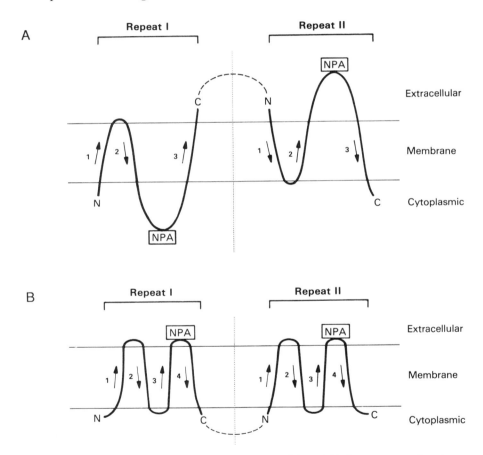

FIGURE 2   MIP family transmembrane domains models. (A) Each repeat contains three transmembrane domains. The second repeat is oriented with a 180° rotation. Each NPA box is oriented toward opposite sides of the membrane (Preston and Agre, 1991; Reizer *et al.*, 1993). (B) Each repeat contains an even number of transmembrane domains (four transmembrane domains are shown). Both repeats are inserted in the membrane with the same orientation. Both NPA boxes are oriented toward the same side of the membrane (Wistow *et al.*, 1991). Each transmembrane domain is indicated with a numbered arrow. The axis of symmetry is indicated with a dotted line. The junction between both repeats is indicated by a broken line.

amino acids per repeat in the MIP family; the first one, 16 amino acids long, is too short for an α helix but would fit a β strand.

When Nod26 was cloned, a model including some β-sheet structure was proposed (Fortin *et al.*, 1987). However, practically all publications that followed describing the new members of the MIP family presented the six-transmembrane-domain model (see Reizer *et al.*, 1993) based on the one proposed for MIP (Gorin *et al.*, 1984). Other inconsistencies of this model were raised by Revel and Yancey (1985). Comparison of protein structure predicted by Kite and Doolittle analysis and X-ray diffraction data of crystallized protein indicated that short β-sheet structures found by crystallographic studies were missed by hydropathy plots in some proteins (Fasman and Gilbert, 1990; Jahnig, 1990).

MIP, CHIP28, and Nod26 are known to be inserted in the membrane with their N and C termini localized to the cytoplasmic side of the membrane (Paul and Goodenough, 1983b; Takemoto *et al.*, 1983; Johnson *et al.*, 1985; Smith and Agre, 1991; Miao *et al.*, 1992). CHIP28 and Nod26 are glycosylated (Denker *et al.*, 1988; Smith and Agre, 1991; Miao *et al.*, 1992; Zhang *et al.*, 1993), indicating that those sites are on the extracellular side of the membrane. Although mutational analysis of CHIP28 indicated that both NPA boxes are important for water transport (Preston *et al.*, 1992b), no information was provided about the orientation of each NPA repeat with respect to the membrane. These biochemical data provide some information about the insertion of the protein molecule in the membrane. However, determining the true structure of the members of the MIP family awaits crystallographic studies of these proteins.

II. MIP Family in Vertebrates

A. MIP

1. Localization

The major intrinsic protein of the ocular lens, MIP, constitutes approximately 60% of the integral membrane protein of the lens and is specifically expressed in the lens fibers (Broekhuyse *et al.*, 1979; Paul and Goodenough, 1983a; Fitzgerald *et al.*, 1985; Alcala and Maisel, 1985). The lens is composed of an anterior layer of epithelial cells that continually differentiate into fiber cells at the equatorial epithelium (Fig. 3; McAvoy, 1980; Maisel *et al.*, 1981; Harding and Crabbe, 1984). Since the differentiation of epithelial cells into fiber cells entails extensive cell elongation with a dramatic increase in the synthesis of new

plasma membrane by elongating cells, proper membrane biosynthesis and physiology are of utmost importance in maintaining the transparent state of the lens (Duncan and Jacob, 1984; Harding and Crabbe, 1984; Kistler and Bullivant, 1989). The large concentration of crystallin proteins associated with the lens fibers, as well as the unique cytoarchitecture of the mature fiber cells, which lack cellular organelles, underlies the ability of the lens to refract and focus incoming light onto the retina (Bloemendal, 1982; Piatigorsky, 1981; Harding and Crabbe, 1984; Bettelheim, 1985). The intercellular space between fibers is extremely narrow, a feature that is essential to maintaining the transparency and correct refractive index of the lens (Duncan and Jacob, 1984; Ehring *et al.*, 1990). The lens, which lacks vasculature and innervation, maintains an extensive network of intercellular transmembrane pathways. Numerous electrophysiological and dye passage studies indicate that lens fiber cells are thoroughly coupled both *in situ* and *in vitro* (Johnson *et al.*, 1988; Mathias and Rae, 1989; Cooper *et al.*, 1991; Lea and Duncan, 1991). MIP immunolocalizes to lens fiber cell membranes, including gap junctional membranes (see Section I,A). Lens fiber gap junctions show a square array of oligomers instead of the hexagonal array observed in classical gap junctions of other tissues (Zampighi *et al.*, 1982; Costello *et al.*, 1989; Kistler and Bullivant, 1989; Benedetti *et al.*, 1990) (see Chapter 26). MIP, arranged in tetragonal arrays of 6–7 nm with two different crystallization patterns, gives a structure of thin and wavy junctions that do not contain coaxially aligned channels. However, thick junctions of 16–17 nm that contain MP70 also contain isolated clusters of MIP inside the plaques of MP70 (Zampighi *et al.*, 1989). Junctions containing MP70 are abundant in the lens cortex but not in the nucleus (Gruitjers *et al.*, 1987; Kistler and Bullivant, 1989; Zampighi *et al.*, 1989). MIP has also been proposed to be involved in close membrane apposition during lens gap junction formation (Costello *et al.*, 1989; Gruitjers, 1989; Zampighi *et al.*, 1989).

MIP gene expression is temporally and spatially regulated during lens development. MIP mRNA and protein are first detectable in the primary lens fibers at the early lens vesicle stage, and are subsequently found in secondary lens fibers as they differentiate from epithelial cells (Waggoner and Maisel, 1978; Yancey *et al.*, 1988; Watanabe *et al.*, 1989). MIP is synthesized by a class of polyribosomes associated with the lens fiber plasma membrane–cytoskeleton complex (Ramaekers *et al.*, 1980). Moreover, *in vitro* translation of MIP mRNA requires the co-translational presence of microsomal membranes and signal recognition particles (SRP) for insertion of the peptide into

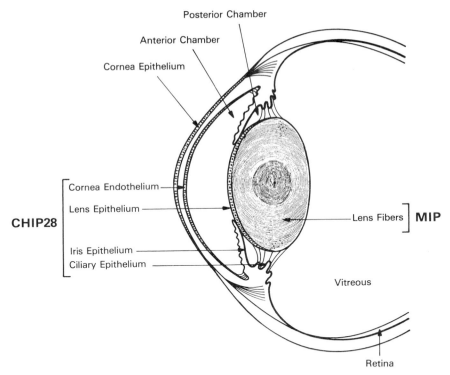

FIGURE 3 Localization of MIP and CHIP28 in the eye. Eye and lens diagram based on Kronfeld (1969) and Paterson and Delamere (1992). CHIP28 localization in cornea endothelium, iris epithelium, ciliary epithelium, and lens epithelium based on Bondy *et al.* (1993), Hasegawa *et al.* (1993), and Nielsen *et al.* (1993b). MIP localization in lens fibers based on Waggoner and Maisel (1978), Broekhuyse *et al.* (1979), Yancey *et al.* (1988), and Watanabe *et al.* (1989).

the membrane, indicating that MIP has at least one unprocessed signal sequence that directs membrane protein integration (Anderson *et al.*, 1983; Paul and Goodenough, 1983b). MIP synthesis is induced as retinal cells transdifferentiate into lens cells in culture, localizing to the membranes of lentoids, a characteristic feature of differentiating lens cells in culture (Moscona *et al.*, 1983; Sas *et al.*, 1985). MIP mRNA levels are drastically decreased in cataracts, in which normal lens fibers are replaced by balloon cells with loss of organized fiber architecture (Beckor, 1988), and in the lens of transgenic mice with perturbed lens fiber differentiation (Limjoco *et al.*, 1991; C. Egwuagu, D. Parker, and A. B. Chepelinsky unpublished data).

2. Selective Proteolysis

The MIP gene encodes a 263-amino-acid protein that has been highly conserved throughout evolution (Fig. 4). MIP undergoes selective proteolysis during cataractogenesis and aging; the proteolytic product (MIP22) is found at higher levels in the nuclear than in the cortical lens fibers (Ngoc *et al.*, 1985; Takemoto *et al.*, 1985,1986,1988; Takemoto and Takehana, 1986; Rintoul and Takemoto, 1991). This posttranslational

cleavage from a 26-kDa to a 22-kDa protein by selective proteolysis of the C-terminal domain appears to regulate the gating of MIP channels (Gooden *et al.*, 1985b; Peracchia and Girsch, 1985b; Ehring *et al.*, 1988; Peracchia, 1989; Girsch, 1991).

3. Phosphorylation

The signal transducers protein kinase A and C have been implicated in the modification of MIP. Phosphorylation of MIP has been implicated as a principal posttranslational determinant of its biological function as a channel protein. Considerable evidence indicates that MIP in bovine lens fiber membranes is a phosphoprotein. MIP was found to be phosphorylated *in vivo* and *in vitro* by both protein kinase A and protein kinase C at serine and, to a lesser extent, at threonine residues. The predominant phosphorylated sites were located approximately 20–40 amino acid residues from the C terminus of the bovine protein, on the cytoplasmic side of the membrane (Garland and Russell, 1985; Johnson *et al.*, 1986; Lampe and Johnson, 1989,1990). Evidence from electrophysiological studies of purified MIP reconstituted into planar lipid bilayers indicates that phosphorylation of MIP, either

```
Human    MWELRSASFWRAIFAEFFATLFYVFFGLGSSLRWAPGPLHVLQVAMAFGLALATL   55
Bovine   - - - - - - - - - - - - C - - - S - - - - - - - - A - - - - - - - - - - - - - L - - - - - - - -
Frog     - - - F - - F - - - - - V - - - - G - M - - - - A - - K - - A - - AN - - V I - L - - - - V - - - M
Chicken  - R - - - - S - - - - - L - - - LGS - L - TLL -   - A - -         - - - -

Human    VQSVGHISGAHVNPAVTFAFLVGSQMSLLRAFCYMAAQLLGAVAGAAVLYSVTPP  110
Bovine   - - A - - - - - - - - - - - - - - - - - - - - - - - - - - - - - - - - I - - V - - - - - - - - - - - - - -
Frog     - - - I - - V - - - - I - - - - - - - - - - - I - - - - F - - I F - I - - - - - - - - - - - G - - - A

Human    AVRGNLALNTLHPAVSVGQATTVEIFLTLQFVLCIFATYDERRNGQLGSVALAVG  165
Bovine   - - - - - - - - - - - - - - G - - - - - - - I - - - - - - - - - - - - - - - - - - R - - - - - - - - -
Rat                                    R F - - - - - - - - - - - - - - - - RM - - - - - - - -
Frog     - I - - - - - - - - - - G - - L - - - - - - - - - - - - - - - - - - - - - R - - - - S - - I -
Chicken                                                          D - HD - RP - - A - - P - -

Human    FSLALGHLFGMYYTGAGMNPARSFAPAILTG|NFT|NHWVYWVGPIIGGGLGSLLYD  220
Bovine   - - - T - - - - - - - - - - - - - - - - - - R - - - - - - - - - - - V - A - - - - - - - -
Rat      - - - T - - - - - - - - - - - - - - - - - - R - - S - - - - - - - - - - - - - - - - - -
Frog     - - - T - - - - - L - - - - - S - - - - - - V - R - - - - - - - - - - - - - - A - G - V - -
Chicken  - - - - - - - - - - IPF - - - - - - - - - VI - R - - - - F - A - - LL - AA - AA - - - E

Human    FLLFPRLK(S)I(S)ERL(S)VLKGAKPDV(S)NGQPEVTGEPVELNTQAL  263
Bovine   - - - - - - - - V - - - - - I - - SR - SE - - - - - - - - - - K - - - -
Rat      - - - - - - - - V - - - - - I - - - R - SD - - - - - G - - - - - - K - - - -
Frog     - I - - - MRGL - - - - - I - - - - R - AEPE - - Q - A - - - - I - - K - - S -
Chicken  LA - C - - AR - MA - - - A - - R - EP - AAAPP - - PPA - - L - - K - - G -
```

FIGURE 4 MIP, from *Rana pipiens* to *Homo sapiens*. Alignment of MIP amino acid sequences of human (Pisano and Chepelinsky, 1991), bovine (Gorin *et al.*, 1984), partial rat (Shiels *et al.*, 1988), partial chicken (Lampe *et al.*, 1989; Kodama *et al.*, 1990), and frog (Austin *et al.*, 1993) homologs. Amino acid identities are noted by a dashed line. A potential glycosylation site at amino acid 197 is boxed. A putative calmodulin binding site at amino acids 225–241 (Peracchia, 1989) is underlined. Potential phosphorylation sites (Garland and Russell, 1985; Pearson *et al.*, 1985; Lampe and Johnson, 1990) at serines 229, 231, 235, and 245 are circled.

in vivo or *in vitro*, with protein kinase A alters channel voltage dependence (see Fig. 5), suggesting that phosphorylation may be a posttranslational determinant of MIP function as a transmembrane channel protein. Ser 243 and 245 may play a role in this regulation (Ehring *et al.*, 1991; Lampe and Johnson, 1990).

4. Calmodulin Binding

Second messengers and signal transducers such as Ca^{2+} and calmodulin have been proposed as modulators of MIP channel-forming ability in lens fiber cells. Bovine and chicken MIP are known to bind calmodulin in a Ca^{2+}-dependent and -independent manner (Welsh *et al.*, 1982; Girsch and Peracchia, 1985a,b; van den Eijnden- van Raaij *et al.*, 1985; Louis *et al.*, 1990). Moreover, a calmodulin-binding site characterized in bovine MIP by Girsch and Peracchia (1991), at amino acids 225–241, is well conserved throughout evolution; the changes in the amino acid sequence are

largely conservative (see Fig. 4). Evolutionary conservation of the calmodulin-binding site in the MIP protein suggests that the signal transducer calmodulin may play a role in the function of MIP as a lens channel protein. Additional support for this proposition is drawn from studies of purified MIP reconstituted into liposomes; the addition of calmodulin to these reconstituted vesicles renders MIP channels sensitive to Ca^{2+}, with a resultant change in channel permeability (Peracchia and Girsch, 1985a,b; Peracchia, 1989). Moreover, the addition of calmodulin to MIP was shown to induce conformational changes in MIP, suggesting that calmodulin may modulate gating of MIP channels in the lens (Girsch and Peracchia, 1985b; Girsch, 1991).

5. Functional Assays

MIP reconstituted into proteoliposomes increases the permeability of molecules of up to 1.5 kDa in

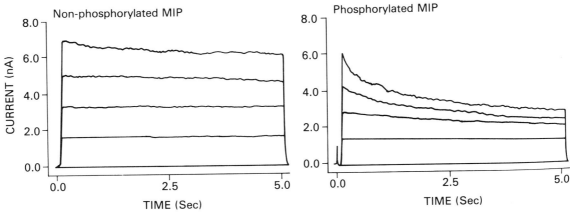

FIGURE 5 Voltage dependence of MIP channels in reconstituted lipid bilayers. (*Left*) Nonphosphorylated MIP. (*Right*) Phosphorylated MIP. *In vivo* phosphorylation or *in vitro* phosphorylation with protein kinase A shows similar patterns. Reprinted with permission from Ehring *et al.* (1991).

molecular mass, indicating a channel diameter of approximately 1.4 nm. Antibodies against MIP block this channel activity (Girsch and Peracchia, 1985a; Gooden *et al.*, 1985a; Nikaido and Rosenberg, 1985; Peracchia and Girsch, 1985a; Scaglione and Rintoul, 1989). Tetragonal structures are observed in reconstituted MIP proteoliposome vesicles that are similar to the arrays observed in lens membranes *in vivo* (Zampighi *et al.*, 1982; Aerts *et al.*, 1990; Benedetti *et al.*, 1990; Ehring *et al.*, 1990). The permeability and gating of these channels is regulated by Ca^{2+} only in the presence of calmodulin; this regulation disappears in the absence of a C-terminal amino acid domain present in MIP26 (Girsch and Peracchia, 1985a; Peracchia and Girsch, 1985a,b).

MIP reconstituted into single planar bilayer membranes forms large, slightly anion-selective channels. The voltage dependence of these channels is modulated by phosphorylation with cAMP-dependent protein kinase. Dephosphorylated MIP forms almost voltage-independent channels whereas phosphorylated MIP forms voltage-dependent channels (Ehring *et al.*, 1991). Proteolytic removal of the C terminus mimics the effects of phosphorylation, suggesting that the C terminus stabilizes the channel in the closed state (Ehring *et al.*, 1991,1993). In the presence of calmodulin, the gating of these MIP channels becomes Ca^{2+} dependent (Ehring and Hall, 1991). Interestingly, a putative calmodulin-dependent protein kinase phosphorylation site is present at MIP Ser 229 (Pearson *et al.*, 1985); several protein kinase C phosphorylation sites are present in the calmodulin-binding domain characterized by Girsch and Peracchia (1991; see Fig. 4). Ser 243 and 245 are phosphorylated by protein kinase A in bovine MIP (Lampe and Johnson, 1990). Peptide phosphorylation by protein kinase C has been shown to disrupt its

calmodulin-binding ability (Hinrichsen and Blackshear, 1993). Therefore, the MIP C-terminal domain may regulate gating by phosphorylation and calmodulin binding. Further, the presence of MIP molecules with shorter C termini (MIP22) in the lens nucleus may result in channels with different regulatory properties in the lens cortex and nucleus. MIP spatial localization in the ocular lens and its properties as a channel suggest its possible role in the maintenance of lens transparency by reducing the interfiber space (Zampighi *et al.*, 1989; Ehring *et al.*, 1990,1993).

B. CHIP28

The CHIP28 protein was isolated from human erythrocyte membranes by Agre and co-workers while searching for an Rh breakdown product. The cDNA was isolated from a human bone marrow cDNA library with oligonucleotides corresponding to the N-terminal CHIP28 sequence (Denker *et al.*, 1988; Preston and Agre, 1991; Smith and Agre, 1991).

Immunohistochemical studies indicated that CHIP28 is also present in kidney. Kidney CHIP28 cDNA was obtained by RT-PCR of rat kidney mRNA with oligonucleotides corresponding to the human erythrocyte CHIP28 sequence and subsequent screening of a rat kidney cDNA library (Deen *et al.*, 1992; Zhang *et al.*, 1993). CHIP28 is expressed in rat kidney cortex and medulla. In the cortex, this protein immunolocalizes to the microvilli of the apical brush border and to basolateral membranes of the convoluted and straight proximal tubules. In the medulla, the protein immunolocalizes to the apical and basolateral membranes of descending thin limbs (Fig. 6; Deen *et al.*, 1992; Sabolic *et al.*, 1992; Agre *et al.*, 1993a; Hasegawa *et al.*, 1993; Nielsen *et al.*, 1993a; Zhang *et al.*, 1993). In these two regions of the nephron, the water

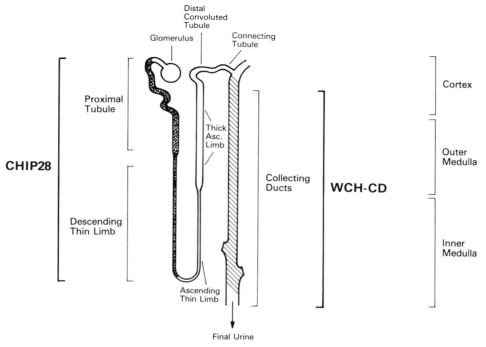

FIGURE 6 Localization of CHIP28 and WCH-CD in the kidney. Dotted area: CHIP28 localization in proximal tubule and descending thin limb (Sabolic *et al.*, 1992; Hasegawa *et al.*, 1993; Nielsen *et al.*, 1993a; Zhang *et al.*, 1993). Hatched area: WCH-CD localization in collecting ducts (Fushimi *et al.*, 1993; Nielsen *et al.*, 1993c) Diagram modified from Nielsen *et al.*, *The Journal of Cell Biology*, **120,** 371–383, by copyright permission of the Rockefeller University Press.

channel genes are expressed constitutively (Harris *et al.*, 1991; Verma, 1992; Nielsen *et al.*, 1993a). In the collecting duct epithelia, where the water channels are regulated by vasopressin (antidiuretic hormone, ADH), Fushimi *et al.*, (1993) found that the gene that is expressed is WCH-CD (Fig. 6), not CHIP28. The exquisitely precise regulation of genes encoding water channels is achieved by the expression of different genes in different cells of the same tissue that are regulated differently.

CHIP28 is expressed in the cornea endothelia, iris epithelia, ciliary epithelia, and lens epithelia of the eye (see Fig. 3; Nielsen *et al.*, 1993b; Hasegawa *et al.*, 1993, 1994; Stamer *et al.*, 1994). CHIP28 has also been found in the trabecular meshwork and Schlemm's canal endothelia of the human eye (Stamer *et al.*, 1994). The localization of MIP in the lens fibers and CHIP28 in the lens epithelia may play a role in maintaining lens transparency, whereas the localization of CHIP28 in the cornea endothelia may play a role in maintaining corneal transparency. Cornea endothelial cells in culture retain the ability to synthesize CHIP28 and to express a CHIP28-like channel (Echevarria *et al.*, 1993b; D. Parker, J. Fischbarg, and A. B. Chepelinsky unpublished data). The expression of CHIP28 in cell

monolayers that surround the anterior and posterior chamber of the eye may be involved in regulating the osmolarity of the aqueous humor.

Unexpectedly, the mouse homolog of the CHIP28 gene was characterized when a delayed early gene was cloned (DER2) and was found to be identical to CHIP28 (Lanahan *et al.*, 1992). CHIP28 is expressed during active cell division, either by stimulation of quiescent fibroblasts with serum, fibroblast growth factor (FGF), or platelet-derived growth factor (PDGF) or in regenerating liver, where active mitosis is occuring; this protein may play an important role in the transition from G_1 to S during the cell cycle (Lau and Nathans, 1991; Lanahan *et al.*, 1992).

CHIP28, a 269-amino-acid protein, is glycosylated *in vivo* and *in vitro* and forms tetramers (Smith and Agre, 1991; Zhang *et al.*, 1993). The CHIP28 amino acid sequence is highly conserved in human, rat, and mouse (see Fig. 7), with 93% amino acid identity between human and mouse or rat and 98% identity between mouse and rat.

CHIP28 behaves as a water channel in functional assays. CHIP28 mRNA is able to induce the appearance of water channels when injected into *Xenopus* oocytes (Preston *et al.*, 1992a). Purified CHIP28 from

Human	M A S E F K K K L F W R A V V A E F L A T T L F V F I S I G S A L G F K Y P V G N N Q T A V Q D N V K V S L A	55
Rat	- - - - I - - - - - - - - - - - - - - - M - - - - - - - - - - - - - - - N - - L E R - - - L - - - - - - - - -	
Mouse	- - - - I - - - - - - - - - - - - - - - M - - - - - - - - - - - - - - - N - - L E R - - - L - - - - - - - - -	

Human	F G L S I A T L A Q S V G H I S G A H L N P A V T L G L L L S C Q I S I F R A L M Y I I A Q C V G A I V A T A	110
Rat	- L - - V - - - - - - - - - - - - - S -	
Mouse	- L - - V - - - - - - - - - - - - - - -	

Human	I L S G I T S S L T G N S L G R N D L A D G V N S G Q G L G I E I I G T L Q L V L C V L A T T D R R R R D L G	165
Rat	- - - - - - - - L E - - - - - - - - R -	
Mouse	- - - - - - - - V D - - - - - - - - H -	

Human	G S A P L A I G L S V A L G H L L A I D Y T G C G I N P A R S F G S A V I T H N F S N H W I F W V G P F I G G	220
Rat	- L - R - - - - - - - - - - - - - S	
Mouse	- L - R - - - - - - - - - - - - - -	

Human	A L A V L I Y D F I L A P R S S D L T D R V K V W T S G Q V E E Y D L D A D D I N S R V E M K P K	269
Rat	- - - - - - - - - - - - - - - - - F - - - M -	
Mouse	- - - - - - - - - - - - - - - - - F - - - M -	

FIGURE 7 Conservation of CHIP28 amino acid sequence. Alignment of human CHIP28 amino acid sequence (Preston and Agre, 1991) with rat (Deen *et al.*, 1992) and mouse (Lanahan *et al.*, 1992) homologs. Amino acid identities are noted by a dashed line. Glycosylation sites are boxed (Preston and Agre, 1991; Preston *et al*, 1992b; Zhang *et al.*, 1993). Cysteine 189 involved in Hg-sensitive water transport is circled (Preston *et al.*, 1993).

red blood cells reconstituted into proteoliposomes results in functional water channels. These channels are water selective and are inhibited by mercury ions; they do not transport urea (Zeidel *et al.*, 1992; van Hoek and Verkman, 1992). Point mutations in the four cysteines potentially responsible for the mercury inhibition of the water channel activity of CHIP28 indicated that only Cys 189 is involved in the water channel activity in *Xenopus* oocytes (Preston *et al.*, 1993). Interestingly, Cys 189 is located two amino acids from the second NPA box repeat (see Table I, Fig. 7).

Although CHIP28 appears to have a broad distribution, its expression is targeted to specific cell types that need its function as a water channel (see Fig. 3,6). This protein is expressed in renal proximal tubules and descending thin limbs, red pulp of the spleen, and membranes of circulating red cells. CHIP28 is also found in several other tissues such as heart, lung, spleen, and prostate (Deen *et al.*, 1992; Agre *et al.*, 1993a; Hasegawa *et al.*, 1994). CHIP28 is expressed in the brain only in the choroid plexus epithelium. CHIP28 functions as a water channel and is involved in water reabsorption in the proximal nephron, in erythrocyte osmoregulation, and in water flow during secretion of cerebrospinal fluid by the choroid plexus (Bondy *et al.*, 1993; Nielsen *et al.*, 1993a,b; Smith *et al.*, 1993). The localization of CHIP28 to secretory and resorptive epithelia and capillary endothelia suggests that it plays a major role in transcellular water movement in a wide range of tissues (Agre *et al.*, 1993a; Nielsen *et al.*, 1993b).

C. WCH-CD

Physiological characterization of water channels indicated that specific channels were responsible for water transport in erythrocytes, kidney proximal tubules, and collecting ducts (van Hoek *et al.*, 1990,1991,1992; Zhang *et al.*, 1990,1991; Harris *et al.*, 1991; Dempster *et al.*, 1992; Verkman, 1992; Echevarria *et al.*, 1993a). The isolation and characterization of CHIP28 as one of those water channels provided insights used in the isolation of a second one, WCH-CD. To isolate WCH-CD, RT-PCR was performed with rat kidney medulla mRNA using oligonucleotides corresponding to the NPA box repeat of the already characterized members of the MIP family (see Section I,B); the RT-PCR product was used to screen a rat kidney cDNA library (Fushimi *et al.*, 1993).

Figure 6 shows the localization of CHIP28 and WCH-CD in the kidney. Although both are expressed in kidney and can be found in the kidney cortex and medulla, they have different localization. CHIP28 is localized in the apical membrane of the proximal tubule and the descending thin limb; no WCH-CD is observed in these cells. However, in the collecting tubule, only WCH-CD is expressed; no CHIP28 expression is observed (Fushimi *et al.*, 1993; Nielsen *et al.*, 1993a,c); whereas CHIP28 (also called aquaporin-1) is constitutively expressed in the kidney, WCH-CD (also called aquaporin-2) is induced by vasopressin. Water deprivation results in a drastic increase in rat kidney WCH-CD mRNA levels, whereas CHIP28 mRNA levels are not affected (Sasaki *et al.*, 1992). WCH-CD mRNA injected into *Xenopus* oocytes induces water channels that are sensitive to Hg^{2+} and similar to the ones induced by CHIP28 (Fushimi *et al.*, 1993).

Figure 8 shows the comparison of MIP, WCH-CD, and CHIP28 amino acid sequences. WCH-CD is 43% identical to CHIP28 and 59% identical to MIP; CHIP28 is 43% identical to MIP (Preston and Agre, 1991; Fushimi *et al.*, 1993). WCH-CD is evolutionarily closer to MIP than CHIP28 (de Jong *et al.*, 1994). Interestingly, MIP and WCH-CD localize to the same chromosome, whereas CHIP28 localizes to a different one (Sparkes *et al.*, 1986; Moon *et al.*, 1993; Deen *et al.*, 1994b,c; Saito *et al.*, 1994; Sasaki *et al.*, 1994). MIP has greater identity to WCH-CD in some regions (mainly in the first repeat) and greater identity to CHIP28 in others (in the second repeat). The first six amino acids are identical in WCH-CD and MIP. CHIP28 has an insert of seven amino acids at position 33 that is found in neither MIP nor WCH-CD. MIP presents a greater degree of similarity to WCH-CD than to CHIP28 in this region. However, Cys 189, located two amino acids upstream from the second NPA box and essential for water transport activity, is conserved in CHIP28 and WCH-CD (Preston *et al.*, 1993; Fushimi *et al.*, 1993). This cysteine residue is not present in MIP nor in other members of the family (see Table I). WCH-CD Arg 187, located next to the second NPA box and highly conserved within MIP family (see Table I) and Ser 216, conserved in MIP and WCH-CD, are required for vasopresin-dependent concentration of urine and for water channel activity in functional assays (Deen *et al.*, 1994b).

A glycosylation site is CHIP28, at Asn 42, is not present in WCH-CD or in MIP. The second potential glycosylation site in CHIP28, at Asn 205, is conserved in MIP but is absent from WCH-CD. The only putative glycosylation site in WCH-CD, at Asn 124, is absent from CHIP28 and MIP. Near the C-terminal end, the tripeptide Val–Glu–Leu is conserved in MIP and WCH-CD. WCH-CD contains several protein kinase A and putative phosphorylation sites near its C terminus (Fushimi *et al.*, 1993), in approximately the same region as in MIP.

FIGURE 8 Alignment of human CHIP28 (Preston and Agre, 1991), rat WCH-CD (Fushimi *et al.*, 1993), human EST2306 (Adams *et al.*, 1992), and *Drosophila bib* (Rao *et al.*, 1990) with human MIP (Pisano and Chepelinsky, 1991). Amino acid identities are indicated by a dashed line. Boxed amino acids indicate the conserved repeats in every member of the MIP family (see Fig. 1).

The similarity between MIP, CHIP28, and WCH-CD is particularly noteworthy in view of the proposed function of MIP in the regulation of volume fluxes between the intercellular space and the cytoplasm (Ehring *et al.*, 1990,1993; Zampighi *et al.*, 1992). An analogous function may exist in the erythrocyte, in which rapid changes in cell shape and volume are required in response to physical constraints imposed by vessels of varying diameters (Smith and Agre, 1991). CHIP28 and WCH-CD function as water channels when their mRNAs are injected into *Xenopus* oocytes. However, when MIP mRNA is injected into *Xenopus* oocytes, functional water channels are not induced (Rice *et al.*, 1993; C. H. van Os, personal communication). Although these experiments suggest that MIP might not be involved in water transport, additional experiments are needed to further characterize the channels formed by MIP.

D. EST02306

EST (expressed sequence tag) 02306 is one of the 2375 cDNAs clones identified from human cDNA libraries (Adams *et al.*, 1992). This partial cDNA sequence (Gene Bank Accession No. M85785) was obtained from a human 17- to 18-week embryo, and has been described as the human homolog of *Drosophila bib* (*big brain*). However, amino acid sequence comparison with available human sequences of the MIP family (see Fig. 8) suggests that EST02306 is a distinct member of the family. Since *bib* and EST02306 are both expressed during embryonic development, Adams *et al.*, (1992) classified this gene as *bib*. However, CHIP28 is also expressed in brain (Bondy *et al.*, 1993; Nielsen *et al.*, 1993b) and is highly conserved throughout evolution (see Fig. 7). At this stage, describing EST02306 as the homolog of *bib* or of any other member of the MIP family is premature; this gene may be a new member of the MIP family.

III. MIP Family in *Drosophila*

The *big brain* gene of *Drosophila*, *bib*, functions to control the number of neuronal precursors in the peripheral nervous system. *bib* was identified as a transcription unit that, when transformed in mutants, rescued the neurogenic phenotype and lethality of *bib* mutants. *bib* begins to be expressed after the cell membrane begins to form around the nuclei in the syncytial blastoderm, and is detectable in all the somatic cells that form in an early stage-5 embryo. *bib* is expressed in ectoderm while the cell membrane is adjacent to its original neighbors. *bib* expression disappears from the neuroblasts after they have segregated from the epidermal layer. Later the mesoderm begins to express *bib*, but the gene product disappears after germ-band shortening (Rao *et al.*, 1990,1992).

The *bib* cDNA encodes a 700 amino acid protein and its N-terminal half shows 40% identity to MIP (Rao *et al.*, 1990). Three initiation codons are possible; if the third AUG is used to initiate translation, the *bib* N terminus is closer in length to that of other members of the family than previously predicted (see Fig. 8). However, the C terminus is much longer than and different in amino acid composition from the other members of the family. The functional role of *bib* during development remains to be established.

IV. MIP Family in Plants

A. Nodulin 26

Nodulin genes are activated during the formation of nitrogen-fixing root nodules on leguminous plants, after infection with the symbiotic rhizobia bacteria. The peribacteroid membrane, which separates the bacteroids from the plant cytoplasm, is involved in the transport of ammonia from the bacteroids to the plant cytoplasm and of nutrients from the cytoplasm to the bacteroids (Verma and Delauney, 1988; Nap and Bisseling, 1990; Brewin, 1991; Verma, 1992). Nod26, also called nodulin 26, is the major intrinsic protein of the peribacteroid membrane in soybean root nodules. This protein is glycosylated, and its N and C termini are located on the host cytoplasmic side of the membrane (Miao *et al.*, 1992).

Nod26 cDNA was isolated by screening a nodule-specific cDNA library, devoid of known nodulin sequences, with a probe obtained from *B. japonicum*-induced nodule polyribosomes precipitated by peribacteroid membrane antisera (Fortin *et al.*, 1987). Nod26 is a 271-amino-acid protein, 35% identical to MIP and approximately 32% identical to other members of the MIP family expressed in plants (Sandal and Marcker, 1988,1990; Pao *et al.*, 1991; Hofte *et al.*, 1992). Purified Nod26 incorporated into liposomes forms voltage sensitive channels in lipid bilayers, similar to those formed by MIP (Shomer *et al.*, 1994).

Nodulin 26 is phosphorylated by a calcium-dependent, calmodulin-independent protein kinase located in the peribacteroid membrane (Weaver *et al.*, 1991; Miao *et al.*, 1992; Weaver and Roberts, 1992). Phosphorylation/dephosphorylation of Nod26 produces significant changes in malate uptake by isolated symbiosomes. Nod26 phosphorylation, in the host cytoplasmic domain at Ser 262, actives malate transport whereas dephosphorylation reversibly inhibits this transport, suggesting that Nod26 behaves as a channel regulated by Ca^{2+} and phosphorylation, and may have important implications for the regulation of symbiotic nitrogen fixation (Ouyang *et al.*, 1991; Weaver and Roberts, 1992).

B. TobRB7

TobRB7 was identified as a root-specific gene because its cDNA was isolated from a *Nicotiana tabacum* root cDNA library by differential screening with root and leaf probes (Conkling *et al.*, 1990; Yamamoto *et al.*, 1990,1991). TobRB7 is specifically expressed in the root vascular system, which is important for absorption and transport of water and nutrients from the root to the aerial parts of the plant. TobRB7 begins to be expressed during development in the root meristem and immature central cylinder regions (Yamamoto *et al.*, 1991). Several *cis*-regulatory elements responsible for the root-specific expression of the TobRB7 gene have been identified in the genomic 5′ flanking sequence by functional assays in transgenic plants (Conkling *et al.*, 1990; Yamamoto *et al.*, 1991). TobRB7 mRNA injected into *Xenopus* oocytes induces water channels similar to those observed with CHIP28 and may also transport other small molecules (M. A. Conkling, personal communication). TobRB7 is a 250-amino-acid protein, 65 and 52% identical to γTIP and αTIP, respectively, two plant tonoplast proteins that are also members of the MIP family (Hofte *et al.*, 1992).

C. AtRB7

AtRB7 was isolated from an *Arabidopsis* root cDNA library screened with a TobRB7 cDNA probe under low stringency conditions (Yamamoto *et al.*, 1990) and is specifically expressed in the *Arabidopsis* root. AtRB7 cDNA encodes a 253-amino-acid protein that is 61% identical to TobRB7. This percentage of identity with TobRB7, the absence of arginine in the second NPA box repeat (see Table I), and different C- and N-terminal sequences (Yamamoto *et al.*, 1990) suggest that AtRB7 is a different member of the NPA gene family, rather than the *Arabidopsis* homolog of TobRB7. AtRB7 has a high degree of identity with *Arabidopsis* γTIP. However, these gene products differ in tissue localization and C-terminal amino acid sequence, suggesting that they are different members of the MIP family (Hofte *et al.*, 1992).

D. Pea R7A

Pea R7A, the product of a turgor-responsive gene, is abundant in roots and is induced in shoots by dehydration and heat shock (Guerrero *et al.*, 1990). Pea R7a cDNA encodes a 289-amino-acid protein that is 69%

identical to RD28, 37% identical to MIP and αTIP, 34% identical to γTIP, 28% identical to Nod26, and 26% identical to *bib* and GlpF amino acid sequences (Pao *et al.*, 1991; Hofte *et al.*, 1992; Yamaguchi-Shinozaki *et al.*, 1992). The regulation of this gene by water deprivation suggests that this member of the MIP family may also be involved in water transport.

E. RD28

RD28 was isolated by differential screening of a cDNA library prepared from desiccated *Arabidopsis thaliana* plants with probes prepared from hydrated and desiccated plants (Yamaguchi-Shinozaki *et al.*, 1992). RD28 cDNA encodes a 285-amino-acid protein that belongs to the MIP family, and is closely related (69% identity) to Pea7A, which is also induced by water deprivation (Hofte *et al.*, 1992).

F. Tonoplast Intrinsic Proteins

In plant cells, the cytoplasm occupies a thin layer between the plasma membrane and the vacuole. The tonoplast is the membrane that separates the vacuole from the cytoplasm. Several tonoplast intrinsic proteins (TIPs) have been characterized and found to be members of the MIP family (Johnson *et al.*, 1990; Maurel *et al.*, 1993; Chrispeels and Maurel, 1994).

αTIP is a seed-specific protein that immunolocalizes to the tonoplast of the protein-storage vacuoles of the embryo and endosperm (Johnson *et al.*, 1989). αTIP has been isolated from soybean and *Arabidopsis thaliana* seeds; 68% amino acid identity exists between both species. αTIP is a phosphoprotein phosphorylated in the N terminus at Ser 7 by a Ca^{2+}-dependent protein kinase (Hofte and Chrispeels, 1992; Hofte *et al.*, 1992; Johnson and Chrispeels, 1992; Reizer *et al.*, 1993; Chrispeels and Maurel, 1994).

γTIP is expressed in the tonoplast of vegetative cells of *A. thaliana*. *In situ* hybridization studies showed that γTIP mRNA was present only in the zone of cell elongation just behind the apical meristem. In seedling shoots, mRNA abundance was also found to be correlated with cell expansion. Functional assays in transgenic plants indicated that the 5′ flanking sequence of the γTIP gene contains *cis*-regulatory elements that are responsible for its expression in differentiating cells (Hofte *et al.*, 1992; Ludevid *et al.*, 1992). γTIP mRNA injected into *Xenopus* oocytes induced formation of water-specific channels (Maurel *et al.*, 1993). The presence of water channels in the tonoplast may allow the plant cells to use the vacuolar space for rapid osmoregulation of the cytoplasm (Chrispeels and Maurel, 1994).

V. MIP Family in Bacteria

The *Escherichia coli* glycerol facilitator (*glpF*) is part of the *glp* regulon, which is inducible by glycerol-3-phosphate and provides a strong growth advantage on glycerol at low external concentrations. This gene is responsible for the facilitated diffusion of glycerol across the *E. coli* inner membrane. GlpF forms a channel that is permeable to neutral molecules smaller than 0.4 nm in diameter, and allows the entry of glycerol and other polyols (pentitols and hexitols), urea, glycine, and D,L-glyceraldehyde (Heller *et al.*, 1980). The glycerol facilitator is the only known example of a plasma membrane transporter in the inner membrane that is believed to catalyze facilitated diffusion in *E. coli* (Nikaido and Saier, 1992).

The glycerol facilitator gene (*glpF*) is part of the *glp* regulon and is the first gene of the *glpFKX* operon (Maramatsu and Mizuno, 1989; Sweet *et al.*, 1990; Truniger *et al.*, 1992; Weissenborn *et al.*, 1992). *glpF* encodes a 281-amino-acid protein that belongs to the MIP family (Baker and Saier, 1990; Sandal and Marcker, 1990; Pao *et al.*, 1991; Reizer *et al.*, 1993). Interestingly, this protein contains two inserts, 14 and 15 amino acids long, that coincide with the junction of exons 1 and 2 and exons 3 and 4, respectively, of the MIP gene (see Fig. 1; Yamamoto *et al.*, 1990; Wistow *et al.*, 1991). The cloned *glpF* gene complements a chromosomal glycerol transport-minus mutation (Sweet *et al.*, 1990). The GlpF protein interacts with glycerol kinase, encoded by the second gene of the operon (Voegele *et al.*, 1993). *glpF* mRNA injected into *Xenopus* oocytes facilitates glycerol uptake without affecting osmotic water permeability (Maurel *et al.*, 1993). The glycerol facilitator from *Streptomyces coelicolor* has been partially characterized and also belongs to the MIP family (Smith and Chater, 1988; Sandal and Marcker, 1990). *Lactococcus lactis* X-PDAP (X-prolyl dipeptidyl aminopeptidase) may be a distant member of the MIP family. It shows similarity to members of this family only in one stretch of amino acids (Mayo *et al.*, 1991).

VI. MIP Family in Yeast

FPS1 was isolated from *Saccharomyces cerevisiae* as a suppressor of the growth defect of the yeast *fdp1* mutant on fermentable sugars. The FPS1 gene encodes a 669-amino-acid protein that shows similarity to members of the MIP family in an internal region, although the NPA boxes are not conserved (Van Aelst *et al.*, 1991).

VII. Summary

The MIP family is an ancient transmembrane channel family that arose by gene duplication of an ancestral gene. The members of this growing family, from such evolutionarily diverse organisms as bacteria and mammals, are specialized for expression of membrane proteins that appear to function as selective channels to regulate osmolarity. γTIP and CHIP28 function as water channels but not as glycerol channels; GlpF functions as a glycerol channel but not as a water channel; αTIP does not function as a water channel; CHIP28 is able to transport water but not urea; MIP does not appear to function as a water channel. CHIP28, WCH-CD, and γTIP have been renamed aquaporins to reflect their specialized function (Agre *et al.*, 1993a,b; Nielsen *et al.*, 1993b; Moon *et al.*, 1993; Smith *et al.*, 1993; Hoffman, 1993; Chrispeels and Maurel, 1994; Deen *et al.*, 1994a,b; Sasaki *et al.*, 1994).

MIP, Nod26, and αTIP are phosphorylated proteins, and phosphorylation regulates the properties of the MIP and Nod26 channels. Since phosphorylation is known to regulate the gating of other channels (West *et al.*, 1991; Busch *et al.*, 1992; Saez *et al.*, 1993), this mechanism may also be a common property of members of the MIP family.

Calmodulin binding modifies the Ca^{2+} dependency of MIP channels. Calmodulin is involved in the gating of other channels (Hsu and Molday, 1993; Hinrichsen and Blackshear, 1993) (see Chapter 29). Calmodulin and Ca^{2+} regulate cell cycle progression (Lu and Means, 1993). Since CHIP28 is also regulated during cell proliferation, calmodulin may have a role in the regulation of the MIP family channels.

The C-terminal domain play an important role in the gating of the MIP channels. Since the C terminus is the most variable region in the members of the MIP family, this domain may play a significant role in the specific properties of each member of the MIP family.

The members of the MIP family are so specialized that different cells of the same tissue express different members of the family. In the kidney, two members of the family are expressed in different cell types, probably to attain different regulation. CHIP28, which is constitutively expressed, is not regulated by vasopressin whereas WCH-CD, which plays an important role in urine concentration during water deprivation, is regulated by vasopressin. In the eye, MIP is expressed in lens fibers, where it may play a role in maintaining the narrow interfiber space. However, CHIP28 expressed in lens epithelia, cornea endothelia, iris epithelia, ciliary epithelia, all of which are cell monolayers that surround the anterior and posterior chamber of the eye, may be involved in regulating the osmolarity of the aqueous humor. The localization of CHIP28 in secretory and resorptive epithelia and capillary endothelia indicates that it plays a major role in transcellular water movement throughout the body. Even when γTIP and αTIP are both localized to the vacuole membrane, they are expressed in different plant tissues with different functions.

The members of the MIP family are expressed when cell volume and shape change during cell differentiation, elongation, or growth. MIP begins to be expressed in the lens when the epithelial cells begin elongating to form the lens fibers. γTIP is expressed in plants during cell elongation, and its presence in the tonoplast regulates cytoplasm volume. CHIP28, expressed during active cell division, may be important in the cell volume changes that accompany mitosis.

How information from physiological, developmental, molecular biological, and evolutionary studies converged to provide new insights into old questions is fascinating. The amazingly rapid growth of the MIP family has provided insights into the structure and function of the members already isolated, as well as clues to searching for new members. The molecular characterization of water channels gives new dimensions to this field, and has opened new approaches to understanding the mechanisms involved in channel regulation at the molecular level.

Acknowledgments

I am grateful to Peter Agre, Martin Chrispeels, Mark Conkling, George Ehring, Jorge Fischbarg, Rick Mathias, Soren Nielsen, Carel H. van Os, Arlyn Garcia Perez, Sei Sasaki, and Daniel Stamer for reprints, preprints, unpublished information, and valuable discussions, and to Graeme Wistow for stimulating discussions, insightful comments, and critical reading of the manuscript.

References

Adams, M. D., Dubnick, M., Kerlavage, A. R., Moreno, R., Kelley, J. M., Utterback, T. R., Nagle, J. W., Fields, C., and Venter, J. C. (1992). Sequence identification of 2,375 human brain genes. *Nature (London)* **355**, 632–634.

Aerts, T., Xia, J.-Z., Slegers, H., de Block, J., and Clauwaert, J. (1990). Hydrodynamic characterization of the major intrinsic protein from the bovine lens fiber membranes. Extraction in *n*-octyl-β-D-glucopyranoside and evidence for a tetrameric structure. *J. Biol. Chem.* **265**, 8675–8680.

Agre, P., Preston, G. M., Smith, B. L., Jung, J. S., Raina, S., Moon, C., Guggino, W. B., and Nielsen, S. (1993a). Aquaporin CHIP: The archetypal molecular water channel. *Am. J. Physiol.* **265**, F463–F476.

Agre, P., Sasaki, S., and Chrispeels, M. J. (1993b). Aquaporins: A family of water channel proteins. *Am. J. Physiol.* **265**, F461.

Alcala, J., and Maisel, H. (1985). Biochemistry of lens plasma membranes and cytoskeleton. *In* "The Ocular Lens: Structure, Function, and Pathology" (H. Maisel, ed.), pp. 169–222. Marcel Dekker, New York.

Anderson, D. J., Mostov, K. E., and Blobel, G. (1983). Mechanisms of integration of *de novo*-synthesized polypeptides into membranes: Signal-recognition particle is required for integration into microsomal membranes of calcium ATPase and of lens MP26 but not of cytochrome b_5. *Proc. Natl. Acad. Sci. USA* **80**, 7249–7253.

Austin, L. R., Rice, S. J., Baldo, G. J., Lange, A. J., Haspel, H. C., and Mathias, R. T. (1993). The cDNA sequence encoding the lens major intrinsic protein of frog lens. *Gene* **124**, 303–304.

Baker, M. E., and Saier, M. H. (1990). A common ancestor for bovine lens fiber major intrinsic protein, soybean nodulin-26 protein, and *E. coli* glycerol facilitator. *Cell* **60**, 185–186.

Bekhor, I. (1988). MP26 messenger RNA sequences in normal and cataractous lens. A molecular probe for abundance and distribution of a fiber cell-specific gene product. *Invest. Ophthamol. Vis. Sci.* **29**, 802–813.

Benedetti, E. L., Dunia, I., Manenti, S., and Bloemendal, H. (1990). Biochemical and structural properties of the protein constituent of junctional domains in eye lens fiber plasma membranes. *In* "Parallels in Cell to Cell Junctions in Plants and Animals" (A. W. Robards, W. J. Lucas, J. D. Pitts, H. J. Jongsma, and D. C. Spray, eds.), pp. 35–52. Springer-Verlag, Berlin.

Bettelheim, F. A. (1985). Physical basis of lens transparency. *In* "The Ocular Lens. Structure, Function, and Pathology" (H. Maisel, ed), pp. 265–300. Marcel Dekker, New York.

Beyer, E. C., Paul, D. L., Goodenough, D. A. (1987). Connexin43: A protein from rat heart homologous to a gap junction protein from liver. *J. Cell Biol.* **105**, 2621–2629.

Beyer, E. C., Goodenough, D. A, and Paul, D. L. (1988). The connexins, a family of related gap junction proteins. *In* "Gap Junctions" (E. L. Hertzberg and R. G. Johnson, eds.), pp. 167–175. Liss, New York.

Beyer, E. C., Kistler, J., Paul, D. L., and Goodenough, D. A. (1989). Antisera directed against connexin43 peptides react with a 43-kD protein localized to gap junctions in myocardium and other tissues. *J. Cell Biol.* **108**, 595–605.

Beyer, E. C., Paul, D. L., and Goodenough, D. A. (1990). Connexin family of gap junction proteins. *J. Membr. Biol.* **116**, 187–194.

Bloemendal, H. (1982). Lens proteins. *CRC Crit. Rev. Biochem.* **12**, 1–38.

Bondy, C., Chin, E., Smith, B. L., Preston, G. M., and Agre, P. (1993). Developmental gene expression and tissue distribution of the CHIP28 water channel protein. *Proc. Natl. Acad. Sci. USA* **90**, 4500–4504.

Brewer, G. J. (1991). Reconstitution of lens channels between two membranes. *In* "Biophysics of Gap Junction Channels" (C. Peracchia, ed.), pp. 301–316. CRC Press, Boca Raton, Florida.

Brewin, N. J. (1991). Development of the legume root nodule. *Annu. Rev. Cell Biol.* **7**, 191–226.

Broekhuyse, R. M., Kuhlmann, E. D., and Stols, A. L. (1976). Lens membranes II. Isolation and characterization of the main intrinsic polypeptide (MIP) of bovine lens fiber membranes. *Exp. Eye Res.* **23**, 365–371.

Broekhuyse, R. M., Kuhlmann, E. D., and Winkens, H. J. (1979). Lens membranes. VII. MIP is an immunologically specific component of lens fibers membranes and is identical with the 26K band protein. *Exp. Eye Res.* **29**, 303–313.

Busch, A. E., Varnum, M. D., North, R. A., and Adelman, J. P. (1992). An amino acid mutation in a potassium channel that prevents inhibition by protein kinase C. *Science* **255**, 1705–1707.

Chrispeels, M. J., and Maurel, C. (1994). Aquaporins: The molecular basis of facilitated water movement between plant cells. *Plant Physiol.* (in press).

Conkling, M. A., Cheng, C-L., Yamamoto, Y. T., and Goodman, H. M. (1990). Isolation of transcriptionally regulated root-specific genes from tobacco. *Plant Physiol.* **93,** 1203–1211.

Cooper, K., Mathias, R. T., and Rae, J. L. (1991). The physiology of lens junctions. *In* "Biophysics of Gap Junction Channels" (C. Peracchia, ed.), pp. 57–74. CRC Press, Boca Raton, Florida.

Costello, M. J., McIntosh, T. J., and Robertson, J. D. (1989). Distribution of gap junctions and square array in the mammalian lens. *Invest. Ophthalmol. Vis. Sci.* **30,** 975–989.

Cowan, S. W., Schirmer, T., Rummel, G., Steiert, M., Ghosh, R., Paupitit, R. A., Jansonius, J. N., and Rosenbusch, J. P. (1992). Crystal structures explain functional properties of two *E. coli* porins. *Nature (London)* **358,** 727–733.

David, L. L., Takemoto, L. J., Anderson, R. S., and Shearer, T. R. (1988). Proteolytic changes in main intrinsic polypeptide (MIP26) from membranes in selenite cataract. *Curr. Eye Res.* **7,** 411–417.

Deen, P. M. T., Dempster, J. A., Wieringa, B., and Van Os, C. H. (1992). Isolation of a cDNA for rat CHIP28 water channel: High mRNA expression in kidney cortex and inner medulla. *Biochem. Biophys. Res. Commun.* **188,** 1267–1273.

Deen, P. M. T., Verdijk, M. A. J., Knoers, N. V. A. M., Wieringa, B., Monnens, L. A. H., van Os, C. H., and van Oost, B. A. (1994a). Requirement of human renal water channel aquaporin-2 for vasopressin-dependent concentration of urine. *Science* **264,** 92–95.

Deen, P. M. T., Weghuis, D. O., van Kessel, A. G., Wieringa, B., and van Os, C. H. (1994b). The human gene for water channel aquaporin 1 (AQP1) is localized on chromosome 7p15 → p14. *Cytogenet Cell Genet.* **65,** 243–246.

Deen, P. M. T., Weghuis, D. O., Sinke, R. J., van Kessel, A. G., Wieringa, B., and van Os, C. H. (1994c). Assignment of the human gene for the water channel of renal collecting duct Aquaporin 2 (AQP2) to chromosome 12 region q12 → q13. *Cytogenet Cell Genet.* **66,** 260–262.

de Jong, W. W., Lubsen, N. H., and Kraft, H. J. (1994). Molecular evolution of the eye lens. *Prog. Retinal Eye Res.* **14** (in press).

Dempster, J. A., Van Hoek, A. N., and Van Os, C. H. (1992). The quest for water channels. *News Physiol. Sci.* **7,** 172–176.

Denker, B. M., Smith, B. L., Kuhajda, F. P., and Agre, P. (1988). Identification, purification, and partial characterization of a novel Mr 28,000 integral membrane protein from erythrocytes and renal tubules. *J. Biol. Chem.* **263,** 15634–15642.

Donaldson, P., and Kistler, J. (1992). Reconstitution of channels from preparations enriched in lens gap junction protein MP70. *J. Membr. Biol.* **129,** 155–165.

Duncan, G., and Jacob, T. J. (1984). The lens as a physicochemical system. *In* "The Eye" (H. Davson, ed.), 3d ed., Vol. 1b, pp. 159–206. Academic Press, New York.

Echevarria, M., Frindt, G., Preston, G. M., Milovanovic, S., Agre, P., Fischbarg, J., and Windhager, E. E. (1993a). Expression of multiple water channel activities in *Xenopus* oocytes injected with mRNA from rat kidney. *J. Gen. Physiol.* **101,** 827–841.

Echevarria, M., Kuang, K., Iserovich, P., Li, J., Preston, G. M., Agre, P., and Fischbarg, J. (1993b). Cultured bovine corneal endothelial cells express CHIP28 water channels. *Am. J. Physiol.* **265,** C1349–C1355.

Ehring, G. R., and Hall, J. E. (1991). Reconstitution into planar lipid bilayers: Application to the study of lens MIP. *In* "Biophysics of Gap Junction Channels" (C. Peracchia, ed.), pp. 333–351. CRC Press, Boca Raton, Florida.

Ehring, G. R., Zampighi, G. A., and Hall, J. E. (1988). Properties of MIP26 channels reconstituted into planar lipid bilayers. *In*

"Gap Junctions" (E. L. Hertzberg and R. G. Johnson, eds.), pp. 335–346. Liss, New York.

Ehring, G. R., Zampighi, G. A., Horwitz, J., Bok, D., and Hall, J. E. (1990). Properties of channels reconstituted from the major intrinsic protein of lens fiber membranes. *J. Gen. Physiol.* **96,** 631–664.

Ehring, G. R., Lagos, N., Zampighi, G. A., and Hall, J. E. (1991). Phosphorylation modulates the voltage dependence of channels reconstituted from the major intrinsic protein of lens fiber membranes. *J. Membr. Biol.* **126,** 75–88.

Ehring, G. R., Zampighi, G. A., and Hall, J. E. (1993). Does MIP play a role in cell–cell communication? *Prog. Cell Res.* **3,** 143–151.

Fasman, G. D., and Gilbert, W. A., (1990). The prediction of transmembrane protein sequences and their conformation: an evaluation. *Trends Biochem. Sci.* **15,** 89–92.

Fitzgerald, P. G., Bok, D., and Horwitz, J. (1983). Immunocytochemical localization of the main intrinsic polypeptide (MIP) in ultrathin frozen sections of rat lens. *J. Cell Biol.* **97,** 1491–1499.

Fitzgerald, P. G., Bok, D., and Horwitz, J. (1985). The distribution of the main intrinsic membrane polypeptide in ocular lens. *Curr. Eye Res.* **4,** 1203–1218.

Fortin, M. G., Morrison, N. A., and Verma, D. P. S. (1987). Nodulin-26, a peribacteroid membrane nodulin is expressed independently of the development of the peribacteroid compartment. *Nucleic Acids Res.* **15,** 813–824.

Fushimi, K., Uchida, S., Hara, Y., Marumo, F., and Sasaki, S. (1993). Cloning and expression of apical membrane water channel of rat kidney collecting tubule. *Nature (London)* **361,** 549–552.

Garland, D., and Russell, P. (1985). Phosphorylation of lens fiber cell membrane proteins. *Proc. Natl. Acad. Sci. USA* **82,** 653–657.

Girsch, S. J. (1991). Reconstitution and spectroscopy as methodologies for probing junction channel protein structure and function. *In* "Biophysics of Gap Junction Channels" (C. Peracchia, ed.), pp. 273–299. CRC Press, Boca Raton, Florida.

Girsch, S. J., and Peracchia, C. (1985a). Lens cell-to-cell channel protein: I. Self-assembly into liposomes and permeability regulation by calmodulin. *J. Membr. Biol.* **83,** 217–225.

Girsch, S. J., and Peracchia, C. (1985b). Lens cell-to-cell channel protein: II. Conformational change in the presence of calmodulin. *J. Membr. Biol.* **83,** 227–233.

Girsch, S. J., and Peracchia, C. (1991). Calmodulin interacts with a C-terminus peptide from the lens membrane protein MIP26. *Curr. Eye Res.* **10,** 839–849.

Gooden, M., Rintoul, D., Takehana, M., and Takemoto, L. (1985a). Major intrinsic polypeptide (MIP26K) from lens membrane: reconstitution into vesicles and inhibition of channel forming activity by peptide antiserum. *Biochem. Biophys. Res. Commun.* **128,** 993–999.

Gooden, M., Takemoto, L. J., and Rintoul, D. A. (1985b). Reconstitution of MIP26 from single human lenses into artificial membranes. I. Differences in pH sensitivity of cataractous vs. normal human lens fiber cell proteins. *Curr. Eye Res.* **4,** 1107–1115.

Gorin, M. B., Yancey, S. B., Cline, J., Revel, J. P., and Horwitz, J. (1984). The major intrinsic protein (MIP) of the bovine lens fiber membrane: Characterization and structure based on cDNA cloning. *Cell* **39,** 49–59.

Gruitjers, W. T. M. (1989). A non-connexon protein (MIP) is involved in eye lens gap-junction formation. *J. Cell Sci.* **93,** 509–513.

Gruijters, W. T. M., Kistler, J., Bullivant, S., and Goodenough, D. A. (1987). Immunolocalization of MP70 in lens fiber 16-17 nm intercellular junctions. *J. Cell Biol.* **104,** 565–572.

Guerrero, F. D., Jones, J. T., and Mullet, J. E. (1990). Turgor-responsive gene transcription and RNA levels increase rapidly when pea shoots are wilted. Sequence and expression of three inducible genes. *Plant Mol. Biol.* **15**, 11–26.

Harding, J. J., and Crabbe, M. J. C. (1984). The lens: Development, proteins, metabolism and cataract. *In* "The Eye" (H. Davson, ed.), 3d Ed., Vol. 1, pp 207–492. Academic Press, New York.

Harris, A. L. (1991). Connexin32 forms ion channels in single artificial membranes. *In* "Biophysics of Gap Junctions Channels" (C. Peracchia, ed.), pp. 373–389. CRC Press, Boca Raton, Florida.

Harris, H. W., Strange, K., and Zeidel, M. L. (1991). Current understanding of the cellular biology and molecular structure of the antidiuretic hormone-stimulated water transport pathway. *J. Clin. Invest.* **88**, 1–8.

Hasegawa, J, Zhang, R., Dohrman, A., and Verkman, A. S. (1993). Tissue-specific expression of mRNA encoding rat kidney water channel CHIP28k by *in situ* hybridization. *Am. J. Physiol.* **264**, C237–C245.

Hasegawa, H., Lian, S- C., Finkbeiner, W. E., and Verkman, A. S. (1994). Extrarenal tissue distribution of CHIP28 water channels by in situ hybridization and antibody staining. *Am. J. Physiol.* **266**, C893–C903.

Heller, K. B., Lin, E. C. C., and Wilson, T. H. (1980). Substrate specificity and transport properties of the glycerol facilitator of *Escherichia coli. J. Bacteriol.* **144**, 274–278.

Hinrichsen, R. D., and Blackshear, P. J. (1993). Regulation of peptide-calmodulin complexes by protein kinase C *in vivo. Proc. Natl. Acad. Sci. USA* **90**, 1585–1589.

Hoffman, J. F. (1993). Aquaporin: A wee burn runs through it. *J. Clin. Invest.* **92**, 1604–1605.

Hofte, H., and Chrispeels, M. J. (1992). Protein sorting to the vacuolar membrane. *Plant Cell* **4**, 995–1004.

Hofte, H, Hubbard, L., Reizer, J., Ludevid, D., Herman, E. M., and Chrispeels, M. J. (1992). Vegetative and seed-specific forms of tonoplast intrinsic protein in the vacuolar membrane of *Arabidopsis thaliana. Plant Physiol.* **99**, 561–570.

Horwitz, J., and Bok, D. (1987). Conformational properties of the main intrinsic polypeptide (MIP26) isolated from lens plasma membranes. *Biochemistry* **26**, 8092–8098.

Hsu, Y-T., and Molday, R. S. (1993). Modulation of the cGMP-gated channel of rod photoreceptor cells by calmodulin. *Nature (London)* **361**, 76–79.

Jahnig, F. (1990). Structure predictions of membrane proteins are not that bad. *Trends Biochem. Sci.* **15**, 93–95.

Jeanteur, D., Lakey, J. H., and Pattus, F. (1991). The bacterial porin superfamily: Sequence alignment and structure prediction. *Mol. Microbiol.* **5**, 2153–2164.

Johnson, K. D., and Chrispeels, M. J. (1992). Tonoplast-bound protein kinase phosphorylates tonoplast intrinsic protein. *Plant Physiol.* **100**, 1787–1795.

Johnson, K. R., Frenzel, E., Johnson, K., Klukas, K., Lampe, P., Sas, D., Sas, M. J., Biegon, R., and Louis, C. (1985). Is lens MP26 a gap junction protein? A phosphoprotein? *In* "Gap Junctions" (M. V. L. Bennett and D. C. Spray, eds.), pp. 91–105. Cold Spring Harbor Laboratory Press, Cold Spring Harbor, New York.

Johnson, K. R., Lampe, P. D., Hur, K. C., Louis, C. F., and Johnson, R. G. (1986). A lens intercellular junction protein, MP26, is a phosphoprotein. *J. Cell Biol.* **102**, 1334–1343.

Johnson, K. D., Herman, E. M., and Chrispeels, M. J. (1989). An abundant, highly conserved tonoplast protein in seeds. *Plant Physiol.* **91**, 1006–1013.

Johnson, K. D., Hofte, H., and Chrispeels, M. J. (1990). An intrinsic tonoplast protein of protein storage vacuoles in seeds is structur-ally related to a bacterial solute transporter (GlpF). *Plant Cell* **2**, 525–532.

Johnson, R. G., Klukas, K. A., Tze-Hong, L., and Spray, D. C. (1988). Antibodies to MP28 are localized to lens junctions, alter intercellular permeability and demonstrate increased expression during development. *In* "Gap Junctions" (E. L. Hertzberg and R. G. Johnson, eds.), pp. 81–98. Liss, New York.

Kistler, J., and Bullivant, S. (1989). Structural and molecular biology of the eye lens membranes. *CRC Crit. Rev. Biochem. Mol. Biol.* **24**, 151–181.

Kistler, J., Kirkland, B., and Bullivant, S. (1985). Identification of a 70,000-D protein in lens membrane junctional domains. *J. Cell Biol.* **101**, 28–35.

Kodama, R., Agata, N., Mochii, M., and Eguchi, G. (1990). Partial amino acid sequence of the major intrinsic protein (MIP) of the chicken lens deduced from the nucleotide sequence of a cDNA clone. *Exp. Eye Res.* **50**, 737–741.

Kronfeld, P. C. (1969). The gross anatomy and embryology of the eye. *In* "The Eye" (H. Davson, ed.), Vol. 1, pp. 1–66. Academic Press, New York.

Lampe, P. D., and Johnson, R. G. (1989). Phosphorylation of MP26, a lens junction protein, is enhanced by activators of protein kinase C. *J. Membr. Biol.* **107**, 145–155.

Lampe, P. D., and Johnson, R. G. (1990). Amino acid sequence of *in vivo* phosphorylation sites in the main intrinsic protein (MIP) of lens membranes. *Eur. J. Biochem.* **194**, 541–547.

Lampe, P. D., Remington, S. G., Jarvis, L. J., Louis, C. F., and Johnson, R. G. (1989). Homologies between chicken and bovine lens MIP. *J. Cell Biol.* **109**, p.46a.

Lanahan, A., Williams, J. B., Sanders, L. K., and Nathans, D. (1992). Growth factor-induced delayed early response genes. *Mol. Cell. Biol.* **12**, 3919–3929.

Lau, L. F., and Nathans, D. (1991). Genes induced by serum growth factors. *In* "The Hormonal Control Regulation of Gene Transcription" (P. Cohen and J. G. Foulkes, eds.), pp. 257–293. Elsevier Science, Amsterdam.

Lea, E. J. A., and Duncan G. (1991). Lens cell communication—From the whole organ to single channels. *In* "Biophysics of Gap Junction Channels" (C. Peracchia, ed.), pp. 353–371. CRC Press, Boca Raton, Florida.

Limjoco, T. I., Carper, D., Bondy, C., and Chepelinsky, A. B. (1991). Accumulation and spatial location of aldose reductase mRNA in a lens tumor of an αA-crystallin/SV40 T antigen transgenic mouse line. *Exp. Eye Res.* **52**, 759–762.

Louis, C. F., Hogan, P., Visco, L., and Strasburg, G. (1990). Identity of the calmodulin-binding proteins in bovine lens plasma membranes. *Exp. Eye Res.* **50**, 495–503.

Lu, K. P., and Means, A. R. (1993). Regulation of the cell cycle by calcium and calmodulin. *Endocrine Rev.* **14**, 40–58.

Ludevid, D., Hofte, H., Himelblau, E., and Chrispeels, M. J. (1992). The expression pattern of the tonoplast intrinsic protein γ-TIP in *Arabidopsis thaliana* is correlated with cell enlargement. *Plant Physiol.* **100**, 1633–1639.

Maisel, H., Harding, C. V., Alcala, J. R., Kuszak, J., and Bradley, R. (1981). The morphology of the lens. *In* "Molecular and Cellular Biology of the Eye Lens" (H. Bloemendal, ed.), pp. 49–84. John Wiley & Sons, Chichester.

Mathias, R. T., and Rae, J. L. (1989). Cell to cell communication in lens. *In* "Cell Interactions and Gap Junctions" (N. Sperelakis and W. C. Cole eds.), Vol. I, pp. 29–50. CRC Press, Boca Raton, Florida.

Maurel, C., Reizer, J., Schroeder, J. I., and Chrispeels, M. J. (1993). The vacuolar membrane protein γ-TIP creates water specific channels in Xenopus oocytes. *EMBO J.* **12**, 2241–2247.

Mayo B., Kok J., Venema K., Bockelmann W., Teuber M., Reinke H., and Venema G. (1991). Molecular cloning and sequence analysis of the X-prolyl dipeptidyl aminopeptidase gene from *Lactococcus lactis* subsp. *cremoiris*. *Appl. Environ. Microbiol.* **57,** 38–44.

McAvoy, J. W. (1980). Induction of the eye lens. *Differentiation* **17,** 137–149.

Miao, G-H., Hong, Z., and Verma, D. P. S. (1992). Topology and phosphorylation of soybean nodulin-26, an intrinsic protein of the peribacteriod membrane. *J. Cell Biol.* **118,** 481–490.

Moon, C., Preston, G. M., Griffin, C. A., Jabs, E. W., and Agre, P. (1993). The human aquaporin-CHIP gene. Structure, organization, and chromosomal localization. *J. Biol. Chem.* **268,** 15772–15778.

Moscona, A. A., Brown, M., Degenstein, L., Fox, L., and Soh, B. M. (1983). Transformation of retinal glia cells into lens phenotype: Expression of MP26, a lens plasma membrane antigen. *Proc. Natl. Acad. Sci. USA* **80,** 7239–7243.

Muramatsu, S., and Mizuno, T. (1989). Nucleotide sequence of the region encompassing the *glpKF* operon and its upstream region containing a bent DNA sequence of *Escherichia coli*. *Nucleic Acids Res.* **17,** 4378.

Musil, L. S., and Goodenough, D. A. (1991). Biochemical analysis of connexin43 intracellular transport, phosphorylation, and assembly into gap junctional plaques. *J. Cell Biol.* **115,** 1357–1374.

Musil, L. S., Beyer, E. C., and Goodenough, D. A. (1990). Expression of the gap junction protein connexin43 in embryonic chick lens: Molecular cloning, ultrastructural localization, and post-translational phosphorylation. *J. Membr. Biol.* **116,** 163–175.

Nap, J-P., and Bisseling, T. (1990). Developmental biology of a plant–prokaryote symbiosis: The legume root nodule. *Science* **250,** 948–954.

Ngoc, L. D., Paroutaud, P., Dunia, I., Benedetti, E. L., and Heobeke, J. (1985). Sequence analysis of peptide fragments from the intrinsic membrane protein of calf lens fiber MP26 and its natural maturation product MP22. *FEBS Lett.* **181,** 74–78.

Nicholson, B. J., Takemoto, L. J., Hunkapiller, M. W., Hood, L. E., and Revel, J-P. (1983). Differences between liver gap junction protein and lens MP 26 from rat: Implications for tissue specificity of gap junctions. *Cell* **32,** 967–978.

Nielsen, S., Smith, B. L., Christensen, E. L., Knepper, M. A., and Agre P. (1993a). CHIP28 water channels are localized in constitutively water-permeable segments of the nephron. *J. Cell Biol.* **120,** 371–383.

Nielsen, S., Smith, B. L., Christensen, E. L., and Agre P. (1993b). Distribution of the aquaporin CHIP in secretory and resorptive epithelia and capillary endothelia. *Proc. Natl. Acad. Sci. USA* **90,** 7275–7279.

Nielsen, S., DiGiovanni, S. R., Christensen, E. I., Knepper, M. A., and Harris, H. W. (1993c). Cellular and subcellular immunolocalization of vasopressin-regulated water channel in rat kidney. *Proc. Natl. Acad. Sci. U.S.A.* **90,** 11663–11667.

Nikaido, H. (1992). Porins and specific channels of bacterial outer membranes. *Mol. Microbiol.* **6,** 435–442.

Nikaido, H., and Rosenberg, E. Y. (1985). Functional reconstitution of lens gap junction proteins into proteoliposomes. *J. Membr. Biol.* **85,** 87–92.

Nikaido, H., and Saier, M. H. (1992). Transport proteins in bacteria: Common themes in their design. *Science* **258,** 936–942.

Ouyang, L-J., Whelan, J., Weaver, C. D., Roberts, D. M., and Day, D. A. (1991). Protein phosphorylation stimulates the rate of malate uptake across the peribacteroid membrane of soybean nodules. *FEBS Lett.* **293,** 188–190.

Pao, G. M., Wu, L-F., Johnson K. D., Hofte, H., Chrispeels M. J., Sweet, G., Sandal, N. N., and Saier, M. H. (1991). Evolution of the MIP family of integral membrane transport proteins. *Mol. Microbiol.* **5,** 33–37.

Paterson, C. A., and Delamere, N. A. (1992). The lens. In ''Adler's Physiology of the Eye: Clinical Application'' (W. M. Hart, Jr., ed.) 9th Ed. pp. 348–390. Mosby-Year Book, St. Louis.

Paul, D. L., and Goodenough, D. A. (1983a). Preparation, characterization, and localization of antisera against bovine MP26, an integral protein from lens fiber plasma membrane *J. Cell Biol.* **96,** 625–632.

Paul, D. L., and Goodenough, D. A. (1983b). *In vitro* synthesis and membrane insertion of bovine MP26, an integral membrane protein from lens fiber plasma membrane. *J. Cell Biol.* **96,** 633–638.

Paul, D. L., Ebihara, L., Takemoto, L. J., Swenson, K. I., and Goodenough, D. A. (1991). Connexin46, a novel gap junction protein, induces voltage gated currents in nonjunctional plasma membrane of *Xenopus* oocytes. *J. Cell Biol.* **115,** 1077–1089.

Pauptit, R. A., Schimer, T., Jansoniuis, J. N., Rosenbusch, J. P., Parker, M. W., Tucker, A. D., Tsernoglou, D., Weiss, M. S., and Schulz, G. E. J. (1991). A common channel-forming motif in evolutionarily distant porins. *Struct. Biol.* **107,** 136–145.

Pearson, R. B., Woodgett, J. R., Cohen, P., and Kemp, B. E. (1985). Substrate specificity of a multifunctional calmodulin-dependent protein kinase. *J. Biol. Chem.* **260,** 14471–14476.

Peracchia, C. (1989). Control of gap junction permeability and calmodulin-like proteins. In ''Cell Interactions and Gap Junctions'' (N. Sperelakis and W. C. Cole, eds.), Vol. I, pp. 125–142. CRC Press, Boca Raton, Florida.

Peracchia, C., and Girsch, S. J. (1985a). Permeability and gating of lens gap junction channels incorporated into liposomes. *Curr. Eye Res.* **4,** 431–439.

Peracchia, C., and Girsch, S. J. (1985b) Is the cleavable C-terminal arm of cell-to cell channel protein the channel gate? *Biochem. Biophsy. Res. Commun.* **133,** 688–695.

Piatigorsky, J. (1981). Lens differentiation in vertebrates. A review of cellular and molecular features. *Differentiation* **19,** 134–153.

Pisano, M. M., and Chepelinsky, A. B. (1991). Genomic cloning, complete nucleotide sequence, and structure of the human gene encoding the major intrinsic protein (MIP) of the lens. *Genomics* **11,** 981–990.

Preston, G. M., and Agre, P. (1991). Isolation of the cDNA for erythrocyte integral membrane protein of 28 kilodaltons: Member of an ancient channel family. *Proc. Natl. Acad. Sci. USA* **88,** 11110–11114.

Preston, G. M., Carroll, T. P., Guggino, W. B., and Agre, P. (1992a). Appearance of water channels in *Xenopus* oocytes expressing red cell CHIP28 protein. *Science* **256,** 385–387.

Preston, G. M., Smith, B. L., Jung, J. S., Guggino, W. B., and Agre, P. (1992b). Structure and function of CHIP28 water channels. *J. Am. Soc. Nephrol.* **3,** 797.

Preston, G. M., Jung, J. S., Guggino, W. B., and Agre, P. (1993). The mercury-sensitive residue at cysteine 189 in the CHIP28 water channel. *J. Biol. Chem.* **268,** 17–20.

Ramaekers, F. C. S., Selten-Versteegen, A. M. E., Benedetti, E. L., Dunia, I., and Bloemendal, H. (1980). *In vitro* synthesis of the major lens membrane protein. *Proc. Natl. Acad. Sci. USA* **77,** 725–729.

Rao, Y., Jan, L. Y., and Jan, Y. N. (1990). Similarity of the product of the *Drosophila* neurogenic gene *big brain* to transmembrane channel proteins. *Nature* (London) **345,** 163–167.

Rao, Y., Bodmer, R., Jan, L. Y., and Jan, Y. N. (1992). The *big brain* gene of *Drosophila* functions to control the number of neuronal precursors in the peripheral nervous system. *Development* **116,** 31–40.

Reizer, J., Reizer, A., and Saier, M. H. (1993). The MIP family of integral channel proteins: Sequence comparisons, evolutionary relationships, reconstructed pathway of evolution, and proposed functional differentiation of the two repeated halves of the proteins. *Crit. Rev. Biochem. Mol. Biol.* **28**, 235–257.

Revel, J. P., and Yancey, S. B. (1985). Molecular conformation of the major intrinsic protein of lens fiber membranes: Is it a junction protein? *In* "Gap Junctions" (V. L. Bennett and D. C. Spray, eds.), pp. 33–48. Cold Spring Harbor Laboratory Press, Cold Spring Harbor, New York.

Rice, S. J., Baldo, G. J., Kushmerick, C., Haspel, H., and Mathias, R. T. (1993). Cloning, expression and functional studies of frog lens MIP. *Invest. Ophthal. Vis. Sci. (Suppl.)* **34**, 1444.

Rintoul, D. A., and Takemoto, L. J. (1991). Reconstitution and molecular processing of MIP26K, a putative gap junctional component of the lens fiber cell membrane. *In* "Biophysics of Gap Junction Channels" (C. Peracchia, ed.), pp. 317–329. CRC Press, Boca Raton, Florida.

Sabolic, I., Valenti, G., Verbavatz, J-M., Van Hoek, A. N., Verkman, A. S., Ausiello, D. A., and Brown, D. (1992). Localization of the CHIP28 water channel in rat kidney. *Am. J. Physiol.* **263**, C1225–C1233.

Saez, J. C., Berthoud, V. M., Moreno, A. P., and Spray, D. C. (1993). Gap junctions. Multiplicity of controls in differentiated and undifferentiated cells and possible functional implications. *In* "Advances in Second Messenger and Phosphoprotein Research" (S. Shenolikar and A. C. Nairn, eds.) Vol 27, pp. 163–198. Raven Press, New York.

Saito, F., Sasaki, S., Chepelinsky, A. B., Fushimi, K., Marumo, F., and Ikeuchi, T. (1994). Human AQP-2 and MIP genes, two members of the MIP family, map within chromosome band 12q13 using two-color FISH. *Cytogen. Cell Genetics (in press)*.

Sandal, N. N., and Marcker, J. A. (1988). Soybean nodulin 26 is homologous to the major intrinsic protein of the bovine lens fiber membrane. *Nucleic Acids Res.* **16**, 9347.

Sandal, N. N., and Marcker, K. A. (1990). Some nodulin and Nod proteins show similarity to specific animal proteins. *In* "Nitrogen Fixation: Achievements and Objectives" (P. M. Gresshoff, L. E. Roth, G. Stacey, and W. E. Newton, eds), pp. 687–692. Chapman and Hall, New York.

Sas, D. F., Sas, M. J., Johnson, K. R., Menko, A. S., and Johnson, R. G. (1985). Junctions between lens fiber cells are labelled with a monoclonal antibody shown to be specific for MP26. *J. Cell Biol.* **100**, 216–225.

Sasaki, S., Fushimi, K., Uchida, S., and Marumo, F. (1992). Regulation and localization of two types water channels in the rat kidney. *J. Am. Soc. Nephrol.* **3**, 798.

Sasaki, S., Fushimi, K., Saito, H., Saito, F., Uchida, S., Ishibashi, K., Kuwabara, M., Ikeuchi, T., Inui, K., Nakajima, K., Watanabe, T. X., and Marumo, F. (1994). Cloning, characterization, and chromosomal mapping of human aquaporin collecting duct. *J. Clin. Invest.* **93**, 1250–1256.

Scaglione, B. A., and Rintoul, D. A. (1989). A fluorescence-quenching assay for measuring permeability of reconstituted lens MIP 26. *Investig. Ophthalmol. Vis. Sci.* **30**, 961–966.

Shen, L., Shrager, P., Girsch, S. J., Donaldson, P. J., and Peracchia, C. (1991). Channel reconstitution in liposomes and planar bilayers with HPLC-purified MIP26 of bovine lens. *J. Membr. Biol.* **124**, 21–32.

Shiels, A., Kent, N. A., McHale, M., and Bangham, J. A. (1988). Homology of MIP26 to Nod26. *Nucleic Acids Res.* **16**, 9348.

Shomer, N. H., Weaver, C. D., Roberts, D. M., and Louis, C. F. (1994). Single channel properties of the nodulin26 protein from soybean nodule symbiosome membranes. *Biophys. J.* **66**, A259.

Smith, B. L., and Agre, P. (1991). Erythrocyte Mr 28,000 transmem-

brane protein exists as a multisubunit oligomer similar to channel proteins. *J. Bio. Chem.* **266**, 6407–6415.

Smith, B. L., Baumgarten, R., Nielsen, S., Raben, D., Zeidel, M. L., and Agre, P. (1993). Concurrent expression of erythroid and renal aquaporin CHIP and appearance of water channel activity in perinatal rats. *J. Clin. Invest.* **92**, 2035–2041.

Smith, C. P., and Chater, K. F. (1988). Structure and regulation of controlling sequences for the *Streptomyces coelicolor* glycerol operon. *J. Mol. Biol.* **204**, 569–580.

Sparkes, R. S., Mohandas, T., Heinzmann, C., Gorin, M. B., Horwitz, J., Law, M. L., Jones, C. A., and Bateman, J. B. (1986). The gene for the major intrinsic protein (MIP) of the ocular lens is assigned to human chromosome 12cen-q14. *Invest. Ophthalmol. Vis. Sci.* **27**, 1351–1354.

Stamer, W. D., Snyder, R. W., Smith, B. L., Agre, P., and Regan, J. W. (1994). Localization of aquaporin CHIP in the human eye: Implications in the pathogenesis of glaucoma and other disorders of ocular fluid balance. *Invest. Ophthalmol. Vis. Sci. (in press)*.

Swamy, M. S., and Abraham, E. C. (1992). Glycation of lens MIP26 affects the permeability in reconstituted liposomes. *Biochem. Biophys. Res. Commun.* **186**, 632–638.

Sweet, G., Gandor, C., Voegele, R., Wittekindt, N., Beuerle, J., Truniger, V., Lin, E. C. C., and Boos, W. (1990). Glycerol facilitator of *Escherichia coli*: Cloning of *glpF* and identification of the *glpF* product. *J. Bacteriol.* **172**, 424–430.

Swenson, K. I., Jordan, J. R., Beyer, E. C., and Paul, D. L. (1989). Formation of gap junctions by expression of connexins in *Xenopus* oocyte pairs. *Cell* **57**, 145–155.

Takemoto, L., and Takehana, M. (1986). Covalent change of major intrinsic polypeptide (MIP26K) of lens membrane during human senile cataractogenesis. *Biochem. Biophys. Res. Commun.* **135**, 965–971.

Takemoto, L. J., Hansen, J. S., Nicholson, B. J., Hunkapiller, M., Revel, J-P., and Horwitz J. (1983). Major intrinsic polypeptide of lens membrane. Biochemical and immunological characterization of the major cyanogen bromide fragment. *Biochim. Biophys. Acta* **731**, 267–274.

Takemoto, L. J., Hansen, J. S., and Horwitz, J. (1985). Antisera to synthetic peptides of lens MIP26K (major intrinsic polypeptide): Characterization and use as site-specific probes of membrane changes in the aging human lens. *Exp. Eye Res.* **41**, 415–422.

Takemoto, L., Takehana, M., and Horwitz, J. (1986). Covalent changes in MIP26K during aging of the human lens membrane. *Invest. Ophthalmol. Vis. Sci.* **27**, 443–446.

Takemoto, L., Kuck, J., and Kuck, K. (1988). Changes in the major intrinsic polypeptide (MIP26K) during opacification of the Emory mouse lens. *Exp. Eye Res.* **47**, 329–336.

Tenbroek, E., Arneson, M., Jarvis, L., and Louis, C. (1992). The distribution of the fiber cell intrinsic membrane proteins MP20 and connexin46 in the bovine lens. *J. Cell Sci.* **103**, 245–257.

Truniger, V., Boos, W., and Sweet, G. (1992). Molecular analysis of the *glpFKX* regions of *Escherichia coli* and *Shigella flexneri*. *J. Bacteriol.* **174**, 6981–6991.

Van Aelst, L., Hohmann, S., Zimmermann, F. K., Jans, A. W. H., and Thevelein, J. M. (1991). A yeast homologue of the bovine lens fiber MIP gene family complements the growth defect of a *Saccharomyces cerevisiae* mutant on fermentable sugars but not its defect in glucose-induced RAS-mediated cAMP signalling. *EMBO J.* **10**, 2095–2104.

Van den Eijnden-van Raaij, A. J. M., de Leeuw, A. L. M., and Broekhuyse, R. M. (1985). Bovine lens calmodulin. Isolation, partial characterization and calcium-independent binding to lens membrane proteins. *Curr. Eye Res.* **4**, 905–912.

Van Hoek, A. N., and Verkman A. S. (1992). Functional reconstitution of the isolate erythrocyte water channel CHIP28. *J. Biol. Chem.* **267**, 18267–18269.

Van Hoek, A. N., de Jong, M. D., and Van Os, C. H. (1990). Effects of dimethylsulfoxide and mercurial sulfhydryl reagents on water and solute permeability of rat kidney brush border membranes. *Biochim. Biophys. Acta* **1030**, 203–210.

Van Hoek, A. N., Hom, M. L., Luthjens, L. H., de Jong, M. D., Dempster, J. A., and Van Os, C. H. (1991). Functional unit of 30kDa for proximal tubule water channels as revealed by radiation inactivation. *J. Biol. Chem.* **266**, 16633–16635.

Van Hoek, A. N., Luthjens, L. H., Hom, M. L., Van Os, C. H., and Dempster, J. A. (1992). A 30 kDa functional size for the erythrocyte water channel determined *in situ* by radiation inactivation. *Biochem. Biophys. Res. Commun.* **184**, 1331–1338.

Verkman, A. S. (1992). Water channels in cell membranes. *Annu. Rev. Physiol.* **54**, 97–108.

Verma, D. P. S. (1992). Signals in root nodule organogenesis and endocytosis of *Rhizobium*. *Plant Cell* **4**, 373–382.

Verma, D. P. S., and Delauney, A. J. (1988). Root nodule symbiosis: Nodulins and nodulin genes. *In* "Temporal and Spatial Regulation of Plant Genes" (D. P. S. Verma and R. B. Goldberg, eds.), pp. 169–199. Springer-Verlag, New York.

Voegele, R. T., Sweet, G. D., and Boos, W. (1993). Glycerol kinase of *Escherichia coli* is activated by interaction with the glycerol facilitator. *J. Bacteriol.* **175**, 1087–1094.

Waggoner, P. R., and Maisel, H. (1978). Immunofluorescent study of a chick lens fiber cell membrane polypeptide. *Exp. Eye Res.* **27**, 151–157.

Watanabe, M., Kobayashi, H., Rutishauser, U., Katar, M., Alcala, J., and Maisel, H. (1989). NCAM in the differentiation of embryonic lens tissue. *Dev. Biol.* **135**, 414–423.

Weaver, C. D., and Roberts, D. M. (1992). Determination of the site of phosphorylation of nodulin 26 by the calcium-dependent protein kinase from soybean nodules. *Biochemistry* **31**, 8954–8959.

Weaver, C. D., Crombie, B., Stacey, G., and Roberts, D. M. (1991). Calcium-dependent phosphorylation of symbiosome membrane proteins from nitrogen-fixing soybean nodules. *Plant Physiol.* **95**, 222–227.

Weissenborn, D. L., Wittekindt, N., and Larson, T. J. (1992). Structure and regulation of the *glpFK* operon encoding glycerol diffusion facilitator and glycerol kinase of *Escherichia coli* K-12. *J. Biol. Chem.* **267**, 6122–6131.

Welsh, M. J., Aster, J. C., Ireland, M., Alcala, J., and Maisel, H. (1982). Calmodulin binds to chick lens gap junction protein in a calcium-independent manner. *Science* **216**, 642–644.

West, J. W., Numann, R., Murphy, B. J., Scheuer, T., and Catterall, W. A. (1991). A phosphorylation site in the Na$^+$ channel required for modulation by protein kinase C. *Science* **254**, 866–868.

White, T. W., Bruzzone, R., Goodenough, D. A., and Paul, D. L. (1992). Mouse Cx50, a functional member of the connexin family of gap junction proteins, is the lens fiber protein MP70. *Mol. Biol. Cell* **3**, 711–720.

Wistow, G., Pisano, M. M., and Chepelinsky, A. B. (1991). Tandem sequence repeats in transmembrane channel proteins. *Trends Biochem. Sci.* **16**, 170–171.

Yamaguchi-Shinozaki, K., Koizumi, M., Urao, S., and Shinozaki, K. (1992). Molecular cloning and characterization of 9 cDNAs for genes that are responsive to desiccation in *Arabidopsis thaliana*: Sequence analysis of one cDNA clone that encodes a putative transmembrane channel protein. *Plant Cell Physiol.* **33**, 217–224.

Yamamoto, Y. T., Cheng, C-L., and Conkling, M. A. (1990). Root-specific genes from tobacco and *Arabidopsis* homologous to an evolutionarily conserved gene family of membrane channel proteins. *Nucleic Acids Res.* **18**, 7449.

Yamamoto, Y. T., Taylor, C. G., Acedo, G. N., Cheng, C-L., and Conkling, M. A. (1991). Characterization of *cis*-acting sequences regulating root specific gene expression in tobacco. *Plant Cell* **3**, 371–382.

Yancey, S. B., Koh, K., Chung, J., and Revel, J. P. (1988). Expression of the gene for main intrinsic polypeptide (MIP): Separate spatial distributions of MIP and β-crystallin gene transcripts in rat lens development. *J. Cell Biol.* **106**, 705–714.

Zampighi, G. A., Simon, S. A., Robertson, J. D., McIntosh, T. J., and Costello, M. J. (1982). On the structural organization of isolated bovine lens fiber junctions. *J. Cell Biol.* **93**, 175–189.

Zampighi, G. A., Hall, J. E., Ehring, G. R., and Simon, S. A. (1989). The structural organization and protein composition of lens fiber junctions. *J. Cell Biol.* **108**, 2255–2275.

Zampighi, G. A., Simon, S. A., and Hall, J. E. (1992). The specialized junctions of the lens. *Int. Rev. Cytol.* **136**, 185–225.

Zeidel, M. L., Ambudkar, S. V., Smith, B. L., and Agre, P. (1992). Reconstitution of functional water channels in liposomes containing purified red cell CHIP28 protein. *Biochemistry* **31**, 7436–7440.

Zhang, R., Logee, K. A., and Verkman, A. S. (1990). Expression of mRNA coding for kidney and red cell water channels in *Xenopus* oocytes. *J. Biol. Chem.* **265**, 15375–15378.

Zhang, R., Alper, S. L., Thorens, B., and Verkman, A. S. (1991). Evidence from oocyte expression that the erythrocyte water channel is distinct from band 3 and the glucose transporter. *J. Clin. Invest.* **88**, 1553–1558.

Zhang, R., Skach, W., Hasegawa, H., van Hoek, A. N., and Verkman, A. S. (1993). Cloning, functional analysis and cell localization of a kidney proximal tubule water transporter homologous to CHIP28. *J. Cell. Biol.* **120**, 359–369.

Calmodulin-Regulated Channels

Calmodulin-Sensitive Channels

YOSHIRO SAIMI, KIT-YIN LING, AND
CHING KUNG

I. Calmodulin and Calmodulin-Binding Targets

Ca^{2+} as a second messenger often works through mediators such as Ca^{2+}-binding proteins. Calmodulin (CaM) is among the most important of such Ca^{2+}-binding protein, because it activates many different enzymes and ion pumps, and participates in many aspects of cell physiology (Cohen and Klee, 1988). CaM is a 17-kDa soluble protein found in all eukaryotic species. Almost no variations are detectable among vertebrate CaMs, and only limited, mostly conservative changes occur among CaM molecules of the diverse species of protists and fungi (Kung *et al.*, 1992). For example, the CaM of *Drosophila* is 98% identical and 99% similar to the vertebrate form (Smith *et al.*, 1987); the *Paramecium* CaM is 88% identical and 94% similar to the vertebrate form (Schaefer *et al.*, 1987a). As do vertebrate CaMs, *Paramecium* CaM can activate enzymes such as phosphodiesterase (Burgess-Cassler *et al.*, 1987), myosin light-chain kinase (Schaefer *et al.*, 1987b), erythrocyte Ca^{2+}–ATPase (D. Kosk-Kosicka, K.-Y. Ling, and C. Kung, unpublished data), and calcineurin (C. Klee, K.-Y. Ling, and C. Kung, unpublished data). The crystal structures of vertebrate and *Paramecium* CaMs have been resolved

to 1.8 Å (Meador *et al.*, 1992; Wu *et al.*, 1992); the two structures are almost identical. This dumbbell-shaped molecule has two lobes, each equipped with a pair of EF hands that include Ca^{2+}-binding sites. These two lobes are connected by a central α helix with seven turns that is exposed to solvent (Strynadka and James, 1989). Each EF hand is a helix–loop–helix structure; residues 1, 3, 5, 7, and 12 in the loop coordinate the Ca^{2+} (Kretsinger, 1980). The loops in EF hands III and IV in the C-terminal lobe bind Ca^{2+} at a higher affinity than loops I and II in the N-terminal lobe. Binding of Ca^{2+} in these EF hands exposes one hydrophobic domain per lobe, enabling these domains to interact with the targets. Activation of target enzymes, such as the myosin light-chain kinase, appears to be by disinhibition, that is, the Ca^{2+}–CaM complex binds to and removes from the catalytic site an autoinhibitory domain that contains a basic amphiphilic α (BAA) helix (Blumenthal *et al.*, 1985; O'Neil and DeGrado, 1990). As revealed by X-ray crystallographic studies (Chattopadhyaya *et al.*, 1992) and NMR (Ikura *et al.*, 1992), the complex of Ca^{2+}–CaM and a BAA helix peptide assumes a very compact form. Apparently, the middle portion of the central α helix of CaM unwinds, and the two lobes clasp onto and engulf the BAA helix. The interaction of these two molecules is largely (80%) due to van der Waals forces; the hydrophobic domains of the two CaM lobes interact with the hydrophobic side of the BAA helix through a multitude of contacts.

The best known targets of Ca^{2+}–CaM are some twenty different enzymes (Cohen and Klee, 1988). In addition, Ca^{2+}–CaM also activates a Ca^{2+} pump in the erythrocyte membrane (Carafoli et al., 1988) that hydrolyses ATP and transports Ca^{2+} outward against the Ca^{2+} electrochemical gradient. Ion channel function is opposite that of pumps: when open, these channels let ions flow down their electrochemical gradients. Evidence suggests that certain ion channels are also activated by Ca^{2+}–CaM, but a distinction must be made between ion channels that are affected by CaM indirectly, for example, by Ca^{2+}–CaM-dependent phosphorylation of the channel protein (e.g., Hoger et al., 1991; Onozuka et al., 1991; Wagner et al., 1991; McCarron et al., 1992; Hell et al., 1993) and ion channels that interact directly with and are thereby regulated by CaM. This chapter reviews only examples of the latter. The reader is referred elsewhere for discussion on two such topics since there are more pertinent chapters (ryanodine-receptor Ca^{2+}-release channels of sarcoplasmic reticulum, Chapter 30; gap junction channels, Chapter 24).

II. Calmodulin-Gated Ion Channels in *Paramecium*

Paramecium is a unicellular protozoan whose motility is powered by typical eukaryotic cilia, the membrane-enclosed "9 + 2" microtubular structures. Cilia normally stroke from anterior to posterior to generate forward power; therefore, normal *Paramecium* locomotion is forward. As do other animals, paramecia respond and adapt to touch, heat, and organic or inorganic chemicals. Various stimuli can elicit a stereotypical "avoidance reaction" in wild-type paramecia, that is, a sequence of reversals of its ciliary beat, swimming backward for a short distance and then resuming forward swimming within 1 sec (Jennings, 1906). This behavior results from depolarizing membrane electrogenesis, the Ca^{2+} action potential, on stimulation. During a Ca^{2+} action potential, a pulse of Ca^{2+} flux enters the cilia (Eckert, 1972). In cells bathed in solutions similar to culture media, this action potential is very complex and is based on at least four membrane currents through two voltage-gated channels (Ca^{2+} and K^+ channels) and two Ca^{2+}-activated channels (Na^+ and K^+ channels). Another set of ion currents can be observed on hyperpolarization of the membrane, including a second Ca^{2+}-dependent K^+ current (Saimi, 1986; Preston, 1990; Preston et al., 1990a,1992). This second type of Ca^{2+}-dependent K^+ channel is pharmacologically distinct from that elicited on depolarization (Preston et al., 1990c). As described subsequently, evidence shows that these three Ca^{2+}-dependent channels, like enzymes, are activated directly by Ca^{2+}–CaM.

Mutations can lead to improper electrogenesis and, therefore, aberrant behavior. Wild-type behavior is apparently not necessary for survival, at least not in the laboratory. For example, "pawn" mutants, which lack the Ca^{2+} current and therefore lack the avoidance reaction, are clearly viable (Kung and Eckert, 1972; Saimi and Kung, 1987). A class of mutants, dubbed "pantophobiacs," overreact to various stimuli, exhibiting exaggerated backward swimming for more than 1 min (Hinrichsen et al., 1985). Experiments with two-electrode voltage clamp showed that pantophobiacs have lost most or all of the Ca^{2+}-dependent K^+ currents activated on depolarization and hyperpolarization (Saimi et al., 1983; Preston et al., 1990b). Curiously, however, some alleles of pantophobiacs show the opposite phenotype, that is, an abbreviated rather than a prolonged avoidance reaction (Van Houten et al., 1977; Kink et al., 1990). Voltage-clamp experiments of these "fast-2" mutants showed that the major defect in these mutants is a loss of the Ca^{2+}-dependent Na^+ current and not the K^+ currents (Saimi, 1986; Kink et al., 1990; Preston et al., 1990b). Through standard genetic analyses, three pantophobiac-type and four fast-2-type mutations in the *cam* gene locus have been reported to date.

Injection of wild-type protein molecules into organisms would be expected to convert mutant phenotypes into wild-type for hours or days. Indeed, injection of wild-type cytoplasm into the cam^1 pantophobiac reverted the mutant phenotype (Hinrichsen et al., 1985). Sequential fractionations of the wild-type cytoplasm eventually traced the "pantophobia-curing activity" to a protein identified as CaM (Hinrichsen et al., 1986). Amino acid residues of the wild-type and cam^1 CaM proteins were sequenced directly using Edman degradation (Schaefer et al., 1987a,b). The cam^1 product was found to have a single serine-to-phenylalanine substitution at amino acid residue 101 (S101F), which is the eighth residue in Ca^{2+}-binding loop III. Subsequently, two more pantophobiac alleles, cam^2 (Wallen-Friedman, 1988) and cam^3, were analyzed and were found to bear an Ile 135 to threonine (I135T) substitution in Ca^{2+}-binding loop IV (by amino acid sequencing; Lukas et al., 1989) and a Met 145 to valine (M145V) substitution (through nucleotide sequencing of the *cam* gene; Kink et al., 1990), respectively. Three fast-2 mutants have been analyzed: cam^{11} has a single glutamate to lysine (E54K) substitution; cam^{12} has two substitutions, glycine to glutamate (G40E) and aspartate to asparagine (D50N);

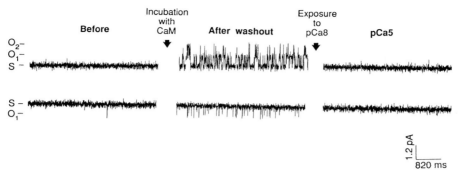

FIGURE 1 Activation of Ca^{2+}-dependent Na^+ channels by Ca^{2+}/CaM in the presence of 100 mM Na^+. When the Na^+ channel became inactive (before, *left*), *Paramecium* CaM (2 μg or 0.4 μM) was added to the bath that contained $10^{-5}M$ free Ca^{2+}. After a 5-min incubation, CaM was washed out of the bath by perfusion with the Na^+ solution without CaM. The Na^+ channel activity increased after incubation with CaM but without ATP (after washout, *middle*). The patch was then exposed to $10^{-8}M$ free Ca^{2+} for 1 min, and returned to $10^{-5}M$ free Ca^{2+}. The Na^+ channel was inactivated after the low Ca^{2+} exposure (pCa5, *right*). Reproduced with permission from Saimi and Ling (1990). Copyright© 1990 by the AAAs.

and *cam*[13] also has two substitutions, valine to isoleucine (V35I) and aspartate to asparagine (D50N) (Kink *et al.*, 1990; Ling *et al.*, 1992).

These mutants clearly showed that mutations in CaM can have profound effects on ion channels in *Paramecium*. Are these effects direct? Results of a set of experiments by Saimi and Ling (1990) suggest that, at least for the Ca^{2+}-dependent Na^+ channel of *Paramecium*, the interaction is direct without any requirement for ATP (Fig. 1). In these experiments, active channels captured in excised inside-out patches were exposed to a bath of low $[Ca^{2+}]$; the channel activities were found to disappear. Replenishing Ca^{2+} did not reactivate the channels until CaM was also added to the bath. After such restoration, the channel could again be inactivated through low $[Ca^{2+}]$ and reactivated with Ca^{2+}–CaM. Because the bath was a simple buffered ion solution with no added ATP, kinase, or other enzymes, the simplest interpretation is that Ca^{2+}–CaM interacts directly with the channel itself or with channel-associated molecule(s).

Since none of the Ca^{2+}-dependent Na^+ and K^+ channel proteins or their corresponding genes have been isolated or cloned, showing that they bind CaM directly or contain BAA helices is not yet possible. However, evidence suggests that an autoinhibitory mechanism similar to that for the Ca^{2+}–CaM-dependent enzymes may be at work in some of these channels. Limited proteolysis activates but does not destroy activity in several CaM target enzymes by removing the inhibitory BAA-helix domain (Cohen and Klee, 1988). The Ca^{2+}-dependent K^+ channel activated by hyperpolarization in *Paramecium* is also activated by proteolysis (Kubalski *et al.*, 1989). After a control treatment of membrane patches with each of three different proteolytic enzymes tested, channels

in the patches became constitutively active and no longer required Ca^{2+}. This limited proteolytic disinhibition did not affect the ion selectivity or the unit conductance of the channel. A similar activation of *Paramecium* Na^+ channel by limited trypsin digestion has been observed (Y. Saimi and K. Y. Ling, unpublished).

The mutations of the first six *Paramecium* CaM mutants, all in evolutionarily conserved residues, were found to have an unexpected but systematic distribution. The three fast-2 mutants—*cam*[11], *cam*[12], and *cam*[13]—defective mainly in their Ca^{2+}-dependent Na^+ current, have amino acid substitutions in the N-terminal lobe. The three pantophobiacs *cam*[1], *cam*[2], and *cam*[3], defective mainly in their Ca^{2+}-dependent K^+ current, have amino acid substitutions in the C-terminal lobe (Kink *et al.*, 1990). Nine additional mutations continue to conform to this pattern (K.-Y. Ling *et al.*, unpublished data). These results suggest that a functional bipartition of the CaM protein reflects its structural bipartition. Apparently each type of Ca^{2+}–CaM-dependent ion channel is specialized and requires specific interaction with a particular lobe of CaM, that is, the Na^+ channel interacts with the N-terminal lobe and the K^+ channels with the C-terminal lobe. Such a functional polarization has been discussed for Ca^{2+}–ATPase as well (Bzdega and Kosk-Kosicka, 1992).

Surprisingly, even the most conserved features of CaM can be altered without lethal effects in *Paramecium* (Kung *et al.*, 1992). This absence of lethality also occurs in budding yeast (Geiser *et al.*, 1991). Perhaps these conserved features of CaM are not for housekeeping or cell-cycle progression functions that are vital to the cell, but for signal transductions in which ion channels play important roles.

III. trp and trpl Gene Products: Photosensory Channels in Drosophila

Photon reception by arthropod compound eyes results in an inward current, thereby causing depolarization in the photoreceptor cell. Although no consensus has been reached in this field of research, investigators generally agree that light excites rhodopsin, which activates phospholipase C through a GTP-binding protein. Inositoltriphosphate (IP_3) generated by the catalytic action of phospholipase C causes a release of Ca^{2+} from an internal store (Ranganathan et al., 1991b). How IP_3 or Ca^{2+} release relates to inward current is not yet understood.

Many Drosophila mutants defective in photoreception were isolated and analyzed behaviorally and by electroretinography (Pak, 1979). The genes corresponding to several of these mutations were then cloned and examined. For example, a blind fly with no receptor potential (norp) in the electroretinogram was found to be mutated in the structural gene for a phospholipase C (Bloomquist et al., 1988). A mutant with neither inactivation nor after-potential (ninaE) in its electroretinogram was found to be mutated in one of the rhodopsins (O'Tousa et al., 1985). Other mutations were not understood as well, even after the corresponding genes were cloned. One mutant that can behaviorally respond properly to dim light and appears to be blind in bright light shows a transient receptor potential (trp) in its electroretinogram (Cosens and Manning, 1969). The trp gene has been cloned by germ-line transformation of the mutant (Montell and Rubin, 1989). Although the hydropathy plot of the predicted peptide clearly indicates that it is an intramembrane protein, researchers realized only recently that this gene is most likely to encode an ion channel.

Voltage clamping of individual photoreceptors has been done using the whole-cell recording technique on dissociated ommatidia (Hardie, 1991; Ranganathan et al., 1991a). The results show that the light-sensitive channels are primarily permeable to Ca^{2+}. Examinations of trp mutants using this technique led to the conclusion that the trp mutation selectively abolishes one class of these Ca^{2+}-passing channels (Hardie and Minke, 1992).

Phillips et al. (1992) used ^{125}I-labeled CaM to probe a lambda expression library of Drosophila head cDNA. Using lambda clones that express CaM-binding peptide fragments, these researchers retrieved a full-length gene from another library. This gene apparently encodes a 127.6-kDa protein. When its sequence was compared with those in the data bank, the protein was found to have a surprising 39% identity of amino acids with the trp protein; hence, the gene locus is now known as trpl for "trp-like." An examination of the trpl sequence shows putative membrane-crossing segments similar to S1 through S6 of voltage-gated ion channels, particularly those of the voltage-gated Ca^{2+} channel (Hui et al., 1991). Even the putative pore-lining SS1–SS2 regions show some similarity, although the S4 segment of the trpl protein is largely devoid of the putative gating charges. Consistent with the fact that trpl was discovered through the binding of its protein fragments to CaM, two regions that can form BAA helices were also identified. Assuming the usual model of six membrane-spanning regions, these BAA helices should face the cytoplasm. Reexamination of the trp sequence shows that it is also like a Ca^{2+} channel and has at least one of the two BAA helices found in trpl. Examinations of trp and trpl transcripts by in situ hybridization showed that these two genes are expressed almost exclusively in the retina of the compound eye. These findings, in conjunction with the voltage-clamp results (Hardie and Minke, 1992), suggest strongly that the products of trp and trpl are Ca^{2+}–CaM-gated ion channels. Since trp mutants lack a sustained component of the receptor current, leaving intact a transient current induced by light, a tentative interpretation is that trpl encodes a channel that inactivates rapidly whereas trp encodes a channel that inactivates very slowly. Binding of CaM to the BAA helices, and consequential conformation changes, may be the mechanism of activation and inactivation (Phillips et al., 1992). Functional expression of these channels in oocytes and isolation of trpl-null mutations will be crucial tests for this hypothesis. Note that trp and trpl encode a peptide that corresponds to a monomeric subunit of a channel that is likely to be tetrameric. Therefore, four or eight BAA helices, that is, multiple CaM-binding sites, should exist in each channel. However, patch-clamp experiments will be necessary for direct testing of the effect(s) of CaM and Ca^{2+} on these channels, as done for the Paramecium Ca^{2+}–CaM-dependent Na^+ channel described earlier (Saimi and Ling, 1990).

IV. Calcium-Dependent K^+ Channels in Kidney and Colon

Potassium ions in the kidney are secreted at the glomerulus, but are then repeatedly reabsorbed and resecreted in the proximal and distal tubules (Jamison and Müller-Suúr, 1987) to retrieve other solutes such as Na^+ and Cl^-, as well as water. Potassium transport in the thick ascending limb of Henle's loop, which is part of the mammalian distal tubule, includes energy-dependent Na^+/K^+ exchange at the basolateral membrane (vascular side), energy-dependent $NaCl/KCl$ co-transport at the apical membrane (lumen side), and a passive pathway(s) at the basolateral and apical membranes of the secretory cell (Giebisch, 1987; Lang and Rehwald, 1992). This leak K^+ conductance is regulated by many factors: pH, Ca^{2+}, depolarization, antidiuretic drugs, and hormones. Ba^{2+} is known to inhibit this passive K^+ transport.

The patch-clamp method has been successfully applied to studies of mammalian renal electrophysiology (for reviews, see Giebisch, 1987; Kolb, 1990; Lang and Rehwald, 1992; *Renal Physiology and Biochemistry* **13**). In the apical membrane of the thick ascending limb in mammals, several K^+ channels have been found (Guggino *et al.*, 1987a,b; Greger *et al.*, 1990). One such channel is a BK-type Ca^{2+}-dependent K^+ channel of 120-pS conductance with charybdotoxin (ChTX) sensitivity, that was found in the cloned A3 cultured cells from rabbit medullary thick ascending limb (Guggino *et al.*, 1987a,b). Although activated at micromolar ranges of Ca^{2+}, this channel cannot be seen *in situ* in thick ascending limbs isolated from rat (Greger *et al.*, 1990) or rabbit (Wang *et al.*, 1990). These Ca^{2+}-dependent K^+ channels are suspected to be required for "housekeeping" functions such as maintenance of cell volume or cellular growth (Lang and Rehwald, 1992). In the A3 cultured cells, activation of the Ca^{2+}-dependent K^+ channel is observed on exposure to a hypotonic solution (Takeguchi and Guggino, 1989; Montrose-Rafizadeh and Guggino, 1991). No reports have been made on Ca^{2+}-dependent K^+ channels in the basolateral membrane of the thick ascending limb (Lang and Rehwald, 1992).

The K^+ conductance has also been studied biochemically. Luminal membrane vesicles from a crude membrane fraction of the outer medulla of pig kidney, where the thick ascending limb is abundant, show $^{86}Rb^+$ uptake in the presence of the Ca^{2+} ionophore A23187 (Burnham *et al.*, 1985,1986; Klaerke *et al.*, 1987a; Klaerke and Jørgensen, 1992). Although half of this uptake is Ca^{2+} independent, the other half is Ca^{2+} dependent with a K_m of 0.2 μM. The Ca^{2+}-dependent component is inhibited by Ba^{2+} (with K_m of 150 μM), by a K^+-channel inhibitor (quinidine), and by low pH. These inhibitions are consistent with those of the passive K^+ conductance in the renal tubule cell (Giebisch, 1987; Lang and Rehwald, 1992). The authors further reconstituted K^+ conductance in vesicles after solubilization with CHAPS, and found most of the Ca^{2+}-sensitive sites to be facing out of the vesicle, since the Ca^{2+} ionophore was no longer needed for $^{86}Rb^+$ flux activation. The Ca^{2+} sensitivity

of the reconstituted K^+ conductance remains the same as in the plasma membrane.

Interestingly, K^+ conductance in plasma membrane vesicles is suppressed by CaM inhibitors (trifluoperazine and calmidazolium), indicating a possible CaM involvement in its activation by Ca^{2+} (Klaerke *et al.*, 1987a). These authors also showed that limited trypsin digestion of the reconstituted vesicle induces K^+ conductance that is inhibited by Ba^{2+} but is no longer Ca^{2+} dependent. This loss of Ca^{2+} sensitivity does not occur without CHAPS solubilization, indicating that an access to the cytoplasmic side of the K^+-conducting molecule is required for trypsin modification. This activation of renal K^+ conductance by limited trypsin digestion resembles that seen for the Ca^{2+}-dependent channels in *Paramecium* (see preceding discussion). The reconstituted K^+ conductance can be activated 2-fold by addition of CaM only in the presence of Ca^{2+}; the K_m for CaM binding is determined to be ~0.1 nM (Klaerke *et al.*, 1987b). These authors presented the idea of direct interaction of CaM with the K^+ conductance in the luminal membrane. The CHAPS-solubilized membrane proteins were chromatographed through a CaM–Sepharose 4B affinity column and eluted with EGTA. This fraction, containing only two major bands of 51 and 36 kDa, was reconstituted in vesicles. The uptake of these vesicles is perfectly Ca^{2+} dependent as well as CaM dependent, although to a lesser degree (10–20% increase), relative to the vesicles prior to solubilization. Phosphorylation of the 51-kDa protein augmented $^{86}Rb^+$ uptake and CaM sensitivity. The 51-kDa band was later identified as β_2-glycoprotein I (Klaerke and Jørgensen, 1992). For some reason, purified K^+ conductance of the kidney is no longer inhibitable by Ba^{2+} (Klaerke *et al.*, 1987b). Perhaps the weakened CaM sensitivity and loss of Ba^{2+} sensitivity of the K^+ conductance after CaM-affinity chromatography are related, and indicate an alteration or partial denaturation of the protein.

Although no electrophysiological evidence exists for the K^+ channel activity in pig kidney outer medulla after solubilization and reconstitution using the planar lipid bilayer or the patch-clamp method, it is difficult to determine whether this kidney K^+ channel is the same as or different from BK-type Ca^{2+}-dependent K^+ channels found in the cloned A3 cultured cells of rabbit thick ascending limb (Guggino *et al.*, 1987a,b).

Adopting the same strategy using CaM-Sepharose column affinity chromatography, Liu *et al.* (1991) purified a fraction of membrane proteins from rabbit colonocytes, that consisted of 120, 60 and 35-kDa proteins as major species. Subsequent reconstitution of these proteins into the planar lipid bilayer revealed

Ca^{2+}-dependent K^+ channels with conductance of approximately 250 pS, indicating BK-type channels. Most interestingly, channel activation required CaM since the channel activity was observed only when CaM was included in the bath at a concentration of 100 nM in the presence of 10 μM Ca^{2+}. Perhaps these channels correspond to those found in rabbit distal colon epithelial cells, shown to have properties of BK-type Ca^{2+}-dependent K^+ channels (Turnheim et al., 1989).

In both examples, they found 35-kDa proteins through CaM–affinity chromatography enriched in fractions containing the channel activity. Note that the BK-type Ca^{2+}-dependent K^+ channel from bovine smooth muscle apparently has a subunit of 35-kDa protein, as judged by ChTX toxin binding and channel reconstitution (Garcia-Calvo et al., 1991; Giangiacomo et al., 1991). Since ChTX binds and blocks the channel pores, it is tempting to speculate that 35-kDa proteins form Ca^{2+}-dependent K^+ channels and have CaM-binding sites for channel activity regulation. However, more recent studies clearly show a deduced molecular weight of more than 100 kDa for Ca^{2+}-dependent K^+ channels in mouse (mSlo channels; Butler et al., 1993). When expressed in Xenopus oocytes, these channels have pharmacological properties shared among BK-type Ca^{2+}-dependent K^+ channels. This apparent discrepancy in molecular weight might suggest multiple families of Ca^{2+}-dependent K^+ channels, one of which may be regulated by CaM. Further molecular analyses would clarify such a possibility.

V. Others

Hsu and Molday (1993) observed a reduction in cGMP sensitivity of more than 1.5 fold for the cGMP-dependent channel in membrane vesicles from the vertebrate rod outer segment (ROS), when CaM was included in vesicle suspension. Through immunoaffinity column chromatography using antibodies against the channel protein, a 240-kDa protein was identified to be associated with the cGMP-dependent channel with molecular weight of 63 kDa. This 240-kDa protein was also shown to bind CaM, experimenting with CaM affinity column chromatography and CaM overlay using ^{125}I-labeled CaM. However, the channel protein itself apparently did not bind CaM in these experiments. In the dark, where internal cGMP of ROS is high, this non-selective channel is open and Ca^{2+} influx maintains a high internal concentration. Therefore, the 240-kDa protein, suspected to be spectrin, is thought to receive a Ca^{2+} signal via CaM and then mediate the signal to the channel, lowering the cGMP-channel sensitivity in the dark. cAMP–cGMP-gated channels of olfactory receptor cells are closely related to the cGMP-gated channel in ROS. In these CaM reduces cAMP–cGMP-channel sensitivity up to 20 fold, although in this case CaM is thought to interact with the channel directly (T.-Y. Chen and K.-Y. Yau, personal communication).

The ryanodine–receptor Ca^{2+}-release channel in the sarcoplasmic reticulum is among the first channels shown to be CaM sensitive (Smith et al., 1989; also see Chapter 30). Cardiac sarcoplasmic reticulum contains chloride channels (116 pS) also inhibited by CaM in a Ca^{2+}-dependent manner (Kawano and Hiraoka, 1993).

The possibility of Ca^{2+}–CaM regulation of Ca^{2+}-dependent K^+ channels has also been explored in the red blood cell (Pape and Kristensen, 1984) and in adipocytes (Pershadsingh et al., 1986). Although Gárdos channels of red cells, which differ from the BK-type Ca^{2+}-dependent K^+ channels, apparently require a soluble factor (Sarkadi et al., 1980; Schwarz and Passow, 1983), this factor appears to be heat labile, unlike CaM.

VI. Conclusions

The apparent functional bipartition of Paramecium CaM in the activation of Ca^{2+}–CaM-dependent K^+ and Na^+ channels not only indicates different sites of interaction between CaM and these channel molecules, but also implies a possible mechanism of controlling the activity of these ion channels. Because the Na^+ current is inward and the K^+ current is outward on depolarization, Ca^{2+}–CaM is likely to coordinate these two currents, probably to sustain a brief depolarization during the action potential (Kung et al., 1992). Similar control of obviously antagonizing factors by Ca^{2+}–CaM can be observed in the interactions of cyclase and phosphodiesterase, both dependent on Ca^{2+}–CaM, by which a pulse of cyclic nucleotide is generated to execute the Ca^{2+} signal. Such orchestrations could be carried out by the differential geometric distribution of the two antagonizing CaM-binding targets, or by differential Ca^{2+} binding of CaM. The two lobes of CaM are known to have different binding affinities for Ca^{2+} and Mg^{2+} (Tsai et al., 1987); therefore, two antagonizing effectors may sense different Ca^{2+}-binding states of CaM. Alternatively, two opposing effectors may recognize CaM only when the proper lobe is filled with Ca^{2+}, thus generating functional bipartition. Our knowledge of how CaM activates individual enzymes in vitro is rather complete.

How it acts *in vivo* to regulate the twenty or more enzymes, channels, and pumps with which it interacts is far from clear. The observation of a functional bipartition and the hypothesis of up–down orchestration may help explore CaM actions in live cells.

Although myriad claims of Ca^{2+}–CaM-dependent channels have been made based only on the effects of anti-CaM drugs, these cases are not reviewed in this chapter because these drugs have other effects on these channels, apparently independent of CaM involvement (e.g., Kihira *et al.*, 1990). Moreover, a potent CaM inhibitor, calmidazolium, has been shown to create "holes" in the membrane (Meissner, 1986), making interpretation of the results of these studies difficult. Ideally, verification that Ca^{2+}–CaM directly regulates any channel requires electrophysiological demonstration as well as molecular confirmation of CaM binding to the channel.

The studies in *Paramecium* and in *Drosophila* were not originally designed to search for Ca^{2+}–CaM-dependent channels. Although findings of already known factors such as CaM could be disappointing to some investigators (Kung *et al.*, 1992), others were able to take advantage of these new systems for study.

Acknowledgments

We thank R. Preston for his comments on this manuscript and for sharing his enthusiasm with us. This work was supported by National Institutes of Health Grants GM22714 and GM26286 and a grant from the Lucille P. Markey Charitable Trust.

References

Bloomquist, B. T., Shortridge, R. D., Schneuwly, S., Perdew, M., Montell, C., Steller, H., Rubin, G., and Pak, W. L. (1988). Isolation of a putative phospholipase C gene of *Drosophila*, *norpA*, and its role in phototransduction. *Cell* **54,** 723–733.

Blumenthal D, K., Takio, K., Edelman, A. M., Charbonneau, H., Titani, K., Walsh, K. A., and Krebs, E. G. (1985). Identification of the calmodulin-binding domain of skeletal muscle myosin light chain kinase. *Proc. Natl. Acad. Sci. USA* **82,** 3187–3191.

Burgess-Cassler, A., Hinrichsen, R. D., Maley, M. E., and Kung, C. (1987). Biochemical characterization of a genetically altered calmodulin in *Paramecium*. *Biochim. Biophys. Acta* **913,** 321–328.

Burnham, C., Karkish, S. J. D., and Jørgensen, P. L. (1985). Identification and reconstitution of a Na/K/Cl cotransporter and K^+ channel from luminal membranes of renal red outer medulla. *Biochim. Biophys. Acta* **821,** 461–469.

Burnham, C., Braw, R., and Karlish, S. J. D. (1986). A Ca^{2+}-dependent K^+ channel in "luminal" membranes from the renal outer medulla. *J. Membr. Biol.* **93,** 177–186.

Butler, A., Tsunoda, S., McCobb, D. P., Wei, A., and Salkoff, L. (1993). *mSlo*, a complex mouse gene encoding "Maxi" calcium-activated potassium channels. *Science* **261,** 221–224.

Bzdega, T., and Kosk-Kosicka, D. (1992). Regulation of the erythrocyte Ca^{2+}-ATPase by mutant calmodulins with Glu→Ala substitutions in the Ca^{2+}-binding domains. *J. Biol. Chem.* **267,** 4394–4397.

Carafoli, E., Krebs, J., and Shiesi, M. (1988). Calmodulin in the transport of calcium across biomembranes. *In* "Calmodulin" (P. Cohen and C. B. Klee, eds.), pp. 297–312. Elsevier, New York.

Chattopadhyaya, R., Meador, W. E., Means, A. R., Quiocho, F. A. (1992). *J. Mol. Biol.* **228,** 1177–1192.

Cohen, P., and Klee, C. B. (1988). "Calmodulin, Molecular Aspects of Cellular Regulation." Elsevier, New York.

Cosens, D. J., and Manning, A. (1969). Abnormal electroretinogram from a *Drosophila* mutant. *Nature (London)* **224,** 285–287.

Eckert, R. (1972). Bioelctric control of ciliary activity. *Science* **176,** 473–481.

Garcia-Calvo, M., Vazquez, J., Smith, M., Kaczorowski, G. J., and Garcia, M. L. (1991). Characterization of the solubilized charibdotoxin receptor from bovine aortic muscle. *Biochemistry* **30,** 11157–11164.

Giangiacomo, K. M., Garcia-Calvo, M., Garcia, M. L., and McManus, O. B. (1991). Functional reconstitution of the large conductance Ca^{2+}-activated K^+ channel purified from bovine aortic smooth muscle. *Biophys. J.* **59,** 214a.

Geiser, J. R., van Tuinen, D., Brockerhoff, S. E., Neff, M. M., and Davis, T. N. (1991). Can calmodulin function without binding calcium? *Cell* **65,** 949–959.

Giebisch, G. H. (1987). Cell models of potassium transport in the renal tubule. *Curr. Top. Membr. Transport* **28,** 133–183.

Greger, R., Bleich, M., and Schlatter, E. (1990). Ion channels in the thick ascending limb of Henle's loop. *Renal Physiol. Biochem.* **13,** 37–50.

Guggino, S. E., Guggino, W. B., Green, N., and Sacktor, B. (1987a). Ca^{2+}-activated K^+ channels in cultured medullary thick ascending limb cells. *Am. J. Physiol.* **252,** C121–127.

Guggino, S. E., Guggino, W. B., Green, N., and Sacktor, B. (1987b). Blocking agents of Ca^{2+}-activated K^+ channels in cultured medullary thick ascending limb cells. *Am. J. Physiol.* **252,** C128–137.

Hardie, R. C. (1991). Whole-cell recordings of the light induced current in dissociated *Drosophila* photoreceptors: Evidence for feedback by calcium permeating the light-sensitive channels. *Proc. R. Soc. London B* **245,** 203–210.

Hardie, R. C., and Minke, R. (1992). The *trp* gene is essential for a light-activated Ca^{2+} channel in *Drosophila* photoreceptors. *Neuron* **8,** 643–651.

Hell, J. W., Yokoyama, C. T., Wong, S. T., Warner, C., Snutch, T. P., and Catterall, W. A. (1993). Differential phosphorylation of two size forms of the neuronal class C L-type calcium channel $\alpha 1$ subunit. *J. Biol. Chem.* **268,** 19451–19457.

Hinrichsen, R. D., Amberger, E., Saimi, Y., Burgess-Cassler, A., and Kung, C. (1985). Genetic analysis of mutants with a reduced Ca^{2+}-dependent K^+ current in *Paramecium tetraurelia*. *Genetics* **111,** 433–445.

Hinrichsen, R. D., Burgess-Cassler, A., Soltvedt, B. C., Hennessey, T., and Kung, C. (1986). Restoration by calmodulin of a Ca^{2+}-dependent K^+ current missing in a mutant of *Paramecium*. *Science* **232,** 503–506.

Hoger, J. H., Walter, A. E., Vance, D., Yu, L., Lester, H. A., and Davidson, N. (1991). Modulation of a cloned mouse brain potassium channel. *Neuron* **6,** 227–236.

Hsu, Y.-T., and Molday, R. S. 1993). Modulation of the cGMP-gated channel of rod photoreceptor cells by calmodulin. *Nature (London)* **361,** 76–79.

Hui, A., Ellinor, P. T., Krizanova, O., Wang, J.-J., Diebold, R. J., and Schwarz, A. (1991). Molecular cloning of multiple subtypes of a novel rat bran isoform of the α_1 subunit of the voltage-dependent calcium channel. *Neuron* **7,** 35–44.

Ikura, M., Clore, G. M., Gronenborn, A. M., Zhu, G., Klee, C. B., and Bax, A. (1992). Solution structure of a calmodulin-target peptide complex by multidimensional NMR. *Science* **256,** 632–638.

Jamison, R. L., and Müller-Suúr, R. (1987). Potassium recycling. *Curr. Top. Membr. Transport* **28,** 115–131.

Jennings, H. S. (1906). "Behavior of Lower Organisms." Indiana University Press, Bloomington.

Kawano, S., and Hiraoka, M. (1993). Protein kinase A-activated chloride channel is inhibited by the Ca^{2+}-calmodulin complex in cardiac sarcoplasmic reticulum. *Circ. Res.* **73,** 751–757.

Kihira, M., Matsuzawa, K., Tokuno, H., and Tomota, T. (1990) Effects of calmodulin antagonists on calcium-activated potassium channels in pregnant rat myometrium. *Br. J. Pharmacol.* **100,** 353–359.

Kink, J. A., Maley, M. E., Preston, R. R., Ling, K.-Y., Wallen-Friedman, M. A., Saimi, Y., and Kung, C. (1990). Mutations in *Paramecium* calmodulin indicate functional differences between the C-terminal and N-terminal lobes *in vivo*. *Cell* **62,** 165–174.

Klaerke, D. A., and Jørgensen, P. L. (1992). Affinity purification and reconstitution of calcium-activated potassium channels. *Meth. Enzymol.* **207,** 564–573.

Klaerke, D. A., Karlish, S. J. D., and Jørgensen, P. L. (1987a). Reconstitution of in phospholipid vesicles of calcium-activated potassium channel from outer renal medulla. *J. Membr. Biol.* **95,** 105–112.

Klaerke, D. A., Petersen, J., and Jørgensen, P. L. (1987b). Purification of Ca^{2+}-activated K^+ channel protein on calmodulin affinity columns after detergent solubilization of luminal membranes from outer renal medulla. *FEBS Lett.* **216,** 211–216.

Kolb, H. A. (1990). Potassium channels is excitable and nonexcitable cells. *Rev. Physiol. Biochem. Pharmacol.* **115,** 51–91.

Kretsinger, R. H. (1980). Structure and evolution of calcium-modulated proteins. *CRC Crit. Rev. Biochem.* **8,** 119–174.

Kubalski, A., Martinac, B., and Saimi, Y. (1989). Proteolytic activation of a hyperpolarization- and calcium-dependent potassium channel in *Paramecium*. *J. Membr. Biol.* **112,** 91–96.

Kung, C., and Eckert, R. (1972. Genetic modification of electric properties in an excitable membrane. *Proc. Natl. Acad. Sci. USA* **69,** 93–97.

Kung, C., Preston, R. R., Maley, M. E., Ling, K.-Y., Kanabrocki, J. A., Seavey, B. R., and Simi, Y. (1992). *In vivo Paramecium* mutants show that calmodulin orchestrates membrane responses to stimuli. *Cell Calcium* **13,** 413–425.

Lang, F., and Rehwald, W. (1992). Potassium channels in renal epithelial transport regulation. *Physiol. Rev.* **72,** 1–32.

Ling, K.-Y., Preston, R. R., Burns, R., Kink, J. A., and Kung, C. (1992). Primary mutations in calmodulin prevent action of the Ca^{2+}-dependent Na^+ channel in *Paramecium*. *Protein Struct. Funct. Genet.* **12,** 365–371.

Liu, S., Dubinsky, W. P., Haddox, M. K., and Schultz, S. G. (1991). Reconstitution of isolated Ca^{2+}-activated K^+ channel proteins from basolateral membranes of rabbit colonocytes. *Am. J. Physiol.* **261,** C713–C717.

Lukas, T. J., Wallen-Friedman, M., Kung, C., and Watterson, D. M. (1989). *In vivo* mutations of calmodulin: A mutant *Paramecium* with altered ion current regulation has an isoleucine-to-threonine change at residue 136 and an altered methylation state at lysine residue 115. *Proc. Natl. Acad. Sci. USA* **86,** 7331–7335.

McCarron, J. G., McGeown, J. G., Reardon, S., Ikebe, M., Fay, F. S., and Walsh, J. V., Jr. (1992). Calcium-dependent enhancement of calcium current in smooth muscle by calmodulin-dependent protein kinase II. *Nature (London)* **357,** 74–77.

Meador, W. E., Means, A. R., and Quiocho, F. A. (1992). Target enzymes recognition by calmodulin: 2.4 Å structure of a calmodulin-peptide complex. *Science* **257,** 1251–1255.

Meissner, G. (1986). Ryanodine activation and inhibition of the Ca^{2+} release channel of sarcoplasmic reticulum. *J. Biol. Chem.* **261,** 6300–6306.

Montell, C., and Rubin, G. (1989). Molecular characterization of the *Drosophila trp* locus: A putative integral membrane protein required for phototransduction. *Neuron* **2,** 1313–1323.

Montrose-Rafizadeh, C., and Guggino, W. B. (1991). Role of intracellular calcium in volume regulation by rabbit medullary thick ascending limb cells. *Am. J. Physiol.* **260,** F402–F409.

O'Neil, K. T., and DeGrado, W. F. (1990). How calmodulin binds its targets: Sequence independent recognition of amphophilic α-helices. *Trends Biochem. Sci.* **15,** 59–64.

Onozuka, M., Furuichi, H., Imai, S., and Ozono, S. (1991). Activation process of calcium-dependent potassium channel in *Euhadra* neurons: Involvement of calcium/calmodulin and subsequent protein phosphorylation. *Comp. Biochem. Physiol. A* **99,** 419–428.

O'Tousa, J. E., Baehr, W., Martin, R. L., Hirsh, J., Pak, W. L., and Applebury, M. L. (1985). The *Drosophila ninaE* gene encodes an opsin. *Cell* **40,** 839–850.

Pak, W. L. (1979). Study of photoreceptor function using *Drosophila* mutants. *In* "Neurogenetics: Genetic Approaches to the Nervous System" (X. O. Breakefield, ed.), pp. 67–99. Elsevier/North Holland, New York.

Pape, L., and Kristensen, B. L. (1984). A calmodulin activated Ca^{2+}-dependent K^+ channel in human erythrocyte membrane inside-out vesicles. *Biochim. Biophys. Acta* **770,** 1–6.

Pershadsingh, H. A., Gale, R. D., Delfert, D. M., and McDonald, J. M. (1986). A calmodulin dpendent Ca^{2+}-activated K^+ channel in the adipocyte plasma membrane. *Biochim. Biophys. Res. Commun.* **135,** 934–941.

Phillips, A. M., Bull, A., and Kelly, L. E. (1992). Identification of a *Drosophila* gene encoding a calmodulin-binding protein with homology to the *trp* phototransduction gene. *Neuron* **8,** 631–642.

Preston, R. R. (1990). A magnesium current in *Paramecium*. *Science* **250,** 285–288.

Preston, R. R., Saimi, Y., and Kung, C. (1990a). Evidence for two K^+ currents activated upon hyperpolarization of *Paramecium tetraurelia*. *J. Membr. Biol.* **115,** 41–50.

Preston, R. R., Wallen-Friedman, M. A., Saimi, Y., and Kung, C. (1990b). Calmodulin defects cause the loss of Ca^{2+}-dependent K^+ currents in two pantophobiac mutants of *Paramecium tetraurelia*. *J. Membr. Biol.* **115,** 51–60.

Preston, R. R., Saimi, Y., Amberger, E., and Kung, C. (1990c). Interactions between mutants with defects in two Ca^{2+}-dependent K^+ currents of *Paramecium tetraurelia*. *J. Membr. Biol.* **115,** 61–69.

Preston, R. R., Saimi, Y., and Kung, C. (1992). Calcium current activated upon hyperpolarization of *Paramecium tetraurelia*. *J. Gen. Physiol.* **100,** 233–251.

Ranganathan, R., Harris, W. A., Stevens, C. F., and Zucker, C. S. (1991a). A *Drosophila* mutant defective in extracellular calcium-dependent photoreceptor deactivation and rapid desensitization. *Nature (London)* **354,** 230–232.

Ranganathan, R., Harris, W. A., and Zuker, C. S. (1991b). The molecular genetics of invertebrate phototransduction. *Trends Neurosci.* **14,** 486–493.

Saimi, Y. (1986). Calcium-dependent sodium currents in *Paramecium*: Mutational manipulations and effects of hyper- and depolarization. *J. Membr. Biol.* **92,** 227–236.

Saimi, Y., and Kung, C. (1987). Behavioral genetics of *Paramecium*. *Ann. Rev. Genet.* **21,** 47–65.

Saimi, Y., and Ling, K.-Y. (1990). Calmodulin activation of calcium-dependent sodium channels in excised membrane patches of *Paramecium*. *Science* **249,** 1441–1444.

Saimi, Y., Hinrichsen, R. D., Forte, M., and Kung, C. (1983). Mutant analysis shows that the Ca^{2+}-induced K^+ current shuts off one type of excitation in *Paramecium*. *Proc. Natl. Acad. Sci. USA* **80,** 5112–5116.

Sarkadi, B., Szebeni, J., and Gádos, G. (1980). Effects of calcium transport processes in inside-out red cell membrane vesicles. *In* "Membrane Transport in Erythrocytes" (U. V. Lassesn, H. H. Ussing, and J. O. Wieth, eds.), pp. 220–235. Munksgaard, Copenhagen.

Schaefer, W. H., Lukas, T. J., Blair, I. A., Schultz, J. E., and Watterson, D. M. (1987a). Amino acid sequence of a novel calmodulin from *Paramecium tetraurelia* that contains dimethyllysine in the first domain. *J. Biol. Chem.* **262,** 1025–1029.

Schaefer, W. H., Hinrichsen, R. D. Burgess-Cassler, A., Kung, C., Blair, L. A., and Watterson, D. M. (1987b). A mutant *Paramecium* with a defective calcium-dependent potassium conductance has an altered calmodulin: A nonlethal selective alteration in calmodulin regulation. *Proc. Natl. Acad. Sci. USA* **84,** 3931–3935.

Schwarz, W., and Passow, H. (1983). Ca^{2+}-activated K^+ channels in erythrocytes and excitable cells. *Annu. Rev. Physiol.* **45,** 359–374.

Smith, J. S., Rousseau, E., and Meissner, G. (1989). Calmodulin modulation of single sarcoplasmic reticulum Ca^{2+}-release channels from cardiac and skeletal muscle. *Circ. Res.* **64,** 352–359.

Smith, V. L., Doyle, K. E., Maune, J. F., Munjaal, R. P., and Beckingham, K. (1987). Structure and sequence of the *Drosophila* calmodulin gene. *J. Mol. Biol.* **196,** 471–485,

Strynadka, N. C. J., and James M. N. G. (1989). Crystal structure of the helix-loop-helix calcium-binding proteins. *Annu. Rev. Biochem.* **58,** 951–998.

Takeguchi, J., and Guggino, W. B. (1989). Membrane stretch: A physiological stimulator of CA^{2+}-activated K^+ channels in thick ascending limb. *Am. J. Physiol.* **275,** F347–352.

Tsai, M.-D., Drakenberg, T., Thulin, E., and Forsén, S. (1987). Is the binding of magnesium(II) to calmodulin significant? An investigation by magnesium-25 nuclear magnetic resonance. *Biochemistry* **26,** 3635–3643.

Turnheim, K., Constantin, J., Chan, S., and Schultz, S. G. (1989). Reconstitution of a calcium-activated potassium channel in basolateral membranes of rabbit colonocytes into planar lipid bilayers. *J. Membr. Biol.* **112,** 247–254.

Van Houten, J., Chang, S. Y., and Kung, C. (1977). Genetic analyses of "Paranoiac" mutants of *Paramecium tetraurelia*. *Genetics* **86,** 113–120.

Wagner, J. A., Cozens, A. L., Schulman, H., Gruenert, D. C., Stryer, L., and Gardner, P. (1991). Activation of chloride channels in normal and cystic fibrosis airway epithelial cells by multifunctional calcium/calmoduli-dependent protein kinase. *Nature* (*London*) **349,** 793–796.

Wallen-Friedman, M. A. (1988). "An Ion-Current Mutant of *Paramecium tetraurelia* with Defects in the Primary Structure and Posttransational *N*-Methylation of Calmodulin." Ph.D. Thesis. University of Wisconsin—Madison.

Wang, W. S., White, S., Geibel, J., and Giebisch, G. H. (1990). A potassium channel in the apical membrane of rabbit thick ascending limb of Henle's loop. *Am. J. Physiol..* **258,** F244–F253.

Wu, S., Rao, S. T., Satyshur, K. A., Ling, K.-Y., Kung, C., and Sundaralingam, M. (1993). Structure of *Paramecium tetraurelia* calmodulin at 1.8Å resolution. *Protein Sci.* **2,** 436–447.

Channels of Unicellular Organisms

29

Microbial Channels

BORIS MARTINAC, XIN-LIANG ZHOU,
ANDRZEJ KUBALSKI, SERGEI SUKHAREV,
AND CHING KUNG

I. Patch-Clamp Investigation of Microbes

Microscopic creatures are vastly more diverse than plants and animals combined. These organisms represent different early evolutionary branches and consist of cells with very different body plans and metabolic styles. A formal classification of microbes is beyond the scope of this chapter. A reminder that the organisms we call microbes include prokaryotes and eukaryotes, among which are protists, algae, and fungi, should suffice. Primarily, this chapter addresses the activities of ion channels discovered in *Paramecium*, *Dictyostelium*, yeasts, and *Escherichia coli*. In addition to microbial biology and channel evolution, researchers have studied channel biology in molecular detail. Thus, most of the microbial species examined with patch-clamp techniques are those that are genetically tractable and have been used extensively in biochemical or molecular biological research.

Prior to the advent of patch-clamp techniques, electrical measurements could only be made on giant cells *in vivo* or by incorporating microbial channels into planar lipid bilayers *in vitro*. The patch-clamp method not only allows examination of individual ion channels (Hamill *et al.*, 1982), but also allows the study of small cells. Little modification of the solutions, electrodes, and amplifiers is needed to record from mi-

crobes. The challenge is in generating appropriate preparations so the microbial membranes can form gigaOhm seals with the patch-clamp pipettes. For example, *Paramecium* is completely covered with cilia; yeast membrane is covered by a cell wall, and different yeasts differ in their wall components; *E. coli* has two membranes, the outer and the inner membrane. No single method can be applied to all these situations. Figure 1 summarizes some of the methods used. Details of these methods can be found in Saimi *et al.* (1992) and in references cited for different organisms in the following discussion.

II. *Paramecium*

Although unicellular, *Paramecium* has many animal-like characteristics. This organism is a heterotroph and relies on dietary amino acids and vitamins. *Paramecium* swims about with 5000 incessantly beating cilia. Like those in higher animals, each cilium consists of a dynein–ATPase powered "9 + 2" assembly of sliding microtubules enclosed in a membrane. The *Paramecium* membrane is excitable, and ciliary motility is governed by membrane electrogenesis. The "avoidance reaction" (Jennings, 1906), for example, correlates with excitation. Depolarization can trigger a Ca^{2+}-based action potential; the resultant intraciliary

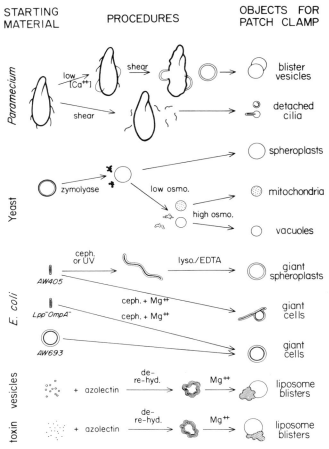

FIGURE 1 Outline of procedures to generate microbial objects suitable for patch-clamp studies. The starting material (*left*)—*Paramecium*, budding yeast, *Escherichia coli*, membrane vesicles, and yeast killer toxin—have dimensions on the order of 100, 10, 1, 0.1, and 0.01 μm, respectively. The procedure (*center*), briefly stated here but detailed in Saimi *et al.* (1992) and references therein, converts them to objects used successfully in patch-clamp experiments. The objects (*right*) are all 5–15 μm in diameter, except detached cilia from *Paramecium* (~2 μm) and yeast mitochondria and vacuoles (~3 μm). *Paramecium* blister vesicles are induced in solutions of high [Na$^+$] and low [Ca^{2+}], and can be detached from the cell mechanically. These blisters have two membranes, the outer cytoplasmic membrane and the inner membrane, probably of alveolar origin. Detached cilia are sheared off paramecia forced through a narrow-bore pipet. Yeast spheroplasts are obtained after cell wall removal with zymolyase, an enzyme mixture from *Athrobacter luteus*. A hypo-osmotic treatment followed by osmotic re-adjustment can open the spheroplasts, each releasing a mitochondrion and a vacuole. These mitochondria and vacuoles are optically distinct and can be selected for patch-clamp studies. *E. coli* wild-type strain AW405 grows as filaments more than 100 μm long in the presence of cephalexin, a penicillin analog, or on exposure to ultraviolet light. These filaments round up into giant spheroplasts when treated with EDTA and lysozyme, which nicks the peptidylglycan cell wall. Giant cells are obtained by growing the wild-type or Lpp$^-$Omp$^-$ double-mutant cells, lacking the major outer-membrane anchors to the cell wall, in the presence of cephalexin and tens of mM [Mg^{2+}]. About 10% of cells of the osmotic-sensitive mutant strain AW693 grown in high [K$^+$] and millimolar [Mg^{2+}] are also giant cells. Vesicles of various microbial membrane fractions and yeast killer toxin are mixed with azolectin which, after a cycle of de- and rehydration, yields large multilamellar liposomes. Liposome blisters grow out of the liposomes after they have been collapsed by tens of mM [Mg^{2+}]. Objects are not drawn to scale. Reproduced with permission from Saimi *et al.* (1992).

increase of Ca^{2+} causes the cilia to beat in a reversed direction to propel the cell backward (Naitoh and Eckert, 1969; Eckert *et al.*, 1972). Hyperpolarization can trigger other ion currents and a more rapid ciliary beat in a near-normal direction, causing the cell to spurt forward (Naitoh and Eckert, 1969,1973). Under certain circumstances, hyperpolarization can become regenerative (Satow and Kung, 1977). Cyclic AMP is thought to play an important role in this spurting response (Hennessey *et al.*, 1985; Bonini *et al.*, 1986; Nakaoka and Machemer, 1990; Preston and Saimi, 1990). Schultz and co-workers (1992) suggested that, in *Paramecium*, the K^+ efflux underlying membrane hyperpolarization and cAMP increase might be affected by the same entity—the ciliary adenylate cyclase. The purified cyclase, on incorporation into planar lipid bilayers, was found to be associated with a 320-pS nonselective cation channel.

Because the *Paramecium* cell is large, some 100 μm in length, it has been investigated with the classical two-electrode voltage clamp as well as the patch clamp. Experiments with these two methods and with planar lipid bilayers revealed at least 13 types of ion currents: $I_{Ca(d)}$, $I_{Ca(h)}$, $I_{Ca(mech)}$, $I_{K(d)}$, $I_{K(h)}$, $I_{K(Ca,d)}$, $I_{K(Ca,h)}$, $I_{K(mech)}$, $I_{Na(Ca)}$, $I_{Mg(Ca)}$, I_{Cl}, I_{M+}, and $I_{M+(ATP)}$. This large number of currents should not be surprising. First, this highly differentiated cell is anatomically complex. Although continuous, the plasma membrane and the ciliary membrane are distinct. Also, organelles are found in or attached to the plasma membrane, including an oral apparatus, a cytoproct, two contractile vacuoles, an alveolar-sac network, thousands of parasomal sacs, and trichocysts (Jurand and Selman, 1969). Second, as a free-living organism, *Paramecium* must move, respond, adapt, endocytose, excrete, secrete, divide, mate, and autogamize (Goertz, 1988). Some of these physiological functions are known or expected to have an ionic or electrical component.

When *Paramecium* is examined bathed in various external solutions under a two-electrode voltage clamp, membrane depolarizations can trigger at least five currents. (1) $I_{Ca(d)}$, a depolarization-activated Ca^{2+} current, activates rapidly by voltage and inactivates in a Ca^{2+}-dependent manner (Oertel *et al.*, 1977; Brehm and Eckert, 1978). Deciliation abolishes this current, indicating that the channels are located in the ciliary membrane and not in the body membrane (Dunlap, 1977; Machemer and Ogura, 1979). "Pawn" mutations abolish or reduce this current (Kung and Eckert, 1972; Oertel *et al.*, 1977; Satow and Kung, 1980a). (2) $I_{K(d)}$, a rapid K^+ current activated by depolarization, does not inactivate (Machemer and Ogura, 1979; Satow and Kung, 1980a). (3) $I_{Na(Ca)}$, a Na^+ current, is activated by internal Ca^{2+}, presumably from the already described Ca^{2+} current (Saimi and Kung,

1980; Saimi, 1986). A subclass of calmodulin mutants called "fast-2" has little or none of this Na^+ current (Kink *et al.*, 1990; Preston *et al.*, 1991; Ling *et al.*, 1992; Kung *et al.*, 1992). (4) $I_{K(Ca,d)}$, a K^+ current, is also activated by internal Ca^{2+} (Satow and Kung, 1980b; Preston *et al.*, 1990b). A different subclass of calmodulin mutants called "pantophobiacs" shows large reductions in this current (Saimi *et al.*, 1983; Kink *et al.*, 1990; Preston *et al.*, 1990b; Kung *et al.*, 1992). This current is greatly enhanced in the *teaA* mutant (Hennessey and Kung, 1987; Preston *et al.*, 1990c). (5) $I_{Mg(Ca)}$, a Mg^{2+}-specific current, is activated by Ca^{2+} (Preston, 1990). This novel current, the first ever reported in any organism, has a selectivity of $Mg^{2+} > Mn^{2+} = Co^{2+} > Sr^{2+} = Ba^{2+} \gg Ca^{2+}$, and is affected in some of the calmodulin mutants (R. R. Preston, unpublished results).

Hyperpolarizations under voltage clamp also reveal at least five currents. (1) $I_{Ca(h)}$ is an unusual Ca^{2+} current. Unlike other Ca^{2+} currents, Ba^{2+} not only cannot carry but blocks this current (Preston *et al.*, 1992a). The current is also blocked by amiloride and shows a Ca^{2+}-dependent inactivation (Preston *et al.*, 1992b). (2) $I_{K(h)}$, a hyperpolarization-activated K^+ current, is blockable by quinidine (Satow and Kung, 1977; Oertel *et al.*, 1978; Preston *et al.*, 1990a). (3) $I_{K(Ca,h)}$, a K^+ current, is activated after $I_{Ca(h)}$. This current is kinetically and pharmacologically distinct from $I_{K(Ca,d)}$ (Preston *et al.*, 1990a). Some calmodulin mutations of the pantophobiac type reduce this current, whereas *rst* causes an early activation of it (Preston *et al.*, 1990b,c). (4) $I_{Na(Ca)}$ is induced by $I_{Ca(h)}$, and is believed to flow through the same Na^+ channel activated by Ca^{2+} from either I_{Ca}, since the Na^+ currents at either polarization are affected by the fast-2 type of calmodulin mutant in an identical manner (Preston *et al.*, 1991). (5) $I_{Mg(Ca)}$ also most likely flows through the Mg^{2+} channel activated by $I_{Ca(d)}$, since the Mg^{2+} currents observed at opposite polarizations have the same ion selectivity and blockage pattern (Preston, 1990).

When *Paramecium* strikes an obstacle in its swim path, it backs up. Mechanical impacts of the anterior end of electrode-pinned cells elicit a depolarization prior to the action potential. This receptor potential was found to be based on Ca^{2+} (Eckert *et al.*, 1972). Although nearly all divalent cations were able to carry $I_{Ca(mech)}$, the corresponding receptor current under voltage clamp, Ca^{2+} presumably is the natural carrier based on the distributions of ions (Ogura and Machemer, 1980; Satow *et al.*, 1983). Touching the poste-rior of a *Paramecium* causes it to spurt forward. This mechanical stimulation induces a hyperpolarization by means of a K^+ efflux, $I_{K(mech)}$ (Naitoh and Eckert, 1973; Machemer, 1976; Ogura and Machemer, 1980).

When a fraction of *Paramecium* ciliary membrane was incorporated into planar lipid bilayers, activities of two types of divalent-cation-passing channel were observed and reported (Ehrlich *et al.*, 1984): (1) a voltage-independent 29-pS channel through which Mg^{2+} is slightly more permeant than Ba^{2+} and Ca^{2+} that might be equivalent to the anterior mechanoreceptor channel and (2) a voltage-dependent 1.6-pS channel that favors Ba^{2+} and Ca^{2+} greatly over Mg^{2+}, which might be the microscopic equivalent of $I_{Ca(d)}$.

The application of patch clamp to surface blisters or detached cilia from *Paramecium* (Fig. 1) allowed recording of activities from many different types of ion channel including at least four types of K^+ channel, a Ca^{2+} channel, a Na^+ channel, a Cl^- channel (I_{Cl}), and two cation-nonspecific channels (I_{M+}), one of which can be activated by ATP ($I_{M+(ATP)}$) (Martinac *et al.*, 1988; Y. Saimi, B. Martinac, and R. R. Preston, unpublished observations). The activities of two channels have been documented more extensively. (1) A Ca^{2+}-dependent K^+ channel of 72-pS conductance, that is more active on hyper- than on depolarization (Saimi and Martinac, 1989), appears to be the microscopic equivalent of $I_{K(Ca,h)}$. Limited proteolytic digestion activates this channel and makes it independent of Ca^{2+} and voltage (Kubalski *et al.*, 1989). (2) A 19-pS Ca^{2+}-dependent Na^+ channel, which depends on Ca^{2+}–calmodulin for its activation (Saimi and Ling, 1990), is apparently the channel responsible for the macroscopic $I_{Na(Ca)}$ described earlier. For details on these two types of channels, see Preston *et al.* (1991) and Chapter 28.

Although space limitations restrict our discussion to work on *Paramecium*, several ciliated protists have also been investigated in terms of their ion currents, including studies with two-electrode voltage clamp on *Stylonychia* (Machemer and Deitmer, 1987), *Fabrea* (Kubalski, 1987), and *Stentor* (Wood, 1982). Also studies of cation channels have been done by incorporating *Tetrahymena* ciliary membrane into planar lipid bilayers (Oosawa and Sokabe, 1985; Fujiwara *et al.*, 1988).

III. *Dictyostelium*

Dictyostelium discoideum is a slime mold that grows as individual amebas. On starvation, the amebas aggregate to form a multicellular fruiting body using cAMP as an intercellular signal. Aggregating amebas, alternating in time and in location with respect to neighbors in a swarm, release and respond to cAMP (Gerisch, 1987). Among other effects, binding of cAMP appears to stimulate a Ca^{2+} uptake (Bumann *et al.*, 1984) and a K^+ release (Aeckerle *et al.*, 1985).

Mueller *et al.* (1986) first reported a patch-clamp study of *Dictyostelium* cells in which they found a 9-pS channel that probably passes K^+. More recently, Mueller and Hartung (1990) reexamined the membranes of amebic cells after induction of differentiation by nutrient removal. In the cell-attached configuration, these researchers consistently observed the activities of three types of unit conductances: DI, an 11-pS K^+ conductance activated on depolarization; DII, a 6-pS K^+ conductance activated on depolarization; and HI, a 3-pS Ca^{2+}-passing conductance activated on hyperpolarization. Based on probability of encounter, channel densities appear low at $\sim 0.1/\mu m^2$ for DI and HI and $\sim 0.02/\mu m^2$ for DII. Like most channels in other microbes, the physiological roles of these channels have yet to be clarified.

IV. Yeasts and Other Fungi

Because of its large cell size, *Neurospora crassa*, a bread mold, was the first fungus studied extensively using electrophysiological methods to investigate its membrane potential and how this potential is regulated (Slayman, 1977).

A. Plasma Membranes

Two species of yeast have been used extensively in contemporary research: the budding yeast *Saccharomyces cerevisiae* and the fission yeast *Schizosaccharomyces pombe*. These organisms can readily be cultured and manipulated in the laboratory. As expected, Ca^{2+}, K^+, Na^+, Mg^{2+}, and several trace metals are required for growth. Ca^{2+}, in particular, is needed for cell-cycle progression (Iida *et al.*, 1990). In *S. cerevisiae*, at least two CDC proteins, products of cell-division cycle genes, have the EF-hand structures indicative of their being Ca^{2+}-binding proteins (Baum *et al.*, 1986; Miyamoto *et al.*, 1987).

The plasma membrane of *S. cerevisiae* can form a gigaseal with pipettes after enzymatic removal of the cell wall (Fig. 1). Patch-clamp investigation showed activities of at least two types of ion channels. (1) A 20-pS channel is activated on depolarization above a threshold of -10 to 0 mV and is not activated by hyperpolarization. This channel is highly specific for K^+ over Na^+ and is blocked by the usual K^+-channel blockers tetraethylammonium and Ba^{2+} (Gustin *et al.*, 1986). Surprisingly, Ramirez *et al.* (1989) assigned an electrical activity to a K^+ channel that appears at $> +100$ mV or < -100 mV in a wild-type. In a mutant defective in the structural gene for the plasma membrane H^+–ATPase, however, these investigators found this activity to appear at lower voltages in the presence of >50 m*M* ATP. The mutation appears to make this K$^+$ channel sensitive to cytoplasmic ATP. Researchers suggested a tight physical coupling between the ATPase and the channel. (2) Stretch forces exerted on the membrane activate a 36-pS channel that is blockable by 10 μM gadolinium applied to the cytoplasmic side. This channel passes a variety of ions including Ca^{2+}, but prefers cations over anions. This channel also shows an adaptation to a sustained stretch force. Interestingly, this adaptation depends on voltage. The adaptation takes place when the force is applied when the membrane is polarized (cytoplasmic negative); a force applied during depolarization (positive) activates these channels but does not inactivate them. In whole-cell recording, previously inactivated channels can apparently recover during depolarization and force application. Experiments with whole spheroplasts of different sizes showed that the channel-activating force is a stretch along the plane of the membrane and not a pressure perpendicular to it (Gustin *et al.*, 1988). Although this stretch force is most conveniently applied by suction to the pipette on the cell, positive pressure or osmotic forces applied by changing bath osmolarity also activate these channels recorded in the whole-cell mode (X. L. Zhou, M. C. Gustin, and B. Martinac, unpublished observation). A failure to elicit the whole-cell macroscopic mechanosensitive current that was anticipated by microscopic currents seen in patches excised from snail neurons, in particular, cautioned against the possible artifactual origin of mechanically induced activities in general (Morris and Horn, 1991). The expected macroscopic currents have been recorded from *S. cerevisiae* spheroplasts (Gustin *et al.*, 1988). As reported by Gustin, and to the relief of Morris and Horn (Gustin *et al.*, 1991), not only can such macroscopic currents be recorded but they saturate at higher stretch forces, as expected of a finite number of open channels in a yeast cell and not expected of breakdown and leakage. Macroscopic currents from "whole spheroplasts" or "whole bleb" have also been observed in two other fungal preparations. Although the function(s) of the mechanosensitive channels is not yet known, they are likely to play a role in osmoregulation, since yeasts show clear physiological responses to osmotic stress (Blomberg and Adler, 1989).

All unicellular fungi are called yeasts. However, unicellularity is not a profound trait; fission yeasts and budding yeasts were apparently separated very early in evolution. Among the many differences are cell shape and wall compositions. Unlike the result for the budding yeast, treatment of the fission yeast *S. pombe* with zymolyase causes "protoplast protuberances" to appear at the pointed end of the cell (Kobori *et al.*, 1989). These blebs can readily be studied with a patch clamp. The more thoroughly documented ion channel from these blebs is a mechanosensitive channel (Zhou and Kung, 1992). Currents through one or two such channels, and ensemble currents of tens of units, have been recorded from excised patches and whole blebs, respectively. These mechanosensitive channels can be activated by slight pressure applied to whole blebs and can be blocked by gadolinium, like those in *S. cerevisiae*. Unlike the *S. cerevisiae* channels, however, the *S. pombe* channels are more conductive (180 pS) and more selective ($P_{K^+}/P_{Cl^-} = 3.6$), and are poorly activated but strongly inactivated on membrane depolarization during force application. Reports have also been made of activities of other ion channels in the *S. pombe* plasma membrane. Larsson *et al.* (1992) noted the activity of a voltage-gated cation channel of four conductance levels; Zhou and Kung (1992) intimated three types of voltage-activated channel in addition to the mechanosensitive one in this membrane. More detailed reports are anticipated.

The bean-rust fungus *Uromyces appendiculatus* is a curious organism of agricultural import. To successfully infect its host, a germ tube emerging from the spore must reach a stoma on the leaf surface. The growth of the germlings is apparently guided to the stomata by the topography of the bean-leaf surface. Differentiation of the germlings into infectious structures is triggered by the topography of the stomatal lips (Hoch *et al.*, 1987). Treating germinated spores with Novozyme® converts them into protoplasts suitable for patch-clamp studies. Whereas voltage activates channels of smaller conductances, suction applied to the patch activates a 600-pS channel. This channel excludes Cl$^-$ but is not very selective among cations. Its permeability to K$^+$ and to Ca^{2+} are nearly the same. Activities can be seen in excised patches as well as in whole protoplasts. The maxima of macroscopic currents indicate about 100 channels in a 4-μm diameter protoplast, or about 2 channels per μm^2 (Zhou *et al.*, 1991). This current could transduce the membrane stress induced by the leaf topography into an influx of Ca^{2+}, which could then act as a second messenger for subsequent differentiation of the infectious structures. However, causal relationships have yet to be established. Note that gadolinium blocks this channel and also prevents germling growth and differentiation.

B. Internal Membranes

As diagrammed in Fig. 1, hypo-osmotic rupturing of the plasma membrane of *S. cerevisiae* releases a giant mitochondrion and a vacuole (Saimi *et al.*, 1992). Both objects can be investigated with patch-clamp techniques.

Although first discovered in mitochondria of *Paramecium* (Schein *et al.*, 1976b), the outer membranes of all mitochondria tested, including those of yeast, are equipped with voltage-dependent anion-selective channels (VDACs). Through these channels, nonprotein molecules traverse the mitochondrial outer membrane. Direct examination of the mitochondria, exposed as shown in Fig. 1, revealed activities of VDAC (Saimi *et al.*, 1992). Wunder and Colombini (1991) studied thoroughly the VDAC of *Neurospora* (a bread mold) reconstituted in liposomes with patch-clamp techniques. Most of the experimentation on VDACs, however, was performed with planar lipid bilayers. VDACs are highly conductive. The open-state conductance is 14 nS in 4 M KCl and is even higher in higher KCl concentrations. Open VDAC is mildly anion selective, with a P_{Cl-}/P_{K+} of 1.8. VDAC also shows several lower-conductance "closed" states of different ion selectivity. The channel tends to open when no voltage is imposed, but closes when the membrane is polarized in either direction. The VDAC of *S. cerevisiae* is a 30-kDa protein consisting of 283 amino acid residues (Mihara and Sato, 1985). The sequence contains no hydrophobic region long enough to span the bilayer as an α helix, except an N-terminal amphipathic α-helical mitochondrial target sequence. However, the VDAC sequence contains 12 stretches of alternating hydrophobic and hydrophilic residues, which could form a β sheet with the former protruding on one side and the latter on the other (Forte *et al.*, 1987). The current model of VDAC is a barrel of a single such β sheet. This model explains why so little protein can enclose such a large pore. Extensive site-directed mutagenesis showed that charge-reversal mutations in sites distributed throughout the length of the molecule can alter ion selectivity. These sites apparently form the filter lining of the pore (Blachly-Dyson *et al.*, 1990). Voltage-induced closure to lower-conductance states is modeled as having some of the β strands slip out of the barrel, causing it to become narrower. At some sites, charge reversals alter the ion selectivity of the open state but not of the closed states (Peng *et al.*, 1992). These sites may be on β strands that are extruded in the closed states. Zimmerberg and Parsegian (1986) found a large decrease in the volume of water within VDAC when it closes. This observation is also consistent with the stave-removal model, and not with a local-constriction model for closure (see also Chapters 34 and 35).

Figure 1 also shows the release of a vacuole from a yeast protoplast after osmotic shock. Patch-clamp examinations of such vacuoles, in vacuole-side-out excised patches, have been carried out. Several types of conductance were noted, including two studies discussed in more detail (Saimi *et al.*, 1992). For example, Minorsky *et al.* (1989) noted activities of a cation conductance that is 140 pS (at cytoplasmic-side positive voltages) and 98 pS (vacuole positive), with subconductance states and a slow kinetics. Bertl and Slayman (1990) made a more thorough study of an ~120-pS channel and showed that it activates at a range of cytoplasmic-side negative voltages. Channel opening is not affected much by cytoplasmic Ca^{2+} until it exceeds 1 mM. Interestingly, however, reducing agents cause the channel to become activated at micromolar levels of cytoplasmic Ca^{2+}.

C. Killer Toxin

Some yeast strains produce toxins that are capable of killing cells of sensitive strains. The K1 killer strains of *S. cerevisiae* harbor an M1 virus with a double-stranded RNA genome. The K1 killer toxin, the product of a viral gene, forms a 118-pS cation channel when incorporated into artificial liposomes (Martinac *et al.*, 1990a; Fig. 1). This channel gates independently of voltage and is capable of passing K^+. It does not discriminate between K^+ and Na^+, but prefers monovalent to divalent cations. These findings are consistent with the notion that K^+ depletion of the sensitive cells during intoxication is responsible for the lethal effect.

V. *Escherichia coli*

Escherichia coli, like other gram-negative bacteria, has two membranes separated by periplasm. Peptidylglycan cell wall in the periplasmic space provides rigidity for the cell. The wall and two membranes are collectively referred to as the cell envelope. The outer membrane is a very specialized structure because the outer monolayer is composed of lipopolysaccharide rather than the usual phospholipids. At least 14 proteins are usually found in this membrane, among which are porins (see subsequent discussion). The inner membrane is the true cytoplasmic membrane, which houses the machinery for electron transport, nutrient uptake, chemoreception, and so on.

Some 10^5 porins constitute a major portion of the outer membrane of an *E. coli* cell. Although other types of inducible porins for special solutes exist, this review covers only the group of major porins comprising OmpC, OmpF, and PhoE, all of which show more than 70% homology in their amino acid sequences (Jap and Walian, 1990; Jeanteur *et al.*, 1991). Although these porins form trimers, each ~30-kDa monomer is an individual aqueous channel that can pass hydrophilic solutes up to 600 daltons in size. The structures of OmpF and PhoE in the open state have been solved by X-ray crystallography (Cowan *et al.*, 1992). Both appear to be trimeric. Each monomeric subunit consists of a 16-stranded antiparallel β barrel enclosing a pore. A loop between β strands 6 and 7 folds into the pore, constricting the pore at about half the height of the barrel to 7×11 Å in an elliptical cross section. This constriction apparently defines the conductance, since selected OmpC mutations that increase the solute-exclusion size limit are substitutions of sites in this loop by residues with smaller side chains (Misra and Benson, 1988). Lakey *et al.* (1991) showed that this constriction apparently also defines selectivity, since a lysine-to-glycine difference at a strategic position of this loop correlates well with the difference in the slight anion preference of PhoE and the slight cation preference of OmpF. Extensive biochemical studies led to the view that porins are responsible for the molecular sieve nature of the outer membrane (Benz, 1988; Nikaido, 1992) and that porins are static pores. However, some studies in planar bilayers showed that porins can be closed by strong membrane polarizations (Schindler and Rosenbusch, 1978; Morgan *et al.*, 1990). Berrier *et al.* (1992) used a patch clamp to examine OmpF and OmpC porins purified from salt-extract reconstituted in liposomes. These porins could be closed by strong voltages in steps of 200 pS, but also showed millisecond flickers in steps of 60–70 pS at lower voltages. Berrier *et al.* (1992) interpreted these results to be the behaviors of the trimers and the monomers, respectively. Using a method modified from that of Criado and Keller (1987), Delcour *et al.* (1989a) incorporated *E. coli* membrane fractions into asolectin liposomes and found three types of channels. One type tends to be open most of the time, but frequently and briefly exists in the closed state. Gating of several units between states appears to be cooperative and is encouraged by depolarization. Strong depolarization irreversibly closes the channels, however (Delcour *et al.*, 1989b). As mentioned earlier, Misra and Benson (1988) generated OmpC mutants that could survive on a substrate larger than the solute-exclusion limit of wild-type OmpC. Lakey *et al.* (1991) showed that this mutant has a larger pore and a greater voltage sensitivity using planar lipid bilayers. Using this mutant, Delcour *et al.* (1991) identified the activity, observed with liposome patches, to be that of OmpC. These investigators showed that the mutant channel has a 9–30% increase in unit conductance, as expected of this mutant, which is known to have a larger solute-exclusion limit. However, the most prominent change is that the mutant channel shows a voltage-dependent closure, mostly in groups of three

rather than in groups of two, as seen in the channels of the parent strain. The differences in porin kinetics observed by Delcour *et al.* (1991) and Berrier *et al.* (1992) may be explained by the use of incorporation of membrane fragments in the former study and the use of purified porins in the latter study. However, why OmpC porins appear to be more active in patch-clamp experiments than in planar bilayer experiments is not clear. Ascribing a physiological function to voltage-gating of porins is also challenging. However, a Donnan potential occurs across the outer membrane. If periplasmic charges are organized and are intimately associated with the inner monolayer of the outer membrane, the potential across this membrane, as experienced by porins, may be larger than can be estimated (Stock *et al.*, 1977). Major contributors to these charges are the membrane-derived oligosaccharides (MDOs) in the periplasm. Interestingly, Delcour *et al.* (1992) showed that MDO not only reduces the OmpC currents but also promotes their cooperative closures.

Osmolarity regulates the synthesis of both porins and MDO. Low osmolarities in the culture medium induce synthesis of MDO, the concentration of which often rises up to 20 mM (Sen *et al.*, 1988). MDOs as fixed negative charges in the periplasm should attract positive charges to the periplasmic space. These molecules may adjust the ionic strength and, therefore, the osmolarity of the periplasm, which then acts as a buffer for the cytoplasmic (inner) membrane against low osmolarity in the external medium. However, changing MDO concentrations in the periplasm does not affect the permeability of cephaloridine as measured by a substrate-conversion method (Sen *et al.*, 1988). Low medium osmolarities also favor higher OmpF/OmpC ratios in the outer membrane of strain K-12. The elaborate molecular mechanism that regulates the OmpF/OmpC ratio has been studied by genetic dissection (Igo *et al.*, 1990). The adaptive value of this regulation is not clear, however, since K-12 mutants lacking OmpF or OmpC and the closely related wild-type strain B lacking OmpC all grow well in the laboratory, regardless of culture osmolarity (see Csonka and Hanson, 1991).

The surfaces of giant spheroplasts or giant cells generated by the method shown in Fig. 1 or by other methods (Buechner *et al.*, 1990) can form gigaOhm seals with patch-clamp pipettes (Martinac *et al.*, 1987). Gigaseal formation is itself surprising, given the conventional view of the outer membrane as a static sieve. Either the outer membrane is dynamic and most porins are closed *in vivo*, or the patch-clamp manipulation induces porin closure. In any event, patch-clamp studies revealed the activity of a mechanosensitive channel in such a preparation (Martinac *et al.*, 1987). Mild suction or pressure greatly increased the open probability of this channel in a manner predictable by a Boltzmann distribution in which the applied mechanical energy partitions the channel between its open and closed conformations. An osmotic change in the solution bathing an inside-out patch can also activate this channel (Martinac *et al.*, 1992). Gadolinium ion at submillimolar concentrations blocks this channel (Martinac *et al.*, 1991). Membrane depolarization favors the open state. This channel has a conductance of about 1 nS in 200 mM KCl. The channel visits substates of lower conductances. The dwell time in these substates is much longer in a mutant lacking the major lipoprotein (Kubalski *et al.*, 1992a). The channel favors the passage of anions slightly over cations, and its kinetic behavior is affected by the ionic species (Martinac *et al.*, 1987). Higher levels of suction applied to spheroplast patches through the pipette often activate a second channel that has an approximately 3-nS conductance and much faster open/close kinetics than the 1-nS channel (Sukharev *et al.*, 1993,1994). Szabo *et al.* (1990) also reported mechanosensitive currents from *E. coli* spheroplasts. These researchers observed a much larger number of conductances in a wild-type than the two studied by Sukharev *et al.* (unpublished data). Differences in strains and procedures may account for some of the differences. *Escherichia coli* mechanosensitive channels in membrane fragments have been reconstituted into asolectin liposomes and were found to retain their functions observed in spheroplasts, including their ability to be activated by mild suction (Delcour *et al.*, 1989a).

The ability to reconstitute the channel activities *in vitro* (Sukharev *et al.*, 1993) allowed enrichment and identification of channel proteins by following the activities through fractionation. Beginning with total *E. coli* membranes, two different series of chromatography enriched the same protein of about 17,000 m.w. corresponding to the 3-nS activity. N-terminal sequence of this protein led to the cloning of its gene, *mscL* (mechano-sensitive *c*hannel of very *l*arge conductance). Insertional disruption of *mscL* (knockout) removed the channel activity. Expressing a plasmid-born *mscL* in the *mscL*-knockout chromosomal background restored the activity. Furthermore, the 17-kD protein was also expressed in a cell-free reticulocyte lysate with *mscL* being the only template and the expressed material containing the channel activity. The open reading frame of *mscL* corresponds to a unique protein of 136 amino-acid residues. The N-terminal $\frac{3}{4}$ of the molecule is highly hydrophobic (Sukharev *et al.*, 1994). Unlike channels gated by ligands or voltage, little is known concerning mechanosensitive channels in terms of molecular structure.

mscL is the first such channel to be cloned and sequenced. Deeper understanding of mechanosensitivity will be reached by dissecting *mscL* as a model molecule. Because this channel can readily be incorporated *in vitro* and function in foreign lipids, it has a very large conductance and therefore a favorable signal-to-noise ratio. Also, it resides in the genetically most amenable biological system, that of *E. coli*.

Although a biochemical and molecular biological approach to studying these channels will be crucial to our understanding in the future, a biophysical approach has been taken to determine how the force is transduced to the channel protein. Cell membranes are asymmetrical; the inner monolayers have more phospholipids with negatively charged head groups than the outer monolayers (Rothman and Lenard, 1977). The outer membrane of *E. coli* is even more asymmetrical, since its outer monolayer is of an unusual composition. An amphipathic molecule is expected to partition differentially into the two monolayers because of this asymmetry. The uneven insertion of amphipaths leads to local membrane buckling, which is readily visible in blood cells, as predicted by the "bilayer couple hypothesis" (Sheetz and Singer, 1974,1976). Such a perturbation is expected to create a mechanical stress force in the membrane. Martinac *et al.* (1990b) showed that a variety of amphipaths activate the 1-nS mechanosensitive channel of *E. coli*. Further, channel activations by different amphipaths correspond in time course and effectiveness to their known solubilities into lipid.

In animal cells, the gating force is believed to be transmitted to the mechanosensitive channels through the cytoskeleton (Sachs, 1988). The amphipath experiment and the reconstitution experiment in *E. coli* indicate that the channel can respond to stress forces in the lipid bilayer instead. Sukharev *et al.* (1993,1994) showed that the *E. coli* mechanosensitive channels, solubilized in a detergent and reconstituted into liposomes of foreign lipids, nonetheless retain their mechanosensitive function. Thus, at least these types of channels are gated by tension transduced exclusively via the lipid bilayer. In *E. coli*, the outer membrane is anchored to the peptidylglycan (cell wall) that provides rigidity. Digestion of this peptidylglycan with lysozyme activates rather than inactivates the channel (Buechner *et al.*, 1990; Martinac *et al.*, 1992). Peptidylglycan, the bacterial equivalent of the cytoskeleton, normally restrains the outer membrane, attenuating rather than transmitting the mechanical forces in the membrane (Martinac, 1992).

Membrane deformation can also be induced by mutations. The most abundant protein in the outer membrane is the 7.2-kDa lipoprotein that is embedded in the inner monolayer and does not protrude into the outer monolayer. This lipoprotein is also a major covalent link between the outer membrane and the peptidylglycan. Kubalski et al. (1993) showed that the outer membrane of a mutant lacking this major lipoprotein is much less able to transmit the stretch force to the mechanosensitive channel. A bulky amphipath (lysolecithin), but not smaller amphipaths, can restore this ability. The forces that gating the channel in the bilayer that result from the relative but simultaneous expansion and compression of the two monolayers have been treated in a theoretical model by Markin and Martinac (1991).

The location of the 1-nS mechanosensitive channel described in the preceding section is controversial. Martinac and co-workers (1987,1992; Buechner et al., 1990) argued for its outer membrane location on several grounds. Giant spheroplasts have an outer membrane as seen by electron microscopy and immunofluorescence microscopy. The patch-clamp pipette is not likely to be able to pass through the outer membrane and still seal onto the inner membrane. Digestion of excised inside-out patches with lysozyme added to the bath activates the channel. Were the channel in the inner membrane, the peptidylglycan would have been facing the pipette side, inaccessible to lysozyme. Also, mutational deletion of lipoproteins clearly located in the outer membrane strongly affects the behavior of these mechanosensitive channels, as described earlier (Kubalski et al., 1993). Finally, patch-clamp experiments on protoplasts with only the cytoplasmic membrane revealed the activities of a new set of channels but not the activities of porins or the 1-nS mechanosensitive channel (Kubalski et al., 1992). Note that these arguments are based on observations made on patches from spheroplasts or protoplasts, preparations more closely resembling live cells than reconstituted liposomes. Even these investigators found mechanosensitive channel activities in asolectin liposomes incorporated with material from either the "inner membrane" or the "outer membrane" fractions, separated by sucrose gradient (Delcour et al., 1989a). A different group of investigators (Berrier et al., 1989) observed mechanosensitive channel activities in 50% of the patches from liposomes incorporated with inner membrane material, but only in 7% of the patches from those incorporated with outer membrane material, and concluded that the channel is localized to the inner membrane. Conclusions must be drawn cautiously because perfect separation of inner and outer membranes by gradient centrifugation cannot be achieved (Osborn et al., 1972) and a direct comparison of different material in liposomes assumes the same efficiency of incorporation and the same degree of channel inactivation in the procedures. Berrier et al. (1989) found the mechanosensitive channel(s) to

be of different unit conductances, from 140 to 950 pS. Events of even higher conductance were observed, although rarely. The differences between the observations by Berrier et al. (1989) and Delcour et al. (1989a), both made on reconstituted liposomes, have not been resolved.

As early as the 1950s, investigators observed that hypotonic shock caused E. coli to jettison its smaller solutes and ions but retain its macromolecules (Britten and McClure, 1962). The pathway for this behavior has not been identified, but it clearly has an exclusion limit. Building on the idea that the mechanosensitive channels observed are in the inner membrane, Berrier et al. (1992) showed that gadolinium ion blocks the channel and also blocks the loss of ATP, glutamate, and lactose and slows down the loss of K^+ from osmotically down-shocked cells. These researchers concluded that this channel is the pathway for the controlled efflux of solutes on hypo-osmotic shock. Although this idea is very attractive, caution is required since gadolinium blocks most if not all mechanosensitive channels.

Based on the methods of Birdsell and Cota-Robles (1967), Kubalski et al. (1992) succeeded in preparing E. coli protoplasts suitable for patch-clamp experimentation. Several types of ion channels, each less than 100 pS in conductance, have consistently been observed. In addition, these investigators have also observed a 600-pS mechanosensitive channel that has kinetics that are completely different from those of the 1-nS or 3-nS channels observed on the outer membrane of E. coli (Kubalski, B. Martinac, and Kung, unpublished observations). Gram-positive bacteria have only one membrane, the cytoplasmic membrane. Note that mechanosensitive channels have also been recorded from the membranes of two such bacteria: Streptococcus faecalis (Zoratti and Petronilli, 1988) and Bacillus subtilis (Zoratti et al., 1990; Martinac et al., 1992).

Using gradient-fractionated inverted vesicles of E. coli plasma membrane fused to planar lipid bilayers, Simon et al. (1989) detected a 115-pS channel at 45 mM potassium glutamate. This channel prefers anions over cations and was more active on negative voltages. Researchers believe it is a channel that conducts protein to the periplasm.

VI. Conclusion

The application of a variety of techniques, especially patch clamp, has clearly shown that microbes, like other organisms, have ion channels. To date, microbial ion channels are largely only known by their activities. Except for VDACs, porins, and Shaker-like channel genes of *Paramecium* (T. Jegla and L. Salkoff, personal communication), the proteins and genes that correspond to these electrical activities have yet to be identified. As stated at the beginning of this review, microbes are vastly more diverse than animals and plants. That different types of channels are encountered in different microbial membranes is, therefore, not surprising.

Interestingly, voltage-gated K^+-specific channels seem to have evolved in all branches of eukaryotic organisms, including plants, animals, protists, slime molds, and several fungi. Voltage-gated Ca^{2+}-specific channels are found in *Paramecium* and *Dictyostelium*, but not in yeasts. Voltage-gated Na^+ channels have not been encountered in microbes to date. Such Na^+ channels might represent late evolutionary modification of Ca^{2+} channels, to exploit only the electrical and not the ionic effects. Also no reports have been made on microbial channels directly gated by external ligands. This deficiency may not reflect an absence of such channels from microbes as much as the difficulty in knowing which of the hundreds of possible ligands should be presented to the microbial surfaces. Gating by internal second messengers is exemplified by four types of Ca^{2+}-gated channels in *Paramecium*. Analyses show clearly that these channels are Ca^{2+}–calmodulin gated. This method of gating is apparently not limited to protists, since evidence suggests that this mechanism is also employed by higher animals (see Chapter 28). Determining whether other current types first described in microbes, such as Mg^{2+}-specific current and hypolarization-activated Ca^{2+} current, also have equivalents in higher forms would be interesting. Channel gating by stretch force in the lipid bilayer appears to be an ancient mechanism. Mechanosensitive channels are found in *Paramecium*, yeasts, *E. coli*, and other bacteria. This fact may also reflect the need common to all microbes to deal with osmotic stress.

The discovery of ion channels in the cytoplasmic membranes of bacteria is surprising to many bacteriologists. Intuitively, the cytoplasmic membrane sustaining a proton motive force should not be leaky. This intuition leads to a claim that bacterial cells, unlike eukaryotic cells, do not have ion channels except in the outer membranes (Maloney, 1987). This claim originated from the absence of reports of channels in the inner membrane. However, until recently, no proper technique has been available by which to evaluate this membrane properly, electrophysiologically. The use of patch clamp, in combination with other methods, should provide a better understanding of bacterial electrophysiology. Although *E. coli* clearly has ion channels in its outer and inner membranes, as reviewed here, the relationship of these channels to eukaryotic channels has yet to be established.

Many basic mechanisms, such as how protein conformation can be governed by a stretch force, remain to be elucidated. How microbial channels function in the lives of microbes is also largely unknown. Clearly, future studies on the molecular biology and molecular physiology of microbial channels hold promise that is likely to be fulfilled, since the microbes reviewed in this chapter are amenable to genetic and molecular manipulations.

Acknowledgments

We would like to thank Yoshiro Saimi, who introduced the patch-clamp technology into our laboratory, making it possible for us to examine single-channel activities of *Paramecium*, yeasts, and *E. coli*. We would also like to thank Robin R. Preston for constructive criticism of this paper. This work was supported by National Institutes of Health Grants GM22714 and GM36186 and a grant from the Lucille P. Markey Charitable Trust.

References

Aeckerle, S., Wurster, B., and Malchow, D. (1985). Oscillations and cyclic AMP-induced changes of the K^+ concentration in *Dictyostelium discoideum*. *EMBO J.* **4**, 39–43.

Baum, P., Furlong, C., and Byers, B. (1986). Yeast gene required for spindle pole body duplication: Homology of its product with Ca^{2+}-binding proteins. *Proc. Natl. Acad. Sci. USA* **83**, 5512–5516.

Benz, R. (1988). Structure and function of porins from gram-negative bacteria. *Annu. Rev. Microbiol.* **42**, 359–393.

Berrier, C., Coulombe, A., Houssin, C., and Ghazi, A. (1989). A patch-clamp study of inner and outer membranes and of contact zones of *E. coli*, fused into giant liposomes. Pressure-activated channels are localized in the inner membrane. *FEBS Lett.* **259**, 27–32.

Berrier, C., Coulombe, A., Szabo, I., Zoratti, M., and Ghazi, A. (1992). Gadolinium ion inhibits loss of metabolites induced by osmotic shock and large stretch-activated channels in bacteria. *Eur. J. Biochem.* **206**, 559–565.

Bertl, A., and Slayman, C. L. (1990). Cation-selective channels in the vacuolar membrane of *Saccharomyces*: Dependence on calcium, redox state, and voltage. *Proc. Natl. Acad. Sci. USA* **87**, 7824–7828.

Birdsell, D. C., and Cota-Robles E. H. (1967). Production and ultrastructure of lysozyme and ethylenediamine tetraacetate-lysozyme spheroplasts of *Escherichia coli*. *J. Bacteriol.* **93**, 427–437.

Blachly-Dyson, E., Peng, S.-Z., Colombini, M., and Forte, M. (1990). Selectivity changes in site-directed mutants of the VDAC ion channel: Structural implications. *Science* **247**, 1233–1236.

Blomberg, A., and Adler, L. (1989). Roles of glycerol and glycerol-3-phosphate dehydrogenase (NAD$^+$) in acquired osmotolerance of *Saccharomyces cerevisiae*. *J. Bacteriol.* **171**, 1087–1092.

Bonini, N., Gustin, M., and Nelson, D. L. (1986). Regulation of ciliary motility by membrane potential in *Paramecium*: A role for cyclic AMP. *Cell Motil. Cytosk.* **6**, 256–272.

Brehm, P., and Eckert, R. (1978). Calcium entry leads to inactivation of calcium channel in *Paramecium*. *Science* **202**, 1203–1206.

Britten, R. J., and McClure, F. T. (1962). The amino acid pool of *Escherichia coli*. *Bacteriol. Rev.* **26**, 292–335.

Buechner, M., Delcour, A. H., Martinac, B., Adler, J., and Kung, C. (1990). Ion channel activities in the *Escherichia coli* outer membrane. *Biochim. Biophys. Acta* **1024**, 111–121.

Bumann, J., Malchow, D., and Wurster, B. (1984). Attractant-induced changes and oscillations of the extracellular Ca^{2+} concentration in suspensions of differentiating *Dictyostelium* cells. *J. Cell Biol.* **98,** 173–178.

Cowan, S. W., Schirmer, T., Rummel, G., Steiert, M., Ghosh, R., Pauptit, R. A., Jansonius, J. N., and Rosenbusch, J. P. (1992). Crystal structures explain functional properties of two *E. coli* porins. *Nature (London)* **358,** 727–733.

Criado, M., and Keller, B. (1987). A membrane fusion strategy for single-channel recordings of membranes usually non-accessible to patch-clamp pipette electrodes. *FEBS Lett.* **224,** 172–176.

Csonka, L. N., and Hanson, A. D. (1991). Prokaryotic osmoregulation: Genetics and physiology. *Annu. Rev. Microbiol.* **45,** 569–606.

Delcour, A. H., Martinac, B., Adler, J., and Kung, C. (1989a). Modified reconstitution method used in patch-clamp studies of *Escherichia coli* ion channels. *Biophys J.* **56,** 631–636.

Delcour, A. H., Martinac, B., Adler, J., and Kung, C. (1989b). Voltage-sensitive ion channel of *Escherichia coli.* *J. Membr. Biol.* **112,** 267–275.

Delcour, A. H., Adler, J., and Kung, C. (1991). A single amino acid substitution alters conductance and gating of OmpC porin of *Escherichia coli.* *J. Membr. Biol.* **119,** 267–275.

Delcour, A. H., Adler, J., Kung, C., and Martinac, B. (1992). Membrane-derived oligosaccharides (MDOs) promote closing of a *E. coli* porin channel. *FEBS Lett.* **304,** 216–220.

Dunlap, K. (1977). Localization of calcium channels in *Paramecium caudatum.* *J. Physiol.* **271,** 119–134.

Eckert, R., Naitoh, Y., and Friedman, K. (1972). Sensory mechanisms in *Paramecium.* I. Two-components of the electric response to mechanical stimulation of the anterior surface. *J. Exp. Biol.* **56,** 683–694.

Ehrlich, B., Finkelstein, A., Forte, M., and Kung, C. (1984). Voltage-dependent calcium channels from *Paramecium* cilia incorporated into planar lipid bilayers. *Science* **225,** 427–428.

Forte, M., Guy, H., and Mannella, C. (1987). Molecular genetics of the VDAC ion channel: Structural model and sequence analysis. *J. Bioenerg. Biomembr.* **19,** 341–350.

Fujiwara, C., Anzai, K., Kirino, Y., Nagao, S., Nozawa, Y., and Takahashi, M. (1988). Cation channels from ciliary membrane of *Tetrahymena* reconstituted into planar lipid bilayer. Comparison between the channels from the wild *T. thermophila* and from its mutant which does not show ciliary reversal. *J. Biochem.* **104,** 344–348.

Gerisch, G. (1987). Cyclic AMP and other signals controlling cell development and differentiation in *Dictyostelium. Annu. Rev. Biochem.* **6,** 853–879.

Goertz, H.-D. (1988). "*Paramecium.*" Spring-Verlag, Berlin.

Gustin, M. C., Martinac, B., Saimi, Y., Culbertson, M. R., and Kung, C. (1986). Ion channels in yeast. *Science* **233,** 1195–1197.

Gustin, M. C., Zhou, X.-L., Martinac, B., and Kung, C. (1988). A mechanosensitive ion channel in the yeast plasma membrane. *Science* **242,** 762–765.

Gustin, M. C., Sachs, F., Sigurdson, W., Ruknudin, A., Bowman, C., Morris, C. E., and Horn, R. (1991). Single-channel mechanosensitive currents. *Science* **253,** 800.

Hamill, O. P., Marty, A., Neher, E., Sakmann, B., and Sigworth, F. J. (1981). Improved patch-clamp techniques for high-resolution current recordings from cells and cell-free membrane patches. *Pfluegers Arch.* **391,** 85–100.

Hennessey, T. M., and Kung, C. (1987). A calcium-dependent potassium current is increased by a single-gene mutation in *Paramecium. J. Membr. Biol.* **98,** 145–155.

Hennessey, T. M., Machemer, H., and Nelson, D. L. (1985). Injected cyclic AMP increases ciliary beat frequency in conjunction with membrane hyperpolarization. *Eur. J. Cell Biol.* **36,** 153–156.

Hoch, H., Staples, R. C., Whitehead, B., Comeau, J., and Wolf, E. D. (1987). Signaling for growth orientation and cell differentiation by surface topography in *Uromyces. Science* **235,** 1659–1662.

Igo, M. M., Slauch, J. M., and Silhavy, T. J. (1990). Signal transduction in bacteria: Kinases that control gene expression. *New Biol.* **2,** 5–9.

Iida, H., Sakaguchi, S., Yagawa, Y., and Anraku, Y. (1990). Cell cycle controls by Ca^{2+} in *Saccharomyces cerevisiae. J. Biol. Chem.* **265,** 21216–21222.

Jap, B. K., and Walian, P. J. (1990). Biophysics of the structure and function of porins. *Q. Rev. Biophys.* **23,** 367–403.

Jeanteur, D., Lakey, J. H., and Pattus, F. (1991). The bacterial porin superfamily: Sequence alignment and structure prediction. *Mol. Microbiol.* **5,** 2153–2164.

Jennings, H. S. (1906). *Behavior of Lower Organisms.* Indiana University Press, Bloomington.

Jurand, A., and Selman, G. G. (1969). "The Anatomy of *Paramecium aurelia.*" St. Martin's Press, London.

Kink, J. R., Maley, M. E., Preston, R. R., Ling, K.-Y., Wallen-Friedman, M. A., Saimi, Y., and Kung, C. (1990). Mutations in *Paramecium* calmodulin indicate functional differences between the C-terminal and the N-terminal lobes *in vivo. Cell* **62,** 165–174.

Kobori, H., Yamada, N., Taki, A., and Osumi, M. (1989). Actin is associated with the formation of the cell wall in reverting protoplasts of the fission yeast *Schizosaccharomyces pombe. J. Cell Science* **94,** 635–646.

Kubalski, A. (1987). The effects of tetraethylammonium on the excitability of a marine ciliate *Fabrea salina. Acta Protozool.* **26,** 135–144.

Kubalski, A., Martinac, B., and Y. Saimi (1989). Proteolytic activation of a hyperpolarization- and calcium-dependent potassium channel in *Paramecium. J. Membr. Biol.* **112,** 91–96.

Kubalski, A., Martinac, B., Ling, K.-Y., Adler, J., and Kung, C. (1993). Activities of a mechanosensitive ion channel in an *E. coli* mutant lacking the major lipoprotein. *J. Membr. Biol.* **131,** 151–160.

Kubalski, A. Martinac, B., Adler, J., and Kung, C. (1992). Patch clamp studies on *E. coli* inner membrane *in vivo. Biophys. J.* **61,** A513.

Kung, C., and Eckert, R. (1972). Genetic modifications of electric properties in an excitable membrane. *Proc. Natl. Acad. Sci. USA* **69,** 93–97.

Kung, C., Preston, R. R., Maley, M. E., Ling, K-Y., Kanabrocki, J. A., Seavey, B. R., and Saimi, Y. (1992). *In vivo Paramecium* mutants show that calmodulin orchestrates membrane responses to stimuli. *Cell Calcium* **13,** 413–425.

Lakey, J. H., Lea, E. J. A., and Pattus, F. (1991). ompC mutants which allow growth on maltodextrins show increased channel size and greater voltage sensitivity. *FEBS Lett.* **278,** 31–34.

Larsson, P., Vacata, V., Lecar, H., and Hofer, M. (1992). Multilevel cationic channel in the plasma membrane of *Schizosaccharomyces pombe. Biophys. J.* **61,** A512.

Ling, K.-Y., Preston, R. R., Burns, R., Kink, J. A., Saimi, Y., and Kung, C. (1992). Primary mutations in calmodulin prevent activation of the Ca^{2+}-dependent Na^+ channel in *Paramecium. Protein Struct. Funct. Genet.* **12,** 365–371.

Machemer, H. (1976). Interactions of membrane potential and cations in regulation of ciliary activity in *Paramecium. J. Expt. Biol.* **65,** 427–448.

Machemer, H., and Deitmer, J. W. (1987). From structure to behaviour: *Stylonychia* as a model system for cellular physiology. *Prog. Protistol* **2,** 213–330.

Machemer, H., and Ogura, A. (1979). Ionic conductances of membranes in ciliated and deciliated *Paramecium. J. Physiol.* **296,** 49–60.

Maloney, P. C. (1987). Coupling to an energized membrane: Role of ion-motive gradients in the transduction of metabolic energy.

In "*Escherichia coli* and *Salmonella typhimurium*, Cellular and Molecular Biology" (F. C. Neidhardt, J. L. Ingraham, K. BrookssLow, B. Magasanik, M. Schaechter, and H. E. Umbarger, eds.), Vol. 1, pp. 222–243. American Society of Microbiology, Washington, D.C.

Markin, V. S., and Martinac, B. (1991). Mechanosensitive ion channels as reporters of bilayer expansion: a theoretical model. *Biophys. J.* **60**, 1120–1127.

Martinac, B. (1992). Mechanosensitive ion channels: biophysics and physiology. *In* "CRC Thermodynamics of Membrane Receptors and Channels" (M. B. Jackson, ed.). CRC Press, Boca Raton, Florida, 327–351.

Martinac, B., Buechner, M., Delcour, A. H., Adler, J., and Kung, C. (1987). A pressure-sensitive ion channel in *Escherichia coli*. *Proc. Natl. Acad. Sci. USA* **84**, 2297–2301.

Martinac, B., Saimi, Y., Gustin, M. C., and Kung, C. (1988). Ion channels of three microbes: *Paramecium*, yeast and *Escherichia coli*. *In* "Calcium and Ion Channel Modulation" (A. D. Grinnell, D. Armstrong, and M. B. Jackson, eds.), pp. 415. Plenum Press, New York.

Martinac, B., Zhu, H., Kubalski, A., Zhou, X.-L., Culbertson, M., Bussey, H., and Kung, C. (1990a). Yeast K1 killer toxin forms ion channels in sensitive yeast spheroplasts and in artificial liposomes. *Proc. Natl. Acad. Sci. USA* **87**, 6228–6232.

Martinac, B., Adler, J., and Kung, C. (1990b). Mechanosensitive ion channels of *E. coli* activated by amphipaths. *Nature (London)* **348**, 261–263.

Martinac, B., Delcour, A. H., Buechner, M., Adler, J., and Kung, C. (1992). Mechanosensitive ion channels in bacteria. *In* "Advances in Comparative and Environmental Physiology" (F. Ito, ed.), Vol. 10, pp. 3–18. Springer-Verlag, Berlin.

Mihara, K., and Sato, R. (1985). Molecular cloning and sequencing of cDNA for yeast porin, an outer mitochondrial membrane protein: A search for targeting signal in the primary structure. *EMBO J.* **4**, 769–774.

Minorsky, P. V., Zhou, X.-L., Culbertson, M. R., and Kung, C. (1989). A patch-clamp analysis of a cation current in the vacuolar membrane of the yeast *Saccharomyces*. *Plant Physiol.* **89**, 149.

Misra, R., and Benson, S. A. (1988). Genetic identification of the pore domain of the OmpC porin of *Escherichia coli* K-12. *J. Bacteriol.* **170**, 528–533.

Miyamaoto, S., Okya, Y., Ohsumi, Y., and Anraku, Y. (1987). Nucleotide sequence of the CLS4 (CDC24) gene of *Saccharomyces cerevisiae*. *Gene* **54**, 125–132.

Morgan, H., Lansdale, J. T., and Adler, G. (1990). Polarity-dependent voltage-gated porin channels form *Escherichia coli* in lipid bilayer membrane. *Biochim. Biophys. Acta* **1021**, 175–181.

Morris, C. E., and Horn, R. (1991). Failure to elicit neuronal macroscopic mechanosensitive currents anticipated by single-channel studies. *Science* **251**, 1246–1249.

Mueller, U., and Hartung, K. (1990). Properties of three different ion channels in the plasma membrane of the slime mold *Dictyostelium discoideum*. *Biochim. Biophys. Acta* **1026**, 204–212.

Mueller, U., Malchow, D., and Hartung, K. (1986). Single ion channels in the slime mold *Dictyostelium discoideum*. *Biochim. Biophys. Acta* **857**, 287–290.

Naitoh, Y., and Eckert, R. (1969). Ionic mechanisms controlling behavioral responses of *Paramecium* to mechanical stimulation. *Science* **164**, 945–965.

Naitoh, Y., and Eckert, R. (1973). Sensory mechanisms in *Paramecium*. II. Ionic basis of the hyperpolarizing mechanoreceptor potential *J. Exp. Biol* **59**, 53–65.

Nakaoka, Y., and Machemer, H. (1990). Effects of cyclic nucleotides and intracellular Ca on voltage-activated ciliary beating in *Paramecium*. *J. Comp. Physiol.* **166**, 401–406.

Nikaido, H. (1992). Porins and specific channels of bacterial outer membranes. *Mol. Microbiol.* **6**, 435–442.

Oertel, D., Schein, S. J., and Kung, C. (1977). Separation of membrane currents using a *Paramecium* mutant. *Nature (London)* **268**, 120–124.

Oertel, D., Schein, S. J., and Kung, C. (1978). A potassium conductance activated by hyperpolarization in *Paramecium*. *J. Membr. Biol.* **43**, 169–185.

Ogura, A., and Machemer, H. (1980). Distribution of mechanoreceptor channels in *Paramecium* surface membrane. *J. Comp. Physiol.* **135**, 233–242.

Oosawa, Y., and Sokabe, M. (1985). Cation channels form *Tetrahymena* cilia incorporated into planar lipid bilayers. *Am. J. Physiol.* **249**, C177–C179.

Osborn, M. J., Gander, J. E., Parisi, E., and Carson, J. (1972). Mechanism of assembly of the outer membrane of *Salmonella typhimurium*. *J. Biol. Chem.* **247**, 3962–3972.

Peng, S.-Z., Blackly-Dyson, E., Colombini, M., and Forte, M. (1992). Large scale rearrangement of protein domains is associated with voltage gating of the VDAC channel. *Biophys. J.* **62**, 123–135.

Preston, R. R. (1990). A magnesium current in *Paramecium*. *Science* **250**, 285–287.

Preston, R. R., and Saimi, Y. (1990). Calcium ions and the regulation of motility in *Paramecium*. *In* "Ciliary and Flagellar Membranes" (R. A. Bloodgood, ed.). Plenum Press, New York. pp 173–200.

Preston, R. R., Saimi, Y., and Kung, C., (1990a). Evidence for two K$^+$ currents activated upon hyperpolarization of *Paramecium tetraurelia*. *J. Membr. Biol.* **115**, 41–50.

Preston, R. R., Wallen-Friedman, M. A., Saimi, Y., and Kung, C. (1990b) Calmodulin defects cause the loss of Ca^{2+}-dependent K$^+$ currents in two pantophobiac mutants of *Paramecium tetraurelia*. *J. Membr. Biol* **115**, 51–60.

Preston, R. R., Saimi, Y., Amberger, E., and Kung, C. (1990c). Interactions between mutants with defects in two Ca^{2+}-dependent K$^+$ currents of *Paramecium tetraurelia*. *J. Membr. Biol.* **115**, 61–69.

Preston, R. R., Kink, J. A., Hinrichsen, R. D., Saimi, Y., and Kung, C. (1991). Calmodulin mutants and Ca^{2+}-dependent channels in *Paramecium*. *Annu. Rev. Physiol.* **53**, 309–319.

Preston, R. R., Saimi, Y., and Kung, C. (1992a). Calcium current activated upon hyperpolarization of *Paramecium tetraurelia*. *J. Gen. Physiol.* **100**, 233–251.

Preston, R. R., Saimi, Y., and Kung, C. (1992b). Ca-dependent inactivation of the calcium current activated upon hyperpolarization of *Paramecium tetraurelia*. *J. Gen. Physiol.* **100**, 253–268.

Ramirez, J. A., Vacata, V., McCusker, J. H., Haber, J. E., Mortimer, R. K., Owen, W. G., and Lecar, H. (1989). ATP-sensitive K$^+$ channels in a plasma membrane hyperpolarization-ATPase mutant of the yeast *Saccharomyces cerevisiae*. *Proc. Natl. Acad. Sci. USA* **86**, 7866–7870.

Rothman, J. E., and Lenard, J. (1977). Membrane asymmetry. The nature of membrane asymmetry provides clues to the puzzle of how membranes are assembled. *Science* **195**, 743–753.

Sachs, F. (1988). Mechanical transduction in biological systems. *CRC Crit. Rev. Biomed. Eng.* **16**, 141–169.

Saimi, Y. (1986). Calcium-dependent sodium currents in *Paramecium*: Mutational manipulations and effects of hyper- and depolarization. *J. Membr. Biol.* **92**, 227–236.

Saimi, Y., and Kung, C. (1980). A Ca-induced Na$^+$ current in *Paramecium*. *J. Exp. Biol.* **88**, 305–325.

Saimi, Y., and Kung, C. (1982). Are ions involved in the gating of calcium channels? *Science* **218**, 153–156.

Saimi, Y., Hinrichsen, R. D., Forte, M., and Kung, C. (1983). Mutant analysis shows that the Ca^{2+}-induced K^+ current shuts off one type of excitation in *Paramecium*. *Proc. Natl. Acad. Sci. USA* **80**, 5112–5116.

Saimi, Y., and Martinac, B. (1989). Calcium-dependent potassium channel in *Paramecium* studied under patch clamp. *J. Membr. Biol.* **112**, 79–89.

Saimi, Y., Martinac, B., Delcour, A. H., Minorsky, P. V., Gustin, M. C., Culbertson, M. R., Adler, J., and Kung, C. (1992). Patch clamp studies of microbial ion channels. *Meth. Enzymol.* **207**, 681–691.

Satow, Y., and Kung, C. (1977). A regenerative hyperpolarization in *Paramecium*. *J. Comp. Physiol* **119**, 99–110.

Satow, Y., and Kung, C. (1980a). Membrane currents of pawn mutants of the *pwA* group in *Paramecium tetraurelia*. *J. Exp. Biol.* **84**, 57–71.

Satow, Y., and Kung, C. (1980b). Ca-induced K^+-outward current in *Paramecium tetraureli*. *J. Exp. Biol* **88**, 293–303.

Satow, Y., Murphy, A. D., and Kung, C. (1983). The ionic basis of the depolarizing mechanoreceptor potential of *Paramecium tetraurelia*. *J. Exp. Biol.* **103**, 253–264.

Schein, S. J., Bennett, M. V. L., and Katz, G. M. (1976a). Altered calcium conductance in pawns, behavioral mutants of *Paramecium aurelia*. *J. Exp. Biol.* **65**, 699–724.

Schein, S. J., Colombini, M., and Finkelstein, A. (1976b). Reconstitution in planar lipid bilayers of voltage-dependent anion-selective channel obtained from *Paramecium* mitochondria. *J. Membr. Biol.* **30**, 99–120.

Schindler, H., and Rosenbusch, J. P. (1978). Matrix protein from *Escherichia coli* outer membranes forms voltage-controlled channels in lipid bilayers. *Proc. Natl. Acad. Sci. USA* **75**, 3751–3755.

Schultz, J. E., Klumpp, S., Benz, R., Schurhoff-Goeters, W. J. Ch., and Schmid, A. (1992). Regulation of adenylyl cyclase from *Paramecium* by an interinsic potassium conductance. *Science* **255**, 600–603.

Sen, K., Hellman, J., and Nikaido, H. (1988). Porin channels in intact cells of *Escherichia coli* are not affected by Donnan potentials across the outer membrane. *J. Biol. Chem.* **263**, 1182–1187.

Sheetz, M. P., and Singer, S. J. (1974). Biological membranes as bilayer couples. A molecular mechanism of drug-erythrocyte interaction. *Proc. Natl. Acad. Sci. USA* **71**, 4457–4461.

Sheetz, M. P., Painter, R. G., and Singer, S. J. (1976). Biological membranes as bilayer couples. III. Compensatory shape changes induced in membranes. *J. Cell Biol.* **70**, 193–203.

Simon, S. M., Blobel, G., and Zimmerberg, J. (1989). Large aqueous channels in membrane vesicles derived from the rough edoplasmic reticulum of canine pancreas or the plasma membrane of *Escherichia coli*. *Proc. Natl. Acad. Sci. USA* **86**, 6176–6180.

Slayman, C. L. (1977). Energetics and control of transport in *Neurospora*. *In* "Water Relations in Membrane Transport in Plants and Animals" (A. M. Jungreis, T. K. Hodges, A. Kleinzeller, and S. G. Schultz, eds.), pp. 69–86. Academic Press, New York.

Stock, J. B., Rauch, B., and Roseman, S. (1977). Periplasmic space in *Salmonella typhimurim* and *Escherichia coli*. *J. Biol. Chem.* **252**, 7850–7861.

Sukharev, S. I., Blount, P., Martinac, B., Blattner, F. R., and Kung, C. (1994). *mscL* alone encodes a large-conductance mechanosensitive channel in *E. coli*. *Nature* (*London*) **368**, 265–268.

Sukharev, S. I., Martinac, B., Arshavsky, V. Y., and Kung, C. (1993). Two types of mechanosensitive channels in the *Escherichia coli* cell envelope: solubilization and functional reconstitution. *Biophys. J.* **65**, 177–183.

Szabo, I., Petronilli, V., Guerra, L., and Zoratti, M. (1990). Cooperative mechanosensitive ion channels in *Escherichia coli*. *Biochem. Biophys. Res. Commun.* **171**, 280–186.

Wood, D. C. (1982). Membrane permeabilities determining resting, action and mechanoreceptor potentials in *Stentor coeruleus*. *J. Comp. Physiol.* **146**, 537–550.

Wunder, U. R., and Colombini, M. (1991). Patch clamping VDAC in liposomes containing whole mitochondrial membranes. *J. Membr. Biol.* **123**, 83–91.

Zhou, X.-L., and Kung, C. (1992). A mechanosensitive ion channel in *Schizosaccharomyces pombe*. *EMBO J.* **11**, 2869–2875.

Zhou, X.-L., Stumpf, M. A., Hoch, H. C., and Kung, C. (1991). A mechanosensitive channel in whole cells and in membrane patches of the fungus *Uromyces*. *Science* **253**, 1415–1417.

Zimmerberg, J., and Parsegian, V. (1986). Polymer inaccessible volume changes during opening and closing of a voltage-dependent ionic channel. *Nature* (*London*) **323**, 36–39.

Zoratti, M., and Petronilli, V. (1988). Ion-conducting channels in a gram-positive bacterium. *FEBS Lett.* **240**, 105–109.

Zoratti, M., Petronilli, V., and Szabo, I. (1990). Stretch-activated composite ion channels in *Bacillus subtilis*. *Biochem. Biophys. Res. Commun.* **168**, 443–450.

CHANNELS OF INTRACELLULAR MEMBRANES

Channels of Calcium Stores

1. Ryanodine Receptor
(Chapters 30 and 31)

2. InsP$_3$ Receptor
(Chapters 32 and 33)

Ryanodine Receptor/ Ca²⁺-Release Channel of Excitable Tissues

GERHARD MEISSNER

I. Introduction

In excitable tissues, release of Ca^{2+} ions from intracellular Ca^{2+} stores can be triggered by an excitatory electrical signal (Fabiato, 1983; Endo, 1977) or can occur via a chain of voltage-independent steps that involve agonist-induced formation of inositol 1,4,5-trisphosphate (IP_3) and subsequent activation of an intracellular membrane receptor–channel complex, the IP_3 receptor (Somlyo et al., 1988; van Breemen and Saida, 1989; see Chapters 32 and 33). The voltage-dependent mechanism, commonly referred to as excitation–contraction (E–C) coupling, has been studied most thoroughly in vertebrate striated muscle. In cardiac and skeletal muscle, rapid release of Ca^{2+} from the intracellular compartment, the sarcoplasmic reticulum (SR), is triggered by a surface-membrane action potential that is communicated to the SR at specialized areas where the "junctional" SR comes in close contact with the surface membrane and tubular in-foldings of the surface membrane (T-tubules). At these areas, protein bridges are present that span the gap between the two membrane systems. These structures have been termed "feet" (Franzini-Armstrong, 1970), and are now more generally known as the SR Ca^{2+}-release channel or ryanodine receptor (RyR), because of he presence of an intrinsic Ca^{2+}-channel activity within

the feet structures and their ability to bind the plant alkaloid [³H]ryanodine with high affinity and specificity (Fleischer and Inui, 1989; Lai and Meissner, 1989). Currently, three mammalian cDNAs have been cloned and fully sequenced, encoding a skeletal RyR (Takeshima et al., 1989; Zorzato et al., 1990), a cardiac RyR (Nakai et al., 1990; Otsu et al., 1990), and a brain RyR (Takamata et al., 1992). The RyRs are composed of ~5000 amino acid residues; the predicted amino acid sequences of the three RyRs were found to be 65–70% identical. Northern blot analysis of mRNA from a variety of mammalian tissues has indicated that the skeletal isoform appears to be restricted to fast- and slow-twitch skeletal muscle, whereas the cardiac isoform is apparently expressed, in addition to in the heart, at low levels throughout the brain, with the exception of the pituitary (Nakai et al., 1990; Otsu et al., 1990; Hakamata et al., 1992; Lai et al., 1992a). The brain RyR mRNA was found to be most abundant in certain areas of the brain (corpus striatum, thalamus, and hippocampus) and to be present in tissues containing smooth muscle, such as aorta and uterus (Hakamata et al., 1992). A mammalian, caffeine-insensitive RyR isoform has been shown by Northern blot analysis to be expressed at low abundance in excitable and nonexcitable mink tissues treated with transforming growth factor β (TGF-β), with the exception of the heart (Giannini et al., 1992).

Partial sequence analysis indicated 94% amino sequence identity with the C-terminal one-eighth of the rabbit brain RyR.

The specific mechanism of RyR/Ca^{2+}-release channel activation remains to be fully defined. In skeletal muscle, a unique mechanism referred to as the mechanical coupling mechanism has been formulated that suggests that small transient charge movements observed during T-tubule depolarization represent the physiological link between a surface-membrane action potential and SR Ca^{2+} release (Rios and Pizarro, 1991). An important implication of the mechanical coupling model is that the SR Ca^{2+}-release channel directly interacts with a voltage-sensing molecule in the T-tubule membrane. Biophysical and pharmacological evidence (Rios and Pizarro, 1991) and cDNA expression studies (Tanabe *et al.*, 1988; see Chapter 11) have suggested that the dihydropyridine (DHP) receptor for Ca^{2+} channel antagonists is the voltage sensor in E-C coupling in vertebrate skeletal muscle. Accordingly, protein conformational changes occurring during T-tubule depolarization may be sensed by the SR Ca^{2+}-release channel, leading to its activation and the release of Ca^{2+} from the SR. The involvement of a M_r ~90,000 protein (triadin) in physically linking the DHP receptor and SR feet/Ca^{2+}-release channels has been postulated (Brandt *et al.*, 1992).

Unlike in vertebrate skeletal muscle, in mammalian cardiac muscle a DHP-receptor isoform has been shown to function as a voltage-dependent Ca^{2+} channel that mediates Ca^{2+} influx during a surface-membrane action potential (Tanabe *et al.*, 1990). The rise in intracellular Ca^{2+} concentration has been suggested to trigger the opening of a Ca^{2+}-activated release channel in the SR (Cannell *et al.*, 1987; Nabauer *et al.*, 1989). Good evidence suggests that the Ca^{2+} trigger mechanism also plays a major role in regulating a SR RyR/Ca^{2+}-release channel in crustacean skeletal muscle (Gyorke and Palade, 1992; Seok *et al.*, 1992). Mammalian smooth muscle (Herrmann-Frank *et al.*, 1991) and brain (Ashley, 1989; Lai *et al.*, 1992a) also contain Ca^{2+}-activated RyR/Ca^{2+}-release channels. However, whether Ca^{2+} is the physiological activator in these cells is unclear.

In subsequent sections, this chapter focuses on studies carried out to determine several of the structural and functional features of the RyR/Ca^{2+}-release channel. These studies have shown that excitable tissues express a 30 S ryanodine-sensitive, Ca^{2+}-gated release channel protein complex. The channel complexes share a high-affinity binding site for [^{3}H]ryanodine and a high-conductance pathway for Ca^{2+} and monovalent cations, but display isoform- and species-dependent differences in their *in vitro* regulation by Ca^{2+} and other effector molecules.

II. Microsomal Membrane Fractions Enriched in Ca^{2+}-Release and [^{3}H]Ryanodine Binding Activities

The molecular properties of the RyR/Ca^{2+}-release channel and its regulation by Ca^{2+} and other endogenous and exogenous effectors have been studied most extensively with rabbit skeletal muscle as the source for RyR. Fragmentation of the SR during homogenization and subsequent fractionation by differential and density gradient centrifugation yielded a "heavy" SR vesicle fraction that was enriched in Ca^{2+}-release and [^{3}H]ryanodine-binding activities, and corresponded to the terminal cisternae or junctional regions of the SR (Nagasaki and Kasai, 1983; Meissner, 1984; Fleischer *et al.*, 1985). Another important advance allowing the study of coupling between T-tubule depolarization and SR Ca^{2+} release has been the isolation of membrane fractions enriched in "triads" composed of a T-tubule vesicle sandwiched by two heavy SR vesicles (Ikemoto *et al.*, 1989). Microsomal membrane fractions enriched in Ca^{2+}-release and/or [^{3}H]ryanodine-binding activities have been also isolated from other excitable tissues, including cardiac and smooth muscle and brain. In most cases, however, these membrane fractions were of a lower purity and less well characterized than those from mammalian skeletal muscle.

III. Isolation of 30 S RyR/Ca^{2+}-Release Channel

The neutral plant alkaloid ryanodine binds in media of high ionic strength, with nanomolar affinity and in a Ca^{2+}-dependent manner, to the RyR/Ca^{2+}-release channel (Table I; Lai and Meissner, 1989; see Chapter 31). Because the drug binds with high specificity and dissociates slowly from the high-affinity site of the membrane-bound and detergent-solubilized receptor, [^{3}H]ryanodine has been found to be an ideal probe in the isolation of the RyR from a variety of tissues and species.

The RyR/Ca^{2+}-release channel was first isolated from striated muscle because its role in regulating free cytoplasmic Ca^{2+} levels was originally recognized in muscle. Relatively large amounts of membranes enriched in [^{3}H]ryanodine-binding activity can be obtained from skeletal and cardiac muscle. The detergent-solubilized rabbit skeletal (Pessah *et al.*, 1986; Imagawa *et al.*, 1987; Inui *et al.*, 1987; Hymel *et al.*, 1988; Lai *et al.*, 1988) and canine cardiac (Inui *et al.*, 1987; Anderson *et al.*, 1989; Rardon *et al.*, 1989) RyRs were purified with retention of their

TABLE I Comparison of Properties of 30 S RyR/Ca²⁺ Release Channel Complexes

Property	Rabbit skeletal	Canine cardiac	Mammalian vascular	Bovine brain	Frog skeletal	Lobster skeletal
Apparent sedimentation coefficient(S)	30	30	30	30	30	30
Number of ~560-kDa polypeptides (on SDS gels)	1	1	1	1	2	1
Morphology	quatrefoil	quatrefoil	ND[a]	quatrefoil	ND	ND
Ryanodine binding						
high affinity (K_D)[b] (nM)	5–6	2–3	ND	3.0	4.0	6.5
low affinity	yes	ND	ND	yes	yes	ND
Single-channel conductance						
in 0.25 M K⁺ (pS)	770	745	370	800	ND	775
in 0.5 M Na⁺ (pS)	600	550	ND	ND	600	ND
in 0.05 M Ca²⁺ (pS)	145[c]	145[c]	~100[c]	140[c]	110[d]	125[c]
Regulation						
activation by Ca²⁺	μM	μM	μM	μM	μM	mM
activation by mM ATP and/or mM caffeine	yes	yes	yes	yes	yes	weak
inhibition by mM Mg²⁺ and/or μM ruthenium red	yes	yes	yes	yes	yes	weak
Modification by ryanodine						
to subconducting state	yes	yes	not observed	yes	yes	yes
to closed state	yes	ND	yes	yes	ND	yes

[a] ND, Not determined.

[b] In 1.0 M NaCl, pH 7 buffer containing 0.1–1.0 mM Ca²⁺ and 5 mM AMP.

[c] With 0.25 M KCl in *cis* chamber.

[d] With 0.15 M NaCl in *cis* chamber.

[³H]ryanodine-binding and intrinsic Ca²⁺ channel activities by sequential column chromatography (Inui *et al.*, 1987; Hymel *et al.*, 1988) and, in essentially one step, by immunoaffinity chromatography (Smith *et al.*, 1988) or density gradient centrifugation (Lai *et al.*, 1988). The membrane-bound Ca²⁺-release channel was solubilized in the presence of the channel-specific probe [³H]ryanodine using the zwitterionic detergent CHAPS and high ionic strength (1.0 M NaCl); then the channel was purified as a 30 S protein complex by centrifugation through a linear sucrose gradient (Fig. 1). Density gradient centrifugation resulted in efficient separation of the large 30 S RyR complex from other smaller solubilized SR proteins because of its faster sedimentation rate. The sucrose gradient centrifugation procedure is relatively simple and straightforward and has been used to isolate the 30 S RyR from rabbit (Lai *et al.*, 1988), frog (Lai *et al.*, 1992b), and lobster (Seok *et al.*, 1992) skeletal muscle, canine cardiac muscle (Anderson *et al.*, 1989), and bovine brain (Lai *et al.*, 1992a), as well as to identify a 30 S RyR complex from canine and porcine aorta (Herrmann-Frank *et al.*, 1991) and toad stomach muscle (Xu *et al.*,

1991) (Table I). SDS gel electrophoresis has shown that the RyR peak fractions of mammalian skeletal and cardiac muscle (Fig. 1; Imagawa *et al.*, 1987; Inui *et al.*, 1987; Lai *et al.*, 1988; Anderson *et al.*, 1989), brain (McPherson *et al.*, 1991; Lai *et al.*, 1992a), crustacean skeletal muscle (Seok *et al.*, 1992), and the nematode *Caenorhaloditis elegans* (Kim *et al.*, 1992) contain a single major high molecular weight polypeptide with a calculated M_r~560,000, as determined by cloning and sequencing of the complementary DNA of the mammalian skeletal and cardiac muscle RyR isoforms (Takeshima *et al.*, 1989; Otsu *et al.*, 1990; Nakai *et al.*, 1990; Zorzato *et al.*, 1990). In contrast, the presence of two major immunologically distinct high molecular weight RyR protein bands has been described for chicken, frog, and toadfish skeletal muscle (Airey *et al.*, 1990; Olivares *et al.*, 1991; Lai *et al.*, 1992b). The two receptor isoforms are present in chicken as discrete homo-oligomers (Airey *et al.*, 1990), and were shown in frog to have immunological properties and an electrophoretic mobility characteristic of the mammalian skeletal and cardiac receptor isoforms, respectively (Lai *et al.*, 1992b).

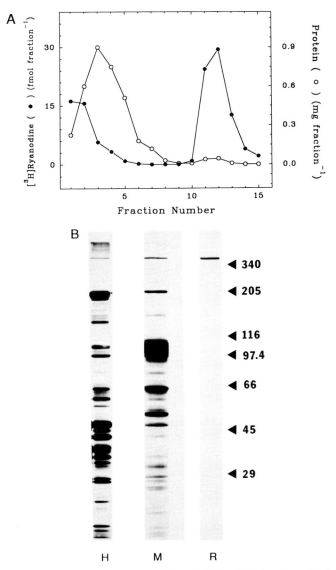

FIGURE 1 Sedimentation profile and SDS gel electrophoresis of rabbit skeletal and canine cardiac ryanodine receptors (RyRs). (A) Heavy skeletal muscle sarcoplasmic reticulum (SR) vesicles were solubilized in CHAPS in the presence of [³H]ryanodine, centrifuged through linear sucrose gradients, and fractionated. Fractions were analyzed for protein and radioactivity. Bound [³H]ryanodine is present in fractions 11–13, comigrating with a small protein peak with an apparent sedimentation coefficient of 30 S. (B) SDS polyacrylamide gel electrophoresis of rabbit skeletal muscle homogenate (H), heavy SR membranes (M), and purified RyR (R), followed by silver staining.

Electron microscopy studies have revealed that the purified RyRs of mammalian skeletal (Lai *et al.*, 1988; Saito *et al.*, 1988) and cardiac (Anderson *et al.*, 1989) muscle and of brain (Lai *et al.*, 1992a) have an overall morphology identical to that of the protein bridges ("feet") that span the T-tubule–SR junctional gap in

isolated SR fractions of vertebrate skeletal muscle (Ferguson *et al.*, 1984). A high apparent sedimentation coefficient and four-leaf-clover-like (quatrefoil) appearance of negative-stained samples (Fig. 2), as well as cross-linking studies (Lai *et al.*, 1989), showed that the skeletal muscle 30 S RyR complex is composed of four M_r ~560,000 polypeptides. Three-dimensional reconstruction of electron micrographs indicated overall projected dimensions of 27 × 27 × 14 nm and suggested the presence of a central channel that connects to four radial channels located in the foot (cytoplasmic) region of the complex (Wagenknecht *et al.*, 1989). Preliminary information regarding the transmembrane arrangement of the receptor has been obtained from amino acid sequence analysis of the mammalian RyRs. Comparison of hydropathy plots of the predicted amino sequences has suggested that the M_r ~560,000 polypeptides consist of two major structural regions, the C-terminal pore region that exhibits a high extent of amino acid sequence similarity and consists of 4 (Takeshima *et al.*, 1989) or 10 (Zorzato *et al.*, 1990) putative transmembrane segments (16 or 40 transmembrane segments per tetrameric RyR, respectively) and a large, more variable extramembrane region that is thought to correspond to the cytoplasmic foot structure. Primary sequence predictions also suggested the presence of several putative cytoplasmic Ca^{2+}, nucleotide, and calmodulin-binding and phosphorylation sites (Takeshima *et al.*, 1989; Zorzato *et al.*, 1990; Hakamata *et al.*, 1992). Of these, however, only one site corresponding to a calmodulin-dependent protein kinase phosphorylation site in the cardiac RyR (Ser 2809) has been confirmed to date (Witcher *et al.*, 1991). The site is present in the cardiac and skeletal RyRs, but only the cardiac site appears to be utilized; on phosphorylation, the release channel is activated by an apparent reversal of inhibition by calmodulin.

IV. RyR/Ca²⁺-Release Channel Conducts Mono- and Divalent Cations

The RyR/Ca²⁺-release channel is a cation-selective channel that displays an unusually large ion conductance for monovalent cations (~600 and ~750 pS with 500 mM Na⁺ and 250 mM K⁺ as the current carriers, respectively) and divalent cations (~100–150 pS with 50 mM Ca²⁺) (Table I). The existence of a large Ca²⁺ conductance was originally demonstrated in single-channel recordings with native skeletal (Smith *et al.*, 1985) and cardiac (Rousseau *et al.*, 1986) muscle SR vesicles fused with planar lipid bilayers, and subsequently in studies with purified RyRs incorporated

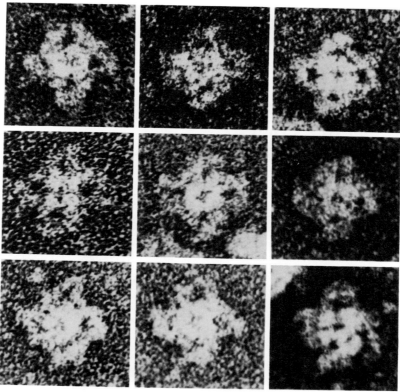

FIGURE 2 Negative-stain electron microscopy of the rabbit skeletal muscle 30 S ryanodine receptor (RyR). A selected panel of particles with the characteristic four-leaf (quatrefoil) structures (Ferguson *et al.*, 1984) is shown. Dimensions of the quatrefoil are 34 nm from the tip of one leaf to the tip of the opposite one; each leaf is 14 nm wide. The central electron-dense region has a diameter of 14 nm with a central hole 1–2 nm in diameter. Reprinted with permission from Lai *et al.* (1988). Copyright © 1988 Macmillan Magazines Ltd.

into lipid bilayers (Hymel *et al.*, 1988; Lai *et al.*, 1988; Smith *et al.*,1988; Anderson *et al.*, 1989; Rardon *et al.*, 1989; Williams, 1992). As shown in Fig. 3, the CHAPS-solubilized 30 S RyR complex, purified in the absence of [^{3}H]ryanodine, can be directly incorporated into planar lipid bilayers. In the presence of a symmetrical 500 mM NaCl buffer, a Ca^{2+}- and ATP-sensitive Na$^+$-conducting single-channel activity with a slope conductance of 595 pS was observed. Perfusion of the *trans* bilayer chamber (corresponding to the SR lumenal side) with a 50 mM Ca^{2+} solution and the *cis* chamber (corresponding to the SR cytoplasmic side) with the slowly permeating cation Tris$^+$ resulted in a Ca^{2+} conductance of 91 pS.

Single-channel recordings have allowed the direct study of the functional consequences of the interaction of ryanodine with its receptor (Imagawa *et al.*, 1987; Rousseau *et al.*, 1987). In turn, the highly characteristic modification of the channel by ryanodine provides a reliable means to distinguish between the Ca^{2+}-release channel and other ion channel currents.

Figure 4 shows that the addition of 10 μM ryanodine to a K$^+$-conducting rabbit skeletal muscle release channel induced the formation of an open channel state of reduced conductance, as originally observed for the native mammalian skeletal and cardiac Ca^{2+}-conducting release channels (Imagawa *et al.*, 1987; Rousseau *et al.*, 1987). Addition of millimolar ryanodine fully closed the channel (Fig. 4), further confirming that the K$^+$-conducting rabbit skeletal channel is a ryanodine-sensitive channel. Table I shows that incorporation into planar bilayers of detergent-solubilized 30 S RyR protein fractions isolated from various tissues and species leads to the formation of a ryanodine-sensitive ion conductance pathway that exhibits a Ca^{2+} and a K$^+$ or Na$^+$ conductance essentially identical to those of the rabbit skeletal muscle Ca^{2+}-release channel. Two exceptions are that the maximal K$^+$ conductance for the mammalian vascular channel is half that of the skeletal and cardiac channels (Herrmann-Frank *et al.*, 1991), and the Ca^{2+}-permeable channel of *C. elegans* has a maximal conductance

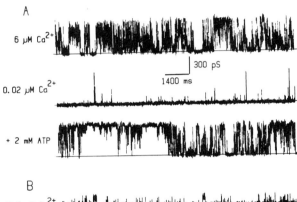

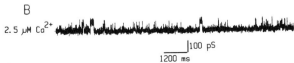

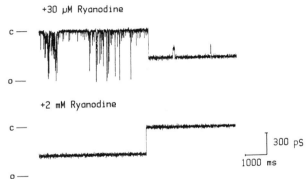

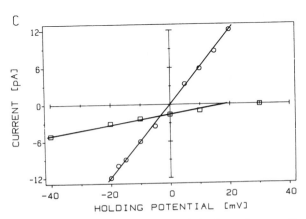

FIGURE 3 Reconstitution of the rabbit skeletal muscle ryanodine receptor (RyR) into planar bilayers. Single-channel currents, shown as upward deflections, were recorded (A) in symmetrical 0.5 M NaCl, 20 mM Na PIPES, pH 7 buffer with the indicated concentrations of free Ca^{2+} and ATP in the *cis* chamber (corresponding to the SR cytoplasmic side) and (B) after perfusion of the *trans* chamber (SR lumenal side) with 50 mM Ca(OH$_2$/250 mM HEPES, pH 7.4 buffer and the *cis* chamber with 125 mM Tris/250 mM HEPES, pH 7.4 buffer plus 2.5 μM free Ca^{2+}. Holding potentials were (A) -15 mV and (B) 0 mV. (C) Current–voltage relationships for recordings in A and B. Values of unit conductances were 595 and 91 pS, with 500 mM Na$^+$ (O) and 50 mM Ca^{2+} ($\square$) as the conducting ion, respectively. Reprinted with permission from Lai *et al.* (1988). Copyright © 1988 Macmillan Magazines Ltd.

FIGURE 4 Effect of ryanodine on a single purified and reconstituted Ca^{2+} release channel. Shown are single-channel currents of the unlabeled purified rabbit skeletal muscle 30 S ryanodine receptor (RyR) incorporated into a planar lipid bilayer in symmetrical 0.25 M KCl buffer containing 50 μM free Ca^{2+}. Holding potential is 20 mV. Upper trace shows appearance of subconducting state with open probability of ~1, several minutes after *cis* addition of 30 μM ryanodine. Lower trace illustrates the sudden transitions from the subconductance state to a fully closed state within 1 min of *cis* addition of 2 mM ryanodine. Bars on left represent the open (o) and closed (c) channel. Reprinted with permission from Lai *et al.* (1989).

of only 218 pS in 250 mM KCl (Kim *et al.*, 1992). However, the appearance of multiple subconductances present within the channel tetramer is a frequently observed phenomenon on reconstitution of the detergent-solubilized, purified channel (Liu *et al.*, 1989). Therefore, further studies may reveal a conductance level of 700–800 pS for these channels, as observed for the skeletal 30 S RyR. Ryanodine modifies the K$^+$ conductance by inducing the formation of an open subconductance state at micromolar concentrations and by fully closing the channel at millimolar concentrations (Table I). The observation of an essentially identical conductance behavior in various RyRs suggests that the pore region of the RyR/Ca^{2+}-release channel complex is highly conserved. In agreement with this prediction, the putative C-terminal pore region shows a high extent of sequence similarity among the four sequenced mammalian RyR isoforms (Giannini *et al.*, 1992; Hakamata *et al.*, 1992).

V. RyR/Ca^{2+}-Release Channel Is a Ligand-Gated Channel

SR Ca^{2+}-release channel activity displays only a weak to moderate voltage dependence in bilayers. Also, the SR of cardiac and skeletal muscle contains, in addition to the Ca^{2+}-release channel, a K$^+$ channel, a Cl$^-$ channel, and an H$^+$ (OH$^-$) permeable pathway (Meissner, 1983). Generation of a large membrane potential and activation of the Ca^{2+}-release channel by a rapid change in SR membrane potential are therefore difficult to visualize.

Vesicle ion flux and single-channel measurements have indicated that RyR/Ca^{2+}-release channels can be activated and inhibited by various effector molecules. Endogenous effector molecules of the mammalian cardiac and skeletal muscle channels include Ca^{2+}, Mg^{2+}, adenine nucleotides, calmodulin (Meissner and Henderson, 1987; and references cited), and possibly IP_3 (Valdivia *et al.*, 1992, and references cited). In addition, a large number of drugs have been found to affect the channel (Palade *et al.*, 1989). The latter include caffeine, which activates the mammalian skeletal and cardiac channels at millimolar concentrations by increasing channel open time without a change in conductance (Rousseau *et al.*, 1988; Rousseau and Meissner, 1989), and ryanodine, which interacts with the release channel by inducing an open subconductance or by fully closing the channel (Fig. 4). Among the inhibitors, ruthenium red is probably the best known because it inhibits the mammalian skeletal release channel at submicromolar concentrations (Miyamoto and Racker, 1982).

Study of Ca^{2+}-release channel activity has been approached by two complementary techniques: measurement of macroscopic $^{45}Ca^{2+}$ fluxes from passively loaded vesicles and recording of microscopic monovalent cation and Ca^{2+} ion currents through single channels incorporated into bilayers using SR vesicles or purified RyRs. These studies have shown that major differences exist in the *in vitro* regulation of the mammalian and crustacean RyRs by Ca^{2+} and other effector molecules. Figure 5 shows that the $^{45}Ca^{2+}$ efflux rate from rabbit skeletal, canine cardiac, and lobster skeletal muscle SR vesicles is dependent on extravesicular (cytoplasmic) $[Ca^{2+}]$. For the mammalian vesicles, a bell-shaped Ca^{2+} activation curve was obtained, with $^{45}Ca^{2+}$ efflux optimal at micromolar $[Ca^{2+}]$; this result suggested that the mammalian skeletal and cardiac release channels possess high-affinity activating and low-affinity inhibitor Ca^{2+}-binding sites. An ~20-fold higher $^{45}Ca^{2+}$ efflux rate from cardiac vesicles indicated that micromolar Ca^{2+} activates the cardiac RyR to a greater extent than the skeletal RyR. The lobster channel appeared to lack corresponding high- and low-affinity regulatory Ca^{2+}-binding sites. $^{45}Ca^{2+}$ efflux was slow at micromolar $[Ca^{2+}]$; an elevation of free Ca^{2+} concentration to 1 mM or greater in the efflux media was required to observe a substantial increase in the $^{45}Ca^{2+}$ efflux rate (Fig. 5). Single-channel measurements with purified RyRs have corroborated vesicle flux measurements by showing that micromolar $[Ca^{2+}]$ activated the mammalian skeletal and cardiac muscle channels (Lai *et al.*, 1988; Anderson *et al.*, 1989), whereas millimolar $[Ca^{2+}]$ were needed to optimally activate the lobster channel (Seok *et al.*, 1992).

Mammalian and crustacean RyRs also exhibit major differences in their sensitivity to *in vitro* activation and inhibition by other effector molecules (Table II). In cardiac and skeletal muscle membrane preparations, the addition of 10 mM caffeine and 5 mM ATP to

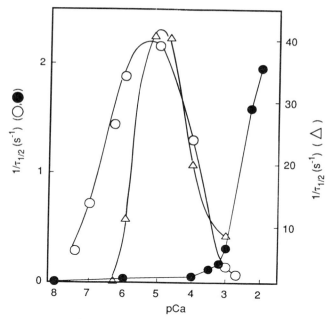

FIGURE 5 Ca^{2+}-induced $^{45}Ca^{2+}$ efflux from rabbit skeletal ($\bigcirc$), canine cardiac ($\triangle$), and lobster skeletal ($\bullet$) sacroplasmic reticulum (SR) vesicles. Vesicles were passively loaded with 1 mM $^{45}Ca^{2+}$ and then diluted into iso-osmolar efflux media containing the indicated concentrations of free Ca^{2+}. $\tau_{1/2}$ indicates the time in which Ca^{2+}-permeable vesicles released half their $^{45}Ca^{2+}$ contents. Reprinted with permission from Seok *et al.* (1992).

nanomolar Ca^{2+} medium or addition of 5 mM adenosine 5' (β,γ-methylene)triphosphate (AMP-PCP) to a micromolar Ca^{2+} medium resulted in nearly complete activation (Smith *et al.*, 1986; Rousseau *et al.*, 1986,1988; Meissner and Henderson, 1987; Rousseau and Meissner, 1989). Intermediate $^{45}Ca^{2+}$ efflux rates were observed when the mammalian channels were activated at nanomolar Ca^{2+} by 5 mM ATP or 10 mM caffeine, or by $10^{-5} M$ Ca^{2+} alone. An important difference was, however, that caffeine and micromolar Ca^{2+} were more effective in activating the cardiac release channel, whereas ATP was more effective in activating the mammalian skeletal Ca^{2+} release channel. In contrast, the addition of neither 5 mM adenine nucleotide or 10 mM caffeine to 10^{-8} or $10^{-3} M$ Ca^{2+} media was effective in giving a more than 2-fold activation of the lobster skeletal muscle release channel. Another significant difference between the mammalian and crustacean release channels was that Ca^{2+}-activated Ca^{2+} release from lobster SR vesicles could not be substantially inhibited by millimolar Mg^{2+} or micromolar ruthenium red concentrations.

Mammalian tissues express, in addition to the skeletal and cardiac muscle isoforms, a brain (Hakamata *et al.*, 1992), TGF-β-regulated caffeine-insensitive

TABLE II Comparison of $^{45}Ca^{2+}$ Efflux from Rabbit Skeletal, Canine Cardiac, and Lobster Skeletal Sarcoplasmic Reticulum Vesicles[a]

Free Ca^{2+} (M)	Additions to efflux medium (M)	$^{45}Ca^{2+}$ efflux ($\tau_{1/2}$) (sec)		
		Rabbit skeletal	Canine cardiac	Lobster skeletal
$<10^{-8}$	—	8	25	130
	5×10^{-3} ATP	0.06	12	90
	10^{-2} caffeine	0.4	0.05	90
	5×10^{-3} ATP + 10^{-2} caffeine	0.01	0.01	60
$\sim10^{-5}$	—	0.6	0.02	110
	5×10^{-3} AMP-PCP	0.01	0.01	70
	10^{-2} caffeine	0.5	0.01	70
	10^{-3} Mg^{2+}	15	0.25	
	10^{-5} ruthenium red	>30	>10	
10^{-3}	—	10	0.1	2.5
	5×10^{-3} Mg^{2+}			3.0
	10^{-5} ruthenium red	>30	>10	5.0

[a] SR vesicles were passively loaded with 1 mM $^{45}Ca^{2+}$ and diluted into efflux media containing the indicated concentrations of free Ca^{2+} and other additions. $\tau_{1/2}$ indicates the time required for Ca^{2+}-permeable vesicles to release half their $^{45}Ca^{2+}$ load.

(Giannini *et al.*, 1992) RyR isoform. Very little is known at present about the regulatory properties of the latter isoform. Its existence has been recognized only recently. Systematic study of the pharmacological properties will be more difficult than study of the skeletal and cardiac muscle isoforms because of the relatively low levels of expression. Planar lipid bilayer reconstitution of partially purified RyR preparations has suggested that mammalian vascular smooth muscle exhibits a Ca^{2+}- and caffeine-sensitive channel activity (Herrmann-Frank *et al.*, 1991). Northern blot analysis (Hakamata *et al.*, 1992) has suggested that this activity is likely to be associated with the brain RyR isoform. On incorporation into bilayers, brain microsomal membrane fractions (Ashley, 1989) and purified RyRs (McPherson *et al.*, 1991; Lai *et al.*, 1992a) also exhibit Ca^{2+}- and caffeine-sensitive channel activities. Because of the expression of cardiac and brain RyR isoforms in brain (Hakamata *et al.*, 1992), however, an assignment of these activities to the correct RyR isoform is not possible. Table I summarizes the currently known properties of the vascular smooth muscle (Herrmann-Frank *et al.*, 1991) and brain (Lai *et al.*, 1992a) RyRs, as well as those of frog skeletal muscle RyR (Lai *et al.*, 1992b).

VI. Conclusion

The identification of [3H]ryanodine as a specific probe has enabled the isolation and subsequent cloning of an intracellular 30 S RyR complex composed of four ~560-kDa polypeptides. Current evidence suggests that at least three distinct RyR genes exist in mammalian cells. The mammalian RyRs show a tissue-specific expression. Those isolated from muscle and neuronal tissue have been shown to share several properties, including an apparent sedimentation coefficient of 30 S, the presence of a high-affinity ryanodine-binding site, activation by Ca^{2+}, and a high-conductance mono- and divalent cation pathway that can be modified by ryanodine in an essentially identical manner. On the other hand, the RyRs exhibit isoform- and species-dependent differences in their sensitivity to activation by Ca^{2+}, ATP, and caffeine and to inhibition by Mg^{2+} and ruthenium red. Further studies are needed to determine the amino acid sequences that are responsible for these differences, as well as to localize the regulatory domains within the large RyR/Ca^{2+}-release channel complex.

Acknowledgments

This work was supported by U.S. Public Health Services grants AR18687 and HL27430.

References

Airey, J. A., Beck, C. F., Murakami, K., Tanksley, S. J., Derrinck, T. J., Ellisman, M. H., and Sutko, J. L. (1990). Identification and localization of two triad junctional foot protein isoforms in mature avian fast twitch skeletal muscle [published erratum appears in *J. Biol. Chem.* (1990) **265(35)**, 22057]. *J. Biol. Chem.* **265**, 14187–14194.

Anderson, K., Lai, F. A., Liu, Q. Y., Rousseau, E., Erickson, H. P., and Meissner, G. (1989). Structural and functional characterization of the purified cardiac ryanodine receptor–Ca²⁺ release channel complex. *J. Biol. Chem.* **264**, 1329–1335.

Ashley, R. H. (1989). Activation and conductance properties of ryanodine-sensitive calcium channels from brain microsomal membranes incorporated into planar lipid bilayers. *J. Membr. Biol.* **111**, 179–189.

Brandt, N. R., Caswell, A. H., Brunschwig, J. P., Kang, J. J., Antoniu, B., and Ikemoto, N. (1992). Effects of anti-triadin antibody on Ca²⁺ release from sarcoplasmic reticulum. *FEBS Lett.* **299**, 57–59.

Cannell, M. B., Berlin, J. R., and Lederer, W. J. (1987). Effect of membrane potential changes on the calcium transient in single rat cardiac muscle cells. *Science* **238**, 1419–1423.

Endo, M. (1977). Calcium release from the sarcoplasmic reticulum. *Physiol. Rev.* **57**, 71–108.

Fabiato, A. (1983). Calcium-induced release of calcium from the cardiac sarcoplasmic reticulum. *Am. J. Physiol.* **245**, C1–C14.

Ferguson, D. G., Schwartz, H. W., and Franzini-Armstrong, C. (1984). Subunit structure of junctional feet in triads of skeletal muscle. A freeze-drying, rotary-shadowing study. *J. Cell Biol.* **99**, 1735–1742.

Fleischer, S., and Inui, M. (1989). Biochemistry and biophysics of excitation–contraction coupling. *Annu. Rev. Biophys. Chem.* **18**, 333–364.

Fleischer, S., Ogunbunmi, E. M., Dixon, M. C., and Fleer, E. A. M. (1985). Localization of calcium release channels with ryanodine in junctional terminal cisternae of sarcoplasmic reticulum of fast skeletal muscle. *Proc. Natl. Acad. Sci. USA* **82**, 7256–7259.

Franzini-Armstrong, C. (1970). Studies of the triad. I. Structure of the junction in frog twitch fibers. *J. Cell Biol.* **47**, 488–499.

Giannini, G., Clementi, E., Ceci, R., Marziali, G., and Sorrentino, V. (1992). Expression of a ryanodine receptor–Ca²⁺ channel that is regulated by TGF-β. *Science* **257**, 91–94.

Gyorke, S., and Palade, P. (1992). Calcium-induced calcium release in crayfish skeletal muscle. *J. Physiol. (London)* **457**, 195–210.

Hakamata, Y., Nakai, J., Takeshima, H., and Imoto, K. (1992). Primary structure and distribution of a novel ryanodine receptor/calcium release channel from rabbit brain. *FEBS Lett.* **312**, 229–235.

Herrmann-Frank, A., Darling, E., and Meissner, G. (1991). Functional characterization of the Ca²⁺-gated Ca²⁺ release channel of vascular smooth muscle sarcoplasmic reticulum. *Pflügers Arch.* **418**, 353–359.

Hymel, L., Inui, M., Fleischer, S., and Schindler, H. (1988). Purified ryanodine receptor of skeletal muscle sarcoplasmic reticulum forms Ca²⁺-activated oligomeric Ca²⁺ channels in planar bilayers. *Proc. Natl. Acad. Sci. USA* **85**, 441–445.

Ikemoto, N., Ronjat, M., and Meszaros, L. G. (1989). Kinetic analysis of excitation-contraction coupling. *J. Bioenerg. Biomembr.* **21**, 247–265.

Imagawa, T., Smith, J. S., Coronado, R., and Campbell, K. P. (1987). Purified ryanodine receptor from skeletal muscle sarcoplasmic reticulum is the Ca²⁺-permeable pore of the calcium release channel. *J. Biol. Chem.* **262**, 16636–16643.

Inui, M., Saito, A., and Fleischer, S. (1987). Isolation of the ryanodine receptor from cardiac sarcoplasmic reticulum and identity with the feet structures. *J. Biol. Chem.* **262**, 15637–15642.

Kim, Y. K., Valdivia, H. H., Maryon, E. B., Anderson, P., and Coronado, R. (1992). High molecular weight proteins in the nematode C. elegans bind [³H]ryanodine and form a large conductance channel. *Biophys. J.* **63**, 1379–1384.

Lai, F. A., and Meissner, G. (1989). The muscle ryanodine receptor and its intrinsic Ca²⁺ channel activity. *J. Bioenerg. Biomembr.* **21**, 227–246.

Lai, F. A., Erickson, H. P., Rousseau, E., Liu, Q. Y., and Meissner, G. (1988). Purification an reconstitution of the calcium release channel from skeletal muscle. *Nature (London)* **331**, 315–319.

Lai, F. A., Misra, M., Xu, L., Smith, H. A., and Meissner, G. (1989). The ryanodine receptor–Ca²⁺ release channel complex of skeletal muscle sarcoplasmic reticulum. Evidence for a cooperatively coupled, negatively charged homotetramer. *J. Biol. Chem.* **264**, 16776–16785.

Lai, F. A., Wickenden, C., Xu, L., Kumari, G., Misra, M., Lee, H. B., Sar, M., and Meissner, G. (1992a). Expression of a cardiac Ca²⁺ release channel isoform in mammalian brain. *Biochem. J.* **288**, 553–564.

Lai, F. A., Liu, Q. Y., Xu, L., El-Hashem, A., Kramarcy, N. R., Sealock, R., and Meissner, G. (1992b). Amphibian ryanodine receptor isoforms are related to those of mammalian skeletal or cardiac muscle. *Am. J. Physiol.* **263**, C365–C372.

Liu, Q. Y., Lai, F. A., Rousseau, E., Jones, R. V., and Meissner, G. (1989). Multiple conductance states of the purified calcium release channel complex from skeletal sarcoplasmic reticulum. *Biophys. J.* **55**, 415–424.

McPherson, P. S., Kim, Y. K., Valdivia, H., Knudson, C. M., Takekura, H., Franzini, A. C., Coronado, R., and Campbell, K. P. (1991). The brain ryanodine receptor: A caffeine-sensitive calcium release channel. *Neuron* **7**, 17–25.

Meissner, G. (1983). Monovalent ion and calcium ion fluxes in sarcoplasmic reticulum. *Mol. Cell. Biochem.* **55**, 65–82.

Meissner, G. (1984). Adenine nucleotide stimulation of Ca²⁺-induced Ca²⁺ release in sarcoplasmic reticulum. *J. Biol. Chem.* **159**, 1365–1374.

Meissner, G., and Henderson, J. S. (1987). Rapid calcium release from cardiac sarcoplasmic reticulum vesicles is dependent on Ca²⁺ and is modulated by Mg²⁺, adenine nucleotide, and calmodulin. *J. Biol. Chem.* **262**, 3065–3073.

Miyamoto, H., and Racker, E. (1982). Mechanism of calcium release from skeletal sarcoplasmic reticulum. *J. Membr. Biol.* **66**, 193–201.

Nabauer, M., Callewaert, G., Cleemann, L., and Morad, M. (1989). Regulation of calcium release is gated by calcium current, not gating charge in cardiac myocytes. *Science* **244**, 800–803.

Nagasaki, K., and Kasai, M. (1983). Fast release of calcium from sarcoplasmic reticulum vesicles monitored by chlortetracycline fluorescence. *J. Biochem. (Tokyo)* **94**, 1101–1108.

Nakai, J., Imagawa, T., Hakamata, Y., Shigekawa, M., Takeshima, H., and Numa, S. (1990). Primary structure and functional expression from cDNA of the cardiac ryanodine receptor/calcium release channel. *FEBS Lett.* **271**, 169–177.

Olivares, E. B., Tanksley, S. J., Airey, J. A., Beck, C. F., Ouyang, Y., Deerinck, T. J., Ellisman, M. H., and Sutko, J. L. (1991). Nonmammalian vertebrate skeletal muscles express two triad junctional foot protein isoforms. *Biophys. J.* **59**, 1153–1163.

Otsu, K., Willard, H. F., Khanna, V. K., Zorzato, F., Green, N. M., and MacLennan, D. H. (1990). Molecular cloning of cDNA encoding the Ca²⁺ release channel (ryanodine receptor) of rabbit cardiac muscle sarcoplasmic reticulum. *J. Biol. Chem.* **265**, 13472–13483.

Palade, P., Dettbarn, C., Brunder, D., Stein, P., and Hals, G. (1989). Pharmacology of calcium release from sarcoplasmic reticulum. *J. Bioenerg. Biomembr.* **21**, 295–320.

Pessah, I. N., Francini, A. O., Scales, D. J., Waterhouse, A. L., and Casida, J. E. (1986). Calcium-ryanodine receptor complex.

Solubilization and partial characterization from skeletal muscle junctional sarcoplasmic reticulum vesicles. *J. Biol. Chem.* **261**, 8643–8648.

Rardon, D. P., Cefali, D. C., Mitchell, R. D., Seiler, S. M., and Jones, L. R. (1989). High molecular weight proteins purified from cardiac junctional sarcoplasmic reticulum vesicles are ryanodine-sensitive calcium channels. *Circ. Res.* **64**, 779–789.

Rios, E., and Pizarro, G. (1991). Voltage sensor of excitation– contraction coupling in skeletal muscle. *Physiol. Rev.* **71**, 849–908.

Rousseau, E., and Meissner, G. (1989). Single cardiac sarcoplasmic reticulum Ca^{2+}-release channel: Activation by caffeine. *Am. J. Physiol.* **256**, H328–H333.

Rousseau, E., Smith, J. S., Henderson, J. S., and Meissner, G. (1986). Single channel and $^{45}Ca^{2+}$ flux measurements of the cardiac sarcoplasmic reticulum calcium channel. *Biophys. J.* **50**, 1009–1014.

Rousseau, E., Smith, J. S., and Meissner, G. (1987). Ryanodine modifies conductance and gating behavior of single Ca^{2+} release channel. *Am. J. Physiol.* **253**, C364–C368.

Rousseau, E., Ladine, J., Liu, Q. Y., and Meissner, G. (1988). Activation of the Ca^{2+} release channel of skeletal muscle sarcoplasmic reticulum by caffeine and related compounds. *Arch. Biochem. Biophys.* **267**, 75–86.

Saito, A., Inui, M., Radermacher, M., Frank, J., and Fleischer, S. (1988). Ultrastructure of the calcium release channel of sarcoplasmic reticulum. *J. Cell Biol.* **107**, 211–219.

Seok, J. H., Xu,, L., Kramarcy, N. R., Sealock, R., and Meissner, G. (1992). The 30 S lobster skeletal muscle Ca^{2+} release channel (ryanodine receptor) has functional properties distinct from the mammalian channel proteins. *J. Biol. Chem.* **267**, 15893–15901.

Smith, J. S., Coronado, R., and Meissner, G. (1985). Sarcoplasmic reticulum contains adenine nucleotide-activated calcium channels. *Nature (London)* **316**, 446–449.

Smith, J. S., Coronado, R., and Meissner, G. (1986). Single channel measurements of the calcium release channel from skeletal muscle sarcoplasmic reticulum. Activation by Ca^{2+} and ATP and modulation by Mg^{2+}. *J. Gen. Physiol.* **88**, 573–588.

Smith, J. S., Imagawa, T., Ma, J., Fill, M., Campbell, K. P., and Coronado, R. (1988). Purified ryanodine receptor from rabbit skeletal muscle is the calcium-release channel of sarcoplasmic reticulum. *J. Gen. Physiol.* **92**, 1–26.

Somlyo, A. P., Walker, J. W., Goldman, Y. E., Trentham, D. R.,

Kobayashi, S., Kitazawa, T., and Somlyo, A. V. (1988). Inositol trisphosphate, calcium and muscle contraction. *Phil. Trans. R. Soc. London* **320**, 399–414.

Takeshima, H., Nishimura, S., Matsumoto, T., Ishida, H., Kangawa, K., Minamino, N., Matsuo, Ueda, M., Hanaoka, M., and Hirose, T. (1989). Primary structure and expression from complementary DNA of skeletal muscle ryanodine receptor. *Nature (London)* **339**, 439–445.

Tanabe, T., Beam, K. G., Powell, J. A., and Numa, S. (1988). Restoration of excitation-contraction coupling and slow calcium current in dysgenic muscle by dihydrophyridine receptor complementary DNA. *Nature (London)* **336**, 134–139.

Tanabe, T., Mikami, A., Numa, S., and Beam, K. G. (1990). Cardiac-type excitation-contraction coupling in dysgenic skeletal muscle injected with cardiac dihydropyridine receptor cDNA. *Nature (London)* **344**, 451–453.

Valdivia, C. V. D., Potter, B. V. L., and Coronado, R. (1992). Fast release of $^{45}Ca^{2+}$ induced by inositol 1,4,5-trisphosphate and Ca^{2+} in the sarcoplasmic reticulum of skeletal muscle: Evidence for two types of Ca^{2+} release channels. *Biophys. J.* **61**, 1184–1193.

van Breemen, C., and Saida, K. (1989). Cellular mechanisms regulating $[Ca^{2+}]_i$ in smooth muscle. *Annu. Rev. Physiol.* **51**, 315–329.

Wagenknecht, T., Grassucci, R., Frank, J., Saito, A,. Inui, M., and Fleischer, S. (1989). Three-dimensional architecture of the calcium channel/foot structure of sarcoplasmic reticulum. *Nature (London)* **338**, 167–170.

Williams, A. J. (1992). Ion conduction and discrimination in the sarcoplasmic reticulum ryanodine receptor/calcium-release channel. *J. Muscle Res. Cell Motil.* **13**, 7–26.

Witcher, D. R., Kovacs, R. J., Schulman, H., Cefali, D. C., and Jones, L. R. (1991). Unique phosphorylation site on the cardiac ryanodine receptor regulates calcium channel activity. *J. Biol. Chem.* **266**, 11144–11152.

Xu, L., Etter, E., Fay, F. S., and Meissner, G. (1991). Evidence for 30 S ryanodine receptor–Ca^{2+} release channel complex for stomach smooth muscle of toad. *Biophys. J.* **59**, 67a.

Zorzato, F., Fujii, J., Otsu, K., Phillips, M., Green, N. M., Lai, F. A., Meissner, G., and MacLennan, D. H. (1990). Molecular cloning of cDNA encoding human and rabbit forms of the Ca^{2+} release channel (ryanodine receptor) of skeletal muscle sarcoplasmic reticulum. *J. Biol. Chem.* **265**, 2244–2256.

31

Pharmacology of Ryanodine-Sensitive Ca²⁺ Release Channels

ILDIKO ZIMANYI AND ISAAC N. PESSAH

I. Introduction

The identity of the functional unit comprising the major Ca^{2+} efflux pathway from sacroplasmic reticulum (SR) that operates in excitation–contraction (E–C) coupling of striated muscle was still undiscovered in the mid-1980s. The naturally occurring plant alkaloid ryanodine provided a much needed research tool. For over half a century, the bark of *Ryania speciosa Vahl*, a woody shrub endemic to the rain forests of Central and South America, has been known to contain active principles that have remarkable biological activity (see Jenden and Fairhurst, 1969, for review). The remarkable toxic potency of ryanodine and its related alkaloids (LD_{50} 20–350 μg/kg in a variety of species; Waterhouse *et al.*, 1987) has inspired many researchers of muscle to understand its mode of action and to identify a specific biological target. The pioneering work of Procita (1958) demonstrated the direct myogenic activity of ryanodine. Jenden and Fairhurst (1969) determined that ryanodine is without an effect on Ca^{2+}–ATPase activity, and clearly demonstrated the existence of an efflux pathway distinct from the Ca^{2+} pump. Localization of the site of ryanodine action required the radiolabeling of the alkaloid. Fairhurst (1971) first attempted labeling the pyrrole ring of ryanodine, but the product did not appear to bind SR in a saturable manner and was of limited use. Identification of 9,21-dehydroryanodine (Waterhouse *et al.*, 1984) as the major active principle in *Ryania* facilitated the synthesis of [³H]-ryanodine. Reduction of 9,21-dehydroryanodine with ³H₂ gas and paladium catalyst produced 9,21-[³H]ryanodine of high specific activity, suitable for receptor binding analyses. This synthetic scheme was soon corroborated (Sutko *et al.*, 1986). Pessah and co-workers (1985) first demonstrated high- and low-affinity binding of [³H]ryanodine to heavy SR membranes of skeletal and cardiac muscle that was influenced by known modulators of SR Ca^{2+} release. Subsequent studies have demonstrated the binding of [³H]ryanodine to high- and low-affinity sites, with K_ds ranging from 1 nM to 3000 nM and binding capacities of 2–64 pmol/mg protein (Fleischer *et al.*, 1985; Seifert and Casida, 1986; Campbell *et al.*, 1987; Lattanzio *et al.*, 1987; Pessah *et al.*, 1987; Inui *et al.*, 1988; Lai *et al.*, 1989; McGrew *et al.*, 1989; Chu *et al.*, 1990; Carroll *et al.*, 1991; Pessah and Zimanyi, 1991; Zimanyi and Pessah, 1991a; Hawkes *et al.*, 1992). The protein that binds ryanodine with high affinity has been successfully solubilized using CHAPS (Pessah *et al.*, 1986; Seifert and Casida, 1986) and has been immunoaffinity purified in the presence of digitonin (Imagawa *et al.*, 1987). Lai and co-workers (1988a) reported the purification of the CHAPS-solubilized ryanodine receptor by density-gradient centrifugation, and demonstrated that the ryanodine receptor migrates as an oligomeric complex with an apparent sedimentation coefficient of 30 S (Lai *et al.*, 1987). The 30 S fraction has been shown to consist primarily of a single high-molecular-weight band of M_r 360,000 on SDS-PAGE (Campbell *et al.*, 1987; Inui *et al.*, 1987a; Lai *et al.*, 1988a). Biochemical, electrophysiological (Fleischer *et al.*, 1985; Pessah *et al.*, 1985,1986,1987; Kawamoto *et al.*, 1986; Campbell *et al.*, 1987; Imagawa *et al.*, 1987; Inui *et al.*, 1987a,b; Lai *et al.*, 1987,1988a,b; Smith *et al.*, 1988; Rardon *et al.*, 1989), and electron microscopic (Somlyo, 1979; Ferguson *et al.*, 1984; Costello *et al.*, 1986; Lai *et al.*, 1988a; Jorgensen *et al.*, 1993) data have all provided corroborating evidence that [³H]ryanodine binds with high affinity to the SR Ca²⁺-release channel, and that the junctional foot protein constitutes a major structural and functional component. Lai *et al.* (1988a) proposed a tetrameric structure of four identical subunits for the native Ca²⁺-release channel. Demonstration of tight associations between the junctional foot protein of skeletal SR and other triadic proteins, such as the immunophilin FKBP12 (Collins, 1991; Jayaraman, 1992; Timerman, *et al.*, 1993) and triadin (Kim *et al.*, 1990; Caswell *et al.*, 1991; Knudson *et al.*, 1993; Liu *et al.*, 1994), suggests a more complex heteromeric structure for the native channel of skeletal muscle. However, cDNA cloning and heterologous expression of the skeletal ryanodine receptor (termed RyR-1) (Marks *et al.*, 1989; Takeshima *et al.*, 1989; Zorzato *et al.*, 1990) have reaffirmed that four copies of a polypeptide with a molecular mass of 565,223 daltons (deduced from cDNA) constitute the major structural and functional element of the "ryanodine receptor." Cloning a gene that encodes a distinct cardiac isoform of the ryanodine receptor, RyR-2, (Nakai *et al.*, 1990; Otsu *et al.*, 1990) and recognition that both cardiac and skeletal isoforms of the ryanodine receptor are expressed in brain (Kuwajima *et al.*, 1992; Lai *et al.*, 1992) suggest a more general distribution for ryanodine-sensitive Ca²⁺ channels in excitable cells. The term "ryanodine receptor" has become synonymous with the calcium-induced Ca²⁺-release (CICR) phenomenon that appears to have a broad occurrence in both excitable and nonexcitable cells. Indeed, the emerging evidence of distinct genes coding for ryanodine receptor-like proteins paralleling the related inositol 1,4,5-trisphosphate (IP₃) receptors, appear to be broadly expressed in a wide variety of cell types (Giannini *et al.*, 1992) and may constitute a major site for convergence for signal transduction involving Ca²⁺ (see Chapter 32, 33).

The availability of [³H]ryanodine, a specific and conformationally sensitive probe, has dramatically improved our understanding of the structure and func-

tion of the Ca²⁺-release channel. However, several perplexing, seemingly contradictory results have plagued researchers trying to draw clear mechanistic conclusions about how ryanodine alters Ca²⁺ channel function. This chapter review has a dual purpose. First, advances in our knowledge of the complex pharmacology of the Ca²⁺-release channel of SR are reviewed. Second, multidisciplinary evidence is offered that discriminating concentrations of ryanodine can stabilize four discrete gating behaviors of the native channel in SR, strongly supporting the existence of multiple interacting effector sites for ryanodine on the channel complex. A sequential mechanism by which ryanodine alters channel function can reconcile discrepancies in the literature and can account for the complex behavior of [³H]ryanodine at its binding sites. The fields of E–C coupling and the pharmacology of ryanodine have grown exponentially in recent years. Any contributions not cited in this review were omitted because of space limitation and reflect no judgment on the quality or significance of the work.

II. Pharmacology of the SR Ca²⁺-Release Channel

A. Inorganic Modulators

1. Calcium

The binding kinetics of [³H]ryanodine to specific sites on muscle SR and brain endoplasmic reticulum (ER) membranes are highly dependent on the extravesicular free Ca²⁺ concentration. Micromolar Ca²⁺ significantly enhances the rate of association of [³H]ryanodine with SR/ER receptors. Ca²⁺ in the presence of physiological levels of Na⁺,K⁺,Mg²⁺, and adenine nucleotides activates [³H]ryanodine equilibrium binding to SR by increasing the density, the apparent affinity, and the apparent cooperativity of the ryanodine-binding sites (Pessah *et al.*, 1987). The apparent K_d for Ca²⁺ activation ranges between 1 and 30 μM, depending on assay conditions and the source of receptor (muscle or brain). The Hill coefficient for Ca²⁺ activation is about 2 for muscle SR membranes indicating the presence of multiple binding sites for Ca²⁺ on the channel complex, whereas brain ER yields Hill coefficients near 1 (Zimanyi and Pessah, 1991a,b). The pharmacological data indicating multiple-cooperative effector sites for calcium on the skeletal-junctional foot protomer is supported by sequence information deduced from cDNA. Three distinct regions having some conformance with the "EF Hand" motif are present within amino acids 4253 to 4264,

4407 to 4416, and 4489 to 4499. The binding of [³H]ryanodine to the junctional SR Ca²⁺-release channels is dependent on the free Ca²⁺ concentration to a greater extent at nanomolar than at micromolar concentrations of the alkaloid (Lai *et al.*, 1989). Reduction of the free Ca²⁺ concentration from 1 mM to 0.1 μM in the presence of 1 M NaCl and adenine nucleotides (see subsequent discussion) results in the loss of high-affinity binding sites; the apparent dissociation constant of the remaining low-affinity site is 2–5 μM. Increasing the Ca²⁺ concentration reveals a population of high-affinity binding sites that represents one-fourth of the total number of binding sites, indicating that the alkaloid binds to multiple, presumably four, sites on the channel complex (Lai *et al.*, 1989).

Greater than 100 μM Ca²⁺ inhibits the high-affinity binding of [³H]ryanodine (Chu *et al.*, 1990; Zimanyi and Pessah, 1991a) when assayed under physiological salt conditions. Excellent agreement exists between the $K_{d/Ca^{2+}}$ and the degree of cooperativity (n_H) measured by radioligand binding assays and the EC$_{50}$ of Ca²⁺ required to induce Ca²⁺ release from actively loaded SR vesicles by spectrophotometric determination using antipyrylazo III in the presence of P$_i$ (Palade, 1987b; Zimanyi and Pessah, 1991a). Lipid bilayers fused with heavy SR vesicles reveal a high-conductance Ca²⁺-selective channel whose open probability increases with 1–100 μM Ca²⁺ in the *cis* chamber (sarcoplasmic side), but decreases with Ca²⁺ > 100 μM (Smith *et al.*, 1986). Ca²⁺ induces these effects without an appreciable change in the lifetime of each open event.

2. Magnesium

Mg²⁺ at physiological concentration (1 mM) slows the rate of association and reduces the affinity and the number of binding sites for [³H]ryanodine when assayed in optimal Ca²⁺, the result of a decrease in apparent affinity of the activator site for Ca²⁺ (Pessah *et al.*, 1987; Chu *et al.*, 1990). The IC$_{50}$ concentration for Mg²⁺ is 0.5–2 mM, depending on the type of SR (i.e., cardiac or skeletal) in the presence of 50 μM Ca²⁺; the logit slope is 1.5–1.9, corresponding to multiple interacting sites for Mg²⁺ on the release channel (Pessah *et al.*, 1985; Zimanyi and Pessah, 1991a). In contrast, inhibition of [³H]ryanodine binding to brain ER is less sensitive to Mg²⁺ (IC$_{50}$ = 10 mM), with a logit slope of 1 (Zimanyi and Pessah, 1991b). Mg²⁺ inhibits single-channel gating and causes flickering of the Ca²⁺-activated channel (Smith *et al.*, 1986).

3. Ionic Strength

High ionic strength, well beyond levels of physiological relevance for KCl or NaCl, has a dramatic in-fluence on the binding of [³H]ryanodine to SR. NaCl (0.3–1 M) enhances the binding of [³H]ryanodine in cardiac SR, whereas it inhibits the binding of [³H]ryanodine to the high-affinity finding state in skeletal SR (Zimanyi and Pessah, 1991a). Michalak and coworkers (1988) found that high NaCl activates the binding of 20 nM [³H]ryanodine to high- and low-affinity states, suggesting that high salt may also alter the binding properties of [³H]ryanodine to lower-affinity states of the channel. The principle action of 1M NaCl on the cardiac receptor is to enhance the rate of [³H]ryanodine association at optimal Ca²⁺ concentration, thereby enhancing the affinity of ryanodine to its binding site without significantly changing the binding capacity of SR. The action of Na⁺ in skeletal muscle is different. Although the rate of association of [³H]ryanodine is also enhanced, high salt (1 M) reduces the maximal binding capacity of skeletal SR. The latter effect is apparently due to an unstable equilibrium, since receptor occupancy drops 37% with 4-hr incubation at 37°C relative to 2-hr incubation (Zimanyi and Pessah, 1991a). Inhibition of [³H]ryanodine binding by high Ca²⁺ is also sensitive to high salt, since the Ca²⁺ inhibition curve shifts to the right in the presence of 1 M NaCl. Single-channel measurements are often performed in the presence of asymmetrical Cs⁺, which serves as the current carrier and inhibits the potassium channels. The high (500 mM in the *cis* chamber) CsCl concentrations used in bilayer experiments have a strong influence on the interaction of ryanodine with its binding sites on the channel complex (see subsequent discussion). The physiological significance, if any, of high salt effects on the kinetic behavior of [³H]ryanodine with its binding site and, more importantly, on channel structure and function remains to be elucidated. KCl or NaCl (>0.5 M) is necessary to demonstrate high affinity [³H]ryanodine-binding sites in brain microsomes (Ashley, 1989; McPherson and Campbell, 1990; Zimanyi and Pessah, 1991b). A modulatory role for the immunophilin FKBP-12 (FK506 binding protein 12 kDa) on skeletal SR Ca²⁺ release channel function has been demonstrated by Timerman *et al.* (1993). The FKBP12/RyR-1 heterocomplex appears to stabilize the closed conformation of the Ca²⁺ release channel, since micromolar of the immunosuppressant FK506 promotes dissociation of FKBP12 to the supernatant of junctional SR membrane preparations and reduces the rates of active Ca²⁺ accumulation in FKBP12-deficient SR vesicles. It is possible that high salt may promote the dissociation of FKBP12 from RyR-1 and stabilize an activated channel conformation, thus stabilizing high-affinity binding sites for [³H]ryanodine.

B. Organic Modulators

1. Adenine Nucleotides

ATP and other adenine nucleotides are physiologically relevant activators of the SR Ca^{2+}-release channel (Meissner, 1984; Meissner *et al.*, 1986; Imagawa *et al.*, 1987; Pessah *et al.*, 1987; Michalak *et al.*, 1988; Zimanyi and Pessah, 1991a). Adenine nucleotides in the presence of 1 mM Mg^{2+} increase the affinity of the activator site for Ca^{2+} and increase the affinity and the maximum binding capacity of the high-affinity site for [³H]ryanodine. Adenine nucleotides restore the association rate of the alkaloid when reduced by 1 mM Mg^{2+}. The rank order of potency of adenine nucleotides for activating the binding of [³H]ryanodine is AMP-PCP[1] > cAMP > ADP = adenosine > AMP (Pessah *et al.*, 1987), which is very similar to the rank order of potency for these agents to stimulate Ca^{2+} release from SR vesicles (Meissner, 1984). In single-channel measurements in bilayers, ATP activates by enhancing the mean open time, but both Ca^{2+} and ATP are necessary to fully activate the channel (Smith *et al.*, 1986; Hymel *et al.*, 1988a). Given the observed potency of AMP-PCP, a nonhydrolyzable ATP analog, the immediate modulatory actions of ATP do not appear to require phosphorylation of the receptor. Supporting this hypothesis, the primary sequence information of the skeletal isoform in the ryanodine receptor reveals at least three putative adenine nucleotide binding sites with the GXGXXG motif. Adenine nucleotides have a biphasic effect on the binding of [³H]ryanodine to rat cortical ER (Zimanyi and Pessah, 1991b).

2. Caffeine

The mechanism by which caffeine activates the Ca^{2+}-release channel and the binding of [³H]ryanodine has been well characterized (Imagawa *et al.*, 1987; Pessah *et al.*, 1987; Michalak *et al.*, 1988). Caffeine induces rapid release of Ca^{2+} from heavy SR vesicles loaded with $^{45}Ca^{2+}$ (Fairhurst and Hasselbach, 1970; Kim *et al.*, 1983) or with Ca^{2+} in the presence of pyrophosphate and ATP (Palade, 1987a). Millimolar caffeine in the presence of 1 mM Mg^{2+} markedly decreases the threshold for Ca^{2+} activation of [³H]ryanodine binding and increases the affinity of Ca^{2+} for its activator site. This effect of caffeine results in the apparent loss of cooperativity for Ca^{2+} binding (Pessah *et al.*, 1987). Caffeine also increases the affinity of the alkaloid for its high-affinity binding site; at low

micromolar Ca^{2+} concentrations, caffeine greatly enhances the rate of association of [³H]ryanodine. At lower concentrations of Ca^{2+} (2 μM), caffeine is more stimulatory than AMP-PCP, and the two agents in combination act synergistically. However, synergism between caffeine and ATP disappears at optimal concentrations of Ca^{2+}, at which AMP-PCP is more stimulatory than caffeine. Caffeine and AMP-PCP together dramatically increase the affinity and the cooperativity of Ca^{2+} binding to the receptor. Millimolar caffeine has been shown to activate single channels from SR membranes fused with planar lipid bilayers (Rousseau *et al.*, 1986; Rousseau and Meissner, 1989; Ondrias *et al.*, 1990). Derivatives of the marine tunicate *Eudistoma olivaceum* have been found to release Ca^{2+} from skeletal SR in a manner similar to caffeine (Seino *et al.*, 1991). Synthesis of 9-[³H]methyl-7-bromoeudistomin D has permitted direct radioligand binding analysis and has revealed apparently competitive inhibition by caffeine (Fang *et al.*, 1993).

The findings of Pessah *et al.* (1987) concerning the modulation of [³H]ryanodine by caffeine and adenine nucleotides suggested that adenine nucleotides exert their stimulation at a site distinct from the caffeine modulatory sites of the receptor. However, at suboptimal concentrations of Ca^{2+}, occupation of both effector sites results in a dramatic increase in the affinity and cooperativity of Ca^{2+} binding, suggesting functional communication between these two effector sites within the receptor complex.

3. Anthraquinones

Zorzato *et al.* (1985) provided the first evidence that doxorubicin, an antineoplastic anthraquinone, induces rapid release of Ca^{2+} from purified junctional SR vesicles from skeletal muscle. Tension measurement of skinned fibers in the presence of doxorubicin and caffeine, singly or in combination, suggested that both agents mobilize Ca^{2+} from the same store. Subsequent experiments by Abramson and colleagues (1988) demonstrated that these effects of anthraquinones are the result of a direct interaction with the ryanodine-receptor complex of skeletal junctional SR. In a manner similar to the mechanism proposed for caffeine-induced Ca^{2+} release, micromolar anthraquinone increases the apparent affinity by which Ca^{2+} activates the [³H]ryanodine binding site and enhances rapid release of Ca^{2+} from cardiac SR vesicles, suggesting an allosteric site for anthraquinone action on the ryanodine receptor complex (Pessah *et al.*, 1990). A good correlation exists between the EC_{50} concentration (5–30 μM) of doxorubicin enhancing the binding of [³H]ryanodine to the high-affinity site, rising tension in skinned muscle fiber, or releasing Ca^{2+} from

---

[1] AMP-PCP is the nonhydrolyzing ATP analog β, γ-methylene-adenosine-5'-triphosphate.

actively filled vesicles (Zorzato *et al.*, 1985; Abramson *et al.*, 1988; Pessah *et al.*, 1990,1992). Doxorubicin increases the affinity of [³H]ryanodine for its high-affinity site by increasing the association rate of ryanodine binding, but it has no effect on the maximal binding capacity. Competition studies with caffeine suggest that doxorubicin and caffeine share the same binding site on the cardiac ryanodine-receptor complex (Pessah *et al.*, 1990). This result supports the finding of Zorzato *et al.* (1986), who demonstrated that photolabeling of junctional SR proteins from rabbit skeletal muscle with doxorubicin is inhibited by caffeine. Doxorubicin, like caffeine, initially activates single channels incorporated into lipid bilayers and Ca^{2+} efflux from SR vesicles, but also inhibits the SR Ca^{2+} channels in a time- and concentration-dependent manner (Ondrias *et al.*, 1990; Pessah *et al.*, 1992). Researchers have suggested that the action of anthraquinones involves modulation of the oxidation state of thiols (Ondrias *et al.*, 1990; Pessah *et al.*, 1990); that are functionally important in gating mechanism of the ryanodine-receptor Ca^{2+}-release channel complex (Abramson and Salama, 1989). The observation that pretreatment of SR vesicles with the thiol-reducing agent dithiothreitol potentiates the doxorubicin-activated binding of [³H]ryanodine and the doxorubicin-induced release of Ca^{2+} (Pessah *et al.*, 1990), and prevents doxorubicin-induced channel inactivation in single-channel recordings (Ondrias *et al.*, 1990), lends support to this hypothesis.

C. Inhibitors of Ryanodine Binding

1. Ruthenium Red

The polycationic dye ruthenium red stimulates Ca^{2+} uptake in skeletal muscle junctional SR preparations (Miyamoto and Racker, 1982). Seiler and co-workers (1984) have shown that ruthenium red enhances the uptake of Ca^{2+} into junctional SR vesicles, an effect not due to direct activation of the Ca^{2+}–ATPase, which supports previous inferences (Jones and Cala, 1981; Miyamoto and Racker, 1982; Jones and Burrows, 1983) that ruthenium red increases net Ca^{2+} sequestration by blocking Ca^{2+} release from the SR vesicles. Ruthenium red has been shown to directly block Ca^{2+} efflux from SR membrane vesicles, (Ohnishi, 1979; Yamamoto and Kasai, 1982; Antoniu *et al.*, 1985; Fleischer *et al.*, 1985; Smith *et al.*, 1985; Imagawa *et al.*, 1987; Lattanzio *et al.*, 1987), Ca^{2+} release from SR of skinned fibers (Volpe *et al.*, 1986), and Ca^{2+} release from SR of intact frog single twitch fibers after microinjection of the inhibitor (Baylor *et al.*, 1989). Ruthenium red has been shown to

block single-channel currents of the purified Ca^{2+} release channel reconstituted into planar lipid bilayer (Lai *et al.*, 1988a). Yamamoto and Kasai (1982) have suggested that ruthenium red inhibits Ca^{2+} release by entering into the channel pore, thus blocking the channel in an open state (occlusion). This hypothesis has been supported by the work of Wyskovsky *et al.* (1990). Binding of [³H]ryanodine to the Ca^{2+}-release channel is blocked by ruthenium red in a competitive manner (Pessah *et al.*, 1985,1986; Imagawa *et al.*, 1987; Chu *et al.*, 1990; Pessah and Zimanyi, 1991; Zimanyi and Pessah, 1991a; Mack *et al.*, 1992). The IC_{50} concentration of ruthenium red inhibition of [³H]ryanodine-binding to the high-affinity site is 10–20 nM. Ruthenium red at low nanomolar concentrations seems to behave as a true competitive inhibitor of [³H]ryanodine at its high-affinity site, by increasing the K_d of ryanodine by slowing the rate of association (Chu *et al.*, 1990). In consonance with a competitive mechanism, ruthenium red inhibits the Ca^{2+}- and the ryanodine-induced Ca^{2+} release from SR vesicles (Palade, 1987a,b; Zimanyi and Pessah, 1991a) with an IC_{50} of 30–60 nM and 11 nM, respectively. However, a discrepancy is detected in the action of high concentrations of ruthenium red. Ruthenium red slows the rate of dissociation induced by 100-fold dilution of [³H]ryanodine (Chu *et al.*, 1990; Pessah and Zimanyi, 1991), indicating a cooperative interaction of multiple binding sites on the receptor complex. A true competitive inhibitor does not change the maximal number of binding sites; however, ruthenium red, even at 25 nM, significantly reduces the maximal binding capacity of ryanodine for the high-affinity site (Mack *et al.*, 1992). Moreover, pretreatment of SR membranes with 25 nM and 1 μM ruthenium red, followed by subsequent elimination of the inhibitor using several wash steps, results in loss of 50 and 96% of the high-affinity binding of [³H]ryanodine, respectively, compared with a control treated in the same way without ruthenium red. These results indicate a persistent effect of ruthenium red at high concentrations.

2. Neomycin

Neomycin has been shown to induce neuromuscular paralysis in experimental animals and during clinical use (Pridgen, 1956; Baeyens *et al.*, 1985). Palade (1987c) showed that neomycin, and several congeners, inhibit Ca^{2+} release from SR vesicles when induced by a number of Ca^{2+} channel agonists at concentrations that are orders of magnitude lower than those required to block phosphoinositide turnover. Neomycin can discriminate or induce multiple affinity states for the binding of nanomolar [³H]ryanodine to the Ca^{2+}-channel complex (Mack *et al.*, 1992). A concentration of neomycin $\leq K_{il}$ (130 nM) reduces the affinity of ryanodine for its skeletal receptor by decreasing the association rate without reducing [³H]ryanodine-binding capacity, whereas high concentrations ($>> K_{il}$, e.g., 30 μM) also reduce the maximal number of binding sites. Neomycin (0.3 μM) significantly reduces the observed association rate constant (k_{obs}) for the binding of 0.5 nM [³H]ryanodine from 0.0244 min^{-1} to 0.016 min^{-1} as expected for a competitive interaction. A true competitive inhibitor would not be expected to alter the dissociation rate of a ligand for its binding site. Indeed, the k_{-1} of nanomolar [³H]ryanodine from its binding site is not significantly different when intiated by a 100-fold dilution of the assay medium in the presence or absence of 0.3 μM neomycin. However, the presence of 30 μM neomycin in the dilution medium significantly slows the k_{-1} of [³H]ryanodine from its high affinity site, in a manner similar to high concentrations of ruthenium red. Kinetic binding studies demonstrate that high concentrations of neomycin induce allosterism among [³H]ryanodine binding sites. As a result, high neomycin concentrations, like ruthenium red and micromolar ryanodine (see subsequent discussion), induce [³H]ryanodine-binding site complexes that are recalcitrant to competitive inhibition. Neomycin ($\geq$0.3 μM) persistently reduces the number of available binding sites for [³H]ryanodine in a manner similar to high concentrations of ruthenium red or ryanodine. Neomycin does not show direct interaction with the Ca²⁺-activator site of the channel complex, but inhibits ryanodine-enhanced release of Ca²⁺ from skeletal SR vesicles in a biphasic manner (Mack *et al.*, 1992) in consonance with its biphasic inhibition of the binding of [³H]ryanodine. The nanomolar potency of neomycin for altering Ca²⁺-release channel function may represent an important mechanism in the etiology of aminoglycoside-induced skeletal muscle paralysis.

3. FLA 365

Chiesi and co-workers (1988) showed that the phenylalkylamine (2,6-dichloro-4-amino-phenyl)isopropylamine, FLA 365, inhibits the release of Ca²⁺ from SR vesicles. This inhibitor of the Ca²⁺-release channel behaves quite differently from ruthenium red and neomycin. FLA 365 inhibits the rate of Ca²⁺ release from SR vesicles, induced by 100 μM CaCl₂, in a monophasic manner with an IC₅₀ of 3.8 μM (Mack *et al.*, 1992). [³H]ryanodine-binding assays demonstrate that FLA 365 inhibits binding with an IC₅₀ of 1.4 μM and a Hill coefficient near 1. FLA 365 also increases the K_d of ryanodine for its high-affinity site, but has no effect on the maximal binding capacity, not even at

high concentration, indicating that FLA 365 is a true competitive inhibitor. Examining possible persistent effects of FLA 365 on the binding of [³H]ryanodine (measured as described for neomycin) reveals that FLA 365 does not persistently alter the ability of ryanodine to interact with its receptor on the Ca²⁺-release channel complex (Mack *et al.*, 1992).

4. Interaction of Ruthenium Red, Neomycin, and FLA 365 on the Ryanodine Receptor

FLA 365 does not influence the inhibitory potency of ruthenium red or neomycin at the [³H]ryanodine-binding site. Neither ruthenium red nor neomycin shifts the inhibitory potency of FLA 365. However, FLA 365 steepens the inhibition curves of neomycin, indicating that FLA 365 modulates the ability of neomycin to induce multiple receptor affinity states for [³H]ryanodine. The interaction of neomycin and ruthenium red seems to be more complex. Activator-induced Ca²⁺ release from passively loaded vesicles has been shown to be triphasic: phase 1 can be inhibited by neomycin, and phase 1 and 2 can be inhibited by ruthenium red, whereas none of these inhibitors is able to inhibit phase 3. These results suggest that the two inhibitors are sensitive to different affinity states of the Ca²⁺-release channel (Wyskovsky *et al.*, 1990). Increasing the concentration of ruthenium red reduces the inhibitory potency of neomycin on the binding of [³H]ryanodine. Graphically, a rightward shift is seen in the neomycin-inhibition curve (Fig. 1A). Inverse experiments show that neomycin also causes a rightward shift in the ruthenium red inhibition curves (Mack *et al.*, 1992).

Neomycin and ruthenium red appear to share a common binding site. Occupancy by either agent is mutually exclusive of ryanodine binding to its high-affinity site. FLA 365, on the other hand, is an uncharged molecule that does not change the K_i for either ruthenium red or neomycin, suggesting that FLA 365 binds to a site separate from that recognized by the cationic inhibitors. The binding of FLA 365 is, however, mutually exclusive with the binding of ryanodine. The conceptual model of Mack and co-workers (1992) is presented in Fig. 1B.

The biphasic character of the inhibition curves and the logit slope significantly less than 1 seen with neomycin suggest multiple interacting sites on each functional channel oligomer. Although at low concentrations neomycin behaves as a true competitive inhibitor with respect to the high-affinity binding site of ryanodine, at high concentrations it slows the dissociation of ryanodine from its receptor, a phenomenon that is also demonstrated with high concentrations of ruthenium red or ryanodine. Further, the behavior of

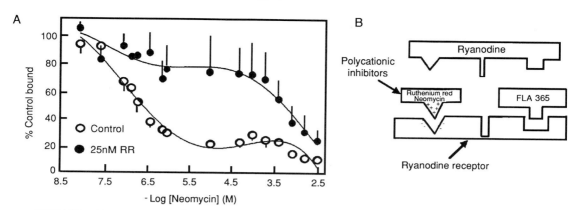

FIGURE 1 (A) Heterologous [³H]ryanodine displacement study with antagonists neomycin and ruthenium red (RR) using junctional sarcoplasmic reticulum (SR) from rabbit skeletal muscle. The presence of 25 nM RR (●) significantly shifts the log–dose inhibition curve for neomycin. The data expressed are normalized as a percentage of the maximal specific [³H]ryanodine bound in the presence of neomycin alone (control, ○) and in the presence of a fixed concentration of RR. (B) Postulated scheme for the relationship of binding site recognizing ryanodine, ruthenium red, neomycin, and FLA 365 on one of four identical subunits. Reprinted with permission from Mack *et al.* (1992).

ruthenium red, neomycin, and ryanodine in mixed-inhibitor binding studies is clearly more complex than predicted by simple competition among these agents at a single (noninteracting) population of binding sites. For example, the presence of 25 nM ruthenium red results in a substantial decrease in the slope of the neomycin inhibition curve, suggesting that ruthenium red modulates the ability of neomycin to induce multiple [³H]ryanodine binding states. Analysis of the binding data reveals an increase in the proportion of [³H]ryanodine binding sites that are recalcitrant to neomycin inhibition in the presence of ruthenium red relative to controls. The proportion of [³H]ryanodine binding sites that are recalcitrant to neomycin inhibition in skeletal SR increases from 20% in controls to 50% in the presence of ruthenium red (Mack *et al.*, 1992). Since ruthenium red and neomycin appear to occupy mutually exclusive sites (Figs. 1A,B), these results could only stem from the presence of multiple interacting sites for these agents on the channel oligomer.

In conclusion, FLA 365 is a true competitive inhibitor of the binding of [³H]ryanodine to its high-affinity state or site, and does not appear to induce complex kinetic behavior with increasing concentration (i.e., does not induce allosterism). This property of FLA 365, as well as its ability to traverse membranes, makes it a valuable probe in receptor binding assays as well as in studies with intact cells. The complex action of ruthenium red and neomycin suggests the presence of several homologous binding sites on the channel oligomer that recognize [³H]ryanodine, ruthenium red, and neomycin, all of which are allosterically cou-

pled and tend to promote multiple conformational and affinity states.

III. Mechanism of Ryanodine Action

A. Does Ryanodine Activate or Inhibit the Ca²⁺ Permeability of SR?

Much controversy is found in the early literature concerning the activating or inhibiting effect of ryanodine on membrane permeability to Ca²⁺ and whether alternative mechanisms of ryanodine action are present in cardiac and skeletal SR (Fairhurst and Hasselbach, 1970; Jones *et al.*, 1979; Chamberlain *et al.*, 1984; Hilgemann *et al.*, 1983; Sutko and Kenyon, 1983; Seiler *et al.*, 1984; Feher and Lipford, 1985; Fleischer *et al.*, 1985). The molecular actions of ryanodine are indeed complex, but these properties of ryanodine have permitted the intricacies of SR Ca²⁺ channel structure and function to unfold. The potency and the qualitative changes in SR permeability induced by ryanodine are greatly influenced by the very ligands that modulate channel function (Meissner, 1986b), as are the kinetics of [³H]ryanodine binding to its receptor sites, as discussed earlier. The conformationally sensitive nature of the [³H]ryanodine-binding sites may help explain observations by Lattanzio and co-workers (1987), who found that, in the presence of low Ca²⁺, ryanodine at or above 100 μM concentration only inhibits Ca²⁺ efflux whereas, in the presence of unphysiological 5 mM Ca²⁺, the alkaloid induces Ca²⁺ release at concentrations ranging from 10 nM to 3 μM;

above 10 μM, it inhibits Ca^{2+} release. Ryanodine at 0.01–10 μM enhances $^{45}Ca^{2+}$ release from skeletal SR vesicles and is characterized by a slow onset, whereas at concentrations greater than 10 μM, ryanodine appears to inhibit channel activation (Meissner, 1986b; Alderson and Feher, 1987; Lattanzio et al., 1987; Lai et al., 1989). Based on such findings, researchers have concluded that ryanodine has two distinct effects on channel activity. However, quantitative discrepancies between binding affinities measured with [^{3}H]ryanodine and the concentrations required to influence membrane permeability to Ca^{2+} have been difficult to reconcile. A quantitative discrepancy also exists between ryanodine-receptor binding and the ability of the alkaloid to influence channel gating in planar lipid bilayers. For example, current records from single Ca^{2+} channels reconstituted in planar lipid bilayers reveal that micromolar ryanodine in the cis chamber (the sarcoplasmic side of the channel) induces a long-lived subconductance state (Imagawa et al., 1987; Rousseau et al., 1987; Hymel et al., 1988a,b; Nagasaki and Fleischer, 1988; Smith et al., 1988; Anderson et al., 1989; Lai and Meissner, 1989) that is rather insensitive to Ca^{2+}, ATP, and Mg^{2+} (Meissner et al., 1986; Lai et al., 1989). This effect of ryanodine has been attributed to occupation of the high-affinity site. The discrepancy between high-affinity binding constants ($K_d = 1–4$ nM) and the amount of ryanodine needed to alter the function of a single channel reconstituted in planar bilayers has been attributed partly to different experimental conditions (Lai and Meissner, 1989), but mainly to slow rates of association of ryanodine with its binding site (Smith et al., 1988).

An insightful study by Jenden and Fairhurst (1969) showed that ryanodine initially inhibited ATP-driven Ca^{2+} uptake. However, after a prolonged interval, spontaneous recovery occurred and eventually the steady-state calcium uptake rate approximated that in control experiments. Feher and Lipford (1985), using micromolar ryanodine, found that ryanodine initially inhibited calcium uptake by SR, but this effect was followed by a subsequent marked stimulation, supporting the initial findings of Jenden and Fairhurst (1969). Could the heterogeneous nature of ryanodine action on SR permeability to Ca^{2+} actually reflect a sequential, perhaps persistent, change in the function of the release channel after exposure to ryanodine either at high concentration or for a long time period? The remainder of this chapter focuses on work that utilizes three distinct approaches to address this question: [^{3}H]ryanodine receptor binding analyses, Ca^{2+}-loading and -release measurements from actively loaded SR vesicles, and single Ca^{2+} channel measurements from planar lipid bilayers fused with SR vesi-

cles. In many cases, the data were gathered under identical experimental conditions to provide direct evidence that sequential occupation of distinct but allosterically coupled ryanodine-binding sites on the Ca^{2+}-release channel complex can adequately account for the seemingly unpredictable kinetic behavior of [^{3}H]ryanodine in binding assays and the multiplicity of ryanodine actions on SR Ca^{2+} transport.

B. Multiple Ryanodine Binding Sites on the Ca^{2+} Release Channel Complex

1. Receptor Binding Studies

Equilibrium binding studies with [^{3}H]ryanodine have indicated the presence of multiple classes of specific, interacting binding sites for [^{3}H]ryanodine on both skeletal and cardiac SR (Pessah et al., 1985; Inui et al., 1987a; Michalak et al., 1988). The binding of [^{3}H]ryanodine (0.1–20 nM) to a saturable site with the highest affinity (K_{d1}-1–5 nM) obeys the law of mass action. Concentrations in excess of 20–30 nM (>10x K_{d1}) result in complex binding kinetics that hint at the presence of sites with lower affinity for the alkaloid. Kinetic analyses of the binding of [^{3}H]ryanodine have shown that introducing 2000-fold excess (10 μM) unlabeled ryanodine, to promote dissociation of specifically bound [^{3}H]ryanodine, results in a polyphasic dissociation with an extremely slow component (half-time of 14.4 hr) representing 75% of all the binding sites (Pessah et al., 1987). These initial results supported the notion, gleaned from equilibrium studies, of multiple binding sites that interact in a complex manner. However, the calculated K_d based on kinetic rate constants is 20-fold lower (i.e., of higher affinity) than the apparent K_d for the site with the highest affinity in equilibrium experiments. This glaring discrepancy has plagued mechanistic interpretations, especially since analysis of equilibrium binding experiments by the method of Hill or Scatchard suggests that presence of sites with low affinity. Detailed studies have also supported the presence of multiple, interacting ryanodine-binding sites with nanomolar and as low as micromolar affinities (Lai et al., 1989; McGrew et al., 1989; Pessah and Zimanyi, 1991a).

Typically, studies of Ca^{2+} transport across SR vesicles, single-channel measurements in planar bilayers, and the binding of [^{3}H]ryanodine to SR membranes are performed under quite different conditions, each optimal for the respective assay. Three buffers routinely used are: (1) binding buffer (A)—20 mM HEPES buffer, pH 7.1, consisting of 250 mM KCl, 15 mM NaCl, and 50 μM $CaCl_2$ at 37°C; (2) transport buffer (B)—95 mM KCl, 20 mM K-MOPS, pH 7.0, 7.5 mM sodium pyrophosphate, 250 μM antipyrylazo III, 1.5 mM MgATP, 25 μg/ml creatine phosphokinase, 5 mM phosphocreatine, and 50 μM CaCl$_2$ at 37°C; and (3) bilayer buffer (C)—500 mM CsCl, 100 μM CaCl$_2$, 5 mM HEPES, pH 7.2 at 25°C. Using buffer A, Scatchard analysis of the binding of 0.5–32 nM, 15–500 nM, 50–5000 nM, and 300–10,000 nM [³H]ryanodine reveals the presence of four affinity states or binding sites on the SR Ca²⁺-release channel. Table I summarizes the corresponding dissociation constants (K_d) and maximal binding capacities (B_{max}). With 0.5–32 nM ryanodine, the Hill coefficient (n_H) is close to 1, indicating the absence of allosterism. However, at higher ryanodine concentrations, the n_H decreases to 0.23, implying that the binding of ryanodine to the low-affinity sites may exhibit negative cooperativity, in agreement with the findings of Lai *et al.* (1989). The shape of the Scatchard curves also suggests negative cooperativity, because in any range of ryanodine concentrations wider than two orders of magnitude, the Scatchard curves are curvilinear and concave. The validity of conclusions drawn from radioligand binding and Ca²⁺ transport assays hinges on the assumption that the characteristics of ryanodine binding to the channel complex are essentially the same under the conditions of the two types of experiments. Binding curves performed in the transport buffer (B) yield similar Scatchard plots for sites 1 and 2 with K_ds of 3.7 and 55.9 nM, and with B_{max}s of 3.9 and 15.7 pmol/ mg protein. [³H]Ryanodine binding in the transport buffer maintains its sensitivity to Mg²⁺, Ca²⁺, adenine nucleotides, and ruthenium red. Equilibrium binding experiments with 0.5–500 nM [³H]ryanodine performed in the same buffer used for single-channel measurements (buffer C, at 25°C) are best fit by two sites with K_ds of 0.7 and 23 nM (n_H, 0.7), whereas binding in the range of 50–5050 nM reveals an addi-

tional site that exhibits significantly lower affinity, with a K_d of 1.5 μM (n_H, 0.4; Table I). The binding curve shows that specific binding reaches plateau at 100–500 nM [³H]ryanodine and rises again above 500 nM up to and beyond 5050 nM, indicating the possibility of a fourth site that is not saturated at 5050 nM and has significantly lower than micromolar affinity. A comparison of the affinities of ryanodine for its sites in buffers A and C implies that 500 mM CsCl markedly increases the affinity of [³H]ryanodine for the higher affinity sites (1 and 2) whereas it reduces the affinity for the lower affinity sites (3 and possibly 4). The maximal binding capacities for the first three detectable sites are in very good agreement with those measured in the absence of CsCl at 37°C. The binding of ryanodine to the three sites detectable (in the presence of 0.5 M CsCl) suggests negative cooperativity with a Hill coefficient less than 1.

The association rate of [³H]ryanodine in the binding buffer (A) increases linearly with the concentration of [³H]ryanodine from 1 nM to 1 μM (Pessah and Zimanyi, 1991). The rate of dissociation, induced by 100-fold dilution of the assay medium, also increases with the concentration of ryanodine from 1 to 50 nM; and the shape of the dissociation curve [plotting time against ln(specific binding at time t/specific binding at time 0)] changes from monophasic to biphasic. The initial fast component represents the low-affinity sites; the slow component represents the high-affinity site. The ratio of the two components at 50 nM [³H]ryanodine is 2.5 : 1, which is in good agreement with the same ratio calculated from B_{max2} and B_{max1} (2.3 : 1). These kinetic results are consistent with an allosteric interaction among ryanodine binding sites that are negatively cooperative. From the kinetic rate constants obtained from association (k_{+1}) and dissociation (k_{-1}) experiments, the dissociation constants

TABLE I   **Binding Constants for [³H]Ryanodine**[a]

| Site | 250 mM KCl, 37°C | | 500 mM CsCl, 25°C | |
	K_d (nM)	B_{max} (pmol/mg prot)	K_d (nM)	B_{max} (pmol/mg prot)
1	4.6 ± 1.9	5.8 ± 0.2	0.7 ± 0.3	3.3 ± 0.8
2	52.8 ± 17.1	12.9 ± 4.1	22.9 ± 11.9	12.2 ± 2.5
3	554.5 ± 66.5	26.1 ± 1.9	1480 ± 880	18.5 ± 7.7
4	2800.0 ± (>100%)	67.2 ± (>100%)	?[b]	?[b]

[a] Adapted from Pessah and Zimanyi (1991) and Buck *et al.* (1992).

[b] The determination of this site has not been performed because of the high cost and the large amount of radioactive waste of high concentration of [³H]ryanodine necessary for that experiment.

$(K_d = k_{-1}/k_{+1})$ can be calculated. The dissociation constants K_{d1} and K_{d2}, calculated from the rate constants (Table II), are in good agreement with those obtained from saturation experiments (Table I). Unexpectedly, the dissociation rate of [³H]ryanodine (1 μM) from its binding sites unpredictably slows ($t_{1/2} = 4.1$ hr). McGrew et al. (1989) have performed kinetic experiments to assess the cooperative nature of the binding of [³H]ryanodine. Ryanodine concentrations 100-fold greater than K_d for the high-affinity site significantly slow the rate of dissociation induced by 100-fold excess of assay medium. Based on this kinetic behavior, the authors have concluded that binding to the low-affinity site decreases the dissociation rate of the high-affinity binding by positive cooperativity. However, measurement of the rate of association and dissociation of 1 μM [³H]ryanodine yields a calculated K_d of 4.8 nM, which is not significantly different from the K_d for the high-affinity site (Pessah and Zinanyi, 1991). Therefore, the slowing of dissociation with micromolar ryanodine is not consistent with the interpretation of full positive cooperativity. Clearly, the significance of cooperativity (positive or negative) depends on the sequence in which labeled and unlabeled ryanodine are added to the receptor.

The apparent conflicting kinetic behavior just described led Pessah and co-workers to examine the possibility that [³H]ryanodine binding and the modulation of the release channel by ryanodine are governed by a sequential mechanism. The association rate of nanomolar [³H]ryanodine with its binding sites is significantly slower at 25°C than at 37°C in buffer A (described earlier) and is attributed to the very high temperature coefficient for activation of [³H]ryanodine-binding sites ($Q_{10} = 3.5$–4; Ogawa and Harafugi, 1990; Carroll et al., 1991). CsCl (500 mM, buffer C) changes the linear relationship between the association rate and increasing ryanodine concentrations to one that is biphasic. Although the apparent association rate constant is a linear function of [³H]ryanodine from 1 to 50 nM, a dramatic drop occurs in the association rate at [³H]ryanodine concentrations >250 nM in the presence of 500 mM Cs$^+$ the drop is also linearly dependent on the [³H]ryanodine concentration (Fig. 2). This decreased association rate (k_{+1}) is in good agreement with the decreased affinity (1/K_d) measured in saturation binding assays in the presence of Cs$^+$ (Table I). When k_{+1} decreases, K_d increases because $K_d = k_{-1}/k_{+1}$, and an increased K_d means lower affinity.

2. Ca²⁺ Transport Measurements

Ca^{2+} release measured from actively loaded skeletal junctional SR vesicles in the presence of pyrophosphate (Palade, 1987a) is enhanced significantly in a concentration-dependent manner by ryanodine and is inhibited by nanomolar ruthenium red. The threshold concentration of ryanodine that is necessary to enhance Ca^{2+}-induced Ca^{2+} release in this assay is between 10 and 25 nM (Pessah and Zimanyi, 1991). Analysis of ryanodine-enhanced Ca^{2+}-induced Ca^{2+}-release rates, according to Lineweaver–Burke, results in a nonlinear fit suggestive of negative cooperativity with respect to the alkaloid effector sites, with a Hill coefficient of 0.52. The estimated K_ds are 41 nM and 3.6 μM, reflecting the combination of the high- and low-affinity sites. The ratio of high- to low-affinity sites is 1:3, which agrees with ratios calculated from equilibrium and kinetic receptor-binding studies. The presence of 500 mM CsCl (used in lipid bilayer studies) does not change the loading capacity of the SR vesicles for Ca^{2+}, and ryanodine enhances the Ca^{2+}-induced Ca^{2+} release with a threshold concentration of 5 nM. A significant observation concerns the time-

TABLE II Kinetic Rate Constants of the Binding of [³H]Ryanodine and the Dissociation Rate Constants Derived from Them[a]

	Kinetic experiment[b]					Calculated dissociation constants	
	k_{obs} (min^{-1})	k_{11} (min^{-1})	k_{-12} (min^{-1})	k_{+11}[c] (nM^{-1}min^{-1})	k_{+12}[c] (nM^{-1}min^{-1})	K_{d1}[d] (nM)	K_{d2} (nM)
Cardiac	0.0637	0.00212	0.0288	0.00123	0.000698	1.7	41.3
Skeletal	0.0623	0.00225	0.0292	0.00120	0.000662	1.9	44.1

[a] Data from Pessah and Zimanyi (1991).
[b] [³H]Ryanodine concentration is 50 nM.
[c] $k_{+1} = (k_{obs} - k_{-1})/$[Ryanodine].
[d] $K_d = k_{-1}/k_{+1}$.

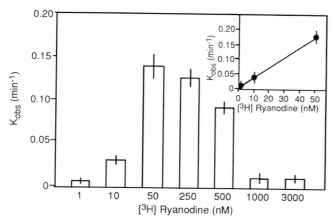

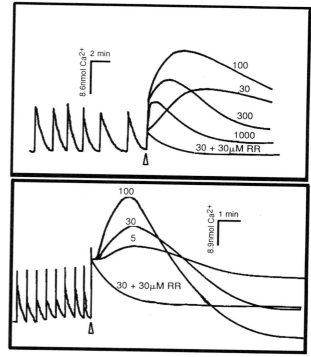

FIGURE 2 The observed association rate constants linearly increase (*inset*) with [³H]ryanodine concentration up to 100 nM, then decrease with >250 nM concentration of the alkaloid in the presence of 500 mM CsCl at 25°C. Binding of [³H]ryanodine to skeletal sarcoplasmic reticulum (SR) is performed under identical conditions as the measurement of single-channel conductance in lipid bilayer membranes. (Reprinted with permission from Zimanyi *et al.* (1992). This kinetic behavior is distinctly different from measurements made in KCl which yields a linear increase in k_{obs} with ryanodine concentrations ranging from 1–1000 nM (Pessah and Zimanyi (1991).

FIGURE 3 Ryanodine-enhanced rapid Ca²⁺ release from cardiac (*top*) and skeletal (*bottom*) (SR) actively loaded in the presence of pyrophosphate. SR vesicles are loaded up to near capacity; then release is initiated (open arrowhead) by addition of ryanodine in the presence of 5 μM (cardiac) and 30 μM (skeletal) CaCl₂. Sequential activation and inactivation of the Ca²⁺ release channel of cardiac and skeletal SR is clearly observed at high concentrations of ryanodine. Ryanodine concentrations are indicated on the traces (μM). RR, ruthenium red. Reprinted with permission from Pessah and Zimanyi (1991).

dependent nature of the action of ryanodine on SR Ca²⁺ permeability, in the absence or presence of 500 mM CsCl. Ryanodine has been shown to inhibit the channel at concentrations greater than 10 μM (Meissner *et al.*, 1986; Alderson and Feher, 1987; Lattanzio *et al.*, 1987; Lai and Meissner, 1989; Lai *et al.*, 1989). The possibility of a sequential, kinetically limited mechanism by which ryanodine receptor occupancy alters SR permeability, as previously inferred by Lattanzio *et al.* (1987) and suggested by Lai *et al.* (1989), has been addressed. The experimental approach of continuously monitoring the impact of ryanodine on Ca²⁺ permeability under active-loading conditions, from the moment the alkaloid is added, has facilitated direct assessment of the possibility of a sequential mechanism. Ryanodine up to 1 mM in concentration initially activates, then inactivates, the release of Ca²⁺ from actively filled SR vesicles, allowing the reaccumulation of Ca²⁺ (Fig. 3). The rate of inactivation depends on the ryanodine concentration, as well as on the extravesicular Ca²⁺ concentration.

3. Measurement of Single-Channel Kinetics in Lipid Bilayer Membranes

Receptor-binding studies confirm the occupancy of ryanodine at 1–50 nM concentrations (to affinity states 1 and 2) on SR vesicles; however, in single-channel measurements, micromolar ryanodine is purportedly

needed to decrease channel conductance and to dramatically increase the fractional open time (Imagawa *et al.*, 1987; Rousseau *et al.*, 1987; Lai *et al.*, 1988; Smith *et al.*, 1986, 1988; Chu *et al.*, 1990). Ryanodine at 10 μM concentration reduces the conductance for Ca²⁺ from 90 to 35 pS; the subconductance state is not reversible by perfusion of the free ryanodine from the *cis* chamber (Smith *et al.*, 1988; Liu *et al.*, 1989). Since the binding of ryanodine is very sensitive to temperature and ionic strength (Pessah *et al.*, 1986; Seifert and Casida, 1986; Ogawa and Harafugi, 1990; Chu *et al.*, 1990; Carroll *et al.*, 1991; Zimanyi and Pessah, 1991a), correlating results obtained at 37°C and low ionic strength (binding experiments) with those obtained at 25°C and high ionic strength (single-channel measurements in 500 mM asymmetrical CsCl) is difficult. [³H]Ryanodine receptor-binding and single-channel current measurements on rabbit skeletal muscle SR were performed under identical conditions (Buck *et al.*, 1992). Rabbit skeletal SR fused with planar lipid bilayers assayed in the presence of asymmetrical CsCl at +20-mV holding potential and activating Ca^{2+} (100 μM, cis) exhibited discrete current fluctuations corresponding to single gating events of the Ca^{2+}-release channel. In the presence of 100 μM Ca^{2+}, ryanodine (5–40 nM) increased the open probability of the channel without changing the unitary conductance of 468 pS (Buck et al., 1992). The open dwell-time constants, which are best fit by a double exponential, increased with a concomitant decrease in the closed dwell-time constants. These results support the insightful work done by Bull and co-workers (1989) using frog skeletal SR, and imply that these actions of nanomolar ryanodine, presumably mediated by occupation of high-affinity sites, are not species specific. The current–voltage relationship of a channel modified by 10 nM ryanodine remains essentially unchanged, and the altered gating kinetics are fully reversible (Buck et al., 1992). Ryanodine (>10 nM) occasionally results in subconductance

states of approximately 50% of the full conductance state (Fig. 4A). When the concentration of ryanodine is raised to 50 nM, the channel makes a transition from the more slowly fluctuating full-conductance state to a long-lived half-conductance state after a lag time of approximately 10 min. Further increases in the concentration of ryanodine result in a more rapid appearance of a long-lived half-conductance state (225 pS). Ryanodine (10 μM) rapidly induces a half-conductance state that persists after perfusion of the cis chamber with Ca^{2+} and ryanodine-free solution (Fig. 4A), in agreement with the findings of Liu et al. (1989). Ryanodine, at concentrations ≥70 μM, causes the channel to fluctuate in half- and quarter-conductance transitions. The quarter-conductances always appear as closures from the half-conductance state and are short-lived (Fig. 4B; Buck et al., 1992).

Both half- and quarter-subconductance states have similar voltage intercepts at zero current and both are inhibited by ruthenium red, indicating that they

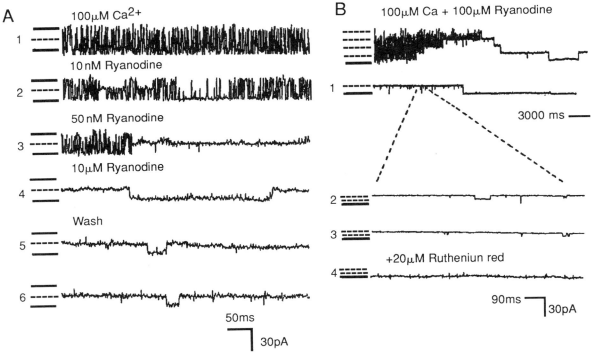

FIGURE 4 Ryanodine stabilizes distinct states of the skeletal Ca^{2+} release channel in planar biolayers. (A) Additions to the cis chamber are (1) 100 μM $CaCl_2$, (2) 10 nM ryanodine, (3) additional 40 nM ryanodine, (4) 10 μM ryanodine, trace (5) and (6) perfusion of the cis side with a Ca^{2+}-free buffer and subsequent addition of 100 μM $CaCl_2$. (B) The top trace shows the effect of 100 μM ryanodine added to the cis chamber in the presence of several channels. (1) Expanded view of a section on trace (2) showing a 1/4 conductance-state transition. (3) Further transitions to the 1/4 state. (4) A dose of 20 μM ruthenium red is added to the cis chamber. The lower solid line to the left of each trace represents the closed state of the channel, whereas the upper solid line represents the full open state; the dashed lines represent the 1/2 and 1/4 conductance states of the channel. All records were made at +20-mV holding potential. Reprinted with permission from Buck et al. (1992).

emanate from the Ca^{2+} channel. A consistent shift to more positive reversal potentials, from -36 ± 17 mV for the full-conductance state to -28 ± 7 mV for the quarter-conductance state, suggests a possible shift in the ion selectivity of the channel pore, but the significance of the observation is questionable because of large standard deviations associated with measurement of reversal potential. Ryanodine, at concentrations ≥ 200 μM, completely closes the channel and no subsequent gating is observed.

The lack of correlation between the binding affinity of [³H]ryanodine to the receptor and the concentration necessary to cause activation (half-conductance state) or inactivation (channel closure) of a single channel in bilayer experiments has been explained by the slow association rate of ryanodine, a conclusion based on experiments performed under entirely different assay conditions. Experiments show that, in the presence of 500 mM CsCl, the binding of [³H]ryanodine to its high-affinity sites proceeds rapidly at 25°C, reaching 8, 12, 22, and 42% of steady-state occupancy within 5 min at 5, 10, 20, and 40 nM concentrations, which is within the time frame of a bilayer experiment (Buck *et al.*, 1992). Ryanodine (10 nM) slows gating transitions of the full-conductance fluctuation, which corresponds well with its binding affinity for the highest affinity site ($K_{d1} \sim 1$ nM). The second high-affinity binding conformation resolved by receptor-binding experiments (K_{d1}-23 nM) is correlated with the onset of occasional half-conductance states that appear from the slower full-conductance fluctuation and rapidly manifest themselves at ryanodine concentrations of 20–50 nM; both effects are readily reversible. Elevating the concentration of ryanodine from nanomolar to micromolar in the same bilayer results in rapid transition to a long-lived half-conductance state, suggesting the sequential mechanism of ryanodine action. The association rate of [³H]ryanodine drops dramatically from 250 nM to 1 μM ryanodine concentrations, suggesting a striking change in channel conformation that is coincident with the transition to the long-lived half-conductance state and with the reduced affinity of [³H]ryanodine to the third low-affinity site (K_{d3}-1.5 μM) in the presence of 500 mM CsCl. The concentrations of ryanodine required to observe quarter-conductance fluctuations (≥ 70 μM) and complete channel closure (>200 μM) do not correlate well with the apparent K_d of the lowest affinity site for ryanodine binding. A possible explanation for this discrepancy is apparent from kinetic experiments with high concentrations of [³H]ryanodine under conditions identical to those used in bilayer experiments. Micromolar ryanodine would occupy only 7% of the binding sites within the time frame of bilayer experi-

ments. The matter is complicated by the fact that the channel transitions induced by micromolar ryanodine are unidirectional and persistent (possibly irreversible).

C. Persistent Effect of Ryanodine on the Ca^{2+} Release Channel

1. Evidence for Persistent Action of Ryanodine

Several lines of evidence suggest that the mechanism of ryanodine action deviates from a simple reversible drug–receptor interaction. The equilibrium binding of 1 μM [³H]ryanodine to terminal cisternae of SR membranes in the presence of 250 mM KCl and 50 μM Ca^{2+} is 11.5 pmol/mg protein, whereas at 50 nM [³H]ryanodine the binding to the same preparation is 7 pmol/mg protein. Assuming the presence of low-affinity binding sites ($K_d = 3$ μM), the expected value at binding of 1 μM [³H]ryanodine should be 25–30 pmol/mg protein. These data suggest a loss of the available binding sites after incubation with 1 μM ryanodine for 1 hr (Pessah and Zimanyi, 1991). Further, the slowing of the rate of dissociation induced by 100-fold excess assay medium by >0.4 μM ryanodine (McGrew *et al.*, 1989; Pessah and Zimanyi, 1991), the slow rate of dissociation induced by 2000-fold excess unlabeled ryanodine (Pessah *et al.*, 1987), and the biphasic association rate in the presence of 500 mM CsCl indicate deviation from a predictable cooperative interaction. The sequential activation and inactivation of the release of Ca^{2+} by increasing concentrations of ryanodine (Lai *et al.*, 1989; Pessah and Zimanyi, 1991), and the persistent effect of 10 μM ryanodine on channel conductance (Smith *et al.*, 1988; Liu *et al.*, 1989; Zimanyi *et al.*, 1992), suggest an irreversible change in the receptor complex induced by ryanodine.

2. Receptor Binding Sites

Junctional SR vesicles from rabbit skeletal muscle pretreated for 1 hr in the absence and presence of ryanodine (0.1–100 μM), followed by several washing steps to remove bound ryanodine, lost the high-affinity binding sites (up to 82%) after exposure to 10 or 100 μM ryanodine (Table III). Scatchard analysis of the binding of 1–5000 nM [³H]ryanodine after pretreatment with 0.1–10 μM ryanodine reveals a significant decrease in the maximal binding capacity of the high-affinity site and a complete loss of the low-affinity sites. These observations suggest that micromolar concentrations of ryanodine cause a persistent change in the receptor. The effects of 1 μM ryanodine persist for at least 48 hr after the treatment. The pharmacological sensitivity of the remaining high-affinity

TABLE III Effect of Ryanodine Pretreatment on the Binding of 1 nM [^{3}H]Ryanodine[a]

[Ryanodine][b] (μM)	Bound (pmol/mg prot.)	n	Inhibition %
0	0.748 ± 0.133	9	—
0.1	0.700 ± 0.253	3	14.05 ± 3.82
1.0	0.365 ± 0.114[c]	6	52.16 ± 8.52[c]
10.0	0.142 ± 0.049[c]	5	82.54 ± 2.23[c]
100.0	0.175 ± 0.027	4	81.47 ± 5.59

[a] Data from Zimanyi et al. (1992).
[b] Ryanodine concentration for pretreatment.
[c] Significantly different from previous concentration. $p < 0.05$, unpaired Student t test.

sites to inhibitors (ruthenium red, neomycin, Mg^{2+}) and activators (caffeine, doxorubicin), after pretreatment with ryanodine, remains unchanged relative to washed control SR (Zimanyi et al., 1992).

3. Ca^{2+} Transport Measurements

Pretreatment with 0.1 or 100 μM ryanodine alters the Ca^{2+} transport properties of SR membrane vesicles. Ryanodine (30 μM) causes rapid efflux of Ca^{2+} from washed control vesicles, whereas the initial rate of Ca^{2+} release from vesicles pretreated with 100 μM ryanodine is completely inhibited, although these vesicles loaded significantly more Ca^{2+} (Fig. 5). Membranes pretreated with 100 nM ryanodine load significantly less Ca^{2+} than the corresponding control vesicles, and exhibit a much smaller release of Ca^{2+} in response to 30 μM ryanodine, even at elevated (15 μM) extravesicular Ca^{2+} (Fig. 5b). The altered Ca^{2+}-uptake characteristics of SR pretreated with ryanodine are not caused by modification of the Ca^{2+} pump protein, since the rate of Ca^{2+}-dependent P$_i$ production shows no significant differences in washed controls and 10 or 100 μM ryanodine-treated samples (Zimanyi et al., 1992).

4. Possible Role of Thiols in the Persistent Action of Ryanodine

An important issue is the role of critical thiols in the gating process of the Ca^{2+}-release channel. Critical thiols on the ryanodine receptor complex are thought to play important roles in gating the Ca^{2+} channel (Abramson and Salama, 1989). Oxidation of thiols seems to correlate with channel opening; reduction correlates with channel closing. To test whether channel thiols play a role in the persistent action of ryanodine, the possible protective effect of dithiothreitol, an effective thiol-reducing agent, has been examined.

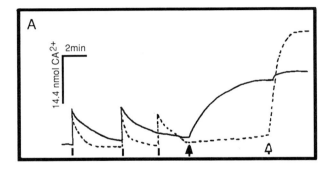

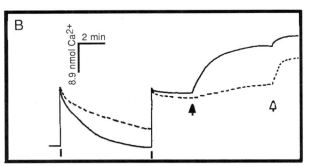

FIGURE 5 Pretreatment of skeletal sarcoplasmic reticulum (SR) with ryanodine results in persistent inhibition of ryanodine-induced Ca^{2+} release. Skeletal SR membranes were pretreated with 100 μM (A, dashed line) or 100 nM (B, dashed line) ryanodine and extensively washed to remove unbound ryanodine. Paired controls were treated identically without ryanodine (solid lines). The vesicles were actively loaded with sequential addition of 12 nmol CaCl$_2$ (vertical bars). Solid arrow indicates the addition of 30 μM ryanodine; open arrow indicates the addition of 2 μg A23187 (Ca^{2+} ionophore). Reprinted with permission from Zimanyi et al. (1992).

Dithiothreitol (1 mM) protects the ryanodine receptors from the action of 1 μM ryanodine by 51%, compared with the appropriate control, taking into account the slight activating effect of dithiothreitol itself. Dithiothreitol increases the affinity of ryanodine for its high-affinity site by enhancement of the association rate and reduction of the rate of dissociation. The protective effect of dithiothreitol has also been assessed at the level of the single channel reconstituted in lipid bilayers. Addition of 0.5 mM dithiothreitol to the cis chamber has little effect on channel gating; further addition of 10 μM ryanodine results in formation of the prolonged half-conductance state (Fig. 6).

In contrast to the persistent (irreversible) effect of 10 μM ryanodine without dithiothreitol, just shown, perfusion with a Ca^{2+}- and ryanodine-free medium, followed by the addition of 50 μM Ca^{2+} and 0.5 mM dithiothreitol, restored the full-conductance state of the channel, showing that dithiothreitol protected the channel from the irreversible action of ryanodine, and that a shift in redox equilibrium of critical thiols of

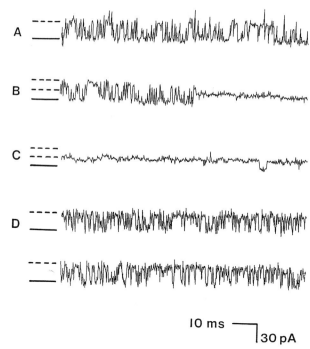

FIGURE 6 Dithiothreitol protects against the irreversible action of micromolar ryanodine concentrations. The following additions were made to the *cis* chamber: (A) 0.5 mM DTT + 50 μM CaCl₂, (B) 10 μM ryanodine, (C) 50 μM CaCl₂, after perfusion of the *cis* chamber with Ca²⁺-free buffer, and (D) 0.5 mM DTT. The solid bar to the left of each trace represents the closed state of the channel. The center dashed line represents the 1/2 conductance state, whereas the upper dashed line represents the full conductance state of the channel. Reprinted with permission from Zimanyi *et al.* (1992).

the channel complex may account, completely or partially, for the hypothesized energy barrier associated with conformational transition resulting in a long-lived half-conductance state, manifested by the sudden reduction of the association rate of micromolar ryanodine in the presence of 500 mM CsCl and the occlusion of the gating pore by ryanodine binding to the low-affinity sites (Zimanyi *et al.*, 1992).

A number of investigators have reported that sulfhydryl-oxidizing reagents including anthraquinones (Abramson *et al.*, 1988; Pessah *et al.*, 1990), reactive disulfides (Pessah *et al.*, 1987; Zaidi *et al.*, 1989), and heavy metal ions (Abramson and Salama, 1989) induce the release of Ca²⁺ from SR vesicles and modify the gating behavior of individual Ca²⁺ channels reconstituted in bilayer lipid membranes (BLM). This suggests that the oxidation state of "critical" sulfhydryl (SH) moieties may be important in regulating the SR Ca²⁺ channel. However, the identity of the proteins targeted by thiol reactive reagents in the SR membrane have not been clearly established. Whether

physiologically relevant modulators of the RyR allosterically alter the thiol redox state of channel-associated proteins is unknown. Liu and coworkers (1994) have provided direct evidence of the existence, location, and essential function of critical thiols within the channel protein complex. The rationale behind these studies argues that if redox cycling of sulfhydryl moieties is essential to normal channel gating, then critical sulfhydryls must surely be chemically reactive to account for the rapidity of channel gating. The nonfluorescent maleimide, 7-diethyl amino-3-(4'-maleimidylphenyl)-4-methylcoumarin (CPM) readily undergoes Michael addition with protein thiols, producing an irreversible adduct with high fluorescent yield (Sipple, 1981). Utilizing CPM at low concentration (1–50 nM, i.e., 0.02–1.0 pmol/μg SR protein), Liu *et al.* (1994) have examined the kinetics of labeling of junctional SR protein from rabbit skeletal or rat cardiac muscle in the presence of physiologically and pharmacologically relevant channel activators or inhibitors. The approach has demonstrated (1) the existence of a discrete class (≤ 1 pmol/μg SR protein) of highly reactive thiol groups on the SR membrane, (2) that their reactivity is markedly influenced by known physiological and pharmacological modulators of the RyR,[2] (3) hyperreactive sulfhydryls are localized primarily on the ryanodine receptor (RyR) protomer and channel associated proteins such as triadin, and (4) that hyperreactive thiols are essential to the normal gating of Ca²⁺ channels. CPM will undoubtedly be a very useful probe for defining the location and role of hyperreactive sulfhydryl chemistry in regulation of the RyR-1 complex.

IV. Concluding Remarks

The pharmacology of the Ca²⁺-release channel of muscle SR has been extensively studied. Ryanodine-sensitive Ca²⁺ channels of SR (and perhaps ER) appear to be highly modulated by a host of chemically heterogeneous physiological and xenobiotic agents. The high degree of chemical modulation and genetic regulation (Marks *et al.*, 1991; Giannini, 1992) to which these intracellular Ca²⁺ channels are subjected undoubtedly reflects an important role, not only in E–C coupling of striated muscle but also in stimulus–secretion coupling and timing of developmental programs.

Ryanodine is a specific conformational probe for the Ca²⁺-release channel. Mechanistically, the coupling between ryanodine-receptor occupancy and channel function is complex. The following questions have been addressed in an attempt understand the action of ryanodine on the channel protein. (1) Do multiple binding sites for the alkaloid exist on the junctional foot tetramer with different affinities, or is a single site transformed to different affinity states? (2) Can an interaction among ryanodine binding sites be demonstrated? (3) Does the binding of ryanodine lead to sequential changes in channel function? (4) What is the relationship between various binding affinities for ryanodine and the concentrations necessary to modulate the gating property of the channel? (5) Does the action of high concentrations of ryanodine persist despite removal of the alkaloid from its binding sites?

Researchers have reached the consensus that ryanodine specifically recognizes the Ca^{2+}-release channel/junctional foot complex of terminal SR. Experimental evidence argues for and against 1, 2, 3, or 4 binding sites per junctional foot tetramer. Undoubtedly, the open channel binds [^{3}H]ryanodine with higher affinity than the closed channel. Solubilization of the SR Ca^{2+}-release channel by Zwittergent 3-14 has been shown to denature high- and low-affinity [^{3}H]ryanodine binding, indicating the segregation of the four subunits of the Ca^{2+}-release channel and the loss of cooperativity among binding sites (Lai et al., 1989). Allosterism and interconversion between low- and high-affinity binding sites is controlled by the Ca^{2+} concentration of the medium, which implies a close relationship between the model describing the action of ryanodine and that describing the modulation of the channel by physiologically relevant agents. Several researchers suggest the presence of one high-affinity (K_d = 1–5 nM under any circumstance) and 2–3 low-affinity binding sites per tetramer (Lai et al., 1989; McGrew et al., 1989; Chu et al., 1990; Carroll et al., 1991; Pessah and Zimanyi, 1991; Hawkes et al., 1992). A discrepancy arises among the exact values of K_d for the low-affinity sites, because of the wide range of experimental conditions used. The experimental variables with the greatest impact on the outcome of in vitro assays include Ca^{2+}, ionic strength, and temperature. As has been suggested by Chu et al. (1990), high ionic strength seems to be necessary to remove an endogenously bound ligand (perhaps FKBP12) from the receptor to facilitate binding of ryanodine. At high ionic strength, the affinity of ryanodine for the high-affinity state increases, whereas the affinity for the low-affinity states decreases. The various affinities of ryanodine can be best explained by the binding of ryanodine to interconvertible sites; therefore, the observed K_d is always a weighted average of ryanodine binding to multiple conformational states. The influence of several modulators and inhibitors on ryanodine binding correlates well with their action on channel gating and underscores a high degree of allosteric control. Examples include the slowing rate of association in response to high concentrations of Ca^{2+} and Mg^{2+} and the slowing rate of dissociation in response to high concentrations of ruthenium red, neomycin, or ryanodine (Chu et al., 1990; Pessah and Zimanyi, 1991; Mack et al., 1992). Currently, more experimental support is available for negative cooperativity (Lai et al., 1989; Chu et al., 1990; Pessah and Zimanyi, 1991; Carroll et al., 1991; Zimanyi et al., 1992) than for positive cooperativity (McGrew et al., 1989; Hawkes et al., 1992) among the ryanodine binding sites. The finding by Carroll et al. (1991) that the covalent stabilization of the ryanodine receptor oligomers by 1,6-bis-maleimidohexane does not change the binding of [^{3}H]ryanodine or the release of Ca^{2+} indicates that the tetramer is the functional form of the ryanodine-receptor channel.

Wyskovsky et al. (1990), using physiological assay conditions, established two three-state models by meticulous kinetic data analyses. The models, based on the inhibitory behavior of ruthenium red and neomycin on SR Ca^{2+} permeability, could not distinguish between Model I, two open states and one closed state, and Model II, one open state and two closed states. The proposed models assume that activation of the channel reduces inactivation, whereas inhibition of the channel accelerates channel inactivation. However, caution should be used when using ruthenium red, neomycin, or ryanodine to gain information about native channel conformations. For example, in the Ca^{2+} transport assays by Wyskovsky et al. (1990), the concentration of neomycin (100 μM) was high enough to induce multiple conformational states of the channel complex. The inherent ability of neomycin not only to antagonize Ca^{2+} efflux from the native channel, but also to significantly and persistently alter channel conformation (and, perhaps, the fundamental unit responsible for channel behavior) must be accounted for in data analyses and kinetic modeling. Liu et al. (1989) demonstrated the presence of four distinct subconductance states of the purified SR channel in single-channel current measurements. Whether four discrete conductances occur in the native channel remains unknown. Could ryanodine stabilize each of these putative channel transitions? One possible model that accounts for the diverse kinetic behavior and actions of the ryanodine on the Ca^{2+}-release channel of SR assumes four allosterically modulated binding sites, for which sequential binding of ryanodine to one site would decrease the affinity of binding to the next site (negative cooperativity). The model should also account for the apparently sequential nature of ryanodine action on channel function.

This interpretation is supported by the biphasic effect of ryanodine on the rate of association and dissociation, by the persistent effect resulting from pretreatment with micromolar ryanodine (including the loss of low-affinity sites and the complete block of Ca^{2+} release), and by the persistent, perhaps irreversible, change in channel conductance. The apparent paradox of slowing dissociation rates of nanomolar [^{3}H]ryanodine by the subsequent addition of micromolar unlabeled alkaloid can also be explained by a sequential mechanism in which the kinetic constants obtained depend on the order in which [^{3}H]ryanodine and unlabeled ryanodine concentrations are raised during kinetic experiments. If the high-affinity site for ryanodine is near the channel pore, sequential occupation of additional sites on the channel oligomer, on elevating ryanodine concentration, could cause partial occlusion of the former and account for the slowing of dissociation. However, unequivocal determination of the stoichiometry of ryanodine binding sites per functional oligomer awaits photoaffinity labeling of the ryanodine receptor.

Acknowledgments

The authors would like to thank Kristy J. Gayman, Mary J. Scheidt, and Emily L. Durie for their excellent technical assistance and Aimee L. Edinger for editing the manuscript. We are indebted to Matthew M. Mack, Edmond Buck, Jonathan J. Abramson, Gouhoa Liu, and Paul D. Pion, for their contribution and insightful discussions. This work has been supported by Grant ES05002 from the National Institutes of Health and a Grant-In-Aid #90-167 from the American Heart Association, California Affiliate, to I. N. Pessah and by Grant GM 44337 from the National Institutes of Health to J. J. Abramson.

References

Abramson, J. J., and Salama, G. (1989). Critical sulfhydryls regulate calcium release from sarcoplasmic reticulum. *J. Bioenerg. Biomembr.* **21**, 283–294.

Abramson, J. J., Buck, E., Salama, G., Casida, J. E., and Pessah, I. N. (1988). Mechanism of anthraquinone-induced calcium release from skeletal muscle sarcoplasmic reticulum. *J. Biol. Chem.* **263**, 18750–18758.

Alderson, B. H., and Feher, J. J. (1987). The interaction of calcium and ryanodine with cardiac sarcoplasmic reticulum. *Biochim. Biophys. Acta* **900**, 221–229.

Anderson, K., Lai, F. A., Liu, Q.-Y., Rousseau, E., Erickson, H. P., and Meissner, G. (1989). Structural and functional characterization of the purified cardiac ryanodine receptor–Ca^{2+} release channel complex. *J. Biol. Chem.* **264**, 1329–1335.

Antoniu, B., Kim, D. H., Morii, M., and Ikemoto, N. (1985). Inhibitors of Ca^{2+} release from the isolated sarcoplasmic reticulum. I. Ca^{2+} channel blockers. *Biochim. Biophys. Acta* **816**, 9–17.

Ashley, R. H. (1989). Activation and conductance properties of ryanodine-sensitive calcium channels from brain microsomal membranes incorporated into planar lipid bilayers. *J. Membr. Biol.* **111**, 179–189.

Baeyens, J. M., and Del Pozo, E. (1985). Comparison of the effect of calcium and channel stimulant Bay K 8644 on neomycin-induced neuromuscular blockade. *Pharmacol. Toxicol.* **65**, 398–401.

Baylor, S. M., Hollingworth, S., and Marshall, M. W. (1989). Effects of intracellular ruthenium red on excitation–contraction coupling in intact frog skeletal muscle fibres. *J. Physiol.* **408**, 617–635.

Buck, E., Zimanyi, I., Abramson, J. J., and Pessah, I. N. (1992). Ryanodine stabilizes multiple conformational states of the skeletal muscle calcium release channel. *J. Biol. Chem.* **267**, 23560–23567.

Bull, R., Marengo, J. J., Suarez-Isla, B. A., Donoso, P., Sutko, J. L., and Hidalgo, C. (1989). Activation of calcium channels in sarcoplasmic reticulum from frog muscle by nanomolar concentrations of ryanodine. *Biophys. J.* **56**, 749–756.

Campbell, K. P., Knudson, C. M., Imagawa, T., Leung, T., Sutko, L., Kahl, D., Raab, C. R., and Madson, L. (1987). Identification and characterization of the high affinity [^{3}H]ryanodine receptor of the junctional sarcoplasmic reticulum Ca^{2+} release channel. *J. Biol. Chem.* **262**, 6460–6463.

Carroll, S., Skarmeta, J. G., Yu, X., Collins, K. D., and Inesi, G. (1991). Interdependence of ryanodine binding, oligomeric receptor interactions, and Ca^{2+} release regulation in junctional sarcoplasmic reticulum. *Arch. Biochem. Biophys.* **260**, 239–247.

Caswell, A. H., Brandt, N. R., Brunschwig, J. P., and Purkerson, S. (1991). Localization and partial characterization of the oligomeric disulfide-linked molecular weight 95,000 protein (triadin) which binds the ryanodine and dihydropyridine receptors in skeletal muscle triadic vesicles. *Biochemistry* **30**, 7507–7513.

Chamberlain, B. K., Volpe, P., and Fleischer, S. (1984). Inhibition of calcium-induced calcium release from purified cardiac sarcoplasmic reticulum vesicles. *J. Biol. Chem.* **259**, 7547–7553.

Chiesi, N., Schwaller, R., and Calviello, G. (1988). Inhibition of rapid Ca^{2+}-release from isolated skeletal and cardiac sarcoplasmic reticulum membranes. *Biochem. Biophys. Res. Commun.* **154**, 1–8.

Chu, A., Diaz-Munoz, M., Hawkes, M. J., Brush, K., and Hamilton, S. L. (1990). Ryanodine as a probe for the functional state of the skeletal muscle sarcoplasmic reticulum calcium release channel. *Mol. Pharmacol.* **37**, 735–741.

Collins, J. H. (1991). Sequence analysis of the ryanodine receptor: Possible association with a 12K, FK506-binding immunophilin/protein kinase C inhibitor. *Biochem. Biophys. Res. Commun.* **17**, 1288–1290.

Costello, B., Chadwick, C., Saito, A., Chu, A., Maurer, A., and Fleischer, S. (1986). Characterization of the junctional face membrane from terminal cisternae of sarcoplasmic reticulum. *J. Cell Biol.* **103**, 741–754.

Fairhurst, A. S. (1971). The preparation of tritium-labeled ryanodine. *J. Labelled Compd.* **7**, 133–136.

Fairhurst, A. S., and Hasselbach, W. (1970). Calcium efflux from a heavy sarcotubular fraction. Effects of ryanodine, caffeine and magnesium. *Eur. J. Biochem.* **13**, 504–509.

Fang, Y-I., Adachi, M., Kobayashi, J., and Ohizumi, Y. (1993). High affinity binding of 9-[^{3}H]methyl-7-bromoeudistomin D to the caffeine-binding site of skeletal muscle sarcoplasmic reticulum. *J. Biol. Chem.* **268**, 18622–18625.

Feher, J. J., and Lipford, G. B. (1985). Mechanism of ryanodine on cardiac sarcoplasmic reticulum. *Biochim. Biophys. Acta* **813**, 77–86.

Ferguson, D. G., Schwartz, H. W., and Franzini-Armstrong, C. (1984). Subunit structure of junctional feet in triads of skeletal muscle: a freeze-drying, rotary-shadowing study. *J. Cell Biol.* **99**, 1735–1742.

Fleischer, S., Ogunbunmi, E. M., Dixon, M. C., and Fleer,

E. A. M. (1985). Localization of Ca^{2+} release channels with ryanodine in junctional terminal cisternae of sarcoplasmic reticulum of fast skeletal muscle. *Proc. Natl. Acad. Sci. USA* **82**, 7256–7259.

Giannini, G., Clementi, E., Ceci, R., Marziali, G., and Sorrentino, V. (1992). Expression of a ryanodine receptor-Ca^{2+} channel that is regulated by TGF-β. *Science* **257**, 91–94.

Hawkes, M. J., Nelson, T. E., and Hamilton, S. L. (1992). [³H]Ryanodine as a probe of changes in the functional state of the Ca^{2+}-release channel in malignant hyperthermia. *J. Biol. Chem.* **267**, 6702–6709.

Hilgemann, D. W., Delay, M. J., and Lager, G. A. (1983). Activation-dependent cumulative depletions of extracellular free calcium in guinea pig atrium measured with antipyrylazo III and tetramethyl murexide. *Circ. Res.* **53**, 779–793.

Hymel, L., Inui, M., Fleischer, S., and Schindler, H. (1988a). Purified ryanodine receptor of skeletal muscle sarcoplasmic reticulum forms Ca^{2+}-activated Ca^{2+} channels in planar bilayers. *Proc. Natl. Acad. Sci. USA* **85**, 441–445.

Hymel, L., Schindler, H., Inui, M., and Fleischer, S. (1988b). Reconstitution of purified cardiac muscle calcium release channel (ryanodine receptor) in planar bilayers. *Biochem. Biophys. Res. Commun.* **152**, 308–314.

Imagawa, T., Smith, J. S., Coronado, R., and Campbell, K. P. (1987). Purified ryanodine receptor from skeletal muscle sarcoplasmic reticulum is the Ca^{2+} permeable pore of the calcium release channel. *J. Biol. Chem.* **262**, 16636–16643.

Inui, M., Saito, A., and Fleischer, S. (1987a). Isolation of the ryanodine receptor from cardiac sarcoplasmic reticulum and identity with the feet structures. *J. Biol. Chem.* **262**, 15637–15642.

Inui, M., Saito, A., and Fleischer, S. (1987b). Purification of the ryanodine receptor and identity with feet structures of junctional terminal cisternae of sarcoplasmic reticulum of fast skeletal muscle. *J. Biol. Chem.* **262**, 1740–1747.

Inui, M., Wang, S., Saito, A., and Fleischer, S. (1988). Characterization of junctional and longitudinal sarcoplasmic reticulum from heart muscle. *J. Biol. Chem.* **263**, 10843–10850.

Jayaraman, T., Brillantes, A-M., Timerman, A. P., Fleischer, S., Erdjument-Bromage, H., Tempst, P., and Marks, A. R. (1992). FK506 binding protein associated with calcium release channel (ryanodine receptor). *J. Biol. Chem.* **267**, 9474–9477.

Jenden, D. J., and Fairhurst, A. S. (1969). The pharmacology of ryanodine. *Pharmacol. Rev.* **21**, 1–25.

Jones, L. R., and Burrows, S. D. (1983). Ruthenium red (RR)-induced stimulation of Ca^{2+} uptake by junctional sarcoplasmic reticulum vesicles (JSRV) isolated from heart. *Fed. Proc.* **42**, 1933 Abstr. 1024.

Jones, L. R., and Cala, S. E. (1981). Biochemical evidence for functional heterogeneity of cardiac sarcoplasmic reticulum vesicles. *J. Biol. Chem.* **256**, 11809–11818.

Jones, L. R., Besch, H. R., Jr., Sutko, J. L., and Willerson, J. T. (1979). Ryanodine-induced stimulation of net Ca^{++} uptake by cardiac sarcoplasmic reticulum vesicles. *J. Pharmacol. Exp. Ther.* **209**, 48–55.

Jorgensen, A. O., Shen, A. C., Arnold, W., McPherson, P. S., and Campbell, K. P. (1993). The Ca^{2+}-release channel/ryanodine receptor is localized in junctional and corbular sarcoplasmic reticulum in cardiac muscle. *J. Cell Biol.* **120**, 969–980.

Kawamoto, R. M., Brunschwig, J. P., Kim, K. C., and Caswell, A. H. (1986). Isolation, characterization, and localization of the spanning protein from skeletal muscle triads. *J. Cell Biol.* **103**, 1405–1414.

Kim, D. H., Ohnishi, S. T., and Ikemoto, N. (1983). Kinetic studies of calcium release from sarcoplasmic reticulum in vitro. *J. Biol. Chem.* **258**, 9662–9668.

Kim, K. C., Caswell, A. H., Talvenheimo, J. A., and Brant,

N. R. (1990). Isolation of a terminal protein which may link the dihydropyridine receptor to the junctional foot protein in skeletal muscle. *Biochemistry* **29**, 9281–9289.

Knudson, C. M., Strong, K. K., Moomaw, C. R., Slaughter, C. A., and Campbell, K. P. (1993). Primary structure and topological analysis of junctional sarcoplasmic reticulum glycoprotein (triadin). *J. Biol. Chem.* **268**, 12646–12654.

Kuwajima, G., Futatsugi, A., Niinobe, M., Nakanishi, S., and Mikoshiba, K. (1992). Two types of ryanodine receptors in mouse brain: skeletal muscle type exclusively in Purkinje cells and cardiac muscle type in various neurons. *Neuron* **9**, 1133–1142.

Lai, F. A., Erickson, H., Block, B. A., and Meissner, G. (1987). Evidence for a junctional feet-ryanodine receptor complex from sarcoplasmic reticulum. *Biochem. Biophys. Res. Commun.* **143**, 704–709.

Lai, F. A., Erickson, H. P., Rousseau, E., Liu, Q.-Y., and Meissner, G. (1988a). Purification and reconstitution of the calcium release channel from skeletal muscle. *Nature (London)* **331**, 315–319.

Lai, F. A., Anderson, K., Rousseau, E., Liu, Q.-Y., and Meissner, G. (1988b). Evidence for a Ca^{2+} channel within the ryanodine receptor complex from cardiac sarcoplasmic reticulum. *Biochem. Biophys. Res. Commun.* **151**, 441–449.

Lai, F. A., and Meissner, G. (1989). The muscle ryanodine receptor and its intrinsic Ca^{2+} channel activity. *J. Bioenerg. Biomembr.* **21**, 227–246.

Lai, F. A., Misra, M., Xu, L., Smith, H. A., and Meissner, G. (1989). The ryanodine receptor-Ca^{2+} release channel complex of skeletal muscle sarcoplasmic reticulum. Evidence for a cooperatively coupled, negatively charged homotetramer. *J. Biol. Chem.* **264**, 16776–16785.

Lai, F. A., Dent, M., Wickenden, C., Xu, L., Kumari, G., Misra, M., Lee, H. B., Sar, M., and Meissner, G. (1992). Expression of a cardiac Ca^{2+}-release channel isoform in mammalian brain. *Biochem. J.* **288**, 553–564.

Lattanzio, F. A. Jr., Schlatterer, R. G., Nicar, M., Campbell, K. P., and Sutko, J. L. (1987). The effects of ryanodine on passive calcium fluxes across sarcoplasmic reticulum membranes. *J. Biol. Chem.* **262**, 2711–2718.

Liu, G., Abramson, J. J., Zable, A. C., and Pessah, I. N. (1994). Direct evidence for existence and functional role of hyperreactive sulfhydryls on ryanodine receptor/triadin complex selectively labeled by the coumarin maleimide CPM. *Mol. Pharmacol.* **45**, 189–200.

Liu, Q.-Y., Lai, F. A., Rousseau, E., Jones, R. V., and Meissner, G. (1989). Multiple conductance states of the purified calcium release channel complex from skeletal sarcoplasmic reticulum. *Biophys. J.* **55**, 415–424.

Mack, M. M., Zimanyi, I., and Pessah, I. N. (1992). Discrimination of multiple binding sites for antagonists of the calcium release channel complex of skeletal and cardiac sarcoplasmic reticulum. *J. Pharmacol. Exp. Ther.* **262**, 1028–1037.

Marks, A. R., Tempst, P., Hwang, K. S., Taubman, M. B., Inui, M., Chadwick, C., Fleischer, S., and Nadal-Ginard, B. (1989). Molecular cloning and characterization of the ryanodine receptor/junctional channel complex cDNA from skeletal muscle sarcoplasmic reticulum. *Proc. Natl. Acad. Sci. USA* **86**, 8683–8687.

McGrew, S. G., Wolleben, C., Siegl, P., Inui, M., and Fleischer, S. (1989). Positive cooperativity of ryanodine binding to the calcium release channel of sarcoplasmic reticulum from heart and skeletal muscle. *Biochemistry* **28**, 1686–1695.

McPherson, P. S., and Campbell, K. P. (1990). Solubilization and biochemical characterization of the high-affinity [³H]ryanodine receptor from rabbit brain membranes. *J. Biol. Chem.* **265**, 18454–18460.

Meissner, G. (1984). Adenine nucleotide stimulation of Ca^{2+}-

induced Ca²⁺ release in sarcoplasmic reticulum. *J. Biol. Chem.* **259**, 2365–2374.

Meissner, G. (1986). Ryanodine activation and inhibition of the Ca²⁺ release channel of sarcoplasmic reticulum. *J. Biol. Chem.* **260**, 6300–6306.

Meissner, G., Darling, E., and Eveleth J. (1986). Kinetics of rapid Ca²⁺ release by sarcoplasmic reticulum. Effects of Ca²⁺, Mg²⁺, and adenine nucleotides. *Biochemistry* **25**, 236–244.

Michalak, M., Dupraz, P., and Soshan-Barmatz, V. (1988). Ryanodine binding to sarcoplasmic reticulum membranes: comparison between cardiac and skeletal muscle. *Biochim. Biophys. Acta* **939**, 587–594.

Miyamoto, H., and Racker, E. (1982). Mechanism of calcium release from skeletal sarcoplasmic reticulum. *J. Membr. Biol.* **66**, 183–201.

Nagasaki, K., and Fleischer, S. (1988). Ryanodine-sensitivity of the calcium release channel of sarcoplasmic reticulum. *Cell Calcium* **9**, 1–7.

Nakai, J., Imagawa, T., Hakamat, Y., Shigekawa, M., Takeshima, H., and Numa, S. (1990). Primary structure and functional expression from cDNA of the cardiac ryanodine receptor/calcium release channel. *FEBS Lett.* **271**, 169–177.

Ogawa, Y., and Harafugi, H. (1990). Effects of temperature on [³H]ryanodine binding to sarcoplasmic reticulum from bullfrog skeletal muscle. *J. Biochem.* **107**, 887–893.

Ohnishi, S. T. (1979). Calcium-induced calcium release from fragmented sarcoplasmic reticulum. *J. Biochem. (Tokyo)* **86**, 1147–1150.

Ondrias, K., Borgatta, L., Kim, D. H., and Ehrlich, B. (1990). Biphasic effects of doxorubicin on the calcium release channel from sarcoplasmic reticulum of cardiac muscle. *Circ. Res.* **67**, 1167–1174.

Otsu, K., Willard, H. F., Khanna, V. K., Zorzato, F., Green, N. M., and MacLennan, D. H. (1990). Molecular cloning of cDNA encoding the Ca²⁺ release channel (ryanodine receptor) of rabbit cardiac muscle sarcoplasmic reticulum. *J. Biol. Chem.* **265**, 13472–13483.

Palade, P. (1987a). Drug-induced Ca²⁺ release from isolated sarcoplasmic reticulum. I. Use of pyrophosphate to study caffeine-induced Ca²⁺ release. *J. Biol. Chem.* **262**, 6135–6141.

Palade, P. (1987b). Drug-induced Ca²⁺ release from isolated sarcoplasmic reticulum. II. Releases involving a Ca²⁺-induced Ca²⁺ release channel. *J. Biol. Chem.* **262**, 6142–6148.

Palade, P. (1987c). Drug-induced Ca²⁺ release from isolated sarcoplasmic reticulum. III. Block of Ca²⁺-induced Ca²⁺ release by organic polyamines. *J. Biol. Chem.* **262**, 6149–6154.

Pessah, I. N., and Zimanyi, I. (1991). Characterization of multiple [³H]ryanodine binding sites on the Ca²⁺ release channel of sarcoplasmic reticulum from skeletal and cardiac muscle: Evidence for a sequential mechanism in ryanodine action. *Mol. Pharmacol.* **39**, 679–689.

Pessah, I. N., Waterhouse, A. L., and Casida, J. E. (1985). The calcium-ryanodine receptor complex of skeletal and cardiac muscle. *Biochem. Biophys. Res. Commun.* **128**, 449–456.

Pessah, I. N., Francini, A. O., Scales, D. J., Waterhouse, A. L., and Casida, J. E. (1986). Calcium-ryanodine receptor complex: Solubilization and partial characterization from skeletal muscle junctional sarcoplasmic reticulum. *J. Biol. Chem.* **262**, 8643–8648.

Pessah, I. N., Stambuk, R. A., and Casida, J. E. (1987). Ca²⁺-activated ryanodine binding: mechanism of sensitivity and intensity modulation by Mg²⁺, caffeine and adenine nucleotides. *Mol. Pharmacol.* **31**, 232–238.

Pessah, I. N., Durie, E., Schiedt, M. J., and Zimanyi, I. (1990). Anthraquinone-sensitizes Ca²⁺ release channel from rat cardiac sarcoplasmic reticulum: Possible receptor-mediated mechanism for doxorubicin cardiomyopathy. *Mol. Pharmacol.* **37**, 503–514.

Pessah, I. N., Schiedt, M. J., Shalaby, M. A., Mack, M., and Giri, S. N. (1992). Etiology of sarcoplasmic reticulum calcium release channel lesions in doxorubicin-induced cardiomyopathy. *Toxicology* **72**, 189–206.

Pridgen, J. E. (1956). Respiratory arrest thought to be due to intra-peritoneal neomycin. *Surgery (St. Louis)* **40**, 571–574.

Procita, L. (1958). Some pharmacological actions of ryanodine in the mammal. *J. Pharmacol. Exp. Ther.* **123**, 296–305.

Rardon, D. P., Cefali, D. C., Mitchell, R. D., Seiler, S. M., and Jones, L. R. (1989). High molecular weight proteins purified from cardiac junctional sarcoplasmic reticulum vesicles are ryanodine-sensitive calcium channels. *Circ. Res.* **64**, 779–789.

Rousseau, E., and Meissner, G. (1989). Single cardiac sarcoplasmic reticulum Ca²⁺ release channel: Activation by caffeine. *Am. J. Physiol.* **256**, H328–H333.

Rousseau, E., Smith, J., Henderson, J., and Meissner, G. (1986). Single channel and calcium flux measurements of the cardiac sarcoplasmic reticulum calcium channel. *Biophys. J.* **50**, 1009–1014.

Rousseau, E., Smith, J. S., and Meissner, G. (1987). Ryanodine modifies conductance and gating behavior of single Ca²⁺ release channel. *Am. J. Physiol.* **253**, C364–C368.

Seifert, J., and Casida, J. E. (1986). Ca²⁺-dependent ryanodine binding site: soluble preparation from rabbit cardiac sarcoplasmic reticulum. *Biochem. Biophys. Acta* **861**, 399–405.

Seiler, S., Wegener, A. D., Whang, D. D., Hathaway, D. R., and Jones, L. R. (1984). High molecular weight proteins in cardiac and skeletal muscle junctional sarcoplasmic reticulum vesicles bind calmodulin, are phosphorylated and are degraded by Ca²⁺-activated protease. *J. Biol. Chem.* **259**, 8550–8557.

Seino, A. Kobayashi, M., Kobayashi, J., Fang, Y-I., Ishibashi, M., Nakamura, H., Momose, K., and Ohizumi, Y. (1991). 9-methyl-7-bromoeudistomin D, a powerful radio-labelable Ca⁺⁺ releaser having caffeine-like properties, acts on Ca⁺⁺-induced Ca⁺⁺ release channels of sarcoplasmic reticulum. *J. Pharmacol. Exp. Ther.* **256**, 861–867.

Sipple, T. O. (1981). New fluorochromes for thiols: Maleimide and iodoacetimide derivatives of 3-phenylcoumarin fluorophore. *J. Histochem. Cytochem.* **29**, 314–321.

Smith, J. S., Coronado, R., and Meissner, G. (1985). Sarcoplasmic reticulum contains adenine-nucleotide-activated calcium channels. *Nature (London)* **316**, 446–449.

Smith, J. S., Coronado, R., and Meissner, G. (1986). Single channel measurements of the calcium release channel from skeletal muscle sarcoplasmic reticulum: Activation by calcium, ATP and modulation by magnesium. *J. Gen. Physiol.* **88**, 573–588.

Smith, J. S., Imagawa, T., Ma, J., Fill, M., Campbell, K. P., and Coronado, R. (1988). Purified ryanodine receptor from rabbit skeletal muscle is the calcium-release channel of the sarcoplasmic reticulum. *J. Gen. Physiol.* **92**, 1–26.

Somlyo, A. V. (1979). Bridging structures spanning the junctional gap at the triad of skeletal muscle. *J. Cell Biol.* **80**, 743–750.

Sutko, J. L., and Kenyon, J. L. (1983). Ryanodine modification of cardiac muscle responses to potassium-free solutions: evidence for inhibition of sarcoplasmic reticulum calcium release. *J. Gen. Physiol.* **82**, 385–404.

Sutko, J. L., Thompson, L. J., Schlatterer, R. G., Lattanzio, F. A., Fairhurst, A. S., Campbell, C., Martin, S. F., Deslongchamps, P., Ruest, L., and Taylor, D. R. (1986). Separation and formation of ryanodine from dehydroryanodine. Preparation of tritium-labeled ryanodine. *J. Labelled Compd. Radiopharmacol.* **23**, 215–222.

Takeshima, H., Nishimura, S., Matsumoto, T., Ishida, H., Kangawa, K., Minamino, N., Matsuo, H., Ueda, M., Hanaoka, M., Hirose, T., and Numa, S. (1989). Primary structure and expression from complementary DNA of skeletal muscle ryanodine receptor. *Nature (London)* **339,** 439–445.

Timerman, A., Ogunbumni, E., Freund, E., Wiederrecht, G., Marks, A. R., and Fleischer, S. (1993). The calcium release channel of sarcoplasmic reticulum is modulated by FK506 binding protein. Dissociation and reconstitution of FKBP-12 to the calcium release channel of skeletal muscle sarcoplasmic reticulum. *J. Biol. Chem.* **268,** 22992–22999.

Volpe, P., Salviati, G., and Chu, A. (1986). Calcium-gated calcium channels in sarcoplasmic reticulum of rabbit skinned skeletal muscle fibers. *J. Gen. Physiol.* **87,** 289–303.

Waterhouse, A. L., Holden, I., and Casida, J. E. (1984). 9,21-Didehydroryanodine: A new principal toxic constituent of the botanical insecticide ryania. *J. Chem. Soc. Chem. Commun.* 1265–1266.

Waterhouse, A. L., Pessah, I. N., Francini, A. O., and Casida, J. E. (1987). Structural aspects of ryanodine action and selectivity. *J. Med. Chem.* **30,** 710–716.

Wyskovsky, W., Hohenegger, M., Plank, B., Hellmann, G., Klein, S., and Suko, J. (1990). Activation and inhibition of the calcium-release channel of isolated skeletal muscle heavy sarcoplasmic reticulum. Models of the calcium-release channel. *Eur. J. Biochem.* **194,** 549–559.

Yamamoto, N., and Kasai, M. (1982). Mechanism and function of the Ca^{2+}-gated cation channel in sarcoplasmic reticulum vesicles. *J. Biochem. (Tokyo)* **92,** 485–496.

Zaidi, N. F., Lagenaur, C. F., Abramson, J. J., Pessah, I. N., and Salama, G. (1989). Reactive disulfides trigger Ca^{2+} release from sarcoplasmic reticulum via an oxidation reaction. *J. Biol. Chem.* **264,** 21725–21736.

Zimanyi, I., and Pessah, I. N. (1991a). Comparison of [^{3}H]ryanodine receptors and Ca^{++} release from rat cardiac and rabbit skeletal muscle sarcoplasmic reticulum. *J. Pharmacol. Exp. Ther.* **256,** 938–946.

Zimanyi, I., and Pessah, I. N. (1991b). Pharmacological characterization of the specific binding of [^{3}H]ryanodine to rat brain microsomal membranes. *Brain Res.* **561,** 181–191.

Zimanyi, I., Buck, E., Abramson, J. J., and Pessah, I. N. (1992). Ryanodine induces persistent activation of the Ca^{2+} release channel from skeletal sarcoplasmic reticulum. *Mol. Pharmacol.* **42,** 1049–1057.

Zorzato, F., Salviati, G., Facchinetti, T., and Volpe, P. (1985). Doxorubicin induces calcium release from terminal cisternae of skeletal muscle. *J. Biol. Chem.* **260,** 7349–7355.

Zorzato, F., Margreth, A., and Volpe, P. (1986). Direct photolabeling of junctional sarcoplasmic reticulum with [^{14}C]doxorubicin. *J. Biol. Chem.* **261,** 13252–13257.

Zorzato, F., Fujii, J., Otsu, K., Phillips, M., Green, N. M., Lai, F. A., Meissner, G., and MacLennan, D. H. (1990). Molecular cloning of cDNA encoding human and rabbit forms of the Ca^{2+} release channel (ryanodine receptor) of skeletal muscle sarcoplasmic reticulum. *J. Biol. Chem.* **265,** 2244–2256.

InsP₃ Receptor and Intracellular Calcium Release

InsP_3 Receptor and Intracellular Calcium Release

TREVOR J. SHUTTLEWORTH

I. Historical Perspective

Over the past decade or so, researchers have determined that many of the essential functions in a variety of cells are regulated by changes in the concentration of intracellular calcium ions ($[Ca^{2+}]_i$). These changes originate, at least in part, from the mobilization of Ca^{2+} from specific intracellular stores as the result of a receptor-enhanced turnover of membrane phosphoinositides. The idea that the activation of receptors for certain neurotransmitters, hormones, and growth factors was associated with the increased turnover of membrane phosphoinositides began with the studies of Hokin and Hokin (1955,1956) in the early 1950s, but its real significance and its relationship to changes in $[Ca^{2+}]_i$ were not fully appreciated until the review by Michell (1975) some 20 years later. In this review, Michell presented the argument that changes in $[Ca^{2+}]_i$ caused by release from intracellular stores were the key intracellular signal (second messenger) linking a variety of neurotransmitters and hormones with their cellular actions and, critically, that these changes were entirely subsequent to (and dependent on) the increased turnover of membrane phosphoinositides. Even then, the views of Michell were very controversial and far from widely accepted. In particular, how a signal generated in the lipid environment of the membrane (phosphoinositide turnover) could be transmitted to Ca^{2+} stores lying within the cytosol remained unclear. Not until the landmark paper by Streb *et al.* (1983) was the identity of inositol (1,4,5)trisphosphate [$Ins(1,4,5)P_3$] as a water-soluble product of phosphoinositide metabolism and its key role in releasing Ca^{2+} from specific intracellular stores revealed.

Over the past 10 years, enormous progress has been made in describing the detailed mechanism of $Ins(1,4,5)P_3$ action and its molecular basis. The mechanism is now known to involve the binding of $Ins(1,4,5)P_3$ to a specific receptor protein in the membranes surrounding certain intracellular Ca^{2+} stores, and the subsequent activation of a Ca^{2+} channel resulting in the mobilization of the calcium from within that store. Evidence from a variety of sources has shown that the $Ins(1,4,5)P_3$-binding properties and Ca^{2+}-channel properties reside in a single protein (Ferris *et al.*, 1989; Mayrleitner *et al.*, 1991), which therefore can be classified as an intracellular member of the group of ligand-gated ion channels. The InsP₃ receptor is now known to act as a Ca^{2+}-release channel in certain nonmitochondrial Ca^{2+} stores in nonmuscle tissues as well as in smooth muscle and, possibly, in the heart. Despite the enormous advances made in recent years, as more and more of the precise molecular mechanisms involved in the functioning of this protein are revealed, it becomes clear that the basic process is far more complicated than first supposed. The main purpose of this chapter is to describe only briefly the current information on the molecular nature of the InsP$_3$ receptor protein and InsP$_3$-induced Ca^{2+} mobilization, much of which has been covered in more detail in excellent reviews by Taylor and Richardson (1991) and Ferris and Snyder (1992). More particularly, the intent is to focus on current areas of controversy regarding the basis of InsP$_3$-activated Ca^{2+} channel function, especially those relating to its modulation and physiological role in cellular signaling.

II. Identification and Characterization of the InsP$_3$ Receptor

A. Radioligand Binding Studies

Specific high-affinity binding sites for InsP$_3$ were first identified in various peripheral tissues, namely adrenal cortex and liver, and in neutrophils (Baukal et al., 1985; Spät et al., 1986; Guillemette et al., 1987). Subsequently, much higher levels of specific binding were demonstrated in the cerebellum (Worley et al., 1987a). This tissue has since been widely used in biochemical studies of receptor function (but see Section III,B). When measured in isolated microsomal membrane preparations or in permeabilized cells, determinations of binding affinity typically demonstrate values for K_d of 5–300 nM. The actual kinetics of binding are far from well understood, however, and few detailed studies have been carried out on this aspect of the receptor.

B. Structure–Activity Relationships

Classically, studies on the binding characteristics and activities of different analogs of receptor ligands have been of considerable importance in obtaining information on the structure of the particular receptor. Such studies assume an additional significance in the case of the InsP$_3$ receptor because of the great diversity of inositol phosphates that exist within the cell and, more particularly, are known to be produced after receptor activation of phosphoinositide turnover. As a result, information on the relative activities of the various inositol phosphates at the InsP$_3$ receptor may be of direct physiological relevance in indicating which inositol phosphates may have Ca^{2+}-releasing activity in the cell. Early studies revealed that inositol phosphate binding to the InsP$_3$ receptor protein and the subsequent activation of Ca^{2+} release showed a high degree of selectivity (Irvine et al., 1986). In particular, vicinal phosphates in the 4 and 5 positions were essential for activity, suggesting that these sites most likely interact in a fairly intimate way with the binding site. The presence of an additional phosphate in the 1 (preferably) or 2 position greatly increases affinity. The hydroxyl groups at the remaining positions also influence activity; their removal, particularly from the 6 position, greatly decreases activity (Safrany et al., 1991). As can be predicted, the enantiomer L-Ins(1,4,5)P$_3$ [equivalent to D-Ins(3,5,6)P$_3$] is essentially inactive (Willcocks et al., 1987; Strupish et al., 1988). From a strictly functional viewpoint, the most critical issue is that the inositol phosphates Ins(1,4)P$_2$, Ins(1,3,4,5)P$_4$, and Ins(1,3,4)P$_3$, the immediate products of Ins(1,4,5)P$_3$ metabolism, are all at least 1000-fold less active than Ins(1,4,5)P$_3$ at the InsP$_3$ receptor.

For many other membrane channels and receptors, the identification and characterization of effective antagonists has proven to be a fruitful approach to the determination of structure–function relationships. Unfortunately, relatively little success has been had along these lines for the InsP$_3$-receptor/Ca^{2+}-release channel protein. The most effective antagonists identified to date are molecules such as heparin (Hill et al., 1987; Ghosh et al., 1988; Guillemette et al., 1989) and other sulfated polysaccharides (e.g., dextran sulfate, polyvinyl sulfate) (Yamamoto et al., 1991), as well as decavanadate (Strupish et al., 1991). However, all these molecules are far from specific and clearly have effects at other sites, including the enzymes that are responsible for the metabolism of Ins(1,4,5)P$_3$ (for example, see Strupish et al., 1991). Despite these problems, the high affinity of heparin for the InsP$_3$-binding site has proven extremely useful in the isolation, purification, and subsequent characterization of the receptor/channel protein (see subsequent discussion).

C. Intracellular Distribution of Receptors

Researchers knew from the earlier studies of InsP$_3$-induced Ca^{2+} release that mitochondrial inhibitors did not affect InsP$_3$-induced release but that inhibitors of the Ca^{2+} pump of the endoplasmic reticulum (ER; e.g., vanadate) did (Streb et al., 1982). However, Ins(1,4,5)P$_3$ only releases Ca^{2+} from a fraction of the total nonmitochondrial pool, the precise proportion varying somewhat in different cell types. Consistent with this finding is the fact that electron microscopic immunocytochemistry localizes InsP$_3$ receptors to components of the ER, but morphologically apparently identical components are not equally labeled (Mignery et al., 1989; Ross et al., 1989; Otsu et al., 1990; Satoh et al., 1990). Consequently, the pool of intracellular Ca²⁺ released by Ins(1,4,5)P₃ appears to be some subcompartment of the ER. By analogy with the sarcoplasmic reticulum of muscle cells, InsP₃-sensitive stores were sought on the basis of the distribution of antibody labeling of the sarcoplasmic reticulum protein calsequestrin. Investigations of nonmuscle cells using this approach revealed reactivity associated with specific membrane-bound structures distributed throughout the cytoplasm. These structures were named "calciosomes" (Volpe *et al.*, 1988) and were initially thought to represent the InsP₃-sensitive stores. However, subsequent investigations have generally failed to support this idea. In particular, immunocytochemical studies with antibodies generated against the InsP₃-receptor protein specifically label structures that are morphologically distinct from the calciosomes (Satoh *et al.*, 1990). A major problem in attempts to resolve the intracellular localization and distribution of InsP₃ receptors and the identity of the InsP₃-sensitive stores is the fact that, apart from cerebellar Purkinje cells, the density of the receptors and, consequently, of the binding sites is generally too low to resolve. In summary, the precise identity of the InsP₃-sensitive stores in nonmuscle cells is unclear, other than that it appears to form some subset of the total ER. Of particular interest is the often high degree of labeling seen in the region of the nuclear membrane (Ross *et al.*, 1989; Satoh *et al.*, 1990). Findings of InsP₃-sensitive stores in the nucleus or perinuclear area (Malviya *et al.*, 1990; Nicotera *et al.*, 1990; Glennon *et al.*, 1992) support this distribution and may provide a basis for the reported effects of the Ins(1,4,5)P₃/[Ca²⁺]ᵢ signaling pathway in modulating gene transcription, as well as cell growth and differentiation.

An area of some discussion over the years has been the question of the relationship between InsP₃ receptors and the plasma membrane. Much of the early interest in this issue arose from a model that offered an explanation for the apparent combined actions of Ins(1,4,5)P₃ in mobilizing Ca²⁺ from intracellular stores and activating Ca²⁺ entry (Putney, 1986). This model proposed that the agonist-sensitive stores were in close association with the plasma membrane, and that emptying of the stores by agonist-generated Ins(1,4,5)P₃ activated a pathway directly linking the open lumen of the stores with the extracellular medium (Putney, 1986). Early evidence for this model centered on an apparent direct refilling of empty stores by Ca²⁺ from the extracellular medium without traversing the cytosol and raising [Ca²⁺]ᵢ. Despite the attractiveness of this model, numerous studies now show that the intracellular stores refill from the cytosol rather than directly from the extracellular medium

(i.e., Ca²⁺ entering the cells passes to the cytosol before entering the intracellular stores) and that a direct pathway such as the one originally proposed does not exist (Takemura and Putney, 1989; Muallem *et al.*, 1990; Shuttleworth, 1990; see Putney, 1990). Nevertheless, several reports have suggested that InsP₃-binding sites may be localized to, or at least associated with, the plasma membrane. Some of the earlier reports claiming a copurification of InsP₃-binding sites and plasma membrane markers in cell fractionation studies (Guillemette *et al.*, 1988) may have resulted from the possible association of the InsP₃ receptor with cytoskeletal elements (Guillemette *et al.*, 1990; Rossier *et al.*, 1991). Similarly, the immunocytochemical studies done by Satoh *et al.* (1990) in Purkinje cells reported labeling of structures immediately beneath, but not actually within, the plasma membrane. Overall, in most cell types studied, little or no conclusive evidence exists to indicate the presence of InsP₃ receptors actually in the plasma membrane. An exception to this rule may be lymphocytes (Kuno and Gardner, 1987) and olfactory cilia (Restrepo *et al.*, 1990), in which plasma membrane ion channels directly gated by Ins(1,4,5)P₃ appear to exist. However, the detailed molecular relationships between these plasma membrane proteins and the InsP₃-receptor protein of intracellular stores is currently unknown.

III. Structural Studies of the Receptor

A. Purification, Molecular Cloning, and Expression

The InsP₃-receptor protein was originally purified from cerebellum (Supattapone *et al.*, 1988a), and subsequently from aortic smooth muscle (Chadwick *et al.*, 1990) and vas deferens (Mourey *et al.*, 1990). Currently, purification protocols generally use a combination of heparin–agarose affinity columns followed by concanavalin A–Sepharose chromatography, the latter relying on the glycosylation of the InsP₃ receptor. An alternative approach is to use an Ins(1,4,5)P₃ affinity column (Prestwich *et al.*, 1991). SDS polyacrylamide gel electrophoresis of the purified protein reveals a subunit polypeptide chain of about 260 kDa (Supattapone *et al.*, 1988a; Maeda *et al.*, 1990; Mourey *et al.*, 1990; but see subsequent discussion). However, size exclusion chromatography indicates a molecular weight of ~1 million, implying that the native molecule is a homotetramer (Supattapone *et al.*, 1988a; Guillemette *et al.*, 1990; Maeda *et al.*, 1991). This configuration is supported by a variety of additional evidence from biochemical studies as well as determinations of subconductance states in the reconstituted receptors (Watras *et al.*, 1991).

The first cloning of the InsP$_3$ receptor was actually based on a specific protein found in Purkinje cells, named P-400, whose relationship to inositol phosphates was originally unsuspected. When it became apparent that the characteristics of the protein P-400 corresponded to those of the isolated InsP$_3$ receptor, the cDNA encoding this protein in mice was cloned and sequenced (Furuichi *et al.*, 1989). Independently, using antibodies based on a partial cDNA that was specific to the Purkinje cells of rat brain (Nordquist *et al.*, 1988), researchers showed that the protein encoded by the cloned P-400 gene corresponded to the InsP$_3$ receptor (Mignery *et al.*, 1990). The full-length transcript is approximately 10 kb and encodes a protein of 2749 amino acids with a molecular weight of 313,000. Confirmation of the functional identity of the cloned genes comes from experiments in which expression of the InsP$_3$-receptor/P-400 protein in a fibroblast cell line has been shown to result in an increased InsP$_3$-dependent intracellular Ca^{2+}-release activity (Miyawaki *et al.*, 1990). Further, reconstitution of the purified receptor in phospholipid vesicles (Ferris *et al.*, 1989) or planar lipid bilayers (Maeda *et al.*, 1991) confirms that Ins(1,4,5)P$_3$-binding and Ca^{2+}-channel activity reside in the same protein.

Evidence from hydropathy analysis, as well as from mutagenesis studies, suggests that the InsP$_3$-binding domain resides in the extreme N-terminal region (last 400 residues) of the protein (Fig. 1). The C-terminal portion, on the other hand, contains 6–8 putative transmembrane domains that are presumed to make up, or at least contain, the Ca^{2+} channel (Mignery and Südhof, 1990; Mignery *et al.*, 1990). Despite initial claims to the contrary (Furuichi *et al.*, 1989), the N and C termini of the protein are now known to reside in the cytoplasm (De Camilli and Takei, 1990). Binding of Ins(1,4,5)P$_3$ to its site induces a large conformational change in the protein (Mignery and Südhof, 1990) that presumably underlies the transduction process responsible for opening the Ca^{2+}-channel portion. Between the InsP$_3$-binding domain and the putative Ca^{2+}-channel domain lies an extensive span, 1400–1500 amino acids in length, that is believed to lie in the cytoplasm of the cell; this region is called the "linking" or "coupling" domain (see Fig. 1). Importantly, within this coupling domain are the various sites for phosphorylation, ATP binding, and alternative splicing events.

B. Receptor Heterogeneity

The original molecular cloning studies on the InsP$_3$ receptor from mammalian brain generally suggested

a high degree of evolutionary conservation in the protein, with less than a 15% difference between rodent and human forms and only a 1% amino acid difference between rat and mouse. Nevertheless, subsequent studies involving other types of tissue have revealed considerable potential heterogeneity within the InsP$_3$-

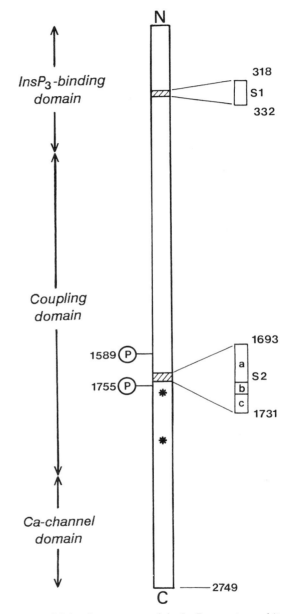

FIGURE 1 Molecular structure of the InsP$_3$ receptor and its functional domains. The amino acid sequence numbering is based on the receptor from rat brain, as reported by Mignery *et al.* (1990). Shaded bars (S1 and S2) represent the known alternative splicing sites; S2 is made up of three separate sequences than can show independent deletion. Sites of phosphorylation by protein kinase A are indicated by P; putative nucleotide binding sites are indicated by asterisks. Adapted from Taylor and Richardson (1991) and Mikoshiba (1993). See text for details.

receptor protein. One source of this heterogeneity arises from the demonstrated existence of multiple genes encoding sequences for the InsP₃ receptor, as first demonstrated by Südhof *et al.* (1991), who obtained full-length sequences of two distinct genes from the cerebellum (InsP₃ receptor types I and II). Both genes encode structurally similar proteins with sequences that are 69% identical. Interestingly, the greatest diversity is seen in the "coupling" domain, raising the possibility of potential differences in allosteric regulation of receptor function between the two types. Subsequently, partial sequences of two (or possibly three) additional genes have been cloned from a mouse placenta cDNA library; the expression of multiple types of InsP₃ receptor within a single cell has been suggested (Ross *et al.*, 1992).

In addition, a major source of heterogeneity in the InsP₃ receptor is the existence of multiple isoforms generated as a result of various alternative splicing events. For example, up to five different subtypes have been identified in mouse brain during development (Nakagawa *et al.*, 1991). Two alternative splicing sites in the InsP₃ receptor are known (see Fig. 1). One lies within the N-terminal InsP₃-binding domain which, although precise functional implications are unknown, clearly raises the potential for modification of the properties of binding. However, the principal site for these alternative splicing events is a 120-nucleotide region in the large "coupling domain" referred to earlier. This region, in fact, comprises three separate domains that are capable of independent removal, giving rise to several potential variants. Interestingly, the region in question lies between two known protein kinase A-dependent phosphorylation sites and close to the two consensus sequences for nucleotide binding. Further, the complete deletion of the 120-nucleotide region actually creates a new nucleotide-binding consensus sequence (Ferris and Snyder, 1992). Clearly, such an arrangement suggests the possibility of significant, and potentially specific, variation in the modulation of receptor function and receptor–channel coupling, the details of which are only just beginning to be revealed (see subsequent discussion). These various alternative InsP₃-receptor proteins can show a highly specific tissue distribution. For example, the deletion of the 120-nucleotide segment within the coupling domain is now known to be characteristic of the InsP₃ receptor of peripheral nonneuronal tissues whereas the "complete form" is typical of neuronal tissues (Danoff *et al.*, 1991). A supposed nuclear InsP₃ receptor has been identified with a molecular mobility indicating a 220-kDa protein (cf., 260-kDa in rat cerebellum and liver; Matter *et al.*, 1993). This protein presumably forms the basis for the reported InsP₃-

induced release of Ca²⁺ in isolated rat liver nuclei (Malviya *et al.*, 1990) referred to earlier.

The InsP₃ receptor is also subject to a variety of posttranslational modifications. For example, the InsP₃ receptor can be phosphorylated by protein kinase A (see Fig. 1; Supattapone *et al.*, 1988b), as well as by protein kinase C and Ca²⁺–CaM-dependent protein kinase II (Ferris *et al.*, 1991), apparently at distinct sites. Interestingly, the InsP₃ receptor itself displays serine protein kinase-like activity. Further, the protein can apparently autophosphorylate (Ferris *et al.*, 1992a) although, curiously, stoichiometric analysis of the autophosphorylation reaction suggests that this behavior is only demonstrated by a subpopulation of the total receptor pool. The functional significance of this activity is, at present, unclear. Not surprisingly, the variations in molecular sequence seen in the various InsP₃ receptors, and referred to earlier, also influence phosphorylation events. In the long form of the InsP₃ receptor, found in the cerebellum, protein kinase A preferentially phosphorylates Ser 1755, whereas in the short form of peripheral smooth muscle, Ser 1589 becomes phosphorylated (Danoff *et al.*, 1991).

C. Relationships with the Ryanodine Receptor Protein

In discussions of the structure of the InsP₃ receptor, many workers have emphasized the close structural and functional relationships between the InsP₃ receptor and the ryanodine receptor. These discussions are not repeated here. Instead, a word of caution is given against overemphasis of such comparisons. The similarities between the respective roles of the InsP₃ receptor and the ryanodine receptor in mediating intracellular Ca²⁺ release are, of course, obvious. With respect to similarities in structure, early studies using antibodies against the C terminus of the InsP₃ receptor concluded that this region was positioned within the lumen of the Ca²⁺ store (Furuichi *et al.*, 1989), an arrangement very different from that of the ryanodine receptor. However, subsequent investigations have favored a cytoplasmic location for this part of the molecule, which is more in accordance with the topology of the ryanodine molecule. With respect to their amino acid sequences, again, some similarities exist although the overall size of the proteins is very different. Specifically, the last four transmembrane domains show the closest sequence homology, particularly within the putative last two spans, but even here only 46% identity is seen. Of course, in both molecules, this region is believed to constitute the Ca²⁺ channel, so a certain amount of similarity is, perhaps, not surprising. Finally, note that the InsP₃ receptor and the ryanodine receptor also show some very marked differences. For example, their respective single-channel Ca^{2+} conductances are very different (26 pS for the $InsP_3$ receptor and 80–100 pS for the ryanodine receptor, see Chapter 33) and, of course, their respective pharmacologies are unique.

IV. Receptor Function

A. Modulatory Factors Affecting Binding and Ca^{2+} Release

A very important point to note when considering the actions of potential modulatory agents at the $InsP_3$ receptor is the distinction between actions at the $InsP_3$-binding site, actions at the Ca^{2+}-release site, and the precise relationships between the two. Thus, certain agents may affect the binding of $Ins(1,4,5)P_3$ without altering Ca^{2+} mobilization; alternatively, Ca^{2+} mobilization may be affected without any influence on $Ins(1,4,5)P_3$ binding. Although such considerations can play a part in any ligand-gated channel protein, they are perhaps more significant in the Ca^{2+}-mobilizing $InsP_3$ receptor because of the presence of a very large stretch of the protein separating ligand-binding and "effector" (Ca^{2+}-channel) domains. Of course, the physiologically relevant effect is an influence on Ca^{2+} mobilization, but this effect has been directly studied in only relatively few examples. How several of the reported modulatory influences on $Ins(1,4,5)P_3$-binding relate to this function is unclear. Another complication is the difficulty in demonstrating a *direct* modulatory action on the $InsP_3$ receptor/ Ca^{2+} channel rather than possible indirect effects resulting from changes in the content of the $InsP_3$-mobilizable pool. These complications are elegantly illustrated by the various effects of protein kinase A on $InsP_3$-mediated effects. Thus, consistent with the presence of protein kinase A-dependent phosphorylation sites in the receptor molecule (see previous discussion), purified $InsP_3$ receptors can be phosphorylated by protein kinase A. In hepatocytes, activation of protein kinase A increases the sensitivity of intracellular Ca^{2+} stores to $Ins(1,4,5)P_3$ and increases the maximum amount of Ca^{2+} mobilized (Burgess *et al.*, 1991). In platelets (Enouf *et al.*, 1987) and in the cerebellum (Supattapone *et al.*, 1988b; Volpe and Anderson-Lang, 1990), protein kinase A activation also results in an increase in the amount of Ca^{2+} mobilized, but phosphorylation of the cerebellar receptor uncouples $Ins(1,4,5)P_3$ binding from Ca^{2+} release with a concomitant decline in the sensitivity to $Ins(1,4,5)P_3$. The observed increases in the amount of Ca^{2+} released may well arise from a protein kinase A-induced stimulation of the Ca^{2+} pump in the stores, although such an effect is not always seen (Volpe and Anderson-Lang, 1990). Significantly, in all cases, protein kinase A does not affect the actual binding affinity of $Ins(1,4,5)P_3$ for its receptor.

Other modulators of likely physiological significance include ATP. At low concentrations (1–10 μM), ATP binding markedly potentiates the binding of $Ins(1,4,5)P_3$ (Spät *et al.*, 1992) and $InsP_3$-induced stimulation of Ca^{2+} release in a cooperative fashion. This effect can be duplicated in the purified receptor and is mediated by the adenosine moiety (Ferris *et al.*, 1990; Maeda *et al.*, 1991). At somewhat higher, although still physiological, nucleotide concentrations (0.1–1 mM), a competitive inhibition between the pyrophosphate group of ATP and the $InsP_3$-binding domain occurs (Nunn and Taylor, 1990). With respect to the physiological relevance of these effects, researchers have suggested that $InsP_3$-mediated Ca^{2+} release would presumably be accompanied by an activation of the Ca^{2+}–ATPase of the Ca^{2+} store (ER), which may result in a local depletion of ATP levels in the vicinity of the $InsP_3$ receptor. At normal concentrations of ATP in the cell (~1 mM), initially this event would result in an increase in Ca^{2+} release as ATP is removed from the low-affinity sites on the $InsP_3$ receptor. Subsequently, as ATP levels fall even further, Ca^{2+} release would decline as the high-affinity sites became affected (Ferris *et al.*, 1990). Whether such a sequence of events actually occurs in the intact cell, and whether it contributes to the transient nature of $InsP_3$-induced Ca^{2+} release and $[Ca^{2+}]_i$ oscillations (see subsequent discussion), as has been suggested, is unknown at this time, but consideration of such possibilities certainly illustrates the potential complexities of the $InsP_3$-mediated Ca^{2+}-mobilization response.

Very clear effects of cytosolic pH also occur at the $InsP_3$ receptor. Binding is profoundly affected by pH, with an optimal level at pH 8–9 (Worley *et al.*, 1987b). When the effects of pH on $InsP_3$-induced Ca^{2+} mobilization have been studied, they show similar effects with an increased Ca^{2+} release at higher values of pH (Brass and Joseph, 1985; Guillemette and Segui, 1988; Joseph *et al.*, 1989; Meyer *et al.*, 1990). Despite the possibility of complications arising from indirect effects, possibly resulting from changes in luminal Ca^{2+} content (see subsequent discussion), the observed effects are of sufficient size to be likely to be of some physiological significance, particularly in view of the often fairly substantial changes in intracellular pH associated with agonist action.

Certainly of considerable importance, and the subject of much current debate, are the modulatory effects of Ca^{2+} itself, both cytosolic and luminal. Such effects are clearly complex and apparently variable among different tissues. Part of such variability probably arises from the multiple forms of the receptor, as discussed earlier, but may also reflect the presence or degree of association of certain intermediary Ca^{2+}-binding proteins. Thus, in the cerebellum, an integral membrane protein named calmedin has been claimed to mediate the inhibitory effects of increases in $[Ca^{2+}]_i$ (within the physiological range) on $Ins(1,4,5)P_3$ binding (Danoff *et al.*, 1988). However, this mediation does not appear to occur in peripheral tissues, where calmedin is rarely detected, although studies suggest that the $InsP_3$ receptors of these tissues apparently retain the potential for calmedin-mediated modulation. For example, the $InsP_3$ receptor of the vas deferens shows no Ca^{2+}-induced inhibition of $Ins(1,4,5)P_3$ binding, but such inhibition is seen when the receptor is mixed with partially purified calmedin (Mourey *et al.*, 1990).

Generally, the effects of increases in cytosolic Ca^{2+} on $InsP_3$-induced Ca^{2+} release show a biphasic response—initially enhancing Ca^{2+} release and then subsequently decreasing it (Iino, 1990; Bezprozvanny *et al.*, 1991; Finch *et al.*, 1991, see Chapter 33). The basis for such effects is presently unclear. In the intact cell, the known protein kinase C or Ca^{2+}–CaM-dependent protein kinase II phosphorylation of the receptor may be involved (Ferris *et al.*, 1991). More recently, investigators have suggested that a demonstrated binding of Ca^{2+} ions by inositol phosphates, including $Ins(1,4,5)P_3$, may also contribute to this phenomenon by a simple process of chelation of Ca^{2+} at appropriate concentrations (Luttrell, 1993). In addition, such effects may be related to the finding that, in hepatocytes, increases in Ca^{2+} reversibly switch the receptor to a nonconducting configuration with a high affinity for $Ins(1,4,5)P_3$ (Pietri *et al.*, 1990). Such conformational switches affecting binding affinity have been suggested by several groups and, although some problems have arisen with definitive proof (see Taylor and Richardson, 1991), considerable evidence has been presented to suggest multiple conformational forms of the receptor with different ligand affinities that can be modulated by changes in $[Ca^{2+}]_i$ (and, consequently, by agonists that induce such changes). Showing that such changes in binding affinity are also associated with changes in the Ca^{2+} conductance of the release channel is, however, more problematical.

Potential modulatory effects exerted via the Ca^{2+} content *within the lumen* of the agonist-sensitive stores have been generally less thoroughly examined, despite considerable interest in the possibility of such modulation providing the basis for the quantal nature

of $InsP_3$-induced Ca^{2+} release (see subsequent discussion). Current evidence suggests that such effects are more variable than the well-established effects of cytosolic Ca^{2+}. Evidence in support of such modulation has been reported in hepatocytes and smooth muscle cells, in which increases in luminal Ca^{2+} render the receptor more sensitive to $Ins(1,4,5)P_3$ (Missiaen *et al.*, 1991; Nunn and Taylor, 1991). However, this effect could not be detected in certain other cells (Shuttleworth, 1992) or even by other groups working on hepatocytes (Combettes *et al.*, 1992). Similar evidence contrary to a modulatory effect of luminal Ca^{2+} comes from studies on rat basophilic leukemia cells (Meyer *et al.*, 1990). In addition, researchers have shown that the reported effects of luminal Ca^{2+} are themselves affected by the level of cytosolic Ca^{2+} (Missiaen *et al.*, 1992a). Another study of the detailed kinetics of the effects of cytosolic Ca^{2+} on Ca^{2+} release suggests that these behaviors may explain many, if not all, of the apparent effects of luminal Ca^{2+} (Iino and Endo, 1992). In any event, the effects of luminal Ca^{2+} appear to be relatively minor compared with the rather pronounced effects of cytosolic Ca^{2+} although, again, considerable potential exists for complex interactions, rendering definitive conclusions difficult.

Various other modulatory factors have also been described, although generally their bases or roles are unknown. Of potential physiological significance, and certainly worthy of further study, is the reported inhibitory effect of Mg^{2+} (Volpe *et al.*, 1990). The precise mechanism involved is unclear, but could involve multiple actions on binding and channel conductance, as well as effects on the metabolism of $Ins(1,4,5)P_3$.

B. Properties of InsP₃-Induced Ca²⁺ Release

The first indication of subtle complexities inherent in the mobilization of internal Ca^{2+} stores by $Ins(1,4,5)P_3$ came from studies on permeabilized preparations of pancreatic acinar cells (Muallem *et al.*, 1989) and from an investigation utilizing flash photolysis of caged $Ins(1,4,5)P_3$ in frog oocytes (Parker and Miledi, 1989). Specifically, researchers showed that increasing submaximal concentrations of $Ins(1,4,5)P_3$ resulted in the release of an increasing *fraction* of the total agonist-sensitive Ca^{2+} pool, rather than an increasing *rate* of release of the total pool. This phenomenon can be shown to arise from a biphasic response of Ca^{2+} efflux to submaximal concentrations of $Ins(1,4,5)P_3$—an initial very rapid efflux followed by a much slower efflux (Muallem *et al.*, 1989; Taylor and Potter, 1990; see Fig. 2A). Such a response might be expected to result from a simple desensitization of the $InsP_3$ receptor, but this does not appear to be the case. For example, stores

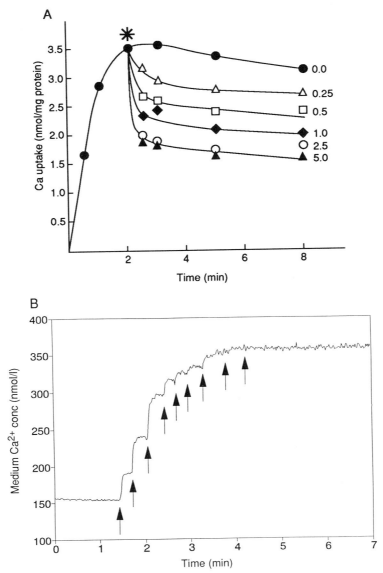

FIGURE 2   (A) The effect of different concentrations of exogenous Ins(1,4,5)P$_3$ on Ca^{2+} release in permeabilized pancreatic acini. Permeabilized acini were allowed to accumulate Ca^{2+} in intracellular stores in the presence of ATP. At the point indicated (*), the acini were exposed to the different concentrations of Ins(1,4,5)P$_3$ (in μM) indicated at left. Data are from Muallem *et al.* (1989); adapted with permission. (B) The effect of sequential increase in Ins(1,4,5)P$_3$ concentration on the release of Ca^{2+} from intracellular stores in permeabilized cells from the exocrine avian nasal gland. Additions of Ins(1,4,5)P$_3$ (at the arrows) raised the concentration to 0.1, 0.2, 0.4, 0.6, 0.8, 1.0, 2.0, 3.0, and 5.0 μM Ins(1,4,5)P$_3$ respectively. Reprinted with permission from Shuttleworth (1992).

that have responded in this way to low concentrations of Ins(1,4,5)P$_3$ retain their ability to respond similarly to further additions of the inositol phosphate (Meyer and Stryer, 1990; Taylor and Potter, 1990) and no intrinsic desensitization of the receptor can be detected, even with prolonged exposure to Ins(1,4,5)P$_3$ (Oldershaw *et al.*, 1992). The net effect can be readily illustrated in permeabilized cell preparations, in which

each addition of a submaximal concentration of Ins(1,4,5)P$_3$ results in a transient step-like release of Ca^{2+} (Fig. 2B). This phenomenon has now been demonstrated for a variety of different cell types and has been termed "incremental" (Meyer and Stryer, 1990) or, more commonly, "quantal" release (Muallem *et al.*, 1989). Two basic models have been proposed to account for this phenomenon. (1) At submaximal concentrations, $Ins(1,4,5)P_3$ results in a partial emptying of all the stores ("steady-state" model; Irvine, 1990). (2) A complete emptying of only some of the stores is the result ("all-or-nothing" response; Muallem *et al.*, 1989; Parker and Ivorra, 1990). Debate concerning these two proposals continues, but the major issue that arises is the basis for the required heterogeneity and/or modulation of the response that is necessary to explain the observed quantal nature of $InsP_3$-induced Ca^{2+} release. Of course, such a process could result from the stores of different cells having different sensitivities to $Ins(1,4,5)P_3$ (i.e., an *inter*cellular rather than *intra*cellular heterogeneity), although researchers generally assume that this is a property of the stores within individual cells. Evidence from the few studies on this phenomenon in individual single cells supports the notion that this is indeed the case (Jacob *et al.*, 1988; Parker and Ivorra, 1990; Bootman *et al.*, 1992; Shuttleworth, 1993).

Clearly, one way in which the sensitivity of $InsP_3$-induced Ca^{2+} release of different stores within an individual cell could vary is as a result of heterogeneity of the $InsP_3$ receptor. As discussed earlier, such heterogeneity of receptor function could arise from differences in basic receptor structure or from various post-translational modifications (e.g., phosphorylation). An alternative explanation that has been proposed is that the sensitivity of the $InsP_3$ receptor to $Ins(1,4,5)P_3$ is subject to modulation by the Ca^{2+} content of the agonist-sensitive store (Irvine, 1990; Tregear *et al.*, 1991). This proposal is undoubtedly attractive, made all the more so by the possibility it raises of explaining the disparate data relating to the control of Ca^{2+} entry (see subsequent discussion; Irvine, 1991). Evidence in support of such a proposal has been produced, at least in certain cell types (Missiaen *et al.*, 1991,1992b; Nunn and Taylor, 1991,1992). However, as discussed earlier, in other cell types such modulation could not be discerned (Combettes *et al.*, 1992; Shuttleworth, 1992). Further, the model proposed does not entirely explain all the features of $InsP_3$-induced Ca^{2+} release. In its original form (Irvine, 1990), $Ins(1,4,5)P_3$ acted only indirectly by modulating Ca^{2+} binding to a luminal site which, in turn, determined the gating of the Ca^{2+} channel. Although such a model reproduced many of the features observed, it failed to explain the very narrow range of concentrations over which $Ins(1,4,5)P_3$ exerts its effects. Alternative versions proposed that luminal Ca^{2+} modulates $Ins(1,4,5)P_3$ binding, which opens the Ca^{2+} channel, or that channel opening is directly dependent on the binding of *both* $Ins(1,4,5)P_3$ and luminal Ca^{2+}. However, simulations based on these versions did not produce a sufficiently rapid release of Ca^{2+} at low $Ins(1,4,5)P_3$ concentra-

tions, and failed to mimic quantal responses adequately (Tregear *et al.*, 1991). The only way the original version of the model, with its reasonable replication of quantal responses, can operate within the very narrow concentration range over which $Ins(1,4,5)P_3$ exerts its effects is by the incorporation of a strong positive cooperativity between the four monomers believed to form the channel. Whether such cooperativity actually exists is an area of considerable current dispute (see Meyer *et al.*, 1990; Watras *et al.*, 1991). An additional problem with the model is the fact that no obvious Ca^{2+}-binding domains are seen on the internal (luminal) side of the $InsP_3$ receptor that might provide a molecular or structural basis for such an effect. To overcome this deficiency, researchers have suggested that such modulation may be exerted indirectly via a luminal accessory protein such as calreticulin or calsequestrin. However, such arguments are difficult to reconcile with the clear evidence that a quantal release response is seen in liposomes in which only the purified $InsP_3$ receptor has been incorporated (Ferris *et al.*, 1992b), indicating that this property is intrinsic to the protein itself. In conclusion, despite the obvious attractiveness of the idea that $InsP_3$-induced Ca^{2+} release is subject to significant modulation by luminal Ca^{2+}, and the subsequent explanation of the quantal nature of the Ca^{2+} release, evidence in support of such a model is, at best, contradictory and its physical basis is still unclear.

C. $[Ca^{2+}]_i$ Oscillations

Experiments in which agonist-induced changes in $[Ca^{2+}]_i$ (or $[Ca^{2+}]_i$-activated membrane currents) have been studied in *individual* cells frequently reveal that the resulting signal is far more complicated than a simple increase in $[Ca^{2+}]_i$. In particular, $[Ca^{2+}]_i$ is often seen to oscillate, sometimes over a considerable range, in response to application of an appropriate agonist (Fig. 3). This result is particularly common at low concentrations of agonist. Various experiments have shown that, although the oscillations depend on extracellular Ca^{2+} entry for their maintenance, the basic oscillatory phenomenon is a property of the intracellular Ca^{2+} stores. The various ideas that have been proposed relative to the physiological significance of such oscillatory behavior is beyond the scope of this review (see, for example, Berridge and Galione, 1988; Goldbeter *et al.*, 1990), but the possible role of certain properties of the $InsP_3$ receptors and the Ca^{2+} stores they control will be discussed.

Several different models have been proposed to explain the origin of agonist-induced $[Ca^{2+}]_i$ oscillations (Woods *et al.*, 1987; Meyer and Stryer, 1988;

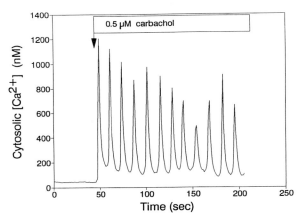

FIGURE 3 Oscillations in $[Ca^{2+}]_i$ in an individual cell from the exocrine avian nasal gland in response to activation of muscarinic receptors by the agonist carbachol. $[Ca^{2+}]_i$ was determined ratiometrically by dual-emission single-cell photon counting of an indo-1-loaded cell.

Goldbeter *et al.*, 1990; Cuthbertson and Chay, 1991). Basically, these models fall into two groups—one in which the oscillations result from cyclical changes in levels of $Ins(1,4,5)P_3$ (Cobbold *et al.*, 1991; Harootunian *et al.*, 1991) and another in which oscillations result from a cyclical release and reuptake of Ca^{2+} in the face of constant levels of $Ins(1,4,5)P_3$ (for review, see Berridge and Galione, 1988; Petersen and Wakui, 1990). Most extensively discussed in the literature is a particular variant of the latter model in which the oscillatory signal is based on a cyclical uptake and release of Ca^{2+} from an $InsP_3$-*insensitive* pool subsequent to the initial release of Ca^{2+} from the $InsP_3$-sensitive pool (Berridge and Galione, 1988; Goldbeter *et al.*, 1990). Specifically, this model proposes that the $InsP_3$-insensitive pool that is the key to the oscillatory response displays the phenomenon of Ca^{2+}-induced Ca^{2+} release and, as such, is sensitive to caffeine and inhibited by ryanodine. Evidence in support of such a model is extensive, at least in cells such as chromaffin cells in which clear evidence exists of separate $InsP_3$-sensitive and $InsP_3$-insensitive (but caffeine-sensitive) pools, the latter showing the property of Ca^{2+}-induced Ca^{2+} release (Cheek *et al.*, 1991). However, evidence also suggests that such a two-pool model is not absolutely necessary to obtain oscillations. In other words, most of the essential features of $[Ca^{2+}]_i$ oscillations can be modeled with a single $InsP_3$-sensitive pool, provided a biphasic $[Ca^{2+}]_i$-dependent activation and inhibition of $InsP_3$-mediated Ca^{2+} release is incorporated (i.e., positive and negative feedback of $[Ca^{2+}]_i$ on $InsP_3$-induced Ca^{2+} release; Somogyi and Stucki, 1991).

Now convincing evidence exists of such a biphasic response from a variety of different cells (see Pietri *et al.*, 1990; Carter and Ogden, 1992; Yao and Parker, 1992; Zhang and Muallem, 1992). Further, some cell types that display pronounced $[Ca^{2+}]_i$ oscillations clearly do not contain an $InsP_3$-insensitive Ca^{2+}-induced Ca^{2+}-release pool (e.g., *Xenopus* oocytes, Parys *et al.*, 1992; hamster eggs, Miyazaki *et al.*, 1992), suggesting that such a single-pool model is entirely sufficient to explain $[Ca^{2+}]_i$ oscillations in these cells. In various other cell types, detailed spatial examination of agonist-induced $[Ca^{2+}]_i$ oscillations reveals that these fluctuations occur as series of discrete waves that traverse the cytosol from a distinct point source (Thomas *et al.*, 1991). In *Xenopus* oocytes, even more complex patterns of changes in $[Ca^{2+}]_i$ are seen that take the form of spiraling waves (Lechleiter *et al.*, 1991). Of possible relevance to such phenomena are studies in which measurements were made of the diffusion coefficients and lifetimes for $Ins(1,4,5)P_3$, free Ca^{2+}, and buffered Ca^{2+} in a cytosolic extract (Allbritton *et al.*, 1992). Calculations based on these measurements indicated that typical effective ranges of these messengers in the cytosol were 24 μm, 0.1 μm, and 5 μm, respectively. These data suggest that, for a transient point source, $Ins(1,4,5)P_3$ would represent essentially a "global" messenger with the potential to induce effects over large distances. Ca^{2+}, on the other hand, would likely be restricted in its actions to a very limited region within the cell.

D. Ins(1,4,5)P₃ Action and Ca²⁺ Entry

The $InsP_3$-induced release of intracellular Ca^{2+} can itself result only in a transient rise in $[Ca^{2+}]_i$ because of the limited capacity of the $InsP_3$-sensitive stores within the cell. However, in the majority of so-called "nonexcitable" cells, the elevation in $[Ca^{2+}]_i$ following neurotransmitter or hormone action is known to be generally sustained as long as receptor activation continues. Indeed, precisely this *sustained* elevation in $[Ca^{2+}]_i$ is usually the necessary physiological signal for the response of the cell. This sustained component of the overall $[Ca^{2+}]_i$ signal is now known to arise from an enhanced entry of Ca^{2+} from the extracellular medium. In contrast to the situation in many nerve and muscle cells, this entry of Ca^{2+} does not occur via the well-characterized voltage-activated Ca^{2+} channels but through a different class of channels that are dihydropyridine insensitive. Other than this, almost nothing is known of this receptor-activated Ca^{2+} entry pathway, but its possible relationships with the $InsP_3$ receptor and the phenomenon of $InsP_3$-

induced Ca^{2+} release have been the subject of much speculation and considerable controversy in recent years.

An early model involving a proposed close physical association of the stores with the plasma membrane, as well as a pathway possibly directly linking the lumen of the empty stores with the extracellular medium (Putney, 1986), has already been mentioned. However, as discussed, this model as originally conceived is not correct (see Putney, 1990). Full discussion of all the current arguments relating to this important unresolved question are beyond the scope of this review, but the key models can be summarized (see Fig. 4). One viewpoint holds that the activation of Ca^{2+} entry across the plasma membrane occurs as a result of emptying the InsP$_3$-sensitive stores per se, and that neither Ins(1,4,5)P$_3$ or any other inositol phosphate is *directly* involved in the activation. In support of this model, for example, is the observation that the Ca^{2+} entry pathway can be activated by addition of the drug thapsigargin, which empties the InsP$_3$-sensitive stores by inhibiting the Ca^{2+}-uptake pump present in their membrane but produces no elevation of either Ins(1,4,5)P$_3$ or Ins(1,3,4,5)P$_4$. The key question of how the "relative emptiness" of the InsP$_3$-sensitive stores is sensed and the message is transmitted to the plasma membrane remains unanswered. The other viewpoint is that Ca^{2+} entry results from a combination of Ins(1,4,5)P$_3$ acting at its receptor on the intracellular stores and Ins(1,3,4,5)P$_4$ [the phosphorylation product of 3-kinase activity on Ins(1,4,5)P$_3$] acting at a separate receptor or binding site that is either in, or associated with, the plasma membrane. Contradictory data have been produced by different groups regarding the requirement for Ins(1,3,4,5)P$_4$; generally, no clear consensus has emerged from the studies performed to date. Clearly, more experiments, particularly involving additional cell types, are required.

Despite the obvious difficulties in reconciling these very different viewpoints, an attempt to do so was made by Irvine (1991); he proposed that Ca^{2+} entry depended on the interaction between two proteins—one the InsP$_3$ receptor in the membrane surrounding the intracellular Ca^{2+} stores and the other a protein in the plasma membrane regulating Ca^{2+} entry, which may be the putative "InsP$_4$ receptor". This model proposed that Ca^{2+} entry depended on the dissociation of these two proteins, a process that was itself modulated by the binding of Ins(1,4,5)P$_3$ and Ins(1,3,4,5)P$_4$, as well as, critically, by the Ca^{2+} content within the stores. This latter effect was suggested to act via a modulatory Ca^{2+}-binding site on the luminal side of the InsP$_3$ receptor, as discussed

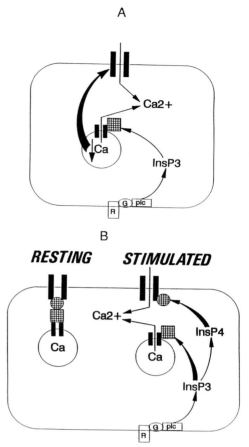

FIGURE 4 Diagrams representing current models describing the relationship between Ins(1,4,5)P$_3$-induced release of intracellular Ca^{2+} stores and the activation of Ca^{2+} entry. (A) The "capacitative" model developed by Putney (see Putney, 1990, for details). (B) Models invoking a combined action of Ins(1,4,5)P$_3$ and Ins(1,3,4,5)P$_4$ in the activation of Ca^{2+} entry. The hatched box represents the InsP$_3$ receptor; the hatched sphere represents the putative InsP$_4$ receptor. See text for further details.

extensively earlier. The argument is that, under normal circumstances, generated Ins(1,4,5)P$_3$ and Ins(1,3,4,5)P$_4$ would bind at their respective sites to cause dissociation of the proteins, and the activation of both Ca^{2+} release and Ca^{2+} entry. Because of the modulatory effect of luminal Ca^{2+}, Ca^{2+} entry would remain activated until the stores were replenished, even if inositol phosphate concentrations returned to resting levels after removal of the agonist. However, under conditions in which the intracellular Ca^{2+} stores were severely depleted (such as following treatment with thapsigargin), the removal of Ca^{2+} from the luminal modulatory site would, itself, result in a dissociation of the two proteins and the activation of Ca^{2+}

entry. The appeal of this model is obvious because it offers an immediate reconciliation of what appear to be directly contradictory data. Unfortunately, direct evidence in support of such a model is lacking and, to date, even the existence (or, at least, the importance) of a modulatory effect of luminal Ca^{2+} on the $InsP_3$ receptor remains controversial (see preceding discussion).

V. Conclusions

In the several years since the discovery of the key role of $Ins(1,4,5)P_3$ in initiating Ca^{2+} release from intracellular stores, enormous advances have undoubtedly been made. The specific protein responsible for the binding of $Ins(1,4,5)P_3$ and Ca^{2+} release has been identified in a variety of tissues; it has been purified, and its properties have been studied in reconstituted systems. The relevant genes have been cloned and sequenced, and the important parts of the molecule have been at least tentatively identified. However, the detailed events following binding of $Ins(1,4,5)P_3$ to its receptor and the complex relationships that exist between $Ins(1,4,5)P_3$ binding and the opening of the Ca^{2+} channel are far from fully understood, as are the actions of the multiple factors that undoubtedly modulate the response in the intact cell. Consequently, researchers have progressed from a "simple" explanation of an apparently "simple" mystery—how does a receptor-activated increase of phosphoinositide turnover in the cell membrane act to initiate release of Ca^{2+} from stores within the cytosol—to exposure of a complex pattern of processes and interrelated control mechanisms that are only just beginning to be appreciated, and that will undoubtedly occupy the attention and efforts of researchers for some considerable time to come.

Acknowledgments

Work from the author's laboratory was supported by National Institutes of Health Grant GM-40457.

References

Allbritton, N. L., Meyer, T., and Stryer, L. (1992). Range of messenger action of calcium ion and inositol 1, 4, 5-trisphosphate. *Science* **258**, 1812–1815.

Baukal, A. J., Guillemette, G., Rubin, R., Spat, A., and Catt, K. J. (1985). Binding sites for inositol trisphosphate in the bovine adrenal cortex. *Biochem. Biophys. Res. Commun.* **133**, 532–538.

Berridge, M. J., and Galione, A. (1988). Cytosolic calcium oscillators. *FASEB J.* **2**, 3074–3082.

Bezprozvanny, I., Watras, J., and Ehrlich, B. E. (1991). Bell-shaped calcium-response curves of $Ins(1,4,5)P_3$- and calcium-gated channels from endoplasmic reticulum of cerebellum. *Nature (London)* **351**, 751–754.

Bootman, M. D., Berridge, M. J., and Taylor, C. W. (1992). All-or-nothing Ca^{2+} mobilization from the intracellular stores of single histamine-stimulated HeLa cells. *J. Physiol. (London)* **450**, 163–178.

Brass, L. F., and Joseph, S. K. (1985). A role for inositol triphosphate in intracellular Ca^{2+} mobilization and granule secretion in platelets. *J. Biol. Chem.* **260**, 15172–15179.

Burgess, G. M., Bird, G. S. J., Obie, J. F., and Putney, J. W., Jr. (1991). The mechanism for synergism between phospholipase C- and adenylylcyclase-linked hormones in liver. Cyclic AMP-dependent kinase augments inositol trisphosphate-mediated Ca^{2+} mobilization without increasing the cellular levels of inositol polyphosphates. *J. Biol. Chem.* **266**, 4772–4781.

Carter, T. D., and Ogden, D. (1992). Kinetics of intracellular calcium release by inositol 1,4,5-trisphosphate and extracellular ATP in porcine cultured aortic endothelial cells. *Proc. R. Soc. London* **250**, 235–241.

Chadwick, C. C., Saito, A., and Fleischer, S. (1990). Isolation and characterization of the inositol trisphosphate receptor from smooth muscle. *Proc. Natl. Acad. Sci. USA* **87**, 2132–2136.

Cheek, T. R., Barry, V. A., Berridge, M. J., and Missiaen, L. (1991). Bovine adrenal chromaffin cells contain an inositol 1,4,5-trisphosphate-insensitive but caffeine-sensitive Ca^{2+} store that can be regulated by intraluminal free Ca^{2+}. *Biochem. J.* **275**, 697–701.

Cobbold, P. H., Sanchez-Bueno, A., and Dixon, C. J. (1991). The hepatocyte calcium oscillator. *Cell Calcium* **12**, 87–95.

Combettes, L., Claret, M., and Champeil, P. (1992). Do submaximal $InsP_3$ concentrations only induce the partial discharge of permeabilized hepatocyte calcium pools because of the concomitant reduction of intraluminal Ca^{2+} concentration. *FEBS Lett.* **301**, 287–290.

Cuthbertson, K. S. R., and Chay, T. R. (1991). Modelling receptor-controlled intracellular calcium oscillators. *Cell Calcium* **12**, 97–109.

Danoff, S. K., Supattapone, S., and Snyder, S. H. (1988). Characterization of a membrane protein from brain mediating the inhibition of inositol 1,4,5-trisphosphate receptor binding by calcium. *Biochem. J.* **254**, 701–705.

Danoff, S. K., Ferris, C. D., Donath, C., Fischer, G. A., Munemitsu, S., Ullrich, A., Snyder, S. H., and Ross, C. A. (1991). Inositol 1,4,5-trisphosphate receptors: Distinct neuronal and nonneuronal forms derived by alternative splicing differ in phosphorylation. *Proc. Natl. Acad. Sci. USA* **88**, 2951–2955.

De Camilli, P., and Takei, K. (1990). $InsP_3$ receptor turnaround. *Nature (London)* **344**, 495.

Enouf, J., Giraud, F., Bredoux, R., Bourdeau, N., and Levy-Toledano, S. (1987). Possible role of a cAMP-dependent phosphorylation in the calcium release mediated by inositol 1,4,5-trisphosphate in human platelet membrane vesicles. *Biochim. Biophys. Acta* **928**, 76–82.

Ferris, C. D., and Snyder, S. H. (1992). Inositol 1,4,5-trisphosphate-activated calcium channels. *Annu. Rev. Physiol.* **54**, 469–488.

Ferris, C. D., Huganir, R. L., Supattapone, S., and Snyder, S. H. (1989). Purified inositol 1,4,5-trisphosphate receptor mediates calcium flux in reconstituted lipid vesicles. *Nature (London)* **342**, 87–89.

Ferris, C. D., Huganir, R. L., and Snyder, S. H. (1990). Calcium flux mediated by purified inositol 1,4,5-trisphosphate receptor in reconstituted lipid vesicles is allosterically regulated by adenine nucleotides. *Proc. Natl. Acad. Sci. USA* **87**, 2147–2151.

Ferris, C. D., Huganir, R. L., Bredt, D. S., Cameron, A. M., and Snyder, S. H. (1991). Inositol trisphosphate receptor: Phosphorylation by protein kinase C and calcium calmodulin-dependent protein kinases in reconstituted lipid vesicles. *Proc. Natl. Acad. Sci. USA* **88**, 2232–2235.

Ferris, C. D., Cameron, A. M., Bredt, D. S., Huganir, R. L., and Snyder, S. H. (1992a). Autophosphorylation of inositol 1,4,5-trisphosphate receptors. *J. Biol. Chem.* **267**, 7036–7041.

Ferris, C. D., Cameron, A. M., Huganir, R. L., and Snyder, S. H. (1992b). Quantal calcium release by purified reconstituted inositol 1,4,5-trisphosphate receptors. *Nature (London)* **356**, 350–352.

Finch, E. A., Turner, T. J., and Goldin, S. M. (1991). Calcium as a coagonist of inositol 1,4,5-trisphosphate-induced calcium release. *Science* **252**, 443–446.

Furuichi, T., Yoshikawa, S., Miyawaki, A., Wada, K., Maeda, N., and Mikoshiba, K. (1989). Primary structure and functional expression of the inositol 1,4,5-trisphosphate-binding protein P₄₀₀. *Nature (London)* **342**, 32–38.

Glennon, M. C., Bird, G., St. J., Takemura, H., Thastrup, O., Leslie, B. A., and Putney, J. W., Jr. (1992). *In situ* imaging of agonist-sensitive calcium pools in AR4-2J pancreatoma cells. Evidence for an agonist- and inositol 1,4,5-trisphosphate-sensitive calcium pool in or closely associated with the nuclear envelope. *J. Biol. Chem.* **267**, 25568–25575.

Ghosh, T. K., Eis, P. S., Mullaney, J. M., Ebert, C. L., and Gill, D. L. (1988). Competitive, reversible, and potent antagonism of inositol 1,4,5-trisphosphate-activated calcium release by heparin. *J. Biol. Chem.* **263**, 11075–11079.

Goldbeter, A., Dupont, G., and Berridge, M. J. (1990). Minimal model for signal-induced Ca^{2+} oscillations and for their frequency encoding through protein phosphorylation. *Proc. Natl. Acad. Sci. USA* **87**, 1461–1465.

Guillemette, G., and Segui, J. A. (1988). Effects of pH, reducing and alkylating reagents on the binding and Ca^{2+} release activities of inositol 1,4,5-triphosphate in the bovine adrenal cortex. *Mol. Endocrinol.* **2**, 1249–1255.

Guillemette, G., Balla, T., Baukal, A. J., Spät, A., and Catt, K. J. (1987). Intracellular receptors for inositol 1,4,5-trisphosphate in angiotensin II target tissues. *J. Biol. Chem.* **262**, 1010–1015.

Guillemette, G., Balla, T., Baukal, A. J., and Catt, K. J. (1988). Characterization of inositol 1,4,5-trisphosphate receptors and calcium mobilization in a hepatic plasma membrane fraction. *J. Biol. Chem.* **263**, 4541–4548.

Guillemette, G., Lamontagne, S., Boulay, G., and Mouillac, B. (1989). Differential effects of heparin on inositol 1,4,5-trisphosphate binding, metabolism, and calcium release activity in the bovine adrenal cortex. *Mol. Pharmacol.* **35**, 339–344.

Guillemette, G., Favreau, I., Boulay, G., and Potier, M. (1990). Solubilization and partial characterization of inositol 1,4,5-trisphosphate receptor of bovine adrenal cortex reveal similarities with the receptor of rat cerebellum. *Mol. Pharmacol.* **38**, 841–847.

Harootunian, A. T., Kao, J. P. Y., Paranjape, S., and Tsien, R. Y. (1991). Generation of calcium oscillations in fibroblasts by positive feedback between calcium and IP₃. *Science* **251**, 75–78.

Hill, T. D., Berggren, P.-O., and Boynton, A. L. (1987). Heparin inhibits inositol trisphosphate-induced calcium release from permeabilized rat liver cells. *Biochem. Biophys. Res. Comm.* **149**, 897–901.

Hokin, L. E., and Hokin, M. R. (1955). Effects of acetylcholine on the turnover of phosphoryl units in individual phospholipids of pancreas slices and brain cortex slices. *Biochim. Biophys. Acta* **18**, 102–110.

Hokin, L. E., and Hokin, M. R. (1956). Phosphoinositides and protein secretion in pancreas slices. *J. Biol. Chem.* **233**, 805–810.

Iino, M. (1990). Biphasic Ca^{2+} dependence of inositol 1,4,5-trisphos-phate-induced Ca release in smooth muscle cells of the guinea pig *Taenia caeci*. *J. Gen. Physiol.* **95**, 1103–1122.

Iino, M., and Endo, M. (1992). Calcium-dependent immediate feedback control of inositol 1,4,5-trisphosphate-induced Ca^{2+} release. *Nature (London)* **360**, 76–78.

Irvine, R. F. (1990). "Quantal" Ca^{2+} release and the control of Ca^{2+} entry by inositol phosphates—A possible mechanism. *FEBS Lett.* **263**, 5–9.

Irvine, R. F. (1991). Inositol tetrakisphosphate as a second messenger: Confusions, contradictions, and a potential resolution. *BioEssays* **13**, 419–427.

Irvine, R. F., Letcher, A. J., Lander, D. J., and Berridge, M. J. (1986). Specificity of inositol phosphate-stimulated Ca^{2+} mobilization from Swiss-mouse 3T3 cells. *Biochem. J.* **240**, 301–304.

Jacob, R., Merritt, J. E., Hallam, T. J., and Rink, T. J. (1988). Repetitive spikes in cytoplasmic calcium evoked by histamine in human endothelial cells. *Nature (London)* **335**, 40–45.

Joseph, S. K., Rice, H. L., and Williamson, J. R. (1989). The effect of external calcium and pH on inositol trisphosphate-mediated calcium release from cerebellum microsomal fractions. *Biochem. J.* **258**, 261–265.

Kuno, M., and Gardner, P. (1987). Ion channels activated by inositol 1,4,5-trisphosphate in plasma membrane of human T-lymphocytes. *Nature (London)* **326**, 301–304.

Lechleiter, J., Girard, S., Peralta, E., and Clapham D. (1991). Spiral calcium wave propagation and annihilation in *Xenopus laevis* oocytes. *Science* **252**, 123–126.

Luttrell, B. M. (1993). The biological relevance of the binding of calcium ions by inositol phosphates. *J. Biol. Chem.* **268**, 1521–1524.

Maeda, N., Niinobe, M., and Mikoshiba, K. (1990). A cerebellar Purkinje cell marker P₄₀₀ protein is an inositol 1,4,5-trisphosphate (InsP₃) receptor protein. Purification and characterization of InsP₃ receptor complex. *EMBO J.* **9**, 61–67.

Maeda, N., Kawasaki, T., Nakade, S., Yokota, N., Taguchi, T., Kasai, M., and Mikoshiba, K. (1991). Structural and functional characterization of inositol 1,4,5-trisphosphate receptor channel from mouse cerebellum. *J. Biol. Chem.* **266**, 1109–1116.

Malviya, A. N., Rogue, P., and Vincendon, G. (1990). Steriospecific inositol 1,4,5-[³²P]trisphosphate binding to isolated rat liver nuclei: Evidence for inositol trisphosphate receptor-mediated calcium release from the nucleus. *Proc. Natl. Acad. Sci. USA* **87**, 9270–9274.

Matter, N., Ritz, M.-F., Freyermuth, S., Rogue, P., and Malviya, A. N. (1993). Stimulation of nuclear protein kinase C leads to phosphorylation of nuclear inositol 1,4,5-trisphosphate receptor and accelerated calcium release by inositol 1,4,5-trisphosphate from isolated rat liver nuclei. *J. Biol. Chem.* **268**, 732–736.

Mayrleitner, M., Chadwick, C. C., Timerman, A. P., Fleischer, S., and Schindler, H. (1991). Purified IP₃ receptor from smooth muscle forms an IP₃ gated and heparin sensitive Ca^{2+} channel in planar bilayers. *Cell Calcium* **12**, 505–514.

Meyer, T., and Stryer, L. (1988). Molecular model for receptor-stimulated calcium spiking. *Proc. Natl. Acad. Sci. USA* **85**, 5051–5055.

Meyer, T., and Stryer, L. (1990). Transient calcium release induced by successive increments of inositol 1,4,5-trisphosphate. *Proc. Natl. Acad. Sci. USA* **87**, 3841–3845.

Meyer, T., Wensel, T., and Stryer, L. (1990). Kinetics of calcium channel opening by inositol 1,4,5-trisphosphate. *Biochemistry* **29**, 32–37.

Michell, R. H. (1975). Inositol phospholipids and cell surface receptor function. *Biochim. Biophys. Acta* **415**, 81–147.

Mignery, G. A., and Südhof, T. C. (1990). The ligand binding site and transduction mechanism in the inositol-1,4,5-triphosphate receptor. *EMBO J.* **9**, 3893–3898.

Mignery, G. A., Südhof, T. C., Takei, K., and De Camilli, P. (1989). Putative receptor for inositol 1,4,5-trisphosphate similar to ryanodine receptor. *Nature (London)* **342**, 192–195.

Mignery, G. A., Newton, C. L., Archer, B. T., III, and Südhof, T. C. (1990). Structure and expression of the rat inositol 1,4,5-trisphosphate receptor. *J. Biol. Chem.* **265**, 12679–12685.

Mikoshiba, K. (1993). Inositol 1,4,5-trisphosphate receptor. *Trends Pharmacol. Sci.* **14**, 86–89.

Missiaen, L., Taylor, C. W., and Berridge, M. J. (1991). Spontaneous calcium release from inositol trisphosphate-sensitive calcium stores. *Nature (London)* **352**, 241–244.

Missiaen, L., De Smedt, H., Droogmans, G., and Casteels, R. (1992a). Luminal Ca^{2+} controls the activation of the inositol 1,4,5-trisphosphate receptor by cytosolic Ca^{2+}. *J. Biol. Chem.* **267**, 22961–22966.

Missiaen, L., Taylor, C. W., and Berridge, M. J. (1992b). Luminal Ca^{2+} promoting spontaneous Ca^{2+} release from inositol trisphosphate-sensitive stores in rat hepatocytes. *J. Physiol. (London)* **455**, 623–640.

Miyawaki, A., Furuichi, T., Maeda, N., and Mikoshiba, K. (1990). Expressed cerebellar-type inositol 1,4,5-trisphosphate receptor, P_{400}, has calcium release activity in a fibroblast L cell line. *Neuron* **5**, 11–18.

Miyazaki, S., Yuzaki, M., Nakada, K., Shirakawa, H., Nakanishi, S., Nakade, S., and Mikoshiba, K. (1992). Block of Ca^{2+} wave and Ca^{2+} oscillation by antibody to the inositol 1,4,5-trisphosphate receptor in fertilized hamster eggs. *Science* **257**, 251–255.

Mourey, R. J., Verma, A., Supattapone, S., and Snyder, S. H. (1990). Purification and characterization of the inositol 1,4,5-trisphosphate receptor protein from rat vas deferens. *Biochem. J.* **272**, 383–389.

Muallem, S., Pandol, S. J., and Beeker, T. G. (1989). Hormone-evoked calcium release from intracellular stores is a quantal process. *J. Biol. Chem.* **264**, 205–212.

Muallem, S., Khademazad, M., and Sachs, G. (1990). The route of Ca^{2+} entry during reloading of the intracellular Ca^{2+} pool in pancreatic acini. *J. Biol. Chem.* **265**, 2011–2016.

Nakagawa, T., Okano, H., Furuichi, T., Aruga, J., and Mikoshiba, K. (1991). The subtypes of the mouse inositol 1,4,5-trisphosphate receptor are expressed in a tissue-specific and developmentally specific manner. *Proc. Natl. Acad. Sci. USA* **88**, 6244–6248.

Nicotera, P., Orrenius, S., Nilsson, T., and Berggren, P.-O. (1990). An inositol 1,4,5-trisphosphate-sensitive Ca^{2+} pool in liver nuclei. *Proc. Natl. Acad. Sci. USA* **87**, 6858–6862.

Nordquist, D. T., Kozak, C. A., and Orr, H. T. (1988). cDNA cloning and characterization of three genes uniquely expressed in cerebellum by Purkinje neurons. *J. Neurosci.* **8**, 4780–4789.

Nunn, D. L., and Taylor, C. W. (1990). Liver inositol 1,4,5-trisphosphate-binding sites are the Ca^{2+}-mobilizing receptors. *Biochem. J.* **270**, 227–232.

Nunn, D. L., and Taylor, C. W. (1991). Regulation of Ca^{2+} mobilization by Ins(1,4,5)P_3 and intraluminal Ca^{2+} in permeabilized hepatocytes. *Biochem. Soc. Trans.* **19**, 206S.

Nunn, D. L., and Taylor, C. W. (1992). Luminal Ca^{2+} increases the sensitivity of Ca^{2+} stores to inositol 1,4,5-trisphosphate. *Mol. Pharmacol.* **41**, 115–119.

Oldershaw, K. A., Richardson, A., and Taylor, C. W. (1992). Prolonged exposure to inositol 1,4,5-trisphosphate does not cause intrinsic desensitization of the intracellular Ca^{2+}-mobilizing receptor. *J. Biol. Chem.* **267**, 16312–16316.

Otsu, H., Yamamoto, A., Maeda, N., Mikoshiba, K., and Tashiro, Y. (1990). Immunogold localization of inositol 1,4,5-trisphosphate (InsP$_3$) receptor in mouse cerebellar Purkinje cells using three monoclonal antibodies. *Cell Struct. Funct.* **15**, 163–173.

Parker, I., and Ivorra, I. (1990). Localized all-or-none calcium liberation by inositol trisphosphate. *Science* **250**, 977–979.

Parker, I., and Miledi, R. (1989). Nonlinearity and facilitation in phosphoinositide signaling studied by the use of caged inositol trisphosphate in *Xenopus* oocytes. *J. Neurosci.* **9**, 4068–4077.

Parys, J. B., Sernett, S. W., DeLisle, S., Snyder, P. M., Welsh, M. J., and Campbell, K. P. (1992). Isolation, characterization, and localization of the inositol 1,4,5-trisphosphate receptor protein in *Xenopus laevis* oocytes. *J. Biol. Chem.* **267**, 18776–18782.

Petersen, O. H., and Wakui, M. (1990). Oscillating intracellular Ca^{2+} signals evoked by activation of receptors linked to inositol lipid hydrolysis: Mechanism of generation. *J. Membr. Biol.* **118**, 93–105.

Pietri, F., Hilly, M., and Mauger, J.-P. (1990). Calcium mediates the interconversion between two states of the liver inositol 1,4,5-trisphosphate receptor. *J. Biol. Chem.* **265**, 17478–17485.

Prestwich, G. D., Marecek, J. F., Mourey, R. J., Theibert, A. B., Ferris, C. D., Danoff, S. K., and Snyder, S. H. (1991). Tethered IP$_3$. Synthesis and biochemical applications of the 1-O-(3-aminopropyl) ester of inositol 1,4,5-trisphosphate. *J. Am. Chem. Soc.* **113**, 1822–1825.

Putney, J. W., Jr. (1986). A model for receptor-regulated calcium entry. *Cell Calcium* **7**, 1–12.

Putney, J. W., Jr. (1990). Capacitative calcium entry revisited. *Cell Calcium* **11**, 611–624.

Restrepo, D., Miyamoto, T., Bryant, B. P., and Teeter, J. H. (1990). Odor stimuli trigger influx of calcium into olfactory neurones of the channel catfish. *Science* **249**, 1166–1168.

Ross, C. A., Meldolesi, J., Milner, T. A., Satoh, T., Supattapone, S., and Snyder, S. H. (1989). Inositol 1,4,5-trisphosphate receptor localized to endoplasmic reticulum in cerebellar Purkinje neurons. *Nature (London)* **339**, 468–470.

Ross, C. A., Danoff, S. K., Schell, M. J., Snyder, S. H., and Ullrich, A. (1992). Three additional inositol 1,4,5-trisphosphate receptors: Molecular cloning and differential localization in brain and peripheral tissues. *Proc. Natl. Acad. Sci. USA* **89**, 4265–4269.

Rossier, M. F., Bird, G. S. J., and Putney, J. W., Jr. (1991). Subcellular distribution of the calcium-storing inositol 1,4,5-trisphosphate-sensitive organelle in rat liver. Possible linkage to the plasma membrane through the actin microfilaments. *Biochem. J.* **274**, 643–650.

Safrany, S. T., Wojcikiewicz, R. J. H., Strupish, J., Nahorski, S. R., Dubreuil, D., Cleophax, J., Gero, S. D., and Potter, B. V. L. (1991). Interaction of synthetic D-6-deoxy-*myo*-inositol 1,4,5-trisphosphate with the Ca^{2+}-releasing D-*myo*-inositol 1,4,5-trisphosphate receptor, and the metabolic enzymes 5-phosphatase and 3-kinase. *FEBS Lett.* **278**, 252–256.

Satoh, T., Ross, C. A., Villa, A., Supattapone, S., Pozzan, T., Snyder, S. H., and Meldolesi, J. (1990). The inositol 1,4,5-trisphosphate receptor in cerebellar purkinje cells: quantitative immunogold labeling reveals concentration in an ER subcompartment. *J. Cell Biol.* **111**, 615–624.

Shuttleworth, T. J. (1990). Receptor-activated calcium entry in exocrine cells does not occur via agonist-sensitive intracellular pools. *Biochem. J.* **266**, 719–726.

Shuttleworth, T. J. (1992). Ca^{2+} release from inositol trisphosphate-sensitive stores is not modulated by intraluminal [Ca^{2+}]. *J. Biol. Chem.* **267**, 3573–3576.

Shuttleworth, T. J. (1993). "Quantal" IP$_3$-induced Ca^{2+} release in individual intact exocrine cells. *FASEB J.* **7**, A215.

Somogyi, R., and Stucki, J. W. (1991). Hormone-induced calcium oscillations in liver cells can be explained by a simple one pool model. *J. Biol. Chem.* **266**, 11068–11077.

Spät, A., Bradford, P. G., McKinney, J. S., Rubin, R. P., and Putney, J. W., Jr. (1986). A saturable receptor for ^{32}P-inositol-1,4,5-trisphosphate in hepatocytes and neutrophils. *Nature (London)* **319**, 514–516.

Spät, A., Eberhardt, I., and Kiesel, L. (1992). Low concentrations of adenine nucleotides enhance the receptor binding of inositol 1,4,5-trisphosphate. *Biochem. J.* **287**, 335–336.

Streb, H., Irvine, R. F., Berridge, M. J., and Schulz, I. (1983). Release of Ca^{2+} from a nonmitochondrial intracellular store in pancreatic acinar cells by inositol-1,4,5-trisphosphate. *Nature (London)* **306**, 67–69.

Strupish, J., Cooke, A. M., Potter, B. V. L., Gigg, R., and Nahorski, S. R. (1988). Stereospecific mobilization of intracellular Ca^{2+} by inositol 1,4,5-trisphosphate. Comparison with inositol 1,4,5-trisphosphorothioate and inositol 1,3,4-trisphosphate. *Biochem. J.* **253**, 901–905.

Strupish, J., Wojcikiewicz, R. J. H., Challiss, R. A. J., Safrany, S. T., Willcocks, A. L., Potter, B. V. L., and Nahorski, S. R. (1991). Is decavanadate a specific inositol 1,4,5-trisphosphate receptor antagonist. *Biochem. J.* **277**, 294.

Südhof, T. C., Newton, C. L., Archer, B. T., III, Ushkaryov, Y. A., and Mignery, G. A. (1991). Structure of a novel InsP$_3$ receptor. *EMBO J.* **10**, 3199–3206.

Supattapone, S., Danoff, S. K., Theibert, A., Joseph, S. K., Steiner, J., and Snyder, S. H. (1988a). Cyclic AMP-dependent phosphorylation of a brain inositol trisphosphate receptor decreases its release of calcium. *Proc. Natl. Acad. Sci. USA* **85**, 8747–8750.

Supattapone, S., Worley, P. F., Baraban, J. M., and Snyder, S. H. (1988b). Solubilization, purification, and characterization of an inositol trisphosphate receptor. *J. Biol. Chem.* **263**, 1530–1534.

Takemura, H., and Putney, J. W., Jr. (1989). Capacitative calcium entry in parotid acinar cells. *Biochem. J.* **258**, 409–412.

Taylor, C. W., and Potter, B. V. L. (1990). The size of inositol 1,4,5-trisphosphate-sensitive Ca^{2+} stores depends on inositol 1,4,5-trisphosphate concentration. *Biochim. J.* **266**, 189–194.

Taylor, C. W., and Richardson, A. (1991). Structure and function of inositol trisphosphate receptors. *Pharmacol. Ther.* **51**, 97–137.

Thomas, A. P., Renard, D. C., and Rooney, T. A. (1991). Spatial and temporal organization of calcium signalling in hepatocytes. *Cell Calcium* **12**, 111–126.

Tregear, R. T., Dawson, A. P., and Irvine, R. F. (1991). Quantal release of Ca^{2+} from intracellular stores by InsP$_3$: Tests of the concept of control of Ca^{2+} release by intraluminal Ca^{2+}. *Proc. R. Soc. London* **243**, 263–268.

Volpe, P., and Alderson-Lang, B. H. (1990). Regulation of inositol 1,4,5-trisphosphate-induced Ca^{2+} release. II. Effect of cAMP-dependent protein kinase. *Am. J. Physiol.* **258**, C1086–C1091.

Volpe, P., Krause, K-H., Hashimoto, S., Zorzato, F., Pozzan, T., Meldolesi, J., and Lew, D. P. (1988). ''Calciosome'', a cytoplasmic organelle: The inositol 1,4,5-trisphosphate-sensitive Ca^{2+} store of nonmuscle cells? *Proc. Natl. Acad. Sci. USA* **85**, 1091–1095.

Volpe, P., Alderson-Lang, B. H., and Nickols, G. A. (1990). Regulation of inositol 1,4,5-trisphosphate-induced Ca^{2+} release. I. Effect of Mg^{2+}. *Am. J. Physiol.* **258**, C1077–C1085.

Watras, J., Bezprozvanny, I., and Ehrlich, B. E. (1991). Inositol 1,4,5-trisphosphate-gated channels in cerebellum: Presence of multiple conductance states. *J. Neurosci.* **11**, 3239–3245.

Willcocks, A. L., Cooke, A. M., Potter, B. V. L., and Nahorski, S. R. (1987). Stereospecific recognition sites for [^{3}H]inositol (1,4,5)-trisphosphate in particulate preparations of rat cerebellum. *Biochem. Biophys. Res. Commun.* **146**, 1071–1078.

Woods, N. M., Cuthbertson, K. S. R., and Cobbold, P. H. (1987). Agonist-induced oscillations in cytoplasmic free calcium concentration in single rate hepatocytes. *Cell Calcium* **8**, 79–100.

Worley, P. F., Baraban, J. M., Colvin, J. S., and Snyder, S. H. (1987a). Inositol trisphosphate receptor localization in brain: Variable stoichiometry with protein kinase C. *Nature (London)* **325**, 159–161.

Worley, P. F., Baraban, J. M., Supattapone, S., Wilson, V. S., and Synder, S. H. (1987b). Characterization of inositol trisphosphate receptor binding in brain. Regulation by pH and calcium. *J. Biol. Chem.* **262**, 12132–12136.

Yamamoto, H., Kanaide, H., and Nakamura, M. (1991). Dextran sulfate inhibits the inositol 1,4,5-trisphosphate-induced Ca^{2+} release from skinned and cultured smooth muscle cells. *Eur. J. Pharmacol.* **206**, 175–179.

Yao, Y., and Parker, I. (1992). Potentiation of inositol trisphosphate-induced Ca^{2+} mobilization in *Xenopus* oocytes by cytosolic Ca^{2+}. *J. Physiol. (London)* **458**, 319–338.

Zhang, B.-X., and Muallem, S. (1992). Feedback inhibition of Ca^{2+} release by Ca^{2+} is the underlying mechanism of agonist-evoked intracellular Ca^{2+} oscillations in pancreatic acinar cells. *J. Biol. Chem.* **267**, 24387–24393.

InsP₃ Receptor: Functional Properties and Regulation

ILYA BEZPROZVANNY AND
BARBARA E. EHRLICH

I. Introduction

Intracellular calcium (Ca^{2+}) acts as a second messenger in many cell types. In response to stimulation of the cell, the cytoplasmic concentration of Ca^{2+} can rise dramatically, often 10- to 100-fold from its resting level. This increase in intracellular Ca^{2+} can trigger a variety of cellular processes including muscle contraction, secretion of hormones and neurotransmitters, gene expression, and cell proliferation.

What is the source of this Ca^{2+}? Ca^{2+} can be elevated by influx of Ca^{2+} through voltage-gated or receptor-operated ion channels in the plasma membrane of the cell. The properties of these plasma membrane channels have been discussed elsewhere (see Chapters 10, 11, 12, and 13; Kostyuk, 1986; Tsien et al., 1988; Tsien and Tsien, 1990; Bertolino and Llinas, 1992). Ca^{2+} can also be released from nonmitochondrial intracellular stores that contain large amounts of accumulated Ca^{2+} (Carafoli, 1987). These stores appear to be the endoplasmic (ER) or sarcoplasmic (SR) reticular membrane structures, although the existence of special Ca^{2+}-containing organelles, called "calciosomes," also has been suggested (Krause et al., 1989).

The Ca^{2+} in these stores can be released by an increase in the intracellular concentration of specific second messengers. Some portion of the accumulated Ca^{2+} can be released by production of inositol 1,4,5-trisphosphate (InsP₃), a compound generated by enzymatic breakdown of the lipid precursor phosphatidylinositol 4,5-bisphosphate after activation of phospholipase C (PLC; Berridge, 1993). Another pool of accumulated Ca^{2+} is not sensitive to InsP₃, but releases Ca^{2+} via a Ca^{2+}-induced Ca^{2+}-release mechanism, in response to caffeine application (see Chapter 30) or via a direct mechanical connection with the plasma membrane dihydropyridine receptor (see Chapter 11). Some cell types have predominantly InsP₃-sensitive [e.g., *Xenopus* oocytes, rat basophilic leukemia (RBL) cells; Meyer et al., 1988; Parys et al., 1992] or InsP₃-insensitive (skeletal and cardiac muscle; Lai et al., 1988; Smith et al., 1988; Anderson et al., 1989) Ca^{2+} stores, but most have both types (smooth muscle, neurons, liver, sea urchin oocytes; Lipscombe et al., 1988; Supattapone et al., 1988; Chadwick et al., 1990; Iino, 1990; Marks et al., 1990; McPherson and Campbell, 1990; Shoshan-Barmatz et al., 1990; Galione et al., 1991). The differences in Ca^{2+} release characteristics of these two types of stores can be explained by the differences in the properties of the two known types of intracellular Ca^{2+}-release channels: the InsP₃ receptor (InsP₃R) and the ryanodine receptor (RyR).

Both receptor types have been purified to homogeneity and cloned. Specific binding of the alkaloid ryanodine to one type of intracellular Ca^{2+} channel made it possible to purify this protein (therefore termed the ryanodine receptor; see Chapters 30, and 31) from skeletal muscle (Lai *et al.*, 1988), cardiac muscle (Anderson *et al.*, 1989), and brain (McPherson and Campbell, 1990). Use of $InsP_3$ binding assays has allowed the purification of the $InsP_3R$ from cerebellum (Supattapone *et al.*, 1988), aortic smooth muscle (Chadwick *et al.*, 1990), and vas deferens (Mourey *et al.*, 1990). The mRNA sequences encoding the RyR from skeletal (Takeshima *et al.*, 1989) and cardiac (Otsu *et al.*, 1990) muscle and the $InsP_3R$ from cerebellum (Furuichi *et al.*, 1989; Mignery *et al.*, 1990) were later determined. In Fig. 1, the main molecular characteristics of these receptors are represented schematically. Several reviews have covered the properties of both receptors in depth (Berridge and Irvine, 1989; Taylor and Richardson, 1991; Berridge, 1993). In this chapter, a brief description of the molecular structure of the $InsP_3R$ is given, but the primary focus is on its functional properties and regulation of its activity. The properties of the RyR will be discussed mainly for comparison with the properties of the $InsP_3R$.

II. Identification and Purification of InsP₃R

After the role of $InsP_3$ as a Ca^{2+}-mobilizing agent was discovered (Streb *et al.*, 1983), several laboratories began to use radiolabeled $InsP_3$ in binding assays to find the molecular target of this messenger. Specific $InsP_3$ binding sites with an affinity in the nanomolar range were first found in liver and adrenal cortex (Baukal *et al.*, 1985; Spat *et al.*, 1986), but the density of these receptors was low. A detailed biochemical characterization of the $InsP_3$-binding sites became possible after an unusually high density of these sites was found in cerebellum (Worley *et al.*, 1987a). A single class of specific $InsP_3$-binding sites ($K_d = 40\ nM$) was found in cerebellar microsomes. The binding of $InsP_3$ to its receptor was dramatically increased by alkalinization and was inhibited by submicromolar concentrations of free Ca^{2+} (Worley *et al.*, 1987b). This inhibitory effect of Ca^{2+} was later attributed to the presence of a Ca^{2+}-binding protein, calmedin (Danoff *et al.*, 1988). Alternatively, the inhibitory effect could be explained by Ca^{2+} activation of PLC, a Ca^{2+}-dependent process that would produce excess $InsP_3$ and subsequently would interfere with the binding assay (Mignery *et al.*, 1992). Interestingly, in the liver, an increase in the Ca^{2+} concentration was found to switch the $InsP_3R$ to a desensitized state with a higher affinity for $InsP_3$ (Pietri *et al.*, 1990; Pietri Rouxel *et al.*, 1992). Use of a rapid ion-exchange binding assay revealed the presence of two classes of $InsP_3$ binding sites ($K_d = 4\ nM$ and $K_d = 120\ nM$) in detergent-solubilized extracts of cerebellar membranes (Hingorani and Agnew, 1991). $InsP_3$ binding to both classes of sites was inhibited by Ca^{2+} (Hingorani and Agnew, 1991). More detailed analyses of $InsP_3$ binding, especially measurements of $InsP_3$ binding and dissociation kinetics and studies of the mechanisms of Ca^{2+}-induced effects, may explain the apparent contradictions among the different sets of experiments.

The development of $InsP_3$-binding assays provided an assay for $InsP_3R$ purification. Because the highest density of $InsP_3$-binding sites was found in cerebellar homogenates (Worley *et al.*, 19887b), this tissue was initially used as the starting material. The use of heparin–agarose and concanavalin A–sepharose chromatography of Triton-solubilized cerebellar extracts yielded the first successful purification of the $InsP_3R$ (Supattapone *et al.*, 1988). More recently, extensive purification of $InsP_3R$ from cerebellum in a single step was demonstrated using a wheat germ agglutinin column (Hingorani and Agnew, 1992) and ligand-based affinity chromatography (Prestwich *et al.*, 1991).

Reconstitution of purified $InsP_3R$ into proteoliposomes established that a single protein mediates recognition of $InsP_3$ and efflux of Ca^{2+} (Ferris *et al.*, 1989), suggesting that the protein is a receptor for $InsP_3R$ as well as a Ca^{2+} channel. This suggestion was confirmed in planar lipid bilayer experiments with purified $InsP_3R$ from cerebellum (Maeda *et al.*, 1991) and smooth muscle (Mayrleitner *et al.*, 1991).

Electrophoretic analysis of the purified $InsP_3R$

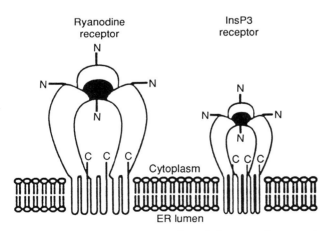

FIGURE 1 Schematic representation of the ryanodine receptor (RyR) and the inositol 1,4,5-trisphosphate receptor ($InsP_3R$), illustrating their relative size and location in the reticular membrane. Both receptors are tetramers. Subunits of the RyR and $InsP_3R$ have molecular masses calculated from cDNA of 565 kDa and 313 kDa, respectively. Reproduced from Henzi and MacDermott (1992) with kind permission from Pergamon Press Ltd., Headington Hill Hall, Oxford CX3 OBW, UK.

preparation revealed a single protein band with an M_r of 260 kDa (Supattapone *et al.*, 1988). If solubilized cerebellar membrane proteins were fractionated on sucrose density gradients, the InsP₃R migrated to a density compatible with an apparent molecular mass of over 1000 kDa (Mignery *et al.*, 1989), suggesting a tetrameric structure for the channel complex. Support for the hypothesis that the InsP₃R is a tetramer was obtained later in experiments in which the purified cerebellar InsP₃R was cross-linked with *bis*(sulfosuccinimidyl)suberate (Maeda *et al.*, 1991) and in electron microscopic analysis of the purified receptor that showed that the complex has the appearance of a pinwheel with four spokes radiating from a common center (Chadwick *et al.*, 1990).

III. Molecular Cloning of InsP₃R

Use of antibodies specific for the InsP₃R revealed that the Purkinje cells of cerebellum contained a very high density of these receptors (Ross *et al.*, 1989). Antibodies against one of the proteins expressed specifically in Purkinje cells (PCD6; Nordquist *et al.*, 1988) cross-reacted with the purified InsP₃R (Mignery *et al.*, 1989). These observations allowed the determination of the sequence encoding the C terminus of InsP₃R (Mignery *et al.*, 1989) and, subsequently, the complete primary structure of the receptor from rat cerebellum (Mignery *et al.*, 1990). At the same time, the coding sequence of P-400, a Purkinje cell-specific protein from mouse, was determined by another laboratory; the primary structure of this protein suggested that it was an InsP₃R (Furuichi *et al.*, 1989). The InsP₃R from both species contains 2749 amino acids, corresponding to an estimated molecular mass of 313 kDa. A very high degree of conservation between rat and mouse forms of the receptor was observed (Furuichi *et al.*, 1989; Mignery *et al.*, 1990).

Soon after the first clone was obtained (InsP₃R type I), several additional isoforms of InsP₃R (type II and type III) were found in rat and human cDNA libraries using polymerase chain reaction (PCR) techniques (Sudhof *et al.*, 1991; Ross *et al.*, 1992). Researchers also found that the InsP₃Rs from *Drosophila melanogaster* (Yoshikawa *et al.*, 1992) and *Xenopus* oocytes (Kume *et al.*, 1993) have extensive sequence similarity to the mouse and rat isoforms of the receptor. Alternative splicing of RNA encoding the InsP₃R was also revealed (Mignery *et al.*, 1990). Interestingly, different splice variants of the receptor were expressed preferentially in specific regions of the brain and at different stages of neuronal development (Nakagawa *et al.*, 1991a,b). PCR analysis also revealed that a 39-amino-

acid sequence was present in the brain but was spliced out in all peripheral tissues tested (Danoff *et al.*, 1991), thus providing a potential marker for neuronal and nonneuronal forms of the receptor.

Cloning of the InsP₃R has facilitated probing the link between structure and function of this protein. In an elegant series of experiments, COS cells were transfected with expression vectors containing truncated constructs of the InsP₃R gene. The biochemical properties of the expressed proteins were evaluated (Mignery and Sudhof, 1990). From the results of these experiments, a domain structure for InsP₃R was proposed (Mignery and Sudhof, 1990). In this model, the InsP₃R consists of three domains, each with a distinct functional role: the N terminus is the InsP₃-binding domain, the C terminal domain forms the Ca^{2+} channel and is important for tetramerization of the receptor complex, and a very long coupling domain exists between the ligand binding and channel-forming domains (Fig. 2). A prediction from this model is that binding of ligand to the InsP₃-binding domain will cause a dramatic change in the conformation of the coupling domain which will, in turn, induce Ca^{2+} channel opening. Indeed, a large-scale conformational change of the coupling domain on binding of InsP₃ was directly observed (Mignery and Sudhof, 1990). This model was later independently supported using truncated constructs of the mouse InsP₃R expressed in NG 108-15 cells (Miyawaki *et al.*, 1991). Analysis of the InsP₃R sequence revealed the presence of several putative regulatory sites in the coupling domain of the receptor (Furuichi *et al.*, 1989; Mignery *et al.*, 1990). These sites include, but are not limited to, two consensus sequences for phosphorylation by cAMP-dependent protein kinase, two putative ATP-binding sites, and a potential Ca^{2+}-binding region (Taylor and Richardson, 1991; Ferris *et al.*, 1992; Mignery *et al.*, 1992). In addition, the alternative splicing event that creates neuronal and nonneuronal forms of the receptor (see preceding discussion) also occurs in the coupling domain of the InsP₃R. Since interaction with the coupling domain appears to be an ideal way to modulate InsP₃-gated channel activity, more sites of regulation are likely to be discovered in this region of the sequence.

FIGURE 2 Domain structure of the InsP₃ receptor. The coupling domain contains two regions with consensus sequences for phosphorylation by protein kinase A, putative binding sites for Ca^{2+} and ATP, and alternative spicing sites. Modified from Sudhof *et al.* (1991) by permission of Oxford University Press.

IV. Functional Studies of InsP₃R/CA²⁺ Channel

The main function of the InsP₃R is to allow the flow of Ca²⁺ from intracellular stores to the cytoplasm in response to InsP₃ binding. Evidence that the receptor and the ion channel reside in the same protein has been obtained from reconstitution of purified receptor (Ferris *et al.*, 1989). In this manner, the InsP₃R is analogous to other receptor-operated ion channels, except that the InsP₃R is primarily associated with intracellular membranes.

How has the function of the InsP₃R/Ca²⁺ channel been studied? When ion channels are relatively inaccessible, as are intracellular channels, electrophysiological techniques such as patch clamping and two-electrode voltage clamping are difficult to use. Observations of the function of intracellular channels such as the InsP₃R can be made using two other methods: Ca²⁺ release from microsomes or permeabilized cells and single-channel current measurements of receptor incorporated into bilayers. As discussed subsequently, each of these methods has been used to assess the functional properties and regulation of the InsP₃R.

A. Channel Properties of InsP₃-Gated Ca²⁺ Channels

InsP₃-gated Ca²⁺ channels were first recorded in bilayers using aortic smooth muscle SR vesicles (Ehrlich and Watras, 1988). These channels had a slope conductance of 10 pS at 0 mV with Ca²⁺ as the current carrier, but when the conductance over the whole voltage range (0 to −50 mV) was calculated, the conductance was 20 pS. When purified cerebellar InsP₃R was incorporated into bilayers, the single-channel conductance observed was 26 pS with Ca²⁺ as the current carrier (Maeda *et al.*, 1991). The similarity of the conductances suggests that the channels formed by the receptors from the two different tissues are similar. When cerebellar microsomes were incorporated into bilayers, multiple subconductance states (3–4 current levels) of the InsP₃-gated channels were observed (Watras *et al.*, 1991). The conductance of each step was 20 pS with Ca²⁺ as the current carrier; the steps were equivalent. In view of the initial observations of the receptor from smooth muscle and of the purified reconstituted receptor, the multiple conductance levels could have been interpreted to indicate that several channels are active in the membrane. Instead, two observations support the hypothesis that the receptor forms a single channel with multiple conductance levels. First, the distribution of current am-

plitudes cannot be fitted by a binomial distribution as expected for multiple channels opening independently (Watras *et al.*, 1991). Second, when optimal recording conditions are used (defined subsequently), the channel spends more than 90% of the open time in one of the subconductance states (Level 3, 60 pS). Openings to this level are more frequent and much longer than openings to the other levels. Thus, in the analysis of open-channel probability and kinetic characteristics of the channels described in this chapter, the focus of discussion is the main subconductance state because the remaining substates represent a small fraction of the total open time.

The activity of InsP₃-gated channels is never observed without InsP₃ present on the cytoplasmic side of the channels (Fig. 3A,B; traces 1 and 2). Addition of 2 μM InsP₃ to the cytoplasmic side of the membrane activated the channel (Fig. 3A,B; traces 3 and 4). Even when optimal conditions are used (2 μM InsP₃, 0.2 μM free Ca²⁺, and millimolar adenine nucleotide; see subsequent discussion), the gating of the channel is characterized by brief (2–5 msec) openings and long (50–100 msec) closings (Fig. 3A,B; traces 3 and 4). Measurements of channel activity at different InsP₃ concentrations imply that InsP₃ acts as a partial agonist of its receptor, because even at very high InsP₃ concentrations the maximum channel open probability observed is ~10% (Watras *et al.*, 1991). Activation of the channels with InsP₃ is reversible; channels stop opening if InsP₃ is washed out (Fig. 3B; traces 5 and 6); re-addition of InsP₃ reactivates the channels (Fig. 3A; traces 7 and 8). Channels do not seem to desensitize, because the activity of the InsP₃-gated channels can be observed for long periods of time (tens of minutes, sometimes hours) in the bilayer without a decrease in channel activity (Fig. 3B; traces 3–6). The addition of heparin, a potent inhibitor of InsP₃ binding (Ghosh *et al.*, 1988), suppresses channel activity (Fig. 3B; traces 7 and 8).

The selectivity of the InsP₃-gated channels among different divalent cations was determined (Bezprozvanny and Ehrlich, 1993b) to be, in order of permeability: Ba²⁺ > Sr²⁺ > Ca²⁺ > Mg²⁺. The same sequence with similar permeability ratios was reported for the cardiac RyR (Tinker and Williams, 1992), suggesting that the "selectivity filters" of these channels are similar. This hypothesis is in good agreement with the high degree of homology in the amino acid sequences in the transmembrane regions of these proteins.

B. InsP₃ Dependence of InsP₃-Gated Ca²⁺ Channels

Clearly binding of InsP₃ is absolutely necessary for these channels to open (Fig. 3A). How many InsP₃

A

0.2 μM Ca
0.5 mM ATP

B

0.2 μM Ca
0.5 mM ATP

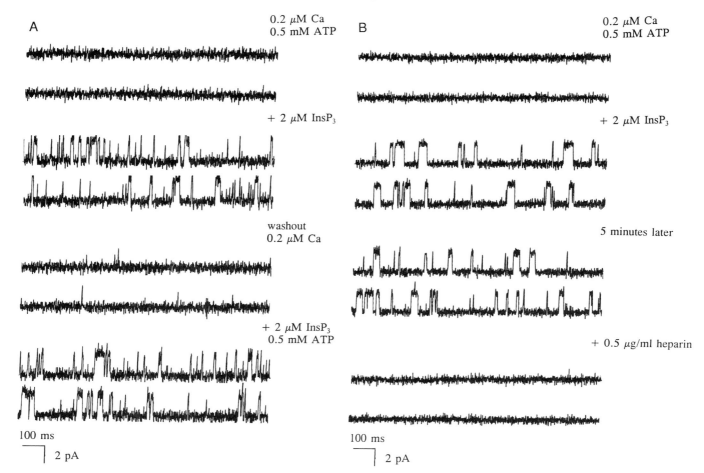

+ 2 μM InsP₃

+ 2 μM InsP₃

washout
0.2 μM Ca

5 minutes later

+ 2 μM InsP₃
0.5 mM ATP

+ 0.5 μg/ml heparin

100 ms

2 pA

100 ms

2 pA

FIGURE 3 Functional properties of the InsP₃ receptor in planar lipid bilayers. (A) InsP₃ is absolutely required for channel activation. Addition of optimal concentrations of Ca²⁺ (0.2 μM) and ATP (0.5 mM) in the absence of InsP₃ were insufficient to open the channel (traces 1 and 2). InsP₃ (2 μM) activated a channel of 2 pA (traces 3 and 4). Further additions of InsP₃ did not increase the open probability of the channel. After washout of InsP₃, channel activity halts, showing that InsP₃ action is reversible (traces 5 and 6). Readdition of InsP₃, ATP, and Ca²⁺ returns channel activity to its maximum (traces 7 and 8). (B) InsP₃-gated channel currents do not inactivate with time, but can be inhibited by heparin. Without InsP₃ present, channel openings are not observed (traces 1 and 2). Addition of InsP₃ activates the channel (traces 3 and 4); this activity is maintained (traces 5 and 6). The activity can be maintained for hours (not shown). Addition of heparin (0.5 μg/ml) completely inhibits channel activity (traces 7 and 8). Heparin is a competitive antagonist of InsP₃ binding (Supattapone *et al.*, 1988). **Methods:** Endoplasmic reticulum vesicles derived from canine cerebellum were made as described previously (Watras *et al.*, 1991). Bilayers made from phosphatidylethanolamine and phosphatidylserine (75% : 25%; Avanti Polar Lipids) were formed by painting a solution of lipid in decane across a 100 μm hole in a Teflon sheet that bisected a Lucite chamber. Channel incorporation into the bilayer was done as described elsewhere (Bezprozvanny *et al.*, 1991). The *trans* chamber contained 53 mM Ba(OH)₂, 250 mM HEPES, pH 7.35 and was held at virtual ground; the *cis* chamber contained 250 mM HEPES-Tris, pH 7.35, 1 mM EGTA. The *cis* and *trans* chambers represent the cytoplasmic and intralumenal sides of the reticular membrane, respectively. Channel openings indicating divalent cation movement from the *trans* to the *cis* side are shown as upward deflections from zero current. Addition of all compounds was made to the *cis* chamber; the transmembrane potential was maintained at 0 mV. Traces are representative of at least 3 min of continuous recording under each condition.

molecules must bind to open the channel? Because of the tetrameric structure of the receptor, four binding sites are found in each channel complex. Experiments with permeabilized RBL cells suggested that all four InsP₃-binding sites must be occupied by InsP₃ to open the channel (Meyer *et al.*, 1988). However, in planar

lipid bilayer experiments, the channel open probability depended on the InsP₃ concentration in accordance with the Michaelis–Menten equation (Watras *et al.*, 1991), which implies that binding to a single site is sufficient to open the channel. Also, no cooperativity in InsP₃ action was reported in measurements of InsP₃-

induced Ca^{2+} efflux from brain synaptosomes (Finch *et al.*, 1991). Binding of one InsP$_3$ molecule to any of the receptor subunits was suggested to be sufficient to open the channel (Watras *et al.*, 1991). In the proposed model, a conformational change of the subunit with InsP$_3$ bound was suggested to change the conformation of the two adjacent subunits and cause the channel to open completely. Alternatively, binding of InsP$_3$ to all four subunits of the channel complex is necessary to open the channel, but it is much harder for the last molecule of InsP$_3$ to bind. This model will also lead to a noncooperative concentration dependence of InsP$_3$ action, as observed in the experiments (Finch *et al.*, 1991; Watras *et al.*, 1991). To explain the cooperativity observed in experiments with RBL cells (Meyer *et al.*, 1988), positive feedback in the process of InsP$_3$-induced Ca^{2+} release due to the Ca^{2+} dependence of these channels (described subsequently) was hypothesized (Iino, 1990; Watras *et al.*, 1991). Because no Ca^{2+} buffer was present in the epxeriments with permeabilized cells, an elevation in the cytoplasmic Ca^{2+} level could make the InsP$_3$ dependence of the release appear to be steeper.

This latter suggestion was examined quantitatively by Iino and Endo (1992). In this study, the InsP$_3$ dependence of Ca^{2+} release from intracellular stores of permeabilized smooth muscles was determined. When the cytoplasmic Ca^{2+} concentration was allowed to change during Ca^{2+} release, a steep InsP$_3$ dependence was observed with a Hill coefficient of 3.82, a value close to that obtained earlier in the experiments with RBL cells (Meyer *et al.*, 1988). However, if the free Ca^{2+} concentration was fixed at 300 nM, a level at which positive feedback from Ca^{2+} would not occur (see subsequent discussion) the InsP$_3$ dependence was less steep, yielding a Hill coefficient of only 2.03 (Iino and Endo, 1992). Although these data indicate that immediate Ca^{2+} feedback may explain the very sharp InsP$_3$ dependence reported earlier (Meyer *et al.*, 1988), some degree of cooperativity in InsP$_3$ action was also found, contrary to observations of InsP$_3$-gated channel activity in bilayers (Watras *et al.*, 1991) and InsP$_3$-induced Ca^{2+} release in brain synaptosomes (Finch *et al.*, 1991). Note that cooperativity in InsP$_3$ action was observed at low (<100 nM) InsP$_3$ concentrations (Meyer *et al.*, 1988; Iino and Endo, 1992), whereas noncooperative dose–response curves were obtained at higher (>100 nM) concentrations of InsP$_3$ (Finch *et al.*, 1991; Watras *et al.*, 1991). Measurements of single-channel activity in the presence of very low InsP$_3$ concentrations would be useful in resolving this discrepancy. Unfortunately, these experiments are difficult to perform because channel activity at InsP$_3$ concentrations below 100 nM is very low, and extremely long recording times are necessary to collect sufficient channel openings for data analysis. Nonetheless, additional experiments are necessary to distinguish among the models of channel activation by InsP$_3$.

V. Regulation of InsP$_3$R Function

The InsP$_3$R is one of the key elements in cell signaling. The importance of this receptor is highlighted by the number and variety of mechanisms that regulate its function. The following sections outline the main regulatory pathways of InsP$_3$R function known to date.

A. Regulation by Cytoplasmic Ca^{2+}

The relationship between Ca^{2+} and InsP$_3$ has turned out to be astonishingly complex (Taylor and Marshall, 1992). These interactions probably provide the basis for a variety of phenomena including the observed spatiotemporal characteristics of Ca^{2+} signaling (Berridge, 1993). Some features of the relationship between Ca^{2+} and InsP$_3$ are discussed in the following two sections.

Increases in cytoplasmic Ca^{2+} concentration were shown to inhibit InsP$_3$ binding to the receptor with half-inhibition at 300 nM Ca^{2+} (Supattapone *et al.*, 1988). These data suggest a monotonic inhibition of InsP$_3$-gated channels with increases in the cytoplasmic Ca^{2+} concentration in the submicromolar range. However, when the Ca^{2+} dependence of InsP$_3$-induced Ca^{2+} release was determined in permeabilized smooth muscle, a biphasic effect of Ca^{2+} was observed with a maximum rate of release at 300 nM free Ca^{2+} (Iino, 1990). The mechanism for activation by Ca^{2+} below 300 nM in these experiments could be a Ca^{2+}-dependent alteration in InsP$_3$ metabolism or a direct effect on the channel gating mechanism. Although Iino suggested that the latter possibility was responsible for the biphasic Ca^{2+} effect (Iino, 1990), use of permeabilized cells did not allow elimination of the other possibilities.

Use of a rapid superfusion system allowed studies that focused on analysis of the subsecond kinetics of InsP$_3$-induced Ca^{2+} release from brain synaptosomal vesicles (Finch *et al.*, 1991). These investigators found that extravesicular Ca^{2+} acted as a coagonist of InsP$_3$-induced release, with a maximum amount of Ca^{2+} released at 500 nM free Ca^{2+}. Further increases in extravesicular Ca^{2+} inhibited InsP$_3$-induced Ca^{2+} release. Although Ca^{2+} potentiation of InsP$_3$-induced release occurred faster than the detection limit of the rapid superfusion system, a more slowly developing Ca^{2+}-induced inhibition was measured with a time constant of 580 msec at 10 μM free Ca^{2+} (Finch *et al.*, 1991). In another report, use of caged Ca^{2+} further improved the time resolution of the experimental system; experiments on permeabilized smooth muscle cells showed that both activating and inhibitory effects of Ca^{2+} on the InsP$_3$R were extremely rapid in the presence of InsP$_3$, which would allow for "immediate feedback control" of InsP$_3$-induced Ca^{2+} release (Iino and Endo, 1992).

To test whether a direct effect of Ca^{2+} on the gating of the channel exists, the effect of cytoplasmic Ca^{2+} on InsP$_3$-gated channel activity was monitored in planar lipid bilayers (Bezprozvanny *et al.*, 1991). The open probability of the channel increased as the free Ca^{2+} concentration was elevated from 10 nM to 250 nM and decreased at Ca^{2+} concentrations above 250 nM (Fig. 4A). Both activating and inhibitory effects of Ca^{2+} on channel activity were reversible, ruling out the possibility of Ca^{2+}-induced chemical modification of the receptors. A bell-shaped Ca^{2+}-dependence curve was obtained with the maximum open probabil-

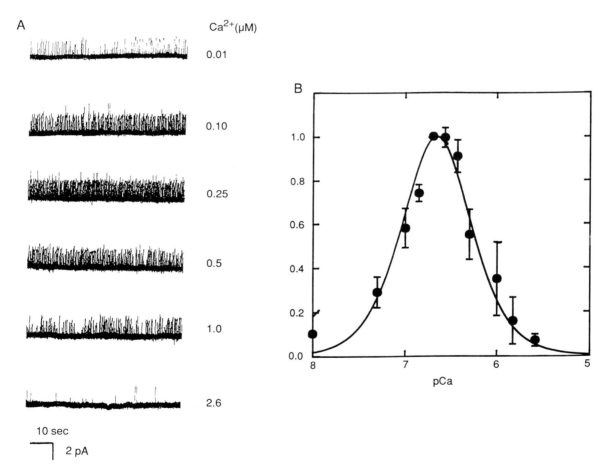

FIGURE 4 Calcium dependence of the InsP$_3$ receptor in planar lipid bilayers. (A) Channel activity was monitored in the presence of 2 μM InsP$_3$ and 330 μM AMP-PCP. To change the free Ca^{2+} concentration in the range from 0.01 to 5 μM, aliquots of a calibrated 20 mM CaCl$_2$ solution were added to a mixture of 1 mM EGTA and 1 mM HEEDTA. The free Ca^{2+} concentration, indicated on the right, was calculated using published binding constants (Fabiato, 1988). The effect of Ca^{2+} was reversible. (B) Open probability of the InsP$_3$-gated channel plotted against the free Ca^{2+} concentration. Using a threshold level of 1 pA, the open probability was calculated from recordings at least 2.5 min long at each Ca^{2+} concentration. To compare four different experiments, values were normalized to the maximum channel activity observed in each experiment, which never exceeded 15% assuming one channel in the bilayer. Methods are described in Fig. 3, except that Ca^{2+} was the current carrier. Modified with permission from Bezprozvanny *et al.* (1991). Copyright © 1991 Macmillan Magazines Ltd.

ity at 200–300 nM free Ca^{2+} (Fig. 4B). The single-channel open probability decreased sharply on both sides of the maximum; the entire curve fell within the physiological range of Ca^{2+} concentrations.

Despite some differences among the reports, a general conclusion can be drawn based on the experimental data just described. Evidently the function of the $InsP_3R$ is regulated by cytoplasmic Ca^{2+} in a biphasic manner with the maximum activity of the channels at Ca^{2+} levels of 200–300 nM. This regulation does not seem to be mediated by some sort of enzymatic modification, because it occurs in an *in vitro* reconstitution system, it is very fast, and it is reversible. Whether these changes in channel function are mediated by Ca^{2+} interactions with the $InsP_3R$ itself, or with Ca^{2+}-binding proteins associated with the receptor, is not clear. The purified $InsP_3R$ binds Ca^{2+} (Danoff *et al.*, 1988); at least one Ca^{2+}-binding site has been found in the transducing domain of the receptor (Mignery *et al.*, 1992). These observations provide evidence in favor of a direct effect of Ca^{2+} on $InsP_3R$ function. The existence of a Ca^{2+}-binding protein, "calmedin," was proposed (Danoff *et al.*, 1988) to explain Ca^{2+}-induced inhibition of $InsP_3$ binding to the receptor (but see Mignery *et al.*, 1992). Inhibition of $InsP_3R$ function by high Ca^{2+} may also develop from Ca^{2+}-induced desensitization of the receptor, rather than from a decrease in $InsP_3$ binding. For example, Ca^{2+}-mediated interconversion between functional and nonfunctional states of the $InsP_3R$ in liver was reported (Pietri *et al.*, 1990; Pietri Rouxel *et al.*, 1992); a similar mechanism might be important for Ca^{2+}-induced inactivation of the $InsP_3R$ in the other tissues as well. The exact location of the "positive" and "negative" Ca^{2+} sensors is not yet known. Experiments with the purified and reconstituted $InsP_3R$ will be necessary to address this issue.

How important is the regulation of $InsP_3R$ function by cytoplasmic Ca^{2+}? Implications of Ca^{2+}-dependent enhancement of channel activity include the observation that the $InsP_3$-gated channel can act as a Ca^{2+}-gated channel in the presence of a fixed concentration of $InsP_3$ if cytoplasmic Ca^{2+} is less than 300 nM. Ca^{2+}-induced Ca^{2+} release was considered to be the functional marker for the presence of RyR in a cell, but this postulate must be reconsidered in light of the data described earlier. Clearly some form of positive feedback in the process of Ca^{2+} release is also necessary to explain the spatiotemporal aspects of Ca^{2+} signaling in the cell. Early reports proposed that Ca^{2+} activation of PLC would generate the periodic surges in $InsP_3$ concentration during each spike, providing the necessary positive feedback in the course of the release (Meyer and Stryer, 1988; Harootunian *et al.*,

1991; Keizer and DeYoung, 1992), but evidence to support this proposal has been difficult to obtain. A two-pool model was also proposed (Berridge, 1990,1991; Goldbeter *et al.*, 1990), in which Ca^{2+}-induced Ca^{2+} release through RyR would provide the positive feedback needed to sustain Ca^{2+} oscillations. Immediate Ca^{2+}-dependent positive feedback during $InsP_3$-induced Ca^{2+} release may replace the other proposed pathways, thus providing a simple mechanism for the generation of Ca^{2+} spikes and oscillations in cells (Berridge, 1993; Berridge and Irvine, 1989). Similarly, Ca^{2+}-induced inhibition can provide the negative feedback necessary for Ca^{2+} oscillations. Although the "feedback by $InsP_3$ generation" model and the "two-pool" model might play a role in some cell types, they are not able to explain the generation of spiral Ca^{2+} waves and Ca^{2+} oscillations in *Xenopus* oocytes (Lechleiter *et al.*, 1991; Parker and Yao, 1991), because *Xenopus* oocytes lack RyR (Parys *et al.*, 1992) and because the nonmetabolizable analog $InsPS_3$ also induces the same pattern of Ca^{2+} signaling seen with $InsP_3$ (Lechleiter and Clapham, 1992). Moreover, monoclonal antibodies against the $InsP_3R$ are able to block Ca^{2+} waves and Ca^{2+} oscillations in fertilized hamster eggs (Miyazaki *et al.*, 1992), thus showing that $InsP_3R$ functional properties constitute the primary and essential mechanism for the spatiotemporal aspects of Ca^{2+} signaling in several systems.

Can Ca^{2+} oscillations be modeled based solely on the known information about the functional properties of the $InsP_3Rs$? Indeed, a quantitative model that simulates the observed characteristics of Ca^{2+} oscillations in many cell types could be generated (De Young and Keizer, 1992). This model was based solely on a fit to the bell-shaped Ca^{2+} dependence of $InsP_3R$ activity (Fig. 4B; Bezprozvanny *et al.*, 1991). Therefore, although other factors might, and probably do, affect the characteristics of Ca^{2+} signaling in the cell, investigations into the molecular basis of the bell-shaped regulation of $InsP_3R$ function by cytoplasmic Ca^{2+} are of primary importance in our efforts to understand the underlying mechanisms of Ca^{2+} oscillations in cells.

B. Regulation by Intraluminal Ca^{2+}

Addition of submaximal $InsP_3$ concentrations causes partial discharge of Ca^{2+} from the intracellular stores. This phenomenon is often called "quantal release" (Muallem *et al.*, 1989). One possible explanation for these observations is the presence within the cell of $InsP_3R$ isoforms with different affinities for $InsP_3$ (Muallem *et al.*, 1989; Kindman and Meyer, 1993). An alternative hypothesis suggested that regulation of InsP₃R sensitivity to InsP₃ by intraluminal Ca^{2+} concentration can explain "quantal release" (Irvine, 1990). According to this model, the affinity of the InsP₃R decreases over the course of Ca^{2+} release because of depletion of the Ca^{2+} stores and a lowering of the intraluminal Ca^{2+} concentration. Thus, only partial discharge of Ca^{2+} is possible in the presence of low InsP₃ concentrations.

Although very attractive, this model required experimental evidence to support the hypothesis of InsP₃R regulation by intraluminal Ca^{2+} levels. Several laboratories have addressed this issue using very similar experimental approaches. In these experiments, Ca^{2+} stores were partially depleted by incubation of permeabilized cells for short times with low concentrations of Ca^{2+} ionophore of the Ca^{2+}–ATPase inhibitor thapsigargin; then the characteristics of InsP₃-induced Ca^{2+} release were studied. Using permeabilized rat hepatocytes (Nunn and Taylor, 1992) and smooth muscle cells (Missiaen *et al.*, 1992a,b), a decrease in InsP₃R sensitivity to InsP₃ on depletion of the Ca^{2+} stores was demonstrated. Interestingly, this effect was evident even in the presence of InsP₃ concentrations as high as 3 μM (Missiaen *et al.*, 1992b), which is a concentration of InsP₃ so high that "quantal release" is no longer observed. In contrast, no changes in InsP₃ sensitivity due to the depletion of Ca^{2+} stores were observed by two other laboratories in their experiments with permeabilized rat hepatocytes (Combettes *et al.*, 1992) and cells from the avian supraorbital nasal gland (Shuttleworth, 1992; see Chapter 32), despite the use of much lower InsP₃ concentrations (100–200 nM).

Unfortunately, evaluation of these data is difficult because the absolute levels of intraluminal free Ca^{2+} and their changes over the course of the experiment are not measurable and cannot be controlled. Most of the Ca^{2+} accumulated inside the reticulum is bound to various Ca^{2+}-binding proteins such as calsequestrin, and the free Ca^{2+} concentration in the lumen may vary depending on the buffering capacity of these proteins. Different levels of intraluminal Ca^{2+} concentrations may have been reached by different laboratories during the depletion process, making it difficult to compare the studies. In at least one study (Combettes *et al.*, 1992), this problem was addressed quantitatively, but more precise measurements of intraluminal Ca^{2+} concentrations are necessary to clarify the issue. Thus, the conclusions are controversial. Investigations of the effect of intraluminal Ca^{2+} on the function of InsP₃-gated channels incorporated into planar lipid bilayer will be useful in this respect, because the free Ca^{2+} concentration on both cytoplasmic and intraluminal sides of the membrane can be easily controlled during such an experiment.

C. Regulation by ATP

The initial studies of InsP₃-mediated signal transduction mechanisms found that the presence of ATP or nonhydrolyzable ATP analogs was essential for InsP₃-induced Ca^{2+} release (Smith *et al.*, 1985). These authors came to the conclusion that ATP was a necessary cofactor for the activation of the then hypothetical InsP₃R (Smith *et al.*, 1985). The role of ATP as an allosteric activator of the InsP₃R was proposed later, based on experiments with receptor that was purified and reconstituted into liposomes (Ferris *et al.*, 1990). These investigators found that 10 μM ATP or nonhy-drolyzable ATP analogs dramatically potentiated InsP₃-mediated Ca^{2+} flux into vesicles containing purified InsP₃R. The existence of a specific ATP-binding site on the InsP₃R was also demonstrated in the same report. In agreement with earlier data (Smith *et al.*, 1985), AMP and GTP could not substitute for ATP in the activation of the InsP₃Rs, but nonhydrolyzable ATP analogs were equally potent (Ferris *et al.*, 1990), ruling out the possibility of channel phosphorylation. Interestingly, further increases in ATP concentration inhibited InsP₃-mediated Ca^{2+} efflux (Ferris *et al.*, 1990), implying the presence of an additional site of ATP action that was inhibitory.

Using permeabilized vascular smooth muscle cells, the effects of adenine nucleotides on the characteristics of InsP₃-induced Ca^{2+} release were studied in greater detail (Iino, 1991). In accordance with the data described earlier, potentiation of InsP₃-induced Ca^{2+} release by submillimolar concentrations of ATP and nonhydrolyzable ATP analogs was observed. AMP also was able to potentiate this release, but much larger concentrations were needed. High concentrations of ATP were able to inhibit InsP₃-induced Ca^{2+} release, but only if low InsP₃ concentrations (0.1 μM) were used in the experiment. The latter observation suggested a competitive antagonism between InsP₃ and high concentrations of ATP.

Despite the general agreement that submillimolar ATP acts as an allosteric activator of InsP₃R function, the mechanism of ATP action is not yet known. Addition of ATP was reported to cause an increase in the open probability of InsP₃-gated Ca^{2+} channels from aortic smooth muscle SR or cerebellar ER reconstituted into planar lipid bilayers (Ehrlich and Watras, 1988; Bezprozvanny *et al.*, 1991). In experiments with the purified and reconstituted InsP₃R from cerebellum, both the open probability and the channel conductance were affected by ATP (Maeda *et al.*, 1991). The modulation of the InsP$_3$-gated channel by ATP has been investigated in more detail (Bezprozvanny and Ehrlich, 1993a). The effects of ATP on the function of the InsP$_3$-gated channel were characterized at the single-channel level and the mechanistic details of the regulation of this channel were described.

In these experiments, addition of 2 μM InsP$_3$ to the cytoplasmic side of the membrane activated the channels even in the absence of ATP (Fig. 5A; second pair of traces), but the open probability was low despite the use of a high InsP$_3$ concentration (2 μM). Subsequent addition of 1 mM ATP in the presence of InsP$_3$ dramatically increased the probability of finding the channel open (Fig. 5A, third pair of traces), although ATP in the absence of InsP$_3$ did not activate the channels (Bezprozvanny et al., 1991). Inhibition of channel activity by heparin (10 μg/ml) shows that the ATP-enhanced channel activity is not the result of opening of the RyR, but is indeed enhancing the activity of the InsP$_3$R (Fig. 5A; bottom pair of traces). Although heparin is known to inhibit the InsP$_3$R (Ghosh et al., 1988; Kobayashi et al., 1988), it activates the RyR (Ritov et al., 1985; Bezprozvanny et al., 1993). Single-channel data analysis revealed that ATP activated the InsP$_3$-gated channels by increasing the average duration of channel openings *and* the frequency of openings (Fig. 5B; *left, center*). The evidence for a nucleotide-dependent alteration in current amplitude (Maeda et al., 1991) was lacking in these experiments (Fig. 5B; *right*). Similar effects were observed when nonhydrolyzable ATP analogs were used in the experiments. Therefore, profound effects of ATP and ATP analogs on the gating of InsP$_3$-gated channels were observed, but open and closed times of the channels were affected, but the conductance of the channels remained unchanged. These effects were only seen in the presence of InsP$_3$; ATP is not an agonist for the channel.

Where must ATP bind to cause these effects? Biochemical studies have identified a specific binding site for ATP on each subunit of the InsP$_3$R with an affinity of 17 μM (Maeda et al., 1991). This site was found to be quite selective for ATP relative to AMP and GTP. Interestingly, the predicted amino acid sequence of the InsP$_3$R contains two consensus sequences identified as ATP-binding sites (Furuichi et al., 1989; Mignery et al., 1990; Mignery and Sudhof, 1990); both sequences are located in the coupling domain of the receptor (see Fig. 2). ATP binding to this region is likely to affect the coupling between InsP$_3$ binding and channel gating, causing the functional effects described earlier. ATP does not seem likely to produce its effect by increasing the affinity of InsP$_3$R for InsP$_3$ because the affinity of InsP$_3$ for its receptor is not increased by adenine nucleotides (Worley et al., 1987b). More recently, an effect of ATP on InsP$_3$ binding was observed (Spat et al., 1992), but the degree of potentiation was small (30%) relative to the dramatic increase in channel activity caused by addition of ATP. ATP binding to the "coupling domain" of the InsP$_3$ receptor is likely to increase the intrinsic efficacy of InsP$_3$ several-fold, that is, ATP improves the ability of InsP$_3$ to open the channel when both compounds are bound to the receptor.

When much higher concentrations of ATP were used, inhibition of InsP$_3$ channel activity was also observed. The ability of ATP and other polyphosphate compounds to inhibit InsP$_3$ binding to the receptor was directly demonstrated in biochemical experiments (Maeda et al., 1991). In these experiments, an increase in the InsP$_3$ concentration was able to restore activity of the channels inhibited by high ATP concentration (Bezprozvanny and Ehrlich, 1993a). These data are interpreted as evidence for competitive antagonism between InsP$_3$ and ATP, in agreement with an earlier proposal (Iino, 1991).

Does InsP$_3$R modulation by ATP have any physiological relevance? Although the precise physiological significance of adenine nucleotide regulation of the InsP$_3$-gated channel in intact cells is unclear, allosteric regulation by ATP is a common motif used by cells to maintain viability (Katz, 1992). This form of regulation can play an important role in the preservation of the cell under conditions of energy starvation, which occurs in diseases such as heart failure and occlusive stroke. For example, after several minutes of ischemia, levels of ATP in the brain fell below 0.1 mM (Abe et al., 1987), a concentration at which changes in ATP alter InsP$_3$-gated channel behavior. Decreased levels of ATP will reduce Ca^{2+} release into the cytoplasm from intracellular stores, which ultimately will avoid Ca^{2+} overload. The dual regulation of InsP$_3$ receptors by ATP was also proposed to be important for switching the response mode of the InsP$_3$R (Iino, 1991) and for generating Ca^{2+} oscillations (Ferris et al., 1990).

D. Regulation by Phosphorylation

P-400 protein (Yamamoto et al., 1989) or PCPP-260 protein (Walaas et al., 1986) (names given to the InsP$_3$R protein before its final identification) is one of the best substrates for endogenous and exogenous cAMP-dependent protein kinase (PKA) in the cerebellum. After biochemical purification of the InsP$_3$R was achieved (Supattapone et al., 1988), a more detailed and quantitative analysis of InsP$_3$R phosphorylation became possible. These investigators initially thought

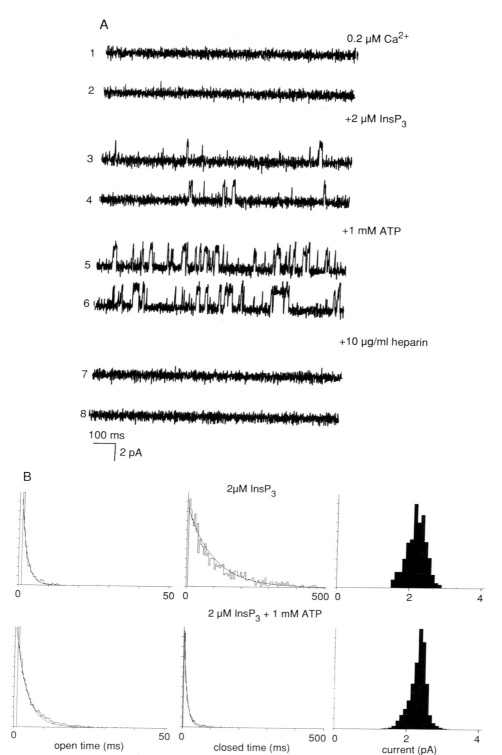

FIGURE 5 The effect of ATP on the InsP$_3$ receptor in planar lipid bilayers. (A) In the absence of ATP and InsP$_3$ channel openings were not observed (traces 1 and 2). Addition of InsP$_3$ (2 μM) activates the channel, but the open probability was submaximal (traces 3 and 4). Subsequent addition of ATP (1 mM) to the chamber containing optimal levels of InsP$_3$ and Ca^{2+} increased the channel open probability (traces 5 and 6). Inhibition of the channel activity by heparin (10 μg/ml) showed that ATP indeed enhanced the activity of the InsP$_3$R (traces 7 and 8). (B) Effect of ATP on the mean open and closed times and on the single channel-conductance of the InsP$_3$-gated channel. Data are from the experiment shown in A. Open time distribution (*left*), observed closed time distribution (*center*), and current amplitude histogram at 0 mV (*right*) shown without ATP (*top*) and in the presence of 1 mM ATP (*bottom*). Histograms were constructed from 470 events in the absence of ATP and 1979 events at 1 mM ATP. Events longer than 1 msec were used for data analysis. Open and closed time distributions were fit with single exponential functions. Openings to subconducting levels of the InsP$_3$-gated channel were infrequent under the conditions used in this series of experiments.

that purified InsP$_3$R was phosphorylated exclusively by PKA (Supattapone *et al.*, 1988), but later experiments revealed that the InsP$_3$R is stoichiometrically (one phosphate per receptor subunit) phosphorylated by PKA, PKC, and Ca^{2+}–calmodulin-dependent protein kinase II (CAM-K-II; Ferris *et al.*, 1991b). Phosphorylation by these three enzymes was additive (Ferris *et al.*, 1991b), indicating that each acts at a different site. Further, serine residues are phosphorylated by all three enzymes, but two-dimensional phosphopeptide maps of InsP$_3$R trypsin digest products showed that, in each case, different sites on the InsP$_3$R peptide were phosphorylated (Ferris *et al.*, 1991b).

Analysis of the InsP$_3$R primary structure revealed the presence of two putative PKA phosphorylation sites (Furuichi *et al.*, 1989; Mignery *et al.*, 1990), both in the coupling domain of the receptor (see Fig. 2). If low PKA concentrations were used in the experiment, only one of these sites in the cerebellar InsP$_3$R (Ser 1756) was phosphorylated. However, at much higher PKA concentrations, another site (Ser 1589) was also phosphorylated, resulting in transfer of two phosphate residues to each subunit of the InsP$_3$R (Ferris *et al.*, 1991a). Note that the amino acid sequence between these two phosphorylation sites is alternatively spliced; a stretch of 39 amino acid residues is deleted in the form of the receptor expressed in nonneuronal tissues (Danoff *et al.*, 1991). What effect, if any, does this splicing event have on the phosphorylation of InsP$_3$R by PKA? To address this question, InsP$_3$R was purified to homogeneity from cerebellum (the long, neuronal form) and from vas deferens (the short, nonneuronal form); phosphorylation of these proteins by PKA was monitored. The InsP$_3$R from cerebellum was phosphorylated by low concentrations of PKA primarily on Ser 1756, whereas the receptor from vas deferens was phosphorylated almost exclusively on Ser 1589 (Danoff *et al.*, 1991). Potentially, this difference may form the basis for tissue-specific regulation of InsP$_3$R function by cAMP-mediated signaling pathways.

What effect does phosphorylation have on the functional characteristics of the InsP$_3$R? This question has been addressed for cAMP-dependent phosphorylation using cerebellar microsomes (Supattapone *et al.*, 1988; Volpe and Alderson-Lang, 1990) and permeabilized hepatocytes (Burgess *et al.*, 1991). Phosphorylation of the InsP$_3$R from cerebellum using the catalytic subunit of PKA caused a 10-fold (Supattapone *et al.*, 1988) or 2-fold (Volpe and Alderson-Lang, 1990) shift to the right in the concentration dependence of InsP$_3$-induced Ca^{2+} release without any effect on InsP$_3$ binding to the receptor, that is, higher concentrations of InsP$_3$ are needed to achieve the same response as that

obtained before phosphorylation. Thus, the ability of InsP$_3$ to open the channel while bound (the efficacy of InsP$_3$) is altered by InsP$_3$R phosphorylation, as predicted from the location of the putative PKA phosphorylation sites in the coupling domain of the receptor (see Fig. 2). Interestingly, phosphorylation of liver InsP$_3$R by PKA caused a 4-fold shift to the left in the concentration dependence of InsP$_3$-induced Ca^{2+} release in permeabilized hepatocytes (lower InsP$_3$ concentrations are needed to achieve the same response) (Burgess *et al.*, 1991). This observation may be the functional manifestation of the differences in PKA phosphorylation sites between neuronal and nonneuronal splice forms of InsP$_3$R, as discussed earlier. Some increase in the total amount of Ca^{2+} released by InsP$_3$ as a result of phosphorylation was also observed in the experiments with cerebellar microsomes and hepatocytes, probably due to stimulation of the Ca^{2+} pump (Supattapone *et al.*, 1988; Burgess *et al.*, 1991; but see Volpe and Alderson-Lang, 1990).

Phosphorylation of the InsP$_3$R by PKA provides a means by which the hormones that affect adenylate cyclase activity could regulate the function of the inositol phospholipid pathway and Ca^{2+} mobilization, and therefore could create the basis for cross-talk between the InsP$_3$- and cAMP-mediated second messenger pathways. Phosphorylation of the InsP$_3$R by PKC and CAM-K-II may also provide additional pathways for feedback regulation of the phosphoinositide transduction mechanism. Activation of PLC in the phosphoinositide cycle signaling pathway generates hydrophobic diacylglycerol (DAG) and the cytoplasmic diffusible messenger InsP$_3$. DAG stimulates the activity of PKC. Production of InsP$_3$ causes an increase in intracellular Ca^{2+} levels due to InsP$_3$-induced Ca^{2+} release; this elevation in the intracellular Ca^{2+} provides additional activation of PKC and turns on CAM-K-II. If the activity of the InsP$_3$R is indeed regulated by PKC- and/or CAM-K-II-mediated phosphorylation, this aspect of receptor phosphorylation could function as a slow feedback mechanism that operates on a different time scale than the immediate and direct regulation of InsP$_3$R function by intracellular Ca^{2+}, as described in previous sections. Interestingly, immunohistochemical experiments demonstrated that the cerebellum-specific form of CAM-K-II is highly concentrated in Purkinje cells (Ouimet *et al.*, 1984; Fukunaga *et al.*, 1988) where the density of InsP$_3$ is the highest (Ross *et al.*, 1989). A variable stoichiometry between PKC and InsP$_3$R in different parts of the brain was reported (Worley *et al.*, 1987a), suggesting the possibility of regional specificity in this form of regulation. Despite the fact that all these suggestions seem very plausible, serious consideration is not possible until the functional consequences of InsP$_3$R phosphorylation are clearly demonstrated.

Autophosphorylation of the purified and reconstituted InsP$_3$R from cerebellum also has been demonstrated (Ferris *et al.*, 1992). This autophosphorylation is magnesium dependent with a maximum stoichiometry of about 0.4 mol phosphate per InsP$_3$R subunit, implying that only a fraction of the receptors is capable of autophosphorylation. In experiments with purified InsP$_3$R or with a synthetic substrate, a serine residue was phosphorylated. This protein appears to be the first example of a receptor with serine autophosphorylation activity; most receptors have tyrosine protein kinase activity (Hanks *et al.*, 1988). The ability of InsP$_3$R to phosphorylate itself does not appear to be regulated by any kind of intracellular messenger (such as Ca^{2+}, InsP$_3$, cAMP, or cGMP); the physiological significance of this process is not yet clear.

VI. Summary

In this review the functional properties and regulation of the InsP$_3$R have been described. This protein is a ligand-gated Ca^{2+}-permeable channel associated with intracellular membranes, predominantly the endoplasmic reticulum. Activation of this protein opens a pore for Ca^{2+}, which can then diffuse from the lumen of the endoplasmic reticulum to the cytoplasm.

Each channel is a tetramer consisting of four subunits of 313 kDa each. Each subunit contains a ligand-binding domain, a transmembrane (channel forming) region, and a coupling domain. Multiple splice variants and isoforms of the receptor exist, probably with different tissue distributions and functional properties.

Binding InsP$_3$ to its receptor is essential for the channel to open. Even at very high InsP$_3$ concentrations, the channels do not open more than 10–15% of the time. This result leads to the conclusion that InsP$_3$ is a partial agonist of these channels. Although InsP$_3$ must bind to the receptor to open the channel, how many InsP$_3$ molecules must bind to the tetrameric channel complex for channel activation is not clear.

The activity of these channels is regulated by cytoplasmic Ca^{2+}. A bell-shaped dependence of InsP$_3$-gated channel activity and Ca^{2+} release was observed, with a maximum activity between 200 and 300 nM free Ca^{2+}. Ca^{2+} binding to the receptor itself or to some other protein may be necessary to cause these effects. Both positive and negative feedback effects of Ca^{2+} on the InsP$_3$R are reversible. The immediate and biphasic regulation of InsP$_3$R function by cytoplasmic Ca^{2+} level appears to be very important in determining spatiotemporal characteristics of Ca^{2+} signaling in the cell.

Submillimolar concentrations of ATP and nonhydrolyzable ATP analogs enhance the activity of InsP$_3$-gated channels about 10-fold. The addition of ATP increases the frequency and the average duration of channel openings, but channel conductance and InsP$_3$ binding are unchanged. ATP presumably acts by binding to an allosteric regulatory site in the coupling domain of the receptor.

These channels may be regulated by intraluminal Ca^{2+} levels, where the degree of depletion of the Ca^{2+} store causes a decrease in the sensitivity of the receptors to InsP$_3$. However, these data are controversial, and are difficult to interpret because the level of luminal Ca^{2+} has not been quantitated.

The InsP$_3$R is stoichiometrically phosphorylated in an additive fashion by PKA, PKC, and CAM-K-II on serine residues. Phosphorylation of InsP$_3$R may act as a mechanism of cross-talk between phosphoinositide- and cAMP-mediated messenger pathways, and may provide a mechanism for slow feedback control over the course of InsP$_3$-induced Ca^{2+} release. The functional consequences of these phosphorylation events *in vivo* are presently under investigation by several laboratories.

Information about the structure, heterogeneity, functional properties, and regulation of the InsP$_3$R is useful in attempting to understand the spatiotemporal aspects of Ca^{2+} signaling by cells. The combination of biochemical, biophysical, and molecular biological techniques has revealed the intricacies of the InsP$_3$R over the past decade. Future investigations will uncover additional facets of this vital receptor.

Acknowledgments

We thank S. Bezprozvannaya for expert technical assistance, and Ed Kaftan for comments on the manuscript. This work was supported by National Institutes of Health Grants HL 33026 and GM 39029.

References

Abe, K., Kogure, K., Yamamoto, H., Imazawa, M., and Miyamoto, K. (1987). Mechanism of arachidonic acid liberation during ischemia in gerbil cerebral cortex. *J. Neurochem.* **48,** 503–509.

Anderson, K., Lai, F. A., Liu, Q.-Y., Rousseau, E., Erickson, H. P., and Meissner, G. (1989). Structural and functional characterization of the purified cardiac ryanodine receptor–Ca^{2+} release channel complex. *J. Biol. Chem.* **264,** 1329–1335.

Baukal, A. J., Guillemette, G., Rubin, R., Spat, A., and Catt, K. J. (1985). Binding sites for inositol trisphosphate in the bovine adrenal cortex. *Biochem. Biophys. Res. Commun.* **133,** 532–538.

Berridge, M. J. (1990). Calcium oscillations. *J. Biol. Chem.* **265**, 9583–9586.

Berridge, M. J. (1991). Cytoplasmic calcium oscillations: A two pool model. *Cell Calcium* **12**, 63–72.

Berridge, M. J. (1993). Inositol trisphosphate and calcium signalling. *Nature (London)* **361**, 315–325.

Berridge, M. J., and Irvine, R. F. (1989). Inositol phosphates and cell signalling. *Nature (London)* **341**, 197–205.

Bertolino, M., and Llinas, R. R. (1992). The central role of voltage-activated and receptor-operated calcium channels in neuronal cells. *Annu. Rev. Pharmacol. Toxicol.* **32**, 399–421.

Bezprozvanny, I., and Ehrlich, B. E. (1993a). ATP modulates the function of inositol 1,4,5-trisphosphate-gated channels at two sites. *Neuron* **10**, 1175–1184.

Bezprozvanny, I., and Ehrlich, B. E. (1993b). Divalent cation conduction of the inositol 1,4,5-trisphosphate gated calcium channels of canine cerebellum. *Biophys. J.* **64**, A328.

Bezprozvanny, I., Watras, J., and Ehrlich, B. E. (1991). Bell-shaped calcium-response curves of Ins(1,4,5)P$_3$- and calcium-gated channels from endoplasmic reticulum of cerebellum. *Nature (London)* **351**, 751–754.

Bezprozvanny, I. B., Ondrias, K., Kaftan, E., Stoyanovsky, D. A., and Ehrlich, B. E. (1993). Activation of the calcium release channel (ryanodine receptor) by heparin and other polyanions is calcium dependent. *Mol. Biol. Cell* **4**, 347–352.

Burgess, G. M., Bird, G. S. J., Obie, J. F., and Putney, J. W. (1991). The mechanism for synergism between phospholipase-C-linked and adenylylcyclase-linked hormones in liver—Cyclic AMP-dependent kinase augments inositol trisphosphate-mediated Ca^{2+} mobilization without increasing the cellular levels of inositol polyphosphates. *J. Biol. Chem.* **266**, 4772–4781.

Carafoli, E. (1987). Intracellular calcium homeostasis. *Annu. Rev. Biochem.* **56**, 395–433.

Chadwick, C. C., Saito, A., and Fleischer, S. (1990). Isolation and characterization of the inositol trisphosphate receptor from smooth muscle. *Proc. Natl. Acad. Sci. USA* **87**, 2132–2136.

Combettes, L., Claret, M., and Champeil, P. (1992). Do submaximal InsP$_3$ concentrations only induce the partial discharge of permeabilized hepatocyte calcium pools because of the concomitant reduction of intraluminal Ca^{2+} concentration? *FEBS Lett.* **301**, 287–290.

Danoff, S. K., Supattapone, S., and Snyder, S. H. (1988). Characterization of a membrane protein from brain mediating the inhibition of inositol 1,4,5-trisphosphate receptor binding by calcium. *Biochem. J.* **254**, 701–705.

Danoff, S. K., Ferris, C. D., Donath, C., Fischer, G. A., Munemitsu, S., Ullrich, A., Snyder, S. H., and Ross, C. A. (1991). Inositol 1,4,5-trisphosphate receptors: Distinct neuronal and nonneuronal forms derived by alternative splicing differ in phosphorylation. *Proc. Natl. Acad. Sci. USA* **88**, 2951–2955.

De Young, G. W., and Keizer, J. (1992). A single-pool inositol 1,4,5-trisphosphate-receptor-based model for agonist-stimulated oscillations in Ca concentration. *Proc. Natl. Acad. Sci. USA* **89**, 9895–9899.

Ehrlich, B. E., and Watras, J. (1988). Inositol 1,4,5-trisphosphate activates a channel from smooth muscle sarcoplasmic reticulum. *Nature (London)* **336**, 583–586.

Fabiato, A. (1988). Computer programs for calculating total from specified free or free from specified total ionic concentrations in aqueous solutions containing multiple metals and ligands. *Meth. Enzymol.* **157**, 378–417.

Ferris, C. D., Huganir, R. L., Supattapone, S., and Snyder, S. H. (1989). Purified inositol 1,4,5-trisphosphate receptor mediates calcium flux in reconstituted lipid vesicles. *Nature (London)* **342**, 87–89.

Ferris, C. D., Huganir, R. L., and Snyder, S. H. (1990). Calcium flux mediated by purified inositol 1,4,5-trisphosphate receptor in reconstituted lipid vesicles is allosterically regulated by adenine nucleotides. *Proc. Natl. Acad. Sci. USA* **87**, 2147–2151.

Ferris, C. D., Cameron, A. M., Bredt, D. S., Huganir, R. L., and Snyder, S. S. (1991a). Inositol 1,4,5-trisphosphate receptor is phosphorylated by cyclic AMP-dependent protein kinase at serines 1755 and 1589. *Biochem. Biophys. Res. Commun.* **175**, 192–198.

Ferris, C. D., Huganir, R. L., Bredt, D. S., Cameron, A. M., and Snyder, S. H. (1991b). Inositol trisphosphate receptor—Phosphorylation by protein kinase-C and calcium calmodulin-dependent protein kinases in reconstituted lipid vesicles. *Proc. Natl. Acad. Sci. USA* **88**, 2232–2235.

Ferris, C. D., Cameron, A. M., Bredt, D. S., Huganir, R. L., and Snyder, S. H. (1992). Autophosphorylation of inositol 1,4,5-trisphosphate receptors. *J. Biol. Chem.* **267**, 7036–7041.

Finch, E. A., Turner, T. J., and Goldin, S. M. (1991). Calcium as a coagonist of inositol 1,4,5-trisphosphate-induced calcium release. *Science* **252**, 443–446.

Fukunaga, K., Goto, S., and Miyamoto, E. (1988). Immunocytochemical localization of calcium/calmodulin-dependent protein kinase II in rat brain and various tissues. *J. Neurochem.* **51**, 1070–1078.

Furuichi, T., Yoshikawa, S., Miyawaki, A., Wada, K., Maeda, N., and Mikoshiba, K. (1989). Primary structure and functional expression of the inositol 1,4,5-trisphosphate-binding protein P400. *Nature (London)* **342**, 32–38.

Galione, A., Lee, H. C., and Busa, W. B. (1991). Ca^{2+}-induced Ca^{2+} release in sea urchin egg homogenates—modulation by cyclic ADP-ribose. *Science* **253**, 1143–1146.

Ghosh, T. K., Eis, P. S., Mullaney, J. M., Ebert, C. L., and Gill, D. L. (1988). Competitive, reversible, and potent antagonism of inositol 1,4,5-trisphosphate-activated calcium release by heparin. *J. Biol. Chem.* **263**, 11075–11079.

Goldbeter, A., Cupont, G., and Berridge, M. J. (1990). Minimal model for signal-induced Ca oscillations and for their frequency encoding through protein phosphorylation. *Proc. Natl. Acad. Sci. USA* **87**, 1461–1465.

Hanks, S. K., Quinn, A. M., and Hunter, T. (1988). The protein kinase family: Conserved features and deduced phylogeny of the catalytic domains. *Science* **241**, 42–52.

Harootunian, A. T., Kao, J. P. Y., Paranjape, S., and Tsien, R. Y. (1991). Generation of calcium oscillations in fibroblasts by positive feedback between calcium and IP$_3$. *Science* **251**, 75–78.

Henzi, V., and MacDermott, A. B. (1992). Characteristics and function of Ca^{2+}- and inositol 1,4,5-trisphosphate-releasable stores of Ca^{2+} in neurons. *Neuroscience* **46**, 251–273.

Hingorani, S., and Agnew, W. S. (1991). A rapid ion-exchange assay for detergent-solubilized inositol 1,4,5-trisphosphate receptors. *Anal. Biochem.* **194**, 204–213.

Hingorani, S. R., and Agnew, W. S. (1992). Assay and purification of neuronal receptors for inositol 1,4,5-trisphosphate. *Meth. Enzymol.* **207**, 573–591.

Iino, M. (1990). Biphasic Ca^{2+} dependence of inositol 1,4,5-trisphosphate-induced Ca release in smooth muscle cells of the guinea pig *Taenia caeci*. *J. Gen. Physiol.* **95**, 1103–1122.

Iino, M. (1991). Effects of adenine nucleotides on inositol 1,4,5-trisphosphate-induced calcium release in vascular smooth muscle cells. *J. Gen. Physiol.* **98**, 681–698.

Iino, M., and Endo, M. (1992). Calcium-dependent immediate feedback control of inositol 1,4,5-trisphosphate-induced Ca release. *Nature (London)* **360**, 76–78.

Irvine, R. F. (1990). "Quantal" Ca release and the control of Ca entry by inositol phosphates—A possible mechanism. *FEBS Lett.* **263**, 5–9;.

Katz, A. M. (1992). "Physiology of the Heart," 2d Ed. Raven Press, New York.

Keizer, J., and Deyoung, G. W. (1992). Two roles for Ca²⁺ in agonist stimulated Ca²⁺ oscillations. *Biophys. J.* **61,** 649–660.

Kindman, L. A., and Meyer, T. (1993). Use of intracellular Ca²⁺ stores from rat basophilic leukemia cells to study the molecular mechanism leading to quantal Ca²⁺ release by inositol 1,4,5-trisphosphate. *Biochemistry* **32,** 1270–1277.

Kobayashi, S., Somlyo, A. V., and Somlyo, A. P. (1988). Heparin inhibits the inositol 1,4,5-triphosphate-dependent, but not the independent, calcium release induced by guanine nucleotide in vascular smooth muscle. *Biochem. Biophys. Res. Commun.* **153,** 625–631.

Kostyuk, P. G. (1986). "Calcium and Cell Excitability." Visshaya Schola, Moscow. (in Russian)

Krause, K. H., Pittet, D., Volpe, P., Pozzan, T., Meldolesi, J., and Lew, D. P. (1989). Calciosome, a sarcoplasmic reticulum-like organelle involved in intracellular Ca²⁺ handling by non-muscle cells: Studies in human neutrophils and HL-60 cells. *Cell Calcium* **10,** 351–361.

Kume, S., Muto, A., Aruga, J., Nakagawa, T., Michikawa, T., Furuichi, T., Nakade, S., Okano, H., and Mikoshiba, K. (1993). The *Xenopus* IP₃ receptor: Structure, function, and localization in oocytes and eggs. *Cell* **73,** 555–570.

Lai, F. A., Erickson, H. P., Rousseau, E., Liu, Q.-Y., and Meissner, G. (1988). Purification and reconstitution of the calcium release channel from skeletal muscle. *Nature (London)* **331,** 315–319.

Lechleiter, J. D., and Clapham, D. E. (1992). Molecular mechanisms of intracellular calcium excitability in *X. laevis* oocytes. *Cell* **69,** 283–294.

Lechleiter, J., Girard, S., Peralta, E., and Clapham, D. (1991). Spiral calcium wave propagation and annihilation in *Xenopus laevis* oocytes. *Science* **252,** 123–126.

Lipscombe, D., Madison, D. V., Poenie, M., Reuter, H., Tsien, R. Y., and Tsien, R. W. (1988). Imaging of cytosolic Ca transients arising from Ca stores and Ca channels in sympathetic neurons. *Neuron* **1,** 355–365.

Maeda, N, Kawasaki, T., Nakade, S., Yokota, N., Taguchi, T., Kasai, M., and Mikoshiba, K. (1991). Structural and functional characterization of inositol 1,4,5-trisphosphate receptor channel from mouse cerebellum. *J. Biol. Chem.* **266,** 1109–1116.

Marks, A. R., Tempst, P., Chadwick, C. C., Riviere, L., Fleischer, S., and Nadal-Ginard, B. (1990). Smooth muscle and brain inositol 1,4,5-trisphosphate receptors are structurally and functionally similar. *J. Biol. Chem.* **265,** 20719–20722.

Mayrleitner, M., Chadwick, C. C., Timerman, A. P., Fleischer, S., and Schindler, H. (1991). Purified IP₃ receptor from smooth muscle forms an IP₃ gated and heparin sensitive Ca²⁺ channel in planar bilayers. *Cell Calcium* **12,** 505–514.

McPherson, P. S., and Campbell, K. P. (1990). Solubilization and biochemical characterization of the high affinity [³H]ryanodine receptor from rabbit brain membranes. *J. Biol. Chem.* **265,** 18454–18460.

Meyer, T., and Stryer, L. (1988). Molecular model for receptor-stimulated calcium spiking. *Proc. Natl. Acad. Sci. USA* **85,** 5051–5055.

Meyer, T., Holowka, D., and Stryer, L. (1988). Highly cooperative opening of calcium channels by inositol 1,4,5-trisphosphate. *Science* **240,** 653–656.

Mignery, G. A., and Sudhof, T. C. (1990). The ligand binding site and transduction mechanism in the inositol-1,4,5-triphosphate receptor. *EMBO J.* **9,** 3893–3898.

Mignery, G., Sudhof, T. C., Takei, K., and De Camilli, P. (1989). Putative receptor for inositol 1,4,5-trisphosphate similar to ryanodine receptor. *Nature (London)* **342,** 192–195.

Mignery, G. A., Newton, C. L., Archer, B. T., and Sudhof, T. C. (1990). Structure and expression of the rat inositol 1,4,5-trisphosphate receptor. *J. Biol. Chem.* **265,** 12679–12685.

Mignery, G. A., Johnston, P. A., and Sudhof, T. C. (1992). Mechanism of Ca²⁺-inhibition of InsP₃ binding to the cerebellar InsP₃ receptor. *J. Biol. Chem.* **267,** 7450–7455.

Missiaen, L., De Smedt, H., Droogmans, G., and Casteels, R. (1992a). Ca release induced by inositol 1,4,5-trisphosphate is a steady-state phenomenon controlled by luminal Ca in permeabilized cells. *Nature (London)* **357,** 599–602.

Missiaen, L., De Smedt, H., Droogmans, G., and Casteels, R. (1992b). Luminal Ca controls the activation of the inositol 1,4,5-trisphosphate receptor by cytosolic Ca. *J. Biol. Chem.* **267,** 22961–22966.

Miyawaki, A., Furuichi, T., Ryou, Y., Yoshikawa, S., Nakagawa, T., Saitoh, T., and Mikoshiba, K. (1991). Structure-function relationships of the mouse inositol 1,4,5-trisphosphate receptor. *Proc. Natl. Acad. Sci. USA* **88,** 4911–4915.

Miyazaki, S., Yuzaki, M., Nakada, K., Shirakawa, H., Nakanishi, S., Nakade, S., and Mikoshiba, K. (1992). Block of Ca wave and Ca oscillation by antibody to the inositol 1,4,5-trisphosphate receptor in fertilized hamster eggs. *Science* **257,** 251–255.

Mourey, R. J., Verma, A., Supattapone, S., and Snyder, S. H. (1990). Purification and characterization of the inositol 1,4,5-trisphosphate receptor protein from rat vas deferens. *Biochem. J.* **272,** 383–389.

Muallem, S., Pandol, S. J., and Beeker, T. G. (1989). Hormone-evoked calcium release from intracellular stores is a quantal process. *J. Biol. Chem.* **264,** 205–212.

Nakagawa, T., Okano, H., Furuichi, T., Aruga, J., and Mikoshiba, K. (1991a). The subtypes of the mouse Inositol 1,4,5-trisphosphate receptor are expressed in a tissue-specific and developmentally specific manner. *Proc. Natl. Acad. Sci. USA* **88,** 6244–6248.

Nakagawa, T., Shiota, C., Okano, H., and Mikoshiba, K. (1991b). Differential localization of alternative spliced transcripts encoding inositol 1,4,5-trisphosphate receptors in mouse cerebellum and hippocampus—*in situ* hybridization study. *J. Neurochem.* **57,** 1807–1810.

Nordquist, D. T., Kozak, C. A., and Orr, H. T. (1988). cDNA cloning and characterization of three genes uniquely expressed in cerebellum by Purkinje neurons. *J. Neurosci.* **8,** 4780–4789.

Nunn, D. L., and Taylor, C. W. (1992). Luminal Ca²⁺ increases the sensitivity of Ca²⁺ stores to inositol 1,4,5-trisphosphate. *Mol. Pharmacol.* **41,** 115–119.

Otsu, K., Willard, H. F., Khanna, V. K., Zorzato, F., Green, N. M., and MacLennan, D. H. (1990). Molecular cloning of cDNA encoding the Ca²⁺ release channel (ryanodine receptor) of rabbit cardiac muscle sarcoplasmic reticulum. *J. Biol. Chem.* **265,** 13472–13483.

Quimet, C. C., McGuinness, T. L., and Greengard, P. (1984). Immunocytochemical localization of calcium/calmodulin-dependent protein kinase II in rat brain. *Proc. Natl. Acad. Sci. USA* **81,** 5604–5608.

Parker, I., and Yao, Y. (1991). Regenerative release of calcium from functionally discrete subcellular stores by inositol trisphosphate. *Proc. R. Soc. London B* **246,** 269.

Parys, J. B., Sernett, S. W., DeLisle, S., Snyder, P. M., Welsh, M. J., and Campbell, K. P. (1992). Isolation, characterization and localization of the inositol 1,4,5-trisphosphate receptor protein in *Xenopus laevis* oocytes. *J. Bio. Chem.* **267,** 18776–18782.

Pietri, F., Hilly, M., and Mauger, J.-P. (1990). Calcium mediates the interconversion between two states of the liver inositol 1,4,5-trisphosphate receptor. *J. Biol. Chem.* **265,** 17478–17485.

Pietri Rouxel, F., Hilly, M., and Mauger, J.-P. (1992). Characterization of a rapidly dissociating inositol 1,4,5-trisphosphate-binding site in liver membranes. *J. Biol. Chem.* **267**, 20017–20023.

Prestwich, G. D., Marecek, J. F., Mourey, R. J., Theibert, A. B., Ferris, C. D., Danoff, S. K., and Snyder, S. H. (1991). Tethered IP$_3$. Synthesis and biochemical applications of the 1-*O*-(3-aminopropyl) ester of inositol 1,4,5-trisphosphate. *J. Am. Chem. Soc.* **113**, 1822–1825.

Ritov, V. B., Men'shikova, E. V., and Kozlov, Y. P. (1985). Heparin induces Ca$^+$ release from the terminal cysterns of skeletal muscle sarcoplasmic reticulum. *FEBS Lett.* **188**, 77–80.

Ross, C. A., Meldolesi, J., Milner, T. A., Satoh, T., Supattapone, S., and Snyder, S. H. (1989). Inositol 1,4,5-trisphosphate receptor localized to endoplasmic reticulum in cerebellar Purkinje neurons. *Nature* (*London*) **339**, 468–470.

Ross, C. A., Danoff, S. K., Schell, M. J., Snyder, S. H., and Ullrich, A. (1992). Three additional inositol 1,4,5-trisphosphate receptors: Molecular cloning and differential localization in brain and peripheral tissues. *Proc. Natl. Acad. Sci. USA* **89**, 4265–4269.

Shoshan-Barmatz, V., Zhang, G. H., Garretson, L., and Krauss-Friedmann, N. (1990). Distinct ryanodine- and inositol 1,4,5-trisphosphate-binding sites in hepatic microsomes. *Biochem. J.* **268**, 699–705.

Shuttleworth, T. J. (1992). Ca^{2+} release from inositol trisphosphate-sensitive stores is not modulated by intraluminal Ca^{2+}. *J. Biol. Chem.* **267**, 3573–3576.

Smith, J. B., Smith, L., and Higgins, B. L. (1985). Temperature and nucleotide dependence of calcium release by myo-inositol 1,4,5-trisphosphate in cultured vascular smooth muscle cells *J. Biol. Chem.* **260**, 14413–14416.

Smith, J. S., Imagawa, T., Ma, J., Fill, M., Campbell, K. P., and Coronado, R. (1988). Rurified ryanodine receptor from rabbit skeletal muscle is the calcium release-channel of sarcoplasmic reticulum. *J. Gen. Physiol.* **92**, 1–26.

Spat, A., Bradford, P. G., McKinney, J. S., Rubin, R. P., and Putney, J. W., Jr (1986). A saturable receptor for ^{32}P-inositol-1,4,5-trisphosphate in hepatocytes and neutrophils. *Nature* (*London*) **319**, 514–516.

Spat, A., Eberhardt, I., and Kiesel, L. (1992). Low concentrations of adenine nucleotides enhance the receptor binding of inositol 1,4,5-trisphosphate. *Biochem. J. Lett.* **287**, 335–336.

Streb, H., Irvine, R. F., Berridge, M. J., and Shulz, I. (1983). Release of Ca^{2+} from a nonmitochondrial store in pancreatic acinar cells by inositol-1,4,5-trisphosphate. *Nature* (*London*) **306**, 67–69.

Sudhof, T. C., Newton, C. L., Archer, B. T., Ushkaryov, Y. A., and Mignery, G. A. (1991). Structure of a novel InsP$_3$ receptor. *EMBO J.* **10**, 3199–3206.

Supattapone, S., Worley, P. F., Baraban, J. M., and Snyder, S. H. (1988). Solubilization, purification, and characterization of an inositol trisphosphate receptor. *J. Biol. Chem.* **263**, 1530–1534.

Takeshima, H., Nishimura, S., Matsumoto, T., Ishida, H., Kangawa, K., Minamino, N., Matsuo, H., Ueda, M., Hanaoka, M., Hirose, T., and Numa, S. (1989). Primary structure and expression from complementary DNA of skeletal muscle ryanodine receptor. *Nature* (*London*) **339**, 439–445.

Taylor, C. W., and Marshall, C. B. (1992). Calcium and inositol 1,4,5-trisphosphate receptors: a complex relationship. *Trends Biochem. Sci.* **17**, 403–407.

Taylor, C. W., and Richardson, A. (1991). Structure and function of inositol trisphosphate receptors. *Pharmacol. Therapeut.* **51**, 97–137.

Tinker, A., and Williams, A. J. (1992). Divalent cation conduction in the ryanodine receptor of sheep cardiac muscle sarcoplasmic reticulum. *J. Gen. Physiol.* **100**, 479–493.

Tsien, R. W., and Tsien, R. Y. (1990). Calcium channels, stores and oscillations. *Annu. Rev. Cell. Biol.* **6**, 715–760.

Tsien, R. W., Lipscombe, D., Madison, D. V., Bley, K. P., and Fox, A. P. (1988). Multiple types of neuronal calcium channels and their selective modulation. *Trends Neurosci.* **11**, 431–438.

Volpe, P., and Alderson-Lang, B. H. (1990). Regulation of inositol 1,4,5-trisphosphate-induced Ca^{2+} release. II. Effect of cAMP-dependent protein kinase. *Am. J. Physiol.* **258**, C1086–C1091.

Walaas, S. I., Nairn, A. C., and Greengard, P. (1986). PCPP-260, a Purkinje cell-specific cyclic AMP-regulated membrane phosphoprotein of Mr 260,000. *J. Neurosci.* **6**, 954–961.

Watras, J., Bezprozvanny, I., and Ehrlich, B. E. (1991). Inositol 1,4,5-trisphosphate-gated channels in cerebellum—Presence of multiple conductance states. *J. Neurosci.* **11**, 3239–3245.

Worley, P. F., Baraban, J. M., Colvin, J. S., and Snyder, S. H. (1987a). Inositol trisphosphate receptor localization in brain: variable stoichiometry with protein kinase C. *Nature* (*London*) **325**, 159–161.

Worley, P. F., Baraban, J. M., Supattapone, S., Wilson, V., and Snyder, S. H. (1987b). Characterization of inositol trisphosphate receptor binding in brain. *J. Biol. Chem.* **262**, 12132–12136.

Yamamoto, H., Maeda, N., Niinobe, M., Miyamoto, E., and Mikoshiba, K. (1989). Phosphorylation of P400 protein by cyclic AMP-dependent protein kinase and Ca/calmodulin-dependent protein kinase II. *J. Neurochem.* **53**, 917–923.

Yoshikawa, S., Tanimura, T., Miyawaki, A., Nakamura, M., Yuzaki, M., Furuichi, T., and Mikoshiba, K. (1992). Molecular cloning and characterization of the inositol 1,4,5-trisphosphate receptor in *Drosophila melanogaster*. *J. Biol. Chem.* **267**, 16613–16619.

B

Mitochondrial Channels

34

Mitochondrial Membrane Channels

HENRY TEDESCHI AND
KATHLEEN W. KINNALLY

I. Historical Perspective

The key reactions of the energy metabolism of most eukaryotes occur in mitochondria. The free energy from the oxidation of substrates drives the synthesis of ATP. These reactions take place in the inner mitochondrial membrane and require the maintenance of ionic gradients. These considerations, and direct studies of mitochondrial permeability, indicate that the inner mitochondrial membrane is highly impermeable. Metabolites and ions are transferred by special transport mechanisms. The outer mitochondrial membrane provides an interface between the inner membrane and cytoplasm, and is generally thought to act as a sieve allowing the passage of solutes smaller than 2 kDa. Channels have been identified in the inner and outer mitochondrial membranes. Although their function is still poorly understood, clearly they are highly regulated and must play an important role in the physiology of the cell.

Studies of the inner membrane were performed similarly to studies of the plasma membrane. Changes in permeability of the plasma membrane led to the postulate of channels gated by voltage or by ligands. This hypothesis was followed by the demonstration of channels as discrete biochemically distinguishable units by pharmacology. The introduction of patch-clamping allowed the detailed study of individual channel activities. The isolation of channel proteins, their reconstitution, and their study with the powerful tools of molecular biology rapidly followed. Except for the putative channels involved in energy transduction [i.e., components of the ATP-synthetase system (see Section III,A,4) or the uncoupling protein (UCP; see Section III,A,3)], the molecular biology of inner membrane channels is in its infancy. In contrast, study of the outer membrane channels has followed an entirely different path. The isolation and reconstitution of the voltage-dependent anion channel (VDAC; mitochondrial porin) into planar bilayers was introduced from the very beginning as a primary research tool, so the molecular biological studies are well advanced.

In recent years, patch clamping has permitted the direct demonstration of channel activity in the outer and inner mitochondrial membranes. Some of the mitochondrial channels are voltage gated, and many respond to nanomolar concentrations of substances, some of which have biological significance, suggesting ligand gating as well.

Although much new information is now available, its interpretation and evaluation in terms of the nature of the channels and their significance in mitochondrial function is far from clear. Further, except for the notable case of VDAC and the proteins related to the transducing functions alluded to earlier, isolation and molecular studies of mitochondrial channels await further work.

II. Channels of the Outer Membrane

A. Voltage-Dependent Anion Channel Behavior, Structure, and Molecular Biology

After the initial discovery of VDAC in extracts of *Paramecium* (Schein *et al.*, 1976), the channel was localized to the outer mitochondrial membrane (Colombini, 1979). VDAC appears to be ubiquitous and has been detected in mitochondria from mammals, fish, insects, plants, fungi, and protozoans. VDAC has generally been found to be the major outer membrane protein. The channel is constituted by a polypeptide of 31 kDa. The VDAC polypeptides of yeast (Mihara and Sato, 1985; Forte *et al.*, 1987), *Neurospora* (Kleene *et al.*, 1987), and human cells (Kayser *et al.*, 1989) have been cloned and sequenced. These molecules correspond to polypeptides with 282–283 amino acids. The general properties and function of VDAC have been reviewed (see Colombini, 1986; Benz, 1990; Mannella, 1992).

The unit conductance of VDAC is large (~650 pS in 150 mM KCl) after reconstitution in planar bilayers following detergent extraction. The open channel is slightly anion selective and voltage dependent, exhibiting a half-closure to ~300 pS with voltage. Although the voltage dependence is variable, it is symmetrical or nearly symmetrical because half-closure occurs at positive and negative voltages. Unlike the fully open state, the half-closed state of ~300 pS is slightly cation selective.

VDAC contains domains of 10–14 amino acids that alternate polar and nonpolar residues. Based on this sequence, models have been proposed in which the amphipathic sequences are viewed as strands in a β-sheet configuration with the polar residues on one face and the nonpolar residues on the other (see Forte *et al.*, 1987; Blachly-Dyson *et al.*, 1990). The β sheet is folded into a cylinder with the polar residues on the inner and the nonpolar residues on the outer face. The inner lining would then be hydrophilic, and the hydrophobic exterior would be ideal for interacting with the hydrocarbon chains of the phospholipids. Another version of a yeast VDAC model is shown in Fig. 1. This model includes sites altered by site-directed mutagenesis. The boxed residues changed the ion selectivity in the open state, whereas those circled did not. The shaded boxed residues correspond to

those that have an effect on the half closed substate (see subsequent discussion). Originally, the number of transmembrane strands was thought to be as high as 19. However, the accessibility to proteases of mammalian VDAC (DePinto and Palmieri, 1992) and the selectivity effects of site-directed mutagenesis on yeast VDAC (Blachly-Dyson *et al.*, 1990) suggest that less of the protein comprises the walls of the lumen. The model in Fig. 1 is made up of only 12 β strands with a thirteenth region forming an amphipathic helix of 20 amino acids. However, forming a pore of the observed diameter (see subsequent discussion) with so few strands necessitates imposing a severe tilt (60°) on the strands, in which case they would not span the lipid bilayer (see Mannella *et al.*, 1992). These considerations suggest the need for further modification of the model.

Possible mechanisms for the voltage gating of VDAC have been considered (Mannella *et al.*, 1992). One model of channel closing involves the entry of a segment of VDAC into the lumen. An alternative model proposes removal of several strands from the barrel. Interestingly, the channel can be closed when reconstituted in planar bilayers by the addition of polymers to either side of the bilayer (Zimmerberg and Parsegian, 1986, 1987). If this effect is purely osmotic, the first mechanism would be unlikely. Site-directed mutagenesis (Peng *et al.*, 1992) identifies domains that affect the ion selectivity of the open state but not that of the half closed state. A possible explanation is that these sites are removed from the hydrophilic inner wall of the channel in the half closed conformation, although other interpretations are possible (see discussion by Peng *et al.*, 1992). For example, the residues may have become shielded by a large conformational change instead of being removed.

Much structural information has been provided by high resolution electron microscopic studies of two-dimensional VDAC crystals using computer image processing (Mannella, 1994). In *Neurospora*, crystals form when the VDAC polypeptides are concentrated by lipid depletion with phospholipase A_2 (Mannella, 1990). VDACs appear in ordered arrays, as shown in Fig. 2 obtained from a density map of a frozen-hydrated two-dimensional crystal. The lumen of the pore appears white, the protein appears dark (higher density), and the phospholipid has an intermediate density. Protein arms (inset) extend laterally between channel hexamers and are thought to correspond to the 20-amino acid α helices. Assuming the VDAC lumen is a β barrel, this density map is best fit by one with a diameter of 3.8 nm at the carbon backbone. Calculations from crystal packing (Mannella, 1987)

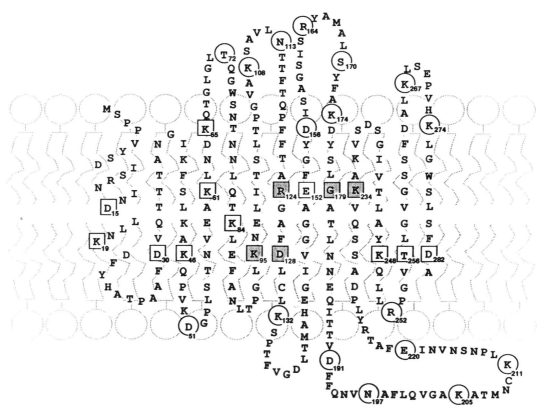

FIGURE 1 Schematic model of the voltage-dependent ionic channel (VDAC). The amino-terminal α helix is shown on the left. The other 12 strands are antiparallel β sheets. The marked amino acids corresponds to the locations of mutations. The boxed residues represent residues at which mutations alter selectivity in the open state. The circles residues indicate mutations that do not affect selectivity. Shaded residues indicate sites at which mutation affects selectivity in the closed state. Reproduced from the *Biophysical Journal*, (1992), **62**, 123–135, by copyright permission of the *Biophysical* Society.

and mass density measurements (Thomas *et al.*, 1991) indicate that each channel is formed by a single 31-kDa polypeptide.

B. Other Channels and/or Plasticity of VDAC Behavior

Several studies have provided evidence for the presence of channels other than VDAC. The evidence is conclusive only for the cases in which channels were demonstrated in VDAC-less mutants. A cation-selective, voltage-insensitive channel with a conductance of 300 pS was found in VDAC-less yeast mitochondria (see Dihanich *et al.*, 1989). Thieffry *et al.* (1988) observed a mitochondrial cation-selective channel incorporated into lipid bilayers formed at the tip of the micropipette. In 150 mM NaCl, the channel exhibits three conductance levels separated by 220 pS. The channel is probably in the outer membrane since it

is sensitive to trypsin digestion of intact mitochondria (Henry *et al.*, 1989). This channel is blocked by a 13-residue peptide from the N-terminal region of subunit IV of cytochrome *c* oxidase and, hence, is called the peptide sensitive channel (PSC; also referred to as voltage-dependent cation channel, VDCC). Since cytochrome oxidase is an inner membrane protein synthesized in the cytoplasm, a role for PSC in protein translocation has been suggested. PSC is distinct from VDAC since activity basically equivalent to that of PSC was demonstrated in membranes from yeast VDAC-less mutants (Thieffry *et al.*, 1992, 1994). In this study, proteoliposomes derived from mitochondrial membranes were incorporated into large liposomes and planar bilayers, either conventional or formed at the tip of the patching pipette.

Other observed channel activities could correspond to VDAC modified by modulators. The activity of VDAC can be modified by a synthetic polyanion that

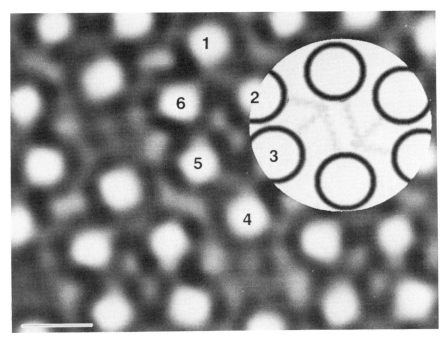

FIGURE 2 Density map of a frozen-hydrated two-dimensional crystal of *Neurospora* VDAC obtained by cryo-electron microscopy. The six channels in the central unit cell are numbered. In the circular inset, the channels are modeled as projections of circular β barrels with C_α backbone diameters of 3.8 nm. The arms in the interhexamer region are represented in the inset as 20-amino-acid α helices extending from each channel into the surrounding bilayer. The amino terminal of the VDAC polypeptide has been localized to this region by immunoelectron microscopy. However, other information suggests that it is part of the wall of the lumen (as represented in Fig. 1). Reproduced with permission from Mannella (1992).

is thought to correspond to a biological modulator. Further, a protein modulator has been isolated from mitochondria (Holden and Colombini, 1988; Liu and Colombini, 1991,1992). These two modulators increase the response of VDAC to voltage when reconstituted in planar bilayers. However, they can produce other effects as well. The addition of the polyanion to the *cis* side of a bilayer produces an asymmetrical voltage dependence, that is, closing will occur only at one polarity (Colombini *et al.*, 1987; Benz, 1990). Also, addition of the polyanion in large concentrations before incorporating VDAC into the bilayers has been reported to close VDAC even in the absence of applied voltage (Colombini *et al.*, 1987). In addition, VDAC has the potential for many substates (Ermishkin and Mirzabekov, 1990). Asymmetrical voltage dependence and several conductances (see subsequent discussion) have been shown for the outer membrane channels and could result from modulation of VDAC.

Early studies of the native outer membrane (see Tedeschi *et al.*, 1989, for review), were carried out at very low ionic strength. The rationale for this choice was that the high density of VDAC in the outer membrane predicted a very low resistance. These studies observed low resistances and changes in whole-patch

conductance that were consistent with the behavior of VDAC at (pipette) positive voltages, that is, a decrease in conductance with voltage to a partially closed level. VDAC was also implicated by the response, at positive voltage, to polyanion addition and the effect of succinic anhydride treatment. No discrete channel behavior was observed as expected from the high conductance, implying that several hundred channels were involved in each patch. At negative voltages, the conductance decreased only rarely and most frequently increased. These latter results were consistent with a biphasic effect: an initial activation of channels followed by an assembly into still larger channels. The involvement of protein channels in these effects is supported by experiments in which the patch is exposed to proteases that decrease the rectification and increase the resistance (B. Popp, K. W. Kinnally, and H. Tedeschi, unpublished results).

Many uncertainties have been introduced by the observation that patch clamping intact mitochondria has resulted in the occasional observation of patch resistances in the range of 0.5 GΩ and, at times, as high as 10 GΩ (Moran *et al.*, 1992; Kinnally and Tedeschi, 1994). Calculations of density of VDAC in mitochondria suggest several hundred in each patch (calculated, for example, from data of Freitag *et al.*, 1982; Lindén *et al.*, 1984; DePinto *et al.*, 1987). These observations suggest that, at least in these mitochondria, the majority of the VDAC are closed completely. The previous reports of much lower patch resistances (e.g., Tedeschi *et al.*, 1987; Kinnally *et al.*, 1987) suggest that VDAC can be activated under some conditions.

Wunder and Colombini (1991) reported patch-clamp results using whole *Neurospora* mitochondrial membranes incorporated into phosphatidylcholine and phosphatidic acid liposomes. These researchers observed a channel of 650 pS exhibiting the responses characteristic of VDAC. In contrast, Moran *et al.* (1992) observed several conductance levels ranging from less than 50 pS to about 300 pS (in 0.15 *M* KCl) when patching intact mitochondria; the same conductance levels were also detected when isolated outer membranes were incorporated into L-α-phosphatidylcholine liposomes. An additional conductance of 530 pS was also reported for the latter case. The voltage dependence of the channel activity was generally asymmetrical unlike that observed for purified VDAC incorporated into bilayers for which closure occurs at both positive and negative potentials.

Kinnally and Tedeschi (1994) have obtained results similar to reports with native outer membranes in intact mitochondria or with outer membranes reconstituted into liposomes. In most cases (~50%), the results resembled those of Moran *et al.* (1992). However, in ~10% of the cases, the results were very similar to those reported by Wunder and Colombini (1991), although the voltage dependence was less pronounced in many patches. A sample current trace of reconstituted 650-pS activity with a half-closed state is shown in Fig. 3.

Many of these observations could be explained by a degree of modulation of VDAC that is currently not recognized.

III. Channels of the Inner Membrane

As in the case of the channels of the plasma membrane, permeability studies of the inner mitochondrial membrane led to the postulate that channels underlie permeability transitions. These studies provided an enormous amount of information and generally preceded the more direct demonstration using patch-clamp techniques. For these reasons the study of permeability transitions is presented first.

A. Evidence from Permeability Studies

Permeability increases can be induced by a variety of means including alkaline pH, elevated Ca^{2+} level, or Mg^{2+}-depletion. These permeability transitions are most readily explained by assuming the activation of a channel.

1. Inner Membrane Anion Channel

The permeability transition induced by alkaline pH is consistent with the activation of the inner membrane anion-selective channel (IMAC; Brierley, 1970; Selwynn *et al.*, 1979; Beavis and Garlid, 1987). Patch clamping has also demonstrated the activation of discrete channels by alkaline pH (see below, Section III,B,2). Further, the conductance increase calculated for one mitochondrion using a few simple and rough approximations shows that a channel could be involved (see Table I). These approximations are correct in order of magnitude. The literature on IMAC has been reviewed elsewhere (Garlid and Beavis, 1986; Beavis, 1992).

IMAC apparently has a regulatory site on the matrix side of the membrane. Its pH activation is reversible and is inhibited by endogenous Mg^{2+}. Other than its anion over cation selectivity, IMAC lacks specificity and can transport many anions, although it favors smaller and multivalent anions. IMAC is present in a

FIGURE 3 Current trace of outer membrane channel activity obtained with patch-clamp techniques. Current was recorded from an excised patch from a phosphatidylcholine liposome fused with *Neurospora* outer membranes by the dehydration–rehydration method of Criado and Keller (1987), as previously reported (Costa *et al.*, 1991). Patch-clamp data were obtained under voltage-clamp conditions, as previously described (Campo *et al.*, 1992). Current data bandwidth, limited to 2 kHz (sampled at 5 kHz), was analyzed using the PAT program (courtesy J. Dempster, University of Strathclyde, UK). Transitions of 650 pS predominate, but also show a half-closed state in 150 m*M* KCl, 5 m*M* HEPES, 1 m*M* EGTA, 1.05 m*M* CaCl, pH 7.4 at −80 mV. *Neurospora crassa* outer membranes were the gift of C. A. Mannella (Wadsworth Center, New York State Dept. of Health) and were prepared as described by Mannella (1982). Voltages are relative to the matrix.

TABLE I Conductance Expected from Various Transport Mechanisms:
Feasible Application of Patch-Clamp Techniques to Native Systems[a]

Pore[b]	J (nmol min^{-1}/ mg protein)	mV	C (pS)	Feasible
IMAC	500[c]	15[d]	53	yes
K$^+$ uniport	4500[e]	200 (extrapolated)	36	yes
Ca^{2+} uniport	60[f]	80[d]	1.2	?
F$_0$	—	100 (pH 2.2)[g]	12[g]	yes
UCP	16[h]	1	2.7[i]	yes

[a] The calculations are approximate. They assume a mitochondrion 2 μm in diameter containing 1×10^{-9} mg protein. Using mitochondria of larger size would increase these values accordingly (e.g., 18× for a giant mitochondrion 5 μm in diameter). The calculations also assume whole mitoplast configuration.

[b] Abbreviations: IMAC, inner membrane anion channel; UCP, uncoupling protein.

[c] Beavis (1992).

[d] Calculated from K$^+$ diffusion potential from our data.

[e] Nicolli et al. (1991).

[f] Scarpa and Azzone (1970). Similar values were obtained using data of Gunter and Pfeiffer (1990).

[g] Schindler and Nelson (1982); single-channel conductance.

[h] Nicholls et al. (1986).

[i] Klitsch and Siemen (1991), using different assumption. Our calculation corresponds to ~25 pS.

variety of tissues including those of plants. Pharmacological studies of IMAC are extensive; a variety of chemicals block the channels including local anesthetics, benzodiazepines, and cationic amphiphiles (see Beavis, 1992, for review).

2. Permeability Transition Pore

A permeability transition of the inner mitochondrial membrane induced by Ca^{2+} has been reviewed by Gunter and Pfeiffer (1990). The transition requires Ca^{2+} accumulation in the matrix and is facilitated by the presence of an inducing agent. These agents are chemically unrelated, suggesting an indirect and complex process. The transition permits the passage of rather large molecules such as sucrose (e.g., Al-Nasser and Crompton, 1986), pyridine nucleotides (e.g., Gazzotti, 1975), and matrix proteins (Igbavboa et al. 1989). Since the activation of a permeability transition pore (PTP) is probably progressive (see Section III,B,3), the increases in permeability may be graded and its magnitude may depend on a variety of conditions. Researchers have proposed that the permeability increase results from the activation of a channel or pore (e.g., Haworth and Hunter, 1980). The process can be inhibited or reversed by a variety of agents including the chelation of Ca^{2+} or the presence of cyclosporin A, ADP, Mg^{2+}, reduced pyridine nucleo-

tides, or low pH. Evidence suggests that the channels act cooperatively (Crompton and Costi, 1988) and may be formed by assembly involving cross-linking by disulfide bridges (Fagian et al., 1990). The possible physiological role of this channel is not clear, but it has been implicated in ischemia–reperfusion injury.

Studies carried out using patch clamping (see Section III,B) show a strong correlation between the properties of PTP and the multiple conductance channel (MCC; Szabó and Zoratti, 1992), suggesting that the two are the same. Other studies (Moran et al., 1990) with preparations enriched in mitochondrial contact sites between inner and outer membranes suggest an involvement of MCC at these sites.

3. Uncoupling Protein

UCP (also called thermogenin) is present in the inner membrane of brown adipose tissue where it constitutes as much as 15% of the protein. The literature concerning UCP is reviewed broadly by Nicholls et al. (1986); other reviews of UCP are available (Klingenberg, 1990; Klaus et al., 1991; Ricquier et al., 1991). UCP activity is very significant in mammals at birth and during cold adaptation or arousal from hibernation. The protein uncouples oxidation from phosphorylation, thereby generating heat. The uncoupling effect can be blocked by purine di- and trinucleotides and the block can be overcome by fatty acids (e.g., see Nicholls *et al.*, 1986).

UCP can be induced by cold acclimation *in vivo* (e.g., Sinohara *et al.*, 1991) or by exposure of brown fat cultured cells to noradrenaline (e.g., Puigserver *et al.*, 1992). Removal of the noradrenaline results in rapid specific degradation of UCP and UCP mRNA (Bianco *et al.*, 1992; Rehnmark *et al.*, 1992). The production of UCP is under transcriptional and posttranscriptional control by noradrenaline and thyroid hormones.

UCP appears to confer a high degree of permeability of H^+ (or perhaps OH^-) and possibly to halides on the inner mitochondrial membrane (see Nicholls *et al.*, 1986). Therefore, its uncoupling effect is attributed to its ability to collapse the electrochemical gradient needed for energy coupling. When UCP is reconstituted into liposomes, the system accumulates H^+ in the presence of valinomycin when the liposomes are preloaded with K^+ (Klingenberg and Winkler, 1985). This system, however, is unable to transport Cl^-. Garlid (1990) reexamined the question and found a GDP-inhibitable anion transport in proteoliposomes containing UCP. However, in the intact mitochondrial system, only the permeability of H^+ is affected by fatty acids. Therefore, the presence of two closely related proteins cannot be disproved at this time.

Nicholls *et al.* (1986) consider the UCP pathway to correspond to a channel. In contrast, Klingenberg and Winkler (1985) and Klingenberg (1990) have argued against this alternative because its turnover is too low when reconstituted in the liposome system (see also subsequent discussion). However, although a high turnover is *prima facie* evidence of a channel, the opposite is not true. A H^+ translocation model in which fatty acids function as proton acceptors and donors has been proposed (Klingenberg and Winkler, 1994).

Nicholls *et al.* (1986) calculate the presence of 0.8 nmol channels/mg protein and a flow of protons of 16 nmol/min mg protein mV. The conductance of an individual channel would be 10 ions/min/mV, or a conductance of 0.03 fS per channel. The halide conductance would be of the same order of magnitude. Such a low conductance would make studying individual channel activity using currently available patch-clamp methods impossible. However, studying transitions in an intact mitochondrion (e.g., in a whole mitoplast configuration) should be possible, since the conductance would reach at least 2.7 pS (Klitsch and Siemen, 1991). Mirzabekov and Akhemerov (1987) have observed a high-conductance channel activity in planar bilayers after the addition of an extract from brown fat containing UCP. The channel was found to be slightly anion selective and to resemble VDAC.

However, these experiments are difficult to evaluate since the report is not sufficiently detailed; even a slight contamination by VDAC could account for these results.

UCP is a protein of 32 kDa that has been isolated from brown fat after affinity labeling with 8-azido ATP (Heaton *et al.*, 1978). The protein is distinct from the adenine nucleotide translocator, as shown immunologically (Ricquier *et al.*, 1983) and from the amino acid sequences (e.g., Aquila *et al.*, 1985; Bouillaud *et al.*, 1986). Immunological procedures were also used to demonstrate its absence from a variety of mammalian tissues other than brown fat (Ricquier and Bouillaud, 1986; see also Ricquier *et al.*, 1992).

UCP has been sequenced (e.g., Aquila *et al.*, 1985; Bouillaud *et al.*, 1986) and is closely related to the adenine nucleotide and the P_i translocators (Aquila *et al.*, 1987; Runswick *et al.*, 1987). All three proteins contain a similar 3-fold repeat of approximately 100 amino acids, as shown in Fig. 4. The physical properties of UCP (e.g., its interaction with detergent, Lin and Klingenberg, 1980) and its amino acid sequence suggest a molecule that may be entirely embedded in the membrane. A variety of approaches suggest two α helices spanning the membrane for each of the three domains (see Klingenberg, 1990). Each α helix is separated by relatively hydrophilic sectors of 35–40 amino acids. The N and C termini of the protein would be on the cytosolic side. Part of the central region could easily provide a hydrophilic path, perhaps corresponding to a channel spanning the membrane.

4. F_0 Complex of ATP Synthase

ATP synthase (F_1F_0) is the enzyme complex responsible for the phosphorylation of ADP. Protons driven by the electrochemical gradient through the synthase provide the energy for the phosphorylation. Acting in reverse, F_0F_1 functions as an ATPase and can transport protons against an electrochemical gradient. The water-soluble part of the complex, F_1, contains the catalytic centers, whereas F_0 contains the proton channel. These topics have been reviewed elsewhere (Schneider and Altendorf, 1987; Futai *et al.*, 1989; Senior, 1990). In submitochondrial particles, F_0 becomes an open channel when F_1 is removed from the complex. Eukaryotic F_0 contains several polypeptides. In *Escherichia coli*, F_0 is composed of the three polypeptides a,b, and c (with a stoichiometry of $1:2:10\pm1$ of a:b:c). All organisms studied have a proteolipid subunit (corresponding to subunit c of *E. coli*) of 70–85 amino acid residues that is soluble in organic solvents. The *E. coli* system has been the focus of most studies since it is simpler and more amenable to genetic manipulation. Analyses of deletion strains (Friedl *et al.*,

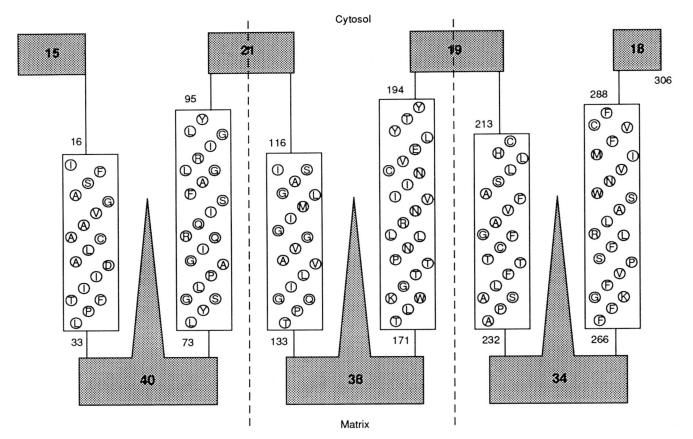

FIGURE 4 Schematic model of the uncoupling protein (UCP) folding through the membrane. The amino and carboxy termini have been localized to the cytoplasmic side. UCP is constituted of three repeat domains, each of about 100 residues, with significant internal homologies. Each domain contains two membrane-spanning α helices separated by relatively hydrophilic stretches of 35–40 residues. Reproduced with permission from Klingenberg (1990).

1983) show that all three subunits are required to establish ATP-dependent proton translocation. Other experiments (Schneider and Altendorf, 1985) show that the three subunits are needed in the native stoichiometry for proton translocation and F_1 binding.

Attempts to determine the proton conductance of F_0 reconstituted into liposomes have provided figures in the range of 10–100 $H^+/sec/F_0$ (for a review, see Schneider and Altendorf, 1987). This rate is insufficient for the functioning of ATP synthase. Assuming 3 H^+/ATP and a maximal rate of 400 ATP/sec for chloroplasts (Junesch and Gräber, 1985), a conductance of 1 fS would be needed (e.g., see Althoff *et al.*, 1989). The lower values found with proteoliposomes are probably attributable to a large number of inactive F_0 subunits. Studies using partially F_0-depleted thylakoid membranes reveal very high conductances, on the order of 1 pS (Lill *et al.*, 1987; Althoff *et al.*, 1989). Incorporation of the proteolipid from the F_0 of yeast into planar bilayers results in an increase in conduc-

tance corresponding to discrete events (Schindler and Nelson, 1982) resolvable at pH 2–3. The channels open and close independently from each other. The open-state conductances and lifetimes are affected by applied voltage. The conductances correspond to 12 pS at pH 2.2 and 100 mV. The assembly into channels is second order, suggesting that the functioning channel is a dimer.

The three polypeptides of F_0 have been sequenced and subjected to genetic analysis. Figure 5 shows models of the F_0 subunits. Sites of missense mutations that inhibit H^+ translocation, without affecting F_0 assembly, are indicated in the diagram.

At this time no evidence exists of a direct involvement of subunit b in proton movement. All mutations of b that impair function also disrupt assembly. Further, part of the subunit can be digested after assembly without affecting the translocation.

In contrast, the a and c subunits are thought to be involved in H^+ translocation, since several missense

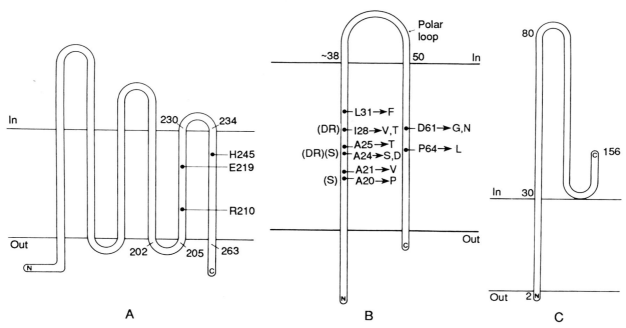

FIGURE 5 Schematic model of the F_0 subunits of *Escherichia coli* showing the residues at which a mutation disrupts H^+ translocation without affecting overall F_0 structure. *A* corresponds to the *a* subunit, *B* to the *c* subunit, and *C* to the *b* subunit. Reproduced with permission from the *Annual Review of Biophysics and Biophysical Chemistry*, Vol. 19, 1990 by Annual Reviews, Inc.

mutations of a and c affect translocation without affecting F_0 assembly. In addition, considerable other evidence is available of c being involved, including the channel behavior discussed earlier. Dicyclohexylcarbodiimide (DCCD) blocks H^+ movement after reacting with Arg61 of this subunit. Incubation of everted submitochondrial vesicles stripped of F_1 with antibodies against the c subunit block proton translocation (Decker-Heberstreit and Altendorf, 1992). Further, as indicated in the figure, several missense mutations interfere with the passage of H^+.

The model currently favored is that of a water-filled channel to accommodate the finding of a Na^+ transport catalyzed by ATPase of *Proprionigenium modestum* (Laubinger and Dimroth, 1988), in which Na^+ takes the role of H^+. Other models involving the so-called proton wire would have difficulty providing an explanation for the Na^+ transport (see Senior, 1990). Chemical studies of the interaction between F_0 and F_1 in providing gating of the channel have implicated a subunit of F_0, the γ subunit of F_1, and possibly other components (Papa *et al.*, 1994).

B. Patch-Clamp Studies

The basic properties of five inner mitochondrial membrane channels are summarized in Table II. Several reviews are available (e.g., Kinnally *et al.*, 1992; Mannella, 1992; Kinnally and Tedeschi, 1994; Sorgato and Moran, 1993). Sample current traces for most inner membrane channel activities are shown in Figs. 6 and 7.

1. Mitochondrial Centum-picoSiemen Channel Activity

The mitochondrial centum-picoSiemen (mCS) channel activity first described by Sorgato *et al.* (1987) is voltage dependent, closing at matrix negative and opening at matrix positive potentials, as shown in Fig. 6. This activity is slightly anion selective. Incorporation of an F_0 fraction into bilayers (Stühmer *et al.*, 1988) or ATP-synthase fractions into liposomes (Sorgato *et al.*, 1989) resulted in the characteristic mCS activity. The channel activity is thought to correspond to a protein that copurifies with F_0 components rather than a protein of the complex, since the activity is not inhibited by oligomycin or DCCD.

mCS has been found to be normally quiescent, but can be activated by washing mitoplasts with EGTA or EDTA (Kinnally *et al.*, 1991). Also, the channel displays bursting behavior (Sorgato *et al.*, 1987). Because of these findings, mCS has been suggested to have multiple closed states (Campo *et al.*, 1992). Sorgato *et al.* (1994) have presented an analysis of open

TABLE II **Summary of Channel Classes**

Class	Size (pS)	Voltage dependence	Selectivity	Effectors	Possible counterparts
Multiconductance (MCC)	~40 to > 1000	slight	slightly varied	Ca^{2+}, voltage, Mg^{2+}, ADP[a]	MMC[b], PTP[c]
mCS	107	yes	slightly anionic	Ca^{2+}, voltage	IMM[d] channel
AAA	~45	no	slightly anionic	pH, Mg^{2+}	IMAC[e]
ACA	~15	no	slightly cationic	Mg^{2+}	K^+ uniport[f]
ATP-sensitive K^+	~15		K^{+g}	ATP	

[a] Szabó and Zoratti (1992).
[b] MMC, Mitochondrial megachannel (Szabó and Zoratti, 1991, 1992).
[c] PTP, Permeability transition pore.
[d] IMM, Inner mitochondrial membrane (Sorgato *et al.*, 1987).
[e] IMAC, Inner membrane anion channel (Garlid and Beavis, 1986).
[f] Nicolli *et al.* (1991).
[g] Inoue *et al.* (1991).

and closed time distribution that supports the existence of multiple closed states. Klitsch and Siemen (1991) patch clamped brown adipose tissue mitochondria treated osmotically to remove the outer membrane. These investigators observed occasional channel openings at one-third, one-half, and two-thirds of the most frequent conductance level of 108 pS. These researchers propose that the 50% conductance level is a substate. A conductance as high as 145 pS was

observed when patches exhibiting the mCS activity were treated with amiodarone (Antonenko *et al.*, 1991a). The higher apparent conductance was proposed (e.g., Kinnally *et al.*, 1992) to result from a change in the apparent closed state, generally not reached without inhibitor. Further, a substate of approximately 50-pS was also encountered (Campo *et al.*, 1992). The 140-pS channel activity and the 50-pS channel activity have now been observed in the

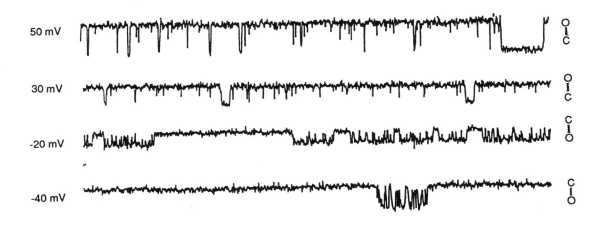

FIGURE 6 Sample current traces of the mitochondrial centum-picoSiemen channel (MCS) activity show the strong voltage dependence of the MCS activity. Current traces were recorded from an attached patch from a mouse liver mitoplast in 150 m*M* KCl, 5 m*M* HEPES, 1 m*M* EGTA, 2 m*M* $MgCl_2$, 2.5 μ*M* rotenone, pH 7.4. French-press mitoplasts were prepared as described by Kinnally *et al.* (1991). Other conditions are as in Fig. 3.

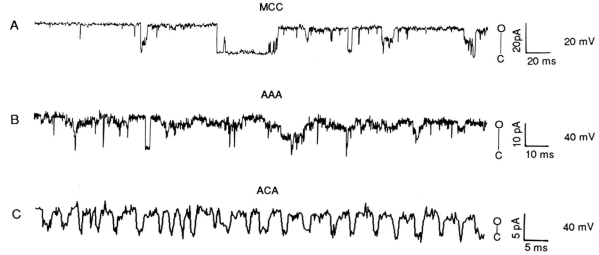

FIGURE 7 Sample current traces of inner membrane channel activities. (A) Multiple conductance channel (MCC) activity was recorded from an excised patch from rat heart mitoplast in symmetrical 150 mM KCl, 5 mM HEPES, 1 mM EGTA, 0.75 mM CaCl$_2$ (~10^{-7}M free Ca^{2+}), pH 7.4. (B) Alkaline pH-induced anion-selective activity (AAA) was recorded from an excised patch from mouse liver mitoplast in symmetrical 1 M KCl, 5 mM HEPES, 10 μM CaCl$_2$ with a pipette medium of pH 6.8 and a bath of pH 8.2. (C) Alkaline pH-induced cation-selective activity (ACA) was recorded from an excised patch from a mouse liver mitoplast in symmetrical 1 M KCl, 5 mM HEPES, 10 μM CaCl$_2$ with the pipette medium at pH 6.8 and the bath at pH 8.2. Other details are as in Fig. 3.

absence of the 100 pS activity or inhibitors (Kinnally and Tedeschi, 1994). These conductances have the same voltage dependence as the more frequently observed conductance of ~100 pS. In addition, at least the 50-pS conductance level has the same selectivity as the ~100-pS channel.

Part of the study by Klitsch and Siemen (1991) includes that of the "whole-mitoplast" configuration, that is, a configuration in which the pipette has access to the matrix space. With this latter design, a sensitivity to purine nucleotides is reported. This configuration was reported to be obtained by exposing the patches to high voltage. Evidence for this configuration was deduced from a reversal of polarity (seen in current–voltage curves) after the high-voltage treatment and a slight increase in capacitance. This interpretation should be treated with caution, since activation of the other inner membrane channel, MCC activity, by high voltage has been observed in *excised* patches. In general, MCC activity was found to have a voltage dependence opposite that of the mCS channel (e.g., Fig. 1B in Kinnally *et al.*, 1989). Further, Inoue *et al.* (1991) claim an insensitivity of the mCS channel to ATP, whereas Szabó and Zoratti (1992) report an inhibition of MCC activity by ADP, although in both studies the adenine nucleotides were introduced on the matrix face of the membrane.

2. Channel Activity Induced by Alkaline pH

During a previous study of the effect of alkaline pH, the activation of a ~15-pS channel was observed. On occasion, a larger channel was also observed (e.g., see Fig. 1 in Antonenko *et al.*, 1991b). Subsequent studies revealed the additional activation of a ~45-pS channel (Antonenko *et al.*, 1993). A study of the ion selectivity of the two alkaline pH-induced channels using asymmetric ion substitutions and reversal potential techniques (e.g., Andersen *et al.*, 1986) was also carried out. The ~15-pS channel was found to be cation selective and to be insensitive to the presence of anions; this is referred to as the alkaline-induced cation-selective activity (ACA). Conversely, the ~45-pS activity was found to be anion selective, and is referred to as the alkaline-induced anion-selective activity (AAA) (see Fig. 7 for sample current traces of ACA and AAA). However, selectivity among the ions of the same charge was very slight.

In view of the observations that alkaline pH or Mg^{2+} depletion induces an anion-selective permeability transition (see previous discussion), one could postulate that the ~45-pS channel activity corresponds to IMAC. A previous study (Antonenko *et al.*, 1991b) examined the pharmacology and dependence on Mg^{2+} or pH of the whole-patch current transition brought about by alkaline conditions, and found it to parallel closely that of IMAC (e.g., Beavis, 1989). An argument against considering IMAC and AAA reflections of the same channel activity comes from selectivity data. The ratio of permeability of Cl^- over glucuronate was reported to be higher than 400 for IMAC (Beavis, 1992), whereas Antonenko et al. (1993) found it to be approximately 3 for AAA.

Hayman and Ashley (1993) and Hayman et al. (1993) recently characterized two multistate anion channels from mitochondria in planar bilayer studies. While many characteristics are similar to those of mCS and AAA, differences exist which may be sorted out with further studies.

The ~15-pS channel could be considered to correspond to the K^+ uniport. A K^+ uniport has been found to be activated by depletion of Mg^{2+} (e.g., Bernardi et al., 1989; Nicolli et al., 1991) and appears to be relatively unselective, transporting Rb^+, Na^+, and Li^+ (Nicolli et al., 1991). The possibility must be considered that the latter transitions also result from the opening of a hitherto inactive channel. The conductance, calculated from the fluxes reported, is in the range expected for a channel (Table I). The correspondence between the uniport and the ~15-pS channel is supported by the fact that the two are relatively unselective in relation to cations. However, current information is not sufficient to draw a firm conclusion.

3. Multiple Conductance Channel Activity

MCC activity, first described in 1988 (Kinnally et al., 1988), exhibits a variety of conductance states (e.g., see Kinnally et al., 1989; Petronilli et al., 1989; Zorov et al., 1992b) with a peak of 1–1.5 nS, as illustrated by a current trace in Fig. 7A. The channel can be induced by Ca^{2+} (e.g., Kinnally et al., 1991; Szabó and Zoratti, 1992; Szabó et al., 1992) or high voltage, generally greater than ±60 mV (e.g., Zorov et al., 1992b). Generally, the activity is voltage dependent: opening occurs at negative potentials and lower conductance levels are occupied at positive potentials. The subconductance levels of MCC activity are often revealed as steps during voltage-induced opening or closing, during the initial voltage activation, or on the addition of inhibitors such as amiodarone (Zorov et al., 1992a) or cyclosporin A (Zorov et al., 1992b). The channel also can remain in the open or closed state for prolonged periods of time (many minutes).

The various conductance levels have been attributed to a single class referred to as MCC activity. The evidence, although persuasive, is not entirely decisive. The various conductance states have many features in common (Kinnally et al., 1992; Zorov et al., 1992a) and, as already mentioned, the activation or

inhibition often proceeds in steps, suggesting that these stages truly represent subconductance levels. On the other hand, some of these activities differ in ion selectivity (see Kinnally et al., 1989). However, subconductance states have been shown to have different selectivities, at least in VDAC (Ludwig et al., 1989; Ermishkin and Mirzabekov, 1990).

Moran et al. (1990) demonstrated activity that we attribute to MCC in preparations enriched with contact sites (Sandri et al., 1988). Studies of MCC activity (Szabó and Zoratti, 1991,1992, Szabó et al., 1992; called the mitochondrial megachannel, MMC, in these reports) and parallel experiments of the PTP using mitochondrial suspensions (Bernardi et al., 1992) have presented convincing evidence that the two are equivalent. Common characteristics include the dependence on divalent cations, ADP, pH, and the effect of cyclosporin A.

4. ATP-sensitive K^+ Channel

Inoue et al. (1991) have described a K^+-selective channel in fused mitoplasts. The channel is inactivated reversibly by ATP on the matrix face of an excised patch and is also inhibited by 4-aminopyridine and glybenclamide plus ATP. In this case, patch-clamp experiments preceded permeability studies on mitochondrial suspensions in defining channel activity. The ATP-sensitive K^+ channel is discussed in detail in Chapter 35.

C. Sensitivity to Drugs

Mitochondrial inner membrane channels appear to be sensitive to a wide spectrum of drugs (e.g., see Beavis, 1992; Kinnally and Tedeschi, 1994), as witnessed by permeability studies using mitochondrial suspensions or patch-clamp studies.

In mitochondrial suspensions, IMAC is affected by a broad range of drugs including amphiphilic drugs (such as quinine or the cardiac drugs propranolol and amiodarone), local anesthetics (such as dibucaine and dupivacaine), and the benzodiazepines Ro 5-4864 and clonazepam (Beavis, 1992). Experiments using patch clamping have generated similar results (see Kinnally et al., 1992). MCC and mCS channel activity are inhibited by amiodarone and propranolol as well as by antimycin A. The amphiphilic drugs also block the whole-patch current transitions induced by alkaline pH (probably ACA). The inhibition of the mCS channel in cardiac mitochondria by the isoquinoline carboxamide PK11195, the benzodiazepine Ro 5-4864, and protoporphyrin IX, an intermediate of heme synthesis, has been reported (Kinnally et al., 1993). MCC activity was also inhibited by Ro 5-4864 and protoporphyrin IX.

One of the unique features of the effect of these inhibitors on inner mitochondrial membrane channels is the activation of channels by high concentrations of the drugs. This effect has been shown for antimycin A (Campo *et al.*, 1992) and ligands of the peripheral benzodiazepine receptors (Kinnally *et al.*, 1993). Results obtained with amiodarone are consistent with this conclusion. Although other interpretations are possible, the results can be explained most simply by assuming the presence of binding receptor sites of lower affinity than the inhibitor sites. At high concentrations, these agents can allosterically block the inhibitory effect. Addition of activating concentrations of a ligand were found to be blocked by inhibitory concentrations of a second higher affinity ligand (K. W. Kinnally and D. B. Zorov, unpublished experiments).

The sensitivity of various mitochondrial channels to the same inhibitors raises an important question. Are the various observed activities manifestations of the same channel protein? For example, the mCS channel could be one of the substates of MCC. Similarly, why are so many supposedly unrelated drugs effective in inhibiting channel activity? Several answers are possible.

(1) The effect of the drugs may be indirect. Since these drugs are highly hydrophobic, they might exert their effect of complexing to the hydrocarbon chains in the membrane. Common effects of various so-called stabilizers, which include tranquilizers and anesthetics, on plasma membranes have been recognized for some time (e.g., Seeman, 1966). However, a common mechanism is not likely since some of the drugs have opposite effects. For example, quinidine, an isomer of quinine, has been found to decrease the fluidity of liver plasma membranes (Needham *et al.*, 1987) whereas propranolol decreases the order of the hydrocarbon chains of the platelet membrane (Dash and Rao, 1990), but both inhibit MCC activity. Further, some indication exists that specific receptors are involved. For example, the benzodiazepine Ro 5-4864, which binds to the mitochondrial benzodiazepine receptor (mBzR) with high affinity, is very effective. In contrast, clonazepam, which is structurally closely related but has a much lower affinity, is inactive or almost inactive (Kinnally *et al.*, 1993).

(2) The channels may be capable of responding to various drugs because they share various binding proteins that are free to move in the plane of the membrane.

(3) A single receptor might be capable of responding to a variety of unrelated compounds. This occurs for mBzR (see Hirsch *et al.*, 1988; Verma and Snyder, 1988), which can bind a broad spectrum of drugs including many that affect either IMAC (Beavis, 1992) or channels observed in patch-clamp studies (Kinnally and Tedeschi, 1994).

(4) Open channels could be restricted to specific structures, assembled or disassembled in response to a broad spectrum of signals and under a complex control system. Contact sites, that is, the sites at which the inner and outer membrane are in contact, could be such structures (see Kinnally and Tedeschi, 1994). Some evidence exists for dynamic regulatory effects in the formation of these sites (e.g., presence of Ca^{2+}, Sandri *et al.*, 1988; metabolism, Knoll and Brdiczka, 1983; for a review of the possible roles of contact sites, see Brdiczka, 1991). This model could allow the coordinated regulation of channel activity. Contact sites are generally present in mitoplasts, in which outer membrane pieces generally persist even after mitoplast preparation.

IV. Interaction between Inner and Outer Membranes

An entirely new perspective has been introduced by evidence of interactions between inner and outer membrane elements. Such interactions are likely to be of great physiological significance since they involve the energy transducing system of mitochondria. As originally described by Hackenbrock (1968), inner and outer membranes make contact at specific sites. Various experiments have suggested a special organization and role for these sites. Patch clamping of liposomes containing membrane preparations from brain mitochondria enriched in contact sites (Sandri *et al.*, 1988) demonstrated channel activity that we consider to correspond to the MCC (Moran *et al.*, 1990). This finding suggests that outer and inner membrane channels are in register at contact sites to provide patent channels between the medium and the patch pipette (Kinnally *et al.*, 1992).

Results from many laboratories seem to implicate complexes between various inner and outer mitochondrial membrane components, including the adenine nucleotide translocator (ANT), VDAC, hexokinase, creatine kinase, and other components possibly occurring at contact sites (see Table III for literature citations). These observations have led to the suggestion of a functional complex of VDAC and ANT. In the appropriate tissues, hexokinase (in the cytoplasmic side; e.g., McEnery, 1992) or creatine kinase (at the surface of the inner membrane; Schlegel *et al.*, 1990) are also thought to form part of this complex. The effect of a synthetic polyanion (for review, see Benz and Brdiczka, 1992) or an endogenous protein regulator (Liu and Colombini, 1992) on membrane function

TABLE III   **Interaction between Various Metabolically Significant Elements**

Interaction	References
Contact sites with hexokinase, creatine kinase, Ca^{2+}, VDAC	Adams *et al.* (1989), Benz and Brdiczka (1992), Brdiczka *et al.* (1986), Kottke *et al.* (1988), Biermans *et al.* (1989)
VDAC with hexokinase	Fiek *et al.* (1982), Lindén *et al.* (1982), Krause *et al.* (1986), Nakashima *et al.* (1986)
glutathione transferase	Krause *et al.* (1986)
Outer membrane with Glycerol kinase	Ostlund *et al.* (1983)
Mitochondrial ATP favored metabolically hexokinase activity	Gots and Bessman (1974), Arora and Pedersen, (1988), BeltrandelRio and Wilson (1991).
creatine phosphokinase	Yang *et al.* (1977), Erickson-Viitanen *et al.* (1982), Gellerich and Saks (1982).
adenine nucleotide translocator	Saks *et al.* (1980,1987), Erickson-Viitanen *et al.* (1982), Moreadith and Jacobus (1982), Barbour *et al.* (1984)

in intact mitochondria supports these views. The polyanion has been shown to interact with VDAC and to affect mitochondrial functions associated with ATP utilization. The modulator has been shown to reduce state 3 respiration and respiratory control ratios (RCR) in intact mitochondria. However, the fact that both polyanions and modulator are effective when placed on the outside of the mitochondrion remains a puzzle. *In situ*, the modulator is supposedly in the intermembrane space.

In agreement with the proposed models, a functioning mBzR complex has been isolated and was found to be constituted by an 18-kDa protein as well as ANT and VDAC, the latter two identified by Western blot analysis (McEnery *et al.*, 1992). These findings suggest the possibility of examining the relationship between outer and inner mitochondrial membrane channels and the mBzR through an examination of the sensitivity of channel activity to benzodiazepines. In accordance with these expectations, these drugs were found to block inner mitochondrial membrane channels (Kinnally *et al.*, 1993).

V. Channel Behavior of Translocators in Reconstituted Systems

Ion translocators have been shown to behave as channels when incorporated into bilayers. This result

is hardly surprising, since these proteins are thought to function as alternating access channels (e.g., Tanford, 1983). Unfortunately, without additional information, ascertaining whether the transport system operates as a channel in the native system is frequently difficult. This difficulty can be illustrated for the case of the Na^+/K^+–ATPase which, in the native system, does not function as a conventional channel. This translocase can exhibit channel behavior when incorporated into planar bilayers (Last *et al.*, 1983; Mironova *et al.*, 1986). The channel behavior is relevant to the transport mechanisms since ouabain and vanadate inhibit channel conductance. However, in other reconstitution experiments, the Na^+/K^+–ATPase behaves as expected from its *in situ* behavior (e.g., Eisenrauch *et al.*, 1991). The currents generated by activation of the translocase are a small fraction of those observed in "channel" experiments already discussed. Further, the voltage dependence is consistent with that of a conventional translocase and not a channel. For example, this protein exhibits a reversal potential in the presence of ATP under symmetrical conditions (i.e., when the concentration of ions is the same on both sides of the planar bilayer), indicating its nonpassive behavior.

In the case of the reconstituted uniports, the significance of channel behavior is difficult to evaluate. Transport is in the direction of the electrochemical gradient, as expected from a channel. Mitochondrial translocases have been studied after incorporation into planar bilayers or liposomes. Mironova *et al.* (1982) detected channel behavior after incorporating the 40-kDa glycoprotein or the 2-kDa polypeptide associated with the Ca^{2+}-transporting system into planar bilayers. The activity was blocked by ruthenium red and thiol reagents. Similarly, a K^+-transporting protein of approximately 60 kDa was found to exhibit channel behavior when incorporated into bilayers (Mironova *et al.*, 1981). More recently, partially purified ion transport proteins were found to exhibit channel behavior when reconstituted into liposomes (Costa *et al.*, 1991; Paliwal *et al.*, 1992; Diwan and Costa, 1993). In the native system, a K^+ uniport could correspond to a channel (see preceding discussion; Table I). However, the corresponding calculations for the Ca^{2+} uniport show rather low conductances. Note, however, that mitochondrial channels may associate and exhibit cooperative behavior (e.g., see discussion of PTP, Section III,A,2), which could lead to much greater conductances in the reconstituted system. In accordance with this view, in some experiments the unit conductance of the channels exhibits several levels (e.g., from 24 to 175 pS in the case of the K^+ translocase studied by Mironova *et al.*, 1981), which could correspond to multiples of a base conductance.

VI. Possible Physiological Roles

The physiological role of mitochondrial channels is still an open question that has been subject to speculation. Some possibilities are summarized in Table IV. The role of VDAC and its association with various enzymes in regulating energy metabolism is obviously important. Inner membrane channels have been argued also to be involved in these interactions, since they may exist in tandem with VDAC (e.g., Kinnally and Tedeschi, 1994). Beavis (1992) proposed that IMAC plays an important role in mitochondrial volume regulation; this role could be extended to several other channels as well. The outer mitochondrial channel, PSC, described by Thieffry *et al.* (1994), is likely to be involved in the transport of newly synthesized polypeptides from the cytoplasm. Since such transfer is thought to occur at contact sites, this finding suggests the presence of this channel at these sites. An mitochondrial inner membrane channel component is also probably involved in some of these transfers.

A strong argument can be made for a role for some of these channels in thermogenic regulation, similar to that of UCP which is limited to brown adipose tissue. Like UCP, the channels could be involved in short-circuiting the H^+-ATPase, thereby increasing thermogenesis.

The possible physiological roles of PTP have been addressed by Gunter and Pfeiffer (1990). One possible function is in the degradation of mitochondria following cell injury. However, a physiological role other than self-destruction is likely since the permeability transition is reversible. A possible role is that of releasing Ca^{2+} previously sequestered to avoid a cytoplasmic Ca^{2+} overload.

The possible metabolic significance of the channels is also suggested by recent experiments of Zizi *et al.* (1994) who found NADH to regulate the gating of VDAC.

Acknowledgments

Our research was funded by National Science Foundation Grant MCB 9117658.

TABLE IV **Some Postulated Physiological Functions of Membrane Channels**

Channel[a]	Function	Selected reference
VDAC	Metabolic regulation	Benz and Brdiczka (1992), Liu and Colombini (1992)
IMAC	Volume regulation	Beavis (1992)
PSC	Protein import	Thieffry *et al.* (1993).
UCP	Thermogenesis	Nicholls *et al.* (1986)
K^+ Uniport	Thermogenesis	G.D. Mironova (personal communication)
PTP	[Ca^{2+}] regulation, mitochondrial turnover	Gunter and Pfeiffer (1990)

[a] Abbreviations: VDAC, voltage-dependent anion channel; IMAC, inner membrane anion channel; PSC, peptide sensitive channel, UCP, uncoupling protein; PTP, permeability transition pore.

References

Adams, V., Bosch, W., Schlegel, J., Walliman, T., and Brdicska, D. (1989). Further characterization of contact sites from mitochondria of different tissues: Topology of peripheral kinases. *Biochim. Biophys. Acta* **981**, 213–225.

Al-Nasser, I., and Crompton, M. (1986). The reversible Ca^{2+}-induced permeabilization of rat liver mitochondria. *Biochem. J.* **239**, 19–29.

Althoff, G., Lill, H., and Junge, W. (1989). Proton channel of the chloroplast ATP synthase, CF_0: Its time-averaged single-channel conductance as a function of pH, temperature, isotopic and ionic medium composition. *J. Membr. Biol.* **108**, 263–271.

Andersen, O. S., Green, W. N., and Urban, B. W. (1986). Ion conduction through sodium channels in planar lipid bilayers. *In* "Ion Channel Reconstitution" (C. Miller, ed.), pp. 385–404. Plenum, New York.

Antonenko, Y. N., Kinnally, K. W., Perini, S., and Tedeschi, H. (1991a). Selective effect of inhibitors on inner mitochondrial membrane channels. *FEBS Lett.* **285**, 89–93.

Antonenko, Y. N., Kinnally, K. W., and Tedeschi, H. (1991b). Identification of anion and cation pathways in the inner mitochondrial membrane by patch clamping of mouse liver mitoplasts. *J. Membr. Biol.* **124**, 151–158.

Antonenko, Y. N., Kinnally, K. W., and Tedeschi, H. (1993). Single channel activity induced in mitoplasts by alkaline pH. *Biophys. J.* 64:343a.

Aquila, H., Link, T. A., and Klingenberg, M. (1985). The uncoupling protein from brown fat mitochondria is related to the mitochondrial ADP/ATP carrier. Analogy of sequence homologies and folding of the protein in the membrane. *EMBO J.* **4**, 2369–2376.

Aquila, H., Link, T. A., and Klingenberg, M. (1987). The sequences of three mitochondrial carriers involved in energy transfer of mitochondria form a homologous protein family. *FEBS Lett.* **212**, 1–9.

Arora, K. K., and Pedersen, P. L. (1988). Functional significance of mitochondrial bound hexokinase in tumor cell metabolism. Evidence for preferential phosphorylation of glucose by intramitochondrially generated ATP. *J. Biol. Chem.* **263**, 17422–17428.

Barbour, R. L., Ribaudo, J., and Chan, S. H. P. (1984). Effect of creatine kinase activity on mitochondrial ADP/ATP transport. *J. Biol. Chem.* **259**, 8246–8251.

Beavis, A. D. (1989). On the inhibition of the mitochondrial inner membrane anion uniporter by cationic amphiphiles and other drugs. *J. Biol. Chem.* **264**, 1508–1518.

Beavis, A. D. (1992). Properties of the inner membrane anion channel in intact mitochondria. *J. Bioenerg. Biomembr.* **24**, 77–90.

Beavis, A. D., and Garlid, K. D. (1986). Evidence for the existence of an inner membrane anion channel in mitochondria. *Biochim. Biophys. Acta* **853**, 187–204.

Beavis, A. D., and Garlid, K. D. (1987). The mitochondrial inner membrane anion channel. *J. Biol.Chem.* **262**, 15085–15093.

BeltrandelRio, H., and Wilson, J. E. (1991). Hexokinase of rat brain mitochondria: Relative importance of adenylate kinase and oxidative phosphorylation as sources of substrate ATP, and interaction with intramitochondrial compartments of ATP and ADP. *Arch. Biochem. Biophys.* **286**, 183–194.

Benz, R. (1990). Biophysical properties of porin pores from mitochondrial outer membrane of eukaryotic cells. *Experientia* **46**, 131–137.

Benz, R., and Brdiczka, D. (1992). The cation-selective substate of the mitochondrial membrane pore: Single channel conductance and influence on intermembrane and peripheral kinases. *J. Bioenerg. Biomembr.* **24**, 33–39.

Bernardi, P., Angrilli, A., Ambrosin, V., and Azzone, G. F. (1989). Activation of latent K^+ uniport in mitochondria treated with the ionophore A23187. *J. Biol. Chem.* **264**, 18902–18906.

Bernardi, P., Vassanelli, S., Veronese, P., Colonna, R., Szabó, I., and Zoratti, M. (1992). Modulation of the mitochondrial transition pore: Effect of protons and divalent cations. *J. Biol. Chem.* **267**, 2934–2939.

Bianco, A. C., Kieffer, J. D., and Silva, J. E. (1992). Adenosine 3′,5′-monophosphate and thyroid hormone control of uncoupling protein messenger ribonucleic acid in freshly dispersed brown adipocytes. *Endocrinology* **130**, 2625–2633.

Biermans, W., De Bie, M., Nijs, B., and Jacob, W. (1989). Ultrastructural localisation of creatine kinase activity in contact sites between inner and outer mitochondrial membranes of rat myocardium. *Biochim. Biophys. Acta* **974**, 74–80.

Blachly-Dyson, E., Peng, S. Z., Colombini, M., and Forte, M. (1990). Alteration of the selectivity of the VDAC ion channel by site-directed mutagenesis. *Science* **247**, 1233–1236.

Bouillaud, F., Weissenbach, J., and Ricquier, D. (1986). Complete cDNA-derived amino acid sequence of brown adipose tissue uncoupling protein. *J. Biol. Chem.* **261**, 1487–1490.

Brdiczka, D. (1991). Contact sites between mitochondrial envelope membranes. Structure and function in energy- and protein-transfer. *Biochim. Biophys. Acta* **1071**, 291–312.

Brdiczka, D., Knoll, G., Riesinger, I., Wieler, U., Klug, G., Benz, R., and Krause, J. (1986). Microcompartmentation at the mitochondrial surface: Its function in metabolic regulation. *In* "Myocardial and Skeletal Muscle Bioenergetics" (N. Brautbar, ed.), pp. 55–69. Plenum Press, New York.

Brierley, G. P. (1970). Energy-linked alteration of the permeability of heart mitochondria to chloride and other ions. *Biochemistry* **9**, 697–707.

Campo, M. L., Kinnally, K. W., and Tedeschi, H. (1992). The effect of antimycin A on mouse liver inner mitochondrial membrane channel activity. *J. Biol. Chem.* **267**, 8123–8127.

Colombini, M. (1979). A candidate for the permeability pathway of the outer mitochondrial membrane. *Nature (London)* **279**, 643–645.

Colombini, M. (1986). Voltage gating in VDAC: Toward a molecular mechanism. *In* "Ion Channel Reconstruction" (C. Miller, ed.), pp. 533–552. Plenum Press, New York.

Colombini, M., Yeung, C. L., Tung, J., and König, T. (1987). The mitochondrial outer membrane VDAC is regulated by a synthetic polyanion. *Biochim. Biophys. Acta* **905**, 279–286.

Costa, G., Kinnally, K. W., and Diwan, J. J. (1991). Patch-clamp analysis of a partially purified ion channel from rat liver mitochondria. *Biochim. Biophys. Res. Commun.* **175**, 305–310.

Criado, M., and Keller, B. U. (1987). A membrane fusion strategy for single channel recording of membranes usually non-accessible to patch-clamp pipette electrodes. *FEBS Lett.* **224**, 172–176.

Crompton, M., and Costi, A. (1988). Kinetic evidence for a heart mitochondrial pore activated by Ca^{2+}, inorganic phosphate and oxidative stress. A potential mechanism of mitochondrial dysfunction during cellular Ca^{2+} overload. *Eur. J. Biochem.* **178**, 488–501.

Dash, D., and Rao, G. R. (1990). Characterization of the effects of propranolol on the physical state of platelet membrane. *Arch. Biochem. Biophys.* **276**, 343–347.

Deckers-Heberstreit, G., and Altendorf, K. (1992). Influence of subunit-specific antibodies on the activity of F_0 complex of the ATP synthase of *Escherichia coli* II Effects of subunit c-specific polyclonal antibodies. *J. Biol. Chem.* **267**, 12370–12374.

De Pinto, V., and Palmieri, F. (1992). Transmembrane arrangement of mitochondrial porin or voltage dependent anion channel (VDAC). *J. Bioenerg. Biomembr.* **24,** 21–26.

De Pinto, V., Ludwig, O., Krause, J., Benz, R., and Palmieri, F. (1987). Porin pores of mitochondrial outer membranes from high and low eukaryotic cells: Biochemical and biophysical characterization. *Biochim. Biophys. Acta* **894,** 109–119.

Dihanich, M., Schmid, A., Oppliger, W., and Benz, R. (1989). Identification of a pore in the mitochondrial outer membrane of a porin deficient yeast mutant. *Eur. J. Biochem.* **181,** 703–708.

Diwan, J. J., and Costa, G. (1993). Purification and patch clamp analysis of a 40 pS mitochondrial channel. *In* "Molecular Biology of Mitochondrial Transport Systems" (M. Forte and M. Colombini, eds.). pp. 199–208. Springer Verlag, Berlin.

Eisenrauch, A., Grell, E., and Bamberg, E. (1991). Voltage dependence of the Na,K-ATPase incorporated into planar bilayers. *In* "The Sodium Pump: Structure, Mechanism, and Regulation" (J. H. Kaplan and P. De Weer, eds.), pp. 317–326. Rockefeller University Press, New York.

Erickson-Viitanen, S., Viitanen, P., Geiger, P. J. Yang, W. C. T., and Bessman, S. P. (1982). Compartmentation of mitochondrial creatine phosphokinase. I. Direct demonstration of compartmentation with the use of labeled precursors. *J. Biol. Chem.* **257,** 14395–14404.

Ermishkin, L. N., and Mirzabekov, T. A. (1990). Redistribution of the electric field within the pore contributes to the voltage dependence of mitochondrial porin channel. *Biochim. Biophys. Acta* **1021,** 161–168.

Fagian, M. M., Pereira da Silva, L., Martins, I. S., and Vercesi, A. E. (1990). Membrane protein thiol cross-linking associated with the permeabilization of the inner mitochondrial membrane by Ca^{2+} plus prooxidants. *J. Biol. Chem.* **265,** 19955–19960.

Fiek, C., Benz, R., Roos, N., and Brdiczka, D. (1982). Evidence for identity between hexokinase-binding protein and the mitochondrial porin of the outer membrane of rat liver mitochondria. *Biochim. Biophys. Acta* **688,** 429–440.

Forte, M., Guy, R., and Mannella, C. (1987). Molecular genetics of the VDAC ion channel: Structural model and sequence analysis. *J. Bioenerg. Biomembr.* **19,** 341–350.

Freitag, H., Neupert, W., and Benz, R. (1982). Purification and characterisation of a pore protein of the outer membrane from *Neurospora crassa. Eur. J. Biochem.* **123,** 619–636.

Friedl, P., Hoppe, J., Gunsulus, R. P., Michelsen, O., von Meyenburg, K., and Schaier, H. U. (1983). Membrane integration and function of the three F_0 subunits of the ATP synthase of *Escherichia coli* K12. *EMBO J.* **2,** 99–103.

Futai, M., Noumi, T., and Maeda, M. (1989). ATP synthase (H^+-ATPase): Results from combined biochemical and molecular biological approaches. *Annu. Rev. Biochem.* **58,** 111–136.

Garlid, K. D. (1990). New insights into mechanisms of anion uniport through the uncoupling protein of brown adipose tissue mitochondria. *Biochim. Biophys. Acta* **1018,** 151–154.

Garlid, K. D., and Beavis, A. D. (1986). Evidence for the existence of an inner membrane anion channel in mitochondria. *Biochim. Biophys. Acta* **853,** 187–204.

Gazzotti, P. (1975). The effect of Ca^{2+} on the oxidation of exogenous NADH in rat liver mitochondria. *Biochem. Biophys. Res. Commun.* **67,** 634–638.

Gellerich, F., and Saks, V. A. (1982). Control of heart mitochondrial oxygen consumption by creatine kinase: The importance of enzyme localization. *Biochem. Biophys. Res. Commun.* **105,** 1473–1481.

Gots, R. E., and Bessman, S. P. (1974). The functional compartmentation of mitochondrial hexokinase. *Arch. Biochem. Biophys.* **163,** 7–14.

Gunter, T. E., and Pfeiffer, D. R. (1990). Mechanisms by which mitochondria transport calcium. *Am. J. Physiol. Cell Physiol.* **27,** C755–786.

Hackenbrock, C. R. (1968). Chemical and physical fixation of isolated mitochondria in low and high energy states. *Proc. Natl. Acad. Sci. USA* **61,** 598–605.

Haworth, R. A., and Hunter, D. R. (1980). Allosteric inhibition of Ca^{2+}-activated hydrophilic channel of the mitochondrial inner membrane by nucleotides. *J. Membr. Biol.* **54,** 231–236.

Hayman, K. A., and Ashley, R. H. (1993). Structural features of multistate cardiac mitoplast anion channel: Inferences from single channel recording. *J. Membr. Biol.* **136,** 191–197.

Hayman, K. A., Spurway, T. D., and Ashley, R. H. (1993). Single anion channels reconstituted from cardiac mitoplasts. *J. Membr. Biol.* **136,** 181–190.

Heaton, G. M., Wagenvoord, R. J., Kemp, A., and Nicholls, D. G. (1978). Brown adipose tissue mitochondria: Photoaffinity labelling of the regulatory site for energy dissipation. *Eur. J. Biochem.* **82,** 515–521.

Henry, J. P., Chich, J.-F., Goldschmidt, D., and Thieffry, M. (1989). Blockade of a mitochondrial cationic channel by an addressing peptide: An electrophysiological study. *J. Membr. Biol.* **112,** 139–147.

Hirsch, J. D., Beyer, C. F., Malkowitz, L., Loullis, C. C., and Blume, A. J. (1988). Characterization of ligand binding to mitochondrial benzodiazepine receptors. *Mol. Pharmacol.* **34,** 164–172.

Holden, M. J., and Colombini, M. (1988). The mitochondrial outer membrane channel, VDAC, is modulated by a soluble protein. *FEBS Lett.* **241,** 105–109.

Igbavboa, U., Zwizinski, C. W., and Pfeiffer, D. R. (1989). Release of mitochondrial matrix protein through a Ca^{2+}-requiring, cyclosporin-sensitive pathway. *Biochem. Biophys. Res. Commun.* **161,** 619–625.

Inoue, I., Nagase, H., Kishi, K., and Higuti, T. (1991). ATP-sensitive K^+ channel in the mitochondrial inner membrane. *Nature (London)* **352,** 244–247.

Junesch, U., and Gräber, P. (1985). The rate of ATP synthesis as a function of ΔpH in normal and dithiothreitol-modified chloroplasts. *Biochim. Biophys. Acta* **809,** 429–434.

Kayser, H., Kratzin, H. D., Thinnes, F. P., Götz, H., Schmidt, W. E., Eckart, K., and Hilschmann, N. (1989). Zur Kenntis der Porine des Menschen. II. Charakterisierung und Primästruktur eines 31-kDa-Porins aus menlischen B-lymphozyten (Porin 31HL)lymphocytes. *Biol. Chem. Hoppe-Seyler* **370,** 1265–1278.

Kinnally, K. W., and Tedeschi, H. (1994). Mitochondrial channels: An integrated view. *In* "Molecular Biology of Mitochondrial Transport Systems" (M. Forte and M. Colombini, eds.). pp. 169–198. Springer Verlag, Berlin.

Kinnally, K. W., Tedeschi, H., and Mannella, C. A. (1987). Evidence for a novel voltage activated channel in the outer mitochondrial membrane. *FEBS Lett.* **226,** 83–87.

Kinnally, K. W., Campo, M. L., and Tedeschi, H. (1988). Evidence for multiple channels in mitochondrial inner membranes: Results from patch-clamping. *Abstr. 14th Int. Cong. Biochem.*, 168.

Kinnally, K. W., Campo, M. L., and Tedeschi, H. (1989). Mitochondrial channel activity studied by patch clamping mitoplasts. *J. Bioenerg. Biomembr.* **21,** 497–506.

Kinnally, K. W., Zorov, D. B., Antonenko, Y. N., and Perini, S. (1991). Calcium modulation of inner mitochondrial channel activity. *Biochem. Biophys. Res. Commun.* **176,** 1183–1188.

Kinnally, K. W., Antonenko, Y. N., and Zorov, D. B. (1992). Modulation of inner mitochondrial membrane channel activity. *J. Bioenerg.* **24,** 99–110.

Kinnally, K. W., Zorov, D. B., Antonenko, Y. N., Snyder, S. H., McEnery, M. W., and Tedeschi, H. (1993). Mitochondrial benzodiazepine receptor linked to inner mitochondrial channels by nanomolar actions of ligands. *Proc. Natl. Acad. Sci.* **90,** 1374–1378.

Klaus, S., Casteilla, L., Bouillaud, F., and Ricquier, D. (1991). The uncoupling protein UCP: A membranous mitochondrial ion carrier exclusively expressed in brown adipose tissue. *Int. J. Biochem.* **23,** 791–801.

Kleene, R., Pfanner, N., Pfaller, R., Lindk, T. A., Sebald, W., Neupert, W., and Tropschug, M. (1987). Mitochondrial porin of *Neurospora crassa*: cDNA cloning, *in vitro* expression and import into mitochondria. *EMBO J.* **6,** 2627–2633.

Klingenberg, M. (1990). Mechanism and evolution of the uncoupling protein of brown adipose tissue. *Trends Biochem. Sci.* **15,** 108–112.

Klingenberg, M., and Winkler, E. (1985). The reconstituted isolated uncoupling protein is a membrane potential driven H$^+$ translocator. *EMBO J.* **4,** 3087–3092.

Klingenberg, M., and Winkler, E. (1994). On the mechanism of H$^+$ transfer in the uncoupling protein of brown adipose tissue. *In* "Molecular Biology of Mitochondrial Transport Systems" (M. A. Forte and M. Colombini, eds.). pp. 11–18. Springer Verlag, Berlin.

Klitsch, T., and Siemen, D. (1991). Inner mitochondrial membrane anion channel is present in brown adipocytes but is not identical with the uncoupling protein. *J. Membr. Biol.* **122,** 69–75.

Knoll, G., and Brdiczka, D. (1983). Changes in freeze-fractured mitochondrial membranes correlated to their energy state. Dynamic interactions of the boundary membranes. *Biochim. Biophys. Acta* **733,** 102–110.

Kottke, M., Adam, V., Riesinger, I., Bremm, G., Bosch, W., Brdiczka, D., Sandri, G., and Panfili, E. (1988). Mitochondrial boundary membrane contact sites in brain: Points of hexokinase and creatine kinase location, and control of Ca^{2+} transport. *Biochim. Biophys. Acta* **935,** 87–102.

Krause, J., Hay, R., Kowollik, C., and Brdiczka, D. (1986). Cross-linking analysis of yeast mitochondria outermembrane. *Biochim. Biophys. Acta* **860,** 690–698.

Last, T. A., Gantzer, M. L., and Tyler, C. D. (1983). Ion-gated channel induced in planar bilayers by incorporation of (Na$^+$,K$^+$)-ATPase. *J. Biol. Chem.* **258,** 2399–2404.

Laubinger, W., and Dimroth, P. (1988). Characterization of the ATP synthase of *Propionigenium modestum* as a primary sodium pump. *Biochemistry* **27,** 7531–7537.

Lill, H., Althoff, G., and Junge, W. (1987). Analysis of ionic channels by flash spectrophotometric technique applicable to thylakoid membranes: CF$_0$, the proton channel of the chloroplast ATP synthase, and, for comparison gramicidin. *J. Membr. Biol.* **98,** 69–78.

Lin, C. S., and Klingenberg, M. (1980). Isolation of uncoupling protein from brown adipose tissue mitochondria. *FEBS Lett.* **113,** 299–303.

Lindén, M., Gellerfors, P., and Nelson, B. D. (1982). Pore protein and the hexokinase-binding protein of rat liver mitochondria are identical. *FEBS Lett.* **141,** 189–192.

Lindén, M., Andersson, G., Gellefors, P., and Nelson, B. D. (1984). Subcellular distribution of rat liver porin. *Biochim. Biophys. Acta* **770,** 93–96.

Liu, M. Y., and Colombini, M. (1991). Voltage gating of the mitochondrial outer membrane channel VDAC is regulated by a very conserved protein. *Am. J. Physiol.* **260,** C371–C374.

Liu, M. Y., and Colombini, M. (1992). Regulation of mitochondrial respiration by controlling the permeability of the outer membrane through the mitochondrial channel VDAC. *Biochim. Biophys. Acta* **1098,** 255–260.

Ludwig, O., Benz, R., and Schulz, J. E. (1989). Porin of *Paramecium*

mitochondria. Isolation, characterization and ion-selectivity of the closed state. *Biochim. Biophys. Acta* **978,** 319–327.

Mannella, C. A. (1982). Structure of the outer mitochondrial membrane: Ordered arrays of pore-like subunits in the outer mitochondrial membrane fractions form *Neurospora crassa. J. Cell Biol.* **94,** 680–687.

Mannella, C. A. (1987). Electron microscopy and image analysis of the mitochondrial outer membrane channel, VDAC. *J. Bioenerg. Biomembr.* **19,** 329–340.

Mannella, C. A. (1990). Structural analysis of mitochondrial pores. *Experientia* **46,** 137–145.

Mannella, C. A. (1992). The "ins" and "outs" of mitochondrial membrane channels. *Trends Biochem. Sci.* **17,** 315–320.

Mannella, C. A. (1994). Insights into the structure of the mitochondrial channel, VDAC, provided by electron microscopy. *In* "Molecular Biology of Mitochondrial Transport Systems" (M. A. Forte and M. Colombini, eds.). pp. 249–264. Springer Verlag, Berlin.

Mannella, C. A., Forte, M., and Colombini, M. (1992). Toward the molecular structure of the mitochondrial channel, VDAC. *J. Bioenerg. Biomembr.* **24,** 7–19.

McEnery, M. W. (1992). The mitochondrial benzodiazepine receptor: Evidence for association with voltage-dependent anion channel (VDAC). *J. Bioenerg. Biomembr.* **24,** 63–70.

McEnery, M. W., Snowman, A. M., Trifiletti, R. R., and Snyder, S. H. (1992). Isolation of the mitochondrial benzodiazepine receptor: Association with the voltage dependent anion channel and the adenine nucleotide carrier. *Proc. Natl. Acad. Sci. USA* **89,** 3170–3174.

Mihara, K., and Sato, R. (1985). Molecular cloning and sequencing of cDNA for yeast porin, an outer mitochondrial membrane protein: A search for targeting signal in the primary structure. *EMBO J.* **4,** 769–774.

Mironova, G. D., Fedotcheva, N. I., Makarov, P. R., Provenich, L. A., and Mironov, G. P. (1981). A protein of the bovine heart mitochondria inducing channel potassium conductivity in bilayer lipid membranes. *Biofizika* **26,** 451–457.

Mironova, G. D., Sirota, T. V., Pronevich, L. A., Trofimenko, N., Mironov, G. P., Gregorjev, P. A., and Kondrashova, M. N. (1982). Isolation and properties of Ca^{2+}-transporting glycoprotein and peptide from beef heart mitochondria. *J. Bioenerg. Biomembr.* **14,** 213–225.

Mironova, G. D., Bocharnikova, N. I., Mirsalikhova, N. M., and Mironov, G. P. (1986). Ion-transporting properties and ATPase activity of Na$^+$/K$^+$)-ATPase large subunit into bilayer lipid membranes. *Biochim. Biophys. Acta* **861,** 224–236.

Mirzabekov, T. A., and Akhemerov, R. N. (1987). Channel forming membrane protein (32kDa) of brown fat mitochondria. *Biofizika* **32,** 345–346.

Moran, O., Sandri, G., Panfili, E., Stühmer, W., and Sorgato, M. C. (1990). Electrophysiological characterization of contact sites in brain mitochondria. *J. Biol. Chem.* **265,** 908–913.

Moran, O., Sciancalepore, M., Sandri, G., Panfili, E., Bassi, R., Ballarin, C., and Sorgato, M. C. (1992). Ionic permeability of the mitochondrial outer membrane. *Eur. Biophys. J.* **21,** 311–319.

Moreadith, R. W., and Jacobus, W. E. (1982). Creatine kinase of heart mitochondria. Functional coupling of ADP transfer to the adenine nucleotide translocase. *J. Biol. Chem.* **257,** 899–905.

Nakashima, R. A., Mangan, P. S., Colombini, M., and Pedersen, P. L. (1986). Hexokinase receptor complex in hepatoma mitochondria: Evidence from N,N'-dicyclohexylcarbodiimide-labeling. Studies for the involvement of VDAC. *Biochemistry* **25,** 1015–1021.

Needham, L., Dodd, N. J. F., and Houslay, M. D. (1987). Quinidine and mellitin both decrease the fluidity of liver plasma membranes and both inhibit hormone-stimulated adenylate cyclase activity. *Biochim. Biophys. Acta* **899**, 44–50.

Nicholls, D. G., Cunningham, S. A., and Rial, E. (1986). The bioenergetic mechanisms of brown adipose tissue mitochondria. *In* "Brown Adipose Tissue" (P. Trayhurn and D. G. Nicholls, eds.), pp. 52–85. Edward Arnold, London.

Nicolli, A., Redetti, A., and Bernardi, P. (1991). The K$^+$ conductance of the inner mitochondrial membrane. *J. Biol. Chem.* **266**, 9465–9470.

Ostlund, A. K., Gohring, U., Krause, J., and Brdiczka, D. (1983). The binding of glycerol kinase to the outer membrane of rat liver mitochondria: Its importance in metabolic regulation. *Biochem. Med.* **30**, 231–245.

Paliwal, R., Costa, G., and Diwan, J. J. (1992). Purification and patch clamp analysis of a 40-pS channel from rat liver mitochondria. *Biochemistry* **31**, 2223–2229.

Papa, S., Guerrieri, F., Zanotti, F., and Capozza, G. (1994). F$_0$ and F$_1$ subunits of the gating and coupling function of mitochondrial H$^+$-ATP synthase. *In* "Molecular Biology of Mitochondrial Transport Systems" (M. A. Forte and M. Colombini, eds.). pp. 19–40. Springer Verlag, Berlin.

Peng, S., Blachly-Dyson, E., Forte, M., and Colombini, M. (1992). Large scale rearrangement of protein domains is associated with voltage gating of the VDAC channel. *Biophys. J.* **62**, 123–135.

Petronilli, V., Szabó, I., and Zoratti, M. (1989). The inner mitochondrial membrane contains ion-conducting channels similar to those found in bacteria. *FEBS Lett.* **259**, 137–143.

Puigserver, O., Herron, D., Gianotti, M., Palou, A., Cannon, B., and Nedergaard, J. (1992). Induction and degradation of uncoupling protein thermogenin in brown adipocytes in vitro and in vivo. Evidence for a rapidly degradable pool. *Biochem. J.* **284**, 393–398.

Rehnmark, S., Bianco, A. C., Kiefer, J. D., and Silva, J. E. (1992). Transcriptional and posttranscriptional mechanisms in uncoupling protein. *Am. J. Physiol.* **262**, E58–67.

Ricquier, D., and Bouillaud, F. (1986). The brown adipose tissue mitochondrial uncoupling protein. *In* "Brown Adipose Tissue" (P. Trayhurn and D. G. Nicholls, eds.), pp. 86–104. Edward Arnold, London.

Ricquier, D., Barlet, J. P., Garel, J. M., Combes-George, M., and Dubois, M. P. (1983). An immunological study of the uncoupling protein of brown adipose tissue mitochondria. *Biochem. J.* **210**, 859–866.

Ricquier, D., Casteilla, L., and Bouillaud, F. (1991). Molecular studies of uncoupling protein. *FASEB J.* **5**, 2237–2242.

Ricquier, D., Raimbault, S., Champigny, O., Miroux, P., and Bouillaud, F. (1992). The coupling protein is not expressed in rat liver. *FEBS Lett.* **303**, 103–106.

Runswick, M. J., Powell, J. T., Nyren, P., and Walker, J. E. (1987). Sequence of the bovine mitochondrial phosphate carrier protein; Structural relationship to ADP/ATP translocase and the brown fat mitochondria uncoupling protein. *EMBO J.* **6**, 1367–1373.

Saks, V. A., Kupriyanov, V. V., and Elizarove, G. V. (1980). Studies of energy transport in heart cells. The importance of creatine kinase localization for the coupling of mitochondrial phosphoryl creatine production to oxidative phosphorylation. *J. Biol. Chem.* **255**, 755–763.

Saks, V. A., Khychua, Z. A., and Kuznetsov, A. V. (1987). Specific inhibition of ATP–ADP translocase in cardiac mitoplasts by antibodies against creatine kinase. *Biochim. Biophys. Acta* **891**, 138–144.

Sandri, G., Siagri, M., and Panfili, E. (1988). Influence of Ca^{2+} on the isolation from rat-brain mitochondria of a fraction enriched in boundary membrane contact sites. *Cell Calcium* **9**, 159–165.

Scarpa, A., and Azzone, G. F. (1970). The mechanism of ion translo-

cation in mitochondria. 4. Coupling of K$^+$ efflux with Ca^{2+} uptake. *Eur. J. Biochem.* **12**, 328–335.

Schein, S. J., Colombini, M., and Finkelstein, A. (1976). Reconstitution in planar lipid bilayers of voltage-dependent anion-selective channel obtained from *Paramecium* mitochondria. *J. Membr. Biol.* **30**, 99–120.

Schindler, H., and Nelson, N. (1982). Proteolipid of adenosinetriphosphatase from yeast mitochondria forms proton-selective channels in planar lipid bilayers. *Biochemistry* **21**, 5787–5794.

Schlegel, J., Wyss, M., Eppenberger, H. M., and Walliman, T. (1990). Functional studies with the octameric and dimeric form of mitochondrial creatine kinase. *J. Biol. Chem.* **265**, 9221–9227.

Schneider, E., and Altendorf, K. (1985). All three subunits are required for reconstitution of an active proton channel F$_0$ of *Escherichia coli* ATP synthase (F$_1$F$_0$). *EMBO J.* **4**, 515–518.

Schneider, E., and Altendorf, K. (1987). Bacterial adenosine 5′-triphosphate synthase (F$_1$F$_0$): Purification and reconstitution of F$_0$ complexes and functional characterization of their subunits. *Microbiol. Rev.* **51**, 477–497.

Seeman, P. M. (1966). Membrane stabilization by drugs: Tranquilizers, steroids and anesthetics. *Int. Rev. Neurobiol.* **9**, 146–221.

Selwyn, M. J., Dawson, A. P., and Fulton, D. V. (1979). An anion-conducting pore in the mitochondrial inner membrane. *Biochem. Soc. Trans.* **7**, 216–219.

Senior, A. E. (1990). The proton-translocating ATPase from *Escherichia coli*. *Annu. Rev. Biophys. Biophys. Chem.* **19**, 7–41.

Sinohara, Y., Shima, A., Kamida, N., and Terada, H. (1991). Uncoupling protein is expressed in liver mitochondria of cold-exposed and newborn rats. *FEBS Lett.* **293**, 173–174.

Sorgato, M. C., and Moran, O. (1993). Channels in mitochondrial membranes: knowns, unknowns, and prospects for the future. *Critical Rev. Biochem. Mol. Biol.* **18**, 127–171.

Sorgato, M. C., Keller, B. U., and Stühmer, W. (1987). Patch-clamping of the inner mitochondrial membrane reveals a voltage-dependent ion channel. *Nature (London)* **330**, 498–500.

Sorgato, M. C., Moran, O., DePinto, V., Keller, B. U., and Stühmer, W. (1989). Further investigation on the conductance ion channel of the inner mitochondrial membrane. *J. Bioenerg. Biomembr.* **21**, 485–496.

Sorgato, M. C., Ballarin, C., and Moran, O. (1994). A minimal kinetic model of the activity of the 107 pS channel of the inner membrane of mitochondria. *In* "Molecular Biology of Mitochondrial Transport Systems" (M. Forte and M. Colombini, eds.). pp. 131–136. Springer Verlag, Berlin.

Stühmer, W., Keller, B., Lippe, G., and Sorgato, M. C. (1988). The outer and inner membranes of mice liver giant mitochondria can be patch clamped: high conductance ion channels are present in both membranes. *In* "Hormones and Cell Regulation" (J. Nunez, J. E. Dumont, and E. Carafoli, eds.), pp. 89–99. INSERM-Libby Eurotext Ltd., London.

Szabó, I., and Zoratti, M. (1991). The giant channel of the inner mitochondrial membrane is inhibited by cyclosporin A. *J. Biol. Chem.* **266**, 3376–3379.

Szabó, I., and Zoratti, M. (1992). The mitochondrial megachannel is the permeability transition pore. *J. Bioenerg. Biomembr.* **24**, 111–117.

Szabó, I., Bernardi, P., and Zoratti, M. (1992). Modulation of the mitochondrial megachannel by divalent cations and protons. *J. Biol. Chem.* **267**, 2940–2946.

Tanford, C. (1983). Mechanism of free energy coupling in active transport. *Annu. Rev. Biochem.* **52**, 379–409.

Tedeschi, H., Mannella, C. A., and Bowman, C. L. (1987). Patch clamping the outer mitochondrial membrane. *J. Membr. Biol.* **97**, 21–29.

Tedeschi, H., Kinnally, K. W., and Mannella, C. A. (1989). Properties of channels in the mitochondrial outer membrane. *J. Bioenerg. Biomembr.* **21,** 451–459.

Thieffry, M., Chich, J. K., Goldsmith, D., and Henry, J. P. (1988). Incorporation in lipid bilayers of a large conductance cationic channel from mitochondrial membranes. *EMBO J.* **7,** 1449–1454.

Thieffry, M., Neyton, J., Pelleschi, M., Fèvre, F., and Henry, J.-P. (1992). Properties of the mitochondrial peptide-sensitive cationic channel studied in planar bilayers and patches of giant liposomes. *Biophys. J.* **63,** 333–339.

Thieffry, M., Fèvre, F., Pelleschi, M., and Henry, J. P. (1994). Characterization of a cationic peptide sensitive channel from mitochondrial outer membrane. *In* "Molecular Biology of Mitochondrial Transport Systems" (M. A. Forte and M. Colombini, eds.). pp. 209–220. Springer Verlag, Berlin.

Thomas, L., Kocsis, E., Colombini, M., Erbe, E., Trus, B. L., and Steven, A. C. (1991). Surface topography and molecular stoichiometry of the mitochondrial channel, VDAC, in crystalline arrays. *J. Struct. Biol.* **106,** 161–171.

Verma, A., and Snyder, S. H. (1988). Characterization of porphyrin interactions with peripheral type benzodiazepine receptors. *Mol. Pharmacol.* **34,** 800–805.

Wunder, U. R., and Colombini, M. (1991). Patch clamping VDAC in liposomes containing whole mitochondrial membranes. *J. Membr. Biol.* **123,** 83–91.

Yang, W. C. T., Geiger, P. J., Bessman, S. P., and Borrebaek, B. (1977). Formation of creatine phosphate and ^{32}P-labelled ATP by isolated rabbit heart mitochondria. *Biochem. Biophys. Res. Commun.* **76,** 882–887.

Zimmerberg, J., and Parsegian, V. A. (1986). Polymer inaccessible volume changes during opening and closing of voltage-dependent ion channel. *Nature (London)* **323,** 36–39.

Zimmerberg, J., and Parsegian, V. A. (1987). Water movement during channel opening and closing. *J. Bioenerg. Biomembr.* **19,** 351–358.

Zizi, M., Forte, M., Blachly-Dyson, E., and Colombini, M. (1994). NADH regulates the gating of VDAC, the mitochondrial outer membrane channel. *J. Biol. Chem.* **269,** 1614–1616.

Zorov, D. B., Kinnally, K. W., Perini, S., and Tedeschi, H. (1992a). Multiple conductance levels in rat heart inner mitochondrial membranes studied by patch clamping. *Biochim. Biophys. Acta* **1105,** 263–270.

Zorov, D. B., Kinnally, K. W., and Tedeschi, H. (1992b). Voltage activation of the heart inner mitochondrial membrane channels. *J. Bioenerg. Biomembr.* **24,** 119–124.

ATP-Sensitive K⁺ Channels in the Inner Membrane of Rat Liver Mitochondria

I. INOUE AND T. HIGUTI

I. Introduction

Mitochondria transport various inorganic and organic ions as well as larger substances such as proteins (cf., Hay et al., 1984; Diwan, 1987). To understand the dynamic processes involved in mitochondrial physiology, the spatial and the time resolution of measurements of ion transport across the mitochondrial membranes must be improved. The patch-clamp technique has achieved these goals. Several attempts have been made to detect microscopic currents from the mitochondrial membranes by applying this technique, and several ion conducting pores (channels) with conductances of 5–1000 pS have been demonstrated in outer and inner membranes (Sorgato et al., 1987,1989; Thieffry et al., 1988; Henry et al., 1989; Kinnally et al., 1989; Moran et al., 1990) (see Chapter 34). Using fused giant mitoplasts, a K⁺-selective channel in the inner mitochondrial membrane has been demonstrated.

II. Patch Clamp to Giant Mitoplast

The outer membranes of isolated rat liver mitochondria were removed by digitonin treatment to make mitoplasts with their inner membrane matrix intact. These mitoplasts were then converted into a spherical configuration by hypotonic swelling, and were fused in a low pH solution (pH 6.5) containing Ca^{2+}. Fusion was stopped by switching the bathing solution to a low Ca^{2+} (10^{-7} M) buffered medium (pH 7.2). Figure 1 shows fused mitoplasts that adhered tightly to a coverslip.

Giant mitoplasts of 7- to 20-μm diameter were patch clamped after switching the bathing solution to one containing 33.3 mM KCl, 66.7 mM NaCl, 7.5 mM MOPS-Na, and 2 mM EGTA at pH 7.2. Patch pipettes were pulled from borosilicate tubing (1.5-μm outer diameter), the tip was slightly heat-polished, and the pipette was filled with a solution containing 100 mM KCl, 7.5 mM MOPS-Na, 1 mM EGTA, and 0.55 mM $CaCl_2$ (pH 7.2). The portion near the tip was then silicone-coated with a drop of N,N-dimethyl-trimethyl-silylamide. The electrode resistance was 5–7 MΩ. Cell-attached patches were made by giving a small negative pressure (0–0.5 cm H_2O) relative to atmospheric pressure. A gigaseal patch was normally formed in 20–30 min. The seal resistance was 5–20 GΩ. Excised inside-out patches could be obtained simply by separating the electrode tip from the mitoplast. All experiments were conducted at a temperature of 35 ± 2°C.

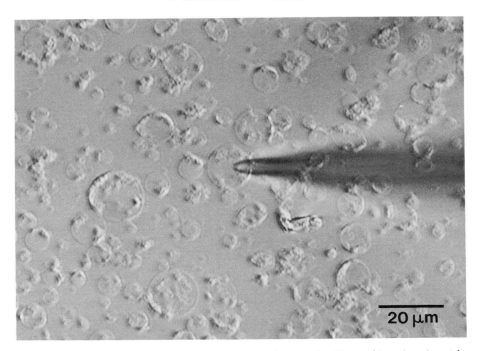

FIGURE 1 Light micrograph of fused giant mitoplasts under Nomarski optics. A patch pipette was introduced onto a giant mitoplast 12 μm in diameter. Methods: Mitochondria were isolated from fresh liver of male Wistar rat (about 250 g) by the method of Schnaitman and Greenawalt (1968). Mitoplasts were prepared by a modified method of Chan *et al.* (1970) as follows. The isolated mitochondria were suspended in H-medium containing 220 mM D-mannitol, 70 mM sucrose, 2.5 mM HEPES-Na, and 0.5 mg/ml crystalline bovine serum albumin at pH 7.4, to give 100-mg mitochondrial protein per ml. Then the mitochondria were treated with the same volume of H-medium containing 10 mg/ml digitonin, purified as described previously (Higuti *et al.*, 1980) for 20 min while stirring gently on ice to remove the outer membrane and to generate mitoplasts with their inner membrane matrix intact. Immediately, the reaction was stopped by adding three volumes of H-medium (DH-medium) diluted 7.5-fold with deionized water and centrifuging the mixture at 10,000 g for 10 min at 4°C. The precipitates obtained (swollen mitoplasts) were suspended with a half volume of DH-medium and centrifuged at 10,000 g for 10 min at 4°C. The resulting precipitate was suspended with a small volume of DH-medium. Electron micrographs have shown that the mitoplasts contain virtually no outer membrane (Schnaitman *et al.*, 1967; Chan *et al.*, 1970). We also confirmed that the resulting mitoplasts had almost no monoamine oxidase activity (an enzyme located in the outer membrane; Schnaitman *et al.*, 1967). Giant mitoplasts were prepared by a modified method of Chazotte *et al.* (1985) as follows. The swollen mitoplasts were suspended in DH-medium at pH 6.5 at a concentration of 0.4 mg protein per ml, and then mixed with the same volume of a Ca^{2+} solution containing 20 mM $CaCl_2$, 5 mM HEPES-K at pH 6.5. This suspension was aliquoted (4 ml) into plastic dishes and incubated at 37°C for 30 min. Coverslips were placed on the bottom of the dishes, to which the giant mitoplasts tightly adhered.

III. Effect of 4-Aminopyridine

Several types of single-channel currents could be recorded from inside-out membrane patches of the mitoplasts. Most channels had large conductances and showed no selectivity to particular ions, as previously reported. Figure 2 shows that a current that seemed to be background noise was further blocked by 4-aminopyridine (4-AP), suggesting that K$^+$-selective channels exist in the mitochondrial inner membrane (Meves and Pichon, 1977). With a higher sweep velocity, the noise was determined to consist of many small-amplitude currents, the smallest of which was -0.75 pA at $V_m = -50$ mV. Nevertheless, current suppression took place in an abrupt manner. Such a highly cooperative openclosed behavior was one of the remarkable characteristics of these channels. With longer recordings (>1 hr), the channel activity was found to run down gradually.

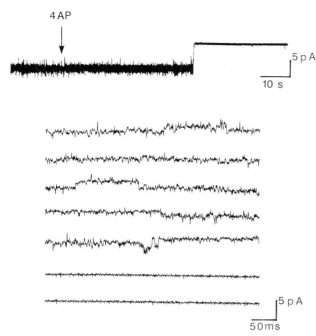

FIGURE 2 (*Top*) Current trace under inside-out patch mode demonstrating that channel activity that seemed to be background noise was inhibited further by 5 mM 4-aminopyridine (4-AP) applied to the bath (matrix side). $V_m = -50$ mV. (*Bottom*) The current record at a higher sweep velocity, showing that the current noise consisted of a number of small channel currents and that channel activities were completely suppressed by 4-AP, suggesting that these channels are K⁺ channels. The membrane potential (V_m) under inside-out patch mode represents the electrical potential of the bath (matrix side) referenced to that of the pipette solution. Transmembrane currents directed from the pipette to the bath have a negative sign, and are displayed as downward deflections in the current records.

IV. ATP Sensitivity of the Channel

A peculiar characteristic of this channel is that the channel activity can be reversibly inhibited by ATP applied to the matrix side. Because of the run down, the channel activity could not be fully restored by removal of ATP from the bath. Current traces in Fig. 3 demonstrate such a reversible ATP effect on channel activity. These records were obtained 40 min after the inside-out patch was made. Since only a few channels were still active in the patch, the current traces demonstrate clear single-channel events. Single-channel currents of -0.85 pA at $V_m = -70$ mV can be seen in the amplitude histograms (Fig. 3B). However, when two channels were active in the patch, the one open state was unstable and behaved like a substate of the dual open state. A similar phenomenon is seen in Fig. 2.

Figure 4 plots the relative open probability as a function of ATP concentration. The channel activity

was half-inhibited at about 0.8 mM ATP and almost blocked above 2 mM. In contrast, ADP or GTP had no significant effect on the ATP-sensitive channel activity. Mg^{2+} was not required for ATP blocking of the channel.

Figure 5 shows ATP-sensitive currents at different membrane potentials, and large currents that appeared only at positive membrane potentials, recorded from the same inside-out patch. The current–voltage (I–V) relationship for the ATP-sensitive currents shows that current amplitude increased linearly in the negative voltage region, and that reversal potential, estimated by linear extrapolation to the voltage axis, was $+27$ mV—close to the equilibrium potential for K⁺ ($+29.5$ mV) and indicating that the ATP-sensitive channel is K⁺ selective. The average value of the slope conductance obtained from six experiments was 9.7 pS with a standard deviation of 1.0 pS. The other voltage-dependent channel had a slope conductance of 105 pS and a reversal potential of 0 mV. The amplitude histograms indicate that the activity of the ATP-sensitive channel is weakly dependent on membrane potential. The histogram at $+70$ mV demonstrates the presence of two ionic channels.

V. Effect of Glibenclamide

Glibenclamide, a sulfonylurea, is a specific and potent blocker of ATP-sensitive K⁺ channels in various types of cells (Ashcroft *et al.*, 1984). Figure 6 shows that the mitochondrial ATP-sensitive K⁺ channel is also blocked by 5 μM glibenclamide applied to the matrix side, in an abrupt manner similar to that shown in Fig. 2. The smallest current amplitude (marked by arrows) corresponds to a 10-pS channel; other current amplitudes were multiples of this unit current amplitude.

The electrophysiological and pharmacological properties suggest that the ATP-sensitive channel of the mitochondrial inner membrane is one of a family of ATP-sensitive K⁺ channels (K_{ATP} channels). These experiments show that the K_{ATP} channel may be only a K⁺-selective channel in the mitochondrial inner membrane. As shown in Figs. 2, 3, and 6, the K_{ATP} channel tends to gate cooperatively rather than independently, suggesting that the K_{ATP} channels exist in clusters and interact with each other. The number of channels in one cluster, estimated from the records in Figs. 2 and 6, is five to seven. Since mitochondria do not have DNA encoding the K_{ATP} channel, the channel protein must be transported from the cytoplasm and be incorporated into the mitochondrial inner membrane.

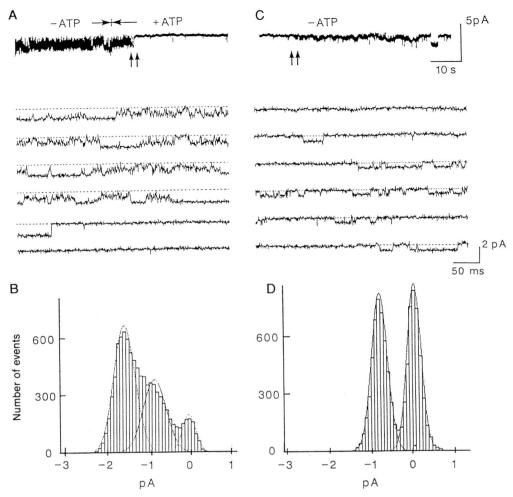

FIGURE 3 (A) Current traces at $V_m = -70$ mV under inside-out patch mode, demonstrating reversible suppression effect of ATP on the channel activity. ATP (2 mM) was applied to the bath solution. Lower traces show the currents in the upper traces (between the two arrows) at a higher sweep velocity; dashed line indicates zero current level. (B) Amplitude histograms of the same current records before application of ATP (*left*) and after reactivation of the channels by removal of ATP (*right*). The channel of -0.85 pA has a slope conductance of 9.0 pS, which was calculated from the current–voltage relationship. Reproduced with permission from Inoue *et al.* (1991).

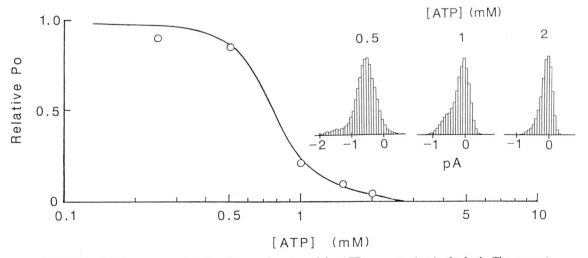

FIGURE 4 Relative open probability (P_o) as a function of the ATP concentration in the bath. The current data were obtained 1 hr after an inside-out patch was made. $V_m = -50$ mV. Amplitude histograms at three ATP concentrations are presented. The mean current amplitude was 0.72 pA, which corresponds to 9.5 pS when the reversal potential is assumed to be $+26$ mV (see Fig. 5). The apparent dissociation constant at the half-maximal inhibition is 8×10^{-4} M. Reproduced with permission from Inoue *et al.* (1991).

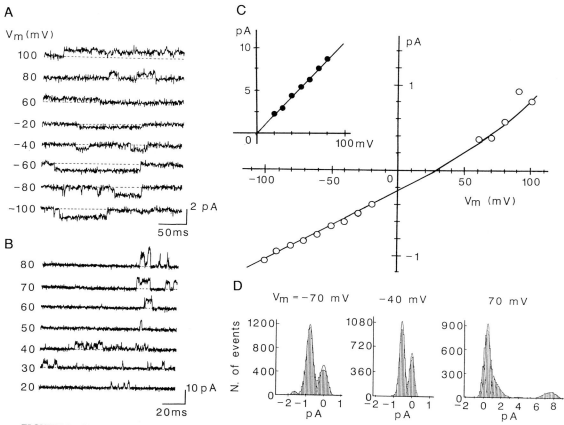

FIGURE 5 (A) ATP-sensitive single-channel currents at various membrane potentials. The records were taken 30–35 min after the inside-out patch was made. The pipette solution contained 100 mM K+; the bath solution contained 33.3 mM K+ and 66.7 mM Na+. (B) ATP-insensitive currents with a larger amplitude, which were recorded only at the positive membrane potentials in the same membrane patch. (C) Current–voltage relationships of the ATP-sensitive channel (○) and the ATP-insensitive channel (●). (D) Amplitude histograms at $V_m = -70$ mV (*left*), at $V_m = -40$ mV (*center*), and at $V_m = +70$ mV (*right*). In negative membrane potentials, only the ATP-sensitive channel was active. P_o at -70 mV was 0.67; P_o at $V_m = -40$ mV was 0.65. Two major peaks at $V_m = +70$ mV, one at 0.41 pA and another at 7.55 pA, represent the ATP-sensitive and -insensitive channels, respectively. Values of P_o of the ATP-sensitive channel and the voltage-dependent channel were 0.58 and 0.11, respectively. Reproduced with permission from Inoue *et al.* (1991).

VI. Possible Role of the K$_{ATP}$ Channel

Currently, the role of the K$_{ATP}$ channel in mitochondrial physiology is unclear. One peculiar property of this channel is that mitochondria generate a K+ conductance if the ATP level of the matrix becomes too low. However, the K$_{ATP}$ channel may also be regulated by other agents, since the mitochondrial ATP concentration does not decrease that much under normal conditions. In any case, once the K$_{ATP}$ channel is activated, the resulting increase in K+ conductance may lead to a depolarization of the inner membrane and may also result in a change in the water content of the mitochondrion. The K$_{ATP}$ channel activity would dissipate the membrane potential of energized mitochondria, which is estimated to be between -150 and -160 mV (Rottenberg, 1979). Mitochondria in living cells are known to swell and contract by changing their water content; some of these changes are related to the respiratory chain (Lehninger, 1964; Garlid, 1988). Low water content is maintained during respiration under phosphorylating conditions, and mitochondria swell when adenine nucleotides or substrates are absent (Lehninger, 1964). Whether the K$_{ATP}$ channel is involved in the swelling–contraction cycle by serving as an ATP sensor, and whether it is related to a Ca^{2+}-dependent regulatory system of oxidative phosphorylation remains to be determined (Harris and Das, 1991; Higuti *et al.*, 1992).

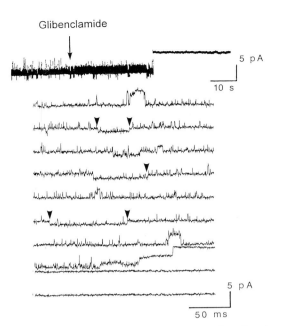

FIGURE 6   (Top) Current trace demonstrating that the channel activity was completely blocked by 5 mM glibenclamide applied to the bath under inside-out configuration. $V_m = -50$ mV. The patch was pretreated by 1 mM ATP, which reduced the channel activity by half. (Bottom) The currents at a higher sweep velocity. Transitions marked by arrows indicate unitary currents with an amplitude of 0.75 pA, which corresponds to 9.9 pS by assuming the reversal potential of +26 mV.

Acknowledgments

I. Inoue was supported by Grant 04266216 and T. Higuti by Grants 02204009, 04680196, and 04266215 from the Ministry of Education and Culture of Japan.

References

Ashcroft, F. M., Harrison, D. E., and Ashcroft, S. J. H. (1984). Glucose induces closure of single potassium channels in isolated rat pancreatic β cells. Nature (London) 312, 446–448.

Chan, T. L., Greenawalt, J. W., and Pedersen, P. L. (1970). Biochemical and ultrastructural properties of a mitochondrial inner membrane fraction deficient in outer membrane and matrix activities. J. Cell Biol. 45, 291–305.

Chazotte, B., Wu, E., Hochli, M., and Hackenbrock, C. R. (1985). Calcium-mediated fusion to produce ultra large osmotically active mitochondrial inner membrane of controlled protein density. Biochim. Biophys. Acta 818, 87–95.

Diwan, J. J. (1987). Mitochondrial transport of K^+ and Mg^{2+}. Biochim. Biophys. Acta 895, 155–165.

Garlid, K. D. (1988). Mitochondrial volume control. In "Integration of Mitochondrial Function" (J. J. Lemasters, C. R. Hackenbrock, R. G. Thuman, and H. V. Westerhoff, eds.), pp. 259–278. Plenum Press, New York.

Harris, D. A., and Das, A. M. (1991). Control of mitochondrial ATP synthesis in the heart. Biochem. J. 280, 561–573.

Hartl, F. U., Pfanner, N., Nicholson, D. W., and Neupert, W. (1989). Mitochondrial protein import. Biochim. Biophys. Acta 988, 1–45.

Hay, R., Bohni, P., and Gasser, S. (1984). How mitochondria import proteins. Biochim. Biophys. Acta 779, 65–87.

Henry, J.-P., Chich, J.-F., Goldschmidt, D., and Thieffry, M. (1989). Blockade of a mitochondrial cationic channel by an addressing peptide: An electrophysiological study. J. Membr. Biol. 112, 139–147.

Higuti, T., Niimi, S., Saito, R., Nakashima, S., Ohe, T., Tani, I., and Yoshimura, T. (1980). Rhodamine 6G, inhibitor of both H^+-ejections from mitochondria energized with ATP and with respiratory substrates. Biochim. Biophys. Acta 593, 1463–1467.

Higuti, T., Kuroiwa, K., Kawamura, Y., and Yoshihara, Y. (1992). The complete amino acid sequence of subunit e of rat liver mitochondrial H^+-ATP synthase. Biochemistry 31, 12451–12454.

Inoue, I., Nagase, H., Kishi, K., and Higuti, T. (1991). ATP-sensitive K^+ channel in the mitochondrial inner membrane. Nature (London) 352, 244–247.

Kinnally, K. W., Campo, M. L., and Tedeschi, H. (1989). Mitochondrial channel activity studied by patch-clamping mitoplasts. J. Bioenerg. Biomembr. 4, 497–506.

Lehninger, A. L. (1964). Energy-coupled changes of volume and structure. In "The Mitochondrion," pp. 180–204. Benjamin, New York.

Meves, H., and Pichon, Y. (1977). The effect of internal and external 4-aminopyridine on the potassium currents in intracellularly perfused squid giant axons. J. Physiol. (London) 268, 511–532.

Moran, O., Sandri, G., Panfili, E., Stuhmer, W., and Sorgato, C. (1990). Electrophysiological characterization of contact sites in brain mitochondria. J. Biol. Chem. 265, 908–913.

Rottenberg, H. (1979). The measurement of membrane potential and ΔpH in cells, organelles, and vesicles. Meth. Enzymol. LV, 547–569.

Schnaitman, C., and Greenawalt, J. W. (1968). Enzymatic properties of the inner and outer membranes of rat liver mitochondria. J. Cell Biol. 38, 158–175.

Schnaitman, C., Erwin, V. G., and Greenawalt, J. W. (1967). The submitochondrial localization of monoamine oxidase. An enzymatic marker for the outer membrane of rat liver mitochondria. J. Cell Biol. 32, 719–735.

Sorgato, M. C., Keller, B. U., and Stuhmer, W. (1987). Patch-clamping of the inner mitochondrial membrane reveals a voltage-dependent ion channel. Nature (London) 330, 498–500.

Sorgato, M. C., Moran, O., De Pinto, V., Keller, B. U., and Stuehmer, W. (1989). Further investigation on the high-conductance ion channel of the inner membrane of mitochondria. J. Bioenerg. Biomembr. 4, 485–496.

Thieffry, M., Chich, J.-F., Goldschmidt, D., and Henry, J.-P. (1988). Incorporation in lipid bilayers of a large conductance cationic channel from mitochondrial membranes. EMBO J. 7, 1449–1454.

Channels
of
Exocytotic
Fusion

Structure and Regulation of the Exocytotic Fusion Pore

JONATHAN R. MONCK,
ANDRES F. OBERHAUSER,
PETER E. R. TATHAM,
AND JULIO M. FERNANDEZ

I. Fusion Pores and Cell Function

The fusion pore is the molecular structure that connects the lumens of two membrane compartments during their fusion. Membrane fusion is a ubiquitous process that occurs in all cells as cellular proteins are shuttled between intracellular compartments or are carried to the cell surface in specialized transport vesicles, which deliver their contents by fusing with the membrane of the targeted compartment. Vesicular traffic is involved in the generation and maintenance of cellular structure. Exocytosis is the specialized case in which a vesicle fuses with the plasma membrane, and is divided into two types. In the first type, which occurs in all eukaryotic cells, membrane proteins are carried to the cell surface in transport vesicles that fuse with the cell membrane in an apparently unregulated constitutive fashion. Many cells, including endocrine, exocrine, and neuronal cells, have evolved a second, more obviously regulated, form in which exocytosis only occurs in response to specific cellular stimuli. Secretory products are packaged in a highly condensed form in the lumen of specialized secretory vesicles (or granules) and are stored in readiness for rapid release. Thus, synaptic vesicles store neurotransmitters, endocrine cells store bioactive peptides and amines, and mast cells store histamine. In re-sponse to the appropriate stimulus, the secretory packages are expelled from the cell without passing through the cytosol.

The nature of the intracellular signals that direct and regulate membrane fusion has been extensively studied. In regulated secretion, an elevation of cytosolic Ca^{2+} concentration and the activation of GTP binding proteins are thought to be the primary stimuli, but the regulation is complex and cell specific, and often involves modulation by cAMP, protein kinase C, and other intracellular signals. The signaling pathways have been studied exhaustively and several excellent reviews cover this topic (Burgoyne, 1991; Lindau and Gomperts, 1991; Plattner *et al.*, 1991). However, little is known about how specific intracellular signals interact with the fusion pore.

Intracellular traffic involves movement of transport vesicles between intracellular compartments and involves pinch-off vesiculation from the donor compartment followed by fusion with the acceptor compartment. Attempts to reconstitute these reactions *in vitro* have revealed a number of proteins required for constitutive vesicular traffic in mammalian cells including NEM sensitive factor (NSF), soluble NSF attachment protein (SNAPS), and small GTP binding proteins from the ADP ribosylation factor (ARF) and *rab* families (Balch, 1989; Rothman and Orci, 1992; Wilson *et al.*, 1992). Genetic studies in yeast have also revealed many proteins necessary for sustained traffic and constitutive exocytosis, some of which are related to the mammalian *rab* and NSF proteins (Bacon *et al.*, 1989; Walworth *et al.*, 1989; Bowser *et al.*, 1992). Although much is known about these proteins, little is known about whether they are involved in regulating membrane fusion per se.

Despite the variety of intracellular reactions in which fusion pores participate, and despite the intensity with which the regulatory mechanisms of membrane fusion have been studied for both intracellular traffic and exocytosis, the nature of the fusion pore is unknown. The principal difficulty has been the lack of techniques to directly monitor the activity of single fusion pores in isolation; the reductionist approach that has proven uniquely suited to the study of ion channels. Patch-clamp techniques have provided the ability to monitor the activity of individual fusion pores in mast cells. These studies have generated a wealth of novel and unexpected observations. The goal of this chapter is to review these findings and to speculate on their significance to our understanding of exocytotic fusion. Although the majority of the results come from mast cells, many of the observations are now being reproduced in other secretory cells, suggesting that the conclusions may have wider significance in other membrane fusion reactions.

II. Rapid-Freeze Electron Microscopy of the Fusion Pore

A. Lipidic Fusion Pore Spans the Gap between Fusing Membranes

Rapid freezing techniques can freeze cells in less than 2 msec and are therefore ideal for observing the transient intermediate states associated with vesicle fusion (Chandler, 1992). In a remarkable series of micrographs, Chandler and Heuser (1980) captured the formation of exocytotic fusion pores in mast cells. Similar fusion pores have since been seen in *Limulus* amebocytes (Ornberg and Reese, 1981), neutrophils (Chandler *et al.*, 1983), chromaffin cells (Schmidt *et al.*, 1983; Nakata *et al.*, 1990), and *Paramecium* (Knoll *et al.*, 1991), thus providing a common picture of the exocytotic event in rapidly frozen cells. The micrographs clearly show membrane lined pores, 20–100 nm in diameter, that provide a water-filled pathway between the lumen of the secretory granule and the extracellular environment. As can be seen in Fig. 1C, these pores appear to be made of a curved bilayer that spans the granule and plasma membranes.

B. Dimple in the Plasma Membrane Is Seen Prior to Fusion Pore Formation

The quick-freeze experiments also provide clues to the events preceding the formation of a fusion pore. In unstimulated mast cells, the plasma membrane and secretory granule membranes are kept at least 100 nm apart (Fig. 1A; Chandler and Heuser, 1980). After stimulation, a new pattern is observed in which the plasma membrane invaginates to form a dimple that approaches the granule membrane (Fig. 1B; Chandler and Heuser, 1980). This pattern of dimple formation after cell stimulation has also been observed in *Limulus* amebocytes (Ornberg and Reese, 1981), chromaffin cells (Schmidt *et al.*, 1983), neutrophils (Chandler *et al.*, 1983), and *Paramecium* (Knoll *et al.*, 1991). The secretory granules are never seen to bulge outward toward the plasma membrane, suggesting that the granule membranes are under tension and the plasma membranes are relatively slack. The micrographs reveal that the dimples typically end in a highly curved tip with a diameter of about 10 nm. Because the dimple creates a focal point at which the plasma and granule membranes could potentially interact at close range, it was suggested that fusion pores form at these focal points of contact. Since the plasma membrane dimples have been seen only on cell stimulation, the dimple forming structures are likely to respond to the intracellular messengers generated by the stimulus–secretion coupling mechanisms.

C. Cytoskeletal Scaffold Bridges the Gap between the Fusing Membranes

Another important observation from the electron micrographs is that filamentous structures are seen spanning the gap between the secretory granule and cell membranes. These structures have been seen in mast cells (Fig. 1A; Chandler and Heuser, 1980) as well as *Limulus* amebocytes (Ornberg and Reese, 1981). These filaments might be made of actin since filamentous networks containing actin have been seen between the cell and granule membranes in many secretory cells (Segawa and Yamashina, 1989; Nielsen, 1990; Nakata and Hirokawa, 1991). Alternatively, they could be made of annexins, which have been shown to form filamentous structures between liposomes that closely resemble the filaments between chromaffin granules and the cell membranes (Nakata *et al.*, 1990). Since the filamentous structures are seen in regions in which fusion pores can form, and given the large amount of biochemical evidence implicating cytoskeletal components in regulation of exocytosis (Burgoyne, 1987, 1991; Linstedt and Kelly, 1987), these fil-

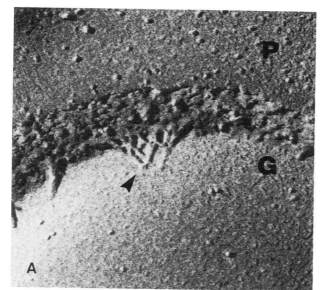

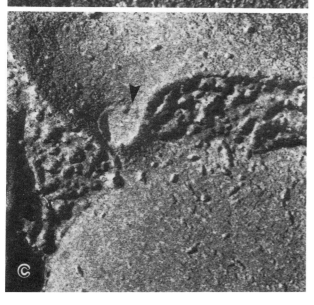

aments may form part of a macromolecular structure that can direct and regulate dimpling and fusion pore formation. This structure can be envisioned as a scaffold which, by bringing about formation of the dimple, causes the two membranes to come into close proximity and promotes conditions that are favorable for membrane fusion.

III. Measurements of Exocytosis and the Fusion Pore

The observations from quick-freeze electron microscopy defined the fusion pore as a discrete entity but provided no kinetic information and are limited to observation of fairly large water-filled structures that can be etched. A glimpse of an earlier smaller fusion pore was provided from patch-clamp studies of exocytosis in the mast cells from the *beige* mouse, a mutant with unusually large secretory granules. As discussed subsequently, admittance measurements with the patch-clamp technique can detect the fusion pore from the instant it conducts ions, and have demonstrated early fusion pores that are much smaller than those seen by electron microscopy. In addition, by measuring the membrane capacitance, the admittance provides a marker of individual exocytotic fusion events and can be used for kinetic studies of the fusion pore as well as for investigating mechanisms of stimulus–secretion coupling. This technique has led to a number of important new discoveries about the nature of the fusion pore and the mechanisms underlying fusion pore formation and development. Although admittance measurements can determine the fusion pore conductance, they cannot measure the current carried by ionic species passing through the fusion pore. Voltammetric methods, which can electrochemically detect the activity of biogenic amines or other redox-sensitive compounds, can pro-

FIGURE 1 Different stages of fusion pore formation and development seen in rapidly frozen rat peritoneal mast cells. (A) Unstimulated cell showing secretory granule clearly separated from the overlying plasma membrane. P, Plasma membrane; G, secretory granule membrane, arrowhead, filamentous structures that appear to span the gap between the plasma and secretory granule membranes. Magnification, 190,000 ×. (B) Cell frozen 15 sec after stimulation with 8 μg/ml compound 48/80. A dimple (arrowhead) forms in the plasma membrane and approaches the secretory granule. Magnification, 225,000 ×. (C) An exocytotic fusion pore (arrowhead) forms at the site of membrane dimpling. Specimen frozen 15 sec after cell stimulation. Magnification, 190,000 ×. Electron micrographs supplied by Douglas E. Chandler (Department of Zoology, Arizona State University, Tempe).

vide a highly sensitive measurement of the secretory products released through a single fusion pore. Before discussing the new findings, we briefly review how these measurements are made.

A. Admittance Measurements with the Patch Clamp Technique

Over the last decade, the patch-clamp technique has created a revolution in biology, making it possible to directly observe the activity of single membrane proteins, namely, ion channels. These studies have revealed a diverse family of ion channels, many novel regulatory mechanisms, and, more recently, in combination with molecular biological techniques, an insight into structure–function relationships (Neher, 1992; Sakmann, 1992). As well as the conventional use to measure transmembrane currents through ion channels, patch clamp can also be used to measure the capacitative currents. The charge carried by the capacitative current is directly proportional to the cell surface area because biological membranes have a fairly constant specific membrane capacitance of 1 μF/cm^2. Therefore, the membrane capacitance can be used to measure the increases in cell surface area that occur on fusion of secretory granules with the cell membrane during exocytosis.

Over the years, several techniques have been used to measure membrane capacitance (Lindau and Neher, 1988, Lindau, 1991). A simple method is to measure the cell admittance by applying a sinusoidal voltage and using a phase-sensitive detector. The earliest measurements of exocytosis with this technique were made over half a century ago (see Cole, 1972). As is the conductance of transmembrane ion currents, the admittance is determined from the measured current using Ohm's Law: $I(\omega) = V(\omega) \cdot Y(\omega)$, where $Y(\omega)$ is the admittance, $\omega = 2\pi f$ and f is the frequency of the sine wave. For a patch-clamped cell, the cell admittance has two orthogonal components, in phase with and 90° out of phase from the membrane potential. If the cell has no ionic conductances and a good seal, the equivalent circuit of the cell simplifies to a capacitor (the cell membrane) in series with the access resistance. When a secretory granule fuses with the cell membrane, an extra capacitor is switched into the equivalent circuit and a change in cell admittance occurs. If the phase detector is properly aligned, the change in admittance is proportional to the capacitance of the secretory granule. Figure 2 shows a staircase of capacitance steps representing the sequential fusion of individual secretory granules in a mast cell stimulated with GTPγS, a hydrolysis-resistant GTP analog.

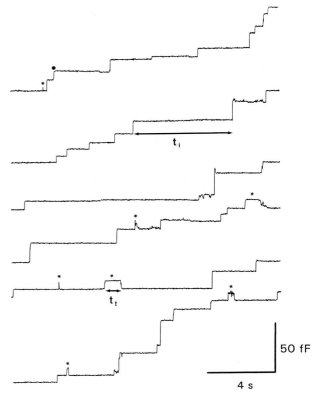

FIGURE 2 Step changes in cell membrane capacitance that occur during the course of mast cell degranulation induced by intracellular GTPγS. All these events correspond to the fusion of single granules with the plasma membrane. The traces are from a continuous record (total of 90 sec, from bottom to top) that occurred after 18 min of cell perfusion with pipette solutions containing 10 μM GTPγS. This record shows two types of events: (1) irreversible fusion events (dots) and (2) transient fusion events (asterisks). From these data, we can measure the time between irreversible fusion events (interstep interval, t_i) and the duration or dwell time of the transient fusions events (t_t). Reprinted with permission from Oberhauser *et al.* (1992a).

This approach has been used successfully to make continuous measurements of cell admittance from patch-clamped cells undergoing exocytosis using both analog (Neher and Marty, 1982; Lindau and Neher, 1988) and digital (Joshi and Fernandez, 1988; Fidler and Fernandez, 1989) phase detectors. Analog lock-in amplifiers have the advantage of the best sensitivity (up to 1 fF for a 10-μm diameter cell), but at this sensitivity the dynamic range is too small to follow the complete degranulation of a mast cell (from 5 to 20 pF). On the other hand, a digital phase detector has slightly lower sensitivity (2–3 fF at best), but maintains high sensitivity over a larger dynamic range. The digital method also avoids the need for extra instrumentation, since the phase detection is done by the software (Joshi and Fernandez, 1988).

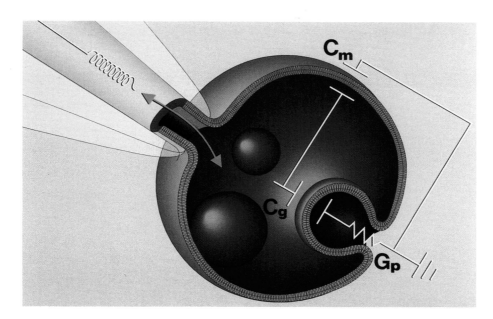

FIGURE 3 Patch-clamp of a mast cell. Schematic representation of whole-cell mode showing electrical elements for the series resistance (G_s), the plasma membrane capacitance (C_m), the secretory granule membrane capacitance (C_g), and the fusion pore conductance (G_p).

A problem with these measurements is that the phase of the cell admittance becomes misaligned from the phase detector as the result of capacitance changes, due to exocytosis or endocytosis, or because of changes in series resistance. However, by inserting a resistor between the cell and the ground, the series resistance can be changed by a known amount and the consequent change in admittance can be used to calculate the new phase angle and to realign the phase detector (Fidler and Fernandez, 1989). In this way, single fusion events can be detected throughout the complete degranulation of a mast cell (Alvarez de Toledo and Fernandez, 1990a; Oberhauser *et al.*, 1992a).

B. Measurement of Fusion Pore Conductance

When a secretory granule fuses with the plasma membrane, a water-filled "fusion pore" forms an electrical connection between the lumen of the secretory vesicle and the extracellular environment. As illustrated in Fig. 3, the membrane capacitance of the secretory granule (C_g) is in series with the conductance of the fusion pore (G_p). Assuming that the secretory granule matrix is freely conducting, this series combination creates a well-defined signature in the electrical admittance of the patch-clamped cell, which can then be used to make time-resolved measurements of the conductance of the fusion pore (Breckenridge and Almers, 1987a; Alvarez de Toledo and Fernandez, 1988). A fusion event, as represented by the equivalent circuit in Fig. 3, will change the cell admittance by an amount equal to

$$\Delta Y = \frac{\omega^2 C_g^2 / G_p}{1 + \omega^2 C_g^2 / G_p^2} + \frac{C_g}{1 + \omega^2 C_g^2 / G_p^2}. \quad (1)$$

The real [Re(ΔY)] and imaginary [Im(ΔY)] components of changes in admittance can be continuously recorded from a patch-clamped cell undergoing exocytosis. Fusion of a secretory granule proceeds as follows. Before fusion, the pore conductance is zero ($G_p = 0$), so both Re(ΔY) and Im(ΔY) are zero. On fusion, the pore conductance becomes finite ($G_p > 0$); thus, both Re(ΔY) and Im(ΔY) change, depending on the fusion pore conductance and vesicle membrane size. During this period, the pore conductance can be calculated using

$$G_p = \omega C_g \, \text{Im}(\Delta Y)/\text{Re}(\Delta Y) \quad (2)$$

(Alvarez de Toledo and Fernandez, 1988; see also Breckenridge and Almers, 1987a, for an alternative method). An example of how the fusion pore conductance changes during fusion of a secretory granule in a mast cell from a *beige* mouse is shown in Fig. 4,

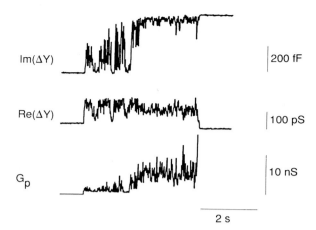

FIGURE 4 Time course of the conductance of an exocytotic fusion pore in a degranulating mast cell from the *beige* mouse. Admittance measurements were made in patch-clamped mast cells, as described in the text. The figure shows, from top to bottom, a continuous recording of the capacitance [Im(ΔY)], the AC conductance [Re(ΔY)], and the calculated fusion pore conductance [Gp = (ωC_g) Im(ΔY)/Re(ΔY)]. The fusion pore opens abruptly and the conductance fluctuates for several seconds before the pore expands irreversibly. The final expansion of the fusion pore is marked by a sudden decrease of the Re(ΔY) trace to the baseline.

with the real and imaginary traces from which it was calculated. As the fusion pore expands, the conductance increases until a limiting conductance is reached, where $G_p \gg \omega C_g$ when Re(ΔY) = 0 and Im(ΔY) = ωC_g. Therefore, Im(ΔY) is a measure of the capacitance contributed by the fusing vesicle when the pore conductance is large. Im(ΔY) is usually referred to as capacitance (in fF), whereas Re(ΔY) is called AC conductance (in pS).

Although the admittance techniques are well suited for continuous measurements of the fusion pore conductance, the time resolution (10 msec at best) is insufficient to measure the fusion pore conductance in the first few milliseconds of its lifetime. Fortunately, an alternative method, developed by Almers and co-workers (Breckenridge and Almers, 1987a; Spruce *et al.*, 1990) allows such a measurement. When the pore forms, a brief current transient occurs corresponding to the current necessary to bring the vesicle membrane to the same potential as the plasma membrane. Analysis of the magnitude and time course of these current transients can be used to determine the conductance of the fusion pore with submillisecond resolution, but only for the few milliseconds that constitute the lifetime of the transient. In contrast, the admittance measurements can be used to follow the pore for hundreds of milliseconds or seconds. By combining the charging transient measurements with admittance measurements, it is possible to study the formation and fast evolution of exocytotic fusion pores from their initial structure to their final irreversible expansion (see Spruce et al., 1990).

C. Fusion of Secretory Vesicles Causes Step Increases in Membrane Capacitance

As predicted by Cole (1972), long before the patch-clamp techniques were available, the capacitance of a secretory cell should reflect increases in surface membrane area because of the addition of membrane on fusion of a secretory granule. Indeed, with the improved resolution afforded by patch-clamp measurements, step-wise increases in the cell membrane capacitance as single secretory granules fuse with the plasma membrane have been demonstrated in a variety of secretory cells undergoing exocytosis (Neher and Marty, 1982; Fernandez et al., 1984; Nusse and Lindau, 1988; Nusse et al., 1990). Since fusion of a secretory vesicle with the plasma membrane increases the surface area of the cell membrane by an amount equal to the vesicle membrane area, the steps have different amplitudes corresponding to different granule sizes (see Fig. 2). For example, a 10-fF capacitance step corresponds to the addition of 1 μm^2 of membrane, equivalent to a spherical granule with a diameter of about 0.5 μm. The size distribution of secretory granules measured from the capacitance has been shown to be in good agreement with the size distributions determined from electron micrographs, not only for mast cells (Fernandez et al., 1984; Alvarez de Toledo and Fernandez, 1990b), but also for chromaffin cells (Neher and Marty, 1982), neutrophils (Nusse and Lindau, 1988), and eosinophils (Nusse et al., 1990), and has provided several new findings about the cell biology of exocytosis (Alvarez de Toledo and Fernandez, 1990a,b).

In addition to step increases in capacitance, step decreases or backsteps are also seen (Neher and Marty, 1982; Fernandez et al., 1984). These events are sometimes due to endocytotic events or membrane blebbing, but more often represent closure of the fusion pore. This surprising finding, that fusion pores can close, was the first in a series of discoveries that have revolutionized our view of the fusion pore (see Section IV). Moreover, the step increases and decreases in membrane capacitance provide markers of fusion pore activity, and allows the study of the kinetics of opening and closing of fusion pores (see Section V).

D. Pipette Biochemistry

A patch pipette typically accesses the cytosol of the cell through a large micrometer-sized opening in the cell membrane through which rapid diffusional exchange of soluble molecules between the pipette and the cytosol occurs. Since the volume of the patch pipette (greater than 10 μl) dwarfs that of the cell (about 1 pl), the soluble contents of the cytosol, including proteins, nucleotides, metabolites, and ions, will be lost into the infinitely larger volume of the patch pipette and are replaced rapidly by the pipette solution. Calculations from the rates of diffusional exchange between cells and a patch pipette predict that, within the first 15 min of recording, virtually all freely soluble proteins of molecular mass equal to or less than 45 kDa will be lost from a typical small cell (Pusch and Neher, 1988; Table I). Investigators have demonstrated that, in some cells, this "washout" uncouples receptor-mediated responses from the secretory response.

The washout of the cytosol is fortuitous because it challenges the investigator to identify the soluble components mediating the sought response. These manipulations, collectively named "pipette biochemistry," have led to several unanticipated discoveries. For example, a long-standing paradigm of the stimulus–secretion field was that an elevation of cytosolic calcium into the micromolar range was sufficient and necessary to elicit exocytotic fusion. Surprisingly, patch-clamped mast cells failed to undergo exocytotic degranulation in response to perfusion with pipette solutions that contained an elevated calcium concentration. In contrast, mast cells undergo a complete and spontaneous degranulation in response to the hydrolysis-resistant GTP analog GTPγS (0.1–50 μM) when Ca^{2+} concentration is clamped below 50 nM with strong calcium buffers (Fernandez et al., 1984; Neher, 1988). These results demonstrate that an elevated concentration of cytosolic calcium is neither suf-

TABLE I Estimated Time for the Wash-Out of Soluble Cytosolic Components through a Patch Pipette[a]

Solute[b]	MW	τ (63%) (sec)	5τ (99%) (min)
K^+	39	16	1.3
GTP	530	39	3.2
rab3a	25,000	140	11.7
Gsα	42,000	167	13.9
βCOP	110,000	230	19.1

[a] Data are for 15-μm diameter cell with 8 MΩ access resistance.

[b] Abbreviations: Gsα, alpha subunit of G_s, a heterotrimeric G-protein; βCOP, coat protein, β polypeptide.

ficient nor necessary under the conditions created in a patch-clamped cell, and directly implicate guanine nucleotide binding proteins in the regulation of exocytotic fusion. Similar results have since been seen in patch-clamped neutrophils (Nusse and Lindau, 1988) and eosinophils (Nusse *et al.*, 1990), as well as in various permeabilized cells (Vallar *et al.*, 1987; Lindau and Gomperts, 1991). The measurement of capacitance in patch-clamped cells has become widely used for the study of stimulus–secretion coupling (Mason *et al.*, 1988; Neher, 1988; Penner, 1988; Penner *et al.*, 1988; Zorec *et al.*, 1991).

E. Voltammetric Measurement of the Release of Secretory Products

Whereas assaying secretion from cell suspensions is relatively straightforward, measurement of release from a single cell presents a greater challenge, because the amount of secretory material released from a single cell is usually very small and tends to diffuse away rapidly. Despite these problems, some secreted substances can be detected electrochemically. The method, termed voltammetry, is based on electrochemical techniques that have been used by analytical chemists for many years to detect and assay oxidizable and reducible substances. The principle is very simple: an electrode is placed near a cell and a voltage is applied that matches the redox potential of one or more of the secretory products. An electrochemical current (distinct from a charging current) is then detected when secretion occurs. Oxidations occur at positive potentials; reductions, at negative potentials. Typical electroactive compounds that may be oxidized include neurotransmitter substances such as epinephrine, norepinephrine, serotonin, dopamine, and their metabolites.

There are many forms of voltammetry differing with respect to the nature and size of the working electrode and the voltage profiles employed (Stamford, 1985,1986). For biological studies, carbon fiber electrodes are favored because they are chemically and biologically inert, and their size (8μm in diameter) provides reasonable spatial resolution. These electrodes also give rise to stable charging currents. Voltage-ramp protocols such as fast cyclic voltammetry and fast differential ramp voltammetry provide the opportunity to discriminate between electroactive species with different redox potentials, and have been used *in vivo* to detect and study the properties of neurotransmitters *in situ* in neuronal tissue (Stamford, 1985; Millar and Williams, 1990) as well as to monitor exocytosis from single cells.

Cyclic voltammetry has been used to detect the release of catecholamines from intact bovine adrenal chromaffin cells (Leszczyszyn *et al.*, 1990; Wightman *et al.*, 1991). Recording the current signals during the oxidation or the reduction segment of a voltammogram (amperometric mode) produced sharp concentration spikes as catecholamines were released in a quantal fashion following stimulation of the cells. Fast dual ramp voltammetry has been used to monitor exocytotic release of serotonin from intact rat mast cells activated by compound 48/80 and mastoparan (Tatham *et al.*, 1991). The time-integrated voltammetric signals provided a measure of the progress of secretion. Measurement of the loss of fluorescence from cells with quinacrine-loaded granules was also used to follow the progress of secretion; when both voltammetry and fluorescence were measured simultaneously, the voltammetric signals lagged behind the fluorescence signals by several seconds. This time-lag reflects a feature common to all voltammetric measurements of release from single cells, namely, the effects of distance and diffusion on the signals generated by an electrode that senses its immediate microenvironment. The time course of voltammetric signals depends on the distance of the working electrode from the source of the electroactive species. An electrode distant from a cell will respond later and less strongly than one close to the cell. This feature is particularly important for the detection of quantal secretory events when high time resolution is required.

Voltammetry at a constant voltage, sometimes referred to as amperometry, should provide the maximum sensitivity and time resolution. This technique has been used to detect release of catecholamine and serotonin from single patch-clamped bovine chromaffin cells (Chow *et al.*, 1992) and mast cells (Fernandez-Chacon and Alvarez de Toledo, 1992; Alvarez de Toledo *et al.*, 1993), respectively. Simultaneous changes in membrane capacitance and electrochemical current took place after a delay following cell stimulation. The voltammetric signals consisted of spikes characteristic of quantal release. Another feature of these signals was that many of them were preceded by a low amplitude, slow increase in current (pedestal or foot signal). These phenomena suggest that slow leakage of secretory products may occur through the fusion pore prior to full membrane fusion. Thus, voltammetric techniques offer a highly sensitive means of monitoring the release of chemical species from single cells or even through a single fusion pore.

IV. Fusion Pore Develops in Distinct Stages

Patch-clamp measurements of secretory cells undergoing exocytosis have revealed the exocytotic fusion pore as a remarkably dynamic entity. The evolution of the fusion pore can be divided into several distinct phases. The pore is first detected as an abrupt increase in conductance, followed by an expansion phase characterized by large fluctuations in conductance. These fluctuations, termed flicker, can be brief (milliseconds) or prolonged (seconds). Flicker usually terminates with a rapidly increasing phase as the pore irreversibly expands to a structure akin to that seen in electron micrographs, but sometimes the pore can close again (Fernandez *et al.*, 1984; Spruce *et al.*, 1990). The properties of these different phases are discussed here.

A. Opening of the Fusion Pore

The fusion pore opens abruptly with a conductance similar to that of a large ion channel (Breckenridge and Almers, 1987a; Spruce *et al.*, 1990). This observation is interesting, since several secretory granule membrane proteins, such as synexins, synaptoporin and synaptophysin, have been shown to form ion channels *in vitro* (Thomas *et al.*, 1988; Knaus *et al.*, 1990; Pollard *et al.*, 1990; Sudhof and Jahn, 1991). This observation has led to the hypothesis that the opening of the fusion pore corresponds to the opening of an ion channel (Almers, 1990; Almers and Tse, 1990). Analysis of the transient current discharges through the fusion pores of *beige* mouse mast cells showed that the median initial conductance is about 300 pS, but varies between 80 and 1000 pS (Breckenridge and Almers, 1987a; Spruce *et al.*, 1990). Admittance measurements give similar values for initial fusion pore conductance (Spruce *et al.*, 1990; Nanavati *et al.*, 1992).

The smallest fusion pores seen in the electron micrographs are approximately 20 nm across and 50 nm long. By modeling the pore as a water-filled tube, this pore can be estimated to have a conductance of approximately 10 nS. Thus, the initial pore detected electrophysiologically is considerably smaller and must represent an earlier, more transient stage in the development of the fusion pore. If this early pore is assumed to be merely a smaller version of those seen in electron micrographs, that is, a lipidic tube with a length of about 15 nm (slightly larger than the width of two bilayers), then an initial conductance of 80 pS corresponds to a diameter of 1 nm.

B. Expansion of the Early Fusion Pore

On opening, the fusion pore begins to expand. The expansion phase can be fast or slow and is characterized by large rapid fluctuations in pore conductance. Hundred-fold changes in conductance (0.1–10 nS) can occur on time scales of tens of milliseconds (Alvarez de Toledo and Fernandez, 1988; Spruce *et al.*, 1990; Monck *et al.*, 1991a). These changes usually appear to be continuous (Breckenridge and Almers, 1987a; Spruce *et al.*, 1990; Nanavati *et al.*, 1992), rather than between discrete levels typical of ion channels, although on occasion several semi-stable conductance states have been seen (Spruce *et al.*, 1990).

Flicker usually ends with the irreversible expansion of the fusion pore to a large conductance that is beyond the detectable limit of the admittance measurements. This value is typically about 10 nS, corresponding to a pore with a diameter of more than 20 nm. Once the pore expands beyond this size, it rarely or never closes, suggesting the possibility of a structural change from an early fusion pore that can close to a larger pore that is unable to close. In mast cells from the *beige* mouse, the lifetime of the early fusion pore can be measured from the length of flicker because flicker begins abruptly when the fusion pore opens and terminates equally abruptly when the pore expands. These measurements show that the time that a fusion pore remains in the flicker state is highly variable. The distribution of flicker length is exponential, allowing extraction of a time constant for pore expansion of about 200 msec (Monck *et al.*, 1991a). However, although some fusion pores flicker for hundreds of milliseconds or even several seconds, most expand in less than 50 msec. Pore expansion may be faster in cells with smaller, more highly curved secretory granules such as synaptic vesicles.

Experiments using voltammetry to measure serotonin release from mast cells have shown that, although a small amount of release occurs during flicker, most of it occurs only after pore expansion (Alvarez de Toledo *et al.*, 1993). Figure 5A shows simultaneous admittance measurements of the fusion pore and voltammetric measurements of product release from a mast cell. During flicker, a small release of serotonin occurs whereas after pore expansion, a rapid, much larger release occurs as the secretory granule matrix decondenses.

C. Granule Swelling Regulates Product Release

Secretory granules typically contain bioactive amines or peptides in highly condensed form, for example, histamine with heparin proteoglycans in mast cell granules or noradrenaline with chromogranins in chromaffin granules. Storage in condensed form allows large amounts to be stored without osmotic complications, but requires a mechanism for rapid decondensation to produce the rapid product release that occurs during neurotransmission or mast cell degranulation.

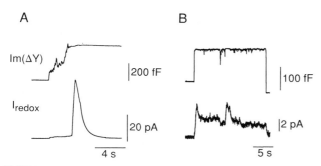

A

Im(ΔY)

| 200 fF

I_redox

| 20 pA

4 s

B

| 100 fF

| 2 pA

5 s

FIGURE 5 Voltammetric measurement of the release of secretory products in mast cells. The figure shows simultaneous measurements of vesicle fusion with the plasma membrane and vesicle content release in *beige* mouse mast cells. Membrane fusion was monitored by measuring the membrane capacitance (top traces). The release of secretory products was monitored by recording the oxidizing currents at a constant voltage using a carbon fiber electrode (bottom traces). The patterns of product release (serotonin) for two types of fusion events are shown: (A) an irreversible exocytotic fusion event and (B) a transient fusion event. (A) The capacitance increases in a fluctuating manner (fusion pore conductance <10 nS) before reaching a stable level (complete expansion of the fusion pore). While the fusion pore is expanding, a small amount of release occurs (Chow *et al.*, 1992; Alvarez de Toledo *et al.*, 1993). Shortly after the final irreversible expansion of the fusion pore, a rapid large increase in the oxidizing current occurs concurrently with the explosive swelling of the granule core. (B) The amperometric signal of a secretory vesicle that fuses transiently with the plasma membrane is shown (indicated by the transient increase in the membrane capacitance). This small amperometric signal follows the fluctuations in pore conductance (not shown) and is due to leak of the granule contents through the fusion pore.

The proteoglycans in mast cells and the mucin glycoproteins in goblet cells of mucosal epithelia share certain important physicochemical properties. These substances form cross-linked gels that condense with bioactive secretory compounds at the acidic pH found in secretory granules. However, on exposure to physiological saline, these structures undergo an explosive decondensation reaction (Verdugo, 1991). Experimentally, the exocytosed core of a mast cell granule can be recondensed to its original volume by exposure to solutions containing histamine at acidic pH, that is, the environment expected in the secretory granule (Curran and Brodwick, 1991; Fernandez *et al.*, 1991). The condensation and decondensation of the granule core proteoglycan is reversible and shows the characteristics of a polymer gel phase transition. These observations led to the proposal that the mechanisms of condensation of histamine with heparin proteoglycans are similar to the coil–globule phase transitions observed in artificial polymer gels (Fernandez *et al.*, 1991). Such artificial gels have been developed to regulate drug release (Kwon *et al.*, 1991). That this process may imitate the biological mechanism is intriguing.

Opening of the fusion pore provides a path for entry of Na^+ and water into the granule matrix. This event triggers the phase transition which, because of its explosive nature, drives the bioactive compounds out of the granule matrix, providing a mechanism for rapid release and dispersal, as is necessary for proper functioning of mast cells and neuronal cells. Proteoglycans are not major components of secretory granules in all secretory cells. However, glycoproteins, such as the highly acidic chromogranins found in neurons and neuroendocrine and endocrine cells (Helle, 1990), could fulfill a similar role—to package secretory product and to facilitate rapid release.

D. Expansion of the Early Fusion Pore Is Not Driven by Granule Swelling

No swelling of the secretory granule has been observed during the periods of fusion pore conductance flicker that occur prior to irreversible pore dilation or closure (Breckenridge and Almers, 1987b; Zimmerberg *et al.*, 1987). These observations have led to the proposition that swelling might provide the driving force for the rapid expansion of the fusion pore (Breckenridge and Almers, 1987b; Zimmerberg *et al.*, 1987; Chandler *et al.*, 1989; Merkle and Chandler, 1989). In support of this hypothesis are electron microscopy studies of sea urchin egg cortical granule exocytosis, showing that fusion pore widening is apparently arrested when matrix swelling and dispersal is inhibited by hyperosmotic solutions and dextran solutions (Zimmerberg *et al.*, 1987; Chandler *et al.*, 1989; Merkle and Chandler, 1989).

The finding that acidic histamine solutions could inhibit swelling of the proteoglycan matrix in mast cells provided a means to directly test the hypothesis that swelling was the driving force for pore expansion (Monck *et al.*, 1991a). As discussed earlier, flicker dwell time provides a convenient estimate of the pore expansion time (at least for the initial phase). Under conditions in which the rate of swelling of the granule matrix was inhibited more than 20-fold, achieved by bathing the cells in an extracellular medium containing histamine at acidic pH, the expansion of the fusion pore was neither prevented nor slowed. Therefore, swelling of the granule matrix due to water influx through the fusion pore cannot be the driving force for the expansion of the early exocytotic fusion pore in mast cells that is seen with the admittance measurements. Some property of the exocytotic scaffold is likely to be responsible for the rapid pore expansion in exocytosis because, in contrast to the exocytotic fusion pore, the fusion pore that is induced between two plasma membranes by influenza virus hemagglutinin is relatively stable and expands slowly (Spruce *et al.*, 1989,1991).

The fusion pores seen in electron micrographs, although too large to be detected with the patch clamp, are still relatively small structures. Therefore, matrix swelling probably provides the force for a second phase of pore expansion leading to complete incorporation of the secretory granule membrane into the cell membrane, and is necessary for rapid expulsion of the secretory granule contents into the extracellular environment.

E. Closure of the Fusion Pore

One of the first new findings provided by patch-clamp measurements of rat mast cells was that a fusion pore does not always expand irreversibly but sometimes closes, leaving an intact secretory vesicle inside the cell (Fernandez *et al.*, 1984). The ability of fusion pores to close was quite unexpected. Transient fusion events have since been observed in normal and *beige* mouse mast cells (Breckenridge and Almers, 1987b; Zimmerberg *et al.*, 1987; Monck *et al.*, 1990; Spruce *et al.*, 1990). Several examples of transient fusion events are shown in Figs. 3, 6, and 7. Sometimes a vesicle can undergo many transient fusions before a final irreversible fusion occurs (Breckenridge and Almers, 1987b; Alvarez de Toledo and Fernandez, 1988; Spruce *et al.*, 1990).

1. Can a Fusion Pore Allow Secretion and Then Close?

Ceccarelli and colleagues proposed that, contrary to the conventional view of neurotransmission, a fusion pore might open, allow discharge of the contents of the vesicle, and then close, obviating the need for membrane merging and subsequent recycling (Valtorta *et al.*, 1990). In mast cells, a delay appears to occur between membrane fusion and release of fluorescent secretory granule markers (Breckenridge and Almers, 1987b). As discussed earlier, low concentrations of biogenic amines can be measured using voltammetric techniques, which allowed direct measurement of catecholamine release from single chromaffin granules (Leszczyszyn *et al.*, 1990; Chow *et al.*, 1992) and serotonin release from single mast cell granules (Tatham *et al.*, 1991; Alvarez de Toledo *et al.*, 1993). Interestingly, small amounts of release were seen to precede the main phase of release (Chow *et al.*, 1992; Alvarez de Toledo *et al.*, 1993). Alvarez de Toledo *et al.* (1993) showed that, in mast cells, this "foot" phase occurs during capacitance flicker before pore expansion (Fig. 5A) and during transient fusion events (Fig. 5B). This result provides support for Ceccarelli's proposal that secretion can occur through a fusion pore that subsequently closes.

2. Fusion Pore Supports a Large Lipid Flux

When a fusion pore connects a mast cell secretory granule to the plasma membrane, a net flow of lipid into the secretory granule occurs (Monck *et al.*, 1990). Analysis of transient fusion events has revealed that the backsteps are usually larger than the onsteps (Fig. 6A,B), indicating a net movement of phospholipid membrane from the cell membrane to the secretory granule during flicker, as illustrated in Fig. 6C. The quantities of the phospholipids moved are large, equivalent to almost 1 million molecules per sec—enough to replace all the lipids in a small lipidic fusion pore every millisecond. Since biological membranes contain a mixture of lipids, the composition of the pore could change dramatically over short time intervals and could provide an explanation for the rapid fusion pore conductance changes if, as seems likely from energetic considerations, some lipids favor pore expansion and others promote closure (Kozlov *et al.*, 1989; Nanavati *et al.*, 1992). The flux of lipid through the fusion pore also indicates that the pore, when it closes, is partially or totally lipidic, an observation that limits the types of molecular models that can be proposed for the structure of the fusion pore.

The rate of movement of membrane through the fusion pore is surprisingly constant throughout its lifetime; the same rate occurs regardless of stimulus strength or a number of other chemical and physical perturbations (Monck *et al.*, 1990,1991b). When the fusion pore opens, a constant driving force for lipid flow exists. A difference in lateral membrane tension between the secretory granule membrane and cell membrane has been proposed to be the driving force for the membrane flow. Because this tension difference is always the same when membranes fuse, a critical tension may be required for membrane fusion. Such a proposal is very compelling since many of the perturbations used to fuse phospholipid bilayers and vesicles in model fusion systems are conditions that have been shown to increase the membrane tension (see Section VI,A). Teleologically, it is pleasing to think that the cell has evolved the ability to utilize an intrinsic property of phospholipid membranes; two membranes, when under tension and close (<2nm) together, can and will fuse spontaneously.

An obvious candidate for increasing the membrane tension is swelling of the proteoglycan core of the secretory granules after membrane fusion. However, swelling of the granule matrix as a consequence of water influx through the fusion pore is unlikely to be

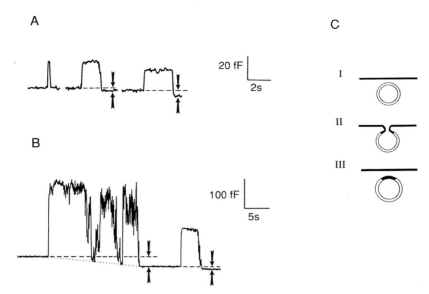

FIGURE 6 Lipid flux through the fusion pore during transient openings. (A) Three granules that underwent transient fusions of different durations recorded in a mouse mast cell. Note that the magnitude of the backstep is larger than the initial step, indicating that the area of the plasma membrane is decreased after a transient fusion event. (B) An example of a transient fusion for a giant secretory granule from a *beige* mouse mast cell where the fusion pore closes to reveal membrane transfer and then reopens. (C) A scheme showing a possible interpretation for the decreased plasma membrane area after a transient fusion event. Before fusion, the capacitance gives a measure of the cell surface area (I). On fusion, the granule is connected to the plasma membrane by a narrow-necked fusion pore (II) and the extra membrane composing the granule membrane contributes to the measured capacitance. At this stage, some of the plasma membrane is drawn into the granule because the granule has a higher membrane tension (see text). When the pore is disrupted at the end of the transient fusion event, the membrane that has been incorporated into the granule membrane remains there and the cell surface area is decreased (III).

the mechanism because the rate of membrane flux through the fusion pore (assumed to be a measure of the tension difference between the two membranes) was not altered in experiments in which the cells were bathed in acidic histamine media (Monck *et al.,* 1990). This experiment does not rule out microswelling caused by ion transport across the granule membrane, but the tension would have to be maintained for the duration of flicker (up to 10 sec) since the rate of membrane flux is constant throughout flicker. Alternatively, an interaction of the granule membrane with the proteoglycan core or with extragranular cytoskeletal elements could be responsible for generating increased membrane tension in the granule membrane.

V. Kinetic Analysis of Fusion Pore Activity

Kinetic analysis of the opening and closing of ion channels led to many discoveries about ion channel function. A similar analysis of the opening and closing of fusion pores can be used to investigate fusion pore kinetics. Figure 7A shows a capacitance record

of a normal mast cell to demonstrate the stepwise changes in capacitance caused by the exocytotic fusion of individual secretory granules, induced by the hydrolysis-resistant guanine nucleotide GTPγS. Two types of fusion events can be seen: (1) irreversible step increases in capacitance on membrane fusion and (2) transient increases (e.g., asterisks) on fusion and subsequent closure of the fusion pore. From such recordings, three parameters can be measured that define the opening and closing of fusion pores: (1) the interval between irreversible fusion events (interstep interval, t_i); (2) the duration or dwell time of transient fusion events (t_t); and (3) the fraction of the total fusion events that are transient. Since several hundred fusion events can be resolved in a normal mast cell, these parameters can be used to estimate the rates of fusion pore opening and closing.

The rate of fusion pore opening can be obtained from the mean interstep interval, in the region where the rate of degranulation is linear, that is, between 20 and 80% of the degranulation (Alvarez de Toledo and Fernandez, 1990c; Oberhauser *et al.,* 1992a). As Fig. 7A shows, the interstep interval covers a wide range

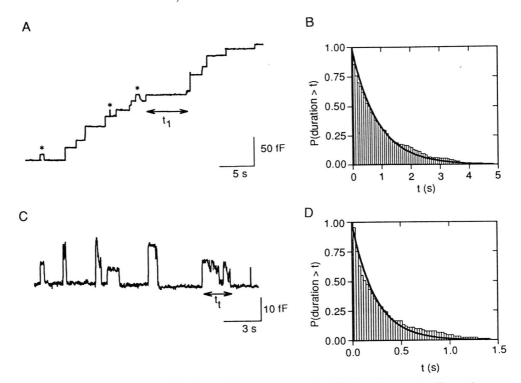

FIGURE 7 Kinetic analysis of the fusion of secretory granules. (A) Step changes in cell membrane capacitance that occur during the course of mast cell degranulation induced by intracellular GTPγS (10 μM). Reversible increases in membrane capacitance are indicated by asterisks. (B) Example of a probability distribution function of the interval between steps for a mast cell stimulated with 10 μM GTPγS. The time constant is 941 ± 15 msec. The data were obtained by measuring time between fusion events (interstep interval, t_i) as indicated in A. The probability distribution function, $P_{(t)}$, is the probability of finding events separated by a time ≥ t. The time constant of the fitted exponential (continuous line) can be used to calculate a rate of fusion pore opening. (C) Example of a membrane capacitance recording showing several transient fusion events. The transient fusion event dwell time (t_t) is obtained by measuring the time between the initial increase in capacitance and the return to the basal level, indicative of complete closure of the fusion pore. (D) Example of a probability distribution function of the dwell time of transient fusion events. The time constant was 230 ± 7 msec (90 events measured in the linear region of the degranulation). The probability distribution function, $P(t)$, gives the probability of a transient fusion event having a lifetime ≥ t. The time constant of the fitted exponential (continuous line) can be used to calculate a rate constant for the events leading to fusion pore opening.

of values. A practical representation of the interstep intervals measured in a single cell can be obtained by plotting these times as probability distribution functions (Fig. 7B), that is, the probability of finding events separated by a time greater than or equal to time t. Since the probability distribution functions are well represented by single exponentials (fitted lines; Alvarez de Toledo and Fernandez, 1990a,b,c), the characteristic time constant (mean interstep interval) can be considered to be a measure of the rate-limiting step leading to the opening of fusion pores. Consequently, the rate of formation of the fusion pore is defined as the inverse of the mean interstep interval.

The rate of fusion pore closure can be obtained by measuring the mean dwell time of transient fusion events, and the fraction of the total fusion events in the cell that are transient. Transient fusion events can be faster than the time resolution of the recordings (10 msec) or can last up to several seconds (Fig. 7C). Like the interstep interval, the duration of transient fusion events also conforms to a Poisson distribution. Figure 7D shows the probability distribution function for the length of time that the fusion pore exists before closure. As for the interstep interval, the probability distribution function is well fitted by a single exponential. The characteristic time constant of this exponential distribution (mean dwell time) and the fraction of transient fusion events can be used to determine a rate constant for the closure of the fusion pore. The inverse of the mean dwell time is a measure of the rate at which a fusion pore leaves the open configuration. The fraction of reversible fusion events represents the probability that a fusion pore will close. Therefore, the rate of closure of the fusion pore can be calculated as the fraction of transient fusion events multiplied by the inverse of the mean transient fusion dwell time.

These methods are analogous to those used in the kinetic analysis of single ion channel data, and have been used to investigate the effects of guanine nucleotides and Ca^{2+} on the kinetics of fusion pore opening and closing (Alvarez de Toledo and Fernandez, 1990c). The dependence of the rates of fusion pore opening and closing on intracellular GTPγS and Ca^{2+} is shown in Fig. 8. The rate of fusion pore opening depends strongly on the concentration of GTPγS (Fig. 8A): a 6-fold increase in the rate occurs on increasing the GTPγS concentration from 0.2 to 50 μM. A significant increase in the rate of fusion pore opening is also observed when the Ca^{2+} concentration is raised above 1 μM in the presence of 0.2 μM GTPγS (Fig. 8A). At Ca^{2+} concentrations above 1 μM, the sensitivity to GTPγS is increased so less GTPγS is needed to reach the same rate of pore opening (data not shown). In contrast, Fig. 8B shows that the rate of fusion pore closure is not significantly affected by cytosolic GTPγS (0.2 to 50 μM) or Ca^{2+} (30 nM to 10 μM).

Oberhauser *et al.* (1992a) used a similar kinetic analysis to examine the temperature dependence of the fusion of secretory granules. Figure 9 shows the rate constants for pore opening and closing plotted as Arrhenius plots. The rate of pore opening is strongly dependent on temperature, suggesting that the rate-limiting step leading to pore formation depends on a conformational change in a protein and probably reflects the activity of a regulatory protein. In contrast, the rate of closure is only weakly dependent on temperature, except for the abrupt discontinuity between 13 and 14°C. These discontinuous temperature dependencies are typical of diffusional processes in homogeneous phospholipid bilayers and occur at the gel–liquid phase transition temperature of the lipid (Krasne *et al.*, 1971; Hoffmann *et al.*, 1980), but had never before been seen in biological membranes. The different temperature dependencies of pore opening and closing strongly suggest that substantial differences exist in the molecular events leading to the opening and closing of fusion pores. Of particular interest, the mechanism of pore closure appears to be sensitive to diffusion of a specific lipid. As discussed in more detail later, theoretical modeling of lipidic pores shows that only certain lipids favor stable pores,

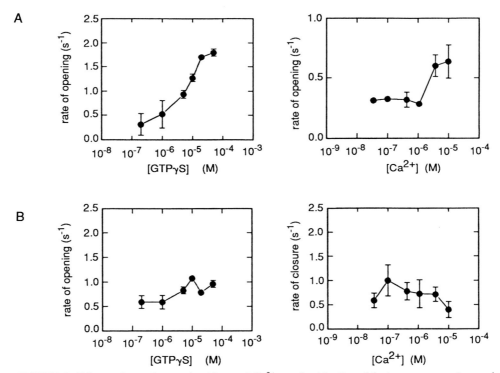

FIGURE 8 Effects of guanine nucleotides and Ca^{2+} on the kinetics of fusion pore opening and closure. (A) Dependence on cytosolic GTPγS (*left;* 0.2 to 50 μM at 30 nM Ca^{2+}) and Ca^{2+} (*right;* 30 nM to 10 μM at 0.2 μM GTPγS) for the rate of fusion pore opening. (B) Dependence on intracellular GTPγS (*left;* 30 nM Ca^{2+}) and Ca^{2+} (*right;* 0.2 μM GTPγS) for the rate of fusion pore closure.

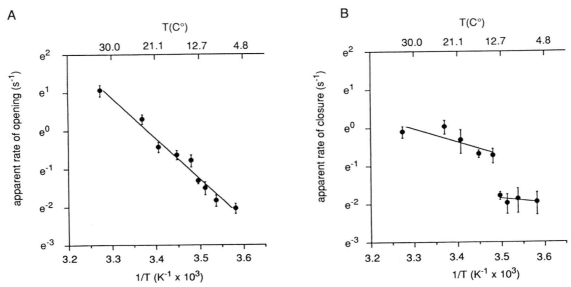

FIGURE 9 Arrhenius plots for the rate of opening (A) and closure (B) of the fusion pore. (A) The activation energy for the rate of formation of the fusion pore is 23 kcal/mol, suggesting that, unlike diffusion, the rate limiting step in fusion pore formation has a large activation energy. (B) Below 13°C, the rate of closure of the fusion pore is independent of temperature; above 14°C, the rate of closure is weakly temperature dependent with an activation energy of 7 kcal/mol. The rate of closure changes abruptly by a factor of three between 13° and 14°C. Reprinted with permission from Oberhauser *et al.* (1992a).

whereas others favor expansion or closure. Since a large lipid flux occurs through the fusion pore, entry of specific lipids may disrupt its structure and favor closure. Below the phase transition temperature, pore-closing lipid molecules may become segregated into crystalline microdomains and be no longer available to induce pore closure (Oberhauser *et al.*, 1992a).

VI. Molecular Mechanisms of Fusion Pore Formation

Fusion pores must exist whenever two membrane compartments fuse, whether in artificial bilayer systems, in virus-mediated fusion, in intracellular traffic, in fertilization, or in exocytosis. These fusion pores are likely to share certain physical principles and molecular structures. Consequently, information learned from the study of other types of fusion pores should aid our understanding of the exocytotic fusion pore.

A. Mechanisms of Fusion Pore Formation in Artificial Bilayer Systems

The structure of the fusion pore and the mechanism of fusion are interrelated. The structure can provide information about the mechanism and vice versa. The fusion of two planar bilayers, or of a vesicle with a

bilayer, has been extensively studied in model lipid systems. Investigators have well established that, when two bilayers approach each other, several attractive and repulsive forces determine whether the membranes will mutually repel and be kept apart or attract and either adhere or fuse (Rand, 1981; Israelachvili and McGuiggan, 1988). At large separations, the balance among the attractive van der Waals forces, the repulsive electrostatic double layer forces, and the fluctuation (or undulation) forces describes the interaction between two membranes. However, as the separation is reduced below 3 nm, repulsive hydration and steric forces begin to dominate and act to keep the bilayers apart. For an approach close enough to allow fusion pore formation, these short-range repulsive forces must be overcome.

Experimentally, several protocols have been used to induce bilayer adhesion and fusion. Protocols that dehydrate the bilayers and reduce the repulsive hydration force cause adhesion and, often, fusion. Evidence is mounting that increased lateral tension, that is, an increased separation of the phospholipid head groups, allows interaction between the hydrophobic interiors of the bilayers and makes fusion energetically favorable (Ohki, 1987; Woodbury and Hall, 1988). Fusogenic protocols that increase membrane tension include subjection to osmotic and hydrostatic pressure, temperature, Ca^{2+} and other divalent cations, pH, and polyethylene glycols (Finkelstein *et al.*, 1986; Holz, 1986; Ohki, 1987; Helm *et al.*, 1989; Yamazaki and Ito, 1990). Likewise, electrical fields can induce fusion because they induce electromechanical stress, increasing the membrane tension and hydrophobicity (Zimmerman, 1986; Needham and Hochmuth, 1989).

Israelachvili and co-workers (Helm *et al.*, 1989,1992) argued that merely reducing the repulsive forces is insufficient to cause membrane fusion and that exposure to a strongly attractive hydrophobic force is necessary for membrane fusion. They showed that phospholipid bilayers applied to mica surfaces could be induced to spontaneously fuse at a separation of 1–2 nm if they were "depleted" by a technique that reduces the density of phospholipid head groups per unit area of membrane, equivalent to increasing the lateral tension. The fusion occurs as the result of localized exposure of the hydrophobic interior, which allows a strong attractive hydrophobic force to bypass the repulsive hydration forces. This conclusion explains why some protocols can overcome the repulsive hydration forces and cause membrane adhesion, but do not induce membrane fusion. Thus, exposure of the hydrophobic force will cause hemifusion of two bilayers to a single bilayer if the separation is less than 2 nm. Once the hemifusion breakthrough has occurred, local stresses in the common bilayer favor pore formation and subsequent full fusion (Helm *et al.*, 1992).

Bilayer fusion, pore formation, and pore expansion can be very fast. For example, fusion of vesicles with planar bilayers can be measured as incorporation of ion channels into the bilayer. The full step in single-channel conductance occurs in less than 200 μsec (Cohen *et al.*, 1980), indicating that, within this time interval, the fusion pore has both formed and expanded to a size where the resistance is negligible, equivalent to a cylindical pore 10 nm long and 30 nm in diameter (Zimmerberg, 1987). Similar pores in a single bilayer induced by electroporation can form on submicrosecond time scales (Hibino *et al.*, 1991). Therefore, such a mechanism is fast enough to account for even the most rapid exocytotic responses such as those occuring during neurotransmission.

B. Virus-Induced Fusion Pores

The ability of viruses to induce fusion pores is particularly instructive since this fusion reaction is the only one in which the proteins that induce fusion have been identified (reviewed by Bentz *et al.*, 1990; Hoekstra, 1990; White, 1990). The fusogenic activity resides in virally encoded membrane glycoproteins, of which the influenza virus hemagglutinin is the best

studied. The hemagglutinin fusion proteins have a transmembrane anchoring segment, which acts to hold the two membranes together, and a fusogenic domain, which forms an amphiphilic peptide with bulky hydrophobic residues on one side and acidic groups on the other. Putative fusion peptides for other viruses have been assigned, but fusogenic activity has been demonstrated only for hemagglutinin and human immunodeficiency virus (HIV) fusion peptides. The fusogenic domains of most viral fusion proteins form sided α helices, with most of the hydrophobic residues on one face (White, 1990).

Models for viral fusion propose that several hemagglutinin trimers assemble to form a "collar" within which the fusion pore forms (Bentz *et al.*, 1990; White, 1990). The membrane region within this collar is approximately 5 nm in diameter. The hydrophobic side of the fusion domain, on the inside of the collar, is proposed to form a bridge that allows lipids from the two membranes to come into contact. However, given that two closely apposed bilayers fuse when under tension or when a local perturbation of the bilayer structure occurs, the arrangement of the viral proteins could draw the two membranes into close apposition within the collar and could induce fusion by increasing the membrane tension within the collar or inducing appropriate local perturbations in the physical structure of the bilayer.

The viral fusion pore has also been examined using electron microscopy and patch-clamp techniques. The lipid pores seen in the electron micrographs appear identical to the exocytotic fusion pore, with a diameter of about 50 nm (Knoll *et al.*, 1988). Patch-clamp measurements of the fusion between two cell membranes, induced by influenza virus hemagglutinin that has been transfected into one of the cells, reveal a fusion pore that opens abruptly and undergoes fluctuations in pore conductance, much like the exocytotic fusion pore, but expands relatively slowly (Spruce *et al.*, 1989,1991). These similarities between the properties of the viral and exocytotic fusion pores suggest similar structures and mechanisms of formation. A sperm glycoprotein implicated in sperm–egg fusion during fertilization has been proposed to contain a sided helical domain like that of the viral fusion peptides (Blobel *et al.*, 1992), suggesting that this mechanism might be widespread.

C. Models for Exocytotic Fusion Pores

Many models for the exocytotic fusion pore have been proposed. One model proposes that the initial pore is a proteinaceous ion channel-like structure traversing the bilayers of the secretory granule and plasma membranes (Rahamimoff *et al.*, 1989; Almers, 1990; Almers and Tse, 1990). According to this hypothesis, the exocytotic fusion pore results from a preassembled ion channel-like structure that opens in response to a cellular messenger. The rationale for this model is that opening of an ion channel would be readily subject to the rapid regulation necessary, for example, in neurosecretion. In addition, how such a fusion pore would close is easy to explain. In analogy with the gap junction, the only known channel to span two bilayers, the initial pore (conductance of 80 pS) would correspond to an oligomeric channel with an internal diameter of about 1 nm (assuming the 15-nm length of the gap junction). The ion channel would show multiple initial conductance levels, like the gap junction, and would expand by intercalation of lipid molecules between the subunits of the structure, leading to a mostly lipidic fusion pore that expands irreversibly, completing exocytosis (Almers, 1990; Almers and Tse, 1990).

Several models have proposed a fusion pore that contains proteins and lipids from its earliest moments, in which the proteins serve as hydrophobic bridges between the membranes. The "hydrophobic bridge" model has been reviewed in two different flavors (Pollard *et al.*, 1991; Zimmerberg *et al.*, 1991). According to this model, proteins form a bridge across which phospholipids can migrate, their hydrophobic tails interacting with the hydrophobic surface of the proteins. The rationale is that the repulsive hydration forces do not have to be overcome *en masse*. In one form, synexin molecules are suggested as the bridge proteins (Pollard *et al.*, 1991). These models have many similarities to models for viral fusion (see Section VI,B). The fusion pore, however, must quickly become mainly lipidic, because otherwise the model could not readily explain the large fluctuations in pore conductance.

Alternatively, the initial fusion pore could be an entirely lipidic structure and the fusion mechanism could be similar to that discussed earlier for artificial bilayer fusion. In the early 1980s, the idea that osmotic swelling might drive exocytotic fusion was proposed after it was found that hyperosmotic solutions could inhibit exocytosis (Orci and Malaisse, 1980; Hampton and Holz, 1983; Pollard *et al.*, 1984; Zimmerberg and Whitaker, 1985; Zimmerberg *et al.*, 1985). However, the inhibitory effects were due to artifactual changes in cell structure that physically prevented the secretory granules from interacting with the plasma membrane (Chandler *et al.*, 1989; Merkle and Chandler, 1989). Subsequently, it has been shown that hyperosmotic solutions do not inhibit exocytotic membrane fusion (Holz and Senter, 1986; Breckenridge and Alm-

ers, 1987b; Zimmerberg *et al.*, 1987; Whitaker and Zimmerberg, 1987) and that fusion precedes visible swelling of the secretory granules in *beige* mouse mast cells (Breckenridge and Almers, 1987b; Zimmerberg *et al.*, 1987). Evidence is now available that the membrane of fusing secretory granules is under tension and that this tension is due to a mechanism independent of matrix swelling (see Section IV,E,2). Since two bilayers will fuse spontaneously at separations below 2 nm when the membranes are under tension, a tension-driven mechanism for formation of a lipidic fusion pore during exocytosis is the favored hypothesis. The role of proteins is likely to be to respond to intracellular signals, to draw the plasma membrane into close proximity with the secretory granule membranes, and to favor fusion by increasing membrane tension or inducing local perturbations in the bilayer structure (Monck and Fernandez, 1992).

VII. Exocytotic Fusion Pore as a Lipidic Structure Supported by a Protein Scaffold

A. Lipid Nature of the Fusion Pore

Comparison of electron micrographs of exocytotic fusion pores with those of virus-induced pores and pores induced by physical perturbation of the bilayer structure shows that all are of similar size with a diameter of 4–50 nm. All these pores are clearly made predominantly of lipids. The question of whether they have developed from smaller lipidic precursors or from pores with a different structure lacks a definitive answer. Nevertheless, several lines of evidence suggest that the smaller reversible fusion pores seen with patch-clamp measurements are also lipidic from an early stage. First, large amounts of membrane flow through the fusion pore (Monck *et al.*, 1990), second, the temperature dependence of fusion pore closure suggests that the closure mechanism involves lipids (Oberhauser *et al.*, 1992a).

Studies of membrane fusion in model bilayer systems have demonstrated that purely lipidic fusion pores do occur (see Section VI,A). But can such lipidic pores show the dynamic properties of the exocytotic fusion pore, namely, multiple conductance levels, rapid fluctuations, and reversibility? A lipidic fusion pore can be modeled by assuming that the free energy of the pore is the sum of the elastic energy associated with the bilayer curvature (Markin *et al.*, 1984; Chernomordic *et al.*, 1987; Kozlov *et al.*, 1989) and the isotropic tension of the bilayer (Nanavati *et al.*, 1992). Figure 10A shows a model of a pore through a single bilayer. The curvature energy depends on the bending rigidity

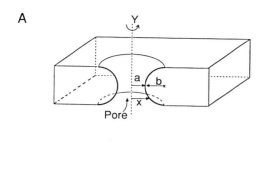

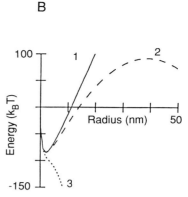

FIGURE 10 Theoretical model of a lipidic pore through a single bilayer. (A) The geometry of the pore was generated by revolving a semicircle (boldfaced) about the axis of revolution, Y. The parameters a, b, and x are defined as a, the radius of the narrowest part of the pore; b, the radius of the generated semicircles; and x, the shortest distance between the axis of revolution and a point on the semicircle. The behavior of such a lipidic pore is governed by the energy that is the sum of the elastic bending energy, or curvature energy, and the work done by tension (see Nanavati *et al.*, 1992, for details). The curvature energy is $W_{bending} = 0.5\,k_c\,(J_m + J_p - J_s)^2$, where k_c is the bending rigidity or elastic modulus, J_s is the spontaneous curvature of the lipids, and J_m and J_p are the meridienal and parallel curvatures defined by the geometry of the pore [$J_m = -1/b$ and $J_p = -(1/b)(1 - (b + a)/x)$]. The tension component is $W_{tension} = \gamma\delta A = -2\,\pi\gamma\,(a^2 - (\pi - 3)b^2 - (\pi - 2)\,a\,b)$. (B) The energy of a lipidic pore as a function of radius in the absence (solid lines) and presence (dashed and dotted lines) of tension. The spontaneous curvature of the lipids was -0.475 nm^{-1}; tensions used were zero (curve 1), 0.1 mN/m (curve 2), and 1.0 mN/m (curve 3).

or elastic modulus of the bilayer (K_c), the spontaneous curvature of the phospholipids in the bilayer (J_s), and the geometry of the pore (defined by J_m and J_p; see legend and Nanavati *et al.*, 1992, for details). Using experimentally determined values for the spontaneous curvature of biological membranes and estimates of the membrane tension present in the secretory granule, free energy profiles for the fusion pore can be calculated (Fig. 10B). From these profiles, the pore radius with the minimum energy as well as the variance of the fluctuations around this value can be determined. Strikingly, the range of conductances predicted for such lipidic pores was indistinguishable from those measured for exocytotic fusion pores in the *beige* mouse mast cell (Nanavati *et al.*, 1992). The model also shows an energy barrier that must be overcome for the fusion pore to expand irreversibly. Changes in the lipid composition due to flux of different lipids through the pore can change the pore from a relatively stable structure to one that expands rapidly or closes. Thus, these theoretical considerations predict the existence of lipidic pores that can expand irreversibly or open for several seconds and then close, much like those seen with the patch-clamp measurements.

Pores in the patch-clamped secretory granule membrane can be induced by a combination of electrical and mechanical stresses. Admittance measurements of such "electromechanical" pores show patterns strikingly similar to those for exocytotic fusion pores: abrupt opening, gradual expansion of the pore, and occasional closure (Oberhauser and Fernandez, 1993). Figure 11 shows an example of admittance measurements of an electromechanical pore; the similarities to measurements of the exocytotic fusion pore (see Fig. 4) are readily apparent. Moreover, the distribution of pore conductances is the same as that for the exocytotic fusion pore and for the theoretical model of a pore in a bilayer (Oberhauser and Fernandez, 1993). Thus, an electromechanical pore in a secretory granule membrane exhibits the properties expected of the fusion pore. Further evidence that such pores through a single bilayer can occur in biological membranes is provided by the observation that electrically induced fusion of eyrthrocytes proceeds through hemifusion (Song *et al.*, 1991).

B. Protein Scaffold Directs Exocytotic Membrane Fusion

The events leading to the formation of the fusion pore can now be considered. Electron micrographs have shown that the secretory granule is positioned approximately 100 nm from the cell membrane, well beyond the range of any of the bilayer–bilayer forces. The first step is dimpling of the plasma membrane toward the secretory granule membrane. According to this model, the macromolecular scaffold of proteins

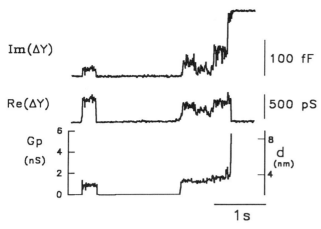

FIGURE 11 Time course of the conductance of an "electromechanical" pore in a secretory granule membrane patch. Electromechanical pores were measured in patch-clamped giant secretory granules from *beige* mouse mast cells using the admittance technique, as for exocytotic fusion pores (for details, see Oberhauser and Fernandez, 1993). The figure shows, from top to bottom, a continuous recording of the capacitance [Im(ΔY)], the AC conductance [Re(ΔY)], and the calculated pore conductance (Eq. 2). Shortly after obtaining a gigaseal, the pore conductance increased spontaneously and abruptly, and fluctuated before returning to zero (transient opening). A second abrupt increase in pore conductance followed; again, further fluctuations and finally an increase to an effectively infinite conductance occurred. The sudden increase in both components of the admittance corresponds to the opening of a pore(s) in the patch membrane, connecting the lumen of the secretory granule with the pipette interior. Right ordinate is the pore diameter calculated from Eq. 8-1 of Hille (1984), for a resistivity of 100 Ω cm, a pore length of 4 nm, and assuming a single pore.

is responsible for directing the dimpling of the cell membrane in response to intracellular signals (Fig. 12). The filamentous structures seen between the two membranes may be part of this scaffold. As the invaginated cell membrane approaches the secretory granule membrane, it will experience the repulsive hydration force. This repulsive force will, at first, be overcome by the process drawing the two membranes together, but will become larger as the separation decreases. The dimple is seen to end in a small, highly curved tip, which will provide a localized region of high tension. When the tip of the dimple approaches the granule membrane, which in the stimulated cell also has a high tension, hemifusion of the two bilayers will occur spontaneously.

Pore formation will follow because stresses in the common bilayer, comprising the intragranular leaflet of the granule membrane and the extracellular leaflet of the plasma membranes, will induce membrane breakdown. Thus, the early fusion pore seen with patch-clamp measurements, which can be as small as 80 pS, is a 1-nm diameter pore in a single bilayer. As

discussed earlier, models of the free energy profiles of such pores predict that they will behave similarly to the measured excytotic fusion pore. These pores should fluctuate around a certain size before expanding irreversibly or closing. An advantage of progressing through hemifusion is that potentially injurious secretory products cannot leak into the cytosol. Pore closure can be explained easily by changes in the spontaneous curvature of the lipids that make up the fusion pore (Nanavati *et al.*, 1992). As the pore expands, the hemifused bilayer retreats and the full fusion pore typical of the electron micrographs is formed irreversibly.

C. Structural and Regulatory Components of the Fusion Pore Scaffold

According to the model just described, the fusion pore is a lipidic structure and the role of the protein scaffold is to present the two membranes in close proximity at the appropriate time (Monck and Fernandez, 1992). The identity of the proteins that might form the scaffold is currently unknown. However, several proteins have locations and/or functional properties that suggest a role associated with exocytosis. Broadly, the putative functions of the scaffold can be divided into three categories: responding to intracellular signals, movement of the two membranes together, and fusogenic activity. In this section, possible proteins involved in these functions are suggested.

Immediately prior to fusion, the plasma membrane must be pulled across the 100-nm gap separating the secretory granules from the cell surface. Similar membrane dimpling occurs prior to pinch-off in endocytosis and in vesiculation of the Golgi. Clathrin structures and scaffolds of βCOPs and ARFs, a family of small GTP binding proteins, are thought to regulate these processes (Rothman and Orci, 1992; Schmid, 1992). Thus, macromolecular scaffolds have precedent in processes involving membrane reorganization. The scaffold involved in plasma membrane dimpling must include some structural elements. Several lines of evidence have indicated a role for the cytoskeleton in exocytosis. Exocytosis is modulated by several drugs that affect the organization of actin and is inhibited by antibodies against spectrin, and reorganization of actin networks during exocytosis has been shown in a number of different cell types (Burgoyne, 1987,1991; Linstedt and Kelly, 1987). The ability of annexins to form filaments similar to those seen in the electron micrographs of chromaffin granules (Nakata *et al.*, 1990) suggests that annexins play a role in plasma membrane dimpling. The actual movement of the

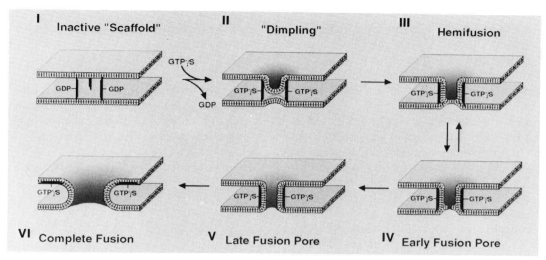

FIGURE 12 Model describing exocytotic fusion as a lipidic fusion event directed by a protein scaffold. (I) Membranes separated by a protein scaffold. Since GTPγS is the trigger for exocytosis in most fusion pore experiments, GDP is depicted as bound to the scaffold to represent the inactive state of a GTP-binding protein, which is the cellular switch for fusion pore formation and forms part of the scaffold. (II) In response to exchange of GDP for GTPγS on the GTP-binding protein, the scaffold directs a dimple in the plasma membrane toward the secretory granule. (III) Hemifusion occurs spontaneously because of local tension in the two membranes. (IV) A small pore forms in the stressed common bilayer of the hemifusion structure. (V) The fusion pore develops into the hourglass structure seen in the electron micrographs. (VI) The pore further expands to allow release of secretory products.

membranes toward each other may result from an interaction between the plasma membrane and charged proteins on the inside of a tubular scaffold, which could draw the plasma membrane into the structure, like a patch pipette draws in membrane during seal formation.

In most cells, Ca^{2+} is the primary intracellular trigger for exocytosis, in many cells, a role for a GTP-binding protein has been implicated in regulation of exocytosis. Therefore, the exocytotic scaffold probably contains GTP- and Ca^{2+}-binding proteins as sensor elements. In patch-clamped mast cells, the response to GTPγS does not wash out (Oberhauser et al., 1992b), suggesting that one GTP-binding protein or more is closely associated with the exocytotic machinery and could be the molecular switch for triggering exocytosis. These could be heterotrimeric G-proteins (Gomperts, 1990; Lindau and Gomperts, 1991) or small GTP-binding proteins of the ras superfamily. Several lines of evidence suggest that a member of the rab3 subfamily might be involved. rab3a has been found only in secretory cells (Mizoguchi et al., 1989), is localized to synaptic vesicles (Darchen et al., 1990; Mizoguchi et al., 1990; Fischer von Mollard et al., 1990), and undergoes cellular redistribution during exocytosis (Fischer von Mollard et al., 1991). Moreover, synthetic oligopeptides homologous to the effector domain of rab3a can stimulate exocytosis in pancreatic

acinar cells (Padfield et al., 1992), chromaffin cells (Senyshyn et al., 1992), and mast cells (Oberhauser et al., 1992b). Several rab proteins have been shown to be involved in controlling fusion between different membrane compartments: rab5 in fusion of late endosomes, Ypt1 (the yeast homolog of rab1) in fusion of transport vesicles with the Golgi in yeast, and Sec4 (the yeast homolog of rab3) in constitutive exocytosis in yeast (Bacon et al., 1989; Walworth et al., 1989; Gorvel et al., 1991). Because different rab proteins are localized to different intracellular compartments (Balch, 1990; Goud and McCaffrey, 1991), different members of the rab3 subfamily might regulate different types of exocytosis.

Several proteins have been proposed to play a role in regulation of Ca^{2+}-dependent exocytosis, including annexin II, exo1, and p145, which prevent washout of Ca^{2+}-dependent exocytosis from adrenal cells (Ali et al., 1989; Morgan and Burgoyne, 1992; Walent et al., 1992); synexin (annexin VII); and synaptotagmins. The annexins are $Ca2+$-binding proteins that bind to secretory granule membranes and to cytoskeletal proteins, and in vitro can form filaments and facilitate chromaffin granule–granule fusion, so these proteins could serve both structural and regulatory roles (Burgoyne and Geisow, 1989; Nakata et al., 1990). The synaptotagmins are Ca^{2+} and phospholipid-binding proteins that undergo conformational changes on binding Ca^{2+} (Brose *et al.*, 1992). Since they contain a sequence that is predicted to form an amphiphilic helix with positively charged residues on one side and bulky hydrophobic groups on the other (Perin *et al.*, 1991), synaptotagmins may fulfill a role similar to that of the amphiphilic helical domains of the viral fusion proteins (Bentz *et al.*, 1990; White, 1990).

In searching for proteins that might form part of the exocytotic scaffold, several other approaches can be taken. First, a number of proteins found in synaptic vesicles, including synapsins, synatophysin, synaptobrevin, and synaptoporin, as well as synaptotagmins, have been proposed to play a role in regulating secretion (Sudhof *et al.*, 1989; Knaus *et al.*, 1990; Pollard *et al.*, 1990; Sudhof and Jahn, 1991); some of these associate to form macromolecular complexes (Bennett *et al.*, 1992). Second, the regulation of intracellular fusion in the Golgi has been proposed to involve a macromolecular complex of proteins, including SNAPS and NSF (Rothman and Orci, 1992; Wilson *et al.*, 1992). Third, genetic studies in yeast have revealed a number of proteins necessary for constitutive exocytosis, some of which are related to those involved in intra-Golgi traffic, including several plasma membrane proteins that bind to each other and to *sec*4, suggesting that they might be part of a scaffold (Bacon *et al.*, 1989; Walworth *et al.*, 1989; Bowser *et al.*, 1992). Given the many similarities between the putative fusion machinery in different cellular fusion reactions, some of these proteins or related proteins may form the exocytotic scaffold and may direct fusion of secretory granules with the plasma membrane.

VIII. Conclusions

The patch-clamp data suggest a fusion pore that is predominantly lipidic from its earliest moments. The case for exocytotic fusion being a spontaneous fusion between two locally tense bilayers that are brought into close proximity by a macromolecular scaffold of proteins has been argued in this chapter. Although the scaffold proteins have not yet been identified, many proteins that have potential structural and regulatory roles in such a scaffold have been discovered. Identification of relevant proteins and characterization of how their interactions promote fusion will provide many challenges in the years ahead. The mast cell is an excellent model system for studying the fusion pore. Modern molecular biological techniques such as the polymerase chain reaction may facilitate identification of mRNAs encoding some of the proteins already discussed (e.g., *rab*s, annexins, synaptotagmins) or closely related proteins in mast cells. The

sequences can then be used to make synthetic oligopeptides for functional studies and to make antipeptide antibodies that can be used for functional studies and immunocytochemistry. The sequences can also be used for designing antisense oligonucleotide probes to try to modify activity of the fusion pore scaffold. We hope that, in the near future, the combination of molecular biological techniques with biophysical measurements, which has proved so successful for study of ion channels, will help define detailed structure–function relationships between the proteins of the scaffold and formation of the fusion pore.

Acknowledgments

We would like to thank Douglas Chandler for supplying the electron micrographs of the fusion pore shown in Fig. 1.

References

Ali, S. M., Geisow, M. J., and Burgoyne, R. D. (1989). A role for calpactin in calcium-dependent exocytosis in adrenal chromaffin cells. *Nature (London)* **340**, 313–315.

Almers, W. (1990). Exocytosis. *Annu. Rev. Physiol.* **52**, 607–624.

Almers, W., and Tse, F. W. (1990). Transmitter release from synapses: Does a preassembled fusion pore initiate exocytosis? *Neuron* **4**, 813–818.

Alvarez de Toledo, G., and Fernandez, J. M. (1988). The events leading to secretory granule fusion. *In* "Cell Physiology of Blood," (R. B. Gunn and J. C. Parker, eds.). Society of General Physiologists Series, Volume 43. pp. 334–344. Rockefeller University Press, New York.

Alvarez de Toledo, G., and Fernandez, J. M. (1990a). Patch-clamp measurements reveal multimodel distribution of granule sizes in rat mast cells. *J. Cell Biol.* **110**, 1033–1039.

Alvarez de Toledo, G., and Fernandez, J. M. (1990b). Compound versus multigranular exocytosis in peritoneal mast cells. *J. Gen. Physiol.* **95**, 397–409.

Alvarez de Toledo, G., and Fernandez, J. M. (1990c). The effect of GTPγS and Ca^{++} on the kinetics of exocytosis of single secretory granules in peritoneal mast cells. *Biophys. J.* **57**, 495a.

Alvarez de Toledo, G., Fernandez-Chacon, R., and Fernandez, J. M. (1993). Evidence of secretion during transient vesicle fusion in beige mouse peritoneal mast cells. *Nature (London)* **363**, 551–557.

Bacon, R. A., Salminen, A., Ruohola, H., Novick, P., and Ferro-Novick, S. (1989). The GTP-binding protein Ypt1 is required for transport in vitro: The Golgi apparatus is defective in ypt1 mutants. *J. Cell Biol.* **109**, 1015–1022.

Balch, W. E. (1989). Biochemistry of interorganelle transport. A new frontier in enzymology emerges from versatile *in vitro* model systems. *J. Biol. Chem.* **264**, 16965–16968.

Balch, W. E. (1990). Small GTP-binding proteins in vesicular transport. *Trends Biochem. Sci.* **15**, 473–477.

Bennett, M. K., Calakos, N., Kreiner, T., and Scheller, R. H. (1992). Synaptic vesicle membrane proteins interact to form a multimeric complex. *J. Cell Biol.* **116**, 761–775.

Bentz, J., Ellens, H., and Alford, D. (1990). An architecture for the fusion site of influenza hemagglutinin. *FEBS Lett.* **276**, 1–5.

Blobel, C. P., Wolfsberg, T. G., Turck, C. W. Myles, D. G., Primakoff, P., and White, J. M. (1992). A potential fusion peptide and an integrin ligand domain in a protein active in sperm-egg fusion. *Nature (London)* **356**, 248–252.

Bowser, R., Muller, H., Govindan, B., and Novick, P. (1992). Sec8p and Sec15p are components of a plasma membrane-associated 19.5 S particle that may function downstream of Sec4p to control exocytosis. *J. Cell Biol.* **118**, 1041–1056.

Breckenridge, L. J., and Almers, W. (1987a). Currents through the fusion pore that forms during exocytosis of a secretory vesicle. *Nature (London)* **328**, 814–817.

Breckenridge, L. J., and Almers, W. (1987b). Final steps in exocytosis observed in a cell with giant secretory granules. *Proc. Natl. Acad. Sci. USA* **84**, 1945–1949.

Brose, N., Petrenko, A. G., Sudhof, T. C., and Jahn, R. (1992). Synaptotagmin: a calcium sensor on the synaptic vesicle surface. *Science* **256**, 1021–1025.

Burgoyne, R. D. (1987). Control of exocytosis. *Nature (London)* **328**, 112–113.

Burgoyne, R. D. (1991). Control of exocytosis in adrenal chromaffin cells. *Biochim. Biophys. Acta.* **1071**, 174–202.

Burgoyne, R. D., and Geisow, M. J. (1989). The annexin family of calcium-binding proteins. *Cell Calcium* **10**, 1–10.

Chandler, D. E. (1992). Membrane fusion as seen in rapidly frozen secretory cells. *Ann. N.Y. Acad. Sci.* **635**, 234–245.

Chandler, D. E., and Heuser, J. E. (1980). Arrest of membrane fusion events in mast cells by quick-freezing. *J. Cell Biol.* **86**, 666–674.

Chandler, D. E., Bennett, J. P., and Gomperts, B. (1983). Freeze-fracture studies of chemotactic peptide-induced exocytosis in neutrophils: Evidence for two patterns of secretory granule fusion. *J. Ultrastruct. Res.* **82**, 221–232.

Chandler, D. E. Whitaker, M., and Zimmerberg, J. (1989). High molecular weight polymers block cortical granule exocytosis in sea urchin eggs at the level of granule matrix dissassembly. *J. Cell Biol.* **109**, 1269–1278.

Chaudhury, M. K., and Ohki, S. (1981). Correlation between membrane expansion and temperature-induced membrane fusion. *Biochim. Biophys. Acta* **642**, 365–374.

Chernomordik, L. V., Melikyan, G. B., and Chizmadzhev, Y. A. (1987). Biomembrane fusion: A new concept derived from model studies using two interacting planar lipid bilayers. *Biochim. Biophys. Acta* **906**, 309–352.

Chow, R. H., von Ruden, L., and Neher, E. (1992). Delay in vesicle fusion revealed by electrochemical monitoring of single secretory events in adrenal chromaffin cells. *Nature (London)* **356**, 60–63.

Cohen, F. S., Zimmerberg, J., and Finkelstein, A. (1980). Fusion of phospholipid vesicles with planar phospholipid bilayer membranes. Incorporation of a vesicular membrane marker into the planar membrane. *J. Gen. Physiol.* **75**, 251–270.

Cohen, F. S., Niles, W. D., and Akabas, M. H. (1989). Fusion of phospholipid vesicles with a planar membrane depends on the membrane permeability of the solute used to create the osmotic pressure. *J. Gen. Physiol.* **93**, 201–210.

Cole, K. S. (1972). "Membranes, Ions and Impulses" (C. A. Tobias, ed.). University of California Press, Berkeley, California.

Curran, M. J., and Brodwick, M. S. (1991). Ionic control of the size of the vesicle matrix of beige mouse mast cells. *J. Gen. Physiol.* **98**, 771–790.

Darchen, F., Zahraoui, A., Hammel, F., Monteils, M. P., Tavitian, A., and Scherman, D. (1990). Association of the GTP-binding protein Rab3A with bovine adrenal chromaffin granules. *Proc. Natl. Acad. Sci. USA* **87**, 5692–5696.

Fernandez, J. M., Neher, E., and Gomperts, B. D. (1984). Capacitance measurements reveal stepwise fusion events in degranulating mast cells. *Nature (London)* **312**, 453–455.

Fernandez, J. M., Lindau, M., and Eckstein, F. (1987). Intracellular stimulation of mast cells with guanine nucleotides antigenic stimulation. *FEBS Lett.* **216**, 89–93.

Fernandez, J. M., Villalon, M., and Verdugo, P. (1991). Reversible condensation of mast cell secretory products *in vitro. Biophys. J.* **59**, 1022–1027.

Fernandez-Chacon, R., and Alvarez de Toledo, G. (1992). In vitro voltammetry of single isolated cells: A new approach to study late exocytosis. *Biophys. J.* **61**, A226.

Fidler, N., and Fernandez, J. M. (1989). Phase tracking: an improved phase detection technique for cell membrane capacitance measurements. *Biophys. J.* **56**, 1153–1162.

Finkelstein, A., Zimmerberg, J., and Cohen, F. S. (1986). Osmotic swelling of vesicles: Its role in the fusion of vesicles with planar phospholipid bilayer membranes and its possible role in exocytosis. *Annu. Rev. Physiol.* **48**, 163–174.

Fischer von Mollard, G., Mignery, G., Baumert, M., Perin, M. S., Hanson, J. J., Burger, P. M., Jahn, R., and Sudhoff, T. C. (1990). rab3 is a small GTP-binding protein exclusively localized to synaptic vesicles. *Proc. Natl. Acad. Sci. USA* **87**, 1988–1992.

Fischer von Mollard, G., Sudhof, T. C., and Jahn, R. (1991). A small GTP-binding protein dissociates from synaptic vesicles during exocytosis. *Nature (London)* **349**, 79–81.

Gomperts, B. D. (1990). G_E: A GTP-binding protein mediating exocytosis. *Annu. Rev. Physiol.* **52**, 591–606.

Gorvel, J.-P., Chavrier, P., Zerial, M., and Gruenberg, J. (1991). rab5 controls early endosome fusion in vitro. *Cell* **64**, 915–925.

Goud, B., and McCaffrey, M. (1991). Small GTP-binding proteins and their role in transport. *Curr. Opin. Cell Biol.* **3**, 626–633.

Hampton, R. Y., and Holz, R. W. (1983). Effects of changes in osmolality on the stability and function of cultured chromaffin cells and the possible role of osmotic forces in exocytosis. *J. Cell Biol.* **96**, 1082–1088.

Helle, K. B. (1990). Chromogranins: Universal proteins in secretory organelles from paramecium to man. *Neurochem. Int.* **17**, 165–175.

Helm, C. A., Israelachvili, J. N., and McGuiggan, P. M. (1989). Molecular mechanisms and forces involved in the adhesion and fusion of amphiphilic bilayers. *Science* **246**, 919–922.

Helm, C. A., Israelachvili, J. N., and McGuiggan, P. M. (1992). Role of hydrophobic forces in bilayer adhesion and fusion. *Biochemistry* **31**, 1794–1805.

Hibino, M., Shigemori, M., Itoh, H., Nagayama, K., and Kinosita, K., Jr. (1991). Membrane conductance of an electroporated cell analyzed by submicrosecondimaging of transmembrane potential. *Biophys. J.* **59**, 209–220.

Hille, B. (1984). Ionic Channels of Excitable Membranes. Sinauer Associates. Sunderland, Massachusetts.

Hoekstra, D. (1990). Membrane fusion of enveloped viruses: Especially a matter of proteins. *J. Bioenerg. Biomem.* **22**, 121–155.

Hoffmann, W., Sarzala, M. G., Gomez-Fernandez, J. C., Goni, F. M., Restall, C. J., and Chapman, D. (1980). Protein rotational diffusion and lipid structure of reconstituted systems of Ca^{2+}-activated adenosine triphosphatase. *J. Mol. Biol.* **141**, 119–132.

Holz, R. W. (1986). The role of osmotic forces in exocytosis from adrenal chromaffin cells. *Annu. Rev. Physiol.* **48**, 175–190.

Holz, R. W., and Senter, R. A. (1986). Effects of osmolality and ionic strength on secretion from adrenal chromaffin cells permeabilized with digitonin. *J. Neurochem.* **46**, 1835–1842.

Israelachvili, J. N., and McGuiggan, P. M. (1988). Forces between surfaces in liquids. *Science* **241**, 795–800.

Joshi, C., and Fernandez, J. M. (1988). Capacitance measurements. An analysis of the phase detector technique used to study exocytosis and endocytosis. *Biophys. J.* **53**, 885–892.

Knaus, P., Marqueze-Pouey, B., Scherer, H., and Betz, H. (1990).

Synaptoporin, a novel putative channel protein of synaptic vesicles. *Neuron* **5**, 453–462.

Knoll, G., Burger, K. N. J., Bron, R., van Meer, G., and Verkleij, A. J. (1988). Fusion of liposomes with the plasma membrane of epithelial cells: Fate of incorporated lipids as followed by freeze fracture and autoradiography of plastic sections. *J. Cell Biol.* **107**, 2511–2521.

Knoll, G., Braun, C., and Plattner, H. (1991). Quenced flow analysis of exocytosis in *Paramecium* cells: Time course, changes in membrane structure, and calcium requirements revealed after rapid mixing and rapid freezing of intact cells. *J. Cell Biol.* **113**, 1295–1304.

Kozlov, M. M., Leikin, S. L., Chernomordik, L. V., Markin, V. S., and Chizmadzhev, Y. A. (1989). Stalk mechanism of vesicle fusion. Intermixing of aqueous contents. *Eur. Biophys. J.* **17**, 121–129.

Krasne, S., Eisenman, G., and Szabo, G. (1971). Freezing and melting of lipid bilayers and the mode of action of nonactin, valinomycin, and gramicidin. *Science* **174**, 412–415.

Kwon, I. C., Bae, Y. H., and Kim, S. W. (1991). Electrically erodible ploymer gel for controlled release of drugs. *Nature (London)* **354**, 291–293.

Leszczyszyn, D. J., Jankowski, J. A., Humberto Viveros, O., Diliberto, E. J., Jr., Near, J. A., and Wightman, R. M. (1990). Nicotinic receptor-mediated catecholamine secretion from individual chromaffin cells. *J. Biol. Chem.* **265**, 14736–14737.

Lindau, M. (1991). Time-resolved capacitance measurements: monitoring exocytosis in single cells. *Q. Rev. Biophys.* **24**, 75–101.

Lindau, M., and Gomperts, B. D. (1991). Techniques and concepts in exocytosis: focus on mast cells. *Biochim. Biophys. Acta* **1071**, 429–471.

Lindau, M., and Neher, E. (1988). Patch-clamp techniques for time-resolved capacitance measurements in single cells. *Pflügers Arch.* **411**, 137–146.

Linstedt, A. D., and Kelly, R. B. (1987). Overcoming barriers to exocytosis. *Trends Neurosci.* **10**, 446–448.

Markin, V. S., Kozlov, M. M., and Borovjagin, V. L. (1984). On the theory of membrane fusion. The stalk mechanism. *Gen. Physiol. Biophys.* **5**, 361–377.

Mason, W. T., Rawlings, S. R., Cobbett, P., Sikdar, S. K., Zorec, R., Akerman, S. N., Benham, C. D., Berridge, M., Cheek, T., and Moreton, R. B. (1988). Control of secretion in anterior pituitary cells—Linking ion channels, messengers and exocytosis. *J. Exp. Biol.* **139**, 287–316.

Merkle, C. J., and Chandler, D. E. (1989). Hyperosmolality inhibits exocytosis in sea urchin eggs by formation of a granule-free zone and arrest of pore widening. *J. Membr. Biol.* **112**, 223–232.

Millar, J., and Williams, G. V. (1990). East differential ramp voltammetry: A new voltammetric technique designed specifically for use in neuronal tissue. *J. Electroanal. Chem.* **282**, 33–49.

Mizoguchi, A., Kim, S., Ueda, T., and Takai, Y. (1989). Tissue distribution of *smg* p25A. A *ras*-like GTP-binding protein, studied by use of a specific monoclonal antibody. *Biochem. Biophys. Res. Commun.* **162**, 1438–1445.

Mizoguchi, A., Kim, S., Ueda, T., Kikuchi, A., Yorifuji, H., Hirokawa, N., and Takai, Y. (1990). Localization and subcellular distribution of *smg* p25A, a *ras* p21-like GTP-binding protein in rat brain. *J. Biol. Chem.* **265**, 11872–11879.

Monck, J. R., and Fernandez, J. M. (1992). The exocytotic fusion pore. *J. Cell Biol.* **119**, 1395–1404.

Monck, J. R., Alvarez de Toledo, G., and Fernandez, J. M. (1990). Tension in secretory granule membranes causes extensive membrane transfer through the exocytotic fusion pore. *Proc. Natl. Acad. Sci. USA* **87**, 7804–7808.

Monck, J. R., Oberhauser, A. F., Alvarez de Toledo, G., and Fernandez, J. M. (1991a). Is swelling of the secretory granule matrix the force that dilates the exocytotic fusion pore? *Biophys. J.* **59**, 39–47.

Monck, J. R., Oberhauser, A. F., Alvarez de Toledo, G., and Fernandez, J. M. (1991b). Tension in secretory granule membranes causes extensive membrane transfer through the exocytotic fusion pore. *Biophys. J.* **59**, 207a.

Morgan, A., and Burgoyne, R. D. (1992). Exo1 and Exo2 proteins stimulate calcium-dependent exocytosis in permeabilized adrenal chromaffin cells. *Nature (London)* **355**, 833–836.

Nakata, T., and Hirokawa, N. (1991). Organization of cortical cytoskeleton of cultured chromaffin cells and involvement in secretion as revealed by quick-freeze, deep-etching and double-label and immunoelectron microscopy. *J. Neurosci.* **12**, 2186–2197.

Nakata, T., Sobue, K., and Hirokawa, N. (1990). Conformational change and localization of calpactin I complex involved in exocytosis as revealed by quick-freeze, deep-etch electron microscopy and immunocytochemistry. *J. Cell Biol.* **110**, 13–25.

Nanavati, C., Markin, V. S., Oberhauser, A., and Fernandez, J. M. (1992). The exocytotic fusion pore as a protein-supported lipidic structure. *Biophys. J.* **63**, 1118–1132.

Needham, D., and Hochmuth, R. M. (1989). Electro-mechanical permeabilization of lipid vesicles. Role of membrane tension and compressibility. *Biophys. J.* **55**, 1001–1009.

Neher, E. (1988). The influence of intracellular calcium concentration on degranulation of dialysed mast cells from rat peritoneum. *J. Physiol.* **395**, 193–214.

Neher, E. (1992). Ion channels for communication between and within cells. *Science* **256**, 498–502.

Neher, E., and Marty, A. (1982). Discrete changes of cell membrane capacitance observed under conditions of enhanced secretion in bovine adrenal chromaffin cells. *Proc. Natl. Acad. Sci. USA* **79**, 6712–6716.

Nielsen, E. H. (1990). A filamentous network surrounding secretory granules from mast cells. *J. Cell Sci.* **96**, 43–46.

Nusse, O., and Lindau, M. (1988). The dynamics of exocytosis in human neutrophils. *J. Cell Biol.* **107**, 2117–2123.

Nusse, O., Lindau, M., Cromwell, O., Kay, A. B., and Gomperts, B. D. (1990). Intracellular application of guanosine-5'-*O*-(3-thiotriphosphate) induces exocytotic granule fusion in guinea pig eosinophils. *J. Exp. Med.* **171**, 775–786.

Oberhauser, A. F., and Fernandez, J. M. (1993). Patch-clamp studies of intact secretory granules. *Biophys. J.* **65**, 1844–1852.

Oberhauser, A. F., Monck, J. R., and Fernandez, J. M. (1992a). Events leading to the opening and closing of the exocytotic fusion pore have markedly different temperature dependencies. Kinetic analysis of single fusion events in patch-clamped mouse mast cells. *Biophys. J.* **61**, 800–809.

Oberhauser, A. F., Monck, J. R., Balch, W. E., and Fernandez, J. M. (1992b). Exocytotic fusion is directly activated by Rab3AL peptides. *Nature (London)* **360**, 270–273.

Ohki, S. (1987). Surface tension, hydration energy and membrane fusion. *In* "Molecular Mechanisms of Membrane Fusion" (S. Ohki, D. Doyle, T. D. Flanagan, S. Withers, and E. Mayhew, eds.), pp. 123–137. Plenum Press, New York.

Orci, L., and Malaisse, W. (1980). Single and chain release of insulin secretory granules is related to anionic transport at exocytotic sites. *Diabetes* **29**, 943–944.

Ornberg, R. L., and Reese, T. S. (1981). Beginning of exocytosis captured by rapid-freezing of *Limulus* amebocytes. *J. Cell Biol.* **90**, 40–54.

Padfield, P. J., Balch, W. E., and Jamieson, J. D. (1992). A synthetic peptide of the rab3a effector domain stimulates amylase release from permeabilized pancreatic acini. *Proc. Natl. Acad. Sci. USA* **89,** 1656–1660.

Penner, R. (1988). Multiple signaling pathways control stimulus-secretion coupling in rat peritoneal mast cells. *Proc. Natl. Acad. Sci. USA* **85,** 9856–9860.

Penner, R., Matthews, G., and Neher, E. (1988). Regulation of calcium influx by second messengers in rat mast cells. *Nature (London)* **334,** 499–504.

Perin, M. S., Brose, N., Jahn, R., and Sudhof, T. C. (1991). Domain structure of synaptotagmin (p65). *J. Biol. Chem.* **266,** 623–629.

Plattner, H., Lumpert, C. J., Knoll, G., Kissmehl, R., Hohne, B., Momayezi, M., and Glas-Albrecht, R. (1991). Stimulus-secretion coupling in *Paramecium* cells. *Eur. J. Cell Biol.* **55,** 3–16.

Pollard, H. B., Pazoles, C. J., Creutz, C. E., Scott, J. H., Zinder, O., and Hotchkiss, A. (1984). An osmotic mechanism for exocytosis from dissociated chromaffin cells. *J. Biol. Chem.* **259,** 1114–1121.

Pollard, H. B., Burns, A. L., and Rojas, E. (1990). Synexin (annexin VII): A cytosolic calcium-binding protein which promotes membrane fusion and forms calcium channels in artificial bilayer and natural membranes. *J. Membr. Biol.* **117,** 101–112.

Pollard, H. B., Rojas, E., Pastor, R. W., Rojas, E. M., Guy, H. R., and Burns, A. L. (1991). Synexin: Molecular mechanism of calcium-dependent membrane fusion and voltage-dependent calcium-channel activity. *Ann. N. Y. Acad. Sci.* **635,** 328–351.

Pusch, M., and Neher, E. (1988). Rates of diffusional exchange between small cells and a measuring patch pipette. *Pflügers Arch.* **411,** 204–211.

Rahamimoff, R., DeReimer, S. A., Ginsburg, S., Kaiserman, I., Sakmann, B., Shapira, R., Stadler, H., and Yakir, N. (1989). Ionic channels in synaptic vesicles: Are they involved in neurotransmitter release? *Q. J. Exp. Physiol.* **74,** 1019–1031.

Rand, R. P. (1981). Interacting phospholipid bilayers: Measured forces and induced structural changes. *Annu. Rev. Biophys. Bioeng.* **10,** 277–314.

Rothman, J. E., and Orci, L. (1992). Molecular dissection of the secretory pathway. *Nature (London)* **355,** 409–415.

Sakmann, B. (1992). Elementary steps in synaptic transmission revealed by currents through single ion channels. *Science* **256,** 503–512.

Schmid, S. L. (1992). The mechanism of receptor-mediated endocytosis: more questions than answers. *BioEssays* **16,** 589–596.

Schmidt, W., Patzak, A., Lingg, G., and Winkler, H. (1983). Membrane events in adrenal chromaffin cells during exocytosis: A freeze-etching analysis after rapid cryofixation. *Eur. J. Cell Biol.* **32,** 31–37.

Segawa, A., and Yamashina, S. (1989). Role of microfilaments in exocytosis: A new hypothesis. *Cell Struct. Funct.* **14,** 531–544.

Senyshyn, J., Balch, W. E., and Holz, R. W. (1992). Synthetic peptides of the effector-binding domain of *rab* enhance secretion from digitonin-permeabilized chromaffin cells. *FEBS Lett.* **309,** 41–46.

Song, L., Ahkong, Q. F., Georgescauld, D., and Lucy, J. A. (1991). Membrane fusion without cytoplasmic fusion (hemi-fusion) in erythrocytes that are subjected to electrical breakdown. *Biochim. Biophys. Acta* **1065,** 54–62.

Spruce, A. E., Iwata, A., White, J. M., and Almers, W. (1989). Patch clamp studies of single cell-fusion events mediated by a viral fusion protein. *Nature (London)* **342,** 555–558.

Spruce, A. E., Breckenridge, L. J., Lee, A. K., and Almers, W. (1990). Properties of the fusion pore that forms during exocytosis of a mast cell secretory vesicle. *Neuron* **4,** 643–654.

Spruce, A. E., Iwata, A., and Almers, W. (1991). The first milliseconds of the pore formed by a fusogenic viral envelope protein

during membrane fusion. *Proc. Natl. Acad. Sci. USA* **88,** 3623–3627.

Stamford, J. A. (1985). *In vivo* voltammetry: Promise and perspective. *Brain Res. Rev.* **10,** 119–135.

Stamford, J. A. (1986). *In vivo* voltammetry: Some methodological considerations. *J. Neurosci. Meth.* **17,** 1–29.

Sudhof, T. C., and Jahn, R. (1991). Proteins of synaptic vesicles involved in exocytosis and membrane recycling. *Neuron* **6,** 665–677.

Sudhof, T. C., Czernik, A. J., Kao, H. T., Takei, K., Johnston, P. A., Horiuchi, A., Kanazir, S. D., Wagner, M. A., Perin, M. S., De Camilli, P., and Greengard, P. (1989). Synapsins: Mosaics of shared and individual domains in a family of synaptic vesicle phosphoproteins. *Science* **245,** 1474–1480.

Tatham, P. E. R., Duchen, M. R., and Millar, J. (1991). Monitoring exocytosis from single mast cells by fast voltammetry. *Eur. J. Physiol. (Pflügers Arch.)* **419,** 409–414.

Thomas, L., Hartung, K., Langosch, D., Rehm, H., Bamberg, E., Franke, W. W., and Bertz, H. (1988). Identification of synaptophysin as a hexameric channel protein of the synaptic vesicle membrane. **242,** 1050–1053.

Vallar, L., Biden, T. J., and Wollheim, C. B. (1987). Guanine nucleotides induce Ca²⁺-independent insulin secretion from permeabilized RINm5F cells. *J. Biol. Chem.* **262,** 5049–5056.

Valtorta, F., Fesce, R., Grohovaz, F., Haimann, C., Hurlbut, W. P., Iezzi, N., Torri Tarelli, F., Villa, A. and Ceccarelli, B. (1990). Neurotransmitter release and synaptic vesicle recycling. *Neurosci.* **35,** 477–489.

Verdugo, P. (1991). Mucin exocytosis. *Am. Rev. Respir. Dis.* **144,** S33–S37.

Walent, J. H., Porter, B. W., and Martin, T. F. J. (1992). A novel 145 kd brain cytosolic protein reconstitutes Ca²⁺-regulated secretion in permeable neuroendocrine cells. *Cell* **70,** 765–775.

Walworth, N. C., Goud, B., Kabcenell, K., and Novick, P. J. (1989). Mutational analysis of *SEC4* suggests a cyclical mechanism for the regulation of vesicular traffic. *EMBO J.* **8,** 1685–1693.

Whitaker, M., and Zimmerberg, J. (1987). Inhibition of secretory granule discharge during exocytosis in sea urchin eggs by polymer solutions. *J. Physiol.* **369,** 527–539.

White, J. M. (1990). Viral and cellular membrane fusion proteins. *Annu. Rev. Physiol.* **52,** 675–697.

Wightman, R. M., Jankowski, J. A., Kennedy, R. T., Kawagoe, K. T., Schroeder, T. J., Leszczyszyn, D. J., Near, J. A., Diliberto, E. J., and Viveros, O. H. (1991). Temporally resolved catecholamine spikes correspond to single vesicle release from individual chromaffin cells. *Proc. Nat. Acad. Sci. USA* **88,** 10754.

Wilson, D. W., Whiteheart, S. W., Wiedmann, M., Brunner, M., and Rothman, J. E. (1992). A multisubunit particle implicated in membrane fusion. *J. Cell Biol.* **117,** 531–538.

Woodbury, D. J., and Hall, J. E. (1988). Role of channels in the fusion of vesicles with a planar bilayer. *Biophys. J.* **54,** 1053–1063.

Yamazaki, M., and Ito, T. (1990). Deformation and instability in membrane structure of phospholipid vesicles caused by osmophobic association: Mechanical stress model for the mechanism of poly(ethylene glycol)-induced membrane fusion. *Biochemistry* **29,** 1309–1314.

Zimmerberg, J. (1987). Molecular mechanisms of membrane fusion: Steps during phospholipid and exocytotic membrane fusion. *Biosci. Rep.* **7,** 251–268.

Zimmerberg, J., and Whitaker, M. (1985). Irreversible swelling of secretory granules during exocytosis caused by calcium. *Nature (London)* **315,** 581–584.

Zimmerberg, J., Sardet, C., and Epel, D. (1985). Exocytosis of sea urchin egg cortical vesicles in vitro is retarded by hyperosmotic sucrose: Kinetics of fusion monitored by quantitative light-scattering microscopy. *J. Cell Biol.* **101,** 2398–2410.

Zimmerberg, J., Curran, M., Cohen, F. S., and Brodwick, M. (1987). Simultaneous electrical and optical measurements show that membrane fusion precedes secretory granule swelling during exocytosis of beige mouse mast cells. *Proc. Natl. Acad. Sci. USA* **84,** 1585–1589.

Zimmerberg, J., Curran, M., and Cohen, F. S. (1991). A lipid/

protein complex hypothesis for exocytotic fusion pore formation. *Ann. N.Y. Acad. Sci.* **635,** 307–317.

Zimmermann, U. (1986). Electrical breakdown, electropermeabilization and electrofusion. *Rev. Physiol. Biochem. Pharmacol.* **105,** 175–256.

Zorec, R., Sikdar, S. K., and Mason, W. T. (1991). Increased cytosolic calcium stimulates exocytosis in bovine lactotrophs. *J. Gen. Physiol.* **97,** 473–497.

Index